Problem Solving Guide for DC/AC

Gary D. Snyder

Prentice Hall

Boston Columbus Indianapolis New York San Francisco Upper Saddle River

Amsterdam Cape Town Dubai London Madrid Milan Munich Paris Montreal Toronto

Delhi Mexico City Sao Paulo Sydney Hong Kong Seoul Singapore Taipei Tokyo

Editorial Director: Vern Anthony
Acquisitions Editor: Wyatt Morris
Editorial Assistant: Yvette Schlarman
Director of Marketing: David Gesell
Marketing Manager: Harper Coles
Marketing Assistant: Chrystal Gonzales
Senior Managing Editor: JoEllen Gohr
Senior Project Manager: Rex Davidson
Senior Operations Supervisor: Pat Tonneman
Operations Specialist: Deidra Skahill
Art Director: Jayne Conte
Cover Designer: Central Design Group
Printer/Binder: Edwards Brothers, Inc.
Cover Printer: Lehigh-Phoenix

10 9 8 7 6 5 4 3 2 1

Prentice Hall
is an imprint of
PEARSON

www.pearsonhighered.com

ISBN-13: 978-0-13-510386-9
ISBN-10: 0-13-510386-X

Contents

Preface

About This Problem Solving Guide

Background

The study of electronics, as with most technical courses, is highly mathematical. Many students find that the greatest challenge to understanding and doing well in electronics is becoming proficient with the math that circuit analysis requires. Although textbooks provide examples and chapter problems to help students develop this proficiency, these can fall short even in the best textbooks. Space constraints often limit the number of examples that the author can include to fully demonstrate practical problem solving techniques. Similarly, solutions to chapter problems (if provided) show only the final results and not the methods of reaching these results. These answers by themselves provide little help for students who arrive at a different answer. Worse, some textbook answers lead only to frustration when they are wrong and the student arrives at the correct (and necessarily different) answers. The instructor typically must address these issues by spending additional class time on specific topics, individually working with students who require additional assistance, or preparing and then checking additional assignments.

Purpose

The intent of the *Problem Solving Guide for DC/AC* is to assist both instructors and students by providing supplemental problems that represent and clearly illustrate essential concepts and techniques in circuit analysis. Each problem falls into one of three categories:

- **Basic.** *Basic* problems deal with the mechanics of electronics and circuit analysis. Typically these will involve problems with specific values and have a specific numerical answer.
- **Advanced.** *Advanced* problems help students gain greater insight into the theory behind electronics and the behavior and operation of electronic components and circuits. These consist of problems that require application of multiple problem solving techniques, finding general solutions to a class of circuits, and similar types of problems.
- **Just for Fun.** As the name suggests, *Just for Fun* problems are problems for those who wish to investigate some additional and often interesting special topics. These problems are generally not covered in electronics courses and allow students to explore some topics that are illustrative, challenging, and (hopefully) enjoyable.

Each problem provides both a detailed, step-by-step procedure for working through and solving the problem and the solution itself.

What the Problem Solving Guide Is

The *Problem Solving Guide for DC/AC* is a representative collection of dc and ac electronic problems and circuit analysis techniques that the reader is likely to encounter in his or her future electronics classes and career. As such, its intent is to be a convenient electronics reference as well as an instructional electronics work. The topics in this guide cover generally parallel topics covered in Tom Floyd's *Principles of Electric Circuits*, *9th Edition*, but are generally applicable to other introductory electronics textbooks. Readers should consider this a companion volume to supplement the information provided in standard basic dc and ac electronics textbooks rather than a substitute.

What the Problem Solving Guide Is Not

The *Problem Solving Guide for DC/AC* is not an answer key or "cheat sheet" for any existing textbook. Nor is it a "cookbook" for designing or analyzing circuits, beyond its role of presenting representative circuits and demonstrating associated circuit analysis techniques. In particular, this guide will not cover electronics theory, other than to provide some context for the problems it contains. It also does not (and cannot) provide examples of every circuit or type of circuit that you may encounter. The problems and solutions in this manual are intended to help you develop problem-solving skills and better understand fundamental electronics concepts, so that you can better analyze circuits that may be unfamiliar to you, as well as those that are familiar.

Problem Solving Guide Organization

This guide consists of 22 sections, each of which generally corresponds to a chapter in Floyd's *Principles of Electric Circuits*, plus an extra chapter devoted to complex numbers and phasors. The first two sections deal with the basic electronic concepts and illustrate how to set up and work through problems in general, with a special emphasis on good problem solving practices (observing significant figures, converting units, ensuring unit consistency, and breaking down larger problems into smaller, more manageable steps). Later sections deal with more concrete examples of circuits, analysis techniques, and electronic topics.

How to Use This Problem Solving Guide

Each example in this guide consists of a problem statement followed immediately by its solution. Before reading the solution, you should attempt to analyze the problem and independently arrive at a solution. You should study the solution only after first attempting to solve the problem. Note that many problems will not have a unique solution method, so you may arrive at the correct answer using a different method than the one shown.

Using Calculators with Circuits

A special note here concerns the use of calculators with ac circuits. This manual works through each problem in detail and does not assume or require the use of any specific model or type of calculator, although some problems assume that the user can solve linear systems and convert between rectangular and polar forms of complex quantities. Section 9 covers the mechanics of solving linear systems, and Section 15 shows how to manually convert between rectangular and polar forms, but most scientific calculators support these types of calculations so that the solutions to many problems will not show the details of working through the equations.

In the case of complex calculations, readers should be aware that calculators support complex calculations in one of two ways: complex form and complex mode.

A calculator that supports complex form can solve complex equations only if the complex values are in a specific format (typically rectangular form). These calculators provide rectangular-to-polar and polar-to-rectangular functions so that the user can convert the values to and from the required format for complex calculations as needed. The TI-36 and HP-33 calculators support complex form.

A calculator that supports complex mode works directly with complex numbers as data objects. These calculators allow users to select the calculator display mode and view numbers in either rectangular or polar form, but process complex numbers as a specific type of data, just as they process real numbers, vectors, matrices, and text as specific types of data. Calculators that support complex mode greatly simplify ac calculations by allowing the user to work with complex values in much the same way as for other numeric values. The HP-48gII and TI-89 are calculators that support complex mode.

If you have a calculator that supports complex calculations, many of the ac calculations in this manual may seem needlessly long and complicated. However, the intent of this manual is not to give just the answer to a problem, but to help you better understand the nature of circuits and the underlying mathematics for analyzing and solving them.

Acknowledgments

I would like to express my great appreciation to David Buchla and Thomas Floyd for the benefit of their considerable authoring talents and experience and for the great opportunity of having worked with them in the past. I also wish to thank Wyatt Morris, Rex Davidson, Lois Porter, and the many talented production personnel of Pearson Education for their invaluable assistance in helping me to prepare this book.

1. Quantities and Units

1.1 Basic

1.1.1 Estimation

Most students today are not familiar with estimation. Modern electronic calculators and software greatly simplify calculations so that correcting a calculation error is not as tedious and time-consuming as it once was. Even so, estimation provides an additional check on the final result and greater confidence that a calculated result is correct.

1. Problem: What is the estimated answer for $664 + 882$?

   Solution: $664 + 882 \approx 700 + 900 = 1600$. Both the augend* (first addition term) and addend (second addition term) were rounded up so that the sum should be somewhat less than 1600.

   *Because the order of addition does not matter, both terms are often called *addends*.

2. Problem: What is the estimated answer for $552 + 45$?

   Solution: $552 + 45 \approx 600 + 40 = 640$. The augend was rounded up more than the addend was rounded down so that the sum is less than 640.

3. Problem: What is the estimated answer for $759 - 1037$?

   Solution: $759 - 1037 \approx 800 - 1000 = -200$. The first difference term (minuend) was rounded up more than the second difference term (subtrahend) was rounded down, so the difference should be somewhat less than -200.

4. Problem: What is the estimated answer for $828 - 339$?

   Solution: $828 - 339 \approx 800 - 300 = 500$. The minuend was rounded down more than the subtrahend was rounded down so that the difference should be somewhat less than 500.

5. Problem: What is the estimated answer for 123×729?

   Solution: $123 \times 729 \approx 100 \times 700 = 70{,}000$. Both the multiplicand (first factor) and multiplier** (second factor) were rounded down so that the product should be somewhat greater than 70,000.

   **Because the order of multiplication does not matter, both factors are often called *multiplicands*.

6. Problem: What is the estimated answer for 36×117?

   Solution: $36 \times 117 \approx 40 \times 100 = 4{,}000$. The multiplicand was proportionally rounded up less than the multiplier was rounded down (+10% vs. –17%) so that the product should be somewhat more than 4,000.

7. Problem: What is the estimated answer for $471 \div 26$?

   Solution: $471 \div 26 \approx 471 \div 30$. $30 \times 16 = 480$ and $30 \times 17 = 510$ so that the quotient should be approximately 16 or 17.

8. Problem: What is the estimated answer for $63 \div 277$?

   Solution: $63 \div 277 \approx 60 \div 300$. $300 \times 0.2 = 60$ so that the quotient should be about 0.2.

1.1.2 Number Representation

1. Problem: Express the floating point value 723,461 in scientific notation.

   Solution: First, express the number as the product of a mantissa and a power of ten:

   $723{,}461 = 723{,}461 \times 10^0$ — The mantissa is **723,461** and the exponent is **0**.

   Next, move the decimal point to the correct position. Scientific notation requires you to move the decimal point until the mantissa is greater than or equal to 1 and less than 10. In this case you will move the decimal point to the **left** so that the exponent will **increase** by one with each move.

   $7\,2\,3\,,4\,6\,1\,. \times 10^0 \rightarrow 7\,2\,,3\,4\,6\,.\,1 \times 10^1$ — 72,346.1 > 10. Keep moving the decimal point.

   $7\,2\,,3\,4\,6\,.\,1 \times 10^1 \rightarrow 7\,,2\,3\,4\,.\,6\,1 \times 10^2$ — 7,234.61 > 10. Keep moving the decimal point.

   $7\,,2\,3\,4\,.\,6\,1 \times 10^2 \rightarrow 7\,2\,3\,.\,4\,6\,1 \times 10^3$ — 723.461 > 10. Keep moving the decimal point.

$7\,2\,3\,.\,4\,6\,1 \times 10^3 \rightarrow 7\,2\,.\,3\,4\,6\,1 \times 10^4$ 72.3461 > 10. Keep moving the decimal point.

$7\,2\,.\,3\,4\,6\,1 \times 10^4 \rightarrow 7\,.\,2\,3\,4\,6\,1 \times 10^5$ $1 \le 7.23461 < 10$. Done.

723,461 in scientific notation is therefore **7.23461 × 10⁵**.

2. Problem: Express the floating point value 0.0009833 in scientific notation.

   Solution: First, express the number as the product of a mantissa and a power of ten:

   $0.0009833 = 0.0009833 \times 10^0$ The mantissa is **0.0009833** and the exponent is **0**.

   Next, move the decimal point to the correct position. Scientific notation requires you to move the decimal point until the mantissa is greater than or equal to 1 and less than 10. In this case you will move the decimal point to the **right** so that the exponent will **decrease** by one with each move.

   $0\,.\,0\,0\,0\,9\,8\,3\,3 \times 10^0 \rightarrow 0\,.\,0\,0\,9\,8\,3\,3 \times 10^{-1}$ 0.009833 < 1. Keep moving the decimal point.

   $0\,.\,0\,0\,9\,8\,3\,3 \times 10^{-1} \rightarrow 0\,.\,0\,9\,8\,3\,3 \times 10^{-2}$ 0.09833 < 1. Keep moving the decimal point.

   $0\,.\,0\,9\,8\,3\,3 \times 10^{-2} \rightarrow 0\,.\,9\,8\,3\,3 \times 10^{-3}$ 0.9833 < 1. Keep moving the decimal point.

   $0\,.\,9\,8\,3\,3 \times 10^{-3} \rightarrow 9\,.\,8\,3\,3 \times 10^{-4}$ $1 \le 9.833 < 10$. Done.

   0.0009833 in scientific notation is therefore **9.833×10^{-4}**.

3. Problem: Express the floating point value 4.0075 in scientific notation.

   Solution: First, express the number as the product of a mantissa and a power of ten:

   $4.0075 = 4.0075 \times 10^0$ The mantissa is **4.0075** and the exponent is **0**.

   Next, move the decimal point to the correct position. Scientific notation requires you to move the decimal point until the mantissa is greater than or equal to 1 and less than 10. In this case the decimal point is already in the correct position, as $1 \le 4.0075 < 10$. Therefore 4.0075 in scientific notation is **4.0075×10^0**.

4. Problem: Express the floating point value 802,993 in engineering notation.

   Solution: First, express the number as the product of a mantissa and a power of ten:

   $802{,}993 = 802{,}993 \times 10^0$ The mantissa is **802,993** and the exponent is **0**.

   Next, move the decimal point to the correct position. Engineering notation requires you to move the decimal point until the mantissa is greater than or equal to 1 and less than 1000 and the exponent is an exact multiple of 3. In this case you will move the decimal point to the **left** so that the exponent will **increase** by one with each move.

   $8\,0\,2\,,\,9\,9\,3\,. \times 10^0 \rightarrow 8\,0\,,\,2\,9\,9\,.\,3 \times 10^1$ 80,299.3 > 1000. Keep moving the decimal point.

   $8\,0\,,\,2\,9\,9\,.\,3 \times 10^1 \rightarrow 8\,,\,0\,2\,9\,.\,9\,3 \times 10^2$ 8,029.93 > 1000. Keep moving the decimal point.

   $8\,,\,0\,2\,9\,.\,9\,3 \times 10^2 \rightarrow 8\,0\,2\,.\,9\,9\,3 \times 10^3$ $1 \le 802.993 < 1000$ and the exponent is an exact multiple of 3, as $3 \times 1 = 3$. Done.

   802,993 in engineering notation is therefore **802.993×10^3**.

5. Problem: Express the floating point value 1.2345 in engineering notation.

   Solution: First, express the number as the product of a mantissa and a power of ten:

   $1.2345 = 1.2345 \times 10^0$ The mantissa is **1.2345** and the exponent is **0**.

   Next, move the decimal point to the correct position. Engineering notation requires you to move the decimal point until the mantissa is greater than or equal to 1 and less than 1000 and the exponent is an exact multiple of 3. In this case the decimal point is already in the correct position, as $1 \le 1.2345 < 1000$ and the exponent of 0 an exact multiple of 3, as $3 \times 0 = 0$. Therefore 1.2345 in engineering notation is **1.2345×10^0**.

6. Problem: Express the floating point value 0.000022 in engineering notation.

   Solution: First, express the number as the product of a mantissa and a power of ten:

$0.000022 = 0.000022 \times 10^0$	The mantissa is **0.000022** and the exponent is **0**.

Next, move the decimal point to the correct position. Engineering notation requires you to move the decimal point until the mantissa is greater than or equal to 1 and less than 1000 (or, alternatively, the exponent is an exact multiple of 3). In this case you will move the decimal point to the **right** so that the exponent will **decrease** by one with each move.

$0.000022 \times 10^0 \rightarrow 0.00022 \times 10^{-1}$	$0.00022 < 1$. Keep moving the decimal point.
$0.00022 \times 10^{-1} \rightarrow 0.0022 \times 10^{-2}$	$0.0022 < 1$. Keep moving the decimal point.
$0.0022 \times 10^{-2} \rightarrow 0.022 \times 10^{-3}$	The exponent of -3 is an exact multiple of 3 (3 x $-1 = -3$) but $0.022 < 1$. Keep moving the decimal point.
$0.022 \times 10^{-3} \rightarrow 0.22 \times 10^{-4}$	$0.22 < 1$. Keep moving the decimal point.
$0.22 \times 10^{-4} \rightarrow 2.2 \times 10^{-5}$	$1 \le 2.2 < 1000$, but the exponent of -5 is not an exact multiple of 3. Keep moving the decimal point.
$2.2 \times 10^{-5} \rightarrow 22 \times 10^{-6}$	$1 \le 22 < 1000$ and the exponent is an exact multiple of 3, as $3 \times -2 = -6$. Done.

0.000022 in engineering notation is therefore **22×10^{-6}**.

7. Problem: Express the value 3.1416×10^2 as a floating point value.

   Solution: To convert a number from an exponential notation to a floating point value, move the decimal point and adjust the exponent until its value is zero. Because the exponent 2 is greater than zero, move the decimal point to the **right** so that the exponent will **decrease** by one with each move.

$3.1416 \times 10^2 \rightarrow 31.416 \times 10^1$	The exponent is still greater than zero. Keep moving the decimal point.
$31.416 \times 10^1 \rightarrow 314.16 \times 10^0$	The exponent is zero so that the mantissa is the floating point value. Done.

3.1416×10^2 as a floating point value is therefore **314.16**.

8. Problem: Express the value 271.83×10^{-3} as a floating point value.

   Solution: To convert a number from an exponential notation to a floating point value, move the decimal point and adjust the exponent until its value is zero. Because the exponent -3 is less than zero, move the decimal point to the **left** so that the exponent will **increase** by one with each move.

$271.83 \times 10^{-3} \rightarrow 27.183 \times 10^{-2}$	The exponent is still less than zero. Keep moving the decimal point.
$27.183 \times 10^{-2} \rightarrow 2.7183 \times 10^{-1}$	The exponent is still less than zero. Keep moving the decimal point.
$2.7183 \times 10^{-1} \rightarrow 0.27183 \times 10^0$	The exponent is zero so that the mantissa is the floating point value. Done.

271.83×10^{-3} as a floating point value is therefore **0.27183**.

1.1.3 Significant Figures and Rounding

Significant figures and rounding are related concepts. Significant figures indicate which digits, or figures, in a number are exact and which only approximate the actual values. As an example, consider the number 127,000 which has three significant figures. The first two digits, **1** and **2**, are exact. The remaining digits, **7,000**, are an approximation of the residual value, which lies somewhere between 6501 and 7500. Rounding determines how the approximate figures are determined. Several rounding methods exist, but this guide will use the "round to even" rule. If your textbook or instructor uses another rounding method, use that method for the rounding step in the solution to each problem.

1. Problem: How many significant figures does the number 0.004927 contain?

Solution: Although there are 7 digits in the number, the first three digits are placeholders only that determine the magnitude and not the actual value of the number. Therefore, only the digits 4927 are significant, so 0.004927 contains four significant figures.

2. Problem: How many significant figures does the number 1000.001 contain?

   Solution: The five zeros in the number 1000.001 are all essential to the value of the number, so they are all significant figures. Therefore, the number 1000.001 contains seven significant figures.

3. Problem: How many significant figures does the number 0.001200 contain?

   Solution: The first three zeros in the number are placeholders only that determine the magnitude and not the actual value of the number. The trailing two zeros are significant digits, as they would not otherwise be shown. Therefore, the number 0.001200 has four significant figures.

4. Problem: How many significant figures does the number 128,000 contain?

   Solution: The number 128,000 contains at least 3 significant figures, but it is not possible to determine from inspection alone exactly how many figures in this floating point number are significant. There is no way to determine how many of the trailing zeros are significant and how many are placeholders. This is one of the weaknesses of floating point number representation.

5. Problem: Round the value 1,735,462 to three significant figures.

   Solution: The place value of the third digit is 10,000. The value to the right of the third digit is 5,462, which is more than half of 10,000. Therefore, round up the third digit to 4 and drop the remaining digits so that 1,735,462 → **1,740,000**.

6. Problem: Round the value 2,150 to two significant figures.

   Solution: The place value of the second digit is 100. The value to the right of the second digit is 50, which is exactly half of 100. Therefore, round the second digit to 2, which is the nearest even value, and drop the remaining digits so that 2,150 → **2,200**.

7. Problem: Round the value 0.00235742 to four significant figures.

   Solution: The first two zeros to the right of the decimal point are placeholders that determine the magnitude rather than value of the number, so they are not significant figures. The place value of the fourth non-zero digit is therefore 0.000001. The value of the remaining digits, 0.00000042, is less than half of 0.000001. Therefore, leave the fourth digit unchanged and drop the remaining digits so that 0.00235742 → **0.002357**.

8. Problem: Round the value 10.0068313 to five significant figures.

   Solution: The first two zeros to the right of the decimal point are integral to the value of the number, so they are significant figures. The place value of the fifth digit is therefore 0.001. The value of the remaining digits, 0.0008313, is more than half of 0.001. Therefore, round up the fifth digit and drop the remaining digits so that 10.0068313 → **10.007**.

1.1.4 Calculations

1. Problem: What is the sum of 2.13 + 0.1073?

   Solution: The basic calculation gives 2.13 + 0.1073 = 2.2373. The least significant figure of the augend (2.1**3**) is hundredths of a unit and the least significant figure of the addend (0.107**3**) is ten thousandths of a unit. The sum must be rounded to the less precise of the two, which is hundredths of a unit. Using the "round to even" rule gives 2.2373 → 2.24.

   Therefore, 2.13 + 0.1073 = **2.24**.

2. Problem: What is the sum of 4735 + 0.009654?

   Solution: The basic calculation gives 4735 + 0.009654 = 4735.009654. The least significant figure of the first addend (473**5**) is units and the least significant figure of the second addend (0.00965**4**) is millionths of a unit. The sum must be rounded to the less precise of the two, which is units. Using the "round to even" rule gives 4735.009654 → 4735.

   Therefore, 4735 + 0.009654 = **4735**.

3. Problem: What is the difference of 11 – 0.0050?

   Solution: The basic calculation gives 11 – 0.0050 = 10.9950. The least significant figure of the minuend (1**1**) is units and the least significant figure of the subtrahend (0.005**0**) is ten thousandths of a unit. The difference must be rounded to the less precise of the two, which is units. Using the "round to even" rule gives 10.995 → 11.

   Therefore, 11 – 0.005 = **11**.

4. Problem: What is the difference of 48.9 – 62.75?

   Solution: The basic calculation gives 48.9 – 62.75 = −13.85. The least significant figure of the minuend (48.**9**) is tenths of a unit and the least significant figure of the subtrahend (62.7**5**) is hundredths of a unit. The difference must be rounded to the less precise of the two, which is tenths of a unit. Using the "round to even" rule gives −13.85 → −13.8.

   Therefore, 48.9 – 62.75 = **–13.8**.

5. Problem: What is the product of 3.8492 × 6.25?

   Solution: The basic calculation gives 3.8492 × 6.25 = 24.0575. The first multiplicand has five significant figures and the second multiplicand has three significant figures. The product must be rounded to the less precise of the two, which is three significant figures. Using the "round to even" rule gives 24.0575 → 24.1.

   Therefore, 3.8492 × 6.25 = **24.1**.

6. Problem: What is the product of 0.0001250 × 24,917,581?

   Solution: The basic calculation gives 0.0001250 × 24,917,581 = 3,114.697625. The first multiplicand has four significant figures and the second multiplicand has eight significant figures. The product must be rounded to the less precise of the two, which is four significant figures. Using the "round to even" rule gives 3,114.697625 → 3,115.

   Therefore, 0.0001250 × 24,917,581 = **3,115**.

7. Problem: What is the quotient of 1.00 ÷ 3.0?

   Solution: The basic calculation gives 1.00 ÷ 3.0 ≈ 0.333333…, which is a non-terminating repeating value. The dividend has three significant figures and the divisor has two significant figures. The quotient must be rounded to the less precise of the two, which is two significant figures. Using the "round to even" rule gives 0.333333… → 0.33.

   Therefore, 1.00 ÷ 3.0 = **0.33**.

8. Problem: What is the quotient of 80.0 ÷ 81.0000?

   Solution: The basic calculation gives 80.00 ÷ 81.0000 ≈ 0.987654321. The dividend has four significant figures and the divisor has six significant figures. The quotient must be rounded to the less precise of the two, which is four significant figures. Using the "round to even" rule gives 0.987654321 → 0.9876.

   Therefore, 80.00 ÷ 81.0000 = **0.9876**.

1.1.5 Metric Prefixes

1. Problem: Express 0.0001435 W using a metric prefix.

   Solution: First, express the number in engineering notation:

   $0.0001435\ \text{W} = 143.5 \times 10^{-6}\ \text{W}$

   The exponent 10^{-6} corresponds to the metric prefix "micro", which uses the symbol "µ". Therefore:

   $143.5 \times 10^{-6}\ \text{W} =$ **143.5 µW**

2. Problem: Express 12,470 V using a metric prefix. Assume that the number contains four significant figures.

   Solution: First, express the number in engineering notation:

   $12{,}470\ \text{V} = 12.47 \times 10^{3}\ \text{V}$. The final zero is dropped because the number contains only four significant figures.

The exponent 10^3 corresponds to the metric prefix "kilo", which uses the symbol "k". Therefore:

12.47×10^3 V = **12.47 kV**

3. Problem: Express 8.20 A using a metric prefix.

Solution: First, express the number in engineering notation:

8.20 A $= 8.20 \times 10^0$ A

The exponent 10^0 does not have a metric prefix. Therefore

8.20×10^0 A = **8.20 A**.

4. Problem: Express 0.47 F using a metric prefix.

Solution: First, express the number in engineering notation:

0.47 F $= 470 \times 10^{-3}$ F

The exponent 10^{-3} corresponds to the metric prefix "milli", which uses the symbol "m". Therefore,

470×10^{-3} F = **470 mF**

5. Problem: Express 6.20 MΩ in terms of the base unit Ω using floating point notation.

Solution: First, express the number in engineering notation. The metric prefix "M" stands for "mega" and corresponds to the exponent 10^6. Therefore,

6.20 MΩ $= 6.20 \times 10^6$ Ω

Next, convert the engineering value to a floating point value:

$6.20 \times 10^6 = 6{,}200{,}000$

Therefore, 6.20 MΩ $= 6.20 \times 10^6$ Ω = **6,200,000 Ω**

6. Problem: Express 0.22 μF in terms of the base unit F using scientific notation.

Solution: First, express the number in engineering notation. The metric prefix "μ" stands for "micro" and corresponds to the exponent 10^{-6}. Therefore,

0.22 μF $= 0.22 \times 10^{-6}$ F

Strictly speaking this value is not in engineering notation as the mantissa is not between 1 and 1000, but the exponent is in engineering notation form. Next, convert the engineering notation value to a floating point value:

$0.22 \times 10^{-6} = 2.2 \times 10^{-7}$

Therefore, 0.22 μF $= 0.22 \times 10^{-6}$ F = $\mathbf{2.2 \times 10^{-7}}$ **F**

7. Problem: Express 1.21 GW in terms of the basic unit W using floating point notation.

Solution: First, express the number in engineering notation. The prefix "G" stands for "giga" and corresponds to the exponent 10^9. Therefore,

1.21 GW $= 1.21 \times 10^9$ W

Next, convert the engineering notation value to a floating point value:

$1.21 \times 10^9 = 1{,}210{,}000{,}000$

Therefore, 1.21 GW $= 1.21 \times 10^9$ W = **1,210,000,000 W**

8. Problem: Express 1.2 nm in terms of the base unit m using floating point notation.

Solution: First, express the number in engineering notation. The prefix "n" stands for "nano" and corresponds to the exponent 10^{-9}. Therefore,

1.2 nm $= 1.2 \times 10^{-9}$ m.

Next, convert the engineering notation value to a floating point value:

$1.2 \times 10^{-9} = 0.0000000012$

Therefore, 1.2 nm $=1.2 \times 10^{-9}$ m = **0.0000000012 m**

1.2 Advanced

1.2.1 Estimation

1. Problem: If light travels at 186,282 miles per second, estimate how many miles light travels in one year.

   Solution: The estimate is

   Miles per minute = (186,282 miles / second) × (60 seconds / minute)
   ≈ (190,000 miles / second) × (60 seconds / minute)
   = (190,000 × 60) [(miles / second) × (seconds / minute)]
   = 11,400,000 miles / minute

   Miles per hour = (11,400,000 miles / minute) × (60 minutes / hour)
   ≈ (11,000,000 miles / minute) × (60 minutes / hour)
   = (11,000,000 × 60) [(miles / minute) × (minutes / hour)]
   = 660,000,000 miles / hour

   Miles per day = (660,000,000 miles / hour) × (24 hours / day)
   ≈ (700,000,000 miles / hour) × (20 hours / day)
   = (700,000,000 × 20) [(miles / hour) × (hours / day)
   = 14,000,000,000 miles / day

   Miles per year = (14,000,000,000 miles / day) × (365 days / year)
   ≈ (14,000,000,000 miles / day) × (400 days / year)
   = (14,000,000,000 × 400 days) [(miles / day) × (days / year)]
   = 5,600,000,000,000 miles per year

   Therefore, light travels an estimated distance of **5,600,000,000,000 miles per year**.

2. Problem: If you could save $5 per day, for about how many years would you need to save to have $1,000,000? Assume that you earn no interest on the money.

   Solution: First, estimate the number of days to save $1,000,000:

   Number of days = (1,000,000 dollars) / (5 dollars / day)
   = (1,000,000 / 5) × [dollars / (dollars /day)]
   = 200,000 [dollars × (day / dollars)]
   = 200,000 days

   Next, estimate the number of years:

   Number of years = 200,000 days / (365 days / year)
   ≈ 200,000 days / (400 days / year)
   = (200,000 / 400) × [days / (days / year)]
   = 500 [days × (year / days)]
   = 500 years

   Because you rounded up the number of days per year by 10%, the estimate is about 10% less than the actual time you would require.

1.2.2 Number Representation

1. Problem: Other than allowing you to express very large and very small numbers in a compact format, what is an advantage of scientific notation and engineering notation over floating point notation?

   Solution: Consider the floating point number 681,000. Without marking the least significant digit in some manner (such as using an overbar or underline) you cannot determine by inspection how many of the trailing zeros are significant digits and the number of significant figures in calculations using this value. With scientific and engineering notation, however, all digits of the mantissa are significant. If the number 681,000 has five significant digits, for example, the scientific and engineering notations representations are 6.8100×10^5 and 68.100×10^3, respectively, so that you can determine the number of significant digits by inspection.

2. Problem: What is an advantage of engineering notation over scientific notation?

Solution: Because engineering notation exponents are always multiples of three, it is much simpler to associate and express numbers in engineering notation with metric prefixes.

1.2.3 Significant Figures and Rounding

1. Problem: What is an advantage of the "round to even" method compared to always rounding up or always rounding down a number ending in 5?

Solution: Consider the following set of values:

0.5, 1.0, 1.5, 2.0, 2.5, 3.0, 3.5, 4.0, 4.5, 5.0, 5.5, 6.0, 6.5, 7.0, 7.5, 8.0, 8.5, 9.0, 9.5, 10.0

The sum of the data set is 105 and the average of the data set is 5.25.

Always rounding up the value to the nearest unit gives

1.0, 1.0, 2.0, 2.0, 3.0, 3.0, 4.0, 4.0, 5.0, 5.0, 6.0, 6.0, 7.0, 7.0, 8.0, 8.0, 9.0, 9.0, 10.0, 10.0

The sum of the rounded up data set is 110 and the average of the rounded up numbers is 5.5. By always rounding up numbers ending in 5, the rounded data set is skewed upwards from the original data set.

Similarly, always rounding down the value to the nearest 10 gives

0.0, 1.0, 1.0, 2.0, 2.0, 3.0, 3.0, 4.0, 4.0, 5.0, 5.0, 6.0, 6.0, 7.0, 7.0, 8.0, 8.0, 9.0, 9.0, 10.0

The sum of the rounded down data set is 100 and the average of the rounded down numbers is 5.0. By always rounding down numbers ending in 5, the rounded data set is skewed down from the original data set.

Using the "round to even" method gives

0.0, 1.0, 2.0, 2.0, 2.0, 3.0, 4.0 4.0, 4.0, 5.0, 6.0, 6.0, 6.0, 7.0, 8.0, 8.0, 8.0, 9.0, 10.0, 10.0

The sum of the "round to even" data set is 105 and average of the "round to even" numbers is 5.25. By effectively rounding up half the numbers and rounding down half the numbers, the rounded data set better corresponds to the original data set. Number distribution and analysis of data populations are more complicated than this, but this example helps to show how some rounding techniques bias or skew the data.

2. Problem: Manufacturing processes introduce variation in component values is so that the actual value of a component deviates from its specified, or **nominal**, value by some fixed percentage. Why would manufacturers not use significant figures to indicate the possible variation in component values from their nominal values?

Solution: Significant figures specify an absolute amount of uncertainty, or deviation, from an indicated value. This uncertainty is a smaller percentage of the indicated value for larger values than for smaller values. Consider as an example the three resistor values 1.0 kΩ, 2.4 kΩ, and 9.1 kΩ. Each value shows two significant figures, which indicate an uncertainty of ±0.050 kΩ.

For the 1.0 kΩ resistor the deviation as a percentage of the stated value is

$$\begin{aligned}\%\text{ deviation} &= (\pm 0.050\text{ k}\Omega / 1.0\text{ k}\Omega) \times 100\% \\ &= (\pm 0.050\text{ k}\Omega / 1.0\text{ k}\Omega) \times 100\% \\ &= \pm 0.050 \times 100\% \\ &= \pm 5.0\ \%\end{aligned}$$

For the 2.4 kΩ resistor the deviation as a percentage of the stated value is

$$\begin{aligned}\%\text{ deviation} &= (\pm 0.050\text{ k}\Omega / 2.4\text{ k}\Omega) \times 100\% \\ &= (\pm 0.050\text{ k}\Omega / 2.4\text{ k}\Omega) \times 100\%\end{aligned}$$

$= \pm 0.021 \times 100\%$

$= \pm 2.1\ \%$

For the 9.1 kΩ resistor the deviation as a percentage of the stated value is

% deviation $= (\pm 0.050\ \text{k}\Omega\ /\ 9.1\ \text{k}\Omega) \times 100\%$

$= (\pm 0.050\ \text{k}\Omega\ /\ 9.1\ \text{k}\Omega) \times 100\%$

$= \pm 0.0055 \times 100\%$

$= \pm 0.55\ \%$

As you can see, the same number of significant figures represents different percent deviations for different component values. Since the manufacturing process determines the percent deviation, manufacturers would require a different manufacturing process for each component value to be able to represent the possible deviation from the nominal value with significant figures, rather than using a common process for all values.

1.2.4 Calculations

1. Problem: What are the sum and difference of the values 7.362×10^3 and 2.56×10^5 expressed in scientific notation?

Solution: To add or subtract numbers in scientific or engineering notation, first ensure that the exponents of the numbers are the same. Although you can choose any exponent, it is typically helpful initially to convert the larger exponent to the smaller exponent.

$7.362 \times 10^3 = 7.362 \times 10^3$

$2.56 \times 10^5 = 256 \times 10^3$

You can see from these common representations that the less precise value is 256×10^3. Consequently the final answers must be rounded to the nearest 1×10^3.

Adding the two values gives

$(7.362 \times 10^3) + (2.56 \times 10^5) = (7.362 \times 10^3) + (256 \times 10^3)$

$= (7.362 + 256) \times 10^3$

$= 263.62 \times 10^3$

$= 263 \times 10^3$ rounded to the nearest 1×10^3

$=$ **2.63×10^5** using proper scientific notation

Subtracting the two values gives

$(7.362 \times 10^3) - (2.56 \times 10^5) = (7.362 \times 10^3) - (256 \times 10^3)$

$= (7.362 - 256) \times 10^3$

$= -248.638 \times 10^3$

$= -249 \times 10^3$ rounded to the nearest 1×10^3

$=$ **-2.49×10^5** using proper scientific notation

2. Problem: What are the product and quotient of the numbers 41.692×10^3 and 61.5×10^{-6} expressed in engineering notation?

Solution: The first number has five significant figures and the second number has three significant figures. For multiplication and division the final answer must be rounded to the lesser number of significant figures, so the final product and quotient for this problem will have three significant figures.

The product of the numbers is

$(41.692 \times 10^3) \times (61.5 \times 10^{-6}) = (41.692 \times 61.5) \times (10^3 \times 10^{-6})$

$= 2{,}564.058 \times 10^{3+(-6)}$

$= 2{,}560 \times 10^{-3}$ rounded to three significant digits

$=$ **2.56×10^0** using proper engineering notation

The quotient of the numbers is

$$(41.692 \times 10^{3}) / (61.5 \times 10^{-6}) = (41.692 / 61.5) \times (10^{3} / 10^{-6})$$
$$= 0.6779187 \times 10^{3-(-6)}$$
$$= 0.678 \times 10^{3} \text{ rounded to three significant digits}$$
$$= \mathbf{678 \times 10^{0}} \text{ using proper engineering notation}$$

3. Problem: Refer to Problems 1 through 4 of Section 1.1.4. Why do you round a sum (or difference) of two numbers to the position of the less precise least significant digit of the two numbers you are adding (or subtracting)?

Solution: The least significant digit in a number indicates the greatest accuracy with which you can express that number. Consider the numbers 634 and 12.53 as an example.

In the first number the digits "6" and "3" are exact, while the least significant digit "4" only approximates the rest of the number to the nearest unit. Similarly, in the second number the digits "1", "2", and "5" are exact, while the least significant digit "3" only approximates the rest of the number to the nearest hundredth of a unit. When you add or subtract these two numbers, you do not know the values for the tenths and hundredths of a unit for the first number, so that adding or subtracting the corresponding tenths and hundredths of a unit for the second number is meaningless. In real-world terms, these digits of the more precise second number are "lost in the noise". Consequently, the less precise least significant digit determines the precision to which you can round the final sum or difference of two numbers.

4. Problem: What are the maximum and minimum values for a resistor with a nominal (ideal) value of 560 Ω if its tolerance is 5% (e.g., the actual value is within 5% of the nominal value)?

Solution: The resistor value can vary by 5% of 560 Ω. This variation is equal to

$$\boldsymbol{R_{VAR}} = 560\ \Omega \times 5\%$$
$$= 560\ \Omega \times (5 / 100)$$
$$= [(560 \times 5) / 100]\ \Omega$$
$$= (2800 / 100)\ \Omega$$
$$= 28\ \Omega$$

The minimum resistance value is

$$\boldsymbol{R_{MIN}} = 560\ \Omega - 28\ \Omega$$
$$= (560 - 28)\ \Omega$$
$$= 532\ \Omega$$

Similarly, the maximum resistor value is

$$\boldsymbol{R_{MAX}} = 560\ \Omega + 28\ \Omega$$
$$= (560 + 28)\ \Omega$$
$$= 588\ \Omega$$

Therefore, the minimum resistance value is **532 Ω** and the maximum resistance value is **588 Ω**.

5. Problem: 10% tolerance resistors have standard values of 3.3 kΩ and 3.6 kΩ. Explain why there are no 10% tolerance resistors with standard values of 3.4 kΩ or 3.5 kΩ.

Solution: The maximum value of a 3.3 kΩ 10% tolerance resistor is

$$\boldsymbol{R_{MAX}} = 3.3\ \text{k}\Omega + (3.3\ \text{k}\Omega \times 10\%)$$
$$= (3.3\ \text{k}\Omega \times 1) + (3.3\ \text{k}\Omega \times 10\%)$$
$$= 3.3\ \text{k}\Omega \times (1 + 10\%)$$
$$= 3.3\ \text{k}\Omega \times [1 + (10 / 100)]$$
$$= 3.3\ \text{k}\Omega \times (1 + 0.1)$$
$$= 3.3\ \text{k}\Omega \times 1.1$$

$= (3.3 \times 1.1)\ \text{k}\Omega$

$= 3.63\ \text{k}\Omega$

The minimum value of a 3.6 kΩ 10% tolerance resistor is

$$\begin{aligned} \boldsymbol{R_{MIN}} &= 3.6\ \text{k}\Omega - (3.6\ \text{k}\Omega \times 10\%) \\ &= (3.6\ \text{k}\Omega \times 1) - (3.6\ \text{k}\Omega \times 10\%) \\ &= 3.6\ \text{k}\Omega \times (1 - 10\%) \\ &= 3.6\ \text{k}\Omega \times [1 - (10 / 100)] \\ &= 3.6\ \text{k}\Omega \times (1 - 0.1) \\ &= 3.6\ \text{k}\Omega \times 0.9 \\ &= (3.6 \times 0.9)\ \text{k}\Omega \\ &= 3.24\ \text{k}\Omega \end{aligned}$$

The 10% tolerance of the 3.3 kΩ and 3.6 kΩ resistors completely covers the resistance range between them. Therefore, there is no practical reason for producing standard 10% tolerance resistors with values in this range.

6. Problem: A certified mechanical chronometer has a specified accuracy of ±2 seconds per day. What accuracy in ppm (parts per million) must the quartz oscillator of a digital watch have to match this accuracy?

Solution: First, find the number of seconds in a day:

$$\begin{aligned} \text{Seconds per day} &= (24\ \text{hours/day}) \times (60\ \text{minutes/hour}) \times (60\ \text{seconds/minute}) \\ &= (24 \times 60 \times 60)\ [(\text{hours/day}) \times (\text{minutes/hour}) \times (\text{seconds/minute})] \\ &= = 86{,}400\ \text{seconds/day} \\ \text{Mechanical accuracy} &= (\pm 2\ \text{seconds/day}) / (86{,}400\ \text{seconds/day}) \\ &= (\pm 2 / 86{,}400)\ [(\text{seconds/day}) / (\text{seconds/day})] \\ &= \pm 0.000023148\ \text{parts / unit} \\ &= \pm 23.148 \times 10^{-6}\ \text{parts / unit} \\ &= (\pm 23.148 \times 10^{-6}\ \text{parts / unit}) \times (1{,}000{,}000\ \text{units / million}) \\ &= (\pm 23.148 \times 10^{-6} \times 1{,}000{,}000)\ [(\text{parts / unit}) \times (\text{units / million})] \\ &= \pm 23.148\ \text{parts per million} \end{aligned}$$

Therefore, a quartz oscillator must have an accuracy of at least **±23.148 ppm** for a digital watch to match the accuracy of a mechanical chronometer with a certified accuracy of ±2 seconds per day.

1.2.5 Metric Prefixes

1. Problem: What is the quotient of "kilo" divided by "milli" expressed as a metric prefix?

Solution: The metric prefix "kilo" corresponds to the exponent 10^3 and the metric prefix "milli" corresponds to the exponent 10^{-3}. The quotient of "kilo" divided by "milli" is then

$$\begin{aligned} 10^3 / 10^{-3} &= 10^{[3-(-3)]} \\ &= 10^6 \end{aligned}$$

The exponent 10^6 corresponds to the metric prefix "mega". Therefore, the quotient of "kilo" divided by "milli" is "**mega**".

2. Problem: What is the product of "micro" multiplied by "milli" expressed as a metric prefix?

Solution: The metric prefix "micro" corresponds to the exponent 10^{-6} and the metric prefix "milli" corresponds to the exponent 10^{-3}. The product of "micro" multiplied by "milli" is then

$$\begin{aligned} 10^{-6} \times 10^{-3} &= 10^{[(-6)+(-3)]} \\ &= 10^{-9} \end{aligned}$$

The exponent 10^{-9} corresponds to the metric prefix "nano". Therefore, the product of "micro" and "milli" is "**nano**".

3. Problem: What is the sum of 6.93 GW + 13.8 MW + 475 kW + 82 W in megawatts? Assume that each value is exact.

Solution: First, convert each of the values to the final unit of megawatts so that you can add them together.

6.93 GW	$= 6.93 \times 10^9$ W
	$= 6{,}930 \times 10^6$ W
	= 6,930 MW
13.8 MW	=13.8 MW
475 kW	$= 475 \times 10^3$ W
	$= 0.475 \times 10^6$ W
	= 0.475 MW
82 W	$= 82 \times 10^0$ W
	$= 0.000082 \times 10^6$ W
	= 0.000082 MW

Now sum the values:

6,930 MW + 13.8 MW + 0.475 MW + 0.000082 MW = 6,944.275082 MW

Therefore, the sum of the individual power values in megawatts is **6,944.275082 MW**.

1.3 Just for Fun

Dimensions are fundamental properties, such as mass, length (or distance), and time, with which physical properties are associated. Many scientific fields use a technique called dimensional analysis to analyze problems. Engineers often use it to verify that their solution to a complex calculation is correct. If, for example, the solution should be a velocity (distance / time) but the engineer's solution is an acceleration (distance / time^2), then dimensional analysis of the two will quickly show that the engineer's proposed solution is incorrect. Another practical application of dimensional analysis is to help determine how one should proceed to obtain an answer to a problem. If you know that the final answer must be an area (distance × distance), and are starting with a volume (distance × distance × distance), then you know you must divide the volume by distance to obtain an area. Note that units and dimensions, while related, are not the same. Units (such as kilograms) are accepted standards by which you can quantify dimensions (such as mass).

One useful application of dimensional analysis is to determine how quantities relate to each other. Consider the following dimensions for current, power, and voltage:

current = charge / time
power = energy / time
voltage = energy / charge

What is the relationship between power, current, and voltage?

First, use an unknown factor, X, to relate two of the known quantities. For this example, choose power and current as the known quantities and assume that the unknown factor times the power will give the current.

X × power = current

X × (energy / time) = charge / time

Now, solve for X:

X × (energy / time) = charge / time

X × (energy / time) × (time / energy) = (charge / time) × (time / energy)

X × (energy / time) × (time / energy) = (charge / time) × (time / energy)

Therefore,

X = charge / energy

The dimensions of the unknown factor are (charge / energy). These dimensions are the inverse of (energy / charge), which are the dimensions for voltage. From this, the unknown factor X must be the inverse of voltage so that

(1 / voltage) × power = current

voltage × (1 / voltage) × power = voltage × current

Therefore, from dimensional analysis,

power = voltage × current

As you continue your studies in electronics you will learn that this relationship is called Watt's Law.

1. Problem: Given that

   voltage = energy / charge

   current = charge / time

   voltage = current × resistance

   What are the dimensions of resistance?

   Solution: First, start with the relationship between voltage, current, and resistance and solve for resistance.

   voltage = current × resistance

   voltage / current = (current × resistance) / current

   = resistance

   Next, substitute in the dimensions for voltage and current:

   resistance = voltage / current

   = (energy / charge) / (charge / time)

   Dividing by (charge / time) is the same as multiplying by (time / charge) so

   resistance = (energy / charge) × (time / charge)

   = (energy × time) / charge^2

   Therefore, the dimensions for resistance are **(energy × time) / $\textbf{charge}^2$**.

2. Problem: Given that capacitance is the ratio of charge to the voltage used to store the charge, what is the product of resistance and capacitance? Use the dimensions of resistance from Problem 1.

   Solution: From the problem statement,

   capacitance = charge / voltage

   From Problem 1,

   voltage = energy / charge

   so

   capacitance = charge / (energy / charge)

   = charge × (charge / energy)

   = charge^2 / energy

   Also from the Problem 1 above,

   resistance = (energy × time) / charge^2

   so the product of resistance and capacitance is

   resistance × capacitance = [(energy × time) / charge^2] × (charge^2 / energy)

   = (energy × time × charge^2) / (charge^2 × energy)

   = time

   Therefore, the dimension for the product of resistance and capacitance is **time**.

3. Problem: Given that

   inductance / resistance = time

   what are the dimensions of inductance? Use the dimensions of resistance from Problem 1.

   Solution: From the problem statement,

inductance / resistance = time

so

(inductance / resistance) × resistance = time × resistance

inductance = time × resistance

From Problem 1,

resistance = (energy × time) / charge^2

so

inductance = (time × energy × time) / charge^2

= (energy × time × time)] / charge^2

= (energy × time^2) / charge^2

= energy × (time^2 / charge^2)

= energy × $(\text{time / charge})^2$

Therefore, **inductance = energy × $(\text{time / charge})^2$**.

4. Problem: How does energy relate to inductance and current? Use the dimensions of inductance from Problem 3.

Solution: From Problem 3,

inductance = energy × $(\text{time / charge})^2$

From this,

inductance / $(\text{time / charge})^2$ = [energy × $(\text{time / charge})^2$] / $(\text{time / charge})^2$

inductance / $(\text{time / charge})^2$ = energy

Dividing inductance by $(\text{time/charge})^2$ is the same as multiplying by $(\text{charge/time})^2$ so

energy = inductance / $(\text{time / charge})^2$

= inductance × $(\text{charge / time})^2$

Since current = (charge / time), it follows that

energy = inductance × $(\text{charge / time})^2$

= inductance × current^2

Therefore, **energy = inductance × current^2**.

2. Voltage, Current, and Resistance

2.1 Basic

2.1.1 Atomic Structure

1. Problem: Which of the elements whose atoms are represented in Figure 2-1 would probably conduct electricity best?

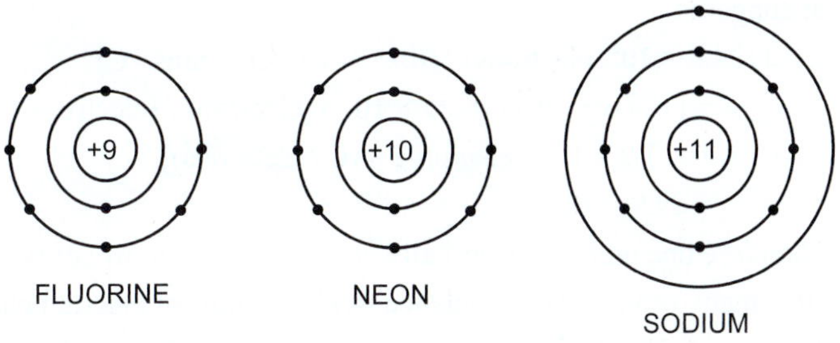

Figure 2-1: Atomic Structures for Problem 1

Solution: The fluorine atom lacks one electron to have a completely filled second shell and would have a greater tendency to capture an electron than to lose electrons, so that fluorine will have few free electrons. The neon atom has a completely filled second shell, so it will not capture nor release electrons and is chemically inert. The sodium atom has a single valence electron in its third shell that requires little energy to become a free electron. Therefore, the element that would conduct electricity best is **sodium**.

2. Problem: Which of the atoms represented in Figure 2-2 is a positive ion?

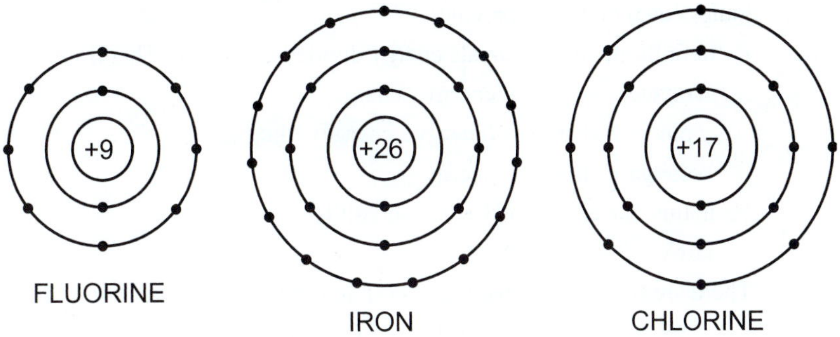

Figure 2-2: Atomic Structures for Problem 2

Solution: By definition, a positive ion is an atom with a net positive charge.

The fluorine atom has 10 electrons and nucleus with 9 positive charges for a net charge of

$$10 \times (-1) + (+9) = (-10) + (+9) = -1$$

The iron atom has 25 electrons and nucleus with 26 positive charges for a net charge of

$$25 \times (-1) + (+26) = (-25) + (+26) = +1$$

The chlorine atom has 17 electrons and nucleus with 17 positive charges for a net charge of

$$17 \times (-1) + (+17) = (-17) + (+17) = 0$$

Therefore, the **iron** atom is a positive ion.

2.1.2 Electrical Charge

1. Problem: A "mole" is a quantity approximately equal to 6.022×10^{24} units. An electrically neutral copper atom contains 29 protons, 35 neutrons, and 29 electrons. How many coulombs of negative charge would one mole of electrically neutral copper atoms contain?

   Solution: An electrically neutral copper atom contains 29 negatively charged electrons. Therefore, one mole of copper atoms contains

   $(6.022 \times 10^{24}$ Cu atoms) × (29 electrons / Cu atom) = 174.6×10^{24} electrons

   One coulomb of charge consists of 6.25×10^{18} electrons, so the number of coulombs in one mole of copper is

   $(174.6 \times 10^{24}$ electrons) / $(6.25 \times 10^{18}$ electrons / C)
   $= (174.6 \times 10^{24} / (6.25 \times 10^{18})$ [electrons / (electrons / C)]
   $= (27.9 \times 10^{6})$ [electrons × (C / electrons)]
   $= 27.9 \times 10^{6}$ C

   Therefore one mole of electrically neutral copper atoms contains **27.9×10^{6} coulombs**.

2. Problem: How many protons would balance five coulombs of electron charge?

   Solution: There are 6.25×10^{18} electrons per coulomb so in five coulombs there are

   (5 C) × $(6.25 \times 10^{18}$ electrons / C) $= (5 \times 6.25 \times 10^{18})$ [C × (electrons / C)]
   $= 31.3 \times 10^{18}$ electrons

   The charge on a proton is equal in magnitude to the charge on an electron so that **31.3×10^{18} protons** would balance five coulombs of electron charge.

2.1.3 Voltage

1. Problem: An electron volt (symbol eV) is the energy required to change the potential difference of an electron (or a charge e that is equivalent to the charge on an electron) by one volt. How many joules equal one electron volt?

   Solution: By definition, voltage equals energy divided by charge. Therefore,

   voltage = energy / charge
   voltage × charge = (energy / charge) × charge
   voltage × charge = energy

   From this, the energy of 1 electron volt is

   1 eV = 1 e × 1 V

   There are 6.25×10^{18} electrons / coulomb = 6.25×10^{18} e/C, so

   1 eV $= [1\ e / (6.25 \times 10^{18}\ e / \text{C})] \times 1$ V
   $= [(1 / 6.25 \times 10^{18}) \times 1]\ [(e \times \text{V}) / (e / \text{C})]$
   $= (160 \times 10^{-21})\ [(e \times \text{V}) \times (\text{C} / e)]$
   $= 160 \times 10^{-21}$ V•C

   (Note that the symbol "•" is often used to represent a product of units to prevent confusing a hyphen with a minus sign.)

   Since one volt equals one joule per one coulomb,

   1 V = 1 J / 1 C
   1 V•C = 1 V × 1 C
   = (1 J / C) × 1 C
   = (1 × 1) [(J / C) × C]
   = 1 J

Therefore,

160×10^{-21} V•C = 160×10^{-21} J

1 eV of energy equals **160×10^{-21} J.**

2. Problem: Twenty coulombs of charge require how much energy to develop a potential difference of four volts?

Solution: By definition, voltage equals energy divided by charge. Therefore

voltage = energy / charge

voltage × charge = (energy / charge) × charge

voltage × charge = energy

Therefore

energy = voltage × charge

= 4 V × 20 C

= (4 × 20) (V × C)

= 80 V•C

1 V•C = 1 J (from Problem 1), so

80 V•C = 80 J

Twenty coulombs of charge require **80 joules** of energy to develop 4 volts of potential difference.

3. Problem: A capacitor is an electronic component that stores charge. If a 6 V battery uses 100 millijoules of energy to store charge on a capacitor, how much charge does the capacitor store?

Solution: By definition, voltage equals energy divided by charge. Therefore,

voltage = energy / charge

voltage × charge = (energy / charge) × charge

(voltage × charge) / voltage = energy / voltage

charge = energy / voltage

Therefore,

charge = energy / voltage

= 100 mJ / 6 V

= $(100 \times 10^{-3}$ J) / 6 V

= $[(100 \times 10^{-3}) / 6]$ J / V

= 16.7×10^{-3} J / V

= 16.7×10^{-3} C

= 16.7 mC

A 6 volt battery using 100 millijoules will store **16.7 mC** of charge on a capacitor.

4. Problem: A student acquires 1 μC of static charge walking across a carpet on a dry winter day. When he touches a doorknob, the spark from the ESD (electrostatic discharge) dissipates 8 mJ of energy. What was the voltage between the student and the doorknob just before the ESD spark?

Solution: By definition, voltage equals energy divided by charge. Therefore,

voltage = energy / charge

= 8 mJ / 1 μC

= $(8 \times 10^{-3}$ J) / $(1 \times 10^{-6}$ C)

= $[(8 \times 10^{-3}) / (1 \times 10^{-6})]$ (J / C)

= 8×10^{3} J / C

= 8×10^{3} V

= 8 kV

Eight kilovolts of potential difference existed between the student and the doorknob just prior to the electrostatic discharge.

2.1.4 Current

1. Problem: A charge counter records the data in Table 2-1 for eight recording intervals.

Table 2-1: Recorded Charge Data

Interval	Charge	Interval	Charge
0.00 s to 0.25 s	1.62 C	1.01 s to 1.25 s	0.01 C
0.26 s to 0.50 s	0.97 C	1.26 s to 1.50 s	2.23 C
0.51 s to 0.75s	1.20 C	1.51 s to 1.75s	0.94 C
0.76 s to 1.00 s	1.15 C	1.76 s to 2.00 s	1.28 C

What is the average current for the eight recorded intervals?

Solution: By definition current equals charge divided by time. The average current will equal the total charge divided by the total elapsed time for the recording intervals.

The total charge $\boldsymbol{Q_T}$ for the recorded intervals is

$$\begin{aligned} Q_T &= 1.62\text{ C} + 0.97\text{ C} + 1.20\text{ C} + 1.15\text{ C} + 0.01\text{ C} + 2.23\text{ C} + 0.94\text{ C} + 1.28\text{ C} \\ &= (1.62 + 0.97 + 1.20 + 1.15 + 0.01 + 2.23 + 0.94 + 1.28)\text{ C} \\ &= 9.40\text{ C} \end{aligned}$$

The elapsed time $\boldsymbol{t}$ over which the data was recorded is the difference between the end time and start time, so

$$\begin{aligned} t &= 2.00\text{ s} - 0.00\text{ s} \\ &= (2.00 - 0.00)\text{ s} \\ &= 2.00\text{ s} \end{aligned}$$

Therefore,

$$\begin{aligned} \text{average current} &= Q_T / t \\ &= 9.40\text{ C} / 2.00\text{ s} \\ &= (9.40 / 2.00)\ (\text{C} / \text{s}) \\ &= 4.70\text{ A} \end{aligned}$$

The average current for the recorded intervals is **4.70 A**.

2. Problem: A fully charged backup capacitor in a calculator stores 150 μC of charge to retain data in the calculator when the batteries are changed. If the calculator draws a constant 120 nA of backup current from the capacitor when the batteries are removed, for how many minutes can the capacitor supply current to the calculator?

Solution: By definition, current equals charge divided by time. Therefore,

$$\begin{aligned} \text{current} &= \text{charge} / \text{time} \\ \text{current} \times \text{time} &= (\text{charge} / \text{time}) \times \text{time} \\ \text{current} \times \text{time} &= \text{charge} \\ (\text{current} \times \text{time}) / \text{current} &= \text{charge} / \text{current} \\ \text{time} &= \text{charge} / \text{current} \end{aligned}$$

From the problem statement, the charge is 150 μC and the current is 120 nA so

$$\begin{aligned} \text{time} &= 150\ \mu\text{C} / 120\text{ nA} \\ &= (150 \times 10^{-6}\text{ C}) / (120 \times 10^{-9}\text{ A}) \end{aligned}$$

$= (150 \times 10^{-6} / 120 \times 10^{-9})$ C / A

$= 1.25 \times 10^{3}$ s

The fully charged capacitor can supply current for $1.25 \times 10^{3} = 1{,}250$ seconds. Converting this to minutes gives

1,250 seconds = 1,250 seconds × (1 minute / 60 seconds)

= [(1,250 × 1) / 60] × [seconds × (minute / seconds)]

= (1,250 /60) minutes

= 20.8 minutes

The fully charged backup capacitor with 150 μC of charge can therefore supply 120 nA of backup current for **20.8 minutes**.

3. Problem: A dynamic RAM (DRAM) cell stores a "1" bit as 30 μC charge on a capacitor. This charge is continually drained by 8 nA of leakage current so that the cell must be refreshed every 1 ms to replace the lost charge. How much charge on the cell is lost between each memory refresh operation?

Solution: By definition, current equals charge divided by time. Therefore,

current = charge / time

current × time = (charge / time) × time

current × time = charge

Charge is drained from the cell by an 8 nA leakage current for 1 ms so

charge = current × time

= 8 nA × 1 ms

$= (8 \times 10^{-9}$ A$) \times (1 \times 10^{-3}$ s$)$

$= [(8 \times 10^{-9}) \times (1 \times 10^{-3})]$ (A × s)

$= 8 \times 10^{-12}$ C

= 8 pC

Therefore, the cell loses **8 pC** of charge between memory refresh operations.

2.1.5 Resistance

1. Problem: The resistivity of gold is 14.67 CM•Ω / ft at 20 °C. How long in inches must an AWG 40 gold wire be to have a resistance of 1 Ω?

Solution: Refer to Table 2-2 on the next page, which shows the cross-sectional areas in CM (circular mils) for AWG 0000 to AWG 40. From the table AWG 40 corresponds to a cross-sectional area of 9.89 CM. By definition, $\boldsymbol{R} = (\boldsymbol{\rho} \times \boldsymbol{l}) / \boldsymbol{A}$, where $\boldsymbol{R}$ is resistance, $\boldsymbol{\rho}$ is resistivity, $\boldsymbol{l}$ is length, and $\boldsymbol{A}$ is cross-sectional area, so

$\boldsymbol{R} = (\boldsymbol{\rho} \times \boldsymbol{l}) / \boldsymbol{A}$

$\boldsymbol{R} \times \boldsymbol{A} = [(\boldsymbol{\rho} \times \boldsymbol{l}) / \boldsymbol{A}] \times \boldsymbol{A}$

$(\boldsymbol{R} \times \boldsymbol{A}) / \boldsymbol{\rho} = (\boldsymbol{\rho} \times \boldsymbol{l}) / \boldsymbol{\rho}$

$(\boldsymbol{R} \times \boldsymbol{A}) / \boldsymbol{\rho} = \boldsymbol{l}$

For $\boldsymbol{R} = 1$ Ω, $\boldsymbol{A} = 9.89$ CM, and $\boldsymbol{\rho} = 14.67$ CM•Ω / ft

$\boldsymbol{l} = (\boldsymbol{R} \times \boldsymbol{A}) / \boldsymbol{\rho}$

= (1 Ω × 9.89 CM) / (14.67 CM•Ω / ft)

= [(1 × 9.89) / 14.67] {(Ω × CM) / [(CM × Ω) / ft]}

= (9.89 / 14.67) {(Ω × CM) × [ft / (CM × Ω)]}

$= 674 \times 10^{-3}$ ft

$= (674 \times 10^{-3}$ ft$) \times (12$ in / ft$)$

$= (674 \times 10^{-3} \times 12)$ [ft × (in / ft)]

= 8.09 in

Therefore, an AWG 40 gold wire must be **8.09 inches** long to have a resistance of 1 Ω.

Table 2-2: American Wire Gauge Cross-Sectional Areas

AWG	Area (CM)	AWG	Area (CM)	AWG	Area (CM)
0000 (4/0)	211,600	12	6,259.0	27	201.50
000 (3/0)	167,810	13	5,178.4	28	159.79
00 (2/0)	133,080	14	4,106.8	29	126.72
0 (1/0)	105,530	15	3,256.7	30	100.50
1	83,694	16	2,582.9	31	79.70
2	66,373	17	2,048.2	32	63.21
3	52,634	18	1,624.3	33	50.13
4	41,742	19	1,288.1	34	39.75
5	33,102	20	1,021.5	35	31.52
6	26,250	21	810.10	36	25.00
7	20,816	22	642.40	37	19.83
8	16,509	23	509.45	38	15.72
9	13,094	24	404.01	39	12.47
10	10,381	25	320.40	40	9.89
11	8,234.0	26	254.10		

2. Problem: What is the conductance of a 1.2 kΩ resistor?

Solution: By definition, conductance is equal to the reciprocal of resistance. Therefore

$$\begin{aligned}\text{conductance} &= 1 / \text{resistance}\\ &= 1 / (1.2\ \text{k}\Omega)\\ &= 1 / (1.2 \times 10^3\ \Omega)\\ &= 833 \times 10^{-3}\ \text{S}\\ &= 833\ \text{mS}\end{aligned}$$

The conductance of a 1.2 kΩ resistor is **833 mS.**

3. Problem: The marking bands on a resistor are green–blue–silver–gold. What are the value and tolerance of the resistor?

Solution: Because the resistor has four bands, the bands have the following meanings:

Band 1: First digit of resistor value

Band 2: Second digit of resistor value

Band 3: Power of 10 multiplier value

Band 4: Tolerance

The colors of the first three bands correspond to the resistor values in Table 2-3.

Table 2-3: Resistor Value Color Codes for 4-Band Markings

Color	Value	Color	Value	Color	Value
Black	0	Yellow	4	Grey	8
Brown	1	Green	5	White	9
Red	2	Blue	6		
Orange	3	Violet	7		

The green–blue colors of Band 1 and Band 2 give a resistor value of **56**.

The color of Band 3 corresponds to the power of 10 multiplier values in Table 2-4.

Table 2-4: Power of 10 Color Codes for 4-Band Markings

Color	Value	Color	Value	Color	Value
Black	0	Yellow	4	Grey	8
Brown	1	Green	5	White	9
Red	2	Blue	6	Silver	-2
Orange	3	Violet	7	Gold	-1

The silver color of Band 3 corresponds to the power of 10 value of −2. Therefore, the multiplier value is $\mathbf{10^{-2}} = \mathbf{0.01}$.

The color of Band 4 corresponds to the tolerance values in Table 2-5.

Table 2-5: Tolerance Color Codes for 4-Band Markings

Color	Tolerance
Silver	10%
Gold	5%

The gold color of Band 4 gives a tolerance of **5%**.

The value and tolerance of a resistor with green–blue–silver–gold marking bands are 56 Ω × 0.01 ± 5% = 0.56 Ω ± 5% = **560 mΩ ± 5%.**

4. Problem: The marking bands on a resistor are red–yellow–brown–red–red. What are the value and tolerance of the resistor?

Solution: Because the resistor has five bands, the bands have the following meanings:

Band 1: First digit of resistor value

Band 2: Second digit of resistor value

Band 3: Third digit of resistor value

Band 4: Power of 10 multiplier value

Band 5: Tolerance

The colors of the first three bands correspond to the resistor values in Table 2-6.

Table 2-6: Resistor Value Color Codes for 5-Band Markings

Color	Value	Color	Value	Color	Value
Black	0	Yellow	4	Grey	8
Brown	1	Green	5	White	9
Red	2	Blue	6		
Orange	3	Violet	7		

The red–yellow–brown colors of Band 1, Band 2, and Band 3 give a resistor value of **241**.

The color of Band 4 corresponds to the power of 10 multiplier values in Table 2-7.

Table 2-7: Power of 10 Color Codes for 5-Band Markings

Color	Value	Color	Value	Color	Value
Black	0	Yellow	4	Grey	8
Brown	1	Green	5	White	9
Red	2	Blue	6	Silver	-2
Orange	3	Violet	7	Gold	-1

The red color of Band 4 corresponds to the power of 10 value of 2. Therefore, the multiplier value is $\mathbf{10^2} = \mathbf{100}$.

The color of Band 5 corresponds to the tolerance values in Table 2-8.

Table 2-8: Tolerance Color Codes for 5-Band Markings

Color	Tolerance	Color	Tolerance
Red	2%	Blue	0.25%
Brown	1%	Violet	0.1%
Green	0.5%		

The red color of Band 5 gives a tolerance of **2%**.

The value and tolerance of a resistor with red–yellow–brown–red–red marking bands are 241 Ω × 100 ± 2% = 24100 Ω ± 2% = **24.1 kΩ ± 2%**.

5. Problem: The marking on a surface-mount chip resistor is 6M8. What is the value of the resistor?

   Solution: The base numeric value of the resistor is 68. The position of the alphabetic marking "M" indicates that the decimal point for the value is between the "6" and "8" and that the metric prefix for the value is "mega". Therefore, the value of the resistor is **6.8 MΩ**.

6. Problem: What are two possible markings for a 13 Ω surface-mount chip resistor?

   Solution: Although not shown, the decimal point for the resistor value 13 is after the "3" so that the alphabetic marking will be after the numeric marking "3". As there is no metric prefix for the value, the alphabetic marking will be "R". Therefore one possible marking for a 13 Ω surface-mount chip resistor is **13R**.

   An alternative marking for a 13 Ω for a surface-mount chip resistor uses only numbers. The first two digits will be the base resistor value, or "13". The third digit will be the value of 10 multiplier for the resistor. $13\ \Omega = 13 \times 1 = 13 \times 10^0$ so that the value of 10 multiplier for the value is 0. The alternative marking for a 13 Ω surface-mount chip resistor is then **130**.

2.1.6 The Electric Circuit

1. Problem: Refer to the circuit for a car headlight system in Figure 2-3.

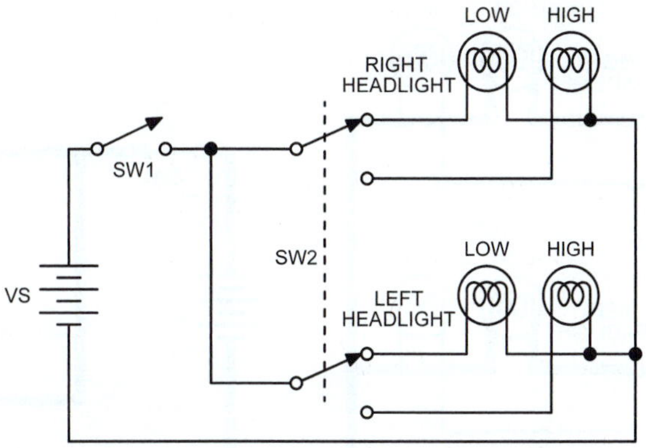

Figure 2-3: Example Electric Circuit

Determine the current path for each of the SW_1 and SW_2 switch settings shown in Table 2-9.

Table 2-9: Example Electric Circuit Switch Settings

SW_1 Setting	SW_2 Setting
Open	Up
Open	Down
Closed	Up
Closed	Down

Solution: For the cases in which switch SW_1 is open, refer to Figure 2-4.

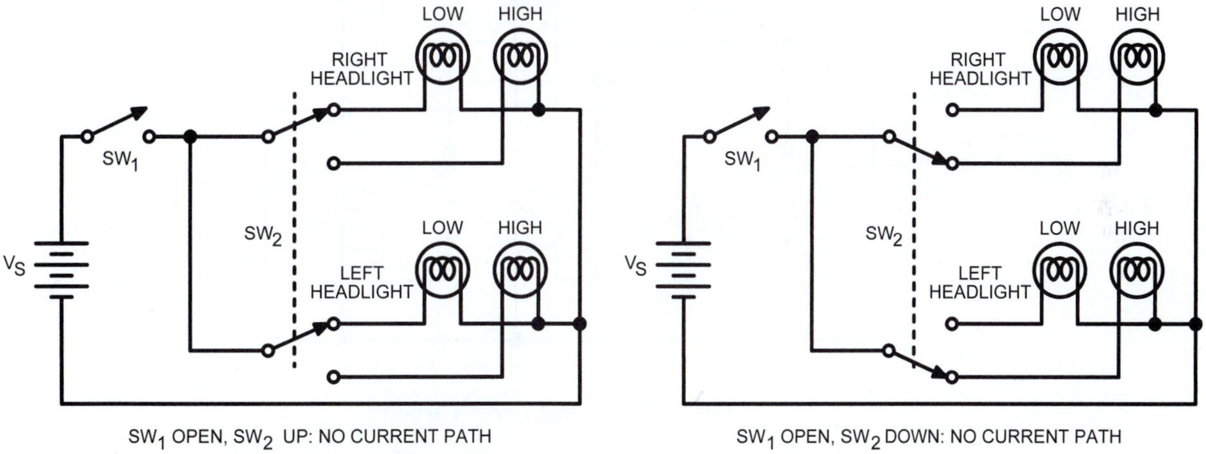

Figure 2-4: Current Paths for Switch *SW1* Open

When switch SW_1 is open, the setting for switch SW_2 does not matter, as there is no path for current from the voltage source V_S to flow.

For the cases in which switch SW_1 is closed, refer to Figure 2-5.

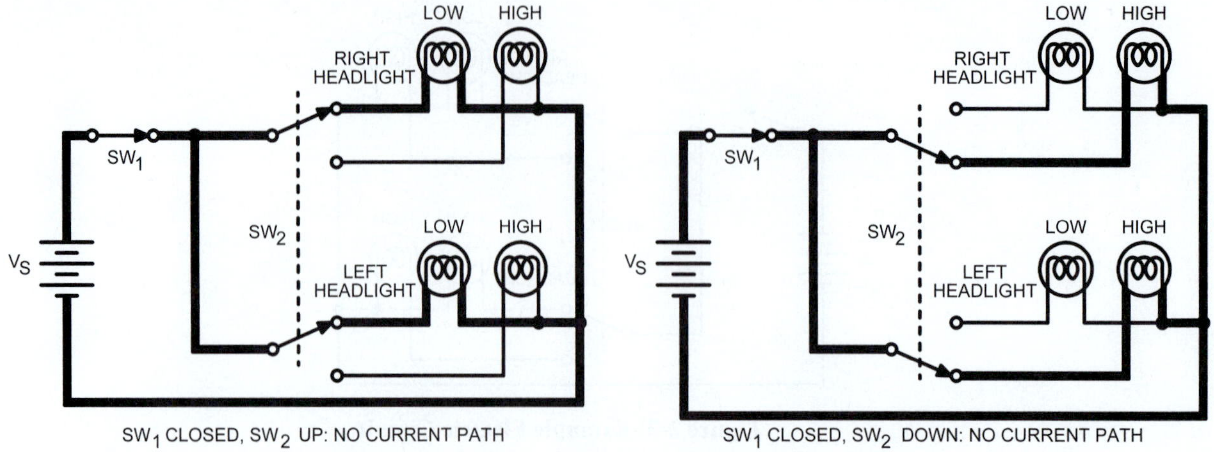

Figure 2-5: Current Paths for Switch *SW1* Closed

When switch SW_1 is closed and switch SW_2 is in the "Up" position, current flows from the voltage source V_S, through the closed switch SW_1, through SW_2 to the "LOW" bulbs of both the right and left headlights, and back to the voltage source.

When switch SW_1 is closed and switch SW_2 is in the "Down" position, current flows from the voltage source V_S, through the closed switch SW_1, through the "HIGH" bulbs of both the right and left headlights, and back to the voltage source.

2. Problem: Refer to the fused lamp circuit in Figure 2-6.

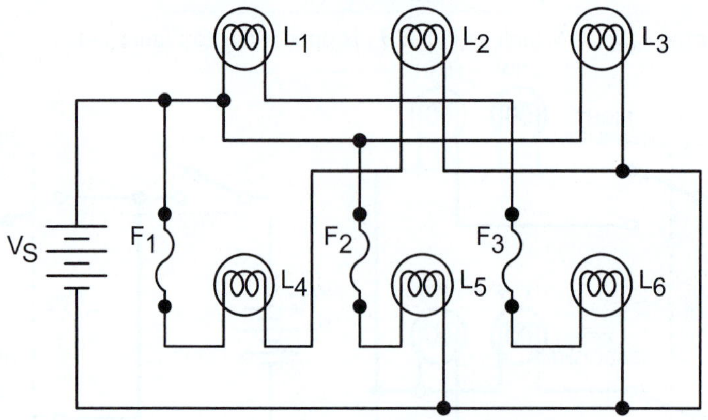

Figure 2-6: Fused Lamp Circuit

Determine which lamp(s) each of the fuses F_1, F_2, and F_3 protects. Will any lamp remain on if all three fuses open?

Solution: Refer to the fusing paths for each of the fuses shown in Figure 2-7.

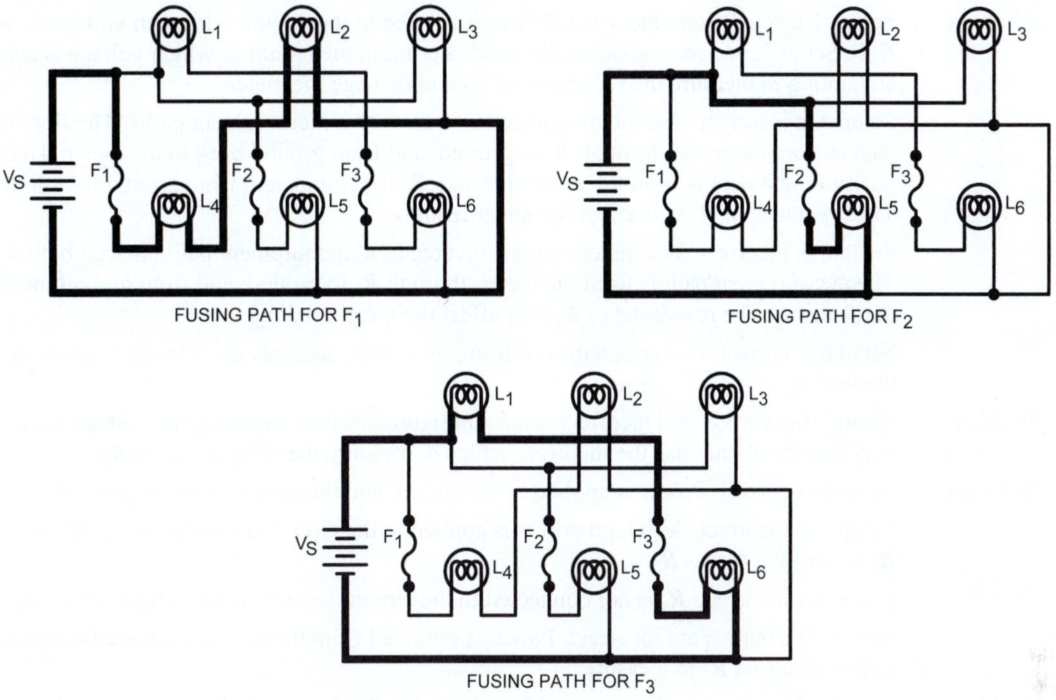

Figure 2-7: Fusing Paths for Lamp Circuit

As Figure 2-7 shows, fuse F_1 protects lamps L_2 and L_4, fuse F_2 protects lamp L_5, and fuse F_3 protects lamps L_1 and L_6. The only lamp that is not fused is lamp L_3, so L_3 will remain on if all three fuses open.

2.1.7 Basic DC Measurements

1. Problem: Identify the correct and incorrect setups in Figure 2-8 for measuring the resistance of R_2. Explain why you should not use the incorrect setups to measure the resistance of R_2.

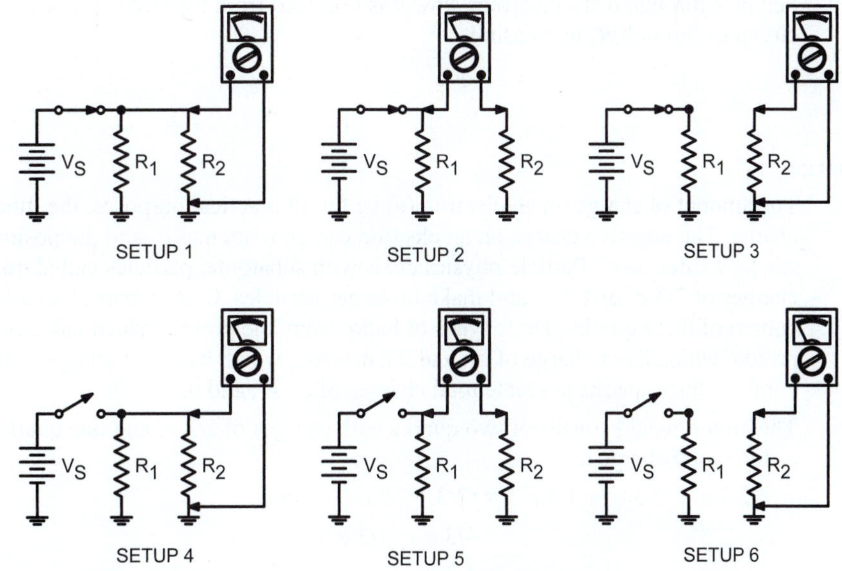

Figure 2-8: Meter Measurement Setups

Solution: Setups 1 through 3 are incorrect. Power is applied to the circuit. Although voltage is not applied to R_2 in Setup 3, you may accidentally touch a point in the circuit to which voltage is applied when attempting to measure the resistance of R_2 and damage the meter.

Setup 4 is incorrect. The meter connection creates two measurement paths. The first measurement path is from the meter, through R_1 to ground, and from ground back to the meter. The second measurement path is from the meter, through R_2 to ground, and from ground back to the meter. The resistance of R_1 will affect the meter reading.

Setup 5 is incorrect. The meter connection creates a measurement path through both R_1 and R_2. The measurement path is from the meter, through R_1 to ground, and from ground through R_2 back to the meter. The resistance of R_1 will affect the meter reading.

Setup 6 is correct. Power is removed from the circuit, and only R_2 is in the measurement path of the meter.

2. Problem: Identify the correct and incorrect setups in Figure 2-8 for measuring the voltage across R_2. Explain why you should not use the incorrect setups to measure the voltage across R_2.

Solution: Setup 1 is correct. Power is applied to the circuit, and the meter connects across R_2.

Setup 2 is incorrect. Although power is applied to the circuit, the meter connects between R_1 and R_2 rather than across R_2.

Setup 3 is incorrect. R_2 is not connected to the circuit, so there is no voltage across R_2 to measure.

Setups 4 through 6 are incorrect. Power is removed from the circuit, so there is no voltage in the circuit or across R_2 to measure.

3. Problem: Identify the correct and incorrect setups in Figure 2-8 for measuring the current through R_2. Explain why you should not use the incorrect setups to measure the current through R_2.

Solution: Setup 1 is incorrect. Power is applied to the circuit, but the meter creates a current path from the voltage source V_S and directly through the low resistance of the meter to ground. The high current may damage the meter.

Setup 2 is correct. Power is applied to the circuit. The meter creates a current path from the voltage source V_S, through the meter, and through resistor R_2 to ground. The current through the meter is the same as the current through R_2.

Setup 3 is incorrect. R_2 is not connected to the circuit, so there is no current through R_2 to measure.

Setups 4 through 6 are incorrect. Power is removed from the circuit, so there is no current in the circuit or through R_2 to measure.

2.2 Advanced

2.2.1 Electrical Charge

1. Problem: The amount of charge on an electron (e) is, for all practical purposes, the fundamental unit of charge. The negative charge on an electron can be written as e^- and the positive charge on a proton can be written as e^+. Particle physics deals with subatomic particles called **quarks**. Quarks have charges of $2/3\ e^+$ or $1/3\ e^-$ and make up larger particles. Certain particles called **hadrons** each consist of three quarks. Three types of hadrons are the proton, which has a charge of e^+, the anti-proton, which has a charge of e^-, and the neutron, which has a net charge of 0. Determine how to combine three quarks to create total charges of e^+, e^-, and 0.

Solution: The proton would consist of two quarks with charges of $2/3\ e^+$ and one quark with a charge of $1/3\ e^-$ for a total charge of

$$2/3\ e^+ + 2/3\ e^+ + 1/3\ e^- = (2/3 + 2/3)\ e^+ + 1/3\ e^-$$
$$= 4/3\ e^+ + 1/3\ e^-$$
$$= 3/3\ e^+$$

$= e^+$

The anti-proton would consist of three quarks with charges of 1/3 e^- for a total charge of

$1/3\ e^- + 1/3\ e^- + 1/3\ e^- = (1/3 + 1/3 + 1/3)\ e^-$

$= 3/3\ e^-$

$= e^-$

The neutron would consist of one quark with a charge of 2/3 e^+ and two quarks with charges of 1/3 e^- for a total charge of

$2/3\ e^+ + 1/3\ e^- + 1/3\ e^- = 2/3\ e^+ + (1/3 + 1/3)\ e^-$

$= 2/3\ e^+ + 2/3\ e^-$

$= 0$

2. Problem: Hydrogen (atomic symbol H) consists of 1 positively charged proton and 1 negatively charged electron and weighs approximately 1 gram per mole (1 g/mol). If 1 kilogram is equal to 2.20 pounds mass, how much negative charge would result from completely ionizing 1 pound of hydrogen?

Solution: From the problem statement, 1 kilogram equals 2.20 pounds.

2.20 lbm = 1 kg

2.20 lbm / 2.20 = 1 kg / 2.20

1 lbm = 0.454 kg

(Note that the above calculation uses "lbm" for "pounds mass" to distinguish it from "lbf", or "pounds force".)

Hydrogen weighs 1 gram per mole.

0.454 kg H $= 0.454 \times 10^3$ g H

= 454 g H

= (454 g H) / (1 g H / mol)

= (454 / 1) [g H / (g H / mol)]

= (454) [g H × (mol / g H)]

= 454 mol

Each mole is equal to 6.022×10^{24} hydrogen atoms.

454 mol $= (454 \text{ mol}) \times (6.022 \times 10^{24} \text{ H / mol})$

$= (454 \times 6.022 \times 10^{24})$ [mol × (H / mol)]

$= 2.73 \times 10^{24}$ hydrogen atoms

Each hydrogen atom releases one electron when ionized.

2.73×10^{24} H atoms $= (2.73 \times 10^{24}$ H atoms$) \times$ (1 electron / H atom)

$= (2.73 \times 10^{24} \times 1)$ [H atoms × (1 electron / H atom)]

$= 2.73 \times 10^{24}$ electrons

Each coulomb of charge contains 6.25×10^{18} electrons.

2.73×10^{24} electrons $= 2.73 \times 10^{24}\ e / (6.25 \times 10^{18}\ e / \text{C})$

$= (2.73 \times 10^{24} / 6.25 \times 10^{18})\ [e / (e / \text{C})]$

$= (2.73 \times 10^{24} / 6.25 \times 10^{18})\ [e \times (\text{C} / e)]$

$= 437 \times 10^6$ C

= 437 MC

Therefore, 1 pound of completely ionized hydrogen would produce **437 megacoulombs** of negative charge.

2.2.2 Voltage

1. Problem: A collection of charge has a potential difference of 5.0 volts. If the amount of charge decreases by 20%, and the total energy in the charge increases by 10%, what is the potential difference of the remaining charge?

Solution: Let $\boldsymbol{Q_I}$ be the initial collection of charge, $\boldsymbol{E_I}$ be the total energy of the initial charge, and V_F be the final potential difference of the remaining charge. By definition, $\boldsymbol{V = E / Q}$ so

$$5.0\text{ V} = E_I / Q_I$$

If the amount of charge decreases by 20% and total energy increases by 10%, then the final potential difference of the remaining charge is

$$\begin{aligned} V_F &= [E_I \times (1 + 10\%)] / [Q_I \times (1 - 20\%)] \\ &= [E_I \times (1 + 10/100)] / [Q_I \times (1 - 20/100)] \\ &= (E_I \times 1.1) / (Q_I \times 0.8) \\ &= (E_I / Q_I) \times (1.1 / 0.8) \\ &= (E_I / Q_I) \times 1.375 \end{aligned}$$

Since $E_I / Q_I = 5$ V,

$$\begin{aligned} V_F &= 5.0\text{ V} \times 1.375 \\ &= 6.875\text{ V} \end{aligned}$$

Therefore, the final potential difference of the remaining charge is **6.9 V**.

2. Problem: The breakdown voltage of air is about 8 kV/in, which means that the potential difference between two points one inch apart must be 4 kV for current (in the form of a spark) to flow. An average lightning bolt (a very large spark) transfers five coulombs of charge with 500 megajoules of energy. What is the average voltage and length in feet of an average lightning bolt?

Solution: By definition, $\boldsymbol{V = E / Q}$ so

$$\begin{aligned} V &= 500\text{ MJ} / 5\text{ C} \\ &= (500 \times 10^6\text{ J}) / (5\text{ C}) \\ &= (500 \times 10^6 / 5)\ (\text{J} / \text{C}) \\ &= 100 \times 10^6\text{ V} \\ &= 100\text{ MV} \end{aligned}$$

The approximate breakdown voltage of air is 8 kV/in, so a 100 MV lightning bolt can discharge over a distance $\boldsymbol{d}$ of

$$\begin{aligned} d &= (100\text{ MV}) / (8\text{ kV} / \text{in}) \\ &= (100 \times 10^6\text{ V}) / (8 \times 10^3\text{ V} / \text{in}) \\ &= (100 \times 10^6 / 8 \times 10^3)\ [\text{V} / (\text{V} / \text{in})] \\ &= 12.5 \times 10^3\ [\text{V} \times (\text{in} / \text{V})] \\ &= 12.5 \times 10^3\text{ in} \end{aligned}$$

One foot equals twelve inches, so

$$\begin{aligned} d &= (12.5 \times 10^3\text{ in}) \times (1\text{ ft} / 12\text{ in}) \\ &= [(12.5 \times 10^3 \times 1) / 12]\ [\text{in} \times (\text{ft} / \text{in})] \\ &= 1.04 \times 10^3\text{ ft} \end{aligned}$$

Therefore, an average lightning bolt has a potential of **100 MV** and length of about **1000 feet**.

2.2.3 Current

1. Problem: Table 2-10 shows the recorded charge data for one period of a sine wave. The positive values for the first half of the sine wave indicate that charge is moving in one direction, and the negative values for the second half sine wave indicate that charge is moving in the opposite direction.

Table 2-10: Recorded Sine Wave Charge Data

Interval	Charge	Interval	Charge
2.500 s to 2.625 s	+76.1 mC	3.500 s to 3.625 s	−76.1 mC
2.626 s to 2.750 s	+217 mC	3.626 s to 3.750 s	−217 mC
2.751 s to 2.875s	+324 mC	3.751 s to 3.875s	−324 mC
2.876 s to 3.000 s	+383 mC	3.876 s to 4.000 s	−383 mC
3.001 s to 3.125 s	+383 mC	4.001 s to 4.125 s	−383 mC
3.126 s to 3.250 s	+324 mC	4.126 s to 4.250 s	−324 mC
3.251 s to 3.375 s	+217 mC	4.251 s to 4.375 s	−217 mC
3.376 s to 3.500 s	+76.1 mC	4.376 s to 4.500 s	−76.1 mC

What is the average current for the first half of the sine wave? What is the average current for the full period of the sine wave?

Solution: By definition, current equals charge divided by time. The average current will equal the total charge divided by the total elapsed time for the recording intervals.

The total current $\boldsymbol{Q_T}$ for the first half of the sine wave is

$$\begin{aligned}\boldsymbol{Q_T} &= (+76.1\text{ mC}) + (+217\text{ mC}) + (+324\text{ mC}) + (+383\text{ mC}) + (+383\text{ mC}) + (+324\text{ mC}) + \\ &\quad (+217\text{ mC}) + (+76.1\text{ mC}) \\ &= [(+76.1) + (+217) + (+324) + (+383) + (+383) + (+324) + (+217) + (+76.1)]\text{ mC} \\ &= +2000 \times 10^{-3}\text{ C} \\ &= +2.00\text{ C}\end{aligned}$$

The elapsed time $\boldsymbol{t}$ for the first half of the sine wave is the difference between the end time and start time for the first half of the period.

$$\begin{aligned}\boldsymbol{t} &= 3.500\text{ s} - 2.500\text{ s} \\ &= (3.500 - 2.500)\text{ s} \\ &= 1.000\text{ s}\end{aligned}$$

The average current is the total charge divided by the elapsed time.

$$\begin{aligned}\text{average current} &= \boldsymbol{Q_T} / \boldsymbol{t} \\ &= 2.00\text{ C} / 1.000\text{ s} \\ &= (2.00 / 1.000)\ (\text{C} / \text{s}) \\ &= 2.00\text{ A}\end{aligned}$$

The average current for the first half period of the sine wave is **2.00 A**.

The total current $\boldsymbol{Q_T}$ for the full period of the sine wave is

$$\begin{aligned}\boldsymbol{Q_T} &= (+76.1\text{ mC}) + (+217\text{ mC}) + (+324\text{ mC}) + (+383\text{ mC}) + (+383\text{ mC}) + (+324\text{ mC}) + \\ &\quad (+217\text{ mC}) + (+76.1\text{ mC}) + (-76.1\text{ mC}) + (-217\text{ mC}) + (-324\text{ mC}) + (-383\text{ mC}) + \\ &\quad (-383\text{ mC}) + (-324\text{ mC}) + (-217\text{ mC}) + (-76.1\text{ mC}) \\ &= [(+76.1) + (+217) + (+324) + (+383) + (+383) + (+324) + (+217) + (+76.1) + (-76.1) + \\ &\quad (-217) + (-324) + (-383) + (-383) + (-324) + (-217) + (-76.1)]\text{ mC} \\ &= +0.00\text{ mC} \\ &= +0.00\text{ C}\end{aligned}$$

The elapsed time $\boldsymbol{t}$ for the full period of the sine wave is the difference between the end time and start time for the full period.

$$\begin{aligned}\boldsymbol{t} &= 4.500\text{ s} - 2.500\text{ s} \\ &= (4.500 - 2.500)\text{ s} \\ &= 2.000\text{ s}\end{aligned}$$

The average current is the total charge divided by the elapsed time.

average current $= \boldsymbol{Q_T / t}$
$= 0.00 \text{ C} / 2.000 \text{ s}$
$= (0.00 / 2.000) \text{ (C / s)}$
$= 0.00 \text{ A}$

Note from this that, although charge is moving (and that current is flowing) for the full period of the sine wave, the amount of charge moving in one direction for the first half of the sine wave is equal to the same amount of charge moving in the opposite direction. As a result, the *net* movement of charge (and therefore current) is zero.

2. Problem: The rest mass of an electron (m_e) is 9.11×10^{-28} grams. Eccentric inventor Dr. Lou D'Cruss decides to invent a machine that will allow dieters to lose weight by extracting electrons from their bodies. If one pound is equal to 0.454 kilograms, how much continuous current must flow for someone to lose one pound of electron mass in one week?

Solution: From the problem statement electron mass is

$m_e = 9.11 \times 10^{-28}$ g / electron

The number of electrons $\boldsymbol{N}$ in one pound is

$\boldsymbol{N} = (1 \text{ lb}) / (9.11 \times 10^{-28} \text{ g / electron})$
$= (1 / 9.11 \times 10^{-28}) \text{ [lb / (g / electron)]}$
$= (1.10 \times 10^{27}) \text{ [(lb} \times \text{electron) / g]}$

One pound of mass is equal to 0.454 kilograms.

$\boldsymbol{N} = (1.10 \times 10^{27}) \text{ [(lb} \times \text{electron) / g]} \times (0.454 \text{ kg / 1 lb})$
$= (1.10 \times 10^{27}) \text{ [(lb} \times \text{electron) / g]} \times (0.454 \times 10^3 \text{ g / 1 lb})$
$= \{[(1.10 \times 10^{27}) \times (0.454 \times 10^3)] / 1\} \text{ [(lb} \times \text{electron} \times \text{g) / (g} \times \text{lb)]}$
$= 498 \times 10^{27}$ electrons

Each electron carries one fundamental charge (e), so the total charge $\boldsymbol{Q_T}$ in one pound of electrons is the number of electrons times the charge per electron.

$\boldsymbol{Q_T} = (498 \times 10^{27} \text{ electrons}) \times (1\ e \text{ / electron})$
$= (498 \times 10^{27} \times 1) \text{ [(electrons} \times e) \text{ / electron]}$
$= 498 \times 10^{27}\ e$

One coulomb is equal to 6.25×10^{18} fundamental charges.

$\boldsymbol{Q_T} = (498 \times 10^{27}\ e) \times (1 \text{ C} / 6.25 \times 10^{18}\ e)$
$= [(498 \times 10^{27} \times 1) / 6.25 \times 10^{18}] \text{ [}(e \times \text{C}) / e]$
$= 79.7 \times 10^9 \text{ C}$

By definition, current is equal to charge divided by time. If the electrons are transferred over a period of one week, the total current $\boldsymbol{I_T}$ is the total charge divided by one week.

$\boldsymbol{I_T} = \boldsymbol{Q_T / t}$
$= 79.7 \times 10^9 \text{ C / 1 week}$
$= (79.7 \times 10^9 / 1) \text{ (C / week)}$
$= 79.7 \times 10^9 \text{ C / week}$

One week equals seven days.

$\boldsymbol{I_T} = (79.7 \times 10^9 \text{ C / week}) \times (1 \text{ week / 7 days})$
$= [(79.7 \times 10^9 \times 1) / 7] \text{ [(C} \times \text{week) / (week} \times \text{days)]}$
$= 11.4 \times 10^9 \text{ C / day}$

One day equals twenty-four hours.

I_T = (11.4 × 10^9 C / day) × (1 day / 24 hr)
= [(11.4 × 10^9 × 1) / 24] [(C × day) / (day × hr)]
= 475 × 10^6 C / hr

One hour equals sixty minutes.

I_T = (475 × 10^6 C / hr) × (1 hr / 60 min)
= [(475 × 10^6 × 1) / 60] (C × hr) / (hr × min)
= 7.91 × 10^6 C / min

One minute equals sixty seconds.

I_T = (7.91 × 10^6 C / min) × (1 min / 60 s)
= [(7.91 × 10^6 × 1) / 60] [(C × min) / (min × s)
= 132 × 10^3 C / s
= 132 × 10^3 A
= 132 kA

Therefore, people who wish to lose one pound of electrons in one week must subject themselves to a constant current of **132 kA**. Just for reference, electrons make up such a small part of an atom's mass that a 200-pound person contains less than one ounce of electrons.

2.2.4 Resistance

1. Problem: A resistor manufacturer decides to cut the costs of a line of wirewound resistor by changing the winding wire from copper (ρ_{Cu} = 10.35 CM•Ω / ft) to aluminum (ρ_{Al} = 16.96 CM•Ω / ft). To minimize the costs of retooling the factory, the length of wire wound around each resistor will be the same so that the wire-cutting and core fabrication machinery will be the same. What must the cross-sectional area of the aluminum wire be compared to the copper wire?

Solution: The resistor values for both the copper and aluminum wirewound resistors will be the same, so

$R_{Al} = R_{Cu}$

Begin with the resistance equation for R.

$R = (\rho \times l) / A$

Substitute the specific values for each element into the resistance equation.

$(\rho_{Cu} \times l_{Cu}) / A_{Cu} = (\rho_{Al} \times l_{Al}) / A_{Al}$
$A_{Al} \times [(\rho_{Cu} \times l_{Cu}) / A_{Cu}] = A_{Al} \times [(\rho_{Al} \times l_{Al}) / A_{Al}]$
$(A_{Al} \times \rho_{Cu} \times l_{Cu}) / A_{Cu} = \rho_{Al} \times l_{Al}$
$[(A_{Al} \times \rho_{Cu} \times l_{Cu}) / A_{Cu}] \times A_{Cu} = (\rho_{Al} \times l_{Al}) \times A_{Cu}$
$A_{Al} \times \rho_{Cu} \times l_{Cu} = \rho_{Al} \times l_{Al} \times A_{Cu}$
$(A_{Al} \times \rho_{Cu} \times l_{Cu}) / l_{Cu} = (\rho_{Al} \times l_{Al} \times A_{Cu}) / l_{Cu}$
$A_{Al} \times \rho_{Cu} = (\rho_{Al} \times l_{Al} \times A_{Cu}) / l_{Cu}$
$(A_{Al} \times \rho_{Cu}) / \rho_{Cu} = [(\rho_{Al} \times l_{Al} \times A_{Cu}) / l_{Cu}] / \rho_{Cu}$
$A_{Al} = (\rho_{Al} \times l_{Al} \times A_{Cu}) / (l_{Cu} \times \rho_{Cu})$

From the problem statement, $l_{Al} = l_{Cu}$. Therefore,

$A_{Al} = (\rho_{Al} \times l_{Al} \times A_{Cu}) / (l_{Cu} \times \rho_{Cu})$
$= (\rho_{Al} \times l_{Cu} \times A_{Cu}) / (l_{Cu} \times \rho_{Cu})$
$= (\rho_{Al} \times A_{Cu}) / \rho_{Cu}$
$= (\rho_{Al} / \rho_{Cu}) \times A_{Cu}$
= [(16.96 CM•Ω / ft) / (10.35 CM•Ω / ft)] × A_{Cu}
= 1.639 × A_{Cu}

Therefore, the cross-sectional area of the aluminum winding wire must be **1.639 times** the area of the copper winding wire for the resistors to have the same resistance.

2. Problem: The manufacturer in Problem 1 discovers that his machinery cannot tin the aluminum leads of his new wirewound resistors so that they will solder properly. A cost analysis shows that it is cheaper to purchase and solder standard-sized pre-tinned leads to the resistor bodies than to purchase new equipment to tin the aluminum leads. Each lead is AWG 22, 1.00 inch long, and composed of pure tin ($\rho_{Sn} = 65.57$ CM•Ω / ft). How much extra resistance will the leads add to each resistor, compared to aluminum leads ($\rho_{Al} = 16.96$ CM•Ω / ft) of the same dimensions?

Solution: From Table 2-2 the cross-sectional area of an AWG 22 lead is 642.40 CM. Begin with the resistance equation for ***R***.

$$\boldsymbol{R} = (\boldsymbol{\rho} \times \boldsymbol{l}) / \boldsymbol{A}$$

Calculate the resistance for each aluminum lead.

$$\begin{aligned}
\boldsymbol{R_{Al}} &= (\boldsymbol{\rho_{Al}} \times \boldsymbol{l_{Al}}) / \boldsymbol{A_{Al}} \\
&= [(16.96 \text{ CM}\bullet\Omega / \text{ft}) \times (1.00 \text{ in})] / 642.40 \text{ CM} \\
&= [(16.96 \times 1.00) / 642.40]\ [(\text{CM} \times \Omega \times \text{in}) / (\text{CM} \times \text{ft})] \\
&= (26.4 \times 10^{-3})\ [(\Omega \times \text{in}) / \text{ft}]
\end{aligned}$$

There are twelve inches in one foot.

$$\begin{aligned}
\boldsymbol{R_{Al}} &= (26.4 \times 10^{-3})\ \{[(\Omega \times \text{in}) / \text{ft}] \times (1 \text{ ft} / 12 \text{ in})\} \\
&= [(26.4 \times 10^{-3} \times 1) / 12]\ [(\Omega \times \text{in} \times \text{ft}) / (\text{ft} \times \text{in})] \\
&= 2.20 \times 10^{-3}\ \Omega \\
&= 2.20 \text{ m}\Omega
\end{aligned}$$

Next, calculate the resistance of each tin lead.

$$\begin{aligned}
\boldsymbol{R_{Sn}} &= (\boldsymbol{\rho_{Sn}} \times \boldsymbol{l_{Sn}}) / \boldsymbol{A_{Sn}} \\
&= [(65.57 \text{ CM}\bullet\Omega / \text{ft}) \times (1.00 \text{ in})] / 642.40 \text{ CM} \\
&= [(65.57 \times 1.00) / 642.40]\ \{[(\text{CM} \times \Omega / \text{ft}) \times \text{in}] / \text{CM}\} \\
&= (102 \times 10^{-3})\ [(\Omega \times \text{in}) / \text{ft}]
\end{aligned}$$

There are twelve inches in one foot.

$$\begin{aligned}
\boldsymbol{R_{Sn}} &= (102 \times 10^{-3})\ \{[(\Omega \times \text{in}) / \text{ft}] \times (1 \text{ ft} / 12 \text{ in})\} \\
&= [(102 \times 10^{-3} \times 1) / 12]\ [(\Omega \times \text{in} \times \text{ft}) / (\text{ft} \times \text{in})] \\
&= 8.51 \times 10^{-3}\ \Omega \\
&= 8.51 \text{ m}\Omega
\end{aligned}$$

Determine the difference in resistance for each lead by finding the difference in resistance to tin and aluminum.

$$\begin{aligned}
\boldsymbol{R_{DIFF}} &= \boldsymbol{R_{Sn}} - \boldsymbol{R_{Al}} \\
&= 8.51 \text{ m}\Omega - 2.20 \text{ m}\Omega \\
&= (8.51 - 2.20) \text{ m}\Omega \\
&= 6.31 \text{ m}\Omega
\end{aligned}$$

Each resistor has two leads, so the added resistance for each resistor is twice the resistance difference per lead.

$$\begin{aligned}
\boldsymbol{R_{TOTAL}} &= 2 \times \boldsymbol{R_{DIFF}} \\
&= 2 \times 6.31 \text{ m}\Omega \\
&= (2 \times 6.31) \text{ m}\Omega \\
&= 12.6 \text{ m}\Omega
\end{aligned}$$

Therefore, using pure tin leads in place of aluminum leads will add **12.6 mΩ** to each resistor.

2.2.5 The Electric Circuit

1. Problem: The cross-sectional areas of American wire gauges are often specified in circular mils. What is the area of one circular mil (CM) in square mils?

Solution: By definition, one circular mil is the area of a circle with a diameter of one mil. The area of a circle in general is

$$A = \pi \times r^2$$

The radius r is half the diameter d.

$$A = \pi \times (d / 2)^2$$
$$= (\pi \times d^2) / 4$$

The diameter of one CM is equal to 1 mil.

$$1 \text{ CM} = [\pi \times (1 \text{ mil})^2] / 4$$
$$= [(\pi \times 1) / 4] \text{ mil}^2$$
$$= [(3.14 \times 1) / 4] (\text{mil}^2)$$
$$= 0.785 \text{ mil}^2$$

Therefore, the area of one circular mil in square mils is **0.785 mil^2**.

2. Problem: One switch that you may encounter in digital circuits is the crosspoint switch. As its name suggests, a crosspoint switch provides an electrical path between rows and columns of conductive wires or traces by connecting the point where the row and column conductors cross. A common application for the crosspoint switch is a keypad, in which each key will connect a specific row to a specific column.

Figure 2-9 illustrates how the crosspoint switch for a hexadecimal numeric keypad operates. A keypad controller applies a voltage to Column 0, scans Row 0 through Row 3 for the applied voltage, and repeats the process for Columns 1 through 3. If no key is pressed, there is no path between the columns and rows, so the row scan will detect no voltage on any row. If a key is pressed, the pressed key will provide a path between one column and one row. When the keypad controller applies a voltage to that column and scans the rows, it will detect the voltage on one of the rows. If key "9" is pressed, for example, the keypad controller will detect the voltage applied to Column 2 on Row 1. Each key will uniquely connect one column to one row. The keypad controller uses the unique row and column combination to determine which key is pressed.

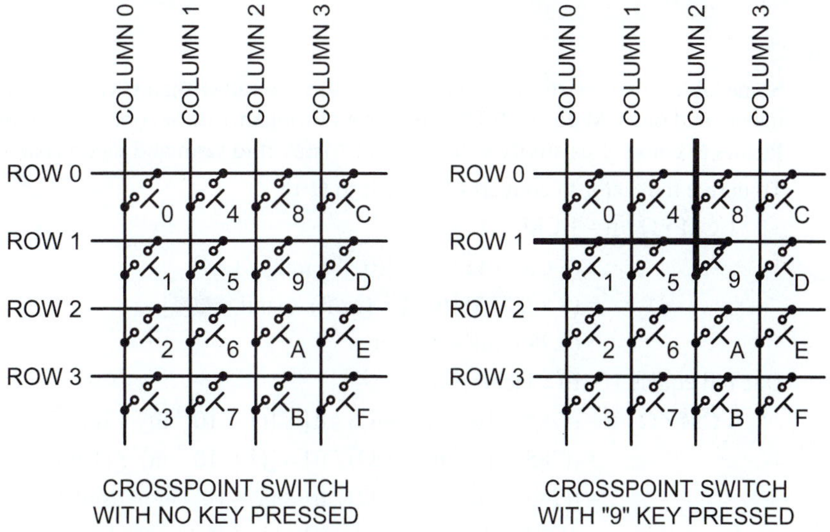

Figure 2-9: Crosspoint Switch Operation

Which column and row combination corresponds to each key for the crosspoint switch in Figure 2-9?

Solution: Table 2-11 shows the column and row combination that corresponds to each key. To verify that the table values are correct, verify that each key provides a path between its indicated column and row.

Table 2-11: Hexadecimal Numeric Keypad Column and Row Combinations

Key	Column	Row	Key	Column	Row
0	0	0	8	2	0
1	0	1	9	2	1
2	0	2	A	2	2
3	0	3	B	2	3
4	1	1	C	3	0
5	1	2	D	3	1
6	1	3	E	3	2
7	1	4	F	3	3

2.2.6 Basic DC Measurements

1. Problem: You are making DC voltage measurements of a circuit with an analog DC voltmeter. What range should you first use for each measurement, and why?

   Solution: Because measurements that exceed the selected range of an analog meter can "peg" the meter and potentially damage the meter movement, you should select the highest range for each initial measurement. Based on the reading, you can then decrease the range one setting at a time as needed until the meter reading is near the center of the meter scale.

2. Problem: What are the first two steps to follow when making circuit resistance measurements and why?

   Solution: Two steps to follow when making circuit resistance measurements are

   - Remove power from the circuit to prevent circuit current from damaging the meter.
   - Disconnect at least one end of the resistor from the circuit to ensure that the meter measures only the resistance of the component of interest.

2.3 Just for Fun

1. Problem: Some tables give resistivity values in units of Ω•m rather than CM•Ω / ft. If one foot equals 0.304 meters and one CM equals 0.785 mil^2 (from Problem 1 of Section 2.2.5 above), derive conversion factors to convert resistivity values from CM•Ω / ft to Ω•m and vice versa.

   Solution: To derive the factor to convert CM•Ω / ft to Ω•m

$$1\text{ CM}\cdot\Omega/\text{ft} = 1\text{ CM}\cdot\Omega/\text{ft}$$
$$= (1\text{ CM}\times\Omega/\text{ft})\times[(0.785\text{ mil}^2)/\text{CM}]$$
$$= (1\times 0.785)\,[(\text{CM}\times\Omega/\text{ft})\times(\text{mil}^2/\text{CM}]$$
$$= (0.785\text{ mil}^2\times\Omega)/\text{ft}$$

One mil equals 1×10^{-3} inches, so

$$1\text{ CM}\cdot\Omega/\text{ft} = [(785\times10^{-3}\text{ mil}^2\times\Omega)/\text{ft}]\times[(1\times10^{-3}\text{ in}/1\text{ mil})^2]$$
$$= [(785\times10^{-3}\text{ mil}^2\times\Omega)/\text{ft}]\times[(1\times10^{-3}\text{ in})^2/(1\text{ mil})^2]$$
$$= [(785\times10^{-3}\text{ mil}^2\times\Omega)/\text{ft}]\times(1\times10^{-6}\text{ in}^2/1\text{ mil}^2)$$
$$= \{(785\times10^{-3})\times(1\times10^{-6})]/1\}\;\{[(\text{mil}^2\times\Omega)/\text{ft}]\times(\text{in}^2/\text{mil}^2)]\}$$
$$= (785\times10^{-9}\text{ in}^2\times\Omega)/\text{ft}$$

One foot equals twelve inches so

$$1 \text{ CM} \cdot \Omega / \text{ft} = [(785 \times 10^{-9} \text{ in}^2 \times \Omega) / \text{ft}] \times [(1 \text{ ft} / 12 \text{ in})^2]$$
$$= [(785 \times 10^{-9} \text{ in}^2 \times \Omega) / \text{ft}] \times [(1 \text{ ft})^2 / (12 \text{ in})^2]$$
$$= [(785 \times 10^{-9} \text{ in}^2 \times \Omega) / \text{ft}] \times (1 \text{ ft}^2 / (144 \text{ in}^2)$$
$$= [(785 \times 10^{-9} \times 1) / 144] \{[(\text{in}^2 \times \Omega) / \text{ft}] \times (\text{ft}^2 / \text{in}^2)\}$$
$$= (785 \times 10^{-9} / 144) [(\text{ft}^2 \times \Omega) / \text{ft}]$$
$$= (785 \times 10^{-9} / 144) [(\text{ft} \times \text{ft} \times \Omega) / \text{ft}]$$
$$= 5.45 \times 10^{-9} \text{ ft} \times \Omega$$

One foot equal 0.304 meters so

$$1 \text{ CM} \cdot \Omega / \text{ft} = (5.45 \times 10^{-9} \text{ ft} \times \Omega) \times (0.304 \text{ m} / \text{ft})$$
$$= (5.45 \times 10^{-9} \times 0.304) [(\text{ft} \times \Omega) \times (\text{m} / \text{ft})]$$
$$= 1.66 \times 10^{-9} \, \Omega \times \text{m}$$

From this, 1 CM•Ω / ft = 1.66×10^{-9} Ω•m. Since 1.66×10^{-9} Ω•m = 1 CM•Ω / ft,

$$1.66 \times 10^{-9} \, \Omega \times \text{m} = (1 \text{ CM} \times \Omega) / \text{ft}$$
$$(1.66 \times 10^{-9} \, \Omega \times \text{m}) / (1.66 \times 10^{-9}) = (1 \text{ CM} \times \Omega / \text{ft}) / (1.66 \times 10^{-9})$$
$$(1.66 \times 10^{-9} / 1.66 \times 10^{-9}) (\Omega \times \text{m}) = (1 / 1.66 \times 10^{-9}) [(\text{CM} \times \Omega) / \text{ft}]$$
$$1 \, \Omega \times \text{m} = 602 \times 10^{6} \text{ CM} \times \Omega / \text{ft}$$

From this, 1 Ω•m = 602×10^6 CM•Ω / ft.

Therefore, **1 CM•Ω / ft = 1.66×10^{-9} Ω•m** and **1 Ω•m = 602×10^6 CM•Ω / ft**.

2. Problem: Reference tables are useful but can be inconvenient if you cannot find the table or a table does not cover the complete range of values you require. In some cases a single equation or function (such as the built-in functions or user-defined programs on a modern calculator) can provide an alternative to a reference table. One reference table that you can replace with a function is the American Wire Gauge table. The cross-sectional areas defined by the American Wire Gauge values form a geometric progression. In other words, the ratio of any two adjacent values in the table is always the same. For example, the ratio of the cross-sectional areas given by AWG 11 / AWG 10 is the same as the ratio of the cross sectional areas given by AWG 30 / AWG 29. Derive a formula that gives the cross-sectional area as a function of the AWG number (e.g., AWG 6 = 6). For simplicity, let the gauges below AWG 0 correspond to negative numbers (i.e., AWG 00 = −1, AWG 000 = −2, etc.).

Solution: Let $\mathbf{A_N}$ be the area in circular mils for value AWG N in the table and **k** be the constant ratio between two adjacent values in the AWG table. As an example, $\mathbf{A_7}$ is the area in circular mils for AWG 7, and $\mathbf{k} = \mathbf{A_8} / \mathbf{A_7}$ (as well as $\mathbf{A_9} / \mathbf{A_8}$, $\mathbf{A_{10}} / \mathbf{A_9}$, etc.). For convenience, let AWG 0 be the reference gauge value from which you will calculate the cross-sectional areas of all other gauge values.

By definition, the cross-sectional area of AWG 0 is $\mathbf{A_0}$. For AWG 0

$$\mathbf{A_0} = \mathbf{A_0}$$
$$= 1 \times \mathbf{A_0}$$

Since any number raised to 0 equals 1, $k^0 = 1$ and so

$$\mathbf{A_0} = \mathbf{k}^0 \times \mathbf{A_0}$$

Therefore, $\mathbf{A_N} = \mathbf{k}^N \times \mathbf{A_0}$ for N = 0.

For values above reference gauge AWG 0,

$$\mathbf{k} = \mathbf{A}_{X+1} / \mathbf{A}_X$$
$$\mathbf{k} \times \mathbf{A}_X = (\mathbf{A}_{X+1} / \mathbf{A}_X) \times \mathbf{A}_X$$
$$\mathbf{k} \times \mathbf{A}_X = \mathbf{A}_{X+1}$$

Therefore, the cross-sectional area of AWG 1, the gauge one value higher than AWG 0, is

$$\mathbf{A_1} = \mathbf{k} \times \mathbf{A_0}$$

$= \mathbf{k}^1 \times \mathbf{A}_0$

Similarly, $\mathbf{A}_2 / \mathbf{A}_1 = \mathbf{k} \rightarrow \mathbf{A}_2 = \mathbf{k} \times \mathbf{A}_1$, so the cross-sectional area of AWG 2, the gauge two values higher than AWG 0, is

$\mathbf{A}_2 \quad = \mathbf{k} \times \mathbf{A}_1$

Since $\mathbf{A}_1 = \mathbf{k} \times \mathbf{A}_0$, then

$$\begin{aligned}\mathbf{A}_2 &= \mathbf{k} \times \mathbf{A}_1 \\ &= \mathbf{k} \times \mathbf{k} \times \mathbf{A}_0 \\ &= \mathbf{k}^2 \times \mathbf{A}_0\end{aligned}$$

Again, $\mathbf{A}_3 / \mathbf{A}_2 = \mathbf{k} \rightarrow \mathbf{A}_3 = \mathbf{k} \times \mathbf{A}_2$, so the cross-sectional area of AWG 3, the gauge three values higher than AWG 0, is

$\mathbf{A}_3 \quad = \mathbf{k} \times \mathbf{A}_2$

Since $\mathbf{A}_2 = \mathbf{k} \times \mathbf{A}_1$, then

$$\begin{aligned}\mathbf{A}_3 &= \mathbf{k} \times \mathbf{A}_2 \\ &= \mathbf{k} \times \mathbf{k} \times \mathbf{A}_1 \\ &= \mathbf{k} \times \mathbf{k} \times \mathbf{k} \times \mathbf{A}_0 \\ &= \mathbf{k}^3 \times \mathbf{A}_0\end{aligned}$$

Repeating this process shows in general that the cross-sectional area for a gauge N values higher than reference gauge AWG 0 will be

$\mathbf{A}_N \quad = \mathbf{k}^N \times \mathbf{A}_0$

Therefore, $\mathbf{A}_N = \mathbf{k}^N \times \mathbf{A}_0$ for $N > 0$.

For gauges below reference gauge AWG 0

$$\begin{aligned}\mathbf{k} &= \mathbf{A}_X / \mathbf{A}_{X-1} \\ 1 / \mathbf{k} &= 1 / (\mathbf{A}_X / \mathbf{A}_{X-1}) \\ &= (\mathbf{A}_{X-1} / \mathbf{A}_X) \times \mathbf{A}_X \\ (1 / \mathbf{k}) \times \mathbf{A}_X &= \mathbf{A}_{X-1} \\ \mathbf{k}^{-1} \times \mathbf{A}_X &= \mathbf{A}_{X-1}\end{aligned}$$

Therefore, the cross-sectional area of AWG 00 = AWG −1, the gauge one value lower than AWG 0, is

$\mathbf{A}_{-1} \quad = \mathbf{k}^{-1} \times \mathbf{A}_0$

Similarly, $\mathbf{A}_{-1} / \mathbf{A}_{-2} = \mathbf{k} \rightarrow \mathbf{A}_{-2} = \mathbf{k}^{-1} \times \mathbf{A}_{-1}$, so the cross-sectional area of AWG −2, the gauge two values lower than AWG 0, is

$\mathbf{A}_{-2} \quad = \mathbf{k}^{-1} \times \mathbf{A}_{-1}$

Since $\mathbf{A}_{-1} = \mathbf{k}^{-1} \times \mathbf{A}_0$, then

$$\begin{aligned}\mathbf{A}_{-2} &= \mathbf{k}^{-1} \times \mathbf{A}_{-1} \\ &= \mathbf{k}^{-1} \times \mathbf{k}^{-1} \times \mathbf{A}_0 \\ &= \mathbf{k}^{-2} \times \mathbf{A}_0\end{aligned}$$

Again, $\mathbf{A}_{-2} / \mathbf{A}_{-3} = \mathbf{k} \rightarrow \mathbf{A}_{-3} = \mathbf{k}^{-1} \times \mathbf{A}_{-2}$, so the cross-sectional area of AWG –3, the gauge three values lower than AWG 0, is

$\mathbf{A}_{-3} \quad = \mathbf{k}^{-1} \times \mathbf{A}_{-2}$

Since $\mathbf{A}_{-2} = \mathbf{k} \times \mathbf{A}_{-1}$, then

$$\begin{aligned}\mathbf{A}_{-3} &= \mathbf{k}^{-1} \times \mathbf{A}_{-2} \\ &= \mathbf{k}^{-1} \times \mathbf{k}^{-1} \times \mathbf{A}_{-1} \\ &= \mathbf{k}^{-1} \times \mathbf{k}^{-1} \times \mathbf{k}^{-1} \times \mathbf{A}_0\end{aligned}$$

$= \mathbf{k}^{-3} \times \mathbf{A}_0$

Repeating this process shows in general that the cross-sectional area for a gauge N values lower than reference gauge AWG 0 will be

$$\mathbf{A}_N = \mathbf{k}^N \times \mathbf{A}_0$$

Therefore $\mathbf{A}_N = \mathbf{k}^N \times \mathbf{A}_0$ for $N < 0$.

From the above, the general equation for any AWG value N is $\mathbf{A}_N = \mathbf{k}^N \times \mathbf{A}_0$ regardless of whether the gauge number is positive, negative, or zero. From Table 2-2, $\mathbf{A}_0 = 105{,}530$ CM and $\mathbf{A}_1 = 83{,}694$ CM so

$$\mathbf{k} = \mathbf{A}_{X+1} / \mathbf{A}_X$$

$$= \mathbf{A}_1 / \mathbf{A}_0$$

$$= 83{,}694 \text{ CM} / 105{,}530 \text{ CM}$$

$$= 0.79308$$

Therefore, the equation for determining the cross-sectional area in circular mils for any AWG value N is

$$\mathbf{A}_N = (0.79308)^N \times 105{,}530 \text{ CM}$$

As a check, the cross-sectional area for AWG 27 is

$$\mathbf{A}_{27} = (0.79308)^{27} \times 105{,}530 \text{ CM}$$

$$= (1.9125 \times 10^{-3}) \times 105{,}530 \text{ CM}$$

$$= 201.82 \text{ CM}$$

This is in close agreement with the Table 2-2 value for AWG 27.

3. Ohm's Law

3.1 Basic

3.1.1 The Relationship of Current, Voltage, and Resistance

1. Problem: Ohm's Law for current states that current is equal to voltage divided by resistance ($I = V / R$). From this, derive Ohm's Law for voltage that expresses voltage in terms of current and resistance.

   Solution: The derivation of Ohm's Law for voltage is

   $$I = V / R$$
   $$I \times R = (V / R) \times R$$
   $$I \times R = V$$

   Therefore, Ohm's Law states that **voltage is equal to current times resistance** ($V = I \times R$).

2. Problem: Ohm's Law for current states that current is equal to voltage divided by resistance ($I = V / R$). From this, derive Ohm's Law for resistance that expresses resistance in terms of voltage and current.

   Solution: The derivation of Ohm's Law for resistance is

   $$I = V / R$$
   $$I \times R = (V / R) \times R$$
   $$(I \times R) / I = V / I$$
   $$R = V / I$$

   Therefore, Ohm's Law states that **resistance is equal to voltage divided by current** ($R = V / I$).

3.1.2 Current Calculations

1. Problem: The starter motor of a car has an effective starting resistance of 160 mΩ. How much current does the motor draw from a 12 V battery?

   Solution: Ohm's Law for current states that current is equal to voltage divided by resistance. Therefore,

   $$I = V / R$$
   $$= 12\ \text{V} / 160\ \text{m}\Omega$$
   $$= 12\ \text{V} / (160 \times 10^{-3}\ \Omega)$$
   $$= [12 / (160 \times 10^{-3})]\ (\text{V} / \Omega)$$
   $$= 75\ \text{A}$$

   Therefore, a starter motor with an effective starting resistance of 160 mΩ draws **75 A** from a 12 V battery.

2. Problem: A 6 V lantern battery powers an electric light with a cold filament resistance of 3.0 Ω. After 30 minutes the battery voltage decreases from 6.00 V to 5.75 V and the resistance of the hot filament increases from 3 Ω to 4.5 Ω. By what percent does the current change after 30 minutes?

   Solution: Ohm's Law for current states that the initial current is

   $$I_{INIT} = V_{INIT} / R_{INIT}$$
   $$= 6.00\ \text{V} / 3.0\ \Omega$$
   $$= (6.00 / 3.0)\ (\text{V} / \Omega)$$
   $$= 2.00\ \text{A}$$

   The current after 30 minutes is

   $$I_{30\,MIN} = V_{30\,MIN} / R_{30\,MIN}$$
   $$= 5.75\ \text{V} / 4.5\ \Omega$$
   $$= (5.75 / 4.5)\ (\text{V} / \Omega)$$

$= 1.28\ \text{A}$

The percentage change of the current after 30 minutes compared to the initial current is equal to

$[(\boldsymbol{I_{30\,MIN}} - \boldsymbol{I_{INIT}}) / \boldsymbol{I_{INIT}}] \times 100\% = [(1.28\ \text{A} - 2.00\ \text{A}) / 2.00\ \text{A}] \times 100\ \%$

$= (-0.72\ \text{A} / 2.00\ \text{A}) \times 100\%$

$= -0.36 \times 100\%$

$= -36\%$

Therefore, current changes by **–36%** after 30 minutes.

3. Problem: The voltage of a fixed dc power supply is specified to be 5 V ± 5%. What are the minimum and maximum currents that the power supply will deliver to a 100 Ω ± 10% resistor?

Solution: From the specifications for the power supply, the supply voltage $\boldsymbol{V_{DC}}$ is

$\boldsymbol{V_{DC}} = 5\ \text{V} \pm 5\%$

$= 5\ \text{V} \pm (5\ \text{V} \times 5\%)$

$= 5\ \text{V} \pm [5\ \text{V} \times (5 / 100)]$

$= 5\ \text{V} \pm [5\ \text{V} \times 0.05]$

$= 5\ \text{V} \pm 0.25\ \text{V}$

Therefore, the minimum and maximum voltages are

$\boldsymbol{V_{MIN}} = 5\ \text{V} - 0.25\ \text{V}$

$= (5 - 0.25)\ \text{V}$

$= 4.75\ \text{V}$

$\boldsymbol{V_{MAX}} = 5\ \text{V} + 0.25\ \text{V}$

$= (5 + 0.25)\ \text{V}$

$= 5.25\ \text{V}$

Similarly, the tolerance of resistor $\boldsymbol{R}$ gives

$\boldsymbol{R} = 100\ \Omega \pm 10\%$

$= 100\ \Omega \pm (100\ \Omega \times 10\%)$

$= 100\ \Omega \pm [100\ \Omega \times (10/100)]$

$= 100\ \Omega \pm [100\ \Omega \times 0.1]$

$= 100\ \Omega \pm 10\ \Omega$

Therefore, the minimum and maximum resistances are

$\boldsymbol{R_{MIN}} = 100\ \Omega - 10\ \Omega$

$= (100 - 10)\ \Omega$

$= 90\ \Omega$

$\boldsymbol{R_{MAX}} = 100\ \Omega + 10\ \Omega$

$= (100 + 10)\ \Omega$

$= 110\ \Omega$

Ohm's Law for current states that current is equal to voltage divided by resistance.

$\boldsymbol{I} = \boldsymbol{V} / \boldsymbol{R}$

This relationship indicates that the current will increase when the voltage increases and/or the resistance decreases, and decrease when the voltage decreases and/or the resistance increases. Therefore,

$\boldsymbol{I_{MIN}} = \boldsymbol{V_{MIN}} / \boldsymbol{R_{MAX}}$

$= 4.75\ \text{V} / 110\ \Omega$

$= (4.75 / 110)\ (\text{V} / \Omega)$

$= 43.2 \times 10^{-3}$ A

$= 43.2$ mA

$$\begin{aligned} \boldsymbol{I_{MAX}} &= \boldsymbol{V_{MAX} / R_{MIN}} \\ &= 5.25 \text{ V} / 90\ \Omega \\ &= (5.25 / 90)\ (\text{V} / \Omega) \\ &= 58.3 \times 10^{-3} \text{ A} \\ &= 58.3 \text{ mA} \end{aligned}$$

The minimum and maximum currents that the supply will deliver to a 100 Ω ± 10% resistor are **43.2 mA** and **58.3 mA**, respectively.

4. Problem: How many coulombs of charge flow from a 1.5 V battery connected to a 1.2 kΩ resistor for 10 seconds?

Solution: From Ohm's Law for current, the current would be

$$\begin{aligned} \boldsymbol{I} &= \boldsymbol{V / R} \\ &= 1.5 \text{ V} / 1.2 \text{ k}\Omega \\ &= 1.5 \text{ V} / 1.2 \times 10^{3}\ \Omega \\ &= [1.5 / (1.2 \times 10^{3})]\ (\text{V} / \Omega) \\ &= 1.25 \times 10^{-3} \text{ A} \\ &= 1.25 \text{ mA} \end{aligned}$$

By definition current is equal to charge divided by time.

$$\begin{aligned} \boldsymbol{I} &= \boldsymbol{Q / t} \\ \boldsymbol{I \times t} &= (\boldsymbol{Q / t}) \times \boldsymbol{t} \\ &= \boldsymbol{Q} \end{aligned}$$

Use this to calculate the charge from the current and the time the current flows.

$$\begin{aligned} \boldsymbol{Q} &= \boldsymbol{I \times t} \\ &= 1.25 \text{ mA} \times 10 \text{ s} \\ &= 1.25 \times 10^{-3} \text{ A} \times 10 \text{ s} \\ &= (1.25 \times 10^{-3} \text{ C} / \text{s}) \times 10 \text{ s} \\ &= (1.25 \times 10^{-3} \times 10)\ [(\text{C} / \text{s}) \times \text{s}] \\ &= 12.5 \times 10^{-3} \text{ C} \\ &= 12.5 \text{ mC} \end{aligned}$$

Thirteen millicoulombs of charge flow from a 1.5 V battery connected to a 1.2 kΩ resistor for 10 seconds.

3.1.3 Voltage Calculations

1. Problem: How much voltage does a solar cell generate if it can supply 12.5 mA to a 120 Ω load?

Solution: Ohm's Law for voltage states that voltage is equal to the product of current and resistance. Therefore,

$$\begin{aligned} \boldsymbol{V} &= \boldsymbol{I \times R} \\ &= 12.5 \text{ mA} \times 120\ \Omega \\ &= 12.5 \times 10^{-3} \text{ A} \times 120\ \Omega \\ &= (12.5 \times 10^{-3} \times 120)\ (\text{A} \times \Omega) \\ &= 1.50 \text{ V} \end{aligned}$$

A solar cell that supplies 12.5 mA to a 120 Ω load generates **1.50 V**.

2. Problem: Three current sources supply current to a 220 Ω load. The first current source supplies 50 mA, the second current source supplies 75 mA, and the third current source supplies 33 mA. What voltage would supply the same total current to the load?

Solution: The total current supplied to the load is

$$\begin{aligned} I_T &= 50\text{ mA} + 75\text{ mA} + 33\text{ mA} \\ &= (50 + 75 + 33)\text{ mA} \\ &= 158\text{ mA} \end{aligned}$$

From Ohm's Law for voltage, a voltage that would supply the same current would be

$$\begin{aligned} V &= \boldsymbol{I} \times \boldsymbol{R} \\ &= 158\text{ mA} \times 220\ \Omega \\ &= 158 \times 10^{-3}\text{ A} \times 220\ \Omega \\ &= (158 \times 10^{-3} \times 220)\ (\text{A} \times \Omega) \\ &= 34.8\text{ V} \end{aligned}$$

The voltage that would supply the same current to the load as the three current sources is **34.8 V**.

3. Problem: A fast-acting fuse that protects a sensitive 330 Ω load will open if the current exceeds 1/8 A. What is the maximum voltage that will not open the fuse?

Solution: From Ohm's Law for voltage, the voltage would be

$$\begin{aligned} V &= \boldsymbol{I} \times \boldsymbol{R} \\ &= 1/8\text{ A} \times 330\ \Omega \\ &= (1/8 \times 330)\ (\text{A} \times \Omega) \\ &= (330 / 8)\text{ V} \\ &= 41.25\text{ V} \end{aligned}$$

The maximum voltage that would not open the 1/8-A fuse is **40 V**.

4. Problem: The absolute maximum rated voltage for the input of a specific static-sensitive part is 5 kV. When an ungrounded technician picks up the part, 2.5 μA of current flows through the 5 GΩ input resistance of the part. Has the technician damaged the part?

Solution: Ohm's Law for voltage states that the input voltage required to produce 2.5 μA of current through 5 GΩ of resistance is

$$\begin{aligned} V &= \boldsymbol{I} \times \boldsymbol{R} \\ &= 2.5\ \mu\text{A} \times 5\text{ G}\Omega \\ &= (2.5 \times 10^{-6}\text{ A}) \times (5 \times 10^{9}\ \Omega) \\ &= [(2.5 \times 10^{-6}) \times (5 \times 10^{9})]\ (\text{A} \times \Omega) \\ &= 12.5 \times 10^{3}\text{ V} \\ &= 12.5\text{ kV} \end{aligned}$$

The 12.5 kV applied voltage exceeds the specified absolute maximum voltage for the part. **The part has been damaged.**

3.1.4 Resistance Calculations

1. Problem: A 120 V compact fluorescent lamp whose light output is equivalent to a 100 W incandescent light draws 200 mA of current. What is the resistance of the lamp?

Solution: Ohm's Law for resistance states that resistance is equal to voltage divided by current. Therefore

$$\begin{aligned} R &= V / I \\ &= 120\text{ V} / 200\text{ mA} \\ &= 120\text{ V} / (200 \times 10^{-3}\text{ A}) \end{aligned}$$

$= [120 / (200 \times 10^{-3})]$ (V / A)

$= 600\ \Omega$

The resistance of a compact fluorescent lamp that draws 200 mA of current is **600 Ω**.

2. Problem: Most ohmmeters operate by applying a known voltage to the leads, measuring the resulting current, and using Ohm's Law to determine the resistance. An analog ohmmeter uses a reference voltage of 1.25 V and has a full-scale current of 125 mA. If the meter reads half-scale when its leads are connected to a resistor, what is the measured resistance?

Solution: If the meter reads half-scale, the meter current is

$$\begin{aligned} \boldsymbol{I_{METER}} &= \boldsymbol{I_{FULL\text{-}SCALE}} / 2 \\ &= 125 \text{ mA} / 2 \\ &= (125 / 2) \text{ mA} \\ &= 62.5 \text{ mA} \end{aligned}$$

From Ohm's Law for resistance, the measured resistance is then

$$\begin{aligned} \boldsymbol{R_{MEAS}} &= \boldsymbol{V_{REF}} / \boldsymbol{I_{METER}} \\ &= 1.25 \text{ V} / 62.5 \text{ mA} \\ &= 1.25 \text{ V} / (62.5 \times 10^{-3} \text{ A}) \\ &= [1.25 / (62.5 \times 10^{-3})] \text{ (V / A)} \\ &= 20.0\ \Omega \end{aligned}$$

The measured resistance is **20.0 Ω**.

3. Problem: A hiker in the mountains encounters an unexpected electrical storm. She immediately crouches with her feet together on her non-conductive backpack to present a minimum target for a potential lightning strike and to insulate herself from the ground, but she still feels a mild tingle between her head and feet. She later determines that the potential difference between her shoes and the ground was about 250 kV. If the tingle she felt was due to 400 μA of current passing through her body, what was the approximate resistance of her backpack?

Solution: From Ohm's Law for resistance, the resistance of the backpack between her feet and the ground was

$$\begin{aligned} \boldsymbol{R} &= \boldsymbol{V} / \boldsymbol{I} \\ &= 250 \text{ kV} / 400\ \mu\text{A} \\ &= (250 \times 10^{3} \text{ V}) / (400 \times 10^{-6} \text{ A}) \\ &= [(250 \times 10^{3}) / (400 \times 10^{-6})] \text{ (V / A)} \\ &= 625 \times 10^{6}\ \Omega \\ &= 625 \text{ M}\Omega \end{aligned}$$

Therefore, the approximate resistance of her backpack was about **600 MΩ**.

4. Problem: A test engineer notes that a unit under test draws 25% more current from a fixed 12 V power supply than it drew at the start of a test. If the final current was 100 mA and the supply voltage was 12.0 V throughout the test, what was the initial resistance of the unit under test?

Solution: If the 100 mA final current $\boldsymbol{I_{FINAL}}$ is 25% higher than the initial current $\boldsymbol{I_{INIT}}$, then

$$\begin{aligned} \boldsymbol{I_{FINAL}} &= \boldsymbol{I_{INIT}} + \boldsymbol{I_{INIT}} \times 25\% \\ &= \boldsymbol{I_{INIT}} \times 1 + \boldsymbol{I_{NIT}} \times 25\% \\ &= \boldsymbol{I_{INIT}} \times (1 + 25\%) \\ &= \boldsymbol{I_{INIT}} \times [1 + 25 / 100] \\ &= \boldsymbol{I_{INIT}} \times 1.25 \end{aligned}$$

Therefore,

$$\begin{aligned} 100 \text{ mA} &= \boldsymbol{I_{INIT}} \times 1.25 \\ 100 \text{ mA} / 1.25 &= (\boldsymbol{I_{INIT}} \times 1.25) / 1.25 \end{aligned}$$

$(100 / 1.25)$ mA $= \boldsymbol{I_{INIT}} \times (1.25 / 1.25)$

80 mA $= \boldsymbol{I_{INIT}}$

From Ohm's Law of resistance,

$$\begin{aligned}\boldsymbol{R_{INIT}} &= \boldsymbol{V_{SUPPLY}} / \boldsymbol{I_{INIT}} \\ &= 12.0 \text{ V} / 80 \text{ mA} \\ &= 12.0 \text{ V} / 80 \times 10^{-3} \text{ A} \\ &= [12.0 / (80 \times 10^{-3})] \text{ (V / A)} \\ &= 150\ \Omega\end{aligned}$$

The initial resistance of the unit under test was **150 Ω**.

3.2 Advanced

3.2.1 The Relationship of Current, Voltage, and Resistance

1. Problem: An ***I-V***, or current-voltage, curve shows the voltage across a device affects the current through it. The graph in Figure 3-1 shows the current-voltage curves for three resistors. Does Resistor 2 or Resistor 3 have more resistance than Resistor 1?

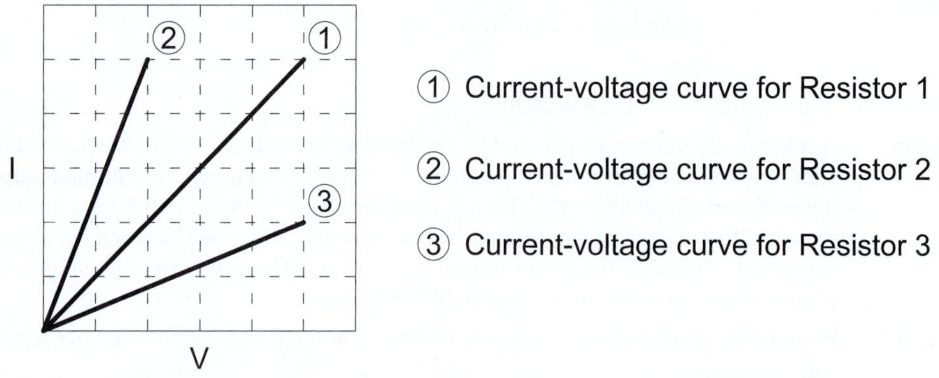

Figure 3-1: Example Current-Voltage Curves

Solution: Let $\boldsymbol{R}_1$, $\boldsymbol{R}_2$, and $\boldsymbol{R}_3$ be the values for Resistors 1, 2, and 3, respectively. The graph shows that the same voltage will produce more current for Resistor 2 than for Resistor 1, so that

$$\boldsymbol{V} / \boldsymbol{I_{R2}} < \boldsymbol{V} / \boldsymbol{I_{R1}}$$

Ohm's Law for resistance states that

$$\boldsymbol{R} = \boldsymbol{V} / \boldsymbol{I}$$

Therefore,

$$\boldsymbol{V} / \boldsymbol{I_{R2}} < \boldsymbol{V} / \boldsymbol{I_{R1}} \rightarrow \boldsymbol{R}_2 < \boldsymbol{R}_1$$

Similarly, the same voltage produces less current for Resistor 3 than for Resistor 1.

$$\boldsymbol{V} / \boldsymbol{I_{R3}} > \boldsymbol{V} / \boldsymbol{I_{R1}} \rightarrow \boldsymbol{R}_3 > \boldsymbol{R}_1$$

Resistor 3 has more resistance than Resistor 1.

2. Problem: The current through a variable resistor in an automatic voltage control circuit increases by 40%. What must the new resistance value be if the voltage across the resistor does not change?

Solution: Let $\boldsymbol{V_{INIT}}$, $\boldsymbol{I_{INIT}}$, and $\boldsymbol{R_{INIT}}$ be the initial voltage, current, and resistance values and $\boldsymbol{V_{NEW}}$, $\boldsymbol{I_{NEW}}$, and $\boldsymbol{R_{NEW}}$ be the new voltage, current, and resistance values. Ohm's Law for voltage states that

$$\boldsymbol{V_{INIT}} = \boldsymbol{I_{INIT}} \times \boldsymbol{R_{INIT}}$$

$$\boldsymbol{V_{NEW}} = \boldsymbol{I_{NEW}} \times \boldsymbol{R_{NEW}}$$

From the problem statement,

$$\boldsymbol{I_{NEW}} = \boldsymbol{I_{INIT}} + \boldsymbol{I_{INIT}} \times 40\%$$
$$= (\boldsymbol{I_{INIT}} \times 1) + (\boldsymbol{I_{INIT}} \times 40\%)$$
$$= \boldsymbol{I_{INIT}} \times (1 + 40\%)$$
$$= \boldsymbol{I_{INIT}} \times (1 + 40 / 100)$$
$$= \boldsymbol{I_{INIT}} \times (1 + 0.4)$$
$$= \boldsymbol{I_{INIT}} \times 1.4$$

and

$$\boldsymbol{V_{NEW}} = \boldsymbol{I_{INIT}} \times 1.4 \times \boldsymbol{R_{NEW}}$$

Since the voltage does not change $\boldsymbol{V_{NEW}} = \boldsymbol{V_{INIT}}$.

$$\boldsymbol{I_{INIT}} \times 1.4 \times \boldsymbol{R_{NEW}} = \boldsymbol{I_{INIT}} \times \boldsymbol{R_{INIT}}$$
$$(\boldsymbol{I_{INIT}} \times 1.4 \times \boldsymbol{R_{NEW}}) / \boldsymbol{I_{INIT}} = (\boldsymbol{I_{INIT}} \times \boldsymbol{R_{INIT}}) / \boldsymbol{I_{INIT}}$$
$$1.4 \times \boldsymbol{R_{NEW}} = \boldsymbol{R_{INIT}}$$
$$(1.4 \times \boldsymbol{R_{NEW}}) / 1.4 = \boldsymbol{R_{INIT}} / 1.4$$
$$\boldsymbol{R_{NEW}} = \boldsymbol{R_{INIT}} / 1.4$$

From this

$$\boldsymbol{R_{NEW}} = \boldsymbol{R_{INIT}} \times (1 / 1.4)$$
$$= \boldsymbol{R_{INIT}} \times 0.714$$
$$= \boldsymbol{R_{INIT}} \times [(0.714 \times 100) / 100]$$
$$= \boldsymbol{R_{INIT}} \times (71.4 / 100)$$
$$= \boldsymbol{R_{INIT}} \times 71.4\%$$

Therefore, the new resistance value must decrease to **71% of the initial resistance value.**

3.2.2 Current Calculations

1. Problem: The initial current through a resistor connected across a battery is 15.0 mA at 25 °C. After 30 minutes the battery voltage drops 10% and the resistor temperature increases to 60 °C. If the resistor's temperature coefficient is +250 ppm / 1 °C (the resistor value changes by +250 parts per million for every 1 °C increase in temperature), what is the new current?

   Solution: Let $\boldsymbol{V_{INIT}}$, $\boldsymbol{I_{INIT}}$, and $\boldsymbol{R_{INIT}}$ be the initial voltage, current, and resistance and $\boldsymbol{V_{30MIN}}$, $\boldsymbol{I_{30MIN}}$, and $\boldsymbol{R_{30MIN}}$ be the voltage, current, and resistance after 30 minutes, respectively. From Ohm's Law for current

$$\boldsymbol{I_{INIT}} = \boldsymbol{V_{INIT}} / \boldsymbol{R_{INIT}}$$
$$15.0 \text{ mA} = \boldsymbol{V_{INIT}} / \boldsymbol{R_{INIT}}$$

and

$$\boldsymbol{I_{30MIN}} = \boldsymbol{V_{30MIN}} / \boldsymbol{R_{30MIN}}$$

From the problem statement the battery voltage after 30 minutes is

$$\boldsymbol{V_{30MIN}} = \boldsymbol{V_{INIT}} - \boldsymbol{V_{INIT}} \times 10\%$$
$$= (\boldsymbol{V_{INIT}} \times 1) - (\boldsymbol{V_{INIT}} \times 10\%)$$
$$= \boldsymbol{V_{INIT}} \times (1 - 10\%)$$
$$= \boldsymbol{V_{INIT}} \times [1 - 10 / 100]$$
$$= \boldsymbol{V_{INIT}} \times (1 - 0.1)$$
$$= \boldsymbol{V_{INIT}} \times 0.9$$

The change in resistance ($\Delta\boldsymbol{R}$) that is due to change in temperature (ΔT) is

$$\Delta\boldsymbol{R} = \boldsymbol{R_{INIT}} \times (+250 \text{ ppm} / 1\ °\text{C}) \times \Delta \text{T}$$
$$= \boldsymbol{R_{INIT}} \times (+250 \text{ ppm})/ 1\ °\text{C}) \times (60\ °\text{C} - 25\ °\text{C})$$

$= \boldsymbol{R_{INIT}} \times (+250\text{ ppm} / 1\text{ °C}) \times 35\text{ °C}$

$= \boldsymbol{R_{INIT}} \times [(+250 \times 35) / 1]\,[(\text{ppm} \times \text{°C}) / \text{°C}]$

$= \boldsymbol{R_{INIT}} \times 8{,}750\text{ ppm}$

so the resistance after 30 minutes is

$\boldsymbol{R_{30MIN}} = \boldsymbol{R_{INIT}} + \boldsymbol{R_{INIT}} \times 8{,}750\text{ ppm}$

$= (\boldsymbol{R_{INIT}} \times 1) + (\boldsymbol{R_{INIT}} \times 8{,}750\text{ ppm})$

$= (\boldsymbol{R_{INIT}} \times 1) + (\boldsymbol{R_{INIT}} \times 8{,}750 / 1{,}000{,}000)$

$= \boldsymbol{R_{INIT}} \times [1 + (8{,}750 / 1{,}000{,}000)]$

$= \boldsymbol{R_{INIT}} \times (1 + 0.00875)$

$= \boldsymbol{R_{INIT}} \times 1.00875$

From Ohm's Law for current

$\boldsymbol{I_{30MIN}} = \boldsymbol{V_{30MIN}} / \boldsymbol{R_{30MIN}}$

$= (\boldsymbol{V_{INIT}} \times 0.9) / (\boldsymbol{R_{INIT}} \times 1.00875)$

$= (\boldsymbol{V_{INIT}} / \boldsymbol{R_{INIT}}) \times (0.9 / 1.00875)$

$= 15.0\text{ mA} \times 0.892193$

$= 13.3829\text{ mA}$

The current through the resistor after 30 minutes is **13.4 mA**.

2. Problem: The voltage from a power supply can vary by ±5%. If the power supply connects to a 5% tolerance resistor, how much can the current vary from the ideal value?

Solution: Ohm's Law for current states the ideal current value $\boldsymbol{I_{IDEAL}}$ will be

$\boldsymbol{I_{IDEAL}} = \boldsymbol{V_{IDEAL}} / \boldsymbol{R_{IDEAL}}$

The actual voltage and resistance $\boldsymbol{V_{ACTUAL}}$ and $\boldsymbol{R_{ACTUAL}}$ will be

$\boldsymbol{V_{ACTUAL}} = \boldsymbol{V_{IDEAL}} + (\boldsymbol{V_{IDEAL}} \times \pm 5\%)$

$\boldsymbol{R_{ACTUAL}} = \boldsymbol{R_{IDEAL}} + (\boldsymbol{R_{IDEAL}} \times \pm 5\%)$

so the actual current $\boldsymbol{I_{ACTUAL}}$ will be

$\boldsymbol{I_{ACTUAL}} = \boldsymbol{V_{ACTUAL}} / \boldsymbol{R_{ACTUAL}}$

$= [\boldsymbol{V_{IDEAL}} + (\boldsymbol{V_{IDEAL}} \times \pm 5\%)] / [\boldsymbol{R_{IDEAL}} + (\boldsymbol{R_{IDEAL}} \times \pm 5\%)]$

The maximum current $\boldsymbol{I_{MAX}}$ occurs when the voltage is maximum and resistance is minimum, so

$\boldsymbol{I_{MAX}} = [\boldsymbol{V_{IDEAL}} + (\boldsymbol{V_{IDEAL}} \times 5\%)] / [\boldsymbol{R_{IDEAL}} + (\boldsymbol{R_{IDEAL}} \times -5\%)]$

$= [(\boldsymbol{V_{IDEAL}} \times 1) + (\boldsymbol{V_{IDEAL}} \times 5\%)] / [(\boldsymbol{R_{IDEAL}} \times 1) + (\boldsymbol{R_{IDEAL}} \times -5\%)]$

$= [\boldsymbol{V_{IDEAL}} \times (1 + 5\%)] / [\boldsymbol{R_{IDEAL}} \times (1 - 5\%)]$

$= \{\boldsymbol{V_{IDEAL}} \times [1 + (5 / 100)]\} / \{\boldsymbol{R_{IDEAL}} \times [1 - (5 / 100)]\}$

$= [\boldsymbol{V_{IDEAL}} \times (1 + 0.05)] / [\boldsymbol{R_{IDEAL}} \times (1 - 0.05)]$

$= (\boldsymbol{V_{IDEAL}} \times 1.05) / (\boldsymbol{R_{IDEAL}} \times 0.95)$

$= (\boldsymbol{V_{IDEAL}} / \boldsymbol{R_{IDEAL}}) \times (1.05 / 0.95)$

$= \boldsymbol{I_{IDEAL}} \times 1.105$

$= \boldsymbol{I_{IDEAL}} \times (1 + 0.105)$

$= (\boldsymbol{I_{IDEAL}} \times 1) + (\boldsymbol{I_{IDEAL}} \times 0.105)$

$= \boldsymbol{I_{IDEAL}} + \{\boldsymbol{I_{IDEAL}} \times [(0.105 \times 100) / 100]\}$

$= \boldsymbol{I_{IDEAL}} + \boldsymbol{I_{IDEAL}} \times 10.5 / 100$

$= \boldsymbol{I_{IDEAL}} + (\boldsymbol{I_{IDEAL}} \times 10.5\%)$

Similarly, the minimum current $\boldsymbol{I_{MIN}}$ occurs when the voltage is minimum and resistance is maximum, so

$$\begin{aligned}
\boldsymbol{I_{MIN}} &= [\boldsymbol{V_{IDEAL}} + (\boldsymbol{V_{IDEAL}} \times -5\%)] / [\boldsymbol{R_{IDEAL}} + (\boldsymbol{R_{IDEAL}} \times 5\%)] \\
&= [(\boldsymbol{V_{IDEAL}} \times 1) + (\boldsymbol{V_{IDEAL}} \times -5\%)] / [(\boldsymbol{R_{IDEAL}} \times 1) + (\boldsymbol{R_{IDEAL}} \times 5\%)] \\
&= [\boldsymbol{V_{IDEAL}} \times (1 - 5\%)] / [\boldsymbol{R_{IDEAL}} \times (1 + 5\%)] \\
&= \{\boldsymbol{V_{IDEAL}} \times [1 - (5 / 100)]\} / \{[\boldsymbol{R_{IDEAL}} \times [1 + (5 / 100)]\} \\
&= [\boldsymbol{V_{IDEAL}} \times (1 - 0.05)] / [\boldsymbol{R_{IDEAL}} \times (1 + 0.05)] \\
&= (\boldsymbol{V_{IDEAL}} \times 0.95) / (\boldsymbol{R_{IDEAL}} \times 1.05) \\
&= (\boldsymbol{V_{IDEAL}} / \boldsymbol{R_{IDEAL}}) \times (0.95 / 1.05) \\
&= \boldsymbol{I_{IDEAL}} \times 0.905 \\
&= \boldsymbol{I_{IDEAL}} \times (1 - 0.095) \\
&= (\boldsymbol{I_{IDEAL}} \times 1) - (\boldsymbol{I_{IDEAL}} \times 0.095) \\
&= \boldsymbol{I_{IDEAL}} - \{\boldsymbol{I_{IDEAL}} \times [(0.095 \times 100) / 100]\} \\
&= \boldsymbol{I_{IDEAL}} - (\boldsymbol{I_{IDEAL}} \times 9.5 / 100) \\
&= \boldsymbol{I_{IDEAL}} - (\boldsymbol{I_{IDEAL}} \times 9.5\%) \\
&= \boldsymbol{I_{IDEAL}} + (\boldsymbol{I_{IDEAL}} \times -9.5\%)
\end{aligned}$$

When the voltage and resistance values can both vary by 5% each, the actual current can vary by **+10.5% / −9.5%**.

3.2.3 Voltage Calculations

1. Problem: Figure 3-2 shows a high-precision potentiometer through which a precision 1.00 mA constant current flows. The resistor body consists of a platinum strip ($\boldsymbol{\rho_{Pt}} = 1.06 \times 10^{-7}\ \Omega\text{•m}$) that is 50.00 mm long, 5.30 mm wide, and 0.10 mm thick and is marked with fifty 1.00 mm graduations. How much voltage does each 1.00 mm graduation represent?

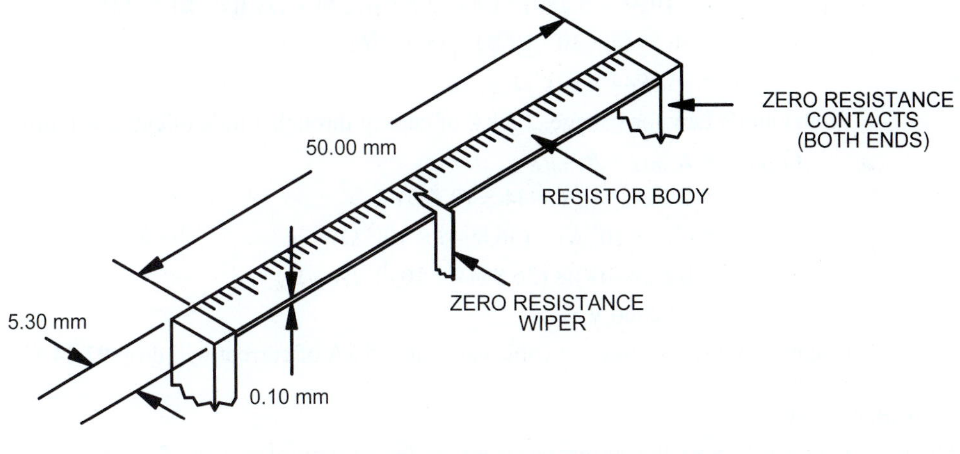

Figure 3-2: High-Precision Platinum Potentiometer

Solution: From Figure 3-2 the cross-sectional area $\boldsymbol{A_{Pt}}$ of the platinum strip between the contacts is

$$\begin{aligned}
\boldsymbol{A_{Pt}} &= 5.30 \text{ mm} \times 0.10 \text{ mm} \\
&= (5.30 \times 10^{-3} \text{ m}) \times (0.10 \times 10^{-3} \text{ m}) \\
&= (5.30 \times 10^{-3} \times 0.10 \times 10^{-3}) (\text{m} \times \text{m}) \\
&= 5.30 \times 10^{-7} \text{ m}^2
\end{aligned}$$

For a 1.00 mm graduation, the length $\boldsymbol{l_{Pt}}$ is 1.00 mm = 1.00×10^{-3} m, so the resistance $\boldsymbol{R_{Pt}}$ between each graduation is

$$\begin{aligned}
\boldsymbol{R_{Pt}} &= (\boldsymbol{\rho_{Pt}} \times \boldsymbol{l_{Pt}}) / \boldsymbol{A_{Pt}} \\
&= [(1.06 \times 10^{-7}\ \Omega\text{•m}) \times (1.00 \times 10^{-3} \text{ m})] / (5.30 \times 10^{-7} \text{ m}^2) \\
&= \{[(1.06 \times 10^{-7}) \times (1.00 \times 10^{-3})] / 5.3 \times 10^{-7}\} [(\Omega \times \text{m} \times \text{m}) / (\text{m} \times \text{m})]
\end{aligned}$$

$= 2.00 \times 10^{-3}\ \Omega$

The current I_{Pt} through the platinum strip is 1.000 mA, so from Ohm's Law for voltage,

$$\begin{aligned}
V_{Pt} &= I_{Pt} \times R_{Pt} \\
&= 1.00\ \text{mA} \times (2.00 \times 10^{-3}\ \Omega) \\
&= (1.00 \times 10^{-3}\ \text{A}) \times (2.00 \times 10^{-3}\ \Omega) \\
&= (1.00 \times 10^{-3} \times 2.00 \times 10^{-3})\ (\text{A} \times \Omega) \\
&= 2.00 \times 10^{-6}\ \text{V} \\
&= 2.00 \times 10^{-6}\ \text{V} \\
&= 2.00\ \mu\text{V}
\end{aligned}$$

Each 1.00 mm graduation on the potentiometer represents **2.00 μV**.

2. Problem: A transmission cable contains a bundle of 7 AWG 0000 copper conductors (ρ_{Cu} = 10.35 CM•Ω/ft). If the cable carries 2.5 kA of current, how much voltage will one mile (5,280 feet) of cable drop?

Solution: From Table 2-2, the cross-sectional area of an AWG 0000 cable is 211,600 CM. The total cross-sectional area of 7 AWG 0000 conductors is

$$\begin{aligned}
A_{CABLE} &= 7 \times 211{,}600\ \text{CM} \\
&= (7 \times 211{,}600)\ \text{CM} \\
&= 1{,}481{,}200\ \text{CM}
\end{aligned}$$

The resistance of one mile of cable is

$$\begin{aligned}
R_{CABLE} &= (\rho_{Cu} \times l_{CABLE}) / A_{CABLE} \\
&= [(10.35\ \text{CM}\cdot\Omega/\text{ft}) \times 1\ \text{mile}] / (1{,}481{,}200\ \text{CM}) \\
&= (10.35\ \text{CM}\cdot\Omega/\text{ft} \times 5{,}280\ \text{ft}) / (1{,}481{,}200\ \text{CM}) \\
&= [(10.35 \times 5{,}280) / 1{,}481{,}200]\ \{[(\text{CM} \times \Omega / \text{ft}) \times \text{ft}] / \text{CM}\} \\
&= 36.8944 \times 10^{-3}\ [(\text{CM} \times \Omega) / \text{CM}] \\
&= 36.8944 \times 10^{-3}\ \Omega
\end{aligned}$$

From Ohm's Law for voltage, 2.5 kA of current through 1 mile of cable will drop

$$\begin{aligned}
V_{CABLE} &= I_{CABLE} \times R_{CABLE} \\
&= 2.5\ \text{kA} \times (36.8944 \times 10^{-3}\ \Omega) \\
&= (2.5 \times 10^{3}\ \text{A}) \times (36.8944 \times 10^{-3}\ \Omega) \\
&= [(2.5 \times 10^{3}) \times (36.8944 \times 10^{-3})]\ (\text{A} \times \Omega) \\
&= 92.2360\ \text{V}
\end{aligned}$$

One mile of the transmission cable carrying 2.5 kA of current will drop **92.24 V**.

3.2.4 Resistance Calculations

1. Problem: Figure 3-3 shows the current sense circuit for the control system of a home wind generator.

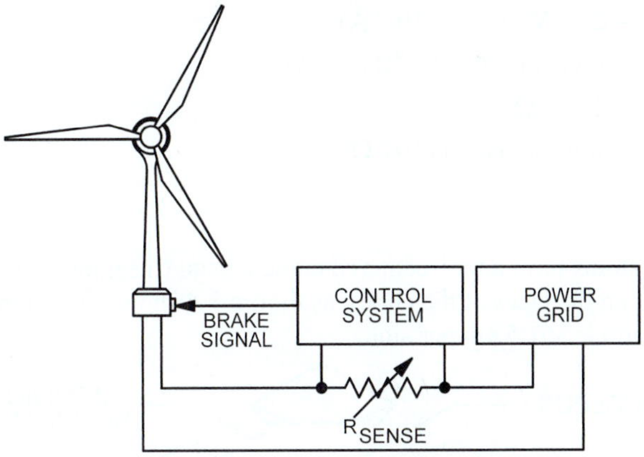

Figure 3-3: Current Sense for Wind Power Control System

The control system will brake the wind generator if the voltage across the sense resistor indicates that the turbine is rotating too rapidly. The wind turbine generates 500 μA/RPM and the control system will trip the brake if the sense voltage V_{SENSE} across the sense resistor R_{SENSE} exceeds 5.0 V. To what resistance should the variable resistor be set to brake the wind generator at 180 RPM?

Solution: The wind turbine generates 500 μA/RPM so the current through R_{SENSE} at 180 RPM is

$$I_{RSENSE} = (500\ \mu A\ /\ RPM) \times 180\ RPM$$
$$= (500 \times 10^{-6}\ A\ /\ RPM) \times 180\ RPM$$
$$= (500 \times 10^{-6} \times 180)\ [(A\ /\ RPM) \times RPM]$$
$$= 90 \times 10^{-3}\ A$$
$$= 90\ mA$$

The brake system will trip when the V_{SENSE} exceeds 5 V. From Ohm's Law for resistance,

$$R_{SENSE} = V_{SENSE}\ /\ I_{SENSE}$$
$$= 5.0\ V\ /\ 90\ mA$$
$$= 5.0\ V\ /\ (90 \times 10^{-3}\ A)$$
$$= [5.0\ /\ (90 \times 10^{-3})]\ (V\ /\ A)$$
$$= 55.6\ \Omega$$

The variable resistor should be set to **56 Ω** to brake the wind generator at 180 RPM.

2. Problem: 750 μC of charge lose 20 mJ of energy in 3 seconds moving through a resistor. What is the value R of the resistor?

Solution: By definition, voltage is energy divided by charge so V_{RES}, the voltage drop across the resistor, is

$$V_{RES} = 20\ mJ\ /\ 750\ \mu C$$
$$= (20 \times 10^{-3}\ J)\ /\ (750 \times 10^{-6}\ C)$$
$$= [(20 \times 10^{-3})\ /\ (750 \times 10^{-6})]\ (J\ /\ C)$$
$$= 26.7\ V$$

Also by definition, current is charge divided by time.

$$I_{RES} = 750\ \mu C\ /\ 3\ s$$
$$= (750 \times 10^{-6}\ C)\ /\ 3\ s$$
$$= (750 \times 10^{-6}\ /\ 3)\ (C\ /\ s)$$
$$= 250 \times 10^{-6}\ A$$

From Ohm's Law for resistance, the resistance is the voltage divided by the current.

$$R = V_{RES}\ /\ I_{RES}$$

$= 26.7\ V / (250 \times 10^{-6}\ A)$
$= [26.7 / (250 \times 10^{-6})]\ (V / A)$
$= 106.7\ \Omega$

The value of the resistor is **107 Ω**.

3.3 Just for Fun

1. Problem: Copper atoms form a face-centered cubic crystal (meaning that the nuclei of the copper atoms are centered on the faces of the cube), as Figure 3-4 shows. The dimensions of the cube that copper atoms form is 361.5 pm per side.

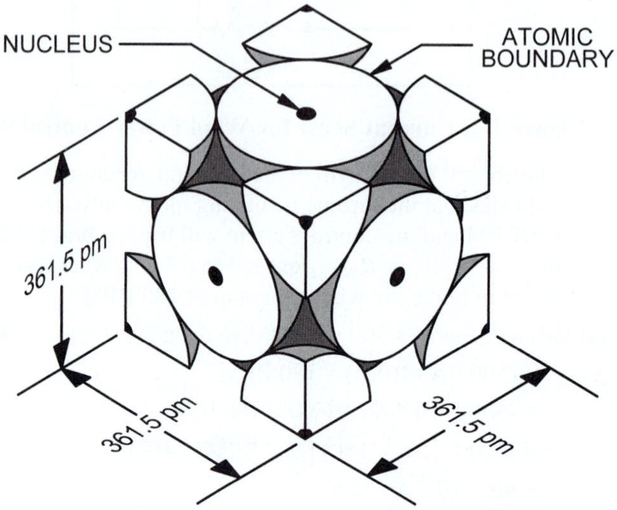

Figure 3-4: Face-Centered Copper Crystal

Calculations for current, voltage, and resistance on an atomic level are not the same as on levels for which Ohm's Law applies. However, if Ohm's Law did apply, how much voltage would be necessary to move one electron through one copper crystal if

1) the resistivity of copper ($\rho_{Cu} = 1.72 \times 10^{-8}\ \Omega{\bullet}m$) applies to one copper crystal, and
2) the velocity of an electron in copper is 4.30 mm/s?

Solution: Ohm's Law for voltage states that voltage is current times resistance. First, find the resistance of the copper crystal.

The cross-sectional area $A_{CRYSTAL}$ of the crystal is

$A_{CRYSTAL} = 361.5\ pm \times 361.5\ pm$
$= (361.5 \times 10^{-12}\ m) \times (361.5 \times 10^{-12}\ m)$
$= [(361.5 \times 10^{-12}) \times (361.5 \times 10^{-12})]\ (m \times m)$
$= 130.7 \times 10^{-21}\ m^2$

The length $l_{CRYSTAL}$ of the copper crystal is just the length of one side of the cube, so that

$l_{CRYSTAL} = 361.5\ pm$

The resistance of the copper crystal is then

$R_{CRYSTAL} = (\rho_{Cu} \times l_{CRYSTAL}) / A_{CRYSTAL}$
$= (1.72 \times 10^{-8}\ \Omega{\bullet}m \times 361.5\ pm) / 130.7 \times 10^{-21}\ m^2$
$= [(1.72 \times 10^{-8}\ \Omega \times m) \times (361.5 \times 10^{-12}\ m)] / (130.7 \times 10^{-21}\ m \times m)$
$= \{[(1.72 \times 10^{-8}) \times (361.5 \times 10^{-12})] / 130.7 \times 10^{-21}\}\ (\Omega \times m \times m) / (m \times m)$
$= 47.58\ \Omega$

Next, find the current. By definition current is equal to charge divided by time, so find the time that it takes the electron to cross the crystal. Velocity $\boldsymbol{v}$ is defined as distance $\boldsymbol{d}$ divided by time $\boldsymbol{t}$, so

$$v = d / t$$
$$v \times (t / v) = (d / t) \times (t / v)$$
$$t = d / v$$

From Figure 3-4, the length of a copper crystal is 361.5 pm, so the time for an electron to pass through the crystal is

$$\begin{aligned} t &= d / v \\ &= 361.5 \text{ pm} / (4.30 \text{ mm} / \text{s}) \\ &= (361.5 \times 10^{-12} \text{ m}) / (4.30 \times 10^{-3} \text{ m} / \text{s}) \\ &= (361.5 \times 10^{-12} / 4.30 \times 10^{-3}) [\text{m} / (\text{m} / \text{s})] \\ &= 230.3 \times 10^{-18} [\text{m} \times (\text{s} / \text{m})] \\ &= 84.07 \times 10^{-9} \text{ s} \end{aligned}$$

The current due to the charge of one electron moving across a copper crystal is

$$\begin{aligned} I_{CRYSTAL} &= Q / t \\ &= (1\ e) / (84.07 \times 10^{-9} \text{ s}) \\ &= [1 / (84.07 \times 10^{-9})] (e / \text{s}) \\ &= 11.86 \times 10^{6}\ e / \text{s} \end{aligned}$$

The charge on one electron equals 6.25×10^{-18} coulombs, so

$$\begin{aligned} I_{CRYSTAL} &= 11.86 \times 10^{6} (e / \text{s}) \\ &= (11.86 \times 10^{6}\ e / \text{s}) \times (6.25 \times 10^{-18} \text{ C} / e) \\ &= [(11.86 \times 10^{6}) \times (6.25 \times 10^{-18})] [(e / \text{s}) \times (\text{C} / e)] \\ &= 74.13 \times 10^{-12} \text{ C} / \text{s} \\ &= 74.13 \times 10^{-12} \text{ A} \end{aligned}$$

From Ohm's Law for voltage,

$$\begin{aligned} V_{CRYSTAL} &= I_{CRYSTAL} \times R_{CRYSTAL} \\ &= (74.13 \times 10^{-12} \text{ A}) \times (47.58\ \Omega) \\ &= [(74.13 \times 10^{-12}) \times (47.58)] (\text{A} \times \Omega) \\ &= 3.526 \times 10^{-9} \text{ V} \\ &= 3.526 \text{ nV} \end{aligned}$$

The theoretical voltage needed to move an electron across one copper crystal is **3.53 nV**.

2. Problem: A concept that you may encounter in electronics is that of **negative resistance**. The term "negative resistance" may suggest that the direction of current is opposite to that of the current in normal resistance, or that it is some unusual property of anti-matter. Actually, negative resistance is a property of some materials in which current through them decreases as the voltage across them increases, and vice versa. Refer to the ***I-V*** curves in Figure 3-5. As the negative resistance graph shows, the current magnitude decreases as the voltage increases but never actually becomes less than zero. Another way to view it is that the change in current, ΔI, is negative when the change in voltage, ΔV, is positive (and vice versa) so that $R = \Delta V / \Delta I$ is negative.

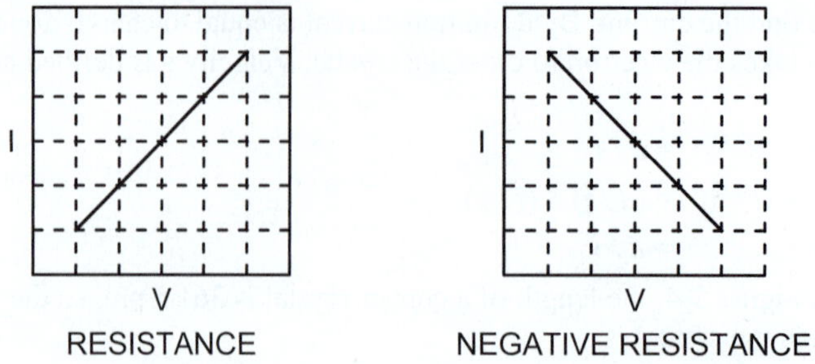

Figure 3-5: Resistance and Negative Resistance I-V Curves

A component that exhibits negative resistance is the tunnel diode. Figure 3-6 shows the measured ***I-V*** curve for a tunnel diode and the indicated initial current (I_0), peak current (I_P), valley current (I_V), and final current (I_F).

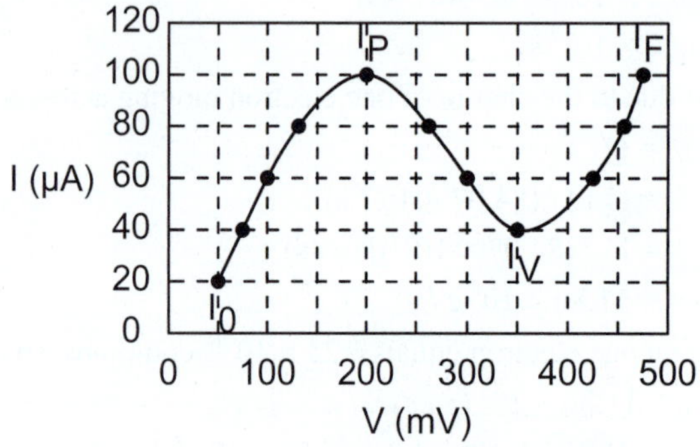

Figure 3-6: Tunnel Diode *I-V* Curve

Use the data in Figure 3-6 to complete Table 3-1. Is the resistance of any data point actually negative? What is the tunnel diode's average resistance between I_P and I_V?

Table 3-1: Tunnel Diode Resistance Calculations

Data Point	V	I	R = V / I
Initial current (I_0)			
Peak current (I_P)			
Valley current (I_V)			
Final current (I_F)			

Solution: From the data in Figure 3-6 the voltage, current, and calculated resistance for each data point are

Initial current (I_0): $V_{I0} = 50\text{ mV}$

$I_{I0} = 20\ \mu\text{A}$

$R_{I0} = V_{I0} / I_{I0}$

$= 50\text{ mV} / 20\ \mu\text{A}$

$= (50 \times 10^{-3}\text{ V}) / (20 \times 10^{-6}\text{ A})$

$= (50 \times 10^{-3} / 20 \times 10^{-6})\ (\text{V} / \text{A})$

$= 2.5 \times 10^{3}\ \Omega$

$= 2.5\text{ k}\Omega$

Peak current (I_P): $V_{IP} = 200\text{ mV}$

$I_{IP} = 100\ \mu\text{A}$

$$\begin{aligned}
R_{IP} &= V_{IP} / I_{IP} \\
&= 200 \text{ mV} / 100 \ \mu\text{A} \\
&= (200 \times 10^{-3} \text{ V}) / (100 \times 10^{-6} \text{ A}) \\
&= (200 \times 10^{-3} / 100 \times 10^{-6}) (\text{V} / \text{A}) \\
&= 2.0 \times 10^{3} \ \Omega \\
&= 2.0 \text{ k}\Omega
\end{aligned}$$

Valley current (I_V): $V_{IV} = 350$ mV

$$\begin{aligned}
I_{IV} &= 40 \ \mu\text{A} \\
R_{IV} &= V_{IV} / I_{IV} \\
&= 350 \text{ mV} / 40 \ \mu\text{A} \\
&= (350 \times 10^{-3} \text{ V}) / (40 \times 10^{-6} \text{ A}) \\
&= (350 \times 10^{-3} / 40 \times 10^{-6}) (\text{V} / \text{A}) \\
&= 8.75 \times 10^{3} \ \Omega \\
&= 8.75 \text{ k}\Omega
\end{aligned}$$

Final current (I_F): $V_{IF} = 475$ mV

$$\begin{aligned}
I_{IF} &= 100 \ \mu\text{A} \\
R_{IF} &= V_{IF} / I_{IF} \\
&= 475 \text{ mV} / 100 \ \mu\text{A} \\
&= (475 \times 10^{-3} \text{ V}) / (100 \times 10^{-6} \text{ A}) \\
&= (475 \times 10^{-3} / 100 \times 10^{-6}) (\text{V} / \text{A}) \\
&= 4.75 \times 10^{3} \ \Omega \\
&= 4.75 \text{ k}\Omega
\end{aligned}$$

Table 3-2 shows the completed table. As the table shows, there are no points for which the resistance of the tunnel diode is negative.

Table 3-2: Solutions for Tunnel Diode Resistance Calculations

Data Point	V	I	R = V / I
Initial current (I_0)	50 mV	20 μA	2.5 kΩ
Peak current (I_P)	200 mV	100 μA	2.0 kΩ
Valley current (I_V)	350 mV	40 μA	8.75 kΩ
Final current (I_F)	475 mV	100 μA	4.75 kΩ

Although the graph between points I_P and I_V is not linear, the average resistance for the region assumes that it is. The average resistance is

$$\begin{aligned}
R_{AVERAGE} (I_V \text{ to } I_P) &= (V_{IV} - V_{IP}) / (I_{IV} - I_{IP}) \\
&= (350 \text{ mV} - 200 \text{ mV}) / (40 \ \mu\text{A} - 100 \ \mu\text{A}) \\
&= [(350 \times 10^{-3} \text{ V}) - (200 \times 10^{-3} \text{ V})] / [(40 \times 10^{-6} \text{ A}) - (100 \times 10^{-6} \text{ A})] \\
&= \{[(350 \times 10^{-3}) - (200 \times 10^{-3})] \text{ V}\} / \{[(40 \times 10^{-6}) - (100 \times 10^{-6})] \text{ A}\} \\
&= (150 \times 10^{-3} \text{ V}) / (-60 \times 10^{-6}) \text{ A} \\
&= (150 \times 10^{-3} / -60 \times 10^{-6}) (\text{V} / \text{A}) \\
&= -2.5 \times 10^{3} \ \Omega \\
&= -2.5 \text{ k}\Omega
\end{aligned}$$

Therefore, although the resistance for each data point is positive, the resistance for the *region* between the peak and valley currents is negative.

4. Energy and Power

4.1 Basic

4.1.1 Energy and Power

1. Problem: A miniature dc motor uses 250 mJ of energy in one minute. How much power does the motor develop?

   Solution: By definition, power is energy divided by time, so

$$\begin{aligned} \boldsymbol{P} &= \boldsymbol{E} / \boldsymbol{t} \\ &= 250 \text{ mJ} / 1 \text{ min} \\ &= (250 \times 10^{-3} \text{ J}) / 1 \text{ min} \\ &= [(250 \times 10^{-3}) / 1) (\text{J} / \text{min}) \\ &= 250 \times 10^{-3} \text{ J} / \text{min} \end{aligned}$$

   One minute equals sixty seconds, so

$$\begin{aligned} \boldsymbol{P} &= 250 \times 10^{-3} (\text{J} /\text{min}) \\ &= 250 \times 10^{-3} (\text{J} /\text{min}) \times (1 \text{ min} / 60 \text{ s}) \\ &= [(250 \times 10^{-3} \times 1) / 60] [(\text{J} /\text{min}) \times (\text{min} / \text{s})] \\ &= 4.167 \times 10^{-3} \text{ J} / \text{s} \\ &= 4.167 \times 10^{-3} \text{ W} \\ &= 4.167 \text{ mW} \end{aligned}$$

   The miniature dc motor develops **4.17 mW** of power.

2. Problem: How many joules are equal to one watt-hour?

   Solution: By definition,

$$1 \text{ Wh} = 1 \text{ W} \times 1 \text{ hr}$$

   One hour equals sixty minutes, so

$$\begin{aligned} 1 \text{ Wh} &= 1 \text{ W} \times 1 \text{ hr} \\ &= (1 \text{ W} \times 1 \text{ hr}) \times (60 \text{ min} / 1 \text{ hr}) \\ &= [(1 \times 1 \times 60) / 1] [(\text{W} \times \text{hr} \times \text{min}) / \text{hr}] \\ &= 60 \text{ W}\bullet\text{min} \end{aligned}$$

   One minute equals sixty seconds, so

$$\begin{aligned} 1 \text{ Wh} &= 60 \text{ W}\bullet\text{min} \\ &= (60 \text{ W} \times \text{min}) \times (60 \text{ s} / 1 \text{ min}) \\ &= [(60 \times 60) / 1] [(\text{W} \times \text{min} \times \text{s}) / \text{min}] \\ &= 3600 \text{ W}\bullet\text{s} \end{aligned}$$

   Since by definition one watt is equal to one joule divided by one second,

$$\begin{aligned} 1 \text{ Wh} &= 3600 \text{ W}\bullet\text{s} \\ &= 3600 \text{ W} \times \text{s} \\ &= 3600 [(\text{J} / \text{s}) \times \text{s}] \\ &= 3600 \text{ J} \end{aligned}$$

   One watt-hour is equal to **3600 J**.

3. Problem: An engineer leaves his work computer continuously powered from Monday morning to Friday afternoon to save time booting up each day rather than turning it off for the fifteen hours he is gone each day. How many joules of energy would he save each week by turning off the computer if the power-save mode of the computer consumes an average of 12.5 watts each hour?

Solution: The computer is on but not used for fifteen hours every Monday, Tuesday, Wednesday, and Thursday so that the computer is on but not used for four days each week. The total idle time per week is

$$\begin{aligned} t &= (15 \text{ hr / day}) \times (4 \text{ days / week}) \times 1 \text{ week} \\ &= (15 \times 4 \times 1)\ [(\text{hr / day}) \times (\text{days / week}) \times \text{week}] \\ &= 60 \text{ hr} \end{aligned}$$

Energy is the product of power and time.

$$\begin{aligned} \boldsymbol{E} &= \boldsymbol{P} \times \boldsymbol{t} \\ &= 12.5 \text{ W} \times 60 \text{ hr} \\ &= (12.5 \times 60)\ (\text{W} \times \text{hr}) \\ &= 750 \text{ Wh} \end{aligned}$$

From Problem 2 above, one watt-hour equals 3600 joules.

$$\begin{aligned} \boldsymbol{E} &= 750 \text{ Wh} \\ &= (750 \text{ Wh}) \times (3600 \text{ J / 1 Wh}) \\ &= [(750 \times 3600) / 1]\ [(\text{Wh} \times \text{J}) / \text{Wh}] \\ &= 2{,}700{,}000 \text{ J} \\ &= 2.7 \times 10^6 \text{ J} \\ &= 2.7 \text{ MJ} \end{aligned}$$

By turning off his computer at the end of each day, the engineer would save **2.7 MJ** of energy.

4.1.2 Power in an Electric Circuit

1. Problem: Derive Watt's Law for voltage and resistance from Watt's Law for voltage and current ($\boldsymbol{P} = \boldsymbol{V} \times \boldsymbol{I}$).

   Solution: Watt's Law for voltage and current states

$$\boldsymbol{P} = \boldsymbol{V} \times \boldsymbol{I}$$

   Using Ohm's Law for current ($\boldsymbol{I} = \boldsymbol{V} / \boldsymbol{R}$) gives

$$\begin{aligned} \boldsymbol{P} &= \boldsymbol{V} \times \boldsymbol{I} \\ &= \boldsymbol{V} \times (\boldsymbol{V} / \boldsymbol{R}) \\ &= (\boldsymbol{V} \times \boldsymbol{V}) / \boldsymbol{R} \\ &= \boldsymbol{V}^2 / \boldsymbol{R} \end{aligned}$$

   Therefore, Watt's Law for voltage and resistance is $\boldsymbol{P} = \boldsymbol{V}^2 / \boldsymbol{R}$.

2. Problem: Derive Watt's Law for current and resistance from Watt's Law for voltage and current ($\boldsymbol{P} = \boldsymbol{V} \times \boldsymbol{I}$).

   Solution: Watt's Law for voltage and current states

$$\mathbf{P} = \boldsymbol{V} \times \boldsymbol{I}$$

   Using Ohm's Law for voltage ($\text{V} = \text{I} \times \text{R}$) gives

$$\begin{aligned} \boldsymbol{P} &= \boldsymbol{V} \times \boldsymbol{I} \\ &= (\boldsymbol{I} \times \boldsymbol{R}) \times \boldsymbol{I} \\ &= (\boldsymbol{I} \times \boldsymbol{I}) \times \boldsymbol{R} \\ &= \boldsymbol{I}^2 \times \boldsymbol{R} \end{aligned}$$

   Therefore, Watt's Law for current and resistance is $\boldsymbol{P} = \boldsymbol{I}^2 \times \boldsymbol{R}$.

3. Problem: You must replace the fuse in a 120 V, 900 W appliance. What current rating should the fuse have, assuming that it must be twice the value of the actual current?

   Solution: Watt's Law for voltage and current states that power is the product of voltage and current so

$$\boldsymbol{P} = \boldsymbol{V} \times \boldsymbol{I}$$
$$\boldsymbol{P} / \boldsymbol{V} = (\boldsymbol{V} \times \boldsymbol{I}) / \boldsymbol{V}$$
$$\boldsymbol{P} / \boldsymbol{V} = \boldsymbol{I}$$

Therefore,

$$\boldsymbol{I} = \boldsymbol{P} / \boldsymbol{V}$$
$$= 900 \text{ W} / 120 \text{ V}$$
$$= (900 / 120) (\text{W} / \text{V})$$
$$= 7.5 \text{ W} / \text{V}$$
$$= 7.5 \text{ A}$$

Since the fuse rating $\boldsymbol{I_{FUSE}}$ should be twice the actual current,

$$\boldsymbol{I_{FUSE}} = 2 \times 7.5 \text{ A}$$
$$= 15.0 \text{ A}$$

The current rating of the replacement fuse should be **15 A**.

4. Problem: What is the resistance of a portable 9 V heating coil that dissipates five watts of power?

Solution: From Watt's Law for voltage and resistance,

$$\boldsymbol{P} = \boldsymbol{V}^2 / \boldsymbol{R}$$
$$\boldsymbol{P} \times \boldsymbol{R} = (\boldsymbol{V}^2 / \boldsymbol{R}) \times \boldsymbol{R}$$
$$= \boldsymbol{V}^2$$
$$(\boldsymbol{P} \times \boldsymbol{R}) / \boldsymbol{P} = \boldsymbol{V}^2 / \boldsymbol{P}$$
$$\boldsymbol{R} = \boldsymbol{V}^2 / \boldsymbol{P}$$

From this,

$$\boldsymbol{R} = (9 \text{ V})^2 / 5 \text{ W}$$
$$= 81 \text{ V}^2 / 5 \text{ W}$$
$$= (81 / 5) (\text{V}^2 / \text{W})$$
$$= 16.2 \ \Omega$$

The resistance of the portable heating coil is **16 Ω**.

4.1.3 Resistor Power Ratings

1. Problem: What is the absolute maximum current that a 1/4 W, 470 Ω resistor can conduct without damage?

Solution: From Watt's Law for current and resistance,

$$\boldsymbol{P} = \boldsymbol{I}^2 \times \boldsymbol{R}$$
$$\text{W} = \text{A}^2 \bullet \Omega$$

so

$$1/4 \text{ W} = \boldsymbol{I}^2 \times 470 \ \Omega$$
$$0.25 \text{ W} / 470 \ \Omega = (\boldsymbol{I}^2 \times 470 \ \Omega) / 470 \ \Omega$$
$$(0.25 / 470) (\text{W} \times \Omega) = [(\boldsymbol{I}^2 \times 470) / 470] (\Omega / \Omega)$$
$$531.0 \times 10^{-6} [(\text{A}^2 / \Omega) \times \Omega] = \boldsymbol{I}^2$$
$$531.0 \times 10^{-6} (\text{A}^2 \times \Omega) / \Omega = \boldsymbol{I}^2$$
$$531.0 \times 10^{-6} \text{ A}^2 = \boldsymbol{I}^2$$

From this

$$\sqrt{\boldsymbol{I}^2} = \sqrt{531.6 \times 10^{-6} \text{ A}^2}$$
$$\sqrt{\boldsymbol{I} \times \boldsymbol{I}} = (\sqrt{531.6 \times 10^{-6}})(\sqrt{(\text{A} \times \text{A})})$$

$I = 23.06 \times 10^{-3}$ A

$I = 23.06$ mA

The absolute maximum current the 1/4 W, 470 Ω resistor can conduct is **23.1 mA.**

2. Problem: A resistor in a circuit must dissipate 0.5 watt with 12 V connected across it. What resistor value and power rating should you choose?

Solution: From Watt's Law for voltage and resistance

$$\boldsymbol{P} = V^2 / \boldsymbol{R}$$
$$\text{W} = \text{V}^2 / \Omega$$

so

$$0.5\text{ W} = (12\text{ V})^2 / \boldsymbol{R}$$
$$0.5\text{ W} \times \boldsymbol{R} = (144\text{ V}^2 / \boldsymbol{R}) \times \boldsymbol{R}$$
$$(0.5\text{ W} \times \boldsymbol{R}) / 0.5\text{ W} = (144\text{ V}^2) / 0.5\text{ W}$$
$$[(0.5 \times 1) / 0.5]\,[(\text{W} \times \boldsymbol{R}) / \boldsymbol{R}] = (144 / 0.5)\,(\text{V}^2 / \text{W})$$
$$1 \times \boldsymbol{R} = 288\,[\text{V}^2 / (\text{V}^2 / \Omega)]$$
$$\boldsymbol{R} = 288\,[\text{V}^2 \times (\Omega / \text{V}^2)]$$
$$\boldsymbol{R} = 288\ \Omega$$

To safely dissipate 0.5 W, the resistor power rating $\boldsymbol{P_R}$ should be twice that value, or

$$\boldsymbol{P_R} = 2 \times 0.5\text{ W}$$
$$= 1.0\text{ W}$$

You should choose a **1 W, 288 Ω** resistor.

3. Problem: The following three resistors are connected across a variable power supply:

$\boldsymbol{R_1}$: 1/16 W, 100 kΩ

$\boldsymbol{R_2}$: 1/2 W, 10 kΩ

$\boldsymbol{R_3}$: 10 W, 100 Ω

As the voltage across the resistors increases from 0 V, which resistor will fail first and at what voltage? Which resistor will fail last and at what voltage?

Solution: Watt's Law for voltage and resistance states

$$\boldsymbol{P} = V^2 / \boldsymbol{R}$$
$$\text{W} = \text{V}^2 / \Omega$$

From this, the maximum voltage a resistor can withstand is

$$\boldsymbol{P} \times \boldsymbol{R} = (V^2 / \boldsymbol{R}) \times \boldsymbol{R}$$
$$= V^2$$

so

$$\sqrt{V^2} = \sqrt{\boldsymbol{P} \times \boldsymbol{R}}$$
$$\sqrt{V \times V} = \sqrt{\boldsymbol{P} \times \boldsymbol{R}}$$
$$V = \sqrt{\boldsymbol{P} \times \boldsymbol{R}}$$

For $\boldsymbol{R_1}$,

$$\boldsymbol{P} \times \boldsymbol{R} = 1/16\text{ W} \times 100\text{ k}\Omega$$
$$= (0.0625\text{ W}) \times (100 \times 10^3\ \Omega)$$
$$= [0.0625 \times (100 \times 10^3)]\,(\text{W} \times \Omega)$$
$$= 6.25 \times 10^3\,[(\text{V}^2 / \Omega) \times \Omega]$$

$= 6.25 \times 10^3\ \mathrm{V}^2$

so

$$V_{MAX}(\boldsymbol{R}_1) = \sqrt{6.25\times 10^3\ \mathrm{V}^2}$$
$$= (\sqrt{6.25\times 10^3})(\sqrt{(\mathrm{V}\times\mathrm{V})})$$
$$= 79.06\ \mathrm{V}$$

For $\boldsymbol{R}_2$,

$\boldsymbol{P}\times\boldsymbol{R} = 1/2\ \mathrm{W} \times 10\ \mathrm{k\Omega}$

$= (0.5\ \mathrm{W}) \times (10 \times 10^3\ \Omega)$

$= [0.5 \times (10 \times 10^3)]\ (\mathrm{W} \times \Omega)$

$= 5.0 \times 10^3\ [(\mathrm{V}^2 / \Omega) \times \Omega]$

$= 5.0 \times 10^3\ \mathrm{V}^2$

so

$$V_{MAX}(\boldsymbol{R}_2) = \sqrt{5.0\times 10^3\ \mathrm{V}^2}$$
$$= (\sqrt{5.0\times 10^3})(\sqrt{(\mathrm{V}\times\mathrm{V})})$$
$$= 70.71\ \mathrm{V}$$

For $\boldsymbol{R}_3$,

$\boldsymbol{P}\times\boldsymbol{R} = 10\ \mathrm{W} \times 100\ \Omega$

$= (10\ \mathrm{W}) \times (100\ \Omega)$

$= [10 \times 100]\ (\mathrm{W} \times \Omega)$

$= 1000\ [(\mathrm{V}^2 / \Omega) \times \Omega]$

$= 1000\ \mathrm{V}^2$

so

$$V_{MAX}(\boldsymbol{R}_3) = \sqrt{1000\ \mathrm{V}^2}$$
$$= (\sqrt{1000})(\sqrt{(\mathrm{V}\times\mathrm{V})})$$
$$= 31.62\ \mathrm{V}$$

$\boldsymbol{R}_3$ has the lowest maximum voltage, so **$\boldsymbol{R}_3$ will be the first resistor to fail at 32 V**. Similarly, $\boldsymbol{R}_1$ has the highest maximum voltage, so **$\boldsymbol{R}_1$ will the last resistor to fail at 79.1 V**. Note that the value of V^2 was enough to determine the first and last resistors to fail, as the largest and smallest values of V^2 will yield the largest and smallest values of V_{MAX}, respectively.

4.1.4 Power Supplies and Batteries

1. Problem: A power supply that delivers 5.0 W of output power $\boldsymbol{P}_{OUT}$ internally dissipates 0.25 W as waste heat $\boldsymbol{P}_{LOSS}$. What is the efficiency of the supply?

   Solution: Power supply efficiency η ("eta") is defined to be

   $$\eta = P_{OUT} / P_{IN}$$

   where

   $$P_{OUT} = P_{IN} - P_{LOSS}$$
   $$P_{OUT} + P_{LOSS} = P_{IN} + -P_{LOSS} + P_{LOSS}$$
   $$P_{OUT} + P_{LOSS} = P_{IN}$$

   From this

   $$\eta = P_{OUT} / P_{IN}$$

$= \boldsymbol{P_{OUT}} / (\boldsymbol{P_{OUT}} + \boldsymbol{P_{LOSS}})$

$= 5.0\ \text{W} / (5.0\ \text{W} + 0.25\ \text{W})$

$= 5.0\ \text{W} / [(5.0 + 0.25)\ \text{W}]$

$= 5.0\ \text{W} / 5.25\ \text{W}$

$= (5.0 / 5.25)\ (\text{W} / \text{W})$

$= 0.952$

The efficiency of the 5.0 W supply is **0.95**.

2. Problem: A 350 mA constant current power supply with an efficiency of 80% is connected to a 75 Ω load $\boldsymbol{R_{LOAD}}$. What is the input power $\boldsymbol{P_{IN}}$ to the supply?

Solution: By definition,

$$\eta = \boldsymbol{P_{OUT}} / \boldsymbol{P_{IN}}$$

$$\eta \times \boldsymbol{P_{IN}} = (\boldsymbol{P_{OUT}} / \boldsymbol{P_{IN}}) \times \boldsymbol{P_{IN}}$$

$$(\eta \times \boldsymbol{P_{IN}}) / \eta = \boldsymbol{P_{OUT}} / \eta$$

$$\boldsymbol{P_{IN}} = \boldsymbol{P_{OUT}} / \eta$$

From Watt's Law for current and resistance

$$\boldsymbol{P} = \boldsymbol{I}^2 \times \boldsymbol{R}$$

$$\text{W} = \text{A}^2 \times \Omega$$

so

$$\boldsymbol{P_{OUT}} = \boldsymbol{I_{OUT}}^2 \times \boldsymbol{R_{LOAD}}$$

$= (350\ \text{mA})^2 \times 75\ \Omega$

$= (350 \times 10^{-3}\ \text{A})^2 \times 75\ \Omega$

$= [(350 \times 10^{-3})^2 \times 75]\ (\text{A}^2 \times \Omega)$

$= [(122.5 \times 10^{-3}) \times 75]\ \text{W}$

$= 9.188\ \text{W}$

Therefore,

$$\boldsymbol{P_{IN}} = \boldsymbol{P_{OUT}} / \eta$$

$= 9.188\ \text{W} / 80\%$

$= 9.188 / (80 / 100)$

$= 9.188 / 0.8$

$= 11.48\ \text{W}$

The input power to the constant current power supply is **11.5** W.

3. Problem: A rechargeable 1.5 V battery has a rating of 1150 mAh. How long can the battery supply current to a 30 Ω load?

Solution: From Ohm's Law for current, a 30 Ω load connected to a 1.5 V battery will draw

$$\boldsymbol{I} = \boldsymbol{V} / \boldsymbol{R}$$

$= 1.5\ \text{V} / 30\ \Omega$

$= (1.5 / 30)\ (\text{V} / \Omega)$

$= 50 \times 10^{-3}\ \text{A}$

By definition, an ampere-hour is the product of current (in amperes) and time (in hours).

$$1150\ \text{mAh} = \boldsymbol{I} \times \boldsymbol{t}$$

$$1150\ \text{mAh} / \boldsymbol{I} = (\boldsymbol{I} \times \boldsymbol{t}) / \boldsymbol{I}$$

$$= \boldsymbol{t}$$

From this,

$t = 1150 \text{ mAh} / \boldsymbol{I}$

$= (1150 \times 10^{-3} \text{ Ah}) / (50 \times 10^{-3} \text{ A})$

$= (1150 \times 10^{-3} / 50 \times 10^{-3}) [(\text{A} \times \text{hr}) / \text{A}]$

$= 23 \text{ hr}$

A 1.5 V battery with a rating of 1150 mAh can supply current to a 30 Ω load for **23 hours**.

4. Problem: An independent testing lab determines that a rechargeable 3.3 V battery pack rated for 2.0 Ah can supply continuous current to a 1.00 kΩ load for four weeks without recharging. Does the battery pack meet its specified ampere-hour rating?

Solution: From Ohm's Law for current,

$\boldsymbol{I} = \boldsymbol{V} / \boldsymbol{R}$

$= 3.3 \text{ V} / 1.00 \text{ k}\Omega$

$= 3.3 \text{ V} / (1.00 \times 10^3 \,\Omega)$

$= [3.3 / (1.00 \times 10^3)] (\text{V} / \Omega)$

$= 3.30 \times 10^{-3} \text{ A}$

The battery rating is equal to current multiplied by time, so

Rating $= (3.30 \times 10^{-3} \text{ A}) \times 4 \text{ weeks}$

$= [(3.30 \times 10^{-3}) \times 4] (\text{A} \times \text{weeks})$

$= 13.2 \times 10^{-3} (\text{A} \times \text{weeks})$

One week is equal to seven days, so

Rating $= (13.2 \times 10^{-3} \text{ A} \times \text{weeks}) \times (7 \text{ days} / \text{week})$

$= [(13.2 \times 10^{-3}) \times 7] [(\text{A} \times \text{weeks} \times \text{days}) / \text{week}]$

$= 92.4 \times 10^{-3} (\text{A} \times \text{days})$

One day is equal to 24 hour, so

Rating $= (92.4 \times 10^{-3} \text{ A} \times \text{days}) \times (24 \text{ hours} / \text{day})$

$= [(92.4 \times 10^{-3}) \times 24] [(\text{A} \times \text{days} \times \text{hours}) / \text{day}]$

$= 2.22 (\text{A} \times \text{hours})$

$= 2.22 \text{ Ah}$

The battery pack **does** meet its specified 2.0 Ah rating.

4.2 Advanced

4.2.1 Energy and Power

1. Problem: An engineer modifies a motor so that it dissipates half the original energy in one-third the original time. How does the power change?

Solution: By definition power is energy divided by time so the original power $\boldsymbol{P_{ORG}}$ is

$\boldsymbol{P_{ORG}} = \boldsymbol{E_{ORG}} / \boldsymbol{t_{ORG}}$

Similarly, the new power $\boldsymbol{P_{NEW}}$ is

$\boldsymbol{P_{NEW}} = \boldsymbol{E_{NEW}} / \boldsymbol{t_{NEW}}$

Since the new energy is half the original energy and the new time is one-third the original time,

$\boldsymbol{P_{NEW}} = (\boldsymbol{E_{ORG}} / 2) / (\boldsymbol{t_{ORG}} / 3)$

$= (\boldsymbol{E_{ORG}} / 2) \times (3 / \boldsymbol{t_{ORG}})$

$= (\boldsymbol{E_{ORG}} \times 3) / (2 \times \boldsymbol{t_{ORG}})$

$= (3 / 2) \times (\boldsymbol{E_{ORG}} / \boldsymbol{t_{ORG}})$

$= 1.5 \times \boldsymbol{P_{ORG}}$

The change in power, $\Delta\boldsymbol{P}$, is the difference between the new and original values.

$$
\begin{aligned}
\Delta P &= P_{NEW} - P_{ORG} \\
&= 1.5 \times P_{ORG} - P_{ORG} \\
&= (1.5 \times P_{ORG}) - (1 \times P_{ORG}) \\
&= (1.5 - 1) \times P_{ORG} \\
&= 0.5 \times P_{ORG} \\
&= (50/100) \times P_{ORG} \\
&= 50\% \times P_{ORG}
\end{aligned}
$$

The power **increases by 50%**.

4.2.2 Power in an Electric Circuit

1. Problem: Show that the product of voltage in volts and current in amperes equals power in watts.

   Solution: By definition,

   $$1\ \text{V} = 1\ \text{J} / \text{C}$$

   and

   $$1\ \text{A} = 1\ \text{C} / \text{s}$$

   so

   $$
   \begin{aligned}
   1\ \text{V} \times 1\ \text{A} &= 1\ \text{V} \times 1\ \text{A} \\
   &= (1\ \text{J} / \text{C}) \times (1\ \text{C} / \text{s}) \\
   &= (1 \times 1)\ [(\text{J} \times \text{C}) / (\text{C} \times \text{s})] \\
   &= 1\ \text{J} / \text{s} \\
   &= 1\ \text{W}
   \end{aligned}
   $$

   The product of voltage and current is energy per time, which by definition is power, and the product of volts and amperes gives watts.

2. Problem: The voltage across a circuit increases by 20%. The voltage change damages the circuit so the current decreases by 30%. By how much does the power change?

   Solution: From Watt's Law for voltage and current

   $$P = V \times I$$

   so the original power P_{ORG} is

   $$P_{ORG} = V_{ORG} \times I_{ORG}$$

   Similarly, the new power P_{NEW} is

   $$P_{NEW} = V_{NEW} \times I_{NEW}$$

   Since the new voltage is 20% more than the original voltage and the new current is 30% less than the original current,

   $$
   \begin{aligned}
   P_{NEW} &= (V_{ORG} + V_{ORG} \times 20\%) \times (I_{ORG} - I_{ORG} \times 30\%) \\
   &= [(V_{ORG} \times 1) + (V_{ORG} \times 20\%)] \times [(I_{ORG} \times 1) - (I_{ORG} \times 30\%)] \\
   &= [V_{ORG} \times (1 + 20\%)] \times [I_{ORG} \times (1 - 30\%)] \\
   &= [V_{ORG} \times (1 + 20/100)] \times [I_{ORG} \times (1 - 30/100)] \\
   &= [V_{ORG} \times (1 + 0.2)] \times [I_{ORG} \times (1 - 0.3)] \\
   &= (V_{ORG} \times 1.2) \times (I_{ORG} \times 0.7) \\
   &= (1.2 \times 0.7) \times (V_{ORG} \times I_{ORG}) \\
   &= 0.84 \times P_{ORG}
   \end{aligned}
   $$

   The change in power ΔP is

   $$\Delta P = P_{NEW} - P_{ORG}$$

$= 0.84 \times P_{ORG} - P_{ORG}$

$= (0.84 \times P_{ORG}) - (1 \times P_{ORG})$

$= (0.84 - 1) \times P_{ORG}$

$= -0.16 \times P_{ORG}$

$= -(16/100) \times P_{ORG}$

$= -16\% \times P_{ORG}$

The power **decreases by 16%**.

4.2.3 Resistor Power Ratings

1. Problem: A 3.3 V switching power supply requires a resistive load that draws 25 mA to remain stable. If the load is a single resistor, what must the resistor's minimum power rating be?

   Solution: From Watt's Law for voltage and current,

   $$\boldsymbol{P} = \boldsymbol{V} \times \boldsymbol{I}$$
   $$\text{W} = \text{V} \times \text{A}$$

   so

   $$\boldsymbol{P} = \boldsymbol{V} \times \boldsymbol{I}$$
   $$= 3.3\ \text{V} \times 25\ \text{mA}$$
   $$= 3.3\ \text{V} \times (25 \times 10^{-3}\ \text{A})$$
   $$= [3.3 \times (25 \times 10^{-3})]\ (\text{V} \times \text{A})$$
   $$= 82.5 \times 10^{-3}\ \text{W}$$
   $$= 82.5\ \text{mW}$$

   Minimum standard power ratings for resistors that are greater than these are 0.1 W and 1/8 W (0.125 W), although most designs would use a 1/4 W (0.25 W) resistor to ensure that the supply operates the resistor well below its power rating and extends the life of the component.

2. Problem: A fixed resistor is at 40.0% of its maximum power rating $\boldsymbol{P_{MAX}}$. How much can the voltage across the resistor $\boldsymbol{R}$ increase before the power exceeds the resistor's maximum power rating?

   Solution: When the resistor is at 40% of its maximum power rating

   $$\boldsymbol{P_{40\%}} = (\boldsymbol{V_{40\%}})^2 / \boldsymbol{R}$$
   $$40.0\% \times \boldsymbol{P_{MAX}} = (\boldsymbol{V_{40\%}})^2 / \boldsymbol{R}$$
   $$(40.0/100) \times \boldsymbol{P_{MAX}} = (\boldsymbol{V_{40\%}})^2 / \boldsymbol{R}$$
   $$0.40 \times \boldsymbol{P_{MAX}} = (\boldsymbol{V_{40\%}})^2 / \boldsymbol{R}$$

   From this

   $$(0.40 \times \boldsymbol{P_{MAX}}) / 0.40 = [(\boldsymbol{V_{40\%}})^2 / \boldsymbol{R}] / 0.40$$
   $$\boldsymbol{P_{MAX}} = [(\boldsymbol{V_{40\%}})^2 / 0.40] / \boldsymbol{R}$$
   $$= [(\boldsymbol{V_{40\%}})^2 / (\sqrt{0.40})^2] / \boldsymbol{R}$$
   $$= [(\boldsymbol{V_{40\%}} / \sqrt{0.40})^2] / \boldsymbol{R}$$
   $$= (\boldsymbol{V_{40\%}} / 0.632)^2 / \boldsymbol{R}$$
   $$= [\boldsymbol{V_{40\%}} \times (1 / 0.632)]^2 / \boldsymbol{R}$$
   $$= (\boldsymbol{V_{40\%}} \times 1.58)^2 / \boldsymbol{R}$$

   The voltage across the resistor can increase **1.58 times** before the power exceeds the resistor's maximum voltage rating.

4.2.4 Power Supplies and Batteries

1. Problem: Determine the efficiency of a power supply in terms of $\boldsymbol{P_{OUT}}$ and $\boldsymbol{P_{LOSS}}$.

Solution: The efficiency of a power supply is defined as

$$\eta = P_{OUT} / P_{IN}$$

and

$$P_{OUT} = P_{IN} - P_{LOSS}$$

From these,

$$P_{OUT} = P_{IN} - P_{LOSS}$$
$$P_{OUT} + P_{LOSS} = P_{IN} + -P_{LOSS} + P_{LOSS}$$
$$P_{OUT} + P_{LOSS} = P_{IN}$$

so

$$\eta = P_{OUT} / P_{IN}$$
$$= P_{OUT} / (P_{OUT} + P_{LOSS})$$

The efficiency of a power supply in terms of P_{OUT} and P_{LOSS} is $\eta = P_{OUT} / (P_{OUT} + P_{LOSS})$.

2. Problem: A fixed 12.0 V power supply dissipates a constant 0.50 W of power as waste heat regardless of the load. What are the minimum and maximum efficiencies of the supply if a 100 Ω resistor with a tolerance of 20% is connected to the output of the supply?

Solution: The actual value for a resistor with a nominal value R_{NOM} with a tolerance of 20% is

$$R_{ACTUAL} = R_{NOM} - (R_{NOM} \times \pm 20\%)$$
$$= (R_{NOM} \times 1) - (R_{NOM} \pm 20\%)$$
$$= R_{NOM} \times (1 \pm 20\%)$$
$$= R_{NOM} \times (1 \pm 20/100)$$
$$= R_{NOM} \times (1 \pm 0.20)$$

The minimum value R_{MIN} is

$$R_{MIN} = R_{NOM} \times (1 - 0.20)$$
$$= 100\ \Omega \times 0.80$$
$$= 80\ \Omega$$

and the maximum value R_{MAX} is

$$R_{MAX} = R_{NOM} \times (1 + 0.20)$$
$$= 100\ \Omega \times 1.20$$
$$= 120\ \Omega$$

From Watt's Law for voltage and resistance

$$P = V^2 / R$$
$$\text{W} = \text{V}^2 / \Omega$$

When voltage is fixed, power will be maximum when resistance is minimum and vice versa, so

$$P_{OUT\,(\text{max})} = (V_{OUT})^2 / R_{MIN}$$
$$= (12.0\ \text{V})^2 / 80\ \Omega$$
$$= 144.0\ \text{V}^2 / 80\ \Omega$$
$$= (144.0 / 80)\ (\text{V}^2 / \Omega)$$
$$= 1.80\ \text{W}$$

and

$$P_{OUT\,(\text{min})} = (V_{OUT})^2 / R_{MAX}$$
$$= (12.0\ \text{V})^2 / 120\ \Omega$$
$$= 144.0\ \text{V}^2 / 120\ \Omega$$
$$= (144.0 / 120)\ (\text{V}^2 / \Omega)$$

$= 1.20\text{ W}$

From Problem 1 above, the efficiency of a power supply in terms of $\boldsymbol{P_{OUT}}$ and $\boldsymbol{P_{LOSS}}$ is

$$\eta = P_{OUT} / (P_{OUT} + P_{LOSS})$$

When $\boldsymbol{P_{OUT}}$ is minimum,

$$\begin{aligned}\eta &= P_{OUT\,(\mathrm{min})} / (P_{OUT\,(\mathrm{min})} + P_{LOSS})\\ &= 1.20\text{ W} / (1.20\text{ W} + 0.5\text{ W})\\ &= 1.20\text{ W} / [(1.20 + 0.50)\text{ W}]\\ &= 1.20\text{ W} / 1.70\text{ W}\\ &= (1.20 / 1.70)\ (\text{W} / \text{W})\\ &= 0.7059\end{aligned}$$

When $\boldsymbol{P_{OUT}}$ is maximum,

$$\begin{aligned}\eta &= P_{OUT\,(\mathrm{max})} / (P_{OUT\,(\mathrm{max})} + P_{LOSS})\\ &= 1.80\text{ W} / (1.80\text{ W} + 0.5\text{ W})\\ &= 1.80\text{ W} / [(1.80 + 0.50)\text{ W}]\\ &= 1.80\text{ W} / 2.30\text{ W}\\ &= (1.80 / 2.30)\ (\text{W} / \text{W})\\ &= 0.7826\end{aligned}$$

The minimum and maximum efficiencies of the power supply are **0.71** and **0.78**, respectively.

3. Problem: In Problem 2 above the efficiency η of the power supply was maximum when $\boldsymbol{P_{OUT}}$ was maximum and minimum when $\boldsymbol{P_{OUT}}$ was minimum. Show that when $\boldsymbol{P_{LOSS}}$ is fixed this will always be true.

Solution: From Problem 1

$$\eta = P_{OUT} / (P_{OUT} + P_{LOSS})$$

so that

$$\begin{aligned}1 / \eta &= 1 / [P_{OUT} / (P_{OUT} + P_{LOSS})]\\ &= (P_{OUT} + P_{LOSS}) / P_{OUT}\\ &= (P_{OUT} / P_{OUT}) + (P_{LOSS} / P_{OUT})\\ &= 1 + (P_{LOSS} / P_{OUT})\end{aligned}$$

When $\boldsymbol{P_{LOSS}}$ is fixed, the ratio $\boldsymbol{P_{LOSS} / P_{OUT}}$ will be maximum when $\boldsymbol{P_{OUT}}$ is minimum and minimum when $\boldsymbol{P_{OUT}}$ is maximum. Therefore, $1 + (P_{LOSS} / P_{OUT})$ will be maximum when P_{LOSS} / P_{OUT} is maximum and minimum when $P_{LOSS}/\mathrm{P_{OUT}}$ is minimum. But $1 + (P_{LOSS} / P_{OUT}) = 1/\eta$, so that $1/\eta$ is also maximum when $\boldsymbol{P_{OUT}}$ is minimum and minimum when $\boldsymbol{P_{OUT}}$ is maximum. Because $1/\eta$ is the reciprocal of the efficiency η, η is maximum when $1/\eta$ is minimum, and minimum when $1/\eta$ is maximum. Therefore, **the efficiency η is maximum when P_{OUT} is maximum and minimum when P_{OUT} is minimum.**

4. Problem: Determine whether or not the ampere-hour is a measure of battery energy.

Solution: Since by definition an ampere-hour is a product of current (in amperes) and time (in hours)

Ampere-hour = current × time

Current is charge divided by time, so

Ampere-hour = current × time
= (charge / time) × time
= charge

The ampere-hour is a unit of battery **charge** rather than battery energy.

4.3 Just for Fun

Every device has a maximum power rating which specifies how much power the device can safely dissipate. This rating is ultimately based on the maximum temperature that the device can withstand before its physical structure is permanently affected. As the device dissipates power, heat builds up in the device and raises its temperature. At some point the heat from the dissipated power will raise the temperature of the device above its maximum temperature. The exact power rating of a device depends on how well the device can eliminate heat and prevent its temperature from increasing above the maximum temperature the device can withstand.

Heat is a particularly important consideration for semiconductor devices such as diodes, transistors, and integrated circuits. One measure of how well a device can eliminate heat is its **thermal resistance**, or $\boldsymbol{\theta}$ ("theta"). The thermal resistance, typically specified in °C / W, determines by how many degrees a specific amount of power will raise a device's temperature. Refer to the simplified diagram of a semiconductor in Figure 4-1.

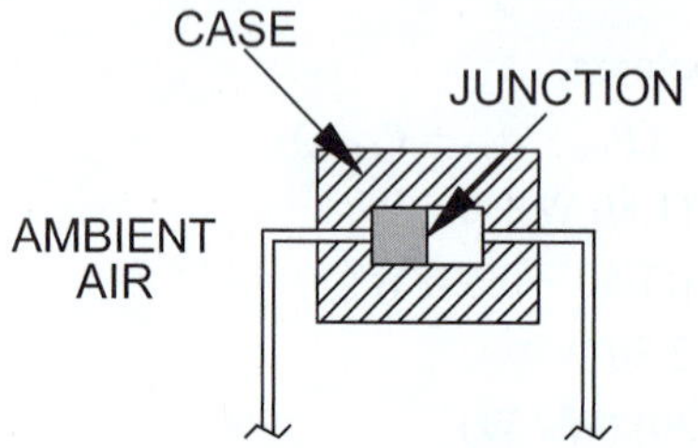

Figure 4-1: Simplified Semiconductor Diagram

As the semiconductor operates, the junction dissipates power and generates heat. The junction-to-case thermal resistance $\boldsymbol{\theta_{JC}}$ determines how effectively the junction transfers heat to the case and how much higher the junction temperature $\boldsymbol{T_J}$ is than the case temperature $\boldsymbol{T_C}$. Similarly, the junction-to-ambient thermal resistance $\boldsymbol{\theta_{JA}}$ determines how effectively the junction transfers heat to the ambient, or surrounding, air and how much higher the junction temperature is than the ambient temperature $\boldsymbol{T_A}$. The case-to-ambient thermal resistance $\boldsymbol{\theta_{CA}}$ determines how effectively the case transfers heat to the ambient air and how much higher the case temperature $\boldsymbol{T_C}$ is than the ambient temperature $\boldsymbol{T_A}$ but is rarely specified or used. Equations that relate temperatures, dissipated power $\boldsymbol{P_D}$, and thermal resistance are

$\boldsymbol{T_J = T_C + (P_D \times \theta_{JC})}$ The junction temperature equals the case temperature plus the product of the dissipated power and junction-to-case thermal resistance.

$\boldsymbol{T_J = T_A + (P_D \times \theta_{JA})}$ The junction temperature equals the ambient temperature plus the product of the dissipated power and junction-to-ambient thermal resistance.

$\boldsymbol{T_C = T_A + (P_D \times \theta_{CA})}$ The case temperature equals the ambient temperature plus the product of the dissipated power and case-to-ambient thermal resistance.

Example 1:

Problem: The ambient temperature surrounding a semiconductor voltage regulator is 25 °C. If $\boldsymbol{\theta_{JA}}$ for the regulator is 60 °C/W, what is the junction temperature when the regulator dissipates 1.5 W?

Solution: From the equation for finding the junction temperature relative to the ambient temperature,

$$\boldsymbol{T_J = T_A + (P_D \times \theta_{JA})}$$

so

$$
\begin{aligned}
T_J &= 25\ °\text{C} + [(1.5\ \text{W}) \times (60\ °\text{C} / \text{W})] \\
&= 25\ °\text{C} + (1.5 \times 60)\ [\text{W} \times (°\text{C} / \text{W})] \\
&= 25\ °\text{C} + 90\ °\text{C} \\
&= (25 + 90)\ °\text{C} \\
&= 115\ °\text{C}
\end{aligned}
$$

Sometimes the intrinsic $\boldsymbol{\theta_{CA}}$ of a semiconductor is not enough to keep the junction of a semiconductor from overheating. In that case, the device requires a heatsink. The heatsink provides a low case-to-ambient thermal resistance that improves the transfer of heat to the ambient air so that the case temperature (and consequently junction temperature) is lower than it would be for the case of the semiconductor device alone.

Example 2:

Problem: The datasheet for a semiconductor device specifies the following thermal data:

$\boldsymbol{T_J}$ (max)= 180 °C

$\boldsymbol{\theta_{JA}}$ = 70 °C

$\boldsymbol{\theta_{JC}}$ = 10 °C

The device must dissipate 1.5 watts when the ambient temperature is 85 °C. Will the device require a heatsink? If so, what must be the maximum thermal resistance of the heatsink?

Solution: From the equation for finding the junction temperature relative to the ambient temperature,

$$\boldsymbol{T_J = T_A + (P_D \times \theta_{JA})}$$

so

$$\begin{aligned} \boldsymbol{T_J} &= 85\ °\text{C} + [1.5\ \text{W} \times (70\ °\text{C} / \text{W})] \\ &= 85\ °\text{C} + (1.5 \times 70)\ [\text{W} \times (°\text{C} / \text{W})] \\ &= 85\ °\text{C} + 105\ °\text{C} \\ &= (85 + 105)\ °\text{C} \\ &= 190\ °\text{C} \end{aligned}$$

This temperature exceeds the $\boldsymbol{T_J}$ (max) specification of 180 °C so the device requires a heatsink. The equation for finding the case temperature relative to the junction temperature is

$$\boldsymbol{T_J = T_C + (P_D \times \theta_{JC})}$$

so if $\boldsymbol{T_J}$ = 180 °C,

$$\begin{aligned} \boldsymbol{T_J} &= \boldsymbol{T_C} + [1.5\ \text{W} \times (10\ °\text{C} / \text{W})] \\ 180\ °\text{C} &= \boldsymbol{T_C} + (1.5 \times 10)\ [\text{W} \times)°\text{C} / \text{W})] \\ &= \boldsymbol{T_C} + 15\ °\text{C} \end{aligned}$$

The case temperature must then be no more than

$$\begin{aligned} 180\ °\text{C} - 15\ °\text{C} &= \boldsymbol{T_C}\ (\text{max}) + 15\ °\text{C} - 15\ °\text{C} \\ (180 - 15)\ °\text{C} &= \boldsymbol{T_C}\ (\text{max}) + (15 - 15)\ °\text{C} \\ 165\ °\text{C} &= \boldsymbol{T_C}\ (\text{max}) + 0\ °\text{C} \\ &= \boldsymbol{T_C}\ (\text{max}) \end{aligned}$$

The equation for finding the case temperature relative to the ambient temperature is

$$\boldsymbol{T_C = T_A + (P_D \times \theta_{CA})}$$

so

$$\begin{aligned} \boldsymbol{T_C - T_A} &= \boldsymbol{T_A + (P_D \times \theta_{CA}) - T_A} \\ &= \boldsymbol{(T_A - T_A) + P_D \times \theta_{CA}} \\ &= \boldsymbol{P_D \times \theta_{CA}} \\ \boldsymbol{[T_C - T_A] / P_D} &= \boldsymbol{(P_D \times \theta_{CA}) / P_D} \\ &= \boldsymbol{\theta_{CA}} \end{aligned}$$

For $\boldsymbol{P_D}$ = 1.5 W, $\boldsymbol{T_C}$ (max) = 165 °C, and $\boldsymbol{T_A}$ = 85 °C, the case-to-ambient thermal resistance of the heatsink must be no more than

$$\begin{aligned} \boldsymbol{\theta_{CA}}\ (\text{max}) &= [\boldsymbol{T_C}\ (\text{max}) - \boldsymbol{T_A}] / \boldsymbol{P_D} \\ &= (165\ °\text{C} - 85\ °\text{C}) / 1.5\ \text{W} \\ &= [(165 - 85)\ °\text{C}] / 1.5\ \text{W} \\ &= 80\ °\text{C} / 1.5\ \text{W} \\ &= (80 / 1.5)\ (°\text{C} / \text{W}) \end{aligned}$$

$= 53.3\ °C / W$

1. Problem: Show that $\theta_{JA} = \theta_{JC} + \theta_{CA}$.

Solution: The basic temperature rise equations for power dissipation are

$$T_J = T_C + (P_D \times \theta_{JC})$$
$$T_J = T_A + (P_D \times \theta_{JA})$$
$$T_C = T_A + (P_D \times \theta_{CA})$$

From these,

$$T_J + T_C = [T_C + (P_D \times \theta_{JC})] + [T_A + (P_D \times \theta_{CA})]$$
$$T_J + T_C + -T_C = T_C + T_A + (P_D \times \theta_{JC}) + (P_D \times \theta_{CA}) + -T_C$$
$$T_J = T_A + (P_D \times \theta_{JC}) + (P_D \times \theta_{CA})$$
$$= T_A + P_D \times (\theta_{JC} + \theta_{CA})$$

Since $T_J = T_A + (P_D \times \theta_{JA})$,

$$T_A + (P_D \times \theta_{JA}) = T_A + P_D \times (\theta_{JC} + \theta_{CA})$$
$$T_A + P_D \times \theta_{JA} + -T_A = T_A + P_D \times (\theta_{JC} + \theta_{CA}) + -T_A$$
$$P_D \times \theta_{JA} = P_D \times (\theta_{JC} + \theta_{CA})$$
$$(P_D \times \theta_{JA}) / P_D = [P_D \times (\theta_{JC} + \theta_{CA})] / P_D$$
$$\theta_{JA} = \theta_{JC} + \theta_{CA}$$

Another approach would be to solve individually for θ_{JC} and θ_{CA} and add the two expressions.

2. Problem: The Engineering department designs a circuit in which a voltage regulator has a voltage drop of 6.0 V between the input and output and draws 250 mA. If T_J (max) = 150 °C and θ_{JA} = 60 °C/W, what is the maximum ambient temperature at which it can operate?

Solution: From Watt's Law for voltage and current, the power dissipation of the regulator is

$$P_D = V \times I$$
$$= 6.0\ V \times 250\ mA$$
$$= 6.0\ V \times (250 \times 10^{-3})\ A$$
$$= [6.0 \times (250 \times 10^{-3})]\ (V \times A)$$
$$= 1.50\ W$$

The equation for the ambient-to-junction temperature rise is

$$T_J = T_A + (P_D \times \theta_{JA})$$

so

$$T_J + -(P_D \times \theta_{JA}) = T_A + (P_D \times \theta_{JA}) + -(P_D \times \theta_{JA})$$
$$= T_A$$

From this,

$$T_A(\text{max}) = T_J(\text{max}) + -(P_D \times \theta_{JA})$$
$$= 150\ °C + -[1.50\ W \times (60\ °C / W)]$$
$$= 150\ °C + -(1.50 \times 60)\ [(W \times °C) / W]$$
$$= 150\ °C + -90\ °C$$
$$= (150 + -90)\ °C$$
$$= 60\ °C$$

The maximum ambient temperature at which the regulator can operate is **60 °C**.

3. Problem: The Marketing department advises the Engineering department that the circuit of Problem 2 above must operate at 70 °C to have an adequate customer base. The Engineering manager decides to add a heatsink to the circuit rather than purchase a more expensive voltage regulator with higher

temperature ratings and assigns you to find one. If $\theta_{JC} = 5$ °C/W for the regulator, what must be the thermal resistance of the heatsink?

Solution: The equation for finding the case temperature relative to the junction temperature is

$$T_J = T_C + (P_D \times \theta_{JC})$$

T_J (max) = 180 °C so

$$T_J\,(\text{max}) = T_C\,(\text{max}) + [1.5\ \text{W} \times (5\ °\text{C / W})]$$

$$150\ °\text{C} = T_C\,(\text{max}) + (1.5 \times 5)\ [(\text{W} \times °\text{C}) / \text{W}]$$

$$= T_C\,(\text{max}) + 7.5\ °\text{C}$$

The maximum case temperature is then

$$180\ °\text{C} + -7.5\ °\text{C} = T_C\,(\text{max}) + 7.5\ °\text{C} + -7.5\ °\text{C}$$

$$(180 - 7.5)\ °\text{C} = T_C\,(\text{max})$$

$$172.5\ °\text{C} = T_C\,(\text{max})$$

The equation to find the case temperature relative to the ambient temperature is

$$T_C\,(\text{max}) = T_A\,(\text{max}) + (P_D \times \theta_{CA})$$

so

$$T_C\,(\text{max}) + -T_A\,(\text{max}) = T_A\,(\text{max}) + (P_D \times \theta_{CA}) + -T_A(\text{max})$$

$$= P_D \times \theta_{CA}$$

$$[T_C\,(\text{max}) - T_A\,(\text{max})] / P_D = (P_D \times \theta_{CA}) / P_D$$

$$= \theta_{CA}$$

For $P_D = 1.5$ W, T_C (max) = 172.5 °C, and T_A (max) = 70 °C, the case-to-ambient thermal resistance of the heatsink must be no more than

$$\theta_{CA} = [T_C\,(\text{max}) - T_A\,(\text{max})] / P_D$$

$$= (172.5\ °\text{C} - 70\ °\text{C}) / 1.5\ \text{W}$$

$$= [(172.5 - 70)\ °\text{C}] / 1.5\ \text{W}$$

$$= 102.5\ °\text{C} / 1.5\ \text{W}$$

$$= (102.5 / 1.5)\ (°\text{C / W})$$

$$= 68.3\ °\text{C / W}$$

The maximum case-to-ambient thermal resistance of the heatsink must be **68.3 °C/W**.

5. Series Circuits

5.1 Basic

5.1.1 Resistors in Series

1. Problem: Which of the resistive circuits in Figure 5-1 are series circuits?

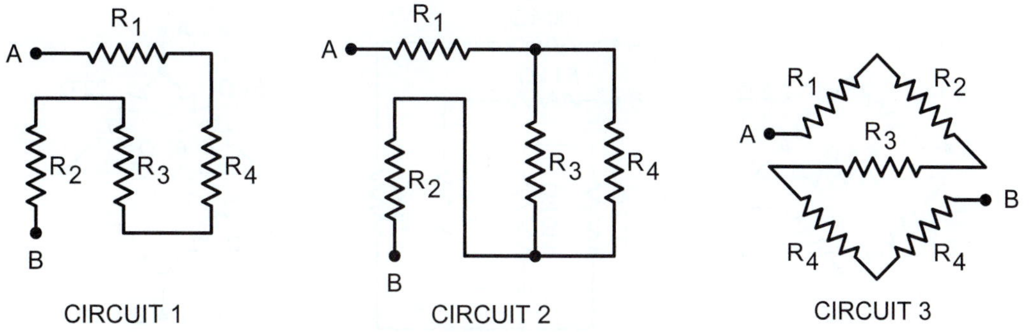

Figure 5-1: Series Circuit Identification

Solution: To determine whether or not each circuit is a series circuit, trace the possible current paths from Point A to Point B.

For Circuit 1, current from Point A must pass through R_1, R_4, R_3, and R_2 to get to Point B. Because the current **must** flow through each of the resistors to get from Point A to Point B, the circuit is a series circuit.

For Circuit 2, current from Point A has two possible paths to get to Point B. One path is from Point A through R_1, R_3, and R_2 to Point B. Another path is from Point A through R_1, R_4, and R_2 to Point B. Because the current has more than one path to get from Point A to Point B, the circuit is not a series circuit. Refer to Figure 5-2.

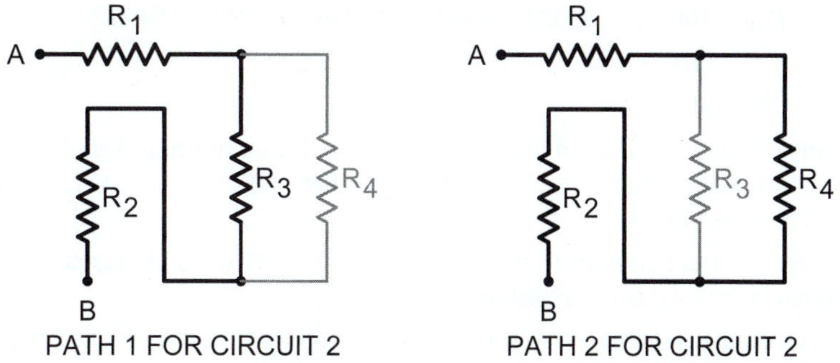

Figure 5-2: Alternate Current Paths for Circuit 2

For Circuit 3, current from Point A must pass through R_1, R_2, R_3, R_4, and R_5 to get to Point B. Because the current **must** flow through each of the resistors to get from Point A to Point B, the circuit is a series circuit.

Circuits 1 and 3 are series circuits.

2. Problem: A printed circuit board has four resistors, labeled R_1 through R_4, connected between test points TP_1 and TP_2. The measured currents for resistors R_1, R_2, R_3, and R_4 are $I_{R1} = 2.50$ mA, $I_{R2} = 2.50$ mA, $I_{R3} = 2.25$ mA, and $I_{R4} = 0.25$ mA, respectively. Assuming that no fault exists in the circuit, are the resistors in series?

Solution: In a series circuit the same current must flow through each resistor so that the measured current for each resistor must be the same. Because I_{R3} and I_{R4} differ from I_{R1}, I_{R2}, and each other, the current through R_3 and R_4 must not be the same current as that through R_1, R_2, and each other. The resistors are not in series.

5.1.2 Total Series Resistance

1. Problem: Determine whether or not each of the circuits in Figure 5-3 is a series circuit. If the circuit is a series circuit, determine the total resistance R_{AB} between Point A and B.

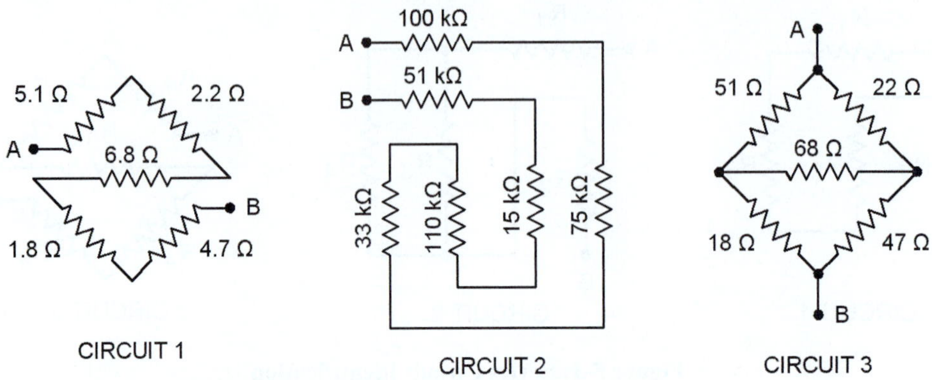

Figure 5-3: Series Resistance Calculation Circuits

Solution: For Circuit 1, current must pass through all the resistors between Point A and Point B, so there is only one circuit path. The circuit is therefore a series circuit. The total resistance R_{AB} is

$$\begin{aligned} R_{AB} &= 5.1\ \Omega + 2.2\ \Omega + 6.8\ \Omega + 1.8\ \Omega + 4.7\ \Omega \\ &= (5.1 + 2.2 + 6.8 + 1.8 + 4.7)\ \Omega \\ &= \mathbf{20.6\ \Omega} \end{aligned}$$

For Circuit 2, current must pass through all the resistors between Point A and Point B, so there is only one circuit path. The circuit is therefore a series circuit. The total resistance R_{AB} is

$$\begin{aligned} R_{AB} &= 100\ k\Omega + 75\ k\Omega + 33\ k\Omega + 110\ k\Omega + 15\ k\Omega + 51\ k\Omega \\ &= (100 + 75 + 33 + 110 + 15 + 51)\ k\Omega \\ &= \mathbf{384\ k\Omega} \end{aligned}$$

For Circuit 3, there are multiple current paths between Point A and Point B (for example, the current leaving Point A can pass through either the 51 Ω or 22 Ω resistor). The circuit is therefore not a series circuit.

2. Problem: A series circuit consists of three 6.2 kΩ resistors, two 1.8 kΩ resistors, and four 9.1 kΩ resistors. What is the total series resistance?

Solution: Because the resistors are in series, you can calculate the resistance for each group of n equal value resistors using the equation

$$R_T = n \times R$$

where

R_T is the total resistance,

n is the number of equal value resistors, and

R is the value of each resistor.

The total value of three 6.2 kΩ resistors in series is

$$\begin{aligned} R_T(6.2\ k\Omega) &= 3 \times 6.2\ k\Omega \\ &= 18.6\ k\Omega \end{aligned}$$

The total value of two 1.8 kΩ resistors in series is

$$\boldsymbol{R_T}\,(1.8\ \text{k}\Omega) = 2 \times 1.8\ \text{k}\Omega = 3.6\ \text{k}\Omega$$

The total value of four 9.1 kΩ resistors in series is

$$\boldsymbol{R_T}\,(9.1\ \text{k}\Omega) = 4 \times 9.1\ \text{k}\Omega = 36.4\ \text{k}\Omega$$

The groups are also in series with each other, so the total series resistance is the sum of the resistance of all the groups. Therefore,

$$\boldsymbol{R_T} = 18.6\ \text{k}\Omega + 3.6\ \text{k}\Omega + 36.4\ \text{k}\Omega = (18.6 + 3.6 + 36.4)\ \text{k}\Omega = \mathbf{58.6\ k\Omega}$$

3. Problem: A series circuit consists of four equal-value resistors. If one resistor shorts, the total series resistance is 5.4 kΩ. What is the total series resistance of the original circuit?

Solution: If one resistor shorts, the total series resistance is due to only three equal-value resistors. From the equation for the total resistance of ***n*** equal-value resistors,

$$\boldsymbol{R_T} = \boldsymbol{n} \times \boldsymbol{R}$$

The total resistance with three resistors is 5.4 kΩ.

$$5.4\ \text{k}\Omega = 3 \times \boldsymbol{R}$$
$$5.4\ \text{k}\Omega / 3 = (3 \times \boldsymbol{R}) / 3$$
$$1.8\ \text{k}\Omega = \boldsymbol{R}$$

The original circuit consists of four equal-value resistors.

$$\boldsymbol{R_T} = \boldsymbol{n} \times \boldsymbol{R} = 4 \times 1.8\ \text{k}\Omega = 7.2\ \text{k}\Omega$$

The total resistance of the original circuit is **7.2 kΩ**.

5.1.3 Application of Ohm's Law

1. Problem: A series resistive circuit connected to a 5 V supply draws a total current of 12.5 mA. How much resistance must be added in series to decrease the current from the 5 V supply to 10.0 mA?

Solution: From Ohm's Law for resistance for the original circuit,

$$\boldsymbol{R_{T\,\text{(org)}}} = \boldsymbol{V_S} / \boldsymbol{I_{T\,\text{(org)}}} = 5\ \text{V} / 12.5\ \text{mA} = 5\ \text{V} / (12.5 \times 10^{-3}\ \text{A}) = [5 / (12.5 \times 10^{-3}]\ (\text{V} / \text{A}) = 400\ \Omega$$

From Ohm's Law for resistance for the new circuit,

$$\boldsymbol{R_{T\,\text{(new)}}} = \boldsymbol{V_S} / \boldsymbol{I_{T\,\text{(new)}}} = 5\ \text{V} / 10.0\ \text{mA} = 5\ \text{V} / (10.0 \times 10^{-3}\ \text{A}) = [5 / (10.0 \times 10^{-3}]\ (\text{V} / \text{A}) = 500\ \Omega$$

Adding a resistor ***R*** in series with $\boldsymbol{R_{T\,\text{(org)}}}$ must increase the resistance to $\boldsymbol{R_{T\,\text{(new)}}}$.

$$\boldsymbol{R_{T\,\text{(org)}}} + \boldsymbol{R} = \boldsymbol{R_{T\,\text{(new)}}}$$
$$\boldsymbol{R_T}\,\text{(org)} + \boldsymbol{R} + -\boldsymbol{R_T}\,\text{(org)} = \boldsymbol{R_{T\,\text{(new)}}} + -\boldsymbol{R_{T\,\text{(org)}}}$$
$$\boldsymbol{R} = \boldsymbol{R_{T\,\text{(new)}}} + -\boldsymbol{R_{T\,\text{(org)}}}$$

$= 500\ \Omega - 400\ \Omega$

$= (500 - 400)\ \Omega$

$= 100\ \Omega$

100 Ω must be added to the circuit to decrease the total current from 12.5 mA to 10.0 mA.

2. Problem: Refer to the circuit and its associated test cases in Figure 5-4.

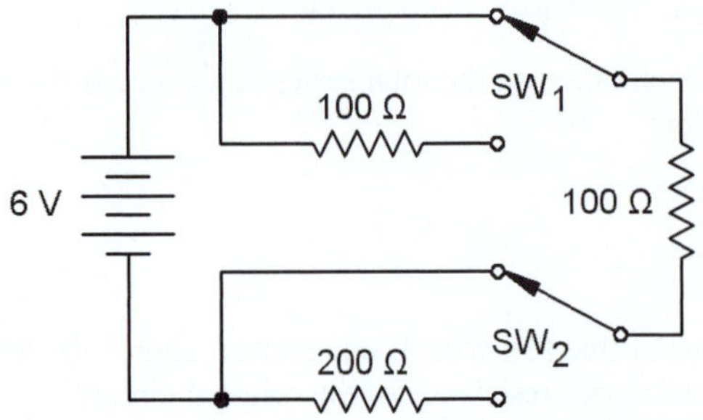

TEST CASE	SW_1	SW_2
1	UP	UP
2	DOWN	UP
3	UP	DOWN
4	DOWN	DOWN

Figure 5-4: Switched Resistance Circuit

Determine the total series resistance and current for each of the four test cases.

Solution: Figure 5-5 shows the current path for each of the test cases.

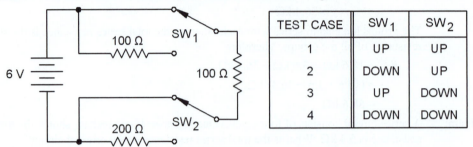

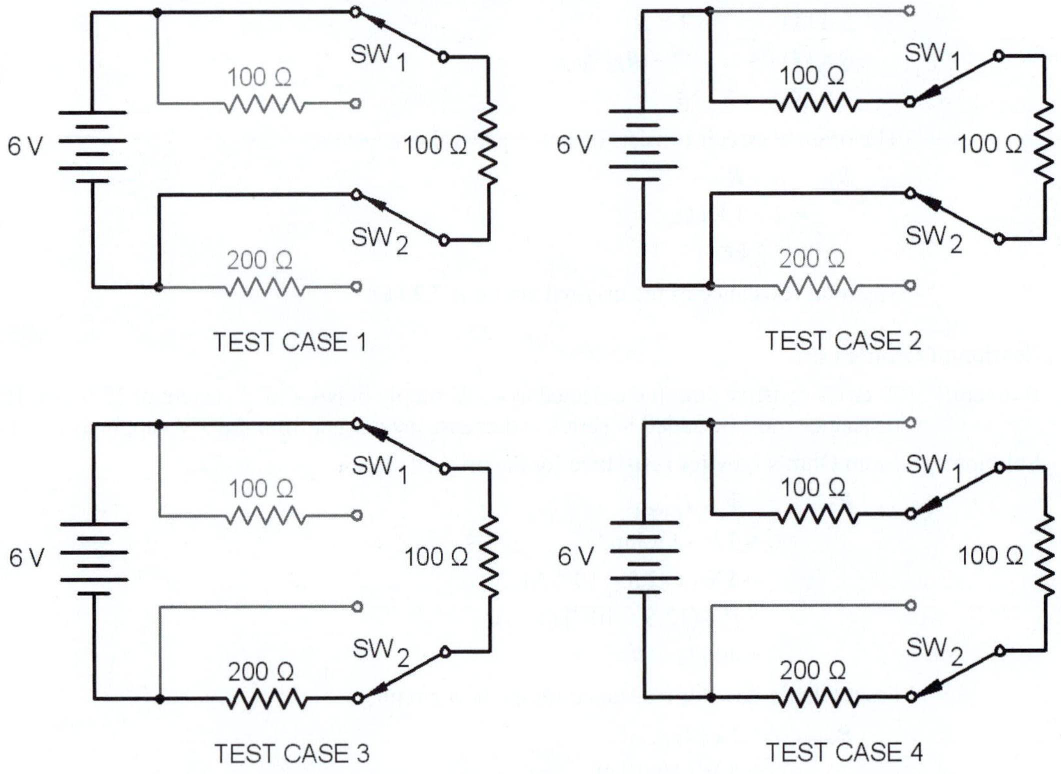

Figure 5-5: Current Paths for Switched Resistance Circuit Test Cases

For Test Case 1, with both SW_1 and SW_2 in the UP position, current from the 6 V battery flows through SW_1, the 100 Ω resistor on the right, SW_2, and back to the 6 V battery. The total resistance is the single 100 Ω resistor, so

$R_T = \mathbf{100\ \Omega}$

The total current is

$$\begin{aligned} \boldsymbol{I_T} &= \boldsymbol{V_S} / \boldsymbol{R_T} \\ &= 6 \text{ V} / 100 \ \Omega \\ &= (6 / 100) (\text{V} / \Omega) \\ &= 60 \times 10^{-3} \text{ A} \\ &= \mathbf{60 \ mA} \end{aligned}$$

For Test Case 2, with $\boldsymbol{SW_1}$ in the DOWN position and $\boldsymbol{SW_2}$ in the UP position, current from the 6 V battery flows through the top 100 Ω resistor, $\boldsymbol{SW_1}$, the 100 Ω resistor on the right, $\boldsymbol{SW_2}$, and back to the 6 V battery. The total resistance is the top and right 100 Ω resistors, so

$$\begin{aligned} \boldsymbol{R_T} &= 100 \ \Omega + 100 \ \Omega \\ &= (100 + 100) \ \Omega \\ &= \mathbf{200 \ \Omega} \end{aligned}$$

Use Ohm's Law to calculate the total current.

$$\begin{aligned} \boldsymbol{I_T} &= \boldsymbol{V_S} / \boldsymbol{R_T} \\ &= 6 \text{ V} / 200 \ \Omega \\ &= (6 / 200) (\text{V} / \Omega) \\ &= 30 \times 10^{-3} \text{ A} \\ &= \mathbf{30 \ mA} \end{aligned}$$

For Test Case 3, with $\boldsymbol{SW_1}$ UP and $\boldsymbol{SW_2}$ DOWN, current from the 6 V battery flows through $\boldsymbol{SW_1}$, the 100 Ω resistor on the right, $\boldsymbol{SW_2}$, the 200 Ω resistor, and back to the 6 V battery. The total resistance is the right 100 Ω resistor and 200 Ω resistor.

$$\begin{aligned} \boldsymbol{R_T} &= 100 \ \Omega + 200 \ \Omega \\ &= (100 + 200) \ \Omega \\ &= \mathbf{300 \ \Omega} \end{aligned}$$

Use Ohm's Law to calculate the total current.

$$\begin{aligned} \boldsymbol{I_T} &= \boldsymbol{V_S} / \boldsymbol{R_T} \\ &= 6 \text{ V} / 300 \ \Omega \\ &= (6 / 300) (\text{V} / \Omega) \\ &= 20 \times 10^{-3} \text{ A} \\ &= \mathbf{20 \ mA} \end{aligned}$$

For Test Case 4, with both $\boldsymbol{SW_1}$ and $\boldsymbol{SW_2}$ in the DOWN position, current from the 6 V battery flows through the top 100 Ω resistor, $\boldsymbol{SW_1}$, the 100 Ω resistor on the right, $\boldsymbol{SW_2}$, the 200 Ω resistor, and back to the 6 V battery. The total resistance is both 100 Ω resistors and the 200 Ω resistor, so

$$\begin{aligned} \boldsymbol{R_T} &= 100 \ \Omega + 100 \ \Omega + 200 \ \Omega \\ &= (100 + 100 + 200) \ \Omega \\ &= \mathbf{400 \ \Omega} \end{aligned}$$

Use Ohm's Law to calculate the total current.

$$\begin{aligned} \boldsymbol{I_T} &= \boldsymbol{V_S} / \boldsymbol{R_T} \\ &= 6 \text{ V} / 400 \ \Omega \\ &= (6 / 400) (\text{V} / \Omega) \\ &= 15 \times 10^{-3} \text{ A} \\ &= \mathbf{15 \ mA} \end{aligned}$$

3. Problem: Five 51 Ω resistors, $\boldsymbol{R_1}$ through $\boldsymbol{R_5}$, are connected in series across a battery. The measured current through the $\boldsymbol{R_1}$ is 47.1 mA. What is the voltage, $\boldsymbol{V_S}$, of the battery?

Solution: The total series resistance of $\boldsymbol{n}$ equal value resistors is

$$\boldsymbol{R_T} = \boldsymbol{n} \times \boldsymbol{R}$$

so

$$R_T = 5 \times 51\ \Omega$$
$$= 255\ \Omega$$

In a series circuit, I_{R1}, the current through resistor R_1 is equal to the total series current I_T, so

$$V_S = I_T \times R_T$$
$$= 47.1\ \text{mA} \times 255\ \Omega$$
$$= (47.1 \times 10^{-3}\ \text{A}) \times 255\ \Omega$$
$$= [(47.1 \times 10^{-3}) \times 255]\ (\text{A} \times \Omega)$$
$$= 12.01\ \text{V}$$

The battery voltage is **12.0 V**.

5.1.4 Voltage Sources in Series

1. Problem: Which of the voltage sources in Figure 5-6 are series-aiding and which are series-opposing with respect to V_{S1}?

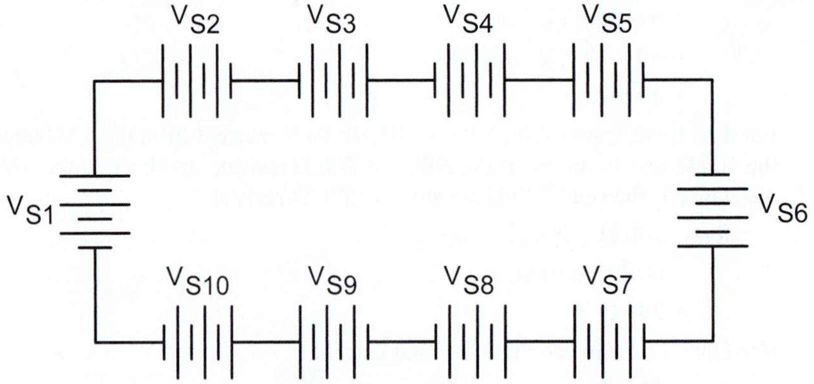

Figure 5-6: Series Voltage Sources

Solution: Voltage sources are **series-aiding** if they are in series and have like polarities in the circuit. Voltage sources are **series-opposing** if they are in series and have opposite polarities in the circuit. A rule to determine whether voltage sources are series-aiding or series-opposing is to trace from one terminal of the reference source around the circuit and compare the polarity of the reference terminal with the polarity of the terminal you enter, or first encounter, for each source. If the polarities are the *same*, the two sources are *series-opposing*. If the polarities are *different*, the two sources are *series-aiding*.

For Figure 5-6, start at the top (positive) terminal of V_{S1} and trace clockwise through the circuit.

The first terminal encountered for V_{S2} is also positive, so V_{S1} and V_{S2} are series-opposing.

The first terminal encountered for V_{S3} is negative, so V_{S1} and V_{S3} are series-aiding.

The first terminal encountered for V_{S4} is negative, so V_{S1} and V_{S4} are series-aiding.

The first terminal encountered for V_{S5} is also positive, so V_{S1} and V_{S5} are series-opposing.

The first terminal encountered for V_{S6} is negative, so V_{S1} and V_{S6} are series-aiding.

The first terminal encountered for V_{S7} is negative, so V_{S1} and V_{S7} are series-aiding.

The first terminal encountered for V_{S8} is also positive, so V_{S1} and V_{S8} are series-opposing.

The first terminal encountered for V_{S9} is negative, so V_{S1} and V_{S9} are series-aiding.

The first terminal encountered for V_{S10} is also positive, so V_{S1} and V_{S10} are series-opposing.

2. Problem: A flashlight uses five 1.5 V D-cell batteries. If one of the batteries is accidentally reversed inside the flashlight, what is the total voltage?

Solution: The total voltage is equal to the algebraic sum of the battery voltages. If four 1.5 V batteries are inserted into the flashlight correctly and one 1.5 V battery is reversed, the total voltage is

$$\begin{aligned} V_S &= (+1.5\text{ V}) + (+1.5\text{ V}) + (+1.5\text{ V}) + (+1.5\text{ V}) + (-1.5\text{ V}) \\ &= [(+1.5) + (+1.5) + (+1.5) + (+1.5) + (-1.5)]\text{ V} \\ &= +4.5\text{ V} \end{aligned}$$

The total voltage is **4.5 V**.

3. Problem: A 12 V car battery consists of six identical lead-acid wet cells connected in series. What is the voltage of each cell?

Solution: The total voltage of $\boldsymbol{n}$ series-aiding (i.e., the voltage polarity is the same) voltage sources is

$$V_S = \boldsymbol{n} \times V$$

so

$$\begin{aligned} V_S / \boldsymbol{n} &= (\text{n} \times V) / \text{n} \\ &= V \end{aligned}$$

From this

$$\begin{aligned} V &= V_S / \boldsymbol{n} \\ &= 12\text{ V} / 6 \\ &= 2\text{ V} \end{aligned}$$

The voltage of each cell is **2 V**.

4. Problem: Two batteries V_1 and V_2 have a total voltage of 24 V when connected series-aiding. The same two batteries have a total voltage of 12 V when connected series-opposing. What are the two battery voltages?

Solution: Batteries connected series-aiding have the same polarity, so the total series-aiding voltage V_{SA} is the sum of the individual voltages.

$$V_{SA} = V_1 + V_2$$

Batteries connected series-opposing have opposite polarities, so the total series-opposing voltage V_{SO} is the difference of the individual voltages.

$$V_{SO} = V_1 - V_2$$

Therefore,

$$\begin{aligned} 24\text{ V} &= V_1 + V_2 \\ 12\text{ V} &= V_1 - V_2 \end{aligned}$$

Adding these two equations gives

$$\begin{aligned} 24\text{ V} + 12\text{ V} &= (V_1 + V_2) + (V_1 - V_2) \\ (24 + 12)\text{ V} &= (V_1 + V_1) + (V_2 - V_2) \\ 36\text{ V} &= 2 \times V_1 \\ 36\text{ V} / 2 &= (2 \times V_1) / 2 \\ 18\text{ V} &= V_1 \end{aligned}$$

Substituting this value into the equation for series-aiding voltage supplies gives

$$\begin{aligned} V_{SA} &= V_1 + V_2 \\ 24\text{ V} &= 18\text{ V} + V_2 \\ 24\text{ V} + -18\text{ V} &= 18\text{ V} + V_2 + -18\text{ V} \\ 6\text{ V} &= V_2 \end{aligned}$$

The voltages of the two batteries are **18 V** and **6 V**.

5.1.5 Kirchhoff's Voltage Law

1. Problem: Determine the value of V_X in the circuit of Figure 5-7.

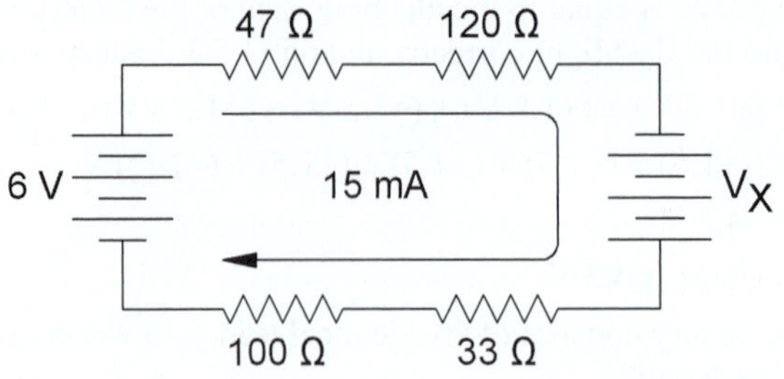

Figure 5-7: Kirchhoff's Voltage Law Example Circuit 1

Solution: From Kirchhoff's Voltage Law, the algebraic sum of the voltages in the circuit must equal zero. The 6 V battery and V_X as shown are series-aiding and are voltage rises, while each resistor voltage is a voltage drop. Ohm's Law for voltage states that the voltage across each resistor is equal to the current times the resistance value. Therefore

$$6\text{ V} - (15\text{ mA} \times 47\ \Omega) - (15\text{ mA} \times 120\ \Omega) + V_X - (15\text{ mA} \times 33\ \Omega) - (15\text{ mA} \times 100\ \Omega) = 0\text{ V}$$

Combining terms gives

$$6\text{ V} - \{(15\text{ mA}) \times [(47\ \Omega + 120\ \Omega + 33\ \Omega + 100\ \Omega)]\} + V_X = 0\text{ V}$$
$$V_X + 6\text{ V} + -[(15 \times 10^{-3}\text{ A}) \times 300\ \Omega] = 0\text{ V}$$
$$V_X + 6\text{ V} + -[(15 \times 10^{-3}) \times 300]\ (\text{A} \times \Omega)] = 0\text{ V}$$
$$V_X + 6\text{ V} + -4.5\text{ V} = 0\text{ V}$$
$$V_X + 1.5\text{ V} = 0\text{ V}$$
$$V_X + 1.5\text{ V} + -1.5\text{ V} = 0\text{ V} + -1.5\text{ V}$$
$$V_X = (0 - 1.5)\text{ V}$$
$$= -1.5\text{ V}$$

V_X is equal to **−1.5 V**. The negative voltage value indicates that although the voltage sources were assumed to be series-aiding and therefore both voltage rises, the polarity of V_X is actually opposite to that shown in the figure.

2. Problem: Determine the value of R_X in the circuit of Figure 5-8.

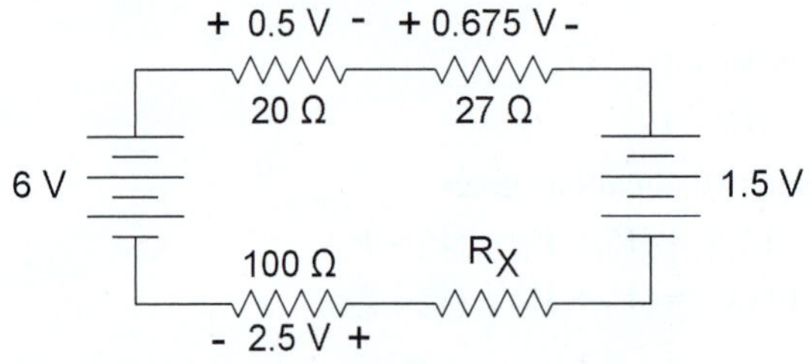

Figure 5-8: Kirchhoff's Voltage Law Example Circuit 2

Solution: From Kirchhoff's Voltage Law, the algebraic sum of the voltages in the circuit must equal zero. The 6 V battery and 1.5 V battery as shown are series-opposing, so the 6 V battery is a voltage rise and the 1.5 V battery is a voltage drop. Each resistor voltage is also a voltage drop, as it decreases the energy of the charge moving through the circuit. Therefore,

$$6\text{ V} - 0.5\text{ V} - 0.675\text{ V} - 1.5\text{ V} - V_{RX} - 2.5\text{ V} = 0\text{ V}$$
$$6\text{ V} + -0.5\text{ V} + -0.675\text{ V} + -1.5\text{ V} + -2.5\text{ V} + -V_{RX} = 0\text{ V}$$
$$(6 + -0.5 + -0.675 + -1.5 + -2.5)\text{ V} + -V_{RX} + V_{RX} = 0\text{ V} + V_{RX}$$

$(6 - 5.175)$ V $= V_{RX}$

0.825 V $= V_{RX}$

For a series circuit, the current I_T through each resistor is the same, so the current through the 100 Ω resistor is equal to the current I_{RX} through R_X. From Ohm's Law for current,

$$I_{RX} = I_T = 2.5 \text{ V} / 100\ \Omega$$
$$= (2.5 / 100)\ (\text{V} / \Omega)$$
$$= 25 \times 10^{-3} \text{ A}$$

Ohm's Law for resistance states that resistance is equal to voltage divided by current so

$$R_X = V_{RX} / I_{RX}$$
$$= 0.825 \text{ V} / (25 \times 10^{-3} \text{ A})$$
$$= [0.825 / (25 \times 10^{-3})]\ (\text{V} / \text{A})$$
$$= 3.3\ \Omega$$

The resistance of R_X is **3.3 Ω**.

5.1.6 Voltage Dividers

1. Problem: You have a fixed 5 V power supply and wish to use a resistor voltage divider to develop n different voltages between 0 V and 5 V. What is the minimum number of fixed resistors you must use for your voltage divider?

   Solution: In a voltage divider, each voltage tap is taken between two adjacent resistors. A single-tap voltage divider requires two resistors, R_1 and R_2, with the tap placed between the two resistors. Each additional resistor adds only one more tap point in the voltage divider so that each voltage divider has one more resistor than the number of tap points. For example, a 2-tap voltage divider requires three resistors, R_1, R_2, and R_3, with the first tap between R_1 and R_2 as for the single-tap voltage divider, and a second tap between R_2 and the additional resistor R_3. Therefore, a voltage divider with n voltage taps will require at least $n + 1$ resistors.

2. Problem: What are the voltages for the voltage divider circuit in Figure 5-9?

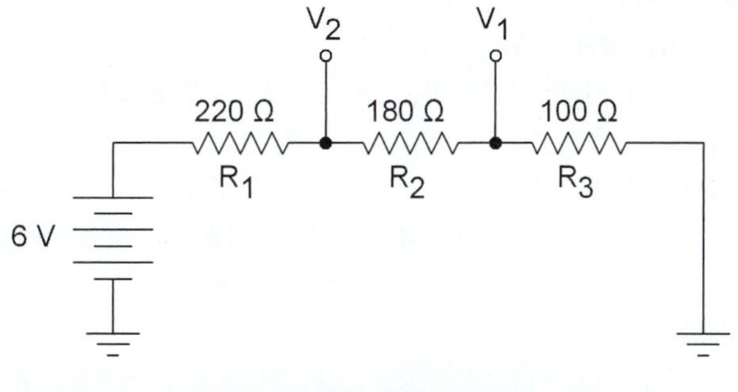

Figure 5-9: Voltage Divider Example Circuit 1

   Solution: From the voltage divider formula,

$$V_X = (R_X / R_T) \times V_S$$

For the series circuit in Figure 5-9,

$$R_T = R_1 + R_2 + R_3$$
$$= 220\ \Omega + 180\ \Omega + 100\ \Omega$$
$$= (220 + 180 + 100)\ \Omega$$
$$= 500\ \Omega$$

For tap voltage V_1, $R_X = R_3$ so

$$V_1 = (R_3 / R_T) \times V_S$$
$$= (100\ \Omega / 500\ \Omega) \times 6 \text{ V}$$

$= [(100 / 500)\ (\Omega / \Omega)] \times 6\ \text{V}$

$= 0.20 \times 6\ \text{V}$

$= 1.20\ \text{V}$

For tap voltage V_2, $\boldsymbol{R_X}$ is $\boldsymbol{R_2}$ and $\boldsymbol{R_3}$ in series, so

$V_2 = (\boldsymbol{R_X} / \boldsymbol{R_T}) \times \boldsymbol{V_S}$

$= [(\boldsymbol{R_2} + \boldsymbol{R_3}) / \boldsymbol{R_T}] \times \boldsymbol{V_S}$

$= [(180\ \Omega + 100\ \Omega) / 500\ \Omega] \times 6\ \text{V}$

$= \{[(180 + 100)\ \Omega] / 500\ \Omega\} \times 6\ \text{V}$

$= (280\ \Omega / 500\ \Omega) \times 6\ \text{V}$

$= [(280 / 500)\ (\Omega / \Omega)] \times 6\ \text{V}$

$= 0.56 \times 6\ \text{V}$

$= 3.36\ \text{V}$

The tap voltages V_1 and V_2 are equal to **1.2 V** and **3.4 V**, respectively.

3. Problem: For the circuit of Figure 5-10, what value of $\boldsymbol{R_2}$ will produce the tap voltage of 1.2 V?

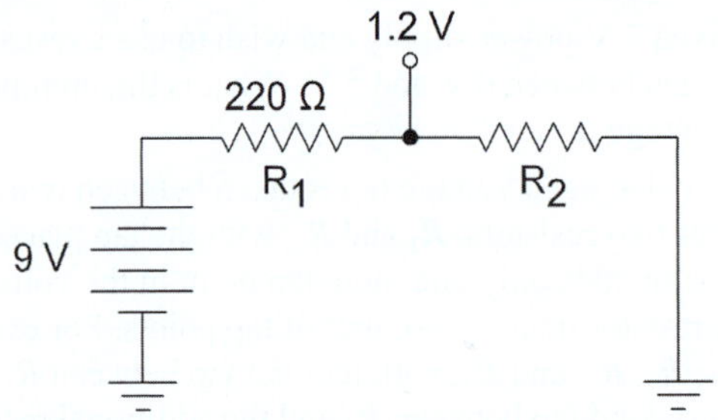

Figure 5-10: Voltage Divider Example Circuit 2

Solution: From the voltage divider formula,

$V_X = (\boldsymbol{R_X} / \boldsymbol{R_T}) \times \boldsymbol{V_S}$

For the circuit in Figure 5-10, $\boldsymbol{R_X}$ is $\boldsymbol{R_2}$ and $\boldsymbol{R_T}$ is $\boldsymbol{R_1}$ and $\boldsymbol{R_2}$ in series so

$1.2\ \text{V} = [\boldsymbol{R_2} / (220\ \Omega + \boldsymbol{R_2})] \times 9\ \text{V}$

$1.2\ \text{V} \times (220\ \Omega + \boldsymbol{R_2}) = [\boldsymbol{R_2} / (220\ \Omega + \boldsymbol{R_2})] \times 9\ \text{V} \times (220\ \Omega + \boldsymbol{R_2})$

$(1.2\ \text{V} \times 220\ \Omega) + (1.2\ \text{V} \times \boldsymbol{R_2}) = 9\ \text{V} \times [\boldsymbol{R_2} / (220\ \Omega + \boldsymbol{R_2})] \times (220\ \Omega + \boldsymbol{R_2})$

$= 9\ \text{V} \times \boldsymbol{R_2}$

From this,

$(1.2\ \text{V} \times 220\ \Omega) + (1.2\ \text{V} \times \boldsymbol{R_2}) + -(1.2\ \text{V} \times \boldsymbol{R_2}) = 9\ \text{V} \times \boldsymbol{R_2} + -(1.2\ \text{V} \times \boldsymbol{R_2})$

$(1.2\ \text{V} \times 220\ \Omega) = (9\ \text{V} + -1.2\ \text{V}) \times \boldsymbol{R_2})$

$1.2\ \text{V} \times 220\ \Omega = 7.8\ \text{V} \times \boldsymbol{R_2}$

$(1.2\ \text{V} \times 220\ \Omega) / 7.8\ \text{V} = (7.8\ \text{V} \times \boldsymbol{R_2}) / 7.8\ \text{V}$

$[(1.2 \times 220) / 7.8]\ [(\text{V} \times \Omega) / \text{V}] = \boldsymbol{R_2}$

$33.8\ \Omega = \boldsymbol{R_2}$

The value of $\boldsymbol{R_2}$ that will produce a tap voltage of 1.2 V is **34 Ω**.

5.1.7 Power in Series Circuits

1. Problem: What is the total power dissipated by the circuit in Figure 5-11?

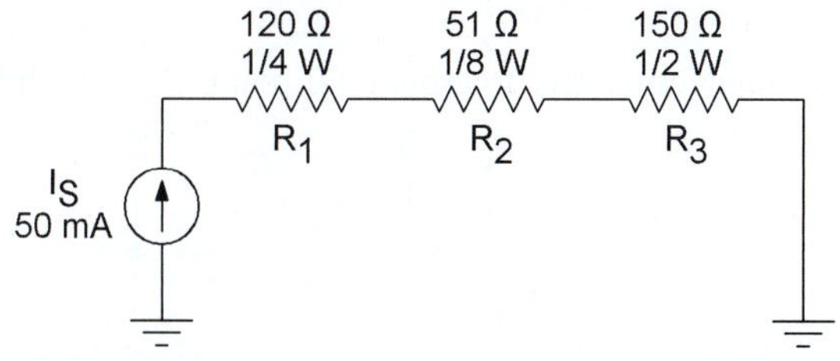

Figure 5-11: Power Dissipation Example Circuit 1

Solution: From Watt's Law for current and resistance,

$$\boldsymbol{P_T} = \boldsymbol{I_T}^2 \times \boldsymbol{R_T}$$
$$\text{W} = \text{A}^2 \times \Omega$$

For the series circuit in Figure 5-11, $\boldsymbol{R_T} = \boldsymbol{R}_1 + \boldsymbol{R}_2 + \boldsymbol{R}_3$.

$$\boldsymbol{R_T} = 120\ \Omega + 51\ \Omega + 150\ \Omega$$
$$= (120 + 51 + 150)\ \Omega$$
$$= 321\ \Omega$$

Therefore,

$$\boldsymbol{P_T} = (50\ \text{mA})^2 \times 321\ \Omega$$
$$= (50 \times 10^{-3}\ \text{A})^2 \times 321\ \Omega$$
$$= [(50 \times 10^{-3})^2 \times 321]\ [(\text{A})^2 \times \Omega]$$
$$= (2500 \times 10^{-6} \times 321)\ (\text{A}^2 \times \Omega)$$
$$= 802.5 \times 10^{-3}\ \text{W}$$
$$= 802.5\ \text{mW}$$

The total power dissipation in the circuit is **800 mW**.

2. Problem: A series circuit consists of a 100 Ω and 1 kΩ resistor. If the 100 Ω resistor safely dissipates 1/8 W during normal circuit operation, how much power must the 1 kΩ resistor safely dissipate during normal circuit operation?

Solution: Because the resistors are in series, the same current flows through both. From Watt's Law for current and resistance, the dissipated power $\boldsymbol{P}$ is

$$\boldsymbol{P} = \boldsymbol{I_T}^2 \times \boldsymbol{R}$$
$$\text{W} = \text{A}^2 \times \Omega$$

From this

$$1/8\ \text{W} = \boldsymbol{I_T}^2 \times 100\ \Omega$$
$$1/8\ \text{W} / 100\ \Omega = (\boldsymbol{I_T}^2 \times 100\ \Omega) / 100\ \Omega$$
$$[1 / (8 \times 100)]\ (\text{W} / \Omega) = \boldsymbol{I_T}^2$$
$$1.25 \times 10^{-3}\ [(\text{A}^2 \times \Omega) / \Omega] = \boldsymbol{I_T}^2$$
$$1.25 \times 10^{-3}\ \text{A}^2 = \boldsymbol{I_T}^2$$

The same current flows through the 1 kΩ resistor, so again from Watt's Law the 1 kΩ resistor must safely dissipate

$$\boldsymbol{P_{1k}} = \boldsymbol{I_T}^2 \times \boldsymbol{R}$$
$$= (1.25 \times 10^{-3}\ \text{A}^2) \times 1\ \text{k}\Omega$$
$$= (1.25 \times 10^{-3}\ \text{A}^2) \times (1 \times 10^3\ \Omega)$$
$$= [(1.25 \times 10^{-3}) \times (1 \times 10^3\ \Omega)]\ (\text{A}^2 \times \Omega)$$
$$= 1.25\ \text{W}$$

The 1 kΩ resistor must safely dissipate **1.25 W** during normal circuit operation.

5.1.8 Voltage Measurements

1. Problem: What are the voltages relative to ground for points A through E for the circuit in Figure 5-12?

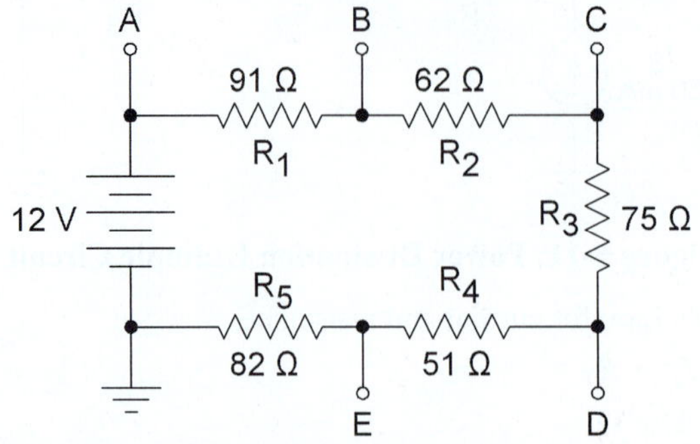

Figure 5-12: Voltage Measurement Example Circuit 1

Solution: The circuit is a series circuit, so the total resistance $\boldsymbol{R_T}$ is

$$\begin{aligned} \boldsymbol{R_T} &= 91\ \Omega + 62\ \Omega + 75\ \Omega + 51\ \Omega + 82\ \Omega \\ &= (91 + 62 + 75 + 51 + 82)\ \Omega \\ &= 361\ \Omega \end{aligned}$$

From Ohm's Law for current, the total current $\boldsymbol{I_T}$ is

$$\begin{aligned} \boldsymbol{I_T} &= \boldsymbol{V_S} / \boldsymbol{R_T} \\ &= 12\ \text{V} / 361\ \Omega \\ &= (12 / 361)\ (\text{V} / \Omega) \\ &= 33.24 \times 10^{-3}\ \text{A} \end{aligned}$$

From Ohm's Law for voltage, each resistor will drop an amount of voltage equal to the product of the resistor value and current through it. The voltages are then

$$\begin{aligned} \boldsymbol{V_A} &= \boldsymbol{V_S} = 12\ \text{V} \\ \boldsymbol{V_B} &= \boldsymbol{V_A} - \boldsymbol{V_{R1}} \\ &= \boldsymbol{V_A} + -(\boldsymbol{I_T} \times \boldsymbol{R_1}) \\ &= 12\ \text{V} + -[(33.24 \times 10^{-3}\ \text{A}) \times (91\ \Omega)] \\ &= 12\ \text{V} + -(33.24 \times 10^{-3} \times 91)\ (\text{A} \times \Omega) \\ &= 12\ \text{V} + -3.025\ \text{V} \\ &= (12 + -3.025)\ \text{V} \\ &= 8.975\ \text{V} \end{aligned}$$

$$\begin{aligned} \boldsymbol{V_C} &= \boldsymbol{V_B} - \boldsymbol{V_{R2}} \\ &= \boldsymbol{V_B} + -(\boldsymbol{I_T} \times \boldsymbol{R_2}) \\ &= 8.975\ \text{V} + -[(33.24 \times 10^{-3}\ \text{A}) \times (62\ \Omega)] \\ &= 8.975\ \text{V} + -(33.24 \times 10^{-3} \times 62)\ (\text{A} \times \Omega) \\ &= 8.975\ \text{V} + -2.061\ \text{V} \\ &= (8.975 - 2.061)\ \text{V} \\ &= 6.914\ \text{V} \end{aligned}$$

$$\begin{aligned} \boldsymbol{V_D} &= \boldsymbol{V_C} - \boldsymbol{V_{R3}} \\ &= \boldsymbol{V_C} + -(\boldsymbol{I_T} \times \boldsymbol{R_3}) \\ &= 6.914\ \text{V} + -[(33.24 \times 10^{-3}\ \text{A}) \times (75\ \Omega)] \\ &= 6.914\ \text{V} + -(33.24 \times 10^{-3} \times 75)\ (\text{A} \times \Omega) \end{aligned}$$

$= 6.914\text{ V} + -2.493\text{ V}$

$= (6.914 + -2.493)\text{ V}$

$= 4.421\text{ V}$

$\boldsymbol{V_E} = \boldsymbol{V_D} - \boldsymbol{V_{R4}}$

$= \boldsymbol{V_D} - (\boldsymbol{I_T} \times \boldsymbol{R_4})$

$= 4.421\text{ V} + -[(33.24 \times 10^{-3}\text{ A}) \times (51\ \Omega)]$

$= 4.421\text{ V} + -(33.24 \times 10^{-3} \times 51)\ (\text{A} \times \Omega)$

$= 4.421\text{ V} + -1.695\text{ V}$

$= (4.421 + -1.695)\text{ V}$

$= 2.726\text{ V}$

The voltages for points A through E are **12 V**, **9.0 V**, **6.9 V**, **4.4 V**, and **2.7 V**, respectively.

2. Problem: What are the voltages relative to ground for points A through D for the circuit of Figure 5-13?

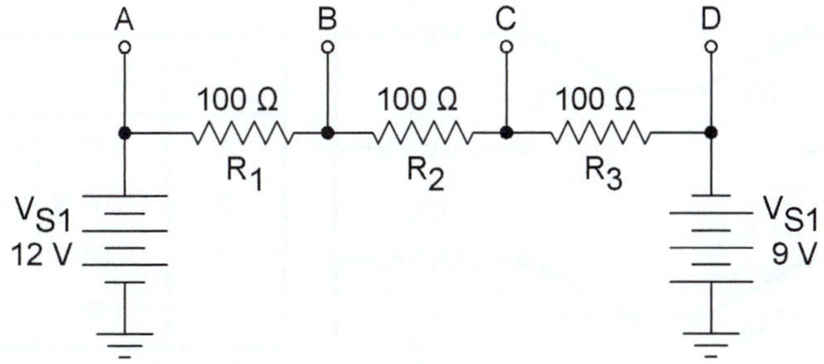

Figure 5-13: Voltage Measurement Example Circuit 2

Solution: The circuit is a series circuit, so the total resistance $\boldsymbol{R_T}$ is

$\boldsymbol{R_T} = 100\ \Omega + 100\ \Omega + 100\ \Omega$

$= (100 + 100 + 100)\ \Omega$

$= 300\ \Omega$

From Ohm's Law for current, the total current $\boldsymbol{I_T}$ is

$\boldsymbol{I_T} = (\boldsymbol{V_{S1}} - \boldsymbol{V_{S2}}) / \boldsymbol{R_T}$

$= [12\text{ V} + -(-9\text{ V})] / 300\ \Omega$

$= 21\text{ V} / 300\ \Omega$

$= (21 / 300)\ (\text{V} / \Omega)$

$= 70 \times 10^{-3}\text{ A}$

From Ohm's Law for voltage, each resistor will drop an amount of voltage equal to the product of the resistor value and current through it. The voltages are then

$\boldsymbol{V_A} = \boldsymbol{V_{S1}} = +12\text{ V}$

$\boldsymbol{V_B} = \boldsymbol{V_A} - \boldsymbol{V_{R1}}$

$= \boldsymbol{V_A} + -(\boldsymbol{I_T} \times \boldsymbol{R_1})$

$= +12\text{ V} + -[(70 \times 10^{-3}\text{ A}) \times (100\ \Omega)]$

$= +12\text{ V} + -(70 \times 10^{-3} \times 100)\ (\text{A} \times \Omega)$

$= +12\text{ V} + -7.0\text{ V}$

$= (+12 + -7.0)\text{ V}$

$= +5.0\text{ V}$

$\boldsymbol{V_C} = \boldsymbol{V_B} - \boldsymbol{V_{R2}}$

$= \boldsymbol{V_B} + -(\boldsymbol{I_T} \times \boldsymbol{R_2})$

$= +5.0\text{ V} + -[(70 \times 10^{-3}\text{ A}) \times (100\ \Omega)]$

$= +5.0\text{ V} + -(70 \times 10^{-3} \times 100)\ (\text{A} \times \Omega)$

$= +5.0\text{ V} + -7.0\text{ V}$

$= (+5.0 + -7.0)\text{ V}$

$= -2.0\text{ V}$

$V_D = V_{S2} = -9\text{ V}$

The voltages for points A through D are **+12 V**, **+5.0 V**, **–2.0 V**, and **–9 V**, respectively.

5.2 Advanced

5.2.1 Resistors in Series

1. Problem: For the circuit of Figure 5-14, determine which resistors are in series for the indicated test cases.

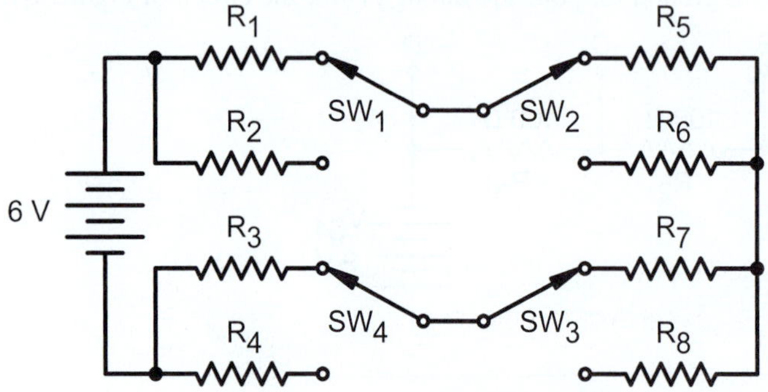

TEST CASE	SW_1	SW_2	SW_3	SW_4
1	UP	DOWN	UP	DOWN
2	DOWN	DOWN	UP	UP
3	DOWN	UP	DOWN	UP
4	UP	DOWN	DOWN	UP

Figure 5-14: Resistors in Series Switched Circuit

Solution: Figure 5-15 shows the current path for each test case.

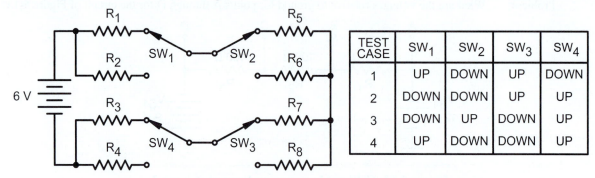

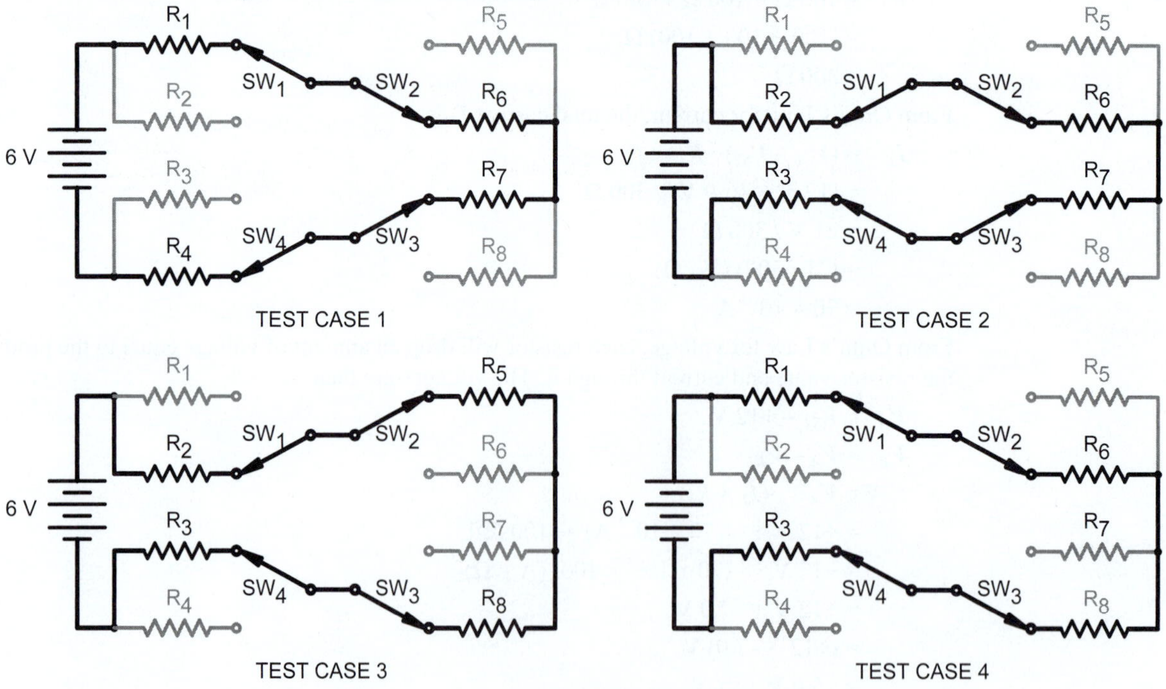

Figure 5-15: Current Paths for Switched Resistor Test Cases

For Test Case 1, the current path shows that R_1, R_6, R_7, and R_4 are in series.

For Test Case 2, the current path shows that R_2, R_6, R_7, and R_3 are in series.

For Test Case 3, the current path shows that R_2, R_5, R_8, and R_3 are in series.

For Test Case 4, the current path shows that R_1, R_6, R_8, and R_3 are in series.

2. Problem: Determine the series resistors between V_S and ground for the switching circuit in Figure 5-16.

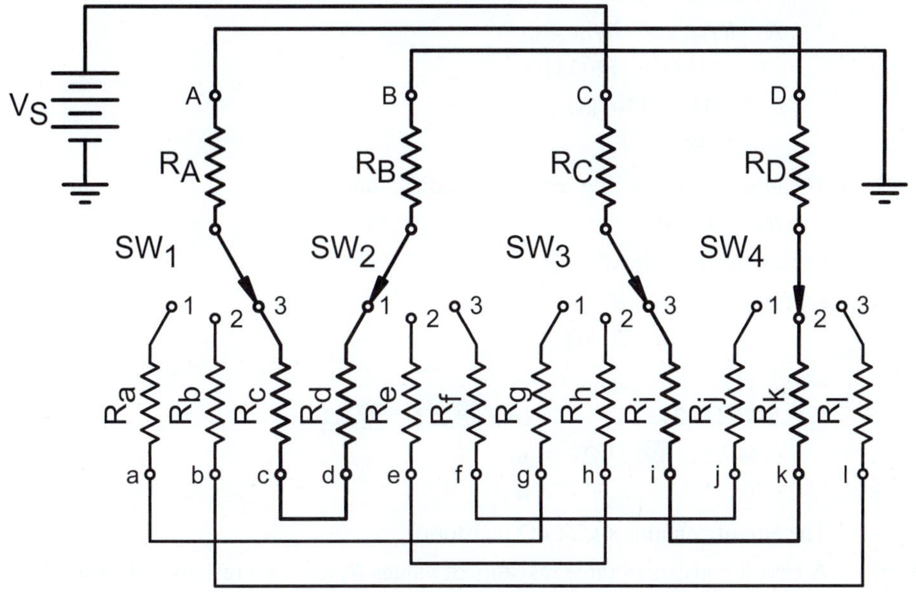

Figure 5-16: Series Resistor Switching Circuit

Solution: Refer to Figure 5-17, which shows the current path for the terminal connections and switch settings. The series resistors and switches between V_S and ground in order are R_C, SW_3, R_i, R_k, SW_4, R_D, R_A, SW_1, R_c, R_d, SW_2, and R_B.

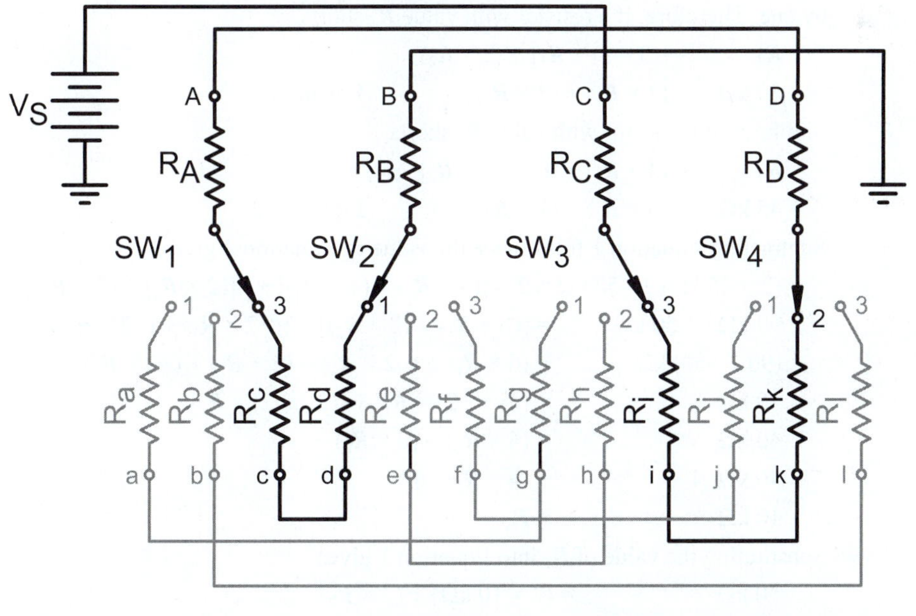

Figure 5-17: Current Path for Series Resistor Switching Circuit of Figure 5-16

5.2.2 Total Series Resistance

1. Problem: A series circuit consists entirely of equal-value resistors and has a total resistance of 144 kΩ. When one resistor shorts, the total resistance decreases to 120 kΩ. How many resistors are in the circuit?

Solution: When one resistor in a series circuit shorts, the change in resistance must be equal to the value of the resistor.

$$\boldsymbol{R} = \boldsymbol{R}_{T\,(\text{normal})} - \boldsymbol{R}_{T\,(\text{shorted})}$$
$$= 144\ \text{k}\Omega + -120\ \text{k}\Omega$$
$$= (144 + -120)\ \text{k}\Omega$$
$$= 24\ \text{k}\Omega$$

The total resistance of $\boldsymbol{n}$ resistors of equal value $\boldsymbol{R}$ is

$$\boldsymbol{R}_T = \boldsymbol{n} \times \boldsymbol{R}$$

For the original circuit

$$144\ \text{k}\Omega = \boldsymbol{n} \times \boldsymbol{R}$$
$$= \boldsymbol{n} \times 24\ \text{k}\Omega$$

so

$$144\ \text{k}\Omega / 24\ \text{k}\Omega = (\boldsymbol{n} \times 24\ \text{k}\Omega) / 24\ \text{k}\Omega$$
$$(144 / 24)\ (\text{k}\Omega / \text{k}\Omega) = \boldsymbol{n}$$
$$6 = \boldsymbol{n}$$

The circuit contains **six** 24 kΩ resistors.

2. Problem: A circuit consists of three resistors of values $\boldsymbol{R}_1$ and two resistors of value $\boldsymbol{R}_2$. If one of the resistors with value $\boldsymbol{R}_1$ shorts the total series resistance is 50 kΩ. If one of the resistors with value $\boldsymbol{R}_2$ shorts the total series resistance is 45 kΩ. What are the values of $\boldsymbol{R}_1$ and $\boldsymbol{R}_2$?

Solution: For the original circuit,

$$\boldsymbol{R}_T = 3 \times \boldsymbol{R}_1 + 2 \times \boldsymbol{R}_2$$

If a resistor shorts, the number of resistors with that value adding to the total resistance decreases by one. Therefore, if a resistor with value $\boldsymbol{R}_1$ shorts,

$$\boldsymbol{R}_T = [(3 - 1) \times \boldsymbol{R}_1] + (2 \times \boldsymbol{R}_2)$$
$$50\ \text{k}\Omega = (2 \times \boldsymbol{R}_1) + (2 \times \boldsymbol{R}_2) \qquad \text{Equation 1}$$

Similarly, if a resistor with value $\boldsymbol{R}_2$ shorts,

$$\boldsymbol{R}_T = (3 \times \boldsymbol{R}_1] + [(2 - 1) \times \boldsymbol{R}_2]$$
$$45\ \text{k}\Omega = (3 \times \boldsymbol{R}_1) + (1 \times \boldsymbol{R}_2) \qquad \text{Equation 2}$$

Subtracting Equation 1 from twice the value of Equation 2 gives

$$(2 \times 45\ \text{k}\Omega) + -50\ \text{k}\Omega = 2 \times [(3 \times \boldsymbol{R}_1) + (1 \times \boldsymbol{R}_2)] + -[(2 \times \boldsymbol{R}_1) + (2 \times \boldsymbol{R}_2)]$$
$$90\ \text{k}\Omega + -50\ \text{k}\Omega = [(6 \times \boldsymbol{R}_1) + (2 \times \boldsymbol{R}_2)] + [(-2 \times \boldsymbol{R}_1) + (-2 \times \boldsymbol{R}_2)]$$
$$(90 + -50)\ \text{k}\Omega = (6 \times \boldsymbol{R}_1) + (-2 \times \boldsymbol{R}_1) + (2 \times \boldsymbol{R}_2) + (-2 \times \boldsymbol{R}_2)$$
$$40\ \text{k}\Omega = [(6 + -2) \times \boldsymbol{R}_1] + [(2 + -2) \times \boldsymbol{R}_2]$$
$$40\ \text{k}\Omega = (4 \times \boldsymbol{R}_1) + (0 \times \boldsymbol{R}_2)$$
$$40\ \text{k}\Omega / 4 = (4 \times \boldsymbol{R}_1) / 4$$
$$10\ \text{k}\Omega = \boldsymbol{R}_1$$

Substituting the value of $\boldsymbol{R}_1$ into Equation 1 gives

$$50\ \text{k}\Omega = (2 \times 10\ \text{k}\Omega) + (2 \times \boldsymbol{R}_2)$$
$$= 20\ \text{k}\Omega + (2 \times \boldsymbol{R}_2)$$

$$
\begin{aligned}
50\ \text{k}\Omega + -20\ \text{k}\Omega &= 20\ \text{k}\Omega + (2 \times \boldsymbol{R_2}) + -20\ \text{k}\Omega \\
(50 + -20)\ \text{k}\Omega &= (2 \times \boldsymbol{R_2}) \\
30\ \text{k}\Omega / 2 &= (2 \times \boldsymbol{R_2}) / 2 \\
15\ \text{k}\Omega &= \boldsymbol{R_2}
\end{aligned}
$$

The values of $\boldsymbol{R_1}$ and $\boldsymbol{R_2}$ in circuit are **10 kΩ** and **15 kΩ**, respectively.

5.2.3 Application of Ohm's Law

1. Problem: Two equal-value resistors are connected in series across a 5.0 V supply. When another resistor of the same value is added in series to the circuit, the current decreases by 10 mA. What is the value of each resistor?

Solution: The total resistance $\boldsymbol{R_T}$ of $\boldsymbol{n}$ resistors in series with the same value $\boldsymbol{R}$ is

$$\boldsymbol{R_T} = \boldsymbol{n} \times \boldsymbol{R}$$

From Ohm's Law for current the total current $\boldsymbol{I_T}$ (org) for the original circuit with two resistors is

$$
\begin{aligned}
\boldsymbol{I_T}\,(\text{org}) &= \boldsymbol{V_S} / \boldsymbol{R_T}\,(\text{org}) \\
&= 5.0\ \text{V} / (2 \times \boldsymbol{R})
\end{aligned}
$$

Similarly, total current $\boldsymbol{I_T}$ (new) for the new circuit with three resistors is

$$
\begin{aligned}
\boldsymbol{I_T}\,(\text{new}) &= \boldsymbol{V_S} / \boldsymbol{R_T}\,(\text{new}) \\
&= 5.0\ \text{V} / (3 \times \boldsymbol{R})
\end{aligned}
$$

From the problem statement $\boldsymbol{I_T}$ (new) is 10 mA less than $\boldsymbol{I_T}$ (org), so

$$
\begin{aligned}
\boldsymbol{I_T}\,(\text{new}) &= \boldsymbol{I_T}\,(\text{org}) - 10\ \text{mA} \\
5.0\ \text{V} / (3 \times \boldsymbol{R}) &= [5.0\ \text{V} / (2 \times \boldsymbol{R})] + -10\ \text{mA} \\
[5.0\ \text{V} / (3 \times \boldsymbol{R})] \times (3 \times \boldsymbol{R}) &= [5.0\ \text{V} / (2 \times \boldsymbol{R}) + -10\ \text{mA}] \times (3 \times \boldsymbol{R}) \\
5.0\ \text{V} &= \{[5.0\ \text{V} / (2 \times \boldsymbol{R})] \times (3 \times \boldsymbol{R})\} + -[10\ \text{mA} \times (3 \times \boldsymbol{R})] \\
5.0\ \text{V} &= \{[5.0\ \text{V} \times (3 \times \boldsymbol{R})] / (2 \times \boldsymbol{R})\} - (10\ \text{mA} \times 3 \times \boldsymbol{R}) \\
5.0\ \text{V} &= \{[(5.0 \times 3) / 2]\ \text{V}\} + -[30\ \text{mA} \times \boldsymbol{R}] \\
5.0\ \text{V} &= 7.5\ \text{V} + -(30\ \text{mA} \times \boldsymbol{R}) \\
5.0\ \text{V} + -7.5\ \text{V} &= 7.5\ \text{V} + -(30\ \text{mA} \times \boldsymbol{R}) + -7.5\ \text{V} \\
-2.5\ \text{V} &= 30\ \text{mA} \times \boldsymbol{R} \\
-2.5\ \text{V} / -30\ \text{mA} &= (-30\ \text{mA} \times \boldsymbol{R}) / (-30\ \text{mA}) \\
-2.5\ \text{V} / (-30 \times 10^{-3}\ \text{A}) &= \boldsymbol{R} \\
[-2.5 / (-30 \times 10^{-3})]\ (\text{V} / \text{A}) &= \boldsymbol{R} \\
83.3\ \Omega &= \boldsymbol{R}
\end{aligned}
$$

The value of each resistor is **83 Ω**.

2. Problem: One common application of a series resistor is that of a sense resistor, which "senses" the current in a circuit by developing a voltage that is proportional to the current. Ideally, the value of this resistor is very small compared to the circuit resistance, so addition of the sense resistor will not significantly affect the circuit resistance or the current. If $\boldsymbol{R_{NOM}}$ is the circuit resistance without a sense resistor and $\boldsymbol{I_{NOM}}$ is the current through $\boldsymbol{R_{NOM}}$ without a sense resistor, what is the maximum value of sense resistor $\boldsymbol{R_{SENSE}}$ compared to $\boldsymbol{R_{NOM}}$ that will change $\boldsymbol{I_{NOM}}$ by less than 0.5%? Assume that the voltage applied to the circuit does not change.

Solution: From Ohm's Law for voltage, the voltage $\boldsymbol{V_S}$ applied to the circuit without a sense resistor is

$$\boldsymbol{V_S} = \boldsymbol{I_{NOM}} \times \boldsymbol{R_{NOM}}$$

From the problem statement, this voltage does not change with the addition of a sense resistor. Addition of a sense resistor $\boldsymbol{R_{SENSE}}$ to the circuit will change the total circuit resistance $\boldsymbol{R_T}$ to

$$\boldsymbol{R_T} = \boldsymbol{R_{NOM}} + \boldsymbol{R_{SENSE}}$$

From Ohm's Law for current, the total current $\boldsymbol{I_T}$ for the circuit with a sense resistor is

$I_T = V_S / R_T$
$= (I_{NOM} \times R_{NOM}) / (R_{NOM} + R_{SENSE})$

If I_T must be within 0.5% of I_{NOM}, then

$I_T = (1 - 0.5\%) \times I_{NOM}$
$= [1 + -(0.5 / 100)] \times I_{NOM}$
$= [1 + -0.005] \times I_{NOM}$
$= 0.995 \times I_{NOM}$

so

$0.995 \times I_{NOM} = (I_{NOM} \times R_{NOM}) / (R_{NOM} + R_{SENSE})$
$(0.995 \times I_{NOM}) \times (R_{NOM} + R_{SENSE}) = [(I_{NOM} \times R_{NOM}) / (R_{NOM} + R_{SENSE})] \times (R_{NOM} + R_{SENSE})$
$(0.995 \times I_{NOM}) \times (R_{NOM} + R_{SENSE}) = (I_{NOM} \times R_{NOM})$

From this.

$[(0.995 \times I_{NOM}) \times (R_{NOM} + R_{SENSE})] / (0.995 \times I_{NOM}) = (I_{NOM} \times R_{NOM}) / (0.995 \times I_{NOM})$
$R_{NOM} + R_{SENSE} = R_{NOM} / 0.995$
$R_{NOM} + R_{SENSE} + -R_{NOM} = [(1 / 0.995) \times R_{NOM}] + -R_{NOM}$
$R_{SENSE} = (1 / 0.995 - 1) \times R_{NOM}$
$R_{SENSE} = 0.00503 \times R_{NOM}$

Therefore,

$R_{SENSE} = R_{NOM} \times 0.00503$
$= R_{NOM} \times [0.00503 \times (100 / 100)]$
$= R_{NOM} \times [(0.00503 \times 100) / 100)]$
$= R_{NOM} \times (0.503 / 100)$
$= R_{NOM} \times 0.503\%$

The maximum value of sense resistor R_{SENSE} that will change the circuit current I_{NOM} by less than 0.5% is **0.503% of the normal circuit resistance R_{NOM}.**

5.2.4 Voltage Sources in Series

1. Problem: V_{S1} consists of n series-aiding voltage sources, each of which has voltage V_X. If one voltage source is reversed, what percentage of V_{S1} is the new voltage V_{S2}?

Solution: The value of V_{S1} from n series-aiding voltage sources each of which has voltage V_X is

$V_{S1} = n \times V_X$

If one voltage source is reversed, there are $n - 1$ series aiding voltage sources and 1 series-opposing voltage source. The new voltage V_{S2} is then

$V_{S2} = [(n - 1) \times V_X] - (1 \times V_X)$
$= (n \times V_X) + -(1 \times V_X) + -(1 \times V_X)$
$= (n + -1 + -1) \times V_X$
$= (n - 2) \times V_X$

The ratio of the new voltage V_{S2} to the original voltage V_{S1} is

$V_{S2} / V_{S1} = [(n - 2) \times V_X] / (n \times V_X)$
$= (n - 2) / n$

Converting this to a percentage gives

$V_{S2} / V_{S1} = [(n - 2) / n] \times (100 / 100)$
$= \{[(n - 2) / n] \times 100\} / 100$

$= \{[(n - 2) / n] \times 100\}$ %

The value of the new voltage as a percentage of the original voltage is **$\{[(n - 2) / n] \times 100\}$ %**. Note that if V_{S1} is due to a single voltage source ($n = 1$) and the source is reversed the new voltage V_{S2} is −100% of V_{S1} (i.e., the magnitude of V_{S2} is the same as that of V_{S1} but the polarity is opposite). Similarly, if there are two voltage sources and one is reversed, V_{S2} is 0% of V_{S1} (i.e., the voltage sources cancel).

2. Problem: Show that if V_{SA} is the series-aiding voltage and V_{SO} is the series-opposing voltage of two batteries V_1 and V_2, then one battery voltage is the average (half the sum) of V_{SA} and V_{SO} and the other battery voltage is half the difference of V_{SA} and V_{SO}.

Solution: When battery voltages V_1 and V_2 are series-aiding, the total voltage V_{SA} is

$$V_{SA} = V_1 + V_2$$

Similarly, when battery voltages V_1 and V_2 are series-opposing, the total voltage V_{SO} is

$$V_{SO} = V_1 + -V_2$$

Taking the sum of V_{SA} and V_{SO} gives

$$V_{SA} + V_{SO} = (V_1 + V_2) + (V_1 + -V_2)$$
$$= 2 \times V_1$$

From this,

$$(2 \times V_1) / 2 = (V_{SA} + V_{SO}) / 2$$
$$V_1 = (V_{SA} + V_{SO}) / 2$$

Similarly, taking the difference of V_{SA} and V_{SO} gives

$$V_{SA} + -V_{SO} = (V_1 + V_2) + -(V_1 + -V_2)$$
$$= (V_1 + V_2) + (-V_1 + V_2)$$
$$= 2 \times V_2$$

From this,

$$(2 \times V_2) / 2 = (V_{SA} + -V_{SO}) / 2$$
$$V_2 = (V_{SA} + -V_{SO}) / 2$$

If V_{SA} is the series-aiding voltage of two batteries and V_{SO} is the series-opposing voltage of the two batteries, then the voltage of one battery is $(V_{SA} + V_{SO}) / 2$ and the voltage of the other battery is $(V_{SA} + -V_{SO}) / 2$.

5.2.5 Kirchhoff's Voltage Law

1. Problem: Determine the value of R_X for the circuit in Figure 5-18.

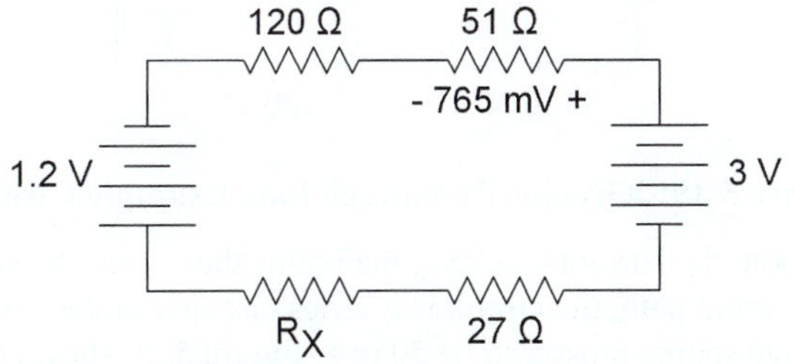

Figure 5-18: Kirchhoff's Voltage Law Example Circuit 3

Solution: From Ohm's Law for current, the current I through the 51 Ω resistor is

$$I = 765\text{ mV} / 51\ \Omega$$
$$= (765 \times 10^{-3}\text{ V}) / (51\ \Omega)$$
$$= (765 \times 10^{-3} / 51)\ (\text{V} / \Omega)$$
$$= 15.0 \times 10^{-3}\text{ A}$$

From the polarity of the voltage across the 51 Ω resistor, conventional current flows counterclockwise. The current is the same for all components in a series circuit, so from Ohm's Law for voltage, the voltages across the 120 Ω and 27 Ω resistors $V_{120\Omega}$ and $V_{27\Omega}$ are

$$V_{120\Omega} = I \times 120\ \Omega$$
$$= (15.0 \times 10^{-3}\ \text{A}) \times (120\ \Omega)$$
$$= (15.0 \times 10^{-3} \times 120)\ (\text{A} \times \Omega)$$
$$= 1.80\ \text{V}$$
$$V_{27\Omega} = I \times 27\ \Omega$$
$$= 15.0 \times 10^{-3}\ \text{A} \times 27\ \Omega$$
$$= (15.0 \times 10^{-3} \times 27)\ (\text{A} \times \Omega)$$
$$= 405 \times 10^{-3}\ \text{V}$$

Kirchhoff's Voltage Law can now find the voltage across R_X. Since conventional current flows counterclockwise, work counterclockwise around the circuit so that all resistor voltages will be voltage drops. Beginning at the negative terminal of the 1.2 V battery gives

$$(+1.2\ \text{V}) + V_{RX} + V_{27\Omega} + (+3\ \text{V}) + (-765\ \text{mV}) + (V_{120\Omega}) = 0\ \text{V}$$
$$(+1.2\ \text{V}) + V_{RX} + (-405 \times 10^{-3}\ \text{V}) + (+3\ \text{V}) + (-765 \times 10^{-3}\ \text{V}) + (-1.80\ \text{V}) = 0\ \text{V}$$
$$[1.2 - (405 \times 10^{-3}) + (+3) + -(765 \times 10^{-3}) + (-1.80)]\ \text{V} + V_{RX} = 0\ \text{V}$$
$$1.23\ \text{V} + V_{RX} + -1.23\ \text{V} = 0\ \text{V} + -1.23\ \text{V}$$
$$V_{RX} = -1.23\ \text{V}$$

As expected, the voltage across R_X is negative, indicating that the voltage across the resistor is in fact a voltage drop. From Ohm's Law for resistance,

$$R_X = V_{RX} / I$$
$$= (1.23\ \text{V}) / (15.0 \times 10^{-3}\ \text{A})$$
$$= [1.23 / (15.0 \times 10^{-3})]\ (\text{V} / \text{A})$$
$$= 82.0\ \Omega$$

The value of R_X is **82 Ω.**

2. Problem: What is the voltage of the 50 mA current source I_S for the circuit of Figure 5-19?

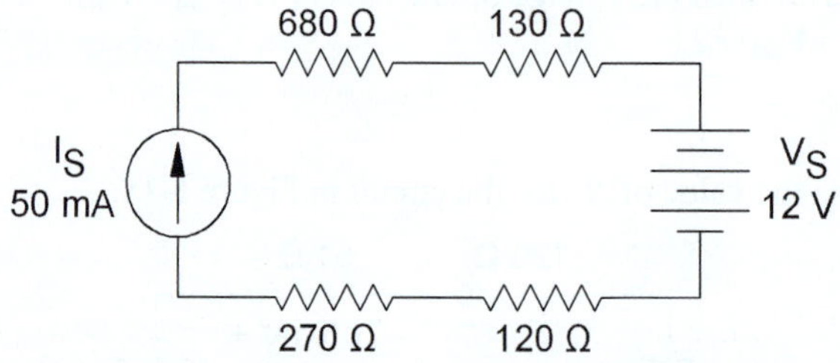

Figure 5-19: Kirchhoff's Voltage Law Example Circuit 4

Solution: By definition, the current source I_S maintains the current through it at 50 mA. Because there is only one current path, the circuit is a series circuit and the current through each resistor and the 12 V voltage source must also be 50 mA. Figure 5-20 shows the polarities of the resistor voltages and that of the 12 V supply.

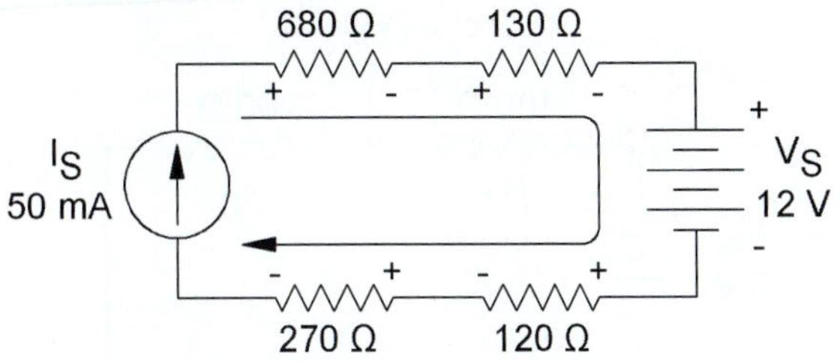

Figure 5-20: Polarity of Voltages for Figure 5-19

From Ohm's Law for voltage, the resistor voltages are

$$\begin{aligned} V_{680\Omega} &= 50 \text{ mA} \times 680\ \Omega \\ &= (50 \times 10^{-3} \text{ A}) \times (680\ \Omega) \\ &= [(50 \times 10^{-3}) \times 680]\ (\text{A} \times \Omega) \\ &= 34.0 \text{ V} \end{aligned}$$

$$\begin{aligned} V_{130\Omega} &= 50 \text{ mA} \times 130\ \Omega \\ &= (50 \times 10^{-3} \text{ A}) \times (130\ \Omega) \\ &= [(50 \times 10^{-3}) \times 130]\ (\text{A} \times \Omega) \\ &= 6.5 \text{ V} \end{aligned}$$

$$\begin{aligned} V_{120\Omega} &= 50 \text{ mA} \times 120\ \Omega \\ &= (50 \times 10^{-3} \text{ A}) \times (120\ \Omega) \\ &= [(50 \times 10^{-3}) \times 120]\ (\text{A} \times \Omega) \\ &= 6.0 \text{ V} \end{aligned}$$

$$\begin{aligned} V_{270\Omega} &= 50 \text{ mA} \times 270\ \Omega \\ &= (50 \times 10^{-3} \text{ A}) \times (270\ \Omega) \\ &= [(50 \times 10^{-3}) \times 270]\ (\text{A} \times \Omega) \\ &= 13.5 \text{ V} \end{aligned}$$

Working through the circuit in the direction of current, Kirchhoff's Voltage Law gives

$$\begin{aligned} (-V_{680\Omega}) + (-V_{130\Omega}) + (-12 \text{ V}) + (-V_{120\Omega}) + (-V_{270\Omega}) + V_{IS} &= 0 \text{ V} \\ (-34.0 \text{ V}) + (-6.5 \text{ V}) + (-12 \text{ V}) + (-6.0 \text{ V}) + (-13.5 \text{ V}) + V_{IS} &= 0 \text{ V} \\ [(-34.0) + (-6.5) + (-12) + (-6.0) + (-13.5) \text{ V}] + V_{IS} &= 0 \text{ V} \\ -72.0 \text{ V} + V_{IS} + 72.0 \text{ V} &= 0 \text{ V} + 72.0 \text{ V} \\ V_{IS} &= 72.0 \text{ V} \end{aligned}$$

The voltage of the current source I_S is *72 V*.

5.2.6 Voltage Dividers

1. Problem: What are the minimum and maximum values of tap voltage V_X for the voltage divider in Figure 5-21 for 5% tolerance resistors and a 5% supply voltage?

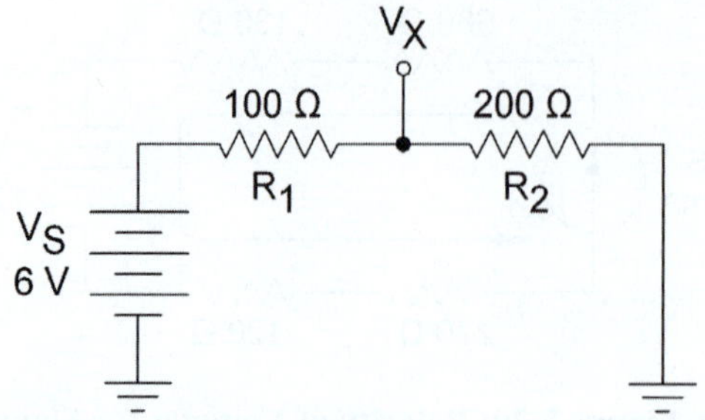

Figure 5-21: Voltage Divider Example Circuit 3

Solution: The value of the tap voltage V_X for the voltage divider of Figure 5-21 is

$$V_X = V_S \times [R_2 / (R_1 + R_2)]$$
$$= (V_S \times R_2) / (R_1 + R_2)$$

From the equation, V_X will be maximum when V_S and R_2 are maximum and R_1 is minimum. Conversely, V_X will be minimum when V_S and R_2 are minimum and R_1 is maximum.

For a 5% voltage supply the variation in V_S will be

$$\Delta V_S = V_S \times 5\%$$
$$= V_S \times 5 / 100$$
$$= 6\text{ V} \times 0.05$$
$$= 0.30\text{ V}$$

Therefore,

$$V_S\text{ (max)} = V_S + \Delta V_S$$
$$= 6\text{ V} + 0.30\text{ V}$$
$$= (6 + 0.30)\text{ V}$$
$$= 6.30\text{ V}$$

and

$$V_S\text{ (min)} = V_S - \Delta V_S$$
$$= 6\text{ V} + -0.30\text{ V}$$
$$= (6 + -0.30)\text{ V}$$
$$= 5.70\text{ V}$$

Similarly, for the resistors,

$$\Delta R_1 = R_1 \times 5\%$$
$$= R_1 \times 5 / 100$$
$$= 100\ \Omega \times 0.05$$
$$= 5\ \Omega$$

so

$$R_1\text{ (max)} = R_1 + \Delta R_1$$
$$= 100\ \Omega + 5\ \Omega$$
$$= (100 + 5)\ \Omega$$
$$= 105\ \Omega$$

and

$$R_1\text{ (min)} = R_1 - \Delta R_1$$
$$= 100\ \Omega + -5\ \Omega$$

$= (100 + -5)\ \Omega$
$= 95\ \Omega$

For resistor $\boldsymbol{R}_2$,

$\Delta \boldsymbol{R}_2 = \boldsymbol{R}_2 \times 5\%$
$= \boldsymbol{R}_2 \times 5 / 100$
$= 200\ \Omega \times 0.05$
$= 10\ \Omega$

so

$\boldsymbol{R}_2\ (\text{max}) = \boldsymbol{R}_2 + \Delta \boldsymbol{R}_2$
$= 200\ \Omega + 10\ \Omega$
$= (200 + 10)\ \Omega$
$= 210\ \Omega$

and

$\boldsymbol{R}_2\ (\text{min}) = \boldsymbol{R}_2 - \Delta \boldsymbol{R}_2$
$= 200\ \Omega + -10\ \Omega$
$= (200 + -10)\ \Omega$
$= 190\ \Omega$

From this,

$V_X\ (\text{max}) = [V_S\ (\text{max}) \times \boldsymbol{R}_2\ (\text{max})] / [\boldsymbol{R}_1\ (\text{min}) + \boldsymbol{R}_2\ (\text{max})]$
$= (6.3\ \text{V} \times 210\ \Omega) / (95\ \Omega + 210\ \Omega)$
$= [(6.3 \times 210)\ (\text{V} \times \Omega)] / [(95 + 210)\ \Omega]$
$= [1323\ (\text{V} \times \Omega)] / (305\ \Omega)$
$= (1323 / 305)\ \text{V}$
$= 4.34\ \text{V}$

and

$V_X\ (\text{min}) = [V_S\ (\text{min}) \times \boldsymbol{R}_2\ (\text{min})] / [\boldsymbol{R}_1\ (\text{max}) + \boldsymbol{R}_2\ (\text{min})]$
$= (5.7\ \text{V} \times 190\ \Omega) / (105\ \Omega + 190\ \Omega)$
$= [(5.7 \times 190)\ (\text{V} \times \Omega)] / [(105 + 190)\ \Omega]$
$= [1083\ (\text{V} \times \Omega)] / (295\ \Omega)$
$= (1083 / 295)\ \text{V}$
$= 3.67\ \text{V}$

The minimum tap voltage V_X (min) is **3.7 V** and the maximum tap voltage V_X (max) is **4.3 V**.

2. Problem: Determine the value of $\boldsymbol{R}_1$ that will develop a tap voltage V_X of 0 V for the circuit of Figure 5-22.

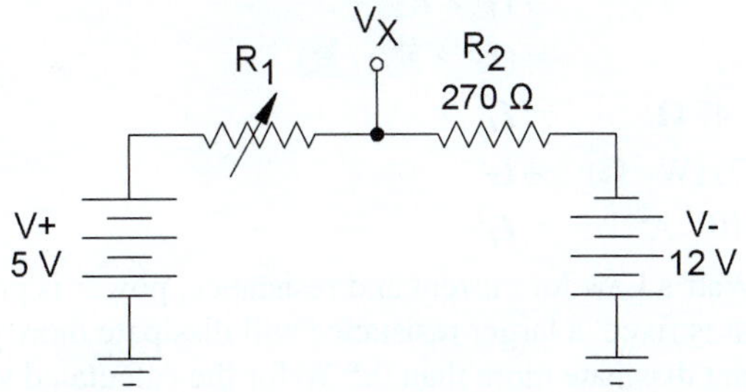

Figure 5-22: Voltage Divider Example Circuit 4

Solution: The voltage divider is a series circuit, so that

$I_{R1} = I_{R2}$

From Ohm's Law for current, the current through R_2 is

$$I_{R2} = [V_X - (V-)] / R_2$$
$$= [(0\ V) + -(-12\ V)] / 270\ \Omega$$
$$= [(0 + 12)\ V] / 270\ \Omega$$
$$= = (12 / 270)\ (V / \Omega)$$
$$= 44.4 \times 10^{-3}\ A$$

From Ohm's Law for resistance,

$$R_1 = [(V+) - V_X] / I_{R1}$$
$$= [(V+) + -V_X] / I_{R2}$$
$$= [(5\ V) + -(0\ V)] / (44.4 \times 10^{-3}\ A)$$
$$= 5\ V / (44.4 \times 10^{-3}\ A)$$
$$= [5 / (44.4 \times 10^{-3})]\ (V / A)$$
$$= 112.5\ \Omega$$

The value of R_1 that will develop a tap voltage V_X of 0 V is **112.5 Ω.**

5.2.7 Power in Series Circuits

1. Problem: The calculated total power P_T for the circuit in Figure 5-23 is 1.5 W. Will this condition exceed the power rating of any of the resistors in the circuit?

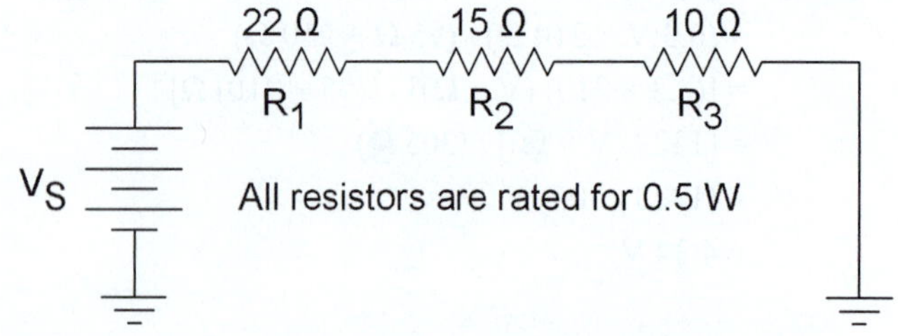

Figure 5-23: Power Dissipation Example Circuit 2

Solution: The total resistance R_T for the series circuit is

$$R_T = R_1 + R_2 + R_3$$
$$= 22\ \Omega + 15\ \Omega + 10\ \Omega$$
$$= (22 + 15 + 10)\ \Omega$$
$$= 47\ \Omega$$

From Watt's Law for current and resistance when the circuit dissipates 1.5 W,

$$P_T = I_T^2 \times R_T$$
$$P_T / R_T = (I_T^2 \times R_T) / R_T$$
$$1.5\ W / 47\ \Omega = I_T^2$$
$$(1.5 / 47)\ (W / \Omega) = I_T^2$$
$$31.9 \times 10^{-3}\ A^2 = I_T^2$$

Also from Watt's Law for current and resistance, power is proportional to the resistance so that when current is fixed, a larger resistance will dissipate more power. The maximum resistance R_{PMAX} that not dissipate more than 0.5 W for the calculated value of I_T^2 is

$$P_{MAX} = I_T^2 \times R_{PMAX}$$
$$P_{MAX} / I_T^2 = (I_T^2 \times R_{PMAX}) / I_T^2$$

$0.5\ \text{W} / (31.9 \times 10^{-3}\ \text{A}^2) = \boldsymbol{R_{PMAX}}$

$[0.5 / (31.9 \times 10^{-3})]\ (\text{W} / \text{A}^2) = \boldsymbol{R_{PMAX}}$

$[0.5 / (31.9 \times 10^{-3})]\ \Omega = \boldsymbol{R_{PMAX}}$

$15.7\ \Omega = \boldsymbol{R_{PMAX}}$

The value of $\boldsymbol{R_1}$ exceeds this value, so **the power dissipated by $\boldsymbol{R_1}$ will exceed its 0.5 W rating**.

2. Problem: What is the maximum current that will not exceed the power rating for any resistor in the series circuit of Figure 5-24?

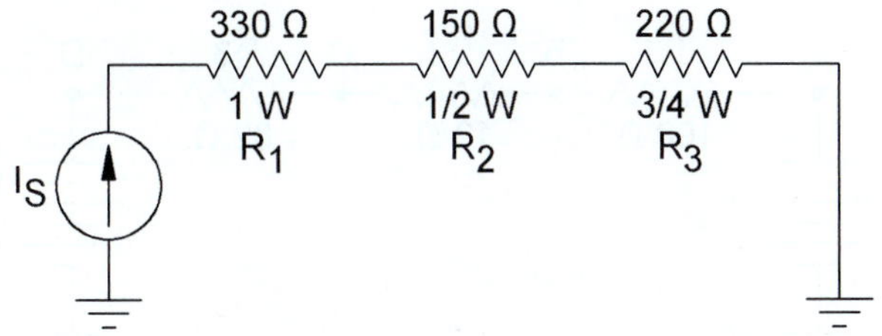

Figure 5-24: Power Dissipation Example Circuit 3

Solution: From Watt's Law for current and resistance,

$\boldsymbol{P_R} = \boldsymbol{I_T}^2 \times \boldsymbol{R}$

$\boldsymbol{P_R} / \boldsymbol{R} = (\boldsymbol{I_T}^2 \times \boldsymbol{R}) / \boldsymbol{R}$

$\boldsymbol{P_R} / \boldsymbol{R} = \boldsymbol{I_T}^2$

$(\boldsymbol{P_R} / \boldsymbol{R})^{1/2} = (\boldsymbol{I_T}^2)^{1/2}$ (Note that $(\boldsymbol{P_R} / \boldsymbol{R})^{1/2}$ is the same as $\sqrt{\boldsymbol{P_R} / \boldsymbol{R}}$)

$= \boldsymbol{I_T}^{(2 \times 1/2)}$

$= \boldsymbol{I_T}$

For $\boldsymbol{R_1}$,

$\boldsymbol{I_T}\ (\text{max}) = (\boldsymbol{P_{R1}} / \boldsymbol{R_1})^{1/2}$

$= (1\ \text{W} / 330\ \Omega)^{1/2}$

$= (1 / 330)^{1/2}\ (\text{W} / \Omega)^{1/2}$

$= (3.03 \times 10^{-3})^{1/2}\ (\text{A}^2)^{1/2}$

$= 55.0 \times 10^{-3}\ \text{A}$

$= 55.0\ \text{mA}$

For $\boldsymbol{R_2}$,

$\boldsymbol{I_T}\ (\text{max}) = (\boldsymbol{P_{R2}} / \boldsymbol{R_2})^{1/2}$

$= (1/2\ \text{W} / 150\ \Omega)^{1/2}$

$= (0.5\ \text{W} / 150\ \Omega)^{1/2}$

$= (0.5 / 150)^{1/2}\ (\text{W} / \Omega)^{1/2}$

$= (3.33 \times 10^{-3})^{1/2}\ (\text{A}^2)^{1/2}$

$= 57.7 \times 10^{-3}\ \text{A}$

$= 57.7\ \text{mA}$

For $\boldsymbol{R_3}$,

$\boldsymbol{I_T}\ (\text{max}) = (\boldsymbol{P_{R3}} / \boldsymbol{R_3})^{1/2}$

$= (3/4\ \text{W} / 220\ \Omega)^{1/2}$

$= (0.75\ \text{W} / 220\ \Omega)^{1/2}$

$= (0.75 / 220)^{1/2}\ (\text{W} / \Omega)^{1/2}$

$= (3.41 \times 10^{-3})^{1/2}\ (\text{A}^2)^{1/2}$

$= 58.4 \times 10^{-3}$ A

$= 58.4$ mA

Because $\boldsymbol{R_1}$ has the lowest value for the maximum value of $\boldsymbol{I_T}$, the maximum current that will not exceed the power rating of a resistor is **55 mA**.

5.2.8 Voltage Measurements

1. Problem: What are the voltages at Points A, B, and D relative to Point C for the circuit in Figure 5-25? Assume that all values are exact.

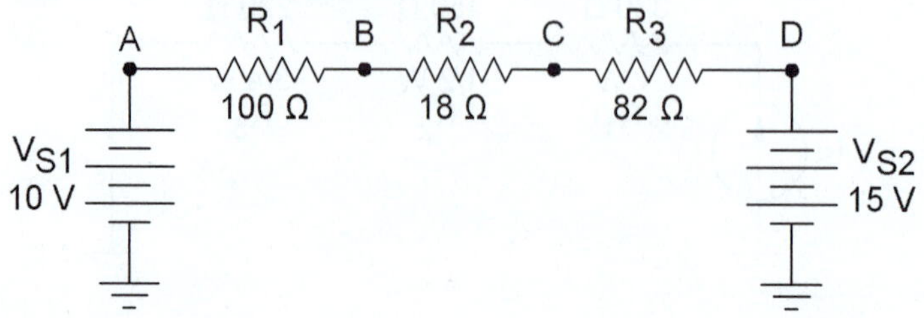

Figure 5-25: Voltage Measurement Example Circuit 3

Solution: Assume that the total current $\boldsymbol{I_T}$ flows clockwise for the circuit so that the voltage polarities are as shown in Figure 5-26.

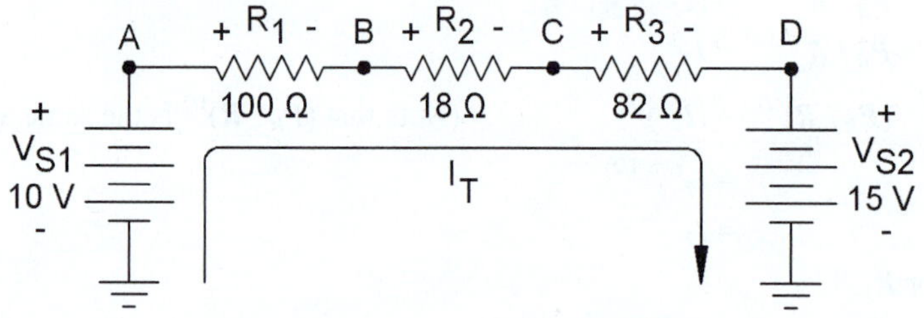

Figure 5-26: Assumed Current Direction and Voltage Polarities for Figure 5-25

By tracing in the direction of the assumed current flow Kirchhoff's Voltage Law gives

$$V_{S1} - V_{R1} - V_{R2} - V_{R3} - V_{S2} = 0 \text{ V}$$

$$V_{S1} + -V_{S2} + -V_{R1} + -V_{R2} + -V_{R3} = 0 \text{ V}$$

From Ohm's Law for voltage,

$$V_{S1} + -V_{S2} + -(I_T \times R_1) + -(I_T \times R_2) + -(I_T \times R_3) = 0 \text{ V}$$

$$V_{S1} + -V_{S2} + -[I_T \times (R_1 + R_2 + R_3)] = 0 \text{ V}$$

$$V_{S1} + -V_{S2} + -[I_T \times (R_1 + R_2 + R_3)] + [I_T \times (R_1 + R_2 + R_3)] = 0 \text{ V} + [I_T \times (R_1 + R_2 + R_3)]$$

$$V_{S1} + -V_{S2} = I_T \times (R_1 + R_2 + R_3)$$

From this,

$$(V_{S1} + -V_{S2}) / (R_1 + R_2 + R_3) = [I_T \times (R_1 + R_2 + R_3)] / (R_1 + R_2 + R_3)$$

$$(V_{S1} + -V_{S2}) / (R_1 + R_2 + R_3) = I_T$$

The total current for the circuit is then

$$I_T = (10 \text{ V} + -15 \text{ V}) / (100\ \Omega + 18\ \Omega + 82\ \Omega)$$

$$= [(10 + -15) \text{ V}] / [(100 + 18 + 82)\ \Omega]$$

$$= -5 \text{ V} / 200\ \Omega$$

$$= (-5 / 200)\ (\text{V} / \Omega)$$

$$= -25 \times 10^{-3} \text{ A}$$

The negative sign for the current means that the actual current direction and resistor voltage polarities are opposite to the assumed direction and polarities. As you will see, this will not affect the calculations as the negative current corrects for the assumed polarity. From Ohm's Law for voltage,

$$\begin{aligned} V_{R1} &= (-25 \times 10^{-3}\ \text{A}) \times (100\ \Omega) \\ &= [(-25 \times 10^{-3}) \times 100]\ (\text{A} \times \Omega) \\ &= -2.5\ \text{V} \end{aligned}$$

$$\begin{aligned} V_{R2} &= (-25 \times 10^{-3}\ \text{A}) \times (18\ \Omega) \\ &= [(-25 \times 10^{-3}) \times 18]\ (\text{A} \times \Omega) \\ &= -450 \times 10^{-3}\ \text{V} \end{aligned}$$

$$\begin{aligned} V_{R3} &= (-25 \times 10^{-3}\ \text{A}) \times (82\ \Omega) \\ &= [(-25 \times 10^{-3}) \times 82]\ (\text{A} \times \Omega) \\ &= -2.05\ \text{V} \end{aligned}$$

If we use Point C as our reference, $V_C = 0$ V, so from Figure 5-26

$$\begin{aligned} V_B &= V_C + V_{R2} \\ &= 0\ \text{V} + (-450 \times 10^{-3}\ \text{V}) \\ &= [0 + (-450 \times 10^{-3})]\ \text{V} \\ &= -450 \times 10^{-3}\ \text{V} \\ &= -450\ \text{mV} \end{aligned}$$

From this,

$$\begin{aligned} V_A &= V_B + V_{R1} \\ &= (-450 \times 10^{-3}\ \text{V}) + (-2.5\ \text{V}) \\ &= [(-450 \times 10^{-3}) + (-2.5)]\ \text{V} \\ &= -2.95\ \text{V} \end{aligned}$$

Similarly,

$$\begin{aligned} V_D &= V_C - V_{R3} \\ &= 0\ \text{V} + -(-2.05\ \text{V}) \\ &= (0 + 2.05)\ \text{V} \\ &= +2.05\ \text{V} \end{aligned}$$

The voltages at Points A, B, and D with respect to Point C are then **−2.95 V**, **−450 mV**, and **+2.05 V** respectively.

2. Problem: What are the voltages at Points A, B, and D relative to Point C for the circuit in Figure 5-25 if the polarity of V_{S1} is reversed? Assume that all values are exact.

Solution: Assume that the total current I_T flows counter-clockwise for the circuit so that the voltage polarities are as shown in Figure 5-27.

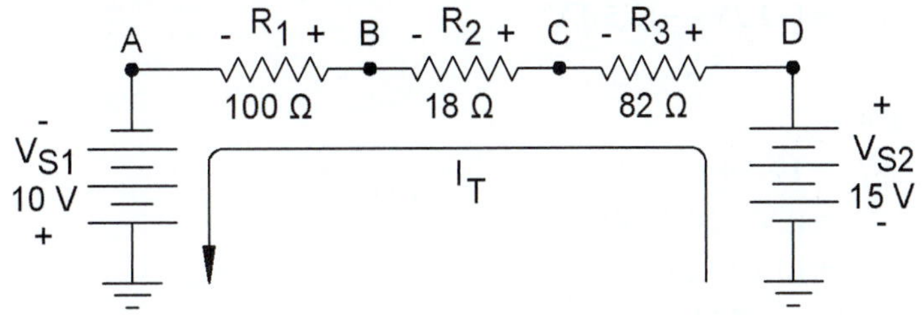

Figure 5-27: Assumed Current Direction and Voltage Polarities for Modified Circuit

By tracing in the direction of the assumed current flow, Kirchhoff's Voltage Law gives

$$V_{S2} - V_{R3} - V_{R2} - V_{R1} + V_{S1} = 0\ \text{V}$$

$$V_{S1} + V_{S2} + -V_{R1} + -V_{R2} + -V_{R3} \quad = 0 \text{ V}$$

From Ohm's Law for voltage,

$$V_{S1} + V_{S2} + -(I_T \times R_1) + -(I_T \times R_2) + -(I_T \times R_3) \quad = 0 \text{ V}$$

$$V_{S1} + V_{S2} + -[I_T \times (R_1 + R_2 + R_3)] \quad = 0 \text{ V}$$

$$V_{S1} + V_{S2} + -[I_T \times (R_1 + R_2 + R_3)] + [I_T \times (R_1 + R_2 + R_3)] \quad = 0 \text{ V} + [I_T \times (R_1 + R_2 + R_3)]$$

$$V_{S1} + V_{S2} \quad = I_T \times (R_1 + R_2 + R_3)$$

Use Ohm's Law to solve for I_T.

$$(V_{S1} + V_{S2}) / (R_1 + R_2 + R_3) \quad = [I_T \times (R_1 + R_2 + R_3)] / (R_1 + R_2 + R_3)$$

$$(V_{S1} + V_{S2}) / (R_1 + R_2 + R_3) \quad = I_T$$

Use the given circuit values to calculate the total current for the circuit.

$$\begin{aligned} I_T &= (10 \text{ V} + 15 \text{ V}) / (100\ \Omega + 18\ \Omega + 82\ \Omega) \\ &= [(10 + 15) \text{ V}] / [(100 + 18 + 82)\ \Omega] \\ &= 25 \text{ V} / 200\ \Omega \\ &= (25 / 200)\ (\text{V} / \Omega) \\ &= 125 \times 10^{-3} \text{ A} \end{aligned}$$

The positive sign for the current means that the actual current direction and resistor voltage polarities are the same as the assumed direction and polarities. From Ohm's Law for voltage,

$$\begin{aligned} V_{R1} &= (125 \times 10^{-3} \text{ A}) \times (100\ \Omega) \\ &= [(125 \times 10^{-3}) \times 100]\ (\text{A} \times \Omega) \\ &= 12.5 \text{ V} \end{aligned}$$

$$\begin{aligned} V_{R2} &= (125 \times 10^{-3} \text{ A}) \times (18\ \Omega) \\ &= [(125 \times 10^{-3}) \times 18]\ (\text{A} \times \Omega) \\ &= 2.25 \text{ V} \end{aligned}$$

$$\begin{aligned} V_{R3} &= (125 \times 10^{-3} \text{ A}) \times (82\ \Omega) \\ &= [(125 \times 10^{-3}) \times 82]\ (\text{A} \times \Omega) \\ &= 10.25 \text{ V} \end{aligned}$$

If we use Point C as our reference, $V_C = 0$ V, so from Figure 5-27

$$\begin{aligned} V_B &= V_C - V_{R2} \\ &= 0 \text{ V} + -2.25 \text{ V} \\ &= (0 + -2.25) \text{ V} \\ &= -2.25 \text{ V} \end{aligned}$$

From this,

$$\begin{aligned} V_A &= V_B - V_{R1} \\ &= (-2.25 \text{ V}) + -(12.5 \text{ V}) \\ &= [-2.25 + -12.5)] \text{ V} \\ &= -14.75 \text{ V} \end{aligned}$$

Similarly

$$\begin{aligned} V_D &= V_C + V_{R3} \\ &= 0 \text{ V} + 10.25 \text{ V}) \\ &= (0 + 10.25) \text{ V} \\ &= +10.25 \text{ V} \end{aligned}$$

The voltages at Points A, B, and D with respect to Point C are then **−14.75 V**, **−2.25 V**, and **+10.25 V** respectively.

5.3 Just for Fun

1. Problem: Verify that the total power dissipated in a series resistive circuit is the sum of the power dissipated by the individual resistors.

Solution: From Watt's Law for current and resistance, the total power dissipated by the resistors in a series circuit is

$$\begin{aligned} \boldsymbol{P_T} &= (\boldsymbol{I_T}^2 \times \boldsymbol{R_1}) + (\boldsymbol{I_T}^2 \times \boldsymbol{R_2}) + \ldots + (\boldsymbol{I_T}^2 \times \boldsymbol{R_N}) \\ &= \boldsymbol{I_T}^2 \times (\boldsymbol{R_1} + \boldsymbol{R_2} + \ldots + \boldsymbol{R_N}) \end{aligned}$$

For a series circuit,

$$\boldsymbol{R_T} = \boldsymbol{R_1} + \boldsymbol{R_2} + \ldots + \boldsymbol{R_N}$$

so

$$\boldsymbol{P_T} = \boldsymbol{I_T}^2 \times \boldsymbol{R_T}$$

The total power $\boldsymbol{P_T}$ in a series circuit is equal to the sum of the power dissipated by the individual resistors.

2. Problem: A common fault in series resistive circuits is a shorted resistor. You can find the shorted resistor by trial and error measurements, but knowing the value of the shorted resistor can simplify the search. When a resistor in a series circuit shorts, the short reduces the total series resistance from $\boldsymbol{R_{NOM}}$ to $\boldsymbol{R_{FAULT}}$ and the increase in current $\Delta\boldsymbol{I}$ is the difference between the nominal current $\boldsymbol{I_{NOM}}$ and the fault current $\boldsymbol{I_{FAULT}}$. Show that the value of the shorted resistor $\boldsymbol{R_X}$ in the circuit is

$$\boldsymbol{R_X} = (\Delta\boldsymbol{I} / \boldsymbol{I_{FAULT}}) \times \boldsymbol{R_{NOM}}$$

Solution: Let $\boldsymbol{V_S}$ be the applied circuit voltage. From the problem statement, the change in current is equal to the difference between the fault current and nominal current.

$$\begin{aligned} \Delta\boldsymbol{I} &= \boldsymbol{I_{FAULT}} - \boldsymbol{I_{NOM}} \\ &= (\boldsymbol{V_S} / \boldsymbol{R_{FAULT}}) + -(\boldsymbol{V_S} / \boldsymbol{R_{NOM}}) \\ &= [\boldsymbol{V_S} / (\boldsymbol{R_{NOM}} + -\boldsymbol{R_X})] + -(\boldsymbol{V_S} / \boldsymbol{R_{NOM}}) \\ &= \{[\boldsymbol{V_S} / (\boldsymbol{R_{NOM}} + -\boldsymbol{R_X})] \times (\boldsymbol{R_{NOM}} / \boldsymbol{R_{NOM}})\} \\ &\qquad + -\{(\boldsymbol{V_S} / \boldsymbol{R_{NOM}}) \times [(\boldsymbol{R_{NOM}} + -\boldsymbol{R_X}) / (\boldsymbol{R_{NOM}} + -\boldsymbol{R_X})]\} \\ &= \{(\boldsymbol{V_S} \times \boldsymbol{R_{NOM}}) / [\boldsymbol{R_{NOM}} \times (\boldsymbol{R_{NOM}} + -\boldsymbol{R_X})]\} \\ &\qquad + -\{[\boldsymbol{V_S} \times (\boldsymbol{R_{NOM}} + -\boldsymbol{R_X})] / [\boldsymbol{R_{NOM}} \times (\boldsymbol{R_{NOM}} + -\boldsymbol{R_X})]\} \\ &= \{(\boldsymbol{V_S} \times \boldsymbol{R_{NOM}}) / [\boldsymbol{R_{NOM}}^2 + -(\boldsymbol{R_{NOM}} \times \boldsymbol{R_X})]\} \\ &\qquad + -\{[\boldsymbol{V_S} \times (\boldsymbol{R_{NOM}} + -\boldsymbol{R_X})] / [\boldsymbol{R_{NOM}}^2 + -(\boldsymbol{R_{NOM}} \times \boldsymbol{R_X})]\} \\ &= \{(\boldsymbol{V_S} \times \boldsymbol{R_{NOM}}) + -[\boldsymbol{V_S} \times (\boldsymbol{R_{NOM}} + -\boldsymbol{R_X})]\} / [\boldsymbol{R_{NOM}}^2 + -(\boldsymbol{R_{NOM}} \times \boldsymbol{R_X})] \\ &= [(\boldsymbol{V_S} \times \boldsymbol{R_{NOM}}) + -(\boldsymbol{V_S} \times \boldsymbol{R_{NOM}}) + (\boldsymbol{V_S} \times \boldsymbol{R_X})] / [\boldsymbol{R_{NOM}}^2 + -(\boldsymbol{R_{NOM}} \times \boldsymbol{R_X})] \\ &= (\boldsymbol{V_S} \times \boldsymbol{R_X}) / [\boldsymbol{R_{NOM}}^2 + -(\boldsymbol{R_{NOM}} \times \boldsymbol{R_X})] \end{aligned}$$

From this,

$$\begin{aligned} &\Delta\boldsymbol{I} \times [\boldsymbol{R_{NOM}}^2 + -(\boldsymbol{R_{NOM}} \times \boldsymbol{R_X})] \\ &\qquad = \{(\boldsymbol{V_S} \times \boldsymbol{R_X}) / [\boldsymbol{R_{NOM}}^2 + -(\boldsymbol{R_{NOM}} \times \boldsymbol{R_X})]\} \times [\boldsymbol{R_{NOM}}^2 + -(\boldsymbol{R_{NOM}} \times \boldsymbol{R_X})] \\ &\Delta\boldsymbol{I} \times [\boldsymbol{R_{NOM}}^2 + -(\boldsymbol{R_{NOM}} \times \boldsymbol{R_X})] = \boldsymbol{V_S} \times \boldsymbol{R_X} \\ &(\Delta\boldsymbol{I} \times \boldsymbol{R_{NOM}}^2) + -(\Delta\boldsymbol{I} \times \boldsymbol{R_{NOM}} \times \boldsymbol{R_X}) = \boldsymbol{V_S} \times \boldsymbol{R_X} \\ &(\Delta\boldsymbol{I} \times \boldsymbol{R_{NOM}}^2) + -(\Delta\boldsymbol{I} \times \boldsymbol{R_{NOM}} \times \boldsymbol{R_X}) + (\Delta\boldsymbol{I} \times \boldsymbol{R_{NOM}} \times \boldsymbol{R_X}) = (\boldsymbol{V_S} \times \boldsymbol{R_X}) + (\Delta\boldsymbol{I} \times \boldsymbol{R_{NOM}} \times \boldsymbol{R_X}) \\ &(\Delta\boldsymbol{I} \times \boldsymbol{R_{NOM}}^2) = (\boldsymbol{V_S} \times \boldsymbol{R_X}) + (\Delta\boldsymbol{I} \times \boldsymbol{R_{NOM}} \times \boldsymbol{R_X}) \end{aligned}$$

so

$$\begin{aligned} &(\Delta\boldsymbol{I} \times \boldsymbol{R_{NOM}}^2) = [\boldsymbol{V_S} + (\Delta\boldsymbol{I} \times \boldsymbol{R_{NOM}})] \times \boldsymbol{R_X} \\ &(\Delta\boldsymbol{I} \times \boldsymbol{R_{NOM}}^2) / [\boldsymbol{V_S} + (\Delta\boldsymbol{I} \times \boldsymbol{R_{NOM}})] = \{[\boldsymbol{V_S} + (\Delta\boldsymbol{I} \times \boldsymbol{R_{NOM}})] \times \boldsymbol{R_X}]\} / [\boldsymbol{V_S} + (\Delta\boldsymbol{I} \times \boldsymbol{R_{NOM}})] \\ &(\Delta\boldsymbol{I} \times \boldsymbol{R_{NOM}}^2) / [\boldsymbol{V_S} + (\Delta\boldsymbol{I} \times \boldsymbol{R_{NOM}})] = \boldsymbol{R_X} \end{aligned}$$

Simplifying this gives

$$\begin{aligned}
\boldsymbol{R_X} &= (\Delta \boldsymbol{I} \times \boldsymbol{R_{NOM}}^2) / [\boldsymbol{V_S} + (\Delta \boldsymbol{I} \times \boldsymbol{R_{NOM}})] \\
&= (\Delta \boldsymbol{I} \times \boldsymbol{R_{NOM}}^2) / \{\boldsymbol{V_S} + [(\boldsymbol{I_{FAULT}} - \boldsymbol{I_{NOM}}) \times \boldsymbol{R_{NOM}}]\} \\
&= (\Delta \boldsymbol{I} \times \boldsymbol{R_{NOM}}^2) / [\boldsymbol{V_S} + (\boldsymbol{I_{FAULT}} \times \boldsymbol{R_{NOM}}) + -(\boldsymbol{I_{NOM}} \times \boldsymbol{R_{NOM}})] \\
&= (\Delta \boldsymbol{I} \times \boldsymbol{R_{NOM}}^2) / [\boldsymbol{V_S} + (\boldsymbol{I_{FAULT}} \times \boldsymbol{R_{NOM}}) + -\boldsymbol{V_S}] \\
&= (\Delta \boldsymbol{I} \times \boldsymbol{R_{NOM}} \times \boldsymbol{R_{NOM}}) / (\boldsymbol{I_{FAULT}} \times \boldsymbol{R_{NOM}}) \\
&= (\Delta \boldsymbol{I} \times \boldsymbol{R_{NOM}}) / \boldsymbol{I_{FAULT}} \\
&= (\Delta \boldsymbol{I} / \boldsymbol{I_{FAULT}}) \times \boldsymbol{R_{NOM}}
\end{aligned}$$

3. Problem: Show that the maximum allowable value current $\boldsymbol{I_{LIMIT}}$ for a series circuit is limited by resistor $\boldsymbol{R_N}$ for which the value of $\boldsymbol{P_{RN}} / \boldsymbol{R_N}$ is minimum.

Solution: Let $\boldsymbol{I_{LIMIT}}$ be the maximum allowable current for the series circuit. Because this current is the same for all resistors in a series circuit, the resistor $\boldsymbol{R_N}$ that has the lowest maximum power rating $\boldsymbol{P_{RN}}$ will limit this current. From the problem statement, $\boldsymbol{P_{RN}}$ is the maximum power dissipation of resistor $\boldsymbol{R_N}$. Watt's Law for current and resistance gives

$$\begin{aligned}
\boldsymbol{P_{RN}} &= (\boldsymbol{I_{LIMIT}})^2 \times \boldsymbol{R_N} \\
\boldsymbol{P_{RN}} / \boldsymbol{R_N} &= (\boldsymbol{I_{LIMIT}})^2 \times \boldsymbol{R_N}) / \boldsymbol{R_N} \\
\boldsymbol{P_{RN}} / \boldsymbol{R_N} &= (\boldsymbol{I_{LIMIT}})^2 \times (\boldsymbol{R_N} / \boldsymbol{R_N}) \\
\boldsymbol{P_{RN}} / \boldsymbol{R_N} &= (\boldsymbol{I_{LIMIT}})^2
\end{aligned}$$

When $\boldsymbol{P_{RN}} / \boldsymbol{R_N}$ is minimum $(\boldsymbol{I_{LIMIT}})^2$ is also minimum, and when $(\boldsymbol{I_{LIMIT}})^2$ is minimum $\boldsymbol{I_{LIMIT}}$ is also minimum. Therefore, the series resistor for which $\boldsymbol{P_{RN}} / \boldsymbol{R_N}$ is minimum determines the maximum allowable value of $\boldsymbol{I_{LIMIT}}$ for the series circuit.

6. Parallel Circuits

6.1 Basic

6.1.1 Resistors in Parallel

1. Problem: For which of the circuits in Figure 6-1 are resistors R_1, R_2, and R_3 in parallel between Points A and B?

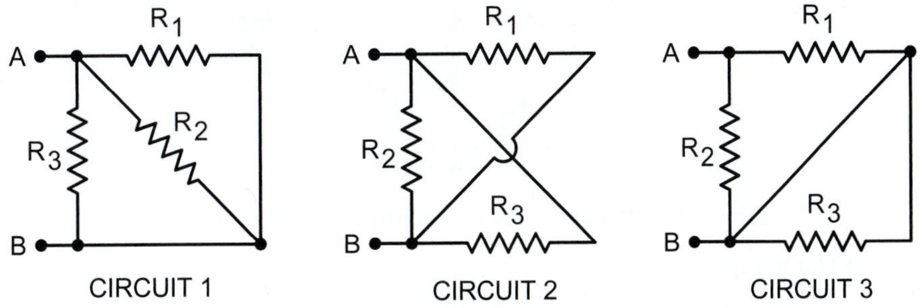

Figure 6-1: Parallel Circuit Identification Example 1

Solution: The resistors are in parallel between Points A and B if each resistor provides a direct current path between Points A and B. Refer to Figure 6-2.

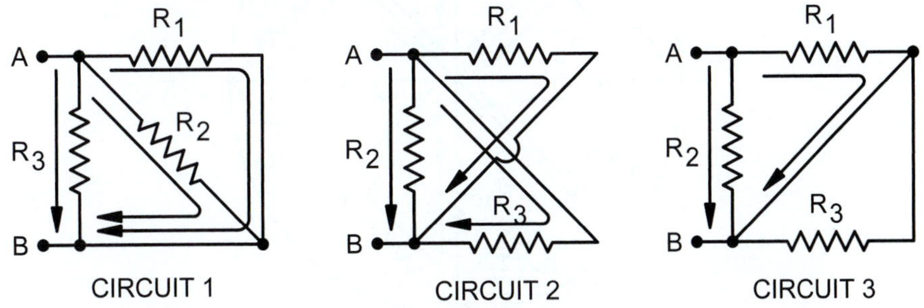

Figure 6-2: Current Paths Between Points A and B for Figure 6-1

In Circuit 1 R_1, R_2, and R_3 each provide a direct current path between Points A and B. Therefore, the resistors in Circuit 1 **are** all in parallel.

In Circuit 2, R_1, R_2, and R_3 each provide a direct current path between Points A and B. Therefore, the resistors in Circuit 2 **are** all in parallel.

In Circuit 3, R_1 and R_2 each provide a direct current path between Points A and B. R_3, however, does not connect directly to Point A so that R_3 does not provide a direct current path between Points A and B. Therefore, the resistors in Circuit 3 **are not** all in parallel.

2. Problem: How many resistors are in parallel between Points A and B of the circuit in Figure 6-3?

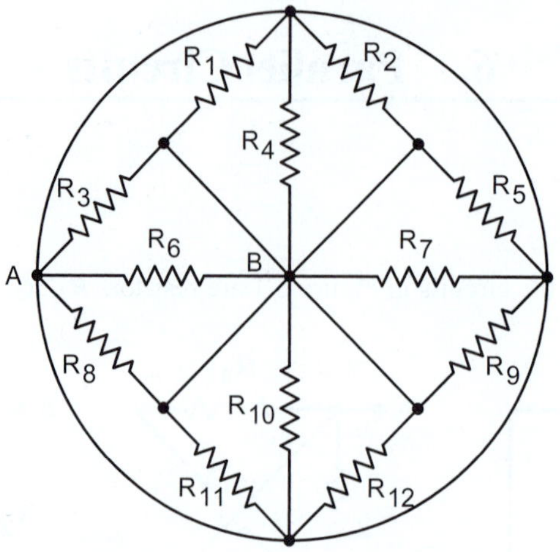

Figure 6-3: Parallel Circuit Identification Example 2

Solution: Refer to Figure 6-4.

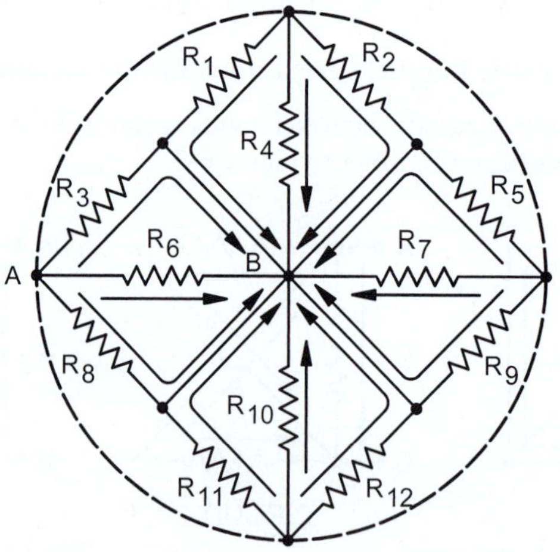

Figure 6-4: Current Paths Between Points A and B for Figure 6-3

As the figure shows, Point A is electrically connected to the junctions of $\boldsymbol{R}_1$ and $\boldsymbol{R}_2$, $\boldsymbol{R}_3$ and $\boldsymbol{R}_8$, $\boldsymbol{R}_5$ and $\boldsymbol{R}_9$, and $\boldsymbol{R}_{11}$ and $\boldsymbol{R}_{12}$ so that each resistor provides a direct current path between Points A and B. Therefore, **all 12 resistors are in parallel.**

6.1.2 Voltage in a Parallel Circuit

1. Problem: A printed circuit board has four resistors, labeled $\boldsymbol{R}_1$ through $\boldsymbol{R}_4$, connected between test points $\boldsymbol{TP}_1$ and $\boldsymbol{TP}_2$. The measured voltages for resistors $\boldsymbol{R}_1$, $\boldsymbol{R}_2$, $\boldsymbol{R}_3$, and $\boldsymbol{R}_4$ are $V_{R1} = 3.3$ V, $V_{R2} = 1.2$ V, $V_{R3} = 3.3$ V, and $V_{R4} = 2.1$ V, respectively. Assuming that no fault exists in the circuit, are the resistors in parallel?

   Solution: By definition, parallel resistors connect between the same two points, so the potential difference between the two points appears across each parallel resistor. The voltages V_{R2} and V_{R4} across $\boldsymbol{R}_2$ and $\boldsymbol{R}_4$ are not equal to each other or to the voltages V_{R1} and V_{R3} across $\boldsymbol{R}_1$ and $\boldsymbol{R}_3$, so the resistors are **not** in parallel.

2. Problem: Three resistors connected in parallel across a 1.5 V battery each have a measured voltage of 1.50 V. When a fourth resistor is added in parallel to the first three resistors, its measured voltage is 1.45 V. What is the measured voltage for the other three resistors?

Solution: By definition, parallel resistors connect between the same two points, so the potential difference between the two points appears across each parallel resistor and the measured voltage across one resistor is equal to the voltage across each resistor in parallel with it. The voltage measured across the fourth resistor is 1.45 V, so the voltage across the other three resistors must also be **1.45 V**.

6.1.3 Kirchhoff's Current Law

1. Problem: What is the current I_{R2} through resistor R_2 for the circuit of Figure 6-5?

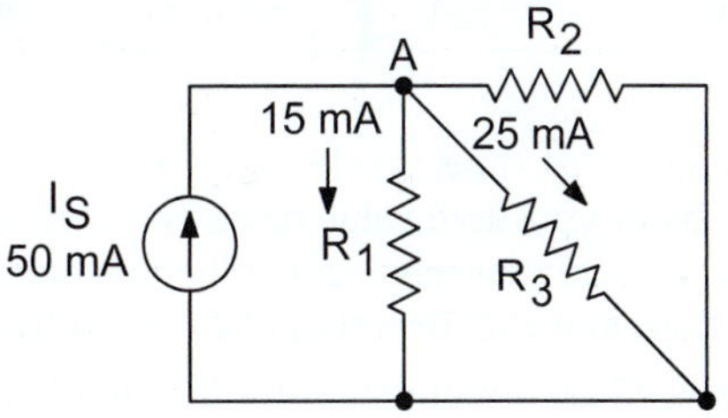

Figure 6-5: Kirchhoff's Current Law Example Circuit 1

Solution: Let currents entering the junction at Point A be positive and those leaving the junction be negative. From Kirchhoff's Current Law

$$+50 \text{ mA} + (-15 \text{ mA}) + (-25 \text{ mA}) + I_{R2} = 0 \text{ mA}$$
$$[+50 + (-15) + (-25)] \text{ mA} + I_{R2} = 0 \text{ mA}$$
$$+10 \text{ mA} + I_{R2} = 0 \text{ mA}$$
$$+10 \text{ mA} + I_{R2} + -10 \text{ mA} = 0 \text{ mA} + -10 \text{ mA}$$
$$I_{R2} = -10 \text{ mA}$$

The current through R_2 is **10 mA**. The negative sign indicates that it exits the junction at Point A.

2. Problem: What are the directions of measured currents I_{R2}, I_{R3}, and I_{R4} for the circuit of Figure 6-6?

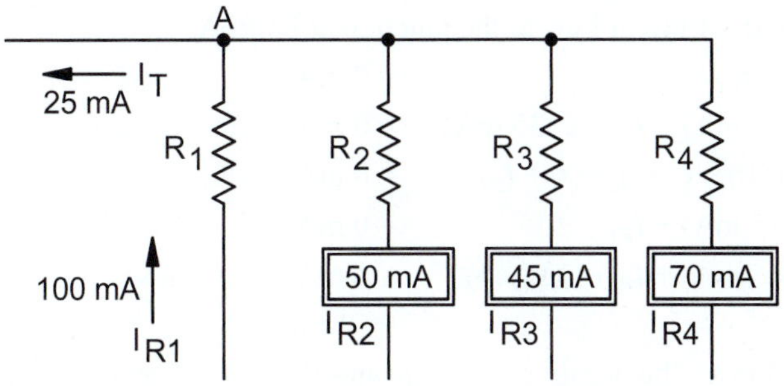

Figure 6-6: Kirchhoff's Current Law Example Circuit 2

Solution: Let currents entering the junction at Point A be positive and currents leaving the junction be negative. From Kirchhoff's Current Law

$$I_T + I_{R1} + I_{R2} + I_{R3} + I_{R4} = 0 \text{ mA}$$
$$(-25 \text{ mA}) + (+100 \text{ mA}) + I_{R2} + I_{R3} + I_{R4} = 0 \text{ mA}$$
$$[(-25) + (+100)] \text{ mA} + I_{R2} + I_{R3} + I_{R4} = 0 \text{ mA}$$
$$(+75 \text{ mA}) + I_{R2} + I_{R3} + I_{R4} = 0 \text{ mA}$$
$$(+75 \text{ mA}) + I_{R2} + I_{R3} + I_{R4} + (-75 \text{ mA}) = 0 \text{ mA} + (-75 \text{ mA})$$
$$I_{R2} + I_{R3} + I_{R4} = -75 \text{ mA}$$

Currents I_{R2}, I_{R3}, and I_{R4} must sum to –75 mA. Table 6-1 shows the sums that result from various combinations of these currents entering and leaving the junction.

Table 6-1: Possible Current Combinations at Point A

I_{R2}	I_{R3}	I_{R4}	Sum	I_{R2}	I_{R3}	I_{R4}	Sum
−50 mA	−45 mA	−70 mA	−165 mA	+50 mA	−45 mA	−70 mA	−65 mA
−50 mA	−45 mA	+70 mA	−25 mA	+50 mA	−45 mA	+70 mA	+75 mA
−50 mA	+45 mA	−70 mA	−75 mA	+50 mA	+45 mA	−70 mA	+25 mA
−50 mA	+45 mA	+70 mA	+65 mA	+50 mA	+45 mA	+70 mA	+165 mA

The only combination that sums to −75 mA is I_{R2} = −50 mA, I_{R3} = +45 mA, and I_{R4} = −70 mA. Since for this problem a positive value represents a current entering the junction at Point A and a negative value represents a current leaving the junction, **the directions of I_{R2} and I_{R4} are down (out of the junction) and the direction of I_{R3} is up (into the junction).**

3. Problem: What are the unspecified current values for the circuit in Figure 6-7?

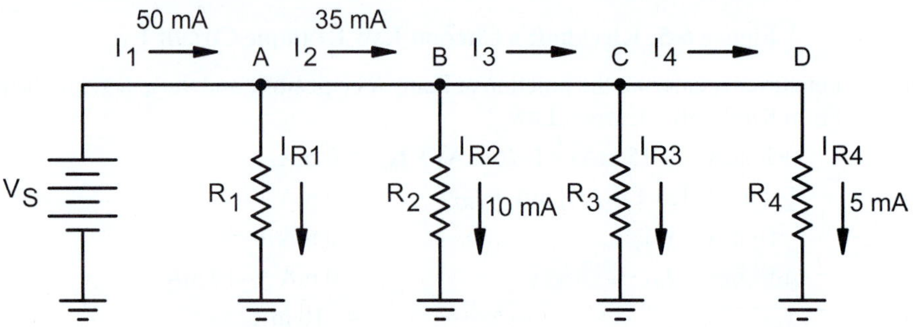

Figure 6-7: Kirchhoff's Current Law Example Circuit 3

Solution: Let currents entering a junction be positive and currents leaving a junction be negative. Apply Kirchhoff's Current Law to the junction at Point A.

$I_1 + I_{R1} + I_2 = 0$ mA

$(+50 \text{ mA}) + I_{R1} + (-35 \text{ mA}) = 0$ mA

$[(+50) + (-35)]$ mA $+ I_{R1} = 0$ mA

$(+15 \text{ mA}) + I_{R1} = 0$ mA

$(+15 \text{ mA}) + I_{R1} + (-15 \text{ mA}) = 0 \text{ mA} + (-15 \text{ mA})$

$I_{R1} = -15$ mA

I_{R1} = 15 mA. The negative sign indicates that I_{R1} is leaving the junction at Point A. Apply Kirchhoff's Current Law to the junction at Point B.

$I_2 + I_{R2} + I_3 = 0$ mA

$(+35 \text{ mA}) + (-10 \text{ mA}) + I_3 = 0$ mA

$[(+35) + (-10)]$ mA $+ I_3 = 0$ mA

$(+25 \text{ mA}) + I_3 = 0$ mA

$(+25 \text{ mA}) + I_3 + (-25 \text{ mA}) = 0 \text{ mA} + (-25 \text{ mA})$

$I_3 = -25$ mA

I_3 = 25 mA. The negative sign indicates that I_3 is leaving the junction at Point B. Apply Kirchhoff's Current Law to the junction at Point D.

$I_4 + I_{R4} = 0$ mA

$I_4 + (-5 \text{ mA}) = 0$ mA

$I_4 + (-5 \text{ mA}) + (+5 \text{ mA}) = 0 \text{ mA} + (+5 \text{ mA})$

$I_4 = +5\text{ mA}$

$\boldsymbol{I_4} = \mathbf{5\ mA}$. The positive sign indicates that I_4 is entering the junction at Point D. Apply Kirchhoff's Current Law to the junction at Point C.

$$I_3 + I_{R3} + I_4 = 0\text{ mA}$$
$$(25\text{ mA}) + I_{R3} + (-5\text{ mA}) = 0\text{ mA}$$
$$[(+25) + (-5)]\text{ mA} + I_{R3} = 0\text{ mA}$$
$$(+20\text{ mA}) + I_{R3} = 0\text{ mA}$$
$$(+20\text{ mA}) + I_{R3} + (-20\text{ mA}) = 0\text{ mA} + (-20\text{ mA})$$
$$I_{R3} = -20\text{ mA}$$

$\boldsymbol{I_{R3}}$ **is 20 mA**. The negative sign indicates that I_{R3} is leaving the junction at Point C.

6.1.4 Total Parallel Resistance

1. Problem: What is the total resistance for the parallel circuit of Figure 6-8?

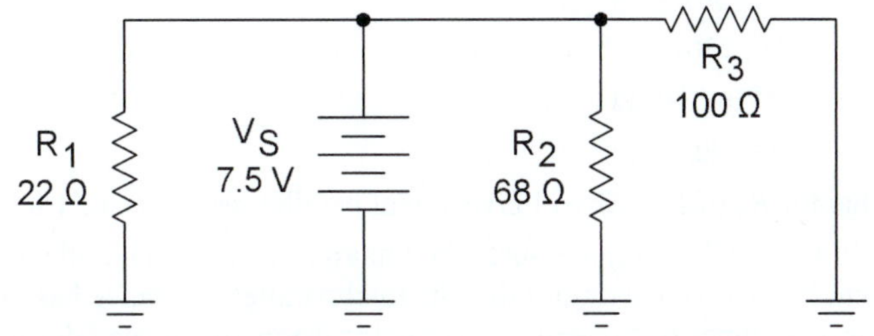

Figure 6-8: Total Parallel Resistance Example Circuit 1

Solution: From the general equation for the resistance of three parallel resistors

$$R_T = 1 / [(1 / R_1) + (1 / R_2) + (1 / R_3)]$$
$$= 1 / [(1 / 22\ \Omega) + (1 / 68\ \Omega) + (1 / 100\ \Omega)]$$
$$= 1 / \{[(1 / 22)(1 / \Omega)] + [(1 / 68)(1 / \Omega)] + [(1 / 100)(1 / \Omega)]\}$$
$$= 1 / [(45.5 \times 10^{-3}\text{ S}) + (14.7 \times 10^{-3}\text{ S}) + (10.0 \times 10^{-3}\text{ S})]$$
$$= 1 / \{[(45.5 \times 10^{-3}) + (14.7 \times 10^{-3}) + (10.0 \times 10^{-3})\text{ S}]\}$$
$$= 1 / (70.2 \times 10^{-3}\text{ S})$$
$$= (1 / 70.2 \times 10^{-3})(1 / \text{S})$$
$$= 14.3\ \Omega$$

The total parallel resistance for the circuit of Figure 6-8 is **14 Ω**.

2. Problem: What is the value of R_X in Figure 6-9 if the total parallel resistance R_T is 1.0 kΩ?

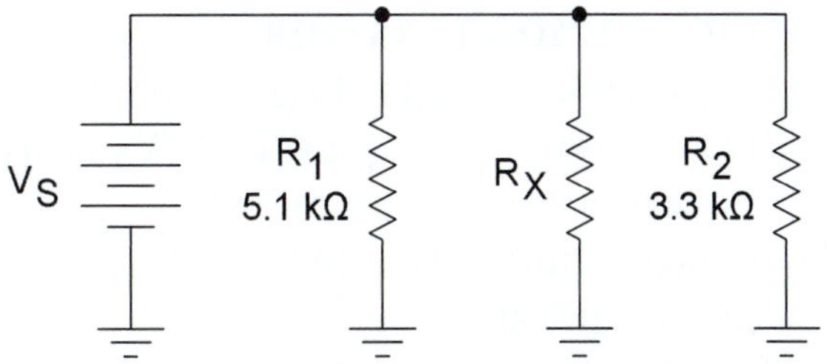

Figure 6-9: Total Parallel Resistance Example Circuit 2

Solution: From the general equation for the resistance of three parallel resistors

$$R_T = 1 / [(1 / R_1) + (1 / R_X) + (1 / R_2)]$$

Taking the reciprocal of both sides gives

$$1 / R_T = (1 / R_1) + (1 / R_X) + (1 / R_2)$$

$$(1 / \boldsymbol{R}_T) - (1/ \boldsymbol{R}_1) - (1 / \boldsymbol{R}_2) = (1 / \boldsymbol{R}_1) + (1 / \boldsymbol{R}_X) + (1 / \boldsymbol{R}_2) - (1/ \boldsymbol{R}_1) - (1 / \boldsymbol{R}_2)$$
$$= [(1 / \boldsymbol{R}_1) - (1/ \boldsymbol{R}_1)] - [(1 / \boldsymbol{R}_2) - (1 / \boldsymbol{R}_2)] + (1 / \boldsymbol{R}_X)$$
$$= 1 / \boldsymbol{R}_X$$

Taking the reciprocal of both sides again gives

$$1 / [(1 / \boldsymbol{R}_T) - (1/ \boldsymbol{R}_1) - (1 / \boldsymbol{R}_2)] = \boldsymbol{R}_X$$

so

$$\boldsymbol{R}_X = 1 / [(1 / 1.0\ \text{k}\Omega) - (1/ 5.1\ \text{k}\Omega) - (1 / 3.3\ \text{k}\Omega)]$$
$$= 1 / \{[1 / (1.0 \times 10^3\ \Omega)] - [1 / (5.1 \times 10^3\ \Omega)] - [1 / (3.3 \times 10^3\ \Omega)]\}$$
$$= 1 / \{[(1 / 1.0 \times 10^3)(1 / \Omega)] - [(1 / 5.1 \times 10^3)(1 / \Omega)] - [(1 / 3.3 \times 10^3)(1 / \Omega)]\}$$
$$= 1 / [(1.00 \times 10^{-3}\ \text{S}) - (196 \times 10^{-6}\ \text{S}) - (303 \times 10^{-6}\ \text{S})]$$
$$= 1 / [(1000 \times 10^{-6}\ \text{S}) - (196 \times 10^{-6}\ \text{S}) - (303 \times 10^{-6}\ \text{S})]$$
$$= 1 / \{[(1000 \times 10^{-6}) - (196 \times 10^{-6}) - (303 \times 10^{-6})\ \text{S}]\}$$
$$= 1 / (501 \times 10^{-6}\ \text{S})$$
$$= [1 / (501 \times 10^{-6})](1 / \text{S})$$
$$= 2.00 \times 10^3\ \Omega$$
$$= 2.00\ \text{k}\Omega$$

A value for $\boldsymbol{R}_X$ of **2.0 kΩ** will give a total parallel resistance of 1.0 kΩ.

3. Problem: ICT (In-Circuit Testing) measures the total resistance of a parallel circuit to be 2.50 kΩ. After burn-in, Final Test determines that the total parallel resistance has increased to 3.33 kΩ. Board Repair determines that one of the parallel resistors overheated during burn-in and opened. What was the original value of the failed resistor?

Solution: Let $\boldsymbol{R}_{FAIL}$ be the resistance of the resistor that failed. From the problem statement, the total parallel resistance, $\boldsymbol{R}_T$, for all the resistors is 2.50 kΩ and the total parallel resistance, $\boldsymbol{R}_{GOOD}$, for the resistors that did not fail is 3.33 kΩ. From the general equation for total parallel resistance,

$$\boldsymbol{R}_T = 1 / [(1 / \boldsymbol{R}_{GOOD}) + (1 / \boldsymbol{R}_{FAIL})]$$

Taking the reciprocal of both sides gives

$$1 / \boldsymbol{R}_T = (1 / \boldsymbol{R}_{GOOD}) + (1 / \boldsymbol{R}_{FAIL})$$
$$(1 / \boldsymbol{R}_T) - (1/ \boldsymbol{R}_{GOOD}) = (1 / \boldsymbol{R}_{GOOD}) + (1 / \boldsymbol{R}_{FAIL}) - (1 / \boldsymbol{R}_{GOOD})$$
$$= [(1 / \boldsymbol{R}_{GOOD}) - (1/ \boldsymbol{R}_{GOOD})] + (1 / \boldsymbol{R}_{FAIL})$$
$$= 1 / \boldsymbol{R}_{FAIL}$$

Taking the reciprocal of both sides again gives

$$1 / [(1 / \boldsymbol{R}_T) - (1/ \boldsymbol{R}_{GOOD})] = \boldsymbol{R}_{FAIL}$$

so

$$\boldsymbol{R}_{FAIL} = 1 / [(1 / 2.50\ \text{k}\Omega) - (1/ 3.33\ \text{k}\Omega)]$$
$$= 1 / \{[1 / (2.50 \times 10^3\ \Omega)] + [1 / (3.33 \times 10^3\ \Omega)]\}$$
$$= 1 / \{[1 / (2.50 \times 10^3)(1 / \Omega)] + [1 / (3.33 \times 10^3)(1 / \Omega)]\}$$
$$= 1 / [(400 \times 10^{-6})\ \text{S} + (300 \times 10^{-6})\ \text{S}]$$
$$= 1 / \{[(400 \times 10^{-6}) - (300 \times 10^{-6})]\ \text{S}\}$$
$$= 1 / (100 \times 10^{-6}\ \text{S})$$
$$= [1 / (100 \times 10^{-6})](1 / \text{S})$$
$$= 10.0 \times 10^3\ \Omega$$
$$= 10.0\ \text{k}\Omega$$

The value of the failed resistor is **10.0 kΩ**.

6.1.5 Application of Ohm's Law

1. Problem: What is the current through each resistor in the circuit of Figure 6-8?

Solution: The applied voltage across a parallel circuit is equal to the voltage across each resistor in the parallel circuit. From Ohm's Law for current

$$\begin{aligned} I_{R1} &= V_S / R_1 \\ &= 7.5\ \text{V} / 22\ \Omega \\ &= (7.5 / 22)\ (\text{V} / \Omega) \\ &= 341 \times 10^{-3}\ \text{A} \\ &= 341\ \text{mA} \end{aligned}$$

$$\begin{aligned} I_{R2} &= V_S / R_2 \\ &= 7.5\ \text{V} / 68\ \Omega \\ &= (7.5 / 68)\ (\text{V} / \Omega) \\ &= 110 \times 10^{-3}\ \text{A} \\ &= 110\ \text{mA} \end{aligned}$$

$$\begin{aligned} I_{R3} &= V_S / R_3 \\ &= 7.5\ \text{V} / 100\ \Omega \\ &= (7.5 / 100)\ (\text{V} / \Omega) \\ &= 75.0 \times 10^{-3}\ \text{A} \\ &= 75.0\ \text{mA} \end{aligned}$$

The currents through R_1, R_2, and R_3 are **340 mA**, **110 mA**, and **75 mA**, respectively.

2. Problem: What is the value of R_4 in the circuit of Figure 6-10 if the total current I_T is 10 mA?

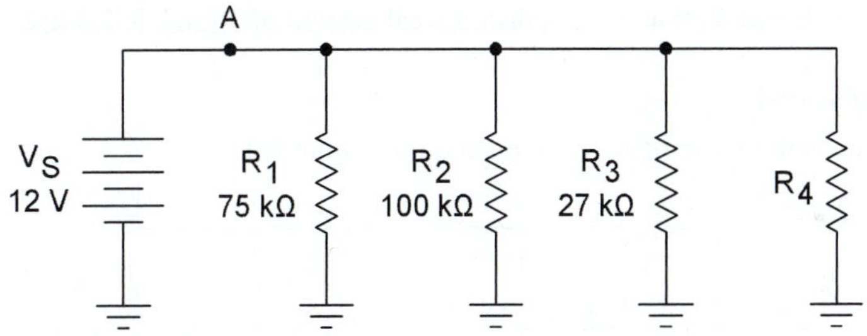

Figure 6-10: Application of Ohm's Law Example Circuit 1

Solution: From Kirchhoff's Current Law, the total current entering the junction at Point A from the voltage source must equal the total current entering the branches. For the total current I_T to equal 10 mA,

$$\begin{aligned} I_T &= I_{R1} + I_{R2} + I_{R3} + I_{R4} \\ I_T + -I_{R1} + -I_{R2} + -I_{R3} &= I_{R1} + I_{R2} + I_{R3} + I_{R4} + -I_{R1} + -I_{R2} + -I_{R3} \\ &= I_{R4} \end{aligned}$$

The voltage across each parallel resistor is equal to the source voltage across the parallel circuit, so from Ohm's Law for current,

$$\begin{aligned} I_{R1} &= V_S / R_1 \\ &= 12\ \text{V} / 75\ \text{k}\Omega \\ &= 12\ \text{V} / (75 \times 10^3\ \Omega) \\ &= [12 / (75 \times 10^3\ \Omega)]\ (\text{V} / \Omega) \\ &= 160 \times 10^{-6}\ \text{A} \end{aligned}$$

$$\begin{aligned} I_{R2} &= V_S / R_2 \\ &= 12\ \text{V} / 100\ \text{k}\Omega \end{aligned}$$

$= 12\text{ V} / (100 \times 10^3\ \Omega)$
$= [12 / (100 \times 10^3\ \Omega)]\ (\text{V} / \Omega)$
$= 120 \times 10^{-6}\text{ A}$

$\boldsymbol{I_{R3}} = \boldsymbol{V_S} / \boldsymbol{R_3}$
$= 12\text{ V} / 27\text{ k}\Omega$
$= 12\text{ V} / (27 \times 10^3\ \Omega)$
$= [12 / (27 \times 10^3\ \Omega)]\ (\text{V} / \Omega)$
$= 444 \times 10^{-6}\text{ A}$

Then,

$\boldsymbol{I_{R4}} = \boldsymbol{I_T} + -\boldsymbol{I_{R1}} + -\boldsymbol{I_{R2}} + -\boldsymbol{I_{R3}}$
$= (10\text{ mA}) + -(160 \times 10^{-6}\text{ A}) + -(120 \times 10^{-6}\text{ A}) + -(444 \times 10^{-6}\text{ A})$
$= (10 \times 10^{-3}\text{ A}) + -(160 \times 10^{-6}\text{ A}) + -(120 \times 10^{-6}\text{ A}) + -(444 \times 10^{-6}\text{ A})$
$= (10{,}000 \times 10^{-6}\text{ A}) + -(160 \times 10^{-6}\text{ A}) + -(120 \times 10^{-6}\text{ A}) + -(444 \times 10^{-6}\text{ A})$
$= (10{,}000 - 160 - 120 - 444) \times 10^{-6}\text{ A}$
$= 9280 \times 10^{-6}\text{ A}$

From Ohm's Law for resistance,

$\boldsymbol{R_4} = \boldsymbol{V_S} / \boldsymbol{I_{R4}}$
$= 12\text{ V} / (9280 \times 10^{-6}\text{ A})$
$= [12 / (9280 \times 10^{-6})]\ (\text{V} / \text{A})$
$= 1.29 \times 10^3\ \Omega$
$= 1.29\text{ k}\Omega$

The value of $\boldsymbol{R_4}$ that will produce a total current of 10 mA is **1.3 kΩ**.

6.1.6 Current Sources in Parallel

1. Problem: What is the value of $\boldsymbol{I_{S3}}$ for the circuit in Figure 6-11?

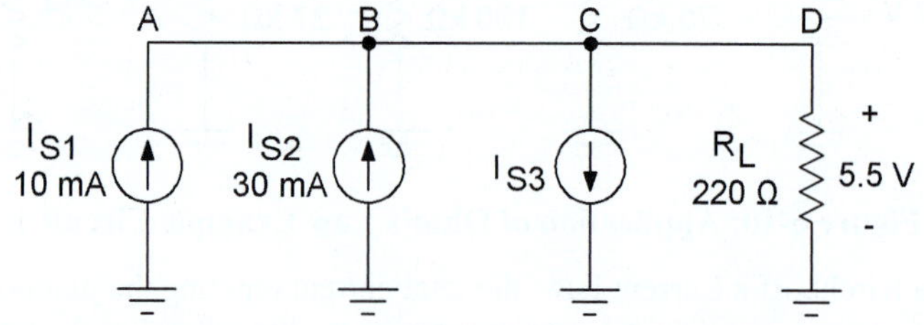

Figure 6-11: Current Sources in Parallel Example Circuit 1

Solution: Refer to Figure 6-12, which shows the currents in the circuit of Figure 6-11. For this problem, let the sign of currents entering a node be positive and the sign of currents leaving a node be negative.

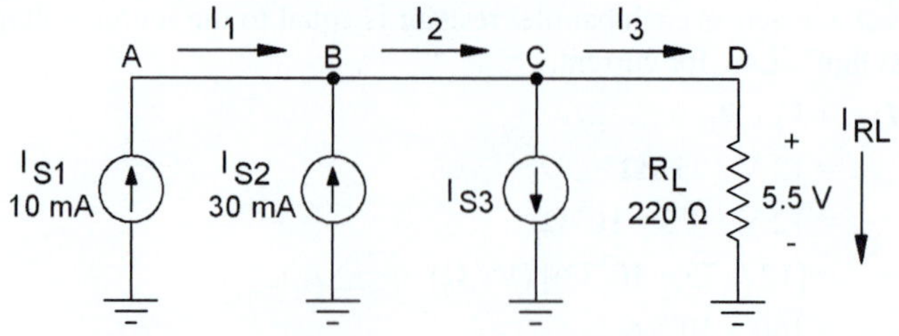

Figure 6-12: Currents for Circuit of Figure 6-11

From Ohm's Law for current,

I_{RL} = 5.5 V / R_L

= 5.5 V / 220 Ω

= (5.5 / 220) (V / Ω)

= 25×10^{-3} A

For the currents at Node D,

$(+I_3) + (-I_{RL}) = 0$ mA

$(+I_3) + (-25\text{ mA}) + (+25\text{ mA}) = 0\text{ mA} + (+25\text{ mA})$

$I_3 = 25$ mA

For the currents at Node A,

$(+I_{S1}) + (-I_1) = 0$ mA

$+10\text{ mA} + (-I_1) + (+I_1) = 0\text{ mA} + (+I_1)$

$10\text{ mA} = I_1$

For the currents at Node B,

$(+I_1) + (+I_{S2}) + (-I_2) = 0$ mA

$(+10\text{ mA}) + (+30\text{ mA}) + (-I_2) = 0$ mA

$[(+10) + (+30)\text{ mA}] + (-I_2) + (+I_2) = 0\text{ mA} + (+I_2)$

$40\text{ mA} = +I_2$

For the currents at Node C,

$(+I_2) + (-I_{S3}) + (-I_3) = 0$ mA

$(+40\text{ mA}) + (-I_{S3}) + (-25\text{ mA}) = 0$ mA

$[(+40) + (-25)\text{ mA}] + (-I_{S3}) + (+I_{S3}) = 0\text{ mA} + (+I_{S3})$

$15\text{ mA} = I_{S3}$

The value of I_{S3} for the circuit in Figure 6-11 is **15 mA**.

2. Problem: What is the value of I_{S3} for the circuit in Figure 6-11 if the polarity of I_{S1} is reversed and the value and polarity of the voltage across R_L is unchanged?

Solution: Refer to Figure 6-13, which shows the circuit currents with I_{S1} reversed. Note that for this problem the figure still shows the direction of I_1 from A to B as in Figure 6-12. Let the sign of currents entering a node be positive and the sign of currents leaving a node be negative.

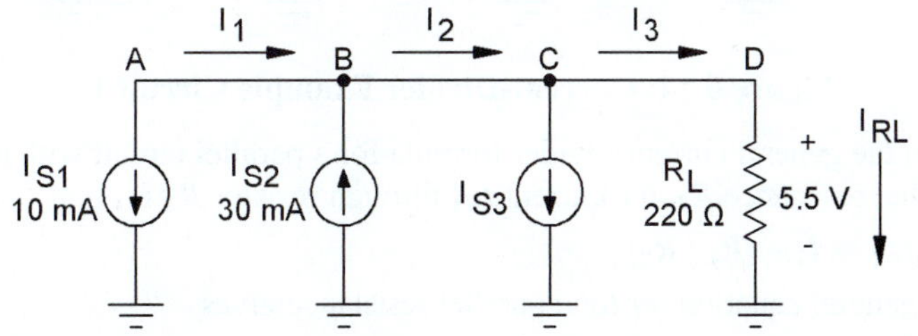

Figure 6-13: Currents for Figure 6-11 with I_{S1} Reversed

From Ohm's Law for current,

I_{RL} = 5.5 V / R_L

= 5.5 V / 220 Ω

= (5.5 / 220) (V / Ω)

= 25×10^{-3} A

For the currents at Node D,

$(+I_3) + (-I_{RL}) = 0$ mA

$$(+I_3) + (-25\text{ mA}) + (+25\text{ mA}) = 0\text{ mA} + (+25\text{ mA})$$
$$I_3 = 25\text{ mA}$$

For the currents at Node A,

$$(-I_{S1}) + (-I_1) = 0\text{ mA}$$
$$-10\text{ mA} + (-I_1) + (+I_1) = 0\text{ mA} + (+I_1)$$
$$-10\text{ mA} = I_1$$

The sign of the current is negative, indicating that the actual direction for I_1 is opposite to that shown in Figure 6-13, so that I_1 is leaving Node B rather than entering it. The currents at Node B will then be

$$(-I_1) + (+I_{S2}) + (-I_2) = 0\text{ mA}$$
$$(-10\text{ mA}) + (+30\text{ mA}) + (-I_2) = 0\text{ mA}$$
$$[-10) + (+30)\text{ mA}] + (-I_2) + (+I_2) = 0\text{ mA} + (+I_2)$$
$$20\text{ mA} = I_2$$

For the currents at Node C

$$(+I_2) + (-I_{S3}) + (-I_3) = 0\text{ mA}$$
$$(+20\text{ mA}) + (-I_{S3}) + (-25\text{ mA}) = 0\text{ mA}$$
$$[(+20) + (-25)\text{ mA}] + (-I_{S3}) + (+I_{S3}) = 0\text{ mA} + (+I_{S3})$$
$$-5\text{ mA} = I_{S3}$$

The sign of the current is negative, indicating that the actual direction for I_{S3} is opposite to that shown by the current source symbol in Figure 6-13. The value of I_{S3} for the circuit in Figure 6-11 is **5 mA flowing from bottom to top**.

6.1.7 Current Dividers

1. Problem: Determine the current through each resistor for the circuit in Figure 6-14.

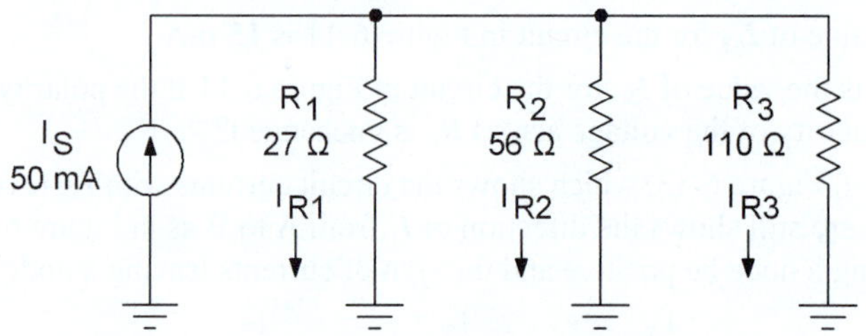

Figure 6-14: Current-Divider Example Circuit 1

Solution: From the general current-divider formula for a parallel circuit with total current I_T and total parallel resistance R_T, the current I_{RX} through resistor R_X is

$$I_{RX} = I_T \times (R_T / R_X)$$

The general equation for total parallel resistance gives

$$\begin{aligned}
R_T &= 1 / [(1 / R_1) + (1 / R_2) + (1 / R_3)] \\
&= 1 / [(1 / 27\ \Omega) + (1 / 56\ \Omega) + (1 / 110\ \Omega)] \\
&= 1 / \{[(1 / 27)(1 / \Omega)] + [(1 / 56)(1 / \Omega)] + [(1 / 110)(1 / \Omega)] \\
&= 1 / [(37.0 \times 10^{-3}\text{ S}) + (17.9 \times 10^{-3}\text{ S}) + (9.09 \times 10^{-3}\text{ S})] \\
&= 1 / \{[(37.0 \times 10^{-3}) + (17.9 \times 10^{-3}) + (9.09 \times 10^{-3})]\text{ S}\} \\
&= 1 / (64.0 \times 10^{-3}\text{ S}) \\
&= [1 / (64.0 \times 10^{-3})]\ (1 / \text{S}) \\
&= 15.6\ \Omega
\end{aligned}$$

For the circuit of Figure 6-14 $I_T = I_S = 50$ mA.

$I_{R1} = I_T \times (R_T / R_1)$
$= 50 \text{ mA} \times (15.6\ \Omega / 27\ \Omega)$
$= 50 \text{ mA} \times [(15.6 / 27)\ (\Omega / \Omega)]$
$= 50 \text{ mA} \times 0.579$
$= 28.9 \text{ mA}$

$I_{R2} = I_T \times (R_T / R_2)$
$= 50 \text{ mA} \times (15.6\ \Omega / 56\ \Omega)$
$= 50 \text{ mA} \times [(15.6 / 56)\ (\Omega / \Omega)]$
$= 50 \text{ mA} \times 0.279$
$= 14.0 \text{ mA}$

$I_{R3} = I_T \times (R_T / R_3)$
$= 50 \text{ mA} \times (15.6\ \Omega / 110\ \Omega)$
$= 50 \text{ mA} \times [(15.6 / 110)\ (\Omega / \Omega)]$
$= 50 \text{ mA} \times 0.142$
$= 7.10 \text{ mA}$

The currents I_{R1}, I_{R2}, and I_{R3} through resistors R_1, R_2, and R_3 are **29 mA, 14 mA**, and **7.1 mA**, respectively.

2. Problem: What value of R_2 for the circuit in Figure 6-15 will produce a current I_{R2} of 15 mA?

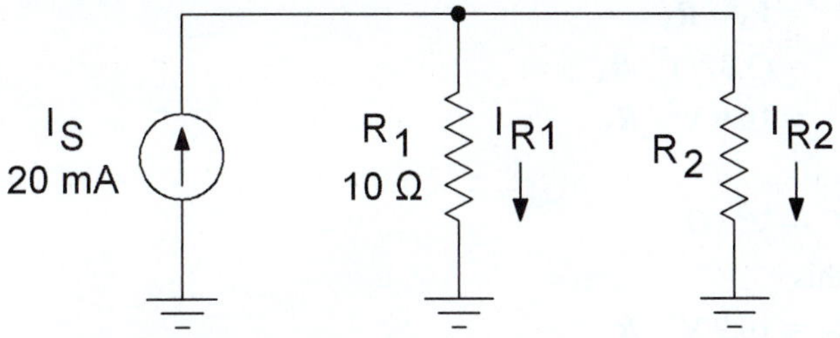

Figure 6-15: Current-Divider Example Circuit 2

Solution: From Kirchhoff's Current Law,

$+I_S - I_{R1} - I_{R2} = 0 \text{ mA}$
$+I_S + -I_{R1} + -I_{R2} + I_{R1} = 0 \text{ mA} + I_{R1}$
$+I_S + -I_{R2} = +I_{R1}$

If $I_{R2} = 15$ mA,

$I_{R1} = I_S - I_{R2}$
$= 20 \text{ mA} + -15 \text{ mA}$
$= (20 + -15) \text{ mA}$
$= 5 \text{ mA}$

From Ohm's Law for voltage, the voltage V_{R1} across resistor R_1 is

$V_{R1} = I_{R1} \times R_1$
$= 5 \text{ mA} \times 10\ \Omega$
$= (5 \times 10^{-3} \text{ A}) \times 10\ \Omega$
$= [(5 \times 10^{-3}) \times 10]\ (\text{A} \times \Omega)$
$= 50 \times 10^{-3} \text{ V}$

Because R_1 and R_2 are in parallel, $V_{R1} = V_{R2}$. From Ohm's Law for resistance,

$R_2 = V_{R2} / I_{R2}$

$= (50 \times 10^{-3}\ \text{V}) / (15\ \text{mA})$

$= (50 \times 10^{-3}\ \text{V}) / (15 \times 10^{-3}\ \text{A})$

$= [(50 \times 10^{-3}) / (15 \times 10^{-3})]\ (\text{V} / \text{A})$

$= 3.33\ \Omega$

The value of $\boldsymbol{R_2}$ necessary to produce a current $\boldsymbol{I_{R2}}$ of 15 mA is **3.3 Ω.**

6.1.8 Power in Parallel Circuits

1. Problem: What is the power dissipation $\boldsymbol{P_{RX}}$ for each resistor $\boldsymbol{R_X}$ in Figure 6-16?

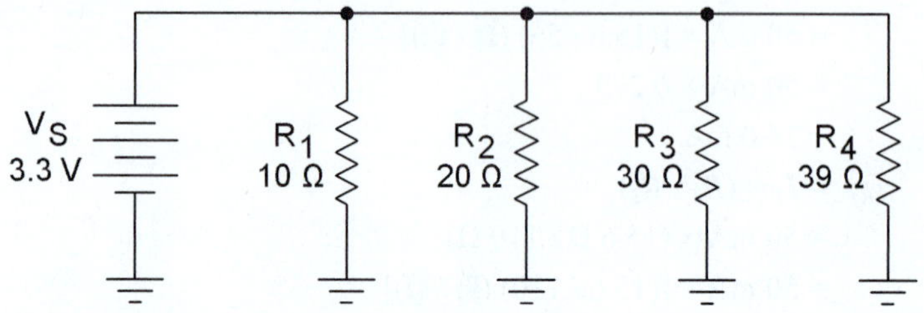

Figure 6-16: Power in Parallel Circuits Example Circuit 1

Solution: Because the resistors are in parallel, the voltage across each resistor $\boldsymbol{V_{RX}}$ is equal to the source voltage $\boldsymbol{V_S}$. From Watt's Law for voltage and resistance

$\boldsymbol{P_{RX}} = \boldsymbol{V_{RX}}^2 / \boldsymbol{R_X}$

$= \boldsymbol{V_S}^2 / \boldsymbol{R_X}$

$= (3.3\ \text{V})^2 / \boldsymbol{R_X}$

$= 10.9\ \text{V}^2 / \boldsymbol{R_X}$

and

$\text{W} = \text{V}^2 / \Omega$

From this,

$\boldsymbol{P_{R1}} = 10.9\ \text{V}^2 / \boldsymbol{R_1}$

$= 10.9\ \text{V}^2 / 10\ \Omega$

$= (10.9 / 10)\ (\text{V}^2 / \Omega)$

$= 1.09\ \text{W}$

$\boldsymbol{P_{R2}} = 10.9\ \text{V}^2 / \boldsymbol{R_2}$

$= 10.9\ \text{V}^2 / 20\ \Omega$

$= (10.9 / 20)\ (\text{V}^2 / \Omega)$

$= 545 \times 10^{-3}\ \text{W}$

$= 545\ \text{mW}$

$\boldsymbol{P_{R3}} = 10.9\ \text{V}^2 / \boldsymbol{R_3}$

$= 10.9\ \text{V}^2 / 30\ \Omega$

$= (10.9 / 30)\ (\text{V}^2 / \Omega)$

$= 363 \times 10^{-3}\ \text{W}$

$= 363\ \text{mW}$

$\boldsymbol{P_{R4}} = 10.9\ \text{V}^2 / \boldsymbol{R_4}$

$= 10.9\ \text{V}^2 / 39\ \Omega$

$= (10.9 / 39)\ (\text{V}^2 / \Omega)$

$= 279 \times 10^{-3}\ \text{W}$

$= 279\ \text{mW}$

The power dissipation $\boldsymbol{P_{R1}}$, $\boldsymbol{P_{R2}}$, $\boldsymbol{P_{R3}}$, and $\boldsymbol{P_{R4}}$ for resistors $\boldsymbol{R_1}$, $\boldsymbol{R_2}$, $\boldsymbol{R_3}$, and $\boldsymbol{R_4}$ are **1.1 W**, **550 mW**, **360 mW**, and **280 mW**, respectively.

2. Problem: What is the power dissipation $\boldsymbol{P_{RX}}$ for each resistor $\boldsymbol{R_X}$ in the circuit of Figure 6-17?

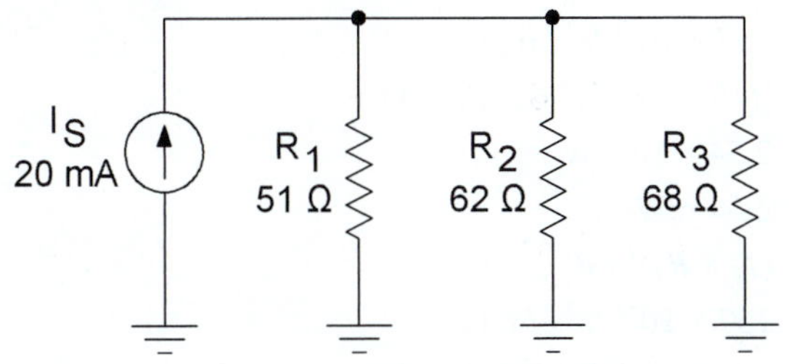

Figure 6-17: Power in Parallel Circuits Example Circuit 2

Solution: From Watt's Law for current and resistance, the power $\boldsymbol{P_{RX}}$ for the current $\boldsymbol{I_{RX}}$ flowing through resistor $\boldsymbol{R_X}$ is

$$\boldsymbol{P_{RX}} = \boldsymbol{I_{RX}}^2 \times \boldsymbol{R_X}$$

From the general current divider formula for a parallel circuit with total current $\boldsymbol{I_T}$ and total parallel resistance $\boldsymbol{R_T}$, the current $\boldsymbol{I_{RX}}$ through resistor $\boldsymbol{R_X}$ is

$$\boldsymbol{I_{RX}} = \boldsymbol{I_T} \times (\boldsymbol{R_T} / \boldsymbol{R_X})$$

so

$$\begin{aligned}
\boldsymbol{P_{RX}} &= \boldsymbol{I_{RX}}^2 \times \boldsymbol{R_X} \\
&= [\boldsymbol{I_T} \times (\boldsymbol{R_T} / \boldsymbol{R_X})]^2 \times \boldsymbol{R_X} \\
&= \boldsymbol{I_T}^2 \times (\boldsymbol{R_T} / \boldsymbol{R_X})^2 \times \boldsymbol{R_X} \\
&= \boldsymbol{I_T}^2 \times [(\boldsymbol{R_T}^2 / (\boldsymbol{R_X} \times \boldsymbol{R_X})] / \boldsymbol{R_X} \\
&= (\boldsymbol{I_T}^2 \times \boldsymbol{R_T}^2) / \boldsymbol{R_X} \\
&= (\boldsymbol{I_T} \times \boldsymbol{R_T})^2 / \boldsymbol{R_X}
\end{aligned}$$

Note that this equation is equivalent to $\boldsymbol{V_T}^2 / \boldsymbol{R_X}$. which is Watt's Law for voltage and resistance. The general equation for total parallel resistance gives

$$\begin{aligned}
\boldsymbol{R_T} &= 1 / [(1 / \boldsymbol{R_1}) + (1 / \boldsymbol{R_2}) + (1 / \boldsymbol{R_3})] \\
&= 1 / [(1 / 51\ \Omega) + (1 / 62\ \Omega) + (1 / 68\ \Omega)] \\
&= 1 / \{[(1 / 51)(1 / \Omega)] + [(1 / 62)(1 / \Omega)] + [(1 / 68)(1 / \Omega)] \\
&= 1 / [(19.6 \times 10^{-3}\ \text{S}) + (16.1 \times 10^{-3}\ \text{S}) + (14.7 \times 10^{-3}\ \text{S})] \\
&= 1 / \{[(19.6 \times 10^{-3}) + (16.1 \times 10^{-3}) + (14.7 \times 10^{-3})]\ \text{S}\} \\
&= 1 / (50.4 \times 10^{-3}\ \text{S}) \\
&= [1 / (50.4 \times 10^{-3})]\ (1 / \text{S}) \\
&= 19.8\ \Omega
\end{aligned}$$

so

$$\begin{aligned}
(\boldsymbol{I_T} \times \boldsymbol{R_T})^2 &= (20\ \text{mA} \times 19.8\ \Omega)^2 \\
&= [(20 \times 10^{-3}\ \text{A}) \times (19.8\ \Omega)]^2 \\
&= [(20 \times 10^{-3}) \times 19.8]\ (\text{A} \times \Omega)]^2 \\
&= (396 \times 10^{-3}\ \text{V})^2 \\
&= 157 \times 10^{-3}\ \text{V}^2
\end{aligned}$$

For the circuit in Figure 6-17, the total current $\boldsymbol{I_T}$ is equal to $\boldsymbol{I_S}$, so

$$\begin{aligned}
\boldsymbol{P_{R1}} &= (\boldsymbol{I_T} \times \boldsymbol{R_T})^2 / \boldsymbol{R_1} \\
&= 157 \times 10^{-3}\ \text{V}^2 / 51\ \Omega
\end{aligned}$$

$= [(157 \times 10^{-3}) / 51] (V^2 / \Omega)$
$= 3.08 \times 10^{-3}$ W
$= 3.08$ mW

$P_{R2} = (I_T \times R_T)^2 / R_2$
$= 157 \times 10^{-3}$ $V^2 / 62\ \Omega$
$= [(157 \times 10^{-3}) / 62] (V^2 / \Omega)$
$= 2.54 \times 10^{-3}$ W
$= 2.54$ mW

$P_{R3} = (I_T \times R_T)^2 / R_3$
$= 157 \times 10^{-3}$ $V^2 / 68\ \Omega$
$= [(157 \times 10^{-3}) / 68] (V^2 / \Omega)$
$= 2.31 \times 10^{-3}$ W
$= 2.31$ mW

The power dissipation P_{R1}, P_{R2}, and P_{R3} for resistors R_1, R_2, and R_3 are **3.1 mW**, **2.5 mW**, and **2.3 mW**, respectively.

6.2 Advanced

6.2.1 Resistors in Parallel

1. Problem: Determine which resistors in Figure 6-18 are in parallel between Points A and B for the indicated settings of SW_1 and SW_2.

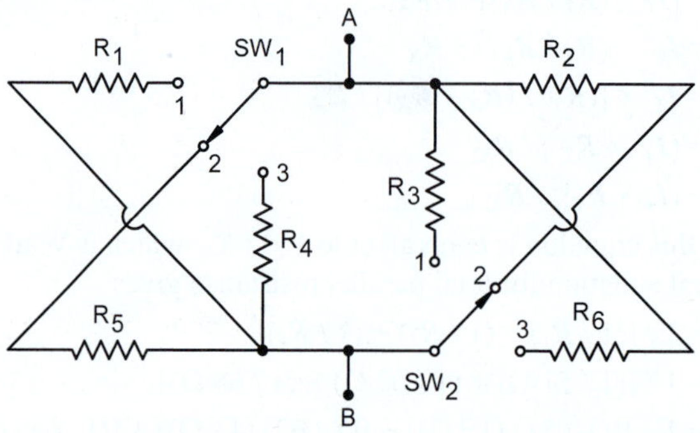

Figure 6-18: Parallel Circuit Identification Example 3

Solution: Resistors are in parallel between Points A and B if each resistor provides a direct current path between Points A and B. Refer to Figure 6-19.

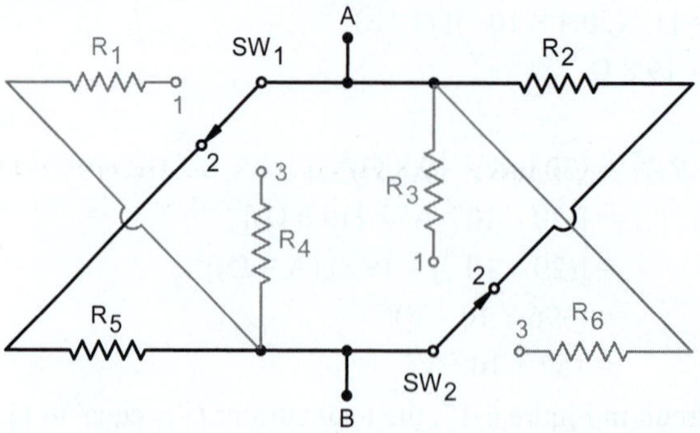

Figure 6-19: Current Paths Between Points A and B for Figure 6-18

As the circuit in Figure 6-19 shows, SW_1 and SW_2 provide current paths between Points A and B through resistors R_5 and R_2, respectively. Therefore, **R_5 and R_2 are in parallel.**

2. Problem: Which switch positions for the circuit in Figure 6-20 will connect resistors of equal value in parallel?

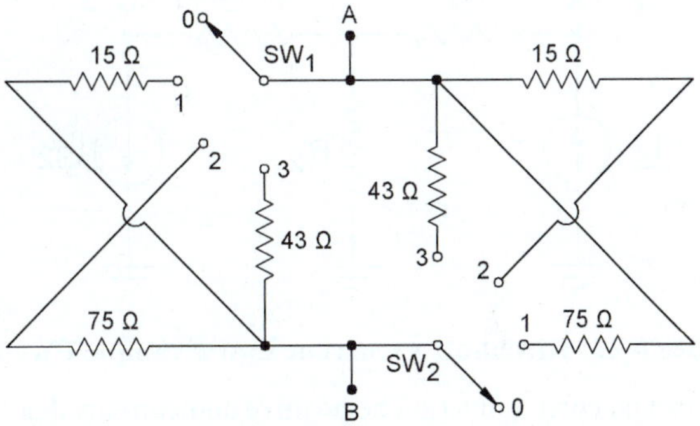

Figure 6-20: Parallel Circuit Identification Example 4

Solution: Refer to the circuits in Figure 6-21.

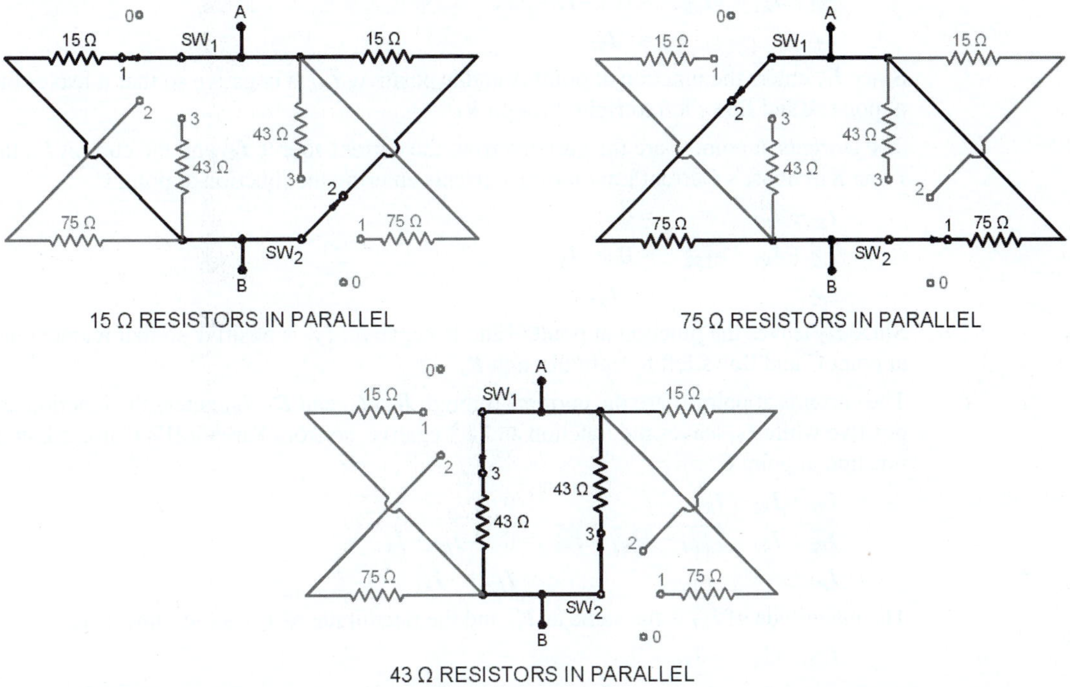

Figure 6-21: Switch Positions and Current Paths for Equal Value Resistors in Parallel

When SW_1 is in position 1, the 15 Ω resistor at the upper left provides a current path between Points A and B. To connect the 15 Ω resistor at the upper right so that it also provides a current path between Points A and B, SW_2 must be in position 2.

When SW_1 is in position 2, the 75 Ω resistor at the lower left provides a current path between Points A and B. To connect the 75 Ω resistor at the lower right so that it also provides a current path between Points A and B, SW_2 must be in position 1.

When SW_1 is in position 3, the 43 Ω resistor at the middle left provides a current path between Points A and B. To connect the 43 Ω resistor at the middle right so that it also provides a current path between Points A and B, SW_2 must be in position 3.

6.2.2 Kirchhoff's Current Law

1. Problem: If I_{S2} decreases for the circuit in Figure 6-22, will the current I_{R2} through R_2 increase, decrease, or stay the same? Assume that $I_{S1} \neq I_{S2}$. **Note: There is more than one answer.**

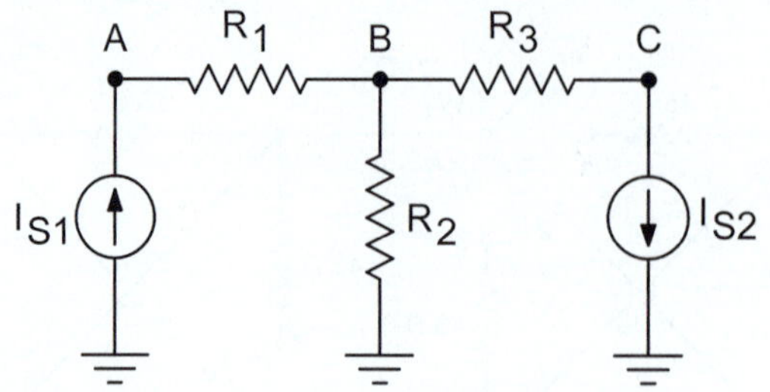

Figure 6-22: Kirchhoff's Current Law Example Circuit 4

Solution: Let currents that enter a junction be positive and currents that leave a junction be negative. The currents at point A are the currents from the current source I_{S1} and the current I_{R1} through R_1. From Kirchhoff's Current Law for the currents entering the junction at point A,

$$I_{S1} + I_{R1} = 0$$
$$I_{S1} + I_{R1} + -I_{S1} = 0 + -I_{S1}$$
$$I_{R1} = -I_{S1}$$

Since I_{S1} enters the junction at point A and is positive, I_{R1} is negative so that it leaves the junction at point A and flows left to right through R_1.

The currents at point C are the currents from the current source I_{S2} and the current I_{R3} through R_3. From Kirchhoff's Current Law for the currents entering the junction at point C

$$I_{S2} + I_{R3} = 0$$
$$I_{S2} + I_{R3} + -I_{S2} = 0 + -I_{S2}$$
$$I_{R3} = -I_{S2}$$

Since I_{S2} leaves the junction at point C and is negative, I_{R3} is positive so that it enters the junction at point C and flows left to right through R_3.

The currents at point B are the currents through R_1, R_2, and R_3. I_{R1} enters the junction and is positive while I_{R3} leaves the junction and is negative, so from Kirchhoff's Current Law for the junction at point C

$$I_{R1} + I_{R2} - I_{R3} = 0$$
$$I_{R1} + I_{R2} + -I_{R3} + -I_{R1} + I_{R3} = 0 + -I_{R1} + I_{R3}$$
$$I_{R2} = I_{R3} + -I_{R1}$$

The magnitude of I_{R1} is the same as I_{S1} and the magnitude of I_{R3} is the same as I_{S2} so

$$I_{R2} = I_{S2} + -I_{S1}$$
$$= I_{S2} - I_{S1}$$

If current source I_{S1} sources more current than current source I_{S2} sinks ($I_{S1} > I_{S2}$), then I_{R2} is negative (leaves the junction at point B) and consists of the "excess" current from current source I_{S1} that I_{S2} cannot sink. If the magnitude of I_{S2} decreases, then the magnitude of I_{R2} must increase to compensate.

If current source I_{S1} sources less current than current I_{S2} sinks ($I_{S1} < I_{S2}$), then I_{R2} is positive (enters the junction at point B) and consists of the "excess" current that I_{S2} must sink. If the magnitude of I_{S2} decreases, then the magnitude of I_{R2} must decrease to compensate.

2. Problem: What are the values and directions for the unknown branch currents X_1 through X_6 in the circuit of Figure 6-23?

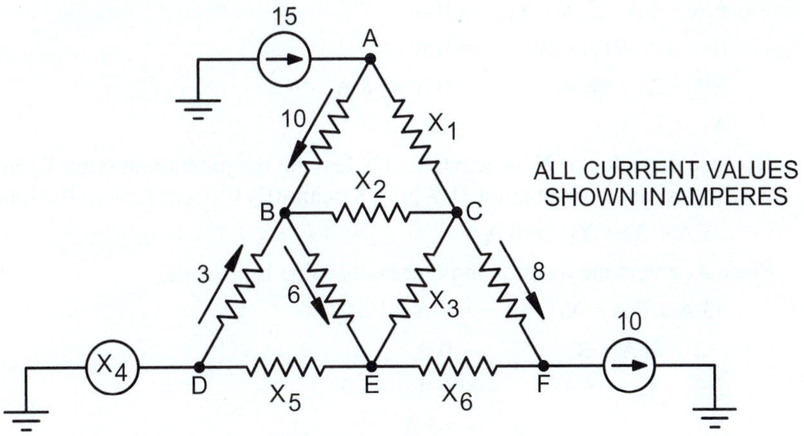

Figure 6-23: Kirchhoff's Current Law Example Circuit 5

Solution: Let currents that enter a junction be positive and currents that leave a current be negative. Apply Kirchhoff's Current Law to the junction at point A.

$$\begin{aligned} 15\text{ A} - 10\text{ A} + \boldsymbol{X}_1 &= 0\text{ A} \\ (15 + -10)\text{ A} + \boldsymbol{X}_1 &= 0\text{ A} \\ 5\text{ A} + \boldsymbol{X}_1 + -5\text{ A} &= 0\text{ A} + -5\text{ A} \\ \boldsymbol{X}_1 &= -5\text{ A} \end{aligned}$$

Because the value of $\boldsymbol{X}_1$ is negative, it is leaving the junction at point A, and consequently entering the junction at point B. Apply Kirchhoff's Current Law to the junction at point B.

$$\begin{aligned} 10\text{ A} + 3\text{ A} - 6\text{ A} + \boldsymbol{X}_2 &= 0\text{ A} \\ (10 + 3 + -6)\text{ A} + \boldsymbol{X}_2 &= 0\text{ A} \\ 7\text{ A} + \boldsymbol{X}_2 + -7\text{ A} &= 0\text{ A} + -7\text{ A} \\ \boldsymbol{X}_2 &= -7\text{ A} \end{aligned}$$

Because the value of $\boldsymbol{X}_2$ is negative, it is leaving the junction at point B, and consequently entering the junction at point C. Apply Kirchhoff's Current Law to the junction at point C.

$$\boldsymbol{X}_1 + \boldsymbol{X}_2 - 8\text{ A} + \boldsymbol{X}_3 = 0\text{ A}$$

Since both $\boldsymbol{X}_1$ and $\boldsymbol{X}_2$ enter the junction at point C, their values are positive.

$$\begin{aligned} 5\text{ A} + 7\text{ A} -8\text{ A} + \boldsymbol{X}_3 &= 0\text{ A} \\ (5 + 7 + -8)\text{ A} + \boldsymbol{X}_3 &= 0\text{ A} \\ 4\text{ A} + \boldsymbol{X}_3 + -4\text{ A} &= 0\text{ A} + -4\text{ A} \\ \boldsymbol{X}_3 &= -4\text{ A} \end{aligned}$$

Because the value of $\boldsymbol{X}_3$ is negative, it is leaving the junction at point C, and consequently entering the junction at point E. Apply Kirchhoff's Current Law to the junction at point F.

$$\begin{aligned} 8\text{ A} - 10\text{ A} + \boldsymbol{X}_6 &= 0\text{ A} \\ (8 + -10)\text{ A} + \boldsymbol{X}_6 &= 0\text{ A} \\ -2\text{ A} + \boldsymbol{X}_6 + 2\text{ A} &= 0\text{ A} + 2\text{ A} \\ \boldsymbol{X}_6 &= +2\text{ A} \end{aligned}$$

Because the value of $\boldsymbol{X}_6$ is positive, it is entering the junction at point F, and consequently leaving the junction at point E. Apply Kirchhoff's Current Law to the junction at point E.

$$6\text{ A} + \boldsymbol{X}_3 + \boldsymbol{X}_6 + \boldsymbol{X}_5 = 0\text{ A}$$

Since $\boldsymbol{X}_3$ enters the junction at point E, its value is positive, and since $\boldsymbol{X}_6$ leaves the junction at point E, its value is negative.

$$6\text{ A} + 4\text{ A} - 2\text{ A} + X_5 = 0\text{ A}$$
$$(6 + 4 + -2)\text{ A} + X_5 = 0\text{ A}$$
$$8\text{ A} + X_5 + -8\text{ A} = 0\text{ A} + -8\text{ A}$$
$$X_5 = -8\text{ A}$$

Because the value of X_5 is negative, it is leaving the junction at point E, and, consequently, entering the junction at point D. Apply Kirchhoff's Current Law to the junction at point D.

$$-3\text{ A} + X_5 + X_4 = 0\text{ A}$$

Since X_5 enters the junction at point D, its value is positive.

$$-3\text{ A} + 8\text{ A} + X_4 = 0\text{ A}$$
$$(-3 + 8)\text{ A} + X_4 = 0\text{ A}$$
$$5\text{ A} + X_4 + -5\text{ A} = 0\text{ A} + -5\text{ A}$$
$$X_4 = -5\text{ A}$$

Since the value of X_4 is negative, it is leaving the junction at point D. Therefore, the direction of the current source is right to left. The circuit in Figure 6-24 summarizes the current values and directions.

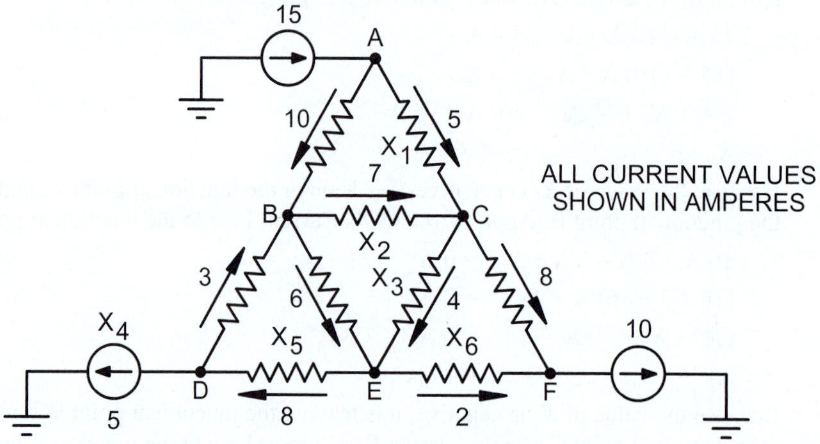

Figure 6-24: Summary of Currents for Figure 6-23

6.2.3 Total Parallel Resistance

1. Problem: A circuit consists of six equal-value resistors in parallel. When you connect two more resistors of the same value in parallel, the total resistance decreases by 100 Ω. What is the value of each resistor?

Solution: The total resistance R_T of n equal-value resistors R is

$$R_T = R / n$$

For the original circuit, $n = 6$ and the original total resistance $R_{T\,(\text{org})}$ is

$$R_{T\,(\text{org})} = R / 6$$

For the modified circuit, $n = 6 + 2 = 8$ and the modified total resistance $R_{T\,(\text{mod})}$ is

$$R_{T\,(\text{mod})} = R / 8$$

Since $R_{T\,(\text{mod})}$ is 100 Ω less than $R_{T\,(\text{org})}$,

$$R_{T\,(\text{mod})} = R / 8$$
$$R_{T\,(\text{org})} - 100\ \Omega = R / 8$$

so

$(\boldsymbol{R} / 6) - 100\ \Omega = \boldsymbol{R} / 8$

$(\boldsymbol{R} / 6) + -100\ \Omega + 100\ \Omega = (\boldsymbol{R} / 8) + 100\ \Omega$

$\boldsymbol{R} / 6 = (\boldsymbol{R} / 8) + 100\ \Omega$

$(\boldsymbol{R} / 6) + -(\boldsymbol{R} / 8) = (\boldsymbol{R} / 8) + 100\ \Omega + -(\boldsymbol{R} / 8)$

$\boldsymbol{R} \times \{[(1 / 6) \times (4 / 4)] + -[(1 / 8) \times (3 / 3)]\} = 100\ \Omega$

$\boldsymbol{R} \times [(4 / 24) + -(3 / 24)] = 100\ \Omega$

$\boldsymbol{R} \times [(4 + -3) / 24] = 100\ \Omega$

$\boldsymbol{R} \times (1 / 24) = 100\ \Omega$

$\boldsymbol{R} \times (1 / 24) \times 24 = 100\ \Omega \times 24$

$\boldsymbol{R} = 2400\ \Omega$

The value of each resistor is **2400 Ω = 2.4 kΩ.**

2. Problem: A circuit consists of seven equal-value resistors in parallel. When you remove three resistors, the total resistance increases by 470 Ω. What is the value of each resistor?

Solution: The total resistance $\boldsymbol{R_T}$ of $\boldsymbol{n}$ equal-value resistors $\boldsymbol{R}$ is

$\boldsymbol{R_T} = \boldsymbol{R} / \boldsymbol{n}$

For the original circuit $\boldsymbol{n} = 7$ and the original total resistance $\boldsymbol{R_{T\,(org)}}$ is

$\boldsymbol{R_{T\,(org)}} = \boldsymbol{R} / 7$

For the modified circuit $\boldsymbol{n} = 7 - 3 = 4$ and the modified total resistance $\boldsymbol{R_{T\,(mod)}}$ is

$\boldsymbol{R_{T\,(mod)}} = \boldsymbol{R} / 4$

Since $\boldsymbol{R_T}$ (mod) is 470 Ω more than $\boldsymbol{R_T}$ (org),

$\boldsymbol{R_{T\,(mod)}} = \boldsymbol{R} / 4$

$\boldsymbol{R_{T\,(org)}} + 470\ \Omega = \boldsymbol{R} / 4$

so

$(\boldsymbol{R} / 7) + 470\ \Omega = \boldsymbol{R} / 4$

$(\boldsymbol{R} / 7) + 470\ \Omega + -(\boldsymbol{R} / 7) = (\boldsymbol{R} / 4) + -(\boldsymbol{R} / 7)$

$470\ \Omega = \boldsymbol{R} \times [(1 / 4) + -(1 / 7)]$

$470\ \Omega = \boldsymbol{R} \times \{[(1 / 4) \times (7 / 7)] + -[(1 / 7) \times (4 / 4)]\}$

$470\ \Omega = \boldsymbol{R} \times [(7 / 28) + -(4 / 28)]$

$470\ \Omega = \boldsymbol{R} \times [(7 + -4) / 28]$

$470\ \Omega = \boldsymbol{R} \times (3 / 28)$

$470\ \Omega \times (28 / 3) = \boldsymbol{R} \times (3 / 28) \times (28 / 3)$

$[(470 \times 28) / 3]\ \Omega = \boldsymbol{R}$

$4387\ \Omega = \boldsymbol{R}$

The value of each resistor is **4400 Ω = 4.4 kΩ.**

6.2.4 Application of Ohm's Law

1. Problem: What values of $\boldsymbol{R_1}$, $\boldsymbol{R_2}$, and $\boldsymbol{R_3}$ will give the branch currents of 10 mA, 20 mA, and 30 mA for the parallel circuit in Figure 6-25? Assume that values are exact.

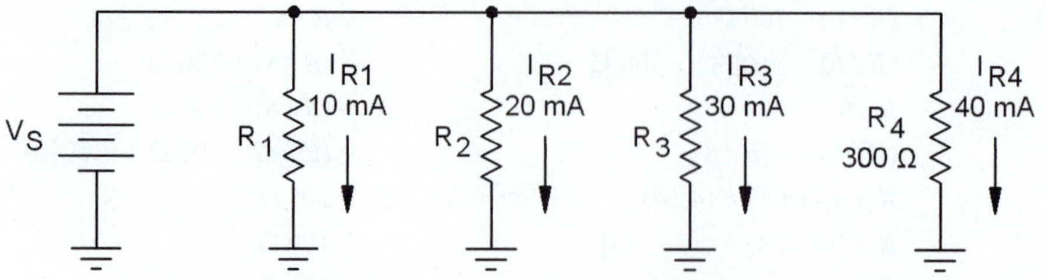

Figure 6-25: Application of Ohm's Law Example Circuit 2

Solution: From Ohm's Law for voltage, the voltage V_{R4} across R_4 is

$$\begin{aligned} V_{R4} &= I_{R4} \times R_4 \\ &= 40 \text{ mA} \times 300\ \Omega \\ &= (40 \times 10^{-3} \text{ A}) \times 300\ \Omega \\ &= [(40 \times 10^{-3}) \times 300]\ (\text{A} \times \Omega) \\ &= 12.0 \text{ V} \end{aligned}$$

Because the resistors are in parallel, the voltages across the resistors must be the same. From Ohm's Law for resistance,

$$\begin{aligned} R_1 &= V_{R1} / I_{R1} \\ &= 12.0 \text{ V} / 10 \text{ mA} \\ &= (12.0 \text{ V}) / (10 \times 10^{-3} \text{ A}) \\ &= [12.0 / (10 \times 10^{-3})]\ (\text{V} / \text{A}) \\ &= 1.2 \times 10^3\ \Omega \\ &= \mathbf{1.2\ k\Omega} \end{aligned}$$

$$\begin{aligned} R_2 &= V_{R2} / I_{R2} \\ &= 12.0 \text{ V} / 20 \text{ mA} \\ &= (12.0 \text{ V}) / (20 \times 10^{-3} \text{ A}) \\ &= [12.0 / (20 \times 10^{-3})]\ (\text{V} / \text{A}) \\ &= \mathbf{600\ \Omega} \end{aligned}$$

$$\begin{aligned} R_3 &= V_{R3} / I_{R3} \\ &= 12.0 \text{ V} / 30 \text{ mA} \\ &= (12.0 \text{ V}) / (30 \times 10^{-3} \text{ A}) \\ &= [12.0 / (30 \times 10^{-3})]\ (\text{V} / \text{A}) \\ &= \mathbf{400\ \Omega} \end{aligned}$$

2. Problem: Refer to the circuit in Figure 6-26. To what value must R_{VAR} be adjusted to produce the voltage V_{SPEC} across R_1?

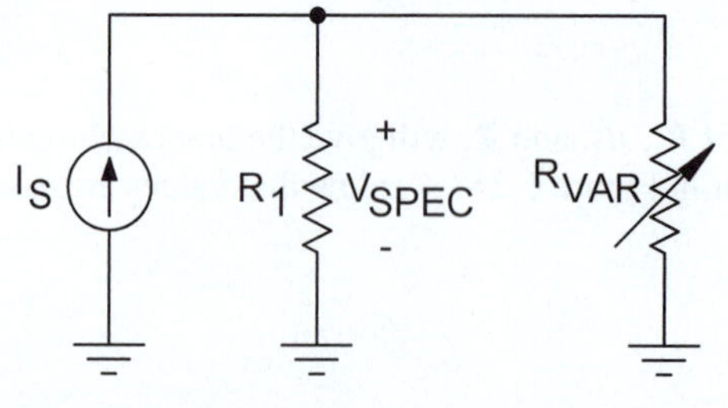

Figure 6-26: Application of Ohm's Law Example Circuit 3

Solution: The total parallel resistance for R_1 and R_{VAR} is

$$1 / R_T = (1 / R_1) + (1 / R_{VAR})$$

so

$$(1 / R_T) + -(1 / R_1) = (1 / R_1) + (1 / R_{VAR}) + -(1 / R_1)$$
$$= 1 / R_{VAR}$$

For the parallel circuit, V_{SPEC} is equal to the total voltage V_T. From Ohm's Law for resistance,

$$R_T = V_T / I_S$$
$$= V_{SPEC} / I_S$$

so

$$1 / R_{VAR} = (1 / R_T) - (1 / R_1)$$
$$= [1 / (V_{SPEC} / I_S)] + -(1 / R_1)$$
$$= (I_S / V_{SPEC}) + -(1 / R_1)$$
$$= [(I_S / V_{SPEC}) \times (R_1 / R_1)] + -[(1 / R_1) \times (V_{SPEC} / V_{SPEC})]$$
$$= [(I_S \times R_1) / (R_1 \times V_{SPEC})] + -[(V_{SPEC} \times 1) / (R_1 \times V_{SPEC})]$$
$$= [(R_1 \times I_S) + -V_{SPEC}] / (R_1 \times V_{SPEC})$$

Therefore,

$$1 / (1 / R_{VAR}) = 1 / \{[(R_1 \times I_S) - V_{SPEC}] / (R_1 \times V_{SPEC})\}$$
$$R_{VAR} = (R_1 \times V_{SPEC}) / [(R_1 \times I_S) - V_{SPEC}]$$

3. Problem: What are the maximum and minimum values of R_{VAR} for the parallel circuit in Figure 6-27?

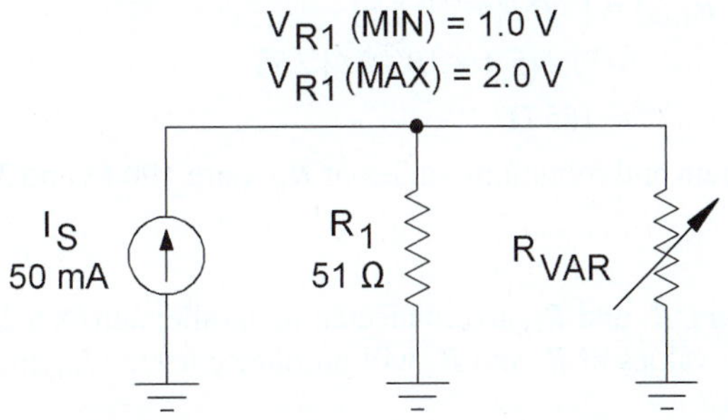

Figure 6-27: Application of Ohm's Law Example Circuit 4

Solution: For the parallel circuit

$$1 / R_T = (1 / R_1) + (1 / R_{VAR})$$

so

$$(1 / R_T) + -(1 / R_1) = (1 / R_1) + (1 / R_{VAR}) + -(1 / R_1)$$
$$= 1 / R_{VAR}$$

For parallel circuits, the voltage V_{R1} across R_1 is equal to the total voltage V_T. Use Ohm's Law for resistance to calculate R_T.

$$R_T = V_T / I_S$$
$$= V_{R1} / I_S$$

Therefore,

$$1 / R_{VAR} = (1 / R_T) - (1 / R_1)$$
$$= [1 / (V_{R1} / I_S)] + -(1 / R_1)$$
$$= (I_S / V_{R1}) + -(1 / R_1)$$

For V_{R1} (min) = 1.0 V,

$$1 / R_{VAR} = (I_S / V_{R1}) - (1 / R_1)$$
$$= (50 \text{ mA} / 1.0 \text{ V}) + -(1 / 51\ \Omega)$$

$= [(50 \times 10^{-3}\ A) / 1.0\ V] + -(19.6 \times 10^{-3}\ S)$
$= [(50 \times 10^{-3}) / (1.0)]\ (A / V) + -(19.6 \times 10^{-3}\ S)$
$= (50 \times 10^{-3}\ S) + -(19.6 \times 10^{-3}\ S)$
$= [(50 \times 10^{-3}) + (-19.6 \times 10^{-3})]\ S$
$= 30.4 \times 10^{-3}\ S$

so

$1 / (1 / \boldsymbol{R_{VAR}}) = 1 / (30.4 \times 10^{-3}\ S)$
$\boldsymbol{R_{VAR}} = [1 / (30.4 \times 10^{-3})]\ (1 / S)$
$= 32.9\ \Omega$

For $\boldsymbol{V_{R1}}$ (max) = 2.0 V,

$1 / \boldsymbol{R_{VAR}} = (\boldsymbol{I_S} / \boldsymbol{V_{R1}}) - (1 / \boldsymbol{R_1})$
$= (50\ mA / 2.0\ V) + -(1 / 51\ \Omega)$
$= [(50 \times 10^{-3}\ A) / 2.0\ V] + -(19.6 \times 10^{-3}\ S)$
$= (50 \times 10^{-3}) / (2.0)\ (A / V) + -(19.6 \times 10^{-3}\ S)$
$= (25 \times 10^{-3}\ S) + -(19.6 \times 10^{-3}\ S)$
$= [(25 \times 10^{-3}) + (-19.6 \times 10^{-3})]\ S$
$= 5.4 \times 10^{-3}\ S$

so

$1 / (1 / \boldsymbol{R_{VAR}}) = 1 / (5.4 \times 10^{-3}\ S)$
$\boldsymbol{R_{VAR}} = [1 / (5.4 \times 10^{-3})]\ (1 / S)$
$= 185\ \Omega$

The maximum and minimum values of $\boldsymbol{R_{VAR}}$ are **190 Ω** and **33 Ω**, respectively.

6.2.5 Current Dividers

1. Problem: Two resistors, $\boldsymbol{R_1}$ and $\boldsymbol{R_2}$, are connected in parallel across a 25 mA supply. If the total voltage $\boldsymbol{V_T}$ is 5.0 V, what values of $\boldsymbol{R_1}$ and $\boldsymbol{R_2}$ will produce currents $\boldsymbol{I_{R1}}$ and $\boldsymbol{I_{R2}}$ of 10 mA and 15 mA, respectively?

   Solution: For a parallel circuit, the current-divider formula is

$\boldsymbol{I_X} = \boldsymbol{I_T} \times (\boldsymbol{R_T} / \boldsymbol{R_X})$
$= (\boldsymbol{I_T} \times \boldsymbol{R_T}) / \boldsymbol{R_X}$

so

$\boldsymbol{I_X} \times \boldsymbol{R_X} = [(\boldsymbol{I_T} \times \boldsymbol{R_T}) / \boldsymbol{R_X}] \times \boldsymbol{R_X}$
$= (\boldsymbol{I_T} \times \boldsymbol{R_T} \times \boldsymbol{R_X}) / \boldsymbol{R_X}$
$= \boldsymbol{I_T} \times \boldsymbol{R_T}$
$(\boldsymbol{I_X} \times \boldsymbol{R_X}) / \boldsymbol{I_X} = (\boldsymbol{I_T} \times \boldsymbol{R_T}) / \boldsymbol{I_X}$
$\boldsymbol{R_X} = (\boldsymbol{I_T} / \boldsymbol{I_X}) \times \boldsymbol{R_T}$

From Ohm's Law for total resistance,

$\boldsymbol{R_T} = \boldsymbol{V_T} / \boldsymbol{I_T}$
$= 5.0\ V / 25\ mA$
$= 5.0\ V / (25 \times 10^{-3}\ A)$
$= [5.0 / (25 \times 10^{-3})]\ (V / A)$
$= 200\ \Omega$

For $\boldsymbol{I_{R1}} = 10$ mA,

$$\begin{aligned}
\boldsymbol{R}_1 &= (\boldsymbol{I_T} / \boldsymbol{I_{R1}}) \times \boldsymbol{R_T} \\
&= (25\text{ mA} / 10\text{ mA}) \times 200\ \Omega \\
&= [(25 \times 10^{-3}\text{ A}) / (10 \times 10^{-3}\text{ A})] \times 200\ \Omega \\
&= [(25 \times 10^{-3}\text{ A}) \times (200\ \Omega)] / (10 \times 10^{-3}\text{ A}) \\
&= \{[(25 \times 10^{-3}) \times (200)] / (10 \times 10^{-3}\text{ A})\}\ [(\text{A} \times \Omega) / \text{A}] \\
&= 500\ \Omega
\end{aligned}$$

For $\boldsymbol{I_{R2}} = 15$ mA,

$$\begin{aligned}
\boldsymbol{R}_2 &= (\boldsymbol{I_T} / \boldsymbol{I_{R2}}) \times \boldsymbol{R_T} \\
&= (25\text{ mA} / 15\text{ mA}) \times 200\ \Omega \\
&= [(25 \times 10^{-3}\text{ A}) / (15 \times 10^{-3}\text{ A})] \times 200\ \Omega \\
&= [(25 \times 10^{-3}\text{ A}) \times (200\ \Omega)] / (15 \times 10^{-3}\text{ A}) \\
&= \{[(25 \times 10^{-3}) \times (200)] / (15 \times 10^{-3}\text{ A})\}\ [(\text{A} \times \Omega) / \text{A}] \\
&= 332\ \Omega
\end{aligned}$$

The values of $\boldsymbol{R}_1$ and $\boldsymbol{R}_2$ are **500 Ω** and **330 Ω**, respectively.

2. Problem: A circuit consists of two equal-value resistors $\boldsymbol{R}_1$ and $\boldsymbol{R}_2$ in parallel with a current source $\boldsymbol{I_S}$. What value of $\boldsymbol{R}_2$ will halve the current through $\boldsymbol{R}_1$?

Solution: For the original circuit, in which $\boldsymbol{R}_1 = \boldsymbol{R}_2$, the current through each resistor is equal.

$$\boldsymbol{I_{R1}}\text{ (org)} = \boldsymbol{I_{R2}}\text{ (org)} = \boldsymbol{I_S} / 2$$

If $\boldsymbol{R}_2$ changes so that $\boldsymbol{I_{R1}}$ is halved, then

$$\begin{aligned}
\boldsymbol{I_{R1}}\text{ (new)} &= \boldsymbol{I_{R1}}\text{ (org)} / 2 \\
&= (\boldsymbol{I_S} / 2) / 2 \\
&= \boldsymbol{I_S} / (2 \times 2) \\
&= \boldsymbol{I_S} / 4
\end{aligned}$$

From Kirchhoff's Current Law, the sum of the resistor currents must equal the current from the current source. Therefore,

$$\begin{aligned}
\boldsymbol{I_S} &= \boldsymbol{I_{R1}}\text{ (new)} + \boldsymbol{I_{R2}}\text{ (new)} \\
&= (\boldsymbol{I_S} / 4) + \boldsymbol{I_{R2}}\text{ (new)} \\
\boldsymbol{I_S} + -(\boldsymbol{I_S} / 4) &= (\boldsymbol{I_S} / 4) + \boldsymbol{I_{R2}}\text{ (new)} + -(\boldsymbol{I_S} / 4) \\
[(4 \times \boldsymbol{I_S}) / 4] + -(\boldsymbol{I_S} / 4) &= \boldsymbol{I_{R2}}\text{ (new)} \\
[(4 \times \boldsymbol{I_S}) + -\boldsymbol{I_S}] / 4 &= \boldsymbol{I_{R2}}\text{ (new)} \\
[(4 + -1) \times \boldsymbol{I_S}] / 4 &= \boldsymbol{I_{R2}}\text{ (new)} \\
(3 \times \boldsymbol{I_S}) / 4 &= \boldsymbol{I_{R2}}\text{ (new)}
\end{aligned}$$

From the current-divider formula for two parallel resistors,

$$\begin{aligned}
\boldsymbol{I_{R1}} &= (\boldsymbol{R_T} / \boldsymbol{R}_1) \times \boldsymbol{I_T} \\
\boldsymbol{I_{R2}} &= (\boldsymbol{R_T} / \boldsymbol{R}_2) \times \boldsymbol{I_T}
\end{aligned}$$

so

$$\begin{aligned}
\boldsymbol{I_{R1}} / \boldsymbol{I_{R2}} &= [(\boldsymbol{R_T} / \boldsymbol{R}_1) \times \boldsymbol{I_T}] / [(\boldsymbol{R_T} / \boldsymbol{R}_2) \times \boldsymbol{I_T}] \\
&= [(\boldsymbol{R_T} \times \boldsymbol{I_T}) / \boldsymbol{R_1}] / [(\boldsymbol{R_T} \times \boldsymbol{I_T}) / \boldsymbol{R_2}] \\
&= (1 / \boldsymbol{R}_1) / (1 / \boldsymbol{R}_2) \\
&= (1 / \boldsymbol{R}_1) \times (\boldsymbol{R}_2 / 1) \\
&= \boldsymbol{R}_2 / \boldsymbol{R}_1
\end{aligned}$$

From this,

$$\begin{aligned}
\boldsymbol{R}_2 / \boldsymbol{R}_1 &= \boldsymbol{I_{R1}} / \boldsymbol{I_{R2}} \\
(\boldsymbol{R}_2 / \boldsymbol{R}_1) \times \boldsymbol{R}_1 &= (\boldsymbol{I_{R1}} / \boldsymbol{I_{R2}}) \times \boldsymbol{R}_1
\end{aligned}$$

$R_2 = (I_{R1} / I_{R2}) \times R_1$

Therefore,

$$
\begin{aligned}
R_2\text{ (new)} &= \{[I_{R1}\text{ (new)}] / [I_{R2}\text{ (new)}]\} \times R_1 \\
&= \{(I_S / 4) / [(3 \times I_S) / 4]\} \times R_1 \\
&= \{(I_S / 4) \times [4 / (3 \times I_S)]\} \times R_1 \\
&= [(I_S \times 4) / (4 \times 3 \times I_S)] \times R_1 \\
&= R_1 / 3 \\
&= R_2\text{ (org)} / 3
\end{aligned}
$$

To halve the current through R_1, **the new value of R_2 must be 1/3 its original value**. This is logical, as the calculated value of I_{R2} (new) is three times the calculated value of I_{R1} (new).

6.2.6 Power in Parallel Circuits

1. Problem: Show that the power dissipation P_{RX} of any resistor R_X in a parallel circuit with total resistance R_T and total power dissipation P_T is equal to $P_T \times (R_T / R_X)$.

   Solution: From Watt's Law for voltage and resistance, the total power dissipation P_T of a circuit with total resistance R_T and applied voltage V_S is

   $$P_T = V_S^2 / R_T$$

   and the power dissipation of resistor R_X is

   $$P_{RX} = V_{RX}^2 / R_X$$

   For a parallel circuit, the applied voltage is equal to the voltage across each resistor.

   $$P_{RX} = V_S^2 / R_X$$

   From this,

   $$
   \begin{aligned}
   P_{RX} / P_T &= (V_S^2 / R_X) / (V_S^2 / R_T) \\
   P_T \times (P_{RX} / P_T) \times P_T &= P_T \times [(V_S^2 / R_X) / (V_S^2 / R_T)] \\
   (P_T \times P_{RX}) / P_T &= P_T \times [(V_S^2 / R_X) / (V_S^2 / R_T)] \\
   P_{RX} &= P_T \times [(V_S^2 / R_X) \times (R_T / V_S^2)] \\
   &= P_T \times (R_T / R_X)
   \end{aligned}
   $$

2. Problem: What is the maximum value of R_3 that will not exceed a total power dissipation of 2.0 W for the circuit in Figure 6-28?

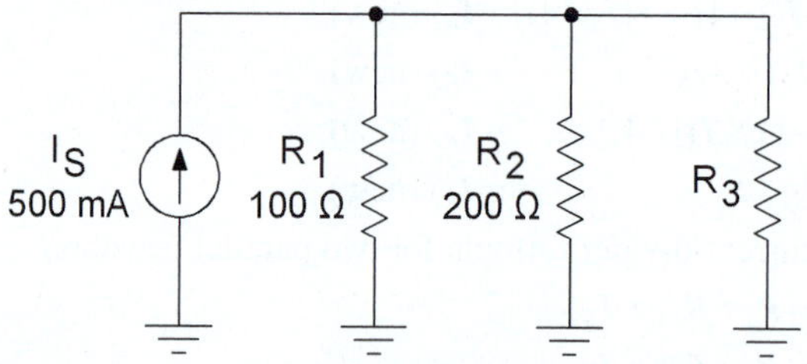

Figure 6-28: Power in Parallel Circuits Example Circuit 3

   Solution: From Watt's Law for current and resistance, the total power dissipation P_T for the circuit is

   $$P_T = I_S^2 \times R_T$$

   $$\text{W} = \text{A}^2 \times \Omega$$

   so

   $$
   \begin{aligned}
   P_T / I_S^2 &= (I_S^2 \times R_T) / I_S^2 \\
   &= R_T
   \end{aligned}
   $$

For $P_{T\,(\max)} = 2$ W, the maximum resistance $R_{T\,(\max)}$ is

$$\begin{aligned}
R_{T\,(\max)} &= P_T / I_S^2 \\
&= (2.0\text{ W}) / (500\text{ mA})^2 \\
&= (2.0\text{ W}) / (500 \times 10^{-3}\text{ A})^2 \\
&= (2.0\text{ W}) / (250 \times 10^{-3}\text{ A}^2) \\
&= [2.0 / (250 \times 10^{-3})]\ (\text{W} / \text{A}^2) \\
&= 8.0\ [(\text{A}^2 \times \Omega) / \text{A}^2] \\
&= 8.0\ \Omega
\end{aligned}$$

For the parallel circuit,

$$\begin{aligned}
1 / R_T &= (1 / R_1) + (1 / R_2) + (1 / R_3) \\
(1 / R_T) + -(1 / R_1) + -(1 / R_2) &= (1 / R_1) + (1 / R_2) + (1 / R_3) + -(1 / R_1) + -(1 / R_2) \\
&= 1 / R_3
\end{aligned}$$

From this,

$$\begin{aligned}
1 / [R_{3\,(\max)}] &= \{1 / [R_T\,(\max)]\} + -(1 / R_1) + -(1 / R_2) \\
1 / [R_{3\,(\max)}] &= [1 / (8.0\ \Omega)] + -[1 / (100\ \Omega)] + -[1 / (200\ \Omega)] \\
&= (1 / 8.0)\,(1 / \Omega) + -(1 / 100)\,(1 / \Omega) + -(1 / 200)\,(1 / \Omega) \\
&= (125 \times 10^{-3}\text{ S}) + -(10 \times 10^{-3}\text{ S}) + -(5 \times 10^{-3}\text{ S}) \\
&= [(125 \times 10^{-3}) + -(10 \times 10^{-3}) + -(5 \times 10^{-3})]\text{ S} \\
&= 110 \times 10^{-3}\text{ S} \\
1 / [1 / R_{3\,(\max)}] &= 1 / (110 \times 10^{-3}\text{ S}) \\
R_{3\,(\max)} &= [1 / (110 \times 10^{-3})]\ (1 / \text{S}) \\
&= 9.09\ \Omega
\end{aligned}$$

The maximum value of R_3 that will not exceed a total power dissipation of 2 .0 W is **9.1 Ω**.

6.3 Just for Fun

1. Problem: Two resistors on a circuit board, R_1 and R_2, have a total series resistance R_S of 100 Ω. An assembler accidentally connects the two resistors in parallel. If the parallel resistance R_P is 15 Ω, what are the values of R_1 and R_2?

Solution: For the resistors R_1 and R_2 in parallel,

$$\begin{aligned}
1 / (15\ \Omega) &= (1 / R_1) + (1 / R_2) \\
1 / (15\ \Omega) + -(1 / R_1) &= (1 / R_1) + (1 / R_2) + -(1 / R_1) \\
1 / (15\ \Omega) + -(1 / R_1) &= (1 / R_2)
\end{aligned}$$

From this

$$\begin{aligned}
1 / R_2 &= [1 / (15\ \Omega) \times (R_1 / R_1)] + -[(1 / R_1) \times (15\ \Omega / 15\ \Omega)] \\
&= [R_1 / (15\ \Omega \times R_1)] + -[15\ \Omega / (15\ \Omega \times R_1)] \\
&= (R_1 - 15\ \Omega) / (15\ \Omega \times R_1) \\
1 / (1 / R_2) &= 1 / [(R_1 + -15\ \Omega) / (15\ \Omega \times R_1)] \\
R_2 &= (15\ \Omega \times R_1) / (R_1 + -15\ \Omega)
\end{aligned}$$

Substituting for R_2 in the series resistance equation gives

$$\begin{aligned}
R_1 + [(15\ \Omega \times R_1) / (R_1 + -15\ \Omega)] &= 100\ \Omega \\
R_1 \times [(R_1 + -15\ \Omega) / (R_1 + -15\ \Omega)] + (15\ \Omega \times R_1) / (R_1 + -15\ \Omega) &= 100\ \Omega \\
\{[R_1^2 + -(15\ \Omega \times R_1)] / (R_1 + -15\ \Omega)\} + (15\ \Omega \times R_1) / (R_1 + -15\ \Omega) &= 100\ \Omega \\
[R_1^2 + -(15\ \Omega \times R_1) + (15\ \Omega \times R_1)] / (R_1 - 15\ \Omega) &= 100\ \Omega \\
R_1^2 / (R_1 + -15\ \Omega) &= 100\ \Omega
\end{aligned}$$

so

$[R_1^2 / (R_1 + -15\ \Omega)] \times (R_1 + -15\ \Omega) = 100\ \Omega \times (R_1 + -15\ \Omega)$

$R_1^2 = (100\ \Omega \times R_1) + -(100\ \Omega \times 15\ \Omega)$

$R_1^2 = (100\ \Omega \times R_1) + -1500\ \Omega$

and

$R_1^2 + -[(100\ \Omega \times R_1) - 1500\ \Omega^2] = [(100\ \Omega \times R_1) + -1500\ \Omega^2] + -[(100\ \Omega \times R_1) - 1500\ \Omega^2]$

$R_1^2 + (-100\ \Omega) \times R_1 + 1500\ \Omega^2 = 0$

This is a quadratic expression of the form $Ax^2 + Bx + C = 0$ which has the general solution

$$R_1 = \frac{-B \pm \sqrt{B^2 - (4 \times A \times C)}}{2 \times A}$$

Note that there are two possible values for R_1, as there are two resistance values and R_1 could be either one (with R_2 taking the other value). Therefore,

$$R_1 = \frac{-(-100\ \Omega) \pm \sqrt{(100\ \Omega)^2 - [4 \times 1 \times (1500\ \Omega)^2]}}{2 \times 1}$$

$$= \frac{100\ \Omega \pm \sqrt{10000\ \Omega^2 - 6000\ \Omega^2}}{2}$$

$$= \frac{100\ \Omega \pm \sqrt{4000\ \Omega^2}}{2}$$

$$= \frac{100\ \Omega \pm 63.2\ \Omega}{2}$$

The first possible value of R_1 is

$R_1 = (100\ \Omega + 63.2\ \Omega) / 2$

$= [(100 + 63.2)\ \Omega] / 2$

$= (163.2\ \Omega) / 2$

$= 81.6\ \Omega$

The second possible value of R_1 is

$R_1 = (100\ \Omega - 63.2\ \Omega) / 2$

$= [(100 - 63.2)\ \Omega] / 2$

$= (36.8\ \Omega) / 2$

$= 18.4\ \Omega$

The resistor values are **82 Ω** and **18 Ω**.

2. Problem: A common fault in parallel resistive circuits is an open resistor. You can find the open resistor by trial and error measurements, but knowing the value of the open resistor can simplify the search. When a voltage source V_S is connected across the circuit and resistor R_X opens, the decrease in total circuit current, ΔI, is the difference between the nominal current I_{NOM} and the fault current I_{FAULT}. Show that the value of the open resistor R_X in the circuit is

$R_X = V_S / \Delta I$

Solution: The total current I_{NOM} for a parallel resistive circuit is equal to the sum of the branch currents, or

$I_{NOM} = I_{R1} + I_{R2} + \ldots + I_{RX} + \ldots + I_{RN}$

$= (V_S / R_1) + (V_S / R_2) + \ldots + (V_S / R_X) + \ldots + (V_S / R_N)$

When R_X opens, $V_S / R_X = I_{RX} = 0$ so the fault current I_{FAULT} is equal to I_{NOM} less the branch current through R_X.

$$\boldsymbol{I}_{NOM} + -\boldsymbol{I}_{RX} = \boldsymbol{I}_{FAULT}$$
$$\boldsymbol{I}_{NOM} + -\boldsymbol{I}_{RX} + \boldsymbol{I}_{RX} = \boldsymbol{I}_{FAULT} + \boldsymbol{I}_{RX}$$
$$\boldsymbol{I}_{NOM} + -\boldsymbol{I}_{FAULT} = \boldsymbol{I}_{FAULT} + \boldsymbol{I}_{RX} + -\boldsymbol{I}_{FAULT}$$
$$\boldsymbol{I}_{NOM} + -\boldsymbol{I}_{FAULT} = \boldsymbol{I}_{RX}$$

By definition the change in current $\boldsymbol{\Delta I}$ is the difference between $\boldsymbol{I}_{NOM}$ and $\boldsymbol{I}_{FAULT}$, so

$$\Delta \boldsymbol{I} = \boldsymbol{I}_{RX}$$
$$= \boldsymbol{V}_S / \boldsymbol{R}_X$$
$$\Delta \boldsymbol{I} \times \boldsymbol{R}_X = (\boldsymbol{V}_S / \boldsymbol{R}_X) \times \boldsymbol{R}_X$$
$$(\Delta \boldsymbol{I} \times \boldsymbol{R}_X) / \Delta \boldsymbol{I} = \boldsymbol{V}_S / \Delta \boldsymbol{I}$$
$$\boldsymbol{R}_X = \boldsymbol{V}_S / \Delta \boldsymbol{I}$$

3. Problem: When a current source $\boldsymbol{I}_S$ is connected across a parallel resistive circuit and resistor $\boldsymbol{R}_X$ opens, the circuit resistance increases from $\boldsymbol{R}_{NOM}$ to $\boldsymbol{R}_{FAULT}$ and the total circuit voltage increases from the nominal voltage $\boldsymbol{V}_{NOM}$ and the fault voltage $\boldsymbol{V}_{FAULT}$. Show that the value of the open resistor $\boldsymbol{R}_X$ in the circuit is given by

$$\boldsymbol{1 / R}_X = \boldsymbol{I}_S \times [(1 / \boldsymbol{V}_{NOM}) + -(1 / \boldsymbol{V}_{FAULT})]$$

Solution: From Ohm's Law for voltage,

$$\boldsymbol{V}_{NOM} = \boldsymbol{I}_S \times \boldsymbol{R}_{NOM}$$

and

$$\boldsymbol{V}_{FAULT} = \boldsymbol{I}_S \times \boldsymbol{R}_{FAULT}$$

For the parallel circuit with no fault,

$$1 / \boldsymbol{R}_{NOM} = (1 / \boldsymbol{R}_1) + (1 / \boldsymbol{R}_2) + \ldots + (1 / \boldsymbol{R}_X) + \ldots + (1/ \boldsymbol{R}_N)$$

When $\boldsymbol{R}_X$ opens, $1 / \boldsymbol{R}_X = 0$ so that

$$1 / \boldsymbol{R}_{FAULT} = (1 / \boldsymbol{R}_1) + (1 / \boldsymbol{R}_2) + \ldots + 0 + \ldots + (1/ \boldsymbol{R}_N)$$
$$= (1 / \boldsymbol{R}_1) + (1 / \boldsymbol{R}_2) + \ldots + 0 + \ldots + (1/ \boldsymbol{R}_N) + [(1 / \boldsymbol{R}_X) - (1 / \boldsymbol{R}_X)]$$
$$= (1 / \boldsymbol{R}_1) + (1 / \boldsymbol{R}_2) + \ldots + (1 / \boldsymbol{R}_X) + \ldots + (1/ \boldsymbol{R}_N) + -(1 / \boldsymbol{R}_X)$$
$$= (1 / \boldsymbol{R}_{NOM}) + -(1 / \boldsymbol{R}_X)$$

From this,

$$(1 / \boldsymbol{R}_{FAULT}) + (1 / \boldsymbol{R}_X) = (1 / \boldsymbol{R}_{NOM}) + -(1 / \boldsymbol{R}_X) + (1 / \boldsymbol{R}_X)$$
$$(1 / \boldsymbol{R}_{FAULT}) + (1 / \boldsymbol{R}_X) + -(1 / \boldsymbol{R}_{FAULT}) = (1 / \boldsymbol{R}_{NOM}) + -(1 / \boldsymbol{R}_{FAULT})$$
$$1 / \boldsymbol{R}_X = (1 / \boldsymbol{R}_{NOM}) + -(1 / \boldsymbol{R}_{FAULT})$$

From Ohm's Law for resistance,

$$1 / \boldsymbol{R}_X = [1 / (\boldsymbol{V}_{NOM} / \boldsymbol{I}_S)] + -[1 / (\boldsymbol{V}_{FAULT} / \boldsymbol{I}_S)]$$
$$= (\boldsymbol{I}_S / \boldsymbol{V}_{NOM}) + -(\boldsymbol{I}_S / \boldsymbol{V}_{FAULT})$$
$$= \boldsymbol{I}_S \times [(1 / \boldsymbol{V}_{NOM}) + -(1 / \boldsymbol{V}_{FAULT})]$$

7. Series-Parallel Circuits

7.1 Basic

7.1.1 Circuit Notation

A standard notation for representing circuit connections uses a "+" operator to represent a series connection and a "||" operator to represent a parallel connection. The "||" operator takes precedence over the "+" operator, just as the multiplication and division operators take precedence over the addition and subtraction operators in mathematical expressions. Just as with mathematical expressions, you can use parentheses to override the standard order of operations. Consider the following expressions:

$$\boldsymbol{R_T} = \boldsymbol{R_1} + \boldsymbol{R_2} \parallel \boldsymbol{R_3}$$

$$\boldsymbol{R_T} = (\boldsymbol{R_1} + \boldsymbol{R_2}) \parallel \boldsymbol{R_3}$$

The first expression represents a circuit in which $\boldsymbol{R_1}$ is in series with the parallel combination of $\boldsymbol{R_2}$ and $\boldsymbol{R_3}$. The second expression represents a circuit in which the series combination of $\boldsymbol{R_1}$ and $\boldsymbol{R_2}$ is in parallel with $\boldsymbol{R_3}$. Refer to the circuits in Figure 7-1, which illustrates the differences between the two expressions.

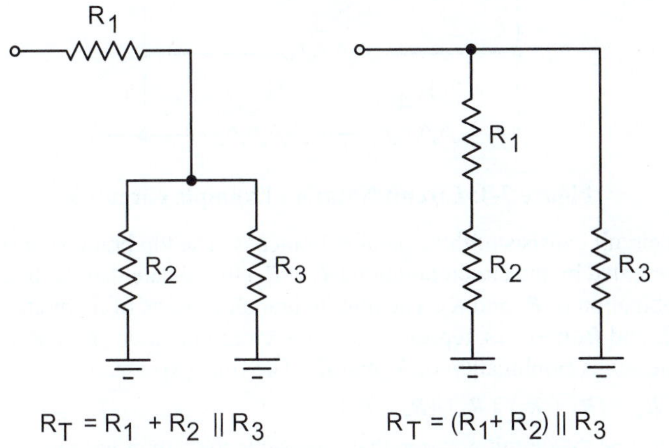

Figure 7-1: Examples of Circuit Notation

1. Problem: Determine the circuit that corresponds to the expression $\boldsymbol{R_T} = (\boldsymbol{R_1} + \boldsymbol{R_2}) \parallel (\boldsymbol{R_3} + \boldsymbol{R_4})$.

   Solution: The parentheses show that the series operators for $\boldsymbol{R_1}$ and $\boldsymbol{R_2}$ and for $\boldsymbol{R_3}$ and $\boldsymbol{R_4}$ take precedence over the parallel operator. Therefore, the circuit consists of the series combination of $\boldsymbol{R_1}$ and $\boldsymbol{R_2}$ in parallel with the series combination of $\boldsymbol{R_3}$ and $\boldsymbol{R_4}$, as Figure 7-2 shows.

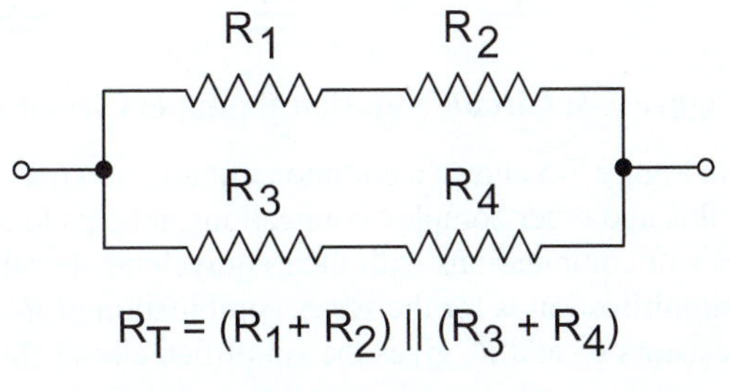

Figure 7-2: Circuit Notation Example Circuit 1

2. Problem: Determine the circuit that corresponds to the expression $\boldsymbol{R_T} = \boldsymbol{R_1} \parallel (\boldsymbol{R_2} + \boldsymbol{R_3} \parallel \boldsymbol{R_4})$.

   Solution: The parentheses show that the expression $\boldsymbol{R_2} + \boldsymbol{R_3} \parallel \boldsymbol{R_4}$ takes precedence and is evaluated before the parallel operator with $\boldsymbol{R_1}$. As there are no other (nested) parentheses inside this expression, the standard order of operations applies so that $\boldsymbol{R_2} + \boldsymbol{R_3} \parallel \boldsymbol{R_4}$ represents $\boldsymbol{R_2}$ in series with the parallel combination of R_3 and R_4. This series-parallel combination is itself in parallel with R_1, as Figure 7-3 shows.

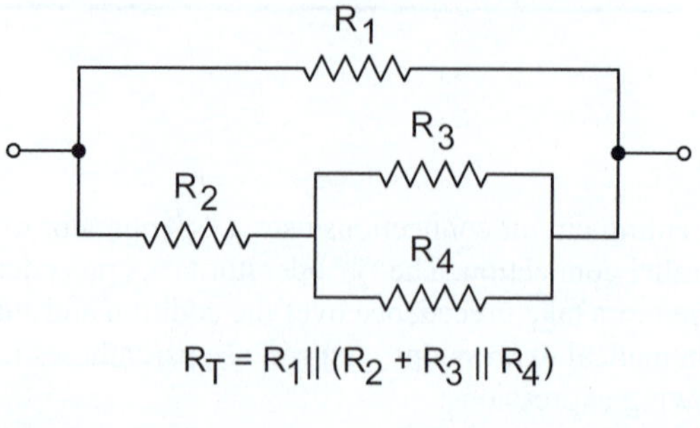

Figure 7-3: Circuit Notation Example Circuit 2

3. Problem: Determine the circuit notation that represents the circuit in Figure 7-4.

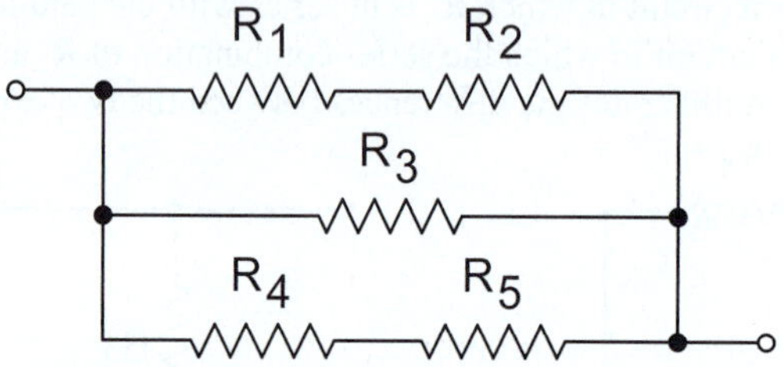

Figure 7-4: Circuit Notation Example Circuit 3

Solution: The circuit consists of three parallel branches. The top branch consists of R_1 and R_2 in series, represented by the circuit notation $(R_1 + R_2)$ to indicate that the branch consists of the series combination of R_1 and R_2. The middle branch consists of R_3 by itself. The bottom branch consists of R_4 and R_5 in series, represented by the circuit notation $(R_4 + R_5)$ to indicate that branch consists of the series combination of R_4 and R_5. The final expression is

$$R_T = (R_1 + R_2) \parallel R_3 \parallel (R_4 + R_5)$$

4. Problem: Determine the circuit notation that represents the circuit in Figure 7-5.

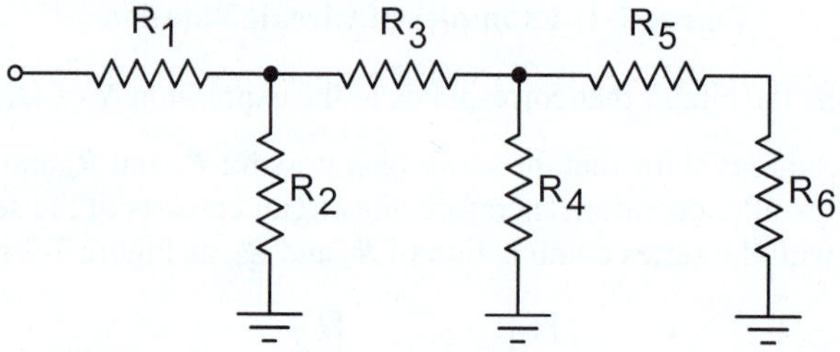

Figure 7-5: Circuit Notation Example Circuit 4

Solution: The circuit in Figure 7-5 shows a common circuit known as a ladder network. To determine the notation for this and other complex connections, it helps to progressively simplify the circuit by replacing resistor combinations with their equivalents. For the original ladder network, the only immediate simplification is for the series combination of R_5 and R_6, represented by $(R_5 + R_6)$. Replacing resistors R_5 and R_6 gives the simplified circuit shown in Figure 7-6.

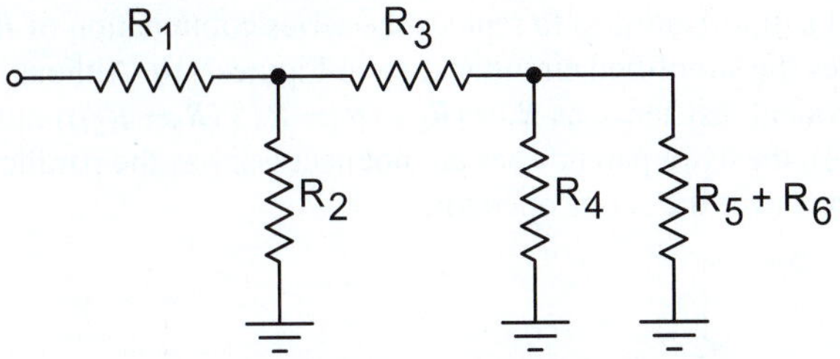

Figure 7-6: First Circuit Simplification

The second simplification is to replace the parallel combination of $\boldsymbol{R}_4$ and $\boldsymbol{R}_5 + \boldsymbol{R}_6$. This gives the simplified circuit shown in Figure 7-7. Note that because $\boldsymbol{R}_4$ is in parallel with the series combination of $\boldsymbol{R}_5$ and $\boldsymbol{R}_6$ and not just $\boldsymbol{R}_5$, the simplified circuit shows $\boldsymbol{R}_4 \parallel (\boldsymbol{R}_5 + \boldsymbol{R}_6)$ rather than $\boldsymbol{R}_4 \parallel \boldsymbol{R}_5 + \boldsymbol{R}_6$.

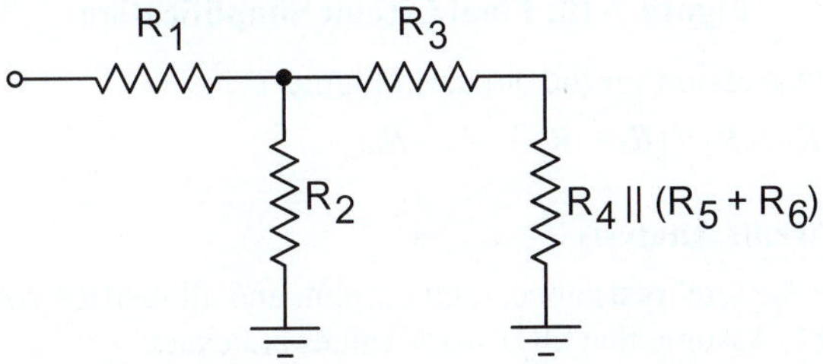

Figure 7-7: Second Circuit Simplification

The third simplification is to replace the series combination of $\boldsymbol{R}_3$ and $\boldsymbol{R}_4 \parallel (\boldsymbol{R}_5 + \boldsymbol{R}_6)$. This gives the simplified circuit shown in Figure 7-8. Although the simplified circuit could show the equivalent resistance as $\boldsymbol{R}_3 + (\boldsymbol{R}_4 \parallel (\boldsymbol{R}_5 + \boldsymbol{R}_6))$ rather than $\boldsymbol{R}_3 + \boldsymbol{R}_4 \parallel (\boldsymbol{R}_5 + \boldsymbol{R}_6)$, the extra parentheses are not necessary as the parallel operator automatically takes precedence over the series operator.

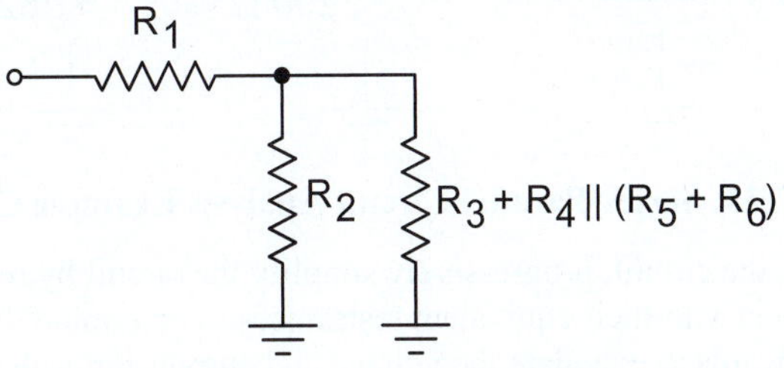

Figure 7-8: Third Circuit Simplification

The fourth simplification is to replace the parallel combination of $\boldsymbol{R}_2$ and $\boldsymbol{R}_3 + \boldsymbol{R}_4 \parallel (\boldsymbol{R}_5 + \boldsymbol{R}_6)$. This gives the simplified circuit shown in Figure 7-9. Note that because $\boldsymbol{R}_2$ is in parallel with the series-parallel combination of $\boldsymbol{R}_3 + \boldsymbol{R}_4 \parallel (\boldsymbol{R}_5 + \boldsymbol{R}_6)$ and not just $\boldsymbol{R}_3$, the simplified circuit shows $\boldsymbol{R}_2 \parallel (\boldsymbol{R}_3 + \boldsymbol{R}_4 \parallel (\boldsymbol{R}_5 + \boldsymbol{R}_6))$ rather than $\boldsymbol{R}_2 \parallel \boldsymbol{R}_3 + \boldsymbol{R}_4 \parallel (\boldsymbol{R}_5 + \boldsymbol{R}_6)$.

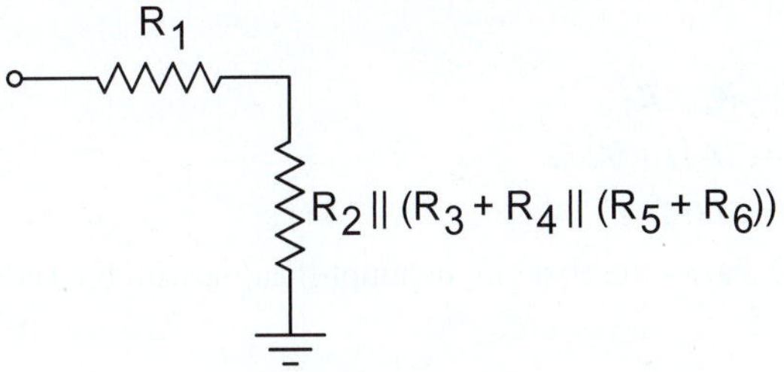

Figure 7-9: Fourth Circuit Simplification

The final simplification is to replace the series combination of $\boldsymbol{R_1}$ and $\boldsymbol{R_2} \parallel (\boldsymbol{R_3} + \boldsymbol{R_4} \parallel (\boldsymbol{R_5} + \boldsymbol{R_6}))$. This gives the simplified circuit shown in Figure 7-10. Although the simplified circuit could show the equivalent resistance as $\boldsymbol{R_1} + (\boldsymbol{R_2} \parallel (\boldsymbol{R_3} + \boldsymbol{R_4} \parallel (\boldsymbol{R_5} + \boldsymbol{R_6})))$ rather than $\boldsymbol{R_1} + \boldsymbol{R_2} \parallel (\boldsymbol{R_3} + \boldsymbol{R_4} \parallel (\boldsymbol{R_5} + \boldsymbol{R_6}))$, the extra parentheses are not necessary as the parallel operator automatically takes precedence over the series operator.

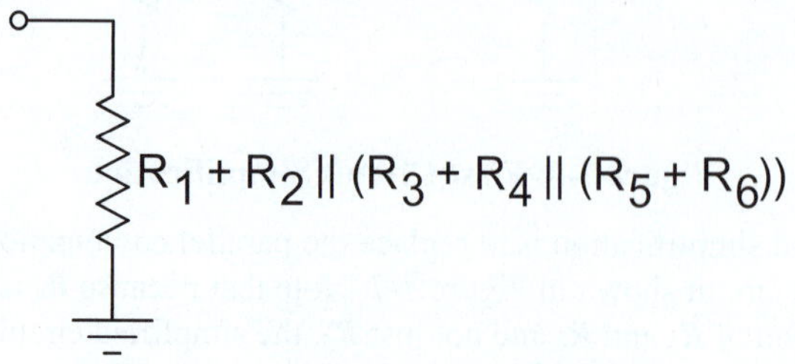

Figure 7-10: Final Circuit Simplification

The final expression for the circuit in Figure 7-5 is

$$\boldsymbol{R_T} = \boldsymbol{R_1} + \boldsymbol{R_2} \parallel (\boldsymbol{R_3} + \boldsymbol{R_4} \parallel (\boldsymbol{R_5} + \boldsymbol{R_6})).$$

7.1.2 Series-Parallel Resistive Circuit Analysis

1. Problem: Determine the total resistance, total current, and all resistor voltages and currents for the circuit in Figure 7-11. Assume that all resistor values are exact.

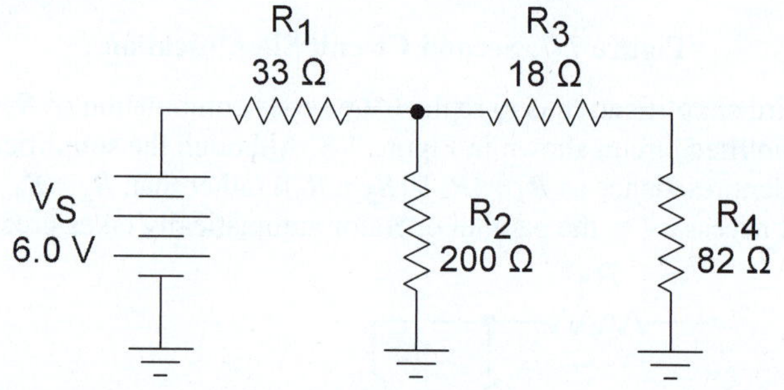

Figure 7-11: Series-Parallel Circuit Analysis Example Circuit 1

Solution: To analyze the circuit, progressively simplify the circuit by replacing series and parallel combinations with their equivalent resistances to determine the total resistance and current. Then, work backwards to calculate the voltage and current for each resistor.

The expression for the circuit in Figure 7-11 is

$$\boldsymbol{R_T} = \boldsymbol{R_1} + \boldsymbol{R_2} \parallel (\boldsymbol{R_3} + \boldsymbol{R_4})$$

From the order of operations for evaluating the expression, the first equivalent resistance $\boldsymbol{R_{EQ1}}$ is the series combination $(\boldsymbol{R_3} + \boldsymbol{R_4})$. From this,

$$\boldsymbol{R_T} = \boldsymbol{R_1} + \boldsymbol{R_2} \parallel \boldsymbol{R_{EQ1}}$$

where

$$\begin{aligned} \boldsymbol{R_{EQ1}} &= \boldsymbol{R_3} + \boldsymbol{R_4} \\ &= 18\ \Omega + 82\ \Omega \\ &= 100\ \Omega \end{aligned}$$

Figure 7-12 shows the first circuit simplification with the series combination of $\boldsymbol{R_3}$ and $\boldsymbol{R_4}$ replaced with $\boldsymbol{R_{EQ1}}$.

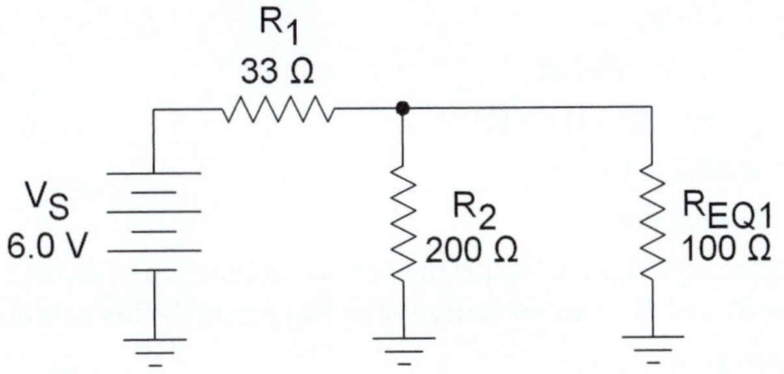

Figure 7-12: First Circuit Simplification for Figure 7-11

The second equivalent resistance R_{EQ2} is the parallel combination $R_2 \parallel R_{EQ1}$. From this

$$R_T = R_1 + R_{EQ2}$$

where

$$R_{EQ2} = R_2 \parallel R_{EQ1}$$
$$= 200\ \Omega \parallel 100\ \Omega$$
$$= 66.7\ \Omega$$

Figure 7-13 shows the second circuit simplification with the parallel combination of R_2 and R_{EQ1} replaced with R_{EQ2}.

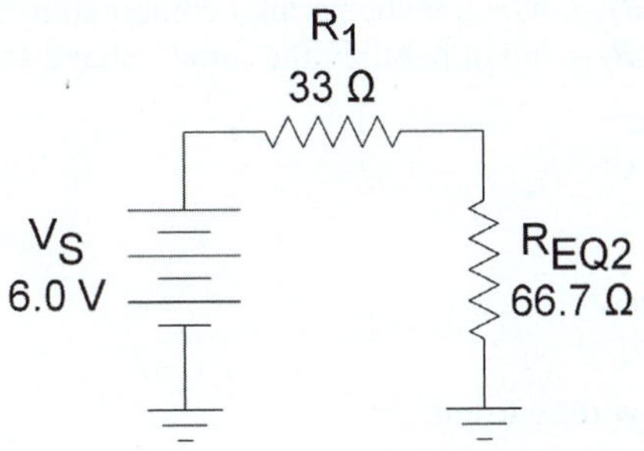

Figure 7-13: Second Circuit Simplification for Figure 7-11

The third (and final) equivalent resistance R_{EQ3} is the series combination $R_1 + R_{EQ2}$. From this

$$R_T = R_{EQ3}$$

where

$$R_{EQ3} = R_1 + R_{EQ2}$$
$$= 33\ \Omega + 66.7\ \Omega$$
$$= 99.7\ \Omega$$

Figure 7-14 shows the final circuit simplification with the series combination of R_1 and R_{EQ2} replaced with R_{EQ3}.

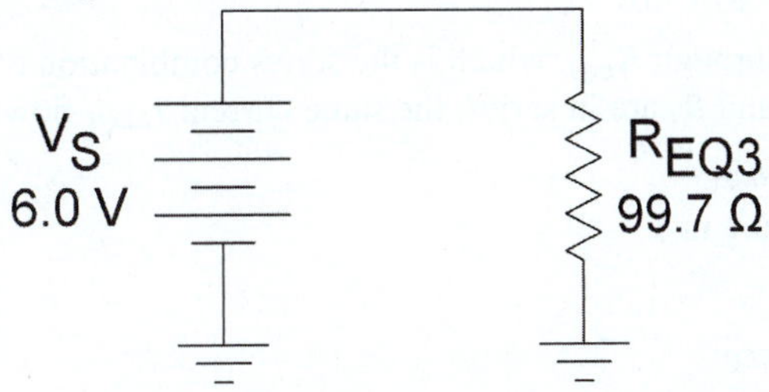

Figure 7-14: Final Circuit Simplification for Figure 7-11

Therefore, $R_T = R_{EQ3} = \mathbf{99.7\ \Omega}$. From Ohm's Law for current, the total current I_T is

$\boldsymbol{I_T} = \boldsymbol{V_S / R_T}$
$= 6.0\ \text{V} / 99.7\ \Omega$
$= (6.0 / 99.7)\ (\text{V} / \Omega)$
$= 60.2 \times 10^{-3}\ \text{A}$
$= \mathbf{60.2\ mA}$

$\boldsymbol{I_T}$ flows through $\boldsymbol{R_{EQ3}}$, which is the series combination of $\boldsymbol{R_1}$ and $\boldsymbol{R_{EQ2}}$ shown in Figure 7-13. Because $\boldsymbol{R_1}$ and $\boldsymbol{R_{EQ2}}$ are in series the same current $\boldsymbol{I_T}$ flows through them. Therefore

$\boldsymbol{I_{R1}} = \boldsymbol{I_T}$
$= \mathbf{60.2\ mA}$

and

$\boldsymbol{I_{REQ2}} = \boldsymbol{I_T}$
$= 60.2\ \text{mA}$

From Ohm's Law for voltage,

$\boldsymbol{V_{REQ1}} = \boldsymbol{I_{REQ2}} \times \boldsymbol{R_{EQ2}}$
$= (60.2 \times 10^{-3}\ \text{A}) \times (66.7\ \Omega)$
$= [(60.2 \times 10^{-3}) \times (66.7)]\ (\text{A} \times \Omega)$
$= 4.01\ \text{V}$

$\boldsymbol{V_{REQ2}}$ is across $\boldsymbol{R_{EQ2}}$, which is the parallel combination of $\boldsymbol{R_2}$ and $\boldsymbol{R_{EQ1}}$ shown in Figure 7-12. Because $\boldsymbol{R_2}$ and $\boldsymbol{R_{EQ1}}$ are in parallel, the same voltage $\boldsymbol{V_{REQ2}}$ is across them. Therefore

$\boldsymbol{V_{R2}} = \boldsymbol{V_{REQ2}}$
$= \mathbf{4.01\ V}$

and

$\boldsymbol{V_{REQ1}} = \boldsymbol{V_{REQ2}}$
$= 4.01\ \text{V}$

From Ohms' Law for current

$\boldsymbol{I_{R2}} = \boldsymbol{V_{R2} / R_2}$
$= 4.01\ \text{V} / 200\ \Omega$
$= (4.01 / 200)\ (\text{V} / \Omega)$
$= 20.1 \times 10^{-3}\ \text{A}$
$= \mathbf{20.1\ mA}$

and

$\boldsymbol{I_{REQ1}} = \boldsymbol{V_{REQ1} / R_{EQ1}}$
$= 4.01\ \text{V} / 100\ \Omega$
$= 40.1 \times 10^{-3}\ \text{A}$
$= 40.1\ \text{mA}$

$\boldsymbol{I_{REQ1}}$ flows through $\boldsymbol{R_{EQ1}}$, which is the series combination of $\boldsymbol{R_3}$ and $\boldsymbol{R_4}$ as shown in Figure 7-11. Because $\boldsymbol{R_3}$ and $\boldsymbol{R_4}$ are in series, the same current $\boldsymbol{I_{REQ1}}$ flows through them. Therefore

$\boldsymbol{I_{R3}} = \boldsymbol{I_{REQ1}}$
$= \mathbf{40.1\ mA}$

and

$\boldsymbol{I_{R4}} = \boldsymbol{I_{REQ1}}$
$= \mathbf{40.1\ mA}$

From Ohm's Law for voltage,

$$\begin{aligned} V_{R3} &= I_{R3} \times R_3 \\ &= (40.1 \times 10^{-3}\ \text{A}) \times (18\ \Omega) \\ &= [(40.1 \times 10^{-3}) \times (18)]\ (\text{A} \times \Omega) \\ &= 723 \times 10^{-3}\ \text{V} \\ &= \mathbf{723\ mV} \end{aligned}$$

and

$$\begin{aligned} V_{R4} &= I_{R4} \times R_4 \\ &= (40.1 \times 10^{-3}\ \text{A}) \times (82\ \Omega) \\ &= [(40.1 \times 10^{-3}) \times (82)]\ (\text{A} \times \Omega) \\ &= \mathbf{3.29\ V} \end{aligned}$$

2. Problem: Determine the total resistance, total current, and all resistor voltages and currents for the circuit in Figure 7-15. Assume that all resistor values are exact.

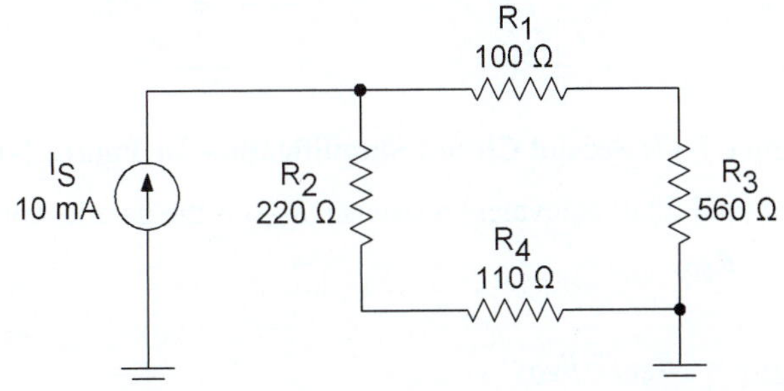

Figure 7-15: Series-Parallel Circuit Analysis Example Circuit 2

Solution: To analyze the circuit, progressively simplify the circuit by replacing series and parallel combinations with their equivalent resistances to determine the total resistance and current. Then, work backwards to calculate the voltage and current for each resistor.

The expression for the circuit in Figure 7-15 is

$$R_T = (R_1 + R_3)\ ||\ (R_2 + R_4)$$

From the order of operations for evaluating the expression, the first equivalent resistance R_{EQ1} is the series combination ($R_1 + R_3$). From this,

$$R_T = R_{EQ1}\ ||\ (R_2 + R_4)$$

where

$$\begin{aligned} R_{EQ1} &= R_1 + R_3 \\ &= 100\ \Omega + 560\ \Omega \\ &= 660\ \Omega \end{aligned}$$

Figure 7-16 shows the first circuit simplification with the series combination of R_1 and R_3 replaced with R_{EQ1}.

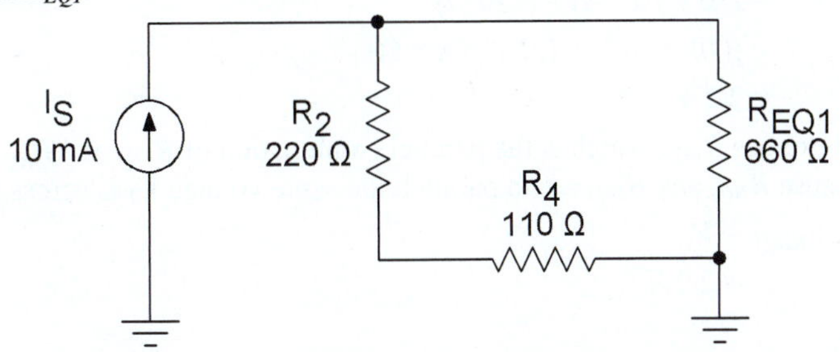

Figure 7-16: First Circuit Simplification for Figure 7-15

The second equivalent resistance R_{EQ2} is the series combination $R_2 + R_4$. From this,

$R_T = R_{EQ1} \parallel R_{EQ2}$

where

$$\begin{aligned} R_{EQ2} &= R_2 + R_4 \\ &= 220\ \Omega + 110\ \Omega \\ &= 330\ \Omega \end{aligned}$$

Figure 7-17 shows the second circuit simplification with the series combination of R_2 and R_4 replaced with R_{EQ2}.

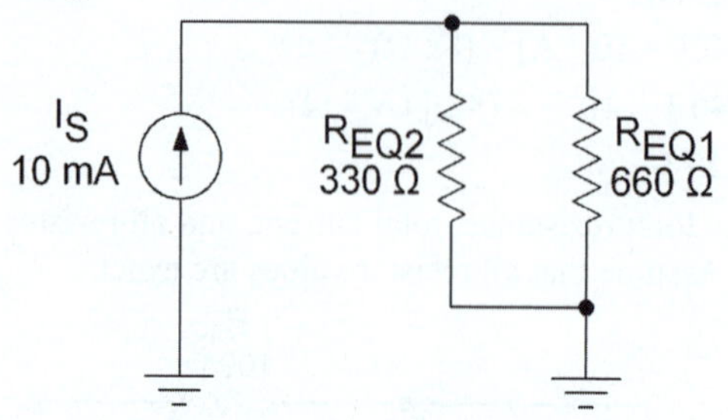

Figure 7-17: Second Circuit Simplification for Figure 7-15

The third (and final) equivalent resistance R_{EQ3} is the parallel combination $R_{EQ1} \parallel R_{EQ2}$. From this,

$R_T = R_{EQ3}$

where

$$\begin{aligned} R_{EQ3} &= R_{EQ1} \parallel R_{EQ2} \\ &= 660\ \Omega \parallel 330\ \Omega \\ &= 220\ \Omega \end{aligned}$$

Figure 7-18 shows the final circuit simplification with the parallel combination of R_{EQ1} and R_{EQ2} replaced with R_{EQ3}.

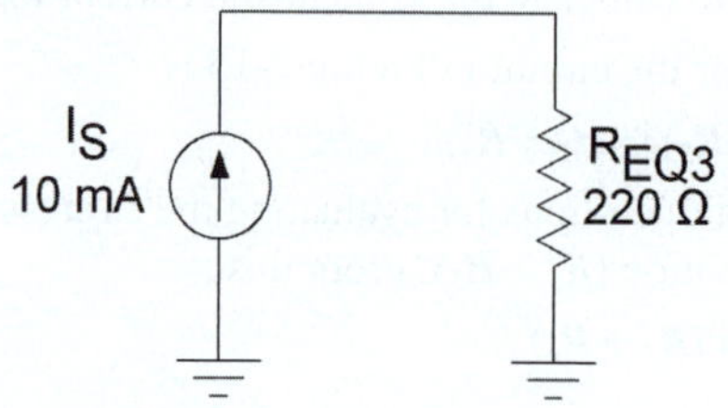

Figure 7-18: Final Circuit Simplification for Figure 7-15

Therefore, $R_T = R_{EQ3} = \mathbf{220\ \Omega}$. From Ohm's Law for voltage, the total voltage V_T is

$$\begin{aligned} V_T &= I_S \times R_T \\ &= 10\text{ mA} \times 220\ \Omega \\ &= (10 \times 10^{-3}\text{ A}) \times (220\ \Omega) \\ &= [(10 \times 10^{-3}) \times (220)]\ (\text{A} \times \Omega) \\ &= \mathbf{2.2\ V} \end{aligned}$$

V_T is across R_{EQ3}, which is the parallel combination of R_{EQ1} and R_{EQ2} shown in Figure 7-17. Because R_{EQ1} and R_{EQ2} are in parallel, the same voltage V_T is across them. Therefore,

$$\begin{aligned} V_{REQ1} &= V_T \\ &= 2.2\text{ V} \end{aligned}$$

and

$V_{REQ2} = V_T$

$= 2.2\ \text{V}$

From Ohm's Law for current,

$$\boldsymbol{I_{REQ1}} = \boldsymbol{V_{REQ1}} / \boldsymbol{R_{EQ1}}$$
$$= (2.2\ \text{V}) / (660\ \Omega)$$
$$= (2.2 / 660)\ (\text{V} / \Omega)$$
$$= 3.33 \times 10^{-3}\ \text{A}$$
$$= 3.33\ \text{mA}$$

and

$$\boldsymbol{I_{REQ2}} = \boldsymbol{V_{REQ2}} / \boldsymbol{R_{EQ2}}$$
$$= (2.2\ \text{V}) / (330\ \Omega)$$
$$= (2.2 / 330)\ (\text{V} / \Omega)$$
$$= 6.67 \times 10^{-3}\ \text{A}$$
$$= 6.67\ \text{mA}$$

$\boldsymbol{I_{REQ2}}$ flows through $\boldsymbol{R_{EQ2}}$, which is the series combination of $\boldsymbol{R_2}$ and $\boldsymbol{R_4}$ shown in Figure 7-16. Because $\boldsymbol{R_2}$ and $\boldsymbol{R_4}$ are in series, the same current $\boldsymbol{I_{REQ2}}$ flows through them.

$$\boldsymbol{I_{R2}} = \boldsymbol{I_{REQ2}}$$
$$= \mathbf{6.67\ mA}$$

and

$$\boldsymbol{I_{R4}} = \boldsymbol{I_{REQ2}}$$
$$= \mathbf{6.67\ mA}$$

From Ohms' Law for voltage,

$$\boldsymbol{V_{R2}} = \boldsymbol{I_{R2}} \times \boldsymbol{R_2}$$
$$= 6.67\ \text{mA} \times 220\ \Omega$$
$$= (6.67 \times 10^{-3}\ \text{A}) \times (220\ \Omega)$$
$$= [(6.67 \times 10^{-3}) \times 220]\ (\text{A} \times \Omega)$$
$$= \mathbf{1.47\ V}$$

and

$$\boldsymbol{V_{R4}} = \boldsymbol{I_{R4}} \times \boldsymbol{R_4}$$
$$= 6.67\ \text{mA} \times 110\ \Omega$$
$$= (6.67 \times 10^{-3}\ \text{A}) \times (110\ \Omega)$$
$$= [(6.67 \times 10^{-3}) \times 110]\ (\text{A} \times \Omega)$$
$$= 733 \times 10^{-3}\ \text{V}$$
$$= \mathbf{733\ mV}$$

$\boldsymbol{I_{REQ1}}$ flows through $\boldsymbol{R_{EQ1}}$, which is the series combination of $\boldsymbol{R_1}$ and $\boldsymbol{R_3}$ as shown in Figure 7-15. Because $\boldsymbol{R_1}$ and $\boldsymbol{R_3}$ are in series, the same current $\boldsymbol{I_{REQ1}}$ flows through them. Therefore,

$$\boldsymbol{I_{R1}} = \boldsymbol{I_{REQ1}}$$
$$= \mathbf{3.33\ mA}$$

and

$$\boldsymbol{I_{R3}} = \boldsymbol{I_{REQ1}}$$
$$= \mathbf{3.33\ mA}$$

From Ohm's Law for voltage,

$$\boldsymbol{V_{R1}} = \boldsymbol{I_{R1}} \times \boldsymbol{R_1}$$
$$= (3.33 \times 10^{-3}\ \text{A}) \times (100\ \Omega)$$

$= [(3.33 \times 10^{-3}) \times (100)]\ (\text{A} \times \Omega)$
$= 333 \times 10^{-3}\ \text{V}$
$= \mathbf{333\ mV}$

and

$V_{R3} = I_{R3} \times R_3$
$= (3.33 \times 10^{-3}\ \text{A}) \times (560\ \Omega)$
$= [(3.33 \times 10^{-3}) \times (560)]\ (\text{A} \times \Omega)$
$= \mathbf{1.87\ V}$

7.1.3 Voltage Divider Loading

1. Problem: What are the unloaded and loaded values of V_{DIV} for the voltage divider circuit of Figure 7-19? Assume that all resistor values are exact.

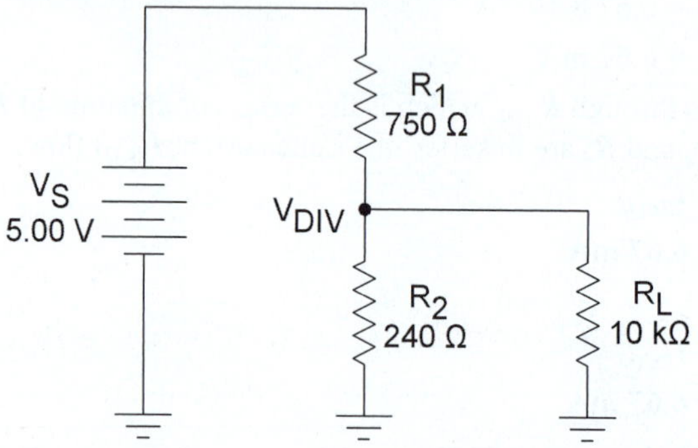

Figure 7-19: Voltage Divider Loading Example Circuit 1

Solution: From the voltage divider formula, the unloaded value V_{DIV} (unloaded) (V_{DIV} with R_L removed) is

V_{DIV} (unloaded) $= V_S \times [R_2 / (R_1 + R_2)]$
$= 5.00\ \text{V} \times [240\ \Omega / (750\ \Omega + 240\ \Omega)]$
$= 5.00\ \text{V} \times \{240\ \Omega / [(750 + 240)\ \Omega]\}$
$= 5.00\ \text{V} \times (240\ \Omega / 990\ \Omega)$
$= 5.00\ \text{V} \times 0.242$
$= \mathbf{1.21\ V}$

With R_L connected the circuit expression is

$R_T = R_1 + R_2 \parallel R_L$

The first step is to simplify the circuit. From the order of operations for evaluating the expression, the first equivalent resistance R_{EQ1} is the parallel combination $R_2 \parallel R_L$. From this,

$R_T = R_1 + R_{EQ1}$

where

$R_{EQ1} = R_2 \parallel R_L$
$= 240\ \Omega \parallel 10\ \text{k}\Omega$
$= 234\ \Omega$

Figure 7-20 shows the first simplified circuit with the parallel combination of R_2 and R_L replaced with R_{EQ1}.

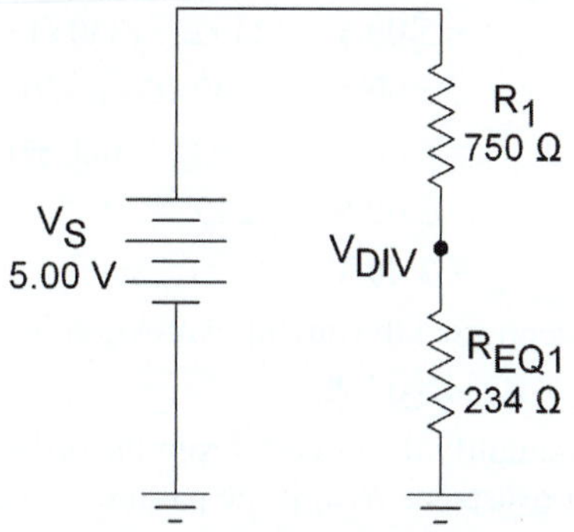

Figure 7-20: Simplified Circuit for Figure 7-19

From the voltage divider formula, the loaded value V_{DIV} (loaded) (V_{DIV} with R_L connected) is

V_{DIV} (loaded) $= V_S \times [R_{EQ1} / (R_1 + R_{EQ1})]$
$= 5.00\text{ V} \times [234\ \Omega / (750\ \Omega + 234\ \Omega)]$
$= 5.00\text{ V} \times \{234\ \Omega / [(750 + 234)\ \Omega]\}$
$= 5.00\text{ V} \times (234\ \Omega / 984\ \Omega)$
$= 5.00\text{ V} \times 0.238$
$= \mathbf{1.19\ V}$

2. Problem: What are the unloaded and loaded values of V_{DIV1} and V_{DIV2} for the voltage divider circuit of Figure 7-21? Assume that all resistor values are exact.

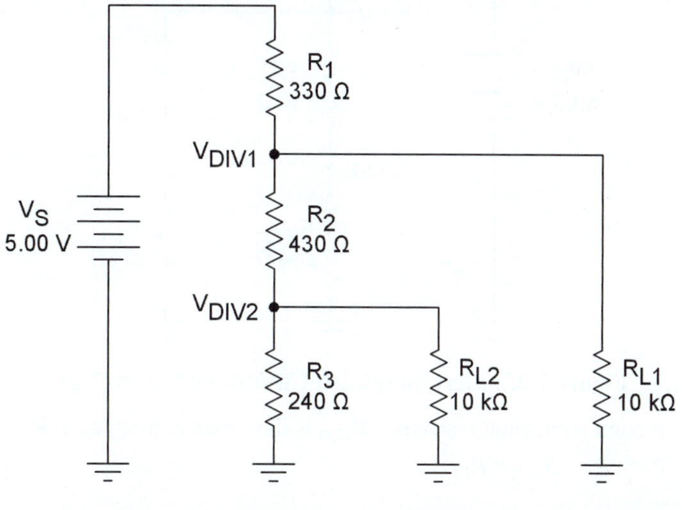

Figure 7-21: Voltage Divider Loading Example Circuit 2

Solution: From the voltage divider formula, the unloaded values V_{DIV1} (unloaded) (V_{DIV1} with R_{L1} removed) and V_{DIV2} (unloaded) (V_{DIV2} with R_{L2} removed) are

V_{DIV1} (unloaded) $= V_S \times [(R_2 + R_3) / (R_1 + R_2 + R_3)]$
$= 5.00\text{ V} \times [(430\ \Omega + 240\ \Omega) / (330\ \Omega + 430\ \Omega + 240\ \Omega)]$
$= 5.00\text{ V} \times \{[(430 + 240)\ \Omega] / [(330 + 430 + 240)\ \Omega]\}$
$= 5.00\text{ V} \times (670\ \Omega / 1000\ \Omega)$
$= 5.00\text{ V} \times 0.670$
$= \mathbf{3.35\ V}$

and

V_{DIV2} (unloaded) $= V_S \times [R_3 / (R_1 + R_2 + R_3)]$

$= 5.00\ \text{V} \times [240\ \Omega / (330\ \Omega + 430\ \Omega + 240\ \Omega)]$

$= 5.00\ \text{V} \times \{(240\ \Omega) / [(330 + 430 + 240)\ \Omega]\}$

$= 5.00\ \text{V} \times (240\ \Omega / 1000\ \Omega)$

$= 5.00\ \text{V} \times 0.240$

$= \mathbf{2.40\ V}$

With R_{L1} and R_{L2} connected the circuit expression is

$R_T = R_1 + (R_2 + R_3 \parallel R_{L2}) \parallel R_{L1}$

The first step is to simplify the circuit. From the order of operations for evaluating the expression, the first equivalent resistance R_{EQ1} is the parallel combination $R_3 \parallel R_{L2}$. From this

$R_T = R_1 + (R_2 + R_{EQ1}) \parallel R_{L1}$

where

$R_{EQ1} = R_3 \parallel R_{L2}$

$= 240\ \Omega \parallel 10\ \text{k}\Omega$

$= 234\ \Omega$

Figure 7-22 shows the first simplified circuit with the parallel combination of R_2 and R_L replaced with R_{EQ1}.

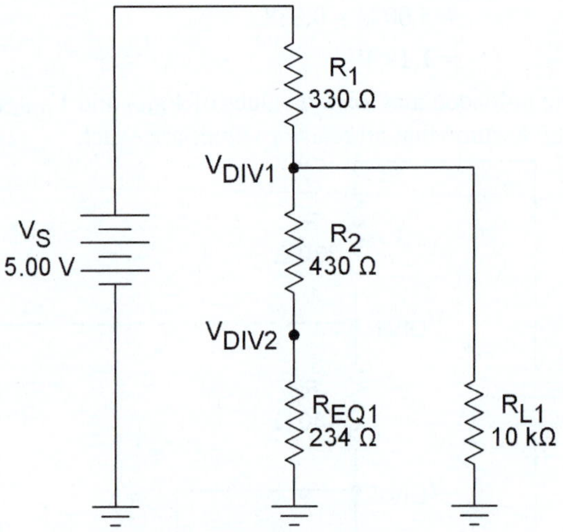

Figure 7-22: First Simplified Circuit for Figure 7-21

The second equivalent resistance R_{EQ2} is the series combination $R_2 + R_{EQ1}$. From this

$R_T = R_1 + R_{EQ2} \parallel R_{L1}$

where

$R_{EQ2} = R_2 + R_{EQ1}$

$= 430\ \Omega + 234\ \Omega$

$= 664\ \Omega$

Figure 7-23 shows the second simplified circuit with the series combination of R_2 and R_{EQ1} replaced with R_{EQ2}.

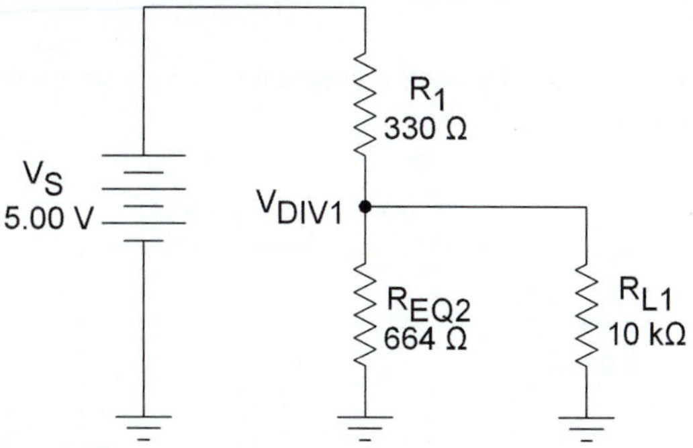

Figure 7-23: Second Simplified Circuit for Figure 7-21

The third equivalent resistance R_{EQ3} is the parallel combination $R_{EQ2} \parallel R_{L1}$. From this

$$R_T = R_1 + R_{EQ3}$$

where

$$\begin{aligned} R_{EQ3} &= R_{EQ2} \parallel R_{L1} \\ &= 664\ \Omega \parallel 10\ \text{k}\Omega \\ &= 623\ \Omega \end{aligned}$$

Figure 7-24 shows the third simplified circuit with the parallel combination of R_{EQ2} and R_{L1} replaced with R_{EQ3}.

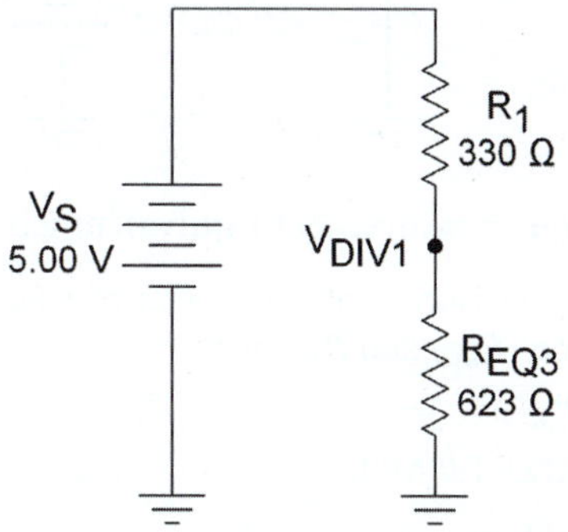

Figure 7-24: Third Simplified Circuit for Figure 7-21

The loaded value V_{DIV1} (loaded) (V_{DIV1} with R_{L1} connected) is

$$\begin{aligned} V_{DIV1}\text{ (loaded)} &= V_S \times [R_{EQ3} / (R_1 + R_{EQ3})] \\ &= 5.00\ \text{V} \times [623\ \Omega / (330\ \Omega + 623\ \Omega)] \\ &= 5.00\ \text{V} \times \{623\ \Omega / [(330 + 623)\ \Omega]\} \\ &= 5.00\ \text{V} \times (623\ \Omega / 953\ \Omega) \\ &= 5.00\ \text{V} \times 0.654 \\ &= \mathbf{3.26\ V} \end{aligned}$$

Figure 7-22 shows that V_{DIV1} (loaded) is across the parallel combination of R_{L1} and the series combination of R_2 and R_{EQ1}. From the voltage divider formula, the loaded value V_{DIV2} (loaded) (V_{DIV2} with R_{L2} connected) is

$$\begin{aligned} V_{DIV2}\text{ (loaded)} &= V_{DIV1}\text{ (loaded)} \times [R_{EQ1} / (R_2 + R_{EQ1})] \\ &= 3.26\ \text{V} \times [234\ \Omega / (430\ \Omega + 234\ \Omega)] \\ &= 3.26\ \text{V} \times \{234\ \Omega / [(430 + 234)\ \Omega]\} \\ &= 3.26\ \text{V} \times (234\ \Omega / 664\ \Omega) \\ &= 3.26\ \text{V} \times 0.353 \\ &= \mathbf{1.15\ V} \end{aligned}$$

7.1.4 Voltmeter Loading

1. Problem: What values of V_{R1} and V_{R2} will a voltmeter with an internal resistance R_M of 1.0 MΩ measure for the circuit in Figure 7-25?

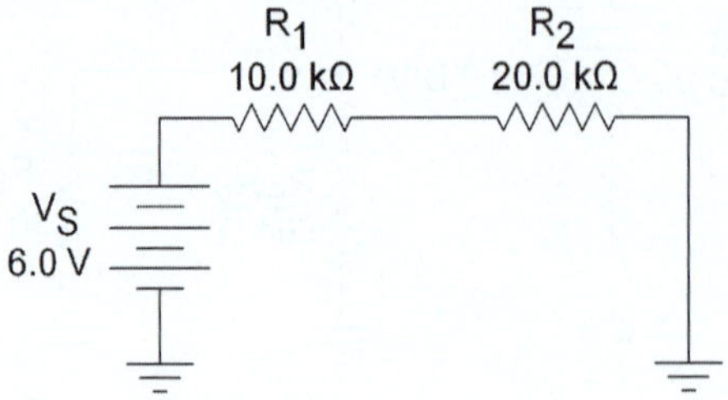

Figure 7-25: Voltmeter Loading Example Circuit 1

Solution: Refer to Figure 7-26, which shows the equivalent circuits for the voltmeter measurements.

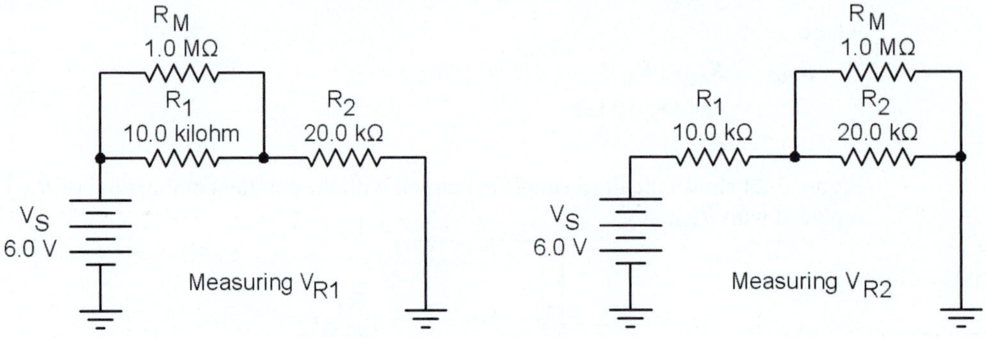

Figure 7-26: Voltmeter Measurement Equivalent Circuits for Figure 7-25

For the V_{R1} and V_{R2} measurements, the meter resistance in parallel with R_1 and R_2 creates equivalent resistances R_{EQ1} and R_{EQ2} where

$$\begin{aligned} R_{EQ1} &= R_1 \parallel R_M \\ &= 10.0\ \text{k}\Omega \parallel 1.0\ \text{M}\Omega \\ &= 9.90\ \text{k}\Omega \end{aligned}$$

and

$$\begin{aligned} R_{EQ2} &= R_2 \parallel R_M \\ &= 20.0\ \text{k}\Omega \parallel 1.0\ \text{M}\Omega \\ &= 19.6\ \text{k}\Omega \end{aligned}$$

Figure 7-27 shows the simplified circuits for the V_{R1} and V_{R2} measurements.

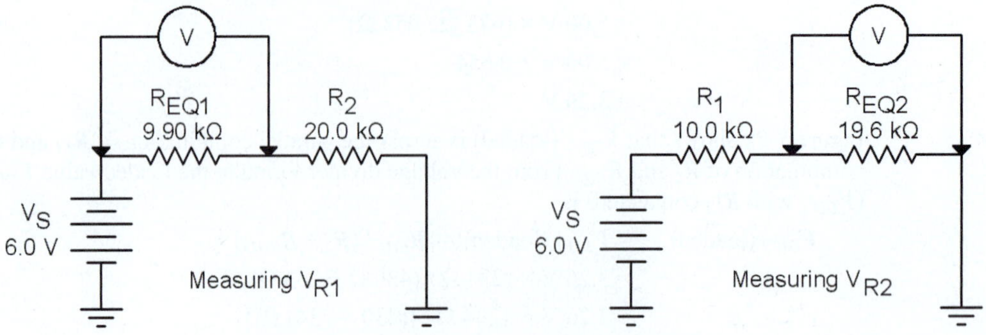

Figure 7-27: Simplified Circuits for Figure 7-26

From the voltage divider formula, the measured values of V_{R1} and V_{R2} with meter loading are

$$V_{R1} = V_S \times [R_{EQ1} / (R_{EQ1} + R_2)]$$
$$= 6.0\text{ V} \times [9.90\text{ k}\Omega / (9.90\text{ k}\Omega + 20.0\text{ k}\Omega)]$$
$$= 6.0\text{ V} \times \{(9.90 \times 10^3\ \Omega) / [(9.90 \times 10^3\ \Omega) + (20.0 \times 10^3\ \Omega)]\}$$
$$=6.0\text{ V} \times \{9.90 \times 10^3\ \Omega / [(9.90 \times 10^3 + 20.0 \times 10^3)\ \Omega]\}$$
$$= 6.0\text{ V} \times (9.90 \times 10^3\ \Omega / 20.9 \times 10^3\ \Omega)$$
$$= 6.0\text{ V} \times 0.331$$
$$= \mathbf{1.99\text{ V}}$$

and

$$V_{R2} = V_S \times [R_{EQ2} / (R_1 + R_{EQ2})]$$
$$= 6.0\text{ V} \times [19.6\text{ k}\Omega / (10.0\text{ k}\Omega + 19.6\text{ k}\Omega)]$$
$$= 6.0\text{ V} \times \{(19.6 \times 10^3\ \Omega) / [(10.0 \times 10^3\ \Omega) + (19.6 \times 10^3\ \Omega)]\}$$
$$=6.0\text{ V} \times \{19.6 \times 10^3\ \Omega / [(10.0 \times 10^3 + 19.6 \times 10^3)\ \Omega]\}$$
$$= 6.0\text{ V} \times (19.6 \times 10^3\ \Omega / 29.6 \times 10^3\ \Omega)$$
$$= 6.0\text{ V} \times 0.662$$
$$= \mathbf{3.97\text{ V}}$$

2. Problem: What will the measured values for V_{R1} and V_{R2} be for the circuit in Figure 7-25 if two voltmeters are used at the same time? Assume that $R_{M1} = R_{M2} = 1.0\text{ M}\Omega$ and that all resistor values are exact.

Solution: Refer to Figure 7-28, which shows the equivalent circuit for the simultaneous voltmeter measurements.

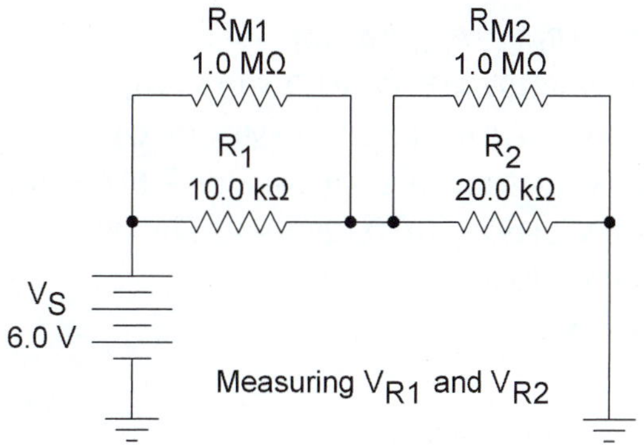

Figure 7-28: Two Voltmeter Measurement Equivalent Circuit for Figure 7-25

The meter resistances in parallel with R_1 and R_2 creates the equivalent resistances R_{EQ1} and R_{EQ2} where

$$R_{EQ1} = R_1 \parallel R_{M1}$$
$$= 10.0\text{ k}\Omega \parallel 1.0\text{ M}\Omega$$
$$= 9.90\text{ k}\Omega$$

and

$$R_{EQ2} = R_2 \parallel R_{M1}$$
$$= 20.0\text{ k}\Omega \parallel 1.0\text{ M}\Omega$$
$$= 19.6\text{ k}\Omega$$

Figure 7-29 shows the simplified circuit for the simultaneous V_{R1} and V_{R2} measurements.

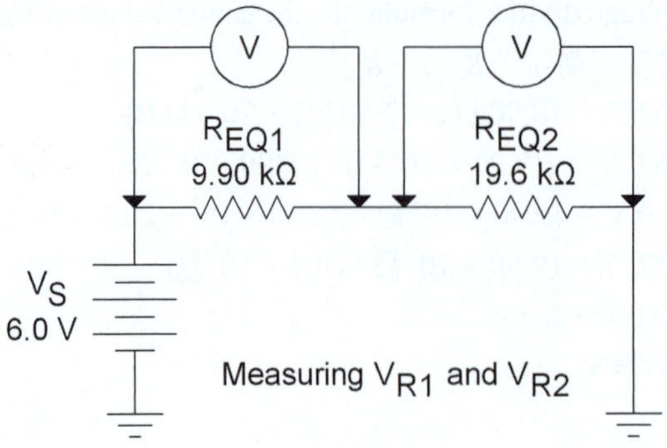

Figure 7-29: Simplified Circuit for Figure 7-28

From the voltage divider formula, the measured values of V_{R1} and V_{R2} with meter loading are

$V_{R1} = V_S \times [R_{EQ1} / (R_{EQ1} + R_{EQ2})]$
$= 6.0\text{ V} \times [9.90\text{ k}\Omega / (9.90\text{ k}\Omega + 19.6\text{ k}\Omega)]$
$= 6.0\text{ V} \times \{(9.90 \times 10^3\ \Omega) / [(9.90 \times 10^3\ \Omega) + (19.6 \times 10^3\ \Omega)]\}$
$=6.0\text{ V} \times \{9.90 \times 10^3\ \Omega / [(9.90 \times 10^3 + 19.6 \times 10^3)\ \Omega]\}$
$= 6.0\text{ V} \times (9.90 \times 10^3\ \Omega / 29.5 \times 10^3\ \Omega)$
$= 6.0\text{ V} \times 0.336$
$=$ **2.01 V**

and

$V_{R2} = V_S \times [R_{EQ2} / (R_{EQ1} + R_{EQ2})]$
$= 6.0\text{ V} \times [19.6\text{ k}\Omega / (9.90\text{ k}\Omega + 19.6\text{ k}\Omega)]$
$= 6.0\text{ V} \times \{(19.6 \times 10^3\ \Omega) / [(9.90 \times 10^3\ \Omega) + (19.6 \times 10^3\ \Omega)]\}$
$=6.0\text{ V} \times \{19.6 \times 10^3\ \Omega / [(9.90 \times 10^3 + 19.6 \times 10^3)\ \Omega]\}$
$= 6.0\text{ V} \times (19.6 \times 10^3\ \Omega / 29.5 \times 10^3\ \Omega)$
$= 6.0\text{ V} \times 0.664$
$=$ **3.99 V**

7.2 Advanced

7.2.1 Series-Parallel Resistive Circuit Analysis

1. Problem: Determine the values of V_{OUT} for the circuit in Figure 7-30 for both positions of switch SW_1.

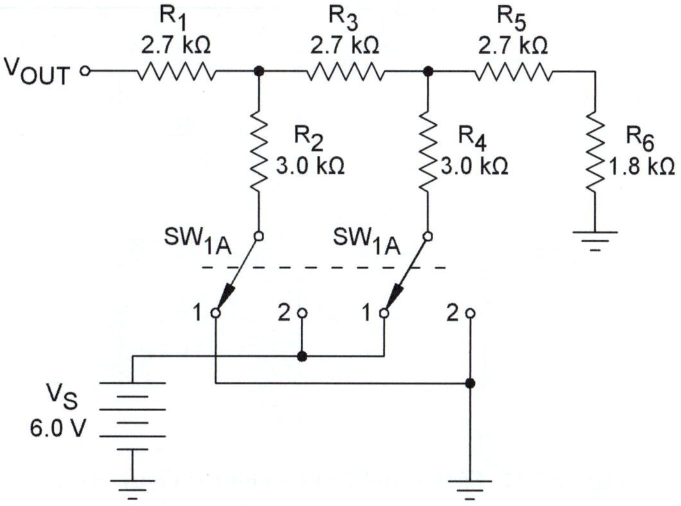

Figure 7-30: Series-Parallel Resistive Circuit Analysis Example Circuit 3

Solution: Figure 7-31 shows the equivalent circuit for Figure 7-30 with SW_1 in position 1 and a redrawn version of the equivalent circuit.

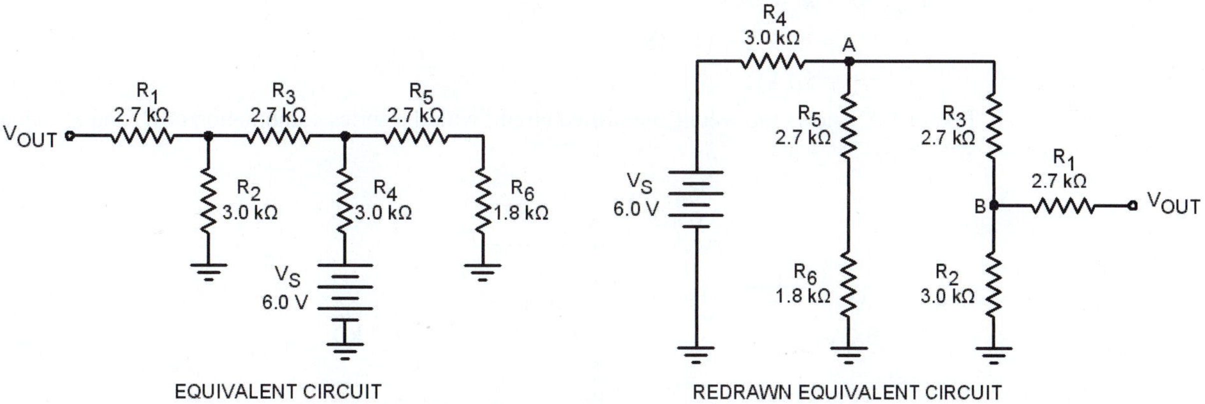

Figure 7-31: Circuit in Figure 7-30 for Switch Position 1

The redrawn circuit in Figure 7-31 shows the series and parallel relationships more clearly. From the redrawn circuit, the expression for the circuit is

$$R_T = R_4 + [(R_5 + R_6) \parallel (R_3 + R_2)]$$

Note that only one end of R_1 connects to the circuit, so that no current flows through R_1. Consequently, R_1 drops no voltage and $V_{OUT} = V_B$.

The first equivalent resistance R_{EQ1} is the series combination ($R_5 + R_6$). From this

$$R_T = R_4 + [R_{EQ1} \parallel (R_3 + R_2)]$$

where

$$R_{EQ1} = R_5 + R_6$$
$$= 2.7\text{ k}\Omega + 1.8\text{ k}\Omega$$
$$= 4.5\text{ k}\Omega$$

Figure 7-32 shows the first simplified circuit with the series combination of R_5 and R_6 replaced with R_{EQ1}. Note that R_1 is not shown as it has no part in the circuit analysis.

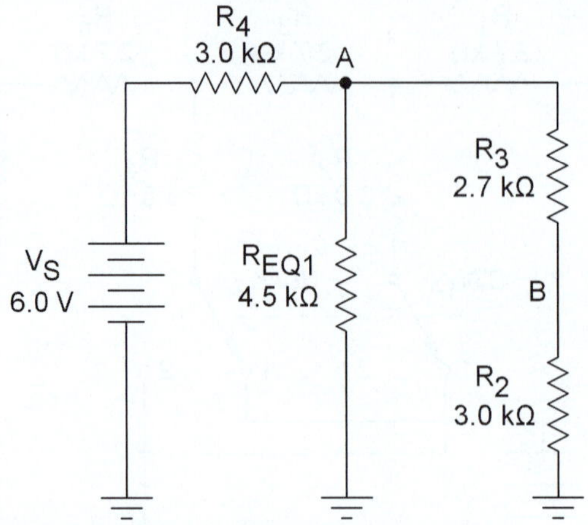

Figure 7-32: First Simplified Circuit for Figure 7-31

The second equivalent resistance R_{EQ2} is the series combination ($R_3 + R_2$). From this

$$R_T = R_4 + R_{EQ1} \parallel R_{EQ2}$$

where

$$R_{EQ2} = R_3 + R_2$$
$$= 2.7\ k\Omega + 3.0\ k\Omega$$
$$= 5.7\ k\Omega$$

Figure 7-33 shows the second simplified circuit with the series combination of R_3 and R_2 replaced with R_{EQ2}.

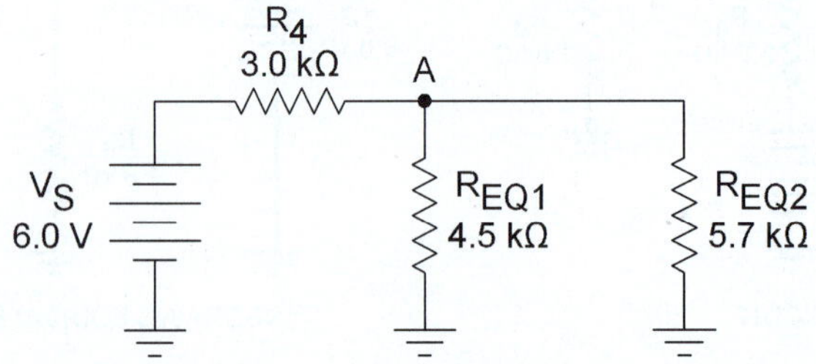

Figure 7-33: Second Simplified Circuit for Figure 7-31

The third equivalent resistance R_{EQ3} is the parallel combination $R_{EQ1} \parallel R_{EQ2}$. From this

$$R_T = R_4 + R_{EQ3}$$

where

$$R_{EQ3} = R_{EQ1} \parallel R_{EQ2}$$
$$= 4.5\ k\Omega \parallel 5.7\ k\Omega$$
$$= 2.51 \times 10^3\ \Omega$$
$$= 2.51\ k\Omega$$

Figure 7-24 shows the third simplified circuit with the parallel combination of R_{EQ1} and R_{EQ2} replaced with R_{EQ3}.

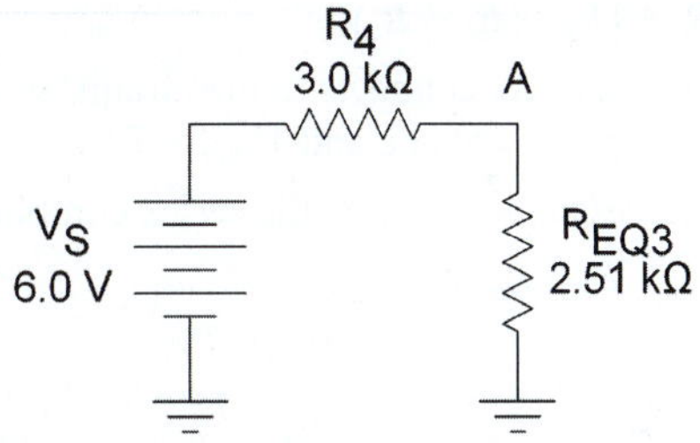

Figure 7-34: Third Simplified Circuit for Figure 7-31

From the voltage divider formula, the voltage V_A at point A is

$$\begin{aligned} V_A &= V_S \times [R_{EQ3} / (R_4 + R_{EQ3})] \\ &= 6.0\text{ V} \times [2.51\text{ k}\Omega / (3.0\text{ k}\Omega + 2.51\text{ k}\Omega)] \\ &= 6.0\text{ V} \times \{(2.51 \times 10^3\ \Omega) / [(3.0 \times 10^3\ \Omega) + (2.51 \times 10^3\ \Omega)]\} \\ &= 6.0\text{ V} \times \{(2.51 \times 10^3\ \Omega) / [(3.0 \times 10^3 + 2.51 \times 10^3)\ \Omega]\} \\ &= 6.0\text{ V} \times (2.51 \times 10^3\ \Omega) / (5.51 \times 10^3\ \Omega) \\ &= 6.0\text{ V} \times 0.456 \\ &= 2.74\text{ V} \end{aligned}$$

Figure 7-32 shows that V_A is across the series combination of R_3 and R_2. From the voltage divider formula, the voltage V_B at point B, which is equal to V_{OUT}, is

$$\begin{aligned} V_B &= V_A \times [R_2 / (R_3 + R_2)] \\ &= 2.74\text{ V} \times [3.0\text{ k}\Omega / (2.7\text{ k}\Omega + 3.0\text{ k}\Omega)] \\ &= 2.74\text{ V} \times \{(3.0 \times 10^3\ \Omega) / [(2.7 \times 10^3\ \Omega) + (3.0 \times 10^3\ \Omega)]\} \\ &= 2.74\text{ V} \times \{(3.0 \times 10^3\ \Omega) / [(2.7 \times 10^3 + 3.0 \times 10^3)\ \Omega]\} \\ &= 2.74\text{ V} \times (3.0 \times 10^3\ \Omega) / (5.7 \times 10^3\ \Omega) \\ &= 2.74\text{ V} \times 0.523 \\ &= 1.44\text{ V} \end{aligned}$$

Figure 7-35 shows the equivalent circuit for Figure 7-30 with SW_1 in position 2 and a redrawn version of the equivalent circuit.

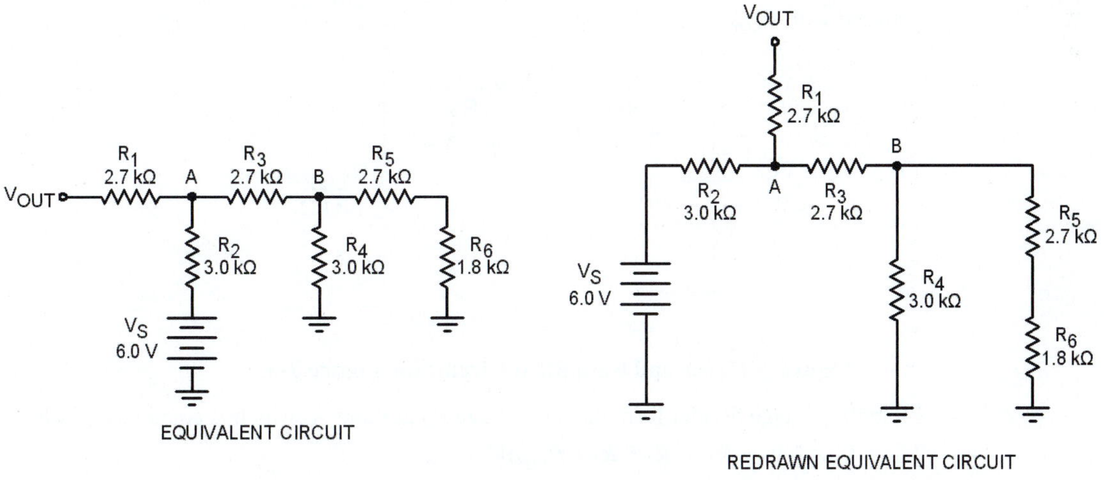

Figure 7-35: Circuit in Figure 7-30 for Switch Position 2

The redrawn circuit in Figure 7-35 shows the series and parallel relationships more clearly. From the redrawn circuit, the expression for the circuit is

$R_T = R_2 + R_3 + [R_4 \parallel (R_5 + R_6)]$

Note that only one end of R_1 connects to the circuit, so that no current flows through R_1. Consequently, R_1 drops no voltage and $V_{OUT} = V_A$.

The first equivalent resistance R_{EQ1} is the series combination $(R_5 + R_6)$. From this,

$R_T = R_2 + R_3 + R_4 \parallel R_{EQ1}$

where

$$\begin{aligned} R_{EQ1} &= R_5 + R_6 \\ &= 2.7\ \text{k}\Omega + 1.8\ \text{k}\Omega \\ &= 4.5\ \text{k}\Omega \end{aligned}$$

Figure 7-36 shows the first simplified circuit with the series combination of R_5 and R_6 replaced with R_{EQ1}. Note that R_1 is not shown, as it has no part in the circuit analysis.

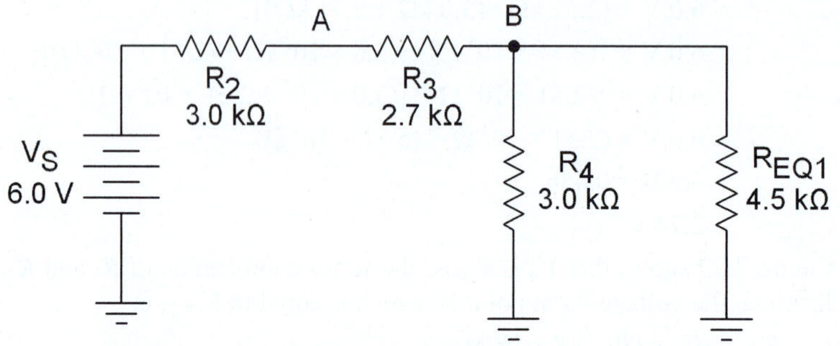

Figure 7-36: First Simplified Circuit for Figure 7-35

The second equivalent resistance R_{EQ2} is the parallel combination $R_4 \parallel R_{EQ1}$. From this,

$R_T = R_2 + R_3 + R_{EQ2}$

where

$$\begin{aligned} R_{EQ2} &= R_4 \parallel R_{EQ1} \\ &= 3.0\ \text{k}\Omega \parallel 4.5\ \text{k}\Omega \\ &= 1.8\ \text{k}\Omega \end{aligned}$$

Figure 7-37 shows the second simplified circuit with the parallel combination of R_4 and R_{EQ1} replaced with R_{EQ2}.

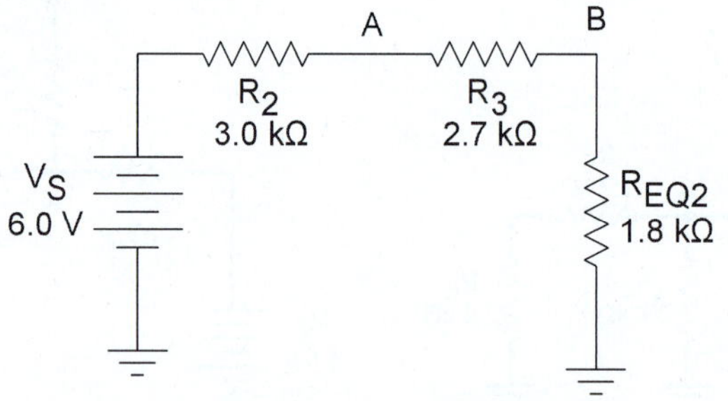

Figure 7-37: Second Simplified Circuit for Figure 7-35

From the voltage divider formula, the voltage V_A at point A, which is equal to V_{OUT}, is

$$\begin{aligned} V_A &= V_S \times [(R_2 + R_3) / (R_2 + R_3 + R_{EQ2})] \\ &= 6.0\ \text{V} \times [(3.0\ \text{k}\Omega + 2.7\ \text{k}\Omega) / (3.0\ \text{k}\Omega + 2.7\ \text{k}\Omega + 1.8\ \text{k}\Omega)] \\ &= 6.0\ \text{V} \times \{(3.0 \times 10^3\ \Omega + 2.7 \times 10^3\ \Omega) / [3.0 \times 10^3\ \Omega + 2.7 \times 10^3\ \Omega + 1.8 \times 10^3\ \Omega]\} \\ &= 6.0\ \text{V} \times \{[(3.0 \times 10^3 + 2.7 \times 10^3)\ \Omega] / [(3.0 \times 10^3 + 2.7 \times 10^3 + 1.8 \times 10^3)\ \Omega]\} \end{aligned}$$

$= 6.0\text{ V} \times (5.7 \times 10^3\ \Omega) / (6.5 \times 10^3\ \Omega)$

$= 6.0\text{ V} \times 0.877$

$= 5.26\text{ V}$

For SW_1 in position 1, V_{OUT} = **1.4 V** and for SW_1 in position 2, V_{OUT} = **5.3 V**.

2. Problem: If R_1, R_2, and R_3 are all 1% tolerance resistors for the Wheatstone bridge in Figure 7-38, what is the maximum possible error for the measured value of R_X when the bridge is balanced (i.e., $V_A = V_B$)?

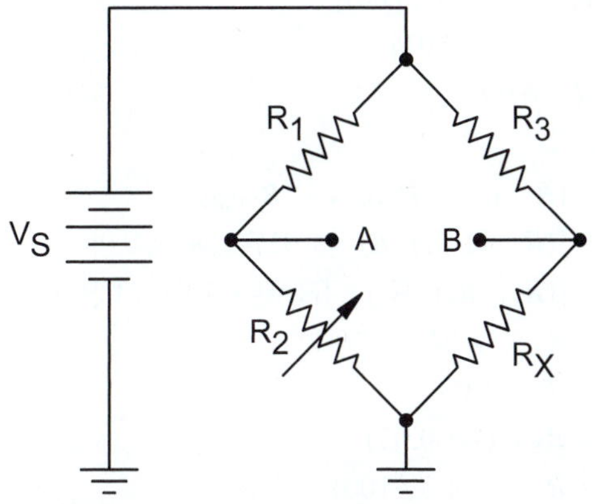

Figure 7-38: Series-Parallel Resistive Circuit Analysis Example Circuit 4

Solution: First, determine the value of R_X. From the voltage divider formula,

$$V_A = V_S \times [R_2 / (R_1 + R_2)]$$

and

$$V_B = V_S \times [R_X / (R_3 + R_X)]$$

When R_2 is adjusted so that the bridge is balanced, $V_A = V_B$.

$$V_A = V_S \times [R_X / (R_3 + R_X)]$$

Equate the V_A terms and solve for R_X.

$$V_S \times [R_2 / (R_1 + R_2)] = V_S \times [R_X / (R_3 + R_X)]$$

$$\{V_S \times [R_2 / (R_1 + R_2)] / V_S\} = \{V_S \times [R_X / (R_3 + R_X)] / V_S\}$$

$$R_2 / (R_1 + R_2) = R_X / (R_3 + R_X)$$

$$[R_2 / (R_1 + R_2)] \times (R_3 + R_X) = [R_X / (R_3 + R_X)]\ (R_3 + R_X)$$

$$[R_2 \times (R_3 + R_X)] / (R_1 + R_2) = R_X$$

$$(R_2 \times R_3 + R_2 \times R_X) / (R_1 + R_2) = R_X$$

$$[(R_2 \times R_3 + R_2 \times R_X) / (R_1 + R_2)] \times (R_1 + R_2) = R_X \times (R_1 + R_2)$$

$$R_2 \times R_3 + R_2 \times R_X = R_X \times R_1 + R_X \times R_2$$

$$(R_2 \times R_3 + R_2 \times R_X) - R_2 \times R_X = (R_1 \times R_X + R_2 \times R_X) - R_2 \times R_X$$

$$(R_2 \times R_3) / R_1 = (R_1 \times R_X) / R_1$$

$$(R_2 \times R_3) / R_1 = R_X$$

The measured value of R_X will be greatest when R_2 and R_3 are maximum and R_1 is minimum. Conversely, the measured value of R_X will be the least when R_2 and R_3 are minimum and R_1 is maximum. For a 1% tolerance resistor

$$R_{MAX} = R + R \times 1\%$$

$$= (R \times 1) + (R \times 1\%)$$

$$= R \times (1 + 1\%)$$

$$= R \times [1 + (1 / 100)]$$

$= \boldsymbol{R} \times [1 + 0.01]$

$= \boldsymbol{R} \times 1.01$

and

$\boldsymbol{R}_{MIN} = \boldsymbol{R} - \boldsymbol{R} \times 1\%$

$= (\boldsymbol{R} \times 1) + -(\boldsymbol{R} \times 1\%)$

$= \boldsymbol{R} \times (1 + -1\%)$

$= \boldsymbol{R} \times [1 + -(1 / 100)]$

$= \boldsymbol{R} \times [1 + -0.01]$

$= \boldsymbol{R} \times 0.99$

Therefore,

$\boldsymbol{R}_{X\,(\mathrm{MAX})} = (\boldsymbol{R}_{2\,(\mathrm{MAX})} \times \boldsymbol{R}_{3\,(\mathrm{MAX})}) / \boldsymbol{R}_{1\,(\mathrm{MIN})}$

$= [(\boldsymbol{R}_2 \times 1.01)\,(\boldsymbol{R}_3 \times 1.01)] / (\boldsymbol{R}_1 \times 0.99)$

$= [(\boldsymbol{R}_2 \times \boldsymbol{R}_3) / \boldsymbol{R}_1] \times [(1.01 \times 1.01) / 0.99]$

$= \boldsymbol{R}_X \times (1.02 / 0.99)$

$= \boldsymbol{R}_X \times 1.03$

$= \boldsymbol{R}_X \times (1 + 0.03)$

$= \boldsymbol{R}_X \times (1 + 3 / 100)$

$= \boldsymbol{R}_X \times (1 + 3\%)$

$= \boldsymbol{R}_X + \boldsymbol{R}_X \times 3\%$

and

$\boldsymbol{R}_{X\,(\mathrm{MIN})} = (\boldsymbol{R}_{2\,(\mathrm{MIN})} \times \boldsymbol{R}_{3(\mathrm{MIN})}) / \boldsymbol{R}_{1\,(\mathrm{MAX})}$

$= [(\boldsymbol{R}_2 \times 0.99)\,(\boldsymbol{R}_3 \times 0.99)] / (\boldsymbol{R}_1 \times 1.01)$

$= [(\boldsymbol{R}_2 \times \boldsymbol{R}_3) / \boldsymbol{R}_1] \times [(0.99 \times 0.99) / 1.01]$

$= \boldsymbol{R}_X \times (0.98 / 1.01)$

$= \boldsymbol{R}_X \times 0.97$

$= \boldsymbol{R}_X \times (1 - 0.03)$

$= \boldsymbol{R}_X \times (1 - 3 / 100)$

$= \boldsymbol{R}_X \times (1 - 3\%)$

$= \boldsymbol{R}_X - \boldsymbol{R}_X \times 3\%$

The measured value of $\boldsymbol{R}_X$ can have a maximum error of **±3%** for a Wheatstone bridge with 1% tolerance resistors.

7.2.2 Voltage Divider Loading

1. Problem: If the voltmeter in Figure 7-39 measures 3.75 V and $\boldsymbol{R}_M = 1\ \mathrm{M\Omega}$, what is the actual (unloaded) value of $\boldsymbol{V}_{DIV}$? Assume that resistor values are exact.

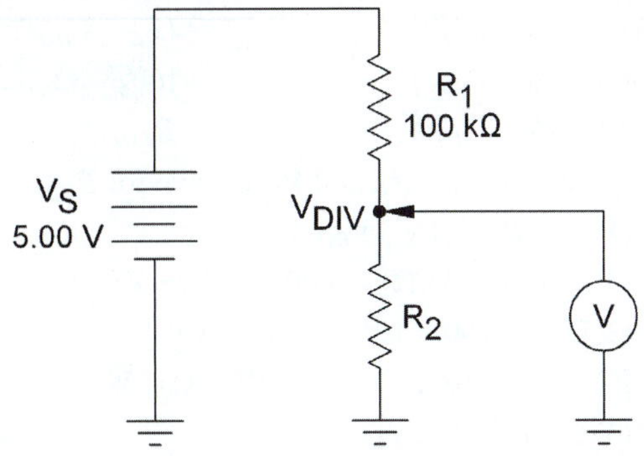

Figure 7-39: Voltage Divider Loading Example Circuit 3

Solution: Refer to Figure 7-40, which shows the equivalent circuit for Figure 7-39.

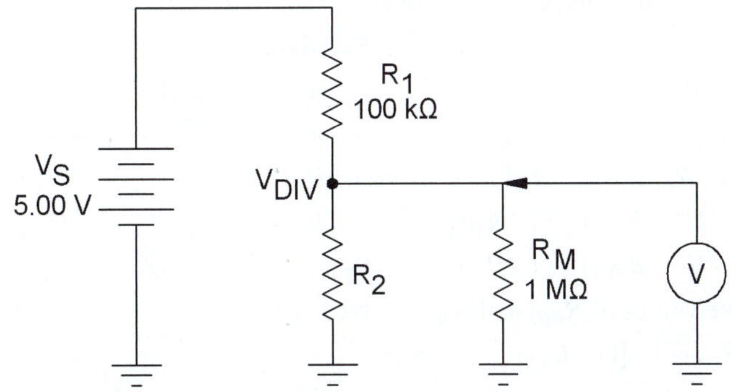

Figure 7-40: Equivalent Circuit for Figure 7-39

The expression for the equivalent circuit is

$\boldsymbol{R_T = R_1 + R_2 \parallel R_M}$

The first equivalent resistance $\boldsymbol{R_{EQ1}}$ is the parallel combination $\boldsymbol{R_2 \parallel R_M}$. From this

$\boldsymbol{R_T = R_1 + R_{EQ1}}$

Figure 7-41 shows the simplified circuit with the parallel combination of $\boldsymbol{R_2}$ and $\boldsymbol{R_M}$ replaced with $\boldsymbol{R_{EQ1}}$.

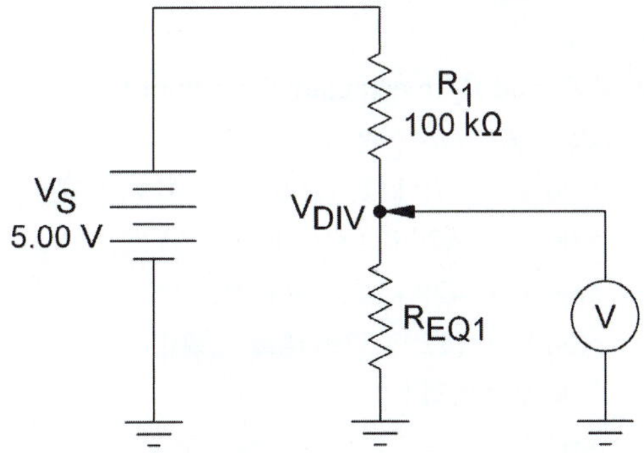

Figure 7-41: Simplified Circuit for Figure 7-40

Use the voltage divider formula to solve for $\boldsymbol{R_{EQ1}}$.

$$V_{DIV} = V_S \times [R_{EQ1} / (R_1 + R_{EQ1})]$$

$$V_{DIV} \times (R_1 + R_{EQ1}) = V_S \times [R_{EQ1} / (R_1 + R_{EQ1})] \times (R_1 + R_{EQ1})$$

$$V_{DIV} \times R_1 + V_{DIV} \times R_{EQ1} = V_S \times R_{EQ1}$$

$$(V_{DIV} \times R_1 + V_{DIV} \times R_{EQ1}) - V_{DIV} \times R_{EQ1} = V_S \times R_{EQ1} - V_{DIV} \times R_{EQ1}$$

$$V_{DIV} \times R_1 = (V_S - V_{DIV}) \times R_{EQ1}$$
$$(V_{DIV} \times R_1) / (V_S - V_{DIV}) = [(V_S - V_{DIV}) \times R_{EQ1}] / (V_S - V_{DIV})$$
$$(V_{DIV} \times R_1) / (V_S - V_{DIV}) = R_{EQ1}$$

Substitute in values for V_{DIV}, R_1, and V_S to solve for R_{EQ1}.

$$\begin{aligned} R_{EQ1} &= (V_{DIV} \times R_1) / (V_S - V_{DIV}) \\ &= (3.75\ \text{V} \times 100\ \text{k}\Omega) / (5.00\ \text{V} - 3.75\ \text{V}) \\ &= (3.75\ \text{V} \times 100 \times 10^3\ \Omega) / (1.25\ \text{V}) \\ &= [(3.75 \times 100 \times 10^3) / 1.25]\ (\text{V} \times \Omega) / \text{V}) \\ &= (375 \times 10^3 / 1.25)\ \Omega \\ &= 300 \times 10^3\ \Omega \\ &= 300\ \text{k}\Omega \end{aligned}$$

Use the value of R_{EQ1} to solve for R_2.

$$R_{EQ1} = R_2 \parallel R_M$$
$$1 / R_{EQ1} = (1 / R_2) + (1 / R_M)$$
$$(1 / R_{EQ1}) + -(1 / R_M) = (1 / R_2) + (1 / R_M) + -(1 / R_M)$$
$$(1 / R_{EQ1}) - (1 / R_M) = (1 / R_2)$$
$$1 / [(1 / R_{EQ1}) - (1 / R_M)] = 1 / (1 / R_2)$$
$$1 / [(1 / R_{EQ1}) - (1 / R_M)] = R_2$$

Use the values of R_{EQ1} and R_M to solve for R_2.

$$\begin{aligned} R_2 &= 1 / [(1 / R_{EQ1}) - (1 / R_M)] \\ &= 1/ [(1 / 300\ \text{k}\Omega) - (1 / 1\ \text{M}\Omega)] \\ &= 1/ \{[1 / (300 \times 10^3\ \Omega)] - [1 / (1 \times 10^6\ \Omega)]\} \\ &= 1/ \{[(1 / 300 \times 10^3)\ (1 / \Omega)] - [(1 / 1 \times 10^6)\ (1 / \Omega)]\} \\ &= 1/ [(3.33 \times 10^{-6}\ \text{S}) - (1 \times 10^{-6}\ \text{S})] \\ &= 1/ [(3.33 \times 10^{-6}) - (1 \times 10^{-6})\ \text{S}] \\ &= 1/ (2.33 \times 10^{-6}\ \text{S}) \\ &= (1 / 2.33 \times 10^{-6})\ (1 / \text{S}) \\ &= 429 \times 10^3\ \Omega \\ &= 429\ \text{k}\Omega \end{aligned}$$

Use the value of R_1 and R_2 to calculate the output for the unloaded voltage divider.

$$\begin{aligned} V_{DIV} &= V_S \times [R_2 / (R_1 + R_2)] \\ &= 5.00\ \text{V} \times [429\ \text{k}\Omega / (100\ \text{k}\Omega + 429\ \text{k}\Omega)] \\ &= 5.00\ \text{V} \times \{429\ \text{k}\Omega / [(100 + 429)\ \text{k}\Omega]\} \\ &= 5.00\ \text{V} \times (429\ \text{k}\Omega / 529\ \text{k}\Omega) \\ &= 5.00\ \text{V} \times [(429 / 529)\ (\text{k}\Omega / \text{k}\Omega)] \\ &= 5.00\ \text{V} \times 0.811 \\ &= 4.05\ \text{V} \end{aligned}$$

The unloaded value of V_{DIV} is **4.05 V**.

2. Problem: For the loaded voltage divider circuit in Figure 7-42, what is the value of R_X? Assume that resistor values are exact.

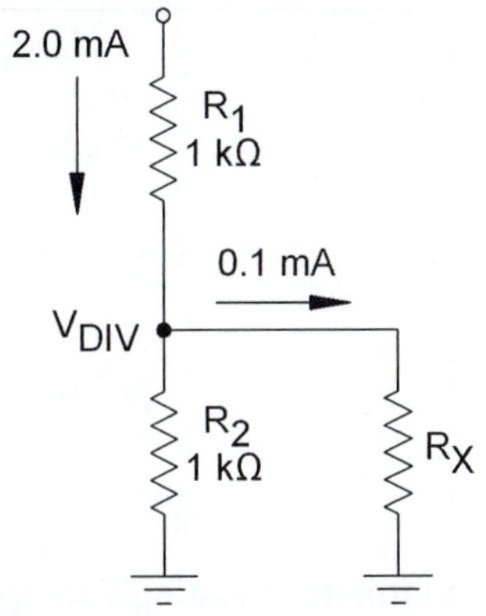

Figure 7-42: Voltage Divider Loading Example Circuit 4

Solution: Let currents entering a junction be positive and currents leaving a junction be negative. From Kirchhoff's Current Law,

$$\begin{aligned}
I_{R1} + I_{R2} + I_{RX} &= 0 \\
I_{R1} + I_{R2} + I_{RX} + -I_{RX} &= 0 + -I_{RX} \\
I_{R1} + I_{R2} + -I_{R1} &= -I_{RX} + -I_{R1} \\
I_{R2} &= -(-\,0.1\text{ mA} + 2.0\text{ mA}) \\
&= -(1.9\text{ mA}) \\
&= -1.9\text{ mA}
\end{aligned}$$

The negative sign indicates that I_{R2} is leaving the junction and flowing downward through R_2, so that the polarity of the voltage V_{R2} across R_2 is positive at the top of R_2. From Ohm's Law for voltage,

$$\begin{aligned}
V_{DIV} = V_{R2} = I_{R2} \times R_2 \\
&= 1.9\text{ mA} \times 1\text{ k}\Omega \\
&= (1.9 \times 10^{-3}\text{ A}) \times (1 \times 10^{3}\ \Omega) \\
&= [(1.9 \times 10^{-3}) \times (1 \times 10^{3})]\ (\text{A} \times \Omega) \\
&= 1.9\text{ V}
\end{aligned}$$

Because R_2 and R_X are in parallel, $V_{R2} = V_{RX}$. From Ohm's Law for resistance,

$$\begin{aligned}
R_X &= V_{RX} / I_{RX} \\
&= (1.9\text{ V}) / (0.1\text{ mA}) \\
&= (1.9\text{ V}) / (0.1 \times 10^{-3}\text{ A}) \\
&= [1.9 / (0.1 \times 10^{-3})]\ (\text{V} / \text{A}) \\
&= 19 \times 10^{3}\ \Omega \\
&= 19\text{ k}\Omega
\end{aligned}$$

The value of R_X is **19 kΩ**.

7.2.3 Voltmeter Loading

1. Problem: For the circuit in Figure 7-43, what must be the minimum voltmeter resistance R_M to ensure that the measured voltage V_{DIV} is within 1% of the actual value? Assume that all resistor values are exact.

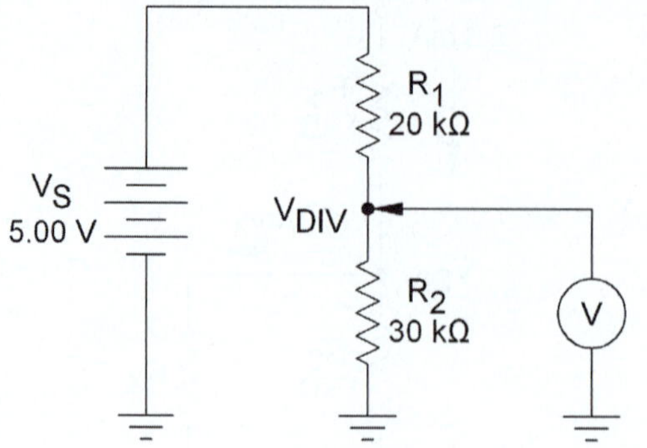

Figure 7-43: Voltmeter Loading Example Circuit 2

Solution: Figure 7-44 shows the equivalent circuit and the simplified circuit for the circuit in Figure 7-43.

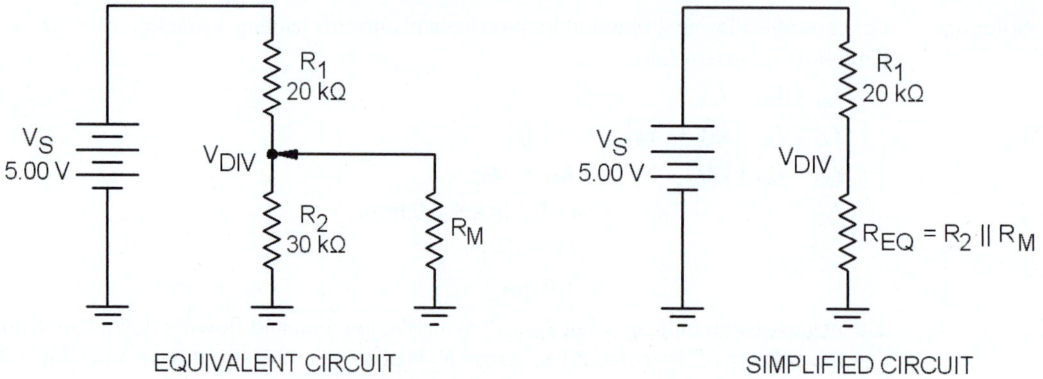

Figure 7-44: Equivalent and Simplified Circuits for Figure 7-43

From the voltage divider formula, the unloaded value of $\boldsymbol{V_{DIV}}$ is

$$\begin{aligned} \boldsymbol{V_{DIV}} &= \boldsymbol{V_S} \times [\boldsymbol{R_2} / (\boldsymbol{R_1} + \boldsymbol{R_2})] \\ &= 5.00\text{ V} \times [30\text{ k}\Omega / (20\text{ k}\Omega + 30\text{ k}\Omega)] \\ &= 5.00\text{ V} \times (30\text{ k}\Omega / 50\text{ k}\Omega) \\ &= 5.00\text{ V} \times [(30 / 50)\ (\text{k}\Omega / \text{k}\Omega)] \\ &= 5.00\text{ V} \times 0.6 \\ &= 3.00\text{ V} \end{aligned}$$

The minimum loaded voltage must be within 1% of this value. Voltmeter loading will reduce the measured voltage, so

$$\begin{aligned} \boldsymbol{V_{DIV}}\ (\text{loaded}) &= \boldsymbol{V_{DIV}} - \boldsymbol{V_{DIV}} \times 1\% \\ &= (\boldsymbol{V_{DIV}} \times 1) + -[\boldsymbol{V_{DIV}} \times (1 / 100)] \\ &= \boldsymbol{V_{DIV}} \times (1 + -0.01) \\ &= \boldsymbol{V_{DIV}} \times 0.99 \\ &= 3.00\text{ V} \times 0.99 \\ &= 2.97\text{ V} \end{aligned}$$

From the voltage divider formula,

$$\boldsymbol{V_{DIV}}\ (\text{loaded}) = \boldsymbol{V_S} \times [\boldsymbol{R_{EQ}} / (\boldsymbol{R_1} + \boldsymbol{R_{EQ}})]$$

where $\boldsymbol{R_{EQ}}$ is the equivalent resistance for the parallel combination of $\boldsymbol{R_2}$ and the meter resistance $\boldsymbol{R_M}$. Use $\boldsymbol{R_2}$ and $\boldsymbol{R_M}$ to solve for $\boldsymbol{R_{EQ}}$.

$$V_{DIV\,\text{(loaded)}} \times (R_1 + R_{EQ}) = V_S \times [R_{EQ} / (R_1 + R_{EQ})] \times (R_1 + R_{EQ})$$
$$[V_{DIV\,\text{(loaded)}} \times R_1] + [V_{DIV\,\text{(loaded)}} \times R_{EQ}] = V_S \times R_{EQ}$$
$$[V_{DIV\,\text{(loaded)}} \times R_1] + [V_{DIV\,\text{(loaded)}} \times R_{EQ}] + -[V_{DIV\,\text{(loaded)}} \times R_{EQ}]$$
$$= V_S \times R_{EQ} + -[V_{DIV\,\text{(loaded)}} \times R_{EQ}]$$
$$[V_{DIV\,\text{(loaded)}} \times R_1] = [V_S + -V_{DIV\,\text{(loaded)}}] \times R_{EQ}$$
$$[V_{DIV\,\text{(loaded)}} \times R_1] / [V_S + -V_{DIV\,\text{(loaded)}}] = \{[V_S + -V_{DIV\,\text{(loaded)}}] \times R_{EQ}\} / [V_S + -V_{DIV\,\text{(loaded)}}]$$
$$\{[V_{DIV\,\text{(loaded)}}] / [V_S - V_{DIV\,\text{(loaded)}}]\} \times R_1 = R_{EQ}$$

Use the circuit voltage values to solve for R_{EQ}.

$$\begin{aligned} R_{EQ} &= [2.97\text{ V} / (5.00\text{ V} - 2.97\text{ V})] \times 20\text{ k}\Omega \\ &= (2.97\text{ V} / 2.03\text{ V}) \times (20\text{ k}\Omega) \\ &= 1.46 \times 20\text{ k}\Omega \\ &= 29.3\text{ k}\Omega \end{aligned}$$

R_{EQ} is R_2 and R_M in parallel, so use R_{EQ} and R_2 to solve for R_M.

$$\begin{aligned} R_{EQ} &= 1 / [(1 / R_2) + (1 / R_M)] \\ 1 / R_{EQ} &= 1 / \{1 / [(1 / R_2) + (1 / R_M)]\} \\ 1 / R_{EQ} &= (1 / R_2) + (1 / R_M) \\ (1 / R_{EQ}) + -(1 / R_2) &= (1 / R_2) + (1 / R_M) + -(1 / R_2) \\ (1 / R_{EQ}) + -(1 / R_2) &= 1 / R_M \\ 1 / [(1 / R_{EQ}) + -(1 / R_2)] &= 1 / (1 / R_M) \\ 1 / [(1 / R_{EQ}) + -(1 / R_2)] &= R_M \end{aligned}$$

Use the values of R_2 and R_{EQ} to solve for R_M.

$$\begin{aligned} R_M &= 1 / [(1 / 29.3\text{ k}\Omega) - (1 / 30\text{ k}\Omega)] \\ &= 1 / \{[1 / (29.3 \times 10^3\ \Omega)] - [(1 / (30 \times 10^3\ \Omega))]\} \\ &= 1 / [(1 / 29.3 \times 10^3)(1 / \Omega) - (1 / 30 \times 10^3)(1 / \Omega)] \\ &= 1 / (34.2 \times 10^{-6}\text{ S} - 33.3 \times 10^{-6}\text{ S}) \\ &= 1 / [(34.2 \times 10^{-6} - 33.3 \times 10^{-6})\text{ S}] \\ &= 1 / (842 \times 10^{-9}\text{ S}) \\ &= (1 / 842 \times 10^{-9})(1 / \text{S}) \\ &= 1.19 \times 10^6\ \Omega \\ &= 1.19\text{ M}\Omega \end{aligned}$$

The minimum meter resistance R_M necessary for the measured value of V_{DIV} to be within 1% of the actual value is **1.19 MΩ.**

2. Problem: Two identical voltmeters are connected to the voltage divider circuit as shown in Figure 7-45. The measured voltage across R_2 is 3.91 V. What are the values of R_1 and R_2?

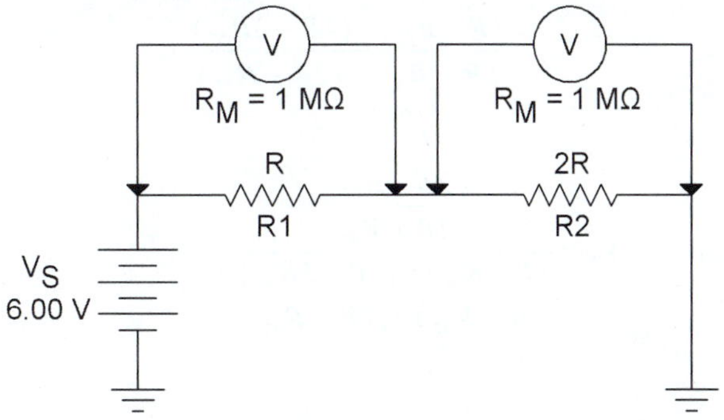

Figure 7-45: Voltmeter Loading Example Circuit 3

Solution: The expression for the voltage divider circuit is

$$R_T = R_{EQ1} + R_{EQ2}$$

where R_{EQ1} and R_{EQ2} are

$$R_{EQ1} = R_M \parallel R_1$$

$$R_{EQ2} = R_M \parallel R_2$$

The product-over-sum expression for two parallel resistors gives

$$R_{EQ1} = \frac{R_1 \times R_M}{R_1 + R_M} = \frac{R \times R_M}{R + R_M}$$

and

$$R_{EQ2} = \frac{2R_2 \times R_M}{2R_2 + R_M} = \frac{2R \times R_M}{2R + R_M}$$

From the voltage divider formula, the voltage V_{REQ2} across R_{EQ2} is

$$V_{REQ2} = V_S \times \frac{R_{EQ2}}{R_{EQ1} + R_{EQ2}}$$

where

$$\begin{aligned}
R_{EQ1} + R_{EQ2} &= \frac{R \times R_M}{R + R_M} + \frac{2R \times R_M}{2R + R_M} \\
&= \frac{(R \times R_M) \times (2R + R_M)}{(R + R_M) \times (2R + R_M)} + \frac{(2R \times R_M) \times (R + R_M)}{(2R + R_M) \times (R + R_M)} \\
&= \frac{(R \times R_M) \times (2R + R_M) + (2R \times R_M) \times (R + R_M)}{(R + R_M) \times (2R + R_M)} \\
&= \frac{(R \times R_M) \times (2R + R_M) + 2 \times (R \times R_M) \times (R + R_M)}{(R + R_M) \times (2R + R_M)} \\
&= \frac{(R \times R_M) \times [(2R + R_M) + 2 \times (R + R_M)]}{(R + R_M) \times (2R + R_M)} \\
&= \frac{(R \times R_M) \times [(2R + R_M) + (2R + 2R_M)]}{(R + R_M) \times (2R + R_M)} \\
&= \frac{(R \times R_M) \times (4R + 3R_M)}{(R + R_M) \times (2R + R_M)}
\end{aligned}$$

so

$$V_{REQ2} = V_S \times \frac{\dfrac{2R \times R_M}{2R + R_M}}{\dfrac{(R \times R_M) \times (4R + 3R_M)}{(R + R_M) \times (2R + R_M)}}$$

$$= V_S \times \frac{2R \times R_M}{2R + R_M} \times \frac{(R + R_M) \times (2R + R_M)}{(R \times R_M) \times (4R + 3R_M)}$$

$$= V_S \times \frac{(2R \times R_M) \times (R + R_M) \times (2R + R_M)}{(R \times R_M) \times (4R + 3R_M) \times (2R + R_M)}$$

$$= V_S \times \frac{2 \times (R \times R_M) \times (R + R_M) \times (2R + R_M)}{(R \times R_M) \times (4R + 3R_M) \times (2R + R_M)}$$

$$= V_S \times \frac{2 \times (R + R_M)}{(4R + 3R_M)}$$

$$= V_S \times \frac{2(R + R_M)}{(4R + 3R_M)}$$

Next, isolate $\boldsymbol{R}$ in terms of the known values V_S, R_M, and V_{REQ}.

$$V_{REQ2} \times (4R + 3R_M) = V_S \times \frac{2(R + R_M)}{(4R + 3R_M)} \times (4R + 3R_M)$$

$$= V_S \times [2(R + R_M)]$$

so

$$(4R \times V_{REQ2}) + (3R_M \times V_{REQ2}) = (2R \times V_S) + (2R_M \times V_S)$$

$$(4R \times V_{REQ2}) + (3R_M \times V_{REQ2}) + -(2R \times V_S)$$
$$= (2R \times V_S) + (2R_M \times V_S) + -(2R \times V_S)$$

$$(4R \times V_{REQ2}) + (3R_M \times V_{REQ2}) + -(2R \times V_S) + -(3R_M \times V_{REQ2})$$
$$= (2R_M \times V_S) + -(3 \times R_M \times V_{REQ2})$$

$$(4R \times V_{REQ2}) + -(2R \times V_S) = R_M \times (2V_S + -3V_{REQ2})$$

$$\frac{R \times (4V_{REQ2} + -2V_S)}{(4V_{REQ2} + -2V_S)} = \frac{R_M \times (2V_S + -3V_{REQ2})}{(4V_{REQ2} + -2V_S)}$$

$$R = \frac{R_M \times (2V_S + -3V_{REQ2})}{(4V_{REQ2} + -2V_S)}$$

From the values for the circuit, $R_1 = R$ and $V_{REQ2} = V_{R2} = 3.91$ V.

$R_1 = R$
$= R_M \times [(2V_S + -3V_{REQ2}) / (4V_{REQ2} + -2V_S)]$
$= 1\ \text{M}\Omega \times [(2 \times 6.00\ \text{V}) + -(3 \times 3.91\ \text{V})] / [(4 \times 3.91\ \text{V}) + -(2 \times 6.00\ \text{V})]$
$= (1 \times 10^6\ \Omega) \times [(12.00\ \text{V} + -11.73\ \text{V}) / (15.64\ \text{V} + -12.00\ \text{V})]$
$= (1 \times 10^6\ \Omega) \times (0.273\ \text{V}) / (3.64\ \text{V})$
$= (1 \times 10^6\ \Omega) \times [(0.273 / 3.64)\ (\text{V} / \text{V})$
$= (1 \times 10^6\ \Omega) \times (75.0 \times 10^{-3})$
$= 75 \times 10^3\ \Omega$
$= 75\ \text{k}\Omega$

and

$R_2 = 2R = 2R_1$
$= 2 \times (75 \times 10^3\ \Omega)$
$= 150 \times 10^3\ \Omega$
$= 150\ \text{k}\Omega$

The values of R_1 and R_2 are **75 kΩ** and **150 kΩ**, respectively.

7.3 Just for Fun

1. Problem: The bridge circuit in Figure 7-46 uses four resistors of equal value, one of which is a potentiometer, to determine the value of an unknown resistor R_X.

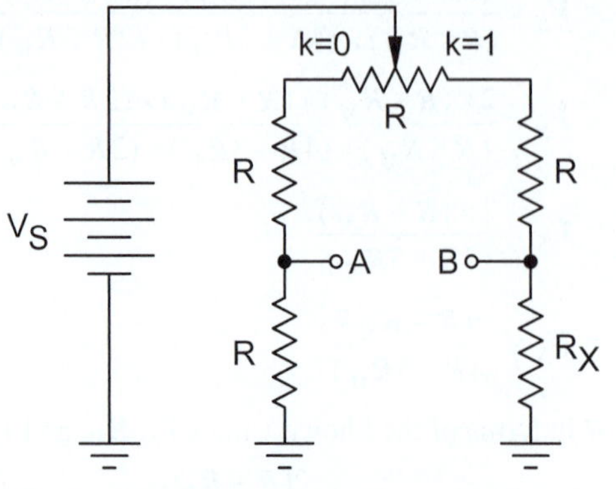

Figure 7-46: Bridge Circuit

The value k represents the position of the potentiometer wiper, where $k = 0$ indicates that the wiper is fully to the left, $k = 0.5$ indicates that the wiper is in the middle of the potentiometer, and $k = 1$ indicates that the wiper is fully to the right. Verify that $R_X = [(2 - k) / (1 + k)] \times R$ when the bridge is balanced (i.e., $V_A = V_B$).

Solution: For any value of k, the potentiometer can be viewed as two fixed resistors R_{LEFT} and R_{RIGHT} whose combined resistance is R. From the definition of k, the value of the left fixed resistor R_{LEFT} is $k \times R$ and the value of the right fixed resistor R_{RIGHT} is $(1 - k) \times R$. Figure 7-47 shows the equivalent circuit for the bridge circuit.

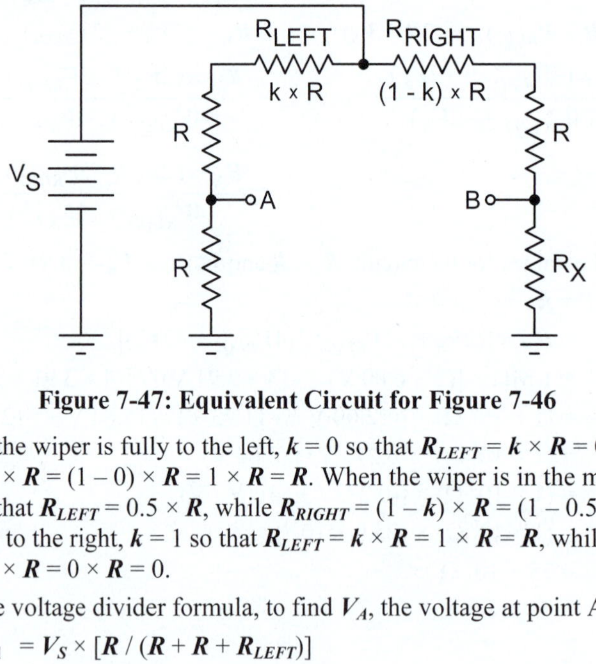

Figure 7-47: Equivalent Circuit for Figure 7-46

When the wiper is fully to the left, $k = 0$ so that $R_{LEFT} = k \times R = 0 \times R = 0$, while $R_{RIGHT} = (1 - k) \times R = (1 - 0) \times R = 1 \times R = R$. When the wiper is in the middle of the potentiometer, $k = 0.5$ so that $R_{LEFT} = 0.5 \times R$, while $R_{RIGHT} = (1 - k) \times R = (1 - 0.5) \times R = 0.5 \times R$. When the wiper is fully to the right, $k = 1$ so that $R_{LEFT} = k \times R = 1 \times R = R$, while $R_{RIGHT} = (1 - k) \times R = (1 - 1) \times R = 0 \times R = 0$.

Use the voltage divider formula, to find V_A, the voltage at point A.

$$
\begin{aligned}
V_A &= V_S \times [R / (R + R + R_{LEFT})] \\
&= (V_S \times R) / (R + R + R_{LEFT}) \\
&= (V_S \times R) / [R + R + (k \times R)] \\
&= (V_S \times R) / [(1 \times R) + (1 \times R) + (k \times R)] \\
&= (V_S \times R) / [(1 + 1 + k) \times R]
\end{aligned}
$$

$= (V_S \times R) / [(2 + k) \times R]\}$

$= V_S / (2 + k)$

Next, use the voltage divider theorem to find V_B, the voltage at point B.

$V_B = V_S \times [R_X / (R_X + R + R_{RIGHT})]$

$= (V_S \times R_X) / \{R_X + R + [(1 - k) \times R]\}$

$= (V_S \times R_X) / \{R_X + (1 \times R) + [(1 - k) \times R)]\}$

$= (V_S \times R_X) / \{R_X + [(1 + 1 - k) \times R]\}$

$= (V_S \times R_X) / \{R_X + [(2 - k) \times R]\}$

When the bridge is balanced, $V_A = V_B$.

$V_A = V_B$

$V_S / (2 + k) = (V_S \times R_X) / \{R_X + [(2 - k) \times R]\}$

$[V_S / (2 + k)] \times \{R_X + [(2 - k) \times R]\}$

$= ((V_S \times R_X) / \{R_X + [(2 - k) \times R]\}) \times \{R_X + [(2 - k) \times R]\}$

$(V_S \times \{R_X + [(2 - k) \times R]\}) / (2 + k)$

$= ((V_S \times R_X \times \{R_X + [(2 - k) \times R]\}) / \{R_X + [(2 - k) \times R]\}$

$[(V_S \times \{R_X + [(2 - k) \times R]\}) / (2 + k)] \times (2 + k)$

$= (V_S \times R_X) \times (2 + k)$

$(V_S \times \{R_X + [(2 - k) \times R] \times (2 + k)\}) / (2 + k)$

$= V_S \times R_X \times (2 + k)$

$V_S \times \{R_X + [(2 - k) \times R]\} = V_S \times R_X \times (2 + k)$

$(V_S \times R_X) + [V_S \times (2 - k) \times R] = [V_S \times (2 + k)] \times R_X$

$(V_S \times R_X) + [V_S \times (2 - k) \times R] - (V_S \times R_X)$

$= \{[V_S \times (2 + k)] \times R_X\} - (V_S \times R_X)$

$(V_S \times R_X) - (V_S \times R_X) + [V_S \times (2 - k) \times R]$

$= \{[V_S \times (2 + k)] - V_S\} \times R_X$

$V_S \times (2 - k) \times R = \{[V_S \times (2 + k)] - (1 \times V_S)\} \times R_X$

$V_S \times (2 - k) \times R = [V_S \times (2 + k - 1)] \times R_X$

$V_S \times (2 - k) \times R = [V_S \times (2 - 1 + k)] \times R_X$

$V_S \times (2 - k) \times R = [V_S \times (1 + k)] \times R_X$

$[V_S \times (2 - k) \times R] / [V_S \times (1 + k)] = \{[V_S \times (1 + k)] \times R_X\} / [V_S \times (1 + k)]$

$[(2 - k) \times R] / (1 + k) = R_X$

$[(2 - k) / (1 + k)] \times R = R_X$

2. Problem: The familiar $R/2R$ ladder network shown in Figure 7-48 consists of two different resistors, one with resistance R and the other with resistance $2 \times R$. An interesting property of the $R/2R$ ladder network is that its equivalent resistance R_{EQ} will always be the same no matter how many $R/2R$ extension sections are added to the base section.

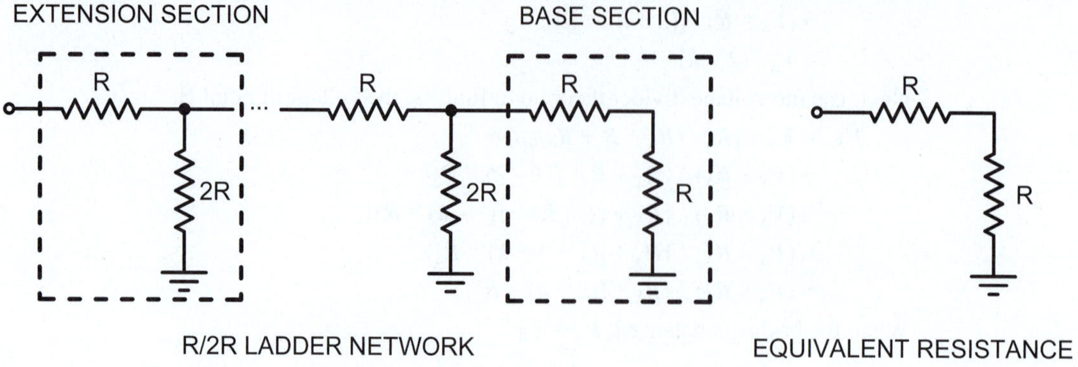

Figure 7-48: The R/2R Ladder Network

A less-familiar network is the ***R/R/R*** ladder network shown in Figure 7-49. This ladder network also has the interesting property that, for a specific value of R_X, its equivalent resistance R_{EQ} is always the same no matter how many ***R/R/R*** sections are added to the base section.

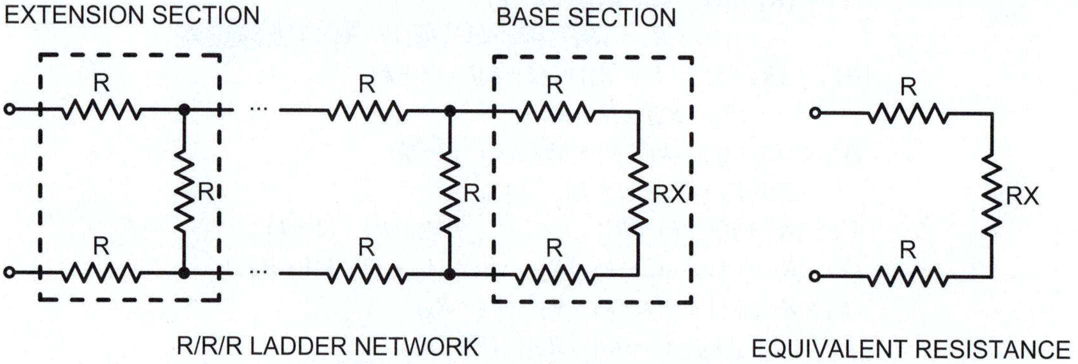

Figure 7-49: The R/R/R Ladder Network

Determine the value of R_X in terms of R that the ***R/R/R*** ladder network requires.

Solution: The key to both the ***R/2R*** and ***R/R/R*** ladder networks is that the equivalent resistance of the base section and the extension section resistor that is in parallel with it must equal to the "key value" of the base section. When this is true, the simplified circuit of the extension section and base section will be identical to the base section, as shown in Figure 7-50 and Figure 7-51. This repeats for each additional extension section, with each repetition giving the same simplified circuit.

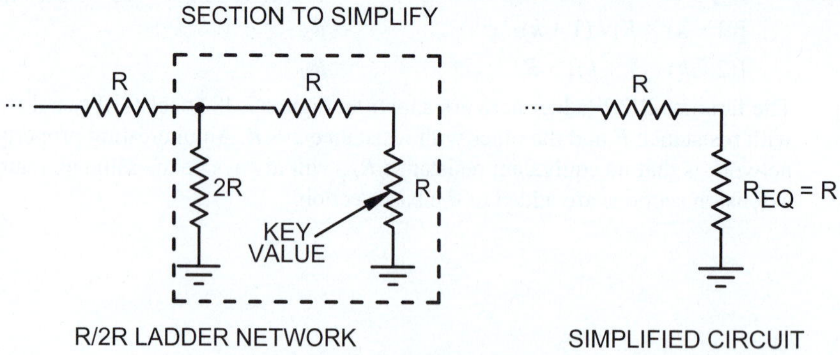

Figure 7-50: Simplification of R/2R Ladder Network

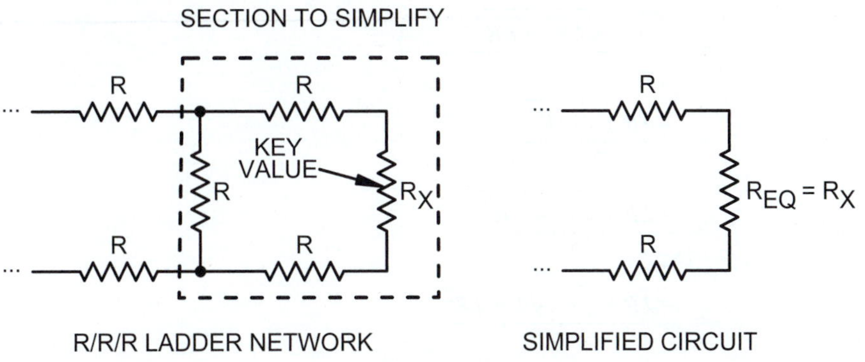

Figure 7-51: Simplification of R/R/R Ladder Network

For the ***R*/*R*/*R*** ladder network, the expression that the section must simplify is

$$R_{EQ} = R \parallel (R + R_X + R)$$

R_{EQ} must equal R_X so

$$R_X = R \parallel (R + R_X + R)$$
$$= R \parallel (R + R + R_X)$$
$$= R \parallel (2R + R_X)$$
$$1 / R_X = (1 / R) + [1 / (2R + R_X)]$$
$$= (1 / R) \times [(2R + R_X) / (2R + R_X)] + [1 / (2R + R_X) \times (R / R)]$$
$$= \{[(2R + R_X) / [R \times (2R + R_X)]\} + \{R / [R \times (2R + R_X)]\}$$
$$= (2R + R_X + R) / [R \times (2R + R_X)]$$
$$= (2R + R + R_X) / [R \times (2R + R_X)]$$
$$= (3R + R_X) / [R \times (2R + R_X)]$$
$$1 / (1 / R_X) = 1 / \{(3R + R_X) / [R \times (2R + R_X)]\}$$
$$R_X = [R \times (2R + R_X)] / (3R + R_X)$$

Isolate and solve for R_X.

$$R_X \times (3R + R_X) = \{[R \times (2R + R_X)] / [(3R + R_X)\} \times (3R + R_X)]$$
$$(R_X \times 3R) + (R_X \times R_X) = [R \times (2R + R_X)] \times (3R + R_X)] / [(3 \times R) + R_X]$$
$$= R \times (2R + R_X)$$
$$= (R \times 2R) + (R \times R_X)$$
$$(3R \times R_X) + R_X^2 = 2R^2 + (R \times R_X)$$
$$(3R \times R_X) + R_X^2 - [2R^2 + (R \times R_X)] = [2R^2 + (R \times R_X)] - [2R^2 + (R \times R_X)]$$
$$R_X^2 + (3R \times R_X) - (R \times R_X) - 2R^2 = 0$$
$$R_X^2 + (3R - R) \times R_X - 2R^2 = 0$$
$$R_X^2 + (2R \times R_X) - 2R^2 = 0$$

This is a quadratic expression of the form $Ax^2 + Bx + C = 0$ and has the solution

$$R_X = \frac{-B \pm \sqrt{B^2 - (4 \times A \times C)}}{2 \times A}$$

where A = 1, B = 2R, and C = $-2R^2$. Substitute for A, B, and C and solve for R_X.

$$R_X = \frac{-2R \pm \sqrt{(2R)^2 - [4 \times 1 \times (-2R^2)]}}{2 \times 1}$$

$$= \frac{-2\boldsymbol{R} \pm \sqrt{4\boldsymbol{R}^2 + 8\boldsymbol{R}^2}}{2}$$

$$= \frac{-2\boldsymbol{R} \pm \sqrt{12\boldsymbol{R}^2}}{2}$$

$$= \frac{-2\boldsymbol{R} \pm \sqrt{3 \times 4 \times \boldsymbol{R}^2}}{2}$$

$$= \frac{-2\boldsymbol{R} \pm \sqrt{3} \times \sqrt{4} \times \sqrt{\boldsymbol{R}^2}}{2}$$

$$= \frac{-2\boldsymbol{R} \pm \sqrt{3} \times 2 \times \boldsymbol{R}}{2}$$

$$= \frac{-2\boldsymbol{R}}{2} \pm \frac{2 \times \sqrt{3} \times \boldsymbol{R}}{2}$$

$$= -\boldsymbol{R} \pm \sqrt{3} \times \boldsymbol{R}$$

Both $\boldsymbol{R_X}$ and $\boldsymbol{R}$ must be positive, so

$$\boldsymbol{R_X} = -\boldsymbol{R} + \sqrt{3} \times \boldsymbol{R}$$

$$= (-1 \times \boldsymbol{R}) + (\sqrt{3} \times \boldsymbol{R})$$

$$= (-1 + \sqrt{3}) \times \boldsymbol{R})$$

$$= (\sqrt{3} - 1) \times \boldsymbol{R})$$

$\boldsymbol{R_X}$ must be equal to about **0.7321** × $\boldsymbol{R}$.

8. Circuit Theorems and Conversions

8.1 Basic

8.1.1 Source Loading

1. Problem: Determine the minimum and maximum loaded values of V_{OUT} for the circuit in Figure 8-1.

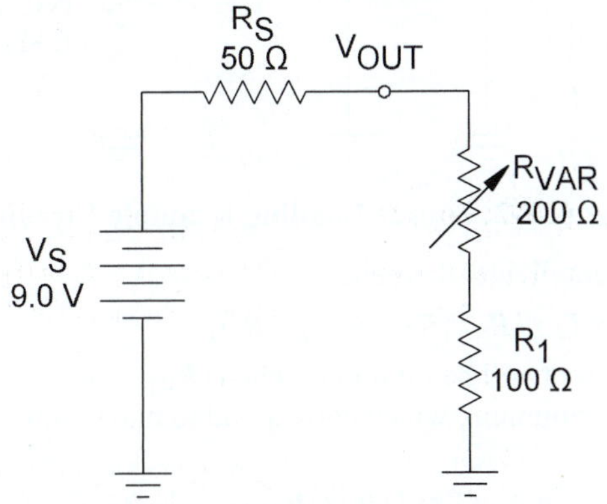

Figure 8-1: Source Loading Example Circuit 1

Solution: From the voltage divider theorem,

$V_{OUT} = V_S \times [(R_{VAR} + R_1) / (R_S + R_{VAR} + R_1)]$

The output voltage will be minimum when $(R_{VAR} + R_1)$ is minimum and maximum when $(R_{VAR} + R_1)$ is maximum, which correspond to minimum and maximum values of R_{VAR}, respectively.

$$V_{OUT\,(min)} = V_S \times [(R_{VAR\,(min)} + R_1) / (R_S + R_{VAR\,(min)} + R_1)]$$
$$= 9.0\text{ V} \times [(0\ \Omega + 100\ \Omega) / (50\ \Omega + 0\ \Omega + 100\ \Omega)]$$
$$= 9.0\text{ V} \times (100\ \Omega / 150\ \Omega)$$
$$= 9.0\text{ V} \times [(100 / 150)\ (\Omega / \Omega)]$$
$$= 9.0\text{ V} \times 0.667$$
$$= 6.00\text{ V}$$

and

$$V_{OUT\,(max)} = V_S \times [(R_{VAR\,(max)} + R_1) / (R_S + R_{VAR\,(max)} + R_1)]$$
$$= 9.0\text{ V} \times [(200\ \Omega + 100\ \Omega) / (50\ \Omega + 200\ \Omega + 100\ \Omega)]$$
$$= 9.0\text{ V} \times (300\ \Omega / 350\ \Omega)$$
$$= 9.0\text{ V} \times [(300 / 350)\ (\Omega / \Omega)]$$
$$= 9.0\text{ V} \times 0.571$$
$$= 7.71$$

The minimum and maximum values of V_{OUT} are **6.0 V** and **7.7 V**, respectively.

2. Problem: Determine the minimum and maximum loaded values of I_{OUT} for the circuit in Figure 8-2.

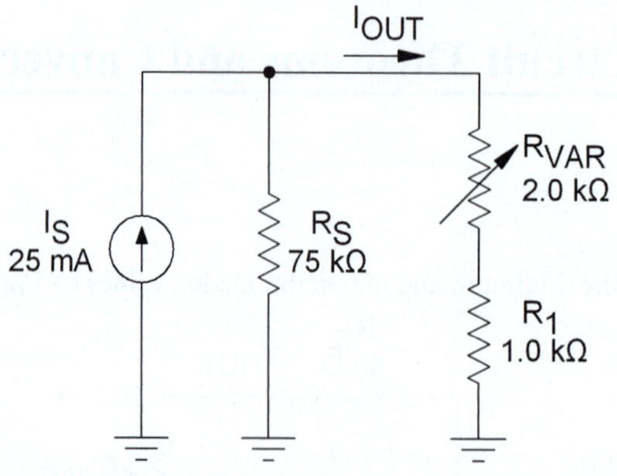

Figure 8-2: Source Loading Example Circuit 2

Solution: From the current divider theorem,

$$I_{OUT} = I_S \times [R_S / (R_S + R_{VAR} + R_1)]$$

The output current will be minimum when ($R_{VAR} + R_1$) is maximum and maximum when ($R_{VAR} + R_1$) is minimum, which correspond to maximum and minimum values of R_{VAR}, respectively.

$$I_{OUT\,(\text{min})} = I_S \times [R_S / (R_S + R_{VAR\,(\text{max})} + R_1)]$$
$$= 25\text{ mA} \times [(75\text{ k}\Omega) / (75\text{ k}\Omega + 2.0\text{ k}\Omega + 1.0\text{ k}\Omega)]$$
$$= 25\text{ mA} \times (75\text{ k}\Omega / 78\text{ k}\Omega)$$
$$= 25\text{ mA} \times [(75 / 78)\ (\text{k}\Omega / \text{k}\Omega)]$$
$$= 25 \times 10^{-3}\text{ A} \times 0.962$$
$$= 24.0 \times 10^{-3}\text{ A}$$
$$= 24.\ 0\text{ mA}$$

and

$$I_{OUT\,(\text{max})} = I_S \times [R_S / (R_S + R_{VAR\,(\text{min})} + R_1)]$$
$$= 25\text{ mA} \times [(75\text{ k}\Omega) / (75\text{ k}\Omega + 0\ \Omega + 1.0\text{ k}\Omega)]$$
$$= 25\text{ mA} \times (75\text{ k}\Omega / 76\text{ k}\Omega)$$
$$= 25\text{ mA} \times [(75 / 76)\ (\text{k}\Omega / \text{k}\Omega)]$$
$$= 25 \times 10^{-3}\text{ A} \times 0.987$$
$$= 24.7 \times 10^{-3}\text{ A}$$
$$= 24.\ 7\text{ mA}$$

The minimum and maximum values of I_{OUT} are **24.0 mA** and **24.7 mA**, respectively.

8.1.2 Superposition Theorem

1. Problem: Determine the currents through each resistor for the circuit in Figure 8-3.

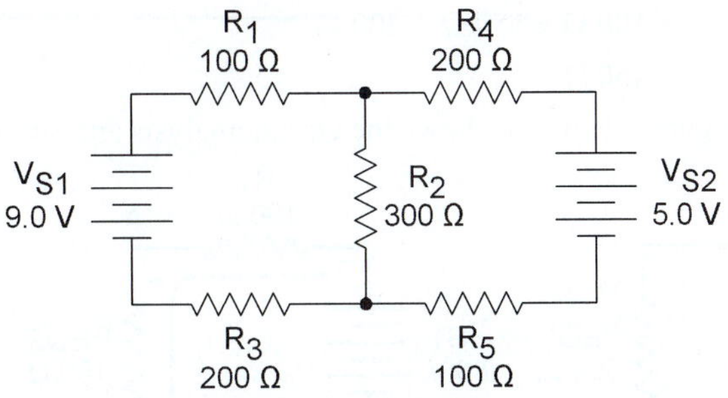

Figure 8-3: Superposition Theorem Example Circuit 1

Solution: To determine the currents through each resistor, separately analyze the series-parallel circuit for each source and algebraically sum the currents. For this example, let currents flowing from left to right and from top to bottom be positive and currents flowing from right to left and from bottom to top be negative. Figure 8-4 shows the circuit for analyzing the effects of V_{S1}.

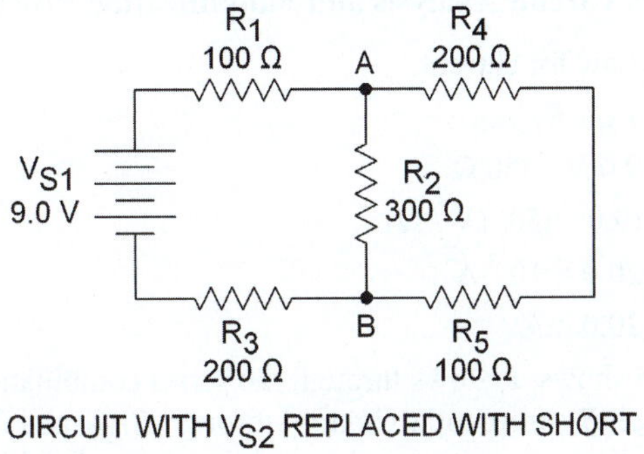

Figure 8-4: Circuit Analysis for V_{S1} for Figure 8-3

After replacing V_{S2} with its internal resistance (a short), the expression for this circuit is

$$R_{T(VS1)} = R_1 + R_2 \parallel (R_4 + R_5) + R_3$$

The first equivalent resistance R_{EQ1} is the series combination $(R_4 + R_5)$.

$$R_{T(VS1)} = R_1 + R_2 \parallel R_{EQ1} + R_3$$

where

$$R_{EQ1} = R_4 + R_5$$
$$= 200\ \Omega + 100\ \Omega$$
$$= 300\ \Omega$$

The second equivalent resistance is the parallel combination $R_2 \parallel R_{EQ1}$.

$$R_{T(VS1)} = R_1 + R_{EQ2} + R_3$$

where

$$R_{EQ2} = R_2 \parallel R_{EQ1}$$

Because R_{EQ2} consists of two equal-value resistors in parallel,

$$R_{EQ2} = R / 2$$
$$= (300\ \Omega) / 2$$
$$= 150\ \Omega$$

The third and final equivalent resistance R_{EQ3} is the series combination $R_1 + R_{EQ2} + R_3$.

$$R_{T(VS1)} = R_{EQ3}$$

where

$$R_{EQ3} = R_1 + R_{EQ2} + R_3$$

$= 100\ \Omega + 150\ \Omega + 200\ \Omega$

$= 450\ \Omega$

Refer to Figure 8-5, which shows the circuit analysis and simplification process for V_{S1}.

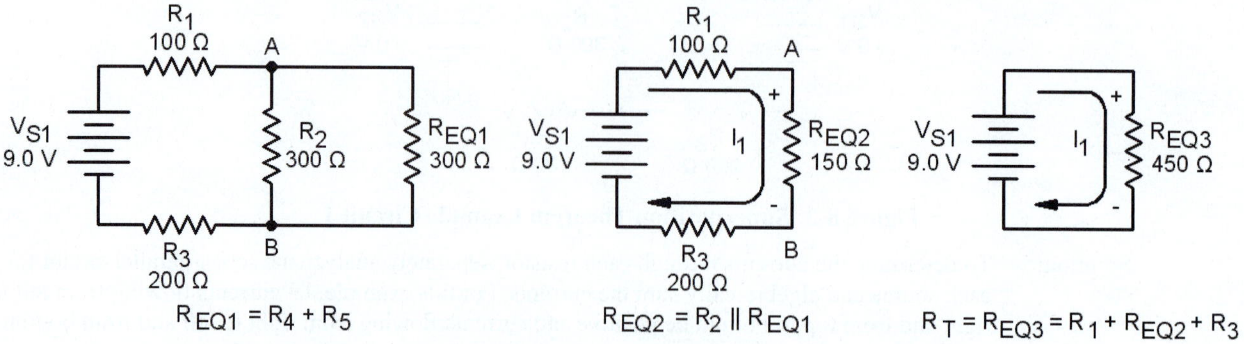

Figure 8-5: Circuit Analysis and Simplification Process for V_{S1}

From Ohm's Law for current,

$$I_{1(VS1)} = V_{S1} / R_{T(VS1)}$$
$$= 9.0\text{ V} / 450\ \Omega$$
$$= (9.0 / 450)\ (\text{V} / \Omega)$$
$$= 20.0 \times 10^{-3}\text{ A}$$
$$= 20.0\text{ mA}$$

As Figure 8-5 shows, I_1 flows through the series combination of $R_1 + R_{EQ2} + R_3$. I_1 flows from left to right through R_1, from top to bottom through R_{EQ2}, and from right to left through R_3. Therefore, from the definition of positive and negative currents for this example,

$$I_{R1\,(VS1)} = +20.0\text{ mA}$$
$$I_{R3\,(VS1)} = -20.0\text{ mA}$$

and from Ohm's Law for voltage,

$$V_{AB} = V_{REQ2}$$
$$= I_1 \times R_{EQ2}$$
$$= 20.0\text{ mA} \times 150\ \Omega$$
$$= (20.0 \times 10^{-3}\text{ A}) \times (150\ \Omega)$$
$$= [(20.0 \times 10^{-3}) \times (150)]\ (\text{A} \times \Omega)$$
$$= 3.00\text{ V}$$

with the polarity shown in Figure 8-5. R_{EQ2} is the parallel combination of R_2 and R_{EQ1}, so this voltage appears across both branches. From Ohm's Law for current,

$$I_{R2\,(VS1)} = V_{AB} / R_2$$
$$= 3.00\text{ V} / 300\ \Omega$$
$$= (3.00 / 300)\ (\text{V} / \Omega)$$
$$= 10.0 \times 10^{-3}\text{ A}$$
$$= 10.0\text{ mA}$$

and

$$I_{REQ1\,(VS1)} = V_{AB} / R_{EQ1}$$
$$= 3.00\text{ V} / 300\ \Omega$$
$$= (3.00 / 300)\ (\text{V} / \Omega)$$

$= 10.0 \times 10^{-3}$ A

$= 10.0$ mA

$I_{REQ1\,(VS1)}$ flows through the series combination of R_4 and R_5, so $I_{R4\,(VS1)} = I_{R5\,(VS1)} = I_{REQ1\,(VS1)}$. The polarity of V_{AB} indicates that $I_{R2\,(VS1)}$ flows from top to bottom, $I_{R4\,(VS1)}$ flows from left to right, and $I_{R5\,(VS1)}$ flows from right to left. By definition,

$I_{R2\,(VS1)} = +10.0$ mA

$I_{R4\,(VS1)} = +10.0$ mA

$I_{R5\,(VS1)} = -10.0$ mA

Figure 8-6 shows the circuit for analyzing the effects of V_{S2}.

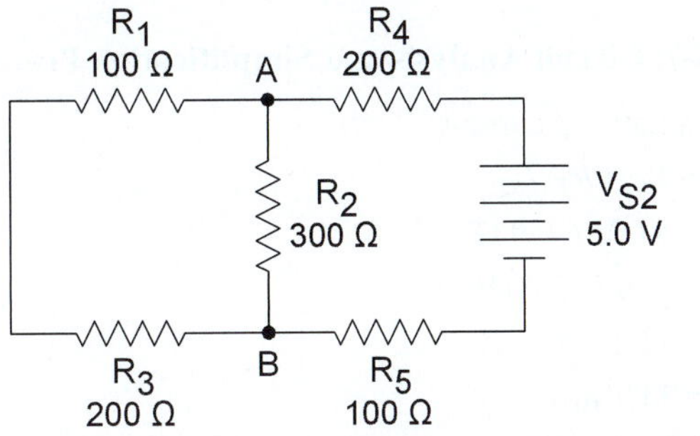

Figure 8-6: Circuit Analysis for V_{S2} for Figure 8-3

After replacing V_{S1} with its internal resistance (a short), the expression for this circuit is

$R_{T\,(VS2)} = R_4 + R_2 \,||\, (R_1 + R_3) + R_5$

The first equivalent resistance R_{EQ1} is the series combination $(R_1 + R_3)$.

$R_T = R_4 + R_2 \,||\, R_{EQ1} + R_5$

where

$R_{EQ1} = R_1 + R_3$

$= 100\ \Omega + 200\ \Omega$

$= 300\ \Omega$

The second equivalent resistance is the parallel combination $R_2 \,||\, R_{EQ1}$.

$R_{T\,(VS2)} = R_4 + R_{EQ2} + R_5$

where

$R_{EQ2} = R_2 \,||\, R_{EQ1}$

Because R_{EQ2} consists of two equal value resistors in parallel

$R_{EQ2} = R / 2$

$= (300\ \Omega) / 2$

$= 150\ \Omega$

The third and final equivalent resistance R_{EQ3} is the series combination $R_4 + R_{EQ2} + R_5$.

$R_{T\,(VS2)} = R_{EQ3}$

where

$R_{EQ3} = R_4 + R_{EQ2} + R_5$

$= 200\ \Omega + 150\ \Omega + 100\ \Omega$

$= 450\ \Omega$

Refer to Figure 8-7, which shows the circuit analysis and simplification process for V_{S2}.

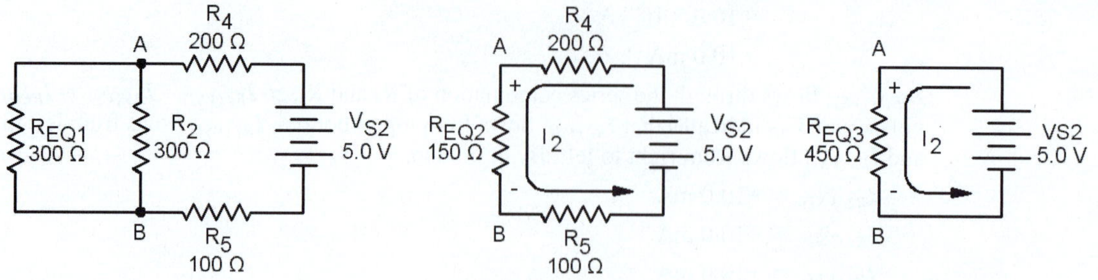

Figure 8-7: Circuit Analysis and Simplification Process for V_{S2}

From Ohm's Law for current,

$$\begin{aligned} I_{2\,(VS2)} &= V_{S2} / R_{T\,(VS2)} \\ &= 5.0\text{ V} / 450\ \Omega \\ &= (5.0 / 450)\ (\text{V} / \Omega) \\ &= 11.1 \times 10^{-3}\text{ A} \\ &= 11.1\text{ mA} \end{aligned}$$

As Figure 8-7 shows, I_2 flows through the series combination of $R_4 + R_{EQ2} + R_5$. I_2 flows from right to left through R_4, from top to bottom through R_{EQ2}, and from left to right through R_5. Therefore from definition of positive and negative currents for this example,

$$I_{R4\,(VS2)} = -11.1\text{ mA}$$

$$I_{R5\,(VS2)} = +11.1\text{ mA}$$

and from Ohm's Law for voltage

$$\begin{aligned} V_{AB} &= V_{REQ2} \\ &= I_{2\,(VS2)} \times R_{EQ2} \\ &= 11.1\text{ mA} \times 150\ \Omega \\ &= (11.1 \times 10^{-3}\text{ A}) \times (150\ \Omega) \\ &= [(11.1 \times 10^{-3}) \times (150)]\ (\text{A} \times \Omega) \\ &= 1.67\text{ V} \end{aligned}$$

with the polarity shown in Figure 8-7. R_{EQ2} is the parallel combination of R_2 and R_{EQ1}, so this voltage appears across both branches. From Ohm's Law for current

$$\begin{aligned} I_{R2\,(VS2)} &= V_{AB} / R_2 \\ &= 1.67\text{ V} / 300\ \Omega \\ &= (1.67 / 300)\ (\text{V} / \Omega) \\ &= 5.56 \times 10^{-3}\text{ A} \\ &= 5.56\text{ mA} \end{aligned}$$

and

$$\begin{aligned} I_{REQ1\,(VS2)} &= V_{AB} / R_{EQ1} \\ &= 1.67\text{ V} / 300\ \Omega \\ &= (1.67 / 300)\ (\text{V} / \Omega) \\ &= 5.56 \times 10^{-3}\text{ A} \\ &= 5.56\text{ mA} \end{aligned}$$

$I_{REQ1\,(VS2)}$ flows through the series combination of R_1 and R_3 so $I_{R1\,(VS2)} = I_{R2\,(VS2)} = I_{REQ1\,(VS2)}$. The polarity of V_{AB} indicates that $I_{R2\,(VS2)}$ flows from top to bottom, $I_{R1\,(VS2)}$ flows from right to left, and $I_{R3\,(VS2)}$ flows from left to right. By definition,

$I_{R1\,(VS2)} = -5.56$ mA

$I_{R2\,(VS2)} = +5.56$ mA

$I_{R3\,(VS2)} = +5.56$ mA

Algebraically sum the currents from V_{S1} and V_{S2} to find the total current through each resistor.

$I_{R1} = I_{R1\,(VS1)} + I_{R1\,(VS2)}$
= (+20.0 mA) + (–5.56 mA)
= [(+20.0) + (–5.56)] mA
= +14.4 mA (flows left to right)

$I_{R2} = I_{R2\,(VS1)} + I_{R2\,(VS2)}$
= (+10.0 mA) + (+5.56 mA)
= [(+10.0) + (+5.56)] mA
= +15.6 mA (flows top to bottom)

$I_{R3} = I_{R3\,(VS1)} + I_{R3\,(VS2)}$
= (–20.0 mA) + (+5.56 mA)
= [(–20.0) + (5.56)] mA
= –14.4 mA (flows right to left)

$I_{R4} = I_{R4\,(VS1)} + I_{R4\,(VS2)}$
= (+10.0 mA) + (–11.1 mA)
= [(+10.0) + (–11.1)] mA
= -1.1 mA (flows right to left)

$I_{R5} = I_{R5\,(VS1)} + I_{R5\,(VS2)}$
= (–10.0 mA) + (+11.1 mA)
= [(–10.0 mA) + (+11.1) mA
= +1.1 mA (flows left to right)

2. Problem: Determine the voltages across each resistor for the circuit in Figure 8-8. Assume that all values are exact.

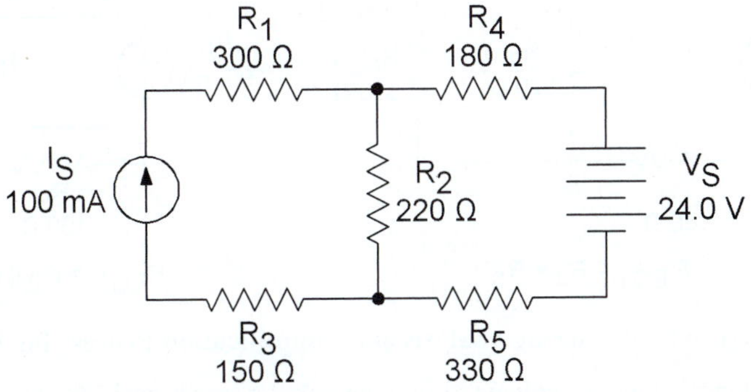

Figure 8-8: Superposition Theorem Example Circuit 2

Solution: To determine the voltages across each resistor, separately analyze the series-parallel circuit for each source and algebraically sum the voltages. For this example, let the voltage polarity on the top or left side determine the sign of the voltage. Figure 8-9 shows the circuit for analyzing the effects of I_S.

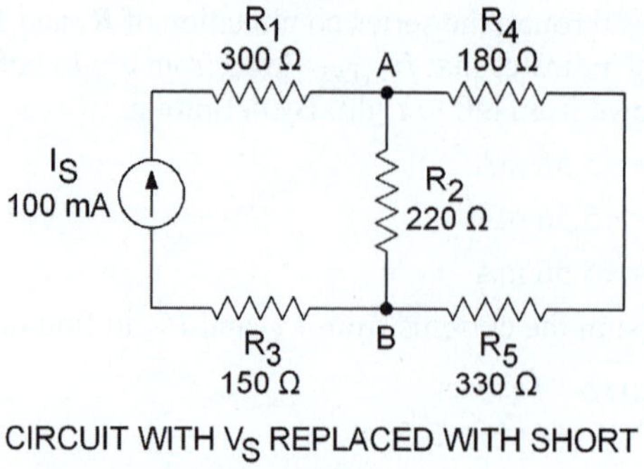

Figure 8-9: Circuit Analysis for I_S for Figure 8-8

After replacing V_S with its internal resistance (a short), the expression for this circuit is

$R_{T(IS)} \quad = R_1 + R_2 \,||\, (R_4 + R_5) + R_3$

The first equivalent resistance, R_{EQ1}, is the series combination ($R_4 + R_5$).

$R_{T(IS)} \quad = R_1 + R_2 \,||\, R_{EQ1} + R_3$

where

$R_{EQ1} \quad = R_4 + R_5$

$= 180\ \Omega + 330\ \Omega$

$= 510\ \Omega$

The second equivalent resistance, R_{EQ2}, is the parallel combination $R_2 \,||\, R_{EQ1}$.

$R_{T(IS)} \quad = R_1 + R_{EQ2} + R_3$

where

$R_{EQ2} \quad = R_2 \,||\, R_{EQ1}$

$= 220\ \Omega \,||\, 510\ \Omega$

$= 154\ \Omega$

Refer to Figure 8-10, which shows the circuit analysis and simplification process for I_S as well as the voltage polarities for R_1, R_{EQ2}, and R_3 due to I_1.

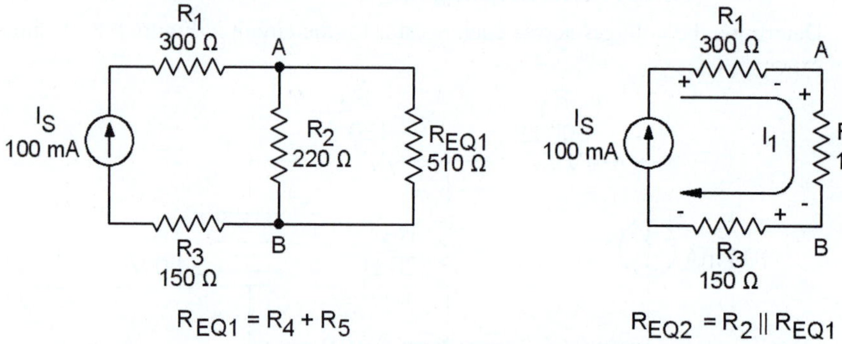

Figure 8-10: Circuit Analysis and Simplification Process for I_S

Use Ohm's Law for voltage to find the voltage across each resistance.

$V_{R1(IS)} \quad = I_1 \times R_1$

$= I_S \times R_1$

$= 100\text{ mA} \times 300\ \Omega$

$= (100 \times 10^{-3}\text{ A}) \times 300\ \Omega$

$= [(100 \times 10^{-3}) \times 300]\ (\text{A} \times \Omega)$

$= +30.0\ \text{V}$ (left end positive)

$\boldsymbol{V_{REQ2\,(IS)}} = \boldsymbol{V_{AB\,(IS)}}$

$= \boldsymbol{I_1} \times \boldsymbol{R_{EQ2}}$

$= \boldsymbol{I_S} \times \boldsymbol{R_{EQ2}}$

$= 100\ \text{mA} \times 154\ \Omega$

$= (100 \times 10^{-3}\ \text{A}) \times 154\ \Omega$

$= [(100 \times 10^{-3}) \times 154]\ (\text{A} \times \Omega)$

$= +15.4\ \text{V}$ (top end positive)

$\boldsymbol{V_{R3\,(IS)}} = \boldsymbol{I_1} \times \boldsymbol{R_3}$

$= \boldsymbol{I_S} \times \boldsymbol{R_3}$

$= 100\ \text{mA} \times 150\ \Omega$

$= (100 \times 10^{-3}\ \text{A}) \times 150\ \Omega$

$= [(100 \times 10^{-3}) \times 150]\ (\text{A} \times \Omega)$

$= -15.0\ \text{V}$ (left end negative)

$\boldsymbol{V_{REQ2\,(IS)}}$ is across the parallel combination of $\boldsymbol{R_2}$ and $\boldsymbol{R_{EQ1}}$, so the resistor voltages are equal.

$\boldsymbol{V_{R2\,(IS)}} = \boldsymbol{V_{REQ2\,(IS)}}$

$= +15.4\ \text{V}$ (top end positive)

$\boldsymbol{V_{REQ1\,(IS)}} = \boldsymbol{V_{REQ2\,(IS)}}$

$= +15.4\ \text{V}$ (top end positive)

$\boldsymbol{R_{EQ1}}$ is the series combination of $\boldsymbol{R_4}$ and $\boldsymbol{R_5}$, as shown in Figure 8-9. From the voltage divider theorem,

$\boldsymbol{V_{R4\,(IS)}} = \boldsymbol{V_{REQ1\,(IS)}} \times [\boldsymbol{R_4} / (\boldsymbol{R_4} + \boldsymbol{R_5})]$

$= 15.4\ \text{V} \times [180\ \Omega / (180\ \Omega + 330\ \Omega)]$

$= 15.4\ \text{V} \times (180\ \Omega / 510\ \Omega)$

$= 15.4\ \text{V} \times [(180 / 510)\ (\Omega / \Omega)]$

$= 15.4\ \text{V} \times 0.353$

$= +5.42\ \text{V}$ (left end positive)

and

$\boldsymbol{V_{R5\,(IS)}} = \boldsymbol{V_{AB\,(IS)}} \times [\boldsymbol{R_5} / (\boldsymbol{R_4} + \boldsymbol{R_5})]$

$= 15.4\ \text{V} \times [330\ \Omega / (180\ \Omega + 330\ \Omega)]$

$= 15.4\ \text{V} \times (330\ \Omega / 510\ \Omega)$

$= 15.4\ \text{V} \times [(330 / 510)\ (\Omega / \Omega)]$

$= 15.4\ \text{V} \times 0.647$

$= -9.95\ \text{V}$ (left end negative)

Figure 8-11 shows the circuits for analyzing the effects of $\boldsymbol{V_S}$.

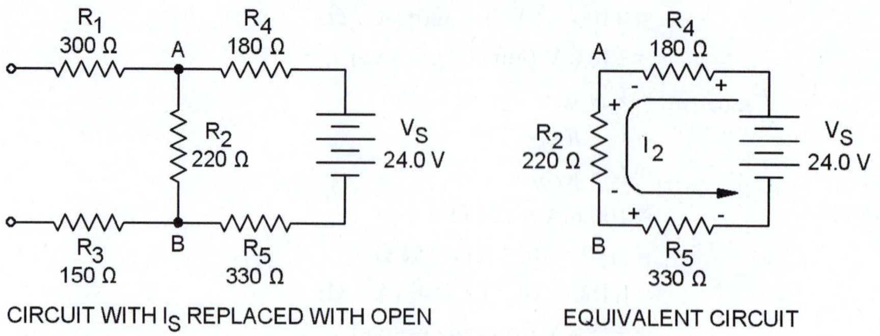

Figure 8-11: Circuit Analysis for V_S for Figure 8-8

After replacing I_S with its internal resistance (an open), no current flows through R_1 or R_3 so the resistors drop no voltage.

$V_{R1\,(VS)} = 0\text{ V}$

$V_{R3\,(VS)} = 0\text{ V}$

Because R_1 and R_3 drop no voltage due to V_S, they can be removed from the equivalent circuit in Figure 8-11. The total resistance seen by V_S are R_4, R_2, and R_5 in series.

$R_{T\,(VS)} = R_4 + R_2 + R_5$
$= 220\ \Omega + 180\ \Omega + 330\ \Omega$
$= 730\ \Omega$

Use the voltage divider theorem to find the voltage drops for the other resistors.

$V_{R2\,(VS)} = V_S \times (R_2 / R_{T\,(VS)})$
$= 24.0\text{ V} \times (220\ \Omega / 730\ \Omega)$
$= 24.0\text{ V} \times [(220 / 730)\ (\Omega / \Omega)]$
$= 24.0\text{ V} \times 0.301$
$= +7.23\text{ V}$ (top end positive)

$V_{R4\,(VS)} = V_S \times (R_4 / R_{T\,(VS)})$
$= 24.0\text{ V} \times (180\ \Omega / 730\ \Omega)$
$= 24.0\text{ V} \times [(180 / 730)\ (\Omega / \Omega)]$
$= 24.0\text{ V} \times 0.247$
$= -5.92\text{ V}$ (left end negative)

$V_{R5\,(VS)} = V_S \times (R_5 / R_{T\,(VS)})$
$= 24.0\text{ V} \times (330\ \Omega / 730\ \Omega)$
$= 24.0\text{ V} \times [(330 / 730)\ (\Omega / \Omega)]$
$= 24.0\text{ V} \times 0.452$
$= +10.8\text{ V}$ (left end positive)

Algebraically sum the voltages from I_S and V_S to find the voltages across each resistor.

$V_{R1} = V_{R1\,(IS)} + V_{R1\,(VS)}$
$= (+30.0\text{ V}) + (0\text{ V})$
$= +30.0\text{ V}$ (left end positive)

$V_{R2} = V_{R2\,(IS)} + V_{R2\,(VS)}$
$= (+15.4\text{ V}) + (+7.23\text{ V})$
$= +22.6\text{ V}$ (top end positive)

$V_{R3} = V_{R3\,(IS)} + V_{R3\,(VS)}$
$= (-15.0\text{ V}) + (0\text{ V})$
$= \mathbf{-15.0\ V}$ **(left end positive)**

$V_{R4} = V_{R4\,(IS)} + V_{R4\,(VS)}$
$= (+5.42\text{ V}) + (-5.92\text{ V})$
$= \mathbf{-0.493\ V}$ **(left end negative)**

$V_{R5} = V_{R5\,(IS)} + V_{R5\,(VS)}$
$= (-9.95\text{ V}) + (+10.8\text{ V})$
$= \mathbf{+0.904\ V}$ **(left end positive)**

8.1.3 Thevenin's Theorem

1. Problem: What are the values of V_{OUT} for the circuit in Figure 8-12 if R_L is halved and doubled?

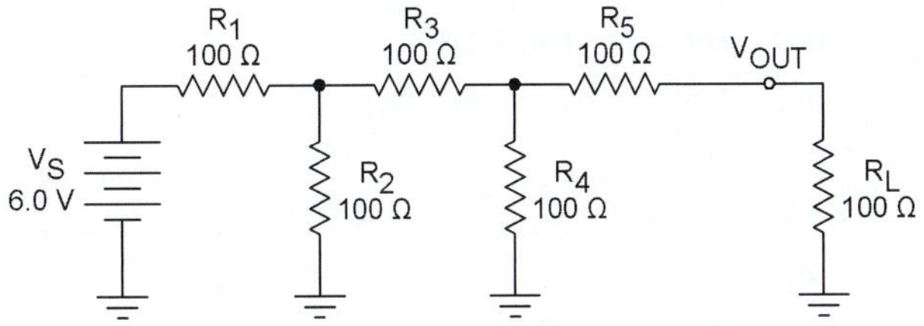

Figure 8-12: Thevenin's Theorem Example Circuit 1

Solution: Rather than change the value of R_L in the circuit to 50 Ω and 200 Ω and analyze each new circuit, it is simpler to Thevenize the circuit relative to R_L to find the two voltages. The first step is to find the open load voltage V_{OL} with R_L removed. Refer to Figure 8-13.

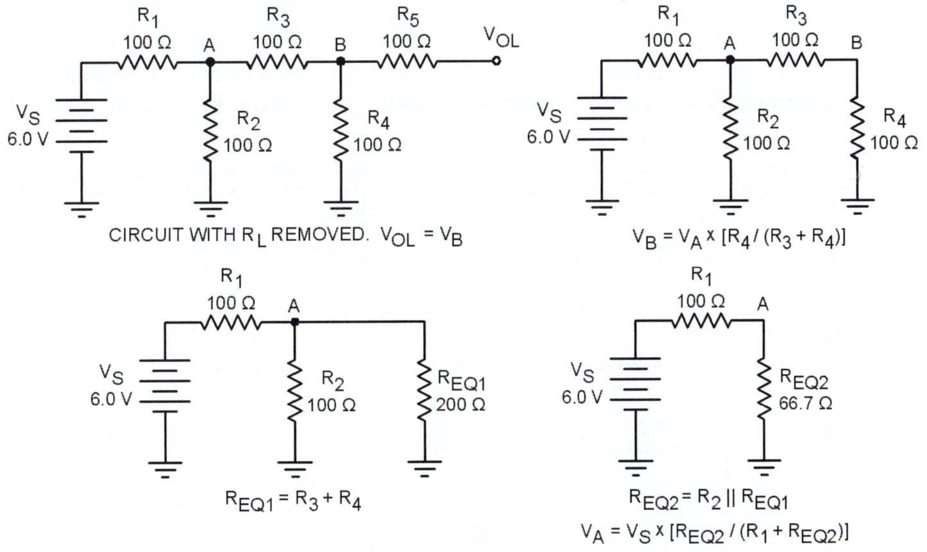

Figure 8-13: Circuit Analysis and Simplification Process for V_{OL}

With R_L removed, no current flows through R_5, so the circuit analysis can ignore R_5 and $V_{OL} = V_B$. From the voltage divider theorem, $V_B = V_A \times [R_4 / (R_3 + R_4)]$, so once V_A is known, V_{OL} can be calculated. The expression for the circuit with R_L removed is

$$R_T = R_1 + R_2 \parallel (R_3 + R_4)$$

The first equivalent resistance R_{EQ1} is the series combination $(R_3 + R_4)$.

$$R_T = R_1 + R_2 \parallel R_{EQ1}$$

where

$$\begin{aligned}
\boldsymbol{R_{EQ1}} &= \boldsymbol{R_3} + \boldsymbol{R_4} \\
&= 100\ \Omega + 100\ \Omega \\
&= (100 + 100)\ \Omega \\
&= 200\ \Omega
\end{aligned}$$

The second equivalent resistance $\boldsymbol{R_{EQ2}}$ is the parallel combination $\boldsymbol{R_2} \parallel \boldsymbol{R_{EQ1}}$.

$$\begin{aligned}
\boldsymbol{R_{EQ2}} &= \boldsymbol{R_2} \parallel \boldsymbol{R_{EQ1}} \\
&= 100\ \Omega \parallel 200\ \Omega \\
&= 66.7\ \Omega
\end{aligned}$$

Use the voltage divider theorem to find $\boldsymbol{V_A}$ and $\boldsymbol{V_{TH}}$.

$$\begin{aligned}
\boldsymbol{V_A} &= \boldsymbol{V_S} \times [\boldsymbol{R_{EQ2}} / (\boldsymbol{R_1} + \boldsymbol{R_{EQ2}})] \\
&= 6.0\ \text{V} \times [66.7\ \Omega / (100\ \Omega + 66.7\ \Omega)] \\
&= 6.0\ \text{V} \times (66.7\ \Omega / 166.7\ \Omega) \\
&= 6.0\ \text{V} \times [(66.7 / 166.7)\ (\Omega / \Omega)] \\
&= 6.0\ \text{V} \times 0.4 \\
&= 2.4\ \text{V}
\end{aligned}$$

$$\begin{aligned}
\boldsymbol{V_{TH}} &= \boldsymbol{V_B} \\
&= \boldsymbol{V_A} \times [\boldsymbol{R_4} / (\boldsymbol{R_3} + \boldsymbol{R_4})] \\
&= 2.4\ \text{V} \times [100\ \Omega / (100\ \Omega + 100\ \Omega)] \\
&= 2.4\ \text{V} \times (100\ \Omega / 200\ \Omega) \\
&= 2.4\ \text{V} \times [(100 / 200)\ (\Omega / \Omega)] \\
&= 2.4\ \text{V} \times 0.5 \\
&= 1.2\ \text{V}
\end{aligned}$$

The second step is to replace the voltage source $\boldsymbol{V_S}$ with its internal resistance (a short) and determine the resistance looking back into the circuit from $\boldsymbol{R_L}$. Refer to Figure 8-14.

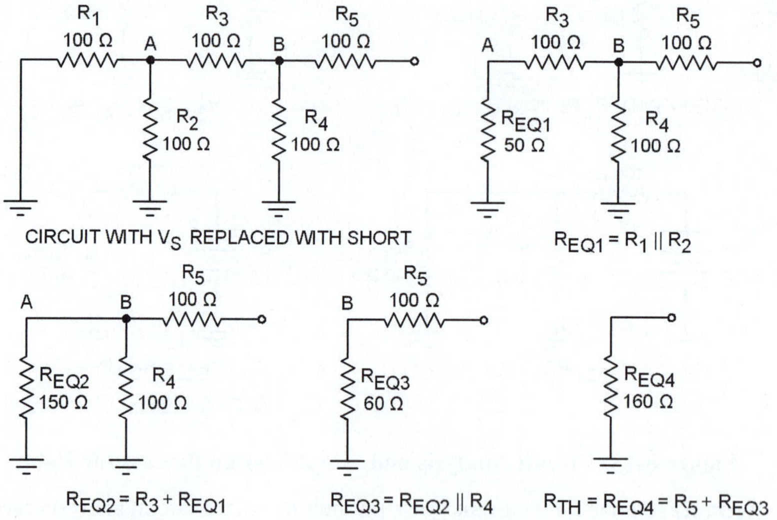

Figure 8-14: Circuit Analysis and Simplification Process for $\boldsymbol{R_{TH}}$

With $\boldsymbol{V_S}$ replaced by a short, the expression for the Thevenin resistance circuit is

$$\boldsymbol{R_{TH}} = \boldsymbol{R_5} + \boldsymbol{R_4} \parallel (\boldsymbol{R_3} + \boldsymbol{R_2} \parallel \boldsymbol{R_1})$$

The first equivalent resistance $\boldsymbol{R_{EQ1}}$ is the parallel combination $\boldsymbol{R_2} \parallel \boldsymbol{R_1}$.

$$\boldsymbol{R_{TH}} = \boldsymbol{R_5} + \boldsymbol{R_4} \parallel (\boldsymbol{R_3} + \boldsymbol{R_{EQ1}})$$

where

$$\begin{aligned} \boldsymbol{R_{EQ1}} &= \boldsymbol{R_2} \parallel \boldsymbol{R_1} \\ &= 100\ \Omega \parallel 100\ \Omega \\ &= 50\ \Omega \end{aligned}$$

The second equivalent resistance $\boldsymbol{R_{EQ2}}$ is the series combination ($\boldsymbol{R_3} + \boldsymbol{R_{EQ1}}$).

$$\boldsymbol{R_{TH}} = \boldsymbol{R_5} + \boldsymbol{R_4} \parallel \boldsymbol{R_{EQ2}}$$

where

$$\begin{aligned} \boldsymbol{R_{EQ2}} &= \boldsymbol{R_3} + \boldsymbol{R_{EQ1}} \\ &= 100\ \Omega + 50\ \Omega \\ &= 150\ \Omega \end{aligned}$$

The third equivalent resistance $\boldsymbol{R_{EQ3}}$ is the parallel combination $\boldsymbol{R_4} \parallel \boldsymbol{R_{EQ2}}$.

$$\begin{aligned} \boldsymbol{R_{EQ3}} &= \boldsymbol{R_4} \parallel \boldsymbol{R_{EQ2}} \\ &= 100\ \Omega \parallel 150\ \Omega \\ &= 60\ \Omega \end{aligned}$$

The fourth (and final equivalent) resistance $\boldsymbol{R_{EQ4}}$ is the series combination $\boldsymbol{R_5} + \boldsymbol{R_{EQ3}}$.

$$\boldsymbol{R_{TH}} = \boldsymbol{R_{EQ4}}$$

where

$$\begin{aligned} \boldsymbol{R_{EQ4}} &= \boldsymbol{R_5} + \boldsymbol{R_{EQ3}} \\ &= 100\ \Omega + 60\ \Omega \\ &= 160\ \Omega \end{aligned}$$

Figure 8-15 shows the Thevenin equivalent circuits in which $\boldsymbol{R_L} = 50\ \Omega$ and $\boldsymbol{R_L} = 200\ \Omega$.

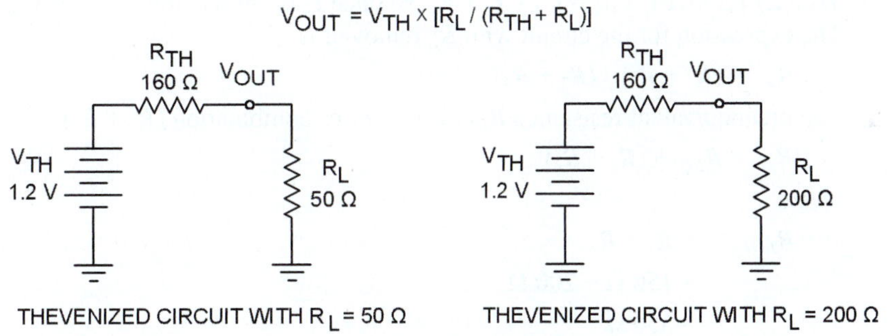

Figure 8-15: Thevenin Equivalent Circuits with 50 Ω and 200 Ω Loads

Use the voltage divider theorem to find $\boldsymbol{V_{OUT}}$ for $\boldsymbol{R_L} = 50\ \Omega$ and $200\ \Omega$.

$$\begin{aligned} \boldsymbol{V_{OUT\,(50\,\Omega)}} &= \boldsymbol{V_{TH}} \times [\boldsymbol{R_L} / (\boldsymbol{R_{TH}} + \boldsymbol{R_L})] \\ &= 1.2\ \text{V} \times [50\ \Omega / (160\ \Omega + 50\ \Omega)] \\ &= 1.2\ \text{V} \times (50\ \Omega / 210\ \Omega) \\ &= 1.2\ \text{V} \times [(50 / 210)\ (\Omega / \Omega)] \\ &= 1.2\ \text{V} \times 0.238 \\ &= 286 \times 10^{-3}\ \text{V} \\ &= \mathbf{286\ mV} \end{aligned}$$

$$\begin{aligned} \boldsymbol{V_{OUT\,(200\,\Omega)}} &= \boldsymbol{V_{TH}} \times [\boldsymbol{R_L} / (\boldsymbol{R_{TH}} + \boldsymbol{R_L})] \\ &= 1.2\ \text{V} \times [200\ \Omega / (160\ \Omega + 200\ \Omega)] \\ &= 1.2\ \text{V} \times (200\ \Omega / 360\ \Omega) \end{aligned}$$

$= 1.2\text{ V} \times [(200 / 360)\ (\Omega / \Omega)]$

$= 1.2\text{ V} \times 0.556$

$= 667 \times 10^{-3}\text{ V}$

$= \mathbf{667\ mV}$

2. Problem: Determine the Thevenin equivalent with respect to $\boldsymbol{R_L}$ for the circuit in Figure 8-16.

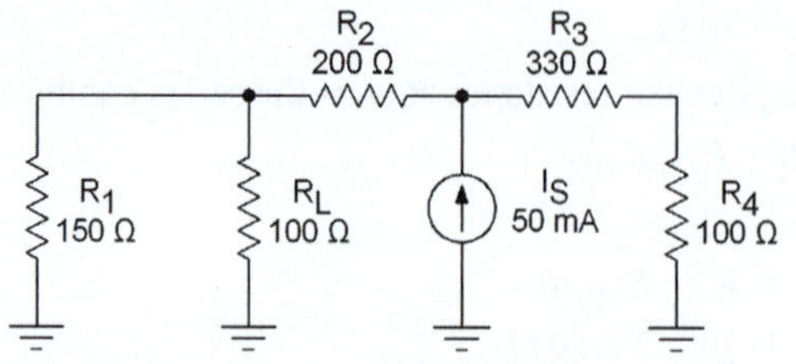

Figure 8-16: Thevenin's Theorem Example Circuit 2

Solution: The first step is to find the open load voltage $\boldsymbol{V_{OL}}$ with $\boldsymbol{R_L}$ removed. Refer to Figure 8-17.

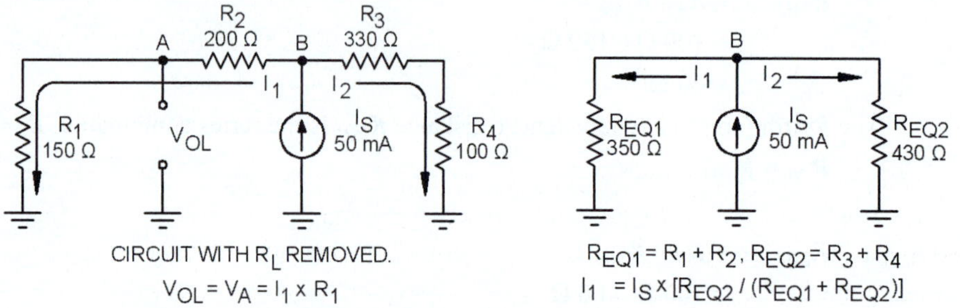

Figure 8-17: Circuit Analysis and Simplification Process for $\boldsymbol{V_{OL}}$

With $\boldsymbol{R_L}$ removed, $\boldsymbol{V_{OL}} = \boldsymbol{V_A} = \boldsymbol{I_1} \times \boldsymbol{R_1}$, so once $\boldsymbol{I_1}$ is known, the value of $\boldsymbol{V_{OL}}$ can be calculated. The expression for the circuit with $\boldsymbol{R_L}$ removed is

$\boldsymbol{R_T} = (\boldsymbol{R_1} + \boldsymbol{R_2}) \parallel (\boldsymbol{R_3} + \boldsymbol{R_4})$

The first equivalent resistance $\boldsymbol{R_{EQ1}}$ is the series combination $(\boldsymbol{R_1} + \boldsymbol{R_2})$.

$\boldsymbol{R_T} = \boldsymbol{R_{EQ1}} \parallel (\boldsymbol{R_3} + \boldsymbol{R_4})$

where

$\boldsymbol{R_{EQ1}} = \boldsymbol{R_1} + \boldsymbol{R_2}$

$= 150\ \Omega + 200\ \Omega$

$= 350\ \Omega$

The second equivalent resistance $\boldsymbol{R_{EQ2}}$ is the series combination $(\boldsymbol{R_3} + \boldsymbol{R_4})$.

$\boldsymbol{R_T} = \boldsymbol{R_{EQ1}} \parallel \boldsymbol{R_{EQ2}}$

where

$\boldsymbol{R_{EQ2}} = \boldsymbol{R_3} + \boldsymbol{R_4}$

$= 330\ \Omega + 100\ \Omega$

$= 430\ \Omega$

Use the current divider theorem to find $\boldsymbol{I_1}$.

$\boldsymbol{I_1} = \boldsymbol{I_S} \times [\boldsymbol{R_{EQ2}} / (\boldsymbol{R_{EQ1}} + \boldsymbol{R_{EQ2}})]$

$= 50\text{ mA} \times [430\ \Omega / (350\ \Omega + 430\ \Omega)]$

$= 50\text{ mA} \times (430\ \Omega / 780\ \Omega)$

$= 50\text{ mA} \times [(430 / 780)\ (\Omega / \Omega)]$

$= (50 \times 10^{-3}\ \text{A}) \times 0.551$

$= 27.6 \times 10^{-3}\ \text{A}$

Use Ohm's Law for voltage to find V_{TH}.

$V_{TH} = V_{OL}$

$= V_A$

$= I_1 \times R_1$

$= (27.6 \times 10^{-3}\ \text{A}) \times (150\ \Omega)$

$= [(27.6 \times 10^{-3}) \times 150]\ (\text{A} \times \Omega)$

$= \mathbf{4.13\ V}$

The second step is to replace the current source I_S with its internal resistance (an open) and determine the resistance looking back into the circuit from R_L. Refer to Figure 8-18.

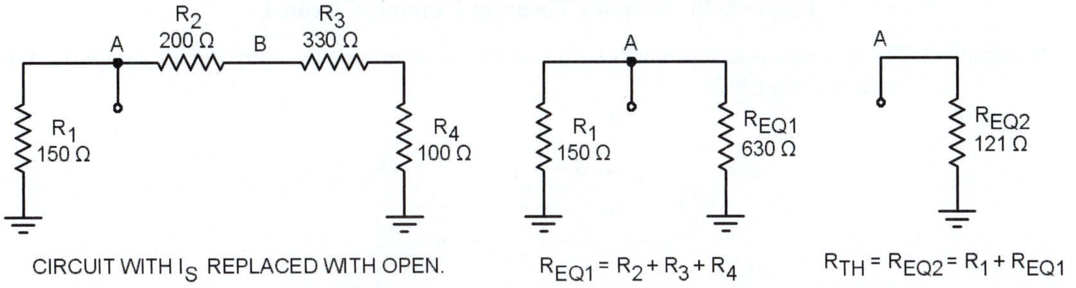

Figure 8-18: Circuit Analysis and Simplification Technique for R_{TH}

With I_S replaced with an open, the expression for the circuit is

$R_{TH} = R_1 \parallel (R_2 + R_3 + R_4)$

The first equivalent resistance R_{EQ1} is the series combination $(R_2 + R_3 + R_4)$.

$R_{TH} = R_1 \parallel R_{EQ1}$

where

$R_{EQ1} = R_2 + R_3 + R_4$

$= 200\ \Omega + 330\ \Omega + 100\ \Omega$

$= 630\ \Omega$

The second (and final) equivalent resistance R_{EQ2} is the parallel combination $R_1 \parallel R_{EQ1}$.

$R_{TH} = R_{EQ2}$

where

$R_{EQ2} = R_1 \parallel R_{EQ1}$

$= 150\ \Omega \parallel 630\ \Omega$

$= \mathbf{121\ \Omega}$

Figure 8-19 shows the Thevenin equivalent for Figure 8-16.

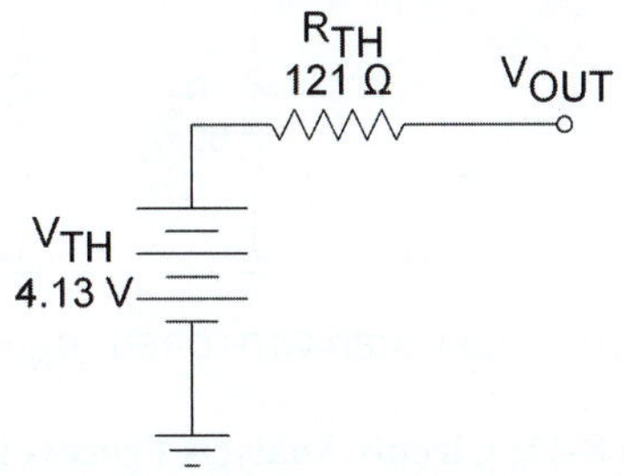

Figure 8-19: Thevenin Equivalent for Figure 8-16

8.1.4 Norton's Theorem

1. Problem: Determine the Norton equivalent with respect to R_L for the circuit in Figure 8-20.

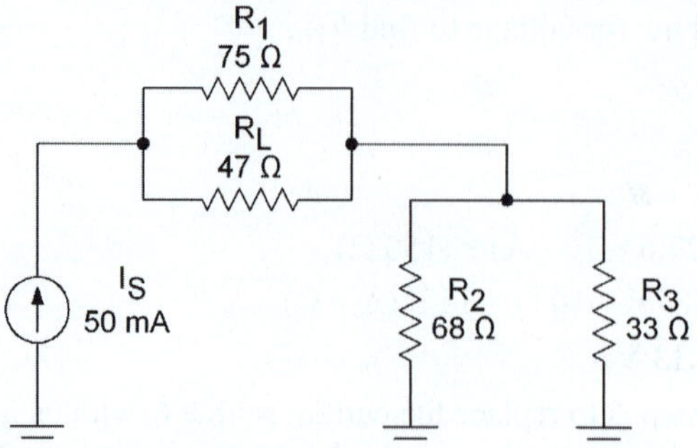

Figure 8-20: Norton's Theorem Example Circuit 1

Solution: The first step is to replace R_L with a short and determine the short circuit current I_{SC} through it. Refer to Figure 8-21.

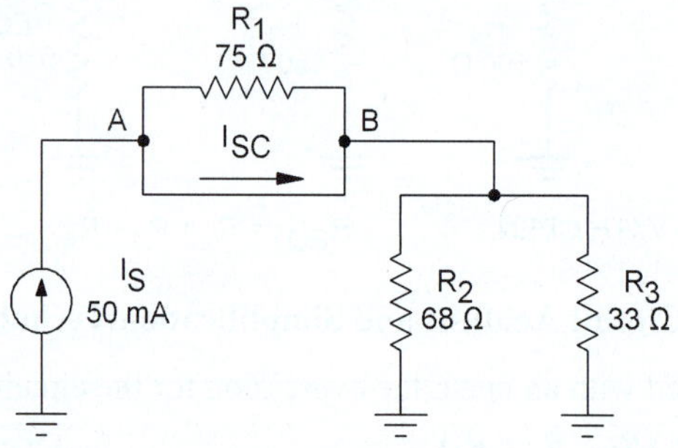

CIRCUIT WITH RL REPLACED WITH SHORT. $I_{SC} = I_S$.

Figure 8-21: Circuit Analysis Process for I_{SC}

With R_L replaced with a short, the short bypasses R_1 so that all the source current I_S passes from point A to point B through the short. Therefore,

$$I_N = I_{SC}$$
$$= I_S$$
$$= 50 \text{ mA}$$

The second step is to replace the current source I_S with its internal resistance (an open) and determine the resistance looking back into the circuit from R_L. Refer to Figure 8-22.

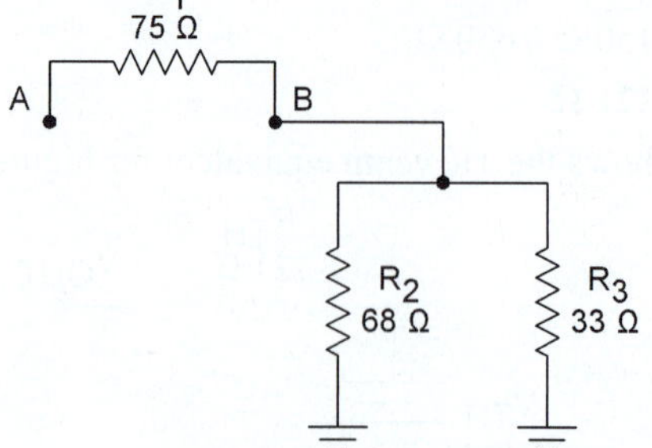

CIRCUIT WITH I_S REPLACED WITH OPEN. $R_N = R_{AB} = R_1$.

Figure 8-22: Circuit Analysis Process for R_N

The resistance between point A and point B is just $\boldsymbol{R_1}$. Therefore,

$$\begin{aligned} \boldsymbol{R_N} &= \boldsymbol{R_{AB}} \\ &= \boldsymbol{R_1} \\ &= 75\ \Omega \end{aligned}$$

Figure 8-23 shows the Norton equivalent for Figure 8-20.

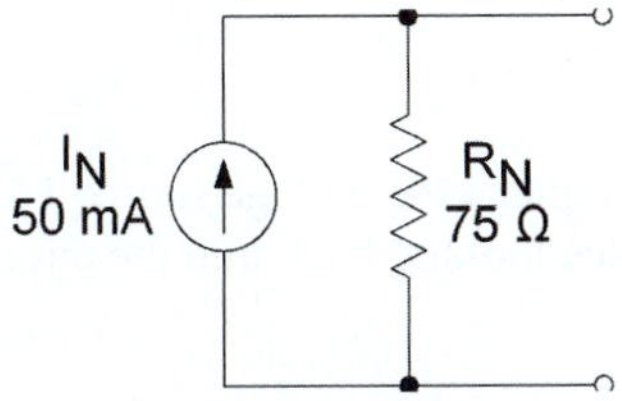

Figure 8-23: Norton Equivalent for Figure 8-20

2. Problem: Determine the Norton equivalent with respect to $\boldsymbol{R_L}$ for the circuit in Figure 8-24.

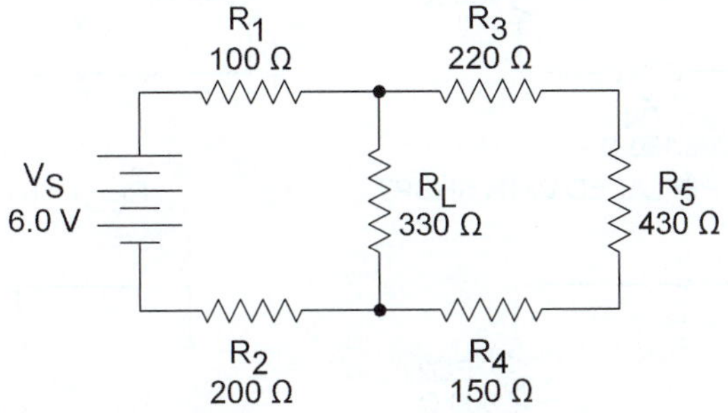

Figure 8-24: Norton's Theorem Example Circuit 2

Solution: The first step is to replace $\boldsymbol{R_L}$ with a short and determine the short circuit current $\boldsymbol{I_{SC}}$ through it. Refer to Figure 8-25.

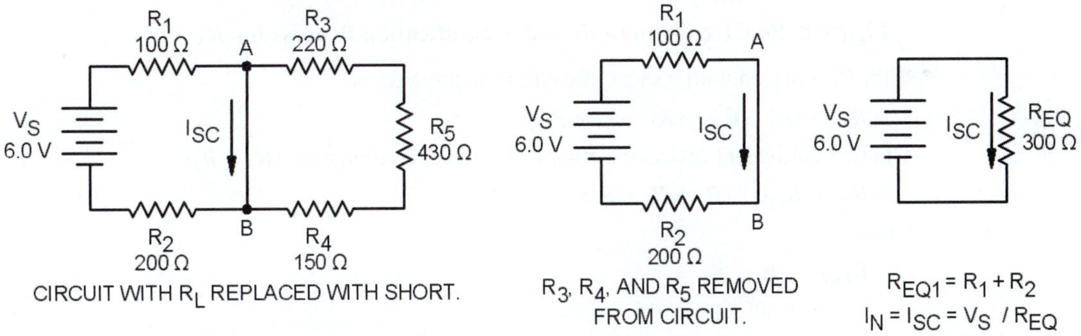

Figure 8-25: Circuit Analysis and Simplification Process for I_N

Replacing $\boldsymbol{R_L}$ with a short bypasses $\boldsymbol{R_3}$, $\boldsymbol{R_4}$, and $\boldsymbol{R_5}$ so that the circuit analysis can ignore them. The expression for the circuit is

$$\boldsymbol{R_T} = \boldsymbol{R_1} + \boldsymbol{R_2}$$

The first (and final) circuit equivalent $\boldsymbol{R_{EQ1}}$ consists of the series combination $\boldsymbol{R_1} + \boldsymbol{R_2}$.

$$\boldsymbol{R_T} = \boldsymbol{R_{EQ}}$$

where

$$\begin{aligned} \boldsymbol{R_{EQ}} &= \boldsymbol{R_1} + \boldsymbol{R_2} \\ &= 100\ \Omega + 200\ \Omega \\ &= 300\ \Omega \end{aligned}$$

From Ohm's Law for current,

$$I_N = I_{SC}$$
$$= V_S / R_{EQ}$$
$$= 6.0\ \text{V} / 300\ \Omega$$
$$= 20 \times 10^{-3}\ \text{V} / \Omega$$
$$= 20 \times 10^{-3}\ \text{A}$$
$$= 20\ \text{mA}$$

The second step is to replace the voltage source V_S with its internal resistance (a short) and determine the resistance looking back into the circuit from R_L. Refer to Figure 8-26.

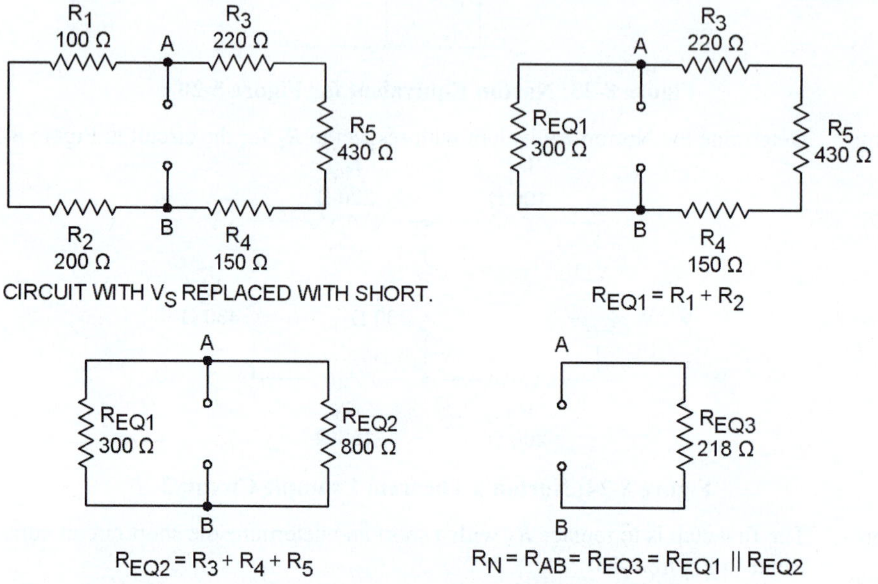

Figure 8-26: Circuit Analysis and Simplification Process for R_N

With V_S replaced with a short, the circuit expression is

$$R_N = (R_1 + R_2) \parallel (R_3 + R_4 + R_5)$$

The first equivalent resistance R_{EQ1} is the series combination $(R_1 + R_2)$.

$$R_N = R_{EQ1} \parallel (R_3 + R_4 + R_5)$$

where

$$R_{EQ1} = R_1 + R_2$$
$$= 100\ \Omega + 200\ \Omega$$
$$= 300\ \Omega$$

The second equivalent resistance R_{EQ2} is the series combination $(R_3 + R_4 + R_5)$.

$$R_N = R_{EQ1} \parallel R_{EQ2}$$

where

$$R_{EQ2} = R_3 + R_4 + R_5$$
$$= 220\ \Omega + 150\ \Omega + 430\ \Omega$$
$$= 800\ \Omega$$

The third (and final) equivalent resistance R_{EQ3} is the parallel combination $R_{EQ1} \parallel R_{EQ2}$.

$$R_N = R_{EQ3}$$

where

$$\boldsymbol{R_{EQ3}} = \boldsymbol{R_{EQ1}} \parallel \boldsymbol{R_{EQ2}}$$
$$= 300\ \Omega \parallel 800\ \Omega)$$
$$= \mathbf{218\ \Omega}$$

Figure 8-27 shows the Norton equivalent for Figure 8-24.

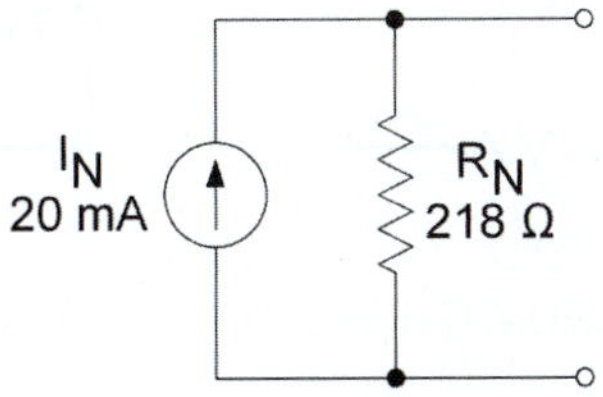

Figure 8-27: Norton Equivalent for Figure 8-24

8.1.5 Maximum Power Transfer Theorem

1. Problem: If $\boldsymbol{R_L}$ for the circuit in Figure 8-28 decreases, will its dissipated power $\boldsymbol{P_{RL}}$ initially increase or decrease?

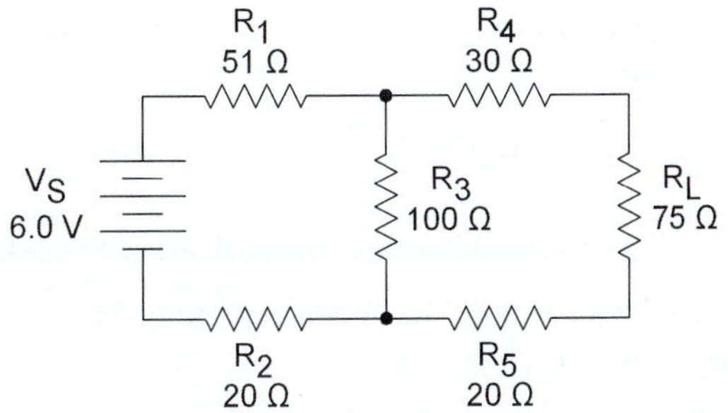

Figure 8-28: Maximum Power Transfer Theorem Example Circuit 1

Solution: $\boldsymbol{R_L}$ dissipates maximum power when $\boldsymbol{R_L} = \boldsymbol{R_{TH}}$. If $\boldsymbol{R_L} > \boldsymbol{R_{TH}}$, then decreasing $\boldsymbol{R_L}$ will initially move it closer to $\boldsymbol{R_{TH}}$ so that power dissipation will initially increase. If $\boldsymbol{R_L} < \boldsymbol{R_{TH}}$, then decreasing $\boldsymbol{R_L}$ will move it further from $\boldsymbol{R_{TH}}$ so that power will decrease. Refer to Figure 8-29.

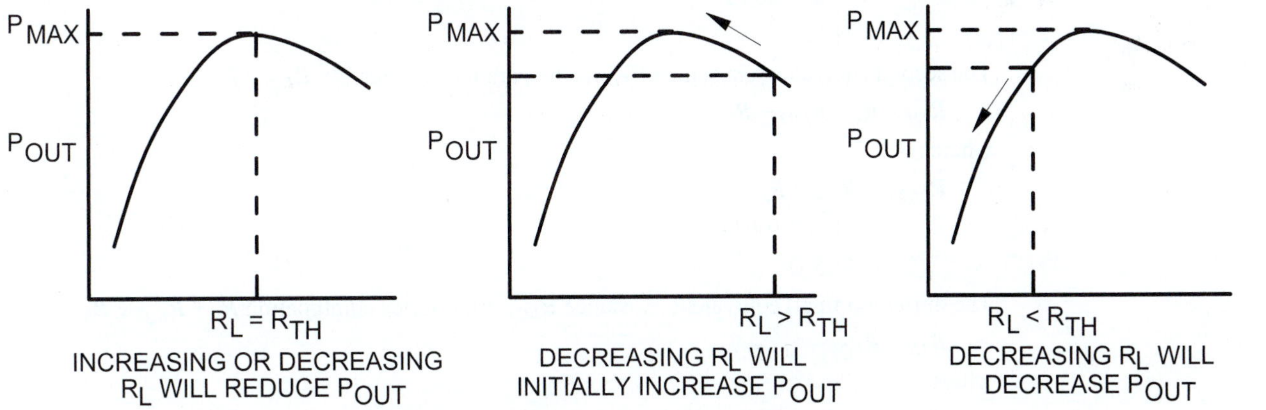

Figure 8-29: Effect on P_{OUT} of Decreasing R_L

To find $\boldsymbol{R_{TH}}$, replace the voltage source V_S with its internal resistance (a short) and determine the resistance looking back into the circuit from $\boldsymbol{R_L}$. Refer to Figure 8-30.

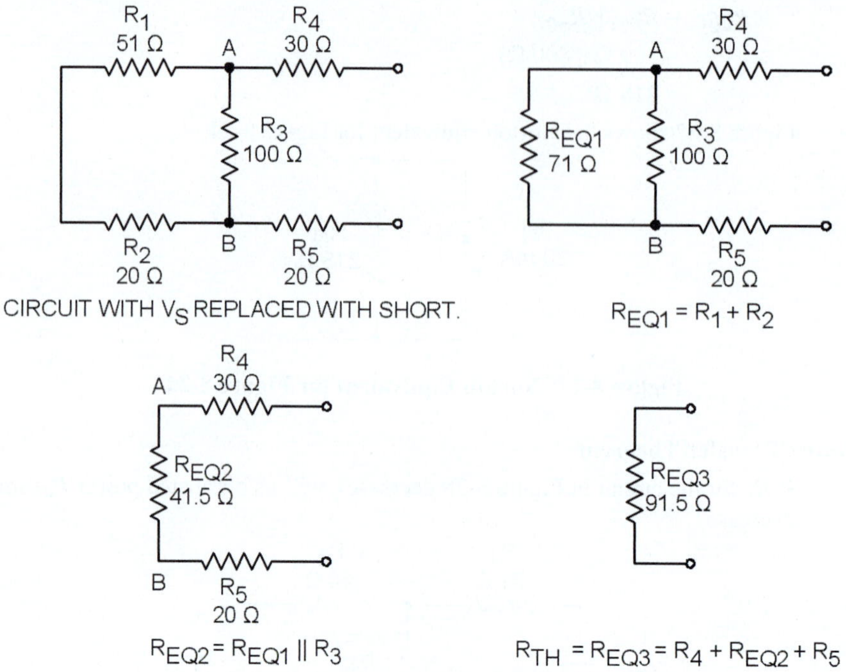

Figure 8-30: Circuit Analysis and Simplification Process for R_{TH}

With V_S replaced with a short, the circuit expression is

$$R_{TH} = R_4 + (R_1 + R_2) \parallel R_3 + R_5$$

The first equivalent resistance R_{EQ1} is the series combination ($R_1 + R_2$).

$$R_{TH} = R_4 + R_{EQ1} \parallel R_3 + R_5$$

where

$$\begin{aligned} R_{EQ1} &= R_1 + R_2 \\ &= 51\ \Omega + 20\ \Omega \\ &= 71\ \Omega \end{aligned}$$

The second equivalent resistance R_{EQ2} is the parallel combination $R_{EQ1} \parallel R_3$.

$$R_{TH} = R_4 + R_{EQ2} + R_5$$

where

$$\begin{aligned} R_{EQ2} &= R_{EQ1} \parallel R_3 \\ &= 71\ \Omega \parallel 100\ \Omega \\ &= 41.5\ \Omega \end{aligned}$$

The third (and final) equivalent resistance R_{EQ3} is the series combination $R_4 + R_{EQ2} + R_5$.

$$R_{TH} = R_{EQ3}$$

where

$$\begin{aligned} R_{EQ3} &= R_4 + R_{EQ2} + R_5 \\ &= 30\ \Omega + 41.5\ \Omega + 20\ \Omega \\ &= 91.5\ \Omega \end{aligned}$$

For $R_L = 75\ \Omega$, $R_L < R_{TH}$ so decreasing R_L will move it further from R_{TH} and move P_{OUT} further from P_{MAX}. Decreasing R_L will initially **decrease** the power dissipation.

2. Problem: Will the maximum power dissipation for variable resistor R_L in Figure 8-31 exceed the 1/8 W power rating?

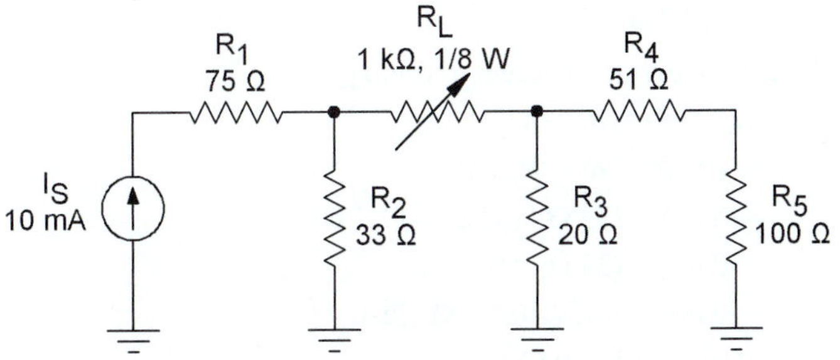

Figure 8-31: Maximum Power Transfer Theorem Example Circuit 2

Solution: To determine the power dissipated by R_L, find the Norton equivalent circuit. If the value of the Norton resistance R_N is between R_L (min) and R_L (max), then R_L will dissipate maximum power when $R_S = R_N$. The first step is to replace R_L with a short and determine the short circuit current $I_{SC} = I_N$ through the short. Refer to Figure 8-32.

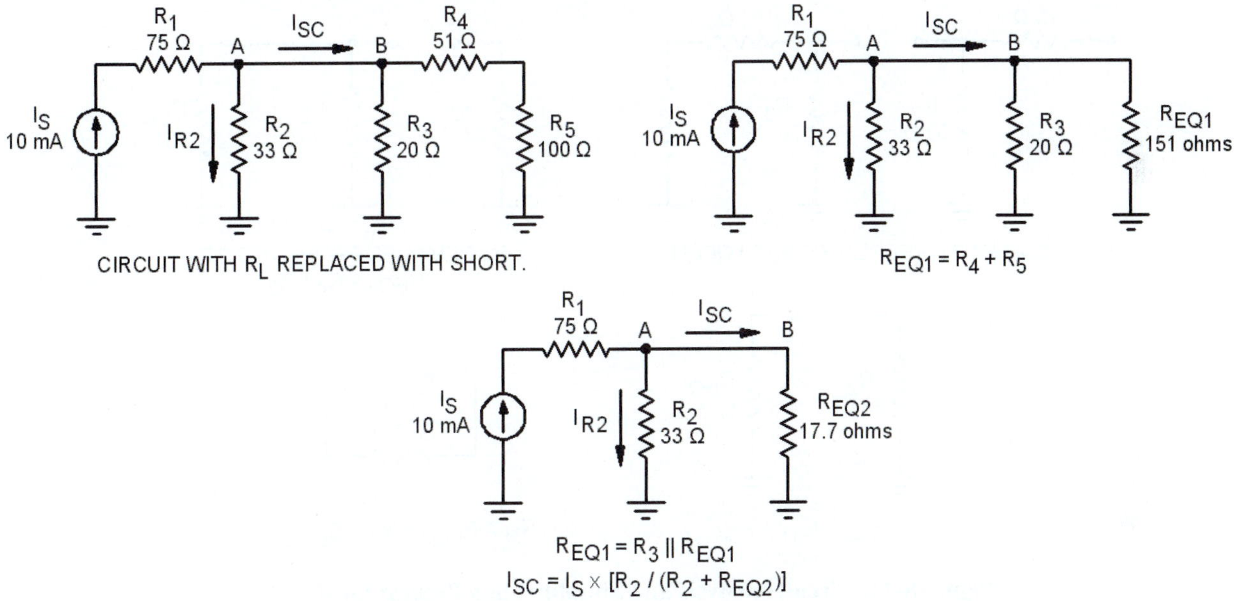

Figure 8-32: Circuit Analysis and Simplification Process for I_N

The source current I_S flows through R_1 and then divides between the current I_{R2} and I_{SC}. The resistance R seen by I_{SC} is

$$R = R_3 \parallel (R_4 + R_5)$$

The first equivalent resistance R_{EQ1} is the series combination ($R_4 + R_5$).

$$R = R_3 \parallel R_{EQ1}$$

where

$$R_{EQ1} = R_4 + R_5$$
$$= 51\ \Omega + 100\ \Omega$$
$$= 151\ \Omega$$

The second (and final) equivalent resistance R_{EQ2} is the parallel combination $R_3 \parallel R_{EQ1}$.

$$R = R_{EQ2}$$

where

$$R_{EQ2} = R_3 \parallel R_{EQ1}$$
$$= 20\ \Omega \parallel 151\ \Omega$$
$$= 17.7\ \Omega$$

Use the current divider theorem to find I_{SC}.

$$I_{SC} = I_S \times [R_2 / (R_2 + R)]$$
$$= I_S \times [R_2 / (R_2 + R_{EQ2})]$$
$$= 10\text{ mA} \times [33\ \Omega / (33\ \Omega + 17.7\ \Omega)]$$
$$= 10\text{ mA} \times (33\ \Omega / 50.7\ \Omega)$$
$$= 10\text{ mA} \times [(33 / 50.7)\ (\Omega / \Omega)]$$
$$= (10 \times 10^{-3}) \times 0.651$$
$$= 6.51 \times 10^{-3}$$
$$= 6.51\text{ mA}$$

To find the effective source resistance, $R_S = R_N$, first replace I_S with its internal resistance (an open) and determine the resistance looking back into the circuit from R_L. Refer to Figure 8-33.

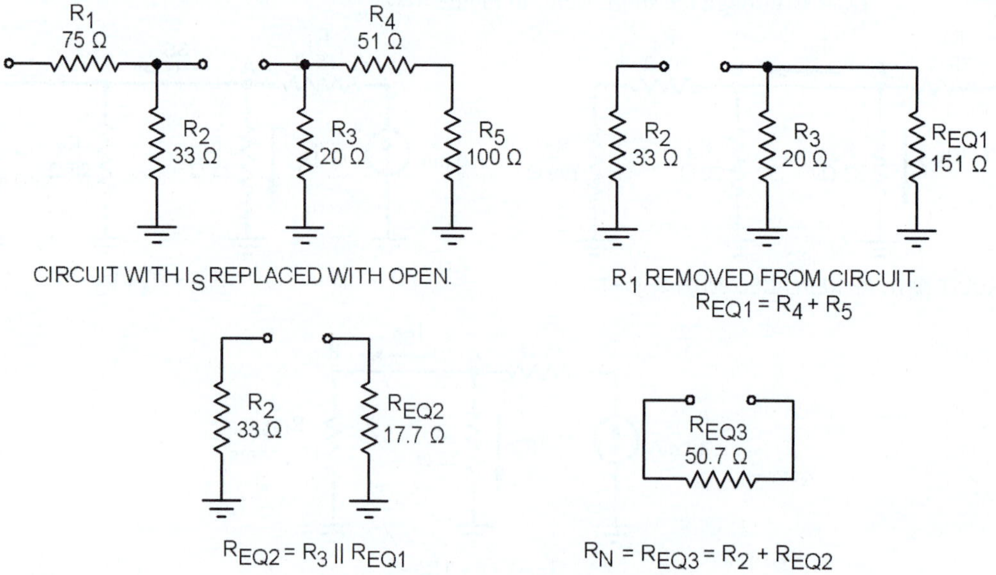

Figure 8-33: Circuit Analysis and Simplification Process for $R_S = R_N$

When I_S is replaced with an open, R_1 connects to the circuit at only one end so that it has no effect and can be ignored for the circuit analysis. The expression for the circuit is

$$R_N = R_2 + R_3 \parallel (R_4 + R_5)$$

The first equivalent resistance R_{EQ1} consists of the series combination $(R_4 + R_5)$.

$$R_N = R_2 + R_3 \parallel R_{EQ1}$$

where

$$R_{EQ1} = R_4 + R_5$$
$$= 51\ \Omega + 100\ \Omega$$
$$= 151\ \Omega$$

The second equivalent resistance R_{EQ2} consists of the parallel combination $R_3 \parallel R_{EQ1}$.

$$R_N = R_2 + R_{EQ2}$$

where

$\boldsymbol{R_{EQ2}} = \boldsymbol{R_3} \parallel \boldsymbol{R_{EQ1}}$
$= 20\ \Omega \parallel 151\ \Omega$
$= 17.7\ \Omega$

The third (and final) equivalent resistance $\boldsymbol{R_{EQ3}}$ consists of the series combination $\boldsymbol{R_2} + \boldsymbol{R_{EQ2}}$.

$\boldsymbol{R_N} = \boldsymbol{R_{EQ3}}$

where

$\boldsymbol{R_{EQ3}} = \boldsymbol{R_2} + \boldsymbol{R_{EQ2}}$
$= 33\ \Omega + 17.7\ \Omega$
$= 50.7\ \Omega$

Figure 8-34 shows the Norton equivalent for the circuit.

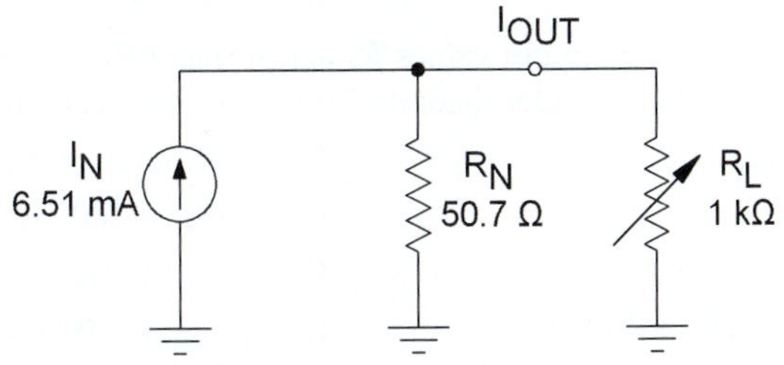

Figure 8-34: Norton Equivalent for Figure 8-31

Because $\boldsymbol{R_N}$ is between $\boldsymbol{R_L}$ (min) and $\boldsymbol{R_L}$ (max), $\boldsymbol{R_L}$ will dissipate its maximum power when it is equal to $\boldsymbol{R_N}$. When $\boldsymbol{R_L} = \boldsymbol{R_N}$, the source current $\boldsymbol{I_N}$ divides equally between $\boldsymbol{R_N}$ and $\boldsymbol{R_L}$.

$\boldsymbol{I_{RL}} = \boldsymbol{I_N} / 2$
$= 6.51\text{ mA} / 2$
$= 3.26\text{ mA}$

Use Watt's Law for current and resistance to calculate the power $\boldsymbol{P_{RL}}$ dissipated by $\boldsymbol{R_L}$.

$\boldsymbol{P_{RL}} = \boldsymbol{I_{RL}}^2 \times \boldsymbol{R_L}$
$= (3.26\text{ mA})^2 \times 1\text{ k}\Omega$
$= (3.26 \times 10^{-3}\text{ A})^2 \times (1 \times 10^3\ \Omega)$
$= [(3.26 \times 10^{-3})^2 \times (1 \times 10^3)]\ (\text{A}^2 \times \Omega)$
$= (10.6 \times 10^{-6}) \times (1 \times 10^3)\text{ W}$
$= 10.6 \times 10^{-3}\text{ W}$
$= 10.6\text{ mW}$

1/8 W = 125 mW so the variable resistor will **not** exceed its maximum power dissipation.

8.2 Advanced

8.2.1 Source Loading

1. Problem: Determine the value of $\boldsymbol{R_S}$ for the circuit in Figure 8-35 for the indicated minimum and maximum values of $\boldsymbol{V_{OUT}}$.

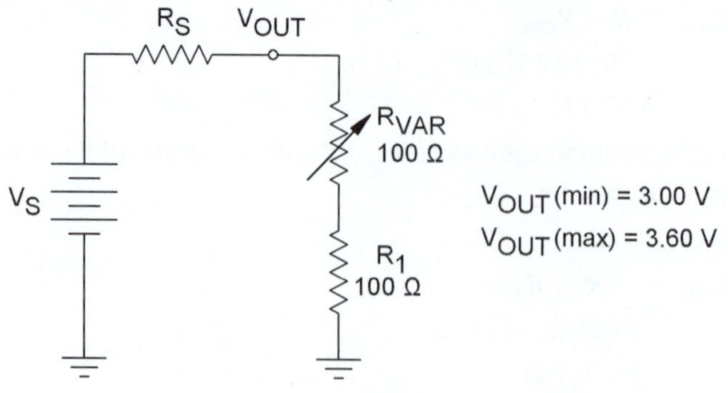

Figure 8-35: Source Loading Example Circuit 3

Solution: The minimum and maximum values R_{VAR} (min) and R_{VAR} (max) are 0 Ω and 100 Ω, respectively. Start with the voltage divider theorem for the minimum and maximum output voltages.

$V_{OUT\,(\text{min})} = V_S \times \{[R_{VAR\,(\text{min})} + R_1] / [R_S + R_{VAR\,(\text{min})} + R_1]\}$

$3.00\text{ V} = V_S \times [(0\ \Omega + 100\ \Omega) / (R_S + 0\ \Omega + 100\ \Omega)]$

$= V_S \times [100\ \Omega / (R_S + 100\ \Omega)]$

$3.00\text{ V} \times (R_S + 100\ \Omega) = V_S \times [100\ \Omega / (R_S + 100\ \Omega)] \times (R_S + 100\ \Omega)$

$[3.00\text{ V} \times (R_S + 100\ \Omega)] / 100\ \Omega = (V_S \times 100\ \Omega) / 100\ \Omega$

$[3.00\text{ V} \times (R_S + 100\ \Omega)] / 100\ \Omega = V_S$

and

$V_{OUT}\,(\text{max}) = V_S \times \{[R_{VAR}\,(\text{max}) + R_1] / [R_S + R_{VAR}\,(\text{max}) + R_1]\}$

$3.60\text{ V} = V_S \times [(100\ \Omega + 100\ \Omega) / (R_S + 100\ \Omega + 100\ \Omega)]$

$= V_S \times [200\ \Omega / (R_S + 200\ \Omega)]$

$3.60\text{ V} \times (R_S + 200\ \Omega) = V_S \times [100\ \Omega / (R_S + 200\ \Omega)] \times (R_S + 200\ \Omega)$

$[3.60\text{ V} \times (R_S + 200\ \Omega)] / 200\ \Omega = (V_S \times 200\ \Omega) / 200\ \Omega$

$[3.60\text{ V} \times (R_S + 200\ \Omega)] / 200\ \Omega = V_S$

Next, set these two equations equal and solve for R_S.

$[3.00\text{ V} \times (R_S + 100\ \Omega)] / 100\ \Omega = [3.60\text{ V} \times (R_S + 200\ \Omega)] / 200\ \Omega$

$\{[3.00\text{ V} \times (R_S + 100\ \Omega)] / 100\ \Omega\} \times 200\ \Omega =$

$\{[3.60\text{ V} \times (R_S + 200\ \Omega)] / 200\ \Omega\} \times 200\ \Omega\}$

$[3.00\text{ V} \times (R_S + 100\ \Omega) \times 2 \times 100\ \Omega] / 100\ \Omega = 3.60\text{ V} \times (R_S + 200\ \Omega)$

$3.00\text{ V} \times (R_S + 100\ \Omega) \times 2 = 3.60\text{ V} \times (R_S + 200\ \Omega)$

$6.00\text{ V} \times (R_S + 100\ \Omega) = 3.60\text{ V} \times (R_S + 200\ \Omega)$

$[6.00\text{ V} \times (R_S + 100\ \Omega)] / 3.60\text{ V} = [3.60\text{ V} \times (R_S + 200\ \Omega)] / 3.60\text{ V}$

$(6.00\text{ V} / 3.60\text{ V}) \times (R_S + 100\ \Omega) = R_S + 200\ \Omega$

$(6.00 / 3.60)\ (\text{V} / \text{V}) \times (R_S + 100\ \Omega) = R_S + 200\ \Omega$

$1.667 \times (R_S + 100\ \Omega) = R_S + 200\ \Omega$

$(1.667 \times R_S) + (1.67 \times 100\ \Omega) = R_S + 200\ \Omega$

$(1.667 \times R_S) + 166.7\ \Omega = R_S + 200\ \Omega$

$(1.667 \times R_S) + 166.7\ \Omega + -166.7\ \Omega = R_S + 200\ \Omega + -166.7\ \Omega$

$1.667 \times R_S + -R_S = R_S + 33.3\ \Omega + -R_S$

$(1.667 + -1) \times R_S = 33.3\ \Omega$

$0.667 \times R_S = 33.3\ \Omega$

$(0.667 \times R_S) / 0.667 \quad = 33.3\ \Omega / 0.667$

$R_S \quad = 50\ \Omega$

The source resistance R_S is **50 Ω**.

2. Problem: Use the indicated maximum and minimum values of V_{OUT} to determine the value of R_S for the circuit in Figure 8-36.

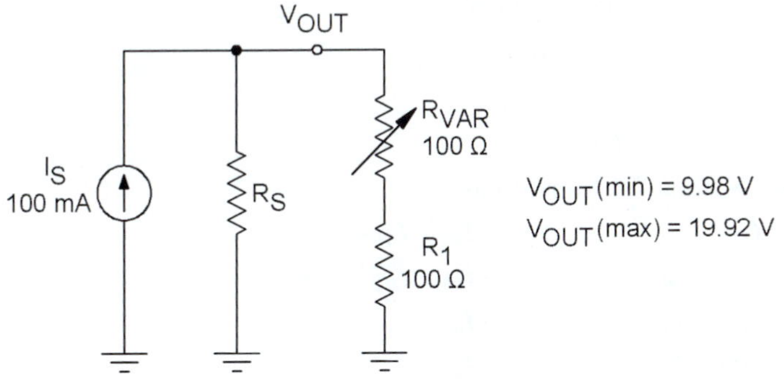

Figure 8-36: Source Loading Example Circuit 4

Solution: I_S divides between R_S and the series combination $R_{VAR} + R_1$. For $R_{VAR\,(\min)} = 0\ \Omega$, the series combination $(R_{VAR\,(\min)} + R_1)$ is minimum and the total parallel resistance $R_T = R_S \parallel (R_{VAR\,(\min)} + R_1)$ is minimum. From Ohm's Law for voltage,

$$V_{OUT} = I_S \times R_T$$

When R_T is minimum, V_{OUT} is the minimum value $V_{OUT\,(\min)}$. Therefore, $V_{OUT\,(\min)}$ corresponds to $R_{VAR\,(\min)}$.

$$\begin{aligned} I_{OUT} &= V_{OUT\,(\min)} / [R_{VAR\,(\min)} + R_1] \\ &= 9.98\ \text{V} / (0\ \Omega + 100\ \Omega) \\ &= 9.98\ \text{V} / 100\ \Omega \\ &= (9.98 / 100)\ (\text{V} / \Omega) \\ &= 99.8 \times 10^{-3}\ \text{A} \\ &= 99.8\ \text{mA} \end{aligned}$$

From Kirchhoff's Current Law,

$$\begin{aligned} I_S &= I_{RS} + I_{OUT} \\ I_S + -I_{OUT} &= I_{RS} + I_{OUT} + -I_{OUT} \\ I_S + -I_{OUT} &= I_{RS} \\ 100\ \text{mA} + -99.8\ \text{mA} &= I_{RS} \\ 0.2\ \text{mA} &= I_{RS} \end{aligned}$$

R_S and $(R_{VAR\,(\min)} + R_1)$ are in parallel, so $V_{OUT\,(\min)}$ is also across R_S. From Ohm's Law for resistance,

$$\begin{aligned} R_S &= V_{OUT\,(\min)} / I_{RS} \\ &= 9.98\ \text{V} / 0.2\ \text{mA} \\ &= 9.98\ \text{V} / (0.2 \times 10^{-3}\ \text{A}) \\ &= [9.98 / (0.2 \times 10^{-3})]\ (\text{V} / \text{A}) \\ &= 49.9 \times 10^{3}\ \Omega \\ &= 49.9\ \text{k}\Omega \end{aligned}$$

For $V_{out\,(\min)}$, the source resistance R_S is **50 kΩ**.

For $R_{VAR\,(\max)} = 100\ \Omega$, the series combination $(R_{VAR\,(\max)} + R_1)$ is maximum and the total parallel resistance $R_T = R_S \parallel (R_{VAR\,(\max)} + R_1)$ is maximum. From Ohm's Law for voltage,

$V_{OUT} = I_S \times R_T$

so when R_T is maximum, V_{OUT} is the maximum value V_{OUT} (max). Therefore, V_{OUT} (max) corresponds to R_{VAR} (max).

$$
\begin{aligned}
I_{OUT} &= V_{OUT}\text{(max)} / [R_{VAR}\text{(max)} + R_1] \\
&= 19.92\text{ V} / (100\ \Omega + 100\ \Omega) \\
&= 19.92\text{ V} / 200\ \Omega \\
&= (19.92 / 200)\ (\text{V} / \Omega) \\
&= 99.6 \times 10^{-3}\text{ A} \\
&= 99.6\text{ mA}
\end{aligned}
$$

From Kirchhoff's Current Law,

$$
\begin{aligned}
I_S &= I_{RS} + I_{OUT} \\
I_S + -I_{OUT} &= I_{RS} + I_{OUT} + -I_{OUT} \\
I_S + -I_{OUT} &= I_{RS} \\
200\text{ mA} + -99.6\text{ mA} &= I_{RS} \\
0.4\text{ mA} &= I_{RS}
\end{aligned}
$$

R_S and ($R_{VAR\,\text{(max)}} + R_1$) are in parallel, so $V_{OUT\,\text{(max)}}$ is also across R_S. From Ohm's Law for resistance,

$$
\begin{aligned}
R_S &= V_{OUT\,\text{(max)}} / I_{RS} \\
&= 19.92\text{ V} / 0.4\text{ mA} \\
&= 19.92\text{ V} / (0.4 \times 10^{-3}\text{ A}) \\
&= [19.92 / (0.4 \times 10^{-3})]\ (\text{V} / \text{A}) \\
&= 49.8 \times 10^{3}\ \Omega \\
&= 49.9\text{ k}\Omega
\end{aligned}
$$

For $V_{out\,\text{(max)}}$, the source resistance R_S is also **50 kΩ**. Therefore, **R_S = 50 kΩ**.

8.2.2 Superposition Theorem

1. Problem: Determine the values of I_{RL} for the values of R_L (min) and R_L (max) for the circuit in Figure 8-37.

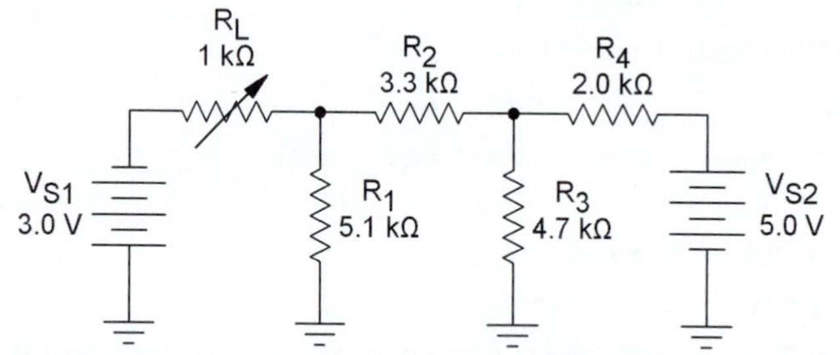

Figure 8-37: Superposition Theorem Example Circuit 3

Solution: For this example, let positive currents through R_L be from left to right and negative currents be from right to left. Refer to Figure 8-38.

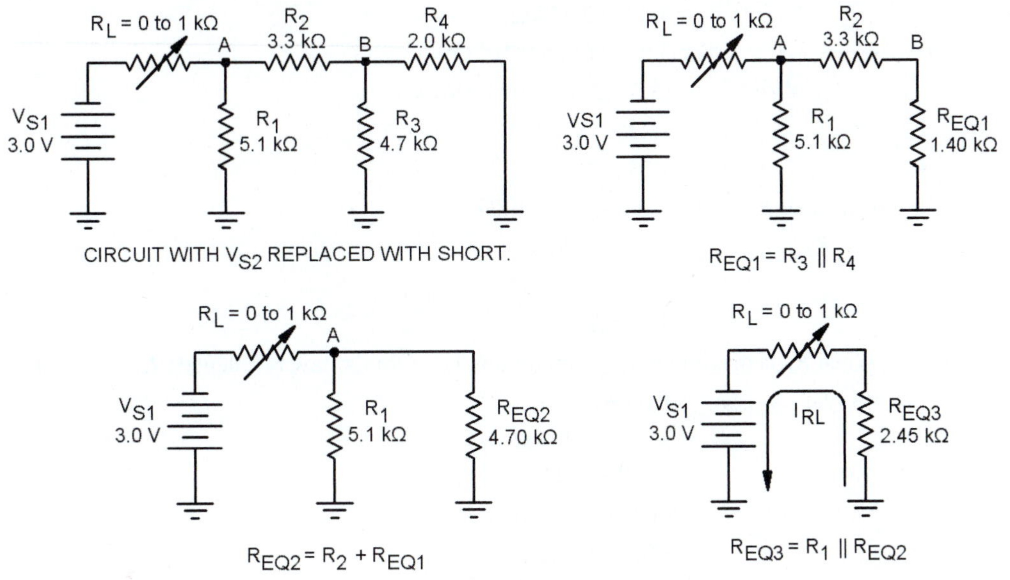

Figure 8-38: Circuit Analysis and Simplification Process for I_{RL} for V_{S1}

To find the current through R_L due to V_{S1}, replace V_{S2} with its internal resistance (a short). With V_{S2} replaced with a short, the expression for the circuit is

$$R_T = R_L + R_1 \parallel (R_2 + R_3 \parallel R_4)$$

The first equivalent resistance R_{EQ1} is the parallel combination $R_3 \parallel R_4$. From this,

$$R_T = R_L + R_1 \parallel (R_2 + R_{EQ1})$$

where

$$R_{EQ1} = R_3 \parallel R_4$$
$$= 4.7\text{ k}\Omega \parallel 2.0\text{ k}\Omega$$
$$= 1.40\text{ k}\Omega$$

The second equivalent resistance R_{EQ2} is the series combination $(R_2 + R_{EQ1})$. From this,

$$R_T = R_L + R_1 \parallel R_{EQ2}$$

where

$$R_{EQ2} = R_2 + R_{EQ1}$$
$$= 3.3\text{ k}\Omega + 1.40\text{ k}\Omega$$
$$= 4.70\text{ k}\Omega$$

The third equivalent circuit R_{EQ3} is the parallel combination $R_1 \parallel R_{EQ2}$. From this,

$$R_T = R_L + R_{EQ3}$$

where

$$R_{EQ3} = R_1 \parallel R_{EQ2}$$
$$= 5.1\text{ k}\Omega \parallel 4.70\text{ k}\Omega$$
$$= 2.45\text{ k}\Omega$$

Calculate R_T for V_{S1} for $R_{L\,(\text{min})} = 0\text{ k}\Omega$.

$$R_{T(VS1)\,(\text{min})} = R_{L\,(\text{min})} + R_{EQ3}$$
$$= 0\text{ k}\Omega + 2.45\text{ k}\Omega$$
$$= 2.45\text{ k}\Omega$$

When R_L is minimum, I_{RL} is maximum. Use Ohm's Law to calculate $I_{RL\,(\text{max})}$ for V_{S1}.

$$I_{RL\,(VS1)\,(\text{max})} = V_{S1} / R_{T(VS1)\,(\text{min})}$$

$= 3.0\ \text{V} / 2.45\ \text{k}\Omega$

$= 3.0\ \text{V} / (2.45 \times 10^3\ \Omega)$

$= [3.0 / (2.45 \times 10^3)]\ (\text{V} / \Omega)$

$= 1.23 \times 10^{-3}\ \text{A}$

$= 1.23\ \text{mA}$

Calculate R_T for V_{S1} for $R_{L\,(\text{max})} = 1\ \text{k}\Omega$.

$$R_{T\,(VS1)\,(\text{min})} = R_{L\,(\text{max})} + R_{EQ3}$$
$$= 1\ \text{k}\Omega + 2.45\ \text{k}\Omega$$
$$= 3.45\ \text{k}\Omega$$

When R_L is maximum, I_{RL} is minimum. Use Ohm's Law to calculate $I_{RL\,(\text{min})}$ for V_{S1}.

$$I_{RL\,(VS1)\,(\text{min})} = V_{S1} / R_{T\,(VS1)\,(\text{min})}$$
$$= 3.0\ \text{V} / 3.45\ \text{k}\Omega$$
$$= 3.0\ \text{V} / (3.45 \times 10^3\ \Omega)$$
$$= [3.0 / (3.45 \times 10^3)]\ (\text{V} / \Omega)$$
$$= 870 \times 10^{-6}\ \text{A}$$
$$= 870\ \mu\text{A}$$

Note that these currents flow from right to left through R_L, so the signed values are $I_{RL\,(VS1)\,(\text{max})} = -1.23$ mA and $I_{RL1\,(\text{min})} = -870\ \mu$A.

To find the current through R_L due to V_{S2}, replace V_{S1} with its internal resistance (a short). Refer to Figure 8-39.

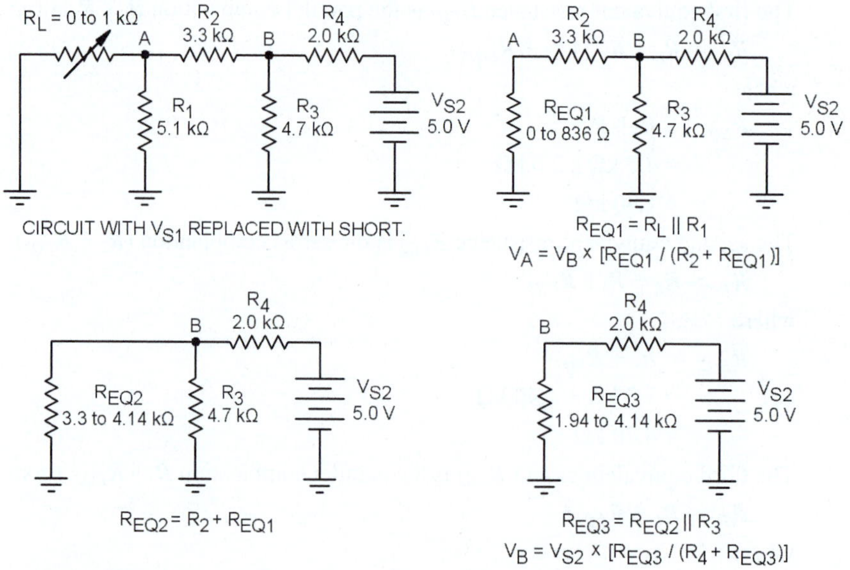

Figure 8-39: Circuit Analysis and Simplification Process for I_{RL} for V_{S2}

With V_{S1} replaced with a short, the circuit expression is

$$R_T = R_4 + R_3 \parallel [R_2 + R_1 \parallel R_L)]$$

The first equivalent resistance R_{EQ1} is the parallel combination $R_1 \parallel R_L$. From this,

$$R_T = R_4 + R_3 \parallel (R_2 + R_{EQ1})$$

where

$$R_{EQ1} = R_1 \parallel R_L$$

For $\boldsymbol{R_{L}}_{(\text{min})} = 0\ \text{k}\Omega$, $\boldsymbol{R_L}$ shorts out $\boldsymbol{R}_1$ so that $\boldsymbol{R_{EQ1}}_{(\text{min})} = 0\ \text{k}\Omega$. For $\boldsymbol{R_L}_{(\text{max})} = 1\ \text{k}\Omega$,

$$\begin{aligned}\boldsymbol{R_{EQ1}}_{(\text{max})} &= \boldsymbol{R}_1 \parallel \boldsymbol{R_L}_{(\text{max})} \\ &= 5.1\ \text{k}\Omega \parallel 1\ \text{k}\Omega \\ &= 836\ \Omega\end{aligned}$$

The second equivalent resistance $\boldsymbol{R_{EQ2}}$ is the series combination $(\boldsymbol{R}_2 + \boldsymbol{R_{EQ1}})$. From this,

$$\boldsymbol{R_T} = \boldsymbol{R}_4 + \boldsymbol{R}_3 \parallel \boldsymbol{R_{EQ2}}$$

where

$$\boldsymbol{R_{EQ2}} = \boldsymbol{R}_2 + \boldsymbol{R_{EQ1}}$$

For $\boldsymbol{R_{EQ1}}_{(\text{min})}$,

$$\begin{aligned}\boldsymbol{R_{EQ2}}_{(\text{min})} &= \boldsymbol{R}_2 + \boldsymbol{R_{EQ1}}\ (\text{min}) \\ &= 3.3\ \text{k}\Omega + 0\ \text{k}\Omega \\ &= 3.3\ \text{k}\Omega\end{aligned}$$

For $\boldsymbol{R_{EQ1}}$ (max),

$$\begin{aligned}\boldsymbol{R_{EQ2}}\ (\text{max}) &= \boldsymbol{R}_2 + \boldsymbol{R_{EQ1}}\ (\text{max}) \\ &= 3.3\ \text{k}\Omega + 836\ \Omega \\ &= 3.3\ \text{k}\Omega + 0.836\ \text{k}\Omega \\ &= 4.14\ \text{k}\Omega\end{aligned}$$

The third equivalent resistance $\boldsymbol{R_{EQ3}}$ is the parallel combination $\boldsymbol{R_3} \parallel \boldsymbol{R_{EQ2}}$. From this,

$$\boldsymbol{R_T} = \boldsymbol{R}_4 + \boldsymbol{R_{EQ3}}$$

where

$$\boldsymbol{R_{EQ3}} = \boldsymbol{R}_3 \parallel \boldsymbol{R_{EQ2}}$$

For $\boldsymbol{R_{EQ2}}_{(\text{min})}$,

$$\begin{aligned}\boldsymbol{R_{EQ3}}_{(\text{min})} &= \boldsymbol{R}_3 \parallel \boldsymbol{R_{EQ2}}\ (\text{min}) \\ &= 4.7\ \text{k}\Omega \parallel 3.3\ \text{k}\Omega \\ &= 1.94\ \text{k}\Omega\end{aligned}$$

For $\boldsymbol{R_{EQ2}}_{(\text{max})}$,

$$\begin{aligned}\boldsymbol{R_{EQ3}}_{(\text{max})} &= \boldsymbol{R}_3 \parallel \boldsymbol{R_{EQ2}}\ (\text{max}) \\ &= 4.7\ \text{k}\Omega \parallel 4.14\ \text{k}\Omega \\ &= 2.20\ \text{k}\Omega\end{aligned}$$

From the voltage divider theorem,

$$\boldsymbol{V_B} = \boldsymbol{V_{S2}} \times [\boldsymbol{R_{EQ3}} / (\boldsymbol{R_{EQ3}} + \boldsymbol{R}_4)]$$

For $\boldsymbol{R_{EQ3}}_{(\text{min})}$,

$$\begin{aligned}\boldsymbol{V_B}_{(\boldsymbol{VS2})\,(\text{min})} &= \boldsymbol{V_{S2}} \times [\boldsymbol{R_{EQ3}}\ (\text{min}) / (\boldsymbol{R_{EQ3}}\ (\text{min}) + \boldsymbol{R}_4)] \\ &= 5.0\ \text{V} \times [1.94\ \text{k}\Omega / (1.94\ \text{k}\Omega + 2.0\ \text{k}\Omega)] \\ &= 5.0\ \text{V} \times (1.94\ \text{k}\Omega / 3.94\ \text{k}\Omega) \\ &= 5.0\ \text{V} \times [(1.94 / 3.94)\ (\text{k}\Omega / \text{k}\Omega)] \\ &= 5.0\ \text{V} \times 0.492 \\ &= 2.46\ \text{V}\end{aligned}$$

For $\boldsymbol{R_{EQ3}}_{(\text{max})}$,

$$\begin{aligned}\boldsymbol{V_B}_{(\boldsymbol{VS2})\,(\text{max})} &= \boldsymbol{V_{S2}} \times [\boldsymbol{R_{EQ3}}\ (\text{max}) / (\boldsymbol{R_{EQ3}}\ (\text{max}) + \boldsymbol{R}_4)] \\ &= 5.0\ \text{V} \times [2.20\ \text{k}\Omega / (2.20\ \text{k}\Omega + 2.0\ \text{k}\Omega)] \\ &= 5.0\ \text{V} \times (2.20\ \text{k}\Omega / 4.20\ \text{k}\Omega) \\ &= 5.0\ \text{V} \times [(2.20 / 4.20)\ (\text{k}\Omega / \text{k}\Omega)]\end{aligned}$$

$= 5.0\ \text{V} \times 0.524$

$= 2.62\ \text{V}$

For $R_{L\,(\text{min})} = 0\ \text{k}\Omega$, the left side of R_2 connects to ground and shorts out R_1 and the maximum value of I_{RL} for V_{S2}, $I_{RL\,(VS2)\,(\text{max})}$, equals $I_{R2\,(\text{max})}$. In addition, the values of V_A and V_B are minimum. Use Ohm's Law for current to calculate $I_{RL\,(\text{max})}$ for V_{S2}.

$$
\begin{aligned}
I_{RL\,(VS2)\,(\text{max})} &= I_{R2\,(\text{max})} \\
&= V_{B\,(VS2)\,(\text{min})} / R_2 \\
&= 2.46\ \text{V} / 3.3\ \text{k}\Omega \\
&= 2.46\ \text{V} / (3.3 \times 10^3\ \Omega) \\
&= [2.46 / (3.3 \times 10^3)]\ (\text{V} / \Omega) \\
&= 746 \times 10^{-6}\ \text{A} \\
&= 746\ \mu\text{A}
\end{aligned}
$$

For $R_{L\,(\text{max})}$, V_A and V_B are maximum. Use the voltage divider theorem to find the $V_{A\,(\text{max})}$ for V_{S2}.

$$
\begin{aligned}
V_{A\,(VS2)\,(\text{max})} &= V_{B\,(VS2)\,(\text{max})} \times [R_{EQ1\,(\text{max})} / (R_{EQ1\,(\text{max})} + R_2)] \\
&= 2.62\ \text{V} \times [836\ \Omega / (836\ \Omega + 3.3\ \text{k}\Omega)] \\
&= 2.62\ \text{V} \times [0.836\ \text{k}\Omega / (0.836\ \text{k}\Omega + 3.3\ \text{k}\Omega)] \\
&= 2.62\ \text{V} \times (0.836\ \text{k}\Omega / 4.14\ \text{k}\Omega) \\
&= 2.62\ \text{V} \times [(0.836 / 4.14)\ (\text{k}\Omega / \text{k}\Omega)] \\
&= 2.62\ \text{V} \times 0.202 \\
&= 529 \times 10^{-3}\ \text{V} \\
&= 529\ \text{mV}
\end{aligned}
$$

When R_L is maximum, I_{RL} is minimum. Use Ohm's Law to $I_{RL\,(\text{min})}$ for V_{S2}.

$$
\begin{aligned}
I_{RL\,(VS2)\,(\text{min})} &= V_{A\,(\text{max})} / R_{L\,(\text{max})} \\
&= 529\ \text{mV} / 1\ \text{k}\Omega \\
&= (529 \times 10^{-3}\ \text{V}) / (1 \times 10^3\ \Omega) \\
&= [(529 \times 10^{-3}) / (1 \times 10^3)]\ (\text{V} / \Omega) \\
&= 529 \times 10^{-6}\ \text{A} \\
&= 529\ \mu\text{A}
\end{aligned}
$$

Note that these currents flow from right to left through R_L so that the signed values are $I_{RL\,(VS2)\,(\text{max})} = -746\ \mu\text{A}$ and $I_{RL\,(VS2)\,(\text{min})} = -529\ \mu\text{A}$. The algebraic sums for these currents are then

$$
\begin{aligned}
I_{RL\,(\text{min})} &= I_{RL\,(VS1)\,(\text{min})} + I_{RL\,(VS2)\,(\text{min})} \\
&= -870\ \mu\text{A} + -529\ \mu\text{A} \\
&= \mathbf{-1.40\ \mu A\ (current\ flows\ from\ right\ to\ left)}
\end{aligned}
$$

$$
\begin{aligned}
I_{RL\,(\text{max})} &= I_{RL\,(VS1)\,(\text{max})} + I_{RL\,(VS2)\,(\text{max})} \\
&= (-1.23\ \text{mA}) + (-746\ \mu\text{A}) \\
&= (-1.23\ \text{mA}) + (-0.746\ \text{mA}) \\
&= \mathbf{-1.97\ mA\ (current\ flows\ from\ right\ to\ left)}
\end{aligned}
$$

2. Problem: Determine the minimum and maximum voltages $V_{RL\,(\text{min})}$ and $V_{RL\,(\text{max})}$ across R_L for the circuit in Figure 8-40.

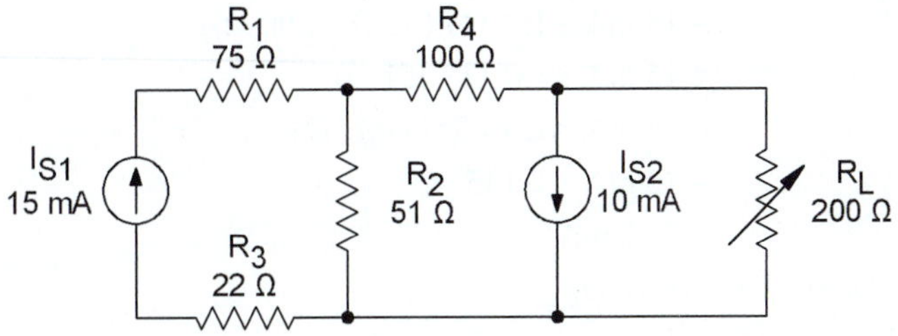

Figure 8-40: Superposition Theorem Example Circuit 4

Solution: For this example, define the sign of the voltage across $\boldsymbol{R_L}$ as that on the top end of $\boldsymbol{R_L}$. Refer to Figure 8-41.

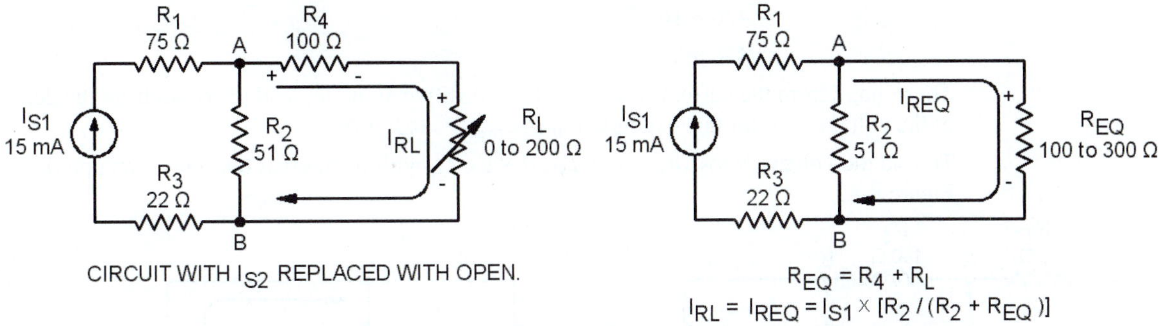

Figure 8-41: Circuit Analysis and Simplification Technique for V_{RL} for I_{S1}

First, note that when $\boldsymbol{R}_{L\,(\mathrm{min})} = 0\ \Omega$, Ohm's Law for voltage gives

$$V_{RL\,(\mathrm{min})} = I_{RL} \times R_{L\,(\mathrm{min})}$$

$$= I_{RL} \times 0\ \Omega$$

$$= 0\ \mathrm{V}$$

Therefore, only the voltages for $\boldsymbol{R}_{L\,(\mathrm{max})} = 200\ \Omega$ need be found. To find the voltage across $\boldsymbol{R_L}$ due to $\boldsymbol{I_{S1}}$, replace $\boldsymbol{I_{S2}}$ with its internal resistance (an open). With $\boldsymbol{I_{S2}}$ replaced with an open, the expression for the circuit is

$$R_{T\,(IS1)} = R_1 + R_2 \parallel (R_4 + R_L) + R_3$$

The first equivalent resistance $\boldsymbol{R_{EQ}}$ (max) is the series combination ($\boldsymbol{R}_4 + \boldsymbol{R}_{L\,(\mathrm{max})}$).

$$R_{T\,(IS1)} = R_1 + R_2 \parallel R_{EQ\,(\mathrm{max})} + R_3$$

where

$$R_{EQ\,(\mathrm{max})} = R_4 + R_{L\,(\mathrm{max})}$$

For $\boldsymbol{R}_{L\,(\mathrm{max})} = 200\ \Omega$,

$$R_{EQ\,(\mathrm{max})} = R_4 + R_{L\,(\mathrm{max})}$$

$$= 100\ \Omega + 200\ \Omega$$

$$= 300\ \Omega$$

From the current divider theorem,

$$I_{REQ} = I_{S1} \times [R_2 / (R_2 + R_{EQ})]$$

Because $\boldsymbol{R_{EQ}}$ is a series resistance containing $\boldsymbol{R_L}$, then $\boldsymbol{I_{RL1}} = \boldsymbol{I_{REQ}}$.

$$I_{RL\,(IS1)} = I_{REQ}$$

$$= I_{S1} \times [R_2 / (R_2 + R_{EQ})]$$

When $\boldsymbol{R_{EQ}}$ is maximum, $\boldsymbol{I_{RL1}} = \boldsymbol{I_{REQ}}$ will be minimum. Therefore,

$$I_{RL\,(IS1)\,(\mathrm{min})} = I_{S1} \times \{[R_2 / [R_2 + R_{EQ\,(\mathrm{max})}]\}$$

$= 15\text{ mA} \times [51\ \Omega / (51\ \Omega + 300\ \Omega)]$

$= 15\text{ mA} \times (51\ \Omega / 351\ \Omega)$

$= 15\text{ mA} \times (51 / 351)\ (\Omega / \Omega)$

$= 15\text{ mA} \times 0.145$

$= 2.18\text{ mA}$

From Ohm's Law for voltage,

$V_{RL\,(IS1)\,(\max)} = I_{RL\,(IS1)\,(\min)} \times R_{L\,(\max)}$

$= 2.18\text{ mA} \times 200\ \Omega$

$= (2.18 \times 10^{-3}\text{ A}) \times 200\ \Omega$

$= [(2.18 \times 10^{-3}) \times 200]\ (\text{A} \times \Omega)$

$= 436 \times 10^{-3}\text{ V}$

$= 436\text{ mV}$

The voltage from the current flow from I_{S1} is positive at the top end of R_L, so from the definition of the voltage sign for this example $V_{RL\,(IS1)\,(\max)} = +436\text{ mV}$.

To find the voltage across R_L due to I_{S2}, replace I_{S1} with its internal resistance (an open). Refer to Figure 8-42.

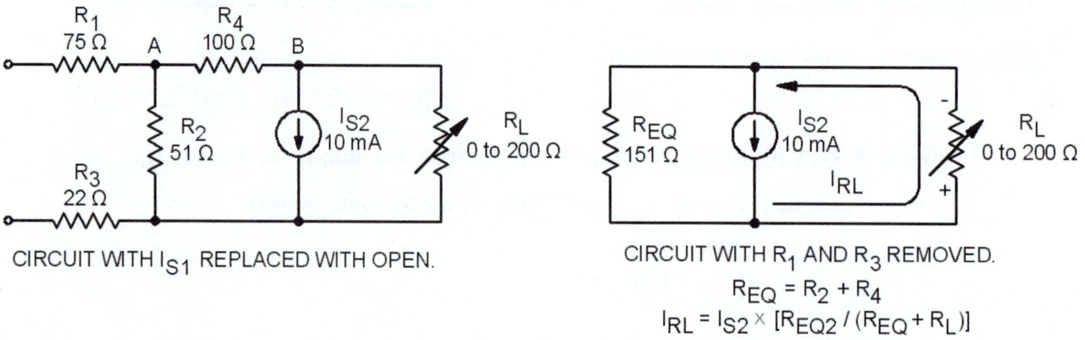

Figure 8-42: Circuit Analysis and Simplification Process for V_{RL} for I_{S2}

With I_{S1} replaced with an open, no current flows through R_1 and R_3 so that they can be ignored when analyzing the circuit. The expression for the circuit is then

$R_{T\,(IS2)} = (R_2 + R_4) \parallel R_L$

The circuit equivalent R_{EQ} is the series combination $(R_2 + R_4)$. From this

$R_{T\,(IS2)} = R_{EQ} \parallel R_L$

where

$R_{EQ} = R_2 + R_4$

$= 51\ \Omega + 100\ \Omega$

$= 151\ \Omega$

From the current divider theorem,

$I_{RL\,(IS2)} = I_{S2} \times [R_{EQ} / (R_{EQ} + R_L)]$

When R_L is maximum, I_{RL2} will be minimum.

$I_{RL\,(IS2)\,(\min)} = I_{S2} \times \{[R_{EQ} / [R_{EQ} + R_{L\,(\max)}]\}$

$= 10\text{ mA} \times [151\ \Omega / (151\ \Omega + 200\ \Omega)]$

$= 10\text{ mA} \times (151\ \Omega / 351\ \Omega)$

$= 10\text{ mA} \times (151 / 351)\ (\Omega / \Omega)$

$= 10 \text{ mA} \times 0.430$

$= 4.30 \text{ mA}$

From Ohm's Law for voltage,

$$V_{RL\ (IS2)\ (\max)} = I_{RL2\ (\min)} \times R_{L\ (\max)}$$
$$= 4.30 \text{ mA} \times 200\ \Omega$$
$$= (4.30 \times 10^{-3} \text{ A}) \times 200\ \Omega$$
$$= [(4.30 \times 10^{-3}) \times 200]\ (\text{A} \times \Omega)$$
$$= 860 \times 10^{-3} \text{ V}$$
$$= 860 \text{ mV}$$

The voltage from the current flow from $I_{S\ (IS2)}$ is negative at the top end of R_L, so from the definition of the voltage sign for this example $V_{RL\ (IS2)\ (\min)} = -860$ mV. The algebraic sum of the superposition voltages is then

$$V_{RL\ (\max)} = V_{RL\ (IS1)\ (\max)} + V_{RL\ (IS2)\ (\max)}$$
$$= (+436 \text{ mV}) + (-860 \text{ mV})$$
$$= \mathbf{-425 \text{ mV (negative polarity at top)}}$$

8.2.3 Thevenin's Theorem

1. Problem: Determine the value of short circuit current I_{SC} through load resistor R_L for the circuit in Figure 8-43. Assume that all values are exact.

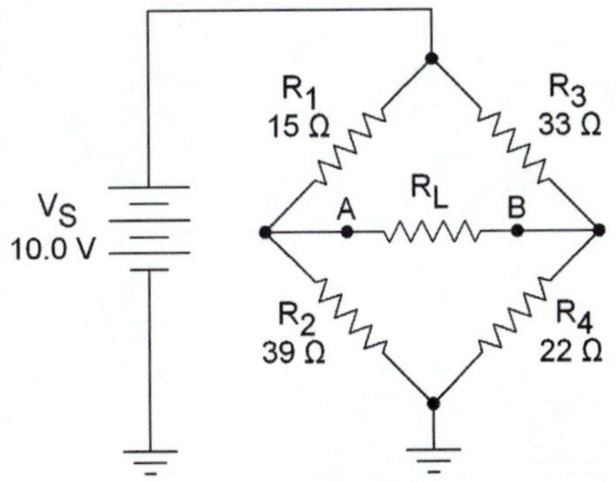

Figure 8-43: Thevenin's Theorem Example Circuit 3

Solution: To determine the value of I_{SC}, Thevenize the circuit relative to R_L and determine the current when $R_L = 0\ \Omega$. The first step is to find the Thevenin voltage V_{TH} for R_L. To do so, first remove R_L and determine the open-load voltage $V_{AB} = V_{TH}$. Refer to Figure 8-44.

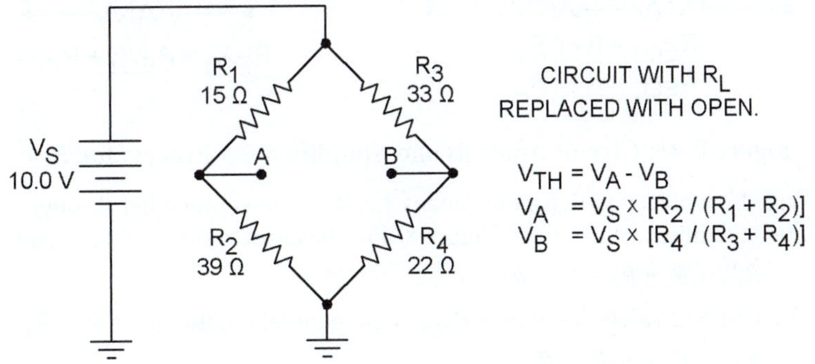

Figure 8-44: Circuit Analysis Process for V_{TH}

By definition, $V_{AB} = V_A - V_B$.

$V_{TH} = V_{AB}$

$= V_A - V_B$

Use the voltage divider theorem to determine V_A and V_B.

$V_A = V_S \times [R_2 / (R_1 + R_2)]$

and

$V_B = V_S \times [R_4 / (R_3 + R_4)]$

Use V_A and V_B to calculate V_{TH}.

$V_{TH} = V_A - V_B$

$= V_S \times [R_2 / (R_1 + R_2)] - V_S \times [R_4 / (R_3 + R_4)]$

$= V_S \times \{[R_2 / (R_1 + R_2)] - [R_4 / (R_3 + R_4)]\}$

$= 10.0\ \text{V} \times \{[39\ \Omega / (15\ \Omega + 39\ \Omega)] + -[22\ \Omega / (33\ \Omega + 22\ \Omega)]\}$

$= 10.0\ \text{V} \times [(39\ \Omega / 54\ \Omega)] + -[(22\ \Omega / 55\ \Omega)]$

$= 10.0\ \text{V} \times [(39 / 54)\ (\Omega / \Omega)] + -[(22 / 55)\ (\Omega / \Omega)]$

$= 10.0\ \text{V} \times (0.722 + -0.400)$

$= 10.0\ \text{V} \times 0.322$

$= 3.22\ \text{V}$

The next step is to find the Thevenin resistance R_{TH}. To do so, replace the voltage source V_S with its internal resistance (a short) and determine the resistance looking back into the circuit from R_L. Refer to Figure 8-45.

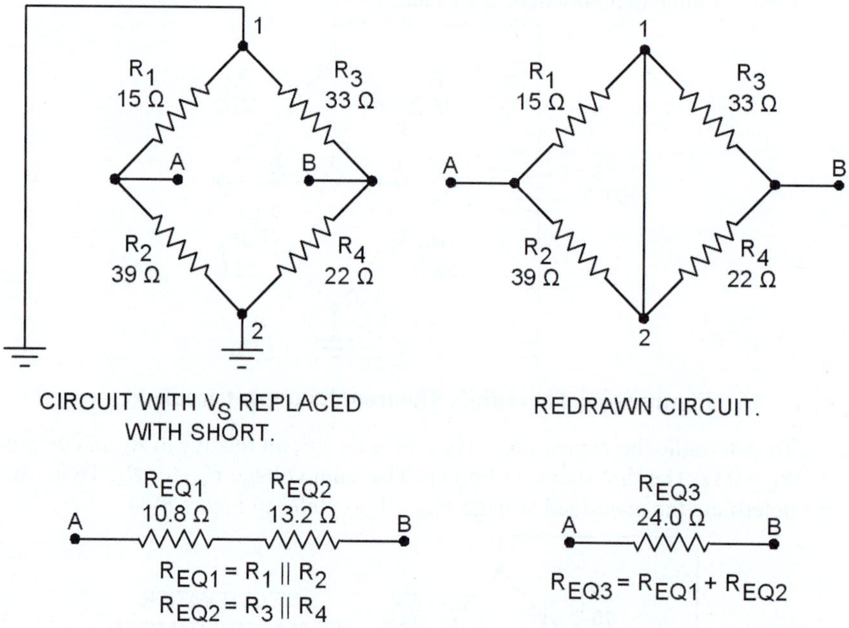

Figure 8-45: Circuit Analysis and Simplification Process for R_{TH}

With V_S replaced with a short, nodes 1 and 2 connect together through ground so that the circuit can be redrawn as shown in Figure 8-45. The expression for this circuit is

$R_{TH} = R_1 \parallel R_2 + R_3 \parallel R_4$

The first equivalent resistance R_{EQ1} is the parallel combination $R_1 \parallel R_2$.

$R_{TH} = R_{EQ1} + R_3 \parallel R_4$

where

$R_{EQ1} = R_1 \parallel R_2$

$= 15\ \Omega \parallel 39\ \Omega$

$= 10.8\ \Omega$

The second equivalent resistance $\boldsymbol{R_{EQ2}}$ is the parallel combination $\boldsymbol{R_3} \parallel \boldsymbol{R_4}$.

$$\boldsymbol{R_{TH}} = \boldsymbol{R_{EQ1}} + \boldsymbol{R_{EQ2}}$$

where

$$\begin{aligned}\boldsymbol{R_{EQ2}} &= \boldsymbol{R_3} \parallel \boldsymbol{R_4}\\ &= 33\ \Omega \parallel 22\ \Omega\\ &= 13.2\ \Omega\end{aligned}$$

The third (and final) equivalent resistance $\boldsymbol{R_{EQ3}}$ is the series combination $\boldsymbol{R_{EQ1}} + \boldsymbol{R_{EQ2}}$.

$$\boldsymbol{R_{TH}} = \boldsymbol{R_{EQ3}}$$

where

$$\begin{aligned}\boldsymbol{R_{EQ3}} &= \boldsymbol{R_{EQ1}} + \boldsymbol{R_{EQ2}}\\ &= 10.8\ \Omega + 13.2\ \Omega\\ &= 24.0\ \Omega\end{aligned}$$

Figure 8-46 shows the Thevenin equivalent circuit with $\boldsymbol{R_L} = 0\ \Omega$.

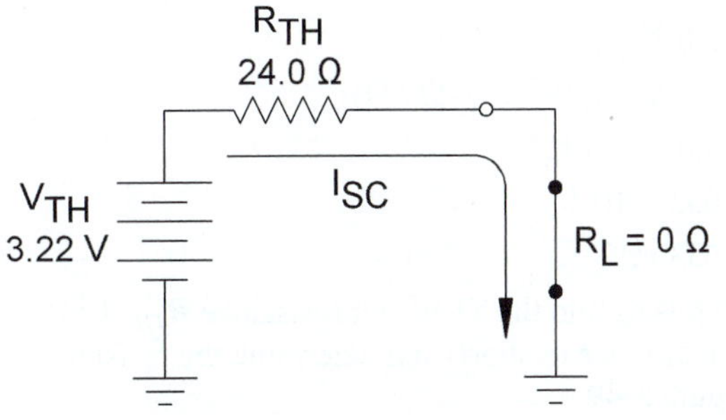

Figure 8-46: Thevenin Equivalent for Figure 8-43 with $\boldsymbol{R_L} = 0\ \Omega$

Use Ohm's Law for current to calculate the short circuit current $\boldsymbol{I_{SC}}$ through $\boldsymbol{R_L}$.

$$\begin{aligned}\boldsymbol{I_{SC}} &= \boldsymbol{V_{TH}} / \boldsymbol{R_{TH}}\\ &= 3.22\text{ V} / 24.0\ \Omega\\ &= (3.22 / 24.0)\ (\text{V} / \Omega)\\ &= 134 \times 10^{-3}\text{ A}\\ &= 134\text{ mA}\end{aligned}$$

The short circuit current through $\boldsymbol{R_L}$ is **134 mA**.

2. Problem: Determine the Thevenin equivalent with respect to $\boldsymbol{R_L}$ for the circuit in Figure 8-47.

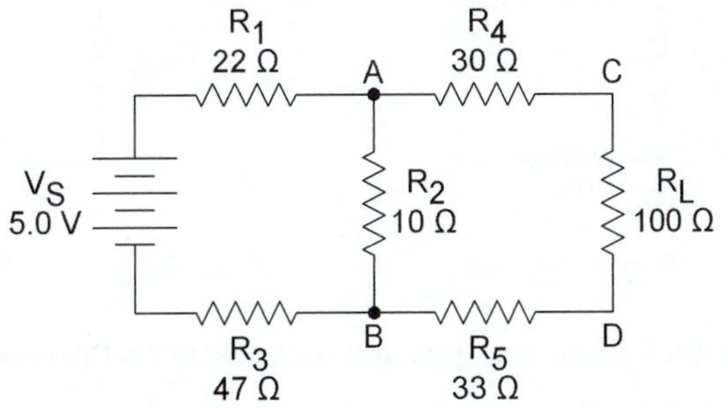

Figure 8-47: Thevenin's Theorem Example Circuit 4

Solution: The first step is to find the Thevenin voltage $\boldsymbol{V_{TH}}$ for $\boldsymbol{R_L}$. To do so, first remove $\boldsymbol{R_L}$ and determine the open-load voltage $\boldsymbol{V_{TH}}$. Refer to Figure 8-48.

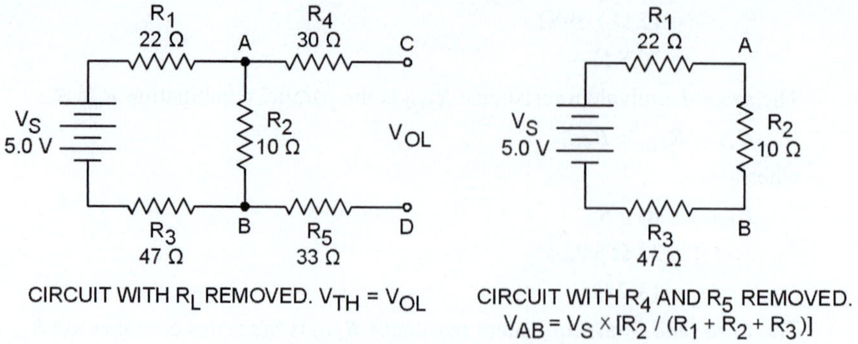

Figure 8-48: Circuit Analysis and Simplification Process for V_{TH}

With R_L removed from the circuit, no current flows through R_4 and R_5, so they can be ignored from the circuit analysis. Use the voltage divider theorem to determine $V_{AB} = V_{OL} = V_{TH}$.

$$
\begin{aligned}
V_{TH} &= V_{AB} \\
&= V_S \times [R_2 / (R_1 + R_2 + R_3)] \\
&= 5.0\text{ V} \times [10\ \Omega / (22\ \Omega + 10\ \Omega + 47\ \Omega)] \\
&= 5.0\text{ V} \times (10\ \Omega / 79\ \Omega) \\
&= 5.0\text{ V} \times [(10 / 79)\ (\Omega / \Omega)] \\
&= 5.0\text{ V} \times 0.127 \\
&= 633 \times 10^{-3} \\
&= 633\text{ mV}
\end{aligned}
$$

The next step is to find the Thevenin resistance R_{TH}. To do so, replace the voltage source V_S with its internal resistance (a short) and determine the resistance looking back into the circuit from R_L. Refer to Figure 8-49.

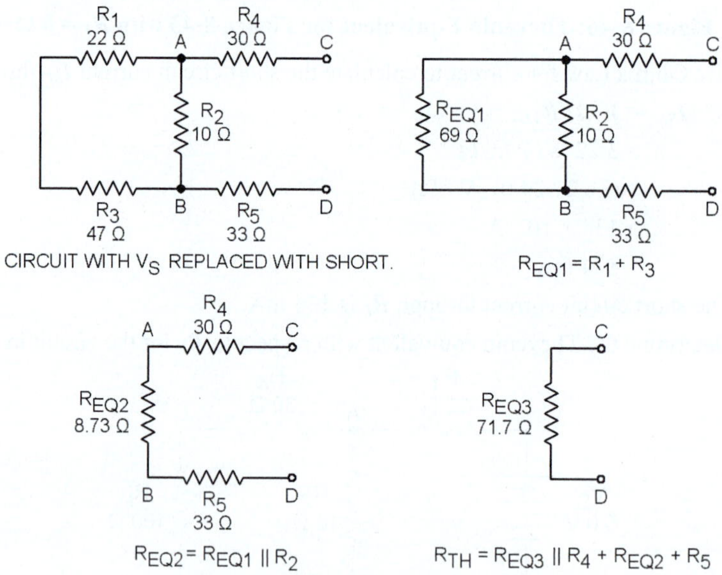

Figure 8-49: Circuit Analysis and Simplification Process for R_{TH}

The expression for $R_{CD} = R_{TH}$ is

$$R_{TH} = R_4 + (R_1 + R_3) \parallel R_2 + R_5$$

The first equivalent resistance R_{EQ1} is the series combination $(R_1 + R_3)$.

$$R_{TH} = R_4 + R_{EQ1} \parallel R_2 + R_5$$

where

$$R_{EQ1} = R_1 + R_3 = 22\ \Omega + 47\ \Omega = 69\ \Omega$$

The second equivalent resistance R_{EQ2} is the parallel combination $R_{EQ1} \parallel R_2$.

$$R_{TH} = R_4 + R_{EQ2} + R_5$$

where

$$R_{EQ2} = R_{EQ1} \parallel R_2 = 69\ \Omega \parallel 10\ \Omega = 8.73\ \Omega$$

The third (and final) equivalent resistance is the series combination $R_4 + R_{EQ2} + R_5$.

$$R_{TH} = R_{EQ3}$$

where

$$R_{EQ3} = R_4 + R_{EQ2} + R_5 = 30\ \Omega + 8.73\ \Omega + 33\ \Omega = 71.7\ \Omega$$

Figure 8-50 shows the Thevenin equivalent for Figure 8-47.

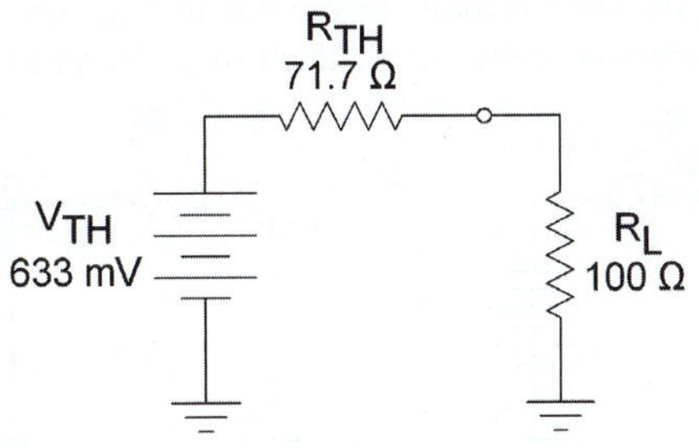

Figure 8-50: Thevenin Equivalent for Figure 8-47

8.2.4 Norton's Theorem

1. Problem: Determine the Norton equivalent with respect to R_L for the circuit in Figure 8-51.

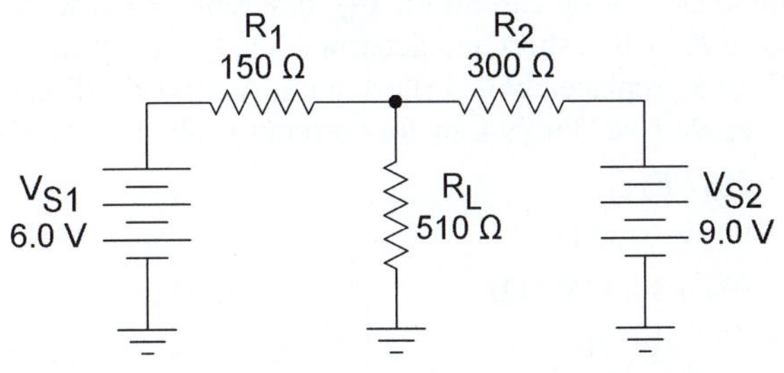

Figure 8-51: Norton's Theorem Example Circuit 3

Solution: The first step is to find the short circuit current I_{SC} through R_L. Because the circuit contains two voltage sources, use the superposition theorem to analyze the individual effects of each source and determine the total short circuit current. Refer to Figure 8-52.

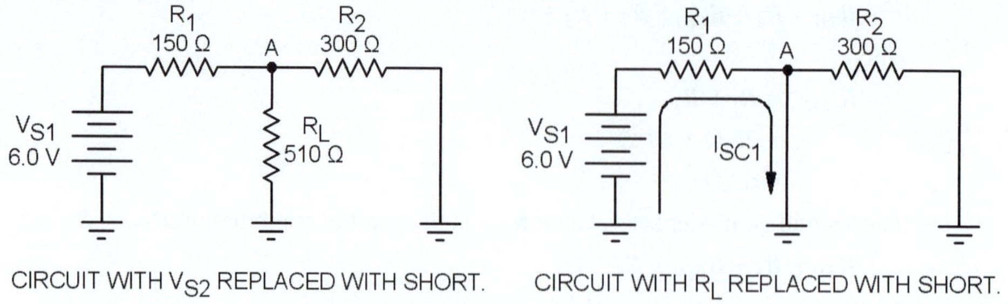

Figure 8-52: Circuit Analysis Process for I_{SC1}

To determine the Norton current for V_{S1}, first replace V_{S2} with its internal resistance (a short). Then replace R_L with a short and determine the short circuit current $I_{SC} = I_N$ between point A and ground. With R_L replaced with a short, point A connects directly to ground so that the current from V_{S1} bypasses R_2. Use Ohm's Law for current to calculate the short circuit current from V_{S1}.

$$\begin{aligned} I_{SC1} &= V_{S1} / R_1 \\ &= 6.0\ \text{V} / 150\ \Omega \\ &= (6.0 / 150)\ (\text{V} / \Omega) \\ &= 40 \times 10^{-3}\ \text{A} \\ &= 40\ \text{mA} \end{aligned}$$

Therefore, I_{SC1}, the short circuit current due to V_{S1} from point A to ground is 40 mA. Next, determine the Norton current for R_L due to V_{S2}. Refer to Figure 8-53.

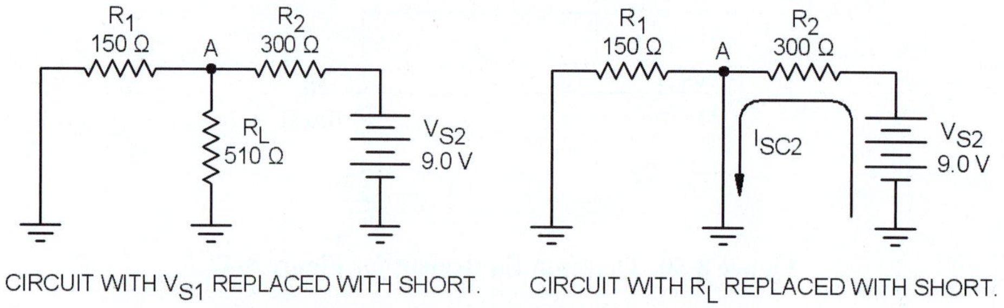

Figure 8-53: Circuit Analysis Process for I_{SC2}

To determine the Norton current for V_{S2}, first replace V_{S1} with its internal resistance (a short). Then replace R_L with a short and determine the short circuit current $I_{SC2} = I_N$ from point A to ground. With R_L replaced with a short, point A connects directly to ground so that the current from V_{S2} bypasses R_1. Use Ohm's Law for current to calculate the short circuit current from V_{S2}.

$$\begin{aligned} I_{SC2} &= V_{S2} / R_2 \\ &= 9.0\ \text{V} / 300\ \Omega \\ &= (9.0 / 300)\ (\text{V} / \Omega) \\ &= 30 \times 10^{-3}\ \text{A} \\ &= 30\ \text{mA} \end{aligned}$$

Therefore, I_{SC2}, the short circuit current due to V_{S2} from point A to ground is 30 mA. Both short circuit currents flow from point A to ground, so the total short circuit current, $I_{SC} = I_N$, is the sum of the individual currents and flows from pint A to ground.

$$\begin{aligned} I_{SC} = I_N &= I_{SC1} + I_{SC2} \\ &= 40\ \text{mA} + 30\ \text{mA} \\ &= 70\ \text{mA} \end{aligned}$$

The second step is to find the Norton resistance R_N by replacing both voltage sources with their internal resistance (a short for each) and determine the resistance looking back into the circuit from R_L. Refer to Figure 8-54.

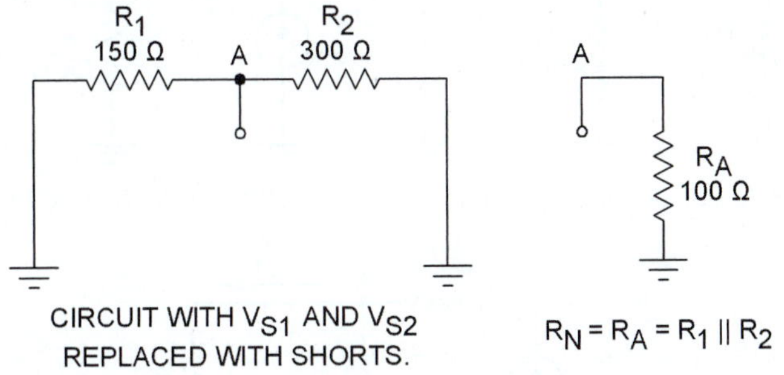

Figure 8-54: Circuit Analysis Process for R_N

With V_{S1} and V_{S2} replaced with shorts, the expression for the circuit resistance at point A is

$$R_N = R_A = R_1 \parallel R_2$$

so

$$\begin{aligned} R_N &= R_1 \parallel R_2 \\ &= 150\ \Omega \parallel 300\ \Omega \\ &= 100\ \Omega \end{aligned}$$

Figure 8-55 shows the Norton equivalent for Figure 8-51.

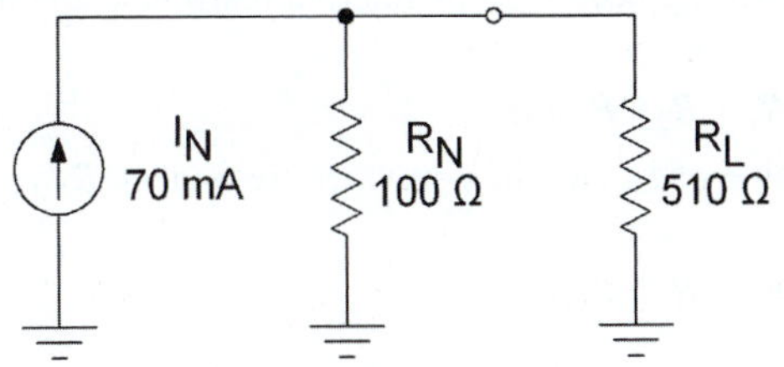

Figure 8-55: Norton Equivalent for Figure 8-51

2. Problem: Determine the Norton equivalent with respect to R_L for the circuit in Figure 8-56.

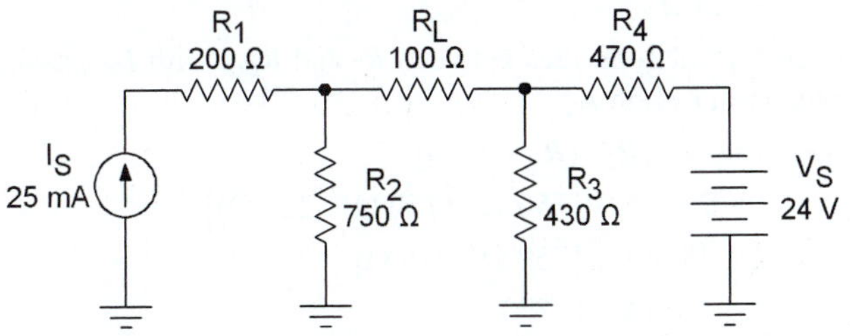

Figure 8-56: Norton's Theorem Example Circuit 4

Solution: For this example define currents flowing from left to right as positive and currents flowing from right to left as negative. The first step is to find the short circuit current I_{SC} through R_L. Because the circuit contains both a current source and a voltage source, use the superposition theorem to determine the current from each source and algebraically sum them to determine the total short circuit current. Refer to Figure 8-57.

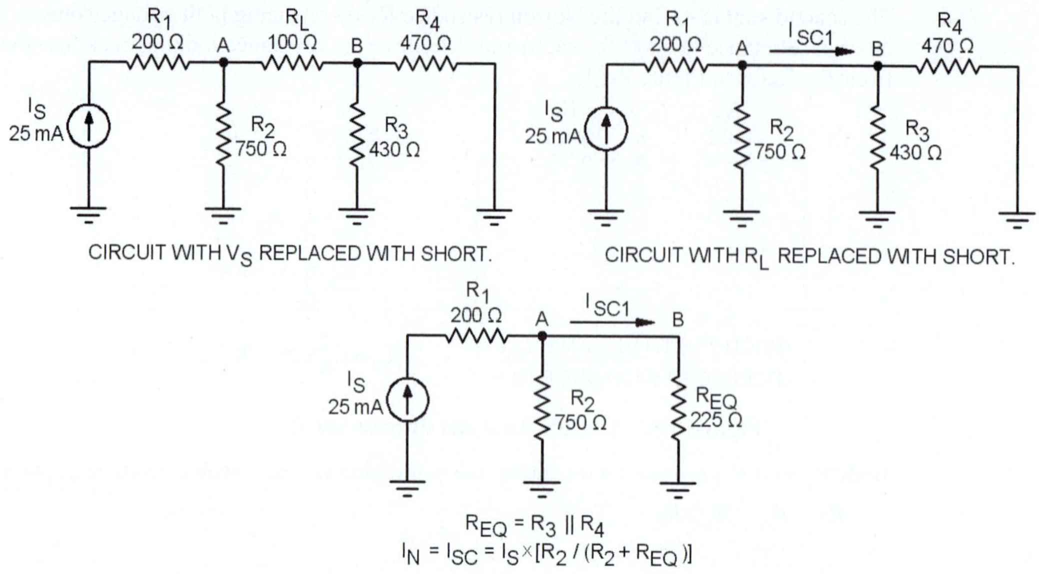

Figure 8-57: Circuit Simplification and Analysis Process for I_{SC1}

To determine the Norton current I_N for I_S, first replace V_S with its internal resistance (a short). Then replace R_L with a short and determine the short circuit current $I_{SC} = I_N$ between point A and point B. With V_S replaced with its internal resistance and R_L replaced with a short the circuit expression is

$$R_{T\,(IS)} = R_1 + R_2 \parallel R_3 \parallel R_4$$

To find the value of I_{SC}, let the equivalent resistance R_{EQ} be the parallel combination $R_3 \parallel R_4$. From this,

$$R_{T\,(IS)} = R_1 + R_2 \parallel R_{EQ}$$

where

$$\begin{aligned} R_{EQ} &= R_3 \parallel R_4 \\ &= 430\ \Omega \parallel 470\ \Omega \\ &= 225\ \Omega \end{aligned}$$

The source current I_S divides between R_2 and R_{EQ}, with I_{SC} passing through R_{EQ}. Use the current divider theorem to find I_{SC}.

$$\begin{aligned} I_{SC\,(IS)} &= I_S \times [R_2 / (R_2 + R_{EQ})] \\ &= 25\text{ mA} \times [750\ \Omega / (750\ \Omega + 225\ \Omega)] \\ &= 25\text{ mA} \times [750\ \Omega / 975\ \Omega] \\ &= 25\text{ mA} \times 0.770 \\ &= 19.2\text{ mA} \end{aligned}$$

$I_{SC\,(IS)}$ flows from left to right so by definition $I_{SC1} = +19.2$ mA.

Refer next to Figure 8-58.

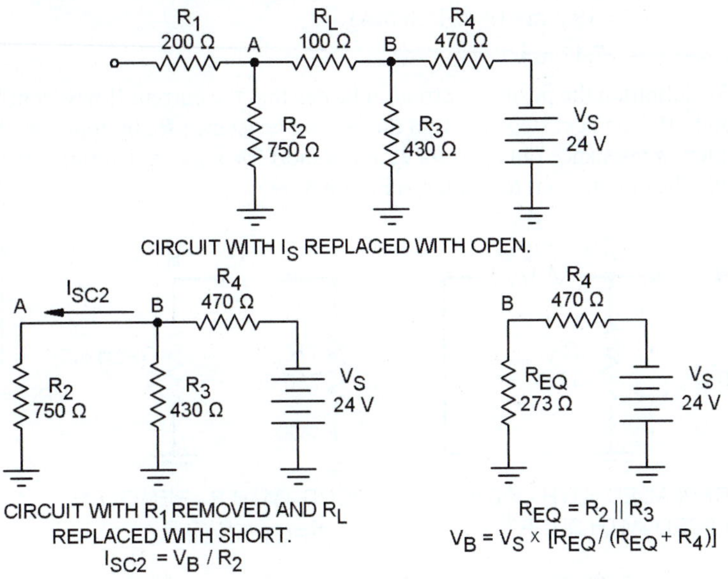

Figure 8-58: Circuit Simplification and Analysis for I_{SC2}

To determine the Norton current for V_S, first replace I_S with its internal resistance (an open). Then replace R_L with a short and determine the short circuit current $I_{SC} = I_N$ between point A and point B. With I_S replaced with its internal resistance and R_L replaced with a short, no current flows through R_1 so that R_1 can be removed for the circuit analysis. The expression for the circuit is

$$R_{T\,(VS)} = R_4 + R_2 \parallel R_3$$

The first (and final) equivalent resistance R_{EQ} is the parallel combination $R_2 \parallel R_3$.

$$R_{T\,(VS)} = R_4 + R_{EQ}$$

where

$$R_{EQ} = R_2 \parallel R_3$$
$$= 750\ \Omega \parallel 430\ \Omega$$
$$=273\ \Omega$$

Use the voltage divider theorem to calculate V_B.

$$V_B = V_S \times [R_{EQ} / (R_{EQ} + R_4)]$$
$$= 24\ \text{V} \times [273\ \Omega / (273\ \Omega + 470\ \Omega)]$$
$$= 24\ \text{V} \times (273\ \Omega / 743\ \Omega)$$
$$= 24\ \text{V} \times 0.368$$
$$= 8.82\ \text{V}$$

Use Ohm's Law to calculate $I_{SC\,(VS)}$.

$$I_{SC\,(VS)} = I_{R2}$$
$$= V_B / R_2$$
$$= 8.82\ \text{V} / 750\ \Omega$$
$$= (8.82 / 750)\ (\text{V} / \Omega)$$
$$= 11.8 \times 10^{-3}\ \text{A}$$
$$= 11.8\ \text{mA}$$

$I_{SC\,(VS)}$ flows from right to left so by definition $I_{SC\,(VS)} = -11.8$ mA. The Norton current I_N for R_L is the sum of the short circuit currents from the two sources.

$$I_N = I_{SC}$$
$$= I_{SC\,(IS)} + I_{SC(VS)}$$

$= (+19.2 \text{ mA}) + (-11.8 \text{ mA})$

$= \mathbf{+7.47 \text{ mA}}$

By definition the positive current indicates that the current flows from left to right from point A to point B. The next step is to find the Norton resistance R_N by replacing both sources with their internal resistance (an open for I_S and a short for V_S) and determine the resistance looking back into the circuit from R_L. Refer to Figure 8-59.

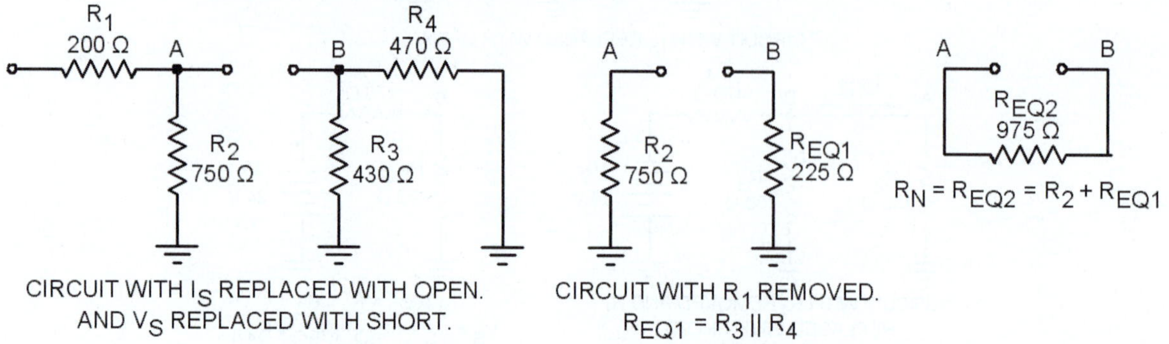

Figure 8-59: Circuit Analysis and Simplification Process for R_N

With I_S replaced by an open, no current flows through R_1 so that R_1 can be removed for the circuit analysis. The expression for the circuit is

$R_N = R_2 + R_3 \,||\, R_4$

The first equivalent resistance R_{EQ1} is for the parallel combination $R_3 \,||\, R_4$.

$R_N = R_2 + R_{EQ1}$

where

$R_{EQ1} = R_3 \,||\, R_4$

$= 430\ \Omega \,||\, 470\ \Omega$

$=225\ \Omega$

The second (and final) equivalent resistance R_{EQ2} is for the series combination $R_2 + R_{EQ1}$.

$R_N = R_{EQ2}$

where

$R_{EQ2} = R_2 + R_{EQ1}$

$= 750\ \Omega + 225\ \Omega$

$= \mathbf{975\ \Omega}$

Figure 8-60 shows the Norton equivalent for Figure 8-56.

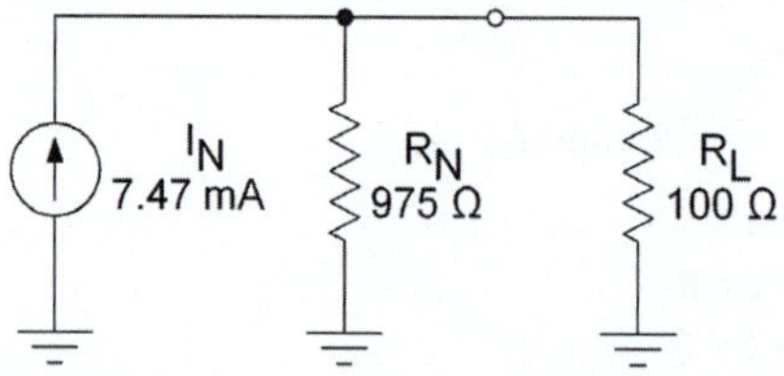

Figure 8-60: Norton Equivalent for Figure 8-56

8.2.5 Maximum Power Transfer Theorem

1. Problem: Determine the maximum value of power P_{RL} dissipated by R_L for the circuit in Figure 8-61. Assume that all values are exact.

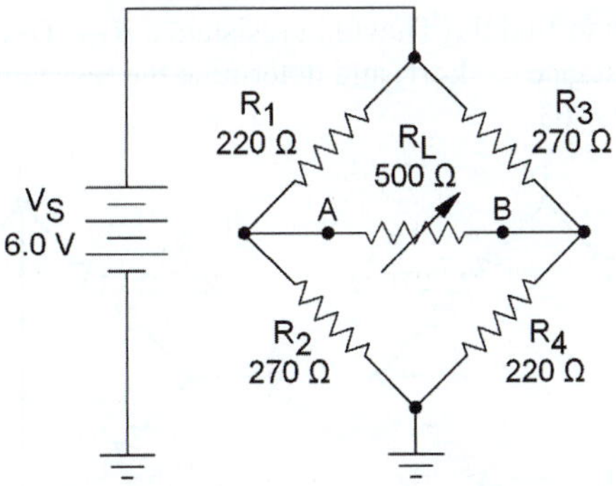

Figure 8-61: Maximum Power Transfer Theorem Example Circuit 3

Solution: To determine the maximum power the circuit will transfer to the load resistor $\boldsymbol{R_L}$, first Thevenize the circuit. The first step is to find the Thevenin voltage $\boldsymbol{V_{TH}}$ for $\boldsymbol{R_L}$. To do so, first remove $\boldsymbol{R_L}$ and determine the open-load voltage $\boldsymbol{V_{TH}} = \boldsymbol{V_{AB}}$. Refer to Figure 8-62.

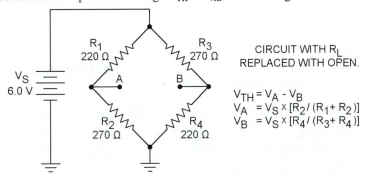

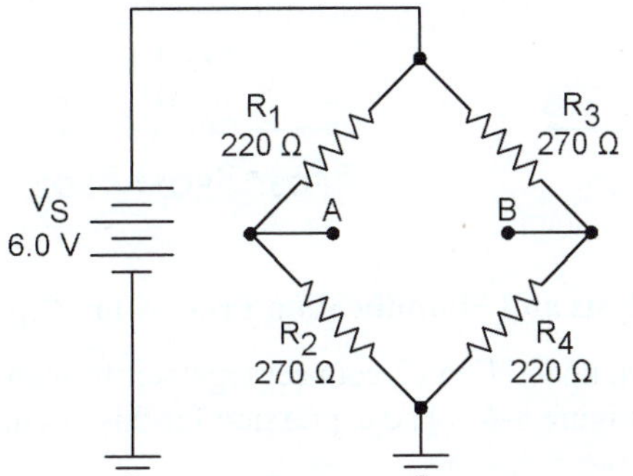

Figure 8-62: Circuit Analysis and Simplification Process for V_{TH}

By definition, $\boldsymbol{V_{AB}} = \boldsymbol{V_A} - \boldsymbol{V_B}$.

$$\boldsymbol{V_{TH}} = \boldsymbol{V_{AB}}$$
$$= \boldsymbol{V_A} - \boldsymbol{V_B}$$

Use the voltage divider theorem to calculate $\boldsymbol{V_A}$ and $\boldsymbol{V_B}$.

$$\boldsymbol{V_A} = \boldsymbol{V_S} \times [\boldsymbol{R_2} / (\boldsymbol{R_1} + \boldsymbol{R_2})]$$

and

$$\boldsymbol{V_B} = \boldsymbol{V_S} \times [\boldsymbol{R_4} / (\boldsymbol{R_3} + \boldsymbol{R_4})]$$

Use $\boldsymbol{V_A}$ and $\boldsymbol{V_B}$ to calculate $\boldsymbol{V_{TH}}$.

$$\boldsymbol{V_{TH}} = \boldsymbol{V_A} - \boldsymbol{V_B}$$
$$= \boldsymbol{V_S} \times [\boldsymbol{R_2} / (\boldsymbol{R_1} + \boldsymbol{R_2})] - \boldsymbol{V_S} \times [\boldsymbol{R_4} / (\boldsymbol{R_3} + \boldsymbol{R_4})]$$
$$= \boldsymbol{V_S} \times \{[\boldsymbol{R_2} / (\boldsymbol{R_1} + \boldsymbol{R_2})] - [\boldsymbol{R_4} / (\boldsymbol{R_3} + \boldsymbol{R_4})]\}$$
$$= 6.0\ \text{V} \times \{[270\ \Omega / (220\ \Omega + 270\ \Omega)] - [220\ \Omega / (270\ \Omega + 220\ \Omega)]\}$$
$$= 6.0\ \text{V} \times [(270\ \Omega / 490\ \Omega)] + -[(220\ \Omega / 490\ \Omega)]$$
$$= 6.0\ \text{V} \times [(270 / 490)\ (\Omega / \Omega)] + -[(220 / 490)\ (\Omega / \Omega)]$$
$$= 6.0\ \text{V} \times (0.551 + -0.449)$$
$$= 6.0\ \text{V} \times 0.102$$
$$= 612 \times 10^{-3}\ \text{V}$$
$$= 612\ \text{mV}$$

The next step is to find the Thevenin resistance R_{TH}. To do so, replace the voltage source V_S with its internal resistance (a short) and determine the resistance looking back into the circuit from R_L. Refer to Figure 8-63.

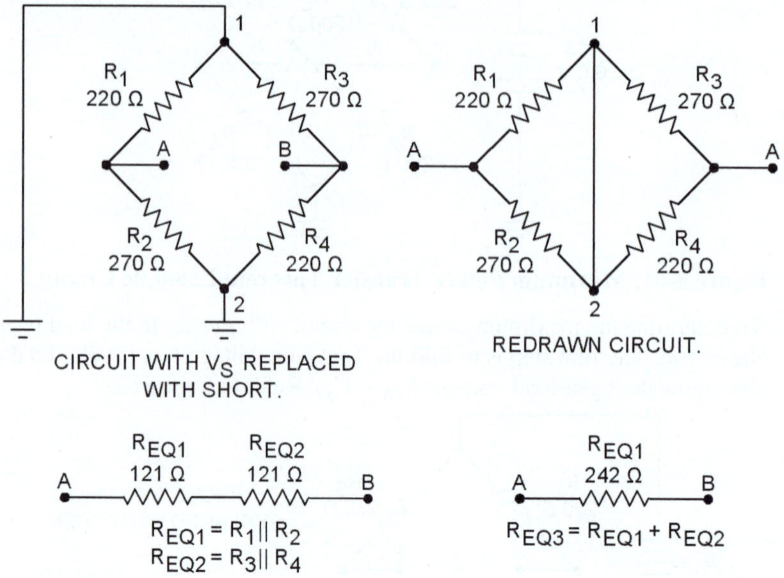

Figure 8-63: Circuit Analysis and Simplification Process for R_{TH}

With V_S replaced with a short, nodes 1 and 2 connect together through ground so that the circuit can be redrawn as shown in Figure 8-63. The expression for this circuit is

$$R_{TH} = R_1 \parallel R_2 + R_3 \parallel R_4$$

The first equivalent resistance R_{EQ1} is the parallel combination $R_1 \parallel R_2$.

$$R_{TH} = R_1 \parallel R_2 + R_3 \parallel R_4$$

where

$$\begin{aligned} R_{EQ1} &= R_1 \parallel R_2 \\ &= 220\ \Omega \parallel 270\ \Omega \\ &= 121\ \Omega \end{aligned}$$

The second equivalent resistance R_{EQ2} is the parallel combination $R_3 \parallel R_4$.

$$R_{TH} = R_{EQ1} + R_{EQ2}$$

where

$$\begin{aligned} R_{EQ2} &= R_3 \parallel R_4 \\ &= 270\ \Omega \parallel 220\ \Omega \\ &= 121\ \Omega \end{aligned}$$

The third (and final) equivalent resistance R_{EQ3} is the series combination $R_{EQ1} + R_{EQ2}$.

$$R_{TH} = R_{EQ3}$$

where

$$\begin{aligned} R_{EQ3} &= R_{EQ1} + R_{EQ2} \\ &= 121\ \Omega + 121\ \Omega \\ &= 242\ \Omega \end{aligned}$$

Figure 8-64 shows the Thevenin equivalent for Figure 8-61.

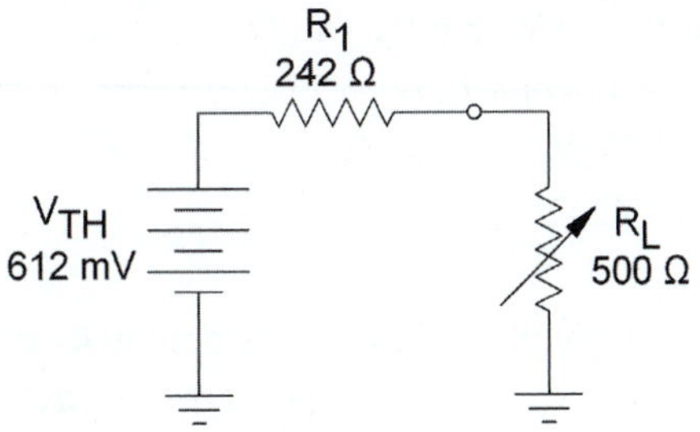

Figure 8-64: Thevenin Equivalent for Figure 8-61

The circuit transfers maximum power when $R_L = R_{TH} = 242\ \Omega$, so that $V_{RL} = V_{TH} / 2$.

$V_{RL} = V_{TH} / 2$
$= 612\ \text{mV} / 2$
$= 306\ \text{mV}$

Use Watt's Law for voltage and resistance to calculate the maximum power dissipated by the load.

$P_{RL\,(\text{max})} = (V_{RL})^2 / R_L$
$= (306 \times 10^{-3}\ \text{V})^2 / 242\ \Omega$
$= [(306 \times 10^{-3})^2 / 242]\ [(\text{V})^2 / \Omega]$
$= [(93.7 \times 10^{-3}) / 242]\ [\text{V}^2 / \Omega]$
$=387 \times 10^{-6}\ \text{W}$
$= 387\ \mu\text{W}$

The maximum power transferred to the load resistance R_L is **387 μW**.

2. Problem: For the circuit in Figure 8-65, $V_{OUT} = 3.6$ V when load resistance $R_L = 12\ \Omega$. What additional resistance R_X must be added in parallel with R_L to maximize the transfer of power to the load?

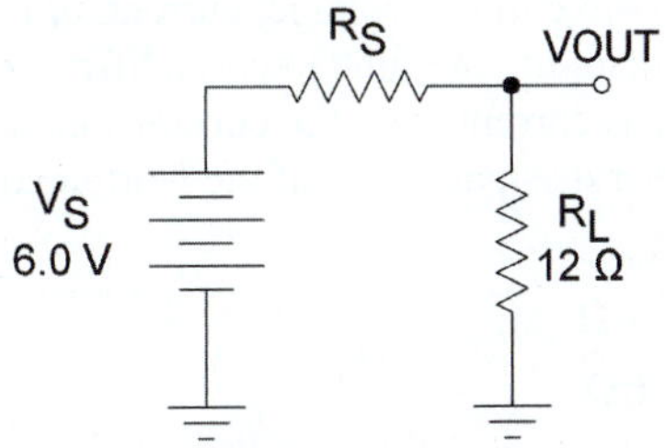

Figure 8-65: Maximum Power Transfer Theorem Example Circuit 4

Solution: When the circuit transfers maximum power to the load $R_S = R_L$. When this is true,

$V_{OUT} = V_S / 2$
$= 6.0\ \text{V} / 2$
$= 3.0\ \text{V}$

Since $V_{OUT} = 3.6$ V for $R_L = 12\ \Omega$, $R_S \neq R_L$, so that the circuit is not transferring maximum power to the load. Use the voltage divider theorem to solve for R_S.

$V_{OUT} = V_S \times [R_L / (R_S + R_L)]$
$V_{OUT} \times (R_S + R_L) = V_S \times [R_L / (R_S + R_L)] \times (R_S + R_L)$
$(V_{OUT} \times R_S) + (V_{OUT} \times R_L) + -(V_{OUT} \times R_L) = (V_S \times R_L) + -(V_{OUT} \times R_L)$
$(V_{OUT} \times R_S) / V_{OUT} = [(V_S + -V_{OUT}) \times R_L] / V_{OUT}$
$R_S = [(V_S + -V_{OUT}) / V_{OUT}] \times R_L$

Substitute values of V_S, V_{OUT}, and R_L to solve for R_1.

$R_S = [(6.0\ \text{V} + -3.6\ \text{V}) / 3.6\ \text{V}] \times 12\ \Omega$

$= [(6.0 + -3.6)\ \text{V} / 3.6\ \text{V}] \times 12\ \Omega$

$= (2.4\ \text{V} / 3.6\ \text{V}) \times 12\ \Omega$

$= (2.4 / 3.6)\ (\text{V} / \text{V}) \times 12\ \Omega$

$= 0.667 \times 12\ \Omega$

$= 8.0\ \Omega$

Use the equation for parallel resistors to determine R_X so that $R_L \parallel R_X = R_S$.

$$1/ R_S = (1 / R_L) + (1 / R_X)$$

$$1 / 8\ \Omega = [(1 / 12)\ \Omega] + (1 / R_X)$$

$$[1 / (8\ \Omega)] + -[1 / (12\ \Omega)] = [1 / (12\ \Omega)] + (1 / R_X) + -[1/ (12\ \Omega)]$$

$$125 \times 10^{-3}\ \text{S} + -83.3 \times 10^{-3}\ \text{S} = 1 / R_X$$

$$(125 \times 10^{-3} + -83.3 \times 10^{-3})\ \text{S} = 1 / R_X$$

$$41.7 \times 10^{-3}\ \text{S} = 1 / R_X$$

$$1 / (41.7 \times 10^{-3}\ \text{S}) = 1 / (1 / R_X)$$

$$[1 / (41.7 \times 10^{-3})]\ (1 / \text{S}) = R_X$$

$$24.0\ \Omega = R_X$$

A **24 Ω** resistor connected across R_L will maximize transfer of power to the load.

8.3 Just for Fun

1. Problem: The following is a question once used in electronics job interviews:

"There are two self-contained 'black boxes' with electrical output contacts before you on a table. One contains a Thevenin equivalent circuit with $V_{TH} = 5$ V and $R_{TH} = 1\ \Omega$. The other is a Norton equivalent circuit with $I_N = 5$ A and $R_N = 1\ \Omega$. How can you identify which box contains the Thevenin equivalent circuit and which box contains the Norton equivalent circuit?"

Solution: Although it is impossible to use voltage, current, or resistance measurement to determine which box contains which circuit, one fundamental difference is that Norton equivalent circuits dissipate power with no load as current from the current source will flow through the parallel source resistance. The power dissipation P_D of the Norton equivalent is

$$P_D = (I_N)^2 \times R_N$$

$$= (5\ \text{A})^2 \times 1\ \Omega$$

$$= 25\ \text{A}^2 \times 1\ \Omega$$

$$= 25\ \text{W}$$

Check the temperature of the boxes, as the power dissipation of the Norton resistance will heat the box that contains it.

2. Problem: Verify Millman's Theorem, which states that K parallel voltage sources (each of which consists of an ideal voltage source and finite series resistance as shown in Figure 8-66) can be replaced by a single voltage source V_M and series resistor R_M, where

$$V_M = [(V_{S1} / R_{S1}) + (V_{S2} / R_{S2}) + \ldots + (V_{SK} / R_{SK})] / [(1 / R_{S1}) + (1 / R_{S2}) + \ldots + (1 / R_{SK})]$$

$$R_M = 1 / [(1 / R_{S1}) + (1 / R_{S2}) + \ldots + (1 / R_{SK})]$$

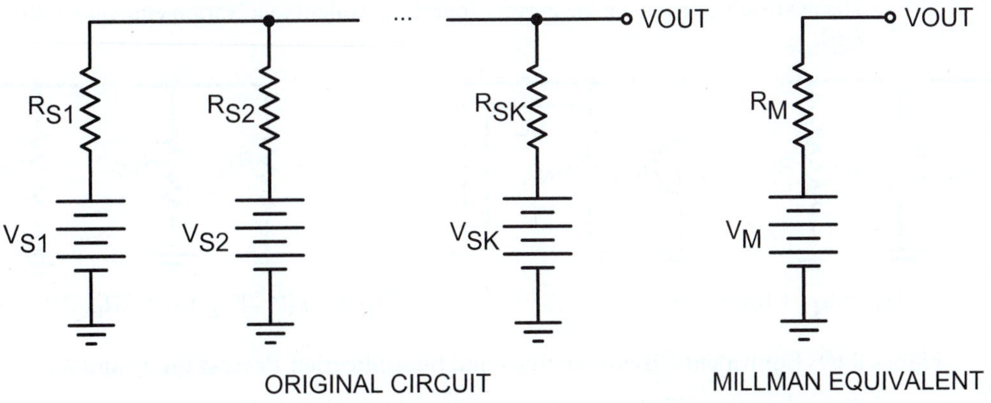

Figure 8-66: Voltage Sources in Parallel and Millman Equivalent

Solution: First, convert each voltage source in the circuit to its Norton equivalent. The first step is to determine the short circuit current for each voltage source. The second step is to replace the voltage source with its internal resistance and determine the resistance looking back into the circuit from the output. Refer to Figure 8-67.

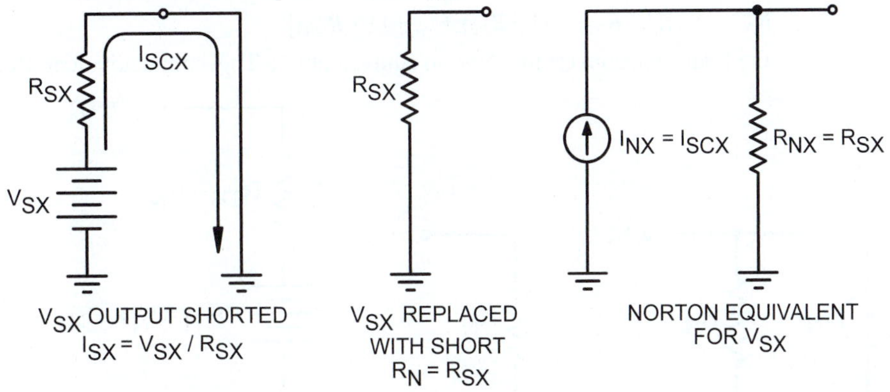

Figure 8-67: Circuit Analysis and Simplification Process for I_N and R_N

From Ohm's Law for current, the short circuit current $I_{SC} = I_N$ of voltage source X is

$$I_{SCX} = V_{SX} / R_{SX}$$

With the ideal voltage source V_{SX} replaced with its internal resistance (a short), the equivalent resistance looking back into the circuit is

$$R_{NX} = R_{SX}$$

The Norton equivalent for voltage source X is then a current source with value $I_{NX} = I_{SCX} = V_{SX} / R_{SX}$, in parallel with resistance $R_{NX} = R_{SX}$. Replacing each voltage source gives K parallel Norton sources as shown in Figure 8-68.

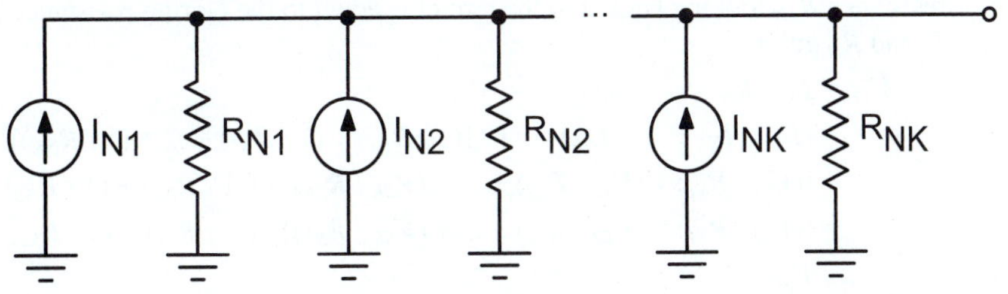

Figure 8-68: Current Source Equivalent of Original Circuit

The next step is to reduce the current source equivalent to a Norton equivalent. Refer to Figure 8-69.

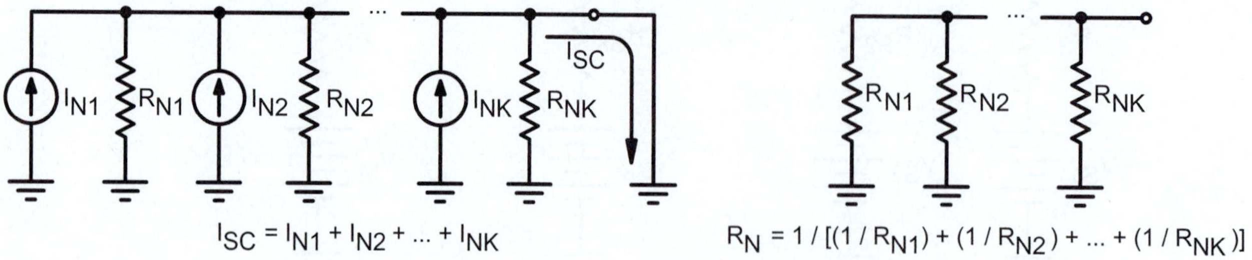

Figure 8-69: Equivalent Circuit Analysis and Simplification Process for I_N and R_N

When the output of the equivalent circuit is shorted, all the Norton resistances are bypassed so that the short circuit current I_{SC} is equal to the sum of the Norton currents I_{N1} through I_{NK}. When each ideal Norton source is replaced with its internal resistance (an open), the resistance looking back into the circuit is equal to the Norton resistances R_{N1} through R_{NK} in parallel. Therefore

$$I_N = I_{SC}$$
$$= I_{N1} + I_{N2} + \ldots + I_{NK}$$
$$R_N = 1 / [(1 / R_{N1}) + (1 / R_{N2}) + \ldots + (1 / R_{NK})]$$

For the final step, convert this Norton equivalent to a Thevenin equivalent. Refer to Figure 8-70.

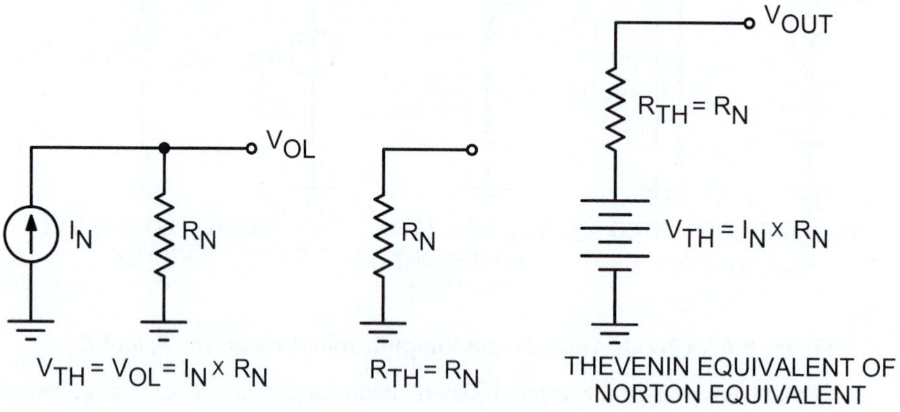

Figure 8-70: Circuit Analysis and Simplification Process for V_{TH} and R_{TH}

The Thevenin voltage V_{TH} is the open load voltage V_{OL} for the circuit. From Ohm's Law for voltage,

$$V_{TH} = V_{OL}$$
$$= I_N \times R_N$$

When the ideal Norton source I_N is replaced with its internal resistance (an open) the Thevenin resistance R_{TH} looking back into the circuit is equal to the Norton resistance R_N. Substituting for I_N and R_N gives

$$V_{TH} = I_N \times R_N$$
$$= (I_{N1} + I_{N2} + \ldots + I_{NK}) \times \{1 / [(1 / R_{N1}) + (1 / R_{N2}) + \ldots + (1 / R_{NK})]\}$$
$$= [(V_{S1} / R_{S1}) + (V_{S2} / R_{S2}) + \ldots + (V_{SK} / R_{SK})] / [(1 / R_{N1}) + (1 / R_{N2}) + \ldots + (1 / R_{NK})]$$
$$= [(V_{S1} / R_{S1}) + (V_{S2} / R_{S2}) + \ldots + (V_{SK} / R_{SK})] / [(1 / R_{S1}) + (1 / R_{S2}) + \ldots + (1 / R_{SK})]$$
$$= V_M$$

and

$$\begin{aligned} R_{TH} &= R_N \\ &= 1 / [(1 / R_{N1}) + (1 / R_{N2}) + \ldots + (1 / R_{NK})] \\ &= 1 / [(1 / R_{S1}) + (1 / R_{S2}) + \ldots + (1 / R_{SK})] \\ &= R_M \end{aligned}$$

This verifies Millman's Theorem.

3. Problem: Prove that, for any number of ladder sections, the circuit in Figure 8-65 transfers maximum power P_{RL} to R_L when $R_L = 2R$.

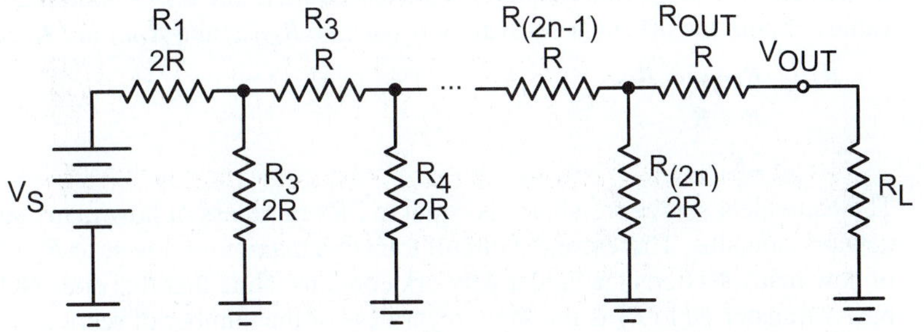

Figure 8-71: *R*/2*R* Maximum Power Transfer Example Circuit

Solution: The circuit will transfer maximum power to the load resistance R_L when R_L is equal to the effective source resistance $R_S = R_{TH} = R_N$. To find the effective source resistance, replace V_S with its internal resistance (a short) and determine the resistance looking back into the circuit from R_L. Refer to Figure 8-72.

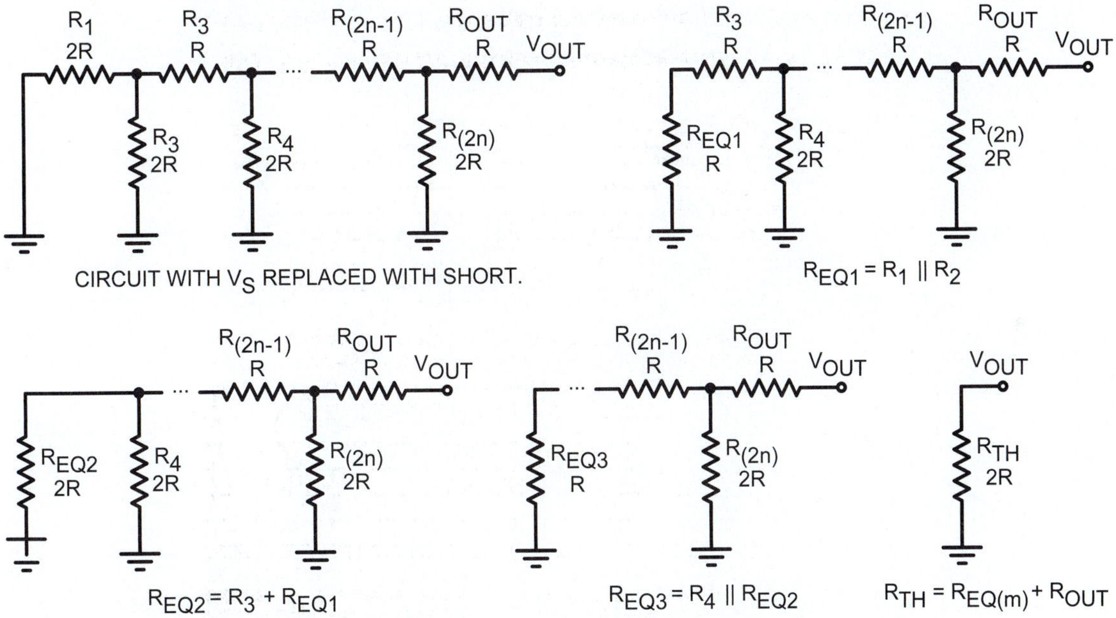

Figure 8-72: Circuit Analysis and Simplification Process for R_{TH}

With V_S replaced with a short, the circuit is that of an $R/2R$ ladder network with n sections. The first equivalent resistance R_{EQ1} is the parallel combination $R_1 \parallel R_2$. The values of R_1 and R_2 are both $2R$, so

$$\begin{aligned} R_{EQ1} &= R_1 \parallel R_2 \\ &= 2R \parallel 2R \\ &= R \end{aligned}$$

The second equivalent resistance $\boldsymbol{R_{EQ2}}$ is the series combination $\boldsymbol{R_{EQ1}} + \boldsymbol{R_3}$, so

$$\begin{aligned} \boldsymbol{R_{EQ2}} &= \boldsymbol{R_{EQ1}} + \boldsymbol{R_3} \\ &= \boldsymbol{R} + \boldsymbol{R} \\ &= 2\boldsymbol{R} \end{aligned}$$

The third equivalent resistance $\boldsymbol{R_{EQ3}}$ is the parallel combination $\boldsymbol{R_{EQ2}} \parallel \boldsymbol{R_4}$. The values of both $\boldsymbol{R_{EQ2}}$ and $\boldsymbol{R_4}$ are $2\boldsymbol{R}$ so, as with the case of $\boldsymbol{R_{EQ1}} = \boldsymbol{R_1} \parallel \boldsymbol{R_2} = \boldsymbol{R}$, $\boldsymbol{R_{EQ3}} = \boldsymbol{R}$. This process repeats for each section of the $\boldsymbol{R}/2\boldsymbol{R}$ ladder network for as many sections as the network contains until the final section for which the equivalent resistance is equal to the series combination $\boldsymbol{R_{EQ(m)}} + \boldsymbol{R_{OUT}}$. The values of both the previous equivalent resistance $\boldsymbol{R_{EQ(m)}}$ and $\boldsymbol{R_{OUT}}$ are $\boldsymbol{R}$, so

$$\begin{aligned} \boldsymbol{R_{TH}} &= \boldsymbol{R_{EQ(m)}} + \boldsymbol{R_{OUT}} \\ &= \boldsymbol{R} + \boldsymbol{R} \\ &= 2\boldsymbol{R} \end{aligned}$$

The equivalent source resistance is equal to $2\boldsymbol{R}$ regardless of how many sections the $\boldsymbol{R}/2\boldsymbol{R}$ ladder network contains. Therefore the circuit transfers maximum power to $\boldsymbol{R_L}$ when $\boldsymbol{R_L} = 2\boldsymbol{R}$ regardless of how many sections the ladder network contains. Note that this does NOT mean that the actual power transferred to $\boldsymbol{R_L}$ is the same regardless of the number of sections in the network, only that the power transferred is maximum when $\boldsymbol{R_L} = 2\boldsymbol{R}$.

9. Branch, Loop, and Node Analyses

This section covers matrices but not the mechanics of solving matrices, as most scientific calculators, spreadsheet programs, and mathematics software have matrix solving features. The section instead discusses setting up and solving the systems of linear equations and matrices that relate to analyzing electronic circuits.

9.1 Basic

9.1.1 Simultaneous Equations

1. Problem: Solve the following linear system of two equations in two unknowns for $\boldsymbol{X}$ and $\boldsymbol{Y}$:

$$\begin{aligned} A:&\quad 3X + 2Y = 20 \\ B:&\quad 4X - Y = 1 \end{aligned}$$

Solution: First, rewrite A and B to contain only sums:

$$\begin{aligned} A:&\quad 3X + 2Y = 20 \\ B:&\quad 4X + -Y = 1 \end{aligned}$$

Next, solve the system by finding multiples of A and B that, when added, will eliminate one of the variables. From the given equations, the sum $A + 2B$ will eliminate variable $\boldsymbol{Y}$ as A contains $2Y$ and $2B$ will contain $-2Y$. The equation $2B$ is

$$2B\text{: } 2 \times (4X + -Y = 1) \rightarrow 8X + -2Y = 2$$

Add the equations A and $2B$ to eliminate the variable Y.

$$\begin{aligned} A:&\quad 3X + 2Y = 20 \\ 2B:&\quad 8X + -2Y = 2 \\ A+2B:&\quad 11X + 0Y = 22 \end{aligned}$$

Use the equation $A + 2B$ to solve for $\boldsymbol{X}$.

$$\begin{aligned} 11X + 0Y &= 22 \\ 11X &= 22 \\ 11X / 11 &= 22 / 11 \\ X &= 2 \end{aligned}$$

Substitute $\boldsymbol{X} = 2$ in equation B and solve for $\boldsymbol{Y}$.

$$\begin{aligned} 4X + -Y &= 1 \\ 4X + -Y + Y &= 1 + Y \\ 4X + -1 &= 1 + Y + -1 \\ 4 \times 2 + -1 &= Y \\ 8 + -1 &= Y \\ 7 &= Y \end{aligned}$$

For the linear system of two equations in two unknowns $\boldsymbol{X} = \mathbf{2}$ and $\boldsymbol{Y} = \mathbf{7}$.

2. Problem: Solve the following linear system of three equations in three unknowns for $\boldsymbol{X}$, $\boldsymbol{Y}$, and $\boldsymbol{Z}$:

$$\begin{aligned} A:&\quad -X + Y + 2Z = -1 \\ B:&\quad 3X + 3Y + 3Z = 0 \\ C:&\quad -2X - 2Y + Z = -9 \end{aligned}$$

Solution: First, rewrite A, B, and C to contain only sums:

$$\begin{aligned} A:&\quad -X + Y + 2Z = -1 \\ B:&\quad 3X + 3Y + 3Z = 0 \\ C:&\quad -2X + -2Y + Z = -9 \end{aligned}$$

Next, solve the system by finding multiples of A, B, and C that, when added, will eliminate one of the variables from two independent equations (i.e., equations that are not multiples of each other). The sum $-3A + B$ will eliminate $\mathbf{Y}$, as $-3A$ will contain $-3\mathbf{Y}$ and B contains $3\mathbf{Y}$. The second sum $2A + C$ will also eliminate $\mathbf{Y}$, as $2A$ will contain $2\mathbf{Y}$ and C contains $-2\mathbf{Y}$. The equations $2A$ and $-3A$ are

$$2A: \quad 2 \times (-\mathbf{X} + \mathbf{Y} + 2\mathbf{Z} = -1) \rightarrow -2\mathbf{X} + 2\mathbf{Y} + 4\mathbf{Z} = -2$$

$$-3A: \quad -3 \times (-\mathbf{X} + \mathbf{Y} + 2\mathbf{Z} = -1) \rightarrow 3\mathbf{X} + -3\mathbf{Y} + -6\mathbf{Z} = 3$$

Add negative $-3A$ to B to create a new equation. For convenience, call it equation D.

$$\begin{aligned} -3A: &\quad 3\mathbf{X} + -3\mathbf{Y} + -6\mathbf{Z} = 3 \\ B: &\quad 3\mathbf{X} + 3\mathbf{Y} + 3\mathbf{Z} = 0 \\ D: &\quad 6\mathbf{X} + 0\mathbf{Y} + -3\mathbf{Z} = 3 \end{aligned}$$

Add $2A$ to C to create a new equation. For convenience, call it equation E.

$$\begin{aligned} 2A: &\quad -2\mathbf{X} + 2\mathbf{Y} + 4\mathbf{Z} = -2 \\ C: &\quad -2\mathbf{X} + -2\mathbf{Y} + \mathbf{Z} = -9 \\ E: &\quad -4\mathbf{X} + 0\mathbf{Y} + 5\mathbf{Z} = -11 \end{aligned}$$

Equation D and equation E can now be solved as a linear system of two equations in two unknowns to eliminate either $\mathbf{X}$ or $\mathbf{Z}$. The sum $2D + 3E$ will eliminate $\mathbf{X}$, as $2D$ will contain $12\mathbf{X}$ and $3E$ will contain $-12\mathbf{X}$. The equations $2D$ and $3E$ are

$$2D: \quad 2 \times (6\mathbf{X} + -3\mathbf{Z} = 3) \rightarrow 12\mathbf{X} + -6\mathbf{Z} = 6$$

$$3E: \quad 3 \times (-4\mathbf{X} + 5\mathbf{Z} = -11) \rightarrow -12\mathbf{X} + 15\mathbf{Z} = -33$$

Add $2D$ to $3E$ to create a new equation. For convenience, call it F.

$$\begin{aligned} 2D: &\quad 12\mathbf{X} + -6\mathbf{Z} = 6 \\ 3E: &\quad -12\mathbf{X} + 15\mathbf{Z} = -33 \\ F: &\quad 0\mathbf{X} + 9\mathbf{Z} = -27 \end{aligned}$$

Use equation F to solve for $\mathbf{Z}$.

$$\begin{aligned} 0\mathbf{X} + 9\mathbf{Z} &= -27 \\ 9\mathbf{Z} &= -27 \\ 9\mathbf{Z} / 9 &= -27 / 9 \\ \mathbf{Z} &= -3 \end{aligned}$$

Substitute $\mathbf{Z} = -3$ into equation D and solve for $\mathbf{X}$.

$$\begin{aligned} 6\mathbf{X} + -3\mathbf{Z} + 3\mathbf{Z} &= 3 + 3\mathbf{Z} \\ 6\mathbf{X} / 6 &= (3 + 3\mathbf{Z}) / 6 \\ \mathbf{X} &= [3 + (3 \times -3)] / 6 \\ &= [3 + (-9)] / 6 \\ &= -6 / 6 \\ &= -1 \end{aligned}$$

Substitute $\mathbf{X} = -1$ and $\mathbf{Z} = -3$ into equation A (or B or C) and solve for $\mathbf{Y}$.

$$\begin{aligned} -\mathbf{X} + \mathbf{Y} + 2\mathbf{Z} + -2\mathbf{Z} &= -1 + -2\mathbf{Z} \\ -\mathbf{X} + \mathbf{Y} + \mathbf{X} &= -1 + -2\mathbf{Z} + \mathbf{X} \\ \mathbf{Y} &= -1 + (-2 \times -3) + -1 \\ &= -1 + 6 + -1 \\ &= 4 \end{aligned}$$

For the linear system of three equations in three unknowns, $\mathbf{X} = \mathbf{-1}$, $\mathbf{Y} = \mathbf{4}$, and $\mathbf{Z} = \mathbf{-3}$.

3. Problem: Represent the following linear systems of equations in $\boldsymbol{A}\bullet\boldsymbol{X} = \boldsymbol{B}$ matrix-vector form:

a. $-2X + 5Y = 4$

$2X - 4Y = -2$

b. $3X - 7Y - 2Z = 10$

$2X + 3Z = -2$

$2Y - Z = 2$

Solution: For a linear system of N equations in N unknowns $X_1, X_2, \ldots X_N$ with solutions $B_1, B_2, \ldots B_N$,

$$A_{11} X_1 + A_{12} X_2 + \ldots + A_{1N} X_N = B_1$$
$$A_{21} X_1 + A_{22} X_2 + \ldots + A_{2N} X_N = B_2$$
$$\ldots$$
$$A_{N1} X_1 + A_{N2} X_2 + \ldots + A_{NN} X_N = B_N$$

the coefficient matrix $\boldsymbol{A}$ is

$$\begin{vmatrix} A_{11} & A_{12} & \cdots & A_{1N} \\ A_{21} & A_{22} & \cdots & A_{2N} \\ \vdots & \vdots & & \vdots \\ A_{N1} & A_{N2} & \cdots & A_{NN} \end{vmatrix}$$

the column vector $\boldsymbol{X}$ is

$$\begin{vmatrix} X_1 \\ X_2 \\ \vdots \\ X_N \end{vmatrix}$$

and the solution matrix $\boldsymbol{B}$ is

$$\begin{vmatrix} B_1 \\ B_2 \\ \vdots \\ B_N \end{vmatrix}$$

The maxtrix-vector form is

$$\begin{vmatrix} A_{11} & A_{12} & \cdots & A_{1N} \\ A_{21} & A_{22} & \cdots & A_{2N} \\ \vdots & \vdots & \vdots & \vdots \\ A_{N1} & A_{N2} & \cdots & A_{NN} \end{vmatrix} \bullet \begin{vmatrix} X_1 \\ X_2 \\ \vdots \\ X_N \end{vmatrix} = \begin{vmatrix} B_1 \\ B_2 \\ \vdots \\ B_N \end{vmatrix}$$

a. A linear system of two equations in two unknowns ($N = 2$) has the form

$$A_{11} X_1 + A_{12} X_2 = B_1$$
$$A_{21} X_1 + A_{22} X_2 = B_2$$

The matrix-vector form $\boldsymbol{A}\bullet\boldsymbol{X} = \boldsymbol{B}$ is

$$\begin{vmatrix} A_{11} & A_{12} \\ A_{21} & A_{22} \end{vmatrix} \bullet \begin{vmatrix} X_1 \\ X_2 \end{vmatrix} = \begin{vmatrix} B_1 \\ B_2 \end{vmatrix}$$

First, rewrite the equations to contain only sums.

$$-2X + 5Y = 4$$
$$2X + -4Y = -2$$

Next, identify the coefficients for each equation.

$A_{11} = -2$, $A_{12} = 5$, $A_{21} = 2$ and $A_{22} = -4$

$X_1 = X$ and $X_2 = Y$

$B_1 = 4$ and $B_2 = -2$

The $A \bullet X = B$ matrix-vector form for the linear system is

$$\begin{vmatrix} -2 & 5 \\ 2 & -4 \end{vmatrix} \bullet \begin{vmatrix} X \\ Y \end{vmatrix} = \begin{vmatrix} 4 \\ -2 \end{vmatrix}$$

b. A linear system of three equations in three unknowns ($N = 3$) has the form

$$A_{11}\,X_1 + A_{12}\,X_2 + A_{13}\,X_3 = B_1$$
$$A_{21}\,X_1 + A_{22}\,X_2 + A_{23}\,X_3 = B_2$$
$$A_{31}\,X_1 + A_{32}\,X_2 + A_{33}\,X_3 = B_3$$

so that the matrix-vector form $A \bullet X = B$ is

$$\begin{vmatrix} A_{11} & A_{12} & A_{13} \\ A_{21} & A_{22} & A_{23} \\ A_{31} & A_{32} & A_{33} \end{vmatrix} \bullet \begin{vmatrix} X_1 \\ X_2 \\ X_3 \end{vmatrix} = \begin{vmatrix} B_1 \\ B_2 \\ B_3 \end{vmatrix}$$

First, rewrite the equations so that each equation contains X, Y, and Z terms:

$$\begin{array}{lrcrcrcr} A: & 3X & - & 7Y & - & 2Z & = & 10 \\ B: & 2X & + & 0Y & + & 3Z & = & -2 \\ C: & 0X & + & 2Y & - & Z & = & 2 \end{array}$$

Next, rewrite the equations to contain only sums:

$$\begin{array}{lrcrcrcr} A: & 3X & + & -7Y & + & -2Z & = & 10 \\ B: & 2X & + & 0Y & + & 3Z & = & -2 \\ C: & 0X & + & 2Y & + & -Z & = & 2 \end{array}$$

For the system of equations in the problem statement

$A_{11} = 3$, $A_{12} = -7$, $A_{13} = -2$, $A_{21} = 2$, $A_{22} = 0$, $A_{23} = 3$, $A_{31} = 0$, $A_{32} = 2$, and $A_{33} = -1$

$X_1 = X$, $X_2 = Y$, and $X_3 = Z$

$B_1 = 10$, $B_2 = -2$, and $B_3 = 2$

so the $A \bullet X = B$ matrix vector form of the linear system is

$$\begin{vmatrix} 3 & -7 & -2 \\ 2 & 0 & 3 \\ 0 & 2 & -1 \end{vmatrix} \bullet \begin{vmatrix} X \\ Y \\ Z \end{vmatrix} = \begin{vmatrix} 10 \\ -2 \\ 2 \end{vmatrix}$$

9.1.2 Branch Current Method

For the branch current method:

- A *node* is any point where two or more components connect.
- A *branch* is a circuit path that connects two adjacent nodes.
- A *loop* is any path through the circuit that does not pass through a component more than once and returns to where it began.
- A loop is *non-redundant* if it contains at least one component that no other loop contains.

The steps for applying the branch current method are:

1) Label the nodes.

2) Determine a reference direction for each non-redundant current loop.
3) Assign branch currents and mark the voltage polarity for each resistor based on the direction of the branch current through it. Rules for doing so are
    - The direction of a branch current must be the same as that of the reference loop that contains it. If the calculated current for the branch is negative, the actual current flow is opposite to the assigned branch current.
    - Branch currents can be assigned only to branches that do not already have an assigned branch current.
    - The polarity of resistor voltages is positive where the branch current enters and negative where the branch current exits.
4) Replace any redundant branch currents. A redundant branch current is a branch current that is the same as another branch current, such as currents for branches that are in series.
5) Remove any nonessential branch currents. A nonessential branch current is one that passes through a voltage or current source and is nonessential because the value of the branch current is independent of the value of the branch voltage.
6) Apply Kirchhoff's Voltage Law and Ohm's Law to each loop to generate a set of linear branch current equations. Omit units for simplicity.
7) Apply Kirchhoff's Current Law to complete the set of linear branch current equations. Define currents entering a node and currents leaving a node to have opposite signs.
8) Solve the set of linear branch current equations.

1. Problem: Use the branch current method to determine the currents through each resistor for the circuit in Figure 9-1.

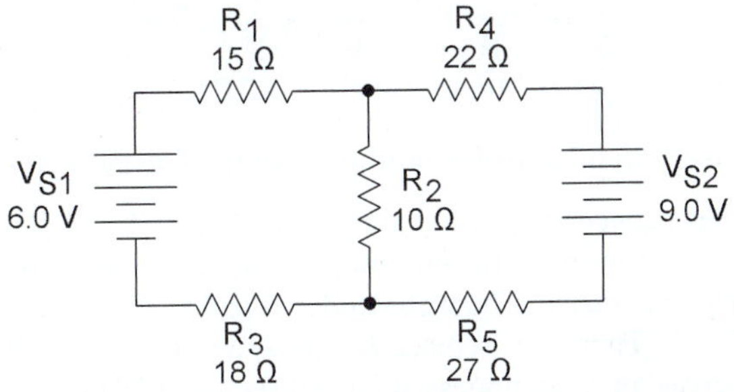

Figure 9-1: Branch Current Method Example Circuit 1

Solution: Label the nodes and determine a reference direction for each non-redundant current loop. The circuit in Figure 9-1 contains six nodes and two non-redundant loops. For this example, choose a clockwise reference direction for each loop. Refer to Figure 9-2.

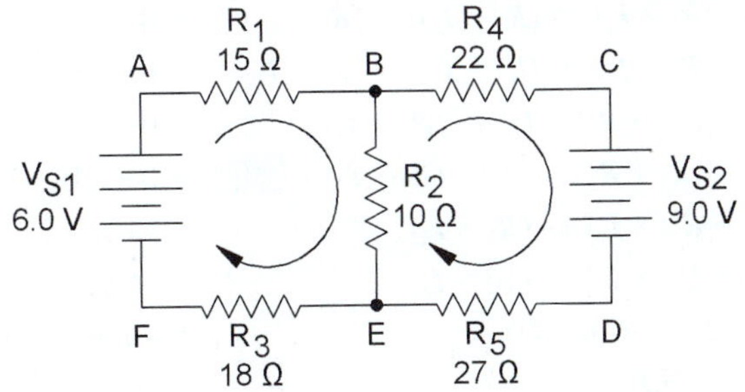

Figure 9-2: Nodes and Reference Directions for Figure 9-1

Assign branch currents and mark the voltage polarity for each resistor based on the direction of the branch current through it. Refer to Figure 9-3.

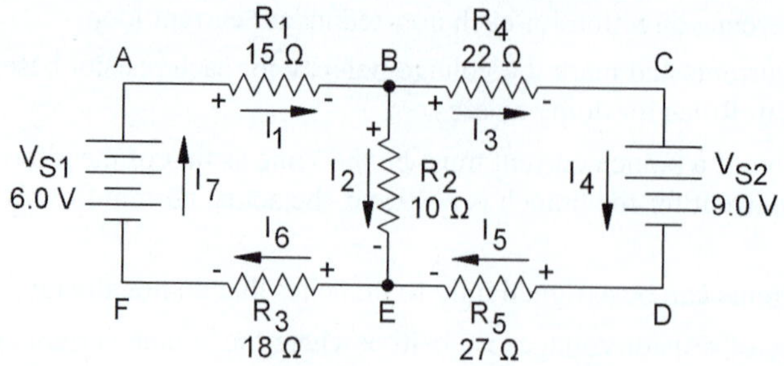

Figure 9-3: Assigned Branch Currents and Resistor Voltage Polarities for Figure 9-1

As Figure 9-3 shows, the branch between nodes B and E exists in both loops in Figure 9-1. The assigned direction for I_2 could be the same for either loop in Figure 9-2, but once you use the left loop to assign the current direction, you ignore the branch for the right loop.

Replace any redundant branch currents and remove any nonessential branch currents. Refer to Figure 9-4.

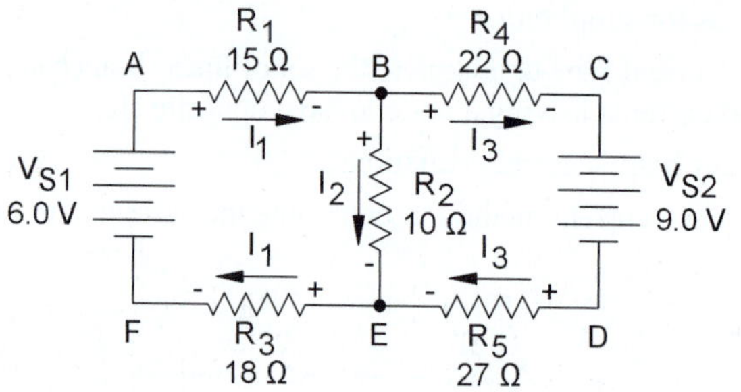

Figure 9-4: Replacement of Redundant Branch Currents for Figure 9-3

The branches between nodes E and F, F and A, and A and B form a series path. From the definition of a series path, the current through each branch must be the same. Therefore $I_1 = I_6 = I_7$. Similarly, the branches between nodes B and C, C and D, and D and E form a series path so that $I_3 = I_4 = I_5$. Therefore, replace I_6 with I_1 and I_5 with I_3. Because I_4 and I_7 both pass through voltage sources, they are nonessential and removed from the circuit.

The next step is to apply Kirchhoff's Voltage Law to each loop. Work in the reference direction for the loop containing V_{S1}, R_1, R_2, and R_3, apply Ohm's Law, and segregate terms.

$$V_{S1} - V_{R1} - V_{R2} - V_{R3} = 0$$

$$V_{S1} + -V_{R1} + -V_{R2} + -V_{R3} = 0$$

$$V_{S1} + -(R_1 \times I_1) + -(R_2 \times I_2) + -(R_3 \times I_1) = 0$$

$$V_{S1} + -(R_1 \times I_1) + -(R_3 \times I_1) + -(R_2 \times I_2) = 0$$

$$V_{S1} + -[(R_1 + R_3) \times I_1] + -(R_2 \times I_2) = 0$$

$$V_{S1} + -[(R_1 + R_3) \times I_1] + -(R_2 \times I_2) + -V_{S1} = 0 + -V_{S1}$$

$$-[(R_1 + R_3) \times I_1] + -(R_2 \times I_2) = -V_{S1}$$

$$-[(15 + 18) \times I_1] + -(10 \times I_2) = -6.0$$

$$-(33 \times I_1) + -(10 \times I_2) = -6.0$$

$$-33I_1 + -10I_2 = -6.0$$

Next, working in the reference direction for the loop containing V_{S2}, R_2, R_4, and R_5, apply Ohm's Law, and segregate terms.

$$-V_{S2} - V_{R5} + V_{R2} - V_{R4} = 0$$

$$-V_{S2} + -V_{R5} + V_{R2} + -V_{R4} = 0$$

$$-V_{S2} + -(R_5 \times I_3) + (R_2 \times I_2) + -(R_4 \times I_3) = 0$$
$$-V_{S2} + -(R_5 \times I_3) + -(R_4 \times I_3) + (R_2 \times I_2) = 0$$
$$-V_{S2} + -[(R_5 + R_4) \times I_3] + (R_2 \times I_2) = 0$$
$$-V_{S2} + -[(R_5 + R_4) \times I_3] + (R_2 \times I_2) + V_{S2} = 0 + V_{S2}$$
$$-[(R_4 + R_5) \times I_3] + (R_2 \times I_2) = V_{S2}$$
$$-[(22 + 27) \times I_3] + (10 \times I_2) = 9.0$$
$$-49I_3 + 10I_2 = 9.0$$
$$10I_2 + -49I_3 = 9.0$$

Combine the two equations to give two equations in three unknowns.

$$-33I_1 + -10I_2 = -6.0\text{ V}$$
$$10I_2 + -49I_3 = 9.0\text{ V}$$

Apply Kirchhoff's Current Law at node B (or node E) for the final equation to construct a system of three equations in three unknowns. Let currents entering the node be positive and currents leaving the node be negative and apply Kirchhoff's Current Law to Node B.

$$I_1 - I_2 - I_3 = 0$$

This gives the linear system of three equations in three unknowns.

$$-33I_1 + -10I_2 = -6.0\text{ V}$$
$$10I_2 + -49I_3 = 9.0\text{ V}$$
$$I_1 - I_2 - I_3 = 0$$

Since the units of the coefficients are in ohms and the solutions on the right-hand side are in volts, the units of I_1, I_2, and I_3 will be in amperes. Rewrite the system of equations so that each equation is a sum and contains each of the variables I_1, I_2, and I_3 with a suitable coefficient to create the system of equations in standard form.

$$\begin{aligned} -33I_1 &+ -10I_2 &+ 0I_3 &= -6.0 \\ 0I_1 &+ 10I_2 &+ -49I_3 &= 9.0 \\ I_1 &- I_2 &- I_3 &= 0 \end{aligned}$$

The matrix-vector form is

$$\begin{vmatrix} -33 & -10 & 0 \\ 0 & 10 & -49 \\ 1 & -1 & -1 \end{vmatrix} \bullet \begin{vmatrix} I_1 \\ I_2 \\ I_3 \end{vmatrix} = \begin{vmatrix} -6.0 \\ 9.0 \\ 0 \end{vmatrix}$$

The solution for this maxtrix equation is

$$\begin{vmatrix} I_1 \\ I_2 \\ I_3 \end{vmatrix} = \begin{vmatrix} 108.3 \times 10^{-3} \\ 242.5 \times 10^{-3} \\ -134.2 \times 10^{-3} \end{vmatrix}$$

Note that the negative value for I_3 indicates that the direction of the assigned branch current is opposite to the actual current direction. The resistor currents are

$I_{R1} = I_1 = 108$ mA (current flows from left to right)

$I_{R2} = I_2 = 243$ mA (current flows from top to bottom)

$I_{R3} = I_1 = 108$ mA (current flows from right to left)

$I_{R4} = I_3 = 134$ mA (current flows from right to left)

$I_{R5} = I_3 = 134$ mA (current flows from left to right)

2. Problem: Use the branch current method to determine the currents through each resistor for the circuit in Figure 9-5.

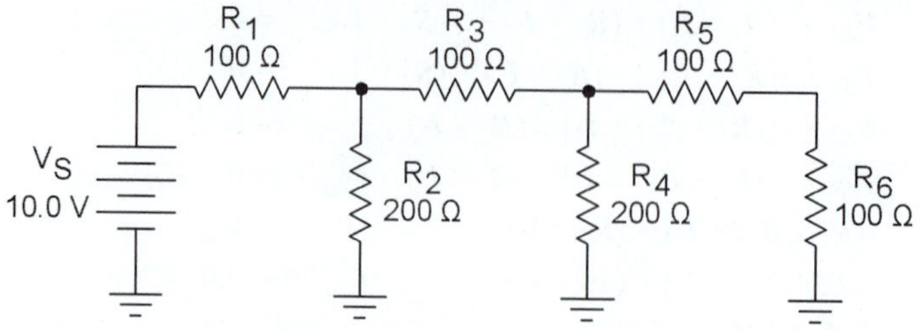

Figure 9-5: Branch Current Method Example Circuit 2

Solution: Label the nodes and determine a reference direction for each non-redundant current loop. The circuit in Figure 9-5 contains five nodes and three non-redundant loops. For this example, choose a clockwise reference direction for each loop. Refer to Figure 9-6.

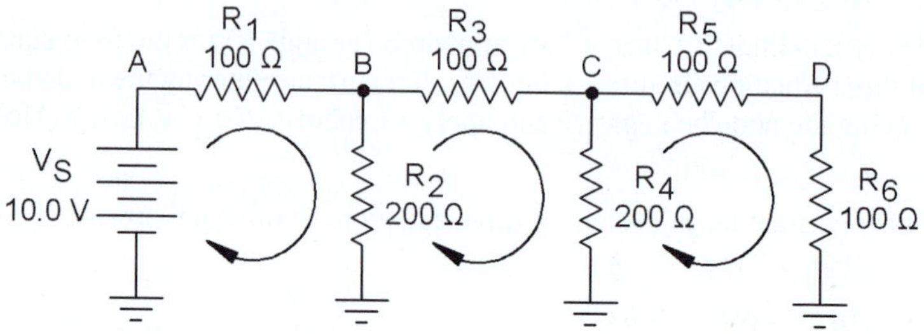

Figure 9-6: Nodes and Reference Directions for Figure 9-5

Note that node 0 represents ground and is shared by all the ground connections in the circuit. Mark the voltage polarity for each resistor based on the direction of the branch current through it. Refer to Figure 9-7.

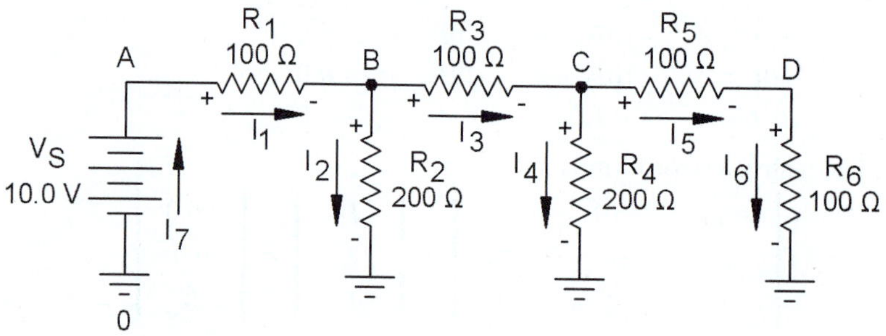

Figure 9-7: Assigned Branch Currents and Resistor Voltage Polarities for Figure 9-5

As Figure 9-7 shows, the branch between node B and ground exists in the left loop that includes nodes A, B, and 0 and in the loop that includes nodes B, C, and 0. The assigned direction for I_2 could be the same as the left or middle loop in Figure 9-6, but once you assign the current direction based on the left loop, you ignore the branch for the middle loop. The same is true for the assigned direction for I_4.

Next, replace any redundant branch currents to remove any nonessential branch currents. The branches between nodes C and D and D and 0 form a series path. From the definition of a series path, the current through each branch must be the same. Therefore $I_5 = I_6$. Similarly, $I_7 = I_1$ but because I_7 passes through the voltage source it is nonessential and removed from the circuit. Refer to Figure 9-8.

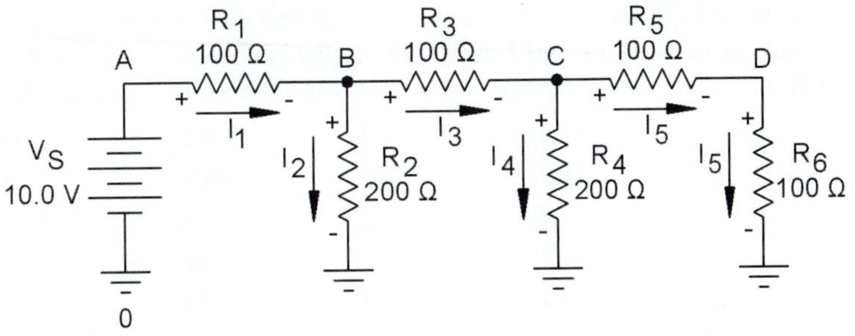

Figure 9-8: Replacement of Redundant Branch Currents for Figure 9-7

Apply Kirchhoff's Voltage Law to each loop. Work in the reference direction for the left-hand loop, apply Ohm's Law, and segregate terms.

$$\begin{aligned}
V_S - V_{R1} - V_{R2} &= 0 \\
V_S + -V_{R1} + -V_{R2} &= 0 \\
V_S + -(R_1 \times I_1) + -(R_2 \times I_2) &= 0 \\
V_S + -(R_1 \times I_1) + -(R_2 \times I_2) + -V_S &= 0 + -V_S \\
-(R_1 \times I_1) + -(R_2 \times I_2) &= -V_S \\
-100I_1 + -200I_2 &= -10.0
\end{aligned}$$

Work in the reference direction for the center loop, apply Ohm's Law, and segregate terms.

$$\begin{aligned}
V_{R2} - V_{R3} - V_{R4} &= 0 \\
V_{R2} + -V_{R3} + -V_{R4} &= 0 \\
(R_2 \times I_2) + -(R_3 \times I_3) + -(R_4 \times I_4) &= 0 \\
(200\ \Omega \times I_2) + -(100\ \Omega \times I_3) + -(200\ \Omega \times I_4) &= 0 \\
200I_2 + -100I_3 + -200I_4 &= 0
\end{aligned}$$

Work in the reference direction for the righthand loop, apply Ohm's Law, and segregate terms.

$$\begin{aligned}
V_{R4} - V_{R5} - V_{R6} &= 0 \\
V_{R4} + -V_{R5} + -V_{R6} &= 0 \\
(R_4 \times I_4) + -(R_5 \times I_5) + -(R_6 \times I_5) &= 0 \\
(R_4 \times I_4) + -[(R_5 + R_6) \times I_5] &= 0 \\
200I_4 + -[(100 + 100) \times I_5] &= 0 \\
200I_4 + -200I_5 &= 0
\end{aligned}$$

Because five unknowns require five equations to solve, the system requires two more equations. Kirchhoff's Current Law at nodes B and C can supply these. Let currents entering a node be positive and currents leaving the node be negative. Apply Kirchhoff's Current Law to node B.

$$I_1 - I_2 - I_3 = 0$$

Similarly, apply Kirchhoff's Current Law to node C.

$$I_3 - I_4 - I_5 = 0$$

Combine the equations to form the system of equations.

$$\begin{aligned}
-100I_1 + -200I_2 &= -10.0\text{ V} \\
200I_2 + -100I_3 + -200I_4 &= 0 \\
200I_4 + -200I_5 &= 0 \\
I_1 - I_2 - I_3 &= 0 \\
I_3 - I_4 - I_5 &= 0
\end{aligned}$$

Since the units of the coefficients are in ohms and the solutions on the right-hand side are in volts, the units of I_1, I_2, I_3, I_4, and I_5 will be in amperes. Rewrite the system of equations so that each equation is a sum and contains each of the variables I_1, I_2, I_3, I_4, and I_5 with a suitable coefficient.

$$\begin{array}{rcrcrcrcrcr}
-100I_1 & + & -200I_2 & + & 0I_3 & + & 0I_4 & + & 0I_5 & = & -10.0 \\
0I_1 & + & 200I_2 & + & -100I_3 & + & -200I_4 & + & 0I_5 & = & 0 \\
0I_1 & + & 0I_2 & + & 0I_3 & + & 200I_4 & + & -200I_5 & = & 0 \\
1I_1 & + & -1I_2 & + & -1I_3 & + & 0I_4 & + & 0I_5 & = & 0 \\
0I_1 & + & 0I_2 & + & 1I_3 & + & -1I_4 & + & -1I_5 & = & 0
\end{array}$$

The matrix-vector form is

$$\left|\begin{array}{rrrrr}
-100 & -200 & 0 & 0 & 0 \\
0 & 200 & -100 & -200 & 0 \\
0 & 0 & 0 & 200 & -200 \\
1 & -1 & -1 & 0 & 0 \\
0 & 0 & 1 & -1 & -1
\end{array}\right| \cdot \left|\begin{array}{c} I_1 \\ I_2 \\ I_3 \\ I_4 \\ I_5 \end{array}\right| = \left|\begin{array}{r} -10.0 \\ 0 \\ 0 \\ 0 \\ 0 \end{array}\right|$$

The solution for this maxtrix equation is

$$\left|\begin{array}{c} I_1 \\ I_2 \\ I_3 \\ I_4 \\ I_5 \end{array}\right| = \left|\begin{array}{c} 50 \times 10^{-3} \\ 25 \times 10^{-3} \\ 25 \times 10^{-3} \\ 12.5 \times 10^{-3} \\ 12.5 \times 10^{-3} \end{array}\right|$$

Note that all currents are positive, so that the reference direction is the same as that for the actual current. The resistor currents are

$I_{R1} = I_1 = 50$ mA (current flows from left to right)

$I_{R2} = I_2 = 25$ mA (current flows from top to bottom)

$I_{R3} = I_3 = 25$ mA (current flows from left to right)

$I_{R4} = I_4 = 12.5$ mA (current flows from top to bottom)

$I_{R5} = I_5 = 12.5$ mA (current flows from left to right)

$I_{R6} = I_5 = 12.5$ mA (current flows from top to bottom)

9.1.3 Loop Current Method

For the loop current method:

- A *loop* is any path through the circuit that does not pass through a component more than once and returns to where it began.
- A loop is *non-redundant* if it contains at least one component that no other loop contains.

The steps for applying the loop current method are:

1) Identify and determine a reference direction for each non-redundant loop current.
2) Assign resistor voltage polarities based on the reference directions for ***each*** current loop. Unlike the branch current method, resistors can have two voltage polarities.
3) Apply Kirchhoff's Voltage Law and Ohm's Law to each loop to generate a set of linear branch current equations. Omit units for simplicity.
4) Solve the set of linear branch current equations.

The loop current method will use the same circuits as those for the branch current method so that you can compare and contrast these methods and see the advantages of each.

1. Problem: Use the loop current method to determine the currents through each resistor for the circuit in Figure 9-9.

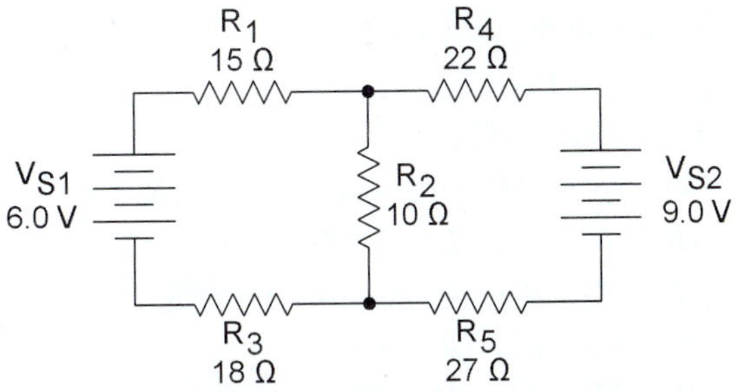

Figure 9-9: Loop Current Method Example Circuit 1

Solution: Identify and determine a reference direction for each non-redundant loop current. For the circuit in Figure 9-9, there are two non-redundant current loops. For this example, let the reference direction for each current be counter-clockwise. Refer to Figure 9-10.

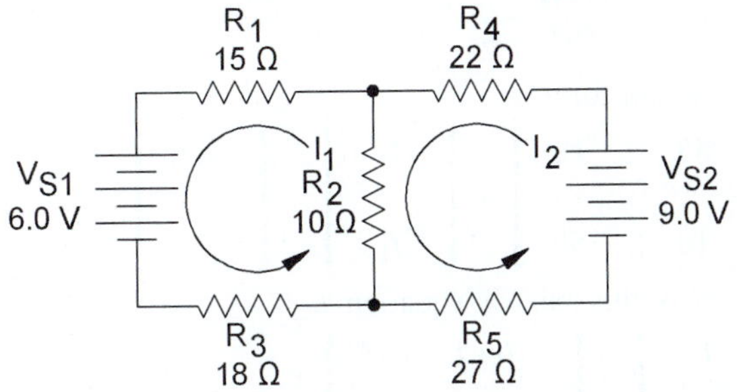

Figure 9-10: Loop Currents for Figure 9-9

Note that another loop exists that contains $\boldsymbol{V_{S1}}$, $\boldsymbol{R_1}$, $\boldsymbol{R_4}$, $\boldsymbol{V_{S2}}$, $\boldsymbol{R_5}$, and $\boldsymbol{R_3}$, but that this loop is redundant because every component in the loop is already included in the loops for $\boldsymbol{I_1}$ and $\boldsymbol{I_2}$. Next, assign resistor voltage polarities based on the reference directions for $\boldsymbol{I_1}$ and $\boldsymbol{I_2}$. Refer to Figure 9-11.

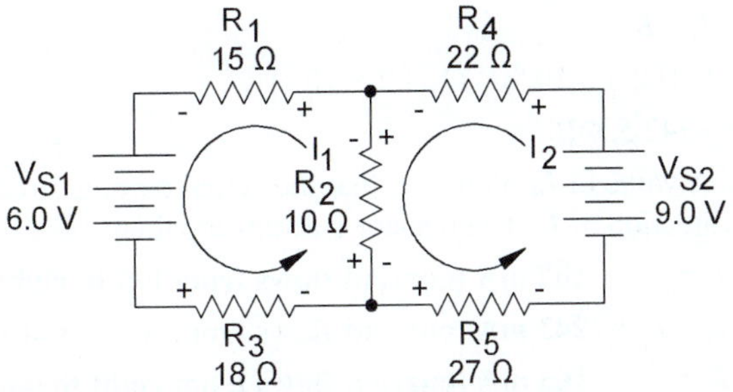

Figure 9-11: Resistor Voltage Polarities for Loop Currents in Figure 9-10

Work in the direction of each loop current to apply Kirchhoff's Voltage Law, apply Ohm's Law, and segregate terms, omitting units for simplicity. Begin with the left loop.

$$-V_{S1} - V_{R3} - V_{R2} - V_{R1} = 0$$

$$-V_{S1} + -V_{R3} + -V_{R2} + -V_{R1} = 0$$

$$-V_{S1} + -(R_3 \times I_1) + -[(R_2 \times I_1) - (R_2 \times I_2)] + -(R_1 \times I_1) = 0$$

$$-V_{S1} + -[(R_3 + R_2 + R_1) \times I_1] + -[-(R_2 \times I_2)] = 0$$

$$-V_{S1} + -[(R_1 + R_2 + R_3) \times I_1] + (R_2 \times I_2) + V_{S1} = 0 + V_{S1}$$

$$-[(R_1 + R_2 + R_3) \times I_1] + (R_2 \times I_2) = V_{S1}$$

$$-(15 + 10 + 18) \times \boldsymbol{I}_1 + 10\boldsymbol{I}_2 = 6.0$$
$$-43\boldsymbol{I}_1 + 10\boldsymbol{I}_2 = 6.0$$

Repeat the process for the right-hand loop.

$$\boldsymbol{V}_{S2} - \boldsymbol{V}_{R4} - \boldsymbol{V}_{R2} - \boldsymbol{V}_{R5} = 0$$
$$\boldsymbol{V}_{S2} + -\boldsymbol{V}_{R4} + -\boldsymbol{V}_{R2} + -\boldsymbol{V}_{R5} = 0$$
$$\boldsymbol{V}_{S2} + -(\boldsymbol{R}_4 \times \boldsymbol{I}_2) + -[(\boldsymbol{R}_2 \times \boldsymbol{I}_2) - (\boldsymbol{R}_2 \times \boldsymbol{I}_1)] + -(\boldsymbol{R}_5 \times \boldsymbol{I}_2) = 0$$
$$\boldsymbol{V}_{S2} + -[(\boldsymbol{R}_4 + \boldsymbol{R}_2 + \boldsymbol{R}_5) \times \boldsymbol{I}_2] - [-(\boldsymbol{R}_2 \times \boldsymbol{I}_1)] = 0$$
$$\boldsymbol{V}_{S2} + -[(\boldsymbol{R}_4 + \boldsymbol{R}_2 + \boldsymbol{R}_5) \times \boldsymbol{I}_2] + (\boldsymbol{R}_2 \times \boldsymbol{I}_1) + -\boldsymbol{V}_{S2} = 0 + -\boldsymbol{V}_{S2}$$
$$-[(\boldsymbol{R}_4 + \boldsymbol{R}_2 + \boldsymbol{R}_5) \times \boldsymbol{I}_2] + (\boldsymbol{R}_2 \times \boldsymbol{I}_1) = -\boldsymbol{V}_{S2}$$
$$-[(22 + 10 + 27) \times \boldsymbol{I}_2] + (10 \times \boldsymbol{I}_1) = -9.0$$
$$-59\boldsymbol{I}_2 + 10\boldsymbol{I}_1 = -9.0$$
$$10\boldsymbol{I}_1 + -59\boldsymbol{I}_2 = -9.0$$

Combine these two equations to give the system of equations.

$$\begin{aligned} -43\boldsymbol{I}_1 &+ -10\boldsymbol{I}_2 = 6.0 \\ 10\boldsymbol{I}_1 &+ -59\boldsymbol{I}_2 = -9.0 \end{aligned}$$

The matrix-vector form is

$$\begin{vmatrix} -43 & 10 \\ 10 & -59 \end{vmatrix} \bullet \begin{vmatrix} \boldsymbol{I}_1 \\ \boldsymbol{I}_2 \end{vmatrix} = \begin{vmatrix} 6.0 \\ -9.0 \end{vmatrix}$$

The solution for this maxtrix equation is

$$\begin{vmatrix} \boldsymbol{I}_1 \\ \boldsymbol{I}_2 \end{vmatrix} = \begin{vmatrix} -108.3 \times 10^{-3} \\ 134.2 \times 10^{-3} \end{vmatrix}$$

Note that the negative value of $\boldsymbol{I}_1$ indicates that the actual direction of $\boldsymbol{I}_1$ is opposite to the assumed reference direction. $\boldsymbol{I}_1$ flows through $\boldsymbol{R}_1$ and $\boldsymbol{R}_3$, so $\boldsymbol{I}_{R1} = \boldsymbol{I}_{R3} = \boldsymbol{I}_1$. Similarly, $\boldsymbol{I}_2$ flows through $\boldsymbol{R}_4$ and $\boldsymbol{R}_5$, so $\boldsymbol{I}_{R4} = \boldsymbol{I}_{R5} = \boldsymbol{I}_2$. The current through $\boldsymbol{R}_2$ is the difference between $\boldsymbol{I}_1$ and $\boldsymbol{I}_2$, so initially assume that the direction of $\boldsymbol{I}_{R2}$ is the same as the reference direction for $\boldsymbol{I}_1$.

$$\begin{aligned} \boldsymbol{I}_{R2} &= \boldsymbol{I}_1 - \boldsymbol{I}_2 \\ &= (-108.3 \times 10^{-3}) - (134.2 \times 10^{-3}) \\ &= -242.5 \times 10^{-3} \end{aligned}$$

The negative value of $\boldsymbol{I}_{R2}$ indicates that the actual current direction is opposite to the assumed reference direction of $\boldsymbol{I}_1$. The resistor currents are then

$\boldsymbol{I}_{R1} = \boldsymbol{I}_1 =$ **108 mA (current flows from left to right)**

$\boldsymbol{I}_{R2} = \boldsymbol{I}_1 - \boldsymbol{I}_2 =$ **243 mA (current flows from top to bottom)**

$\boldsymbol{I}_{R3} = \boldsymbol{I}_1 =$ **108 mA (current flows from right to left)**

$\boldsymbol{I}_{R4} = \boldsymbol{I}_2 =$ **134 mA (current flows from right to left)**

$\boldsymbol{I}_{R5} = \boldsymbol{I}_2 =$ **134 mA (current flows from left to right)**

2. Problem: Use the loop current method to determine the currents through each resistor for the circuit in Figure 9-12.

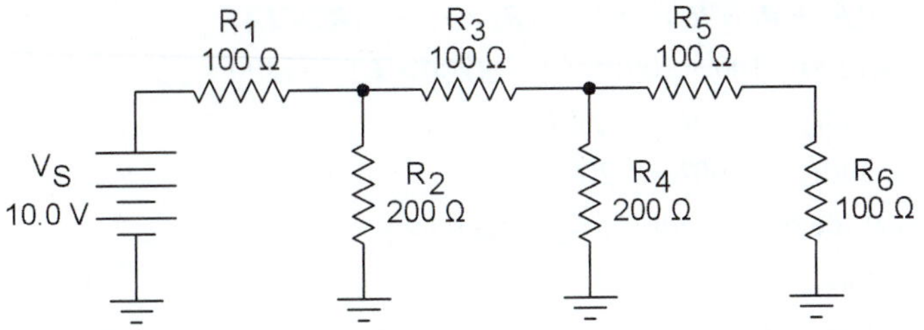

Figure 9-12: Loop Current Method Example Circuit 2

Solution: Identify and determine a reference direction for each non-redundant loop current. For the circuit in Figure 9-12, there are three non-redundant current loops. For this example, let the reference direction for the left-hand and right-hand loop currents be clockwise, and the middle current loop be counter-clockwise. Refer to Figure 9-13.

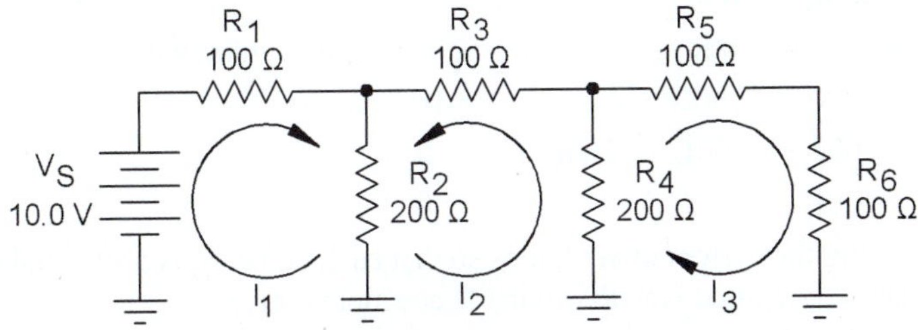

Figure 9-13: Loop Currents for Figure 9-12

Note that another loop exists that contains $\boldsymbol{V_{S1}}$, $\boldsymbol{R_1}$, $\boldsymbol{R_3}$, $\boldsymbol{R_5}$, and $\boldsymbol{R_6}$, but that this loop is redundant because every component in the loop is already included in the loops for $\boldsymbol{I_1}$, $\boldsymbol{I_2}$, and $\boldsymbol{I_3}$. Next, assign resistor voltage polarities based on the reference directions for $\boldsymbol{I_1}$, $\boldsymbol{I_2}$, and $\boldsymbol{I_3}$. Refer to Figure 9-14.

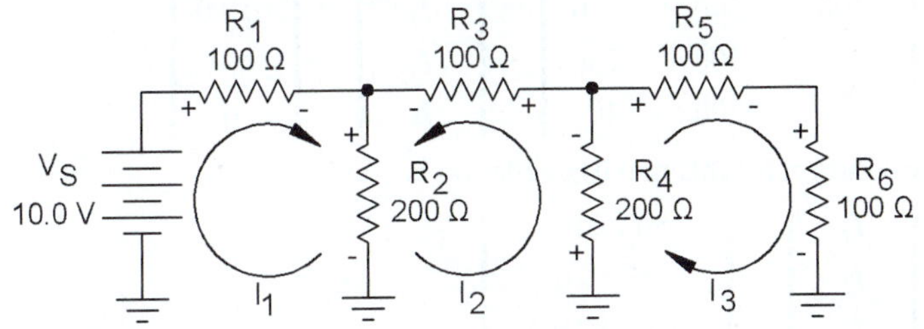

Figure 9-14: Resistor Voltage Polarities for Loop Currents in Figure 9-13

Work in the direction of each loop current to apply Kirchhoff's Voltage Law, apply Ohm's Law, and segregate terms, omitting units for simplicity. Begin with the left-hand loop.

$$\boldsymbol{V_S} - \boldsymbol{V_{R1}} - \boldsymbol{V_{R2}} = 0$$

$$\boldsymbol{V_S} + -\boldsymbol{V_{R1}} + -\boldsymbol{V_{R2}} = 0$$

$$\boldsymbol{V_S} + -(\boldsymbol{R_1} \times \boldsymbol{I_1}) + -[(\boldsymbol{R_2} \times \boldsymbol{I_1}) + (\boldsymbol{R_2} \times \boldsymbol{I_2})] = 0$$

$$\boldsymbol{V_S} + -[(\boldsymbol{R_1} + \boldsymbol{R_2}) \times \boldsymbol{I_1}] + -(\boldsymbol{R_2} \times \boldsymbol{I_2}) = 0$$

$$\boldsymbol{V_S} + -[(\boldsymbol{R_1} + \boldsymbol{R_2}) \times \boldsymbol{I_1}] + -(\boldsymbol{R_2} \times \boldsymbol{I_2}) + -\boldsymbol{V_S} = 0 + -\boldsymbol{V_S}$$

$$-[(\boldsymbol{R_1} + \boldsymbol{R_2}) \times \boldsymbol{I_1}] + -(\boldsymbol{R_2} \times \boldsymbol{I_2}) = -\boldsymbol{V_S}$$

$$-[(100 + 200) \times \boldsymbol{I_1}] + -200\boldsymbol{I_2} = -10.0$$

$$-300\boldsymbol{I_1} + -200\boldsymbol{I_2} = -10.0$$

Repeat the process for the middle loop.

$$-\boldsymbol{V_{R4}} - \boldsymbol{V_{R3}} - \boldsymbol{V_{R2}} = 0$$

$$-\boldsymbol{V_{R4}} + -\boldsymbol{V_{R3}} + -\boldsymbol{V_{R2}} = 0$$

$$[-(\boldsymbol{R_4} \times \boldsymbol{I_2}) + -(\boldsymbol{R_4} \times \boldsymbol{I_3})] + -(\boldsymbol{R_3} \times \boldsymbol{I_2}) + [-(\boldsymbol{R_2} \times \boldsymbol{I_2}) + -(\boldsymbol{R_2} \times \boldsymbol{I_1})] = 0$$

$$-[(R_4 + R_3 + R_2) \times I_2] + -(R_4 \times I_3) + -(R_2 \times I_1) = 0$$
$$-[(200 + 100 + 200) \times I_2] + -(200 \times I_3) + -(200 \times I_1) = 0$$
$$-500I_2 + -200I_3 + -200I_1 = 0$$
$$-200I_1 + -500I_2 + -200I_3 = 0$$

Repeat the process for the right-hand loop.

$$-V_{R4} - V_{R5} - V_{R6} = 0$$
$$-V_{R4} + -V_{R5} + -V_{R6} = 0$$
$$[-(R_4 \times I_3) + -(R_4 \times I_2)] + -(R_5 \times I_3) + -(R_6 \times I_3)] = 0$$
$$-[(R_4 + R_5 + R_6) \times I_3 + -(R_4 \times I_2) = 0$$
$$-[(200 + 100 + 100) \times I_3] + -(200 \times I_2) = 0$$
$$-400I_3 + -200I_2 = 0$$
$$-200I_2 + -400I_3 = 0$$

These give the following three equations in three unknowns:

$$-300I_1 + -200I_2 = -10.0$$
$$-200I_1 + -500I_2 + -200I_3 = 0$$
$$-200I_2 + -400I_3 = 0$$

Rewriting the system of equations so that each equation is a sum and contains each of the variables I_1, I_2, and I_3 with a suitable coefficient gives

$$\begin{array}{rcrcrcr} -300I_1 & + & -200I_2 & + & 0I_3 & = & -10.0 \\ -200I_1 & + & -500I_2 & + & -200I_3 & = & 0 \\ 0I_1 & + & -200I_2 & + & -400I_3 & = & 0 \end{array}$$

The matrix-vector form is

$$\begin{vmatrix} -300 & -200 & 0 \\ -200 & -500 & -200 \\ 0 & -200 & -400 \end{vmatrix} \bullet \begin{vmatrix} I_1 \\ I_2 \\ I_3 \end{vmatrix} = \begin{vmatrix} -10.0 \\ 0 \\ 0 \end{vmatrix}$$

The solution for this maxtrix equation is

$$\begin{vmatrix} I_1 \\ I_2 \\ I_3 \end{vmatrix} = \begin{vmatrix} 50 \times 10^{-3} \\ -25 \times 10^{-3} \\ 12.5 \times 10^{-3} \end{vmatrix}$$

Note that the negative value of I_2 indicates that the actual direction of I_2 is opposite to the assumed reference direction. I_1 flows through R_1, so $I_{R1} = I_1$. Similarly, I_2 flows through R_3 so $I_{R3} = I_2$, and I_3 flows through R_5 and R_6 so $I_{R5} = I_{R6} = I_3$. The current through R_2 is the sum of I_1 and I_2.

$$I_{R2} = I_1 + I_2$$
$$= (50 \times 10^{-3}) + (-25 \times 10^{-3})$$
$$= 25 \times 10^{-3}$$

Similarly, the current through R_4 is the sum of I_2 and I_3.

$$I_{R4} = I_2 + I_3$$
$$= (-25 \times 10^{-3}) + (12.5 \times 10^{-3})$$
$$= -12.5 \times 10^{-3}$$

The negative value of I_{R4} indicates that the actual direction of I_{R4} is opposite to the assumed direction of I_2 and I_3. The resistor currents are

$I_{R1} = I_1 =$ **50 mA (current flows from left to right)**

$I_{R2} = I_1 + I_2 =$ **25 mA (current flows from top to bottom)**

$I_{R3} = I_2 =$	**25 mA (current flows from left to right)**
$I_{R4} = I_2 + I_3 =$	**12.5 mA (current flows from top to bottom)**
$I_{R5} = I_3 =$	**12.5 mA (current flows from left to right)**
$I_{R6} = I_3 =$	**12.5 mA (current flows from top to bottom)**

9.1.4 Node Voltage Method

For the node voltage current method:

- A *node* is any point where two or more components connect.
- A *branch* is a circuit path that connects two adjacent nodes.

The steps for applying the node voltage current method are:

1) Identify and label the nodes and select a reference node.
2) Assign voltage designations for all nodes for which the actual voltage is not known.
3) Assign branch currents at each node for which the actual voltage is not known. The directions of the currents are unimportant, as any negative value will indicate that the actual current direction is opposite to the assigned direction.
4) Apply Kirchhoff's Current Law to each node to which branch currents are assigned to generate a set of linear branch current equations. Define currents entering a node and currents leaving a node to have opposite signs.
5) Use Ohm's Law to express each current in terms of node voltages and resistances. Omit units for simplicity.
6) Solve the set of linear branch current equations.

The node voltage current method will use the same circuits as those for the branch current and loop current methods so that you can compare and contrast these methods and see the advantages of each.

1. Problem: Use the node voltage method to determine the currents through each resistor for the circuit in Figure 9-15.

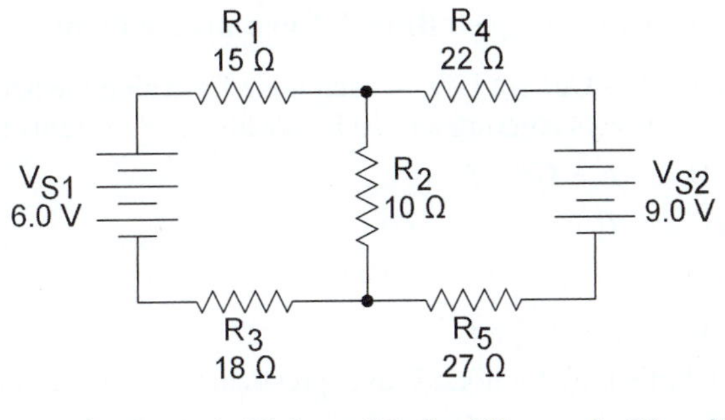

Figure 9-15: Node Voltage Method Example Circuit 1

Solution: Label all the circuit nodes. Refer to Figure 9-16.

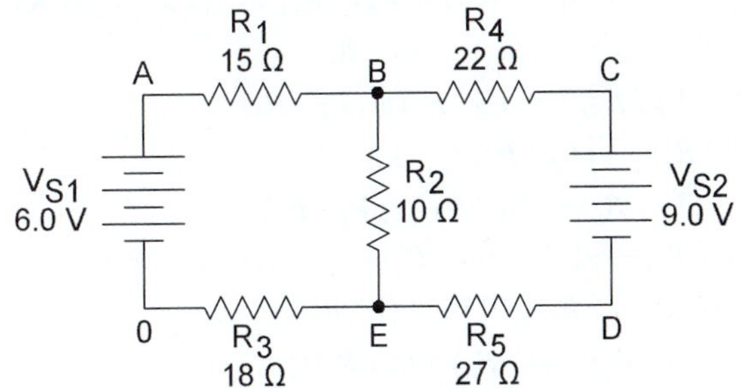

Figure 9-16: Labeled Nodes for Figure 9-15

As Figure 9-16 shows, the circuit contains six nodes. For this example, select the negative terminal of V_{S1} as the reference node, labeled as node 0. Next, assign voltage designations for all nodes for which the voltage is not known. Refer to Figure 9-17.

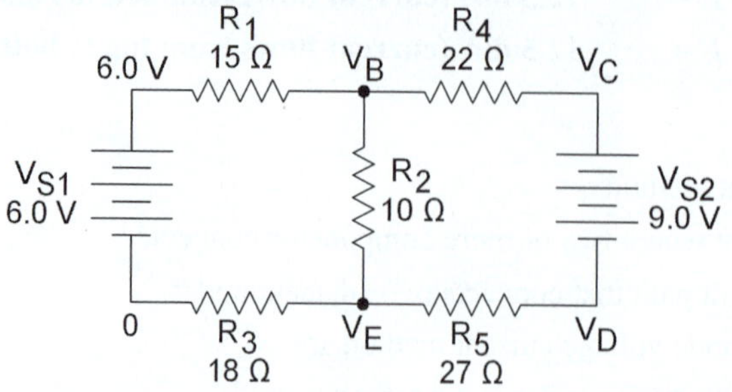

Figure 9-17: Assigned Voltage Designations for Figure 9-15

The voltage at node A relative to the reference node is known, as V_{S1} holds it at 6.0 V more positive than the reference node, so the node is labeled with the actual voltage value. The other nodes are labeled with voltage designations. Next, assign branch currents for all nodes for which the actual voltage is not known. Refer to Figure 9-18.

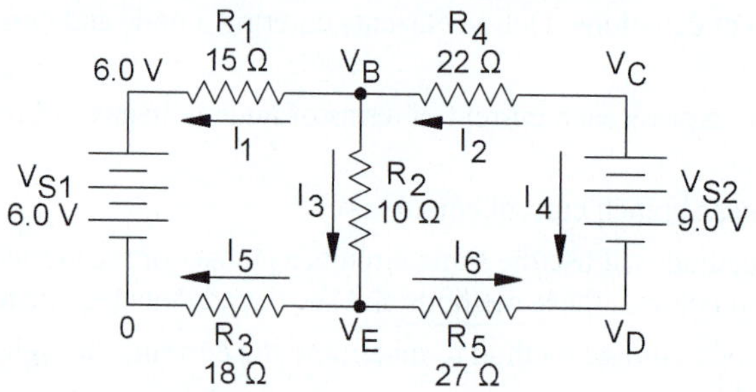

Figure 9-18: Assigned Branch Currents for Figure 9-15

Apply Kirchhoff's Current Law to each node for which the actual voltage is not known. For this example, let currents entering a node be positive and currents leaving a node be negative.

Node B: $-I_1 + I_2 + -I_3 = 0$

Node C: $-I_2 + -I_4 = 0$

Node D: $I_4 + I_6 = 0$

Node E: $I_3 + -I_5 + -I_6 = 0$

Next, use Ohm's Law for node B to express each branch current in terms of the node voltages and resistances.

$$-I_1 + I_2 + -I_3 = 0$$

$$-[(V_B - 6.0\text{ V}) / R_1] + [(V_C - V_B) / R_4] + -[(V_B - V_E) / R_2] = 0$$

$$-(V_B / R_1) + (6.0\text{ V} / R_1) + (V_C / R_4) + -(V_B / R_4) + -(V_B / R_2) + (V_E / R_2) = 0$$

$$(6.0\text{ V} / R_1) + -(V_B / R_1) + -(V_B / R_4) + -(V_B / R_2) + (V_C / R_4) + (V_E / R_2) = 0$$

$$(6.0\text{ V} / R_1) + -\{V_B \times [(1 / R_1) + (1 / R_4) + (1 / R_2)]\} + [V_C \times (1 / R_4)] + [V_E \times (1 / R_2)] = 0$$

$$-\{V_B \times [(1 / R_1) + (1 / R_4) + (1 / R_2)]\} + [V_C \times (1 / R_4)] + [V_E \times (1 / R_2)] + (6.0\text{ V} / R_1) = 0$$

$$-\{V_B \times [(1 / R_1) + (1 / R_2) + (1 / R_2)]\} + [V_C \times (1 / R_4)] + [V_E \times (1 / R_2)] + (6.0\text{ V} / R_1) + -(6.0\text{ V} / R_1) = 0 + -(6.0\text{ V} / R_1)$$

$$-\{V_B \times [(1 / R_1) + (1 / R_2) + (1 / R_4)]\} + [V_C \times (1 / R_4)] + [V_E \times (1 / R_2)] = -(6.0\text{ V} / R_1)$$

$$-\{V_B \times [(1 / 15\ \Omega) + (1 / 10\ \Omega) + (1 / 22\ \Omega)]\} + [V_C \times (1 / 22\ \Omega)] + [V_E \times (1 / 10\ \Omega)] = -(6.0\text{ V} / 15\ \Omega)$$

$$-\{V_B \times [(66.7 \times 10^{-3}\text{ S}) + (100 \times 10^{-3}\text{ S}) + (45.5 \times 10^{-3}\text{ S})]\} + [V_C \times (45.5 \times 10^{-3}\text{ S})] + [V_E \times (100 \times 10^{-3}\text{ S})] = (-6.0 / 15)\ (\text{V}/\Omega)$$

$$-[V_B \times (212 \times 10^{-3}\text{ S})] + [V_C \times (45.5 \times 10^{-3}\text{ S})] + [V_E \times (100 \times 10^{-3}\text{ S})] = -400 \times 10^{-3}\text{ A}$$

$$(-212 \times 10^{-3}\text{ S})V_B + (45.5 \times 10^{-3}\text{ S})V_C + (100 \times 10^{-3}\text{ S})V_E = -400 \times 10^{-3}\text{ A}$$

Repeat the process for node C.

$$-I_2 + -I_4 = 0$$

$$-[(V_C - V_B) / R_4] + -I_4 = 0$$

Repeat the process for node D.

$$I_4 + I_6 = 0$$

$$[(V_E - V_D) / R_5] + I_4 = 0$$

Repeat the process for node E.

$$I_3 + -I_5 + -I_6 = 0$$

$$[(V_B - V_E) / R_2] + -[(V_E - 0) / R_3] + -[(V_E - V_D) / R_5] = 0$$

$$[(V_B + -V_E) / R_2] + -(V_E / R_3) + [(V_D + -V_E) / R_5] = 0$$

$$(V_B / R_2) + -(V_E / R_2) + -(V_E / R_3) + (V_D / R_5) + -(V_E / R_5) = 0$$

$$(V_B / R_2) + (V_D / R_5) + -(V_E / R_2) + -(V_E / R_3) + -(V_E / R_5) = 0$$

$$(V_B / R_2) + (V_D / R_5) + -\{V_E \times [(1 / R_2) + -(1 / R_3) + -(1 / R_5)]\} = 0$$

Nodes C and D deserve special attention because a voltage source, rather than a resistor, separates them. Because of this, the current and voltage between them are independent of each other. Determine the value of I_4 from the currents for node C.

$$-[(V_C - V_B) / R_4] + -I_4 = 0$$

$$-[(V_C - V_B) / R_4] + -I_4 + I_4 = 0 + I_4$$

$$-[(V_C - V_B) / R_4] = I_4$$

Substitute this for I_4 into the expression for node D.

$$[(V_E - V_D) / R_5] + I_4 = 0$$

$$[(V_E - V_D) / R_5] + -[(V_C - V_B) / R_4] = 0$$

Note that this is exactly the same result as for the expression $I_6 + -I_2 = 0$ that would result if the voltage source V_{S2} were replaced with its internal resistance (a short) and nodes C and D were "merged" into a single node. In terms of voltage, V_{S2} maintains 9.0 V between nodes C and D.

$$V_D = V_C - 9.0\text{ V}$$

$$= V_C + -9.0\text{ V}$$

Substitute for V_D in the combined equation for nodes C and D.

$$[(V_E - V_D) / R_5] + -[(V_C - V_B) / R_4] = 0$$

$$[(V_E + -V_D) / R_5] + [(V_B + -V_C) / R_4] = 0$$

$$\{[V_E + -(V_C + -9.0\text{ V})] / R_5\} + [(V_B + -V_C) / R_4] = 0$$

$$\{[V_E + (9.0\text{ V} + -V_C)] / R_5\} + [(V_B + -V_C) / R_4] = 0$$

$$(V_E / R_5) + (9.0\text{ V} / R_5) + -(V_C / R_5) + (V_B / R_4) + (-V_C / R_4) = 0$$

$$(V_B / R_4) + (9.0\text{ V} / R_5) + -(V_C / R_5) + -(V_C / R_4) + (V_E / R_5) = 0$$

$$[V_B \times (1 / R_4)] + \{V_C \times -[(1 / R_5) + (1 / R_4)]\}$$

$$+ [\boldsymbol{V_E} \times (1 / \boldsymbol{R_5})] + (9.0\ \text{V} / \boldsymbol{R_5}) \qquad = 0$$

$$[\boldsymbol{V_B} \times (1 / \boldsymbol{R_4})] + -\{\boldsymbol{V_C} \times [(1 / \boldsymbol{R_5}) + (1 / \boldsymbol{R_4})]\}$$
$$+ [\boldsymbol{V_E} \times (1 / \boldsymbol{R_5})] + (9.0\ \text{V} / \boldsymbol{R_5}) + -(9.0\ \text{V} / \boldsymbol{R_5}) \qquad = 0 + -(9.0\ \text{V} / \boldsymbol{R_5})$$

$$[\boldsymbol{V_B} \times (1 / \boldsymbol{R_4})] + -\{\boldsymbol{V_C} \times [(1 / \boldsymbol{R_4}) + (1 / \boldsymbol{R_5})]\} + [\boldsymbol{V_E} \times (1 / \boldsymbol{R_5})] \qquad = -(9.0\ \text{V} / \boldsymbol{R_5})$$

$$[\boldsymbol{V_B} \times (1 / 22\ \Omega)] + -\{\boldsymbol{V_C} \times [(1 / 22\ \Omega) + (1 / 27\Omega)]\}$$
$$+ [\boldsymbol{V_E} \times (1 / 27\ \Omega)] \qquad = -(9.0\ \text{V} / 27\ \Omega)$$

$$[\boldsymbol{V_B} \times (45.5 \times 10^{-3}\ \text{S})] + -\{\boldsymbol{V_C} \times [(45.5 \times 10^{-3}\ \text{S}) + (37.0 \times 10^{-3}\ \text{S})]\}$$
$$+ [\boldsymbol{V_E} \times (37.0 \times 10^{-3}\ \text{S})] \qquad = -(9.0 / 27)\ (\text{V} / \Omega)$$

$$[\boldsymbol{V_B} \times (45.5 \times 10^{-3}\ \text{S})]$$
$$+ -[\boldsymbol{V_C} \times (82.5 \times 10^{-3}\ \text{S})] + [\boldsymbol{V_E} \times (37.0 \times 10^{-3}\ \text{S})] \qquad = -333 \times 10^{-3}\ \text{A}$$

$$(45.5 \times 10^{-3}\ \text{S})\boldsymbol{V_B} + (-82.5 \times 10^{-3}\ \text{S})\boldsymbol{V_C} + (37.0 \times 10^{-3}\ \text{S})\boldsymbol{V_E} \qquad = -333 \times 10^{-3}\ \text{A}$$

Substitute for $\boldsymbol{V_D}$ in the equation for node E.

$$(\boldsymbol{V_B} / \boldsymbol{R_2}) + (\boldsymbol{V_D} / \boldsymbol{R_5}) + -\{\boldsymbol{V_E} \times [(1 / \boldsymbol{R_2}) + -(1 / \boldsymbol{R_3}) + -(1 / \boldsymbol{R_5})]\} \qquad = 0$$

$$(\boldsymbol{V_B} / \boldsymbol{R_2}) + [(\boldsymbol{V_C} + -9.0\ \text{V}) / \boldsymbol{R_5}]$$
$$+ -\{\boldsymbol{V_E} \times [(1 / \boldsymbol{R_2}) + -(1 / \boldsymbol{R_3}) + -(1 / \boldsymbol{R_5})]\} \qquad = 0$$

$$(\boldsymbol{V_B} / \boldsymbol{R_2}) + (\boldsymbol{V_C} / \boldsymbol{R_5}) + -(9.0\ \text{V} / \boldsymbol{R_5})$$
$$+ -\{\boldsymbol{V_E} \times [(1 / \boldsymbol{R_2}) + -(1 / \boldsymbol{R_3}) + -(1 / \boldsymbol{R_6})]\} \qquad = 0$$

$$(\boldsymbol{V_B} / \boldsymbol{R_2}) + (\boldsymbol{V_C} / \boldsymbol{R_5}) + -\{\boldsymbol{V_E} \times$$
$$[(1 / \boldsymbol{R_2}) + -(1 / \boldsymbol{R_3}) + -(1 / \boldsymbol{R_5})]\} + -(9.0\ \text{V} / \boldsymbol{R_5}) + (9.0\ \text{V} / \boldsymbol{R_5}) \qquad = 0 + (9.0\ \text{V} / \boldsymbol{R_5})$$

$$(\boldsymbol{V_B} / \boldsymbol{R_2}) + (\boldsymbol{V_C} / \boldsymbol{R_5}) + -\{\boldsymbol{V_E} \times [(1 / \boldsymbol{R_2}) + -(1 / \boldsymbol{R_3}) + -(1 / \boldsymbol{R_5})]\} \qquad = (9.0\ \text{V} / \boldsymbol{R_5})$$

$$[\boldsymbol{V_B} \times (1 / \boldsymbol{R_2})] + [\boldsymbol{V_C} \times (1 / \boldsymbol{R_5})]$$
$$+ -\{\boldsymbol{V_E} \times [(1 / \boldsymbol{R_2}) + -(1 / \boldsymbol{R_3}) + -(1 / \boldsymbol{R_5})]\} \qquad = (9.0\ \text{V} / \boldsymbol{R_5})$$

$$[\boldsymbol{V_B} \times (1 / 10\ \Omega)] + [\boldsymbol{V_C} \times (1 / 27\ \Omega)]$$
$$+ -\{\boldsymbol{V_E} \times [(1 / 10\ \Omega) + -(1 / 18\ \Omega) + -(1 / 27\ \Omega)]\} \qquad = (9.0\ \text{V} / 27\ \Omega)$$

$$[\boldsymbol{V_B} \times (100 \times 10^{-3}\ \text{S})] + [\boldsymbol{V_C} \times (37.0 \times 10^{-3}\ \text{S})]$$
$$+ -\{\boldsymbol{V_E} \times [(100 \times 10^{-3}\ \text{S}) + -(56.6 \times 10^{-3}\ \text{S}) + -(37.0 \times 10^{-3}\ \text{S})]\} \qquad = (9.0\ \text{V} / 27\ \Omega)$$

$$[\boldsymbol{V_B} \times (100 \times 10^{-3}\ \text{S})] + [\boldsymbol{V_C} \times (37.0 \times 10^{-3}\ \text{S})] + -[\boldsymbol{V_E} \times (193 \times 10^{-3}\ \text{S})] \qquad = (9.0 / 27)(\text{V} / \Omega)$$

$$(100 \times 10^{-3}\ \text{S})\boldsymbol{V_B} + (37.0 \times 10^{-3}\ \text{S})\boldsymbol{V_C} + (-193 \times 10^{-3}\ \text{S})\boldsymbol{V_E} \qquad = 333 \times 10^{-3}\ \text{A}$$

Omitting the units for simplicity, this gives the following system of three equations in three unknowns:

$$\begin{aligned}
(-212 \times 10^{-3})\boldsymbol{V_B} &+ (45.5 \times 10^{-3})\boldsymbol{V_C} + (100 \times 10^{-3})\boldsymbol{V_E} = -400 \times 10^{-3} \\
(100 \times 10^{-3})\boldsymbol{V_B} &+ (37.0 \times 10^{-3})\boldsymbol{V_C} + (-193 \times 10^{-3})\boldsymbol{V_E} = 333 \times 10^{-3} \\
(45.5 \times 10^{-3})\boldsymbol{V_B} &+ (-82.5 \times 10^{-3})\boldsymbol{V_C} + (37.0 \times 10^{-3})\boldsymbol{V_E} = -333 \times 10^{-3}
\end{aligned}$$

The matrix-vector form is

$$\begin{vmatrix} -212 \times 10^{-3} & 45.5 \times 10^{-3} & 100 \times 10^{-3} \\ 100 \times 10^{-3} & 37.0 \times 10^{-3} & -193 \times 10^{-3} \\ 45.5 \times 10^{-3} & -82.5 \times 10^{-3} & 37.0 \times 10^{-3} \end{vmatrix} \cdot \begin{vmatrix} V_B \\ V_C \\ V_E \end{vmatrix} = \begin{vmatrix} -400 \times 10^{-3} \\ 333 \times 10^{-3} \\ -333 \times 10^{-3} \end{vmatrix}$$

The solution for this maxtrix equation is

$$\begin{vmatrix} V_B \\ V_C \\ V_E \end{vmatrix} = \begin{vmatrix} 4.375 \\ 7.327 \\ 1.950 \end{vmatrix}$$

where the units are volts. From these node voltages, the calculated resistor currents are

$$
\begin{aligned}
\boldsymbol{I_{R1}} &= (\boldsymbol{V_B} - 6.0\text{ V}) / \boldsymbol{R}_1 \\
&= (4.375\text{ V} - 6.0\text{ V}) / 15\ \Omega \\
&= -1.625\text{ V} / 15\ \Omega \\
&= (-1.625 / 15)\ (\text{V} / \Omega) \\
&= -108.3 \times 10^{-3}\text{ A} \\
&= -108.3\text{ mA}
\end{aligned}
$$

$$
\begin{aligned}
\boldsymbol{I_{R2}} &= (\boldsymbol{V_B} - \boldsymbol{V_E}) / \boldsymbol{R}_2 \\
&= (4.375\text{ V} - 1.950\text{ V}) / 10\ \Omega \\
&= 2.425\text{ V} / 10\ \Omega \\
&= (2.425 / 10)\ (\text{V} / \Omega) \\
&= 242.5 \times 10^{-3}\text{ A} \\
&= 242.5\text{ mA}
\end{aligned}
$$

$$
\begin{aligned}
\boldsymbol{I_{R3}} &= (\boldsymbol{V_E} - 0) / \boldsymbol{R}_3 \\
&= (1.950\text{ V} - 0\text{ V}) / 18\ \Omega \\
&= 1.950\text{ V} / 18\ \Omega \\
&= (1.950 / 18)\ (\text{V} / \Omega) \\
&= 108.3 \times 10^{-3}\text{ A} \\
&= 108.3\text{ mA}
\end{aligned}
$$

$$
\begin{aligned}
\boldsymbol{I_{R4}} &= (\boldsymbol{V_C} - \boldsymbol{V_B}) / \boldsymbol{R}_4 \\
&= (7.327\text{ V} - 4.375\text{ V}) / 22\ \Omega \\
&= 2.952\text{ V} / 22\ \Omega \\
&= (2.952 / 22)\ (\text{V} / \Omega) \\
&= 134.2 \times 10^{-3}\text{ A} \\
&= 134.2\text{ mA}
\end{aligned}
$$

$$
\begin{aligned}
\boldsymbol{I_{R5}} &= (\boldsymbol{V_E} - \boldsymbol{V_D}) / \boldsymbol{R}_5 \\
&= [\boldsymbol{V_E} - (\boldsymbol{V_C} - 9.0\text{ V})] / \boldsymbol{R}_5 \\
&= [1.950\text{ V} + -(7.327\text{ V} - 9.0\text{ V})] / 27\ \Omega \\
&= [1.950\text{ V} + 9.0\text{ V} + -7.327\text{ V})] / 27\ \Omega \\
&= 3.623\text{ V} / 27\ \Omega \\
&= (3.623 / 27)\ (\text{V} / \Omega) \\
&= 134.2 \times 10^{-3}\text{ A} \\
&= 134.2\text{ mA}
\end{aligned}
$$

The negative value of $\boldsymbol{I_{R1}}$ indicates that the direction of the actual current is opposite to the direction of the assigned branch current $\boldsymbol{I}_1$. The resistor currents and directions are

I_{R1} = 108 mA (current flows from left to right)

I_{R2} = 243 mA (current flows from top to bottom)

I_{R3} = 108 mA (current flows from right to left)

$I_2 = I_{R4}$ = 134 mA (current flows from right to left)

I_{R5} = 134 mA (current flows from left to right)

2. Problem: Use the node voltage method to determine the currents through each resistor for the circuit in Figure 9-19.

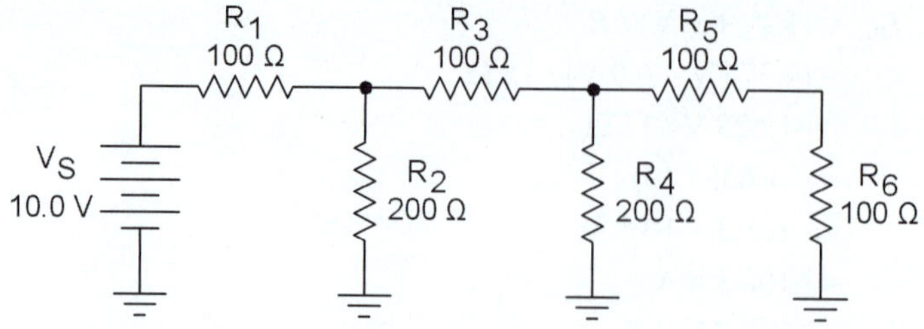

Figure 9-19: Node Voltage Method Example Circuit 2

Solution: First, label all the circuit nodes. Refer to Figure 9-20.

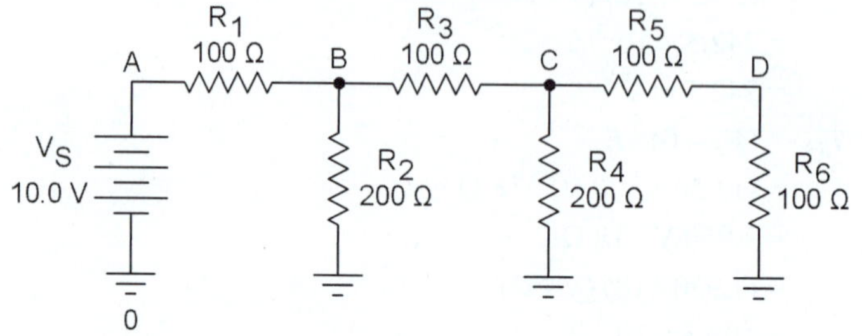

Figure 9-20: Labeled Nodes for Figure 9-19

As Figure 9-20 shows, the circuit contains five nodes. For this example, select the negative terminal of V_S as the reference node, labeled as node 0, which all the ground references share. Next, assign voltage designations for all nodes for which the voltage is not known. Refer to Figure 9-21.

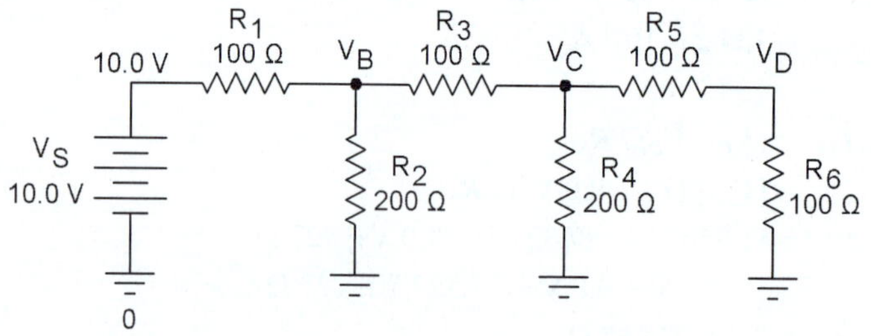

Figure 9-21: Assigned Voltage Designations for Figure 9-19

The voltage at node A relative to the reference node is known, as V_S holds it at 10.0 V more positive than the reference node, so the node is labeled with the actual voltage value. The other nodes are labeled with voltage designations. Next, assign branch currents for all nodes for which the actual voltage is not known. Refer to Figure 9-22.

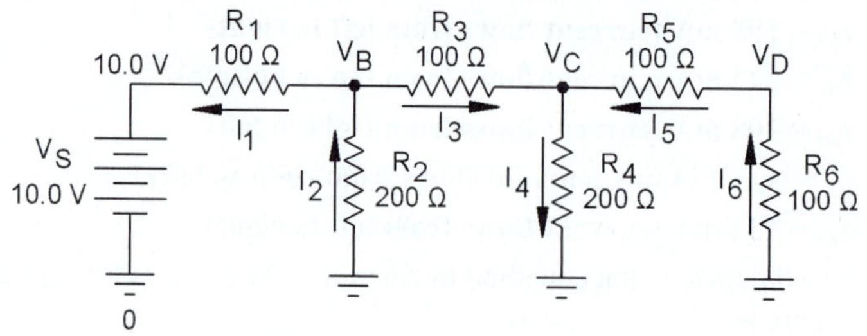

Figure 9-22: Assigned Branch Currents for Figure 9-19

Now, apply Kirchhoff's Current Law to each node for which the actual voltage is not known. For this example, let currents entering a node be positive and currents leaving a node be negative.

Node B: $-I_1 + I_2 + -I_3 = 0$

Node C: $I_3 + -I_4 + I_5 = 0$

Node D: $-I_5 + I_6 = 0$

Next, use Ohm's Law to express each branch current in terms of the node voltages and resistances for node B.

$$-I_1 + I_2 + -I_3 = 0$$
$$-[(V_B - 10.0\text{ V}) / R_1] + [(0 - V_B) / R_2] + -[(V_B - V_C) / R_3] = 0$$
$$-(V_B / R_1) + (10.0\text{ V} / R_1) + -(V_B / R_2) + -(V_B / R_3) + -(V_C / R_3) = 0$$
$$-\{V_B \times [(1 / R_1) + (1 / R_2) + (1 / R_3)]\} + [V_C \times (1 / R_3)] + (10.0\text{ V} / R_1) = 0$$
$$-\{V_B \times [(1 / R_1) + (1 / R_2) + (1 / R_3)]\} + [V_C \times (1 / R_3)] + (10.0\text{ V} / R_1) + -(10.0\text{ V} / R_1) = 0 + -(10.0\text{ V} / R_1)$$
$$-\{V_B \times [(1 / R_1) + (1 / R_2) + (1 / R_3)]\} + [V_C \times (1 / R_3)] = -(10.0\text{ V} / R_1)$$
$$-\{V_B \times [(1 / 100\ \Omega) + (1 / 200\ \Omega) + (1 / 100\ \Omega)]\} + [V_C \times (1 / 100\ \Omega)] = -(10.0\text{ V} / 100\ \Omega)$$
$$-\{V_B \times [(10 \times 10^{-3}\text{ S}) + (5 \times 10^{-3}\text{ S}) + (10 \times 10^{-3}\text{ S})]\} + [V_C \times (10 \times 10^{-3}\text{ S})] = -(10.0 / 100)(\text{V} / \Omega)$$
$$-[V_B \times (25 \times 10^{-3}\text{ S})] + [V_C \times (10 \times 10^{-3}\text{ S})] = -100 \times 10^{-3}\text{ A}$$
$$(-25 \times 10^{-3}\text{ S})V_B + (10 \times 10^{-3}\text{ S})V_C = -100 \times 10^{-3}\text{ A}$$

Repeat the process for node C.

$$I_3 + -I_4 + I_5 = 0$$
$$[(V_B - V_C) / R_3] + -[(V_C - 0) / R_4] + [(V_D - V_C) / R_5] = 0$$
$$(V_B / R_3) + -(V_C / R_3) + -(V_C / R_4) + (V_D / R_5) + -(V_C / R_5) = 0$$
$$[V_B \times (1 / R_3)] + -\{V_C \times [(1 / R_3) + (1 / R_4) + (1 / R_5)] + [V_D \times (1 / R_5)] = 0$$
$$[V_B \times (1 / 100\ \Omega)] + -\{V_C \times [(1 / 100\ \Omega) + (1 / 200\ \Omega) + (1 / 100\ \Omega)] + [V_D \times (1 / 100\ \Omega)] = 0$$
$$[V_B \times (10 \times 10^{-3}\text{ S})] + -\{V_C \times [(10 \times 10^{-3}\text{ S}) + (5 \times 10^{-3}\text{ S}) + (10 \times 10^{-3}\text{ S})]\} + [V_D \times (10 \times 10^{-3}\text{ S})] = 0$$
$$[V_B \times (10 \times 10^{-3}\text{ S})] + -[V_C \times (25 \times 10^{-3}\text{ S})] + [V_D \times (10 \times 10^{-3}\text{ S})] = 0$$
$$(10 \times 10^{-3}\text{ S})V_B + (-25 \times 10^{-3}\text{ S})V_C + (10 \times 10^{-3}\text{ S})V_D = 0$$

Repeat the process for node D.

$$-I_5 + I_6 = 0$$
$$-[(V_D - V_C) / R_5] + [(0 - V_D) / R_6] = 0$$
$$-(V_D / R_5) + (V_C / R_5) + -(V_D / R_6) = 0$$
$$-[V_D \times (1 / R_5)] + [V_C \times (1 / R_5)] + -[V_D \times (1 / R_6)] = 0$$
$$[V_C \times (1 / R_5)] + -[V_D \times (1 / R_5)] + -[V_D \times (1 / R_6)] = 0$$
$$[V_C \times (1 / R_5)] + -\{V_D \times [(1 / R_5) + (1 / R_6)]\} = 0$$
$$[V_C \times (1 / 100\ \Omega)] + -\{V_D \times [(1 / 100\ \Omega) + (1 / 100\ \Omega)]\} = 0$$
$$[V_C \times (10 \times 10^{-3}\text{ S})] + -\{V_D \times [(10 \times 10^{-3}\text{ S}) + (10 \times 10^{-3}\text{ S})]\} = 0$$
$$[V_C \times (10 \times 10^{-3}\text{ S})] + -[V_D \times (20 \times 10^{-3}\text{ S})] = 0$$
$$(10 \times 10^{-3}\text{ S})V_C + (-20 \times 10^{-3}\text{ S})V_D = 0$$

Omitting the units for simplicity and ensuring that all equations have V_B, V_C, and V_D terms, this gives the following system of three equations in three unknowns:

$$
\begin{array}{rcrcrcr}
(-25 \times 10^{-3})\boldsymbol{V_B} & + & (10 \times 10^{-3})\boldsymbol{V_C} & + & (0 \times 10^{-3})\boldsymbol{V_D} & = & -100 \times 10^{-3} \\
(10 \times 10^{-3})\boldsymbol{V_B} & + & (-25 \times 10^{-3})\boldsymbol{V_C} & + & (10 \times 10^{-3})\boldsymbol{V_D} & = & 0 \times 10^{-3} \\
(0 \times 10^{-3})\boldsymbol{V_B} & + & (10 \times 10^{-3})\boldsymbol{V_C} & + & (-20 \times 10^{-3})\boldsymbol{V_E} & = & 0 \times 10^{-3}
\end{array}
$$

The matrix-vector form is

$$
\begin{vmatrix} -25 \times 10^{-3} & 10 \times 10^{-3} & 0 \times 10^{-3} \\ 10 \times 10^{-3} & -25 \times 10^{-3} & 10 \times 10^{-3} \\ 0 \times 10^{-3} & 10 \times 10^{-3} & -20 \times 10^{-3} \end{vmatrix} \bullet \begin{vmatrix} \boldsymbol{V_B} \\ \boldsymbol{V_C} \\ \boldsymbol{V_D} \end{vmatrix} = \begin{vmatrix} -100 \times 10^{-3} \\ 0 \times 10^{-3} \\ 0 \times 10^{-3} \end{vmatrix}
$$

The solution for this maxtrix equation is

$$
\begin{vmatrix} \boldsymbol{V_B} \\ \boldsymbol{V_C} \\ \boldsymbol{V_D} \end{vmatrix} = \begin{vmatrix} 5.000 \\ 2.500 \\ 1.250 \end{vmatrix}
$$

where the units are volts. From these node voltages the calculated branch currents are

$$
\begin{aligned}
\boldsymbol{I_{R1}} &= (\boldsymbol{V_B} - 10.0\text{ V}) / \boldsymbol{R}_1 \\
&= (5.000\text{ V} - 50.0\text{ V}) / 100\ \Omega \\
&= -5.000\text{ V} / 100\ \Omega \\
&= (-5.000 / 100)\ (\text{V} / \Omega) \\
&= -50.0 \times 10^{-3}\text{ A} \\
&= -50.0\text{ mA}
\end{aligned}
$$

$$
\begin{aligned}
\boldsymbol{I_{R2}} &= (0 - \boldsymbol{V_B}) / \boldsymbol{R}_2 \\
&= (0\text{ V} - 5.000\text{ V}) / 200\ \Omega \\
&= -5.000\text{ V} / 200\ \Omega \\
&= (-5.000 / 200)\ (\text{V} / \Omega) \\
&= -25.0 \times 10^{-3}\text{ A} \\
&= -25.0\text{ mA}
\end{aligned}
$$

$$
\begin{aligned}
\boldsymbol{I_{R3}} &= (\boldsymbol{V_B} - \boldsymbol{V_C}) / \boldsymbol{R}_3 \\
&= (5.000\text{ V} - 2.500\text{ V}) / 100\ \Omega \\
&= 2.500\text{ V} / 100\ \Omega \\
&= (2.500 / 100)\ (\text{V} / \Omega) \\
&= 25.0 \times 10^{-3}\text{ A} \\
&= 25.00\text{ mA}
\end{aligned}
$$

$$
\begin{aligned}
\boldsymbol{I_{R4}} &= (\boldsymbol{V_C} - 0) / \boldsymbol{R}_4 \\
&= (2.500\text{ V} - 0\text{ V}) / 200\ \Omega \\
&= 2.500\text{ V} / 200\ \Omega \\
&= (2.500 / 200)\ (\text{V} / \Omega) \\
&= 12.5 \times 10^{-3}\text{ A} \\
&= 12.5\text{ mA}
\end{aligned}
$$

$$
\begin{aligned}
\boldsymbol{I_{R5}} &= (\boldsymbol{V_D} - \boldsymbol{V_C}) / \boldsymbol{R}_5 \\
&= (1.250\text{ V} - 2.500\text{ V}) / 100\ \Omega \\
&= -1.250\text{ V} / 100\ \Omega \\
&= (-1.250 / 100)\ (\text{V} / \Omega) \\
&= -12.5 \times 10^{-3}\text{ A} \\
&= -12.5\text{ mA}
\end{aligned}
$$

$$\begin{aligned} I_{R6} &= (0 - V_D) / R_6 \\ &= (0\text{ V} - 1.250\text{ V}) / 100\ \Omega \\ &= -1.250\text{ V} / 100\ \Omega \\ &= (-1.250 / 100)\ (\text{V} / \Omega) \\ &= -12.5 \times 10^{-3}\text{ A} \\ &= -12.5\text{ mA} \end{aligned}$$

The negative values of I_1, I_2, I_5, and I_6 indicate that the actual directions of I_{R1}, I_{R2}, I_{R5}, and I_{R6} are opposite to the assumed directions of I_1, I_2, I_5, and I_6. The resistor currents are

$I_{R1} =$ **50 mA (current flows from left to right)**

$I_{R2} =$ **25 mA (current flows from top to bottom)**

$I_{R3} =$ **25 mA (current flows from left to right)**

$I_{R4} =$ **12.5 mA (current flows from top to bottom)**

$I_{R5} =$ **12.5 mA (current flows from left to right)**

$I_{R6} =$ **12.5 mA (current flows from top to bottom)**

9.2 Advanced

9.2.1 Simultaneous Equations

1. Problem: How many independent equations would be necessary to solve the following set of equations?

$$\begin{aligned} -5I_2 + 3I_4 &= 2.0 \\ 2I_1 - 2I_2 &= -1.0 \\ -2I_2 + 2I_5 &= 0.5 \\ 4I_2 + 3I_4 &= 3.2 \\ -7I_4 - 4I_5 &= -5.0 \end{aligned}$$

Solution: The equations contain four variables (I_1, I_2, I_4, and I_5) require **four** independent equations.

2. Problem: Determine the matrix-vector form for the following set of equations. Assume that all unknown variables I_N are shown.

$$\begin{aligned} -8I_2 + 6I_4 &= 6.0 \\ -5I_3 + 2I_4 - I_5 &= 0 \\ 3I_1 + 4I_2 &= 0 \\ -4I_1 - 4I_3 &= 0 \\ 2I_2 - 7I_3 &= 0 \end{aligned}$$

Solution: The set of equations include the five unknown variables I_1, I_2, I_3, I_4, and I_5. Include all variable terms in each equation with coefficients.

$$\begin{array}{rcrcrcrcrcr} 0I_1 & + & -8I_2 & + & 0I_3 & + & 6I_4 & + & 0I_5 & = & 6.0 \\ 0I_1 & + & 0I_2 & + & -5I_3 & + & 2I_4 & - & 1I_5 & = & 0 \\ 3I_1 & + & 4I_2 & + & 0I_3 & + & 0I_4 & + & 0I_5 & = & 0 \\ -4I_1 & + & 0I_2 & - & 4I_3 & + & 0I_4 & + & 0I_5 & = & 0 \\ 0I_1 & + & 2I_2 & - & 7I_3 & + & -1I_4 & + & -1I_5 & = & 0 \end{array}$$

Converting all equations to sums gives

$$\begin{array}{rcrcrcrcrcr} 0I_1 & + & -8I_2 & + & 0I_3 & + & 6I_4 & + & 0I_5 & = & 6.0 \\ 0I_1 & + & 0I_2 & + & -5I_3 & + & 2I_4 & + & -1I_5 & = & 0 \\ 3I_1 & + & 4I_2 & + & 0I_3 & + & 0I_4 & + & 0I_5 & = & 0 \\ -4I_1 & + & 0I_2 & + & -4I_3 & + & 0I_4 & + & 0I_5 & = & 0 \\ 0I_1 & + & 2I_2 & + & -7I_3 & + & -1I_4 & + & -1I_5 & = & 0 \end{array}$$

The matrix-vector form is then

$$\begin{vmatrix} 0 & -8 & 0 & 6 & 0 \\ 0 & 0 & -5 & 2 & -1 \\ 3 & 4 & 0 & 0 & 0 \\ -4 & 0 & -4 & 0 & 0 \\ 0 & 2 & -7 & 0 & 0 \end{vmatrix} \bullet \begin{vmatrix} I_1 \\ I_2 \\ I_3 \\ I_4 \\ I_5 \end{vmatrix} = \begin{vmatrix} 6.0 \\ 0 \\ 0 \\ 0 \\ 0 \end{vmatrix}$$

9.2.2 Branch Current Method

Problem: Use the branch current method to determine the Norton equivalent relative to $\boldsymbol{R_L}$ for the circuit in Figure 9-23.

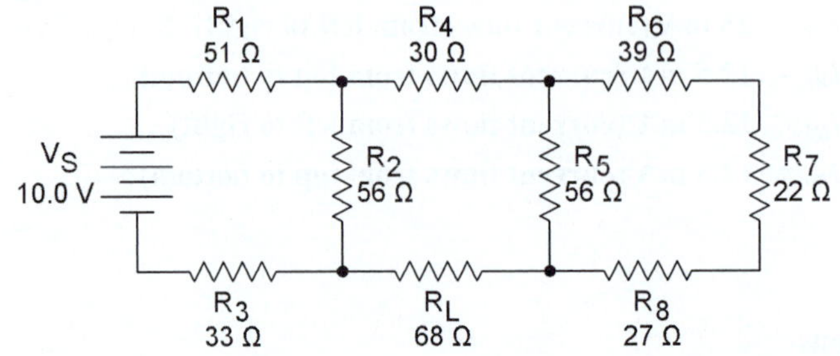

Figure 9-23: Branch Current Method Example Circuit 3

Solution: To find the Norton current, first replace $\boldsymbol{R_L}$ with a short and use the branch current method to determine the short circuit current through it. Then label the nodes and determine a reference direction for each non-redundant current loop. The circuit in Figure 9-23 contains eight nodes and three non-redundant loops. For this example, choose a clockwise reference direction for each loop. Refer to Figure 9-24.

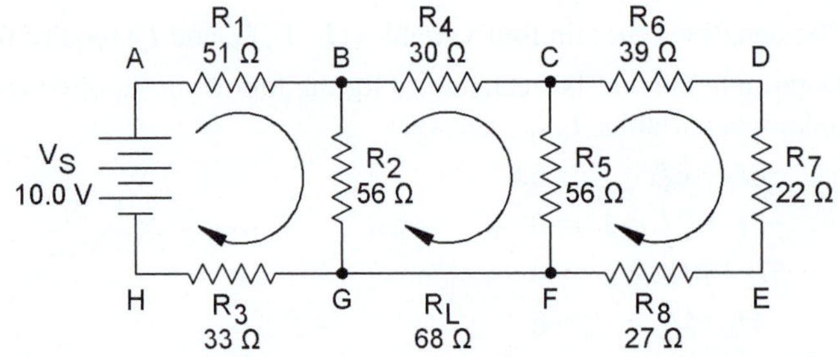

Figure 9-24: Nodes and Reference Directions for Figure 9-23

Assign branch currents and mark the voltage polarity for each resistor based on the direction of the branch current through it. Refer to Figure 9-25.

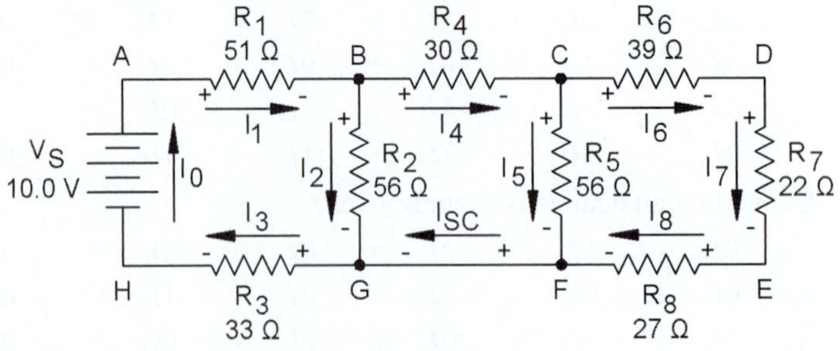

Figure 9-25: Assigned Branch Currents and Resistor Voltage Polarities for Figure 9-23

As Figure 9-25 shows, the branch between nodes B and G exists in the left loop that includes nodes A, B, G, and H and also in the loop that includes nodes B, C, F, and G. The assigned direction for $\boldsymbol{I_2}$ could be the same as the left or middle loop in Figure 9-25, but once current direction is assigned based on the left loop, the branch is ignored for the middle loop. The same is true for the assigned direction for $\boldsymbol{I_5}$.

Now, replace any redundant branch currents and remove any nonessential branch currents. The branches between nodes G and H, H and A, and A and B form a series path. From the definition of a series path, the current through each branch must be the same. Therefore $\boldsymbol{I_0} = \boldsymbol{I_1} = \boldsymbol{I_3}$. Similarly, the branches between nodes C and D, D and E, and E and F form a series path, so that $\boldsymbol{I_6} = \boldsymbol{I_7} = \boldsymbol{I_8}$. Therefore, replace $\boldsymbol{I_3}$ with $\boldsymbol{I_1}$ and $\boldsymbol{I_7}$ and $\boldsymbol{I_8}$ with $\boldsymbol{I_6}$. Because $\boldsymbol{I_0}$ passes through voltage source, it is nonessential and is removed from the circuit. Refer to Figure 9-26.

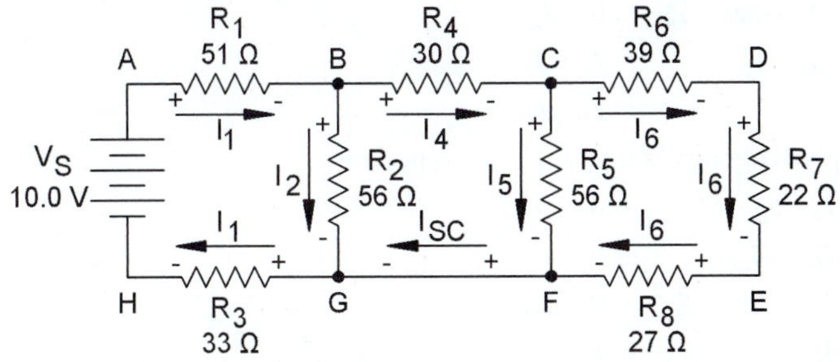

Figure 9-26: Replacement of Redundant Branch Currents for Figure 9-25

The next step is to apply Kirchhoff's Voltage Law to each loop. Work in the reference direction for the loop containing $\boldsymbol{V_S}$, $\boldsymbol{R_1}$, $\boldsymbol{R_2}$, and $\boldsymbol{R_3}$, apply Ohm's Law, and segregate terms.

$$V_S - V_{R1} - V_{R2} - V_{R3} = 0$$
$$V_S + -V_{R1} + -V_{R2} + -V_{R3} = 0$$
$$V_S + -(R_1 \times I_1) + -(R_2 \times I_2) + -(R_3 \times I_1) = 0$$
$$V_S + -(R_1 \times I_1) + -(R_3 \times I_1) + -(R_2 \times I_2) = 0$$
$$V_S + -[(R_1 + R_3) \times I_1] + -(R_2 \times I_2) = 0$$
$$V_S + -[(R_1 + R_3) \times I_1] + -(R_2 \times I_2) + -V_S = 0 + -V_S$$
$$-[(R_1 + R_3) \times I_1] + -(R_2 \times I_2) = -V_S$$
$$-[(51 + 33) \times I_1] + -(56 \times I_2) = -10.0$$
$$-(84 \times I_1) + -(56 \times I_2) = -10.0$$
$$-84I_1 + -56I_2 = -10.0$$

Work in the reference direction for the loop containing $\boldsymbol{R_2}$, $\boldsymbol{R_4}$, $\boldsymbol{R_5}$, and the short for $\boldsymbol{R_L}$, apply Ohm's Law, and segregate terms.

$$V_{R2} - V_{R4} - V_{R5} = 0$$
$$V_{R2} + -V_{R4} + -V_{R5} = 0$$
$$(R_2 \times I_2) + -(R_4 \times I_4) + -(R_5 \times I_5) = 0$$
$$(56 \times I_2) + -(30 \times I_4) + -(56 \times I_5) = 0$$
$$56I_2 + -30I_4 + -56I_5 = 0$$

Work in the reference direction for the loop containing $\boldsymbol{R_5}$, $\boldsymbol{R_6}$, $\boldsymbol{R_7}$, and $\boldsymbol{R_8}$, apply Ohm's Law, and segregate terms.

$$V_{R5} - V_{R6} - V_{R7} - V_{R8} = 0$$
$$V_{R5} + -V_{R6} + -V_{R7} + -V_{R8} = 0$$
$$(R_5 \times I_5) + -(R_6 \times I_6) + -(R_7 \times I_6) + -(R_8 \times I_6) = 0$$
$$(R_5 \times I_5) + -[(R_6 + R_7 + R_8) \times I_6] = 0$$
$$(56 \times I_5) + -[(39 + 22 + 27) \times I_6] = 0$$

$$(56 \times I_5) + -(88 \times I_6) = 0$$
$$56I_5 + -88I_6 = 0$$

These give the following three equations in five unknowns:

$$\begin{aligned} -84I_1 &+ -56I_2 &+ 0I_4 &+ 6I_5 &+ 0I_6 &= -10.0 \\ 0I_1 &+ 56I_2 &+ -30I_4 &+ -56I_5 &+ 0I_6 &= 0 \\ 0I_1 &+ 0I_2 &+ 0I_4 &+ 56I_5 &+ -88I_6 &= 0 \end{aligned}$$

To reduce the number of unknowns, first apply Kirchhoff's Current Law to node C.

$$I_4 - I_5 - I_6 = 0$$
$$I_4 + -I_5 + -I_6 = 0$$
$$I_4 + -I_5 + -I_6 + I_6 = 0 + I_6$$
$$I_4 + -I_5 + I_5 = I_6 + I_5$$
$$I_4 = I_5 + I_6$$

Next, apply Kirchhoff's Current Law to node F.

$$I_5 + I_6 - I_{SC} = 0$$
$$I_5 + I_6 + -I_{SC} = 0$$
$$I_5 + I_6 + -I_{SC} + I_{SC} = 0 + I_{SC}$$
$$I_5 + I_6 = I_{SC}$$

Finally apply Kirchhoff's Current Law to node G.

$$I_2 - I_1 + I_{SC} = 0$$
$$I_2 + -I_1 + I_{SC} + I_1 = 0 + I_1$$
$$I_2 + I_{SC} = I_1$$

Substitute $I_5 + I_6$ for I_{SC}.

$$I_2 + I_{SC} = I_1$$
$$I_2 + I_5 + I_6 = I_1$$

Substitute the expression for I_1 into the equation for the left loop.

$$-84I_1 + -56I_2 + 0I_4 + 0I_5 + 0I_6 = -10.0$$
$$-84(I_2 + I_5 + I_6) + -56I_2 + 0I_4 + 0I_5 + 0I_6 = -10.0$$
$$-84I_2 + -84I_5 + -84I_6 + -56I_2 + 0I_4 + 0I_5 + 0I_6 = -10.0$$
$$-84I_2 + -56I_2 + 0I_4 + -84I_5 + 0I_5 + -84I_6 + 0I_6 = -10.0$$
$$(-84 + -56)I_2 + 0I_4 + (-84 + 0)\, I_5 + (-84 + 0)I_6 = -10.0$$
$$-140I_2 + 0I_4 + -84I_5 + -84I_6 = -10.0$$

Substitute expression for I_4 into the equation for the middle loop.

$$0I_1 + 56I_2 + -30I_4 + -56I_5 + 0I_6 = 0$$
$$0I_1 + 56I_2 + -30(I_5 + I_6) + -56I_5 + 0I_6 = 0$$
$$0I_1 + 56I_2 + -30I_5 + -30I_6 + -56I_5 + 0I_6 = 0$$
$$0I_1 + 56I_2 + -30I_5 + -56I_5 + -30I_6 + 0I_6 = 0$$
$$0I_1 + 56I_2 + (-30 + -56)I_5 + (-30 + 0)I_6 = 0$$
$$0I_1 + 56I_2 + -86I_5 + -30I_6 = 0$$

With I_1 and I_4 eliminated, the set of equations is then one of three equations in three unknowns.

$$\begin{aligned} -140I_2 &+ -84I_5 &+ -84I_6 &= -10.0 \\ 56I_2 &+ -86I_5 &+ -30I_6 &= 0 \\ 0I_2 &+ 56I_5 &+ -88I_6 &= 0 \end{aligned}$$

The matrix-vector form is

$$\begin{vmatrix} -140 & -84 & -84 \\ 56 & -86 & -30 \\ 0 & 56 & -88 \end{vmatrix} \bullet \begin{vmatrix} I_2 \\ I_5 \\ I_6 \end{vmatrix} = \begin{vmatrix} -10.0 \\ 0 \\ 0 \end{vmatrix}$$

The solution for this maxtrix equation is

$$\begin{vmatrix} I_2 \\ I_5 \\ I_6 \end{vmatrix} = \begin{vmatrix} 46.9 \times 10^{-3} \\ 25.0 \times 10^{-3} \\ 15.9 \times 10^{-3} \end{vmatrix}$$

The coefficients are in ohms and the right-hand side is in volts, so the units of the solution are in amperes. The values are all positive, so the directions for the actual currents are the same as the assumed reference directions for the corresponding branch currents. From the calculation for Kirchhoff's Current Law for node F, the value of the Norton current is

$$\begin{aligned} \boldsymbol{I_N} &= \boldsymbol{I_{SC}} \\ &= \boldsymbol{I_5} + \boldsymbol{I_6} \\ &= 25.0 \times 10^{-3}\ \text{A} + 15.9 \times 10^{-3}\ \text{A} \\ &= 40.9 \times 10^{-3}\ \text{A} \\ &= 40.9\ \text{mA} \end{aligned}$$

To determine the Norton resistance $\boldsymbol{R_N}$, replace $\boldsymbol{V_S}$ with its internal resistance (a short), and determine the equivalent resistance $\boldsymbol{R_{EQ}}$ looking back from $\boldsymbol{R_L}$ into the circuit. If necessary, redraw the circuit to determine the relationship of the resistances within the circuit as shown in Figure 9-27. The expression for the circuit is

$$\boldsymbol{R_N} = \boldsymbol{R_2} \,||\, (\boldsymbol{R_1} + \boldsymbol{R_3}) + \boldsymbol{R_4} + \boldsymbol{R_5} \,||\, (\boldsymbol{R_6} + \boldsymbol{R_7} + \boldsymbol{R_8})$$

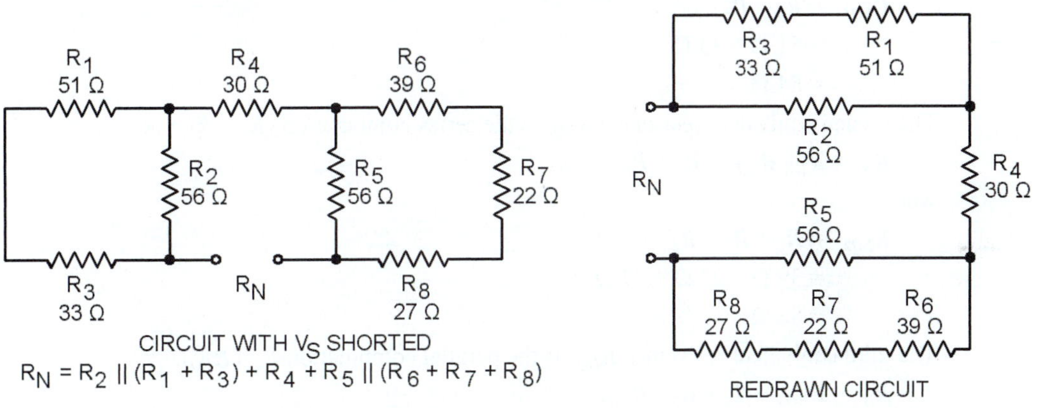

Figure 9-27: Original and Redrawn Circuits with V_S Replaced with Short

Refer to Figure 9-28 for the circuit analysis and simplification process for finding $\boldsymbol{R_N}$.

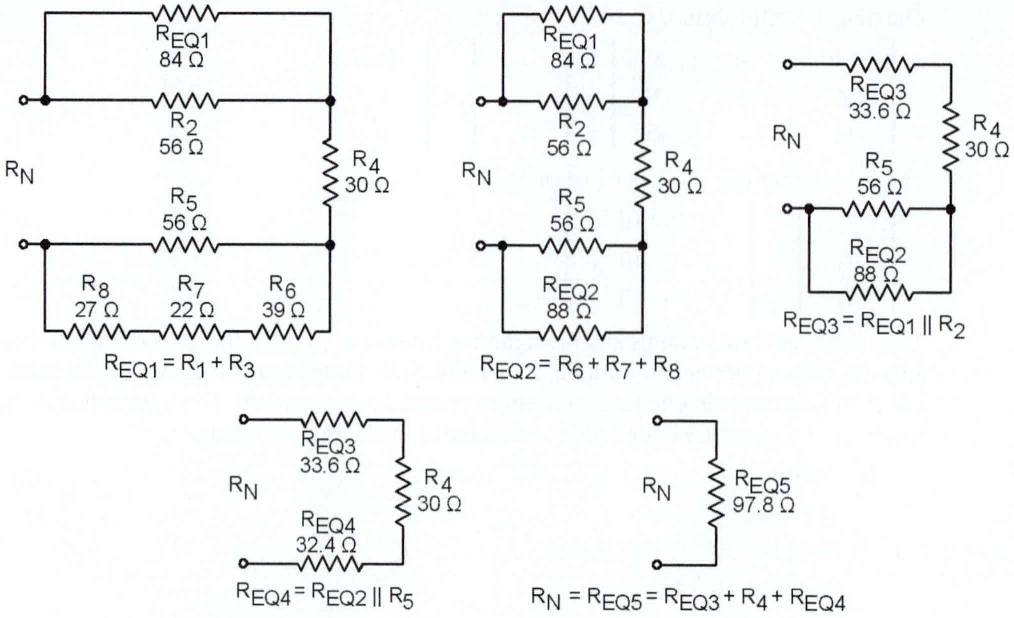

Figure 9-28: Circuit Analysis and Simplification Process for R_N

The first equivalent resistance R_{EQ1} is the series combination ($R_1 + R_3$).

$$R_N = R_2 \parallel R_{EQ1} + R_4 + R_5 \parallel (R_6 + R_7 + R_8)$$

where

$$R_{EQ1} = R_1 + R_3$$
$$= 51\ \Omega + 33\ \Omega$$
$$= 84\ \Omega$$

The second equivalent resistance R_{EQ2} is the series combination ($R_6 + R_7 + R_8$).

$$R_N = R_2 \parallel R_{EQ1} + R_4 + R_5 \parallel R_{EQ2}$$

where

$$R_{EQ2} = R_6 + R_7 + R_8$$
$$= 39\ \Omega + 22\ \Omega + 27\ \Omega$$
$$= 88\ \Omega$$

The third equivalent resistance R_{EQ3} is the parallel combination $R_2 \parallel R_{EQ1}$.

$$R_N = R_{EQ3} + R_4 + R_5 \parallel R_{EQ2}$$

where

$$R_{EQ3} = R_2 \parallel R_{EQ1}$$
$$= 56\ \Omega \parallel 84\Omega$$
$$= 33.6\ \Omega$$

The fourth equivalent resistance R_{EQ4} is the parallel combination $R_5 \parallel R_{EQ2}$.

$$R_N = R_{EQ3} + R_4 + R_{EQ4}$$

where

$$R_{EQ4} = R_5 \parallel R_{EQ2}$$
$$= 56\ \Omega \parallel 88\ \Omega$$
$$= 34.2\ \Omega$$

The fifth (and last) equivalent resistance R_{EQ5} is the series combination ($R_{EQ3} + R_4 + R_{EQ4}$).

$R_N = R_{EQ5}$

where

$$R_{EQ5} = R_{EQ3} + R_4 + R_{EQ4}$$
$$= 33.6\ \Omega + 30\ \Omega + 34.2\ \Omega$$
$$= 97.8\ \Omega$$

Therefore, $R_N = 97.8\ \Omega$. Figure 9-29 shows the Norton equivalent relative to R_L for Figure 9-23.

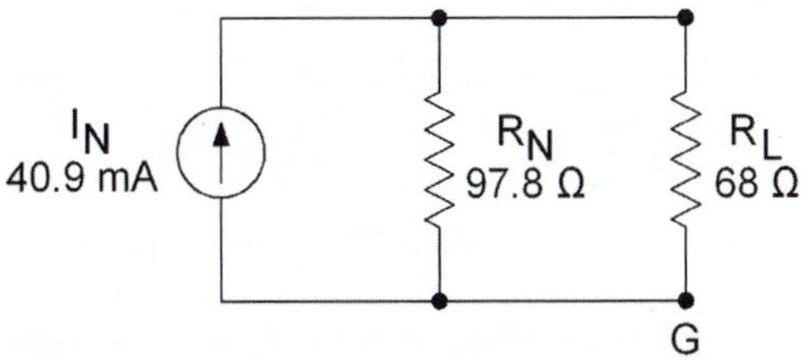

Figure 9-29: Norton Equivalent for Figure 9-23

9.2.3 Loop Current Method

Problem: Use the loop current method to determine the Thevenin equivalent relative to R_L for the circuit in Figure 9-30.

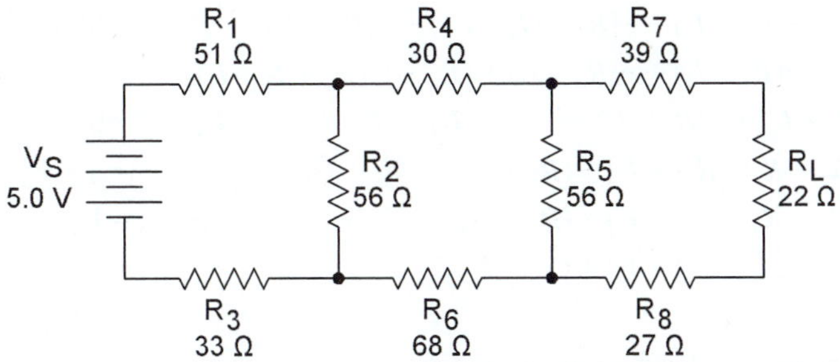

Figure 9-30: Loop Current Method Example Circuit 3

Solution: To find the Thevenin voltage, first remove R_L and then use the loop current method to determine the open load voltage V_{OL}. Identify and determine a reference direction for each non-redundant loop current. For the circuit in Figure 9-30, there are two non-redundant current loops. For this example, let the reference direction for each current be counter-clockwise. Refer to Figure 9-31.

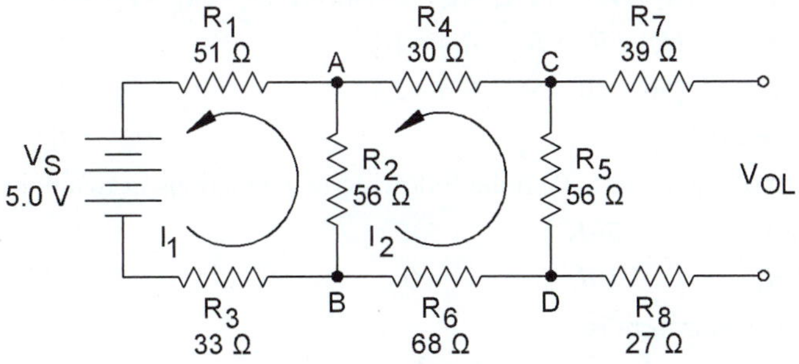

Figure 9-31: Loop Currents for Figure 9-30

Note that another loop exists that contains V_S, R_1, R_4, R_5, R_6, and R_3, but that this loop is redundant because every component in the loop is already included in the loops for I_1 and I_2. Note also that when

R_L is removed from the circuit, no current flows through R_7 and R_8 and the resistors drop no voltage. Consequently, the circuit analysis can ignore R_7 and R_8 so that $V_{TH} = V_{OL} = V_{CD}$. Next, assign resistor voltage polarities based on the reference directions for I_1 and I_2. Refer to Figure 9-32.

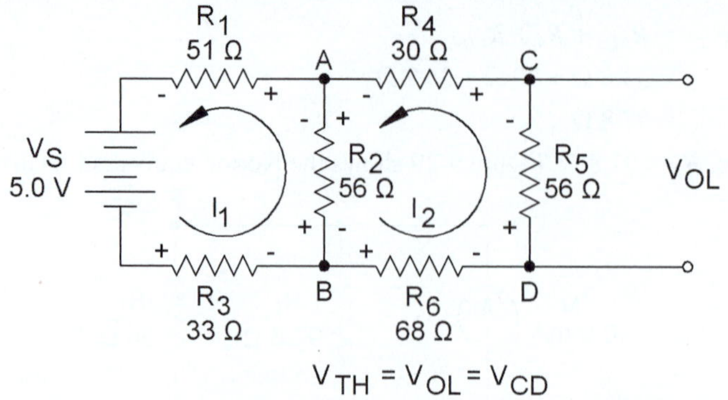

Figure 9-32: Resistor Voltage Polarities for Loop Currents in Figure 9-31

R_7 and R_8 have been removed from the circuit because they have no effect on determining the value of $V_{TH} = V_{OL} = V_{CD}$. Apply Kirchhoff's Voltage law and Ohm's Law to the left-hand loop by working in the direction of the loop current, omitting units for simplicity.

$$-V_S - V_{R1} - V_{R2} - V_{R3} = 0$$
$$-V_S + -V_{R1} + -V_{R2} + -V_{R3} = 0$$
$$-V_S + -(R_1 \times I_1) + -[R_2 \times (I_1 - I_2)] + -(R_3 \times I_1) = 0$$
$$-V_S + -(R_1 \times I_1) + -(R_2 \times I_1) + (R_2 \times I_2) + -(R_3 \times I_1) = 0$$
$$-(R_1 \times I_1) + -(R_2 \times I_1) + -(R_3 \times I_1) + (R_2 \times I_2) + -V_S = 0$$
$$[-(R_1 + R_2 + R_3) \times I_1] + (R_2 \times I_2) + -V_S + V_S = 0 + V_S$$
$$[-(R_1 + R_2 + R_3) \times I_1] + (R_2 \times I_2) = V_S$$
$$[-(51 + 56 + 33) \times I_1] + (56 \times I_2) = 5.0$$
$$-140I_1 + 56I_2 = 5.0$$

Repeat the process for the right-hand loop.

$$-V_{R2} - V_{R6} - V_{R5} - V_{R4} = 0$$
$$-V_{R2} + -V_{R6} + -V_{R5} + -V_{R4} = 0$$
$$-[R_2 \times (I_2 - I_1)] + -(R_6 \times I_2) + -(R_5 \times I_2) + -(R_4 \times I_2) = 0$$
$$(R_2 \times I_1) + -(R_2 \times I_2) + -(R_6 \times I_2) + -(R_5 \times I_2) + -(R_4 \times I_2) = 0$$
$$(R_2 \times I_1) + -(R_2 \times I_2) + -(R_4 \times I_2) + -(R_5 \times I_2) + -(R_6 \times I_2) = 0$$
$$(R_2 \times I_1) + [-(R_2 + R_4 + R_5 + R_6) \times I_2] = 0$$
$$(56 \times I_1) + [-(56 + 30 + 56 + 68) \times I_2] = 0$$
$$56I_1 + -210I_2 = 0$$

The system of equations is then the following two equations in two unknowns:

$$\begin{array}{rcrcl} -140I_1 & + & 56I_2 & = & 5.0 \\ 56I_1 & + & -210I_2 & = & 0 \end{array}$$

The matrix-vector form is

$$\begin{vmatrix} -140 & 56 \\ 56 & -210 \end{vmatrix} \bullet \begin{vmatrix} I_1 \\ I_2 \end{vmatrix} = \begin{vmatrix} 5.0 \\ 0 \end{vmatrix}$$

with the units in amperes. The solution for this maxtrix equation is

$$\begin{vmatrix} I_1 \\ \\ I_2 \end{vmatrix} = \begin{vmatrix} -40.0 \times 10^{-3} \\ \\ -10.7 \times 10^{-3} \end{vmatrix}$$

The negative values for I_1 and I_2 indicate that the directions of the actual currents are opposite to the reference directions. From Ohm's Law for voltage,

$$\begin{aligned} V_{TH} &= V_{OL} \\ &= V_{CD} \\ &= -(I_2 \times R_5) \\ &= -[(-10.7 \times 10^{-3}\ \text{A}) \times (56\ \Omega)] \\ &= -[(-10.7 \times 10^{-3}) \times 56]\ (\text{A} \times \Omega) \\ &= -(-597 \times 10^{-3})\ \text{V} \\ &= 597\ \text{mV} \end{aligned}$$

To determine the Thevenin resistance, replace V_S with its internal resistance (a short) and determine the equivalent resistance R_{EQ} looking back from R_L into the circuit. Refer to Figure 9-33.

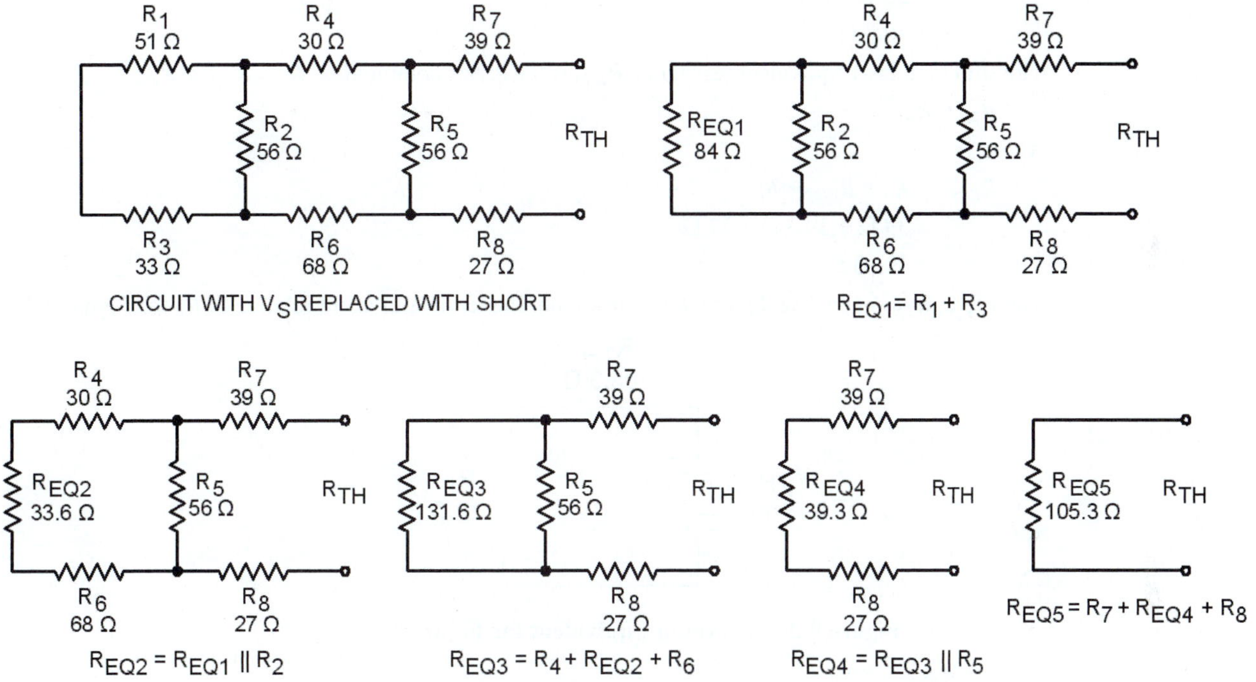

Figure 9-33: Circuit Analysis and Simplification Process for R_{TH}

The expression for the circuit is

$$R_{TH} = R_7 + R_5 \,||\, [R_4 + R_2 \,||\, (R_1 + R_3) + R_6] + R_8$$

The first equivalent resistance R_{EQ1} is the series combination $(R_1 + R_3)$.

$$R_{TH} = R_7 + R_5 \,||\, (R_4 + R_2 \,||\, R_{EQ1} + R_6) + R_8$$

where

$$\begin{aligned} R_{EQ1} &= R_1 + R_3 \\ &= 51\ \Omega + 33\ \Omega \\ &= 84\ \Omega \end{aligned}$$

The second equivalent resistance R_{EQ2} is the parallel combination $R_2 \,||\, R_{EQ1}$.

$$R_{TH} = R_7 + R_5 \,||\, (R_4 + R_{EQ2} + R_6) + R_8$$

where

R_{EQ2} $= R_2 \parallel R_{EQ1}$
$= 56\ \Omega) \parallel 84\ \Omega$
$= 33.6\ \Omega$

The third equivalent resistance R_{EQ3} is the series combination $R_4 + R_{EQ2} + R_6$.

$R_{TH} = R_7 + R_5 \parallel R_{EQ3} + R_8$

where

R_{EQ3} $= R_4 + R_{EQ2} + R_6$
$= 30\ \Omega + 33.6\ \Omega + 68\ \Omega$
$= 131.6\ \Omega$

The fourth equivalent resistance R_{EQ4} is the parallel combination $R_5 \parallel R_{EQ3}$.

$R_{TH} = R_7 + R_{EQ4} + R_8$

where

R_{EQ4} $= R_5 \parallel R_{EQ3}$
$= 56\ \Omega \parallel 131.6\ \Omega$
$= 39.3\ \Omega$

The fifth (and last) equivalent resistance R_{EQ5} is the series combination $R_7 + R_{EQ4} + R_8$.

$R_{TH} = R_{EQ5}$

where

R_{EQ5} $= R_7 + R_{EQ4} + R_8$
$= 39\ \Omega + 39.3\ \Omega + 27\ \Omega$
$= 105.3\ \Omega$

Therefore, $R_{TH} = 105.3\ \Omega$. Figure 9-34 shows the Thevenin equivalent relative to R_L for Figure 9-30.

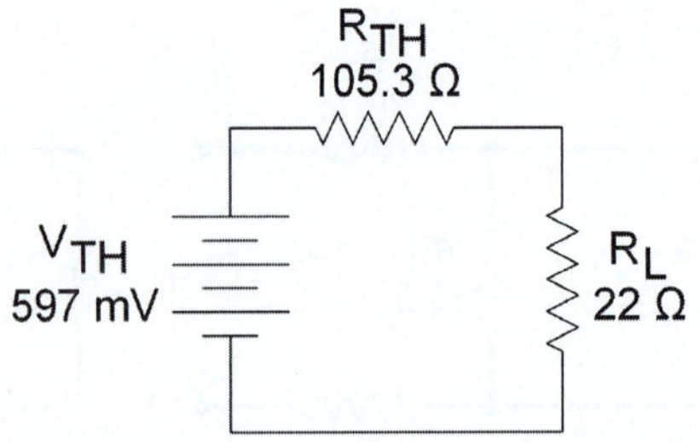

Figure 9-34: Thevenin Equivalent for Figure 9-30

9.2.4 Node Voltage Method

Problem: Use the node voltage method to determine the Norton equivalent relative to R_L for the circuit in Figure 9-35.

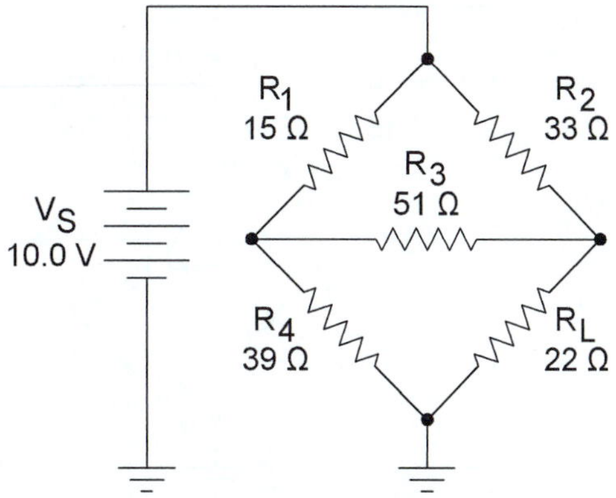

Figure 9-35: Node Voltage Method Example Circuit 3

Solution: To find the Norton current, first remove R_L and then use the node voltage method to determine the short circuit current I_{SC}. Identify and label all circuit nodes. Refer to Figure 9-36.

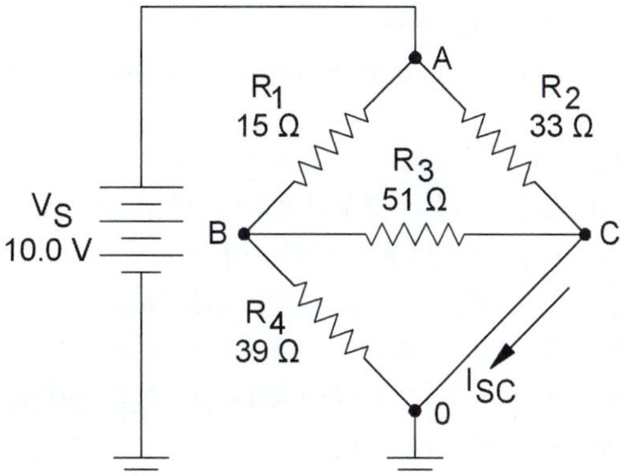

Figure 9-36: Labeled Nodes for Figure 9-35

As Figure 9-36 shows, the circuit contains four nodes. For this example, select the negative terminal of V_S as the reference node, labeled as node 0. Next, assign voltage designations for all nodes for which the voltage is not known. Refer to Figure 9-37.

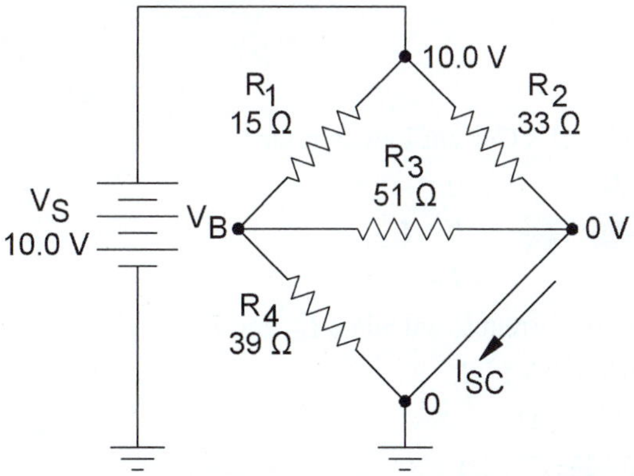

Figure 9-37: Assigned Voltage Designations for Figure 9-15

The voltage at node A relative to the reference node is known, as V_S holds it at 10.0 V more positive than the reference node, so the node is labeled with the actual voltage value. The voltage at node C is also known, because the short to node 0 holds it at 0 V, so the node is labeled with its actual voltage value. Node B is labeled with the voltage designation V_B, because its actual voltage is not known, so it is the only node to which branch currents are assigned. Refer to Figure 9-38.

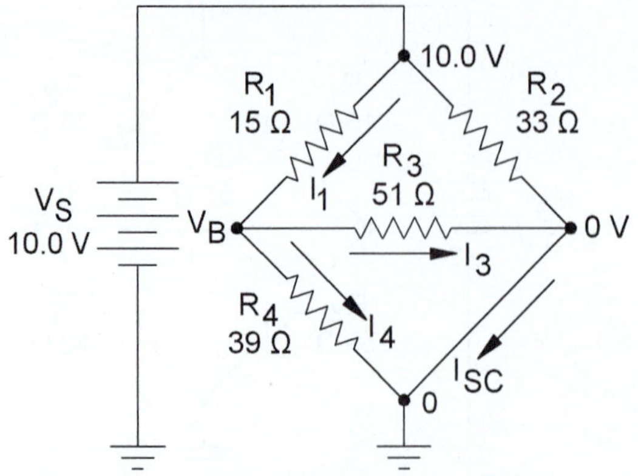

Figure 9-38: Assigned Branch Currents for Figure 9-35

Apply Kirchhoff's Current Law to each node for which the actual voltage is not known. For this example, let currents entering a node be positive and currents leaving a node be negative.

Node B: $I_1 + -I_3 + -I_4 = 0$

Use Ohm's Law to express each branch current in terms of the node voltages and resistances for node B.

$$I_1 + -I_3 + -I_4 = 0$$
$$[(10.0\text{ V} - V_B) / R_1] + -[(V_B - 0\text{ V}) / R_3] + -[(V_B - 0\text{ V}) / R_4] = 0$$
$$[(10.0\text{ V} - V_B) / R_1] + -(V_B / R_3) + -(V_B / R_4) = 0$$
$$(10.0\text{ V} / R_1) + -(V_B / R_1) + -(V_B / R_3) + -(V_B / R_4) = 0$$
$$(10.0\text{ V} / R_1) + -V_B \times [(1 / R_1) + (1 / R_3) + (1 / R_4)] = 0$$
$$(10.0\text{ V} / R_1) + -V_B \times [(1 / R_1) + (1 / R_3) + (1 / R_4)] + -(10.0\text{ V} / R_1) = 0 + -(10.0\text{ V} / R_1)$$
$$-V_B \times [(1 / R_1) + (1 / R_3) + (1 / R_4)] = -(10.0\text{ V} / R_1)$$

Use the resistor values to solve for V_B.

$$-V_B \times [(1 / 15\ \Omega) + (1 / 51\ \Omega) + (1 / 39\ \Omega)] = -(10.0\text{ V} / 15\ \Omega)$$
$$V_B \times -[(66.7 \times 10^{-3}\text{ S}) + (19.6 \times 10^{-3}\text{ S}) + (25.6 \times 10^{-3}\text{ S})] = -(10.0 / 15)(\text{V} / \Omega)$$
$$V_B \times (-112 \times 10^{-3}\text{ S}) / (-112 \times 10^{-3}\text{ S}) = (-667 \times 10^{-3}\text{ A}) / (-112 \times 10^{-3}\text{ S})$$
$$V_B = (-667 \times 10^{-3}\text{ A}) \times (-8.94\ \Omega)$$
$$V_B = (-667 \times 10^{-3} \times -8.94)(\text{A} \times \Omega)$$
$$V_B = 5.96\text{ V}$$

Next, apply Kirchhoff's Current Law to node C.

$$I_{R2} + I_3 + -I_{SC} = 0$$
$$I_{R2} + I_3 + -I_{SC} + I_{SC} = 0 + I_{SC}$$
$$I_{R2} + I_3 = I_{SC}$$

Use Ohm's Law for current to calculate I_{R2} and I_3.

$$I_{R2} = (10.0\text{ V} - 0\text{ V}) / R_2$$
$$= 10.0\text{ V} / 33\ \Omega$$
$$= (10.0 / 33)(\text{V} / \Omega)$$
$$= 303 \times 10^{-3}\text{ A}$$

and

$$I_3 = V_B / R_3$$
$$= 5.96\text{ V} / 51\ \Omega$$

$= 177 \times 10^{-3}$ A

Finally, substitute the values of I_{R2} and I_3 to solve for $I_N = I_{SC}$.

$$I_N = I_{SC} = I_{R2} + I_3$$
$$= (303 \times 10^{-3}\text{ A}) + (177 \times 10^{-3}\text{ A})$$
$$= 480 \times 10^{-3}\text{ A}$$
$$= 480\text{ mA}$$

To determine the Norton resistance R_N, replace V_S with its internal resistance (a short), and determine the equivalent resistance R_{EQ} looking back from R_L into the circuit. If necessary, redraw the circuit to determine the relationship of the resistances within the circuit as shown in Figure 9-39. The expression for the circuit is

$$R_N = R_2 \parallel (R_3 + R_1 \parallel R_4)$$

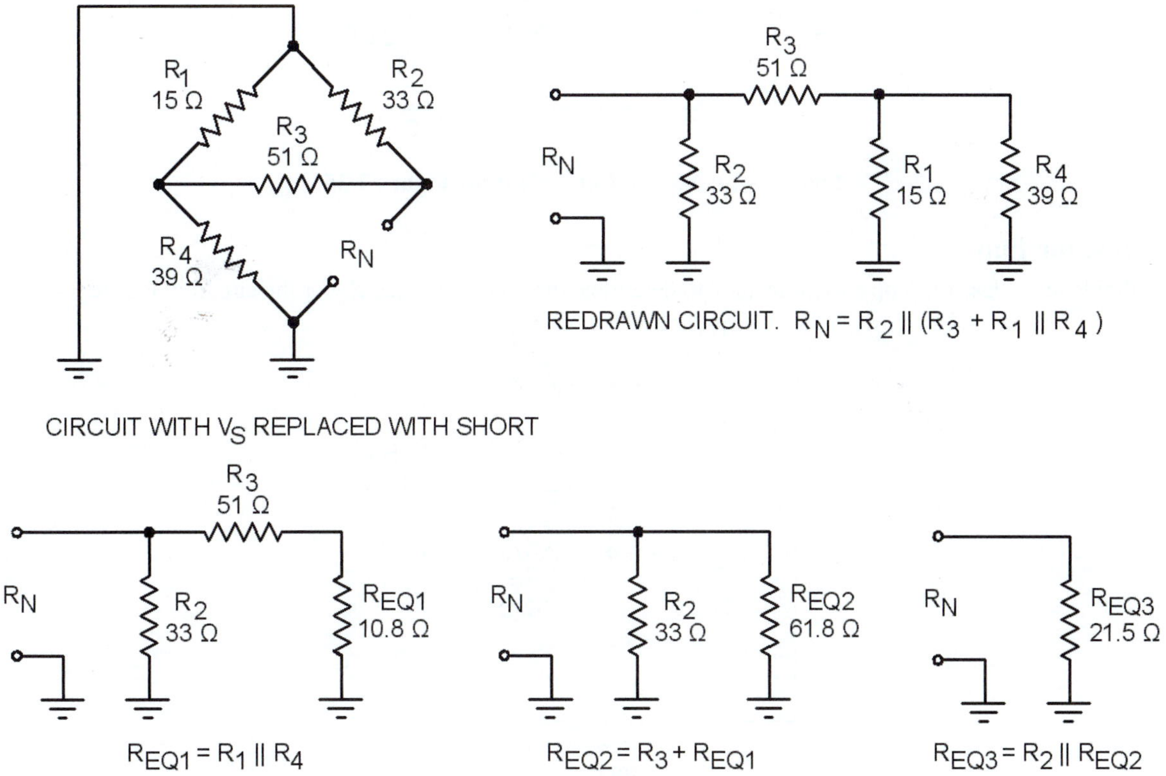

Figure 9-39: Circuit Analysis and Simplification Process for R_N

The first equivalent resistance R_{EQ1} is the parallel combination $R_1 \parallel R_4$.

$$R_N = R_2 \parallel (R_3 + R_{EQ1})$$

where

$$R_{EQ1} = R_1 \parallel R_4$$
$$= 15\ \Omega \parallel 39\ \Omega$$
$$= 10.8\ \Omega$$

The second equivalent resistance R_{EQ2} is the series combination $R_3 + R_{EQ1}$.

$$R_N = R_2 \parallel R_{EQ2}$$

where

$$R_{EQ2} = R_3 + R_{EQ1}$$
$$= 51\ \Omega + 10.8\ \Omega$$

$= 61.8\ \Omega$

The third (and last) equivalent resistance R_{EQ3} is the parallel combination $R_2 \parallel R_{EQ2}$.

$$R_N = R_{EQ3}$$

where

$$R_{EQ3} = R_2 \parallel R_{EQ2}$$
$$= 33\ \Omega \parallel 61.8\ \Omega$$
$$= 21.5\ \Omega$$

Therefore, $R_N = 21.5\ \Omega$. Figure 9-40 shows the Norton equivalent relative to R_L for Figure 9-35.

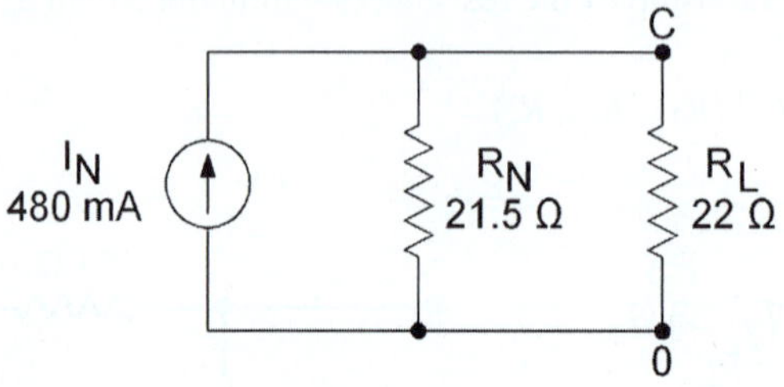

Figure 9-40: Norton Equivalent for Figure 9-35

9.3 Just for Fun

Problem: Use the loop current method to determine the total resistance R_T for the circuit in Figure 9-41.

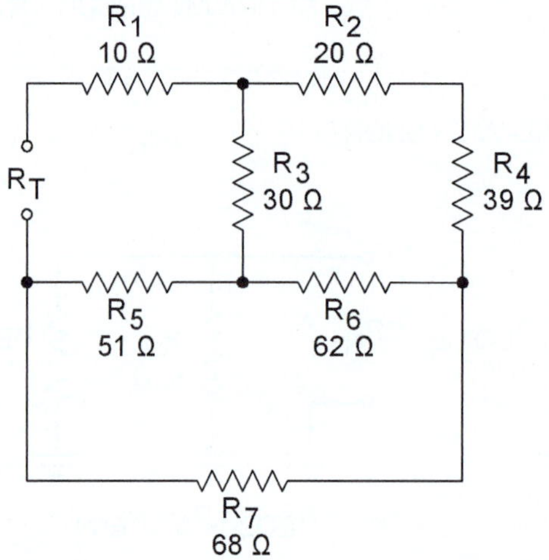

Figure 9-41: Loop Current Method Circuit Resistance Example

Solution: Ohm's Law for resistance can calculate the total circuit resistance R_T from

$$R_T = V_S / I_T$$

For this example, connect a 1 V test voltage source V_S across the circuit, assign current loops as shown in Figure 9-42, and use the loop current method to determine I_T.

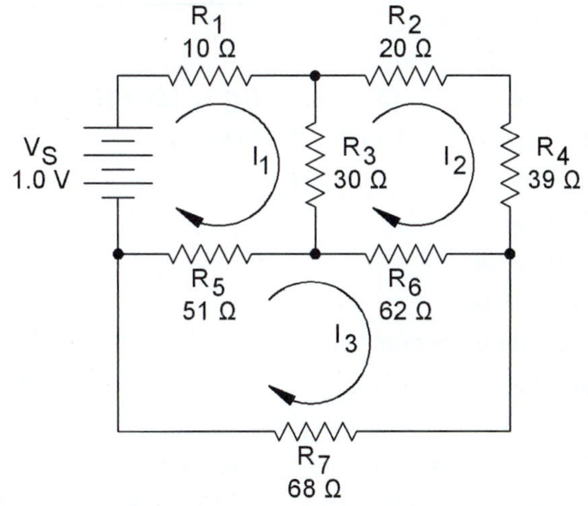

Figure 9-42: Circuit with Test Voltage Source and Loop Currents for Figure 9-41

Figure 9-42 shows that the circuit has three non-redundant loops. Next, assign resistor voltage polarities based on the reference directions for I_1 and I_2. Refer to Figure 9-43.

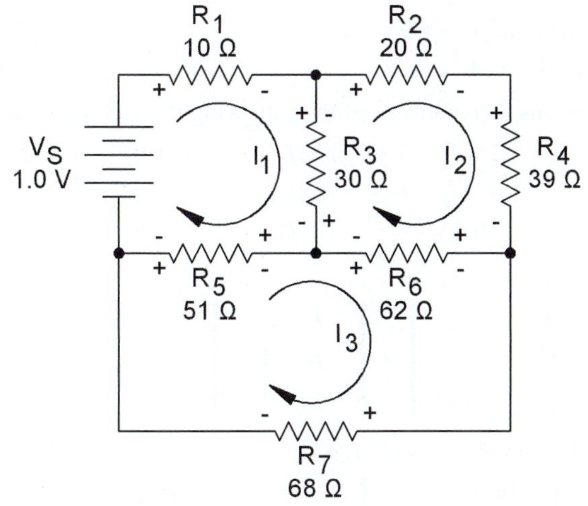

Figure 9-43: Resistor Voltage Polarities for Loop Currents in Figure 9-42

Apply Kirchhoff's Voltage law and Ohm's Law to the top left-hand loop by working in the direction of the loop current, omitting units for simplicity.

$$V_S - V_{R1} - V_{R3} - V_{R5} = 0$$

$$V_S + -V_{R1} + -V_{R3} + -V_{R5} = 0$$

$$V_S + -(R_1 \times I_1) + -[R_3 \times (I_1 + -I_2)] + -[R_5 \times (I_1 + -I_3)] = 0$$

$$V_S + -(R_1 \times I_1) + -(R_3 \times I_1) + (R_3 \times I_2) + -(R_5 \times I_1) + (R_5 \times I_3) = 0$$

$$V_S + -[(R_1 + R_3 + R_5) \times I_1] + (R_3 \times I_2) + (R_5 \times I_3) = 0$$

$$V_S + -[(R_1 + R_3 + R_5) \times I_1] + (R_3 \times I_2) + (R_5 \times I_3) + -V_S = 0 + -V_S$$

$$-[(R_1 + R_3 + R_5) \times I_1] + (R_3 \times I_2) + (R_5 \times I_3) = -V_S$$

$$-[(10 + 30 + 51) \times I_1] + (30 \times I_2) + (51 \times I_3) = -1.0$$

$$-(91 \times I_1) + (30 \times I_2) + (51 \times I_3) = -1.0$$

$$-91I_1 + 30I_2 + 51I_3 = -1.0$$

Repeat the process for the top right-hand loop.

$$-V_{R2} - V_{R4} - V_{R6} - V_{R3} = 0$$
$$-V_{R2} + -V_{R4} + -V_{R6} + -V_{R3} = 0$$
$$-(R_2 \times I_2) + -(R_4 \times I_2) + -[R_6 \times (I_2 + -I_3)] + -[R_3 \times (I_2 + -I_1)] = 0$$
$$-[(R_2 + R_4 + R_6 + R_3) \times I_2] + (R_6 \times I_3) + (R_3 \times I_1) = 0$$
$$-[(20 + 39 + 62 + 30) \times I_2] + (62 \times I_3) + (30 \times I_1) = 0$$
$$-(151 \times I_2) + (62 \times I_3) + (30 \times I_1) = 0$$
$$-151I_2 + 62I_3 + 30I_1 = 0$$
$$30I_1 + -151I_2 + 62I_3 = 0$$

Repeat the process for the bottom loop.

$$-V_{R5} - V_{R6} - V_{R7} = 0$$
$$-V_{R5} + -V_{R6} + -V_{R7} = 0$$
$$-[R_5 \times (I_3 + -I_1)] + -[R_6 \times (I_3 + -I_2)] + -(R_7 \times I_3) = 0$$
$$-(R_5 \times I_3) + (R_5 \times I_1) + -(R_6 \times I_3) + (R_6 \times I_2) + -(R_7 \times I_3) = 0$$
$$-(R_5 \times I_3) + -(R_6 \times I_3) + -(R_7 \times I_3) + (R_5 \times I_1) + (R_6 \times I_2) = 0$$
$$-[(R_5 + R_6 + R_7) \times I_3] + (R_5 \times I_1) + (R_6 \times I_2) = 0$$
$$-[(51 + 62 + 68) \times I_3] + (51 \times I_1) + (62 \times I_2) = 0$$
$$-(181 \times I_3) + (51 \times I_1) + (62 \times I_2) = 0$$
$$(51 \times I_1) + (62 \times I_2) + -(181 \times I_3) = 0$$
$$51I_1 + 62I_2 + -181I_3 = 0$$

The set of equations then consists of the following three equations in three unknowns:

$$\begin{array}{rcrcrcr} -91I_1 & + & 30I_2 & + & -84I_3 & = & -1.0 \\ 30I_1 & + & -151I_2 & + & 62I_3 & = & 0 \\ 51I_1 & + & 62I_5 & + & -181I_3 & = & 0 \end{array}$$

The matrix-vector form is

$$\begin{vmatrix} -91 & 30 & 51 \\ 30 & -151 & 62 \\ 51 & 62 & -181 \end{vmatrix} \cdot \begin{vmatrix} I_1 \\ I_2 \\ I_3 \end{vmatrix} = \begin{vmatrix} -1.0 \\ 0 \\ 0 \end{vmatrix}$$

The solution for this maxtrix equation is

$$\begin{vmatrix} I_1 \\ I_2 \\ I_3 \end{vmatrix} = \begin{vmatrix} 16.9 \times 10^{-3} \\ 6.17 \times 10^{-3} \\ 6.87 \times 10^{-3} \end{vmatrix}$$

where units are in amperes. The values are all positive, so the directions for the actual currents are the same as the assumed reference directions for the corresponding loop currents. For the circuit $I_T = I_1$, so apply Ohm's Law for resistance to calculate R_T.

$$\begin{aligned} R_T &= V_S / I_T \\ &= 1.0\text{ V} / 16.9\text{ mA} \\ &= 1.0\text{ V} / (16.9 \times 10^{-3}\text{ A}) \\ &= [1.0 / (16.9 \times 10^{-3})]\ (\text{V} / \text{A}) \\ &= 59.3\ \Omega \end{aligned}$$

The total resistance R_T is **59.3 Ω**.

10. Magnetism and Electromagnetism

10.1 Basic

10.1.1 The Magnetic Field

1. Problem: An electromagnet creates 10 lines of magnetic force. If the flux density $\boldsymbol{B}$ is 4 μT, what is the cross-sectional area $\boldsymbol{A}$ of the magnetic field?

   Solution: 1 Wb of magnetic flux φ is equal to 1×10^8 LMF* (lines of magnetic force). The flux in Wb for 10 LMF is equal to

   $$\varphi = 10 \text{ LMF} \times [1 \text{ Wb} / (1 \times 10^8 \text{ LMF})]$$
   $$= [(10 \times 1) / (1 \times 10^8)] [(\text{LMF} \times \text{Wb}) / \text{LMF}]$$
   $$= 1 \times 10^{-7} \text{ Wb}$$

   The magnetic flux density of a magnetic field is equal to

   $$\boldsymbol{B} = \varphi / \boldsymbol{A}$$

   so

   $$\boldsymbol{B} \times \boldsymbol{A} = (\varphi / \boldsymbol{A}) \times \boldsymbol{A}$$
   $$(\boldsymbol{B} \times \boldsymbol{A}) / \boldsymbol{B} = \varphi / \boldsymbol{B}$$
   $$\boldsymbol{A} = \varphi / \boldsymbol{B}$$

   For $\boldsymbol{B}$ = 4 μT and $\varphi = 1 \times 10^{-7}$ Wb

   $$\boldsymbol{A} = \varphi / \boldsymbol{B}$$
   $$= (1 \times 10^{-7} \text{ Wb}) / 4\ \mu\text{T}$$
   $$= (1 \times 10^{-7} \text{ Wb}) / (4 \times 10^{-6} \text{ T})$$
   $$= [(1 \times 10^{-7}) / (4 \times 10^{-6})] (\text{Wb} / \text{T})$$
   $$= 25 \times 10^{-3} [\text{Wb} / (\text{Wb} / \text{m}^2)]$$
   $$= 25 \times 10^{-3} [\text{Wb} \times (\text{m}^2 / \text{Wb})]$$
   $$= 25 \times 10^{-3} \text{ m}^2$$

   The cross sectional area of the magnetic field is $\mathbf{25 \times 10^{-3}\ m^2}$.

   (*Note: LMF is not an official acronym and is used in this section for convenience).

2. Problem: The magnetic flux density $\boldsymbol{B}$ of a magnetic field is 3 T. A magnetic lens constricts the field so that the cross-sectional area $\boldsymbol{A}$ through which the flux φ passes is halved to $\boldsymbol{A_{MOD}}$. What is the flux density $\boldsymbol{B_{MOD}}$ of the constricted field?

   Solution: The original magnetic flux density $\boldsymbol{B}$ of a magnetic field with cross-sectional area $\boldsymbol{A}$ and flux φ is equal to

   $$\boldsymbol{B} = \varphi / \boldsymbol{A}$$
   $$= 3 \text{ T}$$

   The magnetic flux density $\boldsymbol{B_{MOD}}$ of the constricted magnetic field with cross-sectional area A_{MOD} is

   $$\boldsymbol{B_{MOD}} = \varphi / \boldsymbol{A_{MOD}}$$

   The field flux is unchanged and $\boldsymbol{A_{MOD}} = A / 2$, so

   $$\boldsymbol{B_{MOD}} = \varphi / \boldsymbol{A_{MOD}}$$
   $$= \varphi / (\boldsymbol{A} / 2)$$
   $$= \varphi \times (2 / \boldsymbol{A})$$
   $$= 2 \times (\varphi / \boldsymbol{A})$$
   $$= 2 \times \boldsymbol{B}$$
   $$= 2 \times 3 \text{ T}$$

$= 6\ \text{T}$

The magnetic flux density of the constricted field is **6 T**.

10.1.2 Electromagnetism

1. Problem: A science project uses an electromagnet that consists of a steel nail wrapped with 10 turns of copper wire. When 100 mA of current $\boldsymbol{I}$ flows through the windings, the electromagnet creates 1.2 μWb of magnetic flux φ. What is the reluctance $\mathfrak{R}$ of the steel nail?

   Solution: The reluctance $\mathfrak{R}$ is

$$\mathfrak{R} = \text{mmf} / \varphi$$

where the mmf in ampere-turns is

$$\begin{aligned}\text{mmf} &= \boldsymbol{I} \times \boldsymbol{N}\\ &= 100\ \text{mA} \times 10\ \text{turns}\\ &= (100 \times 10^{-3}\ \text{A})\ (10\ \text{t})\\ &= [(100 \times 10^{-3}) \times 10]\ (\text{A} \times \text{t})\\ &= 1.0\ \text{At}\end{aligned}$$

For $\varphi = 1.2\ \mu\text{Wb}$

$$\begin{aligned}\mathfrak{R} &= \text{mmf} / \varphi\\ &= (1.0\ \text{At}) / (1.2\ \mu\text{Wb})\\ &= (1.0\ \text{At}) / (1.2 \times 10^{-6}\ \text{Wb})\\ &= [1.0 / (1.2 \times 10^{-6})]\ (\text{At} / \text{Wb})\\ &= 833 \times 10^{3}\ \text{At} / \text{Wb}\end{aligned}$$

The reluctance of the steel nail is $\mathbf{833 \times 10^{3}\ At / Wb}$.

2. Problem: The science student in Problem 1 above decides to double the magnetic flux of her science project by doubling the length $\boldsymbol{l}$ of the nail so that she can wrap twice as many turns of wire around it. Show that, if all other factors remain the same as before, the flux φ will not change.

   Solution: The reluctance of a material with length $\boldsymbol{l}$, permeability μ, and cross-sectional area $\boldsymbol{A}$ is

$$\mathfrak{R} = \boldsymbol{l} / (\mu \times \boldsymbol{A})$$

so

$$\begin{aligned}\mathfrak{R} &= \text{mmf} / \varphi\\ \mathfrak{R} \times \varphi &= (\text{mmf} / \varphi) \times \varphi\\ (\mathfrak{R} \times \varphi) / \mathfrak{R} &= \text{mmf} / \mathfrak{R}\\ \varphi &= \text{mmf} / [\boldsymbol{l} / (\mu \times \boldsymbol{A})]\\ &= \text{mmf} \times [(\mu \times \boldsymbol{A}) / \boldsymbol{l}]\end{aligned}$$

The magnetomotive force mmf for $\boldsymbol{N}$ turns and current $\boldsymbol{I}$ is

$$\text{mmf} = \boldsymbol{N} \times \boldsymbol{I}$$

so

$$\begin{aligned}\varphi &= (\boldsymbol{N} \times \boldsymbol{I}) \times [(\mu \times \boldsymbol{A}) / \boldsymbol{l}]\\ &= (\boldsymbol{N} \times \boldsymbol{I} \times \mu \times \boldsymbol{A}) / \boldsymbol{l}\end{aligned}$$

Calculate the new flux φ_{NEW} when the length $\boldsymbol{l}_{NEW}$ of the new nail is $2 \times \boldsymbol{l}_{ORG}$ of the original nail and the new number of turns $\boldsymbol{N}_{NEW}$ is $2 \times \boldsymbol{N}_{ORG}$.

$$\begin{aligned}\varphi_{NEW} &= (\boldsymbol{N}_{NEW} \times \boldsymbol{I} \times \mu \times \boldsymbol{A}) / \boldsymbol{l}_{NEW}\\ &= [(2 \times \boldsymbol{N}_{ORG}) \times \boldsymbol{I} \times \mu \times \boldsymbol{A}] / (2 \times \boldsymbol{l}_{ORG})\\ &= (\boldsymbol{N}_{ORG} \times \boldsymbol{I} \times \mu \times \boldsymbol{A}) / \boldsymbol{l}_{ORG}\end{aligned}$$

$= \varphi_{ORG}$

Doubling the length of the nail and number of windings will not increase the flux φ.

3. Problem: If doubling the magnetomotive force F_m increases the magnetic flux φ by 50%, what is the change in reluctance $\mathfrak{R}$?

Solution: From Hopkinson's Law (Ohm's Law for magnetic devices),

$$\varphi = F_m / \mathfrak{R}$$
$$\varphi \times \mathfrak{R} = (F_m / \mathfrak{R}) \times \mathfrak{R}$$
$$(\varphi \times \mathfrak{R}) / \varphi = F_m / \varphi$$
$$\mathfrak{R} = F_m / \varphi$$

From the problem statement,

$$F_{m\,(new)} = 2 \times F_{m\,(org)}$$

and

$$\begin{aligned}\varphi_{(new)} &= \varphi + (\varphi \times 50\%)\\ &= \varphi + [\varphi \times (50 / 100)]\\ &= (\varphi \times 1) + (\varphi \times 0.5)\\ &= \varphi \times (1 + 0.5)\\ &= \varphi \times 1.5\end{aligned}$$

so

$$\begin{aligned}\mathfrak{R}_{(new)} &= F_{m\,(new)} / \varphi_{(new)}\\ &= [2 \times F_{m\,(org)}] / [1.5 \times \varphi_{(org)}]\\ &= [F_{m\,(org)}] / [\varphi_{(org)}] \times (2 / 1.5)\\ &= \mathfrak{R}_{(org)} \times 1.33\\ &= \mathfrak{R}_{(org)} \times (1 + 0.33)\\ &= [\mathfrak{R}_{(org)} \times 1] + [\mathfrak{R}_{(org)} \times 0.33]\\ &= \mathfrak{R}_{(org)} + [\mathfrak{R}_{(org)} \times (33 / 100)]\\ &= \mathfrak{R}_{(org)} + [\mathfrak{R}_{(org)} \times 33\%]\end{aligned}$$

The reluctance has **increased by 33%**.

10.1.3 Magnetic Hysteresis

1. Problem: The locking catch to a wind generator is released by actuating a solenoid. If the solenoid consists of 100 turns of wire wound around a 3.0 cm steel rod and is actuated by 10 mA of current, what is the magnetic field intensity H when the solenoid is actuated?

Solution: The magnetic field intensity H is

$$H = F_m / l$$

where magnetomotive force F_m is the product of the number of winding turns N and the winding current I. From this,

$$\begin{aligned}H &= F_m / l\\ &= (N \times I) / l\\ &= (100 \text{ turns} \times 10 \text{ mA}) / 3.0 \text{ cm}\\ &= [\,(10 \times 10^{-3} \text{ A}) \times 100 \text{ turns}] / (3.0 \times 10^{-2} \text{ m})\\ &= \{[100 \times (10 \times 10^{-3})] / (3.0 \times 10^{-2})\}\ [(\text{A} \times \text{t})] / \text{m}]\\ &= 33.3 \text{ At / m}\end{aligned}$$

The magnetic field intensity for the solenoid is **33 At / m**.

2. Problem: The memory cell of a prototype magnetic core memory consists of a single wire loop wrapped around a rod of magnetic material. The material requires a minimum magnetizing force (magnetic field intensity) $H_{MIN} = 5.0 \times 10^{-3}$ At / m to store data. If the length l of the rod that forms the cell is 25 nm long, how much current I must flow through the wire loop to store information?

Solution: Magnetic field intensity H is

$$H = F_m / l$$
$$= (N \times I) / l$$

so

$$H \times l = [(N \times I) / l] \times l$$
$$(H \times l) / N = (N \times I) / N$$
$$(H \times l) / N = I$$

From this

$$I = [(5 \times 10^{-3} \text{ At / m}) \times (25 \text{ nm})] / (1 \text{ turn})$$
$$= [(5 \times 10^{-3} \text{ At / m}) \times (25 \times 10^{-9} \text{ m})] / (1 \text{ turn})$$
$$= \{[(5 \times 10^{-3}) \times (25 \times 10^{-9})] / (1)\} \{[(\text{At / m}) \times \text{m}] / \text{turn}\}$$
$$= 125 \times 10^{-12} [(\text{A} \times \text{turn}) / \text{turn}]$$
$$= 125 \times 10^{-12} \text{ A}$$
$$= 125 \text{ pA}$$

The magnetic memory cell requires ***125 pA*** of current to store information.

10.1.4 Electromagnetic Induction

1. Problem: Figure 10-1 shows the end view of a current-carrying conductor in a magnetic field. If the direction of the current is out of the page as shown, in which direction will the conductor move?

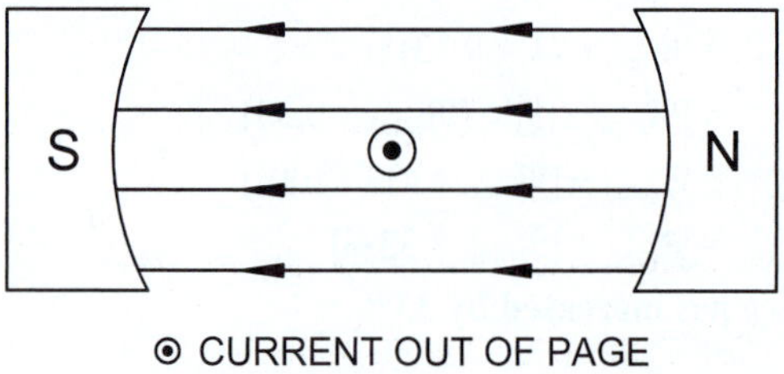

Figure 10-1: Electromagnetic Induction Example

Solution: The first step is to use the right-hand rule to determine the direction of the magnetic field created by the current in the conductor. Refer to Figure 10-2.

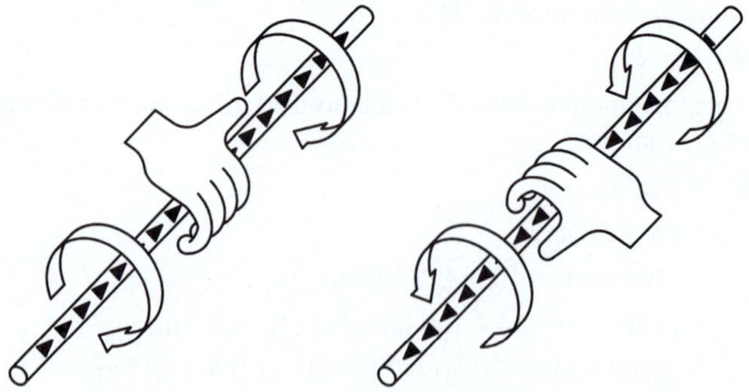

Figure 10-2: Application of the Right-Hand Rule

To use the right-hand rule to find the direction of the magnetic field created by current flowing through a conductor, point the thumb of your right hand in the direction of the current. The direction of the fingers of your right hand then show the direction (north to south) of the magnetic field. As Figure 10-2 shows, current flowing away from you (into the page) sets up a clockwise (CW) magnetic field, while current flowing towards you (out of the page) sets up a counter-clockwise (CCW) magnetic field. Because the current in Figure 10-1 is flowing out of the page, it sets up a CCW magnetic field. Refer to Figure 10-3.

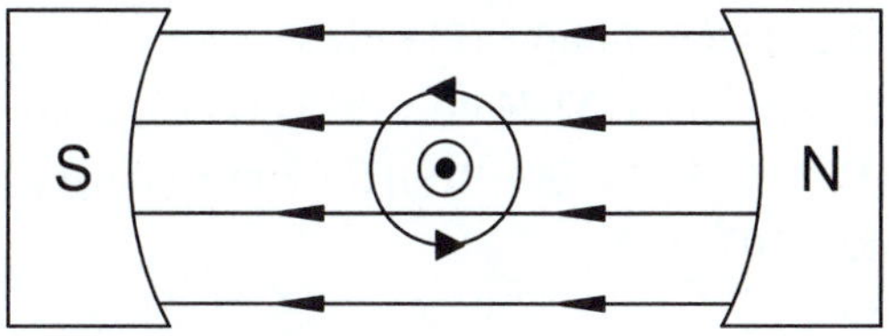

Figure 10-3: Direction of Magnetic Field for Figure 10-1

As Figure 10-3 shows, the polarity of the top of the magnetic field created by the current is the same as the polarity of the field in which the conductor is located and so the two fields repel each other. Conversely, the polarity of the bottom is opposite to the polarity of the field in which the conductor is located and so the two fields attract each other. Because the fields repel at the top of the conductor and attract at the bottom, the force will cause the conductor to move **down**.

Note: There is another way to apply the right-hand rule to problems of this type. To do so, extend the fingers of your right hand in the direction of the current and curl them in the direction of the magnetic field (north to south). Your thumb will point in the direction in which the conductor will move.

2. Problem: A magnetic field is 2.5 cm wide with a magnetic field density $\boldsymbol{B}_\perp = 1.25$ T ($\boldsymbol{B}_\perp$ is the magnetic flux density that is normal, or perpendicular, to the direction of motion). How fast must a conductive wire move through the field to induce an instantaneous voltage $\boldsymbol{v}_{ind}$ of 1.0 V?

Solution: Induced voltage $\boldsymbol{v}_{\mathrm{ind}}$ is equal to

$$\boldsymbol{v}_{ind} = \boldsymbol{B}_\perp \times \boldsymbol{l} \times \mathrm{v} \qquad \text{where } \boldsymbol{v}_{ind} \text{ is in V, } \boldsymbol{B}_\perp \text{ is in T, } \boldsymbol{l} \text{ is in m, and } \mathbf{v} \text{ is in m / s}$$

so

$$\boldsymbol{v}_{ind} / \boldsymbol{B}_\perp = (\boldsymbol{B} \times \boldsymbol{l} \times \mathbf{v}) / \boldsymbol{B}$$

$$(\boldsymbol{v}_{ind} / \boldsymbol{B}_\perp) / \boldsymbol{l} = (\boldsymbol{l} \times \mathbf{v}) / \boldsymbol{l}$$

$$\boldsymbol{v}_{ind} / (\boldsymbol{B}_\perp \times \boldsymbol{l}) = \mathbf{v}$$

From this,

$$\begin{aligned}\mathbf{v} &= (1.0\ \mathrm{V}) / [(1.25\ \mathrm{T}) \times (2.5\ \mathrm{cm})] \\ &= (1.0\ \mathrm{V}) / [(1.25\ \mathrm{T}) \times (2.5 \times 10^{-2}\ \mathrm{m})] \\ &= [1.0\ \mathrm{V} / [1.25 \times (2.5 \times 10^{-2})]\ [\mathrm{V} / (\mathrm{T} \times \mathrm{m})] \\ &= [1.0\ \mathrm{V} / [1.25 \times (2.5 \times 10^{-2})]\ [\mathrm{V} / (\mathrm{T} \times \mathrm{m})] \\ &= 32.0\ \mathrm{m} / \mathrm{s}\end{aligned}$$

The wire must move through the field at **32 m /s** for an induced voltage of 1.0 V.

3. Problem: An SR-71 Blackbird on a reconnaissance mission is flying at Mach 3 (three times the speed of sound, or about 3310 kph). What is the induced voltage, $\boldsymbol{v}_{ind}$, for the plane if the length, $\boldsymbol{l}$, of the SR-71 is 32.74 m, and the earth's magnetic field, $\boldsymbol{B}_\perp$, at the plane's altitude is 15 μT?

Solution: Induced voltage $\boldsymbol{v}_{ind}$ is equal to

$$\boldsymbol{v}_{\mathrm{ind}} = \boldsymbol{B}_\perp \times \boldsymbol{l} \times \mathbf{v} \qquad \text{where } \boldsymbol{v}_{ind} \text{ is in V, } \boldsymbol{B}_\perp \text{ is in T, } \boldsymbol{l} \text{ is in m, and } \mathbf{v} \text{ is in m / s}$$

Converting velocity from kph to m / s gives

$$\begin{aligned}\mathbf{v} &= 3310 \text{ kph}\\ &= (3310 \text{ km} / 1 \text{ hr}) \times (1 \text{ hr} / 60 \text{ min}) \times (1 \text{ min} / 60 \text{ sec})\\ &= [(3310 \times 10^3 \text{ m}) / 1 \text{ hr}] \times (1 \text{ hr} / 60 \text{ min}) \times (1 \text{ min} / 60 \text{ sec})\\ &= \{[(3310 \times 10^3) \times 1 \times 1] / (1 \times 60 \times 60)\} [(\text{m} \times \text{hr} \times \text{min}) / (\text{hr} \times \text{min} \times \text{sec})]\\ &= 919 \text{ m} / \text{s}\end{aligned}$$

From this,

$$\begin{aligned}v_{ind} &= (15\ \mu\text{T}) \times (32.74 \text{ m}) \times (919 \text{ m} / \text{s})\\ &= (15 \times 10^{-6} \text{ T}) \times (32.74 \text{ m}) \times (919 \text{ m} / \text{s})\\ &= [(15 \times 10^{-6}) \times 32.74 \times 919)] [(\text{T} \times \text{m} \times \text{m}) / \text{s}]\\ &= 452 \times 10^{-3} \text{ V}\\ &= 452 \text{ mV}\end{aligned}$$

The induced voltage for an SR-71 travelling at Mach 3 is **450 mV**.

4. Problem: A hospital maintenance worker accidentally leaves a plastic ball-point pen inside an MRI (magnetic resonance imaging) machine. The pen contains a metal spring that is effectively an 8-turn winding. If the MRI machine creates a 3.0 T magnetic flux density, $\boldsymbol{B}$, over a 1.5 m^2 area, $\boldsymbol{A}$, in 2.5 seconds when first powered up, what is the induced voltage, v_{ind}, for the spring when the MRI machine powers up?

Solution: From Faraday's Law,

$v_{ind} = N \times (d\varphi / dt)$ where v_{ind} is in V, N is the number of turns, and $d\varphi / dt$ is in Wb/ s

When the machine is turned on, the magnetic field $\boldsymbol{B}$ changes from 0 T to 3 T in 2.5 seconds.

$$\begin{aligned}\boldsymbol{B}\text{ (off)} &= 0 \text{ T}\\ \boldsymbol{B}\text{ (2.5 s)} &= 3.0 \text{ T}\end{aligned}$$

From the definition of magnetic flux density,

$$\begin{aligned}\boldsymbol{B} &= \varphi / \boldsymbol{A}\\ \boldsymbol{B} \times \boldsymbol{A} &= (\varphi / \boldsymbol{A}) \times \boldsymbol{A}\\ \boldsymbol{B} \times \boldsymbol{A} &= \varphi\end{aligned}$$

Use the expression for the flux, φ, to calculate the change in flux.

$$\begin{aligned}d\varphi / dt &= d(\boldsymbol{B} / \boldsymbol{A}) / dt\\ &= \{[\boldsymbol{B}\text{ (2.5 s)} / \boldsymbol{A}] - [\boldsymbol{B}\text{ (off)} / \boldsymbol{A}]\} / 2.5 \text{ s}\\ &= \{[\boldsymbol{B}\text{ (2.5 s)} - \boldsymbol{B}\text{ (off)}] / \boldsymbol{A}\} / 2.5 \text{ s}\\ &= [(3.0 \text{ T} - 0 \text{ T}) / 1.5 \text{ m}^2] / 2.5 \text{ s}\\ &= (3.0 \text{ T} / 1.5 \text{ m}^2) / 2.5 \text{ s}\\ &= [3.0 / (1.5 \times 2.5)] [(\text{T} / \text{m}^2) / \text{s}]\\ &= 800 \times 10^{-3} \text{ Wb} / \text{s}\end{aligned}$$

Calculate the induced voltage from Faraday's Law.

$$\begin{aligned}v_{ind} &= (8 \text{ turns}) \times (800 \times 10^{-3} \text{ Wb} / \text{s})\\ &= [8 \times (800 \times 10^{-3})] [\text{turns} \times (\text{Wb} / \text{s})]\\ &= 6.4 \text{ V}\end{aligned}$$

The induced voltage across the spring is **6.4 V**.

5. Problem: A copper wire is placed between two 1-inch by 1-inch neodymium magnets that set up a magnetic flux density $\boldsymbol{B}_{\perp}$ of 0.45 T between them. How much current, $\boldsymbol{I}$, must flow through the wire to create a force, $\boldsymbol{F}$, of 2.5 mN on the wire?

Solution: The force $\boldsymbol{F}$ on a conductor carrying current $\boldsymbol{I}$ with length $\boldsymbol{l}$ exposed to a magnetic field with flux density $\boldsymbol{B}_{\perp}$ is

$F \;= B_{\perp} \times I \times l$ where F is in N, B is in T, I is in A, and l is in m

Use this equation to solve for the current, I.

$F / B_{\perp} = (B_{\perp} \times I \times l) / B_{\perp}$

$(F / B_{\perp}) / l = (I \times l) / l$

$(F / B_{\perp}) / l = I$

$F / (B_{\perp} \times l) = I$

Use the values of F, $B_{\perp}$, and l to calculate I.

$I = F / (B_{\perp} \times l)$

$= 2.5 \text{ mN} / (0.45 \text{ T} \times 1 \text{ in})$

$= (2.5 \times 10^{-3} \text{ N}) / \{0.45 \text{ T} \times [1 \text{ in} \times (25.4 \text{ mm / in})]\}$

$= (2.5 \times 10^{-3} \text{ N}) / (0.45 \text{ T} \times \{(1 \times 25.4) [\text{in} \times (\text{mm / in})]\})$

$= (2.5 \times 10^{-3} \text{ N}) / (0.45 \text{ T} \times 25.4 \text{ mm})$

$= (2.5 \times 10^{-3} \text{ N}) / [0.45 \text{ T} \times (25.4 \times 10^{-3} \text{ m})]$

$= \{(2.50 \times 10^{-3}) / [0.45 \times (25.4 \times 10^{-3})]\} [\text{N} / (\text{T} \times \text{m})]$

$= 219 \times 10^{-3} \text{ A}$

$= 219 \text{ mA}$

The conductor must carry **220 mA** to create a force of 2.5 mN on it.

10.1.5 The DC Generator

1. Problem: Determine whether V_{AB} is positive or negative for the DC generator in Figure 10-4.

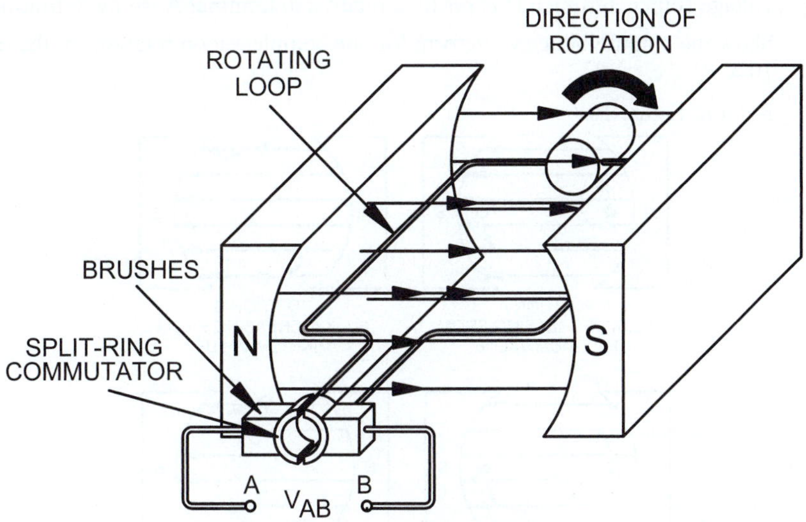

Figure 10-4: DC Generator Example

Solution: Lenz's Law states that an induced voltage will oppose the change that induced it. As the loop rotates in the magnetic field, it opposes the rotation by developing a magnetic field around the conductive loop. There are two different motions that the loop opposes. Refer to Figure 10-5.

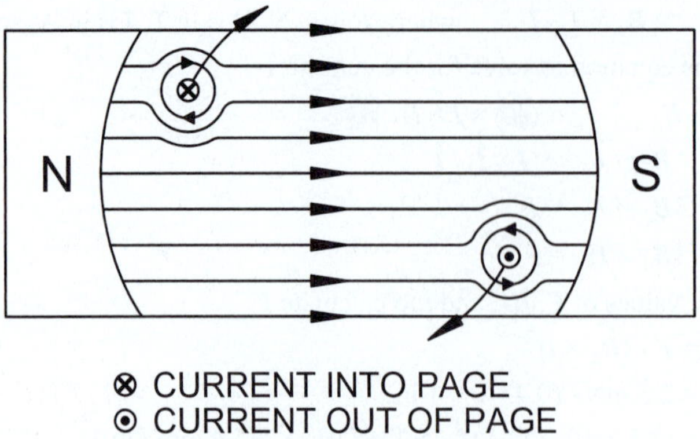

Figure 10-5: View from Commutator of Loop Rotation in the Magnetic Field

At any give time two segments of rotating loop are moving in opposite directions through the magnetic field. Both loop segments oppose the rotation by generating a magnetic field to oppose the magnetic flux in front of the segment and attract the magnetic flux behind the segment. As Figure 10-5 shows, the segment moving upward creates a clockwise field to oppose the field above it and attract the field below it. At the same time, the segment moving downward creates a counter-clockwise field to oppose the field below it and attract the field above it. From the right-hand rule, the direction of the current for the segment moving upward is away from the commutator (into the page) and that for the segment moving downward is towards the commutator (out of the page). Consequently current flows from contact A towards terminal B inside the loop, and from contact B towards contact A external to the loop. Because terminals A and B are for a voltage source, terminal B is positive relative to terminal A, so by definition V_{AB} is **negative**.

2. Problem: Show the output voltage waveform for one complete loop rotation for the dc generator in Figure 10-4.

Solution: Refer to Figure 10-6.

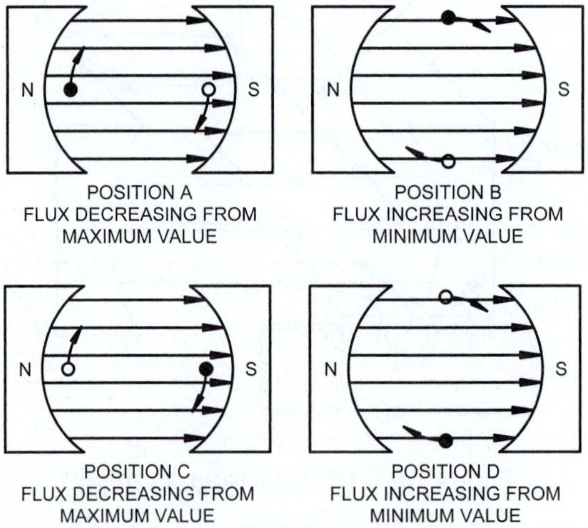

Figure 10-6: Loop Positions for DC Generator of Figure 10-4

When the rotating loop is at position A, the conductors are moving perpendicularly to the magnetic flux so that they cut through the maximum amount of flux per unit time and the induced voltage is maximum. As the loop continues to rotate, the angle between the direction of the conductors and magnetic flux decreases so that the amount of flux cut per unit time decreases.

At position B, the conductors are moving parallel to the lines of magnetic force so that the loop cuts no flux per unit time and the induced voltage is 0. As the loop continues to rotate, the angle between the direction of the conductors and magnetic flux increases so that the amount of flux cut per unit time increases.

At position C, the conductors are moving perpendicularly to the magnetic flux so that they cut through the maximum amount of flux per unit time and the induced voltage is again maximum, although opposite in polarity to the induced voltage at position A. As the loop continues to rotate, the angle between the direction of the conductors and magnetic flux decreases again so that the amount of flux cut per unit time decreases.

At position D, the conductors are moving parallel to the lines of magnetic force so that the loop cuts no flux per unit time and the induced voltage is 0. Although the polarity of the induced loop voltage reverses at positions B and D, the commutator reverses the polarity of the voltage between positions B and D so that the output voltage is a pulsating dc voltage rather an ac voltage. Figure 10-7 shows the output voltage of the generator.

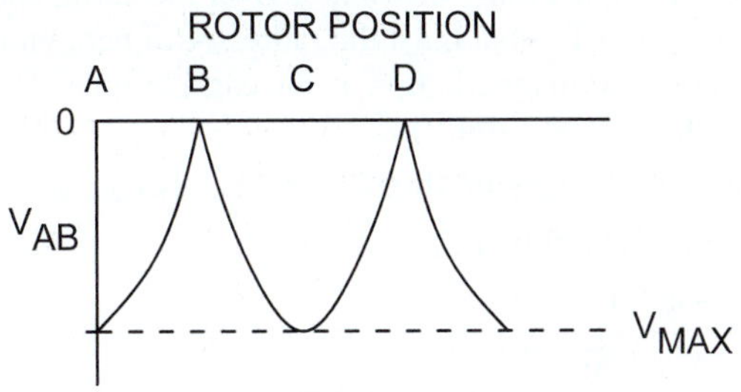

Figure 10-7: Output Voltage Waveform for Figure 10-4

10.1.6 The DC Motor

1. Problem: The torque, T, in newton-meters (N-m) for a dc motor with magnetic flux φ and armature current I is

$$T = K \times \varphi \times I$$

where K, a constant of proportionality, is 25 N-m / Wb-A for the torque rating of the particular dc motor. How much magnetic flux in webers does the motor utilize to generate 2.0 N-m of flux when the motor draws 200 mA of armature current?

Solution: From the formula for torque for a dc motor,

$$T = K \times \varphi \times I$$

$$T / (K \times I) = (K \times \varphi \times I) / (K \times I)$$

$$= \varphi$$

Use the values of T, K, and I to solve for and calculate φ.

$$\varphi = (2.0 \text{ N-m}) / [(25 \text{ N-m / Wb-A}) \times 200 \text{ mA}]$$

$$= (2.0 \text{ N-m}) / [(25 \text{ N-m / Wb-A}) \times (200 \times 10^{-3} \text{ A})]$$

$$= \{2.0 / [25 \times (200 \times 10^{-3})]\} \, ((\text{N} \times \text{m}) / \{[\text{N} \times \text{m} / (\text{Wb} \times \text{A})] \times \text{A}\})$$

$$= (400 \times 10^{-3}) \, [(\text{N} \times \text{m} \times \text{Wb} \times \text{A}) / (\text{N} \times \text{m} \times \text{A})]$$

$$= 400 \times 10^{-3} \text{ Wb}$$

$$= 400 \text{ mWb}$$

The dc motor utilizes **400 mWb** of magnetic flux.

2. Problem: What is the speed, s, of a dc motor that generates 18.9 W of power, P, from 3.0 N-m of torque, T?

Solution: The formula for determining power from torque is

$$P = 0.105 \times T \times s \quad \text{where } P \text{ is in W, } T \text{ is in N-m, and } s \text{ is in rpm}$$

Solve for s.

$P / (0.105 \times T) = (0.105 \times T \times s) / (0.105 \times T)$

$P / (0.105 \times T) = s$

Use the values of P and T to solve for s.

$s = 18.9\ \text{W} / (0.105 \times 3.0\ \text{N-m})$

$= [18.9 / (0.105 \times 3.0)]\ (\text{W} / \text{N-m})$

$= 60.0\ \text{rpm}$

The speed of the dc motor is **60 rpm**.

10.2 Advanced

10.2.1 The Magnetic Field

1. Problem: Assume that the earth's average magnetic flux density, B_E, is about 40 μT and its radius, r_E, is about 6.44 km. If B_E is evenly distributed over the earth, and a neodymium magnet has a magnetic flux density $B_N = 0.5$ T, what must the radius r_N of neodymium magnetic sphere be for the magnet to have the same total magnetic flux as the earth? (Note: The formula for the surface area of a sphere with radius r is $A = 4\pi r^2$.)

Solution: Use the definition of magnetic flux $B = \varphi / A$ to isolate φ.

Solve for the magnetic flux, φ.

$$B = \varphi / A$$
$$B \times A = (\varphi / A) \times A$$
$$= \varphi$$

Set the magnetic flux φ_N of the neodymium sphere with surface area A_N equal to the flux φ_E of the earth with surface area A_E and solve for r_N.

$$\varphi_N = \varphi_E$$
$$B_N \times A_N = B_E \times A_E$$
$$(B_N \times A_N) / B_N = (B_E \times A_E) / B_N$$
$$A_N = (B_E / B_N) \times A_E$$
$$4 \times \pi \times r_N^2 = (B_E / B_N) \times (4 \times \pi \times r_E^2)$$
$$(4 \times \pi \times r_N^2) / (4 \times \pi) = [(B_E / B_N) \times (4 \times \pi \times r_E^2)] / (4 \times \pi)$$
$$r_N^2 = (B_E / B_N) \times r_E^2$$
$$(r_N^2)^{1/2} = [(B_E / B_N) \times r_E^2]^{1/2}$$
$$r_N = [(B_E / B_N) \times r_E^2]^{1/2}$$

Use the values of B_E, B_N, and r_E to solve for r_N.

$$r_N = \{[(40\ \mu\text{T}) / (0.5\ \text{T})] \times (6.44\ \text{km})^2\}^{1/2}$$
$$= \{[(40 \times 10^{-6}\ \text{T}) / (0.5\ \text{T})] \times (6.44 \times 10^3\ \text{m})^2\}^{1/2}$$
$$= \{[(40 \times 10^{-6}\ \text{T}) / (0.5\ \text{T})] \times (41.5 \times 10^6\ \text{m}^2)\}^{1/2}$$
$$= \{[(40 \times 10^{-6}\ \text{T}) \times (41.5 \times 10^6\ \text{m}^2)] / (0.5\ \text{T})]\}^{1/2}$$
$$= \{[(40 \times 10^{-6}) \times (41.5 \times 10^6)] / 0.5\}^{1/2}\ [(\text{T} / \text{T}) \times \text{m}^2]^{1/2}$$
$$= (3.32 \times 10^3)^{1/2}\ (\text{m}^2)^{1/2}$$
$$= 57.6\ \text{m}$$

The radius of a neodymium sphere with the same total magnetic flux as the earth is **57.6 m**.

2. Problem: The radius r_H of a hydrogen atom is about 1.2 angstroms (Å) (1 Å = 100 pm). Use the data from Problem 1 to determine the magnetic flux density B_H of the earth if the earth were the size of a hydrogen atom and the flux remained unchanged.

Solution: From Problem 1, the total magnetic flux of the earth is

$$\varphi_E = \boldsymbol{B}_E \times \boldsymbol{A}_E$$
$$= \boldsymbol{B}_E \times 4 \times \pi \times \boldsymbol{r}_E^2$$
$$= 40\ \mu T \times 4 \times \pi \times (6.44\ km)^2$$
$$= (40 \times 10^{-6}\ T) \times 4 \times \pi \times (6.44 \times 10^3\ m)^2$$
$$= (40 \times 10^{-6}\ T) \times 4 \times \pi \times (41.5 \times 10^6\ m^2)$$
$$= [(40 \times 10^{-6}) \times 4 \times \pi \times (41.5 \times 10^6)]\ (T \times m^2)$$
$$= 20.8 \times 10^3\ [(Wb / m^2) \times m^2]$$
$$= 20.8 \times 10^3\ Wb$$

The surface area of a hydrogen atom $\boldsymbol{A}_H$ is

$$\boldsymbol{A}_H = 4 \times \pi \times r_H^2$$
$$= 4 \times \pi \times (1.2\ Å)^2$$
$$= 4 \times \pi \times [(1.2\ Å) \times (100\ pm / Å)]^2$$
$$= 4 \times \pi \times \{(1.2 \times 100)\ [(Å \times pm) / Å]\}^2$$
$$= 4 \times \pi \times (120\ pm)^2$$
$$= 4 \times \pi \times (120 \times 10^{-12}\ m)^2$$
$$= 4 \times \pi \times (14.4 \times 10^{-21}\ m^2)$$
$$= 181 \times 10^{-21}\ m^2$$

The magnetic flux density $\boldsymbol{B}_H$ is then

$$\boldsymbol{B}_H = \varphi_E / \boldsymbol{A}_H$$
$$= (20.8 \times 10^3\ Wb) / (181 \times 10^{-21}\ m^2)$$
$$= [(20.8 \times 10^3) / (181 \times 10^{-21})]\ (Wb / m^2)$$
$$= 115 \times 10^{21}\ T$$

The magnetic flux density of the earth, if it were shrunk to the size of a hydrogen atom, would be **115×10^{21} T**. By comparison, the most powerful magnet (under construction at the Los Alamos Laboratories in September 2008) had a target strength of 0.1×10^3 T.

10.2.2 Electromagnetism

1. Problem: The inverse of magnetic reluctance $\mathfrak{R}$ is magnetic permeance $\boldsymbol{P} = 1 / \mathfrak{R}$. What is the permeance of a nickel pin with permeability $\mu = 115\ \mu Wb / At\text{-}m$, length $\boldsymbol{l} = 4.0$ cm, and cross sectional area $\boldsymbol{A} = 0.5\ cm^2$?

Solution: From the equation for reluctance,

$$\mathfrak{R} = \boldsymbol{l} / (\mu \times \boldsymbol{A})$$

The permeance is the reciprocal of the reluctance.

$$\boldsymbol{P} = (\mu \times \boldsymbol{A}) / \boldsymbol{l}$$
$$= [(115\ \mu Wb / At\text{-}m) \times (0.5\ cm^2)] / 4.0\ cm$$
$$= \{(115 \times 10^{-6}\ Wb / At\text{-}m) \times (0.5\ cm^2) \times [(1 \times 10^{-2}\ m) / cm]^2\} / (4.0 \times 10^{-2}\ m)$$
$$= \{(115 \times 10^{-6}\ Wb / At\text{-}m) \times (0.5\ cm^2) \times [(1 \times 10^{-2})^2\ (m / cm)^2]\} / (4.0 \times 10^{-2}\ m)$$
$$= \{(115 \times 10^{-6}\ Wb / At\text{-}m) \times [(0.5\ cm^2) \times (1 \times 10^{-4})\ (m^2 / cm^2)]\} / (4.0 \times 10^{-2}\ m)$$
$$= \{[(115 \times 10^{-6}) \times (0.5) \times (1 \times 10^{-4})] / (4.0 \times 10^{-2})\}$$
$$\{[(Wb / At\text{-}m) \times cm^2 \times (m^2 / cm^2)] / m\}$$
$$= 144 \times 10^{-9}\ \{[(Wb / At\text{-}m) \times m^2] / m\}$$
$$= 144 \times 10^{-9}\ [(Wb \times m^2) / (At\text{-}m \times m)]$$
$$= 144 \times 10^{-9}\ [(Wb \times m \times m) / (At \times m \times m)]$$
$$= 144 \times 10^{-9}\ Wb / At$$

The permeance of the nickel pin is $\mathbf{144 \times 10^{-9}}$ **Wb / At**.

2. Problem: For an electromagnet with N turns and length l, let $k_N = N / l$ be the average turns per unit length. Show that the magnetic flux density for an electromagnet with magnetic permeability μ, current I, and average turns per unit length k_N is

$$B = \mu \times I \times k_N$$

Hint: Start by equating the two equations for magnetic reluctance.

Solution: The magnetic reluctance of a material is

$$\mathfrak{R} = l / (\mu \times A)$$

and

$$\mathfrak{R} = F_m / \varphi$$

Set the two equations equal and solve for B.

$$l / (\mu \times A) = F_m / \varphi$$

$$l / (\mu \times A) = (N \times I) / \varphi$$

$$l / (\mu \times A) \times \varphi = [(N \times I) / \varphi] \times \varphi$$

$$(l \times \varphi) / (\mu \times A) = N \times I$$

$$[(\varphi / A) \times (l / \mu)] \times (\mu / l) = (N \times I) \times (\mu / l)$$

$$(\varphi / A) = \mu \times I \times (N / l)$$

$$B = \mu \times I \times k_N$$

10.2.3 Magnetic Hysteresis

1. Problem: Determine the magnetic flux density B for the electromagnet in Figure 10-8 if the core material reluctance is $\mathfrak{R} = 40$ MAt / Wb and the magnetic field intensity is $H = 5.5 \times 10^{-6}$ At / m.

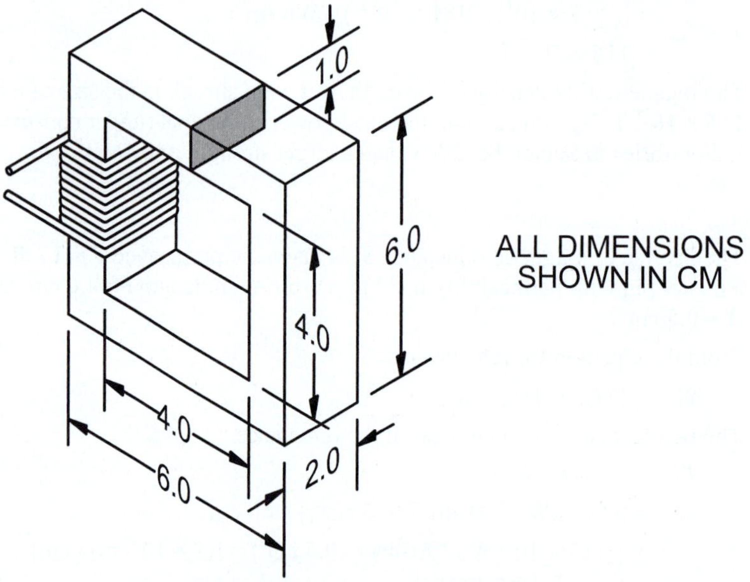

Figure 10-8: Electromagnet for Magnetic Hysteresis Problem

Solution: The solution to this problem will require three separate equations. The first equation is that for magnetic field intensity H for magnetomotive force F_m and core material length l.

$$H = F_m / l$$

The second equation is the definition of magnetic flux density B for magnetic flux φ over area A.

$$B = \varphi / A$$

The third equation is Hopkinson's Law (Ohm's Law for magnetic devices).

$\varphi = F_m / \mathfrak{R}$

Substitute the expression for φ from Hopkinson's Law into the definition for B and solve for F_m.

$$\begin{aligned}
B &= (F_m / \mathfrak{R}) / A \\
&= F_m / (\mathfrak{R} \times A) \\
B \times (\mathfrak{R} \times A) &= [F_m / (\mathfrak{R} \times A)] \times (\mathfrak{R} \times A) \\
B \times \mathfrak{R} \times A &= F_m
\end{aligned}$$

Substitute F_m into the equation for H and solve for B.

$$\begin{aligned}
H &= (B \times \mathfrak{R} \times A) / l \\
H \times l &= [(B \times \mathfrak{R} \times A) / l] \times l \\
(H \times l) / (\mathfrak{R} \times A) &= (B \times \mathfrak{R} \times A) / (\mathfrak{R} \times A) \\
&= B
\end{aligned}$$

Calculate the average path length per side, l_{SIDE}, from the dimensions in Figure 10-8.

$$\begin{aligned}
l_{\mathrm{SIDE}} &= (6.0\ \mathrm{cm} + 4.0\ \mathrm{cm}) / 2 \\
&= (10.0\ \mathrm{cm}) / 2 \\
&= 5.0\ \mathrm{cm} \\
&= 5.0 \times 10^{-2}\ \mathrm{m}
\end{aligned}$$

Use the average length per side to calculate the total average path length l for the core.

$$\begin{aligned}
l &= 4 \times l_{\mathrm{SIDE}} \\
&= 4 \times (5.0 \times 10^{-2}\ \mathrm{m}) \\
&= 20.0 \times 10^{-2}\ \mathrm{m}
\end{aligned}$$

Also from the figure, calculate the cross-sectional area of the core, A.

$$\begin{aligned}
A &= 2.0\ \mathrm{cm} \times 1.0\ \mathrm{cm} \\
&= (2.0 \times 10^{-2}\ \mathrm{m}) \times (1.0 \times 10^{-2}\ \mathrm{m}) \\
&= [(2.0 \times 10^{-2}) \times (1.0 \times 10^{-2})]\ (\mathrm{m} \times \mathrm{m}) \\
&= 2.0 \times 10^{-4}\ \mathrm{m}^2
\end{aligned}$$

Use the values for H, l, $\mathfrak{R}$, and A to calculate B.

$$\begin{aligned}
B &= (H \times l) / (\mathfrak{R} \times A) \\
&= [(5.5 \times 10^{-6}\ \mathrm{At / m}) \times (20.0 \times 10^{-2}\ \mathrm{m})] / [(40\ \mathrm{MAt / Wb}) \times (2.0 \times 10^{-4}\ \mathrm{m}^2)] \\
&= [(5.5 \times 10^{-6}\ \mathrm{At / m}) \times (20.0 \times 10^{-2}\ \mathrm{m})] / [(40 \times 10^{6}\ \mathrm{At / Wb}) \times (2.0 \times 10^{-4}\ \mathrm{m}^2)] \\
&= [(5.5 \times 10^{-6}) \times (20.0 \times 10^{-2})] / [(40 \times 10^{6}) \times (2.0 \times 10^{-4})] \\
&\qquad \{[(\mathrm{At / m}) \times \mathrm{m}] / [(\mathrm{At / Wb}) \times (\mathrm{m}^2)]\} \\
&= [(1.1 \times 10^{-6}) / (80 \times 10^{2})]\ \{[(\mathrm{At / m}) \times \mathrm{m}] / [(\mathrm{At} \times \mathrm{m}^2) / \mathrm{Wb} \times \mathrm{m}]\} \\
&= 8.8 \times 10^{-3}\ \{[(\mathrm{At} \times \mathrm{m}) / \mathrm{m}] / [(\mathrm{At} \times \mathrm{m} \times \mathrm{m}) / \mathrm{Wb}]\} \\
&= 8.8 \times 10^{-3}\ \{[(\mathrm{At} \times \mathrm{m}) / \mathrm{m}] \times [\mathrm{Wb} / (\mathrm{At} \times \mathrm{m} \times \mathrm{m})]\} \\
&= 8.8 \times 10^{-3}\ [(\mathrm{At} \times \mathrm{m} \times \mathrm{Wb}) / (\mathrm{At} \times \mathrm{m} \times \mathrm{m} \times \mathrm{m})] \\
&= 8.8 \times 10^{-3}\ [\mathrm{Wb} / (\mathrm{m} \times \mathrm{m})] \\
&= 8.8 \times 10^{-3}\ \mathrm{Wb / m}^2 \\
&= 8.8 \times 10^{-3}\ \mathrm{T} \\
&= 8.8\ \mathrm{mT}
\end{aligned}$$

The magnetic flux density for the electromagnet is **8.8 mT**.

10.2.4 Electromagnetic Induction

1. Problem: Use dimensional analysis for the equation

$$v_{ind} = N \times (d\varphi / dt)$$

to determine the dimension of φ (lines of magnetic force (LMF)).

Solution: The term N is dimensionless, so the equation reduces to

$$v_{ind} = d\varphi / dt$$

$$v_{ind} \times dt = (d\varphi / dt) \times dt$$

$$v_{ind} \times dt = d\varphi$$

The terms dt and $d\varphi$ represent the changes in the values for t and φ respectively, so their dimensions are the same as for t and φ. Therefore

$$d\varphi = v_{ind} \times dt$$

$$\varphi = v_{ind} \times t$$

LMF = voltage × time
= (energy / charge) × time
= (energy × time) / charge
= energy × (time / charge)
= energy / (charge / time)
= energy / current

The dimension of a line of magnetic force is **energy / current** so that each line of force in a magnetic field represents energy stored in the magnetic field from the flow of current.

2. Problem: An inventor decides to build a hovercraft that uses a magnetic field that repels the earth's magnetic field to levitate. How much current is necessary to negate the effect of gravity on a rod with mass m_{ROD} = 1 g and length l = 1 m, if the earth's magnetic field density is 40 μT and the force due to gravity on a mass m in kilograms is $F_g = m \times 9.8\ \text{m} / \text{s}^2$?

Solution: The force F on a conductor carrying current I with length l exposed to a magnetic field with flux density $B_\perp$ is

$$F = B_\perp \times I \times l$$ where F is in N (kg × m / s²), B is in T, I is in A, and l is in m

This force must be equal to the force of gravity on the 1 g mass.

$$F_g = m_{ROD} \times 9.8\ \text{m} / \text{s}^2$$
$$= 1\ \text{g} \times 9.8\ \text{m} / \text{s}^2$$
$$= (1 \times 10^{-3}\ \text{kg}) \times 9.8\ \text{m} / \text{s}^2$$
$$= [(1 \times 10^{-3}) \times 9.8]\ (\text{kg} \times \text{m} / \text{s}^2)$$
$$= 9.8 \times 10^{-3}\ \text{N}$$

Use the values of F, $B_\perp$, and l to calculate I.

$$F = B_\perp \times I \times l$$
$$9.8 \times 10^{-3}\ \text{N} = (40\ \mu\text{T}) \times I \times (1\ \text{m})$$
$$(9.8 \times 10^{-3}\ \text{N}) / [(40\ \mu\text{T}) \times (1\ \text{m})] = [(40\ \mu\text{T}) \times I \times (1\ \text{m})] / [(40\ \mu\text{T}) \times (1\ \text{m})]$$
$$(9.8 \times 10^{-3}\ \text{N}) / [(40 \times 10^{-6}\ \text{T}) \times (1\ \text{m})] = I$$
$$[(9.8 \times 10^{-3}) / [(40 \times 10^{-6}) \times 1]\ [\text{N} / (\text{T} \times \text{m})] = I$$
$$245\ \text{A} = I$$

A 1-meter rod with a mass of 1 g must carry a current of 245 A to use the earth's magnetic field to negate the effect of gravity. For comparison, a modern US one-cent piece has a mass of 2.5 g.

10.2.5 The DC Generator

Problem: Determine the output voltage waveform V_{AB} for one complete rotation of the two-coil dc generator in Figure 10-9.

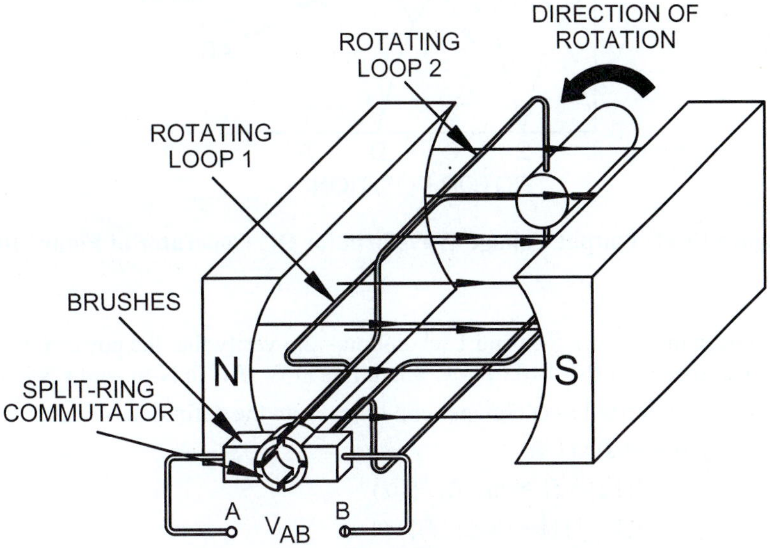

Figure 10-9: Two Coil DC Generator

Solution: The generator is similar in operation to the single-coil dc generator in Figure 10-4. Because the direction of rotation is reversed, however, the polarity of the induced voltage V_{AB} is opposite to that for the single-coil dc generator, because the polarity of the induced magnetic field that opposes the motion of the coils will be reversed. In addition, each coil will produce voltage waveforms that are 90 degrees apart from each other. Refer to Figure 10-10.

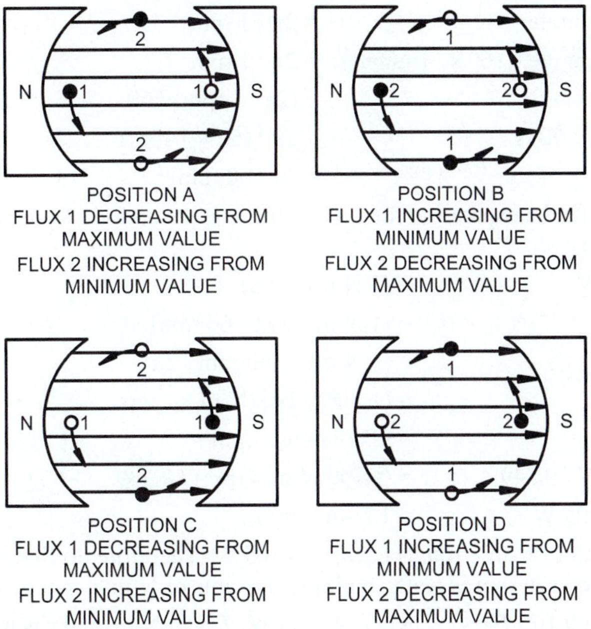

Figure 10-10: Loop Positions for DC Generator of Figure 10-9

As Figure 10-10 shows, when the flux cut by coil 1 is decreasing from its maximum value, the flux cut by coil 2 is increasing from its minimum value and vice versa. Refer to Figure 10-11, which shows the output voltage waveform V_{AB} for the generator.

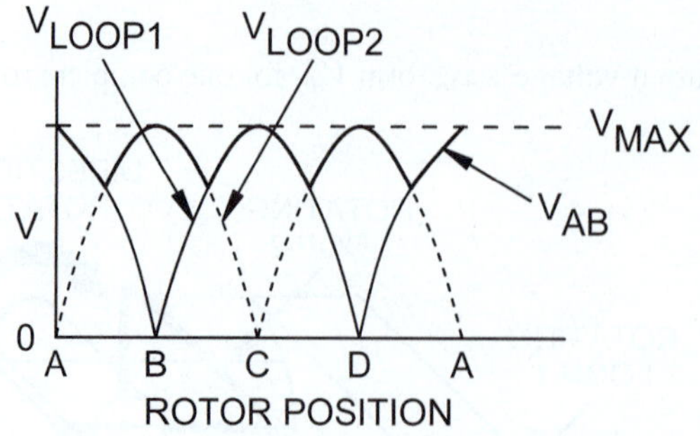

Figure 10-11: Output Voltage Waveform for DC Generator of Figure 10-9

10.2.6 The DC Motor

1. Problem: Given that 1 J = 1 N-m and 1 revolution = 2π, verify that the power $\boldsymbol{P}$ from a dc motor with torque $\boldsymbol{T}$ at speed $\boldsymbol{s}$ is $\boldsymbol{P} = 0.105\ \boldsymbol{T} \times \boldsymbol{s}$, where $\boldsymbol{P}$ is in W, $\boldsymbol{T}$ is in N-m, and $\boldsymbol{s}$ is in rpm.

   Solution: Power is defined as energy per unit time. From the definition of watt,

   1 W = (1 J) / (1 s)
   = [(1 J) × (1 N-m / J)] / (1 s)
   = [(1 × 1) (J × (N-m / J)] / (1 s)
   = (1 N-m) / (1 s)

   Use the definition of rpm to determine 1 s in terms of rpm.

1 rpm	= 1 revolution / 1 minute
	= 2π / 1 minute
1 rpm × 1 minute	= (2π / 1 minute) × 1 minute
(1 rpm × 1 minute) / 1 rpm	= 2π / 1 rpm
1 minute	= 2π / 1 rpm
1 minute × (60 s / 1 minute)	= 2π / 1 rpm
60 s / 60	= (2π / 1 rpm) / 60
(60 / 60) s	= 2π / (60 × 1 rpm)
1 s	= 2π / 60 rpm

   Substitute the expression for 1 s into the expression for 1 W and solve for watts in terms of N-m and rpm.

1 W	= (1 N-m) / (1 s)
	= (1 N-m) / (2π / 60 rpm)
	= (1 N-m) × (60 rpm / 2π)
	= (1 × 60 / 2π) (N-m × rpm)
	= 9.549 N-m × rpm
1 W / 9.549	= (9.549 N-m × rpm) / 9.549
0.105 W	= 1 N-m × rpm

   Each unit of N-m × rpm is equal to 0.105 W. Therefore, to find the product of torque $\boldsymbol{T}$ in N-m and speed $\boldsymbol{s}$ in rpm in watts, multiply the units of N-m × rpm by the conversion factor 0.105.

   $\boldsymbol{P} = 0.105\ \boldsymbol{T} \times \boldsymbol{s}$ where $\boldsymbol{P}$ is in W, $\boldsymbol{T}$ is in N-m, and $\boldsymbol{s}$ is in rpm

2. Problem: What is the efficiency η of a dc motor that delivers 7.5 W of power $\boldsymbol{P_L}$ at a torque $\boldsymbol{T}$ of 2.0 N-m at a speed $\boldsymbol{s}$ of 60 rpm?

   Solution: The generated power $\boldsymbol{P}$ for the dc motor is

$\boldsymbol{P} = 0.105\ \boldsymbol{T} \times \boldsymbol{s}$
$= 0.105 \times 2.5\ \text{N-m} \times 60\ \text{rpm}$
$= (0.105 \times 2.5 \times 60)\ (\text{N-m} \times \text{rpm})$
$= 15.75\ \text{W}$

From this, the efficiency of the motor is

$\eta = (\boldsymbol{P_L} / \boldsymbol{P}) \times 100\%$
$= (7.5\ \text{W} / 15.75\ \text{W}) \times 100\%$
$= 0.476 \times 100\%$
$= 47.6\ \%$

The efficiency of the motor is **48%**.

10.3 Just for Fun

Problem: From the formula for induced voltage,

$\boldsymbol{v}_{\text{ind}} = \boldsymbol{B}_{\perp} \times \boldsymbol{l} \times \mathbf{v}$

show that the energy stored in 1 line of magnetic force (LMF) is equal to 10 nJ / A.

Solution: From the formula for induced voltage, $\boldsymbol{v}_{\text{ind}}$ is in V, $\boldsymbol{B}_{\perp}$ is in Wb / m^2, $\boldsymbol{l}$ is in m, and $\mathbf{v}$ is in m / s.

V	$= (\text{Wb} / \text{m}^2) \times \text{m} \times (\text{m} / \text{s})$
J / C	$= (1 \times 10^8\ \text{LMF} / \text{m}^2) \times \text{m} \times (\text{m} / \text{s})$
	$= (1 \times 10^8)\ [(\text{LMF} \times \text{m} \times \text{m}) / (\text{m}^2 \times \text{s})]$
	$= (1 \times 10^8)\ [(\text{LMF} \times \text{m} \times \text{m}) / (\text{m} \times \text{m} \times \text{s})]$
	$= 1 \times 10^8\ \text{LMF} / \text{s}$
(J / C) × s	$= (1 \times 10^8\ \text{LMF} / \text{s}) \times \text{s}$
J × (s / C)	$= 1 \times 10^8\ \text{LMF}$
J / (C / s)	$= 1 \times 10^8\ \text{LMF}$
$(\text{J} / \text{A}) / (1 \times 10^8)$	$= (1 \times 10^8\ \text{LMF}) / (1 \times 10^8)$
$(\text{J} / \text{A}) \times [1 / (1 \times 10^8)]$	$= 1\ \text{LMF}$
$(\text{J} / \text{A}) \times (10 \times 10^{-9})$	$= 1\ \text{LMF}$
$(10 \times 10^{-9}\ \text{J}) / \text{A}$	$= 1\ \text{LMF}$
10 nJ / A	$= 1\ \text{LMF}$

so that

1 LMF = 10 nJ / A

From this equation, each ampere of current will store 10 nJ in each line of magnetic force it creates.

11. Introduction to Alternating Current and Voltage

11.1 Basic

11.1.1 The Sinusoidal Waveform

For problems in Section 11.1.1, refer to the sinusoidal waveform in Figure 11-1.

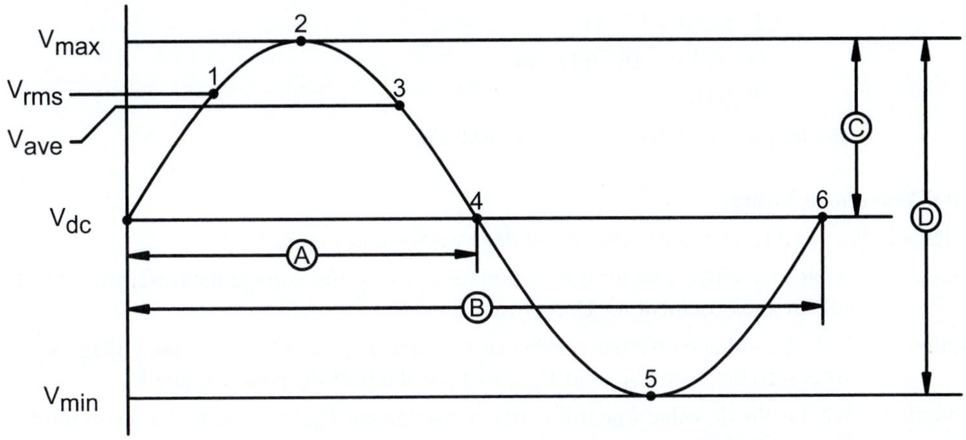

Figure 11-1: Sinusoidal Waveform Example

1. Problem: If t_B, the time for measurement B, is 15 µS, what is the frequency of the waveform?

   Solution: Measurement B is for the full period, T, of the waveform. Frequency f is the reciprocal of the period.

   $$\begin{aligned} f &= 1 / T \\ &= 1 / (15\ \mu\text{s}) \\ &= 1 / (15 \times 10^{-6}\ \text{s}) \\ &= (1 / 15 \times 10^{-6})\ (1/\text{s}) \\ &= 66.7 \times 10^{3}\ \text{Hz} \\ &= 66.7\ \text{kHz} \end{aligned}$$

   The frequency of the waveform is **67 kHz**.

2. Problem: What is t_A, the time for measurement A, if the frequency f of the waveform is 25 kHz?

   Solution: The period T of the waveform is the reciprocal of the frequency f.

   $$\begin{aligned} T &= 1 / f \\ &= 1 / 25\ \text{kHz} \\ &= 1 / (25 \times 10^{3}\ \text{Hz}) \\ &= [1 / (25 \times 10^{3})]\ (1 / \text{Hz}) \\ &= 40 \times 10^{-6}\ \text{s} \\ &= 40\ \mu\text{s} \end{aligned}$$

   Measurement A is half the period.

   $$\begin{aligned} t_A &= (40\ \mu\text{s}) / 2 \\ &= 20\ \mu\text{s} \end{aligned}$$

   The time for measurement A is **20 µs**.

3. Problem: What is the frequency f if point 5 of the waveform occurs 12.5 ms after point 2 of the waveform?

Solution: Points 2 and 5 correspond to V_{max} and V_{min} that, for a sinusoidal waveform occur, one-half period apart.

$$T / 2 = 12.5 \text{ ms}$$

$$(T / 2) \times 2 = (12.5 \text{ ms}) \times 2$$

$$T = 25.0 \text{ ms}$$

The frequency f is the reciprocal of the period T.

$$\begin{aligned} f &= 1 / 25.0 \text{ ms} \\ &= 1 / (25.0 \times 10^{-3} \text{ s}) \\ &= [1 / (25.0 \times 10^{-3})] (1 / \text{s}) \\ &= 40.0 \text{ Hz} \end{aligned}$$

The frequency of the waveform is **40.0 Hz**.

11.1.2 Sinusoidal Waveform Values

For problems in Section 11.1.2, refer to the sinusoidal waveform in Figure 11-1.

1. Problem: What sinusoidal waveform measurements do V_C, the voltage measurement of C, and V_D, the voltage measurement for D, represent?

   Solution: V_C is the voltage difference between V_{max} and V_{DC}, so V_C is the peak voltage V_{pk}. V_D is the voltage difference between V_{max} and V_{min}, so V_D is the peak-to-peak voltage $V_{pk\text{-}pk}$.

2. Problem: What is the dc value V_{DC} if the maximum voltage $V_{max} = +7.0$ V and the minimum voltage $V_{min} = -3.0$ V?

   Solution: The dc value V_{DC} of a sinusoidal waveform is

$$\begin{aligned} V_{DC} &= V_{min} + (V_{max} - V_{min}) / 2 \\ &= (-3.0 \text{ V}) + \{[(+7.0 \text{ V}) - (-3.0 \text{ V})] / 2\} \\ &= (-3.0 \text{ V}) + (+10.0 \text{ V} / 2) \\ &= (-3.0 \text{ V}) + (+5.0 \text{ V}) \\ &= \mathbf{+2.0 \text{ V}} \end{aligned}$$

3. Problem: What is V_C, the voltage for measurement C, if the maximum voltage $V_{max} = +2.5$ V and the minimum voltage $V_{min} = -4.5$ V?

   Solution: V_C is V_{pk}, the peak voltage measurement, and is equal to half the peak-to-peak voltage measurement $V_{pk\text{-}pk}$. The peak-to-peak voltage is equal to the difference between V_{max} and V_{min}, so

$$\begin{aligned} V_C &= V_{pk} \\ &= V_{pk\text{-}pk} / 2 \\ &= (V_{max} - V_{min}) / 2 \\ &= [(+2.5 \text{ V}) - (-4.5 \text{ V})] / 2 \\ &= 7.0 \text{ V} / 2 \\ &= \mathbf{3.5 \text{ V}} \end{aligned}$$

4. Problem: What is the maximum voltage V_{max} if the dc voltage $V_{DC} = 0$ V and the rms voltage $V_{rms} =$ 1.061 V?

   Solution: The rms voltage V_{rms} of a sinusoidal waveform with dc voltage $V_{DC} = 0$ and maximum voltage $V_{max} = V_{pk}$ is

$$\begin{aligned} V_{rms} &= V_{pk} / \sqrt{2} \\ &= V_{pk} \times 0.707 \\ &= V_{max} \times 0.707 \end{aligned}$$

so

$(V_{max} \times 0.707) / 0.707 \quad = V_{rms} / 0.707$

$V_{max} \quad = 1.061 \text{ V} / 0.707$

$= \mathbf{1.500\ V}$

5. Problem: What is the minimum voltage V_{min} if the dc voltage $V_{DC} = -1.2$ V and the maximum voltage $V_{max} = +3.0$ V?

Solution: The dc value V_{DC} of a sinusoidal waveform is

$V_{DC} = V_{min} + (V_{max} - V_{min}) / 2$

so

$V_{DC} \times 2 \quad = [V_{min} + (V_{max} - V_{min}) / 2] \times 2$

$= (2 \times V_{min}) + \{2 \times [(V_{max} + -V_{min}) / 2]\}$

$= (2 \times V_{min}) + (V_{max} + -V_{min})$

$= [(2 \times V_{min}) + (-1 \times V_{min})] + V_{max}$

$= V_{min} + V_{max}$

$V_{DC} \times 2 - V_{max} \quad = V_{min} + V_{max} - V_{max}$

$V_{DC} \times 2 + -V_{max} \quad = V_{min}$

From this,

$V_{min} \quad = (-1.2 \text{ V} \times 2) + (-3.0 \text{ V})$

$= (-2.4 \text{ V}) + (-3.0 \text{ V})$

$= \mathbf{-5.4\ V}$

6. Problem: What is the rms voltage V_{rms} if the average voltage $V_{ave} = 500$ mV and the dc voltage $V_{DC} = 0$?

Solution: For a sinusoidal voltage with dc voltage $V_{DC} = 0$

$V_{rms} \quad = V_{pk} / \sqrt{2}$

$= V_{pk} \times 0.707$

$V_{rms} / 0.707 \quad = (V_{pk} \times 0.707) / 0.707$

$V_{rms} / 0.707 \quad = V_{pk}$

and

$V_{ave} \quad = V_{pk} \times (2 / \pi)$

$V_{ave} / (2 / \pi) \quad = [V_{pk} \times (2 / \pi)] / (2 / \pi)$

$V_{ave} \times (\pi / 2) \quad = V_{pk}$

$V_{ave} \times 1.571 \quad = V_{pk}$

Equating the two gives

$V_{rms} / 0.707 \quad = V_{ave} \times 1.57$

$(V_{rms} / 0.707) \times 0.707 \quad = (V_{ave} \times 1.57) \times 0.707$

$V_{rms} \quad = V_{ave} \times 1.11$

$= 500 \text{ mV} \times 1.11$

$= \mathbf{555\ mV}$

11.1.3 Angular Measurement of a Sine Wave

1. Problem: Given that one complete revolution is 2π radians (rad), verify that $\boldsymbol{D}$ degrees equals

$\boldsymbol{R} = [\boldsymbol{D} \times (\pi / 180)]$ rad

Solution: One complete revolution is 360°, so

$360° \quad = 2\pi$ rad

$360° / 360 \quad = (2\pi / 360)$ rad

$1° \quad = [2\pi / (2 \times 180)]$ rad

$= (\pi / 180)$ rad

From this,

$R = D$ degrees

$= D \times 1°$

$= [D \times (\pi / 180)]$ rad

2. Problem: Given that one complete revolution is 2π radians (rad), how many degrees are in 1 rad?

Solution: One complete revolution is 360°, so

$360° = 2\pi$ rad

$360° / 2\pi = 2\pi \text{ rad} / 2\pi$

$[360 / (2 \times \pi)]° = 1$ rad

$[360 / (2 \times 3.14)]° = 1$ rad

$(360 / 6.28)° = 1$ rad

$57.3° = 1$ rad

One radian equals about **57.3°**.

3. Problem: For the sinusoidal waveforms in Figure 11-2, does Signal 1 lead or lag Signal 2?

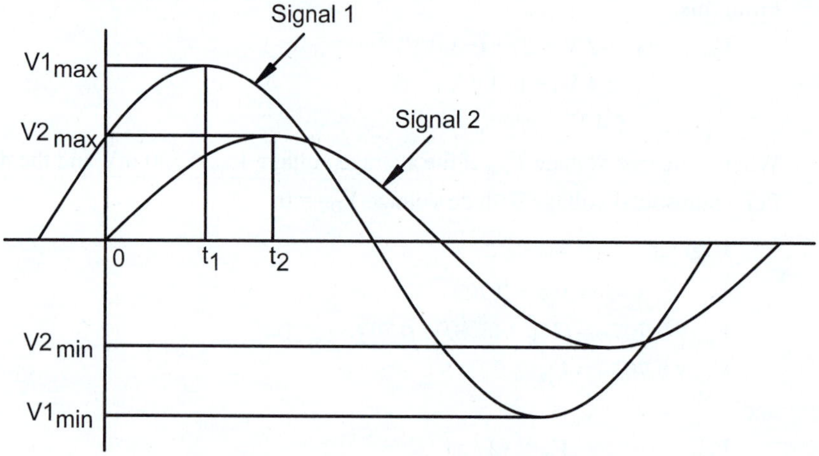

Figure 11-2: Sine Wave Angular Measurement Example Waveforms

Solution: By standard convention the time axis increases from left to right, so that $t_1 < t_2$. This means that $V_{1\,max}$ occurs before $V_{2\,max}$, so that Signal 1 **leads** Signal 2.

4. Problem: If $\mathbf{t_2}$ = 2.0 ms, what is the frequency f of Signal 2?

Solution: The peak amplitude for a sine wave occurs when $\theta = 90°$. Once cycle equals 360°.

$90° = 90° \times (1 \text{ cycle} /360°)$

$= [(90° \times 1) / 360°]$ cycle

$= [90° / 360°]$ cycle

$= 0.25$ cycle

t_2 is therefore one-quarter of the period, T.

$t_2 = 0.25 \times T$

$t_2 / 0.25 = (0.25 \times T) / 0.25$

$t_2 / 0.25 = T$

From this

$T = 2.0 \text{ ms} / 0.25$

$= 8.0\text{ ms}$

The frequency f is the reciprocal to the period T, so

$$
\begin{aligned}
f &= 1 / T \\
&= 1 / 8.0\text{ ms} \\
&= 1 / (8.0 \times 10^{-3}\text{ s}) \\
&= [1 / (8.0 \times 10^{-3})]\,(1 / \text{s}) \\
&= 125\text{ Hz}
\end{aligned}
$$

The frequency of Signal 2 is **125 Hz**.

11.1.4 The Sine Wave Formula

1. Problem: What is the instantaneous voltage $v(\theta)$ at $\theta = 1$ rad for a sine wave with peak voltage $V_{pk} = 1.50$ V?

   Solution: Although you can convert the angle from radians into degrees before calculating the answer, all scientific calculators have a mode option to calculate trigonometric functions in radians so that you can calculate the sine of 1 radian (= 57.3°) directly. From the sine wave formula,

$$
\begin{aligned}
v(\theta) &= V_{pk} \times \sin(\theta) \\
&= 1.50\text{ V} \times \sin(1\text{ rad}) \\
&= 1.50\text{ V} \times 0.841 \\
&= 1.26\text{ V}
\end{aligned}
$$

   The instantaneous voltage is **1.26 V**.

2. Problem: For what angle θ in degrees will a sine wave with peak-to-peak voltage $V_{pk\text{-}pk} = 5.00$ V have an instantaneous voltage $v(\theta) = 1.00$ V?

   Solution: For a sine wave with $V_{pk\text{-}pk} = 5.00$ V,

$$
\begin{aligned}
V_{pk} &= V_{pk\text{-}pk} / 2 \\
&= 5.00\text{ V} / 2 \\
&= 2.50\text{ V}
\end{aligned}
$$

   Use the sine wave formula to solve for θ.

$$
\begin{aligned}
v(\theta) &= V_{pk} \times \sin(\theta) \\
v(\theta) / V_{pk} &= [V_{pk} \times \sin(\theta)] / V_{pk} \\
v(\theta) / V_{pk} &= \sin(\theta) \\
\sin^{-1}[v(\theta) / V_{pk}] &= \sin^{-1}[\sin(\theta)] \\
\sin^{-1}[v(\theta) / V_{pk}] &= \theta
\end{aligned}
$$

   Use the values of $v(\theta)$ and V_{pk} to calculate θ.

$$
\begin{aligned}
\theta &= \sin^{-1}[1.00\text{ V} / 5.00\text{ V}] \\
&= \sin^{-1}[0.200] \\
&= 11.5°
\end{aligned}
$$

   The angle θ = **11.5°** when $v(\theta) = 1.00$ V.

3. Problem: A measured sine wave reaches its peak value $V_{pk} = 2.0$ V when the angle of the reference signal $\theta_{REF} = 112.5°$. What is the phase φ_M in degrees for the measured sine wave?

   Solution: A sine wave reaches its maximum value for $\theta = 90°$. This corresponds to $\theta_{REF} = 112.5°$ for the reference sine wave, so the phase φ_M is the difference between $\theta_M = 90°$ and θ_{REF}.

$$
\begin{aligned}
\theta_M - \theta_{REF} &= \varphi_M \\
90° - 112.5° &= \varphi_M \\
-22.5° &= \varphi_M
\end{aligned}
$$

   The phase of the measured sine wave is **−22.5°**.

11.1.5 Introduction to Phasors

For the problems in Section 11.1.5, refer to the phasor diagram in Figure 11-3.

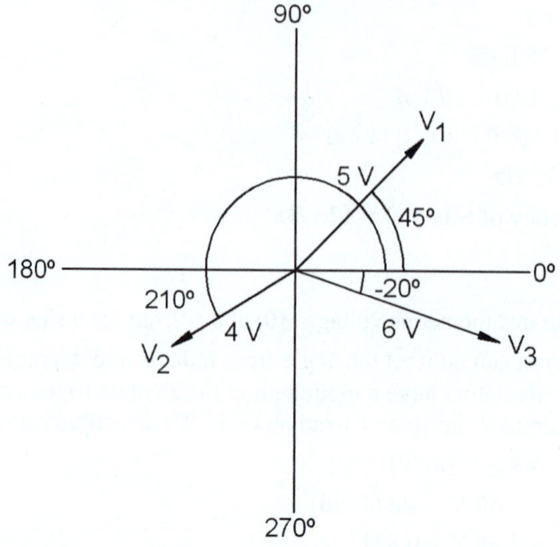

Figure 11-3: Phasor Diagram for Section 11.1.5

1. Problem: What are the peak voltage V_{pk} and phase angle θ of phasor V_1?

   Solution: From the diagram for phasor V_1, the peak voltage is the length of the phasor and the phase angle is the angle between the phasor and the reference (0°) axis. Therefore,

   $V_{pk} = \mathbf{5\ V}$

   $\theta = \mathbf{45°}$

2. Problem: What is the instantaneous voltage v of phasor V_1?

   Solution: The instantaneous voltage v for phasor V_1 is

   $v = V_{pk} \sin(\theta)$

   $= (5\text{ V}) \times \sin(45°)$

   $= 5\text{ V} \times 0.707$

   $= \mathbf{3.53\ V}$

3. Problem: What is the phase relationship between V_1 and V_2?

   Solution: From the phasor diagram, the phase angles θ_1 and θ_2 for phasors V_1 and V_2 are

   $\theta_1 = 45°$

   $\theta_2 = 210°$

   The phase difference of V_2 relative to V_1 is

   $\theta_2 - \theta_1 = 210° - 45°$

   $= +165°$

   Because the answer is positive, V_2 **leads** V_1 **by 165°**.

11.1.6 Analysis of AC Circuits

1. Problem: Determine the total ac current I_T and the currents I_{R1} and I_{R2} through R_1 and R_2, respectively, for the circuit in Figure 11-4. Assume that all values are exact.

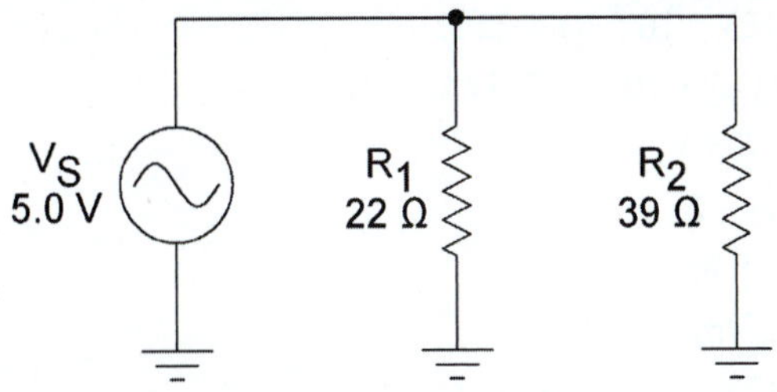

Figure 11-4: Analysis of AC Circuits Example 1

Solution: $\boldsymbol{R_1}$ and $\boldsymbol{R_2}$ are in parallel, so that the ac voltage $\boldsymbol{V_S}$ appears across each. From Ohm's Law for current, the rms currents through $\boldsymbol{R_1}$ and $\boldsymbol{R_2}$ are

$$\begin{aligned} I_{R1} &= V_{R1} / R_1 \\ &= V_S / R_1 \\ &= 5.0\text{ V} / 22\ \Omega \\ &= (5.0 / 22)\ (\text{V} / \Omega) \\ &= 227 \times 10^{-3}\text{ A} \\ &= 227\text{ mA} \end{aligned}$$

$$\begin{aligned} I_{R2} &= V_{R2} / R_2 \\ &= V_S / R_2 \\ &= 5.0\text{ V} / 39\ \Omega \\ &= (5.0 / 39)\ (\text{V} / \Omega) \\ &= 128 \times 10^{-3}\text{ A} \\ &= 128\text{ mA} \end{aligned}$$

From Kirchhoff's Current Law, the current $\boldsymbol{I_T}$ from the source entering the junction to which $\boldsymbol{R_1}$ and $\boldsymbol{R_2}$ connect must equal the total currents $\boldsymbol{I_{R1}}$ and $\boldsymbol{I_{R2}}$ flowing through the resistors.

$$\begin{aligned} I_T &= I_{R1} + I_{R2} \\ &= 227\text{ mA} + 128\text{ mA} \\ &= 355\text{ mA} \end{aligned}$$

The ac currents in the circuit are $\boldsymbol{I_{R1}}$ **= 227 mA**, $\boldsymbol{I_{R2}}$ **= 128 mA**, and $\boldsymbol{I_T}$ **= 355 mA.**

2. Problem: Determine the source voltage $\boldsymbol{V_S}$ and the voltages $\boldsymbol{V_{R1}}$ and $\boldsymbol{V_{R2}}$ across $\boldsymbol{R_1}$ and $\boldsymbol{R_2}$, respectively, for the circuit in Figure 11-5.

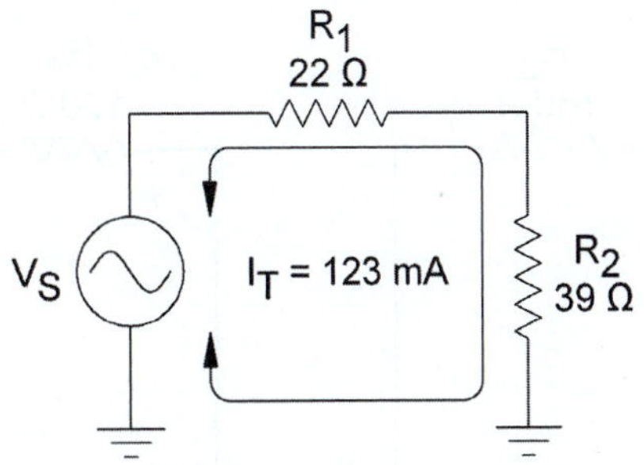

Figure 11-5: Analysis of AC Circuits Example 2

Solution: The circuit is a series circuit so that the total current $\boldsymbol{I_T}$ flows through both $\boldsymbol{R_1}$ and $\boldsymbol{R_2}$. From Ohm's Law for voltage, the voltages $\boldsymbol{V_{R1}}$ and $\boldsymbol{V_{R2}}$ across $\boldsymbol{R_1}$ and $\boldsymbol{R_2}$ are

$$\begin{aligned} V_{R1} &= I_{R1} \times R_1 \\ &= I_T \times R_1 \\ &= 123\text{ mA} \times 22\ \Omega \end{aligned}$$

$= (123 \times 10^{-3}\text{ A}) \times 22\ \Omega$

$= [(123 \times 10^{-3}) \times 22]\ (\text{A} \times \Omega)$

$= 2.706\text{ V}$

$V_{R2} = I_{R2} \times R_2$

$= I_T \times R_2$

$= 123\text{ mA} \times 39\ \Omega$

$= (123 \times 10^{-3}\text{ A}) \times 39\ \Omega$

$= [(123 \times 10^{-3}) \times 39]\ (\text{A} \times \Omega)$

$= 4.797\text{ V}$

From Kirchhoff's Voltage Law, the applied source voltage V_S must equal the total voltages V_{R1} and V_{R2} dropped across the resistors.

$V_S = V_{R1} + V_{R2}$

$= 2.706\text{ V} + 4.797\text{ V}$

$= 7.50\text{ V}$

The ac voltages in the circuit are $V_{R1} = 2.7$ V, $V_{R2} = 4.8$ V, and $V_S = 7.5$ V.

3. Problem: Determine and graph the combined ac and dc voltages V_{R1} and V_{R2} across R_1 and R_2, respectively, for the circuit in Figure 11-6. Use the left side of each resistor for the voltage reference.

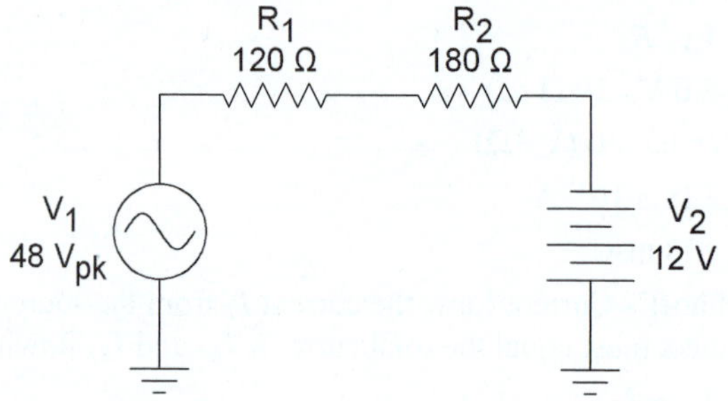

Figure 11-6: Analysis of AC Circuits Example 3

Solution: Superposition is used to determine the combined ac and dc voltages across R_1 and R_2. To determine the ac component, replace V_2 with its internal resistance (a short) and use the voltage divider theorem to find $V_{R1\,(\text{ac})}$ and $V_{R2\,(\text{ac})}$. Similarly, to find the dc component, replace V_1 with its internal resistance (a short) and use the voltage divider theorem to find $V_{R1\,(\text{dc})}$ and $V_{R2\,(\text{dc})}$. Refer to Figure 11-7.

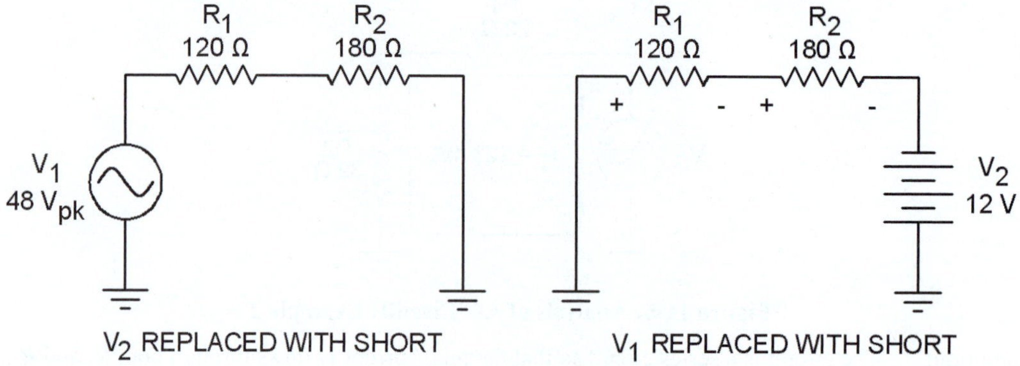

Figure 11-7: Circuits for Superposition Theorem

First, calculate the total series resistance for the circuit.

$$\begin{aligned} \boldsymbol{R_T} &= \boldsymbol{R}_1 + \boldsymbol{R}_2 \\ &= 120\ \Omega + 180\ \Omega \\ &= 300\ \Omega \end{aligned}$$

Next, use the voltage divider theorem to calculate the ac voltages $\boldsymbol{V_{R1}}_{\text{(ac)}}$ and $\boldsymbol{V_{R2}}_{\text{(ac)}}$.

$$\begin{aligned} \boldsymbol{V_{R1}}_{\text{(ac)}} &= \boldsymbol{V}_1 \times (\boldsymbol{R}_1 / \boldsymbol{R_T}) \\ &= 48\ \text{V}_{\text{pk}} \times (120\ \Omega / 300\ \Omega) \\ &= 48\ \text{V}_{\text{pk}} \times 0.4 \\ &= 19.2\ \text{V}_{\text{pk}} \end{aligned}$$

$$\begin{aligned} \boldsymbol{V_{R2}}_{\text{(ac)}} &= \boldsymbol{V}_1 \times (\boldsymbol{R}_2 / \boldsymbol{R_T}) \\ &= 48\ \text{V}_{\text{pk}} \times (180\ \Omega / 300\ \Omega) \\ &= 48\ \text{V}_{\text{pk}} \times 0.6 \\ &= 28.8\ \text{V}_{\text{pk}} \end{aligned}$$

From the polarities shown in Figure 11-7, use the voltage divider theorem to calculate the dc voltages $\boldsymbol{V_{R1}}$ (dc) and $\boldsymbol{V_{R2}}$ (dc).

$$\begin{aligned} \boldsymbol{V_{R1}}_{\text{(dc)}} &= \boldsymbol{V}_2 \times (\boldsymbol{R}_1 / \boldsymbol{R_T}) \\ &= -12\ \text{V}_{\text{dc}} \times (120\ \Omega / 300\ \Omega) \\ &= -12\ \text{V}_{\text{dc}} \times 0.4 \\ &= -4.8\ \text{V}_{\text{dc}} \end{aligned}$$

$$\begin{aligned} \boldsymbol{V_{R2}}_{\text{(dc)}} &= \boldsymbol{V}_2 \times (\boldsymbol{R}_2 / \boldsymbol{R_T}) \\ &= -12\ \text{V}_{\text{dc}} \times (180\ \Omega / 300\ \Omega) \\ &= -12\ \text{V}_{\text{dc}} \times 0.6 \\ &= -7.2\ \text{V}_{\text{dc}} \end{aligned}$$

Finally, combine the ac and dc voltages to determine the voltages for each resistor.

$$\begin{aligned} \boldsymbol{V_{R1}} &= \boldsymbol{V_{R1}}_{\text{(ac)}} + \boldsymbol{V_{R1}}_{\text{(dc)}} \\ \boldsymbol{V_{R1}}_{\text{(max)}} &= 19.2\ \text{V}_{\text{pk}} + -4.8\ \text{V}_{\text{dc}} \\ &= \mathbf{14.4\ V} \\ \boldsymbol{V_{R1}}_{\text{(min)}} &= -19.2\ \text{V}_{\text{pk}} + -4.8\ \text{V}_{\text{dc}} \\ &= \mathbf{-24.0\ V} \end{aligned}$$

$$\begin{aligned} \boldsymbol{V_{R2}} &= \boldsymbol{V_{R1}}_{\text{(ac)}} + \boldsymbol{V_{R1}}_{\text{(dc)}} \\ \boldsymbol{V_{R2}}_{\text{(max)}} &= 28.8\ \text{V}_{\text{pk}} + -7.2\ \text{V}_{\text{dc}} \\ &= \mathbf{21.6\ V} \\ \boldsymbol{V_{R2}}_{\text{(min)}} &= -28.8\ \text{V}_{\text{pk}} + -7.2\ \text{V}_{\text{dc}} \\ &= \mathbf{-36.0\ V} \end{aligned}$$

Figure 11-8 shows the combined voltages $\boldsymbol{V_{R1}}$ and $\boldsymbol{V_{R2}}$.

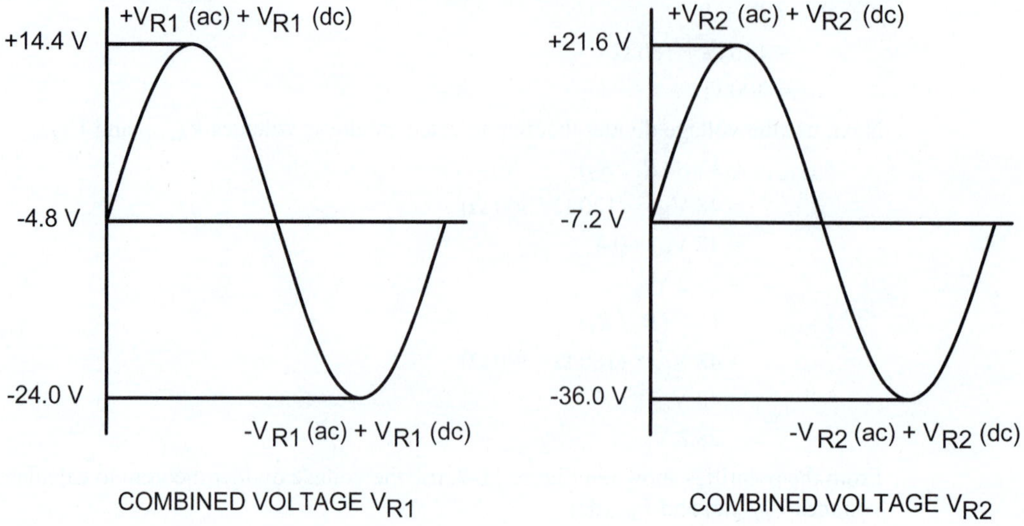

Figure 11-8: Combined Voltages V_{R1} and V_{R2}

4. Problem: Determine the power dissipation P_{RL} for the load resistor R_L for the circuit in Figure 11-9.

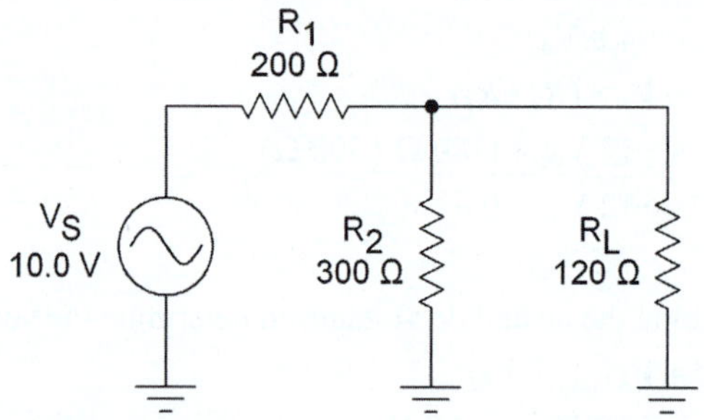

Figure 11-9: Analysis of AC Circuits Example 4

Solution: To determine the power dissipated by R_L, first simplify the circuit. Refer to Figure 11-10.

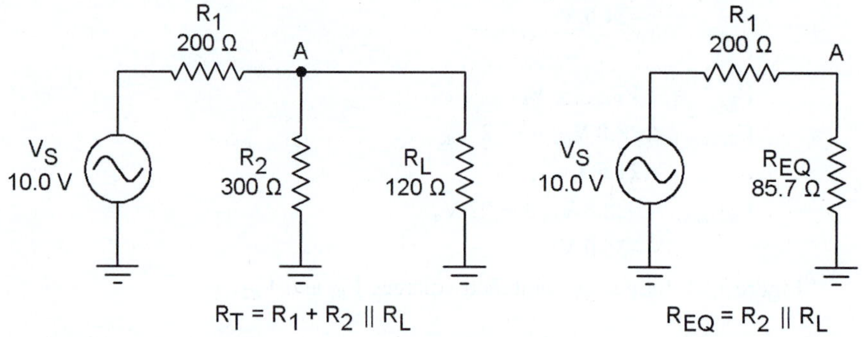

Figure 11-10: Circuit Analysis and Simplification Process for Figure 11-9

The expression for the circuit is

$$R_T = R_1 + R_2 \parallel R_L$$

The first (and only) equivalent resistance is $R_{EQ} = R_2 \parallel R_L$. From this,

$$R_T = R_1 + R_{EQ}$$

where

$\boldsymbol{R_{EQ}} = \boldsymbol{R_2} \parallel \boldsymbol{R_L}$
$= 300\ \Omega \parallel 120\ \Omega$
$= 85.7\ \Omega$

You can now use the voltage divider theorem to find $\boldsymbol{V_A}$, which is the voltage $\boldsymbol{V_{REQ}}$ across $\boldsymbol{R_{EQ}}$. Because $\boldsymbol{R_{EQ}}$ is the parallel combination of $\boldsymbol{R_2} \parallel \boldsymbol{R_L}$, $\boldsymbol{V_A}$ equals both the voltages $\boldsymbol{V_{R2}}$ and $\boldsymbol{V_{RL}}$ across $\boldsymbol{R_2}$ and $\boldsymbol{R_L}$, respectively.

$\boldsymbol{V_{RL}} = \boldsymbol{V_{REQ}}$
$= \boldsymbol{V_A}$
$= \boldsymbol{V_S} \times [\boldsymbol{R_{EQ}} / (\boldsymbol{R_1} + \boldsymbol{R_{EQ}})]$
$= 10.0\ \text{V} \times [85.7\ \Omega / (200\ \Omega + 85.7\ \Omega)]$
$= 10.0\ \text{V} \times (85.7\ \Omega / 285.7\ \Omega)$
$= 10.0\ \text{V} \times 0.3$
$= 3.0\ \text{V}$

Because the ac voltage of $\boldsymbol{V_S}$ is rms by default, use Watt's Law for voltage and resistance.

$\boldsymbol{P_{RL}} = \boldsymbol{V_{RL}}^2 / \boldsymbol{R_L}$
$= (3.0\ \text{V})^2 / 120\ \Omega$
$= 9.0\ \text{V}^2 / 120\ \Omega$
$= (9.0 / 120)\ (\text{V}^2 / \Omega)$
$= 75 \times 10^{-3}\ \text{W}$
$= 75\ \text{mW}$

The power dissipated by $\boldsymbol{R_L}$ is **75 mW**.

11.1.7 The Alternator (AC Generator)

1. Problem: Derive the alternator frequency equation

$\boldsymbol{f} = (\boldsymbol{N} \times \boldsymbol{s}) / 120$

where

$\boldsymbol{f}$ is frequency of the output in Hz,

$\boldsymbol{N}$ is the number of poles, and

$\boldsymbol{s}$ is the rotational speed of the alternator in rpm

Solution: Each set of magnetic poles consists of two poles. From the operation of an alternator, each set of poles generates one alternation (cycle) of ac voltage per revolution.

$\boldsymbol{N}$ poles × 1 rev	$= \boldsymbol{N} / 2$ cycles
($\boldsymbol{N}$ poles × 1 rev) / 1 s	$= (\boldsymbol{N} / 2$ cycles$) / (1$ s$)$
$\boldsymbol{N}$ poles × (1 rev / 1 s)	$= (\boldsymbol{N} / 2)$ (1 cycle / 1 s)
$\boldsymbol{N}$ poles × (1 rev / 1 s)	$= (\boldsymbol{N} / 2)$ Hz
$\boldsymbol{N}$ poles × (1 rev / 1 s) × (60 s / 1 min)	$= (\boldsymbol{N} / 2)$ Hz
$\boldsymbol{N}$ poles × (1 rev × 60 s / (1 s × 1 min)	$= (\boldsymbol{N} / 2)$ Hz
$\boldsymbol{N}$ poles × (60 rev / 1 min)	$= (\boldsymbol{N} / 2)$ Hz
($\boldsymbol{N}$ poles × 60 rpm) × (2 / $\boldsymbol{N}$)	$= (\boldsymbol{N} / 2)$ Hz × (2 / $\boldsymbol{N}$)
($\boldsymbol{N}$ poles × 60 rpm × 2) / $\boldsymbol{N}$	= 1 Hz
1 pole × 120 rpm	= 1 Hz

From this, 120 rpm for each pole will generate a frequency of 1 Hz. Each rpm will therefore generate 1 / 120th of this frequency for each pole.

$\boldsymbol{f} = (\boldsymbol{N} \times \boldsymbol{s}) / 120$

where

f is frequency of the output in Hz,

N is the number of poles, and

s is the rotational speed of the alternator in rpm

2. Problem: An 10-pole alternator generates a 400 Hz ac voltage. What is the speed s of the alternator in rpm?

Solution: From the equation in Problem 1, $f = (N \times s) / 120$. Solve for s.

$$f = (N \times s) / 120$$

$$f / N = [(N \times s) / 120] / N$$

$$(f / N) \times 120 = (s / 120) \times 120$$

$$(f / N) \times 120 = s$$

From the problem statement, $N = 10$ and $f = 400$ Hz.

$$s = (400 \text{ Hz} / 10) \times 120$$

$$= 40 \text{ Hz} \times 120$$

$$= 4{,}800 \text{ rpm}$$

The speed of the alternator is **4,800 rpm**.

11.1.8 Nonsinusoidal Waveforms

1. Problem: Show that the average value V_{ave} of a pulse waveform with baseline V_{base}, amplitude V_A, and duty cycle DC is

$$V_{ave} = V_{base} + (V_A \times \text{DC})$$

Solution: You can use superposition to analyze the pulse and baseline voltages of a general pulse waveform and then sum them for the overall pulse waveform. Refer to Figure 11-11.

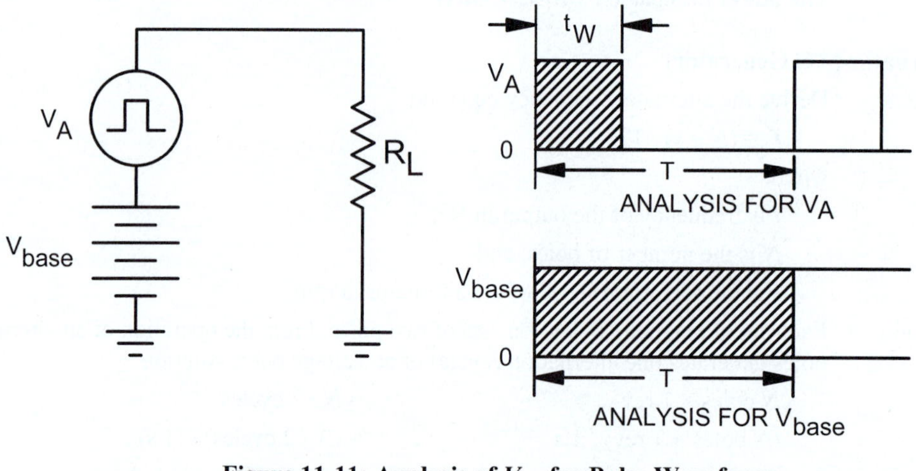

Figure 11-11: Analysis of V_{ave} for Pulse Waveforms

By definition, the average value of a waveform is the area between the waveform and the x-axis (0 V level) divided by its duration. As Figure 11-11 shows, the area for the pulse component with amplitude V_A for one period T is

$$A_{PULSE} = V_A \times t_W$$

Similarly, the area for the baseline component V_{base} for one period T is

$$A_{BASE} = V_{base} \times T$$

Note that if $V_{base} < 0$ then $A_{BASE} < 0$. The total area A_T is the sum of these, so

$$A_T = A_{PULSE} + A_{BASE}$$

$$= (V_A \times t_W) + (V_{base} \times T)$$

$= (V_{base} \times T) + (V_A \times t_W)$

so the average value V_{ave} over one period T is

$V_{ave} = A_T / T$

$= [(V_{base} \times T) + (V_A \times t_W)] / T$

$= [(V_{base} \times T) / T] + [(V_A \times t_W) / T]$

$= V_{base} + [V_A \times (t_W / T)]$

$= V_{base} + (V_A \times DC)$

2. Problem: A pulse waveform has an amplitude voltage $V_A = 5.0$ V and baseline voltage $V_{base} = -2.0$ V. What percent duty cycle DC will result in an average value $V_{ave} = 0$ V?

Solution: Use the equation from Problem 1 to calculate the average value of a pulse waveform and solve for the duty cycle, DC.

$V_{ave} = V_{base} + (V_A \times DC)$

$V_{ave} + -V_{base} = [V_{base} + (V_A \times DC)] + -V_{base}$

$(V_{ave} + -V_{base}) / V_A = (V_A \times DC) / V_A$

$(V_{ave} + -V_{base}) / V_A = DC$

Use the values of V_{ave} and V_{base} to calculate DC.

$DC = \{[0\text{ V} + -(-2.0\text{ V})] / 5.0\text{ V}\} \times 100\%$

$= (2.0\text{ V} / 5.0\text{ V}) \times 100\%$

$= 0.4 \times 100\%$

$= 40\%$

The duty cycle that will produce an average value of 0 V is **40%**.

3. Problem: What is the frequency f of a waveform with a pulse width $t_W = 20$ μs and duty cycle DC = 30%?

Solution: Use the definition of the percent duty cycle of a pulse waveform with period T to solve for f.

$DC = (t_W / T) \times 100\%$

$DC / 100\% = [(t_W / T) \times 100\%] / 100\%$

$(DC / 100\%) / t_W = (t_W / T) / t_W$

$DC / (100\% \times t_W) = 1 / T$

$DC / (100\% \times t_W) = f$

Use the values of t_W and DC to calculate f.

$f = 30\% / (100\% \times 20\text{ μs})$

$= 30\% / [100\% \times (20 \times 10^{-6}\text{ s})]$

$= \{30 / [100 \times (20 \times 10^{-6})]\} [\% / (\% \times s)]$

$= 15 \times 10^3 (1 / s)$

$= 15 \times 10^3$ Hz

$= 15$ kHz

The frequency of the pulse waveform is **15 kHz**.

4. Problem: What is the frequency f of a triangular waveform that has a positive peak of $V_{max} = +2.5$ V, negative peak of $V_{min} = -1.0$ V, positive ramp rate $k_P = 5.0$ V / ms, and negative ramp rate $k_N = -3.0$ V / ms?

Solution: Assume that the waveform starts at the negative peak of −1.0 V. The time t_P to ramp up from $V_{min} = -1.0$ V to the positive peak V_{max} of +2.5 V at the positive ramp rate of $k_P = 5.0$ V / ms is

$t_P = (V_{max} - V_{min}) / k_P$

$= [(+2.5\text{ V}) - (-1.0\text{ V})] / (5.0\text{ V / ms})$

$= 3.5\text{ V} \times (1\text{ ms} / 5.0\text{ V})$

$= 3.5\text{ V} \times [(1 \times 10^{-3}\text{ s}) / 5.0\text{ V}]$
$= \{[3.5 \times (1 \times 10^{-3})] / 5.0\} [(\text{V} \times \text{s}) / \text{V}]$
$= 700 \times 10^{-6}\text{ s}$

The time t_N to ramp down from $V_{max} = +2.5$ V to the negative peak V_{min} of −1.0 V at the negative ramp rate of $k_N = -3.0$ V / ms is

$t_N = (V_{min} - V_{max}) / k_N$
$= [(-1.0\text{ V}) - (+2.5\text{ V})] / (-3.0\text{ V / ms})$
$= -3.5\text{ V} \times (1\text{ ms} / -3.0\text{ V})$
$= -3.5\text{ V} \times [(1 \times 10^{-3}\text{ s}) / -3.0\text{ V}]$
$= \{[-3.5 \times (1 \times 10^{-3})] / -3.0\} [(\text{V} \times \text{s}) / \text{V}]$
$= 1.167 \times 10^{-3}\text{ s}$

The total period for the triangular waveform is the sum of the positive and negative ramp times.

$T = t_P + t_N$
$= 700 \times 10^{-6}\text{ s} + 1.167 \times 10^{-3}\text{ s}$
$= 1.867 \times 10^{-3}\text{ s}$

The frequency is the reciprocal of the period.

$f = 1 / T$
$= 1 / (1.867 \times 10^{-3}\text{ s})$
$= [1 / (1.867 \times 10^{-3})] (1 / \text{s})$
$= [1 / (1.867 \times 10^{-3})] (1 / \text{s})$
$= 536\text{ Hz}$

The frequency of the triangular waveform is **540 Hz.**

11.1.9 The Oscilloscope

For all problems in Section 11.1.9 refer to the oscilloscope display in Figure 11-12.

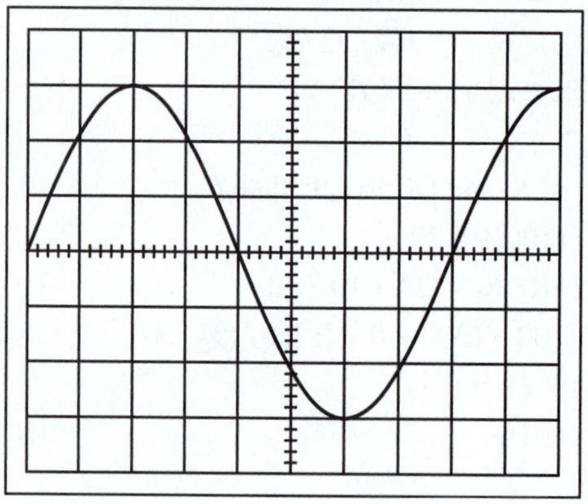

Figure 11-12: Oscilloscope Display for Section 11.1.9

1. Problem: What is the peak-to-peak voltage $V_{pk\text{-}pk}$ for the waveform if the measurement used a 1× probe and the oscilloscope is set for 200 mV / DIV?

   Solution: The peak-to-peak amplitude of the waveform is 6 divisions so for a 1× probe the peak-to-peak voltage is

   $V_{pk\text{-}pk} = (6\text{ DIV}) \times (200\text{ mV / DIV}) \times 1$
   $= (6\text{ DIV}) \times [(200 \times 10^{-3}\text{ V}) / \text{DIV}] \times 1$

$= [6 \times (200 \times 10^{-3}) \times 1]\ [(\text{DIV} \times \text{V}) / \text{DIV}]$

$= 1.2\ \text{V}$

2. Problem: What is the peak voltage V_{pk} for the waveform if the measurement used a 10× probe and the oscilloscope is set for 50 μV / DIV?

   Solution: The peak amplitude of the waveform is 3 divisions, so for a 10× probe the peak voltage is

$$\begin{aligned} V_{pk} &= (3\ \text{DIV}) \times (50\ \mu\text{V} / \text{DIV}) \times 10 \\ &= (3\ \text{DIV}) \times [(50 \times 10^{-6}\ \text{V}) / \text{DIV}] \times 10 \\ &= [3 \times (50 \times 10^{-6}) \times 10]\ [(\text{DIV} \times \text{mV}) / \text{DIV}] \\ &= 1.5 \times 10^{-3}\ \text{V} \\ &= \mathbf{1.5\ mV} \end{aligned}$$

3. Problem: What is the period T of the waveform if the oscilloscope is set for 1 ms / DIV?

   Solution: The measurement from the positive zero crossing for the waveform at the left edge of the waveform to the next positive zero crossing, representing the period T, is 8 horizontal divisions. The period of the waveform is

$$\begin{aligned} T &= (8\ \text{DIV}) \times (1\ \text{ms} / \text{DIV}) \\ &= (8 \times 1)\ [(\text{DIV} \times \text{ms}) / \text{DIV}] \\ &= \mathbf{8\ ms} \end{aligned}$$

4. Problem: What is the frequency of the waveform if the oscilloscope is set for 50 ns / DIV and the horizontal calibration control is set to 10×?

   Solution: The measurement from the positive zero crossing for the waveform at the left edge of the waveform to the next positive zero crossing representing the period T is 8 horizontal divisions. With the horizontal calibration control set to 10×, the measured time is magnified by 10 times so each horizontal division is 1/10th the indicated value, or 50 ns / 10 = 5 ns. The period of the waveform is

$$\begin{aligned} T &= (8\ \text{DIV}) \times (5\ \text{ns} / \text{DIV}) \\ &= (8 \times 5)\ [(\text{DIV} \times \text{ns}) / \text{DIV}] \\ &= 40\ \text{ns} \end{aligned}$$

   From this the frequency of the waveform is

$$\begin{aligned} f &= 1 / T \\ &= 1 / (40\ \text{ns}) \\ &= 1 / (40 \times 10^{-9}\ \text{s}) \\ &= [1 / (40 \times 10^{-9})]\ (1 / \text{s}) \\ &= 25 \times 10^{6}\ \text{Hz} \\ &= \mathbf{25\ MHz} \end{aligned}$$

11.2 Advanced

11.2.1 The Sinusoidal Waveform

1. Problem: Show that the power P_D dissipated in a resistance R by a sinusoidal voltage with peak value V_{pk} is $(V_{pk}^2 / R) / 2$

   Solution: Start with Watt's Law for voltage and resistance for the power P_D dissipated by a dc voltage V_{DC} across a resistance R.

$$P_D = V_{DC}^2 / R$$

   By definition, the rms voltage V_{rms} for a sinusoidal current is equal to the dc voltage that would dissipate the same amount of power.

$$P_D = V_{rms}^2 / R$$

For a sine wave with peak value V_P, $V_{rms} = V_{pk} / \sqrt{2}$. Substitute this value for V_{rms}.

$$P_D = (V_{pk} / \sqrt{2})^2 / R$$
$$= [(V_{pk}{}^2 / R) / (\sqrt{2})^2$$
$$= (V_{pk}{}^2 / R) / 2$$

2. Problem: Show that the power P_D dissipated by a resistor with peak voltage V_{pk} across it and peak current I_{pk} flowing through it is

$$P_D = (V_{pk} \times I_{pk}) / 2$$

Solution: Start with Watt's Law for voltage and current for the power P_D dissipated by a resistor.

$$P_D = V_{DC} \times I_{DC}$$

By definition, the rms voltage V_{rms} for a sinusoidal current is equal to the dc voltage that would dissipate the same amount of power and the rms current I_{rms} is the sinusoidal current equal to the dc current that would dissipate the same amount for power.

$$P_D = V_{rms} \times I_{rms}$$

The rms values for current and voltage for a sinusoidal waveform are

$$V_{rms} = V_{pk} / \sqrt{2}$$
$$I_{rms} = I_{pk} / \sqrt{2}$$

Substitute the values of V_{rms} and I_{rms} into the equation for P_D.

$$P_D = (V_{pk} / \sqrt{2}) \times (I_{pk} / \sqrt{2})$$
$$= (V_{pk} \times I_{pk}) / (\sqrt{2} \times \sqrt{2})$$
$$= (V_{pk} \times I_{pk}) / 2$$

11.2.2 Sinusoidal Waveform Values

1. Problem: The value of point A on a sine wave is 0.629 and the value of point B on the same sine wave is 0.961. What is the difference in degrees for $\Delta\theta = \theta_B - \theta_A$ between the angles θ_A and θ_B for points A and B, respectively? Both θ_A and θ_B are less than 90°.

Solution: Solve for point A on the sine wave from $\sin(\theta_A)$.

$$\sin(\theta_A) = 0.629$$
$$\sin^{-1}(\sin(\theta_A)) = \sin^{-1}(0.629)$$
$$\theta_A = 39.0°$$

Solve for point B on the sine wave from $\sin(\theta_B)$.

$$\sin(\theta_B) = 0.961$$
$$\sin^{-1}(\sin(\theta_B)) = \sin^{-1}(0.961)$$
$$\theta_B = 74.0°$$

Finally calculate the difference $\Delta\theta$ between θ_A and θ_B.

$$\Delta\theta = \theta_B - \theta_A$$
$$= 74.0° - 39.0°$$
$$= \mathbf{35°}$$

2. Problem: Although two angles in a sine wave period can have the same sine value, the arcsine ($\sin^{-1}$) function on most calculators will calculate only the angle that is between ±90°. Refer to Figure 11-13.

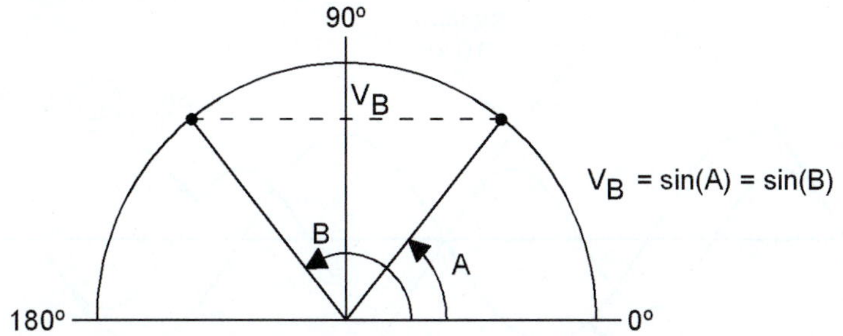

Figure 11-13: Example of Two Angles with the Same Sine Value

Sometimes you will need to know the value of the angle other than that between ±90°. If the arcsine function of your calculator calculates that angle A has sine V_B, what is the value of angle B?

Solution: Although you can prove it using trigonometry, from the symmetry of a circle alone the value of the angle between B and 90° is the same as that of the angle between 90° and A.

$$B - 90° = 90° - A$$
$$B + -90° + 90° = 90° - A + 90°$$
$$B = 180° - A$$

11.2.3 Angular Measurement of a Sine Wave

1. Problem: A point on a sine wave with phase $\varphi = 0$ radians occurs at time $t_1 = 10\ \mu s$. When the sine wave is shifted so that $\varphi = \pi / 12$ radians, the point occurs at time $t_2 = 15\ \mu s$. What is the frequency of the sine wave?

Solution: Let $\Delta\varphi$ represent the angular phase shift and Δt represent the time shift. The ratio of the angular phase shift to a complete cycle (2π) is the same as the ratio of the time shift to the period (T).

$$\Delta t / T = \Delta\varphi / 2\pi$$
$$(15\ \mu s - 10\ \mu s) / T = [(\pi / 12) - 0] / 2\pi$$
$$5\ \mu s / T = (\pi / 12) / 2\pi$$
$$5\ \mu s \times (1 / T) = \pi / (12 \times 2\pi)$$
$$5\ \mu s \times (1 / T) = 1 / 24$$
$$(5\ \mu s \times f) / 5\ \mu s = (1 / 24) / 5\ \mu s$$
$$f = 1 / (24 \times 5\ \mu s)$$
$$= 1 / (120\ \mu s)$$
$$= 1 / (120 \times 10^{-6}\ s)$$
$$= [1 / (120 \times 10^{-6})]\ (1 / s)$$
$$= 8.33 \times 10^3\ Hz$$
$$= 8.33\ kHz$$

The frequency of the sine wave is **8.3 kHz.**

2. Problem: The signals shown in Figure 11-14 all pass through a circuit with a constant phase shift of +45°.

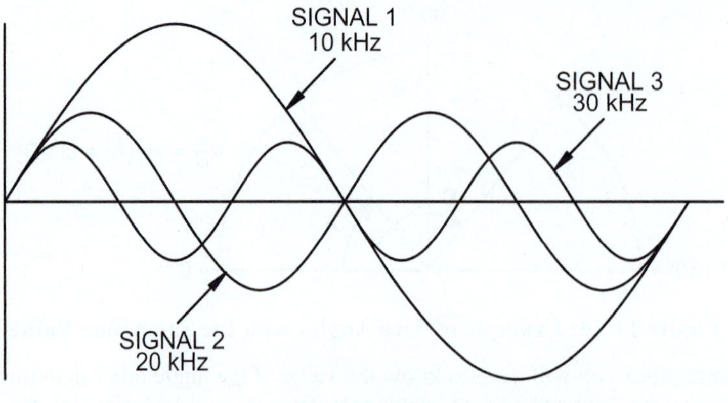

Figure 11-14: Signal Phase Shift Example

What are the relationships of the signals after passing through the circuit?

Solution: 45° is equal to (45° / 360°) = 1/8 of a complete cycle, so each of the signals will shift to the right by 1/8 of its cycle, as shown in Figure 11-15. As you can see, the relationship between the signals differs greatly from the original relationship shown in Figure 11-14. This is because the time that 45° represents differs for each signal. This is the problem with shifting signals with different frequencies by the same angle.

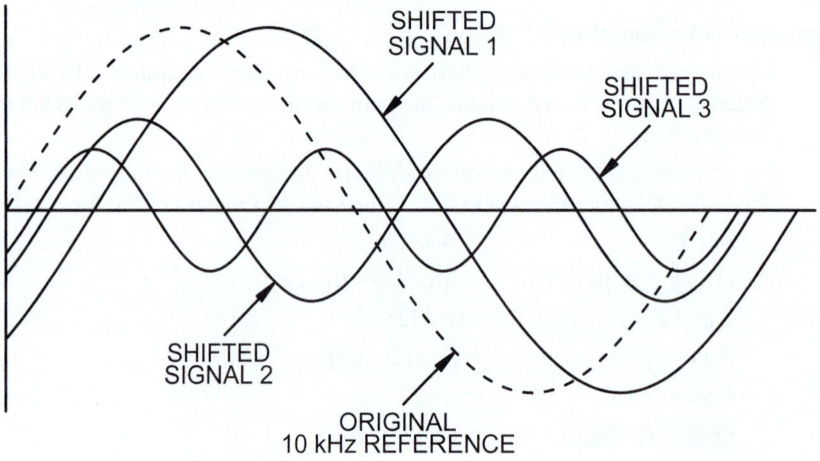

Figure 11-15: Signals Shifted by 45°

11.2.4 The Sine Wave Formula

1. Problem: The positive peaks on two sine waves with frequency $f = 1$ MHz are 150 ns apart. What is the phase angle in degrees between the sine waves?

Solution: The period of the sine waves is

$$\begin{aligned} \boldsymbol{T} &= 1 / \boldsymbol{f} \\ &= 1 / 1\ \text{MHz} \\ &= 1 / (1 \times 10^6\ \text{Hz}) \\ &= [1 / (1 \times 10^6)]\ (1 / \text{Hz}) \\ &= 1 \times 10^{-6}\ \text{s} \end{aligned}$$

The ratio of the phase angle φ between the two sine waves relative to a complete cycle of 360° equals the ratio of the time difference between the corresponding points to the period $\boldsymbol{T}$ of the sine waves.

$$\varphi / 360° = \Delta t / T$$
$$(\varphi / 360°) \times 360° = (\Delta t / T) \times 360°$$
$$\varphi = (\Delta t / T) \times 360°$$
$$\varphi = [150 \text{ ns} / (1 \times 10^{-6} \text{ s})] \times 360°$$
$$= [(150 \times 10^{-9} \text{ s}) / (1 \times 10^{-6} \text{ s})] \times 360°$$
$$= 150 \times 10^{-3} \times 360°$$
$$= 54°$$

The phase angle between the two sine waves is **54°**.

2. Problem: One 100 kHz sine wave leads another 100 kHz sine wave by $\pi / 6$ radians. What is the time delay Δt between a point on the leading sine wave and the corresponding point on the lagging sine wave? Assume all values are exact.

Solution: The period of the sine waves is

$$T = 1 / f$$
$$= 1 / 100 \text{ kHz}$$
$$= 1 / (100 \times 10^3 \text{ Hz})$$
$$= [1 / (100 \times 10^3)] (1 / \text{Hz})$$
$$= 10 \times 10^{-6} \text{ s}$$

The ratio of the phase angle φ relative to a complete cycle of 2π radians equals the ratio of the time difference between the corresponding points on the sine waves to their period T.

$$\Delta t / T = \varphi / 2\pi$$
$$(\Delta t / T) \times T = (\varphi / 2\pi) \times T$$
$$\Delta t = (\varphi / 2\pi) \times T$$
$$\Delta t = [(\pi / 6) / 2\pi] \times (10 \times 10^{-6} \text{ s})$$
$$= [\pi / (6 \times 2\pi)] \times (10 \times 10^{-6} \text{ s})$$
$$= (1 / 12) \times (10 \times 10^{-6} \text{ s})$$
$$= (10 \times 10^{-6} \text{ s}) / 12$$
$$= 833 \times 10^{-9} \text{ s}$$
$$= 833 \text{ ns}$$

The time delay between the two sine waves is **833 ns**.

11.2.5 Introduction to Phasors

1. Problem: What is the angular velocity ω in radians of a phasor if the sine wave it represents has a peak amplitude $V_p = 5.0$ V and the instantaneous voltage $v(t)$ increases from −100 mV to +100 mV in 25 ns?

Solution: Start with the equation for the instantaneous voltage $v(t)$ at time t for a sine wave with peak amplitude V_{pk} and angular frequency ω and solve for t.

$$v(t) = V_{pk} \times \sin(\omega \times t)$$
$$v(t) / V_{pk} = [V_p \times \sin(\omega \times t)] / V_p$$
$$\sin^{-1} [v(t) / V_p] = \sin^{-1}[\sin(\omega \times t)]$$
$$= \omega \times t$$
$$\sin^{-1} [v(t) / V_p] / \omega = (\omega \times t) / \omega$$
$$\sin^{-1} [v(t) / V_p] / \omega = t$$

Let $v(t_1) = -100$ mV at time t_1 and $v(t_2) = +100$ mV at time t_2.

$$t_1 = \sin^{-1} [v(t_1) / V_p] / \omega$$
$$t_2 = \sin^{-1} [v(t_2) / V_p] / \omega$$

Find the difference between the two times and solve for f.

$t_2 - t_1 = \{\sin^{-1}[v(t_2) / V_p] / \omega\} - \{\sin^{-1}[v(t_1) / V_p] / \omega\}$
$= \{\sin^{-1}[v(t_2) / V_p] - \sin^{-1}[v(t_1) / V_p]\} / \omega$

$(t_2 - t_1) \times \omega = (\{\sin^{-1}[v(t_2) / V_p] - \sin^{-1}[v(t_1) / V_p]\} / \omega) \times \omega$

$[(t_2 - t_1) \times \omega] / (t_2 - t_1) = \{\sin^{-1}[v(t_2) / V_p] - \sin^{-1}[v(t_1) / V_p]\} / (t_2 - t_1)$

$f = \{\sin^{-1}[v(t_2) / V_p] - \sin^{-1}[v(t_1) / V_p]\} / (t_2 - t_1)$
$= \{\sin^{-1}[v(t_2) / V_p] - \sin^{-1}[v(t_1) / V_p]\} / (t_2 - t_1)$

From the problem statement, $V_p = 5.0$ V and $t_2 - t_1 = 25$ ns.

$f = [\sin^{-1}(+100 \text{ mV} / 5.0 \text{ V}) - \sin^{-1}(-100 \text{ mV} / 5.0 \text{ V})] / 25 \text{ ns}$
$= \{\sin^{-1}[(+100 \times 10^{-3} \text{ V}) / 5.0 \text{ V}] - \sin^{-1}[(-100 \times 10^{-3} \text{ V}) / 5.0 \text{ V}]\} / 25 \text{ ns}$
$= [\sin^{-1}(+20 \times 10^{-3}) - \sin^{-1}(-20 \times 10^{-3})] / 25 \text{ ns}$
$= [\sin^{-1}(+20 \times 10^{-3}) - \sin^{-1}(-20 \times 10^{-3})] / 25 \text{ ns}$
$= [(+20.0 \times 10^{-3} \text{ rad}) - (-20.0 \times 10^{-3} \text{ rad})] / 25 \text{ ns}$
$= (+40.0 \times 10^{-3} \text{ rad}) / (25 \times 10^{-9} \text{ s})$
$= [(+40.0 \times 10^{-3}) / (25 \times 10^{-9}) (\text{rad} / \text{s})$
$= +1.60 \times 10^{6} \text{ rad} / \text{s}$

The angular velocity of the phasor is **+1.60 × 10⁶ rad / s**. The positive sign indicates that the phasor is rotating in positive (counter-clockwise) direction.

2. Problem: For the phasor of Problem 1, at what times t_1 and t_2 will $v(t_1) = -100$ mV and $v(t_2) = +100$ mV, respectively, if $v(0) = 0$ V?

Solution: From the derivation of Problem 1,

$t_1 = \sin^{-1}[v(t_1) / V_p] / \omega$
$= \sin^{-1}[-100 \times 10^{-3} \text{ V}) / 5.0 \text{ V}] / (1.60 \times 10^{6} \text{ rad} / \text{s})$
$= \sin^{-1}(-20 \times 10^{-3}) / (1.60 \times 10^{6} \text{ rad} / \text{s})$
$= (-20.0 \times 10^{-3} \text{ rad}) / (1.60 \times 10^{6} \text{ rad} / \text{s})$
$= [(-20.0 \times 10^{-3}) / (1.60 \times 10^{6})] [\text{rad} / (\text{rad} / \text{s})]$
$= -12.5 \times 10^{-9} [\text{rad} \times (\text{s} / \text{rad})]$
$= -12.5 \times 10^{-9} \text{ s}$
$= -12.5 \text{ ns}$

The negative sign indicates that this occurs before time $t = 0$ (when the phasor angle is 0 radians).

$t_2 = \sin^{-1}[v(t_2) / V_p] / \omega$
$= \sin^{-1}[+100 \times 10^{-3} \text{ V}) / 5.0 \text{ V}] / (1.60 \times 10^{6} \text{ rad} / \text{s})$
$= \sin^{-1}(+20 \times 10^{-3}) / (1.60 \times 10^{6} \text{ rad} / \text{s})$
$= (+20.0 \times 10^{-3} \text{ rad}) / (1.60 \times 10^{6} \text{ rad} / \text{s})$
$= [(+20.0 \times 10^{-3}) / (1.60 \times 10^{6})] [\text{rad} / (\text{rad} / \text{s})]$
$= +12.5 \times 10^{-9} [\text{rad} \times (\text{s} / \text{rad})]$
$= +12.5 \times 10^{-9} \text{ s}$
$= +12.5 \text{ ns}$

The positive sign indicates that this occurs after time $t = 0$ (when the phase angle is 0 radians). The times for which the instantaneous voltage equal −100 mV and +100 mV are **−12.5 ns** and **+12.5 ns**, respectively, This is expected, due to the symmetry of the sine wave.

11.2.6 Analysis of AC Circuits

1. Problem: Use Ohm's Law to verify that the voltage divider theorem applies to ac resistive circuits.

   Solution: For a series circuit of N resistors and applied voltage V_{ac}, the total series resistance is

   $$R_T = R_1 + R_2 + ... + R_N$$

   From Ohm's Law for ac current, the total ac current is

   $$I_{ac} = V_{ac} / R_T$$

   From Ohm's Law for ac voltage, the ac voltage V_{RK} across any resistor R_K in the series circuit is

   $$V_{RK} = I_{RK} \times R_K$$

   Because the circuit is a series circuit, $I_{RK} = I_{ac}$.

   $$V_{RK} = I_{ac} \times R_K$$
   $$= (V_{ac} / R_T) \times R_K$$
   $$= V_{ac} \times (R_T / R_K)$$

   The voltage divider theorem applies to ac resistive circuits.

2. Problem: Use Ohm's Law to verify that the current divider theorem applies to ac resistive circuits.

   Solution: For a parallel circuit of N resistors and applied voltage V_{ac}, the total parallel resistance is

   $$R_T = 1 / [(1 / R_1) + (1 / R_2) + ... + (1 / R_N)]$$

   From Ohm's Law for ac current, the total ac current is

   $$I_{ac} = V_{ac} / R_T$$

   so

   $$I_{ac} \times R_T = (V_{ac} / R_T) \times R_T$$
   $$= V_{ac}$$

   Also from Ohm's Law for ac current, the ac current I_{RK} through any resistor R_K in the parallel circuit is

   $$I_{RK} = V_{RK} / R_K$$

   Because the circuit is a parallel circuit, $V_{RK} = V_{ac}$, so

   $$I_{RK} = V_{ac} / R_K$$
   $$= (I_{ac} \times R_T) / R_K$$
   $$= I_{ac} \times (R_T / R_K)$$

   The current divider theorem applies to ac resistive circuits.

11.2.7 Nonsinusoidal Waveforms

1. Problem: Show that the average voltage V_{ave} over a half cycle for the ramp waveforms in Figure 11-16 is

   $$V_{ave} = [(V_{max} - V_{min}) / 4] + V_{base}$$

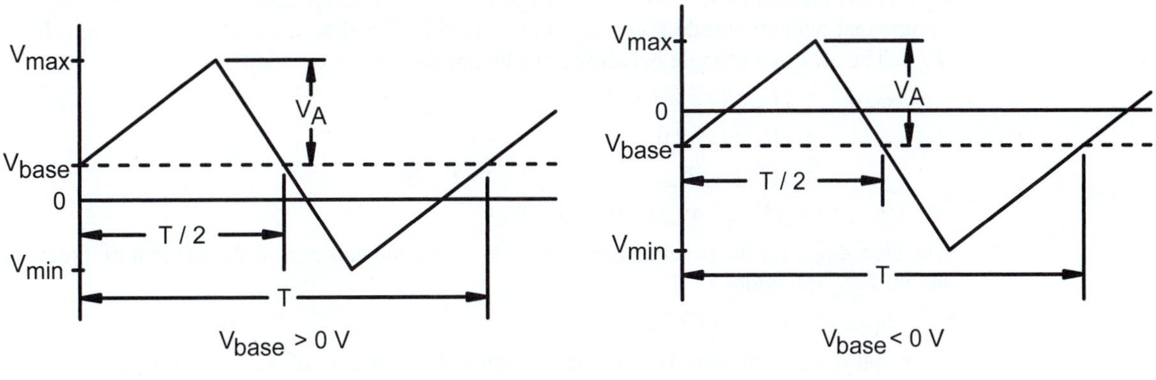

Figure 11-16: V_{ave} for Ramp Waveforms

Solution: First, use the symmetry of the ramp waveforms to determine V_A in terms of V_{max} and V_{min}.

$$\begin{aligned}
V_{max} - V_{base} &= V_{base} - V_{min} \\
V_{max} + -V_{base} &= V_{base} + -V_{min} \\
V_{max} + -V_{base} + V_{base} &= (V_{base} + -V_{min}) + V_{base} \\
V_{max} + V_{min} &= (2 \times V_{base} + -V_{min}) + V_{min} \\
(V_{max} + V_{min}) / 2 &= (2 \times V_{base}) / 2 \\
(V_{max} + V_{min}) / 2 &= V_{base}
\end{aligned}$$

Use this result to solve for $V_A = V_{max} - V_{base}$.

$$\begin{aligned}
V_A &= V_{max} - V_{base} \\
&= V_{max} + -V_{base} \\
&= V_{max} + -[(V_{max} + V_{min}) / 2] \\
&= [(2 \times V_{max}) / 2] + [(-V_{max} + -V_{min}) / 2)] \\
&= [(2 \times V_{max}) + (-V_{max}) + (-V_{min})] / 2 \\
&= [V_{max} + (-V_{min})] / 2 \\
&= (V_{max} - V_{min}) / 2
\end{aligned}$$

Next, use superposition to analyze the ramp and baseline voltages of a general ramp waveform and then sum them for the overall ramp waveform. Refer to Figure 11-17.

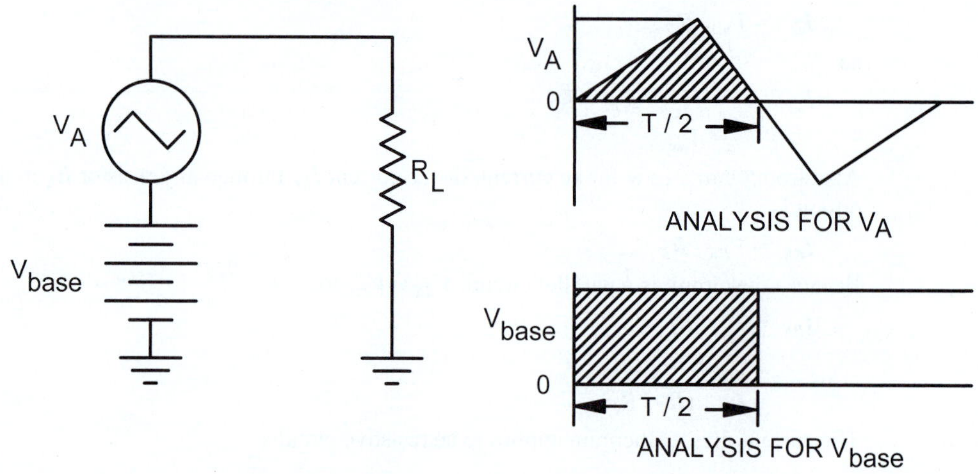

Figure 11-17: Analysis of V_{ave} for Ramp Waveforms

By definition, the average value of a waveform is the area between the waveform and the x-axis (0 V level) divided by its duration. As Figure 11-17 shows, the area A_{RAMP} for the ramp component with amplitude V_A for one half period $T / 2$ is that of a triangle with height V_A and base $T / 2$. The area of a triangle is half the product of the base and height.

$$\begin{aligned}
A_{RAMP} &= [V_A \times (T / 2)] \times (1 / 2) \\
&= [V_A \times (1 / 2)] \times (T / 2) \\
&= \{(V_{max} - V_{min}) / 2] \times (1 / 2)\} \times (T / 2) \\
&= [(V_{max} - V_{min}) / 4] \times (T / 2)
\end{aligned}$$

The area A_{BASE} for the baseline component V_{base} for one half period $T / 2$ is that of a rectangle with height V_{base} and width $T / 2$.

$$A_{BASE} = V_{base} \times (T / 2)$$

Note that if $V_{base} < 0$ then $A_2 < 0$. The total area A_T is the sum of A_{RAMP} and A_{BASE}.

$A_T = \{[(V_{max} - V_{min}) / 4] \times (T / 2)\} + [V_{base} \times (T / 2)]$

$= \{[(V_{max} - V_{min}) / 4] + V_{base}\} \times (T / 2)$

The average value V_{ave} over one half period $T / 2$ is

$V_{ave} = A_T / (T / 2)$

$= (\{[(V_{max} - V_{min}) / 4] + V_{base}\} \times (T / 2)) / (T / 2)$

$= [(V_{max} - V_{min}) / 4] + V_{base}$

2. Problem: Use ramp rates to verify that the average voltage V_{ave} over a half cycle for the ramp waveforms in Figure 11-16 is

$$V_{ave} = [(V_{max} - V_{min}) / 4] + V_{base}$$

given that the positive ramp rate is k_P and negative ramp rate is k_N.

Solution: Refer to Figure 11-18.

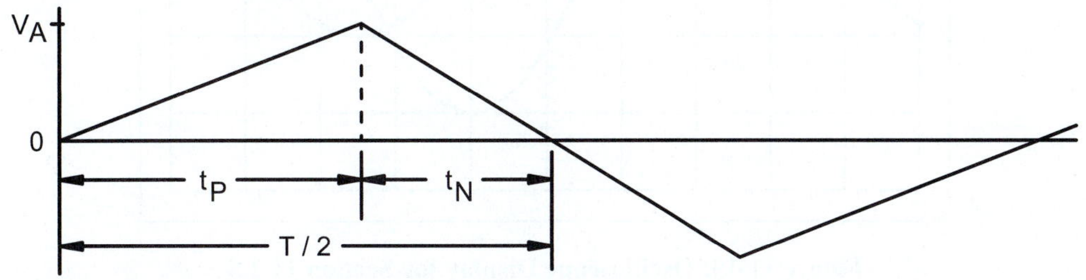

Figure 11-18: Analysis of V_{ave} for Ramp Waveforms Using Ramp Rates

The area of the waveform over the half cycle $T / 2$ is a triangle with height V_A and base $T / 2$. The area of a triangle is half the product of the base and height.

$A_{RAMP} = [V_A \times (T / 2)] \times (1 / 2)$

The area A_{BASE} for the baseline component V_{base} for one half period $T / 2$ in the product of the base and height.

$A_{BASE} = V_{base} \times (T / 2)$

The total area of the sawtooth waveform is the sum of A_{RAMP} and A_{BASE}.

$A_T = A_{RAMP} + A_{BASE}$

$= \{[V_A \times (T / 2)] \times (1 / 2)\} + [V_{base} \times (T / 2)]$

$= \{[V_A \times (1 / 2)] + V_{base}\} \times (T / 2)$

$= [(V_A / 2) + V_{base}] \times (T / 2)$

The average value V_{ave} is the area, A_T, divided by the total time, $T / 2$.

$V_{ave} = A_T / (T / 2)$

$= \{[(V_A / 2) + V_{base}] \times (T / 2)\} / (T / 2)$

$= (V_A / 2) + V_{base}$

From Problem 1, $V_A = (V_{max} - V_{min}) / 2$.

$V_{ave} = [(V_{max} - V_{min}) / 2] / 2) + V_{base}$

$= [(V_{max} - V_{min}) / (2 \times 2)] + V_{base}$

$= [(V_{max} - V_{min}) / 4] + V_{base}$

11.2.8 The Oscilloscope

For problems in Section 11.2.8, refer to the oscilloscope display in Figure 11-19.

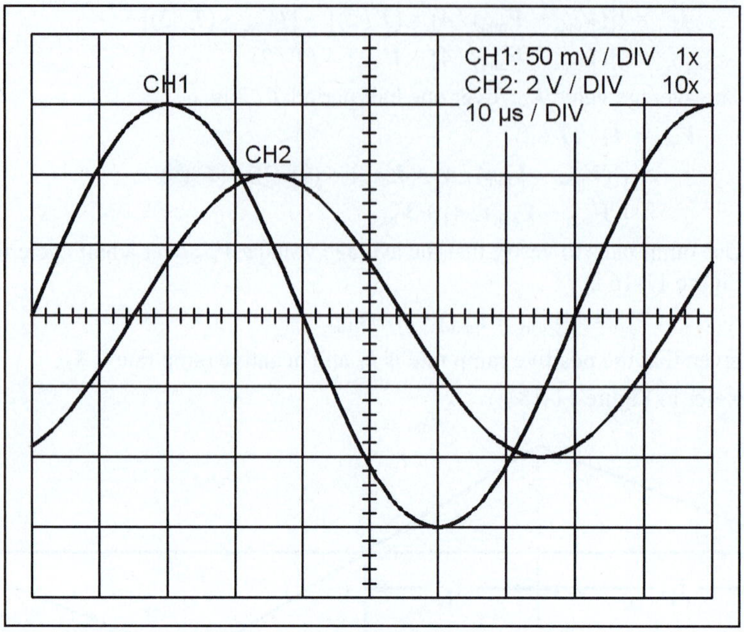

Figure 11-19: Oscilloscope Display for Section 11.2.8

1. Problem: What is the frequency ***f*** for the measured waveforms?

   Solution: The horizontal scale is 10 μs per division and one cycle for each waveform is 8 divisions. The period ***T*** is

$$\boldsymbol{T} = (10\ \mu s / DIV) \times 8\ DIV$$
$$= (10 \times 8)\ [(\mu s / DIV) \times DIV]$$
$$= 80\ \mu s$$

so

$$\boldsymbol{f} = 1 / \boldsymbol{T}$$
$$= 1 / 80\ \mu s$$
$$= 1 / (80 \times 10^{-6}\ s)$$
$$= [1 / (80 \times 10^{-6})\ (1 / s)$$
$$= 12.5 \times 10^{3}\ Hz$$
$$= \mathbf{12.5\ kHz}$$

2. Problem: What are the rms voltages for V_1 and V_2 on Channel 1 and Channel 2, respectively?

   Solution: The vertical scale for Channel 1 is 50 mV per division, Channel 1 is using a 1× probe, and the peak-to-peak amplitude of the Channel 1 signal is 6 divisions, so the Channel 1 voltage is

$$V_1 = (50\ mV_{pk\text{-}pk} / DIV) \times 6\ DIV \times 1$$
$$= (50 \times 6 \times 1)\ [(mV_{pk\text{-}pk} / DIV) \times DIV]$$
$$= 300\ mV_{pk\text{-}pk}$$

For a sine wave, $1\ V_{pk\text{-}pk} = 0.354\ V_{rms}$, so

$$V_1 = (300\ mV_{pk\text{-}pk}) \times (0.354\ V_{rms} / 1\ V_{pk\text{-}pk})$$
$$= (300 \times 10^{-3}\ V_{pk\text{-}pk}) \times (0.354\ V_{rms} / 1\ V_{pk\text{-}pk})$$
$$= \{[(300 \times 10^{-3}) \times 0.354] / 1\}\ [(V_{pk\text{-}pk} \times V_{rms}) / V_{pk\text{-}pk}]$$
$$= 106 \times 10^{-3}\ V_{rms}$$
$$= \mathbf{106\ mV_{rms}}$$

The vertical scale for Channel 2 is 2 V per division, Channel 2 is using a 10× probe, and the peak-to-peak amplitude of the Channel 2 signal is 4 divisions, so the Channel 2 voltage is

$$V_2 = (2\ V_{pk\text{-}pk} / \text{DIV}) \times 4\ \text{DIV} \times 10$$
$$= (2 \times 4 \times 10)\ [(V_{pk\text{-}pk} / \text{DIV}) \times \text{DIV}]$$
$$= 80\ V_{pk\text{-}pk}$$

For a sine wave, 1 $V_{pk\text{-}pk} = 0.354\ V_{rms}$, so

$$V_2 = (80\ V_{pk\text{-}pk}) \times (0.354\ V_{rms} / 1\ V_{pk\text{-}pk})$$
$$= [(80 \times 0.354) / 1]\ [(V_{pk\text{-}pk} \times V_{rms}) / V_{pk\text{-}pk}]$$
$$= \mathbf{28.3\ V_{rms}}$$

3. Problem: What is the phase of the Channel 1 waveform relative to the Channel 2 waveform?

Solution: The negative zero-crossing of the Channel 1 waveform occurs 1.5 divisions before the negative zero crossing of the Channel 2 waveform, and the horizontal scale is 10 µs per division. The time Δt by which Channel 1 leads Channel 2 is

$$\Delta t = (10\ \mu s / \text{DIV}) \times 1.5\ \text{DIV}$$
$$= (10 \times 1.5)\ [(\mu s / \text{DIV}) \times \text{DIV}]$$
$$= 15\ \mu s$$

The ratio of the phase φ to the full cycle of 360° equals the ratio of Δt to the full period T, so

$$\varphi / 360° = \Delta t / T$$
$$\varphi / 360° = (\Delta t / T) \times 360°$$
$$\varphi = (\Delta t / T) \times 360°$$

From this,

$$\varphi = (15\ \mu s / 80\ \mu s) \times 360°$$
$$= [(15 / 80) \times 360]\ [(\mu s / \mu s) \times °]$$
$$= 67.5°$$

The waveform on Channel 1 leads the waveform on Channel 2 by 67.5°.

11.3 Just for Fun

Lissajous figures are graphic representations of the relative amplitude, frequency, and phase of two sinusoidal voltages. Although they are not used as much now as before the introduction of digital oscilloscopes, some areas of signal measurement such as rotor dynamics still use them in some form, and you can sometimes see them rotating on the oscilloscopes in old science-fiction movies.

You can consider an oscilloscope as a device that plots a series of x-y coordinates. For most measurements, the x-coordinates are time values and the y-coordinates are amplitudes of a signal that correspond to the time values. However, most oscilloscopes have an X-Y mode display that allows you to use sinusoidal signals to determine both the x- and y-coordinates. When you do this, you create a Lissajous figure. Refer to Figure 11-20.

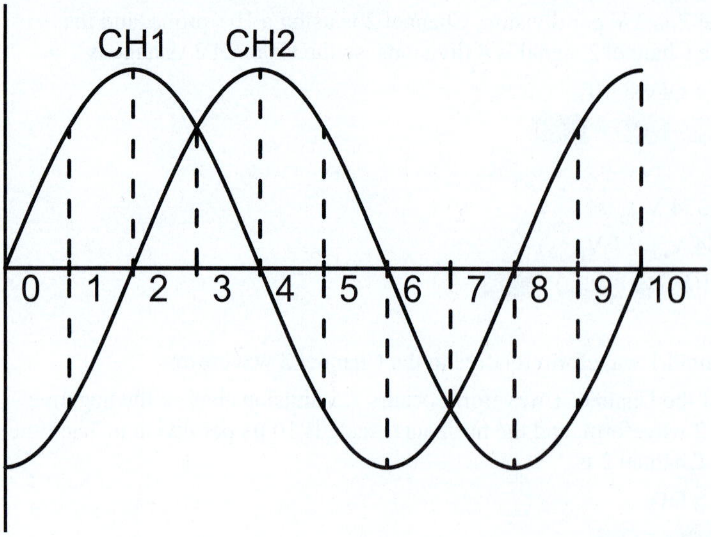

	CHANNEL 1		CHANNEL 2	
SAMPLE	DEGREES	VOLTAGE	DEGREES	VOLTAGE
0	0	0.000	-90	-6.000
1	45	4.242	-45	-4.242
2	90	6.000	0	0.000
3	135	4.242	45	4.242
4	180	0.000	90	6.000
5	225	-4.242	135	4.242
6	270	-6.000	180	0.000
7	315	-4.242	225	-4.242
8	360	0.000	270	-6.000

Figure 11-20: Lissajous Coordinates for Example 1

Figure 11-20 shows two 6.0 V sine waves, for which the signal on CH2 lags the signal on CH1 signal by 90°. The table shows the angles in degrees and the voltage values that correspond to 9 reference sample points 90° apart. At sample point 0, the angle for CH1 is 0° so the channel 1 voltage V_{CH1} for sample 0 is

$$\begin{aligned} V_{\mathrm{CH1}}(0) &= 6.0\text{ V} \times \sin(0°) \\ &= 6.0\text{ V} \times 0.000 \\ &= 0.000\text{ V} \end{aligned}$$

Because CH2 lags CH1 by 90°, the channel 2 voltage V_{CH2} for sample 0 is

$$\begin{aligned} V_{CH2}(0) &= 6.0\text{ V} \times \sin(0° - 90°) \\ &= 6.0\text{ V} \times \sin(-90°) \\ &= 6.0\text{ V} \times -1.000 \\ &= -6.000\text{ V} \end{aligned}$$

The x-y coordinates for sample 0 are then (0.000 V, −6.000 V). For sample 1, 45° after sample 0, the voltages are

$$\begin{aligned} V_{CH1}(1) &= 6.0\text{ V} \times \sin(+45°) \\ &= 6.0\text{ V} \times +0.707 \\ &= +4.242\text{ V} \end{aligned}$$

$$\begin{aligned} V_{CH2}(1) &= 6.0\text{ V} \times \sin(+45° - 90°) \\ &= 6.0\text{ V} \times \sin(-45°) \\ &= 6.0\text{ V} \times -0.707 \\ &= -4.242\text{ V} \end{aligned}$$

The x-y coordinates for sample 1 are then (+4.242 V, −4.242 V). Continuing for the remaining samples gives

$$\begin{aligned} (V_{CH1}(2), V_{CH2}(2)) &= (+6.000\text{ V}, 0.000\text{ V}) \\ (V_{CH1}(3), V_{CH2}(3)) &= (+4.242\text{ V}, +4.242\text{ V}) \\ (V_{CH1}(4), V_{CH2}(4)) &= (+0.000\text{ V}, +6.000\text{ V}) \\ (V_{CH1}(5), V_{CH2}(5)) &= (-4.242\text{ V}, +4.242\text{ V}) \\ (V_{CH1}(6), V_{CH2}(6)) &= (-6.000\text{ V}, 0.000\text{ V}) \\ (V_{CH1}(7), V_{CH2}(7)) &= (-4.242\text{ V}, -4.242\text{ V}) \\ (V_{CH1}(8), V_{CH2}(8)) &= (0.000\text{ V}, -6.000\text{ V}) \end{aligned}$$

These coordinates then define the shape of the Lissajous figure for the waveforms of Figure 11-20. Refer to Figure 11-21.

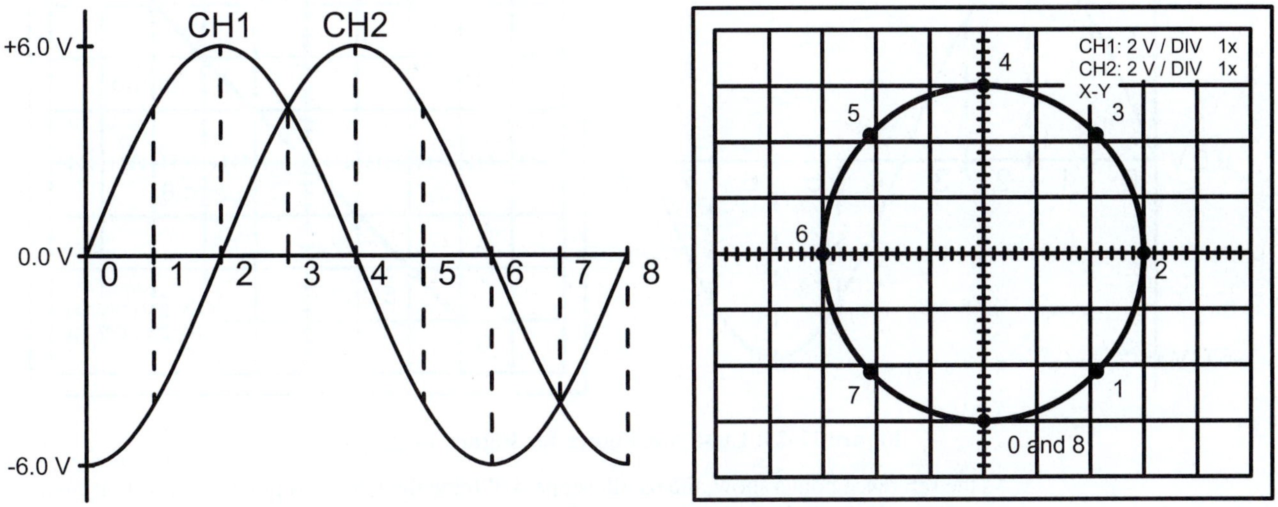

Figure 11-21: Lissajous Figure for Figure 11-20

Figure 11-21 shows that the signals on CH1 and CH2 signals will trace a circle. The reference sample points marked on the display show that the signals trace the circle in a counter-clockwise direction. Note that the Lissajous figure is a circle because the amplitudes for CH1 and CH2 are equal so that the horizontal and vertical axes of the figure are both 12.0 Vpk-pk = 6 divisions. If the amplitudes (or settings) for CH1 and CH2 had not been equal, one axis of the figure would have been longer than the other so that the figure would be an ellipse.

1. Problem: Determine the Lissajous figure for Figure 11-20 if CH1 and CH2 are in phase (i.e., the phase difference $\varphi = 0°$).

   Solution: Refer to Figure 11-22.

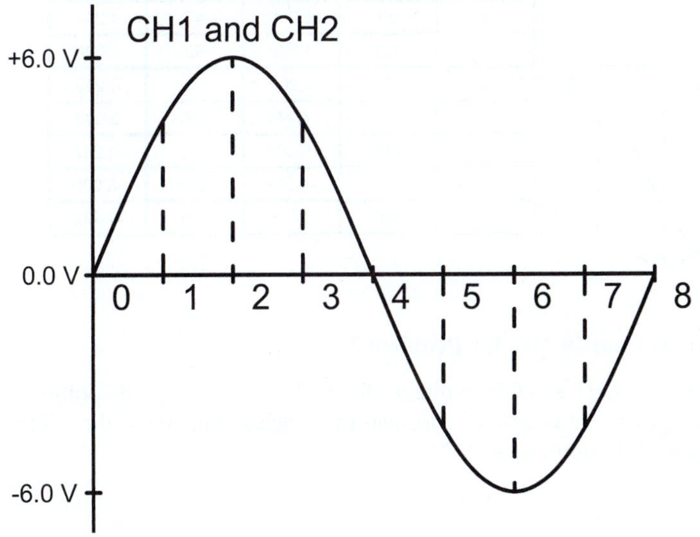

	CHANNEL 1		CHANNEL 2	
SAMPLE	DEGREES	VOLTAGE	DEGREES	VOLTAGE
0	0	0.000	0	0.000
1	45	4.242	45	4.242
2	90	6.000	90	6.000
3	135	4.242	135	4.242
4	180	0.000	180	0.000
5	225	-4.242	225	-4.242
6	270	-6.000	270	-6.000
7	315	-4.242	315	-4.242
8	360	0.000	360	0.000

Figure 11-22: Lissajous Coordinates for Problem 1

The table in Figure 11-22 shows the CH1 and CH2 voltages for each reference sample point. Because the CH1 and CH2 signals are in phase, each voltage for CH1 and CH2 is equal. Consequently, the x- and y- coordinates for each point on the Lissajous figure will be the same and it appears that only one curve exists. The result is the straight line shown in Figure 11-23.

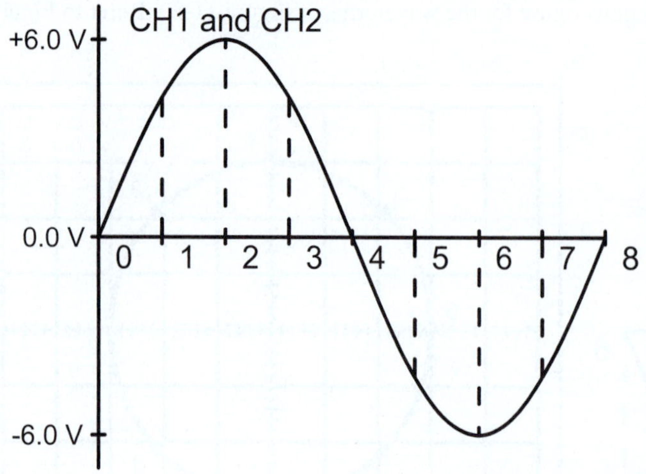

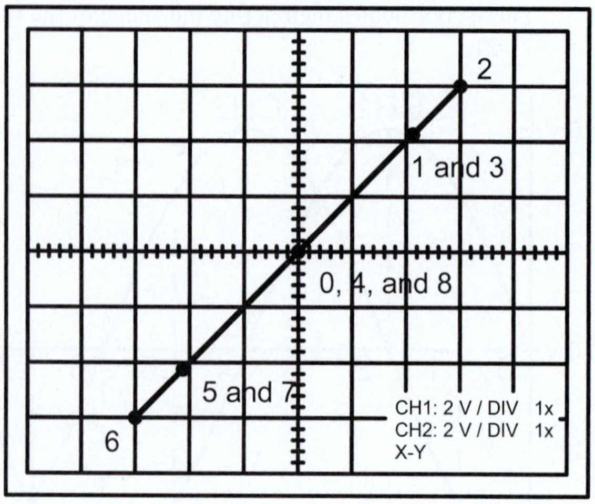

Figure 11-23: Lissajous Figure for Figure 11-22

As the reference points show, the oscilloscope will trace the line from point 0 upward through points 1 and 2, back down through points 3 through 6, and then back up again through points 7 and 8. Note that because the amplitudes of CH1 and CH2 are equal the slope of the line is 1 (i.e., is at a 45° angle).

2. Problem: Determine the Lissajous figure for Figure 11-20 if CH2 lags CH1 by 45° (i.e., the phase φ = 45°).

Solution: Refer to Figure 11-24.

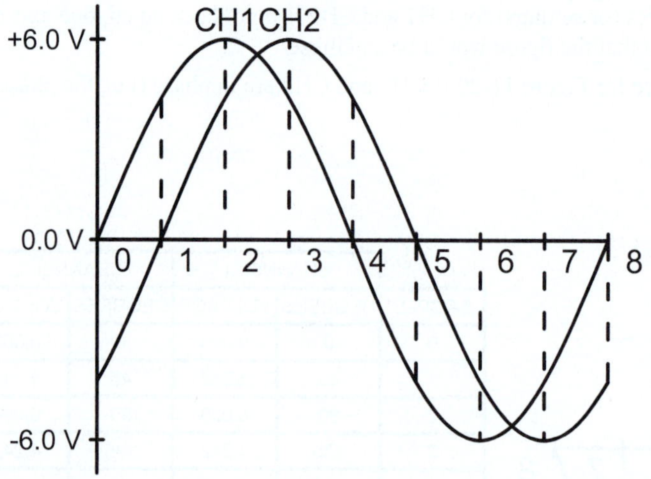

	CHANNEL 1		CHANNEL 2	
SAMPLE	DEGREES	VOLTAGE	DEGREES	VOLTAGE
0	0	0.000	-45	-4.242
1	45	4.242	0	0.000
2	90	6.000	45	4.242
3	135	4.242	90	6.000
4	180	0.000	135	4.242
5	225	-4.242	180	0.000
6	270	-6.000	225	-4.242
7	315	-4.242	270	-6.000
8	360	0.000	315	-4.242

Figure 11-24: Lissajous Coordinates for Problem 2

The table in Figure 11-24 shows the CH1 and CH2 voltages for each reference sample point. Plotting the CH1 and CH2 voltages as the x- and y-coordinates for each sample point then defines the shape of the Lissajous figure. Refer to Figure 11-25.

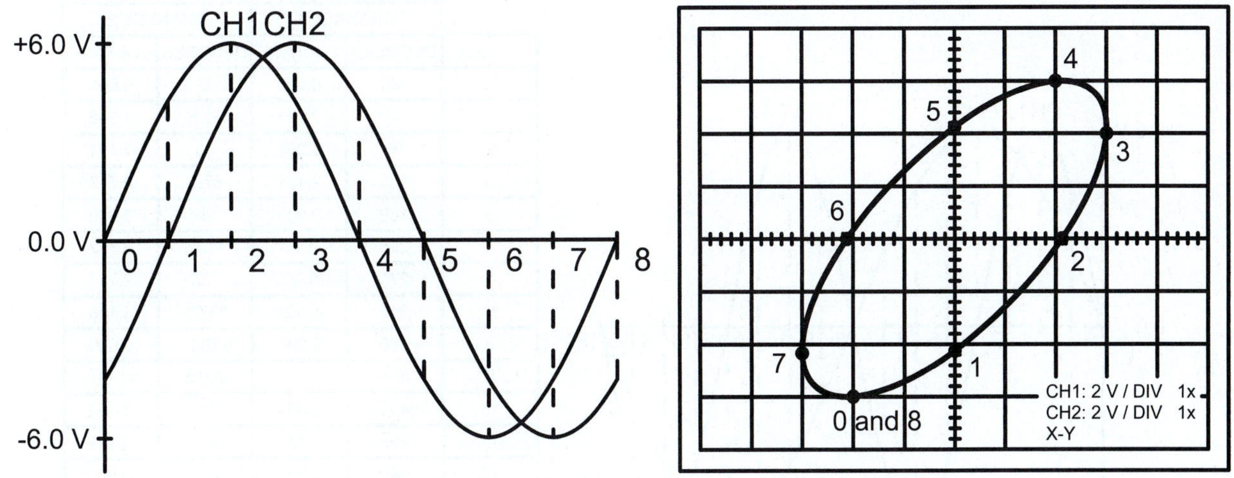

Figure 11-25: Lissajous Figure for Figure 11-24

As the reference points show, the oscilloscope will trace an ellipse for $\varphi = -45°$ (a shape midway between the circle for $\varphi = -90°$ and the line for $\varphi = 0°$). If the amplitudes for CH1 and CH2 were not equal, the ellipse would be tilted at an angle other than 45°.

3. Problem: Use the indicated reference sample points to determine the Lissajous figure for the sine waves shown in Figure 11-26.

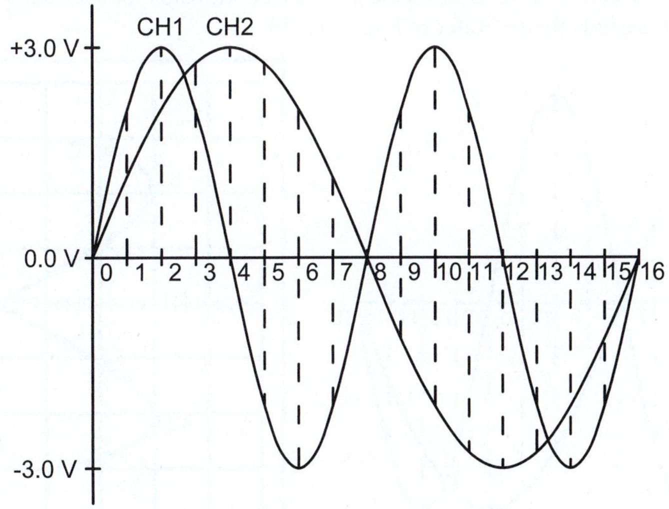

Figure 11-26: Waveforms for Problem 3

Solution: CH1 completes two cycles in the same time that CH2 completes one cycle, so the frequency of the CH1 waveform is twice the frequency of the CH2 waveform. The reference sample points define eight equal intervals for each CH1 cycle and sixteen equal intervals for CH2 so the spacing θ_{CH1} and θ_{CH2} of the sample points are

$$\theta_{CH1} = 360° / 8$$
$$= 45°$$
$$\theta_{CH2} = 360° / 16$$
$$= 22.5°$$

From these angular increments you can determine the angles and voltages that correspond to each sample point. Refer to Figure 11-27.

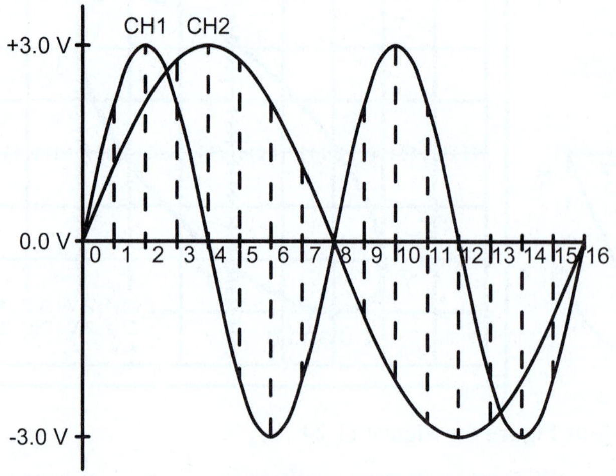

	CHANNEL 1		CHANNEL 2	
SAMPLE	DEGREES	VOLTAGE	DEGREES	VOLTAGE
0	0	0.000	0	0.000
1	45	2.121	22.5	1.148
2	90	3.000	45	2.121
3	135	2.121	67.5	2.772
4	180	0.000	90	3.000
5	225	-2.121	112.5	2.772
6	270	-3.000	135	2.121
7	315	-2.121	157.5	1.148
8	360	0.000	180	0.000
9	405	2.121	202.5	-1.148
10	450	3.000	225	-2.121
11	495	2.121	247.5	-2.772
12	540	0.000	270	-3.000
13	585	-2.121	292.5	-2.772
14	630	-3.000	315.5	-2.121
15	675	-2.121	337.5	-1.148
16	720	0.000	360	0.000

Figure 11-27: Lissajous Coordinates for Problem 3

Plotting the CH1 and CH2 voltages as the x- and y-coordinates for each sample point then defines the shape of the Lissajous figure. Refer to Figure 11-28.

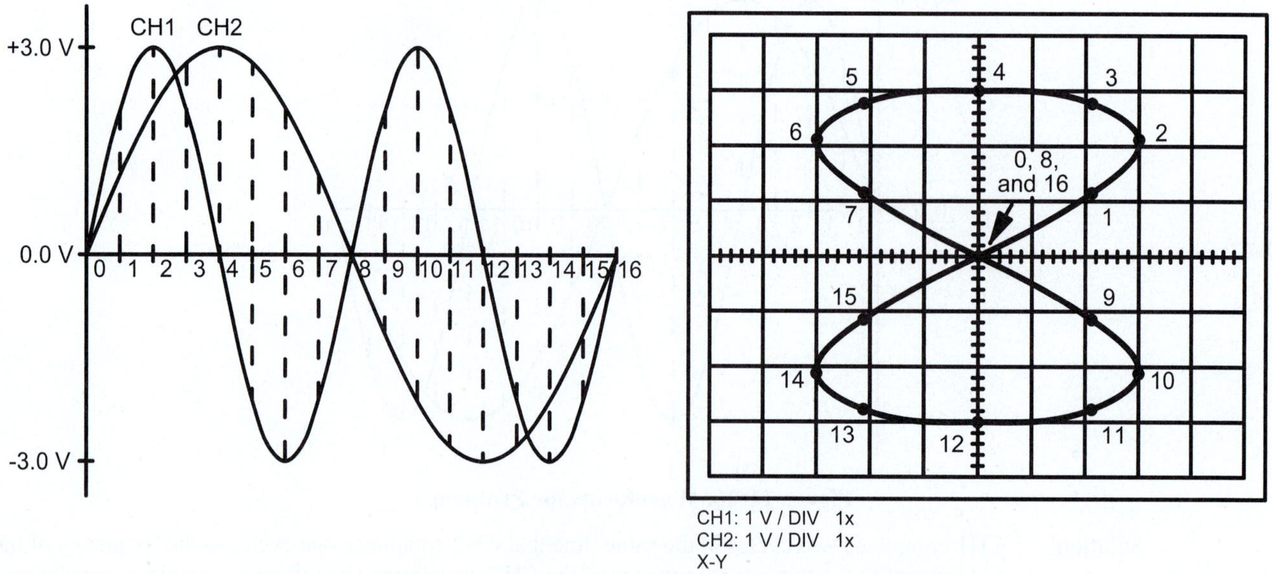

Figure 11-28: Lissajous Figure for Figure 11-26

Because CH1 completes two horizontal cycles in the time that CH2 completes one vertical cycle, the Lissajous figure contains two loops, rather than just one as for when the frequencies were equal. If the frequency of CH1 were three times the frequency of CH2, the figure would contain three vertical loops. Similarly, if the frequency of CH2 were twice the frequency of CH1, the figure would contain two horizontal loops.

12. Capacitors

12.1 Basic

12.1.1 The Basic Capacitor

1. Problem: From the definition of capacitance $\boldsymbol{C}$ in terms of voltage $\boldsymbol{V}$ and charge $\boldsymbol{Q}$, derive the expression for charge $\boldsymbol{Q}$.

   Solution: From the definition of capacitance $\boldsymbol{C}$ in terms of voltage $\boldsymbol{V}$ and charge $\boldsymbol{Q}$,

$$C = Q / V$$
$$C \times V = (Q / V) \times V$$
$$C \times V = Q$$

so

$$Q = C \times V$$

2. Problem: From the definition of capacitance $\boldsymbol{C}$ in terms of voltage $\boldsymbol{V}$ and charge $\boldsymbol{Q}$, derive the expression for voltage $\boldsymbol{V}$.

   Solution: From the definition of capacitance $\boldsymbol{C}$ in terms of voltage $\boldsymbol{V}$ and charge $\boldsymbol{Q}$

$$C = Q / V$$
$$C \times V = (Q / V) \times V$$
$$(C \times V) / C = Q / C$$
$$V = Q / C$$

3. Problem: What is the maximum voltage rating $\boldsymbol{V}_{max}$ of a 4.7 μF capacitor that can store a maximum of 240 μC of charge?

   Solution: Start with the relationship of voltage $\boldsymbol{V}$ in volts in terms of capacitance $\boldsymbol{C}$ in farads and $\boldsymbol{Q}$ in coulombs.

$$V = Q / C$$
$$V_{max} = Q_{max} / C$$
$$= 240\ \mu C / 4.7\ \mu F$$
$$= (240 \times 10^{-6}\ C) / (4.7 \times 10^{-6}\ F)$$
$$= [(240 \times 10^{-6}) / (4.7 \times 10^{-6})]\ (C / F)]$$
$$= 51.1\ V$$

   The maximum voltage rating V_{max} is **51 V**.

4. Problem: A company decides to redesign its 100 pF capacitor so that the new dielectric is air ($\varepsilon_r = 1.0$) and the new plate separation is 2.54 μm. What is the plate area $\boldsymbol{A}$ of the new capacitor?

   Solution: Start with the equation for the capacitance $\boldsymbol{C}$ in farads in terms of plate area $\boldsymbol{A}$ in m^2, dielectric constant ε_r, and plate separation $\boldsymbol{d}$ in m.

$$C = [A \times \varepsilon_r \times (8.85 \times 10^{-12}\ F / m)] / d$$

   Rearrange the equation to solve for the area, $\boldsymbol{A}$.

$$C \times d = \{[A \times \varepsilon_r \times (8.85 \times 10^{-12}\ F / m)] / d\} \times d$$
$$(C \times d) / \varepsilon_r = \{[A \times \varepsilon_r \times (8.85 \times 10^{-12}\ F / m)]\ \} / \varepsilon_r$$
$$[(C \times d) / \varepsilon_r] / (8.85 \times 10^{-12}\ F / m) = A \times (8.85 \times 10^{-12}\ F / m)]\ \} / (8.85 \times 10^{-12}\ F / m)$$
$$(C \times d) / [\varepsilon_r \times (8.85 \times 10^{-12}\ F / m)] = A$$

   Use the given values of $\boldsymbol{C}$, $\boldsymbol{d}$, and ε_r to calculate the value of $\boldsymbol{A}$.

$$A = (100\ pF \times 2.54\ \mu m) / [1.0 \times (8.85 \times 10^{-12}\ F / m)]$$
$$= [(100 \times 10^{-12}\ F) \times (2.54 \times 10^{-6}\ m)] / [1.0 \times (8.85 \times 10^{-12}\ F / m)]$$

$= \{[(100 \times 10^{-12}) \times (2.54 \times 10^{-6})] / [1.0 \times (8.85 \times 10^{-12})]\}\ [(\text{F} \times \text{m})] / (\text{F} / \text{m})]$

$= 28.7 \times 10^{-6}\ [(\text{F} \times \text{m})] \times (\text{m} / \text{F})]$

$= 28.7 \times 10^{-6}\ [(\text{F} \times \text{m} \times \text{m}) / \text{F}]$

$= 28.2 \times 10^{-6}\ \text{m}^2$

The plate area of the new capacitor is **28.7 × 10^{-6} m^2**.

12.1.2 Series Capacitors

1. Problem: How much capacitance C_2 must be added in series to $C_1 = 5.1$ nF to give a total capacitance $C_T = 1.0$ nF?

Solution: Begin with the total capacitance C_T for two capacitors C_1 and C_2 and solve for C_2.

$1 / C_T = (1 / C_1) + (1 / C_2)$

$(1 / C_T) + -(1 / C_1) = (1 / C_1) + (1 / C_2) + -(1 / C_1)$

$(1 / C_T) + -(1 / C_1) = (1 / C_2)$

$1 / [(1 / C_T) + -(1 / C_1)] = 1 / (1 / C_2)$

$1 / [(1 / C_T) + -(1 / C_1)] = C_2$

Use the values of C_T and C_1 to solve for C_2.

$C_2 = 1 / \{[1 / (1.0\ \text{nF})] + -[1 / (5.1\ \text{nF})]\}$

$= 1 / \{[1 / (1.0 \times 10^{-9}\ \text{F})] + -[1 / (5.1 \times 10^{-9}\ \text{F})]\}$

$= 1 / \{[1 \times 10^{9}\ \text{F}^{-1})] + -[1 / (196 \times 10^{6}\ \text{F}^{-1})]\}$

$= 1 / [(1000 \times 10^{6}\ \text{F}^{-1}) + -[(196 \times 10^{6}\ \text{F}^{-1})]$

$= 1 / [(1000 \times 10^{6} + -196 \times 10^{6})\ \text{F}^{-1}]$

$= 1 / (803 \times 10^{6}\ \text{F}^{-1})$

$= 1.24 \times 10^{-9}\ \text{F}$

$= 1.24\ \text{nF}$

The value of added capacitance is 1.2 nF.

2. Problem: What is the value of C_X for the capacitive voltage divider circuit of Figure 12-1?

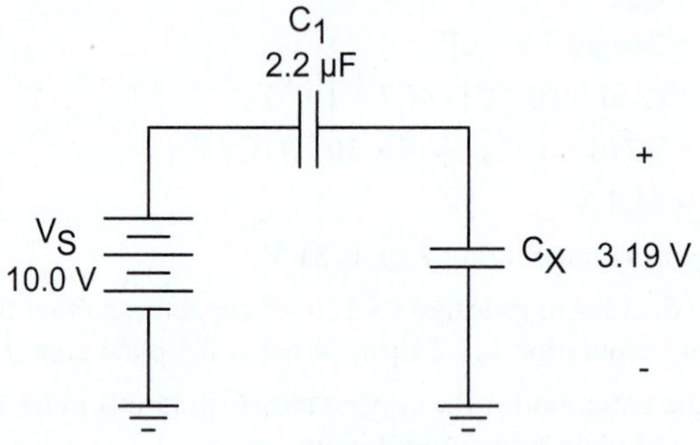

Figure 12-1: Series Capacitors Example Circuit 1

Solution: Use Kirchhoff's Voltage Law to solve for V_{CI}.

$V_S - V_{CI} - V_{CX} = 0\ \text{V}$

$V_S + -V_{CI} + -V_{CX} + V_{CI} = 0\ \text{V} + V_{CI}$

$V_S + -V_{CX} = V_{CI}$

Calculate the value of V_{CI} from the given values of V_S and V_{CX}.

$V_{CI} = 10.0\ \text{V} + -3.19\ \text{V}$

$= 6.81\ \text{V}$

Next, calculate the charge Q_{C1} on C_1.

$$\begin{aligned} Q_{C1} &= C_1 \times V_{C1} \\ &= 2.2\ \mu\text{F} \times 6.81\ \text{V} \\ &= (2.2 \times 10^{-6}\ \text{F}) \times 6.81\ \text{V} \\ &= [(2.2 \times 10^{-6}) \times 6.81]\ (\text{F} \times \text{V}) \\ &= 15.0 \times 10^{-6}\ \text{C} \end{aligned}$$

In a series circuit the charge on all capacitors is equal, so the charge Q_{CX} on C_X is equal to Q_{C1}.

$$\begin{aligned} C_X &= Q_{CX} / V_{CX} \\ &= Q_{C1} / V_{CX} \\ &= (15.0 \times 10^{-6}\ \text{C}) / (3.19\ \text{V}) \\ &= [(15.0 \times 10^{-6}) / 3.19]\ (\text{C} / \text{V}) \\ &= 4.70 \times 10^{-6}\ \text{F} \\ &= 4.70\ \mu\text{F} \end{aligned}$$

The value of C_X is **4.7 μF**.

12.1.3 Parallel Capacitors

1. Problem: Two capacitors C_1 and C_2 are connected in parallel. If $C_1 = 15\ \mu\text{F}$ and $C_2 = 10\ \mu\text{F}$, what capacitance C_3 must be added in parallel to produce a total capacitance C_T of 47 μF?

   Solution: Start with the equation for the total parallel capacitance C_T and solve for C_3.

$$\begin{aligned} C_T &= C_1 + C_2 + C_3 \\ C_T + -C_1 + -C_2 &= C_1 + C_2 + C_3 + -C_1 + -C_2 \\ C_T + -C_1 + -C_2 &= C_3 \end{aligned}$$

   Use the given values of C_T, C_1, and C_2 to calculate C_3.

$$\begin{aligned} C_3 &= C_T + -C_1 + -C_2 \\ &= 47\ \mu\text{F} + -15\ \mu\text{F} + -10\ \mu\text{F} \\ &= 22\ \mu\text{F} \end{aligned}$$

   The value of C_3 is **22 μF**.

2. Problem: The total capacitance of three capacitors of value C_1 and two capacitors of value C_2 in parallel is 36.8 μF. The total capacitance of two capacitors of value C_1 and three capacitors of value C_2 in parallel is 38.2 μF. What are the values of C_1 and C_2?

   Solution: Start by expressing the total capacitance C_T for three C_1 and two C_2 capacitors in parallel.

$$\begin{aligned} C_1 + C_1 + C_1 + C_2 + C_2 &= 36.8\ \mu\text{F} \\ 3C_1 + 2C_2 &= 36.8\ \mu\text{F} \end{aligned}$$

   Solve for C_2 in terms of C_1.

$$\begin{aligned} 3C_1 + 2C_2 + -3C_1 &= 36.8\ \mu\text{F} + -3C_1 \\ 2C_2 / 2 &= (36.8\ \mu\text{F} + -3C_1) / 2 \\ C_2 &= (36.8\ \mu\text{F} + -3C_1) / 2 \end{aligned}$$

   Next, express the total capacitance C_T for three C_1 and two C_2 capacitors in parallel.

$$\begin{aligned} C_1 + C_1 + C_2 + C_2 + C_2 &= 38.2\ \mu\text{F} \\ 2C_1 + 3C_2 &= 38.2\ \mu\text{F} \end{aligned}$$

   Substitute the value of C_2 into the second equation and solve for C_1.

$$\begin{aligned} 2C_1 + 3[(36.8\ \mu\text{F} + -3C_1) / 2] &= 38.2\ \mu\text{F} \\ 2C_1 + [(3 \times 36.8\ \mu\text{F}) / 2] + [(3 \times -3C_1) / 2] &= 38.2\ \mu\text{F} \\ 2C_1 + (110.4\ \mu\text{F} / 2) + (-9C_1 / 2) &= 38.2\ \mu\text{F} \end{aligned}$$

$2C_1 + 55.2\ \mu F + [(-9/2) \times C_1] = 38.2\ \mu F$

$2C_1 + [(-9/2) \times C_1] + 55.2\ \mu F + -55.2\ \mu F = 38.2\ \mu F + -55.2\ \mu F$

$[(2/2) \times 2C_1] + [(-9/2) \times C_1] = -17.0\ \mu F$

$[(4/2) \times C_1] + [(-9/2) \times C_1] = -17.0\ \mu F$

$[(4/2) + (-9/2)] \times C_1 = -17.0\ \mu F$

$(-5/2) \times C_1 = -17.0\ \mu F$

$[(-5/2) \times C_1] \times (-2/5) = -17.0\ \mu F \times (-2/5)$

$C_1 = 6.8\ \mu F$

Use the value of C_1 to calculate C_2.

$$C_2 = [36.8\ \mu F + (-3 \times 6.8\ \mu F)] / 2$$
$$= [36.8\ \mu F + (-20.4\ \mu F)] / 2$$
$$= 16.4\ \mu F / 2$$
$$= 8.2\ \mu F$$

The values of C_1 and C_2 are **6.8 μF** and **8.2 μF**, respectively.

12.1.4 Series-Parallel Capacitors

1. Problem: What is the value of total capacitance C_T for the circuit in Figure 12-2?

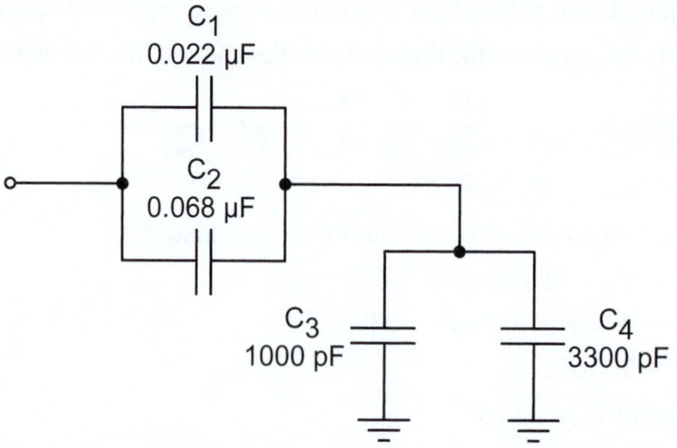

Figure 12-2: Series-Parallel Capacitors Example Circuit 1

Solution: The expression for the circuit is

$$C_T = C_1 \,\|\, C_2 + C_3 \,\|\, C_4$$

Refer to Figure 12-3 for the circuit analysis and simplification process.

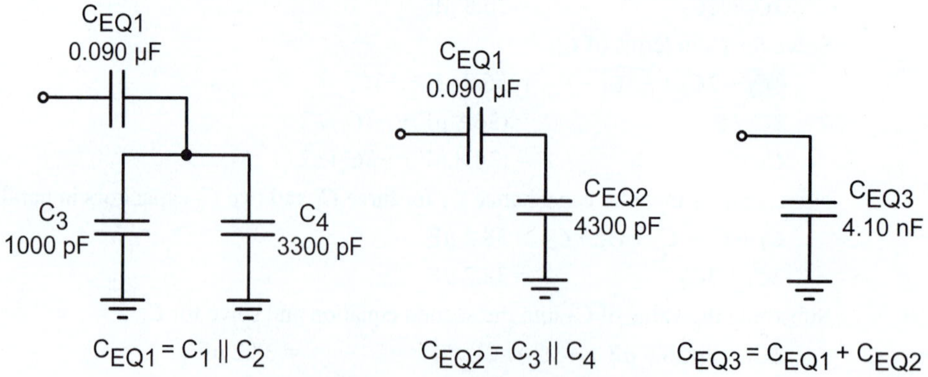

Figure 12-3: Circuit Analysis and Simplification Process for Figure 12-2

The first equivalent capacitance C_{EQ1} is the parallel combination $C_1 \parallel C_2$.

$$C_T = C_{EQ1} + C_3 \parallel C_4$$

where

$$C_{EQ1} = C_1 \parallel C_2$$
$$= 0.022\ \mu F \parallel 0.068\ \mu F$$
$$= 0.022\ \mu F + 0.068\ \mu F$$
$$= 0.090\ \mu F$$

The second equivalent capacitance C_{EQ2} is the parallel combination $C_3 \parallel C_4$.

$$C_T = C_{EQ1} + C_{EQ2}$$

where

$$C_{EQ2} = C_3 \parallel C_4$$
$$= 1000\ pF \parallel 3300\ pF$$
$$= 1000\ pF + 3300\ pF$$
$$= 4300\ pF$$

The third (and last) equivalent capacitance C_{EQ3} is the series combination $C_{EQ1} + C_{EQ2}$.

$$C_T = C_{EQ3}$$

where

$$C_{EQ3} = C_{EQ1} + C_{EQ2}$$
$$= 1 / \{[1 / (0.090\ \mu F)] + [1 / (4300\ pF)]\}$$
$$= 1 / \{[1 / (0.090 \times 10^{-6}\ F)] + [1 / (4300 \times 10^{-12}\ F)]\}$$
$$= 1 / (\{[1 / (0.090 \times 10^{-6})]\ (1/ F)\} + \{[1 / (4300 \times 10^{-12})]\ (1/ F)]\})$$
$$= 1 /[(11.1 \times 10^{6}\ F^{-1}) + (232.5 \times 10^{6}\ F^{-1})]$$
$$= 1 /(243.7 \times 10^{6}\ F^{-1})$$
$$= 4.10 \times 10^{-9}\ F$$
$$= 4.10\ nF$$

The total capacitance C_T is **4.1 nF**.

2. Problem: Determine the value of C_T for the circuit in Figure 12-4.

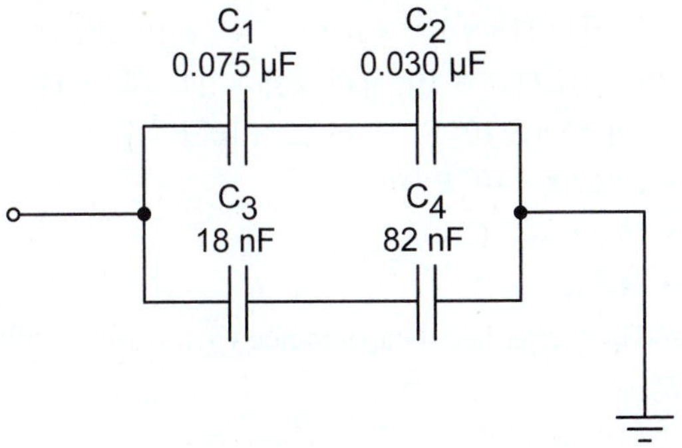

Figure 12-4: Series-Parallel Capacitors Example Circuit 2

Solution: The expression for this circuit is

$$C_T = (C_1 + C_2) \parallel (C_3 + C_4)$$

Refer to Figure 12-5 for the circuit analysis and simplification process.

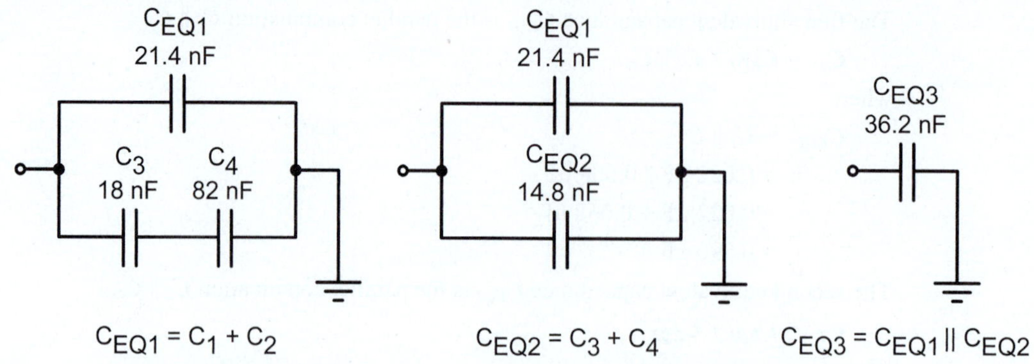

Figure 12-5: Circuit Analysis and Simplification Process for Figure 12-4

The first equivalent capacitance $\boldsymbol{C_{EQ1}}$ is the series combination ($\boldsymbol{C_1 + C_2}$).

$$C_T = C_{EQ1} \parallel (C_3 + C_4)$$

where

$$
\begin{aligned}
C_{EQ1} &= C_1 + C_2 \\
&= 1 / \{[1 / (0.075\ \mu\text{F})] + [1 / (0.030\ \mu\text{F})]\} \\
&= 1 / \{[1 / (0.075 \times 10^{-6}\ \text{F})] + [1 / (0.030 \times 10^{-6}\ \text{F})]\} \\
&= 1 / (\{[1 / (0.075 \times 10^{-6})]\ (1/\ \text{F})\} + \{[1 / (0.030 \times 10^{-6})]\ (1/\ \text{F})]\}) \\
&= 1 / [(13.3 \times 10^{6}\ \text{F}^{-1}) + (33.3 \times 10^{6}\ \text{F}^{-1})] \\
&= 1 / (46.7 \times 10^{6}\ \text{F}^{-1}) \\
&= 21.4 \times 10^{-9}\ \text{F} \\
&= 21.4\ \text{nF}
\end{aligned}
$$

The second equivalent capacitance $\boldsymbol{C_{EQ2}}$ is the series combination ($\boldsymbol{C_3 + C_4}$).

$$C_T = C_{EQ1} \parallel C_{EQ2}$$

where

$$
\begin{aligned}
C_{EQ2} &= C_3 + C_4 \\
&= 1 / \{[1 / (18\ \text{nF})] + [1 / (82\ \text{nF})]\} \\
&= 1 / \{[1 / (18 \times 10^{-9}\ \text{F})] + [1 / (82 \times 10^{-9}\ \text{F})]\} \\
&= 1 / (\{[1 / (18 \times 10^{-9})]\ (1/\ \text{F})\} + \{[1 / (82 \times 10^{-9})]\ (1/\ \text{F})]\}) \\
&= 1 / [(55.6 \times 10^{6}\ \text{F}^{-1}) + (12.2 \times 10^{6}\ \text{F}^{-1})] \\
&= 1 / (67.8 \times 10^{6}\ \text{F}^{-1}) \\
&= 14.8 \times 10^{-9}\ \text{F} \\
&= 14.8\ \text{nF}
\end{aligned}
$$

The third (and last) equivalent capacitance $\boldsymbol{C_{EQ3}}$ is the parallel combination $\boldsymbol{C_{EQ1} \parallel C_{EQ2}}$.

$$C_T = C_{EQ3}$$

where

$$
\begin{aligned}
C_{EQ3} &= C_{EQ1} \parallel C_{EQ2} \\
&= 21.4\ \text{nF} \parallel 14.8\ \text{nF} \\
&= 21.4\ \text{nF} + 14.8\ \text{nF} \\
&= 36.2\ \text{nF}
\end{aligned}
$$

The total capacitance $\boldsymbol{C_T}$ is **36 nF**.

12.1.5 Capacitors in DC Circuits

1. Problem: A capacitor $\boldsymbol{C} = 220$ pF with $V_C = +6.0$ V is discharged to 0 V through resistor $\boldsymbol{R} = 33$ kΩ. What is the time, $\boldsymbol{t}$, required for the capacitor to discharge from +4.0 V to +2.0 V?

   Solution: The general equation for the capacitor voltage $\boldsymbol{v(t)}$ at time $\boldsymbol{t}$ for a circuit with capacitance $\boldsymbol{C}$, resistance $\boldsymbol{R}$, initial capacitor voltage V_0, and final capacitor voltage V_F is

$$v(t) = V_F + (V_0 - V_F)\, e^{-(t/RC)}$$

so

$$v(t) + -V_F = V_F + (V_0 - V_F)\, e^{-(t/RC)} + -V_F$$
$$(v(t) + -V_F) / (V_0 + -V_F) = [(V_0 + -V_F)\, e^{-(t/RC)}] / (V_0 + -V_F)$$
$$\ln [(v(t) + -V_F) / (V_0 + -V_F)] = \ln[e^{-(t/RC)}]$$
$$\ln [(v(t) + -V_F) / (V_0 + -V_F)] \times -(RC) = [-(t / RC)] \times -(RC)$$
$$-(RC) \times \ln [(v(t) + -V_F) / (V_0 + -V_F)] = -(-t)$$
$$-(RC) \times \ln [(v(t) + -V_F) / (V_0 + -V_F)] = t$$

Calculate the elapsed time t_1 when $v(t_1) = 4.0$ V.

$$t_1 = -(RC) \times \ln [(v(t_1) + -V_F) / (V_0 + -V_F)]$$
$$= -(33\ \text{k}\Omega) \times (220\ \text{pF}) \times \ln[(+4.0\ \text{V} + -0\ \text{V}) / (+6.0\ \text{V} + -0)]$$
$$= -(33 \times 10^3\ \Omega) \times (220 \times 10^{-12}\ \text{F}) \times \ln(4.0\ \text{V} / 6.0\ \text{V})$$
$$= -[(33 \times 10^3) \times (220 \times 10^{-12})]\ (\Omega \times \text{F}) \times \ln(0.667)$$
$$= -(7.26 \times 10^{-6}\ \text{s}) \times -0.405$$
$$= 2.94\ \mu\text{s}$$

Next, calculate the elapsed time t_2 when $v(t_2) = 2.0$ V.

$$t_2 = -(RC) \times \ln [(v(t_2) + -V_F) / (V_0 + -V_F)]$$
$$= -(33\ \text{k}\Omega) \times (220\ \text{pF}) \times \ln[(+2.0\ \text{V} + -0\ \text{V}) / (+6.0\ \text{V} + -0)]$$
$$= -(33 \times 10^3\ \Omega) \times (220 \times 10^{-12}\ \text{F}) \times \ln(+2.0\ \text{V} / +6.0\ \text{V})$$
$$= -[(33 \times 10^3) \times (220 \times 10^{-12})]\ (\Omega \times \text{F}) \times \ln(0.333)$$
$$= -(7.26 \times 10^{-6}\ \text{s}) \times -1.110$$
$$= 7.98\ \mu\text{s}$$

Finally, calculate the time t from t_1 to t_2.

$$t = 7.98\ \mu\text{s} - 2.94\ \mu\text{s}$$
$$= 5.03\ \mu\text{s}$$

The time for the capacitor to discharge from +4.0 V to +2.0 V is **5.0 µs**.

2. Problem: What is the time constant $\boldsymbol{\tau}$ for each setting of switches SW_1 and SW_2 for the circuit in Figure 12-6? Assume all values are exact.

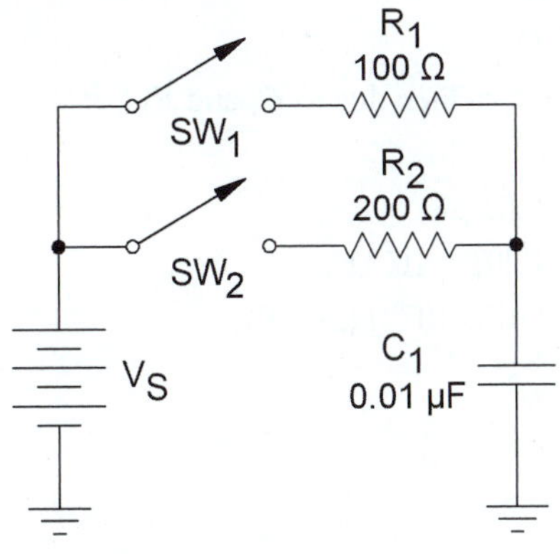

Figure 12-6: Capacitors in DC Circuits Example Circuit 1

Solution: Refer to Figure 12-7 for the circuit switch settings.

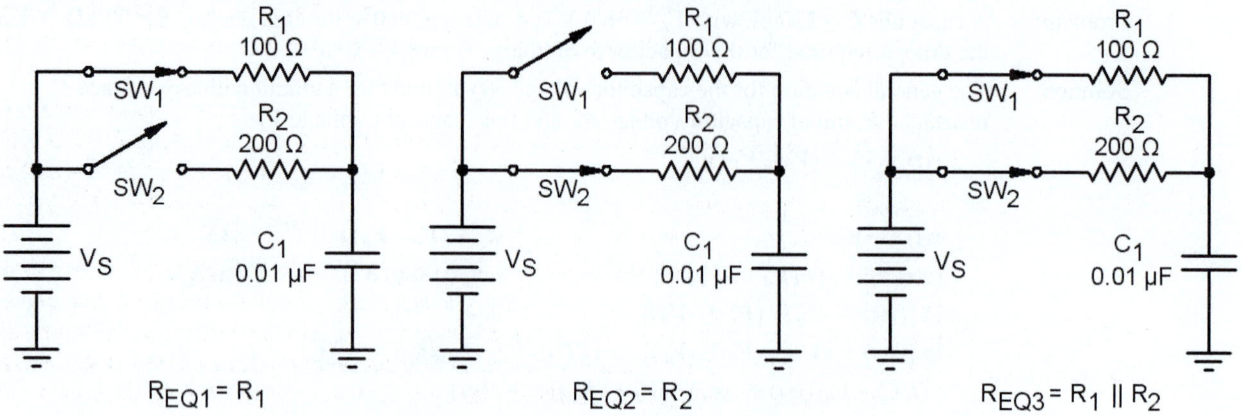

Figure 12-7: Switch Settings for Figure 12-6

For $\boldsymbol{SW_1}$ closed and $\boldsymbol{SW_2}$ open, the equivalent resistance $\boldsymbol{R_{EQ1}}$ in series with $\boldsymbol{C_1}$ is $\boldsymbol{R_1}$ alone. The time constant $\boldsymbol{\tau_1}$ is then due to $\boldsymbol{C_1}$ and $\boldsymbol{R_1}$.

$$\tau_1 = R_{EQ1} \times C_1$$
$$= 100\ \Omega \times 0.01\ \mu F$$
$$= 100\ \Omega \times (0.01 \times 10^{-6}\ F)$$
$$= [100 \times (0.01 \times 10^{-6})]\ (\Omega \times F)$$
$$= 1.0 \times 10^{-6}\ s$$
$$= \mathbf{1.0\ \mu s}$$

For $\boldsymbol{SW_2}$ closed and $\boldsymbol{SW_1}$ open, the equivalent resistance $\boldsymbol{R_{EQ2}}$ in series with $\boldsymbol{C_1}$ is $\boldsymbol{R_2}$ alone. The time constant $\boldsymbol{\tau_2}$ is then due to $\boldsymbol{C_1}$ and $\boldsymbol{R_2}$.

$$\tau_2 = R_{EQ2} \times C_1$$
$$= 200\ \Omega \times 0.01\ \mu F$$
$$= 200\ \Omega \times (0.01 \times 10^{-6}\ F)$$
$$= [100 \times (0.01 \times 10^{-6})]\ (\Omega \times F)$$
$$= 2.0 \times 10^{-6}\ s$$
$$= \mathbf{2.0\ \mu s}$$

For both $\boldsymbol{SW_1}$ and $\boldsymbol{SW_2}$ closed, the equivalent resistance $\boldsymbol{R_{EQ3}}$ in series with $\boldsymbol{C_1}$ is the parallel combination $\boldsymbol{R_1 \parallel R_2}$.

$$R_{EQ3} = R_1 \parallel R_2$$
$$= 100\ \Omega \parallel 200\ \Omega$$
$$= 66.7\ \Omega$$

The time constant $\boldsymbol{\tau_3}$ is then due to $\boldsymbol{C_1}$ and $\boldsymbol{R_1 \parallel R_2}$.

$$\tau_3 = R_{EQ3} \times C_1$$
$$= 66.7\ \Omega \times 0.01\ \mu F$$
$$= 66.7\ \Omega \times (0.01 \times 10^{-6}\ F)$$
$$= [66.7 \times (0.01 \times 10^{-6})]\ (\Omega \times F)$$
$$= 667 \times 10^{-9}\ s$$
$$= \mathbf{66.7\ ns}$$

12.1.6 Capacitors in AC Circuits

1. Problem: Verify that the total capacitive reactance for capacitors in series and parallel are calculated the same as for total resistance for resistors in series and parallel.

Solution: By definition, the value of capacitive reactance X_C for a capacitance C at frequency f is

$$X_C = 1 / [2\pi f C]$$

For two capacitances C_1 and C_2 in series, the total capacitance C_T is

$$C_T = 1 / [(1 / C_1) + (1 / C_2)]$$

The total capacitive reactance X_{CT} is then

$$\begin{aligned} X_{CT} &= 1 / [2\pi f C_T] \\ &= 1 / (2\pi f \{1 / [(1 / C_1) + (1 / C_2)]\}) \\ &= [1 / (2\pi f)] \times [(1 / C_1) + (1 / C_2)] \\ &= \{[1 / (2\pi f)] \times (1 / C_1)\} + \{[1 / (2\pi f)] \times (1 / C_2)]\} \\ &= [1 / (2\pi f C_1)] + [1 / (2\pi f C_2)] \\ &= 1 / [(2\pi f C_1)] + 1 / [(2\pi f C_2)] \\ &= X_{C1} + X_{C2} \end{aligned}$$

Therefore, total series capacitive reactance is the sum of the individual capacitive reactances, just as total series resistance is the sum of the individual resistances. For two capacitances C_1 and C_2 in parallel, the total capacitance C_T is

$$C_T = C_1 + C_2$$

The total capacitive reactance X_{CT} is then

$$\begin{aligned} X_{CT} &= 1 / [2\pi f C_T] \\ &= 1 / [2\pi f (C_1 + C_2)] \\ &= 1 / [2\pi f C_1 + 2\pi f C_2] \\ &= 1 / [(1 / X_{C1}) + (1 / X_{C2})] \end{aligned}$$

Therefore, total parallel capacitive reactance is the reciprocal of the sum of the reciprocals of the individual capacitive reactances, just as total parallel resistance is the reciprocal of the sum of the reciprocal of the individual resistances.

2. Problem: For the voltage divider circuit in Figure 12-8, show that for an ac source voltage

$$V_{OUT} = V_S \times [C_1 / (C_1 + C_2)]$$

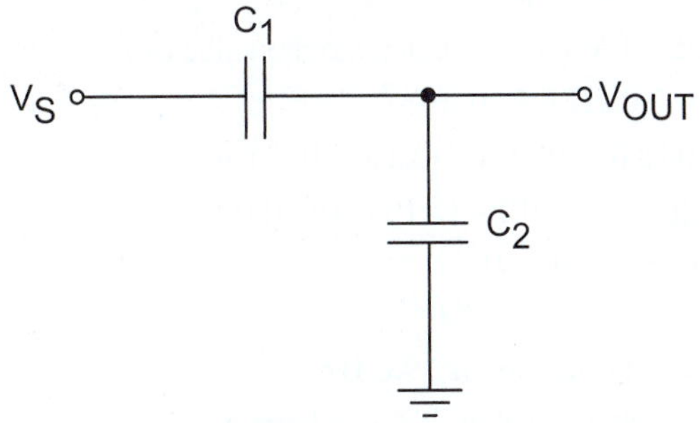

Figure 12-8: Capacitors in AC Circuits Example Circuit 1

Solution: Start with the capacitive reactances X_{C1} and X_{C2} for capacitances C_1 and C_2 at frequency f.

$$X_{C1} = 1 / (2\pi f C_1)$$

$$X_{C2} = 1 / (2\pi f C_2)$$

Next, calculate the total capacitive reactance X_{CT} for a series circuit.

$$\begin{aligned} X_{CT} &= X_{C1} + X_{C2} \\ &= 1 / (2\pi f C_1) + 1 / (2\pi f C_2) \end{aligned}$$

$= [1 / (2\pi f)] \times [(1 / C_1) + (1 / C_2)]$

Apply Ohm's Law for current to determine the total current I_T.

$I_T = V_S / X_{CT}$

Apply Ohm's Law for voltage to determine the voltage V_{OUT}.

$V_{OUT} = V_{C2}$

$= I_{C2} \times X_{C2}$

For a series circuit, the current through C_2 is the same as the total current.

$V_{OUT} = I_T \times X_{C2}$

$= (V_S / X_{CT}) \times X_{C2}$

$= (V_S \times X_{C2}) / X_{CT}$

$= \{V_S \times [1 / (2\pi f C_2)]\} / \{[1 / (2\pi f)] \times [(1 / C_1) + (1 / C_2)]\}$

$= \{V_S \times [1 / (2\pi f)] / [1 / (2\pi f)]\} \times \{(1 / C_2) / [(1 / C_1) + (1 / C_2)]\}$

$= (V_S / C_2) / [(1 / C_1) + (1 / C_2)]$

$= (V_S / C_2) / [(1 / C_1) \times (C_2 / C_2) + (1 / C_2) \times (C_1 / C_1)]$

$= (V_S / C_2) / \{[C_2 / (C_1 \times C_2)] + [C_1 / (C_2 \times C_1)]\}$

$= (V_S / C_2) / [(C_2 + C_1) / (C_1 \times C_2)]$

$= (V_S \times C_1 \times C_2) / [C_2 \times (C_1 + C_2)]$

$= (V_S \times C_1) / (C_1 + C_2)$

$= V_S \times [C_1 / (C_1 + C_2)]$

12.1.7 Switched Capacitor Circuits

1. Problem: A switched capacitor with $C = 5.10\ \mu\text{F}$ emulates an effective resistance $R = 10.0\ \text{k}\Omega$. What is the frequency of operation f?

   Solution: Start with the equation for the effective resistance R of a switched capacitor circuit and solve for f.

   $R = 1 / fC$

   $R \times f = (1 / fC) \times f$

   $(R \times f) / R = (1 / C) / \text{R}$

   $f = (1 / C) \times (1 / R)$

   $= 1 / RC$

   Use the values of R and C to calculate the value of f.

   $f = 1 / (10.0\ \text{k}\Omega \times 5.10\ \mu\text{F})$

   $= 1 / [(10.0 \times 10^3\ \Omega) \times (5.10 \times 10^{-6}\ \text{F})]$

   $= 1 / \{[(10.0 \times 10^3) \times (5.10 \times 10^{-6})]\ (\Omega \times \text{F})\}$

   $= 1 / (51.0 \times 10^{-3}\ \text{s})$

   $= 19.6\ \text{Hz}$

   The frequency of operation is **19.6 Hz**.

2. Problem: Show from unit analysis that, if f is in hertz and C is in farads, the units of $1 / fC$ is ohms.

   Solution: By definition,

   $\text{F} = \text{C} / \text{V}$

   and

   $\text{Hz} = 1 / \text{s}$

   Apply these units for the expression $1 / fC$.

   $1 / fC = 1 / [(1 / \text{s}) \times (\text{C} / \text{V})]$

$= 1 / [(C / s) / V]$
$= 1 / [A / V]$
$= V / A$
$= \Omega$

12.2 Advanced

12.2.1 The Basic Capacitor

1. Problem: A capacitor initially has $Q_0 = 10\ \mu C$ of charge stored on its plates. When the corresponding capacitor voltage V_0 increases by 1 V to V_F, the corresponding charge on the capacitor increases to $Q_F = 12\ \mu C$. What is the value C of the capacitor?

   Solution: Begin with the definition of capacitance.

   $$C = Q / V$$

   Apply the definition of capacitance for the initial condition in the problem statement.

   $$\begin{aligned} C &= Q_0 / V_0 \\ &= 10\ \mu C / V_0 \end{aligned}$$

   Apply the definition of capacitance for the final condition in the problem statement.

   $$\begin{aligned} C &= Q_F / V_F \\ &= 12\ \mu C / (V_0 + 1\ V) \end{aligned}$$

   Equate the two expressions for the value of C.

   $$\begin{aligned}
   10\ \mu C / V_0 &= 12\ \mu C / (V_0 + 1\ V) \\
   V_0 / 10\ \mu C &= (V_0 + 1\ V) / 12\ \mu C \\
   V_0 / 10\ \mu C &= (V_0 / 12\ \mu C) + (1\ V / 12\ \mu C) \\
   (V_0 / 10\ \mu C) + -(V_0 / 12\ \mu C) &= (V_0 / 12\ \mu C) + (1\ V / 12\ \mu C) + -(V_0 / 12\ \mu C) \\
   [V_0 / (1\ \mu C \times 10)] + -[V_0 / (1\ \mu C \times 12)] &= 1\ V / 12\ \mu C \\
   (V_0 / 1\ \mu C)\ [(1 / 10) + -(1 / 12)] &= 1\ V / 12\ \mu C \\
   (V_0 / 1\ \mu C)\ (0.100 + -0.0833) &= 1\ V / 12\ \mu C \\
   (V_0 / 1\ \mu C)\ [0.0167] &= 1\ V / 12\ \mu C \\
   V_0 \times (0.0167 / 1\ \mu C) &= 1\ V / 12\ \mu C \\
   V_0 \times (0.0167 / 1\ \mu C) \times (1\ \mu C / 0.0167) &= (1\ V / 12\ \mu C) \times (1\ \mu C / 0.0167) \\
   V_0 &= [(1 \times 1) / (12 \times 0.0167)]\ [(V \times \mu C) / \mu C] \\
   &= 5.0\ V
   \end{aligned}$$

   Use the values of Q_0 and V_0 to calculate C.

   $$\begin{aligned} C &= 10\ \mu C / 5.0\ V \\ &= (10 \times 10^{-6}\ C) / 5.0\ V \\ &= [(10 \times 10^{-6}) / 5.0)\ (C / V) \\ &= 2.0 \times 10^{-6}\ F \\ &= 2.0\ \mu F \end{aligned}$$

   The value of the capacitor is **2.0 μF**.

2. Problem: A capacitor with $C = 1.0\ \mu F$ is charged to $V_0 = 10.0$ V. A load is connected across the capacitor for time $t = 30$ s and the capacitor voltage decreases to $V_F = 7.5$ V. What average current I_{AVG} flowed through the load? Assume all values are exact.

   Solution: Begin with the definition of capacitance.

   $$C = Q / V$$

   Use the definition of capacitance to calculate the initial charge Q_0 on the capacitor.

$1.0\ \mu F = Q_0 / 10.0\ V$

$1.0\ \mu F \times 10.0\ V = (Q_0 / 10.0\ V) \times 10.0\ V$

$(1.0 \times 10^{-6}\ F) \times 10.0\ V = Q_0$

$[(1.0 \times 10^{-6}) \times 10.0)\ (F \times V) = Q_0$

$10.0 \times 10^{-6}\ C = Q_0$

Repeat the process to calculate the final charge Q_F on the capacitor after 30 seconds.

$1.0\ \mu F = Q_F / 7.5\ V$

$1.0\ \mu F \times 7.5\ V = (Q_F / 7.5\ V) \times 7.5\ V$

$(1.0 \times 10^{-6}\ F) \times 7.5\ V = Q_F$

$[(1.0 \times 10^{-6}) \times 7.5)\ (F \times V) = Q_F$

$7.5 \times 10^{-6}\ C = Q_F$

Current is defined as the net flow of charge per unit time, so the average current I_{AVG} is

$I_{AVG} = \Delta Q / \Delta t$

$= (Q_F - Q_0) / 30\ s$

$= [(7.5 \times 10^{-6}\ C) - (10.0 \times 10^{-6}\ C)] / 30\ s$

$= -2.5 \times 10^{-6}\ C / 30\ s$

$= [(-2.5 \times 10^{-6}) / 30]\ (C / s)$

$= -83.3 \times 10^{-9}\ A$

$= -83.3\ nA$

An average of **–83.3 nA** flowed through the load. Note that the negative sign indicates that the capacitor is discharging.

12.2.2 Series Capacitors

1. Problem: Two capacitors C_1 and C_2 are connected in series across a dc voltage source V_S. Verify that

$V_{C1} = V_S \times [C_2 / (C_1 + C_2)]$

Solution: First, use the definition of capacitance to solve for Q.

$C = Q / V$

$C \times V = (Q / V) \times V$

$C \times V = Q$

Use this expression to express Q_{C1} and Q_{C2}.

$Q_{C1} = C_1 \times V_{C1}$

$Q_{C2} = C_2 \times V_{C2}$

The charge on capacitors in series is equal, so $Q_{C1} = Q_{C2}$. Use this to solve for V_{C1}.

$Q_{C1} = Q_{C2}$

$C_1 \times V_{C1} = C_2 \times V_{C2}$

$(C_1 \times V_{C1}) / C_1 = (C_2 \times V_{C2}) / C_1$

$V_{C1} = (V_{C2} \times C_2) / C_1$

$= V_{C2} \times (C_2 / C_1)$

From Kirchhoff's Voltage Law, the sum of the capacitor voltages equals the source voltage.

$V_S = V_{C1} + V_{C2}$

$V_S + -V_{C1} = V_{C1} + V_{C2} + -V_{C1}$

$V_S + -V_{C1} = V_{C2}$

Substitute this expression for V_{C2} in the equation for V_{C1}.

$$V_{C1} = (V_S + -V_{C1}) \times (C_2 / C_1)$$
$$= [V_S \times (C_2 / C_1)] + -[V_{C1} \times (C_2 / C_1)]$$
$$V_{C1} + [V_{C1} \times (C_2 / C_1)]$$
$$= [V_S \times (C_2 / C_1)] + -[V_{C1} \times (C_2 / C_1)] + [V_{C1} \times (C_2 / C_1)]$$
$$V_{C1} \times [1 + (C_2 / C_1)] = [V_S \times (C_2 / C_1)]$$
$$V_{C1} \times [1 + (C_2 / C_1)] / [1 + (C_2 / C_1)] = [V_S \times (C_2 / C_1)] / [1 + (C_2 / C_1)]$$
$$V_{C1} = V_S \times [(C_2 / C_1) / [1 + (C_2 / C_1)]$$
$$= V_S \times (C_2 / \{C_1 \times [1 + (C_2 / C_1)]\})$$
$$= V_S \times (C_2 / \{[(C_1 \times 1) + [C_1 \times (C_2 / C_1)]\})$$
$$= V_S \times [C_2 / (C_1 + C_2)]$$

2. Problem: Refer to the circuit of N series capacitors with total capacitance C_T in Figure 12-9.

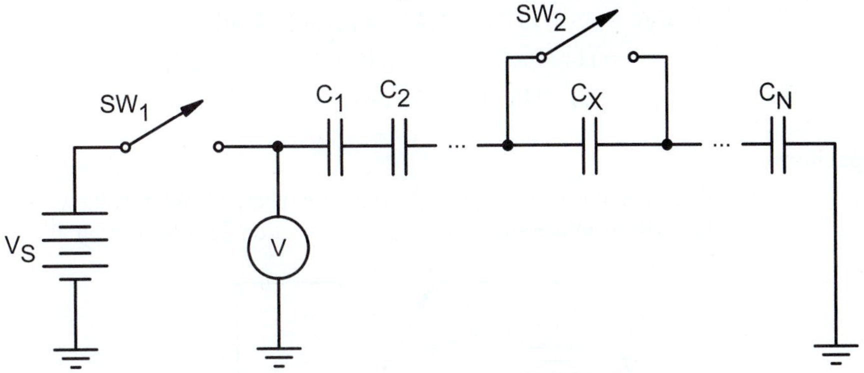

Figure 12-9: Series Capacitors Example Circuit 2

The capacitors are charged to V_S by first closing and then opening SW_1. SW_2 is then closed, shorting capacitor C_X, and the voltage V_{FAULT} measured across the capacitors. Show that the capacitance C_X of the shorted capacitor is

$$C_X = C_T \times [V_S / (V_S - V_{FAULT})]$$

Solution: The total capacitance C_T with no fault is

$$1 / C_T = (1 / C_1) + (1 / C_2) + ... + (1 / C_K) + ... + 1 / C_N$$

and the total capacitance C_{FAULT} with C_X shorted is

$$1 / C_{FAULT} = (1 / C_1) + (1 / C_2) + ... + 0 + ... + 1 / C_N$$
$$= (1 / C_1) + (1 / C_2) + ... + (1 / C_X) + -(1 / C_X) + ... + 1 / C_N$$
$$= (1 / C_1) + (1 / C_2) + ... + (1 / C_X) + ... + 1 / C_N + -(1 / C_X)$$
$$= (1 / C_T) + -(1 / C_X)$$

Use these equations to solve for C_X in terms of C_T and C_{FAULT}.

$$(1 / C_{FAULT}) + (1 / C_X) = (1 / C_T) + -(1 / C_X) + (1 / C_X)$$
$$(1 / C_{FAULT}) + (1 / C_X) + -(1 / C_{FAULT}) = (1 / C_T) + -(1 / C_{FAULT})$$
$$1 / C_X = (1 / C_T) + -(1 / C_{FAULT})$$

Use the definition of capacitance to solve for Q_T.

$$1 / C_T = V_S / Q_T$$
$$(1 / C_T) \times Q_T = (V_S / Q_T) \times Q_T$$
$$(Q_T / C_T) \times C_T = V_S \times C_T$$
$$Q_T = V_S \times C_T$$

The stored charge for the shorted circuit is the same as that for the non-shorted circuit, because the stored charge has no path through which it can dissipate. Therefore, $Q_{FAULT} = Q_T$.

$$1 / C_{FAULT} = V_{FAULT} / Q_{FAULT}$$
$$= V_{FAULT} / Q_T$$
$$= V_{FAULT} / (V_S \times C_T)$$
$$= (V_{FAULT} / V_S) \times (1 / C_T)$$

Substitute this expression for $1 / C_{FAULT}$ into the expression for $1 / C_X$ and solve for C_X.

$$1 / C_X = (1 / C_T) + -(1 / C_{FAULT})$$
$$= (1 / C_T) + -[(V_{FAULT} / V_S) \times (1 / C_T)]$$
$$= (1 / C_T) \times [1 + -(V_{FAULT} / V_S)]$$
$$= (1 / C_T) \times \{[1 \times (V_S / V_S)] + [-(V_{FAULT} / V_S)]\}$$
$$= (1 / C_T) \times [(V_S / V_S)] + -(V_{FAULT} / V_S)]$$
$$= (1 / C_T) \times [(V_S + -V_{FAULT}) / V_S)]$$
$$1 / (1 / C_X) = 1 / \{(1 / C_T) \times [(V_S - V_{FAULT}) / V_S)]\}$$
$$C_X = [1 / (1 / C_T)] \times \{1 / [(V_S - V_{FAULT}) / V_S)]\}$$
$$= C_T \times [V_S / (V_S - V_{FAULT})]$$

12.2.3 Parallel Capacitors

1. Problem: Assume that capacitor C_1 in Figure 12-10 is fully charged. What is the voltage V_C across C_1 and C_2 if the switch SW_1 is moved from position 1 to position 2? Assume that all values are exact.

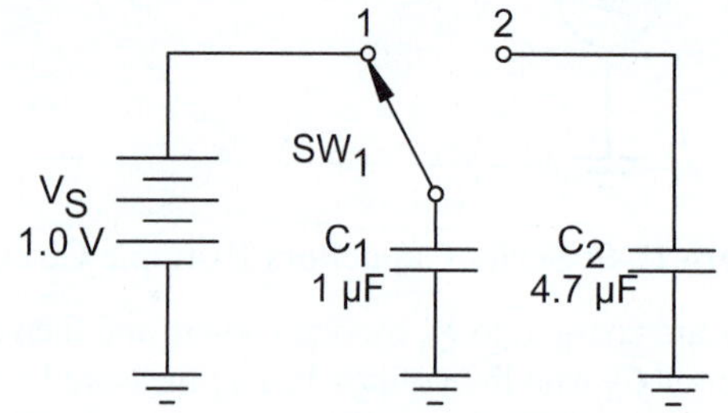

Figure 12-10: Parallel Capacitors Example Circuit 1

Solution: First, calculate the initial total charge Q_0 on the plates of C_1 when it is fully charged.

$$Q_0 = C_1 \times V_S$$
$$= 1\ \mu F \times 1.0\ V$$
$$= (1 \times 10^{-6}\ F) \times 1.0\ V$$
$$= [(1 \times 10^{-6}) \times 1.0]\ (F \times V)$$
$$= 1 \times 10^{-6}\ C$$

When SW_1 is moved from position 1 to position 2, Kirchhoff's Voltage Law requires that the voltages V_{C1} and V_{C2} across C_1 and C_2 be equal.

$$V_{C1} = V_{C2}$$
$$Q_{C1} / C_1 = Q_{C2} / C_2$$
$$(Q_{C1} / C_1) \times C_1 = (Q_{C2} / C_2) \times C_1$$
$$Q_{C1} = Q_{C2} \times (C_1 / C_2)$$

The total charge on both capacitors must equal the initial total charge Q_0 on C_1.

$$Q_0 = Q_{C1} + Q_{C2}$$
$$= Q_{C2} \times (C_1 / C_2) + Q_{C2}$$
$$= Q_{C2} \times [(C_1 / C_2) + 1]$$
$$Q_0 / [(C_1 / C_2) + 1] = \{Q_{C2} \times [(C_1 / C_2) + 1]\} / [(C_1 / C_2) + 1]$$

$\boldsymbol{Q_0} / [(C_1 / C_2) + 1] \quad = \boldsymbol{Q_{C2}}$

Use the value of $\boldsymbol{Q_0}$ to calculate the charge on C_2.

$\boldsymbol{Q_{C2}} = (1 \times 10^{-6}\ \text{C}) / [(1\ \mu\text{F} / 4.7\ \mu\text{F}) + 1]$
$= (1 \times 10^{-6}\ \text{C}) / [(1 / 4.7) + 1]$
$= (1 \times 10^{-6}\ \text{C}) / [0.213 + 1]$
$= (1 \times 10^{-6}\ \text{C}) / 1.213$
$= 825 \times 10^{-9}\ \text{C}$

Finally, use the value of $\boldsymbol{Q_{C2}}$ to calculate $\boldsymbol{V_{C1}} = \boldsymbol{V_{C2}}$.

$\boldsymbol{V_{C1}} = \boldsymbol{V_{C2}} = \boldsymbol{Q_{C2}} / \boldsymbol{C_2}$
$= (825 \times 10^{-9}\ \text{C}) / 4.7\ \mu\text{F}$
$= (825 \times 10^{-9}\ \text{C}) / (4.7 \times 10^{-6}\ \text{F})$
$= [(825 \times 10^{-9}) / (4.7 \times 10^{-6})]\ (\text{C} / \text{F})$
$= 175 \times 10^{-3}\ \text{V}$
$= 175\ \text{mV}$

The voltage across $\boldsymbol{C_1}$ and $\boldsymbol{C_2}$ is **175 mV**.

2. Problem: Refer to the circuit of $\boldsymbol{N}$ parallel capacitors with total capacitance $\boldsymbol{C_T}$ in Figure 12-11.

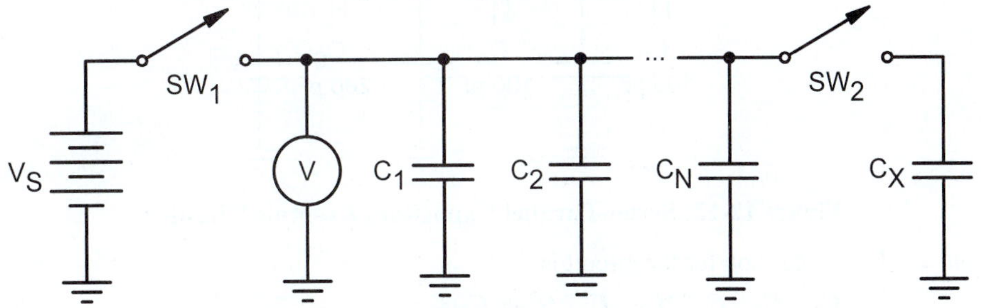

Figure 12-11: Parallel Capacitors Example Circuit 2

The capacitors are charged to $\boldsymbol{V_S}$ by closing and then opening $\boldsymbol{SW_1}$. $\boldsymbol{SW_2}$ is then closed, connecting capacitor $\boldsymbol{C_X}$ in parallel with the circuit and the voltage $\boldsymbol{V_{MOD}}$ measured across the capacitors. Show that the capacitance $\boldsymbol{C_X}$ of the added capacitor is

$\boldsymbol{C_X} = \boldsymbol{C_T} \times [(\boldsymbol{V_S} / \boldsymbol{V_{MOD}}) - 1]$

Solution: The total capacitance $\boldsymbol{C_T}$ without $\boldsymbol{C_X}$ is

$\boldsymbol{C_T} = \boldsymbol{C_1} + \boldsymbol{C_2} + \ldots + \boldsymbol{C_N}$

and the total capacitance $\boldsymbol{C_{MOD}}$ with $\boldsymbol{C_X}$ connected in parallel is

$\boldsymbol{C_{MOD}} = \boldsymbol{C_1} + \boldsymbol{C_2} + \ldots + \boldsymbol{C_N} + \boldsymbol{C_X}$
$= \boldsymbol{C_T} + \boldsymbol{C_X}$

Use this equation to solve for $\boldsymbol{C_X}$ in terms of $\boldsymbol{C_T}$ and $\boldsymbol{C_{MOD}}$.

$\boldsymbol{C_{MOD}} = \boldsymbol{C_T} + \boldsymbol{C_X}$
$\boldsymbol{C_{MOD}} + -\boldsymbol{C_T} = \boldsymbol{C_T} + \boldsymbol{C_X} + -\boldsymbol{C_T}$
$\boldsymbol{C_{MOD}} + -\boldsymbol{C_T} = \boldsymbol{C_X}$

Next, use the definition of capacitance to solve for $\boldsymbol{Q_T}$.

$\boldsymbol{C} = \boldsymbol{Q} / \boldsymbol{V}$
$\boldsymbol{C_T} = \boldsymbol{Q_T} / \boldsymbol{V_S}$
$\boldsymbol{C_T} \times \boldsymbol{V_S} = (\boldsymbol{Q_T} / \boldsymbol{V_S}) \times \boldsymbol{V_S}$
$\boldsymbol{C_T} \times \boldsymbol{V_S} = \boldsymbol{Q_T}$

For the circuit with the C_X connected the stored charge for the modified circuit is the same as for the original circuit, because the stored charge has no path through which to dissipate.

$$\begin{aligned} C_{MOD} &= Q_{MOD} / V_{MOD} \\ &= Q_T / V_{MOD} \\ &= (C_T \times V_S) / V_{MOD} \\ &= C_T \times (V_S / V_{MOD}) \end{aligned}$$

Finally, substitute the value of C_{MOD} into the equation $C_X = C_{MOD} + -C_T$.

$$\begin{aligned} C_X &= [C_T \times (V_S / V_{MOD})] + -C_T \\ &= [C_T \times (V_S / V_{MOD})] + (C_T \times -1) \\ &= C_T \times [(V_S / V_{MOD}) + -1] \\ &= C_T \times [(V_S / V_{MOD}) - 1] \end{aligned}$$

12.2.4 Series-Parallel Capacitors

1. Problem: What is the total capacitance C_T for the $C/2C$ ladder network in Figure 12-12?

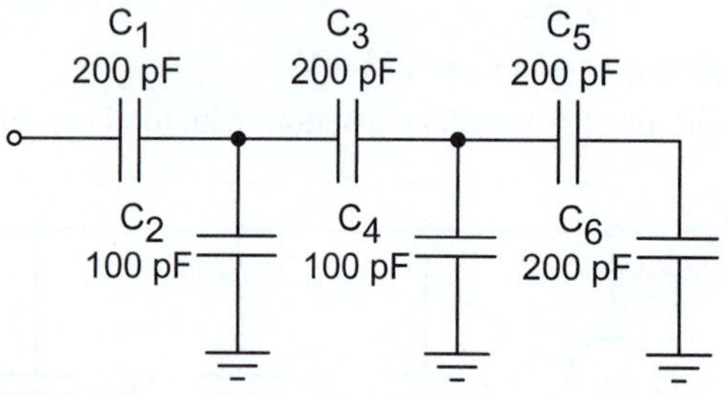

Figure 12-12: Series-Parallel Capacitors Example Circuit 3

Solution: The expression for the circuit is

$$C_T = C_1 + C_2 \parallel [C_3 + C_4 \parallel (C_5 + C_6)]$$

Refer to Figure 12-13 for the circuit analysis and simplification process for the circuit.

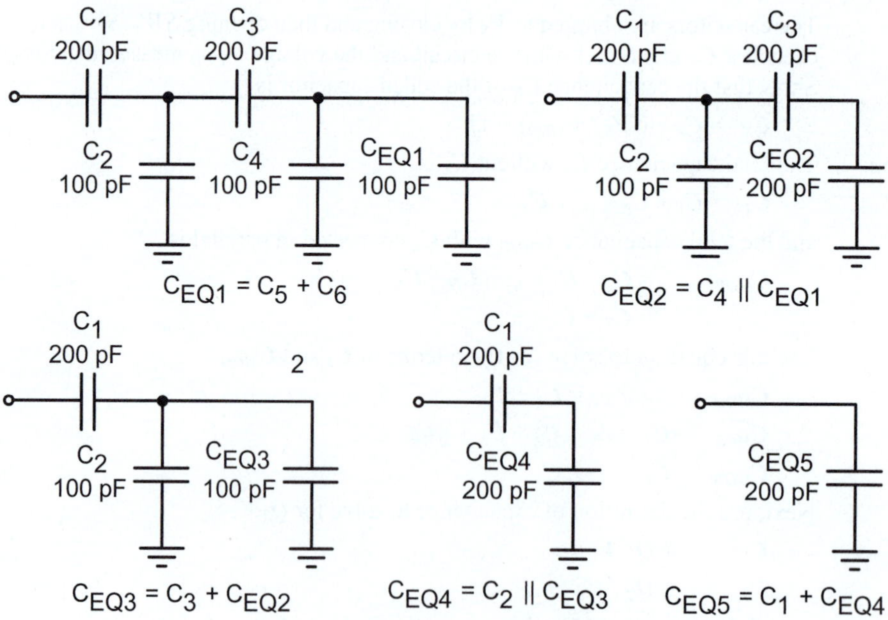

Figure 12-13: Circuit Analysis and Simplification Process for Figure 12-12

The first equivalent capacitance C_{EQ1} is the series combination ($C_5 + C_6$).

$C_T = C_1 + C_2 \parallel (C_3 + C_4 \parallel C_{EQ1})$

where

$C_{EQ1} = C_5 + C_6$

Because $C_5 = C_6$, the value of C_{EQ1} is

$C_{EQ1} = C_5 / 2 = C_6 / 2$
$= 200 \text{ pF} / 2$
$= 100 \text{ pF}$

The second equivalent capacitance C_{EQ2} is the parallel combination $C_4 \parallel C_{EQ1}$.

$C_T = C_1 + C_2 \parallel (C_3 + C_{EQ2})$

where

$C_{EQ2} = C_4 \parallel C_{EQ1}$

Because $C_4 = C_{EQ1}$, the value of C_{EQ2} is

$C_{EQ2} = C_4 \times 2 = C_{EQ1} \times 2$
$= 100 \text{ pF} \times 2$
$= 200 \text{ pF}$

The third equivalent capacitance C_{EQ3} is the series combination ($C_3 + C_{EQ2}$).

$C_T = C_1 + C_2 \parallel C_{EQ3}$

where

$C_{EQ3} = C_3 + C_{EQ2}$

Because $C_3 = C_{EQ2}$, the value of C_{EQ3} is

$C_{EQ3} = C_3 / 2 = C_{EQ2} / 2$
$= 200 \text{ pF} / 2$
$= 100 \text{ pF}$

The fourth equivalent capacitance C_{EQ4} is the parallel combination $C_2 \parallel C_{EQ3}$.

$C_T = C_1 + C_{EQ4}$

where

$C_{EQ4} = C_2 \parallel C_{EQ3}$

Because $C_2 = C_{EQ3}$, the value of C_{EQ4} is

$C_{EQ4} = C_2 \times 2 = C_{EQ3} \times 2$
$= 100 \text{ pF} \times 2$
$= 200 \text{ pF}$

The fifth (and last) equivalent capacitance C_{EQ5} is the series combination $C_1 + C_{EQ4}$.

$C_T = C_{EQ5}$

where

$C_{EQ5} = C_1 + C_{EQ4}$

Because $C_1 = C_{EQ4}$, the value of C_{EQ5} is

$C_{EQ5} = C_1 / 2 = C_{EQ5} / 2$
$= 200 \text{ pF} / 2$
$= 100 \text{ pF}$

The total resistance C_T for the network is **100 pF**.

12.2.5 Capacitors in DC Circuits

1. Problem: What is the capacitor voltage V_C for the circuit in Figure 12-14 at time t = 1.0 ms after switch ***SW*** closes?

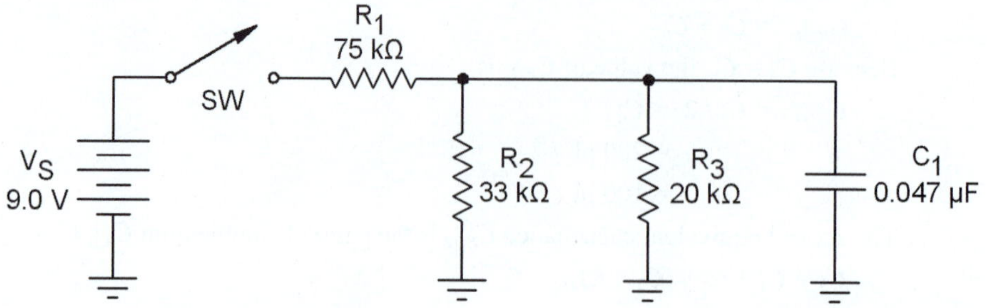

Figure 12-14: Capacitors in DC Circuits Example Circuit 2

Solution: To determine the charging characteristics for the capacitor, you must first reduce the circuit to a simple ***RC*** circuit. To do so, Thevenize the circuit as seen by the capacitor C_1.

To find the Thevenin voltage, find the open load voltage V_{OL} with the capacitor removed from the circuit. Refer to Figure 12-15.

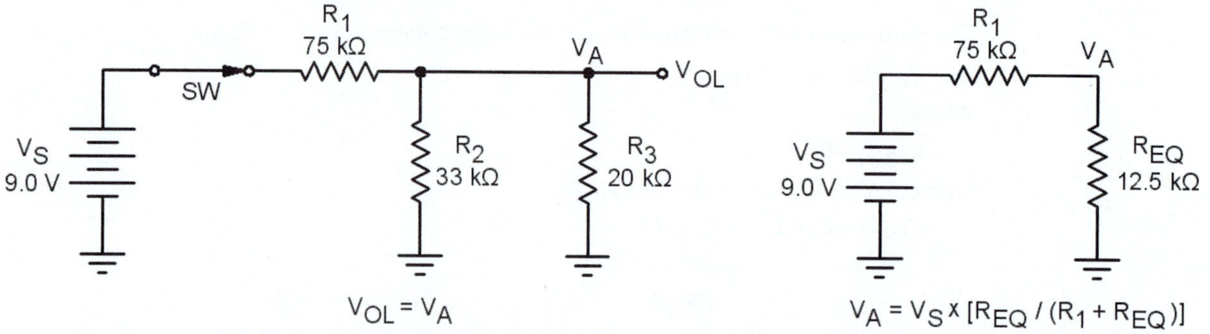

Figure 12-15: Determining the Thevenin Voltage for Figure 12-14

The open load voltage V_{OL} is equal to the voltage V_A which you can determine from standard analysis techniques for series-parallel resistive circuits. The expression for the circuit resistance R_T with C_1 removed is

$$R_T = R_1 + R_2 \parallel R_3$$

The first (and only) equivalent resistance R_{EQ} is the parallel combination $R_2 \parallel R_3$. From this

$$R_T = R_1 + R_{EQ}$$

where

$$\begin{aligned} R_{EQ} &= R_2 \parallel R_3 \\ &= 33\text{ k}\Omega \parallel 20\text{ k}\Omega \\ &= 12.5\text{ k}\Omega \end{aligned}$$

V_{TH} (= V_{OL} = V_A) can now be found from the simplified circuit. Use the voltage divider theorem to find V_A.

$$\begin{aligned} V_{TH} &= V_S \times [R_{EQ} / (R_1 + R_{EQ})] \\ &= 9.0\text{ V} \times [12.5\text{ k}\Omega / (75\text{ k}\Omega + 12.5\text{ k}\Omega)] \\ &= 9.0\text{ V} \times (12.5\text{ k}\Omega / 87.5\text{ k}\Omega) \\ &= 9.0\text{ V} \times 0.142 \\ &= 1.28\text{ V} \end{aligned}$$

To determine the Thevenin resistance, replace the voltage source V_S with its internal resistance (a short) and determine the resistance seen by C_1 looking back into the circuit. Refer to Figure 12-16.

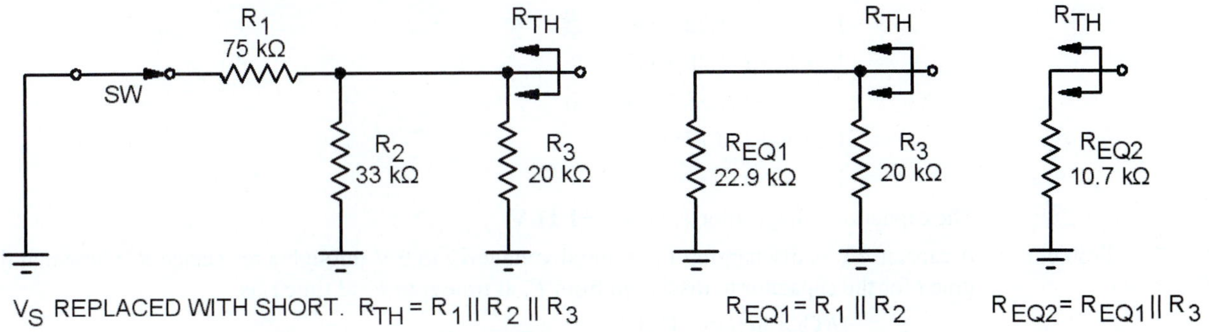

Figure 12-16: Determining the Thevenin Resistance for Figure 12-14

With V_S replaced with a short the Thevenin resistance R_{TH} looking back into the circuit is

$$R_{TH} = R_1 \,||\, R_2 \,||\, R_3$$

The first equivalent resistance R_{EQ1} is the parallel combination $R_1 \,||\, R_2$.

$$R_{TH} = R_{EQ1} \,||\, R_3$$

where

$$\begin{aligned} R_{EQ1} &= R_1 \,||\, R_2 \\ &= 75\text{ k}\Omega \,||\, 33\text{ k}\Omega \\ &= 22.9\text{ k}\Omega \end{aligned}$$

The second equivalent resistance R_{EQ2} is the parallel combination $R_{EQ1} \,||\, R_2$.

$$R_{TH} = R_{EQ2}$$

where

$$\begin{aligned} R_{EQ2} &= R_{EQ1} \,||\, R_2 \\ &= 22.9\text{ k}\Omega \,||\, 20\text{ k}\Omega \\ &= 10.7\text{ k}\Omega \end{aligned}$$

Figure 12-17 shows the Thevenin equivalent for the original circuit.

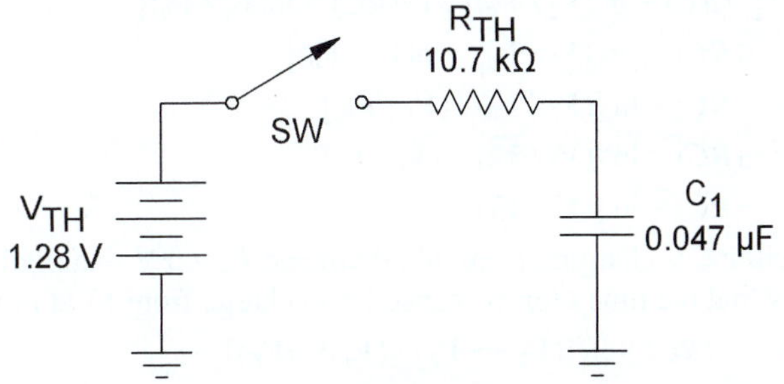

Figure 12-17: Thevenin Equivalent for Figure 12-14

Use the Thevenin equivalent to use the general exponential equation to find $v(t)$.

$$v(t) = V_F + (V_0 - V_F)\, e^{-(t/RC)}$$

where

$$\begin{aligned} RC &= 10.7\text{ k}\Omega \times 0.047\ \mu\text{F} \\ &= (10.7 \times 10^3\ \Omega) \times (0.047 \times 10^{-6}\text{ F}) \\ &= [(10.7 \times 10^3) \times (0.047 \times 10^{-6})]\ (\Omega \times \text{F}) \\ &= 502 \times 10^{-6}\text{ s} \end{aligned}$$

For $V_0 = 0$ V, $V_F = +1.28$ V, and t = 1.0 μs

$$
\begin{aligned}
v(t) &= +1.28\text{ V} + (0\text{ V} - +1.28\text{ V})\,\mathrm{e}^{-(1.0\text{ ms}\,/\,502\text{ μs})} \\
&= +1.28\text{ V} + (-1.28\text{ V})\,\mathrm{e}^{-(1000\text{ μs}\,/\,502\text{ μs})} \\
&= +1.28\text{ V} + (-1.28\text{ V})\,\mathrm{e}^{-1.99} \\
&= +1.28\text{ V} + (-1.28\text{ V} \times 0.136) \\
&= +1.28\text{ V} + (-0.175) \\
&= +1.11\text{ V}
\end{aligned}
$$

The capacitor voltage after 1.0 ms is **+1.11 V**.

2. Problem: A capacitor C is discharging from initial voltage V_0 to 0 V through a resistance R. Show that the time t for the capacitor to discharge from V_1 at time t_1 to V_2 at time t_2 is

$$t = -(RC) \times \ln\,[(V_2 / V_1)]$$

Solution: The general exponential equation for capacitor voltage $v(t)$ is

$$v(t) = V_F + (V_0 - V_F)\,e^{-(t/RC)}$$

so

$$
\begin{aligned}
v(t) + -V_F &= V_F + (V_0 - V_F)\,e^{-(t/RC)} + -V_F \\
(v(t) + -V_F) / (V_0 + -V_F) &= [(V_0 + -V_F)\,e^{-(t/RC)}] / (V_0 + -V_F) \\
\ln\,[(v(t) + -V_F) / (V_0 + -V_F)] &= \ln\,[e^{-(t/RC)}] \\
\ln\,[(v(t) + -V_F) / (V_0 + -V_F)] \times -(RC) &= [-(t / RC)] \times -(RC) \\
-(RC) \times \ln\,[(v(t) + -V_F) / (V_0 + -V_F)] &= -(-t) \\
-(RC) \times \ln\,[(v(t) + -V_F) / (V_0 + -V_F)] &= t
\end{aligned}
$$

Use this equation to express t_1 and t_2.

$$
\begin{aligned}
t_1 &= -(RC) \times \ln\,[(V_1 + -0\text{ V}) / (V_0 + -0\text{ V})] \\
&= -(RC) \times \ln\,(V_1 / V_0) \\
t_2 &= -(RC) \times \ln\,[(V_2 + -0\text{ V}) / (V_0 + -0\text{ V})] \\
&= -(RC) \times \ln\,(V_2 / V_0)
\end{aligned}
$$

The discharge time is the difference between t_1 and t_2.

$$
\begin{aligned}
t &= t_2 - t_1 \\
&= [-(RC) \times \ln\,(V_2 / V_0)] - [-(RC) \times \ln(V_1 / V_0)] \\
&= -(RC) \times [\ln\,(V_2 / V_0) - \ln(V_1 / V_0)] \\
&= -(RC) \times \ln\,[(V_2 / V_0) / (V_1 / V_0)] \\
&= -(RC) \times \ln\,[(V_2 / V_0) \times (V_0 / V_1)] \\
&= -(RC) \times \ln\,[(V_2 / V_1)]
\end{aligned}
$$

3. Problem: A capacitor C is charging from initial voltage $V_0 = 0$ V to a final voltage V_F through a resistance R. Show that the time t for the capacitor to charge from V_1 at time t_1 to V_2 at time t_2 is

$$t = -(RC) \times \ln\,[(V_2 + -V_F) / (V_1 + -V_F)]$$

Solution: The general exponential equation for capacitor voltage v(t) is

$$v(t) = V_F + (V_0 - V_F)\,e^{-(t/RC)}$$

so

$$
\begin{aligned}
v(t) + -V_F &= V_F + (V_0 - V_F)\,e^{-(t/RC)} + -V_F \\
(v(t) + -V_F) / (V_0 + -V_F) &= [(V_0 + -V_F)\,e^{-(t/RC)}] / (V_0 + -V_F) \\
\ln\,[(v(t) + -V_F) / (V_0 + -V_F)] &= \ln\,[e^{-(t/RC)}] \\
\ln\,[(v(t) + -V_F) / (V_0 + -V_F)] \times -(RC) &= [-(t / RC)] \times -(RC) \\
-(RC) \times \ln\,[(v(t) + -V_F) / (V_0 + -V_F)] &= -(-t)
\end{aligned}
$$

$-(\boldsymbol{RC}) \times \ln [(\boldsymbol{v(t)} + -\boldsymbol{V_F}) / (\boldsymbol{V_0} + -\boldsymbol{V_F})] \quad = \boldsymbol{t}$

Use this equation to express t_1 and t_2.

$t_1 = -(\boldsymbol{RC}) \times \ln [(V_1 + -\boldsymbol{V_F}) / (0\text{ V} + -\boldsymbol{V_F})]$

$\quad = -(\boldsymbol{RC}) \times \ln [(V_1 + -\boldsymbol{V_F}) / -\boldsymbol{V_F}]$

$t_2 = -(\boldsymbol{RC}) \times \ln [(V_2 + -\boldsymbol{V_F}) / (0\text{ V} + -\boldsymbol{V_F})]$

$\quad = -(\boldsymbol{RC}) \times \ln [(V_2 + -\boldsymbol{V_F}) / -\boldsymbol{V_F}]$

The time $\boldsymbol{t}$ is the difference between t_1 and t_2.

$\boldsymbol{t} = t_2 - t_1$

$\quad = \{-(\boldsymbol{RC}) \times \ln [(V_2 + -\boldsymbol{V_F}) / -\boldsymbol{V_F}]\} - \{-(\boldsymbol{RC}) \times \ln[(V_1 + -\boldsymbol{V_F}) / -\boldsymbol{V_F}]\}$

$\quad = -(\boldsymbol{RC}) \times (\{\ln [(V_2 + -\boldsymbol{V_F}) / -\boldsymbol{V_F}]\} - \{-(\boldsymbol{RC}) \times \ln[(V_1 + -\boldsymbol{V_F}) / -\boldsymbol{V_F}]\})$

$\quad = -(\boldsymbol{RC}) \times \ln \{[(V_2 + -\boldsymbol{V_F}) / -\boldsymbol{V_F}] / [(V_1 + -\boldsymbol{V_F}) / -\boldsymbol{V_F}]\}$

$\quad = -(\boldsymbol{RC}) \times \ln \{[(V_2 + -\boldsymbol{V_F}) / -\boldsymbol{V_F}] \times [-\boldsymbol{V_F} / (V_1 + -\boldsymbol{V_F})]\}$

$\quad = -(\boldsymbol{RC}) \times \ln [(V_2 + -\boldsymbol{V_F}) / (V_1 + -\boldsymbol{V_F})]$

12.2.6 Capacitors in AC Circuits

1. Problem: What is the total capacitive reactance X_{CT} for the circuit in Figure 12-18?

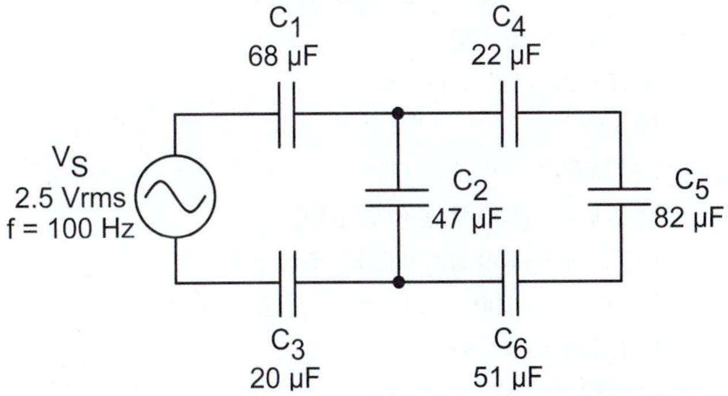

Figure 12-18: Capacitors in AC Circuits Example Circuit 2

Solution: First, determine the capacitive reactance at 100 Hz for each of the capacitors.

$X_{C1} = 1 / [2\pi f C_1]$

$\quad = 1 / [2 \times \pi \times (100\text{ Hz}) \times (68\ \mu\text{F})]$

$\quad = 1 / [2 \times \pi \times (100\text{ Hz}) \times (68 \times 10^{-6}\text{ F})]$

$\quad = 1 / \{[2 \times \pi \times 100 \times (68 \times 10^{-6})] (\text{Hz} \times \text{F})]\}$

$\quad = 1 / (42.7 \times 10^{-3}\text{ S})$

$\quad = [1 / (42.7 \times 10^{-3})] (1 / \text{S})$

$\quad = 23.4\ \Omega$

$X_{C2} = 1 / [2\pi f C_2]$

$\quad = 1 / [2 \times \pi \times (100\text{ Hz}) \times (47\ \mu\text{F})]$

$\quad = 1 / [2 \times \pi \times (100\text{ Hz}) \times (47 \times 10^{-6}\text{ F})]$

$\quad = 1 / \{[2 \times \pi \times 100 \times (47 \times 10^{-6})] (\text{Hz} \times \text{F})]\}$

$\quad = 1 / (29.5 \times 10^{-3}\text{ S})$

$\quad = [1 / (29.5 \times 10^{-3})] (1 / \text{S})$

$\quad = 33.9\ \Omega$

$X_{C3} = 1 / [2\pi f C_3]$

$\quad = 1 / [2 \times \pi \times (100\text{ Hz}) \times (20\ \mu\text{F})]$

$= 1 / [2\times \pi \times (100\text{ Hz}) \times (20 \times 10^{-6}\text{ F})]$
$= 1 / \{[2\times \pi \times 100 \times (20 \times 10^{-6})] (\text{Hz} \times \text{F})]\}$
$= 1 / (12.6 \times 10^{-3}\text{ S})$
$= [1 / (12.6 \times 10^{-3})] (1 / \text{S})$
$= 79.6\ \Omega$

$X_{C4} = 1 / [2\pi f C_4]$
$= 1 / [2\times \pi \times (100\text{ Hz}) \times (22\ \mu\text{F})]$
$= 1 / [2\times \pi \times (100\text{ Hz}) \times (22 \times 10^{-6}\text{ F})]$
$= 1 / \{[2\times \pi \times 100 \times (22 \times 10^{-6})] (\text{Hz} \times \text{F})]\}$
$= 1 / (13.8 \times 10^{-3}\text{ S})$
$= [1 / (13.8 \times 10^{-3})] (1 / \text{S})$
$= 72.3\ \Omega$

$X_{C5} = 1 / [2\pi f C_5]$
$= 1 / [2\times \pi \times (100\text{ Hz}) \times (82\ \mu\text{F})]$
$= 1 / [2\times \pi \times (100\text{ Hz}) \times (82 \times 10^{-6}\text{ F})]$
$= 1 / \{[2\times \pi \times 100 \times (82 \times 10^{-6})] (\text{Hz} \times \text{F})]\}$
$= 1 / (51.5 \times 10^{-3}\text{ S})$
$= [1 / (51.5 \times 10^{-3})] (1 / \text{S})$
$= 19.4\ \Omega$

$X_{C6} = 1 / [2\pi f C_6]$
$= 1 / [2\times \pi \times (100\text{ Hz}) \times (51\ \mu\text{F})]$
$= 1 / [2\times \pi \times (100\text{ Hz}) \times (51 \times 10^{-6}\text{ F})]$
$= 1 / \{[2\times \pi \times 100 \times (51 \times 10^{-6})] (\text{Hz} \times \text{F})]\}$
$= 1 / (32.0 \times 10^{-3}\text{ S})$
$= [1 / (32.0 \times 10^{-3})] (1 / \text{S})$
$= 31.2\ \Omega$

Next, replace the capacitors with their capacitive reactances and analyze the circuit as you would for a resistive circuit. Refer to Figure 12-19, which shows the reactive equivalent and the circuit analysis and simplification process.

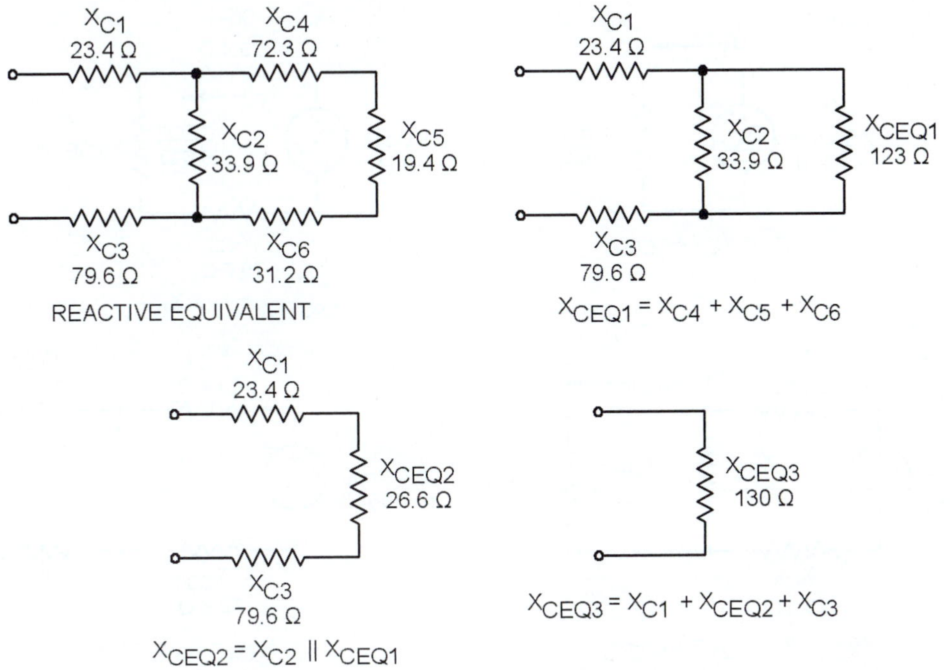

Figure 12-19: Circuit Analysis and Simplification Process for Figure 12-18

The expression for the circuit is

$$X_{CT} = X_{C1} + X_{C2} \,||\, (X_{C4} + X_{C5} + X_{C6}) + X_{C3}$$

The first equivalent reactance X_{CEQ1} is the series combination $(X_{C4} + X_{C5} + X_{C6})$.

$$X_{CT} = X_{C1} + X_{C2} \,||\, X_{CEQ1} + X_{C3}$$

where

$$\begin{aligned} X_{CEQ1} &= X_{C4} + X_{C5} + X_{C6} \\ &= 72.3\ \Omega + 19.4\ \Omega + 31.2\ \Omega \\ &= 123.0\ \Omega \end{aligned}$$

The second equivalent reactance X_{CEQ2} is the parallel combination $X_{C2} \,||\, X_{CEQ1}$.

$$X_{CT} = X_{C1} + X_{CEQ2} + X_{C3}$$

where

$$\begin{aligned} X_{CEQ2} &= X_{C2} \,||\, X_{CEQ1} \\ &= 33.9\ \Omega \,||\, 123.0\ \Omega \\ &= 26.6\ \Omega \end{aligned}$$

The third (and last) equivalent reactance X_{CEQ3} is the series combination $X_{C1} + X_{CEQ2} + X_{C3}$.

$$X_{CT} = X_{CEQ3}$$

where

$$\begin{aligned} X_{CEQ3} &= X_{C1} + X_{CEQ2} + X_{C3} \\ &= 23.4\ \Omega + 26.6\ \Omega + 79.6\ \Omega \\ &= 130.0\ \Omega \end{aligned}$$

The total capacitive reactance is **130 Ω.**

2. Problem: Use the results of Problem 1 to determine the currents through each capacitor for the circuit in Figure 12-18.

   Solution: To determine the currents through the capacitors for the circuit, use Ohm's Law for the circuit equivalents of Problem 1 and work backwards through the circuit. Refer to Figure 12-20.

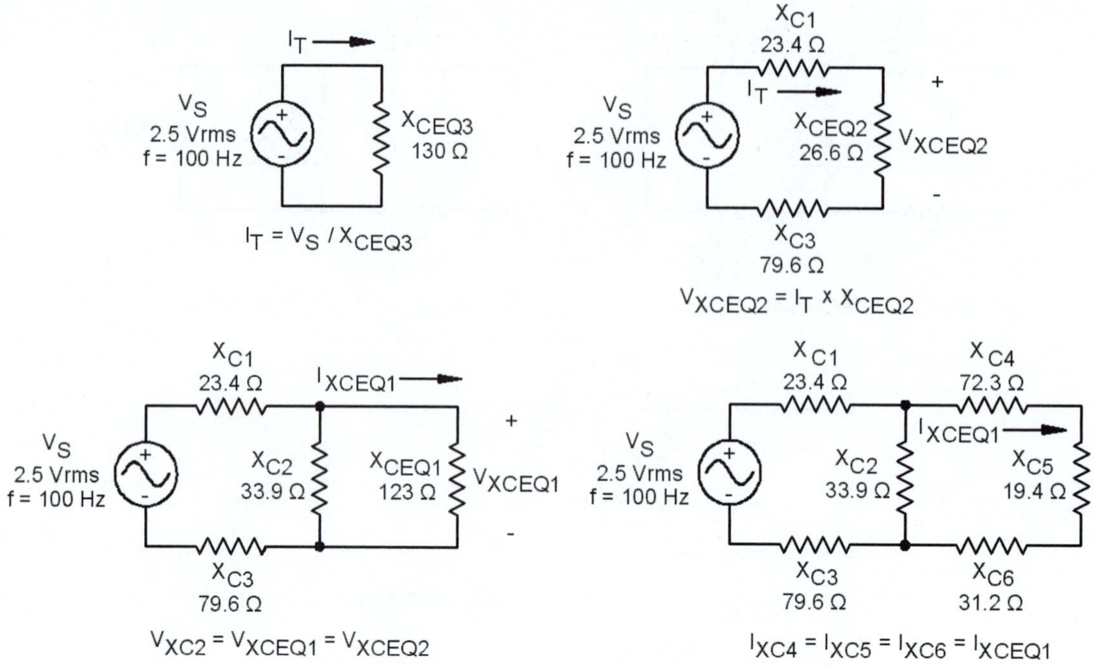

Figure 12-20: Circuit Current Analysis for Figure 12-18

From Ohm's Law for current, the total current I_T is the current through X_{CEQ3}.

$$\begin{aligned} I_T &= V_S / X_{CT} \\ &= V_S / X_{CEQ3} \\ &= 2.5\ \mathrm{V_{rms}} / 130.0\ \Omega \\ &= (2.5 / 130.0)\ (\mathrm{V_{rms}} / \Omega) \\ &= 19.3 \times 10^{-3}\ \mathrm{A_{rms}} \\ &= 19.3\ \mathrm{mA_{rms}} \end{aligned}$$

X_{CEQ3} is the series combination of X_{C1}, X_{CEQ2}, and X_{C3}, so the current through X_{CEQ3} is equal to the current through X_{C1}, X_{CEQ2}, and X_{C3}.

$$\begin{aligned} I_{XC1} = I_{XCEQ2} = I_{XC3} &= I_T \\ &= \mathbf{19.3\ mA_{rms}} \end{aligned}$$

Use Ohm's Law for voltage to calculate the voltage across X_{CEQ2}.

$$\begin{aligned} V_{XCEQ2} &= I_{XCEQ2} \times X_{CEQ2} \\ &= I_T \times X_{CEQ2} \\ &= 19.3 \times 10^{-3}\ \mathrm{A_{rms}} \times 26.6\ \Omega \\ &= [(19.3 \times 10^{-3}) \times 26.6]\ (\mathrm{A_{rms}} \times \Omega) \\ &= 512 \times 10^{-3}\ \mathrm{V_{rms}} \end{aligned}$$

X_{CEQ2} is the parallel combination of X_{C2} and X_{CEQ1} so the voltage across X_{CEQ2} is equal to the voltage across X_{C2} and X_{CEQ1}.

$$\begin{aligned} V_{XC2} = V_{XCEQ1} &= V_{XCEQ2} \\ &= 512 \times 10^{-3}\ \mathrm{V_{rms}} \end{aligned}$$

Use Ohm's Law for current to calculate the currents through X_{C2} and X_{CEQ1}.

$$I_{XC2} = V_{XC2} / X_{C2}$$

$= V_{XCEQ2} / X_{C2}$
$= 512 \times 10^{-3}\ V_{rms} / 33.9\ \Omega$
$= (512 \times 10^{-3} / 33.9)\ (V_{rms} / \Omega)$
$= 15.1 \times 10^{-3}\ A_{rms}$
$= \mathbf{15.1\ mA_{rms}}$

$I_{XCEQ1} = V_{XCEQ1} / X_{CEQ1}$
$= V_{XCEQ2} / X_{CEQ1}$
$= 512 \times 10^{-3}\ V_{rms} / 123.0\ \Omega$
$= (512 \times 10^{-3} / 123.0)\ (V_{rms} / \Omega)$
$= 4.17 \times 10^{-3}\ A_{rms}$
$= 4.17\ mA_{rms}$

X_{CEQ1} is the series combination of X_{C4}, X_{C5}, and X_{C6}, so the current through X_{CEQ1} is equal to the current through X_{C4}, X_{C5}, and X_{C6}.

$I_{XC4} = I_{XC5} = I_{XC6} = I_{XCEQ1}$
$= \mathbf{4.17\ mA_{rms}}$

12.3 Just for Fun

1. Problem: The graph in Figure 12-21 shows the curves for an ***RC*** circuit as it charges from 0 V to V_S and discharges from V_S to 0 V. The two curves intersect at some point, which indicates that the capacitor voltage will be the same regardless of whether the capacitor is charging or discharging. What are the time, t, and the voltage, $v(t)$, where the curves intersect?

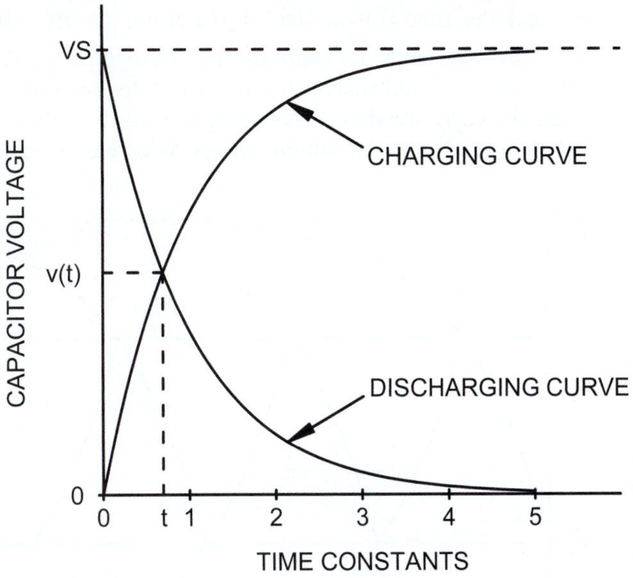

Figure 12-21: *RC* Circuit Charging and Discharging Curves

Solution: The general exponential equation for capacitor voltage $v(t)$ is

$v(t) = V_F + (V_0 - V_F)\, e^{-(t/RC)}$

For the discharging curve $V_0 = V_S$ and $V_F = 0$ V.

$v(t) = 0\ V + (V_S - 0\ V)\, e^{-(t/RC)}$
$= V_S \times e^{-(t/RC)}$

For the charging curve $V_0 = 0$ V and $V_F = V_S$.

$v(t) = V_S + (0\ V - V_S)\, e^{-(t/RC)}$

$= V_S + [(-V_S)\ e^{-(t/RC)}]$

$= V_S \times [1 + -e^{-(t/RC)}]$

Setting these two terms equal gives

$V_S \times e^{-(t/RC)}$	$= V_S \times [1 + -e^{-(t/RC)}]$
$[V_S \times e^{-(t/RC)}] / V_S$	$= V_S \times [1 + -e^{-(t/RC)}] / V_S$
$e^{-(t/RC)} + e^{-(t/RC)}$	$= 1 + -e^{-(t/RC)} + e^{-(t/RC)}$
$[2 \times e^{-(t/RC)}] / 2$	$= 1 / 2$
$e^{-(t/RC)}$	$= 0.5$
$\ln [e^{-(t/RC)}]$	$= \ln (1 /2)$
$RC \times [-(t / RC)]$	$= RC \times \ln (1 / 2)$
$-(-t)$	$= -[RC \times \ln (1 / 2)]$
t	$= RC \times -\ln (1 / 2)$
	$= RC \times \ln (2 / 1)$
	$= RC \times \ln (2)$

Use this value of t in the discharging equation to calculate $v(t)$.

$v(t) = V_S \times e^{-\{[RC \times \ln (2)] / RC\}}$

$= V_S \times e^{-[\ln (2)]}$

$= V_S \times e^{[\ln (1/2)]}$

$= V_S \times (1 / 2)$

$= V_S / 2$

Both curves reach the same voltage $v(t) = V_S / 2$ at time $t = RC \times \ln (2)$.

2. Problem: Refer to the steady-state capacitor charging and discharging graph in Figure 12-22. The capacitor voltage starts at $v(t) = V_L$ and charges towards V_S. After one time constant the capacitor voltage reaches V_H and the capacitor starts discharging towards 0 V. After one time constant the capacitor voltage reaches V_L again and the pattern repeats. What are the values of V_L and V_H?

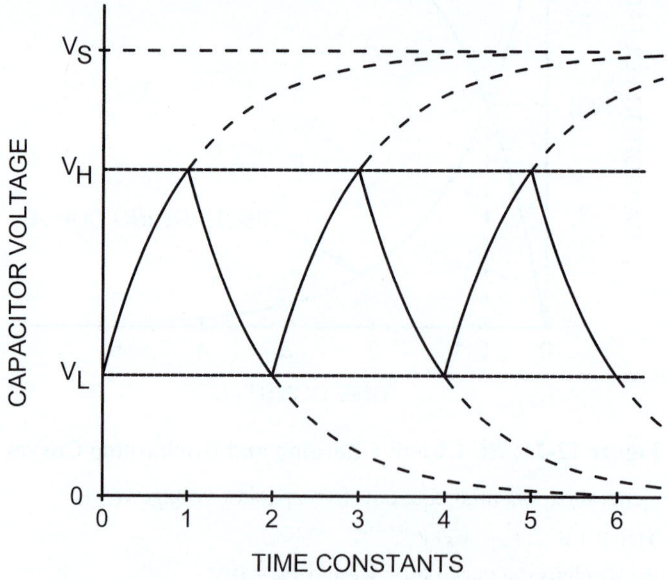

Figure 12-22: Capacitor Steady-State Charging and Discharging Graph

Solution: The general exponential equation for capacitor voltage $v(t)$ is

$$v(t) = V_F + (V_0 - V_F)\, e^{-(t/RC)}$$

For the charging portion of the curve, $v(t) = V_H$, $V_0 = V_L$, $V_F = V_S$, and $t = RC$.

$$V_H = V_S + (V_L - V_S)\, e^{-(RC/RC)}$$
$$= V_S + (V_L - V_S)\, e^{-1}$$

For the discharging portion of the curve, $v(t) = V_L$, $V_0 = V_H$, $V_F = 0$ V, and $t = RC$.

$$V_L = 0\text{ V} + (V_H - 0\text{ V})\, e^{-(RC/RC)}$$
$$= V_H \times e^{-1}$$

Substitute $V_H \times e^{-1}$ for V_L into the charging equation.

$$V_H = V_S + (V_H \times e^{-1} - V_S)\, e^{-1}$$
$$= V_S + (V_H \times e^{-1} \times e^{-1}) + -(V_S \times e^{-1})$$
$$= V_S + (V_H \times e^{-2}) + -(V_S \times e^{-1})$$
$$V_H + -(V_H \times e^{-2}) = V_S + -(V_S \times e^{-1}) + (V_H \times e^{-2}) + -(V_H \times e^{-2})$$
$$V_H \times (1 + -e^{-2}) = V_S \times (1 + -e^{-1})$$
$$V_H \times (1 + -e^{-2}) / (1 + -e^{-2}) = [V_S \times (1 + -e^{-1})] / (1 + -e^{-2})$$
$$V_H = V_S \times [(1 - e^{-1})] / (1 - e^{-2})]$$

Substitute this equation of V_H into the equation for V_L.

$$V_L = V_H \times e^{-1}$$
$$= V_S \times [(1 - e^{-1})] / (1 - e^{-2})] \times e^{-1}$$
$$= V_S \times \{[(1 - e^{-1}) \times e^{-1}] / (1 - e^{-2})\}$$
$$= V_S \times \{[(1 \times e^{-1}) - (e^{-1} \times e^{-1})] / (1 - e^{-2})\}$$
$$= V_S \times [(e^{-1} - e^{-2}) / (1 - e^{-2})]$$

For the steady state waveform, the values of V_L and V_H are

$$\mathbf{V_L = V_S \times [(e^{-1} - e^{-2}) / (1 - e^{-2})]}$$
$$\mathbf{V_H = V_S \times [(1 - e^{-1})] / (1 - e^{-2})]}$$

13. Inductors

13.1 Basic

13.1.1 The Basic Inductor

1. Problem: What is the inductance, $\boldsymbol{L}$, of a coil that develops an induced voltage, $\boldsymbol{v_{ind}}$, of 25.0 μV when the rate of change of the current, $\boldsymbol{dI / dt}$, through it is 532 mA/s?

   Solution: Begin with the equation for the induced voltage $\boldsymbol{v_{ind}}$ and solve for $\boldsymbol{L}$.

   $v_{ind} = L \times (dI / dt)$ where L is in H and dI / dt is in A / s

   $v_{ind} / (dI / dt) = [L \times (dI / dt)] / (dI / dt)$

   $v_{ind} / (dI / dt) = L$

   Use the given values of $\boldsymbol{v_{ind}}$ and $\boldsymbol{dI / dt}$ to calculate $\boldsymbol{L}$.

   $L = (25.0\ \mu\text{V}) / (532\ \text{mA /s})$

   $= (25.0 \times 10^{-6}\ \text{V}) / (532 \times 10^{-3}\ \text{A /s})$

   $= [(25.0 \times 10^{-6}) / (532 \times 10^{-3})]\ [\text{V} / (\text{A /s})]$

   $= 47.0 \times 10^{-6}\ \text{H}$

   The inductor value is **47.0 μH.**

2. Problem: An inductor with $\boldsymbol{L}$ = 5.0 mH is constructed of $\boldsymbol{N}$ = 100 turns of wire on a core with length $\boldsymbol{l}$ = 4.0 cm and cross-sectional area $\boldsymbol{A}$ = 1.0 cm^2. What is the permeability μ of the core material?

   Solution: Begin with the equation for the inductance $\boldsymbol{L}$ of a coil and solve for μ.

   $L = (N^2 \times \mu \times A) / 1$ where $\boldsymbol{L}$ is in H, $\boldsymbol{N}$ is turns, $\boldsymbol{A}$ is in m^2, and $\boldsymbol{l}$ is in m

   $L \times l = [(N^2 \times \mu \times A) / l] \times l$

   $(L \times l) / (N^2 \times A) = (N^2 \times \mu \times A) / (N^2 \times A)$

   $(L \times l) / (N^2 \times A) = \mu$

   Use the given values of $\boldsymbol{L}$, $\boldsymbol{l}$, $\boldsymbol{N}$, and $\boldsymbol{A}$ to calculate μ.

   $\mu = (5.0\ \text{mH} \times 4.0\ \text{cm}) / [(100)^2 \times 1.0\ \text{cm}^2]$

   $= [(5.0 \times 10^{-3}\ \text{H}) \times (4.0 \times 10^{-3}\ \text{m})] / \{(100)^2 \times (1.0\ \text{cm}^2) \times [(0.01\ \text{m}) / (1.0\ \text{cm})]^2\}$

   $= [(5.0 \times 10^{-3}\ \text{H}) \times (4.0 \times 10^{-3}\ \text{m})] / \{10{,}000 \times (1.0\ \text{cm}^2) \times [(0.01\ \text{m})^2 / (1.0\ \text{cm})^2]\}$

   $= [(5.0 \times 10^{-3}\ \text{H}) \times (4.0 \times 10^{-3}\ \text{m})] / [10{,}000 \times (1.0\ \text{cm}^2) \times (0.0001\ \text{m}^2 / 1.0\ \text{cm}^2)]$

   $= [(5.0 \times 10^{-3}) \times (4.0 \times 10^{-3})] / [10{,}000 \times 1.0 \times (0.0001 / 1.0)]\ [(\text{H} \times \text{m}) / [\text{cm}^2 \times (\text{m}^2 / \text{cm}^2)]$

   $= [(20.0 \times 10^{-6}) / 10]\ [(\text{H} \times \text{m}) / (\text{t}^2 \times \text{m} \times \text{m})]$ (Note that turns t^2 have no units.)

   $= 2.0 \times 10^{-6}\ \text{H / m}$

   The core permeability is $\mathbf{2.0 \times 10^{-6}\ H / m}$.

3. Problem: The steady-state magnetic field around a 50.0 mH inductor, $\boldsymbol{L}$, contains 6.25 mJ of energy, $\boldsymbol{E}$. How much current, $\boldsymbol{I}$, is flowing through the inductor?

   Solution: Begin with the equation for power stored by the magnetic field of an inductor and solve for $\boldsymbol{I}$.

   $E = (L \times I^2) / 2$ where $\boldsymbol{E}$ is in J, $\boldsymbol{L}$ is in H, and $\boldsymbol{I}$ is in A

   $E \times 2 = [(L \times I^2) / 2] \times 2$

   $(E \times 2) / L = (L \times I^2) / L$

   $[(E \times 2) / L]^{1/2} = (I^2)^{1/2}$

   $[(E \times 2) / L]^{1/2} = I$

   Use the given values of $\boldsymbol{E}$ and $\boldsymbol{L}$ to calculate $\boldsymbol{I}$.

   $I = [(2 \times 6.25\ \text{mJ}) / 50\ \text{mH}]^{1/2}$

$= \{[2 \times (6.25 \times 10^{-3}\ J)] / (50 \times 10^{-3}\ H)\}^{1/2}$
$= [(12.5 \times 10^{-3}) / (50 \times 10^{-3})]^{1/2}\ (J / H)^{1/2}$
$= (250 \times 10^{-3})^{1/2}\ A$
$= 500 \times 10^{-3}\ A$
$= 500\ mA$

The current flowing through the coil is **500 mA**.

4. Problem: The current for the circuit of Figure 13-1 has reached a steady-state condition. What is the polarity of the induced voltage across the inductor when switch ***SW*** opens?

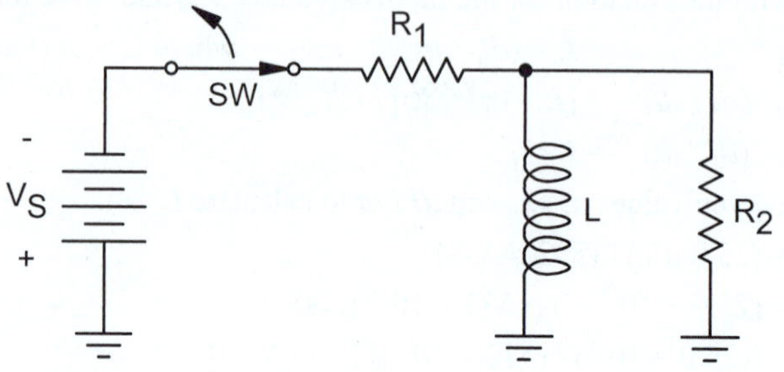

Figure 13-1: Basic Inductor Example Circuit 1

Solution: Refer to Figure 13-2.

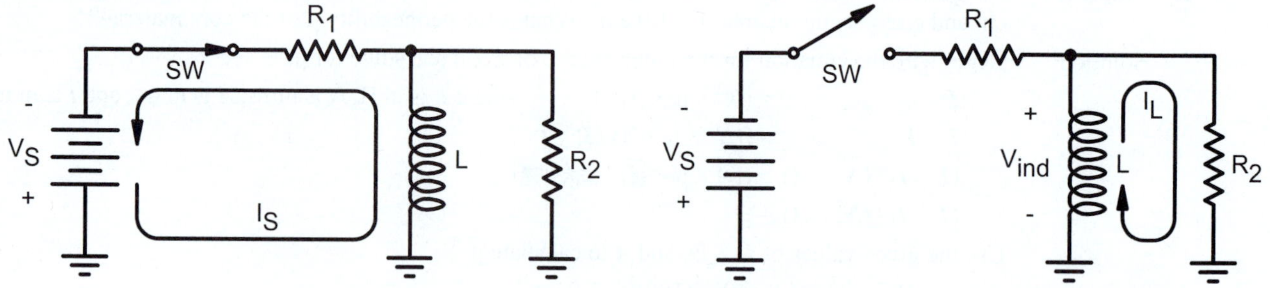

Figure 13-2: Current in Circuit with Switch Closed and Open

When the switch ***SW*** is closed, the voltage source V_S sets up a conventional current I_S that flows counter-clockwise and up through the inductor ***L***. When the switch opens and interrupts the current from V_S, the magnetic field around the inductor collapses. This induces a voltage V_{ind} to maintain the current that is flowing through ***L***. Because the current is flowing up through ***L*** just before the switch opens, V_{ind} must have the polarity shown in Figure 13-2 so that the inductor current continues to flow in the up direction. This creates a clockwise current, I_L, as shown.

13.1.2 Series and Parallel Inductors

1. Problem: Two inductors L_1 and L_2 in series have a total inductance L_T of 80 mH. When another inductor L_3 is added in series, the total inductance increases to $L_{NEW} = 127$ mH. If L_3 has the same value as L_1, what are the values of L_1, L_2, and L_3?

Solution: Begin with the equation for total series inductance for L_T and L_{NEW}.

$$L_T = L_1 + L_2$$
$$L_{NEW} = L_1 + L_2 + L_3$$

Subtract L_T from L_{NEW} to find the change in total inductance due to the addition of L_3.

$$L_{NEW} - L_T = (L_1 + L_2 + L_3) - (L_1 + L_2)$$
$$127\ \text{mH} - 80\ \text{mH} = (L_1 + L_2 + L_3) + -(L_1 + L_2)$$

$$(127 - 80)\text{ mH} = (L_1 + L_2 + L_3) + (-L_1 + -L_2)$$

$$47\text{ mH} = (L_1 + -L_1) + (L_2 + -L_2) + L_3$$

$$= L_3$$

From the problem statement, $L_3 = L_1 = 47$ mH. Use the value of L_1 to calculate L_2.

$$L_T = L_1 + L_2$$

$$80\text{ mH} = 47\text{ mH} + L_2$$

$$= 47\text{ mH} + L_2 + -47\text{ mH}$$

$$33\text{ mH} = L_2$$

The values of L_1, L_2, and L_3 are **47 mH**, **33 mH**, and **47 mH** respectively.

2. Problem: A circuit consists of nine parallel inductors. The value of three inductors is $L_1 = 100$ µH each, the value of two inductors is $L_2 = 51$ µH each, and the value of each of the remaining four inductors is L_3. If the total circuit inductance $L_T = 2.13$ µH, what is the value of L_3?

Solution: Apply the equation for the total inductance L_T of N equal inductances L in parallel and apply it to the number of L_1, L_2, and L_3 inductors in parallel.

$$L_T = L / N$$

$$L_{T(1)} = L_1 / 3$$

$$= 100\text{ µH} / 3$$

$$= 33.3\text{ µH}$$

$$L_{T(2)} = L_2 / 2$$

$$= 51\text{ µH} / 2$$

$$= 25.5\text{ µH}$$

$$L_{T(3)} = L_3 / 4$$

Next, start the general equation for total parallel inductance and use the calculated values to solve for $L_{T(3)}$.

$$\frac{1}{L_T} = \frac{1}{L_{T(1)}} + \frac{1}{L_{T(2)}} + \frac{1}{L_{T(3)}}$$

$$\frac{1}{L_T} + \frac{-1}{L_{T(1)}} + \frac{-1}{L_{T(2)}} = \frac{1}{L_{T(1)}} + \frac{1}{L_{T(2)}} + \frac{1}{L_{T(3)}} + \frac{-1}{L_{T(1)}} + \frac{-1}{L_{T(2)}}$$

$$\frac{1}{L_T} + \frac{-1}{L_{T(1)}} + \frac{-1}{L_{T(2)}} = \frac{1}{L_{T(3)}}$$

$$\frac{1}{L_T} + \frac{-1}{L_{T(1)}} + \frac{-1}{L_{T(2)}} = \frac{1}{L_3 / 4}$$

$$\frac{1}{L_T} + \frac{-1}{L_{T(1)}} + \frac{-1}{L_{T(2)}} = \frac{4}{L_3}$$

$$\left(\frac{1}{L_T} + \frac{-1}{L_{T(1)}} + \frac{1}{L_{T(2)}}\right) \times \frac{1}{4} = \frac{4}{L_3} \times \frac{1}{4}$$

$$\left(\frac{1}{L_T} + \frac{-1}{L_{T(1)}} + \frac{-1}{L_{T(2)}}\right) / 4 = \frac{1}{L_3}$$

Use the values of $L_{T(1)}$ and $L_{T(2)}$ to calculate the value of L_3.

$$\frac{1}{L_3} = \left(\frac{1}{L_T} + \frac{-1}{L_{T(1)}} + \frac{-1}{L_{T(2)}}\right)/4$$

$$= \left(\frac{1}{2.13\ \mu\text{H}} + \frac{-1}{33.3\ \mu\text{H}} + \frac{-1}{25.5\ \mu\text{H}}\right)/4$$

$$= \left[\left(\frac{1}{2.13\times10^{-6}}\right)\left(\frac{1}{\text{H}}\right) + \left(\frac{-1}{33.3\times10^{-6}}\right)\left(\frac{1}{\text{H}}\right) + \left(\frac{-1}{25.5\times10^{-6}}\right)\left(\frac{1}{\text{H}}\right)\right]/4$$

$$= \left[\left(\frac{1}{2.13\times10^{-6}} + \frac{-1}{33.3\times10^{-6}} + \frac{1}{25.5\times10^{-6}}\right)\left(\frac{1}{\text{H}}\right)\right]/4$$

$$= \left[\left(469\times10^3 + -30.0\times10^3 + -39.2\times10^3\right)\left(\frac{1}{\text{H}}\right)\right]/4$$

$$= \left[\left(400\times10^3\right)\left(\frac{1}{\text{H}}\right)\right]/4$$

$$1/\left(\frac{1}{L_3}\right) = 1/\left\{\left[\left(400\times10^3\right)\left(\frac{1}{\text{H}}\right)\right]/4\right\}$$

$$L_3 = 4/(400\times10^3)\ \text{H}$$

$$= 10.0\times10^{-6}\ \text{H}$$

$$= 10.0\ \mu\text{H}$$

The value of the four L_3 inductors is **10 μH** each.

3. Problem: What is the total inductance of the circuit in Figure 13-3?

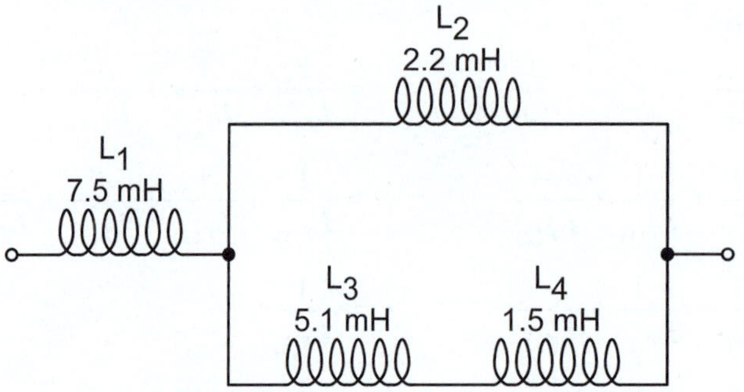

Figure 13-3: Series and Parallel Inductors Example Circuit 1

Solution: The expression for total inductance of the circuit is

$$L_T = L_1 + L_2 \parallel (L_3 + L_4)$$

The first equivalent inductance L_{EQ1} is the series combination ($L_3 + L_4$).

$$L_T = L_1 + L_2 \parallel L_{EQ1}$$

where

$$L_{EQ1} = L_3 + L_4$$

$$= 5.1\ \text{mH} + 1.5\ \text{mH}$$

$$= 6.6\ \text{mH}$$

The second equivalent inductance L_{EQ2} is the parallel combination $L_2 \parallel L_{EQ1}$.

$$L_T = L_1 + L_{EQ2}$$

where

L_{EQ2} $= L_2 \parallel L_{EQ1}$
$= 2.2\text{ mH} \parallel 6.6\text{ mH}$
$= 1 / [(1 / 2.2\text{ mH}) + (1 / 6.6\text{ mH})]$
$= 1 / \{[1 / (2.2 \times 10^{-3}\text{ H})] + [1 / (6.6 \times 10^{-3}\text{ H})]\}$
$= 1 / \{[1 / (2.2 \times 10^{-3})(1/\text{H})] + [1 / (6.6 \times 10^{-3})(1 / \text{H})]\}$
$= 1 / (455\text{ H}^{-1} + 152\text{ H}^{-1})$
$= 1 / (606\text{ H}^{-1})$
$= (1 / 606)(1 / \text{H}^{-1})$
$= 1.65 \times 10^{-3}\text{ H}$
$= 1.65 \times \text{mH}$

The third (and last) equivalent inductance L_{EQ3} is the series combination $L_1 + L_{EQ2}$.

$L_T = L_{EQ3}$

where

L_{EQ3} $= L_1 + L_{EQ2}$
$= 7.5\text{ mH} + 1.65\text{ mH}$
$= 9.15\text{ mH}$

Refer to Figure 13-4 for the circuit analysis and simplification process.

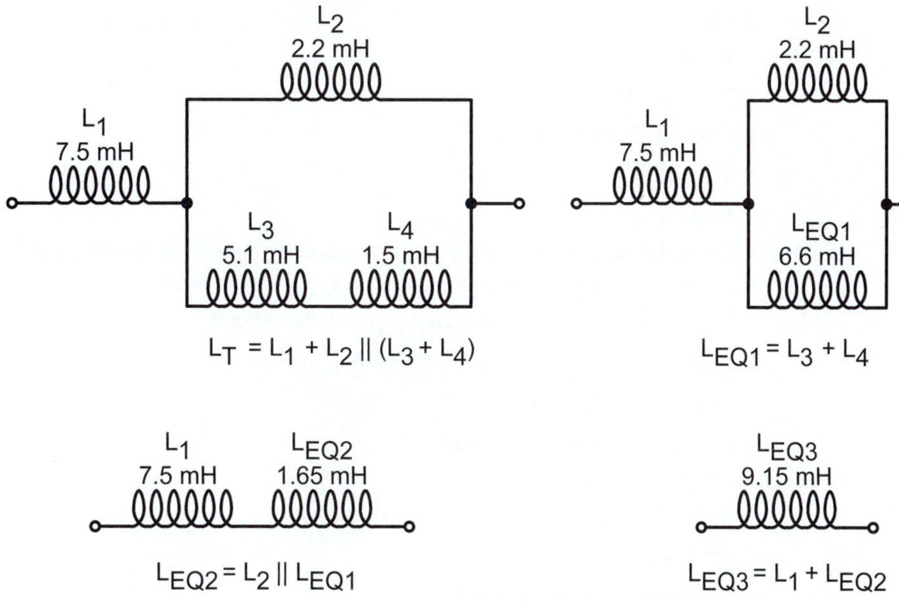

Figure 13-4: Circuit Analysis and Simplification for Figure 13-3

The total inductance for the circuit of Figure 13-3 is **9.2 mH**.

13.1.3 Inductors in DC Circuits

1. Problem: Switch ***SW*** closes at time $t = 0$ for the circuit in Figure 13-5. What is the current $i(t)$ for the circuit in Figure 13-5 at time $t = 12.0\ \mu\text{s}$? Assume that the inductor is ideal with no winding resistance.

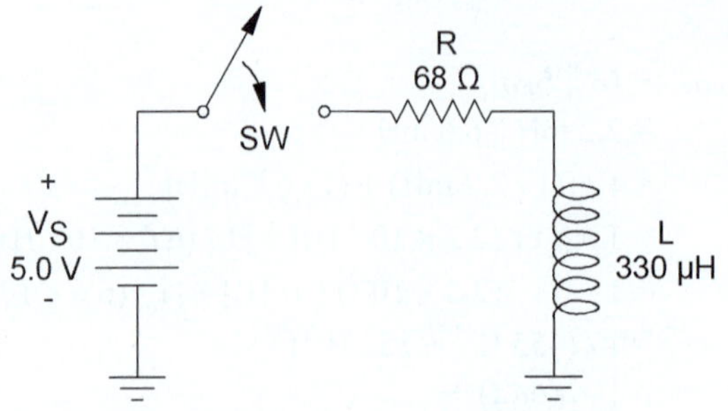

Figure 13-5: Inductors in DC Circuits Example Circuit 1

Solution: Until the switch closes there is no current in the circuit, so the initial current $I_0 = 0$ A. The ideal inductor L has no resistance, so the final current I_F will be

$I_F = V_S / R$

$= 5.0 \text{ V} / 68\ \Omega$

$= (5.0 / 68)\ (\text{V} / \Omega)$

$= 73.5 \times 10^{-3}$ A

The exponential current equation for current $i(t)$ at time t for the inductor circuit is

$i(t) = I_F + (I_0 - I_F)\ e^{-(t/\tau)}$

Use the given values of L and R to calculate the inductive time constant τ.

$\tau = L / R$

$= 330\ \mu\text{H} / 68\ \Omega$

$= (330 \times 10^{-6} \text{ H}) / 68\ \Omega$

$= [(330 \times 10^{-6}) / 68]\ (\text{H} / \Omega)$

$= 4.85 \times 10^{-6}$ s

$= 4.85\ \mu$s

Use the calculated time constant and general exponential equation to solve $i(t)$ for $t = 12.0\ \mu$s.

$i(t) = (73.5 \times 10^{-3} \text{ A}) + [0 \text{ A} - (73.5 \times 10^{-3} \text{ A})]\ e^{-(12.0\ \mu s / 4.85\ \mu s)}$

$= (73.5 \times 10^{-3} \text{ A}) + -(73.5 \times 10^{-3} \text{ A})]\ e^{-(12.0\ \mu s / 4.85\ \mu s)}$

$= (73.5 \times 10^{-3} \text{ A}) \times [1 + -e^{-(12.0\ \mu s / 4.85\ \mu s)}]$

$= (73.5 \times 10^{-3} \text{ A}) \times [1 + -e^{-2.47}]$

$= (73.5 \times 10^{-3} \text{ A}) \times [\ 1 + -0.0844]$

$= (73.5 \times 10^{-3} \text{ A}) \times 0.916$

$= 67.3 \times 10^{-3}$ A

$= 67.3$ mA

The current at time $t = 12\ \mu$s is **67 mA**.

2. Problem: Switch ***SW*** for the circuit in Figure 13-6 is moved from position 1 to position 2 at time $t = 0$. At time $t = 25\ \mu$s ***SW*** is moved back to position 1. Use the exponential current equation to determine the instantaneous induced voltage $v(t)$ across the inductor L immediately after the switch is moved to position 1 at time $t = 25\ \mu$s. Assume that the inductor is ideal with no winding resistance.

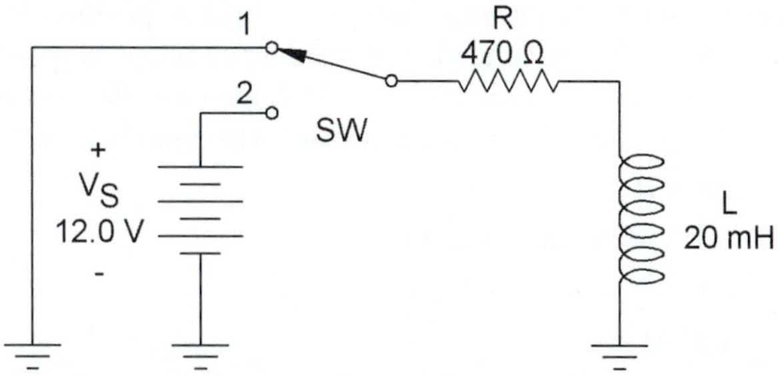

Figure 13-6: Inductors in DC Circuits Example Circuit 2

Solution: Refer to Figure 13-7.

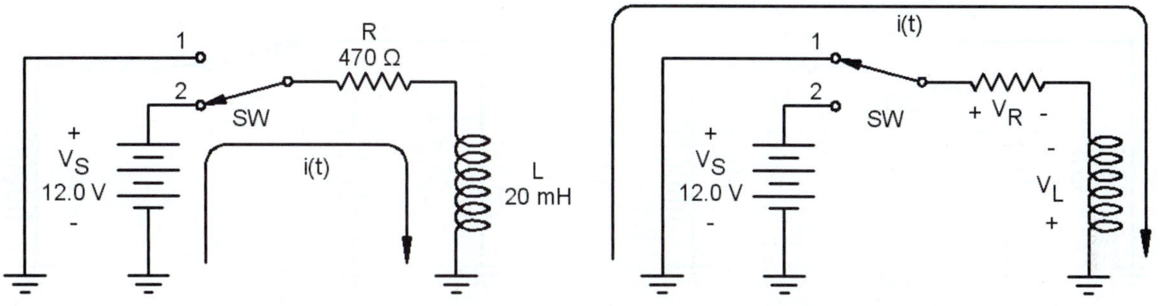

Figure 13-7: Analysis of Circuit Currents and Voltages

When ***SW*** is moved to position 1, the current through the inductor begins to exponentially increase from its initial value of I_0 to its final value of I_F according to the exponential current equation:

$$i(t) = I_F + (I_0 - I_F)\, e^{-(t/\tau)}$$

Use the given values of ***L*** and ***R*** to calculate the inductive time constant **τ**.

$$\begin{aligned}\tau &= L / R \\ &= 20 \text{ mH} / 470\ \Omega \\ &= (20 \times 10^{-3}\text{ H}) / 470\ \Omega \\ &= [(20 \times 10^{-3}) / 470]\ (\text{H} / \Omega) \\ &= 42.6 \times 10^{-6}\text{ s} \\ &= 42.6\ \mu\text{s}\end{aligned}$$

There is no current flowing until ***SW*** is moved to position 2, so $I_0 = 0$ A. The ideal inductor ***L*** has no resistance, so use Ohm's Law for current to calculate the final current value I_F from V_S and ***R***.

$$\begin{aligned}I_F &= V_S / R \\ &= 12.0\text{ V} / 470\ \Omega \\ &= (12.0 / 470)\ (\text{V} / \Omega) \\ &= 25.5 \times 10^{-3}\text{ A}\end{aligned}$$

Use the exponential current equation to calculate the current at time $t = 25$ µs.

$$\begin{aligned}i(t) &= (25.5 \times 10^{-3}\text{ A}) + [0\text{ A} - (25.5 \times 10^{-3}\text{ A})]\, e^{-(25\ \mu\text{s} / 42.6\ \mu\text{s})} \\ &= (25.5 \times 10^{-3}\text{ A}) + -(25.5 \times 10^{-3}\text{ A})]\, e^{-(25\ \mu\text{s} / 42.6\ \mu\text{s})} \\ &= (25.5 \times 10^{-3}\text{ A}) \times [1 + -e^{-(25\ \mu\text{s} / 42.6\ \mu\text{s})}] \\ &= (25.5 \times 10^{-3}\text{ A}) \times [1 + -e^{-0.588}] \\ &= (25.5 \times 10^{-3}\text{ A}) \times [\ 1 + -0.556] \\ &= (25.5 \times 10^{-3}\text{ A}) \times 0.444 \\ &= 11.3 \times 10^{-3}\text{ A}\end{aligned}$$

Immediately after $t = 25$ µs when ***SW*** moves back to position 1, the current through the inductor $\boldsymbol{L}$ and resistor $\boldsymbol{R}$ will be unchanged, as the induced voltage across the inductor will oppose any change in the current. From Kirchhoff's Voltage Law, the voltage V_L across the inductor must equal the voltage drop V_R across the resistor. Use Ohm's Law to calculate V_R from $i(t)$ and $\boldsymbol{R}$.

$$\begin{aligned} V_R &= i(t) \times R \\ &= (11.3 \times 10^{-3}\ \text{A}) \times 470\ \Omega \\ &= [(11.3 \times 10^{-3}) \times 470]\ [\text{A} \times \Omega] \\ &= 5.33\ \text{V} \end{aligned}$$

The voltage across $\boldsymbol{L}$ at $t = 25$ µs is **5.3 V**, with the polarity shown in Figure 13-7.

3. Problem: Solve Problem 2 using the exponential voltage equation for V_L.

Solution: Refer to Figure 13-8.

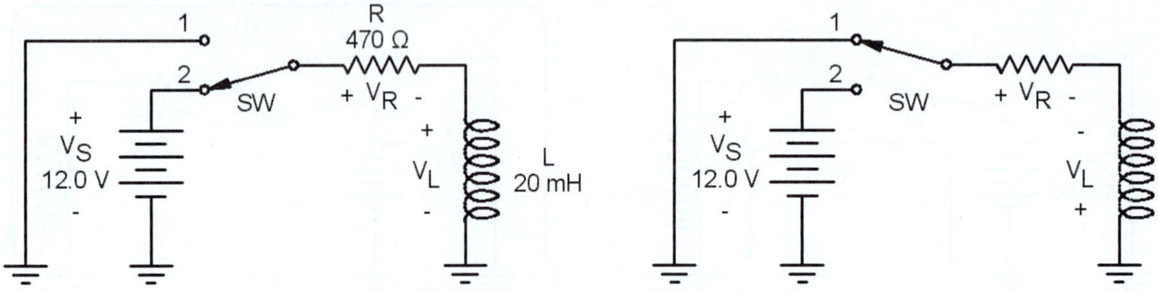

Figure 13-8: Voltage Analysis of Figure 13-6

The exponential voltage equation gives the instantaneous voltage, $v(t)$, at time t.

$$v(t) = V_F + (V_0 - V_F)\, e^{-(t/\tau)}$$

Use the given values of $\boldsymbol{L}$ and $\boldsymbol{R}$ to calculate the time constant τ.

$$\begin{aligned} \tau &= L / R \\ &= 20\ \text{mH} / 470\ \Omega \\ &= (20 \times 10^{-3}\ \text{H}) / 470\ \Omega \\ &= [(20 \times 10^{-3}) / 470]\ (\text{H} / \Omega) \\ &= 42.6 \times 10^{-6}\ \text{s} \\ &= 42.6\ \mu\text{s} \end{aligned}$$

Initially the current in the circuit is 0 A. When the switch ***SW*** is moved from position 1 position 2 at time $t = 0$, the induced voltage V_L across $\boldsymbol{L}$ is equal and opposite to V_S to keep the current at 0 A. Therefore, $V_0 = V_S$. When the current reaches its maximum value the resistor will drop all the source voltage, so that the final value of $V_L = 0$ V. Therefore, $V_F = 0$ V. Use these values of V_0 and V_F to calculate the instantaneous voltage $v(t)$ across the inductor $\boldsymbol{L}$.

$$\begin{aligned} v(t) &= 0\ \text{V} + (12.0\ \text{V} - 0\ \text{V})\, e^{-(t/42.6\ \mu\text{s})} \\ &= (12.0\ \text{V})\, e^{-(t/42.6\ \mu\text{s})} \end{aligned}$$

At time $t = 25$ µs, just before the switch is moved back to position 1, the voltage across $\boldsymbol{L}$ is

$$\begin{aligned} v(t) &= (12.0\ \text{V})\, e^{-(25\ \mu\text{s}/42.6\ \mu\text{s})} \\ &= (12.0\ \text{V}) \times e^{-0.588} \\ &= (12.0\ \text{V}) \times 0.556 \\ &= 6.67\ \text{V} \end{aligned}$$

Use Kirchhoff's Voltage Law to find V_R in terms of V_S and V_L.

$$V_S + -V_R + -V_L \qquad = 0\ \text{V}$$

$V_S + -V_R + -V_L + V_R \quad = 0\text{ V} + V_R$

$V_S + -V_L \quad = V_R$

Use the values of V_S and V_L to calculate V_R at time $t = 25\ \mu s$.

$V_R \quad = 12.0\text{ V} + -6.67\text{ V}$

$\quad = 5.33\text{ V}$

Immediately after ***SW*** is moved back to position 1, the source voltage V_S is removed from the current loop, but the inductor keeps the current through R unchanged so from Ohm's Law the resistor voltage, V_R, is also unchanged. Use Kirchhoff's Voltage Law to calculate V_L.

$V_L + V_R \quad = 0\text{ V}$

$V_L + V_R + -V_R \quad = 0\text{ V} + -V_R$

$V_L \quad = -V_R$

$\quad = -5.33\text{ V}$

The voltage across L at t = 25 μs is **5.3 V** with the polarity shown in Figure 13-8.

13.1.4 Inductors in AC Circuits

1. Problem: Verify that X_{LT} for inductors in series and parallel are calculated the same as for total resistance for resistors in series and parallel.

Solution: By definition, the value of inductive reactance X_L for an inductor L at frequency f is

$X_L \quad = 2\pi f L$

For two inductors L_1 and L_2 in series, the total capacitance L_T is

$L_T \quad = L_1 + L_2$

The total inductive reactance is then

$X_{LT} = 2\pi f L_T$

$= 2\pi f(L_1 + L_2)$

$= 2\pi f L_1 + 2\pi f L_2$

$= X_{L1} + X_{L2}$

Therefore, total series inductive reactance is the sum of the individual inductive reactances, just as total series resistance is the sum of the individual resistances. For two inductors L_1 and L_2 in parallel, the total capacitance L_T is

$L_T \quad = 1 / [(1 / L_1) + (1 / L_2)]$

Substitute this value of L_T into the equation for total inductive reactance, X_{LT}.

$X_{LT} \quad = 2\pi f L_T$

$= 2\pi f\{1 / [(1 / L_1) + (1 / L_2)]\}$

$1 / X_{LT} \quad = 1 / (2\pi f\{1 / [(1 / L_1) + (1 / L_2)]\})$

$= 1 / (2\pi f) \times [(1 / L_1) + (1 / L_2)]$

$= \{[1 / (2\pi f)] \times (1 / L_1)\} + \{[1 / (2\pi f)] \times (1 / L_2)\}$

$= [1 / (2\pi f L_1)] + [1 / (2\pi f L_2)]$

$= (1 / X_{L1}) + (1 / X_{L2})$

$1 / (1 / X_{LT}) \quad = 1 / [(1 / X_{L1}) + (1 / X_{L2})]$

$X_{LT} \quad = 1 / [(1 / X_{L1}) + (1 / X_{L2})]$

Therefore, total parallel inductive reactance is the reciprocal of the sum of the reciprocals of the individual inductive reactances, just as total parallel resistance is the reciprocal of the sum of the reciprocal of the individual resistances.

2. Problem: For the current divider circuit in Figure 13-9, show that for an ac current source I_S

$I_{L1} = I_S \times (L_T / L_1)$

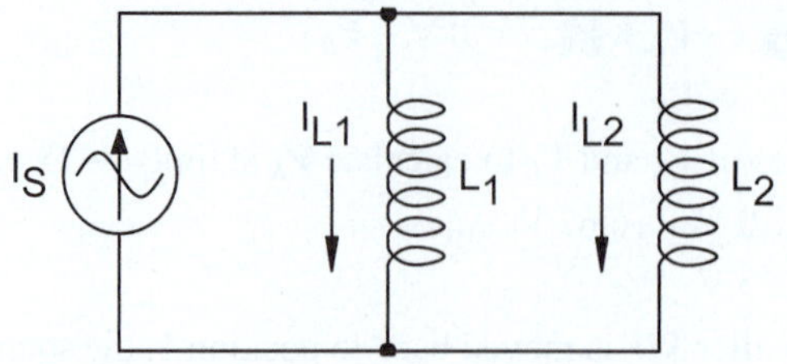

Figure 13-9: Inductor Current Divider Example Circuit for Problem 2

Solution: Let X_{L1} be the inductive reactance of L_1, X_{L2} be the inductive reactance of L_2, and X_{LT} be the total inductive reactance of the parallel inductors. From Ohm's Law for voltage, the source voltage V_S is the source current times the total inductive reactance.

$V_S = I_S \times X_{LT}$

Because the inductors are in parallel, the source voltage is across both inductors. Use Ohm's Law for current to find the current, I_{L1}, through L_1.

$I_{L1} = V_S / X_{L1}$

$= (I_S \times X_{LT}) / X_{L1}$

$= I_S \times (X_{LT} / X_{L1})$

Substitute the equation for inductance, $X_L = 2\pi fL$, for X_{LT} and X_{L1}.

$I_{L1} = I_S \times (2\pi f L_T / 2\pi f L_1)$

$= I_S \times (L_T / L_1)$

3. Problem: What is the Q of the inductor L in Figure 13-10 at frequency f = 12.5 kHz?

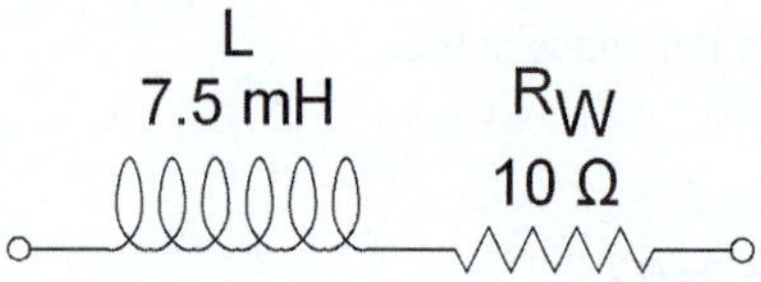

Figure 13-10: Inductor *Q* Example for Problem 3

Solution: Begin with the definition of Q.

$Q = X_L / R$

Use the given values of f and L to calculate X_L.

$X_L = 2\pi fL$

$= 2 \times \pi \times 12.5\text{ kHz} \times 7.5\text{ mH}$

$= 2 \times \pi \times (12.5 \times 10^3\text{ Hz}) \times (7.5 \times 10^{-3}\text{ H})$

$= [2 \times \pi \times (12.5 \times 10^3) \times (7.5 \times 10^{-3})]\ (\text{H} \times \text{Hz})$

$= [2 \times \pi \times (12.5 \times 10^3) \times (7.5 \times 10^{-3})]\ (\text{H} \times \text{Hz})$

$= 589\ \Omega$

The resistance R is the winding resistance R_W.

$R = R_W$

$= 10\ \Omega$

Use the values of X_L and R to calculate Q.

$Q = X_L / R$

$= 589\ \Omega / 10\ \Omega$

$= \mathbf{58.9}$

13.2 Advanced

13.2.1 The Basic Inductor

1. Problem: A company decides to save the material costs of its 15 mH inductor cores. If it reduces the core length by 25%, by how much must it reduce the core diameter if the number of winding turns is not changed?

Solution: Let l_{ORG} be the original core length, A_{ORG} be the original core cross-sectional area, l_{NEW} be the new core length, and A_{NEW} be the new core cross-sectional area. Apply the equation for inductance to the original and core values.

$$L = (N^2 \times \mu \times A) / l$$
$$15\text{ mH} = (N^2 \times \mu \times A_{ORG}) / l_{ORG}$$
$$15\text{ mH} = (N^2 \times \mu \times A_{NEW}) / l_{NEW}$$

Set the two terms equal and solve for A_{NEW}.

$$(N^2 \times \mu \times A_{NEW}) / l_{NEW} = (N^2 \times \mu \times A_{ORG}) / l_{ORG}$$
$$[(N^2 \times \mu \times A_{NEW}) / l_{NEW}] / (N^2 \times \mu) = [(N^2 \times \mu \times A_{ORG}) / l_{ORG}] / (N^2 \times \mu)$$
$$(A_{NEW} / l_{NEW}) \times l_{NEW} = (A_{ORG} / l_{ORG}) / l_{NEW}$$
$$A_{NEW} = A_{ORG} \times (l_{NEW} / l_{ORG})$$

From the problem statement l_{NEW} is 25% less than l_{ORG}.

$$A_{NEW} = A_{ORG} \times [l_{ORG} \times (1 - 25\%)] / l_{ORG}$$
$$A_{NEW} = A_{ORG} \times [1 - (25 / 100)]$$
$$A_{NEW} = A_{ORG} \times [1 - 0.25]$$
$$A_{NEW} = A_{ORG} \times 0.75$$

The formula for the area A of a circle with radius r is $A = 2\pi r^2$.

$$2\pi r_{NEW}^2 = 2\pi r_{ORG}^2 \times 0.75$$
$$(2\pi r_{NEW}^2) / (2\pi) = (2\pi r_{ORG}^2 \times 0.75) / (2\pi)$$
$$r_{NEW}^2 = r_{ORG}^2 \times 0.75$$
$$(r_{NEW}^2)^{1/2} = (r_{ORG}^2 \times 0.75)^{1/2}$$
$$r_{NEW} = r_{ORG} \times 0.75$$
$$r_{NEW} = r_{ORG} \times 0.866$$

The radius r of a circle is half the diameter d.

$$d_{NEW} / 2 = (d_{ORG} / 2) \times 0.866$$
$$(d_{NEW} / 2) \times 2 = [(d_{ORG} \times 0.866) / 2] \times 2$$
$$d_{NEW} = d_{ORG} \times 0.866$$
$$d_{NEW} = d_{ORG} \times (1 - 0.134)$$
$$d_{NEW} = d_{ORG} \times [1 - (13.4 / 100)]$$
$$d_{NEW} = d_{ORG} \times (1 - 13.4\%)$$

The new diameter d_{NEW} is ***13.4%*** less than the original diameter d_{ORG}.

13.2.2 Series and Parallel Inductors

1. Problem: Prove that total series inductance L_T of two coils is the sum of the coil inductances L_1 and L_2.

Solution: Begin with the definition of inductance.

$$v_{ind} = L\,(\mathrm{d}i / \mathrm{d}t)$$

Voltages in series add, so the total induced voltage is the sum of the induced voltages for the individual coils.

$$v_{ind\,(\text{total})} = v_{ind\,(1)} + v_{ind\,(2)}$$

$L_T\,(di_T / dt) \;= L_1\,(di_1 / dt) + L_2\,(di_2 / dt)$

Because the current in a series circuit is the same for each series component, the rate of change for the series current di / dt must also be the same for the circuit and each series component.

$$L_T\,(di / dt) = L_1\,(di / dt) + L_2\,(di / dt)$$
$$L_T \times (di / dt) = (L_1 + L_2) \times (di / dt)$$
$$[L_T \times (di / dt)] / (dt / di) = (L_1 + L_2) \times (di / dt) \times (dt / di)$$
$$L_T = L_1 + L_2$$

The total series inductance L_T is the sum of the individual inductances L_1 and L_2.

2. Problem: Prove that the total parallel inductance L_T is the reciprocal of the sum of the reciprocals of the individual coil inductances L_1 and L_2.

Solution: Begin with the definition of inductance.

$$v_{ind} = L\,(di / dt)$$
$$v_{ind} / L = [L\,(di / dt)] / L$$
$$v_{ind} / L = (di / dt)$$

From Kirchhoff's Current Law, the total current i_T for the parallel combination of the coils is equal to the sum of the individual coil currents i_1 and i_2, so the rate of change for the total current di_T / dt must equal the sum of the rates of change for the individual coil currents di_1 / dt and di_2 / dt.

$$di_T / dt = (di_1 / dt) + (di_2 / dt)$$
$$v_{ind\,(total)} / L_T = [v_{ind\,(1)} / L_1] + [v_{ind\,(2)} / L_2]$$

Because the voltage across a parallel circuit is the same for each component in parallel, the induced voltage, v_{ind}, must be the same for each parallel component.

$$v_{ind} / L_T = (v_{ind} / L_1) + (v_{ind} / L_2)$$
$$v_{ind} / L_T = v_{ind} \times [(1 / L_1) + (1 / L_2)]$$
$$(v_{ind} / L_T) / v_{ind} = \{v_{ind} \times [(1 / L_1) + (1 / L_2)]\} / v_{ind}$$
$$1 / L_T = (1 / L_1) + (1 / L_2)$$

The total parallel inductance L_T is the reciprocal of the sum of the reciprocals of the individual inductances L_1 and L_2.

13.2.3 Inductors in DC Circuits

1. Problem: Refer to Figure 13-11.

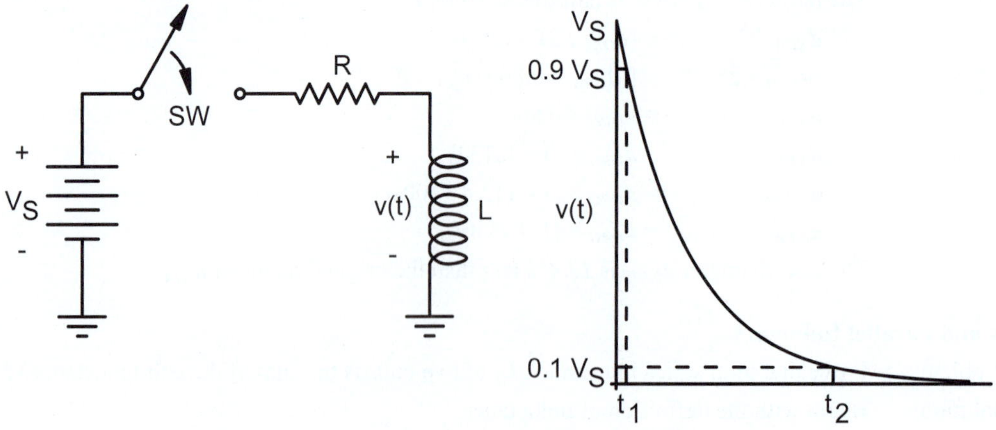

Figure 13-11: Inductors in DC Circuits Example Circuit 3

When the switch ***SW*** closes at time $t = 0$, the inductor voltage $v(t)$ initially equals $v_i = V_S$ and then decays exponentially towards $v_f = 0$ V, as shown in the graph. How many time constants τ does it take the inductor voltage to change from 0.9 V_S to 0.1 V_S?

Solution: As the graph shows, $v(t) = 0.9\ V_S$ at $t = t_1$ and $v(t) = 0.1\ V_S$ at $t = t_2$ so the time for $v(t)$ to change from 0.9 V_S to 0.1 V_S is $t_2 - t_1$.

Use the exponential charging and discharging equation to solve for t_1 when $v(t) = 0.9\ V_S$.

$$v(t) = v_f + (v_0 - v_f)e^{-t/\tau}$$

$$0.9\ V_S = 0\text{ V} + (V_S - 0\text{ V})e^{-t_1/\tau}$$

$$0.9\ V_S = V_S \times e^{-t_1/\tau}$$

$$(0.9\ V_S)\ /\ V_S = (V_S \times e^{-t_1/\tau})/V_S$$

$$0.9 = e^{-t_1/\tau}$$

$$\ln(0.9) = \ln(e^{-t_1/\tau})$$

$$\ln(0.9) = -t_1 / \tau$$

$$\ln(0.9) \times -\tau = (-t_1 / \tau) \times -\tau$$

$$-\ln(0.9) \times \tau = t_1$$

Next, use the exponential charging and discharging equation to solve for t_2 when $v(t) = 0.1\ V_S$.

$$v(t) = v_f + (v_0 - v_f)e^{-t/\tau}$$

$$0.1\ V_S = 0\text{ V} + (V_S - 0\text{ V})e^{-t_2/\tau}$$

$$0.1\ V_S = V_S \times e^{-t_2/\tau}$$

$$(0.1\ V_S)\ /\ V_S = (V_S \times e^{-t_2/\tau})/V_S$$

$$0.1 = e^{-t_2/\tau}$$

$$\ln(0.1) = \ln(e^{-t_2/\tau})$$

$$\ln(0.1) = -t_2 / \tau$$

$$\ln(0.1) \times -\tau = (-t_2 / \tau) \times -\tau$$

$$-\ln(0.1) \times \tau = t_2$$

From these

$$\begin{aligned} t_2 - t_1 &= [-\ln(0.1) \times \tau] - [-\ln(0.9) \times \tau] \\ &= [-\ln(0.1) \times \tau] + [\ln(0.9) \times \tau] \\ &= [-\ln(0.1) + \ln(0.9)] \times \tau \\ &= [\ln(0.9) - \ln(0.1)] \times \tau \\ &= [\ln(0.9 / 0.1)] \times \tau \\ &= \ln(9.0) \times \tau \\ &= 2.2 \times \tau \end{aligned}$$

The time for the inductor voltage to change from 0.9 V_S to 0.1 V_S is **2.2 time constants**.

2. Problem: Although you can use an oscilloscope to directly measure the time constant of an ***RL*** circuit, it is sometimes impractical to view and measure enough of the curve to do so. In these cases, you can use any two points on the curve to find the time constant if you know the final value to which the circuit is charging or discharging. Determine the time constant τ for an ***RL*** circuit if the final current value is i_f and the measured current at times t_1 and t_2 are $i(t_1)$ and $i(t_2)$.

Solution: A property of the ***RL*** exponential charging and discharging curves is that the value of the curve will change by 63% for each time constant, regardless of the current value. This means that you can use any point on the curve for the initial value, provided that you express the time for each point on the curve relative to your chosen initial point. For the ***RL*** circuit, let $i(t_1)$ be the initial point, so the time to point $i(t_2)$ is $t_2 - t_1$. Substitute these values in the general exponential charging and discharging equation for $i(t_2)$.

$$i(t_2) = i_f + (i_0 - i_f)e^{-t_2/\tau}$$

$$i(t_2) = i_f + [i(t_1) - i_f]e^{-(t_2-t_1)/\tau}$$

Use this equation to solve for τ.

$$i(t_2) - i_f = i_f + [i(t_1) - i_f]e^{-(t_2-t_1)/\tau} - i_f$$

$$\frac{i(t_2) - i_f}{i(t_1) - i_f} = \frac{[i(t_1) - i_f]e^{-(t_2-t_1)/\tau} - i_f}{[i(t_1) - i_f]}$$

$$\ln\left(\frac{i(t_2) - i_f}{i(t_1) - i_f}\right) = \ln[e^{-(t_2-t_1)/\tau}]$$

$$\frac{\ln\left(\frac{i(t_2) - i_f}{i(t_1) - i_f}\right)}{-(t_2 - t_1)} = \frac{-(t_2 - t_1)/\tau}{-(t_2 - t_1)}$$

$$1/\left[\frac{\ln\left(\frac{i(t_2) - i_f}{i(t_1) - i_f}\right)}{-(t_2 - t_1)}\right] = 1/(1/\tau)$$

$$\frac{-(t_2 - t_1)}{\ln\left(\frac{i(t_2) - i_f}{i(t_1) - i_f}\right)} = \tau$$

13.2.4 Inductors in AC Circuits

1. Problem: What is the total inductive reactance X_T for the circuit in Figure 13-12?

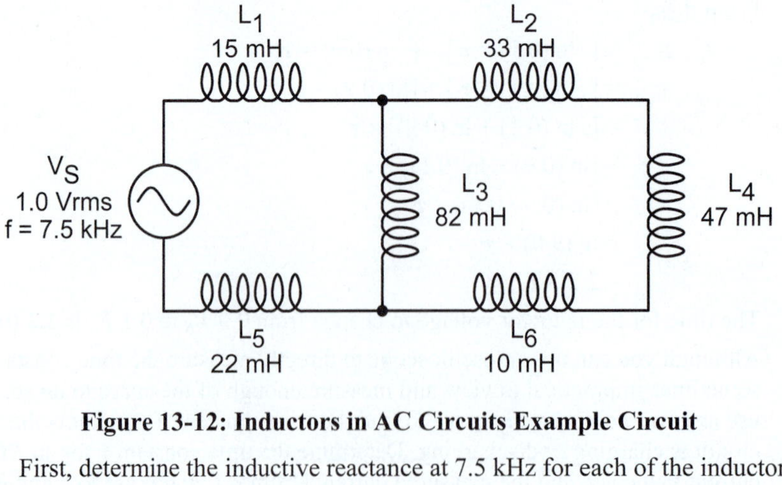

Figure 13-12: Inductors in AC Circuits Example Circuit

Solution: First, determine the inductive reactance at 7.5 kHz for each of the inductors.

$$
\begin{aligned}
X_{L1} &= 2\pi f L_1 \\
&= 2\times \pi \times (7.5\text{ kHz}) \times (15\text{ mH})] \\
&= 2\times \pi \times (7.5 \times 10^3\text{ Hz}) \times (15 \times 10^{-3}\text{ H}) \\
&= [2\times \pi \times (7.5 \times 10^3) \times (68 \times 10^{-6})]\ (\text{Hz} \times \text{H})]\} \\
&= 707\ \Omega
\end{aligned}
$$

$$
\begin{aligned}
X_{L2} &= 2\pi f L_2 \\
&= 2\times \pi \times (7.5\text{ kHz}) \times (33\text{ mH})] \\
&= 2\times \pi \times (7.5 \times 10^3\text{ Hz}) \times (33 \times 10^{-3}\text{ H}) \\
&= [2\times \pi \times (7.5 \times 10^3) \times (33 \times 10^{-3})]\ (\text{Hz} \times \text{H})]\} \\
&= 1.56 \times 10^3\ \Omega \\
&= 1.56\text{ k}\Omega
\end{aligned}
$$

$$
\begin{aligned}
X_{L3} &= 2\pi f L_3 \\
&= 2\times \pi \times (7.5\text{ kHz}) \times (82\text{ mH})] \\
&= 2\times \pi \times (7.5 \times 10^3\text{ Hz}) \times (82 \times 10^{-3}\text{ H}) \\
&= [2\times \pi \times (7.5 \times 10^3) \times (82 \times 10^{-3})]\ (\text{Hz} \times \text{H})]\} \\
&= 3.86 \times 10^3\ \Omega \\
&= 3.86\text{ k}\Omega
\end{aligned}
$$

$$
\begin{aligned}
X_{L4} &= 2\pi f L_4 \\
&= 2\times \pi \times (7.5\text{ kHz}) \times (47\text{ mH})] \\
&= 2\times \pi \times (7.5 \times 10^3\text{ Hz}) \times (47 \times 10^{-3}\text{ H}) \\
&= [2\times \pi \times (7.5 \times 10^3) \times (47 \times 10^{-3})]\ (\text{Hz} \times \text{H})]\} \\
&= 2.21 \times 10^3\ \Omega \\
&= 2.21\text{ k}\Omega
\end{aligned}
$$

$$
\begin{aligned}
X_{L5} &= 2\pi f L_5 \\
&= 2\times \pi \times (7.5\text{ kHz}) \times (22\text{ mH})] \\
&= 2\times \pi \times (7.5 \times 10^3\text{ Hz}) \times (22 \times 10^{-3}\text{ H}) \\
&= [2\times \pi \times (7.5 \times 10^3) \times (22 \times 10^{-3})]\ (\text{Hz} \times \text{H})]\} \\
&= 1.04 \times 10^3\ \Omega \\
&= 1.04\text{ k}\Omega
\end{aligned}
$$

$$
\begin{aligned}
X_{L6} &= 2\pi f L_6 \\
&= 2\times \pi \times (7.5\text{ kHz}) \times (10\text{ mH})] \\
&= 2\times \pi \times (7.5 \times 10^3\text{ Hz}) \times (10 \times 10^{-3}\text{ H}) \\
&= [2\times \pi \times (7.5 \times 10^3) \times (10 \times 10^{-3})]\ (\text{Hz} \times \text{H})]\} \\
&= 471\ \Omega
\end{aligned}
$$

Next, replace the inductors with their inductive reactances and analyze the circuit as for a resistive circuit. Refer to Figure 13-13, which shows the reactive equivalent and the circuit analysis and simplification process.

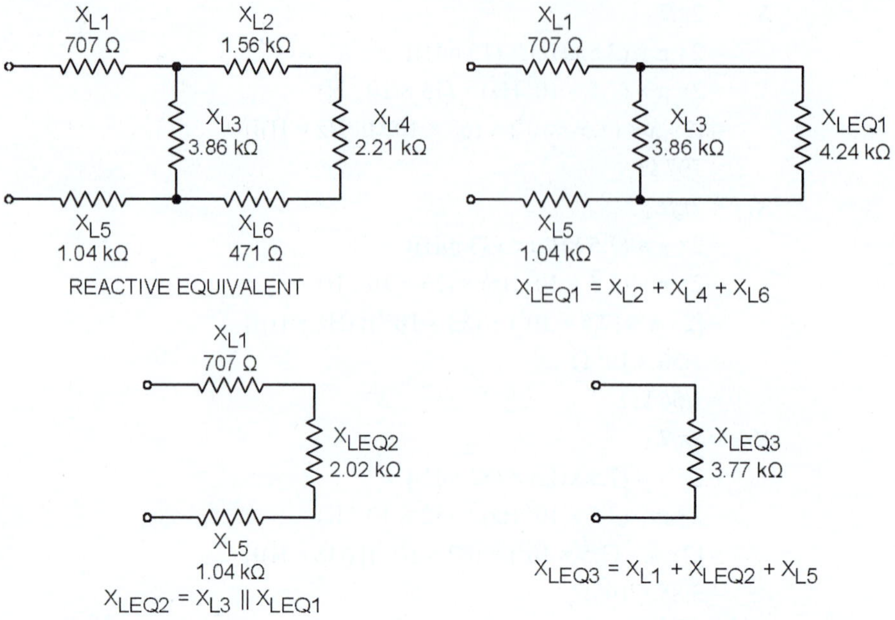

Figure 13-13: Circuit Analysis and Simplification Process for Figure 13-12

The expression for the circuit is

$$X_{LT} = X_{L1} + X_{L3} \parallel (X_{L2} + X_{L4} + X_{L6}) + X_{L5}$$

The first equivalent reactance X_{LEQ1} is the series combination $(X_{L2} + X_{L4} + X_{L6})$.

$$X_{LT} = X_{L1} + X_{L3} \parallel X_{LEQ1} + X_{L5}$$

where

$$\begin{aligned} X_{LEQ1} &= X_{L2} + X_{L4} + X_{L6} \\ &= 1.56 \text{ k}\Omega + 2.21 \text{ k}\Omega + 471 \ \Omega \\ &= 4.24 \text{ k } \Omega \end{aligned}$$

The second equivalent reactance X_{LEQ2} is the parallel combination $X_{L3} \parallel X_{LEQ1}$.

$$X_{LT} = X_{L1} + X_{LEQ2} + X_{L5}$$

where

$$\begin{aligned} X_{LEQ2} &= X_{L3} \parallel X_{LEQ1} \\ &= 3.86 \text{ k}\Omega \parallel 4.24 \text{ k}\Omega \\ &= 2.02 \text{ k}\Omega \end{aligned}$$

The third (and last) equivalent reactance X_{LEQ3} is the series combination $X_{L1} + X_{LEQ2} + X_{L5}$.

$$X_{LT} = X_{LEQ3}$$

where

$$\begin{aligned} X_{LEQ3} &= X_{L1} + X_{LEQ2} + X_{L5} \\ &= 707 \ \Omega + 2.02 \text{ k}\Omega + 1.04 \text{ k}\Omega \\ &= 3.77 \text{ k}\Omega \end{aligned}$$

The total inductive reactance is **3.77 kΩ.**

2. Problem: Use the results of Problem 1 to determine the voltage across each inductor for the circuit in Figure 13-12.

Solution: To determine the voltages across the inductors for the circuit, use Ohm's Law for the circuit equivalents of Problem 1 and work backwards through the circuit. Refer to Figure 13-14.

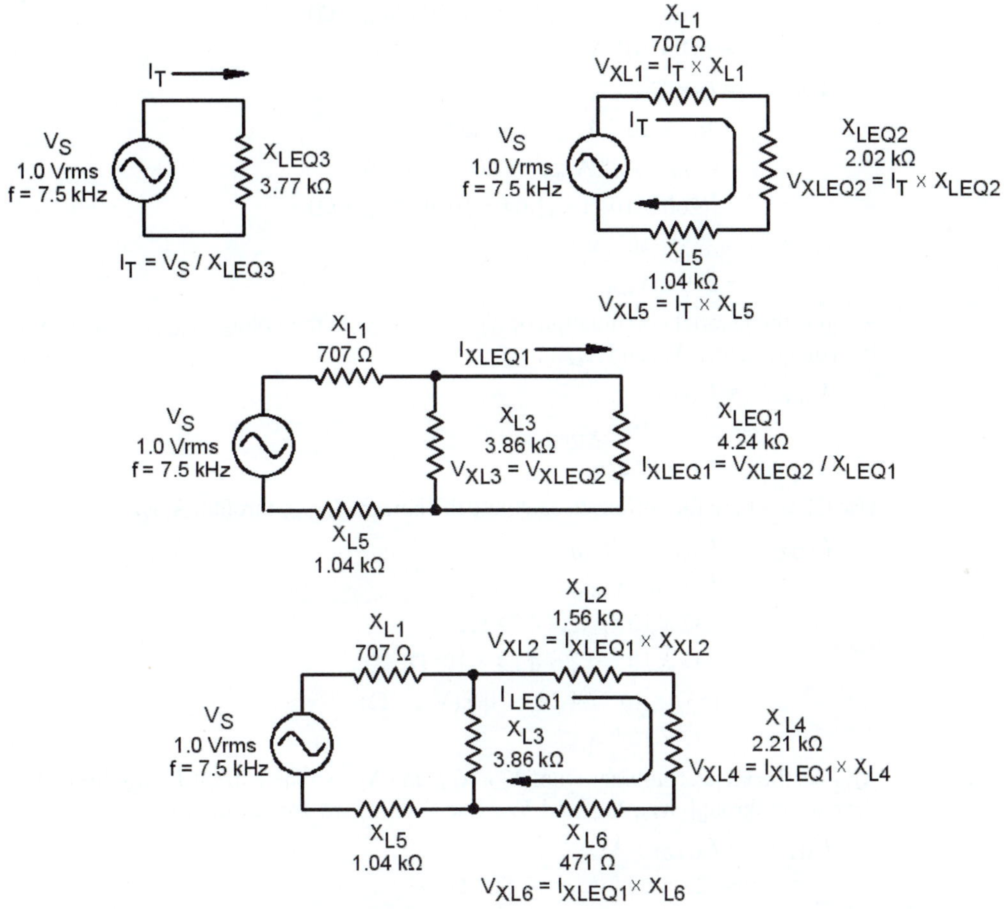

Figure 13-14: Circuit Current Analysis for Figure 13-12

Use Ohm's Law for current to calculate the total current I_T, which is the current through X_{LEQ3}.

$$
\begin{aligned}
I_T &= V_S / X_T \\
&= V_S / X_{LEQ3} \\
&= 1.0\ \text{V}_{\text{rms}} / 3.77\ \text{k}\Omega \\
&= 1.0\ \text{V}_{\text{rms}} / 3.77 \times 10^3\ \Omega \\
&= (1.0 / 3.77 \times 10^3)\ (\text{V}_{\text{rms}} / \Omega) \\
&= 266 \times 10^{-6}\ \text{A}_{\text{rms}}
\end{aligned}
$$

X_{LEQ3} is the series combination of X_{L1}, X_{LEQ2}, and X_{L5}, so the current I_T through X_{LEQ3} is equal to the current through X_{L1}, X_{LEQ2}, and X_{L5}. Use Ohm's Law for voltage to calculate the voltage across each.

$$
\begin{aligned}
V_{XL1} &= I_T \times X_{L1} \\
&= 266 \times 10^{-6}\ \text{A}_{\text{rms}} \times 707\ \Omega \\
&= [(266 \times 10^{-6}) \times 707]\ (\text{A}_{\text{rms}} \times \Omega) \\
&= 188 \times 10^{-3}\ \text{V}_{\text{rms}} \\
&= \mathbf{188\ mV_{rms}}
\end{aligned}
$$

$$
\begin{aligned}
V_{XLEQ2} &= I_T \times X_{L1} \\
&= 266 \times 10^{-6}\ \text{A}_{\text{rms}} \times 2.02\ \text{k}\Omega \\
&= (266 \times 10^{-6}\ \text{A}_{\text{rms}}) \times (2.02 \times 10^3\ \Omega)
\end{aligned}
$$

$= [(266 \times 10^{-6}) \times (2.02 \times 10^{3})]\ (A_{rms} \times \Omega)$

$= 537 \times 10^{-3}\ V_{rms}$

$\boldsymbol{V_{XL5}} \quad \boldsymbol{= I_T \times X_{L5}}$

$= 266 \times 10^{-6}\ A_{rms} \times 1.04\ k\Omega$

$= (266 \times 10^{-6}\ A_{rms}) \times (1.04 \times 10^{3}\ \Omega)$

$= [(266 \times 10^{-6}) \times (1.04 \times 10^{3})]\ (A_{rms} \times \Omega)$

$= 275 \times 10^{-3}\ V_{rms}$

$= \mathbf{275\ mV_{rms}}$

X_{LEQ2} is the parallel combination of X_{L3} and X_{LEQ1} so the voltage V_{XLEQ2} across X_{LEQ2} is equal to the voltage across X_{L3} and X_{LEQ1}.

$\boldsymbol{V_{XL3}} \quad \boldsymbol{= V_{XLEQ2}}$

$= 537 \times 10^{-3}\ V_{rms}$

$= \mathbf{537\ mV_{rms}}$

Use Ohm's Law for current to calculate the current I_{XLEQ1} through X_{LEQ1}.

$\boldsymbol{I_{XLEQ1}} \quad = \boldsymbol{V_{XLEQ1} / X_{LEQ1}}$

$= \boldsymbol{V_{XLEQ2} / X_{LEQ1}}$

$= 537 \times 10^{-3}\ V_{rms} / 4.24\ k\Omega$

$= 537 \times 10^{-3}\ V_{rms} / 4.24 \times 10^{3}\ \Omega$

$= (537 \times 10^{-3} / 4.24 \times 10^{3})\ (V_{rms} / \Omega)$

$= 127 \times 10^{-6}\ A_{rms}$

X_{LEQ1} is the series combination of X_{L2}, X_{L4}, and X_{L6}, so the current I_{XLEQ1} through X_{LEQ1} is equal to the current through X_{L2}, X_{L4}, and X_{L6}. Use Ohm's Law for current to find the voltage across each.

$\boldsymbol{V_{XL2}} \quad = \boldsymbol{I_{XLEQ1} \times X_{L2}}$

$= 127 \times 10^{-6}\ A_{rms} \times 1.56\ k\Omega$

$= (127 \times 10^{-6}\ A_{rms}) \times (1.56 \times 10^{3}\ \Omega)$

$= [(127 \times 10^{-6}) \times (1.56 \times 10^{3})]\ (A_{rms} \times \Omega)$

$= 197 \times 10^{-3}\ V_{rms}$

$= \mathbf{197\ mV_{rms}}$

$\boldsymbol{V_{XL4}} \quad = \boldsymbol{I_{XLEQ1} \times X_{L4}}$

$= 127 \times 10^{-6}\ A_{rms} \times 2.21\ k\Omega$

$= (127 \times 10^{-6}\ A_{rms}) \times (2.21 \times 10^{3}\ \Omega)$

$= [(127 \times 10^{-6}) \times (2.21 \times 10^{3})]\ (A_{rms} \times \Omega)$

$= 280 \times 10^{-3}\ V_{rms}$

$= \mathbf{280\ mV_{rms}}$

$\boldsymbol{V_{XL6}} \quad = \boldsymbol{I_{XLEQ1} \times X_{L6}}$

$= 127 \times 10^{-6}\ A_{rms} \times 471\ \Omega$

$= [(127 \times 10^{-6}) \times 471]\ (A_{rms} \times \Omega)$

$= 59.7 \times 10^{-3}\ V_{rms}$

$= \mathbf{59.7\ mV_{rms}}$

13.3 Just for Fun

Scaling is a method of determining unknown quantities in an equation. For inductors, the reactance is a function of both inductance and frequency. Scaling allows you to determine one value by proportionally adjusting, or scaling, the other values in a reference equation for inductive reactance. For inductors, the reference equation assumes a frequency of 1 Hz and inductance of 1 H. For these values the inductive reactance is

$$\begin{aligned} X_L &= 2\pi f L \\ &= 2\pi\,(1\text{ Hz})\,(1\text{ H}) \\ &= 6.28\ \Omega \end{aligned}$$

or

$$6.28\ \Omega = 2\pi\,(1\text{ Hz})\,(1\text{ H})$$

Suppose now that you wish to determine the value of L that will give an inductive reactance of 15 Ω at 10 kHz. To do this,

1) Find the scale factor K that will change the 6.28 Ω reference value to 15 Ω.

$$\begin{aligned} K &= 15\ \Omega / 6.28\ \Omega \\ &= 2.39 \end{aligned}$$

2) Multiply both sides of the reference equation by K. X_L will be 15 Ω, as K was calculated so that X_L = 15 Ω. You can multiply either quantity on the right-hand side by K, but for this example multiply L by K, because it will simplify scaling f to its desired value of 10 kHz.

$$\begin{aligned} 15\ \Omega &= 2\pi\,(1\text{ Hz})\,(1\text{ H} \times 2.39) \\ &= 2\pi\,(1\text{ Hz})\,(2.39\text{ H}) \end{aligned}$$

3) Now, scale both f and L in the right-hand side of the equation so that f = 10 kHz. Because you must multiply f by 10,000 you must divide L by 10,000 to keep the equation balanced.

$$\begin{aligned} 15\ \Omega &= 2\pi\,(1\text{ Hz} \times 10{,}000)\,(2.39\text{ H} / 10{,}000) \\ &= 2\pi\,(10\text{ kHz})\,(239\ \mu\text{H}) \end{aligned}$$

The value of L required for X_L = 15 Ω for f = 10 kHz is 239 μH. The nearest standard value is 240 μH.

1. Problem: Use scaling to determine the value of X_L for f = 250 Hz and L = 15 mH.

Solution: The basic form for inductive reactance is

$$6.28\ \Omega = 2\pi\,(1\text{ Hz})\,(1\text{ H})$$

For f = 250 Hz, you must scale f by 250. Multiply the left side of the equation by 250 to keep the equation balanced.

$$\begin{aligned} 6.28\ \Omega \times 250 &= 2\pi\,(1\text{ Hz} \times 250)\,(1\text{ H}) \\ 1.571 \times 10^3\ \Omega &= 2\pi\,(250\text{ Hz})\,(1\text{ H}) \end{aligned}$$

For L = 15 mH = 15 × 10^{-3} H, you must scale L by 15 × 10^{-3}. Multiply the left side of the equation by 15 × 10^{-3} to keep the equation balanced.

$$\begin{aligned} (1.571 \times 10^3\ \Omega) \times (15 \times 10^{-3}) &= 2\pi\,(250\text{ Hz})\,[1\text{ H} \times (15 \times 10^{-3})] \\ 23.6\ \Omega &= 2\pi\,(250\text{ Hz})\,(15 \times 10^{-3}\text{ H}) \\ &= 2\pi\,(250\text{ Hz})\,(15\text{ mH}) \end{aligned}$$

The value of X_L is **23.6 Ω.**

2. Problem: Use scaling to determine the value of f for X_L = 5.0 kΩ and L = 470 μH.

Solution: For L = 470 μH = 470 × 10^{-6} H, you must scale L by 470 × 10^{-6}. Multiply the left side of the equation by 470 × 10^{-6} to keep the equation balanced.

$$\begin{aligned} 6.28\ \Omega \times (470 \times 10^{-6}) &= 2\pi\,(1\text{ Hz})\,[1\text{ H} \times (470 \times 10^{-6})] \\ 2.95 \times 10^{-3}\ \Omega &= 2\pi\,(1\text{ Hz})\,(470\ \mu\text{H}) \end{aligned}$$

Next, find the scale factor K that will change the value of X_L from 2.95 × 10^{-3} Ω value to 5.0 kΩ.

$$\begin{aligned} K &= 5.0\text{ k}\Omega / 2.95 \times 10^{-3}\ \Omega \\ &= (5.0 \times 10^3\ \Omega) / (2.95 \times 10^{-3}\ \Omega) \\ &= 1.69 \times 10^6 \end{aligned}$$

Multiply both sides of the reference equation by $\boldsymbol{K}$. X_L will be 5.0 kΩ, as $\boldsymbol{K}$ was calculated so that $X_L = 5.0\ \text{k}\Omega$.

$$5.0\ \text{k}\Omega = 2\pi\ [1\ \text{Hz} \times (1.69 \times 10^6)]\ (470\ \mu\text{H})$$
$$= 2\pi\ (1.69 \times 10^6\ \text{Hz})\ (470\ \mu\text{H})$$
$$= 2\pi\ (1.69\ \text{MHz})\ (470\ \mu\text{H})$$

The value of f is **1.69 MHz**.

14. Transformers

A common transformer specification is that of its turns ratio. Although sources define the turns ratio differently, this manual defines the turns ratio $\boldsymbol{n}$ of a transformer as $\boldsymbol{n} = N_{sec} / N_{pri}$, where N_{pri} is the number of primary windings and N_{sec} is the number of secondary windings. Another convention is to specify a ratio equivalent to $N_{pri} : N_{sec}$.

14.1 Basic

14.1.1 Mutual Inductance

1. Problem: A 56 μWb magnetic flux φ_{1-2} passes from one coil through a second coil. How much flux φ_1 does the first coil produce if the coefficient of coupling $\boldsymbol{k}$ is 0.4?

   Solution: By definition $\boldsymbol{k} = \varphi_{1-2} / \varphi_1$.

$$k = \varphi_{1-2} / \varphi_1$$
$$k \times \varphi_1 = (\varphi_{1-2} / \varphi_1) \times \varphi_1$$
$$(k \times \varphi_1) / k = \varphi_{1-2} / k$$
$$\varphi_1 = \varphi_{1-2} / k$$

   Use the given values of φ_{1-2} and $\boldsymbol{k}$ to calculate φ_1.

$$\varphi_1 = \varphi_{1-2} / k$$
$$= 56\ \mu\text{Wb} / 0.4$$
$$= \mathbf{140\ \mu Wb}$$

2. Problem: Two inductors, L_1 and L_2, with coefficient of coupling $k = 0.75$ have mutual inductance $L_M = 250$ mH. If $L_2 = 250$ mH, what is the value of L_1?

   Solution: By definition $L_M = k\,(L_1L_2)^{1/2}$.

$$L_M^2 = [k\,(L_1L_2)^{1/2}]^2$$
$$L_M^2 = k^2\,[(L_1L_2)^{1/2}]^2$$
$$L_M^2 / (k^2L_2) = (k^2L_1L_2) / (k^2L_2)$$
$$L_M^2 / (k^2L_2) = L_1$$

   Use the given values of L_M, k, and L_2 to calculate L_1.

$$L_1 = (250\text{ mH})^2 / [0.75^2\,(250\text{ mH})]$$
$$= [250^2 / (0.75^2 \times 250)]\,(\text{mH}^2 / \text{mH})$$
$$= 250 / 0.5625\text{ mH}$$
$$= \mathbf{444\ mH}$$

3. Problem: A 10 mH inductor produces 15 μWb of magnetic flux. 10 μWb of this flux pass through a 22.0 mH inductor. What is the mutual inductance of the two inductors?

   Solution: From the problem statement, $L_1 = 10.0$ mH, $L_2 = 22.0$ mH, $\varphi_1 = 15$ μWb, and $\varphi_{1-2} = 10$ μWb. By definition, $k = \varphi_{1-2} / \varphi_1$ and $L_M = k\,(L_1L_2)^{1/2}$.

$$L_M = k\,(L_1L_2)^{1/2}$$
$$= (\varphi_{1-2} / \varphi_1)\,(L_1L_2)^{1/2}$$
$$= (10\ \mu\text{Wb} / 15\ \mu\text{Wb})[(10.0\text{ mH})(22.0\text{ mH})]^{1/2}$$
$$= [(10 / 15)(10.0)^{1/2}(22.0)^{1/2}]\,[(\mu\text{Wb} / \mu\text{Wb})(\text{mH} \times \text{mH})^{1/2}]$$
$$= (1.5)(220)^{1/2}\,(\text{mH}^2)^{1/2}$$
$$= (1.5)(14.8)\text{ mH}$$
$$= \mathbf{22.2\ mH}$$

14.1.2 The Basic Transformer

1. Problem: A transformer has 240 turns in its primary and 12 turns in its secondary. What is its turn ratio $\boldsymbol{n}$?

   Solution: From the problem statement, $N_{pri} = 240$ and $N_{sec} = 12$. By definition, $\boldsymbol{n} = N_{sec} / N_{pri}$.

   $$\begin{aligned} n &= N_{sec} / N_{pri} \\ &= 12 / 120 \\ &= \mathbf{0.1} \end{aligned}$$

2. Problem: A transformer with a turns ratio of 0.25 has 24 turns in its primary. If three primary turns short together, what is the new turns ratio?

   Solution: From the problem statement, $N_{pri} = 24$ and $n_{ORG} = 0.25$. By definition, $\boldsymbol{n} = N_{sec} / N_{pri}$.

   $$\begin{aligned} n &= N_{sec} / N_{pri} \\ n \times N_{pri} &= (N_{sec} / N_{pri}) \times N_{pri} \\ n \times N_{pri} &= N_{sec} \end{aligned}$$

   Use the given values of $\boldsymbol{n}$ and N_{pri} to calculate N_{sec}.

   $$\begin{aligned} N_{sec} &= n \times N_{pri} \\ &= 0.25 \times 24 \\ &= 6 \end{aligned}$$

   After three primary turns short together, $N_{pri(SHORT)} = 24 - 3 = 21$.

   $$\begin{aligned} n_{SHORT} &= N_{sec} / N_{pri(SHORT)} \\ &= 6 / 21 \\ &= \mathbf{0.286} \end{aligned}$$

3. Problem: Refer to the transformer in Figure 14-1.

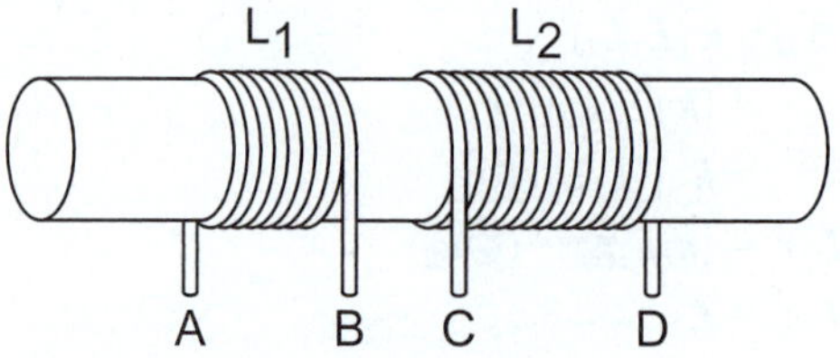

Figure 14-1: Transformer Example 1

   When conventional current flows into terminal $\boldsymbol{A}$ and out of terminal $\boldsymbol{B}$ of L_1, what is the conventional current direction for L_2?

   Solution: From the right-hand rule for conventional current and magnetism, the direction of magnetic flux from I_{L1} is from left to right. To oppose this field, I_{L2} attempts to create a counter-flux from right to left. From the right-hand rule, I_{L2} must flow into terminal $\boldsymbol{C}$ and out of terminal $\boldsymbol{D}$.

4. Problem: Refer to the transformer in Figure 14-2.

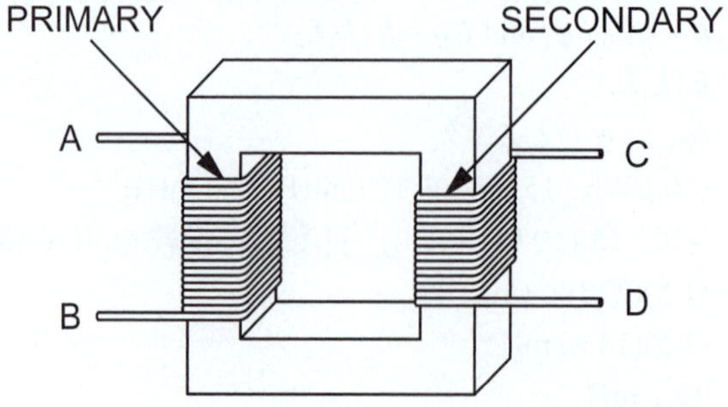

Figure 14-2: Transformer Example 2

When conventional current flows into terminal ***A*** and out of terminal ***B*** of the primary, what is the conventional current direction for the secondary?

Solution: From the right-hand rule for current and magnetism, the primary current sets up a counter-clockwise magnetic flux in the transformer core. To oppose this field, the secondary current attempts to create a clockwise magnetic counter-flux. From the right-hand rule, the secondary current must flow into terminal ***D*** and out of terminal ***C***.

5. Problem: For Figure 14-3, which representations shows the correct dot convention for the transformer?

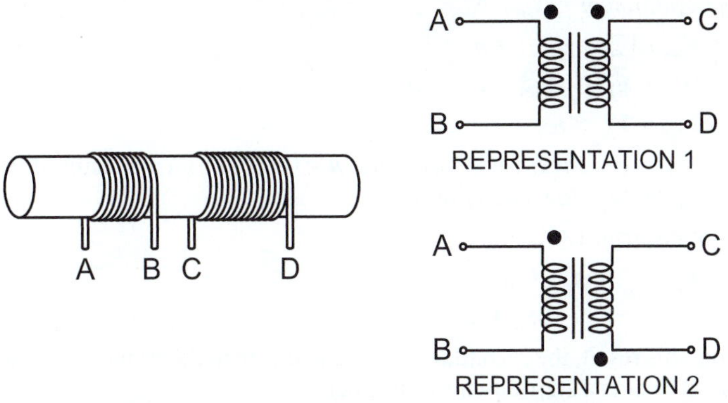

Figure 14-3: Dot Convention Example 1

Solution: Assume that conventional current flows into terminal ***A*** and out of terminal ***B*** of the primary. From the right-hand rule, the direction of magnetic flux is from left to right. The secondary attempts to set up a magnetic counter-flux from right to left, so that by the right-hand rule the secondary current flows into terminal ***D*** and out of terminal ***C*** of the secondary. Because current flows into the dot for the primary, current must flow out of the dot for the secondary. Therefore, the dot on the secondary corresponds to terminal ***C***. **Representation 1 shows the correct dot convention for the transformer.**

6. Problem: For Figure 14-4, which representation shows the correct dot convention for the transformer?

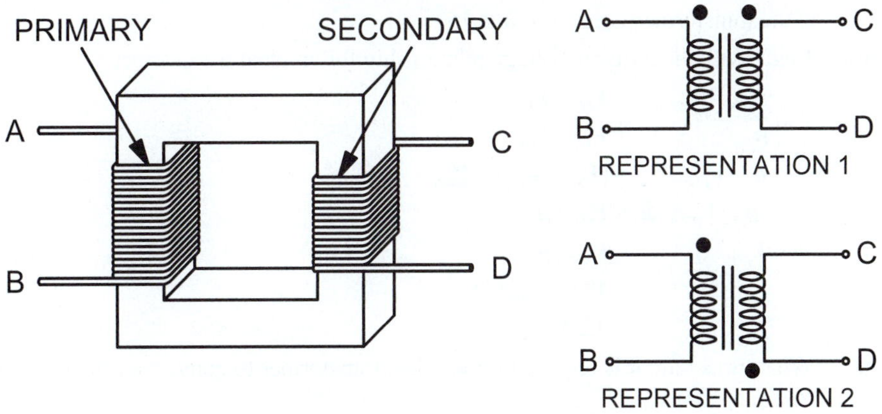

Figure 14-4: Dot Convention Example 2

Solution: Assume that conventional current flows into terminal ***A*** and out of terminal ***B*** of the primary. From the right-hand rule, the direction of magnetic flux is counter-clockwise. The secondary attempts to set up a clockwise magnetic counter-flux, so that by the right-hand rule the secondary current flows into terminal ***C*** and out of terminal ***D*** of the secondary. Because current flows into the dot for the primary, current must flow out of the dot for the secondary. Therefore, the dot on the secondary corresponds to terminal ***D***. **Representation 2 shows the correct dot convention for the transformer.**

14.1.3 Step-Up and Step-Down Transformers

1. Problem: Show that, for an ideal transformer with primary voltage V_{pri}, secondary voltage V_{sec}, and turns ratio n,

   $V_{sec} = V_{pri} \times n$

   Solution: Start with the voltage-to-turns relationship for an ideal transformer.

   $V_{sec} / V_{pri} = N_{sec} / N_{pri}$

   By definition, $n = N_{sec} / N_{pri}$.

   $V_{sec} / V_{pri} = n$

   $(V_{sec} / V_{pri}) \times V_{pri} = V_{pri} \times n$

   $V_{sec} = V_{pri} \times n$

2. Problem: An ideal transformer has a turns ratio $n = 0.166$. From the result of Problem 1, is the transformer a step-up or step-down transformer?

   Solution: From Problem 1, $V_{sec} = V_{pri} \times n$.

   $V_{sec} = n \times V_{pri}$

   $= 0.166 \times V_{pri}$

   From this result, the secondary voltage is less than the primary voltage. By definition, the transformer is a **step-down** transformer.

3. Problem: An ideal transformer steps down voltage from $V_{pri} = 120.0\ V_{rms}$ on the primary to $V_{sec} = 24.0\ V_{rms}$ on the secondary. What is the transformer turns ratio n?

   Solution: Begin with the turns-to-voltage relationship for an ideal transformer.

   $N_{sec} / N_{pri} = V_{sec} / V_{pri}$

   By definition, $n = N_{sec} / N_{pri}$.

   $n = V_{sec} / V_{pri}$

   $= 24.0\ V_{rms} / 120.0\ V_{rms}$

   $= \mathbf{0.2}$

4. Problem: The secondary voltage of an ideal transformer is $V_{sec} = 48.0\ V_{ac}$. What is the primary voltage V_{pri} if the transformer turns ratio n is 15?

   Solution: Begin with the turns-to-voltage relationship for an ideal transformer.

   $N_{sec} / N_{pri} = V_{sec} / V_{pri}$

   $n = V_{sec} / V_{pri}$

   $n \times V_{pri} = (V_{sec} / V_{pri}) \times V_{pri}$

   $(n \times V_{pri}) / n = V_{sec} / n$

   $V_{pri} = V_{sec} / n$

   $= 48.0\ V_{rms} / 15$

   $= \mathbf{3.2\ V_{rms}}$

5. Problem: What turns ratio n is required for an ideal transformer to convert a primary voltage $V_{pri} = 440\ V_{rms}$ to $V_{sec} = 12.0\ V_{rms}$?

   Solution: Begin with the turns-to-voltage relationship for an ideal transformer.

   $N_{sec} / N_{pri} = V_{sec} / V_{pri}$

   $n = V_{sec} / V_{pri}$

   $= 12\ V_{rms} / 440\ V_{rms}$

   $= \mathbf{0.0273}$

14.1.4 Loading the Secondary

For Problems 1 through 3, refer to the ideal transformer circuit in Figure 14-5.

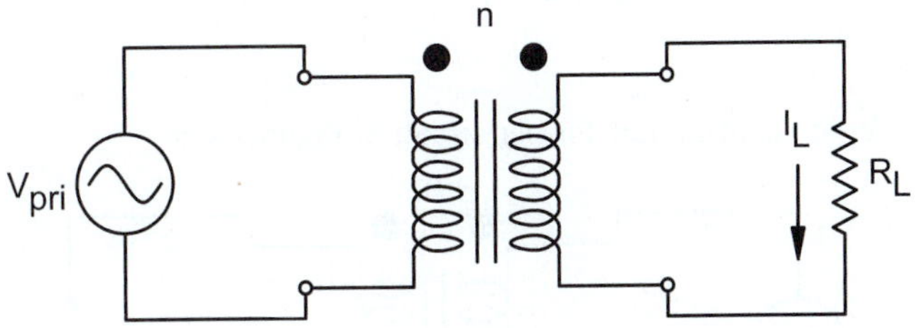

Figure 14-5: Secondary Loading Example Circuit 1

1. Problem: For source voltage $V_S = 2.5\ \text{V}_{\text{rms}}$, turns ratio $n = 12$, and load resistance $R_L = 120\ \Omega$, what is P_L, the power in the load?

   Solution: For an ideal transformer, $V_{sec} = V_{pri} \times n$.

$$\begin{aligned} V_{sec} &= V_{pri} \times n \\ &= V_S \times n \\ &= 2.5\ \text{V}_{\text{rms}} \times 12 \\ &= 3.0\ \text{V}_{\text{rms}} \end{aligned}$$

   Use Watt's Law for voltage and resistance to calculate the load power.

$$\begin{aligned} P_L &= V_{RL}{}^2 / R_L \\ &= V_{sec}{}^2 / R_L \\ &= (3.0\ \text{V}_{\text{rms}})^2 / 120\ \Omega \\ &= \mathbf{75\ mW} \end{aligned}$$

2. Problem: For source voltage $V_S = 5.0\ \text{V}_{\text{rms}}$, load current $I_L = 50\ \text{mA}_{\text{rms}}$, and load resistance $R_L = 600\ \Omega$, what is the turns ratio n?

   Solution: Use Ohm's Law for voltage to calculate the load voltage, V_{RL}.

$$\begin{aligned} V_{RL} &= I_L \times R_L \\ V_{sec} &= I_L \times R_L \\ &= 50\ \text{mA}_{\text{rms}} \times 600\ \Omega \\ &= 30\ \text{V}_{\text{rms}} \end{aligned}$$

   Next, use the definition of turns ratio for an ideal transformer to calculate the turns ratio, n.

$$\begin{aligned} n &= V_{sec} / V_{pri} \\ &= V_{sec} / V_S \\ &= 30\ \text{V}_{\text{rms}} / 5.0\ \text{V}_{\text{rms}} \\ &= \mathbf{6} \end{aligned}$$

3. Problem: For primary power $P_{pri} = 3.5$ W and load resistance $R_L = 330\ \Omega$, what is the secondary current I_L?

   Solution: For an ideal transformer, the primary power P_{pri} equals the secondary power P_{sec}.

$$\begin{aligned} P_{pri} &= P_{sec} \\ P_{pri} &= P_{RL} \end{aligned}$$

   Use Watt's Law for current and resistance to calculate the load current, I_L.

$$\begin{aligned} I_L{}^2 \times R_L &= P_{RL} \\ (I_L{}^2 \times R_L) / R_L &= P_{RL} / R_L \\ \sqrt{I_L{}^2} &= \sqrt{P_{RL} / R_L} \\ I_L &= \sqrt{P_{RL} / R_L} \end{aligned}$$

$= \sqrt{3.5\text{ W} / 330\ \Omega}$

$= \mathbf{103\ mA}$

14.1.5 Reflected Load

For Problems 1 and 2, refer to the ideal transformer circuit in Figure 14-6.

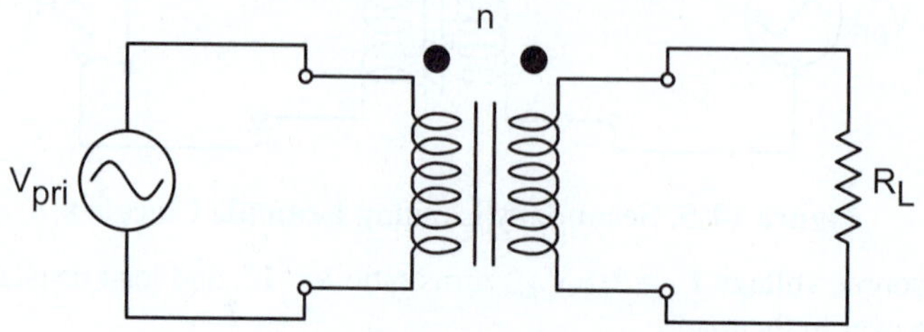

Figure 14-6: Reflected Load Example Circuit 1

1. Problem: Prove that, for the ideal transformer circuit, the reflected load $R_{pri} = R_L / n^2$.

   Solution: For an ideal transformer, $P_{sec} = P_{pri}$. Begin with Watt's Law for voltage and resistance and solve for R_{pri}.

$$\begin{aligned}
P_{sec} &= P_{pri} \\
V_{sec}^2 / R_L &= V_{pri}^2 / R_{pri} \\
(V_{sec}^2 / R_L) \times R_{pri} &= (V_{pri}^2 / R_{pri}) \times R_{pri} \\
(V_{sec}^2 / R_L) \times R_{pri} \times (R_L / V_{sec}^2) &= V_{pri}^2 \times (R_L / V_{sec}^2) \\
R_{pri} &= (V_{pri}^2 / V_{sec}^2) \times R_L \\
&= [1 / (V_{sec} / V_{pri})^2]\, R_L \\
&= (1 / n^2)\, R_L \\
&= R_L / n^2
\end{aligned}$$

2. Problem: For turns ratio $n = 0.2$ and load resistance $R_L = 8\ \Omega$, what is the reflected resistance R_{pri}?

   Solution: For an ideal transformer, the reflected load $R_{pri} = R_L / n^2$.

$$\begin{aligned}
R_{pri} &= R_L / n^2 \\
&= (1/ 0.2)^2\ (8\ \Omega) \\
&= (5)^2 \times (8\ \Omega) \\
&= \mathbf{200\ \Omega}
\end{aligned}$$

14.1.6 Impedance Matching

For Problems 1 through 3, refer to the ideal transformer circuit in Figure 14-7.

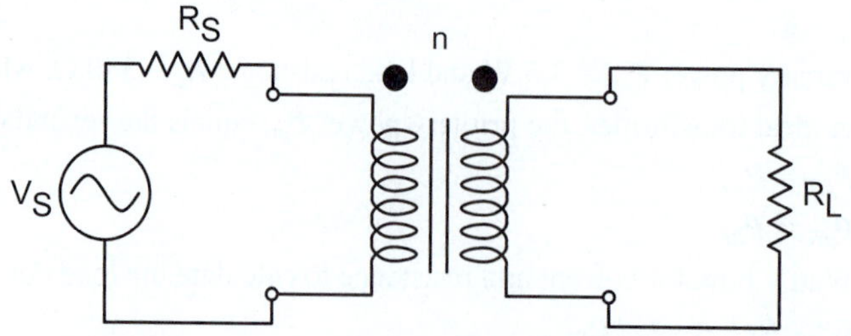

Figure 14-7: Impedance Matching Example Circuit 1

1. Problem: What must be the turns ratio n to match load resistance $R_L = 600\ \Omega$ to source resistance $R_S = 75\ \Omega$?

Solution: The transformer matches load resistance $\boldsymbol{R_L}$ to source resistance $\boldsymbol{R_S}$ when the reflected impedance $\boldsymbol{R_{pri}} = \boldsymbol{R_S}$. For an ideal transformer, the reflected impedance $\boldsymbol{R_{pri}} = \boldsymbol{R_L} / \boldsymbol{n}^2$.

$$\boldsymbol{R_S} = \boldsymbol{R_{pri}}$$
$$\boldsymbol{R_S} = \boldsymbol{R_L} / \boldsymbol{n}^2$$
$$\boldsymbol{R_S} \times \boldsymbol{n}^2 = [\boldsymbol{R_L} / \boldsymbol{n}^2] \times \boldsymbol{n}^2$$
$$(\boldsymbol{R_S} \times \boldsymbol{n}^2) / \boldsymbol{R_S} = \boldsymbol{R_L} / \boldsymbol{R_S}$$
$$\sqrt{\boldsymbol{n}^2} = \sqrt{\boldsymbol{R_L} / \boldsymbol{R_S}}$$
$$\boldsymbol{n} = \sqrt{\boldsymbol{R_L} / \boldsymbol{R_S}}$$
$$= \sqrt{600\ \Omega / 75\ \Omega}$$
$$= \mathbf{2.83}$$

2. Problem: For source resistance $\boldsymbol{R_S} = 60\ \Omega$ and turns ration $\boldsymbol{n} = 0.25$, what must the load resistance $\boldsymbol{R_L}$ be for maximum power transfer?

Solution: Maximum power transfer occurs for $\boldsymbol{R_{pri}} = \boldsymbol{R_S}$. For an ideal transformer, $\boldsymbol{R_{pri}} = \boldsymbol{R_L} / \boldsymbol{n}^2$.

$$\boldsymbol{R_{pri}} = \boldsymbol{R_L} / \boldsymbol{n}^2$$
$$\boldsymbol{R_{pri}} \times \boldsymbol{n}^2 = [\boldsymbol{R_L} / \boldsymbol{n}^2] \times \boldsymbol{n}^2$$
$$\boldsymbol{R_{pri}} \times \boldsymbol{n}^2 = \boldsymbol{R_L}$$

Use the given values of $\boldsymbol{R_S} = 60\ \Omega$ and $\boldsymbol{n} = 0.25$ to calculate $\boldsymbol{R_L}$.

$$\boldsymbol{R_L} = 60\ \Omega \times (0.25)^2$$
$$= 60\ \Omega \times 0.0625$$
$$= \mathbf{3.75\ \Omega}$$

3. Problem: For source voltage $\boldsymbol{V_S} = 1.0\ \text{V}_{\text{rms}}$, source resistance $\boldsymbol{R_S} = 1.0\ \text{k}\Omega$ and load resistance $\boldsymbol{R_L} = 10\ \Omega$, determine the load power, $\boldsymbol{P_{RL}}$, for turns ratios of $\boldsymbol{n} = 1.0$, $\boldsymbol{n} = 0.1$, and $\boldsymbol{n} = 0.01$.

Solution: For $\boldsymbol{n} = 1$, $\boldsymbol{R_{pri}} = \boldsymbol{R_L} = 10\ \Omega$. Use the voltage divider formula to calculate the primary voltage.

$$\boldsymbol{V_{pri}} = \boldsymbol{V_S}\,[\boldsymbol{R_{pri}} / (\boldsymbol{R_S} + \boldsymbol{R_{pri}})]$$
$$= (1.0\ \text{V}_{\text{rms}})\,[10\ \Omega / (1.0\ \text{k}\Omega + 10\ \Omega)]$$
$$= 9.90\ \text{mV}_{\text{rms}}$$

For an ideal transformer, $\boldsymbol{V_{sec}} = \boldsymbol{n} \times \boldsymbol{V_{pri}}$.

$$\boldsymbol{V_{sec}} = 1.0 \times 9.90\ \text{mV}_{\text{rms}}$$
$$= 9.90\ \text{mV}_{\text{rms}}$$

Use Watt's Law for voltage and resistance to calculate the load power.

$$\boldsymbol{P_{RL}} = \boldsymbol{V_{RL}}^2 / \boldsymbol{R_L}$$
$$= (9.90\ \text{mV}_{\text{rms}})^2 / 10\ \Omega$$
$$= \mathbf{9.8\ \mu W}$$

For $n = 0.1$, calculate the reflected resistance, $\boldsymbol{R_{pri}}$.

$$\boldsymbol{R_{pri}} = \boldsymbol{R_L} / \boldsymbol{n}^2$$
$$= [1 / (0.1)^2]\,(10\ \Omega)$$
$$= 1.0\ \text{k}\Omega$$

Use the voltage divider formula to calculate the primary voltage.

$$\boldsymbol{V_{pri}} = \boldsymbol{V_S}\,[\boldsymbol{R_{pri}} / (\boldsymbol{R_S} + \boldsymbol{R_{pri}})]$$
$$= (1.0\ \text{V}_{\text{rms}})\,[1.0\ \text{k}\Omega / (1.0\ \text{k}\Omega + 1.0\ \text{k}\Omega)]$$
$$= 500\ \text{mV}_{\text{rms}}$$

For an ideal transformer, $\boldsymbol{V_{sec}} = \boldsymbol{n} \times \boldsymbol{V_{pri}}$.

$$\boldsymbol{V_{sec}} = 0.1 \times 500\ \text{mV}_{\text{rms}}$$

$= 50.0\ \text{mV}_{\text{rms}}$

Use Watt's Law for voltage and resistance to calculate the load power.

$$\boldsymbol{P_{RL}} = \boldsymbol{V_{RL}}^2 / \boldsymbol{R_L}$$
$$= \boldsymbol{V_{sec}}^2 / \boldsymbol{R_L}$$
$$= (50.0\ \text{mV}_{\text{rms}})^2 / 10\ \Omega$$
$$= \mathbf{250\ \mu W}$$

For $\boldsymbol{n} = 0.01$, calculate the reflected resistance, $\boldsymbol{R_{pri}}$.

$$\boldsymbol{R_{pri}} = \boldsymbol{R_L} / \boldsymbol{n}^2$$
$$= [1 / (0.01)^2]\ (10\ \Omega)$$
$$= 100\ \text{k}\Omega$$

Use the voltage divider formula to calculate the primary voltage.

$$\boldsymbol{V_{pri}} = \boldsymbol{V_S}\ [\boldsymbol{R_{pri}} / (\boldsymbol{R_S} + \boldsymbol{R_{pri}})]$$
$$= (1.0\ \text{V}_{\text{rms}})\ [100\ \text{k}\Omega / (1.0\ \text{k}\Omega + 100\ \text{k}\Omega)]$$
$$= 990\ \text{mV}_{\text{rms}}$$

For an ideal transformer, $\boldsymbol{V_{sec}} = \boldsymbol{n} \times \boldsymbol{V_{pri}}$.

$$\boldsymbol{V_{sec}} = 0.01 \times 990\ \text{mV}_{\text{rms}}$$
$$= 9.90\ \text{mV}_{\text{rms}}$$

Use Watt's Law for voltage and resistance to calculate the load power.

$$\boldsymbol{P_{RL}} = \boldsymbol{V_{RL}}^2 / \boldsymbol{R_L}$$
$$= (9.90\ \text{mV}_{\text{rms}})^2 / 10\ \Omega$$
$$= \mathbf{9.8\ \mu W}$$

14.1.7 Transformer Ratings and Characteristics

1. Problem: For the ideal transformer circuit in Figure 14-8, what is the required apparent power rating for the transformer if $\boldsymbol{V_S} = 120\ \text{V}_{\text{rms}}$, turns ratio $\boldsymbol{n} = 0.1$, and $\boldsymbol{R_L} = 100\ \Omega$?

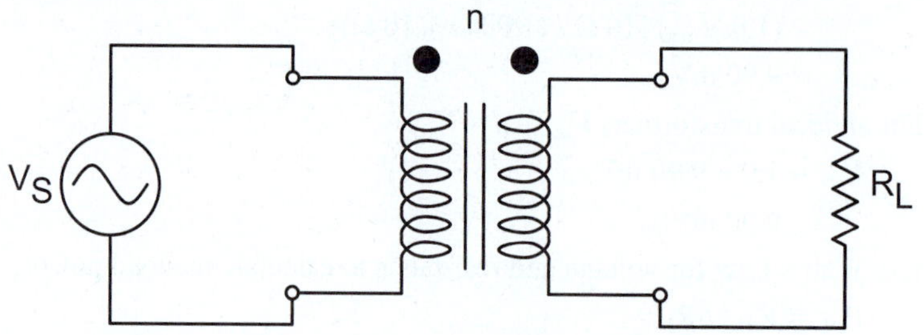

Figure 14-8: Transformer Rating Example Circuit 1

Solution: For an ideal transformer $\boldsymbol{V_{sec}} = \boldsymbol{n} \times \boldsymbol{V_{pri}}$.

$$\boldsymbol{V_{sec}} = \boldsymbol{n} \times \boldsymbol{V_{pri}}$$
$$= \boldsymbol{n} \times \boldsymbol{V_S}$$
$$= 0.1 \times 120\ \text{V}_{\text{rms}}$$
$$= 12.0\ \text{V}_{\text{rms}}$$

Use Ohm's Law for current to calculate $\boldsymbol{I_{RL}} = \boldsymbol{I_{sec}}$.

$$\boldsymbol{I_{RL}} = \boldsymbol{V_{RL}} / \boldsymbol{R_L}$$
$$\boldsymbol{I_{sec}} = \boldsymbol{V_{sec}} / \boldsymbol{R_L}$$
$$= 12.0\ \text{V}_{\text{rms}} / 100\ \Omega$$

$= 120\ mA_{rms}$

For an ideal transformer, $\boldsymbol{P_{pri}} = \boldsymbol{P_{sec}}$, so either can be used to determine the apparent power rating. Use Watt's Law for voltage and current to calculate $\boldsymbol{P_{sec}}$.

$\boldsymbol{P_{sec}} = \boldsymbol{V_{sec}}\,\boldsymbol{I_{sec}}$

$= (12.0\ V_{rms})\ (120\ mA_{rms})$

$= \mathbf{1.44\ VA}$

2. Problem: What is the transformer efficiency η if the $\boldsymbol{V_S} = 48.0$ V, $\boldsymbol{I_{pri}} = 15.0$ mA, $\boldsymbol{I_{sec}} = 25.0$ mA, and $\boldsymbol{R_L} = 910\ \Omega$?

Solution: Use Watt's Law for voltage and current to calculate $\boldsymbol{P_{in}}$.

$\boldsymbol{P_{in}} = \boldsymbol{V_{pri}}\,\boldsymbol{I_{pri}}$

$\boldsymbol{P_{in}} = \boldsymbol{V_S}\,\boldsymbol{I_{pri}}$

$= (48.0\ V)(15.0\ mA)$

$= 720\ mVA$

Next, use Watt's Law for current and resistance to calculate $\boldsymbol{P_{out}}$.

$\boldsymbol{P_{out}} = \boldsymbol{I_{sec}}^2\,\boldsymbol{R_L}$

$= (25.0\ mA)^2\ (910\ \Omega)$

$= [(625 \times 10^{-6})(910)]\ (A^2 \times \Omega)$

$= 569\ mVA$

By definition, the efficiency $\eta = (\boldsymbol{P_{out}} / \boldsymbol{P_{in}}) \times 100\%$.

$\eta = (\boldsymbol{P_{out}} / \boldsymbol{P_{in}}) \times 100\%$

$= (569\ mVA / 720\ mVA) \times 100\%$

$= 0.790 \times 100\%$

$= \mathbf{79.0\%}$

14.1.8 Tapped and Multi-Winding Transformers

For Problems 1 and 2, refer to the ideal tapped transformer circuit in Figure 14-9.

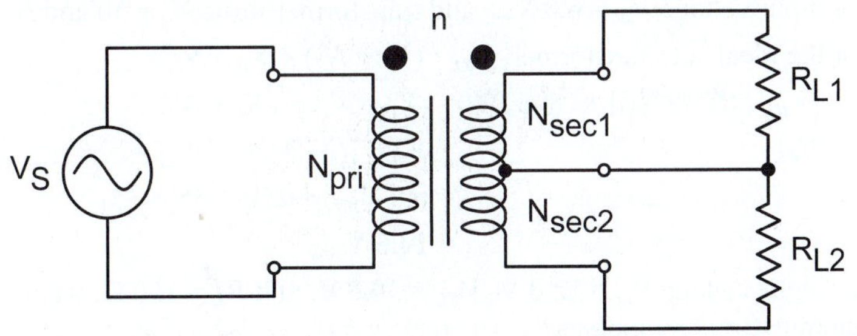

Figure 14-9: Tapped Transformer Example Circuit 1

1. Problem: For source voltage $\boldsymbol{V_S} = 12.0\ V_{rms}$, primary turns $\boldsymbol{N_{pri}} = 75$, and secondary turns $\boldsymbol{N_{sec1}} = 50$ and $\boldsymbol{N_{sec2}} = 175$, what are the load voltages $\boldsymbol{V_{RL1}}$ and $\boldsymbol{V_{RL2}}$?

Solution: For the circuit, $\boldsymbol{V_{pri}} = \boldsymbol{V_S}$. For the load resistor $\boldsymbol{R_{L1}}$, $\boldsymbol{V_{RL1}} = \boldsymbol{V_{sec1}}$.

$\boldsymbol{V_{sec1}} = \boldsymbol{n_1} \times \boldsymbol{V_{pri}}$

$\boldsymbol{V_{RL1}} = (\boldsymbol{N_{sec1}} / \boldsymbol{N_{pri}}) \times \boldsymbol{V_S}$

$= (50 / 75) \times 12.0\ V_{rms}$

$= \mathbf{8.00\ Vrms}$

For the load resistor $\boldsymbol{R_{L2}}$, $\boldsymbol{V_{RL2}} = \boldsymbol{V_{sec2}}$.

$\boldsymbol{V_{sec2}} = \boldsymbol{n_2} \times \boldsymbol{V_{pri}}$

$V_{RL2} = (N_{sec2} / N_{pri}) \times V_S$

$= (175 / 75) \times 12.0\ V_{rms}$

$= \mathbf{28.0\ Vrms}$

2. Problem: For source voltage $V_S = 15.0\ V_{rms}$ and primary turns $N_{pri} = 120$, what secondary turns N_{sec1} and N_{sec2} are required to produce load voltages $V_{RL1} = 6.00\ V_{rms}$ and $V_{RL2} = 18.0\ V_{rms}$?

Solution: For the ideal tapped transformer circuit, load voltage $V_{RL1} = V_{sec1}$.

$N_{sec1} / N_{pri} = V_{sec1} / V_{pri}$

$N_{sec1} / N_{pri} = V_{RL1} / V_{pri}$

$(N_{sec1} / N_{pri}) \times N_{pri} = (V_{RL1} / V_{pri}) \times N_{pri}$

$N_{sec1} = (6.00\ V_{rms} / 15.0\ V_{rms}) \times 120$

$= \mathbf{48.0}$

For the ideal tapped transformer circuit, load voltage $V_{RL2} = V_{sec2}$.

$N_{sec2} / N_{pri} = V_{sec2} / V_{pri}$

$N_{sec2} / N_{pri} = V_{RL2} / V_{pri}$

$(N_{sec2} / N_{pri}) \times N_{pri} = (V_{RL2} / V_{pri}) \times N_{pri}$

$N_{sec2} = (18.0\ V_{rms} / 15.0\ V_{rms}) \times 120$

$= \mathbf{144}$

For Problems 3 and 4, refer to the ideal autotransformer circuit in Figure 14-10.

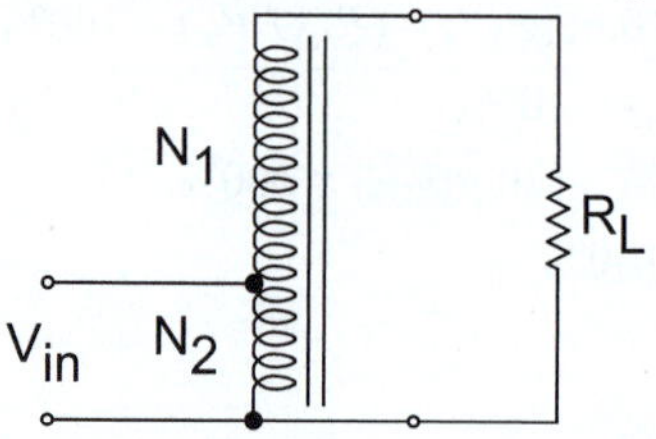

Figure 14-10: Autotransformer Example Circuit 1

3. Problem: For input voltage $V_{in} = 6.0\ V_{rms}$ and transformer turns $N_1 = 50$ and $N_2 = 35$, what is V_{RL}?

Solution: For the ideal autotransformer, $V_{RL} / (N_1 + N_2) = V_{in} / N_2$.

$[V_{RL} / (N_1 + N_2)] \times (N_1 + N_2) = (V_{in} / N_2) \times (N_1 + N_2)$

$V_{RL} = V_{in} \times [(N_1 + N_2) / N_2]$

$= 6.0\ V_{rms} \times [(50 + 35) / 35]$

$= \mathbf{14.6\ V_{rms}}$

4. Problem: For input voltage $V_{in} = 12.0$ V, $V_{RL} = 30.0$ V, and $R_L = 150\ \Omega$, what is the power in the autotransformer windings?

Solution: Refer to Figure 14-11.

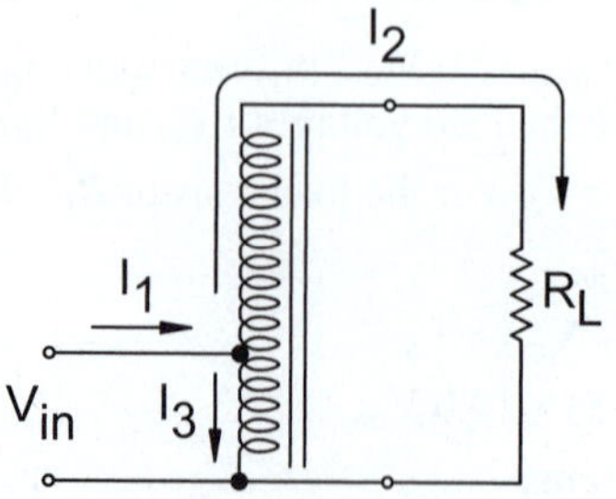

Figure 14-11: Analysis of Autotransformer Winding Power

Use Ohm's Law for voltage and current to calculate $I_2 = I_{RL}$.

$$\begin{aligned} I_2 &= V_{RL} / R_L \\ &= 30.0 \text{ V} / 150\ \Omega \\ &= 200 \text{ mA} \end{aligned}$$

Use Watt's Law for voltage and current to calculate P_{RL}.

$$\begin{aligned} P_{RL} &= V_{RL} \times I_2 \\ &= 30.0 \text{ V} \times 200 \text{ mA} \\ &= 6.0 \text{ VA} \end{aligned}$$

For an ideal autotransformer, $P_{in} = P_{RL}$.

$$\begin{aligned} P_{in} &= P_{RL} \\ V_{in} \times I_1 &= 6.0 \text{ VA} \\ (V_{in} \times I_1) / V_{in} &= 6.0 \text{ VA} / V_{in} \\ I_1 &= 6.0 \text{ VA} / 12.0 \text{ V} \\ &= 500 \text{ mA} \end{aligned}$$

Use Kirchhoff's Current Law to calculate I_3 from I_1 and I_2. Let currents entering the node be positive and currents leaving the node be negative.

$$\begin{aligned} I_1 - I_2 - I_3 &= 0 \\ I_1 - I_2 - I_3 + I_3 &= 0 + I_3 \\ I_1 - I_2 &= I_3 \\ 500 \text{ mA} - 200 \text{ mA} &= I_3 \\ 300 \text{ mA} &= I_3 \end{aligned}$$

From Watt's Law for voltage and current for the power P_{UPPER} in the upper windings,

$$\begin{aligned} P_{UPPER} &= (V_{RL} - V_{in}) \times I_2 \\ &= (30.0 \text{ V} - 12.0 \text{ V}) \times 200 \text{ mA} \\ &= 18.0 \text{ V} \times 200 \text{ mA} \\ &= \mathbf{3.6\ VA} \end{aligned}$$

From Watt's Law for voltage and current for the power P_{LOWER} in the lower windings,

$$\begin{aligned} P_{LOWER} &= V_{in} \times I_3 \\ &= 12.0 \text{ V} \times 300 \text{ mA} \\ &= \mathbf{3.6\ VA} \end{aligned}$$

14.2 Advanced

14.2.1 The Basic Transformer

1. Problem: For an ideal transformer wound on a uniform core, derive the turns ratio n in terms of the primary and secondary self-inductances L_{pri} and L_{sec}.

   Solution: Begin with the equation that relates the value L of an inductor with N turns, core permeability μ, core cross-sectional area A, and core length l and solve for N.

$$\begin{aligned} L &= \frac{N^2 \mu A}{l} \\ L\left(\frac{l}{\mu A}\right) &= \left(\frac{N^2 \mu A}{l}\right)\left(\frac{l}{\mu A}\right) \\ \sqrt{L\left(\frac{l}{\mu A}\right)} &= \sqrt{N^2} \end{aligned}$$

$$\sqrt{L\left(\frac{l}{\mu A}\right)} = N$$

By definition, the turns ratio $n = N_{sec} / N_{pri}$.

$$n = \frac{\sqrt{L_{sec}\left(\frac{l_{sec}}{\mu_{sec} A_{sec}}\right)}}{\sqrt{L_{pri}\left(\frac{l_{pri}}{\mu_{pri} A_{pri}}\right)}}$$

$$= \sqrt{\frac{L_{sec}}{L_{pri}}} \bullet \sqrt{\left(\frac{l_{sec}}{\mu_{sec} A_{sec}}\right)\left(\frac{\mu_{pri} A_{pri}}{l_{pri}}\right)}$$

For a uniform core, $l_{pri} = l_{sec}$, $\mu_{pri} = \mu_{sec}$, and $A_{pri} = A_{sec}$. These variables cancel in the second term to give

$$n = \sqrt{\frac{L_{sec}}{L_{pri}}}$$

2. Problem: For an ideal transformer, use Faraday's Law to show that $N_{sec} / N_{pri} = V_{sec} / V_{pri}$.

Solution: From Faraday's Law, the induced voltage V_{ind} from changing flux $d\varphi / dt$ in a coil of N turns is $v_{ind} = N\,(d\varphi / dt)$.

$$v_{ind} = N\,(d\varphi / dt)$$

$$v_{ind} / N = [N\,(d\varphi / dt)] / N$$

$$v_{ind} / N = d\varphi / dt$$

In an ideal transformer, $d\varphi_{pri} / dt = d\varphi_{sec} / dt$.

$$d\varphi_{pri} / dt = d\varphi_{sec} / dt$$

$$V_{pri} / N_{pri} = V_{sec} / N_{sec}$$

$$(V_{pri} / N_{pri})\,(N_{sec} / V_{pri}) = (V_{sec} / N_{sec})\,(N_{sec} / V_{pri})$$

$$N_{sec} / N_{pri} = V_{sec} / V_{pri}$$

14.2.2 Step-Up and Step-Down Transformers

1. Problem: Determine the load voltage V_{RL} for the ideal transformer circuit in Figure 14-12. Is the circuit a step-up or step-down circuit?

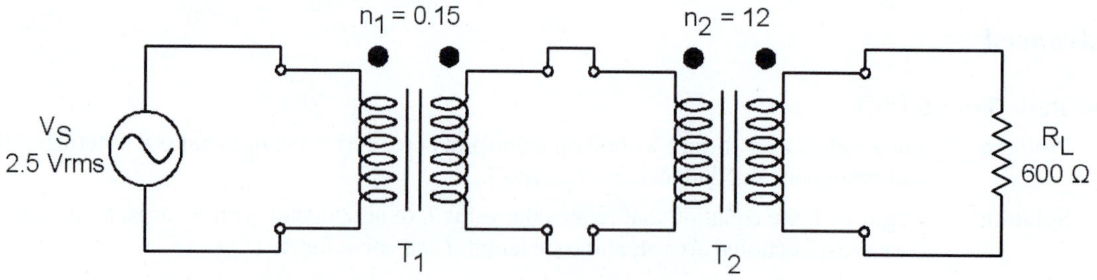

Figure 14-12: Cascaded Transformer Example Circuit 1

Solution: For ideal transformer T_1,

$$V_{sec1} = n_1 \times V_{pri1}$$

$$= n_1 \times V_S$$

$= 0.15 \times 2.5\ V_{rms}$

$= 375\ mV_{rms}$

For ideal transformer T_2,

$V_{sec2} = n_2 \times V_{pri2}$

$V_{RL} = n_2 \times V_{sec1}$

$= 12 \times 375\ mV_{rms}$

$= \mathbf{4.5\ V_{rms}}$

Because $V_{RL} > V_S$, the circuit is a **step-up** circuit.

2. Problem: Determine the effective turns ratio for the ideal transformer circuit in Figure 14-13. What relationship between n_1 and n_2 must exist for the circuit to be a step-up ($V_{RL}/ V_S > 1$), unit ratio ($V_{RL} / V_S = 1$), or step-down ($V_{RL} / V_S < 1$) circuit?

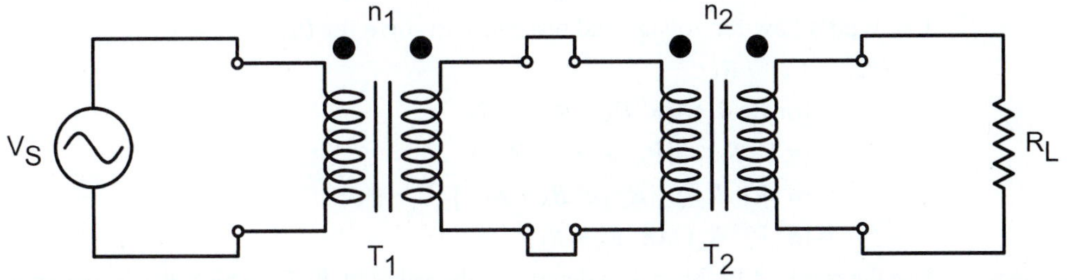

Figure 14-13: Cascaded Transformer Example Circuit 2

Solution: For ideal transformer T_1,

$V_{sec1} = n_1 \times V_{pri1}$

$= n_1 \times V_S$

For ideal transformer T_2,

$V_{sec2} = n_2 \times V_{pri2}$

$V_{RL} = n_2 \times V_{sec1}$

$= n_2 \times n_1 \times V_S$

The effective turns ratio of the circuit is V_{RL} / V_S.

$V_{RL} / V_S = (n_1 \times n_2 \times V_S) / V_S$

$= n_1 \times n_2$

For a step-up circuit, $n_1 \times n_2 > 1$. For an unity ratio circuit, $n_1 \times n_2 = 1$ (n_1 and n_2 must be reciprocals of each other). For a step-down circuit, $n_1 \times n_2 < 1$.

14.2.3 Loading the Secondary

1. Problem: For the ideal transformer circuit in Figure 14-14, determine the load power P_{RL} in terms of source voltage V_S, source resistance R_S, turns ratio n, and load resistance R_L.

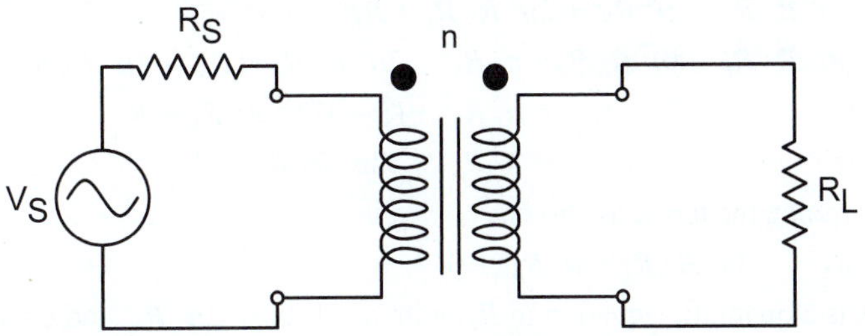

Figure 14-14: Secondary Loading Example Circuit 2

Solution: For an ideal transformer, the reflected load $R_{pri} = R_L / n^2$.

$$R_{pri} = R_L / n^2$$

Use the voltage divider formula to solve for the primary voltage, V_{pri}.

$$\begin{aligned} V_{pri} &= V_S\,[R_{pri} / (R_S + R_{pri})] \\ &= V_S\,\{(R_L / n^2) / [R_S + (R_L / n^2)]\} \\ &= (V_S\,R_L) / \{n^2\,[R_S + (R_L / n^2)]\} \\ &= (V_S\,R_L) / [n^2\,R_S + n^2\,(R_L / n^2)] \\ &= (V_S\,R_L) / (n^2\,R_S + R_L) \end{aligned}$$

The load voltage V_{RL} is equal to the secondary voltage $V_{sec} = n \times V_{pri}$.

$$V_{sec} = n\,[(V_S\,R_L) / (n^2\,R_S + R_L)]$$

$$V_{RL} = (n\,V_S\,R_L) / (n^2\,R_S + R_L)$$

Use Watt's Law for voltage and resistance to solve for P_{RL}.

$$\begin{aligned} P_{RL} &= V_{RL}^{\,2} / R_L \\ &= [(n\,V_S\,R_L) / (n^2\,R_S + R_L)]^2 / R_L \\ &= (n\,V_S\,R_L)^2 / [R_L\,(n^2\,R_S + R_L)^2] \\ &= (n^2\,V_S^2\,R_L^2) / [R_L\,(n^2\,R_S + R_L)^2] \\ &= (n^2\,V_S^2\,R_L) / (n^2\,R_S + R_L)^2 \end{aligned}$$

2. Problem: Use the result of Problem 1 to determine the value of R_L for which P_{RL} is maximum for the ideal transformer circuit in Figure 14-14.

Solution: From the maximum power transfer theorem, the circuit will transfer maximum power to the transformer when the primary resistance $R_{pri} = R_S$, so that from the voltage divider theorem $V_{pri} = V_S / 2$. Use Watt's Law from voltage and resistance to calculate the primary power, P_{pri}.

$$\begin{aligned} P_{pri} &= V_{pri}^{\,2} / R_{pri} \\ &= (V_S / 2)^2 / R_S \\ &= (V_S^2 / 4) / R_S \\ &= V_S^2 / 4R_S \end{aligned}$$

The load power P_{RL} is maximum when the primary power P_{pri} is maximum.

$$\begin{aligned} P_{RL} &= P_{pri} \\ (n^2\,V_S^2\,R_L) / (n^2\,R_S + R_L)^2 &= V_S^2 / 4R_S \\ [(n^2\,V_S^2\,R_L) / (n^2\,R_S + R_L)^2] / V_S^2 &= (V_S^2 / 4R_S) / V_S^2 \\ [(n^2\,R_L) / (n^2\,R_S + R_L)^2]\,4R_S &= (1 / 4R_S)\,4R_S \\ (4n^2\,R_S\,R_L) / (n^2\,R_S + R_L)^2]\,(n^2\,R_S + R_L)^2 &= 1 \times (n^2\,R_S + R_L)^2 \\ 4n^2\,R_S\,R_L &= (n^2\,R_S + R_L)^2 \end{aligned}$$

Expand the right-hand term.

$$\begin{aligned} 4n^2\,R_S\,R_L &= (n^2\,R_S + R_L)^2 \\ 4n^2\,R_S\,R_L &= n^4\,R_S^2 + 2n^2\,R_S\,R_L + R_L^{\,2} \\ 4n^2\,R_S\,R_L - 4n^2\,R_S\,R_L &= n^4\,R_S^2 + 2n^2\,R_S\,R_L + R_L^{\,2} - 4n^2\,R_S\,R_L \\ 0 &= n^4\,R_S^2 + (2 - 4)\,n^2\,R_S\,R_L + R_L^{\,2} \\ 0 &= n^4\,R_S^2 + -2n^2\,R_S\,R_L + R_L^{\,2} \end{aligned}$$

Re-arrange the terms as shown.

$$R_L^{\,2} + -2n^2\,R_S\,R_L + n^4\,R_S^2 = 0$$

This is a quadratic equation in R_L, with a = 1, b = $-2n^2\,R_S$, and c = $n^4\,R_S^2$. Use the quadratic formula to solve for R_L.

$$R_L = \frac{-b \pm \sqrt{b^2 - 4ac}}{2a}$$

$$= \frac{-(-2n^2 R_S) \pm \sqrt{(-2n^2 R_S)^2 - 4(1)(n^4 R_S{}^2)}}{2(1)}$$

$$= \frac{2n^2 R_S \pm \sqrt{4n^4 R_S{}^2 - 4n^4 R_S{}^2}}{2}$$

$$= \frac{2n^2 R_S}{2}$$

$$= n^2 R_S$$

Note that, from an intermediate result for R_{pri} from Problem 1, that this is consistent with $R_{pri} = R_S$.

$$R_L / n^2 = R_S$$

$$n^2 (R_L / n^2) = n^2 R_S$$

$$R_L = n^2 R_S$$

14.2.4 Reflected Load

1. Problem: For the ideal transformer circuit in Figure 14-15, what is the reflected resistance?

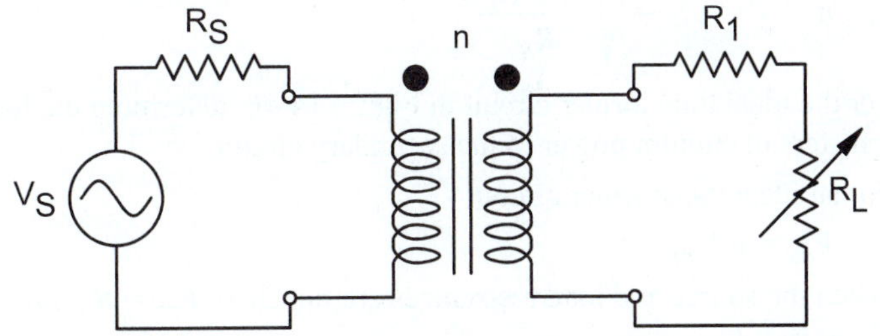

Figure 14-15: Reflected Load Example Circuit 2

Solution: For an ideal transformer circuit with secondary resistance R_{sec} and turns ratio n, the reflected resistance $R_{pri} = R_{sec} / n^2$.

$$R_{pri} = R_{sec} / n^2$$

$$= (R_1 + R_L) / n^2$$

2. Problem: For the ideal transformer circuit in Figure 14-15, determine R_L for which P_{RL} is maximum.

Solution: The transformer will transfer maximum power to the secondary circuit when $R_{pri} = R_S$. The secondary in turn will transfer maximum power to the load when $R_L = R_1$. Begin with the equation for reflected load for an ideal transformer and solve for R_L.

$$R_{pri} = R_S$$

$$(R_1 + R_L) / n^2 = R_S$$

$$[(R_1 + R_L) / n^2]\, n^2 = n^2 R_S$$

$$R_1 + R_L = n^2 R_S$$

$$R_L + R_L = n^2 R_S$$

$$(2R_L) / 2 = (n^2 R_S) / 2$$

$$R_L = (n^2 R_S) / 2$$

14.2.5 Impedance Matching

1. Problem: For the ideal transformer circuit in Figure 14-16, derive the necessary turns ratio to transfer maximum power to the secondary circuit.

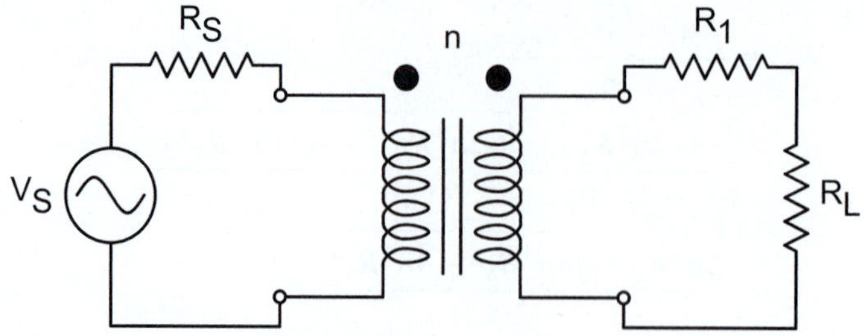

Figure 14-16: Impedance Matching Example Circuit 2

Solution: The ideal transformer circuit will transfer maximum power to the secondary circuit when the source resistance $\boldsymbol{R_S}$ equals the reflected resistance $\boldsymbol{R_{pri}}$.

$$R_S = R_{pri}$$
$$= (R_1 + R_L) / n^2$$
$$n^2 R_S = [(R_1 + R_L) / n^2]\, n^2$$
$$(n^2 R_S) / R_S = (R_1 + R_L) / R_S$$
$$\sqrt{n^2} = \sqrt{\frac{R_1 + R_L}{R_S}}$$
$$n = \sqrt{\frac{R_1 + R_L}{R_S}}$$

2. Problem: For the ideal transformer circuit in Figure 14-16, determine the load power $\boldsymbol{P_{RL}}$ when the circuit transfers maximum power to the secondary circuit.

Solution: For an ideal transformer circuit,

$$V_{sec} = n\, V_{pri}$$

When the source and load resistances are matched, $\boldsymbol{R_{pri}} = \boldsymbol{R_S}$. From the voltage divider theorem, $\boldsymbol{V_{pri}} = \boldsymbol{V_S} / 2$.

$$V_{sec} = n\,(V_S / 2)$$

Use Watt's Law for voltage and resistance to calculate the power transferred to the transformer.

$$P_{pri} = V_{pri}^{\ 2} / R_{pri}$$
$$= (V_S / 2)^2 / R_S$$

For an ideal transformer, $\boldsymbol{P_{sec}} = \boldsymbol{P_{pri}}$.

$$P_{sec} = (V_S / 2)^2 / R_S$$

Use the voltage divider theorem to calculate the power transferred to the load.

$$V_{RL} = V_{sec}\,[R_L / (R_1 + R_L)]$$
$$V_{RL}\, I_{sec} = V_{sec}\, I_{sec}\,[R_L / (R_1 + R_L)]$$
$$V_{RL}\, I_{RL} = V_{sec}\, I_{sec}\,[R_L / (R_1 + R_L)]$$
$$P_{RL} = P_{sec}\,[R_L / (R_1 + R_L)]$$
$$= [(V_S / 2)^2 / R_S]\,[R_L / (R_1 + R_L)]$$
$$= [(V_S^{\ 2} / 4R_S]\,[R_L / (R_1 + R_L)]$$

14.2.6 Transformer Ratings and Characteristics

For Problems 1 and 2, refer to the transformer circuit in Figure 14-17.

1. Problem: For the ideal transformer circuit, the source voltage $V_S = 120$ V, source resistance $R_S = 75\ \Omega$, and turns ratio $n = 0.1$. The maximum apparent power rating on the transformer nameplate is 100 VA. What is the minimum value of R_L that will not exceed this rating?

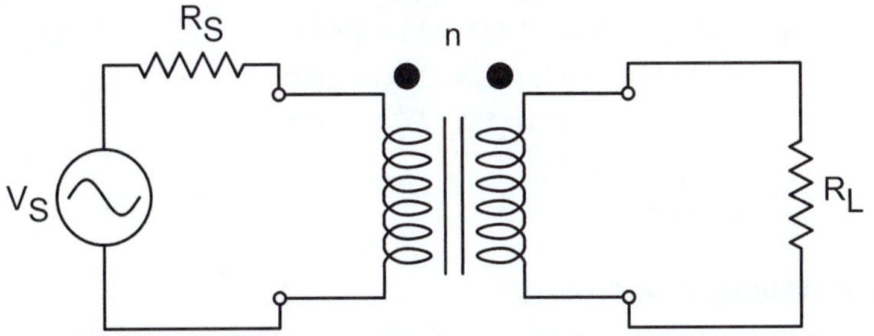

Figure 14-17: Transformer Rating Example Circuit 2

Solution: For an ideal transformer the primary power P_{pri} equals the secondary power P_{sec}, so that the primary and secondary apparent power values are the same. From Watt's Law for voltage and resistance, $P_{pri} = V_S^2 / (R_S + R_{pri})$. Also, the reflected resistance in the primary is $R_{pri} = R_L / n^2$.

$$P_{pri} = V_S^2 / (R_S + R_{pri})$$
$$= V_S^2 / [R_S + (R_L / n^2)]$$
$$P_{pri} \times [R_S + (R_{L\,(\min)} / n^2)] = \{V_S^2 / [R_S + (R_L / n^2)]\} \times [R_S + (R_L / n^2)]$$
$$\{ P_{pri} \times [R_S + (R_{L\,(\min)} / n^2)]\} / P_{pri} = V_S^2 / P_{pri}$$
$$R_S + (R_{L\,(\min)} / n^2) - R_S = (V_S^2 / P_{pri}) - R_S$$
$$(R_{L\,(\min)} / n^2) \times n^2 = [(V_S^2 / P_{pri}) - R_S] \times n^2$$
$$R_{L\,(\min)} = [(V_S^2 / P_{pri}) - R_S] \times n^2$$

Substitute in the given values from the problem statement to calculate $R_{L(\min)}$.

$$R_{L\,(\min)} = \{[(120\text{ V})^2 / 100\text{ VA}] - 75\ \Omega\} \times 0.1^2$$
$$= [(14{,}400\text{ V}^2 / 100\text{ VA}) - 75\ \Omega] \times 0.01$$
$$= [(144\text{ V / A}) - 75\ \Omega] \times 0.01$$
$$= (144\ \Omega - 75\ \Omega) \times 0.01$$
$$= 69\ \Omega \times 0.01$$
$$= \mathbf{0.69\ \Omega}$$

2. Problem: For the ideal transformer circuit, the source voltage $V_S = 24.0$ V, the source resistance $R_S = 50\ \Omega$, the primary current $I_{pri} = 150$ mA, the primary current $I_{sec} = 25.0$ mA, and the load resistance $R_L = 1.3$ kΩ. What is the efficiency η of the transformer?

Solution: For the primary circuit, $V_{pri} = V_S - I_{pri} R_S$.

$$V_{pri} = 24.0\text{ V} - (150\text{ mA})(50\ \Omega)$$
$$= 24.0\text{ V} - (0.150 \times 50)(\text{A} \times \Omega)$$
$$= 24.0\text{ V} - 7.5\text{ V}$$
$$= 16.5\text{ V}$$

From Watt's Law for voltage and current, the transformer primary power $P_{pri} = V_{pri} I_{pri}$.

$$P_{pri} = (16.5\text{ V})(150\text{ mA})$$
$$= (16.5 \times 0.150)(\text{V} \times \text{A})$$
$$= 2.475\text{ VA}$$

From Watt's Law for current and resistance, the transformer secondary power is $P_{sec} = I_{sec}^2 R_L$.

$$P_{sec} = (25.0\text{ mA})^2 \times 1.30\text{ k}\Omega$$
$$= 625 \times 10^{-6}\text{ A}^2 \times 1.30\text{ k}\Omega$$

$= [(625 \times 10^{-6}) \times (1.30 \times 10^{3})]\ (A^2 \times \Omega)$

$= 812.5\ mVA$

By definition the efficiency $\eta = (\boldsymbol{P_{sec}} / \boldsymbol{P_{pri}}) \times 100\%$.

$\eta \quad = (812.5\ mVA / 2.475\ VA) \times 100\%$

$= (0.8125 / 2.475)\ (VA / VA) \times 100\%$

$= (0.8125 / 2.475)\ (VA / VA) \times 100\%$

$= 0.328 \times 100\%$

$=$ **32.8%**

14.2.7 Tapped and Multi-Winding Transformers

1. Problem: For the ideal multi-winding transformer circuit in Figure 14-18, what is the reflected resistance $\boldsymbol{R_{pri}}$?

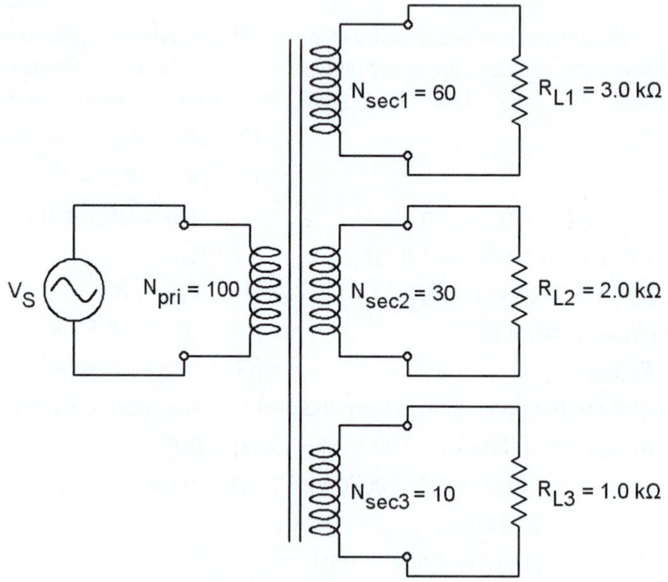

Figure 14-18: Tapped Transformer Example Circuit 2

Solution: Calculate the voltage and power for the top secondary.

$\boldsymbol{V_{sec1}} = \boldsymbol{n_{sec1}} \times \boldsymbol{V_S}$

$= (\boldsymbol{N_{sec1}} / \boldsymbol{N_{pri}}) \times \boldsymbol{V_S}$

$= (60 / 100) \times \boldsymbol{V_S}$

$= 0.60\ \boldsymbol{V_S}$

$\boldsymbol{P_{sec1}} = \boldsymbol{V_{sec1}}^2 / \boldsymbol{R_{L1}}$

$= (0.60\ \boldsymbol{V_S})^2 / 3.0\ k\Omega$

$= [0.36 / (3 \times 10^3)]\ (\boldsymbol{V_S}^2 / \Omega)$

$= 120 \times 10^{-6}\ \boldsymbol{V_S}^2 / \Omega$

Calculate the voltage and power for middle secondary.

$\boldsymbol{V_{sec2}} = \boldsymbol{n_{sec2}} \times \boldsymbol{V_S}$

$= (\boldsymbol{N_{sec2}} / \boldsymbol{N_{pri}}) \times \boldsymbol{V_S}$

$= (30 / 100) \times \boldsymbol{V_S}$

$= 0.30\ \boldsymbol{V_S}$

$$
\begin{aligned}
P_{sec2} &= V_{sec2}^{\;2} / R_{L2} \\
&= (0.30\ V_S)^2 / 2.0\ \text{k}\Omega \\
&= [0.09 / (2 \times 10^3)]\ (V_S^2 / \Omega) \\
&= 45 \times 10^{-6}\ V_S^2 / \Omega
\end{aligned}
$$

Calculate the voltage and power for bottom secondary.

$$
\begin{aligned}
V_{sec3} &= n_{sec3} \times V_S \\
&= (N_{sec3} / N_{pri}) \times V_S \\
&= (10 / 100) \times V_S \\
&= 0.10\ V_S \\
P_{sec3} &= V_{sec3}^{\;2} / R_{L3} \\
&= (0.10\ V_S)^2 / 1.0\ \text{k}\Omega \\
&= [0.01 / (1 \times 10^3)]\ (V_S^2 / \Omega) \\
&= 10 \times 10^{-6}\ V_S^2 / \Omega
\end{aligned}
$$

Sum the individual secondary powers to calculate the total secondary power.

$$
\begin{aligned}
P_{sec} &= P_{sec1} + P_{sec2} + P_{sec3} \\
&= 120 \times 10^{-6}\ V_S^2 / \Omega + 45 \times 10^{-6}\ V_S^2 / \Omega + 10 \times 10^{-6}\ V_S^2 / \Omega \\
&= 175 \times 10^{-6}\ V_S^2 / \Omega.
\end{aligned}
$$

For an ideal transformer $P_{sec} = P_{pri}$. Use Watt's Law for voltage and resistance to calculate R_{pri}.

$$
\begin{aligned}
P_{sec} &= V_{pri}^{\;2} / R_{pri} \\
P_{sec} \times R_{pri} &= (V_S^2 / R_{pri}) \times R_{pri} \\
(P_{sec} \times R_{pri}) / P_{sec} &= V_S^2 / P_{sec} \\
R_{pri} &= V_S^2 / P_{sec} \\
&= V_S^2 / (175 \times 10^{-6}\ V_S^2 / \Omega) \\
&= (1\ V_S^2) / (175 \times 10^{-6}\ V_S^2)\ \Omega \\
&= (1 / 175 \times 10^{-6})\ \Omega \\
&= \mathbf{5.71\ k\Omega}
\end{aligned}
$$

2. Problem: For the ideal autotransformer circuit (sometimes called a **variac**) shown in Figure 14-19, assume that the output voltage $V_{out} = V_{org}$ for the original values of N_1, N_2, and V_{in}. If V_{in} is not changed, what is the new output voltage if the wiper is moved so that the value of N_1 is doubled? What is the new output voltage if the wiper is moved so that the value of N_1 is halved?

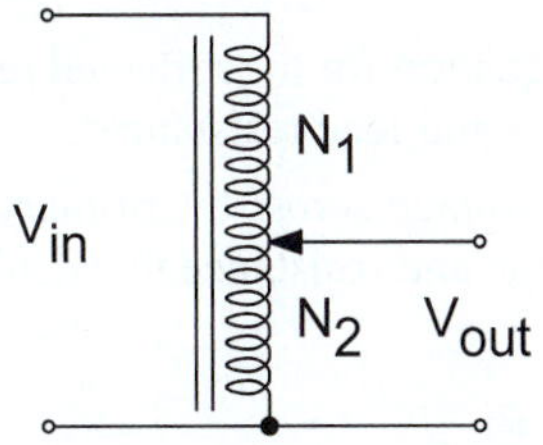

Figure 14-19: Autotransformer Example Circuit 2

Solution: For the autotransformer, the total number of turns $N_T = N_1 + N_2$ and does not change. For an ideal autotransformer, $V_{out} / N_2 = V_{in} / (N_1 + N_2) = V_{in} / N_T$. First, solve for V_{out} in terms of the autotransformer circuit values.

$$
\begin{aligned}
V_{out} / N_2 &= V_{in} / N_T \\
V_{out} / (N_T - N_1) &= V_{in} / N_T \\
[V_{out} / (N_T - N_1)] \times (N_T - N_1) &= (V_{in} / N_T) \times (N_T - N_1) \\
V_{out} &= V_{in}\ [(N_T - N_1) / N_T]
\end{aligned}
$$

Next, use the original circuit values to solve for N_1.

$$V_{org} = V_{in}\,[(N_T - N_1) / N_T]$$
$$N_T \times V_{org} = V_{in}\,[(N_T - N_1) / N_T] \times N_T$$
$$(N_T \times V_{org}) / V_{in} = [V_{in}\,(N_T - N_1)] / V_{in}$$
$$N_T\,(V_{org} / V_{in}) = [V_{in}\,(N_T - N_1)] / V_{in}$$
$$N_T\,(V_{org} / V_{in}) - N_T = N_T - N_1 - N_T$$
$$-[N_T\,(V_{org} / V_{in}) - N_T] = -(-N_1)$$
$$N_T - N_T\,(V_{org} / V_{in}) = N_1$$
$$N_T\,[1 - (V_{org} / V_{in})] = N_1$$

Let V_K be the output voltage for $N_K = kN_1$. For $V_{out} = V_K$,

$$\begin{aligned}V_K &= V_{in}\,[(N_T - N_K) / N_T]\\ &= V_{in}\,[(N_T - kN_1) / N_T]\\ &= V_{in}\,[(N_T - k\{N_T\,[1 - (V_{org} / V_{in})]\}) / N_T]\\ &= V_{in}\,[(N_T - kN_T\,[1 - (V_{org} / V_{in})]\}) / N_T]\\ &= V_{in}\,[(N_T\,\{1 - k[1 - (V_{org} / V_{in})]\}) / N_T]\\ &= V_{in}\,\{1 - k[1 - (V_{org} / V_{in})]\}\\ &= V_{in}\,\{1 - k + k(V_{org} / V_{in})]\}\\ &= V_{in} - kV_{in} + kV_{in}\,(V_{org} / V_{in})\\ &= V_{in} - kV_{in} + kV_{org}\\ &= (1 - k)V_{in} + kV_{org}\end{aligned}$$

When N_1 is doubled, $k = 2$.

$$\begin{aligned}V_{out} &= (1 - 2)V_{in} + 2V_{org}\\ &= -V_{in} + 2V_{org}\\ &= \mathbf{2V_{org} - V_{in}}\end{aligned}$$

When N_1 is halved, $k = 0.5$.

$$\begin{aligned}V_{out} &= (1 - 0.5)V_{in} + 0.5V_{org}\\ &= 0.5V_{in} + 0.5V_{org}\\ &= \mathbf{0.5(V_{org} + V_{in})}\end{aligned}$$

14.3 Just for Fun

1. Problem: Derive the general equation for the reflected resistance in an ideal multi-winding transformer with X secondary windings and load resistances.

Solution: Let V_{pri} be the input voltage across the primary and N_{pri} be the number of primary turns. Use Watt's Law for voltage and resistance to calculate the power for a secondary with N_{sec} turns.

$$\begin{aligned}P_{sec} &= V_{sec}^2 / R_L\\ &= (n\,V_{pri})^2 / R_L\\ &= [(N_{sec} / N_{pri})\,V_{pri}]^2 / R_L\\ &= [(V_{pri} / N_{pri})\,N_{sec}]^2 / R_L\\ &= (V_{pri} / N_{pri})^2\,(N_{sec}^2 / R_L)\end{aligned}$$

The total secondary power $P_{T(sec)}$ for X secondaries and load resistors is

$$\begin{aligned}P_{T(sec)} &= (V_{pri} / N_{pri})^2\,(N_{sec1}^2 / R_{L1}) + \ldots + (V_{pri} / N_{pri})^2\,(N_{secX}^2 / R_{LX})\\ &= (V_{pri} / N_{pri})^2\,[(N_{sec1}^2 / R_{L1}) + \ldots + (N_{secX}^2 / R_{LX})]\end{aligned}$$

For an ideal transformer, $P_{pri} = P_{T(sec)}$.

$$\begin{aligned}
P_{pri} &= (V_{pri} / N_{pri})^2\, [(N_{sec1}^2 / R_{L1}) + \ldots + (N_{secX}^2 / R_{LX})] \\
V_{pri}\, I_{pri} &= (V_{pri}^2 / N_{pri}^2)\, [(N_{sec1}^2 / R_{L1}) + \ldots + (N_{secX}^2 / R_{LX})] \\
(V_{pri}\, I_{pri}) / V_{pri}^2 &= \{(V_{pri}^2 / N_{pri}^2)\, [(N_{sec1}^2 / R_{L1}) + \ldots + (N_{secX}^2 / R_{LX})]\} / V_{pri}^2 \\
I_{pri} / V_{pri} &= (1 / N_{pri}^2)\, [(N_{sec1}^2 / R_{L1}) + \ldots + (N_{secX}^2 / R_{LX})] \\
&= [(N_{sec1}^2 / R_{L1}) + \ldots + (N_{secX}^2 / R_{LX})] / N_{pri}^2 \\
1 / (I_{pri} / V_{pri}) &= 1 / \{[(N_{sec1}^2 / R_{L1}) + \ldots + (N_{secX}^2 / R_{LX})] / N_{pri}^2\} \\
V_{pri} / I_{pri} &= N_{pri}^2 / [(N_{sec1}^2 / R_{L1}) + \ldots + (N_{secX}^2 / R_{LX})] \\
\mathbf{R_{pri}} &= \mathbf{N_{pri}^2 / [(N_{sec1}^2 / R_{L1}) + \ldots + (N_{secX}^2 / R_{LX})]}
\end{aligned}$$

2. Problem: For the ideal transformer circuit in Figure 14-20, determine the required turns ratio for T_2 to double the output voltage from T_1 alone.

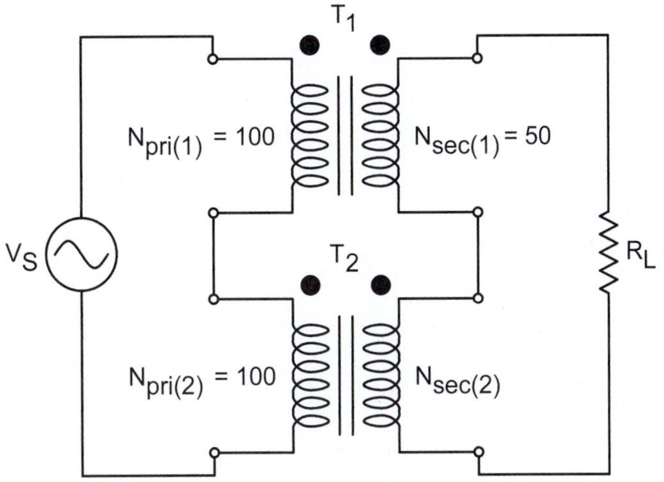

Figure 14-20: Turns Ratio Example Circuit

Solution: Because the same currents flow through the primaries and secondaries of both ideal transformers, they can be viewed as a single ideal transformer. For an ideal transformer, $V_{sec} / N_{sec} = V_{pri} / N_{pri}$. For T_1,

$$\begin{aligned}
V_{sec\,(1)} / 50 &= V_{pri} / 100 \\
(V_{sec\,(1)} / 50) \times 50 &= (V_{pri} / 100) \times 50 \\
V_{sec\,(1)} &= 0.5 V_{pri}
\end{aligned}$$

To double the secondary voltage for T_1 and T_2 combined, $V_{sec} = 2 \times 0.5 V_{pri} = V_{pri}$.

$$\begin{aligned}
V_{sec} / (N_{sec\,(1)} + N_{sec\,(2)}) &= V_{pri} / [N_{pri\,(1)} + N_{pri\,(2)}] \\
V_{sec} / (50 + N_{sec\,(2)}) &= V_{sec} / [100 + 100] \\
V_{sec} / (50 + N_{sec\,(2)}) &= V_{sec} / 200 \\
[V_{sec} / (50 + N_{sec\,(2)})]\, (50 + N_{sec\,(2)}) &= (V_{sec} / 200)\, (50 + N_{sec\,(2)}) \\
V_{sec}\, (200 / V_{sec}) &= [(V_{sec} / 200)\, (50 + N_{sec\,(2)})]\, (200 / V_{sec}) \\
200 - 50 &= 50 + N_{sec\,(2)} - 50 \\
150 &= N_{sec\,(2)}
\end{aligned}$$

By definition $n = N_{sec} / N_{pri}$.

$$\begin{aligned}
n_{T2} &= N_{sec\,(2)} / N_{pri\,(2)} \\
&= 150 / 100 \\
&= \mathbf{1.5}
\end{aligned}$$

15. Complex Numbers and Phasors

15.1 Basic

Complex numbers have a number of equivalent forms. This manual gives polar and rectangular answers in "standard forms", defined as follows:

Rectangular: The "standard rectangular form" is a sum of real and imaginary components.

Polar: The "standard polar form" has a positive magnitude and angle between ±180°.

15.1.1 Standard Form

1. Problem: Convert each of the following complex numbers to "standard rectangular form".

   a. $-2.0 - j\,4.0$

   b. $7.0 + j\,3.0$

   c. $5.0 - j\,(-5.0)$

   d. $-4.0 + j\,(-8.0)$

   Solution: To convert to "standard rectangular form", ensure that the complex number is a sum of real and imaginary components.

   a. $-2.0 - j\,4.0$: The number is a difference of real and imaginary components. To convert to "standard rectangular form", change the sign of the imaginary component and the difference to a sum:

   $$-2.0 - j\,4.0 = \mathbf{-2.0 + j\,(-4.0)}$$

   b. $7.0 + j\,3.0$: The number is a sum of real and imaginary components and already in "standard rectangular form". **No change is necessary**.

   c. $5.0 - j\,(-5.0)$: The number is a difference of real and imaginary components. To convert to "standard rectangular form", change the sign of the imaginary component and the difference to a sum:

   $$5.0 - j\,(-5.0) = 5.0 + j\,[-(-5.0)]$$
   $$= \mathbf{5.0 + j\,5.0}$$

   d. $-4.0 + j\,(-8.0)$: The number is a sum of real and imaginary components and already in "standard rectangular form". **No change is necessary**.

2. Problem: Convert each of the following complex numbers to "standard polar form".

   a. $1.0 \angle 495°$

   b. $-5.0 \angle 240°$

   c. $6.0 \angle -373°$

   d. $-8.0 \angle -540°$

   Solution: To convert to "standard polar form", ensure that the magnitude is a positive value and that the angle is between ±180°.

   a. $1.0 \angle 495°$: The angle is greater than 180°. To convert to "standard polar form", repeatedly subtract 180° (half revolutions) until the angle is between ±180°.

   $$1.0 \angle 495° = 1.0 \angle (495° - 180°)$$
   $$= 1.0 \angle (315° - 180°)$$
   $$= \mathbf{1.0 \angle 135°}$$

   b. $-5.0 \angle 240°$: The magnitude is not a positive value. To convert to "standard polar form", change the sign of the magnitude and adjust the original angle by ±180° (a half revolution). Because the original angle is greater than 180°, subtract 180° so that the new angle will be between ±180°.

$$-5.0 \angle 240° = 5.0 \angle (240° - 180°) = \mathbf{5.0 \angle 60°}$$

c. $6.0 \angle -373°$: The angle is less than −180°. To convert to "standard polar form", repeatedly add 180° (half revolutions) until the angle is between ±180°.

$$\begin{aligned} 6.0 \angle -373° &= 6.0 \angle (-373° + 180°) \\ &= 6.0 \angle (-193° + 180°) \\ &= \mathbf{6.0 \angle (-13°)} \end{aligned}$$

d. $-8.0 \angle -540°$: The magnitude is not a positive value and the angle is not between ±180°. To convert to "standard polar form", first change the sign of the magnitude and adjust the original angle by ±180° (a half revolution). Because the original angle is less than 180°, subtract 180° so that the new angle will be closer to ±180°.

$$\begin{aligned} -8.0 \angle -540° &= -8.0 \angle (-540° + 180°) \\ &= 8.0 \angle (-360°) \end{aligned}$$

The angle is less than −180°. To convert to "standard polar form", repeatedly add 180° (half revolutions) until the angle is between ±180°.

$$\begin{aligned} 8.0 \angle (-360°) &= 8.0 \angle (-360° + 180°) \\ &= 8.0 \angle (-180° + 180°) \\ &= \mathbf{8.0 \angle 0°} \end{aligned}$$

15.1.2 The Complex Number Plane

For Problems 1 and 2, refer to the complex number coordinate systems in Figure 15-1. As Figure 15-1 shows, degrees increase in the counter-clockwise direction (0°, 90°, 180°, 270°, and 360°) and decrease in the clockwise direction (0°, −90°, −180°, −270°, and −360°).

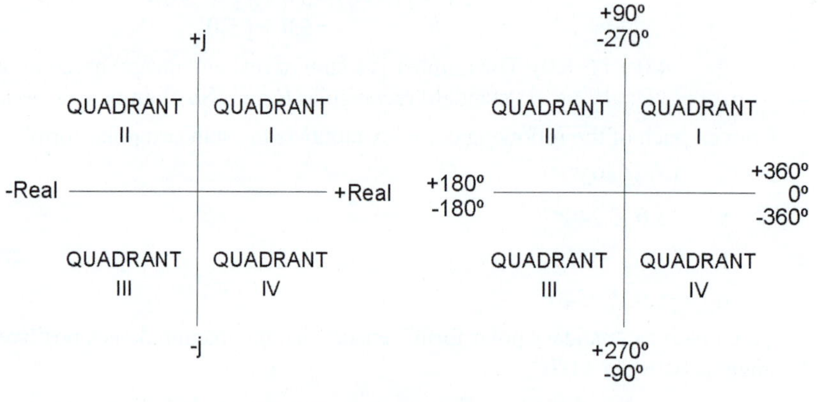

Figure 15-1: Complex Number Coordinate Systems

1. Problem: Identify the quadrant that corresponds to each of the following complex numbers:
   a. $-1 + j\,2$
   b. $4 + j\,3$
   c. $-5 - j\,5$
   d. $2 + j$
   e. $3 - j\,3$

Solution: The quadrant that corresponds to each complex number depends on the signs of the real and imaginary components.

a. $-1 + j\,2$: The real component (-1) is negative and the imaginary component ($+2$) is positive. The imaginary number lies in **Quadrant II**.

b. $4 + j\,3$: The real component ($+4$) is positive and the imaginary component ($+3$) is positive. The imaginary number lies in **Quadrant I**.

c. $-5 - j\,5$: This is the same as $-5 + j\,(-5)$ when rewritten in "standard form" (as the sum of real and imaginary components). The real component (-5) is negative and the imaginary component (-5) is negative. The imaginary number lies in **Quadrant III**.

d. $2 + j$: This is the same as $2 + j\,1$. The real component ($+2$) is positive and the imaginary component ($+1$) is positive. The imaginary number lies in **Quadrant I**.

e. $3 - j\,3$: This is the same as $3 + j\,(-3)$ when rewritten in "standard form" as the sum of real and imaginary components. The real component ($+3$) is positive and the imaginary component (-3) is negative. The imaginary number lies in **Quadrant IV**.

2. Problem: Identify the quadrant that corresponds to each of the following complex numbers:

a. $-1.41 \angle 23°$

b. $0.71 \angle -115°$

c. $4.50 \angle 315°$

d. $-1.0 \angle 210°$

e. $2.3 \angle -290°$

Solution: The quadrant that corresponds to each complex number depends on the signs and values of the magnitude and angle.

a. $-1.41 \angle 23°$: The positive angle ($+23°$) lies in Quadrant 1, but the negative magnitude (-1.41) reverses this (changes the angle by 180°). The imaginary number lies in **Quadrant III**.

b. $0.71 \angle -115°$: The negative angle ($-115°$) lies in Quadrant II, and the positive magnitude ($+0.71$) does not modify this. The imaginary number lies in **Quadrant II**.

c. $4.50 \angle 315°$: The positive angle ($+315°$) lies in Quadrant IV, and the positive magnitude ($+4.50$) does not modify this. The imaginary number lies in **Quadrant IV**.

d. $-1.0 \angle 210°$: The positive angle ($+210°$) lies in Quadrant III, but the negative magnitude (-1.0) reverses this (changes the angle by 180°). The imaginary number lies in **Quadrant I**.

e. $2.3 \angle -290°$: The negative angle ($-290°$) lies in Quadrant I, and the positive magnitude ($+2.3$) does not change this. The imaginary number lies in **Quadrant I**.

15.1.3 Complex Number Conversions

1. Problem: Convert each of the following complex numbers from rectangular to polar form.

a. $-40.0 - j\,9.0$

b. $7.0 + j\,24.0$

c. $-12.0 + j\,5.0$

d. $4.0 - j\,3.0$

Solution: To convert a complex number $\boldsymbol{A} + j\,\boldsymbol{B}$ to polar form $\boldsymbol{R} \angle \varphi$, calculate the magnitude $\boldsymbol{R}$ and angle φ from the equations

$R = \sqrt{A^2 + B^2}$ and $\varphi = \tan^{-1}(B / A)$

Note that the complex number should be in "standard rectangular form" to determine the correct quadrant for the complex number. Note also that because the square root of a number can be either positive or negative (corresponding to a positive or negative magnitude), the quadrant of the original number must be checked to determine whether the calculated angle should be adjusted by ±180° for the final answer.

a. $-40.0 - j\,9.0$: First, convert the number to "standard rectangular form".

$$-40.0 - j\,9.0 = -40.0 + j\,(-9.0)$$

Next, calculate the magnitude R.

$$R = \sqrt{(-40.0)^2 + (-9.0)^2} = \sqrt{1681} = 41.0$$

Next, calculate the angle φ.

$$\varphi = \tan^{-1}(-9.0 / -40.0) = \tan^{-1}(0.225) = 12.7°$$

The complex number $-40.0 + j\,(-9.0)$ is in Quadrant III, but the calculated angle is in Quadrant I. To convert the angle from Quadrant I to Quadrant III, subtract 180° from φ.

$$\varphi - 180° = 12.7° - 180° = -167.3°$$

From this,

$$-40.0 - j\,9.0 = \mathbf{41.0 \angle -167.3°}$$

b. $7.0 + j\,24.0$: The number is already in "standard rectangular form", so first calculate the magnitude R.

$$R = \sqrt{(7.0)^2 + (24.0)^2} = \sqrt{625} = 25.0$$

Next, calculate the angle φ.

$$\varphi = \tan^{-1}(24.0 / 7.0) = \tan^{-1}(3.43) = 73.7°$$

Both the complex number $7.0 + j\,24.0$ and calculated angle are in Quadrant I, so that no adjustment to φ is necessary. From this,

$$7.0 + j\,24.0 = \mathbf{25.0 \angle 73.7°}$$

c. $-12.0 + j\,5.0$: The number is already in "standard rectangular form", so first calculate the magnitude R.

$$R = \sqrt{(-12.0)^2 + (5.0)^2} = \sqrt{169} = 13.0$$

Next, calculate the angle φ.

$$\varphi = \tan^{-1}(5.0 / -12.0)$$

$= \tan^{-1}(-0.417)$

$= -22.6°$

The complex number −12.0 + j 5.0 is in Quadrant II, but the calculated angle is in Quadrant IV. To convert the angle from Quadrant IV to Quadrant II, add 180° to φ.

$\varphi + 180° = -22.6° + 180°$

$= +157.4°$

From this,

$-12.0 + j\,5.0 = \mathbf{13.0 \angle +157.4°}$

d. 4.0 − j 3.0: First, convert the number to "standard rectangular form".

$4.0 - j\,3.0 = 4.0 + j\,(-3.0)$

Next, calculate the magnitude $\boldsymbol{R}$.

$\boldsymbol{R} = \sqrt{(4.0)^2 + (-3.0)^2}$

$= \sqrt{25}$

$= 5.0$

Next, calculate the angle φ.

$\varphi = \tan^{-1}(-3.0 / 4.0)$

$= \tan^{-1}(-0.75)$

$= -36.9°$

Both the complex number 4.0 + j (−3.0) and the calculated angle are in Quadrant IV so no adjustment to φ is necessary. From this,

$4.0 - j\,3.0 = \mathbf{5.0 \angle -36.9°}$

2. Problem: Convert each of the following complex numbers from polar to rectangular form.

a. $1.41 \angle 45°$

b. $-2.50 \angle -120°$

c. $0.707 \angle 585°$

d. $1.00 \angle -200°$

Solution: To convert a complex number $\boldsymbol{R} \angle \varphi$ to rectangular form $\boldsymbol{A} + j\,\boldsymbol{B}$, calculate the real component $\boldsymbol{A}$ and imaginary component $\boldsymbol{B}$ from the equations

$\boldsymbol{A} = \boldsymbol{R} \cos \varphi$ and $\boldsymbol{B} = \boldsymbol{R} \sin \varphi$

Note that the complex number should be in "standard polar form" to determine the correct quadrant for the complex number. Note also that because $\cos(-\varphi) = \cos(\varphi)$ and $\sin(-\varphi) = -\sin(\varphi)$ the quadrant of the original number must be checked to determine the correct signs for $\boldsymbol{A}$ and $\boldsymbol{B}$.

a. $1.41 \angle 45°$: The number is already in "standard polar form", so calculate the value of $\boldsymbol{A}$.

$\boldsymbol{A} = 1.41 \cos(45°)$

$= (1.41)(0.707)$

$= 0.997$

Next, calculate the value of $\boldsymbol{B}$.

$\boldsymbol{B} = 1.41 \sin(45°)$

$= (1.41)(0.707)$

$= 0.997$

The complex number $1.41 \angle 45°$ is in Quadrant I, so both ***A*** and ***B*** should be positive. From this,

$$1.41 \angle 45° = \mathbf{0.997 + j\,0.997}$$

b. $-2.50 \angle -120°$: First, convert the number to "standard polar form".

$$-2.50 \angle -120° = 2.50 \angle [(-120) + 180°]$$
$$= 2.50 \angle 60°$$

Next, calculate the value of ***A***.

$$\boldsymbol{A} = 2.50 \cos (60°)$$
$$= (2.5)(0.50)$$
$$= 1.25$$

Next, calculate the value of *B*.

$$\boldsymbol{B} = 2.50 \sin (60°)$$
$$= (2.50)(0.866)$$
$$= 2.17$$

The complex number $2.50 \angle 60°$ is in Quadrant I, so both ***A*** and ***B*** should be positive. From this,

$$-2.50 \angle -120° = \mathbf{1.25 + j\,2.17}$$

c. $0.707 \angle 585°$: First, convert the number to "standard polar form".

$$0.707 \angle 585° = 0.707 \angle (585° - 360°)$$
$$= 0.707 \angle (225° - 360°)$$
$$= 0.707 \angle -135°$$

Next, calculate the value of ***A***.

$$\boldsymbol{A} = 0.707 \cos (-135°)$$
$$= (0.707)(-0.707)$$
$$= -0.50$$

Next, calculate the value of ***B***.

$$\boldsymbol{B} = 0.707 \sin (-135°)$$
$$= (0.707)(-0.707)$$
$$= -0.50$$

The complex number $0.707 \angle -135°$ is in Quadrant III, so both ***A*** and ***B*** should be negative. From this,

$$0.707 \angle 585° = \mathbf{-0.50 + j\,(-0.50)}$$

d. $1.00 \angle -200°$: First, convert the number to "standard polar form".

$$1.00 \angle -200° = 1.00 \angle (-200° + 360°)$$
$$= 1.00 \angle 160°$$

Next, calculate the value of ***A***.

$$\boldsymbol{A} = 1.00 \cos (160°)$$
$$= (1.00)(-0.940)$$
$$= -0.940$$

Next, calculate the value of ***B***.

$$\boldsymbol{B} = 1.00 \sin (160°)$$
$$= (1.00)(0.342)$$

$= 0.342$

The complex number $1.00 \angle 160°$ is in Quadrant II, so ***A*** should be negative and ***B*** should be positive. From this,

$1.00 \angle -200° = \mathbf{-0.940 + j\,0.342}$

15.1.4 Complex Addition and Subtraction

1. Problem: Perform each of the following addition and subtraction operations and express the answer in "standard rectangular form".

   a. $[-1.25 - j\,(-0.35)] + (0.63 - j\,0.35)$

   b. $(0.50 + j\,0.71) + (-1.25 + j\,0.87)$

   c. $[1.25 + j\,(-0.35)] - (0.87 + j\,0.71)$

   d. $[-0.87 - j\,(-0.71)] - [-1.00 - j\,(-0.50)]$

   Solution: To obtain the answers in "standard rectangular form", first convert the expressions to "standard rectangular form" as needed and then perform the operation.

   a. $[-1.25 - j\,(-0.35)] + (0.63 - j\,0.35) = [-1.25 + j\,0.35] + [(0.63 + j\,(-0.35)]$
   $= (-1.25 + 0.35) + j\,(0.35 + -0.35)$
   $= \mathbf{-0.90 + j\,0.00}$

   b. $(0.50 + j\,0.71) + (-1.25 + j\,0.87) = (0.50 + -1.25) + j\,(0.71 + 0.87)$
   $= \mathbf{-0.75 + j\,1.58}$

   c. $[1.25 + j\,(-0.35)] - (0.87 + j\,0.71) = [1.25 + j\,(-0.35)] + [-(0.87 + j\,0.71)]$
   $= [1.25 + j\,(-0.35)] + [-0.87 + j\,(-0.71)]$
   $= (1.25 + -0.87) + j\,(-0.35 + -0.71)$
   $= \mathbf{0.38 + j\,(-1.06)}$

   d. $[-0.87 - j\,(-0.71)] - [-1.00 - j\,(-0.50)] = [-0.87 - j\,(-0.71)] - [-1.00 - j\,(-0.50)]$
   $= (-0.87 + j\,0.71) - (-1.00 + j\,0.50)$
   $= (-0.87 + j\,0.71) + [-(-1.00 + j\,0.50)]$
   $= (-0.87 + j\,0.71) + [1.00 + j\,(-0.50)]$
   $= (-0.87 + 1.00) + j\,[0.71 + (-0.50)]$
   $= \mathbf{0.13 + j\,0.21}$

2. Problem: Perform each of the following addition and subtraction operations and express the answer in "standard polar form".

   a. $(2.00 \angle -120°) + (-1.00 \angle -30°)$

   b. $(-1.00 \angle 200°) + (-3.50 \angle -100°)$

   c. $(-1.41 \angle -315°) - (1.41 \angle 315°)$

   d. $(-1.00 \angle -330°) - (-0.50 \angle 500°)$

   Solution: To obtain the answers in "standard polar form", first convert the expressions to "standard polar form" as needed, convert to "standard rectangular form", perform the operation, and then convert the answer back to "standard polar form".

   a. $(2.00 \angle -120°) + (-1.00 \angle -30°) = (2.00 \angle -120°) + (1.00 \angle 150°)$
   $= [-1.00 + j\,(-1.73)] + (-0.87 + j\,0.50)$
   $= (-1.00 + -0.87) + j\,(-1.73 + 0.50)$
   $= -1.87 + j\,(-1.23)$
   $= \mathbf{2.24 \angle -147°}$

   b. $(-1.00 \angle 200°) + (-3.50 \angle -100°) = (1.00 \angle 20°) + (3.50 \angle 80°)$
   $= (0.94 + j\,0.34) + (0.61 + j\,3.44)$

$= (0.94 + 0.61) + j\,(0.34 + 3.44)$
$= 1.55 + j\,3.79$
$= \mathbf{4.09 \angle 67.8°}$

c. $(-1.41 \angle -315°) - (1.41 \angle 315°)$ $= (1.41 \angle 135°) - (1.41 \angle -45°)$
$= [-1.00 + j\,(-1.00)] - [1.00 + j\,(-1.00)]$
$= [-1.00 + j\,(-1.00)] + (-1.00 + j\,1.00)$
$= (-1.00 + -1.00) + j\,(-1.00 + 1.00)$
$= -2.00 + j\,0.00$
$= \mathbf{2.00 \angle 180°}$

d. $(-1.00 \angle -330°) - (-0.50 \angle 500°)$ $= (1.00 \angle -150°) - (0.50 \angle 320°)$
$= (1.00 \angle -150°) - (0.50 \angle -40°)$
$= [-0.87 + j\,(-0.50)] - [0.38 + j\,(-0.32)]$
$= [-0.87 + j\,(-0.50)] + (-0.38 + j\,0.32)$
$= (-0.87 + -0.38) + j\,(-0.50 + 0.32)$
$= -1.25 + j\,(-0.18)$
$= \mathbf{1.26 \angle -172°}$

15.1.5 Complex Multiplication and Division

1. Problem: Perform each of the following multiplication and division operations and express the answer in "standard polar form".

   a. $(0.71 \angle 120°)\,(-0.50 \angle -30°)$
   b. $(2.83 \angle -310°)\,(-1.00 \angle 45°)$
   c. $(-0.87 \angle 420°) / (0.87 \angle -1000°)$
   d. $(-1.73 \angle -190°) / (-5.20 \angle -90°)$

   Solution: To obtain the answers in "standard polar form", first convert the expressions to "standard polar form" as needed, perform the operation and then convert the answer back to "standard polar form".

   a. $(0.71 \angle 120°)\,(-0.50 \angle -30°)$ $= (0.71 \angle 120°)\,(0.50 \angle 150°)$
   $= [(0.71)(0.50)] \angle (120° + 150°)$
   $= 0.36 \angle 270°$
   $= \mathbf{0.36 \angle -90°}$

   b. $(2.83 \angle -310°)\,(-1.00 \angle 45°)$ $= (2.83 \angle 50°)\,(1.00 \angle -135°)$
   $= [(2.83)(1.00)] \angle (50° + -135°)$
   $= \mathbf{2.83 \angle -85°}$

   c. $(-0.87 \angle 420°) / (0.87 \angle -1000°)$ $= (0.87 \angle 240°) / (0.87 \angle 80°)$
   $= [(0.87) / (0.87)] \angle (240° - 80°)$
   $= \mathbf{1.00 \angle 160°}$

   d. $(-1.73 \angle -190°) / (-5.20 \angle -90°)$ $= (1.73 \angle -10°) / (5.20 \angle 90°)$
   $= (1.73) / (5.20) \angle (-10° + 90°)$
   $= \mathbf{0.33 \angle 80°}$

2. Problem: Perform each of the following multiplication and division operations and express the answer in rectangular form.

a. $(12.0 + j\ 5.0)\ (-12.0 - j\ 5.0)$
b. $(12.0 - j\ 5.0)\ (4.0 - j\ 3.0)$
c. $(-7.0 + j\ 24.0)\ /\ (3.0 - j\ 4.0)$
d. $(-40.0 - j\ 9.0)\ /\ (-7.0 - j\ 24.0)$

Solution: To obtain the answers in "standard rectangular form", first convert the expressions to "standard rectangular form" as needed, convert the expressions to "standard polar form", perform the operation, and then convert the answer back to "standard rectangular form".

a. $(12.0 + j\ 5.0)\ (-12.0 - j\ 5.0) = (12.0 + j\ 5.0)\ [-12.0 + j\ (-5.0)]$
$= (13.0 \angle 22.6°)\ (13.0 \angle -157°)$
$= [(13.0)(13.0)] \angle (22.6° + -157°)$
$= 169.0 \angle -135°$
$= \mathbf{-119 + j\ (-120)}$

b. $(12.0 - j\ 5.0)\ (4.0 - j\ 3.0) = [12.0 + j\ (-5.0)]\ [4.0 + j\ (-3.0)]$
$= (13.0 \angle -22.6°)\ (5.0 \angle -36.9°)$
$= [(13.0)(5.0)] \angle (-22.6° + -36.9°)$
$= 65.0 \angle -59.5°$
$= \mathbf{-33.0 + j\ (-56.0)}$

c. $(-7.0 + j\ 24.0)\ /\ (3.0 - j\ 4.0) = (-7.0 + j\ 24.0)\ /\ [3.0 + j\ (-4.0)]$
$= (25.0 \angle 106.3°)\ /\ (5.0 \angle -53.1°)$
$= [(25.0)\ /\ (5.0)] \angle [(106.3° - (-53.1°)]$
$= [(25.0)\ /\ (5.0)] \angle (106.3° + 53.1°)$
$= 5.0 \angle 159.4°$
$= \mathbf{-4.68 + j\ 1.76}$

d. $(-40.0 - j\ 9.0)\ /\ (-7.0 - j\ 24.0) = [-40.0 + j\ (-9.0)]\ /\ [-7.0 + j\ (-24.0)]$
$= (-41.0 \angle -167.3°)\ /\ (25.0 \angle -106.3°)$
$= [(-41.0)\ /\ (25.0)] \angle [-167.3° - (-106.3°)]$
$= [(-41.0)\ /\ (25.0)] \angle (-167.3° + 106.3°)$
$= 1.64 \angle -61.1$
$= \mathbf{0.79 + j\ (-1.44)}$

15.1.6 Phasors

For Problems 1 through 3, refer to Figure 15-2.

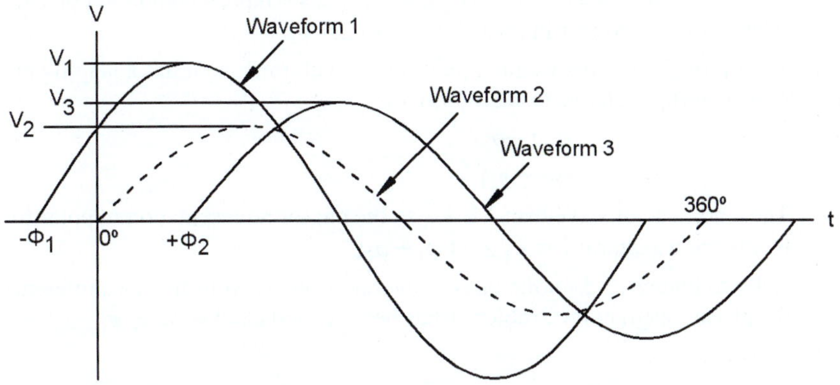

Figure 15-2: Sinusoidal Waveforms

1. Problem: What are the phasor representations in polar form of waveforms 1 and 3 relative to waveform 2? Sketch the phasor representations of the waveforms.

Solution: Waveform 1 crosses the x-axis before waveform 2, so it has a leading (positive) phase angle. The phase of waveform 1 relative to waveform 2 is

$$0° - (-\varphi_1) = +\varphi_1$$

The amplitude of waveform 1 is V_1, so the phasor representation in polar form of waveform 1 relative to waveform 2 is $\boldsymbol{V_1 \angle +\varphi_1}$.

Waveform 3 crosses the x-axis after waveform 2, so it has a lagging (negative) phase angle. The phase of waveform 3 relative to waveform 2 is

$$0° - (+\varphi_2) = -\varphi_2$$

The amplitude of waveform 3 is V_3, so the phasor representation in polar form of waveform 3 relative to waveform 2 in polar form is $\boldsymbol{V_3 \angle -\varphi_2}$.

Refer to Figure 15-3 for the phasor diagram using waveform 2 as the reference.

2. Problem: What are the phasor representations in polar form of waveforms 1 and 2 relative to waveform 3? Sketch the phasor representations of the waveforms.

Solution: Waveform 1 crosses the x-axis before waveform 3, so it has a leading phase angle. The phase of waveform 1 relative to waveform 3 is

$$\begin{aligned} +\varphi_2 - (-\varphi_1) &= +\varphi_2 - (-\varphi_1) \\ &= +\varphi_2 + \varphi_1 \\ &= +(\varphi_1 + \varphi_2) \end{aligned}$$

The amplitude of waveform 1 is V_1, so the phasor representation in polar form of waveform 1 relative to waveform 3 in polar form is $\boldsymbol{V_1 \angle +(\varphi_1 + \varphi_2)}$.

Waveform 2 also crosses the x-axis before waveform 3, so it has a leading phase angle. The phase of waveform 2 relative to waveform 3 is

$$+\varphi_2 - 0° = +\varphi_2$$

The amplitude of waveform 2 is V_2, so the phasor representation in polar form of waveform 2 relative to waveform 3 is $\boldsymbol{V_2 \angle +\varphi_2}$.

Refer to Figure 15-3 for the phasor diagram using waveform 3 as a reference. As the figure shows, the phasor diagram of Problem 1 has been rotated counterclockwise by φ_2.

3. Problem: What are the phasor representations of waveforms 2 and 3 relative to waveform 1? Sketch the phasor representations of the waveforms.

Solution: Waveform 2 crosses the x-axis after waveform 1, so it has a lagging phase angle. The phase of waveform 2 relative to waveform 1 is

$$-\varphi_1 - 0° = -\varphi_1$$

The amplitude of waveform 2 is V_2, so the phasor representation in polar form of waveform 2 relative to waveform 1 in polar form is $\boldsymbol{V_2 \angle -\varphi_1}$.

Waveform 3 also crosses the x-axis after waveform 1, so it has a lagging phase angle. The phase of waveform 3 relative to waveform 1 is

$$\begin{aligned} -\varphi_1 - (+\varphi_2) &= -\varphi_1 + -\varphi_2 \\ &= -(\varphi_1 + \varphi_2) \end{aligned}$$

The amplitude of waveform 3 is V_3, so the phasor representation in polar form of waveform 3 relative to waveform 1 is $\boldsymbol{V_3 \angle -(\varphi_1 + \varphi_2)}$.

Refer to Figure 15-3 for the phasor diagram using waveform 1 as a reference. As the figure shows, the phasor diagram for Problem 1 has been rotated clockwise by φ_1.

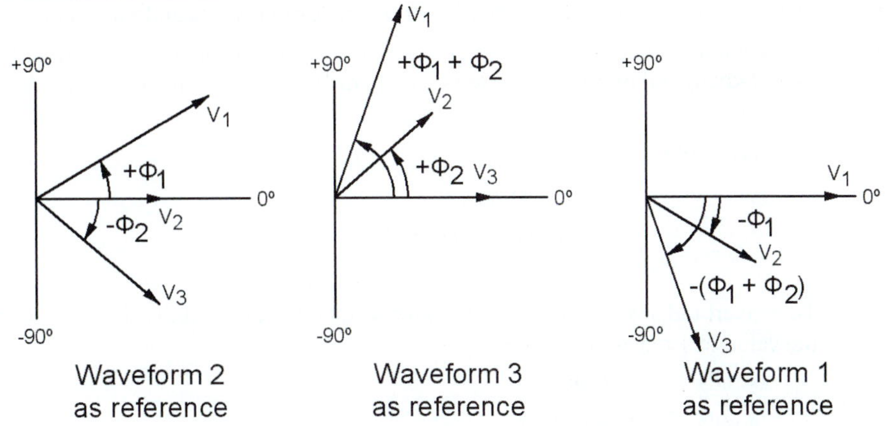

Figure 15-3: Phasor Diagrams for Problems 1 through 3

15.2 Advanced

15.2.1 The Complex Number Plane

Engineers and others who regularly work with complex calculations often use radians (abbreviated *rads*) rather than degrees to measure angles. The fundamental reason for this is that the "degree" is an artificial unit of angular measure, whereas the "radian" is a natural unit for angular measure that reflects an inherent property of circular measurement. Early mathematicians used 360° to represent one complete revolution because 360 is an easy number to divide. 2π radians represents one complete revolution because the ratio of the circumference to the radius of any circle is 2π. 1 radian represents the angle for which the length of the corresponding arc of a circle equals the radius of the circle. Because radians are the ratio of two lengths, you can easily use dimensional analysis to show that the units of length cancel and that the radian is dimensionless.

Radians can be expressed as a number (e.g., "1.50 rads") or in terms of π (e.g., "3π rads" or "$\pi/6$ rads"). Because π is an irrational number approximately equal to 3.1416, radians are usually represented in terms of π unless a numerical value is required, such as when solving problems on some calculators. Scientific calculators invariably support both degrees and radians and have a function that allows you to convert one to the other.

Refer to Figure 15-4 to see how angles are measured and represented in radians.

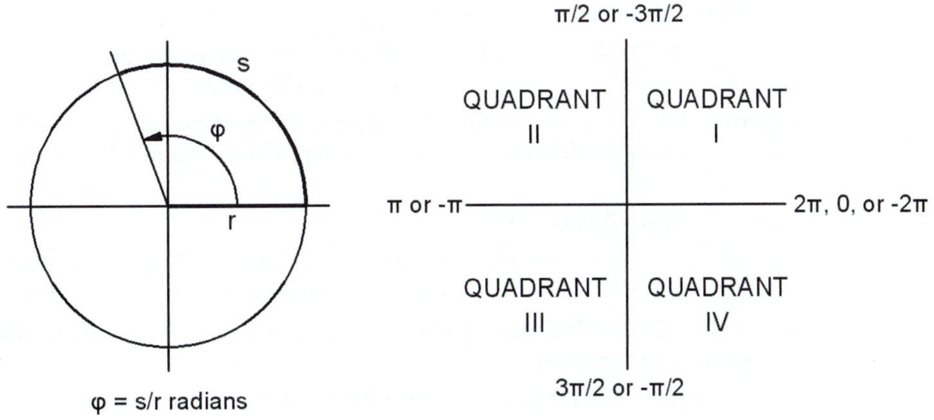

Figure 15-4: Radian Angular Measure

As Figure 15-4 shows, the radian measure φ of an angle is the ratio $\boldsymbol{s} / \boldsymbol{r}$, where $\boldsymbol{s}$ is the length of the arc for the angle and $\boldsymbol{r}$ is the radius of the arc. Radians increase in the counterclockwise direction (0 rads, $\pi/2$ rads, π rads, $3\pi/2$ rads, and 2π rads) and decrease in the clockwise direction (0 rads, $-\pi/2$ rads, $-\pi$ rads, $-3\pi/2$ rads, and -2π rads).

1. Problem: Derive the conversion factor for converting angles to radians and vice versa.

Solution: One complete rotation is equal to 360° or 2π radians. To convert angles to radians, set 360° equal to 2π radians and divide each side by 360 to determine the value of 1° in radians.

$$
\begin{aligned}
360^\circ &= 2\pi \text{ rads} \\
360^\circ / 360 &= 2\pi \text{ rads} / 360 \\
1^\circ &= (2\pi / 360) \text{ rads} \\
&= (2 \times \pi) / (2 \times 180) \text{ rads} \\
&= \boldsymbol{\pi / 180} \textbf{ rads}
\end{aligned}
$$

To convert radians to angles, set 2π radians equal to 360° and divide each side by 2π to determine the value of 1 radian in degrees.

$$
\begin{aligned}
2\pi \text{ rads} &= 360^\circ \\
2\pi \text{ rads} / 2\pi &= 360^\circ / 2\pi \\
1 \text{ rad} &= (2 \times 180^\circ) / (2 \times \pi) \\
&= \boldsymbol{180^\circ / \pi}
\end{aligned}
$$

2. Problem: Determine the smallest positive and negative radian values that are equivalent to the following angular measurements.

a. $17\pi/4$ rads
b. $-9\pi/6$ rads
c. $12\pi/3$ rads
d. -100 rads

Solution: To find the smallest positive and negative radian values, first increase (or decrease) the radian value by 2π radians (one complete rotation) until the value is between 0 and $\pm 2\pi$ radians. An alternate method is to divide the angular measure by 2π and then multiply the fractional portion by 2π. If the equivalent value is not 0, increase (or decrease) the value by 2π one more time to find the smallest positive (or negative) radian value.

a. $17\pi/4 > 2\pi$, so $17\pi/4$ rads first must be decreased by $2\pi = 8\pi/4$ to find the smallest positive radian value.

$17\pi/4 - 8\pi/4 = 9\pi/4$ — $9\pi/4 \geq 2\pi$ so subtract 2π once again.

$9\pi/4 - 8\pi/4 = \pi/4$ — $0 \leq \pi/4 < 2\pi$ so $\boldsymbol{\pi/4}$ **rads** is the smallest positive radian value that is equivalent to $17\pi/4$ rads.

Next, subtract $8\pi/4$ one more time.

$\pi/4 - 8\pi/4 = -7\pi/4$ — $\boldsymbol{-7\pi/4}$ **rads** is the smallest negative radian value that is equivalent to $17\pi/4$ rads.

b. $0 \geq -9\pi/6 > -2\pi$, so $\boldsymbol{-9\pi/6}$ **rads** is already the smallest negative angular value. To find the smallest positive radian value that is equivalent, increase the angular measurement by 2π (= $12\pi/6$).

$-9\pi/6 + 12\pi/6 = 3\pi/6$

$= \pi/2$ — $0 \leq \pi/2 < 2\pi$ so $3\pi/6 = \boldsymbol{\pi/2}$ **rads** is the smallest positive radian value that is equivalent to $-9\pi/6$ rads.

c. $12\pi/3 \geq 2\pi$ so $12\pi/3$ rads first must be decreased by $2\pi = 6\pi/3$ to find the smallest positive radian value.

$12\pi/3 - 6\pi/3 = 6\pi/3$ — $6\pi/6 \geq 2\pi$ so subtract 2π once again.

$6\pi/3 - 6\pi/3 = 0$ — The value is equal to 0, so **0 rads** is both the smallest positive and smallest negative radian values equal to $12\pi/3$.

d. -100 is not a multiple of π, so use the alternate method to find the smallest negative radian value for -100 rads.

$-100 / 2\pi$	$= -15.9155$	The integer portion (–15) represents complete rotations and is discarded. Multiply the fractional rotation by $2\pi = 6.2832$ to find the equivalent angular value.
-0.9155×6.2832	$= -5.752$	**–5.752 rads** is the smallest negative radian value that is equivalent to 100 rads.

Next, add 2π to find the smallest positive radian value.

$-5.752 + 6.2832$	$= 0.5310$	**0.5310 rads** is the smallest positive radian value that is equivalent to 100 rads.

3. Problem: Along which axes or in which quadrants do the following angles lie?
   a. $2\pi/3$ rads
   b. $-\pi/4$ rads
   c. $5\pi/2$ rads
   d. $7\pi/8$ rads

   Solution: To determine the location of each angle, use the fractional value of the angle and compare it to the values that correspond to the coordinate axes.
   a. For $2\pi/3$ rads, the fractional value is 2/3. The value of 2/3 = 4/6 is greater than 3/6 = 1/2 and less than 6/6 = 1, so $\pi/2 < 2\pi/3 < \pi$. **The angle lies between the positive y-axis and negative x-axis in Quadrant II.**
   b. For $-\pi/4$ rads, the fractional value is −1/4. The value of −1/4 is greater than −2/4 = −1/2 and less than 0, so $-\pi/2 < -\pi/4 < 0$. **The angle lies between the negative y-axis and positive x-axis in Quadrant IV.**
   c. For $5\pi/2$ rads, the fractional value is 5/2. This is an improper fraction, so first convert the angle to a proper fractional value.

      $5\pi/2 = 2\pi/2 + 2\pi/2 + 1\pi/2$

      $= 2\pi + \pi/2$

      2π is one complete revolution, so it does not affect the location of the angle and can be discarded. **The effective angle is then $\pi/2$, which lies along the positive y-axis.**
   d. For $7\pi/8$ rads, the fractional value is 7/8. The value of 7/8 is greater than 6/8 = 3/4 and less than 8/8 = 1, so $3\pi4 < 7\pi/8 < 1$. **The angle lies between the negative y-axis and positive x-axis in Quadrant IV.**

15.2.2 Complex Number Conversions

For Problems 1 and 2, refer to Figure 15-5.

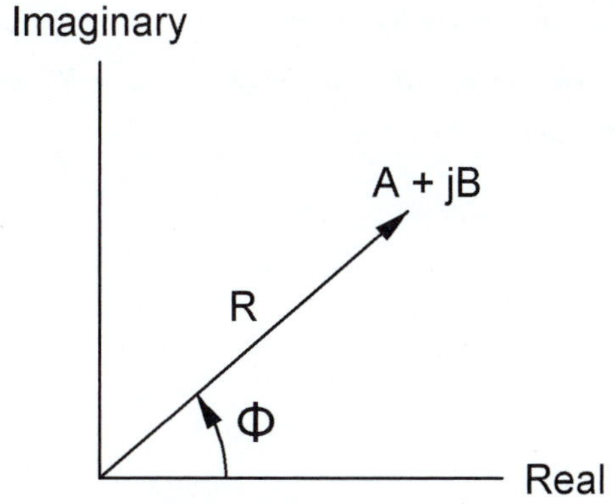

Figure 15-5: Complex Number Diagram for Problems 1 and 2

1. Problem: Verify the formulas $R = \sqrt{A^2 + B^2}$ and $\varphi = \tan^{-1}(B / A)$ for converting complex number from rectangular form $A + j B$ to polar form $R \angle \varphi$.

   Solution: Refer to Figure 15-6.

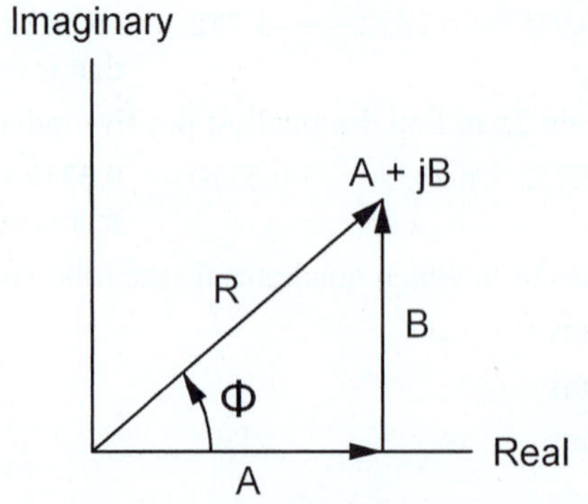

Figure 15-6: Analysis of Complex Number Conversions

As Figure 15-6 shows, the complex number $A + j B$ represents a magnitude A along the x-axis and magnitude B along the y-axis. A, B, and R form right triangle, so using Pythagorean Theorem

$$R^2 = A^2 + B^2$$

$$R = \sqrt{A^2 + B^2}$$

From basic trigonometry,

$$\tan(\varphi) = B / A$$

$$\tan^{-1}[\tan(\varphi)] = \tan^{-1}(B / A)$$

$$\varphi = \tan^{-1}(B / A)$$

2. Problem: Verify the polar-to-rectangular conversion formulas $A = R \cos \varphi$ and $B = R \sin \varphi$.

   Solution: Refer to Figure 15-6. From basic trigonometry

$$A = R \cos(\varphi)$$

and

$$B = R \sin(\varphi)$$

15.2.3 Complex Addition and Subtraction

1. Problem: Show that the magnitude and phase angle of the sum $(R_1 \angle \varphi_1) + (R_2 \angle \varphi_2)$ in polar form are

$$R_T = \sqrt{R_1^{\,2} + R_2^{\,2} + 2R_1R_2\cos(\varphi_1 - \varphi_2)}$$

$$\varphi_T = \tan^{-1}[(R_1 \sin \varphi_1 + R_2 \sin \varphi_2) / (R_1 \cos \varphi_1 + R_2 \cos \varphi_2)]$$

   Solution: Refer to Figure 15-7.

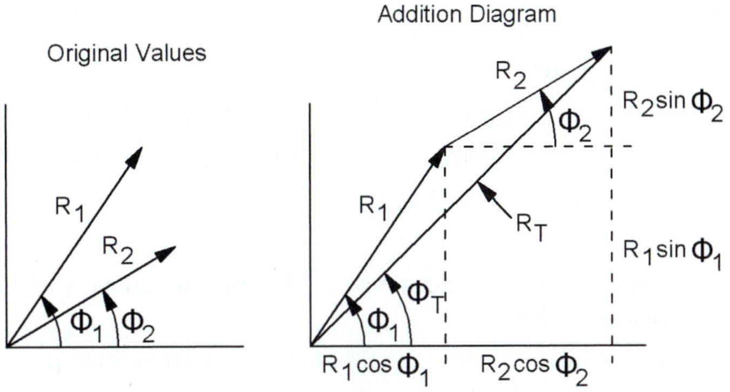

Figure 15-7: Analysis of Complex Sum

The diagram on the left shows the original complex values $\boldsymbol{R}_1 \angle \varphi_1$ and $\boldsymbol{R}_2 \angle \varphi_2$ and the diagram on the right graphically shows the addition of the two. From basic trigonometry, the real and imaginary components of $\boldsymbol{R}_1 \angle \varphi_1$ are $\boldsymbol{R}_1 \cos \varphi_1$ and $\boldsymbol{R}_1 \sin \varphi_1$, respectively, and the real and imaginary components of $\boldsymbol{R}_2 \angle \varphi_2$ are $\boldsymbol{R}_2 \cos \varphi_2$ and $\boldsymbol{R}_2 \sin \varphi_2$, respectively. As the diagram shows, the real component of $\boldsymbol{R}_T \angle \varphi_T$ is the sum of the real components, $\boldsymbol{R}_1 \cos \varphi_1 + \boldsymbol{R}_2 \cos \varphi_2$, and the imaginary component is the sum of the imaginary components, $\boldsymbol{R}_1 \sin \varphi_1 + \boldsymbol{R}_2 \sin \varphi_2$. From the Pythagorean Theorem,

$$\begin{aligned}\boldsymbol{R}_T^2 &= (\boldsymbol{R}_1 \cos \varphi_1 + \boldsymbol{R}_2 \cos \varphi_2)^2 + (\boldsymbol{R}_1 \sin \varphi_1 + \boldsymbol{R}_2 \sin \varphi_2)^2 \\ &= \boldsymbol{R}_1^2 \cos^2 \varphi_1 + 2\boldsymbol{R}_1\boldsymbol{R}_2 (\cos \varphi_1)(\cos \varphi_2) + \boldsymbol{R}_2^2 \cos^2 \varphi_2 + \\ &\quad \boldsymbol{R}_1^2 \sin^2 \varphi_1 + 2\boldsymbol{R}_1\boldsymbol{R}_2 (\sin \varphi_1)(\sin \varphi_2) + \boldsymbol{R}_2^2 \sin^2 \varphi_2 \\ &= \boldsymbol{R}_1^2 (\cos^2 \varphi_1 + \sin^2 \varphi_1) + \boldsymbol{R}_2^2 (\cos^2 \varphi_2 + \sin^2 \varphi_2) + \\ &\quad 2\boldsymbol{R}_1\boldsymbol{R}_2 [(\cos \varphi_1)(\cos \varphi_2) + (\sin \varphi_1)(\sin \varphi_2)]\end{aligned}$$

From standard trigonometric identities, $\cos^2 \boldsymbol{\theta} + \sin^2 \boldsymbol{\theta} = 1$ and $\cos \boldsymbol{\alpha} \cos \boldsymbol{\beta} + \sin \boldsymbol{\alpha} \sin \boldsymbol{\beta} = \cos (\boldsymbol{\alpha} - \boldsymbol{\beta})$, so the magnitude of $\boldsymbol{R}_T \angle \varphi_T$ is

$$\begin{aligned}\boldsymbol{R}_T^2 &= \boldsymbol{R}_1^2 + \boldsymbol{R}_2^2 + 2\boldsymbol{R}_1\boldsymbol{R}_2 [\cos (\varphi_1 - \varphi_2)] \\ \sqrt{\boldsymbol{R}_T^2} &= \sqrt{\boldsymbol{R}_1^2 + \boldsymbol{R}_2^2 + 2\boldsymbol{R}_1\boldsymbol{R}_2[\cos(\varphi_1 - \varphi_2)]} \\ \boldsymbol{R}_T &= \sqrt{\boldsymbol{R}_1^2 + \boldsymbol{R}_2^2 + 2\boldsymbol{R}_1\boldsymbol{R}_2[\cos(\varphi_1 - \varphi_2)]}\end{aligned}$$

Basic trigonometry also gives that the tangent of φ_T is equal to the ratio of the imaginary component divided by the real component of $\boldsymbol{R}_T \angle \varphi_T$. Therefore, the phase angle of $\boldsymbol{R}_T \angle \varphi_T$ is

$$\begin{aligned}\tan \varphi_T &= (\boldsymbol{R}_1 \sin \varphi_1 + \boldsymbol{R}_2 \sin \varphi_2) / (\boldsymbol{R}_1 \cos \varphi_1 + \boldsymbol{R}_2 \cos \varphi_2) \\ \tan^{-1} (\tan \varphi_T) &= \tan^{-1} [(\boldsymbol{R}_1 \sin \varphi_1 + \boldsymbol{R}_2 \sin \varphi_2) / (\boldsymbol{R}_1 \cos \varphi_1 + \boldsymbol{R}_2 \cos \varphi_2)] \\ \varphi_T &= \tan^{-1} [(\boldsymbol{R}_1 \sin \varphi_1 + \boldsymbol{R}_2 \sin \varphi_2) / (\boldsymbol{R}_1 \cos \varphi_1 + \boldsymbol{R}_2 \cos \varphi_2)]\end{aligned}$$

2. Problem: Use the formulas in Problem 1 to calculate the magnitude and phase angles for the following complex sums.

   a. $1.00 \angle 45.0° + 2.50 \angle 45.0°$
   b. $3.00 \angle 0.0° + 4.00 \angle 90.0°$
   c. $1.50 \angle 45.0° + 1.50 \angle -45.0°$
   d. $2.50 \angle 30.0° + 5.00 \angle 60.0°$

Solution: Using the formulas for magnitude and phase angle in Problem 1 for each problem gives:

a. $$\begin{aligned}\boldsymbol{R}_T^2 &= 1.00^2 + 2.50^2 + 2(1.00)(2.50) [\cos (45.0° - 45.0°)] \\ &= 1.00 + 6.25 + (5.00) (\cos 0.0°) \\ &= 1.00 + 6.25 + (5.00)(1) \\ &= 12.25\end{aligned}$$

$R_T = \sqrt{12.25}$

$= \mathbf{3.50}$

$\varphi_T = \tan^{-1}$ [(1.00 sin 45.0° + 2.50 sin 45.0°) / (1.00 cos 45.0° + 2.50 cos 45.0°)]

$= \tan^{-1}$ {[(1.00)(0.707) + (2.50)(0.707)] / [(1.00)(0.707) + (2.50)(0.707)]}

$= \tan^{-1}$ (2.475 / 2.475)

$= \tan^{-1}$ (1.00)

Because both the numerator and denominator are positive, the angle is in Quadrant I so that φ_T= **45.0°**.

b. $R_T^2 = 3.00^2 + 4.00^2 + 2(3.00)(4.00)$ [cos (0.0° – 90.0°)]

= 9.00 + 16.0 + (24.0) [cos (–90.0°)]

= 9.00 + 16.0 + (24.0)(0)

= 25.0

$R_T = \sqrt{25.0}$

$= \mathbf{5.00}$

$\varphi_T = \tan^{-1}$ [(3.00 sin 0.0° + 4.00 sin 90.0°) / (3.00 cos 0.0° + 4.00 cos 90.0°)]

$= \tan^{-1}$ {[(3.00)(0) + (4.00)(1.00)] / [(3.00)(1.00) + (4.00)(0)]}

$= \tan^{-1}$ (4.00 / 3.00)

$= \tan^{-1}$ (1.33)

Because both the numerator and denominator are positive, the angle is in Quadrant I so that φ_T= **53.1°**.

c. $R_T^2 = 1.50^2 + 1.50^2 + 2(1.50)(1.50)$ {cos [45.0° – (–45.0°)]}

= 2.25 + 2.25 + (4.50)[cos (90.0°)]

= 2.25 + 2.25 + (4.50)(0)

= 4.50

$R_T = \sqrt{4.50}$

$= \mathbf{2.12}$

$\varphi_T = \tan^{-1}$ {[1.50 sin 45.0° + 1.50 sin (–45.0°)] / [1.50 cos 45.0° + 1.50 cos (–45°)]}

$= \tan^{-1}$ {[(1.50)(0.707) + (1.50)(–0.707)] / [(1.50)(0.707) + (1.50)(0.707)]}

$= \tan^{-1}$ (0.00 / 2.12)

$= \tan^{-1}$ (0.00)

= **0.00°**

d. $R_T^2 = 2.50^2 + 5.00^2 + 2(2.50)(5.00)$ [cos (30.0° – 60.0°)]

= 6.25 + 25.0 + (25.0) [cos (–30.0°)]

= 6.25 + 25.0 + (25.0)(0.866)

= 52.9

$R_T = \sqrt{52.9}$

$= \mathbf{7.27}$

$\varphi_T = \tan^{-1}$ {[2.50 sin 30.0° + 5.00 sin (60.0°)] / [2.50 cos 30.0° + 5.00 cos (60°)]}

$= \tan^{-1}$ {[(2.50)(0.500) + (5.00)(0.866)] / [(2.50)(0.866) + (5.00)(0.500)]}

$= \tan^{-1}$ (5.58 / 4.67)

$= \tan^{-1}$ (1.20)

Because both the numerator and denominator are positive, the angle is in Quadrant I so that φ_T= **50.1°**.

3. Problem: Verify the results of Problem 2 by first converting the complex values to rectangular form, performing the addition operations, and then converting the answers to polar form.

Solution: Calculating the sums for Problem 2 by using rectangular forms gives:

a. $1.00 \angle 45.0° + 2.50 \angle 45.0° = (0.707 + j\,0.707) + (1.768 + j\,1.768)$
$= 2.475 + j\,2.475$
$= \mathbf{3.50 \angle 45.0°}$

The answer checks.

b. $3.00 \angle 0.0° + 4.00 \angle 90.0° = (3.00 + j\,0.0) + (0 + j\,4.00)$
$= 3.00 + j\,4.00$
$= \mathbf{5.00 \angle 53.1°}$

The answer checks.

c. $1.50 \angle 45.0° + 1.50 \angle -45.0° = (1.06 + j\,1.06) + [1.06 + j\,(-1.06)]$
$= 2.12 + j\,0.0$
$= \mathbf{2.12 \angle 0.0°}$

The answer checks.

d. $2.50 \angle 30.0° + 5.00 \angle 60.0° = (2.17 + j\,1.25) + (2.50 + j\,4.33)$
$= 4.67 + j\,5.58$
$= \mathbf{7.27 \angle 50.1°}$

The answer checks.

4. Problem: Show that the magnitude and phase of the difference $(\boldsymbol{R}_1 \angle \varphi_1) - (\boldsymbol{R}_2 \angle \varphi_2)$ in polar form are

$$\boldsymbol{R}_T = \sqrt{\boldsymbol{R}_1^{\,2} + \boldsymbol{R}_2^{\,2} - 2\boldsymbol{R}_1\boldsymbol{R}_2 \cos(\varphi_1 - \varphi_2)}$$

$$\varphi_T = \tan^{-1}\left[(\boldsymbol{R}_1 \sin\varphi_1 - \boldsymbol{R}_2 \sin\varphi_2) / (\boldsymbol{R}_1 \cos\varphi_1 - \boldsymbol{R}_2 \cos\varphi_2)\right]$$

Solution: Just as you can subtract a real number by adding its negative (e.g., $5 - 3 = 5 + -3$), you can subtract a complex number by adding its negative value (e.g., $\boldsymbol{R}_1 \angle \varphi_1 - [\boldsymbol{R}_2 \angle \varphi_2] = \boldsymbol{R}_1 \angle \varphi_1 + -[\boldsymbol{R}_2 \angle \varphi_2]$). A complex number's negative value has the same magnitude as the complex number, but points in the opposite direction. In other words, $-(\boldsymbol{R} \angle \varphi) = \boldsymbol{R} \angle (\varphi \pm 180°)$. Because the negative value points in the opposite direction to that of the original value, the real and imaginary components are negative values of the original value. This means that adding the negative value subtracts the real and imaginary components of the original number. Refer to Figure 15-8.

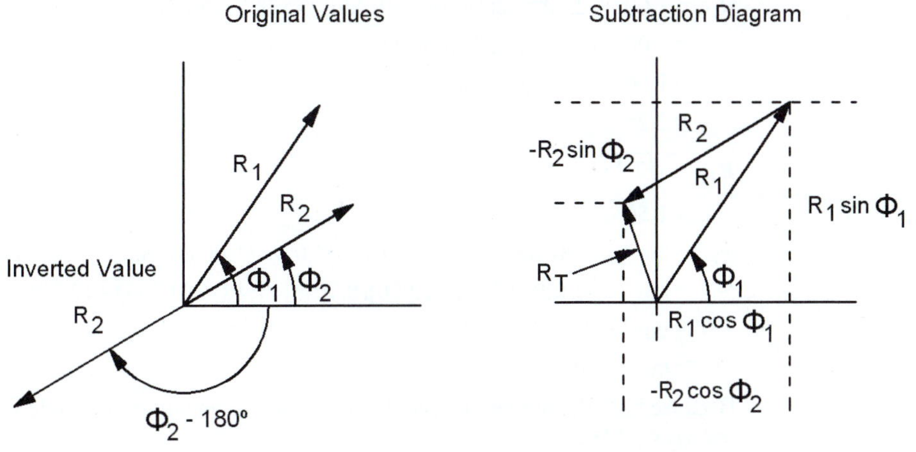

Figure 15-8: Analysis of Complex Difference

The diagram on the left shows the original complex values $\boldsymbol{R}_1 \angle \varphi_1$ and $\boldsymbol{R}_2 \angle \varphi_2$ and the negative value of $-(\boldsymbol{R}_2 \angle \varphi_2) = \boldsymbol{R}_2 \angle (\varphi_2 - 180°)$. The diagram on the right graphically shows the addition of $\boldsymbol{R}_1 \angle \varphi_1$ and $-(\boldsymbol{R}_2 \angle \varphi_2)$. From basic trigonometry, the real and imaginary components of $\boldsymbol{R}_1 \angle \varphi_1$ are $\boldsymbol{R}_1 \cos \varphi_1$ and $\boldsymbol{R}_1 \sin \varphi_1$, respectively, and the real and imaginary components of $-(\boldsymbol{R}_2 \angle \varphi_2)$ are $-\boldsymbol{R}_2 \cos \varphi_2$ and $-\boldsymbol{R}_2 \sin \varphi_2$, respectively. As the diagram shows, the real component of $\boldsymbol{R}_T \angle \varphi_T$ is the sum of the real components, $\boldsymbol{R}_1 \cos \varphi_1 + (-\boldsymbol{R}_2 \cos \varphi_2)$, and the imaginary component is the sum of the imaginary components, $\boldsymbol{R}_1 \sin \varphi_1 + (-\boldsymbol{R}_2 \sin \varphi_2)$. From Pythagorean Theorem

$$\begin{aligned}\boldsymbol{R}_T^2 &= [\boldsymbol{R}_1 \cos \varphi_1 + (-\boldsymbol{R}_2 \cos \varphi_2)]^2 + [\boldsymbol{R}_1 \sin \varphi_1 + (-\boldsymbol{R}_2 \sin \varphi_2)]^2\\ &= \boldsymbol{R}_1^2 \cos^2 \varphi_1 - 2\boldsymbol{R}_1\boldsymbol{R}_2 (\cos \varphi_1)(\cos \varphi_2) + \boldsymbol{R}_2^2 \cos^2 \varphi_2 +\\ &\qquad \boldsymbol{R}_1^2 \sin^2 \varphi_1 - 2\boldsymbol{R}_1\boldsymbol{R}_2 (\sin \varphi_1)(\sin \varphi_2) + \boldsymbol{R}_2^2 \sin^2 \varphi_2\\ &= \boldsymbol{R}_1^2 (\cos^2 \varphi_1 + \sin^2 \varphi_1) + \boldsymbol{R}_2^2 (\cos^2 \varphi_2 + \sin^2 \varphi_2) -\\ &\qquad 2\boldsymbol{R}_1\boldsymbol{R}_2 [(\cos \varphi_1)(\cos \varphi_2) + (\sin \varphi_1)(\sin \varphi_2)]\end{aligned}$$

From standard trigonometric identities, $\cos^2 \theta + \sin^2 \theta = 1$ and $\cos \alpha \cos \beta + \sin \alpha \sin \beta = \cos (\alpha - \beta)$, so the magnitude of $\boldsymbol{R}_T \angle \varphi_T$ is

$$\begin{aligned}\boldsymbol{R}_T^2 &= \boldsymbol{R}_1^2 + \boldsymbol{R}_2^2 - (2\boldsymbol{R}_1\boldsymbol{R}_2)[\cos (\varphi_1 - \varphi_2)]\\ \sqrt{\boldsymbol{R}_T^2} &= \sqrt{\boldsymbol{R}_1^2 + \boldsymbol{R}_2^2 - 2\boldsymbol{R}_1\boldsymbol{R}_2[\cos(\varphi_1 - \varphi_2)]}\\ \boldsymbol{R}_T &= \sqrt{\boldsymbol{R}_1^2 + \boldsymbol{R}_2^2 - 2\boldsymbol{R}_1\boldsymbol{R}_2[\cos(\varphi_1 - \varphi_2)]}\end{aligned}$$

Basic trigonometry also gives that the tangent of φ_T is equal to the ratio of the imaginary component divided by the real component of $\boldsymbol{R}_T \angle \varphi_T$. Therefore the phase angle of $\boldsymbol{R}_T \angle \varphi_T$ is

$$\begin{aligned}\tan \varphi_T &= [\boldsymbol{R}_1 \sin \varphi_1 + (-\boldsymbol{R}_2 \sin \varphi_2)] / [\boldsymbol{R}_1 \cos \varphi_1 + (-\boldsymbol{R}_2 \cos \varphi_2)]\\ \tan^{-1} (\tan \varphi_T) &= \tan^{-1} [(\boldsymbol{R}_1 \sin \varphi_1 - \boldsymbol{R}_2 \sin \varphi_2) / (\boldsymbol{R}_1 \cos \varphi_1 - \boldsymbol{R}_2 \cos \varphi_2)]\\ \varphi_T &= \tan^{-1} [(\boldsymbol{R}_1 \sin \varphi_1 - \boldsymbol{R}_2 \sin \varphi_2) / (\boldsymbol{R}_1 \cos \varphi_1 - \boldsymbol{R}_2 \cos \varphi_2)]\end{aligned}$$

5. Problem: Use the formulas in Problem 4 to calculate the magnitude and phase angles for the following complex subtraction operations.

a. $1.00 \angle 45.0° - 2.50 \angle 45.0°$

b. $3.00 \angle 0.0° - 4.00 \angle 90.0°$

c. $1.50 \angle 45.0° - 1.50 \angle -45.0°$

d. $2.50 \angle 30.0° - 5.00 \angle 60.0°$

Solution: Using the formulas for magnitude and phase angle in Problem 4 for each problem gives:

a. $$\begin{aligned}\boldsymbol{R}_T^2 &= 1.00^2 + 2.50^2 - 2(1.00)(2.50) [\cos (45.0° - 45.0°)]\\ &= 1.00 + 6.25 - (5.00) (\cos 0.0°)\\ &= 1.00 + 6.25 - (5.00)(1)\\ &= 2.25\end{aligned}$$

$$\begin{aligned}\boldsymbol{R}_T &= \sqrt{2.25}\\ &= \mathbf{1.50}\end{aligned}$$

$$\begin{aligned}\varphi_T &= \tan^{-1} [(1.00 \sin 45.0° - 2.50 \sin 45.0°) / (1.00 \cos 45.0° - 2.50 \cos 45.0°)]\\ &= \tan^{-1} \{[(1.00)(0.707) - (2.50)(0.707)] / [(1.00)(0.707) - (2.50)(0.707)]\}\\ &= \tan^{-1} (-1.06 / -1.06)\\ &= \tan^{-1} (1.00)\end{aligned}$$

Because both the numerator and denominator are negative, the angle is in Quadrant III so that $\varphi_T = \mathbf{-135.0°}$.

b. $R_T^2 = 3.00^2 + 4.00^2 - 2(3.00)(4.00)\,[\cos(0.0° - 90.0°)]$
$= 9.00 + 16.0 - (24.0)\,[\cos(-90.0°)]$
$= 9.00 + 16.0 - (24.0)(0)$
$= 25.0$

$R_T = \sqrt{25.0}$
$= \mathbf{5.00}$

$\varphi_T = \tan^{-1}[(3.00 \sin 0.0° - 4.00 \sin 90.0°) / (3.00 \cos 0.0° - 4.00 \cos 90.0°)]$
$= \tan^{-1}\{[(3.00)(0) - (4.00)(1.00)] / [(3.00)(1.00) - (4.00)(0)]\}$
$= \tan^{-1}(-4.00 / 3.00)$
$= \tan^{-1}(-1.33)$

Because the numerator is negative and denominator is positive, the angle is in Quadrant IV so that $\varphi_T = \mathbf{-53.1°}$.

c. $R_T^2 = 1.50^2 + 1.50^2 - 2(1.50)(1.50)\,\{\cos[45.0° - (-45.0°)]\}$
$= 2.25 + 2.25 - (4.50)[\cos(90.0°)]$
$= 2.25 + 2.25 - (4.50)(0)$
$= 4.50$

$R_T = \sqrt{4.50}$
$= \mathbf{2.12}$

$\varphi_T = \tan^{-1}\{[1.50 \sin 45.0° - 1.50 \sin(-45.0°)] / [1.50 \cos 45.0° - 1.50 \cos(-45°)]\}$
$= \tan^{-1}\{[(1.50)(0.707) - (1.50)(-0.707)] / [(1.50)(0.707) - (1.50)(0.707)]\}$
$= \tan^{-1}(2.12 / 0.00)$

Because (2.12 / 0.00) is positive but undefined, $\varphi_T = \mathbf{90.0°}$.

d. $R_T^2 = 2.50^2 + 5.00^2 - 2(2.50)(5.00)\,[\cos(30.0° - 60.0°)]$
$= 6.25 + 25.0 - (25.0)\,[\cos(-30.0°)]$
$= 6.25 + 25.0 - (25.0)(0.866)$
$= 9.60$

$R_T = \sqrt{9.60}$
$= \mathbf{3.10}$

$\varphi_T = \tan^{-1}\{[2.50 \sin 30.0° - 5.00 \sin(60.0°)] / [2.50 \cos 30.0° - 5.00 \cos(60°)]\}$
$= \tan^{-1}\{[(2.50)(0.500) - (5.00)(0.866)] / [(2.50)(0.866) - (5.00)(0.500)]\}$
$= \tan^{-1}(-3.08 / -0.335)$
$= \tan^{-1}(9.20)$

Because both the numerator and denominator are negative, the angle is in Quadrant IV so that $\varphi_T = \mathbf{-96.2°}$.

6. Problem: Verify the results of Problem 5 by first converting the complex values to rectangular form, performing the subtraction operations, and then converting the answers to polar form.

Solution: Calculating the differences for Problem 5 by using rectangular forms gives:

a. $1.00 \angle 45.0° - 2.50 \angle 45.0° = (0.707 + j\,0.707) - (1.768 + j\,1.768)$
$= -1.061 - j\,1.061$
$= \mathbf{1.5 \angle -135.0°}$

The answer checks.

b. $3.00 \angle 0.0° - 4.00 \angle 90.0° = (3.00 + j\,0.0) - (0 + j\,4.00)$
$= 3.00 - j\,4.00$

$= \mathbf{5.00 \angle -53.1°}$

The answer checks.

c. $1.50 \angle 45.0° - 1.50 \angle -45.0° = (1.06 + j\,1.06) - [1.06 + j\,(-1.06)]$
$= 0.00 + j\,2.12$
$= \mathbf{2.12 \angle 90.0°}$

The answer checks.

d. $2.50 \angle 30.0° - 5.00 \angle 60.0° = (2.17 + j\,1.25) - (2.50 + j\,4.33)$
$= -0.335 - j\,3.08$
$= \mathbf{3.10 \angle -96.2°}$

The answer checks.

15.2.4 Complex Multiplication and Division

1. Problem: Show that the real and imaginary components of the product $\boldsymbol{A_T} + j\,\boldsymbol{B_T} = (\boldsymbol{A}_1 + j\,\boldsymbol{B}_1)\,(\boldsymbol{A}_2 + j\,\boldsymbol{B}_2)$ in rectangular form are

$$\boldsymbol{A_T} = \boldsymbol{A}_1\boldsymbol{A}_2 - \boldsymbol{B}_1\boldsymbol{B}_2$$
$$\boldsymbol{B_T} = \boldsymbol{A}_1\boldsymbol{B}_2 + \boldsymbol{A}_2\boldsymbol{B}_1$$

Solution: Expanding the product $(\boldsymbol{A}_1 + j\,\boldsymbol{B}_1)\,(\boldsymbol{A}_2 + j\,\boldsymbol{B}_2)$ using the FOIL (First, Outer, Inner, Last) rule for expansion of terms gives

$$(\boldsymbol{A}_1 + j\,\boldsymbol{B}_1)\,(\boldsymbol{A}_2 + j\,\boldsymbol{B}_2) = \boldsymbol{A}_1\boldsymbol{A}_2 + j\,(\boldsymbol{A}_1\boldsymbol{B}_2) + j\,(\boldsymbol{B}_1\boldsymbol{A}_2) + j^2\,(\boldsymbol{B}_1\boldsymbol{B}_2)$$
$$= \boldsymbol{A}_1\boldsymbol{A}_2 + j^2\,(\boldsymbol{B}_1\boldsymbol{B}_2) + j\,([\boldsymbol{A}_1\boldsymbol{B}_2 + \boldsymbol{B}_1\boldsymbol{A}_2)$$

By definition $j = \sqrt{-1}$, so $j^2 = -1$. Substituting gives

$$(\boldsymbol{A}_1 + j\,\boldsymbol{B}_1)\,(\boldsymbol{A}_2 + j\,\boldsymbol{B}_2) = \boldsymbol{A}_1\boldsymbol{A}_2 + (-1)\,(\boldsymbol{B}_1\boldsymbol{B}_2) + j\,(\boldsymbol{A}_1\boldsymbol{B}_2 + \boldsymbol{B}_1\boldsymbol{A}_2)$$
$$= [\boldsymbol{A}_1\boldsymbol{A}_2 + (-\boldsymbol{B}_1\boldsymbol{B}_2)] + j\,(\boldsymbol{A}_1\boldsymbol{B}_2 + \boldsymbol{A}_2\boldsymbol{B}_1)$$
$$= (\boldsymbol{A}_1\boldsymbol{A}_2 - \boldsymbol{B}_1\boldsymbol{B}_2) + j\,(\boldsymbol{A}_1\boldsymbol{B}_2 + \boldsymbol{A}_2\boldsymbol{B}_1)$$

Therefore, $\boldsymbol{A_T} + j\,\boldsymbol{B_T} = (\boldsymbol{A}_1\boldsymbol{A}_2 - \boldsymbol{B}_1\boldsymbol{B}_2) + j\,(\boldsymbol{A}_1\boldsymbol{B}_2 + \boldsymbol{A}_2\boldsymbol{B}_1)$. Because the real and imaginary terms for equal complex numbers must be equal, $\boldsymbol{A_T} = \boldsymbol{A}_1\boldsymbol{A}_2 - \boldsymbol{B}_1\boldsymbol{B}_2$ and $\boldsymbol{B_T} = \boldsymbol{A}_1\boldsymbol{B}_2 + \boldsymbol{A}_2\boldsymbol{B}_1$.

2. Problem: Use the formulas in Problem 1 to calculate the real and imaginary components for the following complex multiplication problems:

a. $(3.00 + j\,4.00)\,(5.00 - j\,12.0)$
b. $(-2.00 - j\,2.00)\,(-1.00 - j\,1.00)$
c. $(3.00 + j\,2.25)\,(-6.00 + j\,2.50)$
d. $(1.00 - j\,2.00)\,(3.00 - j\,4.00)$

Solution: Using the formulas for the real and imaginary components in Problem 1 for each problem gives:

a. $\boldsymbol{A_T} = (3.00)(5.00) - (4.00)(-12.0)$
$= 15.0 + 48.0$
$= \mathbf{63.0}$

$\boldsymbol{B_T} = (3.00)(-12.0) + (5.00)(4.00)$
$= -36.0 + 20.0$
$= \mathbf{-16.0}$

b. $\boldsymbol{A_T} = (-2.00)(-1.00) - (-2.00)(-1.00)$
$= 2.00 - 2.00$
$= \mathbf{0.00}$

$\boldsymbol{B_T} = (-2.00)(-1.00) + (-2.00)(\,-1.00)$

$= 2.00 + 2.00$

$= \mathbf{4.00}$

c. $A_T = (3.00)(-6.00) - (2.25)(2.50)$

$= -18.0 - 5.63$

$= \mathbf{-23.6}$

$B_T = (3.00)(2.50) + (2.25)(-6.00)$

$= 7.50 - 13.5$

$= \mathbf{-6.00}$

d. $A_T = (1.00)(3.00) - (-2.00)(-4.00)$

$= 3.00 - 8.00$

$= \mathbf{-5.00}$

$B_T = (1.00)(-4.00) + (-2.00)(3.00)$

$= -4.00 - 6.00$

$= \mathbf{-10.0}$

3. Problem: Verify the results of Problem 2 by first converting the complex values to polar form, performing the multiplication operations, and then converting the answers to rectangular form.

Solution: Calculating the products for Problem 2 by using polar forms gives:

a. $(3.00 + j\,4.00)\,(5.00 - j\,12.0) = (5.00 \angle 53.1°)\,(13.0 \angle -67.4°)$

$= (5.00)\,(13.0) \angle [53.1° + (-67.4°)]$

$= 65.0 \angle -14.3°$

$= \mathbf{63.0 - j\,16.0}$

The answer checks.

b. $(-2.00 - j\,2.00)\,(-1.00 - j\,1.00) = (2.83 \angle -135°)\,(1.41 \angle -135°)$

$= (2.83)\,(1.41) \angle [-135° + (-135°)]$

$= 4.00 \angle -270° = 4.00 \angle 90°$

$= \mathbf{0.00 + j\,4.00}$

The answer checks.

c. $(3.00 + j\,2.25)\,(-6.00 + j\,2.50) = (3.75 \angle 36.9°)\,(6.50 \angle 157.4°)$

$= (3.75)\,(6.50) \angle (36.9° + 157.4°)$

$= 24.4 \angle 194° = 24.4 \angle -166°$

$= \mathbf{-23.6 - j\,6.00}$

The answer checks.

d. $(1.00 - j\,2.00)\,(3.00 - j\,4.00) = (2.24 \angle -63.4°)\,(5.00 \angle -53.1°)$

$= (2.24)\,(5.00) \angle [-63.4° + (-53.1°)]$

$= 11.2 \angle -117°$

$= \mathbf{-5.00 - j\,10.0}$

The answer checks.

4. Problem: Show that the real and imaginary components of the quotient $A_T + j\,B_T = (A_1 + j\,B_1)\,/\,(A_2 + j\,B_2)$ in rectangular form are

$$A_T = (A_1A_2 + B_1B_2)\,/\,(A_2^2 + B_2^2)$$

$$B_T = (A_2B_1 - A_1B_2)\,/\,(A_2^2 + B_2^2)$$

Solution: To find the real and imaginary terms of the quotient, first convert the denominator of the expression $(A_1 + j\,B_1)\,/\,(A_2 + j\,B_2)$ to a real value. To do so, multiply the numerator and denominator by the complex conjugate of the denominator. The complex conjugate of a complex value $\boldsymbol{A} + \text{j}\,\boldsymbol{B}$ is $\boldsymbol{A} - \text{j}\,\boldsymbol{B}$.

$$\begin{aligned}
\boldsymbol{A_T} + \text{j}\,\boldsymbol{B_T} &= [(\boldsymbol{A}_1 + \text{j}\,\boldsymbol{B}_1) / (\boldsymbol{A}_2 + \text{j}\,\boldsymbol{B}_2)]\,[(\boldsymbol{A}_2 - \text{j}\,\boldsymbol{B}_2) / (\boldsymbol{A}_2 - \text{j}\,\boldsymbol{B}_2)] \\
&= [(\boldsymbol{A}_1 + \text{j}\,\boldsymbol{B}_1)(\boldsymbol{A}_2 - \text{j}\,\boldsymbol{B}_2)] / [(\boldsymbol{A}_2 + \text{j}\,\boldsymbol{B}_2)(\boldsymbol{A}_2 - \text{j}\,\boldsymbol{B}_2)] \\
&= [\boldsymbol{A}_1\boldsymbol{A}_2 + \text{j}\,(\boldsymbol{B}_1\boldsymbol{A}_2) - \text{j}\,(\boldsymbol{B}_2\boldsymbol{A}_1) - \text{j}^2\,(\boldsymbol{B}_1\boldsymbol{B}_2)] / [(\boldsymbol{A}_2^2 + \text{j}\,(\boldsymbol{B}_2\boldsymbol{A}_2) - \text{j}\,(\boldsymbol{A}_2\boldsymbol{B}_2) - \text{j}^2\,(\boldsymbol{B}_2^2)] \\
&= [\boldsymbol{A}_1\boldsymbol{A}_2 - (-1)(\boldsymbol{B}_1\boldsymbol{B}_2) + \text{j}\,(\boldsymbol{B}_1\boldsymbol{A}_2 - \boldsymbol{B}_2\boldsymbol{A}_1)] / [(\boldsymbol{A}_2^2 + \text{j}\,(\boldsymbol{A}_2\boldsymbol{B}_2 - \boldsymbol{A}_2\boldsymbol{B}_2) - (-1)(\boldsymbol{B}_2^2)] \\
&= [\boldsymbol{A}_1\boldsymbol{A}_2 + \boldsymbol{B}_1\boldsymbol{B}_2 + \text{j}\,(\boldsymbol{A}_2\boldsymbol{B}_1 - \boldsymbol{A}_1\boldsymbol{B}_2)] / [(\boldsymbol{A}_2^2 + \boldsymbol{B}_2^2 + \text{j}\,(0)] \\
&= [\boldsymbol{A}_1\boldsymbol{A}_2 + \boldsymbol{B}_1\boldsymbol{B}_2 + \text{j}\,(\boldsymbol{A}_2\boldsymbol{B}_1 - \boldsymbol{A}_1\boldsymbol{B}_2)] / (\boldsymbol{A}_2^2 + \boldsymbol{B}_2^2) \\
&= (\boldsymbol{A}_1\boldsymbol{A}_2 + \boldsymbol{B}_1\boldsymbol{B}_2) / (\boldsymbol{A}_2^2 + \boldsymbol{B}_2^2) + \text{j}\,[(\boldsymbol{A}_2\boldsymbol{B}_1 - \boldsymbol{A}_1\boldsymbol{B}_2) / (\boldsymbol{A}_2^2 + \boldsymbol{B}_2^2)]
\end{aligned}$$

Therefore, $\boldsymbol{A_T} = (\boldsymbol{A}_1\boldsymbol{A}_2 + \boldsymbol{B}_1\boldsymbol{B}_2) / (\boldsymbol{A}_2^2 + \boldsymbol{B}_2^2)$ and $\boldsymbol{B_T} = (\boldsymbol{A}_2\boldsymbol{B}_1 - \boldsymbol{A}_1\boldsymbol{B}_2) / (\boldsymbol{A}_2^2 + \boldsymbol{B}_2^2)$.

5. Problem: Use the formulas in Problem 4 to calculate the real and imaginary components for the following complex division problems:

a. (3.00 + j 4.00) / (5.00 – j 12.0)
b. (–2.00 – j 2.00) / (–1.00 – j 1.00)
c. (3.00 + j 2.25) / (–6.00 + j 2.50)
d. (1.00 – j 2.00) / (3.00 – j 4.00)

Solution: Using the formulas for the real and imaginary components in Problem 4 for each problem gives:

a.
$$\begin{aligned}
\boldsymbol{A_T} &= [(3.00)(5.00) + (4.00)(-12.0)] / [(5.00)^2 + (-12.0)^2] \\
&= (15.0 - 48.0) / (25.0 + 144) \\
&= -33.0 / 169 \\
&= \mathbf{-0.195}
\end{aligned}$$
$$\begin{aligned}
\boldsymbol{B_T} &= [(5.00)(4.00) - (3.00)(\,-12.0)] / [(5.00)^2 + (-12.0)^2] \\
&= (20.0 + 36.0) / (25.0 + 144) \\
&= 56.0 / 169 \\
&= \mathbf{0.331}
\end{aligned}$$

b.
$$\begin{aligned}
\boldsymbol{A_T} &= [(-2.00)(-1.00) + (-2.00)(-1.00)] / [(-1.00)^2 + (-1.00)^2] \\
&= (2.00 + 2.00) / (1.00 + 1.00) \\
&= 4.00 / 2.00 \\
&= \mathbf{2.00}
\end{aligned}$$
$$\begin{aligned}
\boldsymbol{B_T} &= [(-2.00)(-1.00) - (-2.00)(-1.00)] / [(5.00)^2 + (-12.0)^2] \\
&= (2.00 - 2.00) / (25.0 + 144) \\
&= 0.00 / 169 \\
&= \mathbf{0.00}
\end{aligned}$$

c.
$$\begin{aligned}
\boldsymbol{A_T} &= [(3.00)(-6.00) + (2.25)(2.50)] / [(-6.00)^2 + (2.50)^2] \\
&= (-18.0 + 5.63) / (36.0 + 6.25) \\
&= -12.4 / 42.3 \\
&= \mathbf{-0.293}
\end{aligned}$$
$$\begin{aligned}
\boldsymbol{B_T} &= [(-6.00)(2.25) - (3.00)(2.50)] / [(-6.00)^2 + (2.50)^2] \\
&= (-13.5 - 7.50) / (36.0 + 6.25) \\
&= -21.0 / 42.3 \\
&= \mathbf{-0.497}
\end{aligned}$$

d.
$$\begin{aligned}
\boldsymbol{A_T} &= [(1.00)(\,3.00) + (-2.00)(-4.00)] / [(-3.00)^2 + (-4.00)^2] \\
&= (3.00 + 8.00) / (9.00 + 16.0)
\end{aligned}$$

$= 11.0 / 25.0$

$= \mathbf{0.440}$

$\boldsymbol{B_T} = [(3.00)(-2.00) - (1.00)(-4.00)] / [(-3.00)^2 + (-4.00)^2]$

$= (-6.00 + 4.00) / (9.00 + 16.0)$

$= -2.00 / 25.0$

$= \mathbf{-0.080}$

6. Problem: Verify the results of Problem 5 by first converting the complex values to polar form, performing the division operations, and then converting the answers to rectangular form.

Solution: Calculating the quotients for Problem 5 by using polar forms gives:

a. $(3.00 + j\,4.00) / (5.00 - j\,12.0) = (5.00 \angle 53.1°) / (13.0 \angle -67.4°)$

$= (5.00) / (13.0) \angle [53.1° - (-67.4°)]$

$= 0.385 \angle 121°$

$= \mathbf{-0.195 + j\,0.331}$

The answer checks.

b. $(-2.00 - j\,2.00) / (-1.00 - j\,1.00) = (2.83 \angle -135°) / (1.41 \angle -135°)$

$= (2.83) / (1.41) \angle [-135° - (-135°)]$

$= 2.00 \angle 0.00°$

$= \mathbf{2.00 + j\,0.00}$

The answer checks.

c. $(3.00 + j\,2.25) / (-6.00 + j\,2.50) = (3.75 \angle 36.9°) / (6.50 \angle 157.4°)$

$= (3.75) / (6.50) \angle (36.9° - 157.4°)$

$= 0.577 \angle -121°$

$= \mathbf{-0.293 - j\,0.497}$

The answer checks.

d. $(1.00 - j\,2.00) / (3.00 - j\,4.00) = (2.24 \angle -63.4°) / (5.00 \angle -53.1°)$

$= (2.24) / (5.00) \angle [-63.4° - (-53.1°)]$

$= 0.447 \angle -10.3°$

$= \mathbf{0.440 - j\,0.080}$

The answer checks.

15.3 Just for Fun

In ac electronics, circuits contain both resistance, for which the phase between voltage and current is 0°, and a quantity known as reactance, for which the phase between voltage and current is 90°. Sometimes a circuit design requires that the response be purely resistive (real) or purely reactive (imaginary). Knowing how to find values that will produce a purely real or imaginary result can be useful.

1. Problem: To what value $\boldsymbol{R_2} \angle \varphi_2$ must the complex value $\boldsymbol{R_1} \angle \varphi_1$ be added so that the result $\boldsymbol{R_3} \angle \varphi_3$ is a purely real value (i.e., $\varphi_3 = 0°$)? Test your answer for $\boldsymbol{R_1} \angle \varphi_1 = 1.00 \angle 45.0°$ and $\boldsymbol{R_3} \angle \varphi_3 = 1.00 \angle 0.00°$.

Solution: For the answer to be a purely real value,

$$\boldsymbol{R_1} \angle \varphi_1 + \boldsymbol{R_2} \angle \varphi_2 = \boldsymbol{R_3} \angle 0°$$

Convert the equation to rectangular form.

$$(\boldsymbol{R_1} \cos \varphi_1 + j\,\boldsymbol{R_1} \sin \varphi_1) + (\boldsymbol{R_2} \cos \varphi_2 + j\,\boldsymbol{R_2} \sin \varphi_2) = (\boldsymbol{R_3} \cos 0° + j\,\boldsymbol{R_3} \sin 0°)$$

$$= \boldsymbol{R_3} + j\,0°$$

Determine the real component of $\boldsymbol{R_2} \angle \varphi_2$.

$$\begin{aligned}
\mathbf{R}_1 \cos \varphi_1 + \mathbf{R}_2 \cos \varphi_2 &= \mathbf{R}_3 \\
\mathbf{R}_1 \cos \varphi_1 + \mathbf{R}_2 \cos \varphi_2 + -(\mathbf{R}_1 \cos \varphi_1) &= \mathbf{R}_3 + -(\mathbf{R}_1 \cos \varphi_1) \\
\mathbf{R}_2 \cos \varphi_2 &= \mathbf{R}_3 + -(\mathbf{R}_1 \cos \varphi_1)
\end{aligned}$$

Determine the imaginary component of $\mathbf{R}_2 \angle \varphi_2$.

$$\begin{aligned}
\mathbf{R}_1 \sin \varphi_1 + \mathbf{R}_2 \sin \varphi_2 &= 0.00 \\
\mathbf{R}_1 \sin \varphi_1 + \mathbf{R}_2 \sin \varphi_2 + -(\mathbf{R}_1 \sin \varphi_1) &= 0.00 + -(\mathbf{R}_1 \sin \varphi_1) \\
\mathbf{R}_2 \sin \varphi_2 &= -\mathbf{R}_1 \sin \varphi_1
\end{aligned}$$

Use these results to calculate $\mathbf{R}_2$.

$$\begin{aligned}
(\mathbf{R}_2 \cos \varphi_2)^2 + (\mathbf{R}_2 \sin \varphi_2)^2 &= [\mathbf{R}_3 + -(\mathbf{R}_1 \cos \varphi_1)]^2 + (-\mathbf{R}_1 \sin \varphi_1)^2 \\
\mathbf{R}_2^2 \cos^2 \varphi_2 + \mathbf{R}_2^2 \sin^2 \varphi_2 &= \mathbf{R}_3^2 - 2\mathbf{R}_1\mathbf{R}_3 \cos \varphi_1 + \mathbf{R}_1^2 \cos^2 \varphi_1 + \mathbf{R}_1^2 \sin^2 \varphi_1 \\
(\mathbf{R}_2^2)(\cos^2 \varphi_2 + \sin^2 \varphi_2) &= (\mathbf{R}_3^2 - 2\mathbf{R}_1\mathbf{R}_3 \cos \varphi_1 + \mathbf{R}_1^2)(\cos^2 \varphi_1 + \sin^2 \varphi_1) \\
(\mathbf{R}_2^2)(1) &= (\mathbf{R}_3^2 - 2\mathbf{R}_1\mathbf{R}_3 \cos \varphi_1 + \mathbf{R}_1^2)(1) \\
\mathbf{R}_2^2 &= \mathbf{R}_1^2 + \mathbf{R}_3^2 - 2\mathbf{R}_1\mathbf{R}_3 \cos \varphi_1 \\
\sqrt{\mathbf{R}_2^2} &= \sqrt{\mathbf{R}_1^2 + \mathbf{R}_3^2 - 2\mathbf{R}_1\mathbf{R}_2 \cos\varphi_1} \\
\mathbf{R}_2 &= \sqrt{\mathbf{R}_1^2 + \mathbf{R}_3^2 - 2\mathbf{R}_1\mathbf{R}_2 \cos\varphi_1}
\end{aligned}$$

Next, calculate φ_2.

$$\begin{aligned}
(\mathbf{R}_2 \sin \varphi_2) / (\mathbf{R}_2 \cos \varphi_2) &= -(\mathbf{R}_1 \sin \varphi_1) / [\mathbf{R}_3 + -(\mathbf{R}_1 \cos \varphi_1)] \\
(\sin \varphi_2) / (\cos \varphi_2) &= (\mathbf{R}_1 \sin \varphi_1) / [(\mathbf{R}_1 \cos \varphi_1) - \mathbf{R}_3] \\
\tan \varphi_2 &= (\mathbf{R}_1 \sin \varphi_1) / [(\mathbf{R}_1 \cos \varphi_1) - \mathbf{R}_3] \\
\tan^{-1} (\tan \varphi_2) &= \tan^{-1} \{(\mathbf{R}_1 \sin \varphi_1) / [(\mathbf{R}_1 \cos \varphi_1) - \mathbf{R}_3]\} \\
\varphi_2 &= \tan^{-1} \{(\mathbf{R}_1 \sin \varphi_1) / [(\mathbf{R}_1 \cos \varphi_1) - \mathbf{R}_3]\}
\end{aligned}$$

For $\mathbf{R}_1 \angle \varphi_1 = 1.00 \angle 45.0°$ and $\mathbf{R}_3 \angle \varphi_3 = 1.00 \angle 0.00°$,

$$\begin{aligned}
\mathbf{R}_2 &= \sqrt{(1.00)^2 + (1.00)^2 - 2(1.00)(1.00)\cos 45.0°} \\
&= \sqrt{1.00 + 1.00 - (2.00)(0.707)} \\
&= \sqrt{0.586} \\
&= \mathbf{0.765} \\
\varphi_2 &= \tan^{-1} \{(1.00 \sin 45°) / [(1.00 \cos 45°) - 1.00]\} \\
&= \tan^{-1} \{[(1.00)(0.707)] / [(1.00)(0.707) - 1.00]\} \\
&= \tan^{-1} [(0.707) / (-0.293)] \\
&= \tan^{-1} (-2.41) \\
&= \mathbf{-67.5°}
\end{aligned}$$

To verify this,

$$\begin{aligned}
1.00 \angle 45° + 0.756 \angle -67.5° &= (0.707 + \mathrm{j}\,0.707) + (0.293 - \mathrm{j}\,0.707) \\
&= 1.00 + \mathrm{j}\,0.00 \\
&= 1.00 \angle 0.00°
\end{aligned}$$

The answer checks.

2. Problem: To what value $\mathbf{R}_2 \angle \varphi_2$ must the complex value $\mathbf{R}_1 \angle \varphi_1$ be added so that the result $\mathbf{R}_3 \angle \varphi_3$ is a purely imaginary positive value (i.e., $\varphi_3 = 90°$)? Test your answer for $\mathbf{R}_1 \angle \varphi_1 = 1.00 \angle 45.0°$ and $\mathbf{R}_3 \angle \varphi_3 = 1.00 \angle 90.0°$.

Solution: For the answer to be a purely imaginary value,

$$\boldsymbol{R}_1 \angle \varphi_1 + \boldsymbol{R}_2 \angle \varphi_2 = \boldsymbol{R}_3 \angle 90.0°$$

Convert the equation to rectangular form.

$$(\boldsymbol{R}_1 \cos \varphi_1 + \text{j}\, \boldsymbol{R}_1 \sin \varphi_1) + (\boldsymbol{R}_2 \cos \varphi_2 + \text{j}\, \boldsymbol{R}_2 \sin \varphi_2) = (\boldsymbol{R}_3 \cos 90.0° + \text{j}\, \boldsymbol{R}_3 \sin 90.0°)$$
$$= 0.00 + \text{j}\, \boldsymbol{R}_3$$

Determine the real component of $\boldsymbol{R}_2 \angle \varphi_2$.

$$\boldsymbol{R}_1 \cos \varphi_1 + \boldsymbol{R}_2 \cos \varphi_2 = 0.00$$
$$\boldsymbol{R}_1 \cos \varphi_1 + \boldsymbol{R}_2 \cos \varphi_2 + -(\boldsymbol{R}_1 \cos \varphi_1) = 0.00 + -(\boldsymbol{R}_1 \cos \varphi_1)$$
$$\boldsymbol{R}_2 \cos \varphi_2 = -\boldsymbol{R}_1 \cos \varphi_1$$

Determine the imaginary component of $\boldsymbol{R}_2 \angle \varphi_2$.

$$\boldsymbol{R}_1 \sin \varphi_1 + \boldsymbol{R}_2 \sin \varphi_2 = \boldsymbol{R}_3$$
$$\boldsymbol{R}_1 \sin \varphi_1 + \boldsymbol{R}_2 \sin \varphi_2 + -(\boldsymbol{R}_1 \sin \varphi_1) = \boldsymbol{R}_3 + -(\boldsymbol{R}_1 \sin \varphi_1)$$
$$\boldsymbol{R}_2 \sin \varphi_2 = \boldsymbol{R}_3 + -(\boldsymbol{R}_1 \sin \varphi_1)$$

Use these results to calculate $\boldsymbol{R}_2$.

$$(\boldsymbol{R}_2 \cos \varphi_2)^2 + (\boldsymbol{R}_2 \sin \varphi_2)^2 = [(-\boldsymbol{R}_1 \cos \varphi_1)^2 + [\boldsymbol{R}_3 + -(\boldsymbol{R}_1 \sin \varphi_1)]^2$$
$$\boldsymbol{R}_2^2 \cos^2 \varphi_2 + \boldsymbol{R}_2^2 \sin^2 \varphi_2 = \boldsymbol{R}_1^2 \cos^2 \varphi_1 + \boldsymbol{R}_3^2 - 2\, \boldsymbol{R}_1\boldsymbol{R}_3 \sin \varphi_1 + \boldsymbol{R}_1^2 \sin^2 \varphi_1$$
$$\boldsymbol{R}_2^2 (\cos^2 \varphi_2 + \sin^2 \varphi_2) = \boldsymbol{R}_1^2 (\cos^2 \varphi_1 + \sin^2 \varphi_1) + \boldsymbol{R}_3^2 - 2\, \boldsymbol{R}_1\boldsymbol{R}_3 \sin \varphi_1$$
$$\boldsymbol{R}_2^2 (1) = \boldsymbol{R}_1^2 (1) + \boldsymbol{R}_3^2 - 2\, \boldsymbol{R}_1\boldsymbol{R}_3 \sin \varphi_1$$
$$\boldsymbol{R}_2^2 = \boldsymbol{R}_1^2 + \boldsymbol{R}_3^2 - 2\, \boldsymbol{R}_1\boldsymbol{R}_3 \sin \varphi_1$$
$$\sqrt{\boldsymbol{R}_2^2} = \sqrt{\boldsymbol{R}_1^2 + \boldsymbol{R}_3^2 - 2\boldsymbol{R}_1\boldsymbol{R}_2 \sin \varphi_1}$$
$$\boldsymbol{R}_2 = \sqrt{\boldsymbol{R}_1^2 + \boldsymbol{R}_3^2 - 2\boldsymbol{R}_1\boldsymbol{R}_2 \sin \varphi_1}$$

Next, calculate φ_2.

$$(\boldsymbol{R}_2 \sin \varphi_2) / (\boldsymbol{R}_2 \cos \varphi_2) = [\boldsymbol{R}_3 + -(\boldsymbol{R}_1 \sin \varphi_1)] / (-\boldsymbol{R}_1 \cos \varphi_1)$$
$$(\sin \varphi_2) / (\cos \varphi_2) = [\boldsymbol{R}_3 + -(\boldsymbol{R}_1 \sin \varphi_1)] / (-\boldsymbol{R}_1 \cos \varphi_1)$$
$$\tan \varphi_2 = [\boldsymbol{R}_3 + -(\boldsymbol{R}_1 \sin \varphi_1)] / (-\boldsymbol{R}_1 \cos \varphi_1)$$
$$\tan^{-1} (\tan \varphi_2) = \tan^{-1} \{[\boldsymbol{R}_3 + -(\boldsymbol{R}_1 \sin \varphi_1)] / (-\boldsymbol{R}_1 \cos \varphi_1)\}$$
$$\varphi_2 = \tan^{-1} \{[\boldsymbol{R}_3 + -(\boldsymbol{R}_1 \sin \varphi_1)] / (-\boldsymbol{R}_1 \cos \varphi_1)\}$$

For $\boldsymbol{R}_1 \angle \varphi_1 = 1.00 \angle 45.0°$ and $\boldsymbol{R}_3 \angle \varphi_3 = 1.00 \angle 90.0°$,

$$\boldsymbol{R}_2 = \sqrt{(1.00)^2 + (1.00)^2 - 2(1.00)(1.00)\sin 45.0°}$$
$$= \sqrt{1.00 + 1.00 - (2.00)(0.707)}$$
$$= \sqrt{0.586}$$
$$= \mathbf{0.765}$$

$$\varphi_2 = \tan^{-1} \{1.00 + -(1.00 \cos 45°)] / (-1.00 \cos 45°)\}$$
$$= \tan^{-1} \{[1.00 - (1.00)(0.707)] / (-0.707)\}$$
$$= \tan^{-1} [(0.293) / (-0.707)]$$
$$= \tan^{-1} (-0.414)$$
$$= \mathbf{-22.5°} \text{ or } \mathbf{157.5°}$$

Because both the real and imaginary parts of $\boldsymbol{R}_1 \angle \varphi_1$ are positive, the real part of $\boldsymbol{R}_2 \angle \varphi_2$ must be negative and the imaginary part must be positive so that the real part of $\boldsymbol{R}_3 \angle \varphi_3$ is zero and the imaginary part is positive. Therefore φ_2 must lie in Quadrant II.

$$\varphi_2 = \mathbf{157.5°}$$

To verify this,

$$1.00 \angle 45° + 0.765 \angle 157.5° = (0.707 + j\,0.707) + (-0.707 + j\,0.293)$$
$$= 0.00 + j\,1.00$$
$$= 1.00 \angle 90.0°$$

The answer checks.

3. Problem: To what value $A_2 + j\,B_2$ must the complex value $A_1 + j\,B_1$ be multiplied so that the result $A_3 + j\,B_3$ is a purely real value (i.e., $B_3 = 0$)? Test your answer for $A_1 + j\,B_1 = 1.00 + j\,1.00$ and $A_3 + j\,B_3 = 1.00 + j\,0.00$.

Solution: For the answer to be a purely real value, $B_3 = 0$.

$$(A_1 + j\,B_1)(A_2 + j\,B_2) = A_3 + j\,0.00$$

Use the FOIL rule of expansion.

$$A_1A_2 + j\,A_1B_2 + j\,B_1A_2 + j^2\,B_1B_2 = A_3 + j\,0.00$$
$$A_1A_2 + j\,A_1B_2 + j\,A_2B_1 + (-1)\,B_1B_2 = A_3 + j\,0.00$$
$$(A_1A_2 - B_1B_2) + j\,(A_1B_2 + A_2B_1) = A_3 + j\,0.00$$

From this,

$$A_1A_2 - B_1B_2 = A_3$$
$$A_1B_2 + A_2B_1 = 0.00$$

This system of two equations in two unknowns can be solved using standard techniques.

$$(A_1A_2 - B_1B_2) / A_1 = A_3 / A_1$$
$$(A_1B_2 + A_2B_1) / B_1 = 0.00 / B_1$$

$$A_2 - [(B_1B_2) / A_1] = A_3 / A_1$$
$$A_2 + [(A_1B_2) / B_1] = 0.00$$

$$A_2 - [(B_1 / A_1)\,B_2] = A_3 / A_1$$
$$A_2 + [(A_1 / B_1)\,B_2] = 0.00$$

Subtract the bottom equation from the top equation.

$$\{A_2 - [(B_1 / A_1)\,B_2]\} - \{A_2 + [(A_1 / B_1)\,B_2]\} = (A_3 / A_1) - 0.00$$
$$[-(B_1 / A_1)\,B_2] - [(A_1 / B_1)\,B_2] = A_3 / A_1$$
$$[-(B_1 / A_1) + -(A_1 / B_1)]\,B_2 = A_3 / A_1$$
$$[-(B_1 / A_1)(B_1 / B_1) + -(A_1 / B_1)(A_1 / A_1)]\,B_2 = A_3 / A_1$$
$$\{[-B_1^2 / (A_1B_1)] + [-A_1^2 / (A_1B_1)]\}\,B_2 = A_3 / A_1$$
$$[(-B_1^2 + -A_1^2) / (A_1B_1)]\,B_2 = A_3 / A_1$$
$$\{[(-B_1^2 + -A_1^2) / (A_1B_1)]\,B_2\}\,[(A_1B_1) / (-B_1^2 + -A_1^2)] = (A_3 / A_1)\,[(A_1B_1) / (-B_1^2 + -A_1^2)]$$
$$B_2 = (A_3B_1) / (-B_1^2 + -A_1^2)]$$
$$= -(B_1A_3) / (A_1^2 + B_1^2)$$

Substitute this result for B_2 into $A_1B_2 + A_2B_1 = 0.00$ to solve for A_2.

$$B_1A_2 + A_1[-(B_1A_3) / (A_1^2 + B_1^2)] = 0.00$$
$$B_1A_2 + [-(A_1B_1A_3) / (A_1^2 + B_1^2)] = 0.00$$
$$B_1A_2 + [-(A_1B_1A_3 / (A_1^2 + B_1^2)] + (A_1B_1A_3) / (A_1^2 + B_1^2) = 0.00 + (A_1B_1A_3) / (A_1^2 + B_1^2)$$
$$(B_1A_2) / B_1 = [(A_1B_1A_3) / (A_1^2 + B_1^2)] / B_1$$
$$A_2 = (A_1A_3) / (A_1^2 + B_1^2)$$

For $A_1 + j\,B_1 = 1.00 + j\,1.00$ and $A_3 + j\,B_3 = 1.00 - j\,0.00$,

$$\begin{aligned}\boldsymbol{A}_2 &= (1.00)(1.00) / (1.00^2 + 1.00^2)\\ &= 1.00 / 2.00\\ &= \mathbf{0.500}\\ \boldsymbol{B}_2 &= -(1.00)(1.00) / (1.00^2 + 1.00^2)\\ &= -1.00 / 2.00\\ &= \mathbf{-0.500}\end{aligned}$$

To verify this,

$$\begin{aligned}(1.00 + j\,1.00)\,(0.500 - j\,0.500) &= (1.00)(0.500) + j\,(1.00)(0.500) + j\,(1.00)(-0.500)\\ &\quad + j^2\,(1.00)(-0.500)\\ &= 0.500 + j\,(0.500 - 0.500) + (-1)(-0.500)\\ &= 0.500 + 0.500 + j\,0.00\\ &= 1.00 + j\,0.00\end{aligned}$$

The answer checks.

4. Problem: By what value $\boldsymbol{A}_2 + j\,\boldsymbol{B}_2$ must the complex value $\boldsymbol{A}_1 + j\,\boldsymbol{B}_2$ be multiplied so that the result $\boldsymbol{A}_3 + j\,\boldsymbol{B}_3$ is a purely imaginary value (i.e., $\boldsymbol{A}_3 = 0$)? Test your answer for $\boldsymbol{A}_1 + j\,\boldsymbol{B}_1 = 1.00 + j\,1.00$ and $\boldsymbol{A}_3 + j\,\boldsymbol{B}_3 = 0.00 - j\,1.00$.

Solution: For the answer to be a purely imaginary value

$$(\boldsymbol{A}_1 + j\,\boldsymbol{B}_1) / (\boldsymbol{A}_2 + j\,\boldsymbol{B}_2) = 0.00 + j\,\boldsymbol{A}_3$$

Using the algebraic FOIL (First, Outer, Inner, Last) rule of expansion,

$$\begin{aligned}\boldsymbol{A}_1\boldsymbol{A}_2 + j\,\boldsymbol{A}_1\boldsymbol{B}_2 + j\,\boldsymbol{B}_1\boldsymbol{A}_2 + j^2\,\boldsymbol{B}_1\boldsymbol{B}_2 &= 0.00 + j\,\boldsymbol{B}_3\\ \boldsymbol{A}_1\boldsymbol{A}_2 + j\,\boldsymbol{A}_1\boldsymbol{B}_2 + j\,\boldsymbol{B}_1\boldsymbol{A}_2 + (-1)\boldsymbol{B}_1\boldsymbol{B}_2 &= 0.00 + j\,\boldsymbol{B}_3\\ (\boldsymbol{A}_1\boldsymbol{A}_2 - \boldsymbol{B}_1\boldsymbol{B}_2) + j\,(\boldsymbol{A}_1\boldsymbol{B}_2 + \boldsymbol{A}_2\boldsymbol{B}_1) &= 0.00 + j\,\boldsymbol{B}_3\end{aligned}$$

From this,

$$\begin{aligned}\boldsymbol{A}_1\boldsymbol{A}_2 - \boldsymbol{B}_1\boldsymbol{B}_2 &= 0.00\\ \boldsymbol{A}_1\boldsymbol{B}_2 + \boldsymbol{A}_2\boldsymbol{B}_1 &= \boldsymbol{B}_3\end{aligned}$$

Use standard techniques to solve this system of two equations in two unknowns.

$$\begin{aligned}(\boldsymbol{A}_1\boldsymbol{A}_2 - \boldsymbol{B}_1\boldsymbol{B}_2) / \boldsymbol{A}_1 &= 0.00 / \boldsymbol{A}_1\\ (\boldsymbol{A}_1\boldsymbol{B}_2 + \boldsymbol{A}_2\boldsymbol{B}_1) / \boldsymbol{B}_1 &= \boldsymbol{B}_3 / \boldsymbol{B}_1\end{aligned}$$

$$\begin{aligned}\boldsymbol{A}_2 - (\boldsymbol{B}_1\boldsymbol{B}_2) / \boldsymbol{A}_1 &= 0.00\\ \boldsymbol{A}_2 + (\boldsymbol{A}_1\boldsymbol{B}_2) / \boldsymbol{B}_1 &= \boldsymbol{B}_3 / \boldsymbol{B}_1\end{aligned}$$

$$\begin{aligned}\boldsymbol{A}_2 - [(\boldsymbol{B}_1 / \boldsymbol{A}_1)\,\boldsymbol{B}_2] &= 0.00\\ \boldsymbol{A}_2 + [(\boldsymbol{A}_1 / \boldsymbol{B}_1)\,\boldsymbol{B}_2] &= \boldsymbol{B}_3 / \boldsymbol{B}_1\end{aligned}$$

Subtract the bottom equation from the top equation.

$$\begin{aligned}\{\boldsymbol{A}_2 - [(\boldsymbol{B}_1 / \boldsymbol{A}_1)\,\boldsymbol{B}_2]\} - \{\boldsymbol{A}_2 + [(\boldsymbol{A}_1 / \boldsymbol{B}_1)\,\boldsymbol{B}_2]\} &= 0.00 - (\boldsymbol{B}_3 / \boldsymbol{B}_1)\\ [(\boldsymbol{B}_1 / \boldsymbol{A}_1)\,\boldsymbol{B}_2] - [(\boldsymbol{A}_1 / \boldsymbol{B}_1)\,\boldsymbol{B}_2] &= -(\boldsymbol{B}_3 / \boldsymbol{B}_1)\\ [-(\boldsymbol{B}_1 / \boldsymbol{A}_1) + -(\boldsymbol{A}_1 / \boldsymbol{B}_1)]\,\boldsymbol{B}_2 &= -(\boldsymbol{B}_3 / \boldsymbol{B}_1)\\ [-(\boldsymbol{B}_1 / \boldsymbol{A}_1)\,(\boldsymbol{B}_1 / \boldsymbol{B}_1) + -(\boldsymbol{A}_1 / \boldsymbol{B}_1)\,(\boldsymbol{A}_1 / \boldsymbol{A}_1)]\,\boldsymbol{B}_2 &= -(\boldsymbol{B}_3 / \boldsymbol{B}_1)\\ \{[-\boldsymbol{B}_1^2 / (\boldsymbol{A}_1\boldsymbol{B}_1)] + [-\boldsymbol{A}_1^2 / (\boldsymbol{A}_1\boldsymbol{B}_1)]\}\,\boldsymbol{B}_2 &= -(\boldsymbol{B}_3 / \boldsymbol{B}_1)\\ [(-\boldsymbol{B}_1^2 + -\boldsymbol{A}_1^2) / (\boldsymbol{A}_1\boldsymbol{B}_1)]\,\boldsymbol{B}_2 &= -(\boldsymbol{B}_3 / \boldsymbol{B}_1)\\ \{[(-\boldsymbol{B}_1^2 + -\boldsymbol{A}_1^2) / (\boldsymbol{A}_1\boldsymbol{B}_1)]\,\boldsymbol{B}_2\}\,[(\boldsymbol{A}_1\boldsymbol{B}_1) / (-\boldsymbol{B}_1^2 + -\boldsymbol{A}_1^2)] &= (\boldsymbol{B}_3 / \boldsymbol{B}_1)\,[(\boldsymbol{A}_1\boldsymbol{B}_1) / (-\boldsymbol{B}_1^2 + -\boldsymbol{A}_1^2)]\\ \boldsymbol{B}_2 &= (\boldsymbol{A}_1\boldsymbol{B}_3) / (-\boldsymbol{B}_1^2 + -\boldsymbol{A}_1^2)]\\ &= -(\boldsymbol{A}_1\boldsymbol{B}_3) / (\boldsymbol{A}_1^2 + \boldsymbol{B}_1^2)\end{aligned}$$

Substitute this result for $\boldsymbol{B}_2$ into $\boldsymbol{A}_1\boldsymbol{A}_2 - \boldsymbol{B}_1\boldsymbol{B}_2 = 0.00$ to solve for $\boldsymbol{A}_2$.

$$\boldsymbol{A}_1\boldsymbol{A}_2 - \boldsymbol{B}_1[-(\boldsymbol{A}_1\boldsymbol{B}_3) / (\boldsymbol{A}_1^{\,2} + \boldsymbol{B}_1^{\,2})] = 0.00$$

$$\boldsymbol{A}_1\boldsymbol{A}_2 + [(\boldsymbol{B}_1\boldsymbol{A}_1\boldsymbol{B}_3) / (\boldsymbol{A}_1^{\,2} + \boldsymbol{B}_1^{\,2})] = 0.00$$

$$\boldsymbol{A}_1\boldsymbol{A}_2 + [(\boldsymbol{B}_1\boldsymbol{A}_1\boldsymbol{B}_3) / (\boldsymbol{A}_1^{\,2} + \boldsymbol{B}_1^{\,2})] - (\boldsymbol{B}_1\boldsymbol{A}_1\boldsymbol{B}_3) / (\boldsymbol{A}_1^{\,2} + \boldsymbol{B}_1^{\,2}) = 0.00 - (\boldsymbol{B}_1\boldsymbol{A}_1\boldsymbol{B}_3) / (\boldsymbol{A}_1^{\,2} + \boldsymbol{B}_1^{\,2})$$

$$(\boldsymbol{A}_1\boldsymbol{A}_2) / \boldsymbol{A}_1 = -[(\boldsymbol{B}_1\boldsymbol{A}_1\boldsymbol{B}_3) / (\boldsymbol{A}_1^{\,2} + \boldsymbol{B}_1^{\,2})] / \boldsymbol{A}_1$$

$$\boldsymbol{A}_2 = -(\boldsymbol{B}_1\boldsymbol{B}_3) / (\boldsymbol{A}_1^{\,2} + \boldsymbol{B}_1^{\,2})$$

For $\boldsymbol{A}_1 + \mathrm{j}\,\boldsymbol{B}_1 = 1.00 + \mathrm{j}\ 1.00$ and $\boldsymbol{A}_3 + \mathrm{j}\,\boldsymbol{B}_3 = 0.00 - \mathrm{j}\ 1.00$,

$$\begin{aligned}\boldsymbol{A}_2 &= -(1.00)(-1.00) / [1.00^2 + 1.00^2] \\ &= 1.00 / 2.00 \\ &= \mathbf{0.500}\end{aligned}$$

$$\begin{aligned}\boldsymbol{B}_2 &= -(1.00)(-1.00) / (1.00^2 + 1.00^2) \\ &= 1.00 / 2.00 \\ &= \mathbf{0.500}\end{aligned}$$

To verify this,

$$\begin{aligned}(1.00 + \mathrm{j}\ 1.00)\ (-0.500 + \mathrm{j}\ 0.500) &= (1.00)(0.500) + \mathrm{j}\ (1.00)(0.500) + \mathrm{j}\ (1.00)(0.500) \\ &\quad + \mathrm{j}^2\ (1.00)(0.500) \\ &= 0.500 + \mathrm{j}\ (0.500 + 0.500) + (-1)(0.500) \\ &= (0.500 - 0.500) + \mathrm{j}\ (0.500 + 0.500) \\ &= 0.00 + \mathrm{j}\ 1.00\end{aligned}$$

The answer checks.

16. *RC* Circuits

Note: For all problems in this section, the phase of ac sources is 0° unless otherwise stated.

16.1 Basic

16.1.1 Capacitive Reactance

1. Problem: Calculate the capacitive reactance, X_C, for each value of capacitance, C, and frequency, f.

   a. $C = 1.0\ \mu\text{F}, f = 1.0\ \text{kHz}$

   b. $C = 47\ \text{nF}, f = 60\ \text{Hz}$

   c. $C = 6.8\ \text{pF}, f = 5.0\ \text{MHz}$

   d. $C = 2.2\ \text{F}, f = 2.5\ \text{mHz}$

   Solution: The equation for capacitive reactance, X_C, for frequency, f, and capacitance, C, is

   $$X_C = 1 / (2\pi fC)$$

   For the values of capacitance, C, and frequency, f, in Problem 1, the capacitive reactances, X_C, are:

   a. $X_C = 1 / [2\pi(1\ \text{kHz})(1\ \mu\text{F})]$
   $= 1 / [2 \times 3.14 \times (1.0 \times 10^3\ \text{Hz}) \times (1.0 \times 10^{-6}\ \text{F})]$
   $= 1 / (6.28 \times 10^{-3}\ \text{S})$
   $= \mathbf{159\ \Omega}$

   b. $X_C = 1 / [2\pi(60\ \text{Hz})(47\ \text{nF})]$
   $= 1 / [2 \times 3.14 \times (60\ \text{Hz}) \times (47 \times 10^{-9}\ \text{F})]$
   $= 1 / (17.7 \times 10^{-6}\ \text{S})$
   $= \mathbf{56.4\ k\Omega}$

   c. $X_C = 1 / [2\pi(5.0\ \text{MHz})(6.8\ \text{pF})]$
   $= 1 / [2 \times 3.14 \times (5.0 \times 10^6\ \text{Hz}) \times (6.8 \times 10^{-12}\ \text{F})]$
   $= 1 / (214 \times 10^{-6}\ \text{S})$
   $= \mathbf{4.68\ k\Omega}$

   d. $X_C = 1 / [2\pi(2.5\ \text{mHz})(2.2\ \text{F})]$
   $= 1 / [2 \times 3.14 \times (2.5 \times 10^{-3}\ \text{Hz}) \times (2.2\ \text{F})]$
   $= 1 / (34.6 \times 10^{-3}\ \text{S})$
   $= \mathbf{28.9\ \Omega}$

2. Problem: Determine the capacitance, C, for each value of frequency, f, and capacitive reactance, X_C.

   a. $X_C = 2.2\ \text{k}\Omega, f = 125\ \text{kHz}$

   b. $X_C = 470\ \Omega, f = 500\ \text{Hz}$

   c. $X_C = 1.0\ \Omega, f = 1.0\ \text{Hz}$

   d. $X_C = 15\ \text{M}\Omega, f = 2.5\ \text{kHz}$

   Solution: The equation for capacitance, C, for frequency, f, and capacitive reactance, X_C, is

   $$C = 1 / (2\pi f X_C)$$

   For the values of capacitive reactance, X_C, and frequency, f, in Problem 2, the capacitances, C, are:

   a. $C = 1 / [2\pi(125\ \text{kHz})(2.2\ \text{k}\Omega)]$
   $= 1 / [2 \times 3.14 \times (125 \times 10^3\ \text{Hz}) \times (2.2 \times 10^3\ \Omega)]$
   $= 1 / (1.73 \times 10^9\ \text{F}^{-1})$
   $= \mathbf{579\ pF}$

   b. $C = 1 / [2\pi(500\ \text{Hz})(470\ \Omega)]$

$= 1 / [2 \times 3.14 \times (500\ \text{Hz}) \times (470\ \Omega)]$
$= 1 / (1.48 \times 10^{6}\ \text{F}^{-1})$
$=$ **677 nF**

c. $C = 1 / [2\pi(1.0\ \text{Hz})(1.0\ \Omega)]$
$= 1 / [2 \times 3.14 \times (1.0\ \text{Hz}) \times (1.0\ \Omega)]$
$= 1 / (6.28\ \text{F}^{-1})$
$=$ **159 mF**

d. $C = 1 / [2\pi(2.5\ \text{kHz})(15\ \text{M}\Omega)]$
$= 1 / [2 \times 3.14 \times (2.5 \times 10^{3}\ \text{Hz}) \times (15 \times 10^{6}\ \Omega)]$
$= 1 / (236 \times 10^{9}\ \text{F}^{-1})$
$=$ **4.24 pF**

3. Problem: Determine the frequency, f, for each value of capacitance, C, and capacitive reactance, X_C.

a. $C = 3.3\ \mu\text{F}, X_C = 390\ \Omega$
b. $C = 5.6\ \text{pF}, X_C = 510\ \text{k}\Omega$
c. $C = 820\ \text{nF}, X_C = 68\ \text{k}\Omega$
d. $C = 0.10\ \mu\text{F}, X_C = 1.8\ \text{k}\Omega$

Solution: The equation for frequency, f, for capacitance, C, and capacitive reactance, X_C, is

$$f = 1 / (2\pi C X_C)$$

For the values of capacitive reactance, X_C, and capacitance, C, in Problem 3, the frequencies, f, are:

a. $f = 1 / [2\pi(3.3\ \mu\text{F})(390\ \Omega)]$
$= 1 / [2 \times 3.14 \times (3.3 \times 10^{-6}\ \text{F}) \times (390\ \Omega)]$
$= 1 / (8.09 \times 10^{-9}\ \text{s})$
$=$ **124 Hz**

b. $f = 1 / [2\pi(5.6\ \text{pF})(510\ \text{k}\Omega)]$
$= 1 / [2 \times 3.14 \times (5.6 \times 10^{-12}\ \text{F}) \times (510 \times 10^{3}\ \Omega)]$
$= 1 / (17.9 \times 10^{-6}\ \text{s})$
$=$ **55.7 kHz**

c. $f = 1 / [2\pi(820\ \text{nF})(68\ \text{k}\Omega)]$
$= 1 / [2 \times 3.14 \times (820 \times 10^{-9}\ \text{F}) \times (68 \times 10^{3}\ \Omega)]$
$= 1 / (350 \times 10^{-3}\ \text{s})$
$=$ **2.85 Hz**

d. $f = 1 / [2\pi(0.10\ \mu\text{F})(1.8\ \text{k}\Omega)]$
$= 1 / [2 \times 3.14 \times (0.1 \times 10^{-6}\ \text{F}) \times (1.8 \times 10^{3}\ \Omega)]$
$= 1 / (1.13 \times 10^{-3}\ \text{s})$
$=$ **884 Hz**

16.1.2 Series Circuits

1. Problem: Calculate the total impedance for the circuit in Figure 16-1 in both rectangular and polar forms. Then sketch the impedance phasor diagram for the circuit, identifying the resistance, reactance, and impedance phasors.

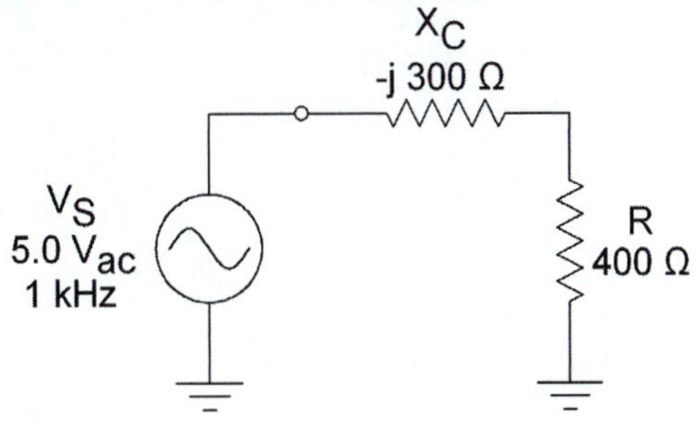

Figure 16-1: Series *RC* Circuit for Problem 1

Solution: The total impedance, Z_T, of the series circuit is

$$Z_T = R - \mathrm{j}\,X_C$$
$$= \mathbf{400\ \Omega - j\ 300\ \Omega}$$

Using rectangular to polar conversion magnitude, Z, of the impedance is

$$Z = \sqrt{R^2 + X_C{}^2}$$
$$= \sqrt{(400\ \Omega)^2 + (-300\ \Omega)^2}$$
$$= \sqrt{160{,}000\ \Omega^2 + 90{,}000\ \Omega^2}$$
$$= \sqrt{250{,}000\ \Omega^2}$$
$$= 500\ \Omega$$

The phase angle, φ, is

$$\varphi = \tan^{-1}(X_C / R)$$
$$= \tan^{-1}(-300\ \Omega / 400\ \Omega)$$
$$= \tan^{-1}(-0.750)$$
$$= -36.9°$$

The polar form of the impedance is **500 Ω ∠ –36.9°**. Figure 16-2 shows the phasor diagram of the circuit impedances.

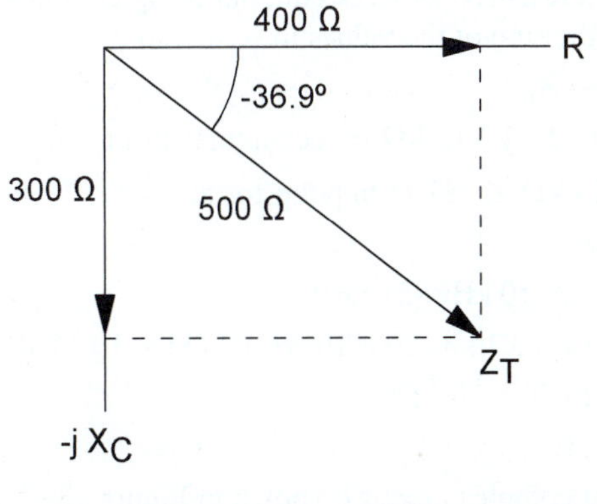

Figure 16-2: Impedance Phasor Diagram for Problem 1

2. Problem: Calculate the total impedance of the circuit in Figure 16-3 for f = 1.0 kHz and f = 10 kHz. Express the answer in both polar and rectangular forms.

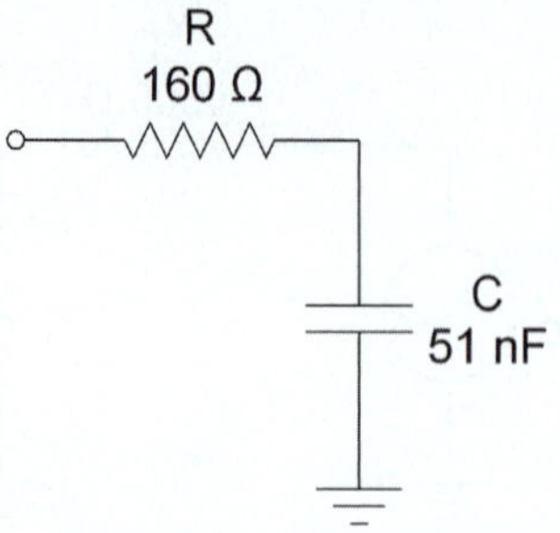

Figure 16-3: Series *RC* Circuit for Problem 2

Solution: First, calculate the reactance X_C. Then use complex arithmetic to find the total impedance $Z_T = R - \text{j}\,X_C$ of the circuit.

For $f = 1.0$ kHz

$$X_C = 1 / [2\pi(1.0\text{ kHz})(51\text{ nF})]$$
$$= 1 / [2 \times 3.14 \times (1.0 \times 10^3\text{ Hz}) \times (51 \times 10^{-9}\text{ F})]$$
$$= 1 / (320 \times 10^{-6}\text{ S})$$
$$= 3.12\text{ k}\Omega$$

The equivalent complex circuit is shown in Figure 16-4.

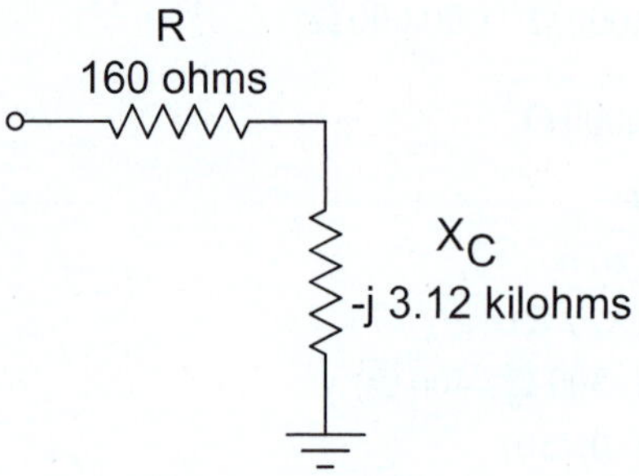

Figure 16-4: Equivalent Complex Series Circuit for $f = 1.0$ kHz

For a complex series circuit, the total impedance is calculated in the same manner as the total resistance for a resistive series circuit, but using complex arithmetic. For a series circuit the total impedance is the sum of the values in series, so

$$Z_T = R - \text{j}\,X_C$$
$$= \mathbf{160\ \Omega - j\ 3.12\ k\Omega} \text{ in rectangular form}$$
$$= \mathbf{3.13\ k\Omega \angle -87.1^\circ} \text{ in polar form.}$$

For $f = 10$ kHz,

$$X_C = 1 / [2\pi(10\text{ kHz})(51\text{ nF})]$$
$$= 1 / [2 \times 3.14 \times (10 \times 10^3\text{ Hz}) \times (51 \times 10^{-9}\text{ F})]$$
$$= 1 / (3.20 \times 10^{-3}\text{ S})$$
$$= 312\ \Omega$$

The equivalent complex circuit is shown in Figure 16-5.

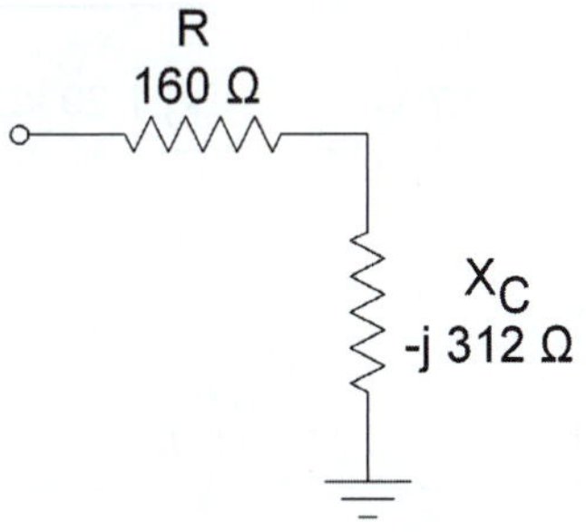

Figure 16-5: Equivalent Complex Series Circuit for *f* = 10 kHz

The total impedance is

$Z_T = R - \text{j}\, X_C$

$= \mathbf{160\ \Omega - j\ 312\ \Omega}$ in rectangular form

$= \mathbf{351\ \Omega \angle -62.9°}$ in polar form.

3. Problem: For the circuit in Figure 16-6, calculate the total impedance, total current, and voltages across each component for *f* = 15 kHz. Express all answers in polar form.

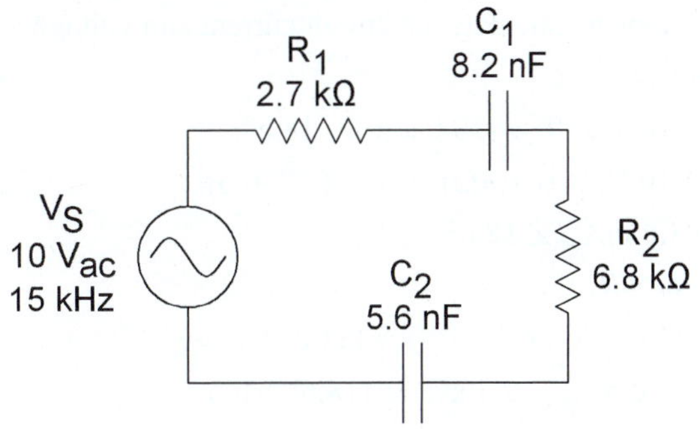

Figure 16-6: Series *RC* Circuit for Problem 3

Solution: First, calculate the reactance, X_C, for each capacitor in the circuit. Then use complex arithmetic to find the total reactance, Z_T and Ohm's Law to calculate the total current and voltage across each component. Express all answers in polar form.

For C_1,

$X_{C1} = 1 / [2\pi(15\text{ kHz})(8.2\text{ nF})]$

$= 1 / [2 \times 3.14 \times (15 \times 10^3\text{ Hz}) \times (8.2 \times 10^{-9}\text{ F})]$

$= 1 / (773 \times 10^{-6}\text{ S})$

$= 1.29\text{ k}\Omega$

For C_2,

$X_{C2} = 1 / [2\pi(15\text{ kHz})(5.6\text{ nF})]$

$= 1 / [2 \times 3.14 \times (15 \times 10^3\text{ Hz}) \times (8.2 \times 10^{-9}\text{ F})]$

$= 1 / (528 \times 10^{-6}\text{ S})$

$= 1.90\text{ k}\Omega$

The equivalent complex circuit is shown in Figure 16-7.

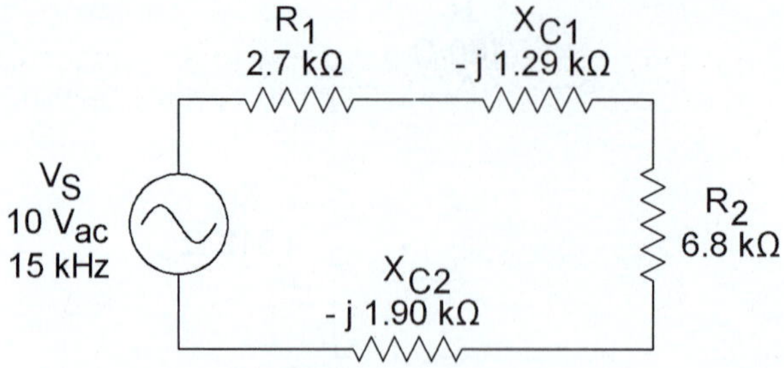

Figure 16-7: Equivalent Complex Series Circuit for Problem 3

For a series circuit, the total impedance is the sum of the series values.

$\boldsymbol{Z_T} = \boldsymbol{R_1} - \text{j}\,\boldsymbol{X_{C1}} + \boldsymbol{R_2} - \text{j}\,\boldsymbol{X_{C2}}$

$= 2.7\text{ k}\Omega - \text{j }1.29\text{ k}\Omega + 6.8\text{ k}\Omega - \text{j }1.90\text{ k}\Omega$

$= (2.7\text{ k}\Omega + 6.8\text{ k}\Omega) - \text{j }(1.29\text{ k}\Omega + 1.90\text{ k}\Omega)$

$= \mathbf{9.5\text{ k}\Omega - j\ 3.19\text{ k}\Omega}$ in rectangular form

$= \mathbf{10.0\text{ k}\Omega \angle -18.6°}$ in polar form

Use Ohm's Law to calculate the circuit current and voltage values.

$\boldsymbol{I_T} = \boldsymbol{V_S} / \boldsymbol{Z_T}$

$= (10\text{ V} \angle 0°) / (10.0\text{ k}\Omega \angle -18.6°)$

$= (10\text{ V}_{ac} / 10.0\text{ k}\Omega) \angle [0° - (-18.6°)]$

$= \mathbf{1.0\text{ mA}_{ac} \angle 18.6°}$

$\boldsymbol{V_{R1}} = \boldsymbol{I_T} \times \boldsymbol{R_1}$

$= (1.0\text{ mA}_{ac} \angle 18.6°)(2.7\text{ k}\Omega \angle 0°)$

$= (1.0\text{ mA}_{ac} \times 2.7\text{ k}\Omega) \angle (18.6° + 0°)$

$= \mathbf{2.7\text{ V}_{ac} \angle 18.6°}$

$\boldsymbol{V_{C1}} = \boldsymbol{I_T} \times (-\text{j}\,\boldsymbol{X_{C1}})$

$= (1.0\text{ mA}_{ac} \angle 18.6°)(-\text{j }1.29\text{ k}\Omega)$

$= (1.0\text{ mA}_{ac} \angle 18.6°)(1.29\text{ k}\Omega \angle -90°)$

$= (1.0\text{ mA}_{ac} \times 1.29\text{ k}\Omega) \angle (18.6° - 90°)$

$= \mathbf{1.29\text{ V}_{ac} \angle -71.4°}$

$\boldsymbol{V_{R2}} = \boldsymbol{I_T} \times \boldsymbol{R_2}$

$= (1.0\text{ mA}_{ac} \angle 18.6°)(6.8\text{ k}\Omega \angle 0°)$

$= (1.0\text{ mA}_{ac} \times 6.8\text{ k}\Omega) \angle (18.6° + 0°)$

$= \mathbf{6.8\text{ V}_{ac} \angle 18.6°}$

$\boldsymbol{V_{C2}} = \boldsymbol{I_T} \times (-\text{j}\,\boldsymbol{X_{C2}})$

$= (1.0\text{ mA}_{ac} \angle 18.6°)(-\text{j }1.90\text{ k}\Omega)$

$= (1.0\text{ mA}_{ac} \angle 18.6°)(1.90\text{ k}\Omega \angle -90°)$

$= (1.0\text{ mA}_{ac} \times 1.90\text{ k}\Omega) \angle (18.6° - 90°)$

$= \mathbf{1.90\text{ V}_{ac} \angle -71.4°}$

4. Problem: For the voltage divider circuit in Figure 16-8, calculate the value of $\boldsymbol{R}$ required for $\boldsymbol{V_{out}} = 0.5\text{ V}_{ac}$. Verify your answer.

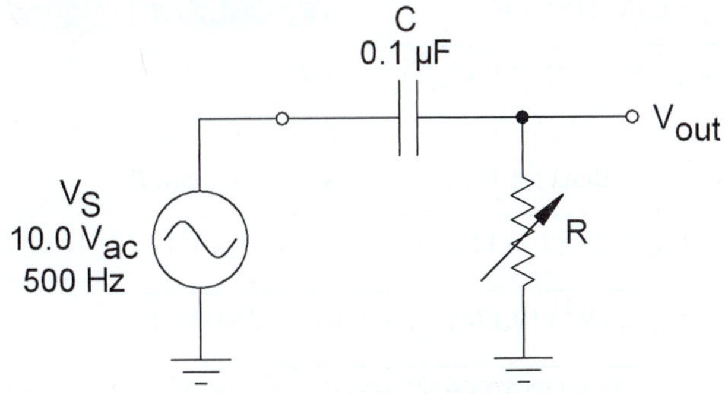

Figure 16-8: Series *RC* Circuit for Problem 4

Solution: First, calculate the reactance X_C. Then use Ohm's Law and substitution to determine what value of R will produce $V_{out} = 5.0\ V_{ac}$.

For C,

$$X_C = 1 / [2\pi(500\text{ Hz})(0.1\ \mu\text{F})]$$
$$= 1 / [2 \times 3.14 \times (500\text{ Hz}) \times (0.1 \times 10^{-6}\text{ F})]$$
$$= 1 / (314 \times 10^{-6}\text{ S})$$
$$= 3.18\text{ k}\Omega$$

For the series circuit in Figure 16-8, the output voltage is taken across R.

$$V_{out} = V_R$$
$$= I_T \times R$$

Use Ohm's Law to find the total current.

$$I_T = V_S / Z_T$$

The magnitude of Z_T is equal to the sum of the squares of R and X_C.

$$Z_T = \sqrt{R^2 + X_C^{\,2}}$$

Substitute the value of Z_T into the equation for I_T.

$$I_T = V_S / \sqrt{R^2 + X_C^{\,2}}$$

Substitute this value of I_T into the equation for V_{out}.

$$V_{out} = I_T \times R$$
$$= (V_S / Z_T) \times R$$
$$= \left(V_S / \sqrt{R^2 + X_C^{\,2}}\right) \times R$$
$$= \left((V_S R) / \sqrt{R^2 + X_C^{\,2}}\right)$$

Finally, solve for R.

$$V_{out}^{\,2} = \left((V_S R) / \sqrt{R^2 + X_C^{\,2}}\right)^2$$
$$V_{out}^{\,2} = [(V_S^2 R^2) / (R^2 + X_C^2)]$$
$$V_{out}^{\,2}(R^2 + X_C^2) = [V_S^2 R^2 / (R^2 + X_C^2)]\,(R^2 + X_C^2)$$
$$V_{out}^{\,2} R^2 + V_{out}^{\,2} X_C^2 = V_S^2 R^2$$
$$V_{out}^{\,2} R^2 + V_{out}^{\,2} X_C^2 - V_{out}^{\,2} R^2 = V_S^2 R^2 - V_{out}^{\,2} R^2$$
$$V_{out}^{\,2} X_C^2 = (V_S^2 - V_{out}^2)\, R^2$$

$$(V_{out}^2 X_C^2) / (V_S^2 - V_{out}^2) = [(V_S^2 - V_{out}^2) R^2] / (V_S^2 - V_{out}^2)$$

$$\sqrt{(V_{out}^2 X_C^2)/(V_S^2 - V_{out}^2)} = \sqrt{R^2}$$

$$= R$$

Substitute the values for V_S, V_{out}, and X_C to calculate R.

$$R = \sqrt{(5\text{ V})^2 (3.18\text{ k}\Omega)^2 / [(10\text{ V})^2 - (5.0\text{ V})^2]}$$

$$= \sqrt{(25\text{ V}^2)(10.1\text{ M}\Omega^2)/(100\text{ V}^2 - 25.0\text{ V}^2)}$$

$$= \sqrt{[(25\text{ V}^2)(10.1\text{ M}\Omega^2/(75\text{ V}^2)])}$$

$$= \sqrt{(0.333)(10.1\text{ M}\Omega^2)}$$

$$= \sqrt{(3.38\text{ M}\Omega^2)}$$

$$= \mathbf{1.84\text{ k}\Omega}$$

Use the values of R and X_C to calculate V_R.

$$Z_T = R - \text{j}X_C$$
$$= 1.84\text{ k}\Omega - \text{j }3.18\text{ k}\Omega$$
$$= 3.67\text{ k}\Omega \angle -60.0°$$

$$I_T = V_S / Z_T$$
$$= (10\text{ V}_{ac} \angle 0°)(3.67\text{ k}\Omega \angle -60.0°)$$
$$= [(10\text{ V}_{ac})(3.67\text{ k}\Omega)] \angle [0 - (-60.0°)]$$
$$= 2.72\text{ mA}_{ac} \angle 60.0°$$

$$V_R = I_T \times R$$
$$= (2.72\text{ mA}_{ac} \angle 60.0°)(1.84\text{ k}\Omega \angle 0°)$$
$$= [(2.72\text{ mA}_{ac})(1.84\text{ k}\Omega) \angle (60.0° + 0°)$$
$$= 5.00\text{ V}_{ac} \angle 60.0°$$

The answer checks.

5. Problem: Draw the phasor voltage diagrams for each circuit in Figure 16-9 using V_S as a reference. Determine from the phasor for V_{out} whether each circuit is a phase lag or phase lead circuit.

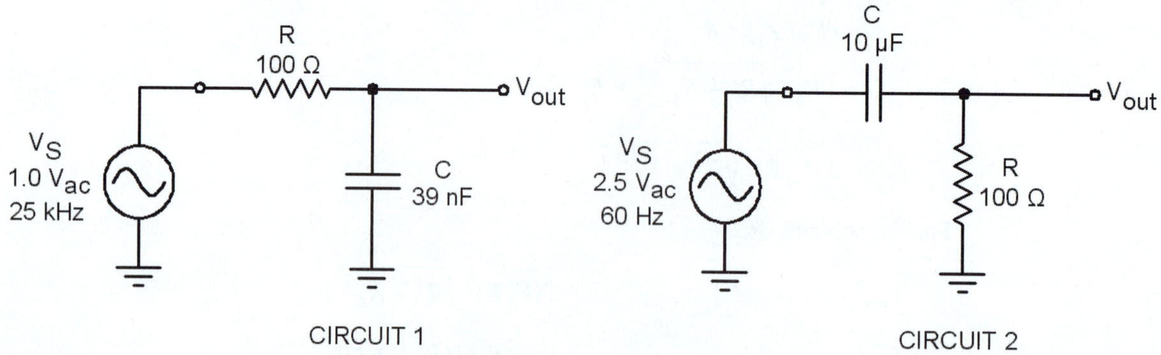

Figure 16-9: Series *RC* Circuits for Problem 5

Solution: For each circuit, first calculate the reactance X_C. Then use Ohm's Law to calculate the total current I_T and circuit voltages V_R and V_C.

For Circuit 1:

$X_C = 1 / [2\pi(25\ \text{kHz})(39\ \text{nF})]$
$= 1 / [2 \times 3.14 \times (25 \times 10^3\ \text{Hz}) \times (39 \times 10^{-9}\ \text{F})]$
$= 1 / (6.13 \times 10^{-6}\ \text{S})$
$= 163\ \Omega$

For Circuit 2:

$X_C = 1 / [2\pi(60\ \text{Hz})(10\ \mu\text{F})]$
$= 1 / [2 \times 3.14 \times (60\ \text{Hz}) \times (10 \times 10^{-6}\ \text{F})]$
$= 1 / (3.77 \times 10^{-3}\ \text{S})$
$= 265\ \Omega$

The equivalent complex circuits are shown in Figure 16-10.

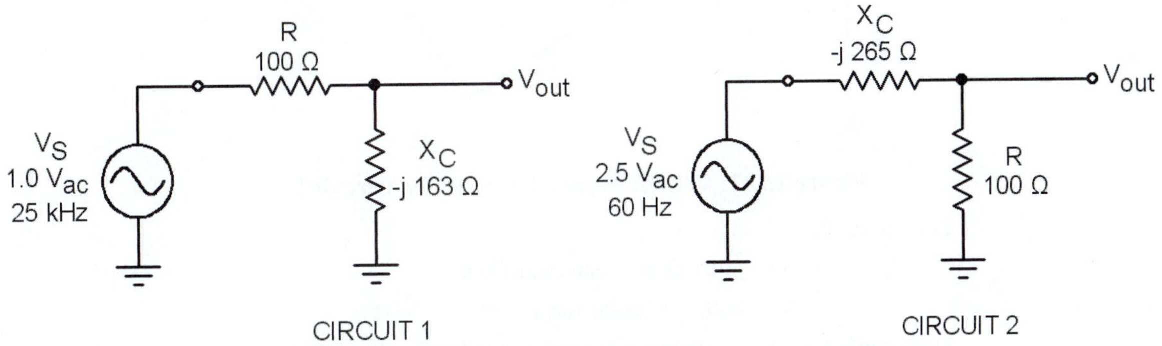

Figure 16-10: Equivalent Complex Phase Circuits for Problem 5

For Circuit 1:

$Z_T = 100\ \Omega - \text{j}\ 163\ \Omega$ in rectangular form
$= 191\ \Omega \angle -58.5°$ in polar form

Use Ohm's Law to calculate the current and voltage values.

$I_T = V_S / Z_T$
$= (1.0\ \text{V}_{ac} \angle 0°) / (191\ \Omega \angle -58.5°)$
$= (1.0\ \text{V}_{ac} / 191\ \Omega) \angle [0° - (-58.5°)]$
$= 5.22\ \text{mA}_{ac} \angle 58.5°$

$V_R = I_T \times R$
$= (5.22\ \text{mA}_{ac} \angle 58.5°)(100\ \Omega)$
$= (5.22\ \text{mA}_{ac} \angle 58.5°)(100\ \Omega \angle 0°)$
$= (5.22\ \text{mA}_{ac} \times 100\ \Omega) \angle [58.5° + (0°)]$
$= 522\ \text{mV}_{ac} \angle 58.5°$

$V_{out} = V_C$
$= I_T \times -\text{j}\ X_C$
$= (5.22\ \text{mA}_{ac} \angle 58.5°)(-\text{j}\ 163\ \Omega)$
$= (5.22\ \text{mA}_{ac} \angle 58.5°)(163\ \Omega \angle -90°)$
$= (5.22\ \text{mA}_{ac} \times 163\ \Omega) \angle [58.5° + (-90°)]$
$= 853\ \text{mV}_{ac} \angle -31.5°$

Figure 16-11 shows the voltage phasor diagram for the circuit. As the diagram shows, V_{out} lags V_S by 31.5° (i.e., $\varphi_{out} = -31.5°$), so the circuit is a ***phase lag*** circuit.

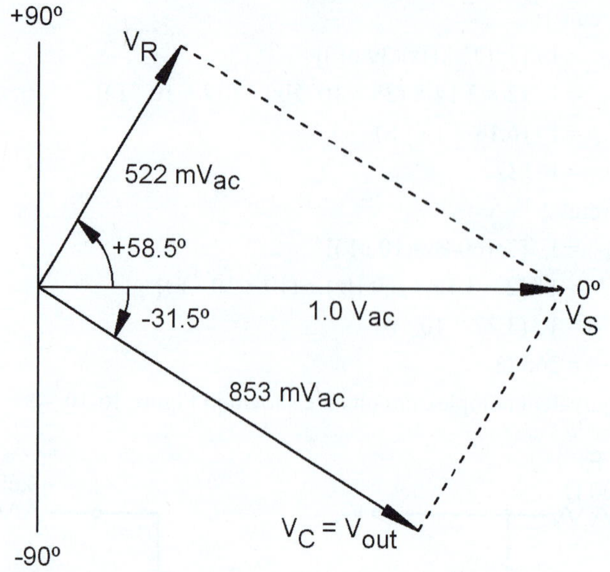

Figure 16-11: Voltage Phasor Diagram for Circuit 1

For Circuit 2:

$\boldsymbol{Z_T}$ = 100 Ω – j 265 Ω in rectangular form

= 284 Ω ∠ –69.3° in polar form

Use Ohm's Law to calculate the current and voltage values.

$\boldsymbol{I_T} = \boldsymbol{V_S} / \boldsymbol{Z_T}$

$= (2.5\ V_{ac} \angle 0°) / (284\ \Omega \angle -69.3°)$

$= (2.5\ V_{ac} / 284\ \Omega) \angle [0° - (-69.3°)]$

$= 8.82\ mA_{ac} \angle 69.3°$

$\boldsymbol{V_C} = \boldsymbol{I_T} \times -j\,\boldsymbol{X_C}$

$= (8.82\ mA_{ac} \angle 69.3°)(-j\ 265\ \Omega)$

$= (35.3\ mA_{ac} \angle 69.3°)(265\ \Omega \angle -90°)$

$= (35.3\ mA_{ac} \times 265\ \Omega) \angle [69.3° + (-90°)]$

$=2.34\ V_{ac} \angle -20.7°$

$\boldsymbol{V_{out}} = \boldsymbol{V_R}$

$= \boldsymbol{I_T} \times \boldsymbol{R}$

$= (8.82\ mA_{ac} \angle 69.3°)(100\ \Omega)$

$= (8.82\ mA_{ac} \angle 69.3°)(100\ \Omega \angle 0°)$

$= (8.82\ mA_{ac} \times 100\ \Omega) \angle (69.3° + 0°)$

$= 882\ mV_{ac} \angle 69.3°$

Figure 16-12 shows the voltage phasor diagram for the circuit. As the diagram shows, $\boldsymbol{V_{out}}$ leads $\boldsymbol{V_S}$ by 69.3° (i.e., $\varphi_{out} = 69.3°$), so the circuit is a ***phase lead*** circuit.

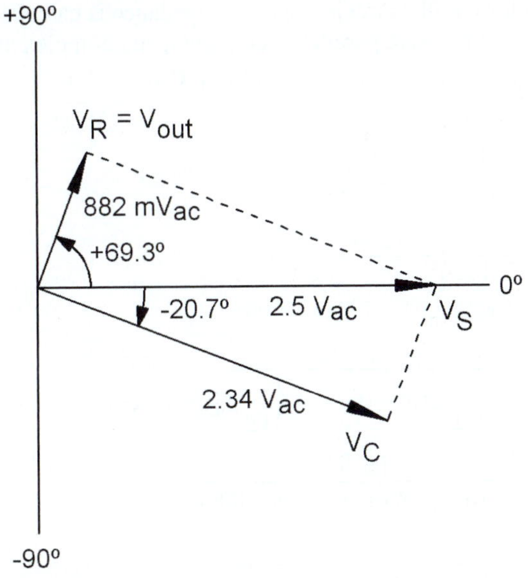

Figure 16-12: Voltage Phasor Diagram for Circuit 2

16.1.3 Parallel Circuits

1. Problem: Refer to the circuit in Figure 16-13. Calculate the total impedance of the circuit for f = 1.0 kHz and 10 kHz. Express the answer in both rectangular and polar forms.

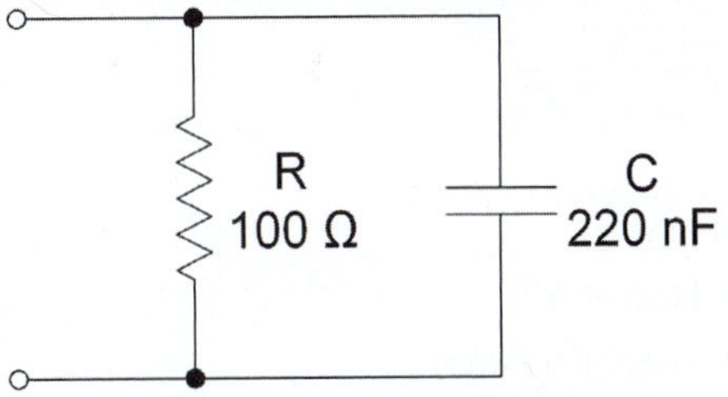

Figure 16-13: Parallel *RC* Circuit for Problem 1

Solution: First, calculate the reactance X_C. Then use complex arithmetic to find the total impedance $Z_T = R \parallel j X_C$ of the circuit. First, calculate X_C for f = 1.0 kHz.

$$X_C = 1 / [2\pi(1.0 \text{ kHz})(220 \text{ nF})]$$
$$= 1 / [2 \times 3.14 \times (1.0 \times 10^3 \text{ Hz}) \times (220 \times 10^{-9} \text{ F})]$$
$$= 1 / (1.38 \times 10^{-3} \text{ S})$$
$$= 723\ \Omega$$

The equivalent complex circuit is shown in Figure 16-14.

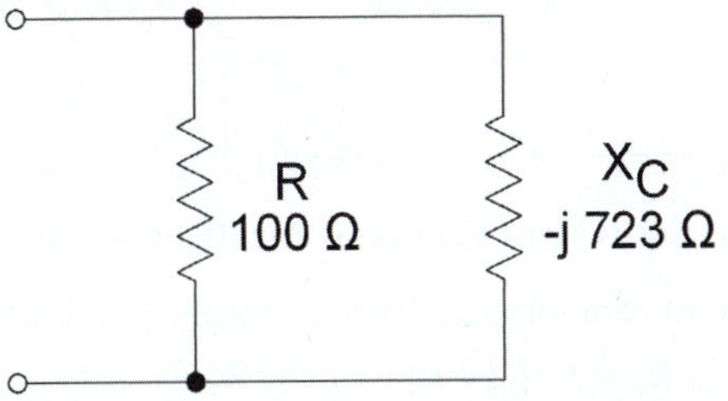

Figure 16-14: Equivalent Complex Parallel Circuit for f = 1.0 kHz

For a complex parallel circuit, the total impedance is calculated in the same manner as the total resistance for a resistive parallel circuit, but using complex arithmetic.

$$\boldsymbol{Z_T} = \boldsymbol{R} \parallel -\text{j}\,\boldsymbol{X_C} = 1 / [(1 / \boldsymbol{R}) + (1 / -\text{j}\,\boldsymbol{X_C})]$$

$$= \frac{1}{\frac{1}{100\,\Omega} + \frac{1}{-\text{j}\,723\,\Omega}}$$

$$= \frac{1\angle 0°}{\frac{1\angle 0°}{100\,\Omega\angle 0°} + \frac{1\angle 0°}{723\,\Omega\angle\text{-}90°}}$$

$$= \frac{1\angle 0°}{\frac{1}{100\,\Omega}\angle(0° - 0°) + \frac{1}{723\,\Omega}\angle[0° \text{ - } (\text{-}90°)]}$$

$$= \frac{1\angle 0°}{(10\text{ mS}\angle 0°) + (1.38\text{mS}\angle 90°)}$$

$$= \frac{1\angle 0°}{(10\text{ mS} + \text{j}\,0) + (0 + \text{j}\,1.38\text{mS})}$$

$$= \frac{1\angle 0°}{(10\text{ mS} + \ 0) + \text{j}\,(0 + 1.38\text{mS})}$$

$$= \frac{1\angle 0°}{10\text{ mS} + \text{j}\,1.38\text{mS}}$$

$$= \frac{1\angle 0°}{10.1\text{ mS}\angle 7.87°}$$

$$= \frac{1}{10.1\text{ mS}}\angle(0° \text{ - } 7.87°)$$

$$= \mathbf{99.1\ \Omega\angle -7.87°}$$

Next, calculate X_C for $f = 10$ kHz.

$$\boldsymbol{X_C} = 1 / [2\pi(10\text{ kHz})(220\text{ nF})]$$

$$= 1 / [2 \times 3.14 \times (10 \times 10^3\text{ Hz}) \times (220 \times 10^{-9}\text{ F})]$$

$$= 1 / (13.8 \times 10^{-3}\text{ S})$$

$$= 72.3\ \Omega$$

The equivalent complex circuit is shown in Figure 16-15.

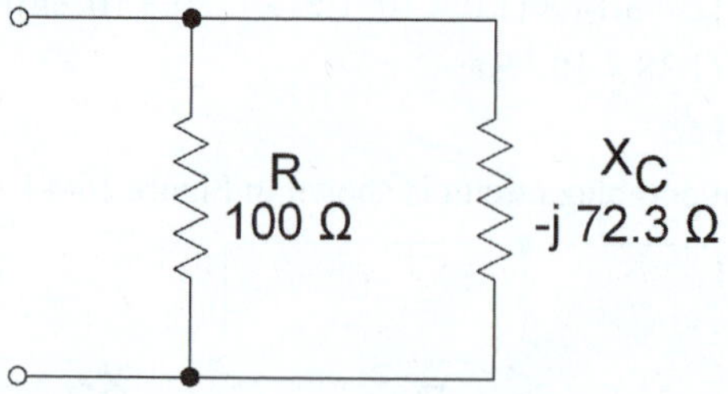

Figure 16-15: Equivalent Complex Parallel Circuit for *f* = 10 kHz

Calculate the total parallel impedance as for the previous circuit.

$$\boldsymbol{Z_T} = \boldsymbol{R} \parallel -\text{j}\,\boldsymbol{X_C} = 1 / [(1 / \boldsymbol{R}) + (1 / -\text{j}\,\boldsymbol{X_C})]$$

$$= \frac{1}{\frac{1}{R} + \frac{1}{-jX_C}}$$

$$= \frac{1}{\frac{1}{100\,\Omega} + \frac{1}{-j72.3\,\Omega}}$$

$$= \frac{1\angle 0°}{\frac{1\angle 0°}{100\,\Omega\angle 0°} + \frac{1\angle 0°}{72.3\,\Omega\angle -90°}}$$

$$= \frac{1\angle 0°}{\frac{1}{100\,\Omega}\angle(0° - 0°) + \frac{1}{72.3\,\Omega}\angle[0° - (-90°)]}$$

$$= \frac{1\angle 0°}{(10\text{ mS}\angle 0°) + (13.8\text{ mS}\angle 90°)}$$

$$= \frac{1\angle 0°}{(10\text{ mS} + j0) + (0 + j13.8\text{ mS})}$$

$$= \frac{1\angle 0°}{(10\text{ mS} + 0) + j(0 + 13.8\text{ mS})}$$

$$= \frac{1\angle 0°}{10\text{ mS} + j13.8\text{ mS}}$$

$$= \frac{1\angle 0°}{17.1\text{ mS}\angle 54.1°}$$

$$= \frac{1}{17.1\text{ mS}}\angle(0° - 54.1°)$$

$$= \mathbf{58.6\,\Omega\angle -54.1°}$$

2. Problem: Refer to the circuit in Figure 16-16. Calculate the total impedance, total current, and currents through each component at f = 2.5 kHz. Express your answers in polar form.

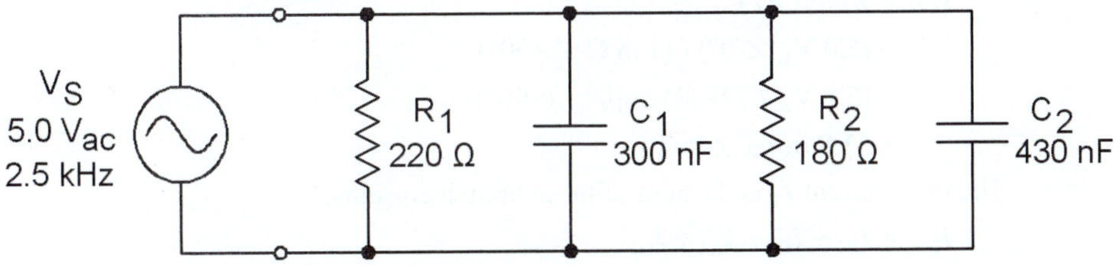

Figure 16-16: Parallel *RC* Circuit for Problem 2

Solution: First, calculate the reactance, X_C, for each capacitor. Then, use Ohm's Law to calculate the current through each branch, sum the branch currents to find the total current, I_T, and use Ohm's Law to find the total impedance, Z_T.

For C_1,

$$X_{C1} = 1 / [2\pi(2.5\text{ kHz})(300\text{ nF})]$$
$$= 1 / [2 \times 3.14 \times (2.5 \times 10^3\text{ Hz}) \times (300 \times 10^{-9}\text{ F})]$$
$$= 1 / (4.71 \times 10^{-3}\text{ S})$$
$$= 212\,\Omega$$

For C_2,

$X_{C2} = 1 / [2\pi(2.5\text{ kHz})(430\text{ nF})]$
$= 1 / [2 \times 3.14 \times (2.5 \times 10^3\text{ Hz}) \times (430 \times 10^{-9}\text{ F})]$
$= 1 / (6.75 \times 10^{-3}\text{ S})$
$= 148\ \Omega$

The equivalent complex circuit is shown in Figure 16-17.

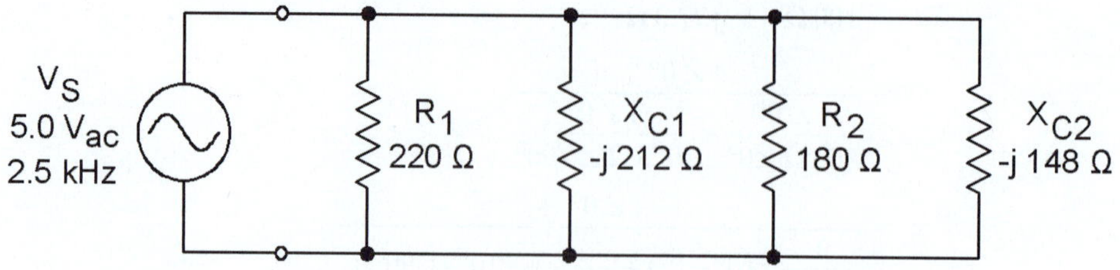

Figure 16-17: Equivalent Complex Parallel Circuit for Problem 2

Because all the components are in parallel, the voltage across each is equal to V_S. Use Ohm's Law to calculate the current through each component.

$I_{R1} = V_S / R_1$
$= (5.0\text{ V}_{ac} \angle 0°) / (220\ \Omega \angle 0°)$
$= (5.0\text{ V}_{ac} / 220\ \Omega) \angle (0° - 0°)$
$= \mathbf{22.7\text{ mA}_{ac} \angle 0°}$

$I_{C1} = V_S / -j X_{C1}$
$= (5.0\text{ V}_{ac} \angle 0°) / (212\ \Omega \angle -90°)$
$= (5.0\text{ V}_{ac} / 212\ \Omega) \angle [0° - (-90°)]$
$= \mathbf{23.6\text{ mA}_{ac} \angle 90°}$

$I_{R2} = V_S / R_2$
$= (5.0\text{ V}_{ac} \angle 0°) / (180\ \Omega \angle 0°)$
$= (5.0\text{ V}_{ac} / 180\ \Omega) \angle (0° - 0°)$
$= \mathbf{27.8\text{ mA}_{ac} \angle 0°}$

$I_{C2} = V_S / -j X_{C2}$
$= (5.0\text{ V}_{ac} \angle 0°) / (148\ \Omega \angle -90°)$
$= (5.0\text{ V}_{ac} / 148\ \Omega) \angle [0° - (-90°)]$
$= \mathbf{33.8\text{ mA}_{ac} \angle 90°}$

The total current I_T is the sum of these branch currents.

$I_T = I_{R1} + I_{C1} + I_{R2} + I_{C2}$
$= (22.7\text{ mA}_{ac} \angle 0°) + (23.6\text{ mA}_{ac} \angle 90°) + (27.8\text{ mA}_{ac} \angle 0°) + (33.8\text{ mA}_{ac} \angle 90°)$
$= (22.7\text{ mA}_{ac} + j\ 0\text{ A}_{ac}) + (0\text{ A}_{ac} + j\ 23.6\text{ mA}_{ac})$
$+ (27.8\text{ mA}_{ac} + j\ 0\text{ A}_{ac}) + (0\text{ A}_{ac} + j\ 33.8\text{ mA}_{ac})$
$= (22.7\text{ mA}_{ac} + 0\text{ A}_{ac} + 27.8\text{ mA}_{ac} + 0\text{ A}_{ac})$
$+ j\ (0\text{ A}_{ac} + 23.6\text{ mA}_{ac} + 0\text{ A}_{ac} + 33.8\text{ mA}_{ac})$
$= 50.5\text{ mA}_{ac} + j\ 57.\ 3\text{ mA}_{ac}$
$= \mathbf{76.4\text{ mA}_{ac} \angle 48.6°}$

Us Ohm's Law for impedance to calculate Z_T from V_S and I_T.

$Z_T = V_S / I_T$

$= (5.0\ V_{ac} \angle 0°) / (76.4\ mA_{ac} \angle 48.6°)$

$= (5.0\ V_{ac} / 76.4\ mA_{ac}) \angle (0° - 48.6°)$

$= \mathbf{65.4\ \Omega \angle -48.6°}$

3. Problem: For the circuit in Figure 16-18, for what value of ***C*** is $I_C = 5.00\ mA_{ac}$? What is the value of V_C for this value of ***C***?

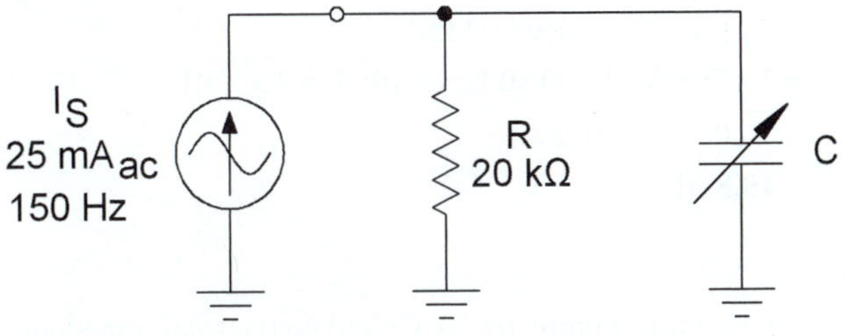

Figure 16-18: Parallel *RC* Circuit for Problem 3

Solution: First, determine the value of I_R when $I_C = 5.00\ mA_{ac}$. The current through ***R*** is in phase with the source voltage, and the current through ***C*** leads the source voltage by 90°. Choose I_R to be the reference current, so that $\varphi_R = 0°$ and $\varphi_C = 90°$. Because the currents through ***R*** and ***C*** are 90° out of phase, the total current, I_T, is the phasor sum of I_R and I_C., as shown in Figure 16-19.

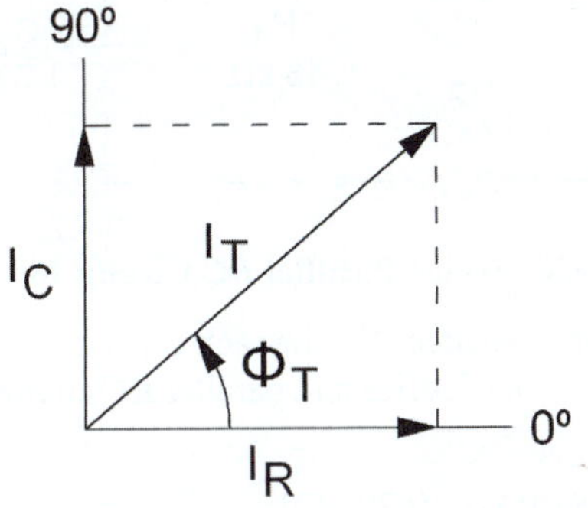

Figure 16-19: Phase Relationships of Currents for Problem 3

Use the Pythagorean Theorem for the diagram in Figure 16-19 to find I_T.

$I_R^2 + I_C^2 = I_T^2$

$I_R^2 + (5.00\ mA_{ac})^2 = (25\ mA_{ac})^2$

$I_R^2 + 25.0\ \mu A_{ac}^2 = 625\ \mu A_{ac}^2$

$I_R^2 + 25.0\ \mu A_{ac}^2 - 25.0\ \mu A_{ac}^2 = 625\ \mu A_{ac}^2 - 25.0\ \mu A_{ac}^2$

$I_R^2 = 600\ \mu A_{ac}^2$

$\sqrt{I_R^2} = \sqrt{600\ \mu A_{ac}^2}$

$I_R = 24.5\ mA_{ac} \angle 0°$

Note that I_R has a phase angle of 0° because it was chosen to be the reference current and the capacitor current leads I_R by 90°. Use Ohm's Law for voltage to find V_R.

$V_R = I_R \times R$

$= (24.5\ mA_{ac} \angle 0°)(20\ k\Omega \angle 0°)$

$= (24.5\ mA_{ac} \times 20\ k\Omega) \angle (0° + 0°)$

$= 490\ V_{ac} \angle 0°$

For parallel circuits the voltage across each component is equal to the source voltage, so that $V_R = V_S = V_C$. Use Ohm's Law for reactance to find X_C.

$$X_C = V_C / I_C$$
$$= (490\ V_{ac} \angle 0°) / (5.0\ mA_{ac} \angle 90°)$$
$$= (490\ V_{ac} / 5.0\ mA_{ac}) \angle (0° - 90°)$$
$$= 98.0\ k\Omega \angle -90°$$

Finally, use $X_C = 98.0$ kΩ and $f = 150$ Hz, to calculate C.

$$C = 1 / [2\pi(150\ Hz)(98.0\ k\Omega)]$$
$$= 1 / [2 \times 3.14 \times (150\ Hz) \times (98.0 \times 10^3\ \Omega)]$$
$$= 1 / (92.3 \times 10^6\ F^{-1})$$
$$= \mathbf{10.8\ nF}$$

16.1.4 Series-Parallel Circuits

1. Problem: Refer to the circuit in Figure 16-20. Calculate the total impedance of the circuit for $f = 1.0$ kHz and 10 kHz. Express the answers in both rectangular and polar forms.

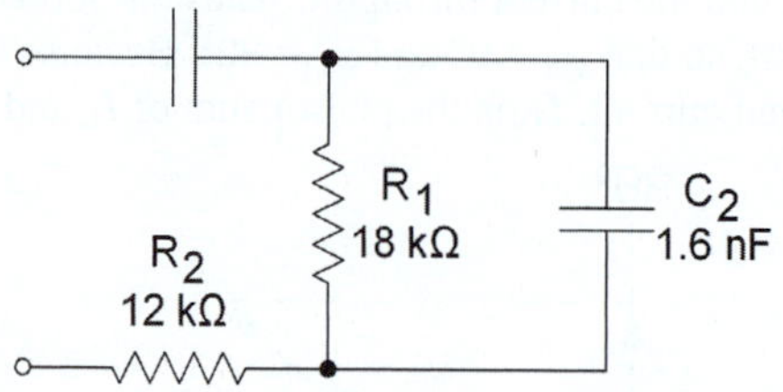

Figure 16-20: Series-Parallel *RC* Circuit for Problem 1

Solution: First, calculate the reactance, X_C, for each capacitor in the circuit. Then use the techniques for finding the impedance of series and parallel ***RC*** circuits to find the total reactance, Z_T.

For $f = 1$ kHz:

$$X_{C1} = 1 / [2\pi(1.0\ kHz)(620\ pF)]$$
$$= 1 / [2 \times 3.14 \times (1.0 \times 10^3\ Hz) \times (620 \times 10^{-12}\ F)]$$
$$= 1 / (3.90 \times 10^{-6}\ S)$$
$$= 257\ k\Omega$$
$$X_{C2} = 1 / [2\pi(1.0\ kHz)(1.6\ nF)]$$
$$= 1 / [2 \times 3.14 \times (1.0 \times 10^3\ Hz) \times (1.6 \times 10^{-9}\ F)]$$
$$= 1 / (10.1 \times 10^{-6}\ S)$$
$$= 99.5\ k\Omega$$

The equivalent complex circuit is shown in Figure 16-21.

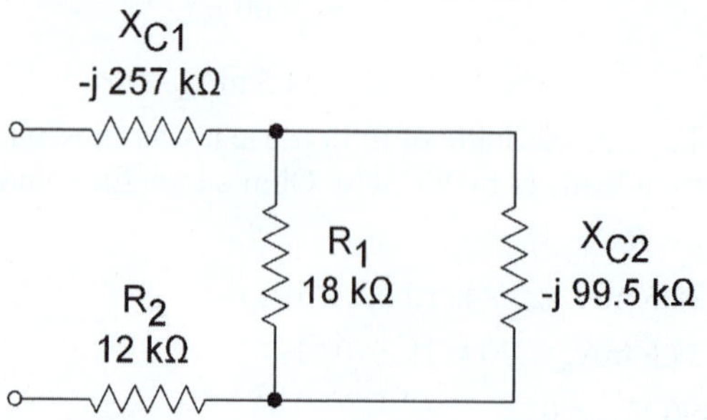

Figure 16-21: Equivalent Complex Series-Parallel Circuit for $f = 1.0$ kHz

For a complex parallel circuit, the total impedance is calculated in the same manner as the total resistance for a resistive series-parallel circuit, but using complex arithmetic. Apply standard analysis of series-parallel circuits to solve for Z_T.

$$Z_T = X_{C1} + R_1 \parallel X_{C2} + R_2$$

From the standard order of operations, the first equivalent impedance Z_{EQ1} is the parallel combination $R_1 \parallel X_{C2}$.

$$Z_T = X_{C1} + Z_{EQ1} + R_2$$

where

$$\begin{aligned}
Z_{EQ1} &= R_1 \parallel X_{C2} \\
&= \frac{1}{\frac{1}{R_1} + \frac{1}{-\mathrm{j}X_{C2}}} \\
&= \frac{1}{\frac{1}{18\ \mathrm{k\Omega}} + \frac{1}{-\mathrm{j}99.5\ \mathrm{k\Omega}}} \\
&= \frac{1\angle 0^\circ}{\frac{1\angle 0^\circ}{18\ \mathrm{k\Omega}\angle 0^\circ} + \frac{1\angle 0^\circ}{99.5\ \mathrm{k\Omega}\angle -90^\circ}} \\
&= \frac{1\angle 0^\circ}{\frac{1}{18\ \mathrm{k\Omega}}\angle(0^\circ - 0^\circ) + \frac{1}{99.5\ \mathrm{k\Omega}}\angle[0^\circ - (-90^\circ)]} \\
&= \frac{1\angle 0^\circ}{(55.6\ \mu\mathrm{S}\angle 0^\circ) + (10.1\ \mu\mathrm{S}\angle 90^\circ)} \\
&= \frac{1\angle 0^\circ}{(55.6\ \mu\mathrm{S} + \mathrm{j}0\ \mathrm{S}) + (0\ \mathrm{S} + \mathrm{j}10.1\ \mathrm{mS})} \\
&= \frac{1\angle 0^\circ}{(55.6\ \mu\mathrm{S} + 0\ \mathrm{S}) + \mathrm{j}(0\ \mathrm{S} + 10.1\ \mu\mathrm{S})} \\
&= \frac{1\angle 0^\circ}{55.6\ \mu\mathrm{S} + \mathrm{j}10.1\ \mu\mathrm{S}} \\
&= \frac{1\angle 0^\circ}{56.5\ \mu\mathrm{S}\angle 10.3^\circ} \\
&= \frac{1}{56.5\ \mu\mathrm{S}}\angle(0^\circ - 10.3^\circ) \\
&= 17.7\ \mathrm{k\Omega}\angle -10.3^\circ
\end{aligned}$$

The second (and final) equivalent impedance is the series combination $X_{C1} + Z_{EQ1} + R_2$. Substitute this value into the equation for Z_T.

$$Z_T = Z_{EQ2}$$

where

$$\begin{aligned}
Z_{EQ2} &= X_{C1} + Z_{EQ1} + R_2 \\
&= (0\ \Omega - \mathrm{j}\ 257\ \mathrm{k\Omega}) + (17.7\ \mathrm{k\Omega}\angle -10.3^\circ) + (12\ \mathrm{k\Omega} + \mathrm{j}\ 0\ \Omega) \\
&= (0\ \Omega - \mathrm{j}\ 257\ \mathrm{k\Omega}) + (17.4\ \mathrm{k\Omega} - \mathrm{j}\ 3.15\ \mathrm{k\Omega}) + (12\ \mathrm{k\Omega} + \mathrm{j}\ 0\ \Omega) \\
&= (0\ \Omega + 17.4\ \mathrm{k\Omega} + 12\ \mathrm{k\Omega}) - \mathrm{j}\ (257\ \mathrm{k\Omega} + 3.15\ \mathrm{k\Omega} + 0\ \mathrm{k\Omega}) \\
&= \mathbf{29.4\ k\Omega - j\ 260\ k\Omega} \text{ in rectangular form}
\end{aligned}$$

$= \mathbf{262\ k\Omega \angle -83.5°}$ in polar form

For f = 10 kHz:

$X_{C1} = 1 / [2\pi(10\ \text{kHz})(620\ \text{pF})]$
$= 1 / [2 \times 3.14 \times (10 \times 10^3\ \text{Hz}) \times (620 \times 10^{-12}\ \text{F})]$
$= 1 / (39.0 \times 10^{-6}\ \text{S})$
$= 25.7\ \text{k}\Omega$

$X_{C2} = 1 / [2\pi(10\ \text{kHz})(1.6\ \text{nF})]$
$= 1 / [2 \times 3.14 \times (10 \times 10^3\ \text{Hz}) \times (1.6 \times 10^{-9}\ \text{F})]$
$= 1 / (101 \times 10^{-6}\ \text{S})$
$= 9.95\ \text{k}\Omega$

The equivalent complex circuit is shown in Figure 16-22.

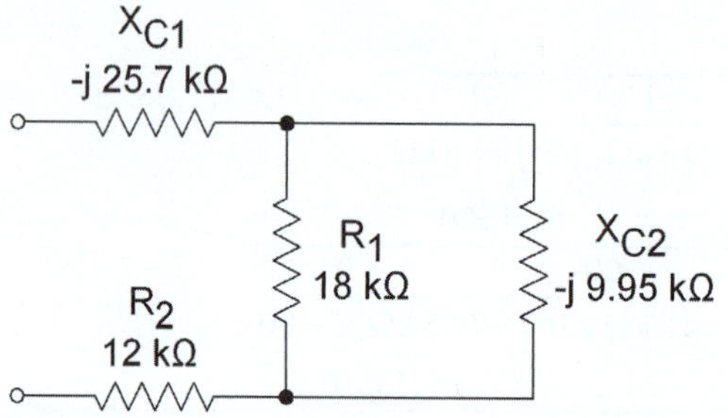

Figure 16-22: Equivalent Complex Series-Parallel Circuit for *f* = 10 kHz

Repeat the analysis process as for f = 1.0 kHz.

$Z_T = X_{C1} + Z_{EQ1} + R_2$

where

$$Z_{EQ1} = R_1 \parallel X_{C2}$$

$$= \frac{1}{\dfrac{1}{R_1} + \dfrac{1}{-\text{j}X_{C2}}}$$

$$= \frac{1}{\dfrac{1}{18\ \text{k}\Omega} + \dfrac{1}{-\text{j}9.95\ \text{k}\Omega}}$$

$$= \frac{1\angle 0°}{\dfrac{1\angle 0°}{18\ \text{k}\Omega \angle 0°} + \dfrac{1\angle 0°}{9.95\ \text{k}\Omega \angle -90°}}$$

$$= \frac{1\angle 0°}{\dfrac{1}{18\ \text{k}\Omega}\angle(0° - 0°) + \dfrac{1}{9.95\ \text{k}\Omega}\angle[0° - (-90°)]}$$

$$= \frac{1\angle 0°}{(55.6\ \mu\text{S} \angle 0°) + (101\ \mu\text{S} \angle 90°)}$$

$$= \frac{1\angle 0°}{(55.6\ \mu\text{S} + \text{j}0\ \text{S}) + (0\ \text{S} + \text{j}101\ \text{mS})}$$

$$= \frac{1\angle 0°}{(55.6\ \mu\text{S} + 0\ \text{S}) + \text{j}(0\ \text{S} + 101\ \mu\text{S})}$$

$$= \frac{1\angle 0°}{55.6\,\mu S + j101\,\mu S}$$

$$= \frac{1\angle 0°}{115\,\mu S\angle 61.1°}$$

$$= \frac{1}{115\,\mu S}\angle(0° - 61.1°)$$

$= 8.71\ k\Omega \angle -61.1°$

$\boldsymbol{Z_T} = \boldsymbol{Z_{EQ2}}$, where

$\boldsymbol{Z_{EQ2}} = \boldsymbol{X_{C1}} + \boldsymbol{Z_{EQ1}} + \boldsymbol{R_2}$

$= (0\ \Omega - j\ 25.7\ k\Omega) + (8.71\ k\Omega \angle -61.1°) + (12\ k\Omega + j\ 0\ \Omega)$

$= (0\ \Omega - j\ 25.7\ k\Omega) + (4.21\ k\Omega - j\ 7.62\ k\Omega) + (12\ k\Omega + j\ 0\ \Omega)$

$= (0\ \Omega + 4.21\ k\Omega + 12\ k\Omega) - j\ (25.7\ k\Omega + 7.62\ k\Omega + 0\ k\Omega)$

= **16.2 kΩ – j 33.3 kΩ** in rectangular form

= **37.0 kΩ ∠–64.0°** in polar form

2. Problem: Refer to the circuit in Figure 16-23. Calculate the total impedance, total current, and the voltages across and currents through each component at $f = 5.0$ kHz. Express all answers in polar form.

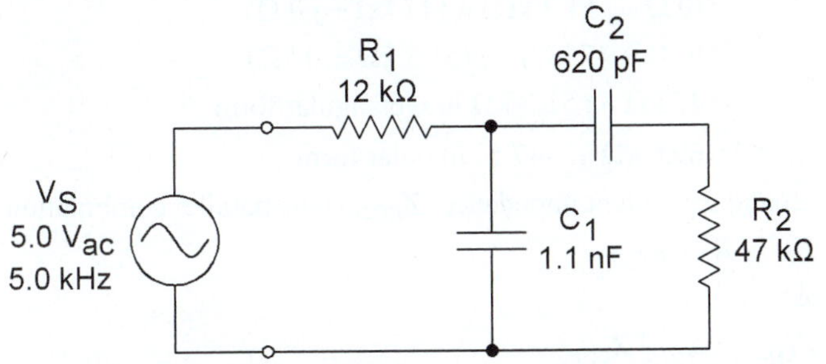

Figure 16-23: Series-Parallel *RC* Circuit for Problem 2

Solution: First, calculate the reactance, X_C, for each capacitor in the circuit. Then use the techniques for analyzing series and parallel ***RC*** circuits to find the impedances, currents, and voltages of the circuit.

For $f = 5$ kHz:

$X_{C1} = 1 / [2\pi(5.0\ kHz)(1.1\ nF)]$

$= 1 / [2 \times 3.14 \times (5.0 \times 10^3\ Hz) \times (1.1 \times 10^{-9}\ F)]$

$= 1 / (31.6 \times 10^{-6}\ S)$

$= 28.9\ k\Omega$

$X_{C2} = 1 / [2\pi(5.0\ kHz)(620\ pF)]$

$= 1 / [2 \times 3.14 \times (5.0 \times 10^3\ Hz) \times (620 \times 10^{-12}\ F)]$

$= 1 / (19.5 \times 10^{-6}\ S)$

$= 51.3\ k\Omega$

The equivalent complex circuit is shown in Figure 16-24.

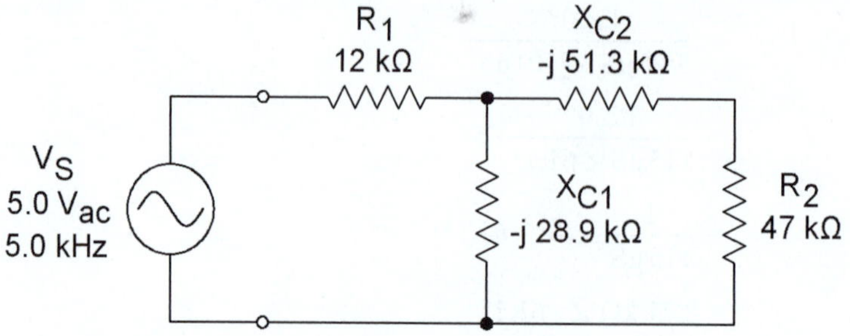

Figure 16-24: Equivalent Complex Series-Parallel Circuit for Problem 2

From standard analysis of series-parallel circuits,

$$\boldsymbol{Z_T} = \boldsymbol{R_1} + \boldsymbol{X_{C1}} \parallel (\boldsymbol{X_{C2}} + \boldsymbol{R_2})$$

From the standard order of operations, the first equivalent impedance, $\boldsymbol{Z_{EQ1}}$, is the series combination $\boldsymbol{X_{C2}} + \boldsymbol{R_2}$.

$$\boldsymbol{Z_T} = \boldsymbol{R_1} + \boldsymbol{X_{C1}} \parallel \boldsymbol{Z_{EQ1}}$$

where

$$\begin{aligned}
\boldsymbol{Z_{EQ1}} &= \boldsymbol{X_{C2}} + \boldsymbol{R_2} \\
&= (0\ \Omega - \text{j}\ 51.3\ \text{k}\Omega) + (47\ \text{k}\Omega + \text{j}\ 0\ \Omega) \\
&= (0\ \Omega + 47\ \text{k}\Omega) - \text{j}\ (51.3\ \text{k}\Omega + 0\ \text{k}\Omega) \\
&= 47\ \text{k}\Omega - \text{j}\ 51.3\ \text{k}\Omega \text{ in rectangular form} \\
&= 69.6\ \text{k}\Omega\ \angle{-47.5^\circ} \text{ in polar form}
\end{aligned}$$

The second equivalent impedance, $\boldsymbol{Z_{EQ2}}$, is the parallel combination $\boldsymbol{X_{C1}} \parallel \boldsymbol{Z_{EQ1}}$.

$$\boldsymbol{Z_T} = \boldsymbol{R_1} + \boldsymbol{Z_{EQ2}}$$

where

$$\begin{aligned}
\boldsymbol{Z_{EQ2}} &= \boldsymbol{X_{C1}} \parallel \boldsymbol{Z_{EQ1}} \\
&= \frac{1}{\dfrac{1}{-\text{j}X_{C1}} + \dfrac{1}{Z_{EQ1}}} \\
&= \frac{1}{\dfrac{1}{-\text{j}\,28.9\ \text{k}\Omega} + \dfrac{1}{47\ \text{k}\Omega - \text{j}\,51.3\ \text{k}\Omega}} \\
&= \frac{1\angle 0^\circ}{\dfrac{1\angle 0^\circ}{28.9\ \text{k}\Omega\angle -90^\circ} + \dfrac{1\angle 0^\circ}{69.6\ \text{k}\Omega\angle -47.5^\circ}} \\
&= \frac{1\angle 0^\circ}{\dfrac{1}{28.9\ \text{k}\Omega}\angle[0^\circ - (-90^\circ)] + \dfrac{1}{69.6\ \text{k}\Omega}\angle[0^\circ - (-47.5^\circ)]} \\
&= \frac{1\angle 0^\circ}{(34.6\ \mu\text{S}\angle 90^\circ) + (14.4\ \mu\text{S}\angle 47.5^\circ)} \\
&= \frac{1\angle 0^\circ}{(0\,\text{S} + \text{j}\,34.6\,\mu\text{S}) + (9.70\,\mu\text{S} + \text{j}\,10.6\,\mu\text{S})} \\
&= \frac{1\angle 0^\circ}{(9.70\,\mu\text{S} + 0\,\text{S}) + \text{j}\,(34.6\,\mu\text{S} + 10.6\,\mu\text{S})}
\end{aligned}$$

$$= \frac{1\angle 0°}{9.70\,\mu S + j\,45.2\,\mu S}$$

$$= \frac{1\angle 0°}{46.2\,\mu S \angle 77.9°}$$

$$= \frac{1}{46.2\,\mu S}\angle(0° - 77.9°)$$

$= 21.7\ k\Omega \angle -77.9°$ in polar form

$= 4.55\ k\Omega - j\ 21.2\ k\Omega$ in rectangular form

The third (and final) equivalent impedance, $\boldsymbol{Z_{EQ3}}$, is the series combination $\boldsymbol{R_1} + \boldsymbol{Z_{EQ2}}$.

$\boldsymbol{Z_T} = \boldsymbol{Z_{EQ3}}$

where

$\boldsymbol{Z_{EQ3}} = \boldsymbol{R_1} + \boldsymbol{Z_{EQ2}}$

$= (12\ k\Omega + j\ 0\ \Omega) + (4.55\ k\Omega - j\ 21.2\ k\Omega)$

$= (12\ k\Omega + 4.55\ k\Omega) - j\ (0\ \Omega + 21.2\ k\Omega)$

$= \mathbf{16.6\ k\Omega - j\ 21.2\ k\Omega}$ in rectangular form

$= \mathbf{26.9\ k\Omega \angle -52.0°}$ in polar form

Use Ohm's Law for current to calculate $\boldsymbol{I_T}$.

$\boldsymbol{I_T} = \boldsymbol{V_S} / \boldsymbol{Z_T}$

$= (5.0\ V_{ac} \angle 0°) / (26.9\ k\Omega \angle -52.0°)$

$= (5.0\ V_{ac} / 26.9\ k\Omega) \angle [0° - (-52.0°)]$

$= \mathbf{186\ \mu A_{ac} \angle 52.0°}$

$\boldsymbol{I_T}$ flows through $\boldsymbol{R_1}$ and $\boldsymbol{Z_{EQ2}}$, so $\boldsymbol{I_{R1}} = \boldsymbol{I_{ZEQ2}} = \boldsymbol{I_T} = \mathbf{186\ \mu A_{ac} \angle 52.0°}$. Use Ohm's Law for voltage to calculate $\boldsymbol{V_{R1}}$ and $\boldsymbol{V_{ZEQ2}}$.

$\boldsymbol{V_{R1}} = \boldsymbol{I_{R1}} \times \boldsymbol{R_1}$

$= (186\ \mu A_{ac} \angle 52.0°)\ (12\ k\Omega \angle 0°)$

$= (186\ \mu A_{ac} \times 12\ k\Omega) \angle (52.0° + 0°)$

$= \mathbf{2.23\ V_{ac} \angle 52.0°}$

$\boldsymbol{V_{ZEQ2}} = \boldsymbol{I_{R1}} \times \boldsymbol{Z_{EQ2}}$

$= (186\ \mu A_{ac} \angle 52.0°)\ (21.7\ k\Omega \angle -77.9°)$

$= (186\ \mu A_{ac} \times 21.7\ k\Omega) \angle (52.0° + -77.9°)$

$= \mathbf{4.03\ V_{ac} \angle -25.9°}$

$\boldsymbol{Z_{EQ2}}$ is the parallel combination of $\boldsymbol{X_{C1}}$ and $\boldsymbol{Z_{EQ1}}$. The voltage across each component in parallel is the same, so $\boldsymbol{V_{C1}} = \boldsymbol{V_{ZEQ1}} = \boldsymbol{V_{ZEQ2}} = \mathbf{4.03\ V_{ac} \angle -25.9°}$. Use Ohm's Law to calculate $\boldsymbol{I_{C1}}$ and $\boldsymbol{I_{ZEQ1}}$.

$\boldsymbol{I_{C1}} = \boldsymbol{V_{C1}} / \boldsymbol{X_{C1}}$

$= (4.03\ V_{ac} \angle -25.9°) / (28.9\ k\Omega \angle -90°)$

$= (4.03\ V_{ac} / 28.9\ k\Omega) \angle [-25.9° - (-90°)]$

$= \mathbf{139\ \mu A_{ac} \angle 64.1°}$

$\boldsymbol{I_{ZEQ1}} = \boldsymbol{V_{ZEQ1}} / \boldsymbol{Z_{EQ1}}$

$= (4.03\ V_{ac} \angle -25.9°) / (69.6\ k\Omega \angle -47.5°)$

$= (4.03\ V_{ac} / 69.6\ k\Omega) \angle [-25.9° - (-47.5°)]$

$= \mathbf{57.9\ \mu A_{ac} \angle 21.6°}$

$\boldsymbol{Z_{EQ1}}$ is the series combination of $\boldsymbol{X_{C2}}$ and $\boldsymbol{R_2}$. The current through each component in series is the same, so $\boldsymbol{I_{C2}} = \boldsymbol{I_{R2}} = \mathbf{57.9\ \mu A_{ac} \angle 21.6°}$. Use Ohm's Law to calculate $\boldsymbol{V_{C2}}$ and $\boldsymbol{V_{R2}}$.

$V_{C2} = I_{C2} \times X_{C2}$

$= (57.9\ \mu A_{ac} \angle 21.6°)\ (51.3\ k\Omega \angle -90°)$

$= (57.9\ \mu A_{ac} \times 51.3\ k\Omega) \angle (21.6° + -90°)$

$= \mathbf{2.97\ V_{ac} \angle -68.4°}$

$V_{R2} = I_{C2} \times R_2$

$= (57.9\ \mu A_{ac} \angle 21.6°)\ (47\ k\Omega \angle 0°)$

$= (57.9\ \mu A_{ac} \times 47\ k\Omega) \angle (21.6° + 0°)$

$= \mathbf{2.72\ V_{ac} \angle 21.6°}$

3. Problem: Refer to the circuit in Figure 16-25. Determine V_{out} for $R_L = 1.00\ k\Omega$ and $R_L = 10.0\ k\Omega$.

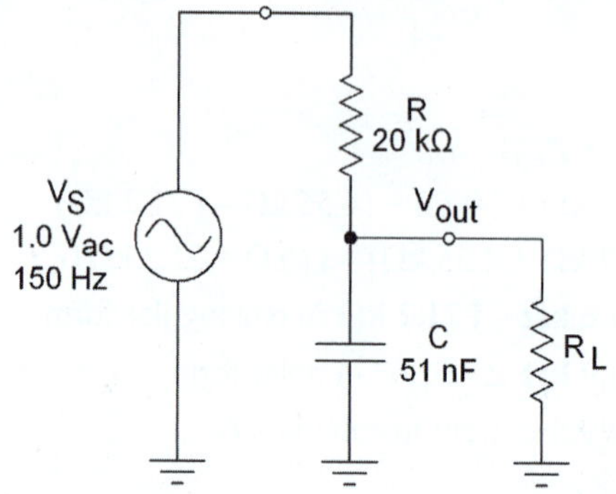

Figure 16-25: Series-Parallel *RC* Circuit for Problem 3

Solution: First, calculate the reactance, X_C, for the capacitor. Then use the techniques for analyzing series and parallel ***RC*** circuits to find the circuit impedances, currents, and voltages to determine V_{out}.

For f = 150 Hz:

$X_C = 1 / [2\pi(150\ Hz)(51\ nF)]$

$= 1 / [2 \times 3.14 \times (150\ Hz) \times (51 \times 10^{-9}\ F)]$

$= 1 / (48.1 \times 10^{-6}\ S)$

$= 20.8\ k\Omega$

The equivalent complex circuit is shown in Figure 16-26.

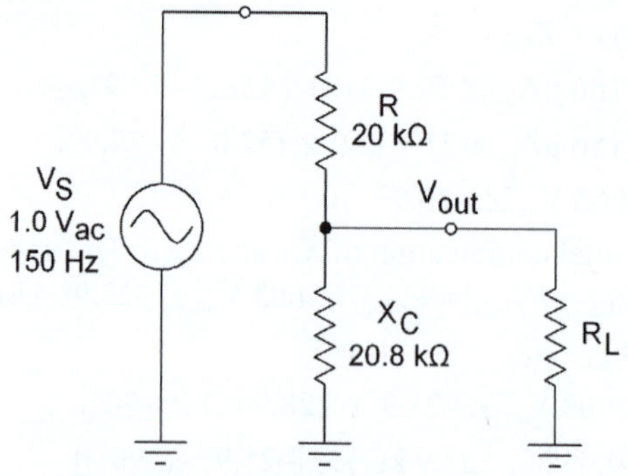

Figure 16-26: Equivalent Complex Circuit for Problem 3

From standard analysis of series-parallel circuits,

$Z_T = R + X_C \parallel R_L$

The first equivalent impedance, Z_{EQ1}, is the parallel combination $X_C \parallel R_L$.

$Z_T = R + Z_{EQ1}$

where

$Z_{EQ1} = X_C \parallel R_L$

For R_L = 1.00 kΩ:

$$Z_{EQ1} = X_C \parallel R_L$$
$$= \frac{1}{\dfrac{1}{-\mathrm{j}X_C} + \dfrac{1}{R_L}}$$
$$= \frac{1}{\dfrac{1}{-\mathrm{j}\,20.8\,\mathrm{k\Omega}} + \dfrac{1}{1.00\,\mathrm{k\Omega}}}$$
$$= \frac{1\angle 0^\circ}{\dfrac{1\angle 0^\circ}{20.8\,\mathrm{k\Omega}\angle -90^\circ} + \dfrac{1\angle 0^\circ}{1.00\angle 0^\circ}}$$
$$= \frac{1\angle 0^\circ}{\dfrac{1}{20.8\,\mathrm{k\Omega}}\angle[0^\circ - (-90^\circ)] + \dfrac{1}{1.00\,\mathrm{k\Omega}}\angle 0^\circ}$$
$$= \frac{1\angle 0^\circ}{(48.1\,\mu\mathrm{S}\angle 90^\circ) + (1.00\,\mathrm{mS}\angle 0^\circ)}$$
$$= \frac{1\angle 0^\circ}{(0\,\mathrm{S} + \mathrm{j}\,48.1\,\mu\mathrm{S}) + (1.00\,\mathrm{mS} + \mathrm{j}\,0\,\mathrm{S})}$$
$$= \frac{1\angle 0^\circ}{(0\,\mathrm{S} + 1.00\,\mathrm{mS}) + \mathrm{j}(48.1\,\mu\mathrm{S} + 0\,\mathrm{S})}$$
$$= \frac{1\angle 0^\circ}{1.00\,\mathrm{mS} + \mathrm{j}\,48.1\,\mu\mathrm{S}}$$
$$= \frac{1\angle 0^\circ}{1.001\,\mathrm{mS}\angle 2.75^\circ}$$
$$= \frac{1}{1.001\,\mathrm{mS}}\angle(0^\circ - 2.75^\circ)$$

= 999 Ω $\angle$ –2.75° in polar form

= 998 Ω – j 48.0 Ω in rectangular form

The second equivalent impedance, Z_{EQ2}, is the series combination $R + Z_{EQ1}$.

$Z_T = Z_{EQ2}$

where

$Z_{EQ2} = R + Z_{EQ1}$

= (20 kΩ + j 0 Ω) + (998 Ω – j 48.0 Ω)

= (20 kΩ + 998 Ω) – j (0 Ω + 48.0 Ω)

= 20.1 kΩ – j 48.0 Ω in rectangular form

= 20.1 kΩ $\angle$ –0.131° in polar form

Use Ohm's Law to calculate the total current I_T.

$I_T = V_S / Z_T$

= (1.0 V_{ac} $\angle$ 0°) / (20.1 kΩ $\angle$ –0.131°)

= (1.0 V_{ac} / 20.1 kΩ) $\angle$ [0° – (–0.131°)]

= 47.6 μA_{ac} $\angle$ 0.131°

Z_T is the series combination of $\boldsymbol{R}$ and Z_{EQ2}. The current through each component in series is the same, so that $I_{ZEQ} = I_T$, and $V_{OUT} = V_{ZEQ1}$. Use Ohm's Law for voltage to calculate V_{out}.

$$\begin{aligned}
V_{out} &= V_{ZEQ1} \\
&= I_{ZEQ1} \times Z_{EQ1} \\
&= I_T \times Z_{EQ1} \\
&= (47.6\ \mu\text{A}_{ac} \angle 0.131°)(999\ \Omega \angle -2.75°) \\
&= (47.6\ \mu\text{A}_{ac} \times 999\ \Omega) \angle (0.131° + -2.75°) \\
&= \mathbf{47.6\ mV_{ac} \angle -2.62°}
\end{aligned}$$

For $R_L = 10.0\ \text{k}\Omega$:

$$\begin{aligned}
Z_{EQ1} &= X_C \parallel R_L \\
&= \frac{1}{\dfrac{1}{-\text{j}X_C} + \dfrac{1}{R_L}} \\
&= \frac{1}{\dfrac{1}{-\text{j}\,20.8\ \text{k}\Omega} + \dfrac{1}{10.0\ \text{k}\Omega}} \\
&= \frac{1\angle 0°}{\dfrac{1\angle 0°}{20.8\ \text{k}\Omega \angle -90°} + \dfrac{1\angle 0°}{10.0 \angle 0°}} \\
&= \frac{1\angle 0°}{\dfrac{1}{20.8\ \text{k}\Omega}\angle[0° - (-90°)] + \dfrac{1}{10.0\ \text{k}\Omega}\angle 0°} \\
&= \frac{1\angle 0°}{(48.1\ \mu\text{S} \angle 90°) + (100\ \mu\text{S} \angle 0°)} \\
&= \frac{1\angle 0°}{(0\,\text{S} + \text{j}\,48.1\ \mu\text{S}) + (100\ \mu\text{S} + \text{j}\,0\,\text{S})} \\
&= \frac{1\angle 0°}{(0\,\text{S} + 100\ \mu\text{S}) + \text{j}(48.1\ \mu\text{S} + 0\,\text{S})} \\
&= \frac{1\angle 0°}{100\ \mu\text{S} + \text{j}\,48.1\ \mu\text{S}} \\
&= \frac{1\angle 0°}{111\ \mu\text{S} \angle 25.7°} \\
&= \frac{1}{111\ \mu\text{S}} \angle (0° - 25.7°) \\
&= 9.01\ \text{k}\Omega \angle -25.7° \text{ in polar form} \\
&= 8.12\ \text{k}\Omega - \text{j}\,3.91\ \text{k}\Omega \text{ in rectangular form}
\end{aligned}$$

The second equivalent impedance, Z_{EQ2}, is the series combination $\boldsymbol{R} + Z_{EQ1}$.

$$Z_T = Z_{EQ2}$$

where

$$\begin{aligned}
Z_{EQ2} &= R + Z_{EQ1} \\
&= (20\ \text{k}\Omega + \text{j}\,0\ \Omega) + (8.12\ \text{k}\Omega - \text{j}\,3.91\ \text{k}\Omega)
\end{aligned}$$

$= (20\ \text{k}\Omega + 8.12\ \text{k}\Omega) - \text{j}\ (0\ \Omega + 3.91\ \text{k}\Omega)$

$= 28.1\ \text{k}\Omega - \text{j}\ 3.91\ \text{k}\Omega$ in rectangular form

$= 28.4\ \text{k}\Omega \angle -7.90°$ in polar form

Use Ohm's Law to calculate the total current $\boldsymbol{I_T}$.

$$\boldsymbol{I_T} = \boldsymbol{V_S} / \boldsymbol{Z_T}$$
$$= (1.0\ \text{V}_{ac} \angle 0°) / (28.4\ \text{k}\Omega \angle -7.90°)$$
$$= (1.0\ \text{V}_{ac} / 28.4\ \text{k}\Omega) \angle [0° - (-7.90°)]$$
$$= 35.2\ \mu\text{A}_{ac} \angle 7.90°$$

$\boldsymbol{Z_T}$ is the series combination of $\boldsymbol{R}$ and $\boldsymbol{Z_{EQ1}}$. The current through each component in series is the same, so that $\boldsymbol{I_{ZEQ1}} = \boldsymbol{I_T}$, and $\boldsymbol{V_{OUT}} = \boldsymbol{V_{ZEQ1}}$, so use Ohm's Law to calculate $\boldsymbol{V_{OUT}}$.

$$\boldsymbol{V_{OUT}} = \boldsymbol{V_{ZEQ1}}$$
$$= \boldsymbol{I_T} \times \boldsymbol{Z_{EQ1}}$$
$$= (35.2\ \mu\text{A}_{ac} \angle 7.90°)\ (9.01\ \text{k}\Omega \angle -25.7°)$$
$$= (47.6\ \mu\text{A}_{ac} \times 9.01\ \text{k}\Omega) \angle (7.90° + -25.7°)$$
$$= \mathbf{317\ mV_{ac} \angle -17.8°}$$

16.1.5 Power in *RC* Circuits

1. Problem: Determine the total real, reactive, and apparent power for the circuit in Figure 16-27.

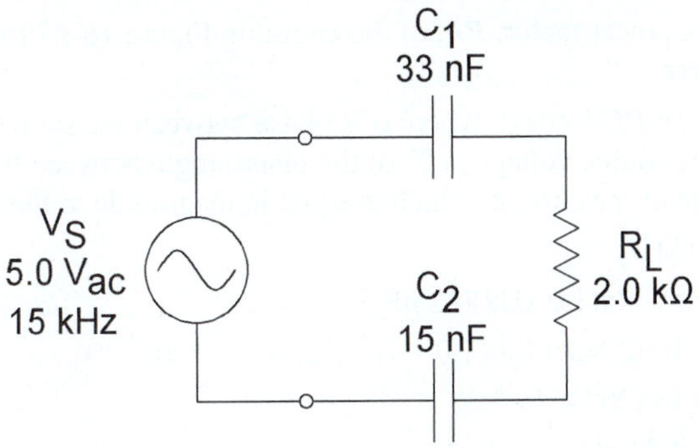

Figure 16-27: Power in *RC* Circuits Example for Problem 1

Solution: First, calculate the reactance, $\boldsymbol{X_C}$, for each capacitor. Then determine the total impedance for the circuit and use Watt's Law and the power triangle to calculate the powers in the circuit.

For $\boldsymbol{f}$ = 15 kHz,

$$\boldsymbol{X_{C1}} = 1 / [2\pi(15\ \text{kHz})(33\ \text{nF})]$$
$$= 1 / [2 \times 3.14 \times (15 \times 10^3\ \text{Hz}) \times (33 \times 10^{-9}\ \text{F})]$$
$$= 1 / (3.11 \times 10^{-3}\ \text{S})$$
$$= 322\ \Omega$$

$$\boldsymbol{X_{C2}} = 1 / [2\pi(15\ \text{kHz})(15\ \text{nF})]$$
$$= 1 / [2 \times 3.14 \times (15 \times 10^3\ \text{Hz}) \times (15 \times 10^{-9}\ \text{F})]$$
$$= 1 / (1.41 \times 10^{-3}\ \text{S})$$
$$= 707\ \Omega$$

For the series circuit,

$$\boldsymbol{Z_T} = \boldsymbol{X_{C1}} + \boldsymbol{R_L} + \boldsymbol{X_{C2}}$$
$$= (0\ \Omega - \text{j}\ 322\ \Omega) + (2.0\ \text{k}\Omega + \text{j}\ 0\ \Omega) + (0\ \Omega - \text{j}\ 707\ \Omega)$$
$$= (0\ \Omega + 2.0\ \text{k}\Omega + 0\ \Omega) - \text{j}\ (322\ \Omega + 0\ \Omega + 707\ \text{k}\Omega)$$

$= 2.0\ k\Omega - j\ 1.03\ k\Omega$ in rectangular form

$= 2.25\ k\Omega \angle -27.2°$ in polar form

The supply voltage is in rms units, so use Watt's Law to calculate the apparent power, $\boldsymbol{P_a}$.

$\boldsymbol{P_a} = \boldsymbol{V_S}^2 / \boldsymbol{Z_T}$

$= (5.0\ V_{ac})^2 / 2.25\ k\Omega$

$= 25.0\ V_{ac}^{\ 2} / 2.25\ k\Omega$

$= \mathbf{11.1\ mVA}$

The total impedance indicates that the phase angle for the circuit is $\varphi = -27.2°$. Use the power triangle to relate the real power, $\boldsymbol{P_r}$, and the reactive power, $\boldsymbol{P_{reac}}$, to the apparent power, $\boldsymbol{P_a}$.

$\boldsymbol{P_r} = \boldsymbol{P_a} \cos \varphi$

$= (11.1\ mVA)[\cos(-27.2°)]$

$= (11.1\ mVA)(0.889)$

$= \mathbf{9.88\ mW}$

and

$\boldsymbol{P_{reac}} = \boldsymbol{P_a} \sin \varphi$

$= (11.1\ mVA)[\sin(-27.2°)]$

$= (11.1\ mVA)(0.458)$

$= \mathbf{5.09\ mVAR}$

2. Problem: What is the power factor, ***PF***, of the circuit in Figure 16-27 for $\boldsymbol{f} = 1.0$ kHz, $\boldsymbol{f} = 5.0$ kHz, and $\boldsymbol{f} = 10.0$ kHz?

Solution: By definition $\boldsymbol{PF} = \cos \varphi$, where φ is phase between the source voltage and current. The phase angle of the source voltage is 0° so the phase angle between the voltage, $\boldsymbol{V_S}$, and current, $\boldsymbol{I_T}$, is the phase angle of the current, which is equal in magnitude to the phase angle of the impedance.

For $\boldsymbol{f} = 1.0$ kHz,

$\boldsymbol{X_{C1}} = 1 / [2\pi(1.0\ kHz)(33\ nF)]$

$= 1 / [2 \times 3.14 \times (1.0 \times 10^3\ Hz) \times (33 \times 10^{-9}\ F)]$

$= 1 / (207 \times 10^{-6}\ S)$

$= 4.82\ k\Omega$

$\boldsymbol{X_{C2}} = 1 / [2\pi(1.0\ kHz)(15\ nF)]$

$= 1 / [2 \times 3.14 \times (1.0 \times 10^3\ Hz) \times (15 \times 10^{-9}\ F)]$

$= 1 / (94.3 \times 10^{-6}\ S)$

$= 10.6\ k\Omega$

For the series circuit,

$\boldsymbol{Z_T} = \boldsymbol{X_{C1}} + \boldsymbol{R_L} + \boldsymbol{X_{C2}}$

$= (0\ \Omega - j\ 4.82\ k\Omega) + (2.0\ k\Omega + j\ 0\ \Omega) + (0\ \Omega - j\ 10.6\ k\Omega)$

$= (0\ \Omega + 2.0\ k\Omega + 0\ \Omega) - j\ (4.82\ k\Omega + 0\ \Omega + 10.6\ k\Omega)$

$= 2.0\ k\Omega - j\ 15.4\ k\Omega$ in rectangular form

$= 15.6\ k\Omega \angle -82.6°$ in polar form

Use the phase angle of the impedance, $\varphi = -82.6°$, to calculate the power factor, ***PF***.

$\boldsymbol{PF} = \cos(-82.6°)$

$= \mathbf{0.129}$

For $\boldsymbol{f} = 5.0$ kHz,

$\boldsymbol{X_{C1}} = 1 / [2\pi(5.0\ kHz)(33\ nF)]$

$= 1 / [2 \times 3.14 \times (5.0 \times 10^3 \text{ Hz}) \times (33 \times 10^{-9} \text{ F})]$
$= 1 / (1.07 \times 10^{-3} \text{ S})$
$= 965\ \Omega$

$X_{C2} = 1 / [2\pi(5.0 \text{ kHz})(15 \text{ nF})]$
$= 1 / [2 \times 3.14 \times (5.0 \times 10^3 \text{ Hz}) \times (15 \times 10^{-9} \text{ F})]$
$= 1 / (471 \times 10^{-6} \text{ S})$
$= 2.12\ \text{k}\Omega$

For the series circuit

$Z_T = X_{C1} + R_L + X_{C2}$
$= (0\ \Omega - j\ 965\ \Omega) + (2.0\ \text{k}\Omega + j\ 0\ \Omega) + (0\ \Omega - j\ 2.12\ \text{k}\Omega)$
$= (0\ \Omega + 2.0\ \text{k}\Omega + 0\ \Omega) - j\ (965\ \Omega + 0\ \Omega + 2.12\ \text{k}\Omega)$
$= 2.0\ \text{k}\Omega - j\ 3.09\ \text{k}\Omega$ in rectangular form
$= 3.68\ \text{k}\Omega \angle -57.1°$ in polar form

Use the phase angle of the impedance, $\varphi = -57.1°$, to calculate the power factor, ***PF***.

$PF = \cos(-57.1°)$
$= \mathbf{0.544}$

For $f = 10.0$ kHz,

$X_{C1} = 1 / [2\pi(10.0 \text{ kHz})(33 \text{ nF})]$
$= 1 / [2 \times 3.14 \times (10.0 \times 10^3 \text{ Hz}) \times (33 \times 10^{-9} \text{ F})]$
$= 1 / (2.07 \times 10^{-3} \text{ S})$
$= 482\ \Omega$

$X_{C2} = 1 / [2\pi(10.0 \text{ kHz})(15 \text{ nF})]$
$= 1 / [2 \times 3.14 \times (10.0 \times 10^3 \text{ Hz}) \times (15 \times 10^{-9} \text{ F})]$
$= 1 / (943 \times 10^{-6} \text{ S})$
$= 1.06\ \text{k}\Omega$

For the series circuit

$Z_T = X_{C1} + R_L + X_{C2}$
$= (0\ \Omega - j\ 482\ \Omega) + (2.0\ \text{k}\Omega + j\ 0\ \Omega) + (0\ \Omega - j\ 1.06\ \text{k}\Omega)$
$= (0\ \Omega + 2.0\ \text{k}\Omega + 0\ \Omega) - j\ (482\ \Omega + 0\ \Omega + 1.06\ \text{k}\Omega)$
$= 2.0\ \text{k}\Omega - j\ 1.54\ \text{k}\Omega$ in rectangular form
$= 2.53\ \text{k}\Omega \angle -37.7°$ in polar form

Use the phase angle of the impedance, $\varphi = -37.7°$, to calculate the power factor, ***PF***.

$PF = \cos(-37.7°)$
$= \mathbf{0.792}$

16.2 Advanced

16.2.1 Series Circuits

1. Problem: Prove that the total capacitive reactance for ***N*** capacitors in series is

$X_{CT} = X_{C1} + X_{C2} + \ldots + X_{CN}$

Solution: For ***N*** capacitors in series, the total capacitance C_T is

$C_T = 1 / [(1 / C_1) + (1 / C_2) + \ldots + (1 / C_N)]$
$1 / C_T = 1 / \{[(1 / C_1) + (1 / C_2) + \ldots + (1 / C_N)]\}$
$1 / C_T = (1 / C_1) + (1 / C_2) + \ldots + (1 / C_N)$

The capacitive reactance, X_C, for a capacitance, C, at frequency, f, is

$$X_C = 1 / (2\pi f C)$$

Use this to calculate the total capacitive reactance, X_{CT}, for total capacitance, C_T.

$$\begin{aligned} X_{CT} &= 1 / (2\pi f C_T) \\ &= (1 / 2\pi f)\,(1 / C_T) \\ &= (1 / 2\pi f)\,[(1 / C_1) + (1 / C_2) + \ldots + (1 / C_N)] \\ &= [1 / (2\pi f C_1)] + [1 / (2\pi f C_2)] + \ldots + [1 / (2\pi f C_N)] \\ &= X_{C1} + X_{C2} + \ldots + X_{CN} \end{aligned}$$

2. Problem: For a series **RC** circuit connected across voltage source V_S, prove that the voltage divider theorem is valid so that the voltage across component K is

$$V_{RK} = V_S\,[R_K / Z_T] \text{ for resistor } R_K$$

$$V_{CK} = V_S\,[X_{CK} / Z_T] \text{ for capacitor } C_K$$

Solution: For a series **RC** circuit with M resistors and N capacitors, the total impedance is Z_T. From Ohm's Law,

$$I_T = V_S / Z_T$$

For a series circuit, the current through any component is equal to I_T. From Ohm's Law, the voltage drop across resistor R_K is

$$\begin{aligned} V_{RK} &= I_T\,R_K \\ &= (V_S / Z_T)\,R_K \\ &= (V_S\,R_K / Z_T) \\ &= V_S\,(R_K / Z_T) \end{aligned}$$

Similarly, the voltage drop across capacitor C_K is

$$\begin{aligned} V_{CK} &= I_T\,X_{CK} \\ &= (V_S\,X_{CK}) / Z_T) \\ &= V_S\,(X_{CK} / Z_T) \end{aligned}$$

16.2.2 Parallel Circuits

1. Problem: Prove that the total reactance of N capacitors in parallel is

$$X_{CT} = 1 / [(1 / X_{C1}) + (1 / X_{C2}) + \ldots + (1 / X_{CN})]$$

Solution: For N capacitors in parallel, the total capacitance C_T is

$$C_T = C_1 + C_2 + \ldots + C_N$$

The capacitive reactance, X_C, for a capacitor, C, and frequency, f, is

$$X_C = 1 / (2\pi f C)$$

From this,

$$\begin{aligned} 1 / X_C &= 1 / [1 / (2\pi f C)] \\ &= 2\pi f C \end{aligned}$$

For total capacitance, C_T,

$$\begin{aligned} 1 / X_{CT} &= 2\pi f C_T \\ &= 2\pi f\,(C_1 + C_2 + \ldots + C_N) \\ &= 2\pi f C_1 + 2\pi f C_2 + \ldots + 2\pi f C_N \\ &= (1 / X_{C1}) + (1 / X_{C2}) + \ldots + (1 / X_{CN}) \\ 1 / (1 / X_{CT}) &= 1 / [(1 / X_{C1}) + (1 / X_{C2}) + \ldots + (1 / X_{CN})] \\ X_{CT} &= 1 / [(1 / X_{C1}) + (1 / X_{C2}) + \ldots + (1 / X_{CN})] \end{aligned}$$

2. Problem: For a parallel ***RC*** circuit with total current $\boldsymbol{I_T}$, prove that the current divider theorem is valid so that the current through component ***K*** is

$$I_{RK} = I_T\,[Z_T / R_K] \text{ for resistor } R_K$$

$$I_{CK} = I_T\,[Z_T / X_{CK}] \text{ for capacitor } C_K$$

Solution: From Ohm's Law,

$$V_T = I_T Z_T$$

For a parallel circuit, the voltage across any component is equal to $\boldsymbol{V_T}$. From Ohm's Law, the current $\boldsymbol{I_{RK}}$ through resistor $\boldsymbol{R_K}$ is

$$\begin{aligned} I_{RK} &= V_T / R_K \\ &= (I_T Z_T) / R_K \\ &= I_T (Z_T / R_K) \end{aligned}$$

and the current through capacitor $\boldsymbol{C_K}$ is

$$\begin{aligned} I_{CK} &= V_T / X_{CK} \\ &= (I_T Z_T) / X_{CK} \\ &= I_T (Z_T / X_{CK}) \end{aligned}$$

3. Problem: For a resistor and capacitor in parallel with resistance, ***R***, and reactance, $\boldsymbol{X_C}$, prove that the product over sum method can be used to calculate the total impedance, $\boldsymbol{Z_T}$, so that

$$Z_T = (-\text{j}\, RX_C) / (R - \text{j}\, X_C)$$

Solution: Assume that a voltage $\boldsymbol{V_S}$ is applied to the parallel combination of ***R*** and $\boldsymbol{X_C}$. Because the components are in parallel the same applied voltage, $\boldsymbol{V_S}$, appears across each. From Ohm's Law,

$$I_R = V_S / R$$

and

$$I_{XC} = V_S / -\text{j}\, X_C$$

From Kirchhoff's Current Law the total current, $\boldsymbol{I_T}$, must equal the sum of these currents.

$$\begin{aligned} I_T &= I_R + I_{XC} \\ &= (V_S / R) + (V_S / -\text{j}\, X_C) \\ &= (V_S / R)\,[(-\text{j}\, X_C) / (-\text{j}\, X_C)] + (V_S / -\text{j}\, X_C)\,(R / R) \\ &= [(-\text{j}\, V_S X_C) / (-\text{j}\, RX_C)] + [V_S R / (-\text{j}\, RX_C)] \\ &= [(-\text{j}\, V_S X_C) + V_S R] / (-\text{j}\, RX_C) \\ &= [V_S (-\text{j}\, X_C + R)] / (-\text{j}\, RX_C) \\ &= [V_S (R - \text{j}\, X_C)] / (-\text{j}\, RX_C) \end{aligned}$$

By definition, the total impedance, $\boldsymbol{Z_T}$, is the applied voltage, $\boldsymbol{V_S}$, divided the total current, $\boldsymbol{I_T}$.

$$\begin{aligned} Z_T &= V_S / I_T \\ &= V_S / \{[V_S (R - \text{j}\, X_C)] / (-\text{j}\, RX_C)\} \\ &= V_S\, \{(-\text{j}\, RX_C) / [V_S (R - \text{j}\, X_C\} \\ &= (-\text{j}\, V_S RX_C) / [V_S (R - \text{j}\, X_C)] \\ &= (-\text{j}\, RX_C) / (R - \text{j}\, X_C) \end{aligned}$$

16.2.3 Series-Parallel Circuits

1. Problem: For the ***RC*** bridge circuit in Figure 16-28, for what circuit values will $\boldsymbol{V_A} = \boldsymbol{V_B}$ so that $\boldsymbol{I_{RL}} = 0$? At what frequency ***f*** will this be true?

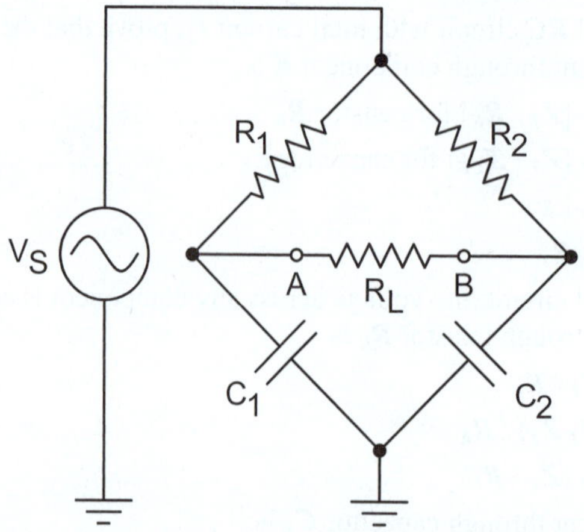

Figure 16-28: *RC* Bridge Circuit for Problem 1

Solution: Begin by using the voltage divider theorem for ***RC*** circuits.

$$V_A = -\text{j}\,X_{C1} / (R_1 - \text{j}\,X_{C1})$$

$$V_B = -\text{j}\,X_{C2} / (R_2 - \text{j}\,X_{C2})$$

Next, set V_A and V_B equal.

$$V_A = V_B$$

$$-\text{j}\,X_{C1} / (R_1 - \text{j}\,X_{C1}) = -\text{j}\,X_{C2} / (R_2 - \text{j}\,X_{C2})$$

$$[-\text{j}\,X_{C1} / (R_1 - \text{j}\,X_{C1})]\,(R_1 - \text{j}\,X_{C1}) = [-\text{j}\,X_{C2} / (R_2 - \text{j}\,X_{C2})]\,(R_1 - \text{j}\,X_{C1})$$

$$(-\text{j}\,X_{C1})\,(R_2 - \text{j}\,X_{C2}) = \{[(-\text{j}\,X_{C2})\,(R_1 - \text{j}\,X_{C1})] / (R_2 - \text{j}\,X_{C2})\}\,(R_2 - \text{j}\,X_{C2})$$

$$-\text{j}\,R_2X_{C1} + \text{j}^2\,X_C X_{C2} = -\text{j}\,R_1X_{C2} + \text{j}^2\,X_{C1}X_{C2}$$

$$-\text{j}\,R_2X_{C1} + (-1)\,X_C X_{C2} = -\text{j}\,R_1X_{C2} + (-1)\,X_{C1} X_{C2}$$

$$-\text{j}\,R_2X_{C1} - X_{C1}X_{C2} + X_{C1}X_{C2} = -\text{j}\,R_1X_{C2} - X_{C1}X_{C2} + X_{C1}X_{C2}$$

$$(-\text{j}\,R_2X_{C1}) / -\text{j} = (-\text{j}\,R_1X_{C2}) / -\text{j}$$

$$(R_2X_{C1}) / (R_1X_{C1}) = (R_1X_{C2}) / (R_1X_{C1})$$

$$R_2 / R_1 = X_{C2} / X_{C1}$$

$$= [1 / (2\pi f C_2)] / [1 / (2\pi f C_1)]$$

$$= (2\pi f C_1)] / (2\pi f C_2)$$

$$= C_1 / C_2$$

V_A will equal V_B when the ratio of C_1 / C_2 is the inverse of the ratio of R_2 / R_1. Because f does not appear in this relationship, if $V_A = V_B$ at one frequency it will be true at all frequencies.

2. Problem: For the ***RC*** bridge circuit in Figure 16-29, for what frequency f will $V_A = V_B$?

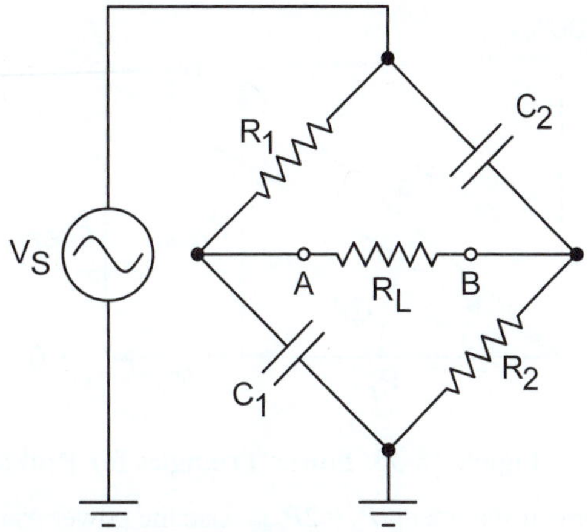

Figure 16-29: *RC* Bridge Circuit for Problem 2

Solution: Begin by using the voltage divider theorem for ***RC*** circuits.

$$V_A = -\text{j}\,X_{C1} / (R_1 - \text{j}\,X_{C1})$$

$$V_B = R_2 / (R_2 - \text{j}\,X_{C2})$$

Setting these terms equal gives

$$\begin{aligned}
V_A &= V_B \\
-\text{j}\,X_{C1} / (R_1 - \text{j}\,X_{C1}) &= R_2 / (R_2 - \text{j}\,X_{C2}) \\
[-\text{j}\,X_{C1} / (R_1 - \text{j}\,X_{C1})]\,(R_1 - \text{j}\,X_{C1}) &= [R_2 / (R_2 - \text{j}\,X_{C2})]\,(R_1 - \text{j}\,X_{C1}) \\
(-\text{j}\,X_{C1})\,(R_2 - \text{j}\,X_{C2}) &= \{[(R_2)\,(R_1 - \text{j}\,X_{C1})] / (R_2 - \text{j}\,X_{C2})\}\,(R_2 - \text{j}\,X_{C2}) \\
-\text{j}\,R_2 X_{C1} + \text{j}^2 X_{C1} X_{C2} + \text{j}\,R_2 X_{C1} &= R_1 R_2 - \text{j}\,R_2 X_{C1} + \text{j}\,R_2 X_{C1} \\
(-1)\,X_{C1} X_{C2} &= R_1 R_2 \\
-(X_{C1} X_C) &= R_1 R_2 \\
-[1 / (2\pi f C_1)]\,[1 / (2\pi f C_2)] &= R_1 R_2 \\
-1 / [(2\pi f C_1)\,(2\pi f C_2)] &= R_1 R_2 \\
-1 / (4\pi^2 f^2 C_1 C_2) &= R_1 R_2 \\
[-1 / (4\pi^2 f^2 C_1 C_2)]\,f^2 &= (R_1 R_2)\,f^2 \\
[-1 / (4\pi^2 C_1 C_2)] / (R_1 R_2) &= [(R_1 R_2)\,f^2] / (R_1 R_2) \\
-1 / (4\pi^2 R_1 R_2 C_1 C_2) &= f^2
\end{aligned}$$

For V_A and V_B to be equal, f^2 must equal $-1 / (4\pi^2 R_1 R_2 C_1 C_2$. But $1 / (4\pi^2 R_1 R_2 C_1 C_2)$ is always positive, so $-1 / (4\pi^2 R_1 R_2 C_1 C_2)$ must always be negative. Because f^2 is always a positive value, **there is no frequency or component values for which $V_A = V_B$.**

16.2.4 Power in *RC* Circuits

1. Problem: The real power, P_r, in an unloaded ***RC*** circuit is twice the reactive power, P_{reac}. When the circuit is loaded, the real power increases by 1.00 W to P'_r, the reactive power is unchanged so that the loaded reactive power $P'_{reac} = P_{reac}$, and the phase angle decreases by 3° to φ'. What is the apparent power, P'_a, in the loaded circuit?

   Solution: Refer to Figure 16-30, which shows the unloaded and loaded power triangles for the circuit.

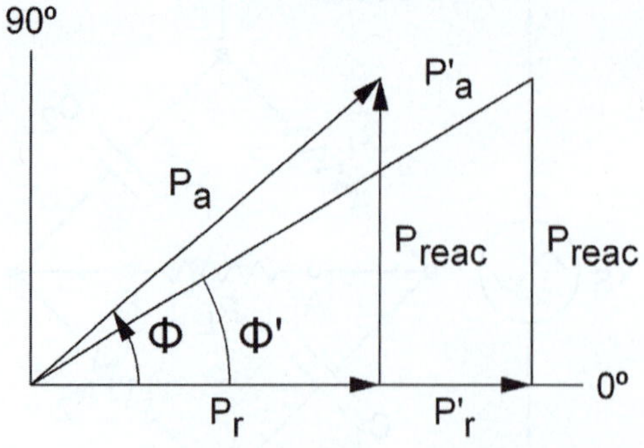

Figure 16-30: Power Triangles for Problem 1

From the problem statement, $P_r = 2P_{reac}$. Use the power triangle for the unloaded circuit to calculate φ.

$$\begin{aligned}\varphi &= \tan^{-1}(P_{reac} / P_r)\\ &= \tan^{-1}(P_{reac} / 2P_{reac})\\ &= \tan^{-1}(1/2)\\ &= 26.6°\end{aligned}$$

From the problem statement, the loaded phase angle 3° less than the unloaded phase angle.

$$\begin{aligned}\varphi' &= \varphi - 3°\\ &= 23.6°\end{aligned}$$

From the power triangle for the loaded circuit,

$$\begin{aligned}P'_a \cos\varphi' &= P'_r\\ (P'_a \cos\varphi') / \cos\varphi' &= P'_r / \cos\varphi'\\ P'_a &= P'_r / \cos\varphi'\\ &= P'_r / \cos 23.6°\\ &= P'_r / 0.917\\ P'^2_a &= (P'_r / 0.917)^2\\ &= P'^2_r / 0.840\end{aligned}$$

From the power triangle, the relationship between apparent, real, and reactive power is $P_a^2 = P_r^2 + P_{reac}^2$, so $P'^2_a = P'^2_r + P'^2_{reac}$. From the problem statement, $P'_{reac} = P_{reac}$, $P_{reac} = P_r / 2$, and $P_r = P'_r - 0.50$ W.

$$\begin{aligned}P'^2_a &= P'^2_r + P_{reac}^2\\ &= P'^2_r + (P_r / 2)^2\\ &= P'^2_r + [(P'_r - 1.00\text{ W}) / 2]^2\\ &= P'^2_r + (P'_r - 1.00\text{ W})^2 / 4\\ &= P'^2_r + [P'^2_r - (2.00\text{ W})P'_r + 1.00\text{ W}^2] / 4\\ &= P'^2_r + 0.25P'^2_r - (.500\text{ W})P'_r + 0.250\text{ W}^2\\ &= 1.25P'^2_r - (0.500\text{ W})P'_r + 0.100\text{ W}^2\end{aligned}$$

Set the two expressions for P'^2_r equal.

$$\begin{aligned}1.25P'^2_r - (0.500\text{ W})P'_r + 0.100\text{ W}^2 &= P'^2_r / 0.840\\ (1.25P'^2_r - (0.500\text{ W})P'_r + 0.100\text{ W}^2)\,0.840 &= (P'^2_r / 0.840)\,0.840\\ 1.050P'^2_r - (0.420\text{ W})P'_r + 0.210\text{ W}^2 - P'^2_r &= P'^2_r - P'^2_r\\ .0502P'^2_r - (0.420\text{ W})P'_r + 0.210\text{ W}^2 &= 0\end{aligned}$$

This is a quadratic formula with $\boldsymbol{a} = 0.0502$, $\boldsymbol{b} = -0.420$, and $\boldsymbol{c} = 0.210$.

$$\begin{aligned} P'_r &= \frac{-b \pm \sqrt{b^2 - 4ac}}{2a}\ \text{W} \\ &= \frac{0.420 \pm \sqrt{(-0.420)^2 - 4(0.0502)(0.210)}}{2(0.0502)}\ \text{W} \\ &= \frac{0.420 \pm \sqrt{0.177 - 0.0422}}{0.100}\ \text{W} \\ &= \frac{0.420 \pm \sqrt{0.134}}{0.100}\ \text{W} \\ &= \frac{0.420 \pm 0.367}{0.100}\ \text{W} \\ &= 0.534\ \text{W or } 7.83\ \text{W} \end{aligned}$$

Because $P_r = P'_r - 1.0$ W, P'_r must equal 7.83 W. $P'_a = P'_r / 0.917$, so for $P'_r = 7.83$ W,

$$\begin{aligned} P'_a &= 7.83\ \text{W} / 0.917 \\ &= \mathbf{8.54\ VA} \end{aligned}$$

2. Problem: Refer to the circuit in Figure 16-31. For what value of R_L is the power factor in the load 0.500?

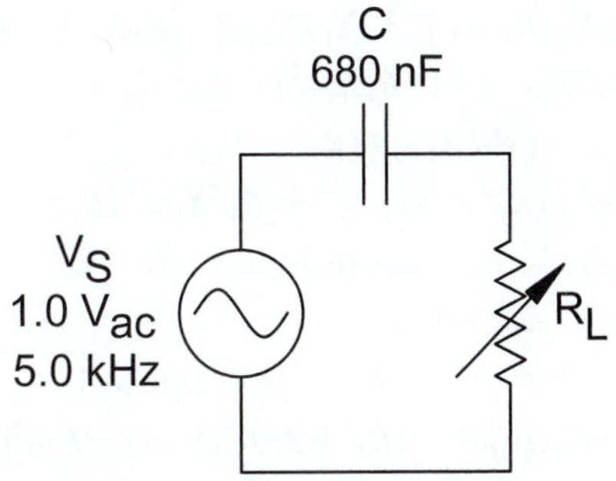

Figure 16-31: Power in *RC* Circuits Example for Problem 2

Solution: The power factor in the load is 0.500 when cos $\varphi = 0.500$. For a series ***RC*** circuit,

$$\begin{aligned} \varphi &= \tan^{-1} (X_C / R) \\ \tan \varphi &= \tan [\tan^{-1} (X_C / R)] \\ (\tan \varphi) / X_C &= (X_C / R) / X_C \\ 1 / [(\tan \varphi) / X_C] &= 1 / (1 / R) \\ X_C / \tan \varphi &= R \end{aligned}$$

For $C = 680$ nF and $f = 5.0$ kHz,

$$\begin{aligned} X_C &= 1 / [2\pi(5.0\ \text{kHz})(680\ \text{nF})] \\ &= 1 / [2 \times 3.14 \times (5.0 \times 10^3\ \text{Hz}) \times (680 \times 10^{-9}\ \text{F})] \\ &= 1 / (21.4 \times 10^{-3}\ \text{S}) \\ &= 46.8\ \Omega \end{aligned}$$

For cos $\varphi = 0.500$,

$$\begin{aligned} \cos^{-1} (\cos \varphi) &= \cos^{-1} (0.500) \\ \varphi &= 60° \end{aligned}$$

From this,

$$R = X_C / \tan \varphi$$

$= 46.8\ \Omega / \tan 60°$

$= 46.8\ \Omega / 1.73$

$= \mathbf{27.0\ \Omega}$

16.3 Just For Fun

1. Problem: For a specific frequency f, what values of resistance R_S and capacitance C_S in a series RC circuit will give the same impedance Z_T as the values of resistance R_P and capacitance C_P in a parallel RC circuit?

Solution: At a specific frequency f, the reactance X_{CS} and X_{CP} of the capacitors C_S and C_P are

$X_{CS} = 1 / (2\pi f C_S)$

$X_{CP} = 1 / (2\pi f C_P)$

Use the product over sum method to find the total parallel impedance, Z_T, for R_P and C_P.

$Z_{PT} = R_P \parallel -\text{j}\, X_{CP}$

$= [(R_P + \text{j}\,0)\,(0 - \text{j}\,X_{CP})] / (R_P - \text{j}\,X_{CP})$

$= [(R_P)(0) - (R_P)(-\text{j}\,X_{CP}) + (\text{j}\,0)(0) + (\text{j}\,0)(-\text{j}\,X_P)] / (R_P - \text{j}\,X_{CP})$

$= -\text{j}\,R_P X_{CP} / (R_P - \text{j}\,X_{CP})$

$= [-\text{j}\,R_P X_{CP} / (R_P - \text{j}\,X_{CP})][(R_P + \text{j}X_{CP}) / (R_P + \text{j}\,X_{CP})]$

$= [(-\text{j}\,R_P X_{CP})\,(R_P + \text{j}\,X_{CP})][(R_P - \text{j}X_{CP})\,(R_P + \text{j}\,X_{CP})]$

$= [-\text{j}\,R_P X_{CP} R_P - (\text{j}^2\,R_P X_{CP} X_{CP})] / (R_P R_P + \text{j}\,X_{CP} R_P - \text{j}\,X_{CP} R_P - \text{j}^2\,X_{CP} X_{CP})$

$= (-\text{j}\,R_P^2 X_{CP} - (-1)\,R_P X_{CP}^2) / (R_P^2 + \text{j}\,0 - (-1)\,X_{CP}^2)$

$= (R_P X_{CP}^2 - \text{j}\,R_P^2 X_{CP}) / (R_P^2 + X_{CP}^2)$

$= [R_P X_{CP}^2 / (R_P^2 + X_{CP}^2)] - \text{j}\,[R_P^2 X_{CP} / (R_P^2 + X_{CP}^2)]$

The real term in Z_{PT} is the series resistance, R_S.

$R_S = R_P X_{CP}^2 / (R_P^2 + X_{CP}^2)$

$= R_P\,[1 / (2\pi f C_P)]^2 / \{R_P^2 + [1 / (2\pi f C_P)^2]\}$

$= R_P\,[1 / (2\pi f C_P)]^2 / \{R_P^2\,[(2\pi f C_P)^2 / (2\pi f C_P)^2] + [1 / (2\pi f C_P)^2]\}$

$= R_P\,[1 / (2\pi f C_P)]^2 / \{[R_P^2\,(2\pi f C_P)^2 + 1] / (2\pi f C_P)^2\}$

$= R_P\,(2\pi f C_P)^2 / \{[R_P^2\,(2\pi f C_P)^2 + 1]\,(2\pi f C_P)^2\}$

$= \mathbf{R_P / [R_P^2\,(2\pi f C_P)^2 + 1]}$

Similarly, the imaginary term is the series capacitance, C_S.

$C_S = R_P^2 X_{CP} / (R_P^2 + X_{CP}^2)$

$= R_P^2\,[1 / (2\pi f C_P)] / \{R_P^2 + [1 / (2\pi f C_P)^2]\}$

$= R_P^2\,[1 / (2\pi f C_P)]^2 / \{R_P^2\,[(2\pi f C_P)^2 / (2\pi f C_P)^2] + [1 / (2\pi f C_P)^2]\}$

$= R_P^2\,[1 / (2\pi f C_P)]^2 / \{[R_P^2\,(2\pi f C_P)^2 + 1] / (2\pi f C_P)^2\}$

$= R_P^2\,(2\pi f C_P)^2 / \{[R_P^2\,(2\pi f C_P)^2 + 1]\,(2\pi f C_P)\}$

$= \mathbf{R_P^2\,(2\pi f C_P) / [R_P^2\,(2\pi f C_P)^2 + 1]}$

2. Problem: For a specific frequency f, what values of resistance R_P and capacitance C_P in a parallel RC circuit will give the same impedance Z_T as the values of resistance R_S and capacitance C_S in a series RC circuit?

Solution: At a specific frequency f, the reactance X_{CS} and X_{CP} of the capacitors C_S and C_P are

$X_{CS} = 1 / (2\pi f C_S)$

$X_{CP} = 1 / (2\pi f C_P)$

The total series impedance Z_{ST} for R_S and C_S is

$$Z_{ST} = R_S - j\,X_S$$

By definition, the total series admittance, Y_{ST}, is the reciprocal of Z_{ST}.

$$\begin{aligned}
Y_{ST} &= 1 / (R_S - j\,X_{CS}) \\
&= [1 / (R_S - j\,X_{CS})]\,[(R_S + j\,X_{CS})\,(R_S + j\,X_{CS})] \\
&= (R_S + j\,X_{CS}) / [(R_S - j\,X_{CS})\,(R_S + j\,X_{CS})] \\
&= (R_S + j\,X_{CS}) / (R_S R_S + j\,R_S X_{CS} - j\,R_S X_{CS} - j^2\,X_{CS} X_{CS}) \\
&= (R_S + j\,X_{CS}) / (R_S^2 + j\,0 - (-1)\,X_{CS}^2) \\
&= (R_S + j\,X_{CS}) / (R_S^2 + X_{CS}^2) \\
&= [R_S / (R_S^2 + X_{CS}^2)] + j\,[X_{CS} / (R_S^2 + X_{CS}^2)]
\end{aligned}$$

The total parallel impedance, Z_{PT}, for R_P and C_P is

$$Z_{PT} = [(1 / R_P) + (1/ -j\,X_P)]^{-1}$$

By definition, the total parallel admittance, Y_{PT}, is the reciprocal of Z_{PT}.

$$\begin{aligned}
Y_{PT} &= (1 / R_P) + (1/ -j\,X_P) \\
&= (1 / R_P) + (1 / -j\,X_P)\,(j / j) \\
&= (1 / R_P) + [j / (-j^2\,X_P)] \\
&= (1 / R_P) + \{j / [-(-1)\,X_P]\} \\
&= (1 / R_P) + j\,(1 / X_P)
\end{aligned}$$

For Z_{ST} and Z_{PT} to be equal, Y_{ST} and Y_{PT} must be equal. The real (resistive) term in Y_{PT} corresponds to the real term in Y_{ST}.

$$\begin{aligned}
1 / R_P &= [R_S / (R_S^2 + X_S^2)] \\
1 / (1 / R_P) &= 1 / [R_S / (R_S^2 + X_S^2)] \\
R_P &= (R_S^2 + X_S^2) / R_S \\
&= \mathbf{[R_S^2 + (1 / 2\pi f C_S)^2] / R_S}
\end{aligned}$$

Similarly, the imaginary (reactive) term in Y_{PT} corresponds to the imaginary term in Y_{ST}.

$$\begin{aligned}
1 / X_{CP} &= [X_{CS} / (R_S^2 + X_{CS}^2)] \\
1 / [1 / (2\pi f C_P)] &= (1 / 2\pi f C_S) / [R_S^2 + (1 / 2\pi f C_S)^2] \\
2\pi f C_P &= 1 / \{2\pi f C_S\,[R_S^2 + (1 / 2\pi f C_S)^2]\} \\
&= 1 / \{R_S^2\,(2\pi f C_S) + [(1 / 2\pi f C_S)^2\,(2\pi f C_S)]\} \\
&= 1 / \{[R_S^2\,(2\pi f C_S)]\,(2\pi f C_S / 2\pi f C_S) + (1 / 2\pi f C_S)\} \\
&= 1 / \{[R_S^2\,(2\pi f C_S)^2 / 2\pi f C_S] + (1 / 2\pi f C_S)\} \\
&= 1 / \{[R_S^2\,(2\pi f C_S)^2 + 1] / 2\pi f C_S\} \\
&= 2\pi f C_S / [R_S^2\,(2\pi f C_S)^2 + 1] \\
(2\pi f C_P) / 2\pi f &= \{2\pi f C_S / [R_S^2\,(2\pi f C_S)^2 + 1]\} / 2\pi f \\
C_P &= \mathbf{C_S / [R_S^2\,(2\pi f C_S)^2 + 1]}
\end{aligned}$$

3. Problem: The circuit analysis techniques for resistive circuits can also be applied to ***RC*** circuits once the capacitors are replaced with capacitive reactances. For the ladder circuit shown in Figure 16-32, use loop current analysis to determine the output voltage V_{out} when $R = X_C$.

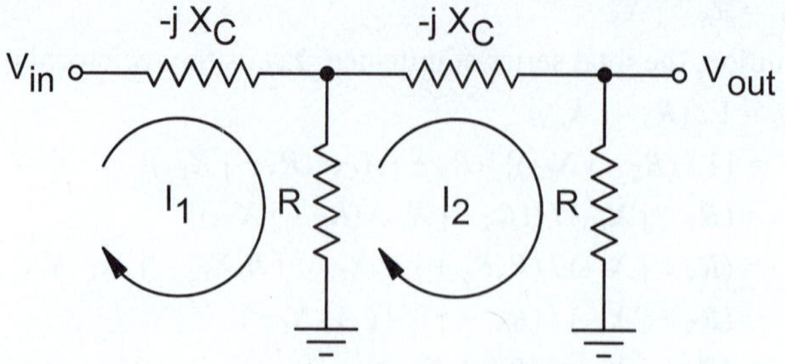

Figure 16-32: *RC* Ladder Network for Problem 3

Solution: From Ohm's Law, $V_{out} = I_2R$. For the reference currents shown, the loop equation for I_1 is

$$V_{in} - I_1(-j\,X_C) - (I_1 - I_2)R = 0$$
$$V_{in} + j\,I_1X_C - I_1R + I_2R = 0$$
$$V_{in} + I_1(j\,X_C - R) + I_2R = 0$$

For the reference currents shown, the loop equation for I_2 is

$$-(I_2 - I_1)R - I_2(-j\,X_C) - I_2R = 0$$
$$-I_2R + I_1R + I_2(j\,X_C) - I_2R = 0$$
$$I_1R - I_2R + j\,I_2X_C - I_2R = 0$$
$$I_1R + I_2(j\,X_C - 2R) = 0$$

Rearranging the loop equations gives

$$(R - j\,X_C)I_1 + -RI_2 = V_{in}$$
$$-RI_1 + (2R - j\,X_C)I_2 = 0$$

From the problem statement, $R = X_C$.

$$(R - j\,R)\,I_1 + -RI_2 = V_{in}$$
$$-RI_1 + (2R - j\,R)I_2 = 0$$

The matrix-vector form is

$$\begin{vmatrix} R - j\,R & -R \\ -R & 2R - j\,R \end{vmatrix} \bullet \begin{vmatrix} I_1 \\ I_2 \end{vmatrix} = \begin{vmatrix} V_{in} \\ 0 \end{vmatrix}$$

Use characteristic matrices to solve this system for I_2. The matrix for I_2 is

$$\begin{vmatrix} R - j\,R & V_{in} \\ -R & 0 \end{vmatrix}$$

with the determinant

$$(R - j\,R)(0) - (-R)(V_{in}) = 0 + V_{in}R$$
$$= V_{in}R$$

The characteristic matrix is

$$\begin{vmatrix} R - j\,R & -R \\ -R & 2R - j\,R \end{vmatrix}$$

with the determinant

$$(R - j\,R)(\,2R - j\,R) - (-R)(-R) = 2RR - j\,RR - 2j\,RR + j^2\,RR - RR$$
$$= 2R^2 - j\,R^2 - 2j\,R^2 + j^2\,R^2 - R^2$$
$$= 2R^2 - j\,3RX_C + (-1)\,R^2 - R^2$$
$$= 2R^2 - R^2 - R^2 - j\,3R^2$$

$= -\text{j}\,3\boldsymbol{R}^2$

$\boldsymbol{I}_2$ is equal to the quotient of determinants for the $\boldsymbol{I}_2$ matrix and characteristic matrix.

$\boldsymbol{I}_2 = (\boldsymbol{V}_{in}\boldsymbol{R}) / (-\text{j}\,3\boldsymbol{R}^2)$

$= \boldsymbol{V}_{in} / (-\text{j}\,3\boldsymbol{R})$

From this,

$\boldsymbol{V}_{out} = \boldsymbol{I}_2\boldsymbol{R}$

$= [\boldsymbol{V}_{in} / (-\text{j}\,3\boldsymbol{R})]\,\boldsymbol{R}$

$= (\boldsymbol{V}_{in} / -\text{j}\,3)\,(\text{j}\,3 / \text{j}\,3)$

$= \boldsymbol{V}_{in}\,[\text{j}\,3 / (-\text{j}^2\,9)]$

$= \boldsymbol{V}_{in}\,\{\text{j}\,3 / [-(-1)\,(9)]\}$

$= \boldsymbol{V}_{in}\,[\text{j}\,(3 / 9)]$

$= \boldsymbol{V}_{in}\,[\text{j}\,(1 / 3)]$

$= \text{j}\,(\boldsymbol{V}_{in} / 3)$

$\mathbf{= (\boldsymbol{V}_{in} / 3) \angle 90°}$

The output voltage has an amplitude 1/3 that of the input voltage and leads the input voltage by 90°.

17. *RL* Circuits

Note: For all problems in this section, the phase of ac sources is 0° unless otherwise stated.

17.1 Basic

17.1.1 Inductive Reactance

1. Problem: Calculate the inductive reactance, X_L, for each value of inductance, L, and frequency, f.

   a. $L = 5.6$ mH, $f = 15$ kHz

   b. $L = 39$ μH, $f = 250$ Hz

   c. $L = 150$ mH, $f = 1.5$ MHz

   d. $L = 5.0$ H, $f = 20$ mHz

   Solution: The equation for capacitive reactance X_L for frequency, f, and inductance, L, is

   $X_L = 2\pi fL$

   For the values of inductance, L, and frequency, f, in Problem 1, the inductive reactances, X_L, are:

   a. $X_L = 2\pi(15 \text{ kHz})(5.6 \text{ mH})$
   $= 2 \times 3.14 \times (15 \times 10^3 \text{ Hz}) \times (5.6 \times 10^{-3} \text{ H})$
   $= \mathbf{528\ \Omega}$

   b. $X_L = 2\pi(250 \text{ Hz})(39\ \mu\text{H})$
   $= 2 \times 3.14 \times (250 \text{ Hz}) \times (39 \times 10^{-6} \text{ H})$
   $= \mathbf{61.3\ m\Omega}$

   c. $X_L = 2\pi(1.5 \text{ MHz})(150 \text{ mH})$
   $= 2 \times 3.14 \times (1.5 \times 10^6 \text{ Hz}) \times (150 \times 10^{-3} \text{ H})$
   $= \mathbf{1.41\ M\Omega}$

   d. $X_L = 2\pi(20 \text{ mHz})(5.0 \text{ H})$
   $= 2 \times 3.14 \times (20 \times 10^{-3} \text{ Hz}) \times (5.0 \text{ H})]$
   $= \mathbf{628\ m\Omega}$

2. Problem: Determine the inductance, L, for each value of frequency, f, and inductive reactance, X_L.

   a. $X_L = 1.0$ kΩ, $f = 60$ Hz

   b. $X_L = 33$ MΩ, $f = 150$ kHz

   c. $X_L = 62$ kΩ, $f = 5.0$ MHz

   d. $X_L = 47$ Ω, $f = 440$ Hz

   Solution: The equation for inductance, L, for frequency, f, and inductive reactance, X_L, is

   $L = X_L / (2\pi f)$

   For the values of capacitive reactance, X_L, and frequency, f, in Problem 2, the inductances, L, are:

   a. $L = (1.0\ \text{k}\Omega) / [2\pi(60 \text{ Hz})]$
   $= (1.0 \times 10^3\ \Omega) / [2 \times 3.14 \times (60 \text{ Hz})]$
   $= (1.0 \times 10^3\ \Omega) / (377 \text{ Hz})$
   $= \mathbf{2.65\ H}$

   b. $L = (33 \times 10^6\ \Omega) / [2\pi(150 \text{ kHz})]$
   $= (33 \times 10^6\ \Omega) / [2 \times 3.14 \times (150 \text{ kHz})]$
   $= (33 \times 10^6\ \Omega) / (943 \times 10^3 \text{ Hz})$
   $= \mathbf{35\ H}$

   c. $L = (62\ \text{k}\Omega) / [2\pi(5.0 \text{ MHz})]$

$= (62\ \text{k}\Omega) / [2 \times 3.14 \times (5.0 \times 10^6\ \text{Hz})]$
$= (62\ \text{k}\Omega) / (31.4 \times 10^6\ \text{Hz})$
$= \mathbf{1.97\ mH}$

d. $L = (47\ \Omega) / [2\pi(440\ \text{Hz})]$
$= (47\ \Omega) / [2 \times 3.14 \times (440\ \text{Hz})]$
$= (47\ \Omega) / (2.77 \times 10^3\ \text{Hz})]$
$= \mathbf{17.0\ mH}$

3. Problem: Determine the frequency, f, for each value of inductance, L, and inductive reactance, X_L.

a. $L = 5.1\ \mu\text{H}, X_L = 2.2\ \text{k}\Omega$
b. $L = 680\ \text{mH}, X_L = 27\ \text{k}\Omega$
c. $L = 91\ \text{mH}, X_L = 1.2\ \text{M}\Omega$
d. $L = 3.3\ \text{nH}, X_L = 1.8\ \Omega$

Solution: The equation for frequency, f, for inductance, L, and capacitive reactance, X_L, is

$f = X_L / (2\pi L)$

For the values of capacitive reactance, X_L, and inductance, L, in Problem 3, the frequencies, f, are:

a. $f = 2.2\ \text{k}\Omega / [2\pi(5.1\ \mu\text{H})]$
$= (2.2 \times 10^3\ \Omega) / [2 \times 3.14 \times (5.1 \times 10^{-6}\ \text{H})]$
$= (2.2 \times 10^3) / (32.0 \times 10^{-6}\ \text{H})$
$= \mathbf{68.7\ MHz}$

b. $f = 27\ \text{k}\Omega / [2\pi(680\ \text{mH})]$
$= (27 \times 10^3\ \Omega) / [2 \times 3.14 \times (680 \times 10^{-3}\ \text{H})]$
$= (27 \times 10^3\ \Omega) / (4.27\ \text{H})$
$= \mathbf{6.32\ kHz}$

c. $f = 1.2\ \text{M}\Omega / [2\pi(91\ \text{mH})]$
$= (1.2 \times 10^6\ \Omega) / [2 \times 3.14 \times (91 \times 10^{-3}\ \text{H})]$
$= (1.2 \times 10^6\ \Omega) / (572 \times 10^{-3}\ \text{H})$
$= \mathbf{2.10\ MHz}$

d. $f = 1.8\ \Omega / [2\pi(3.3\ \text{nH})]$
$= 1.8\ \Omega / [2 \times 3.14 \times (3.3 \times 10^{-9}\ \text{H})]$
$= 1.8\ \Omega / (20.7 \times 10^{-9}\ \text{H})$
$= \mathbf{86.8\ MHz}$

17.1.2 Series Circuits

1. Problem: Calculate the total impedance for the circuit in Figure 17-1 in both rectangular and polar forms. Then sketch the impedance phasor diagram for the circuit, identifying the resistance, reactance, and impedance phasors.

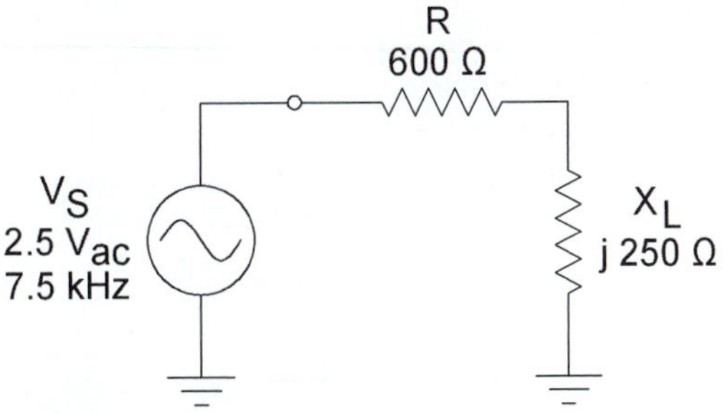

Figure 17-1: Series *RL* Circuit for Problem 1

Solution: The total impedance, Z_T, of the series circuit is

$$Z_T = R + \text{j}\, X_L$$
$$= \mathbf{600\ \Omega + j\ 250\ \Omega}$$

Using rectangular to polar conversion magnitude, Z, of the impedance is

$$Z = \sqrt{R^2 + X_L{}^2}$$
$$= \sqrt{(600\ \Omega)^2 + (250\ \Omega)^2}$$
$$= \sqrt{360{,}000\ \Omega^2 + 62{,}500\ \Omega^2}$$
$$= \sqrt{422{,}500\ \Omega^2}$$
$$= 650\ \Omega$$

The phase angle, φ, is

$$\varphi = \tan^{-1}(X_L / R)$$
$$= \tan^{-1}(250\ \Omega / 600\ \Omega)$$
$$= \tan^{-1}(0.417)$$
$$= 22.6°$$

The polar form of the total impedance is **650 Ω ∠ –22.6°**. Figure 17-2 shows the phasor diagram of the circuit impedances.

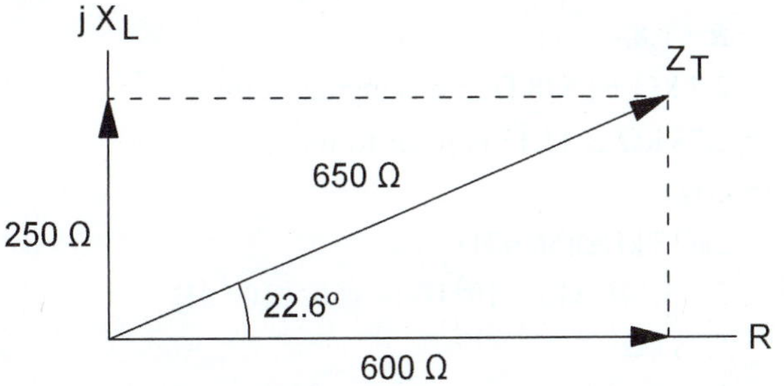

Figure 17-2: Impedance Phasor Diagram for Problem 1

2. Problem: Calculate the total impedance of the circuit in Figure 17-3 for ***f*** = 1.5 kHz and ***f*** = 15 kHz. Express the answer in both polar and rectangular forms.

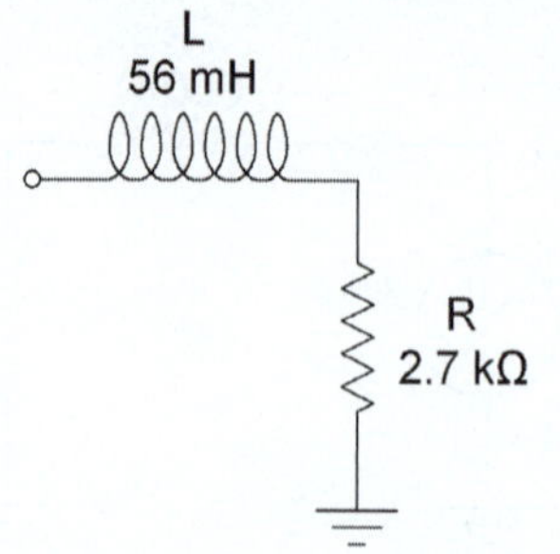

Figure 17-3: Series *RL* Circuit for Problem 2

Solution: First, calculate the reactance X_L. Then use complex arithmetic to find the total impedance $Z_T = R + \mathrm{j}\,X_L$ of the circuit.

For f = 1.5 kHz,

$$X_L = 2\pi(1.5\text{ kHz})(56\text{ mH})$$
$$= 2 \times 3.14 \times (1.5 \times 10^3\text{ Hz}) \times (56 \times 10^{-3}\text{ H})$$
$$= 528\ \Omega$$

The equivalent complex circuit is shown in Figure 17-4.

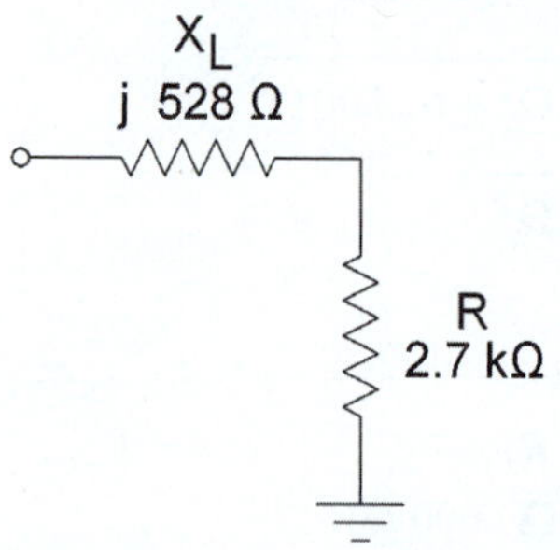

Figure 17-4: Equivalent Complex Series Circuit for f = 1.5 kHz

For a complex series circuit, the total impedance is calculated in the same manner as the total resistance for a resistive series circuit, but using complex arithmetic. For a series circuit the total impedance is the sum of the values in series.

$$Z_T = R + \mathrm{j}\,X_C$$
$$= \mathbf{2.7\ k\Omega + j\ 528\ \Omega} \text{ in rectangular form}$$
$$= \mathbf{2.75\ k\Omega \angle 11.1^\circ} \text{ in polar form.}$$

For f = 15 kHz

$$X_L = 2\pi(15\text{ kHz})(56\text{ mH})$$
$$= 2 \times 3.14 \times (15 \times 10^3\text{ Hz}) \times (56 \times 10^{-3}\text{ H})$$
$$= 5.28\text{ k}\Omega$$

The equivalent complex circuit is shown in Figure 17-5.

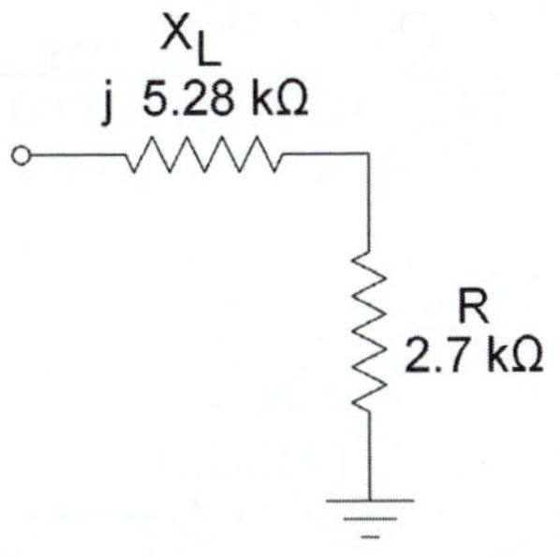

Figure 17-5: Equivalent Complex Series Circuit for *f* = 15 kHz

The total impedance is

$Z_T = R + \text{j}\, X_L$

$= \mathbf{2.7\ k\Omega + j\ 5.28\ k\Omega}$ in rectangular form

$= \mathbf{5.93\ k\Omega \angle 62.9°}$ in polar form.

3. Problem: For the circuit in Figure 17-6, calculate the total impedance, total current, and voltages across each component for *f* = 15 kHz. Express all answers in polar form.

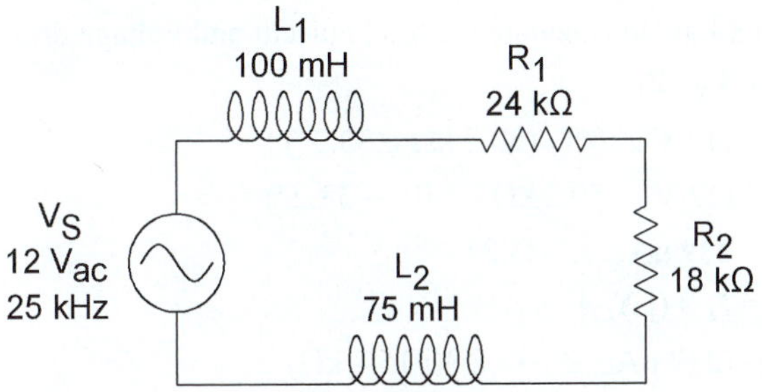

Figure 17-6: Series *RL* Circuit for Problem 3

Solution: First, calculate the reactance, X_L, for each capacitor in the circuit. Then use complex arithmetic to find the total reactance, Z_T and Ohm's Law to calculate the total current and voltage across each component. Express all answers in polar form.

For L_1,

$X_{L1} = 2\pi(25\text{ kHz})(100\text{ mH})$

$= 2 \times 3.14 \times (25 \times 10^3\text{ Hz}) \times (100 \times 10^{-3}\text{ H})$

$= 15.7\text{ k}\Omega$

For L_2,

$X_{L2} = 2\pi(25\text{ kHz})(75\text{ mH})$

$= 2 \times 3.14 \times (25 \times 10^3\text{ Hz}) \times (75 \times 10^{-3}\text{ H})$

$= 11.8\text{ k}\Omega$

The equivalent complex circuit is shown in Figure 17-7.

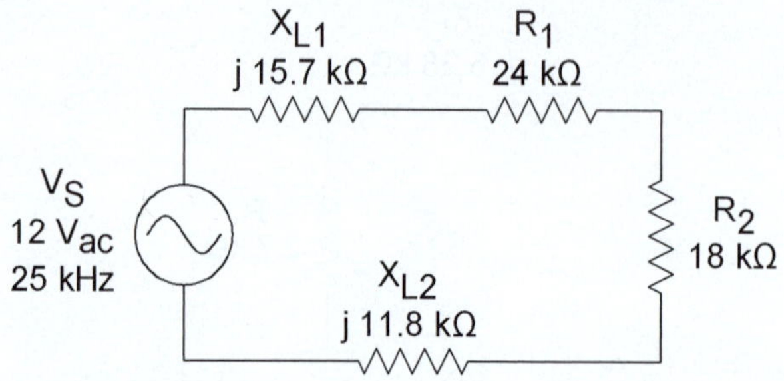

Figure 17-7: Equivalent Complex Series Circuit for Problem 3

For a series circuit, the total impedance is the sum of the series values.

$$\begin{aligned}
\boldsymbol{Z_T} &= \text{j}\,\boldsymbol{X_{L1}} + \boldsymbol{R_1} + \boldsymbol{R_2} + \text{j}\,\boldsymbol{X_{L2}} \\
&= \text{j}\ 15.7\ \text{k}\Omega + 24\ \text{k}\Omega + 18\ \text{k}\Omega + \text{j}\ 11.8\ \text{k}\Omega \\
&= (24\ \text{k}\Omega + 18\ \text{k}\Omega) + \text{j}\ (15.7\ \text{k}\Omega + 11.8\ \text{k}\Omega) \\
&= \mathbf{42\ k\Omega + j\ 27.5\ k\Omega} \text{ in rectangular form} \\
&= \mathbf{50.2\ k\Omega \angle 33.2°} \text{ in polar form}
\end{aligned}$$

Use Ohm's Law to calculate the total current and voltage drop across each component.

$$\begin{aligned}
\boldsymbol{I_T} &= \boldsymbol{V_S} / \boldsymbol{Z_T} \\
&= (12\ \text{V} \angle 0°) / (50.2\ \text{k}\Omega \angle 33.2°) \\
&= (12\ \text{V}_{ac} / 50.2\ \text{k}\Omega) \angle (0° - 33.2°) \\
&= \mathbf{239\ \mu A_{ac} \angle -33.2°}
\end{aligned}$$

$$\begin{aligned}
\boldsymbol{V_{L1}} &= \boldsymbol{I_T} \times (\text{j}\,\boldsymbol{X_{L1}}) \\
&= (239\ \mu\text{A}_{ac} \angle -33.2°)(\text{j}\ 15.7\ \text{k}\Omega) \\
&= (239\ \mu\text{A}_{ac} \angle -33.2°)(15.7\ \text{k}\Omega \angle 90°) \\
&= (239\ \mu\text{A}_{ac} \times 15.7\ \text{k}\Omega) \angle (-33.2° + 90°) \\
&= \mathbf{3.76\ V_{ac} \angle 56.8°}
\end{aligned}$$

$$\begin{aligned}
\boldsymbol{V_{R1}} &= \boldsymbol{I_T} \times \boldsymbol{R_1} \\
&= (239\ \mu\text{A}_{ac} \angle -33.2°)(24\ \text{k}\Omega \angle 0°) \\
&= (239\ \mu\text{A}_{ac} \times 24\ \text{k}\Omega) \angle (-33.2° + 0°) \\
&= \mathbf{5.74\ V_{ac} \angle -33.2°}
\end{aligned}$$

$$\begin{aligned}
\boldsymbol{V_{R2}} &= \boldsymbol{I_T} \times \boldsymbol{R_2} \\
&= (239\ \mu\text{A}_{ac} \angle -33.2°)\ (18\ \text{k}\Omega \angle 0°) \\
&= (239\ \mu\text{A}_{ac} \times 18\ \text{k}\Omega) \angle (-33.2° + 0°) \\
&= \mathbf{4.30\ V_{ac} \angle -33.2°}
\end{aligned}$$

$$\begin{aligned}
\boldsymbol{V_{L2}} &= \boldsymbol{I_T} \times (\text{j}\,\boldsymbol{X_{L2}}) \\
&= (239\ \mu\text{A}_{ac} \angle -33.2°)(\text{j}\ 11.8\ \text{k}\Omega) \\
&= (239\ \mu\text{A}_{ac} \angle -33.2°)(11.8\ \text{k}\Omega \angle 90°) \\
&= (239\ \mu\text{A}_{ac} \times 11.8\ \text{k}\Omega) \angle (-33.2° + 90°) \\
&= \mathbf{2.82\ V_{ac} \angle 56.8°}
\end{aligned}$$

4. Problem: For the voltage divider circuit in Figure 17-8, calculate the value of $\boldsymbol{L}$ required for $\boldsymbol{V_{out}} = 2.5\ \text{V}_{ac}$. Verify your answer.

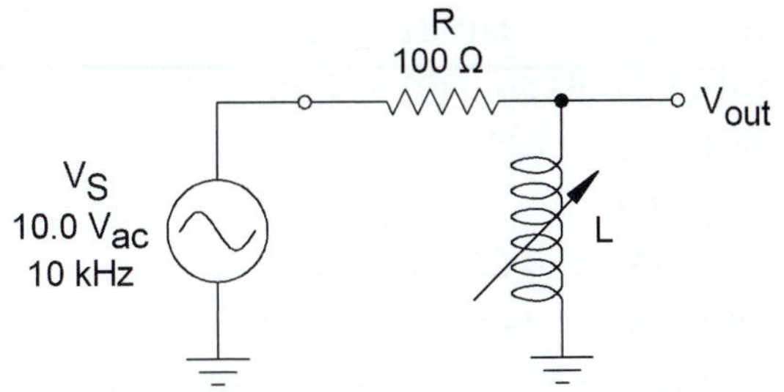

Figure 17-8: Series *RL* Circuit for Problem 4

Solution: First, use the voltage divider formula to determine the value of X_L that will produce $V_{out} = 2.5\ \text{V}_{ac}$.

$V_{out} = 2.5\ \text{V}_{ac} = (10.0\ \text{V}_{ac})\,(X_L / Z_T)$

$2.5\ \text{V}_{ac} / 10.0\ \text{V}_{ac} = [(10.0\ \text{V}_{ac})\,(X_L / Z_T)] / 10.0\ \text{V}_{ac}$

$0.25\ Z_T = (X_L / Z_T)\ Z_T$

$(0.25\ Z_T)^2 = X_L^2$

$0.0625\ Z_T^2 = X_L^2$

For a series ***RL*** circuit, $Z_T^2 = R^2 + X_L^2$.

$X_L^2 = 0.0625\ Z_T^2$

$= 0.0625\ (R^2 + X_L^2)$

$X_L^2 - 0.0625\ X_L^2 = 0.0625\ R^2 + 0.0625\ X_L^2 - 0.0625\ X_L^2$

$(1 - 0.0625)\ X_L^2 = 0.0625\ R^2$

$(0.9375\ X_L^2) / 0.9375 = (0.0625\ R^2) / 0.9375$

$X_L^2 = (0.0625 / 0.9375)\ R^2$

$= 0.0667\ R^2$

Take the square root of both sides.

$\sqrt{X_L^2} = \sqrt{0.0667\ R^2}$

$X_L = 0.258\ R$

$= (0.258)\ (100\ \Omega)$

$= 25.8\ \Omega$

Calculate the inductance ***L*** required for reactance $X_L = 25.8\ \Omega$ at frequency $f = 10$ kHz.

$L = X_L / (2\pi f)$

$= 25.8\ \Omega / (2 \times \pi \times 10\ \text{kHz})$

$= 25.8\ \Omega / [2 \times \pi \times (10 \times 10^3\ \text{Hz})]$

$= 25.8\ \Omega / (62.8 \times 10^3\ \text{Hz})$

= 411 µH

Verify this answer by calculating the value of V_{out}.

$X_L = 2\pi f L$

$= 2\pi\ (10\ \text{kHz})(411\ \mu\text{H})$

$= 2 \times \pi \times (10 \times 10^3\ \text{Hz}) \times (411 \times 10^{-6}\ \text{H})$

$= 25.8\ \Omega$

$Z_T = R + jX_L$

$= 100\ \Omega + j\ 25.8\ \Omega$ in rectangular form

$= 103\ \Omega \angle 14.5°$

$I_T = V_S / Z_T$

$= (10.0\ V_{ac} \angle 0°) / (103\ \Omega \angle 14.5°)$

$= (10.0\ V_{ac} / 103\ \Omega) \angle (0° - 14.5°)$

$= 96.8\ mA \angle -14.5°$

$\boldsymbol{V_{out}} = \boldsymbol{I_T} \times (j\ \boldsymbol{X_L})$

$= (96.8\ mA_{ac} \angle -14.5°)\ (j\ 25.8\ \Omega)$

$= (96.8\ mA_{ac} \angle -14.5°)\ (25.8\ \Omega \angle 90°)$

$= (96.8\ mA_{ac} \times 25.8\ \Omega) \angle (-14.5° + 90°)$

$= 2.50\ V_{ac} \angle 77.5°$

The answer checks.

5. Problem: Draw the phasor voltage diagrams for each circuit in Figure 17-9 using $\boldsymbol{V_S}$ as a reference. Determine from the phasor for $\boldsymbol{V_{out}}$ whether each circuit is a phase lag or phase lead circuit.

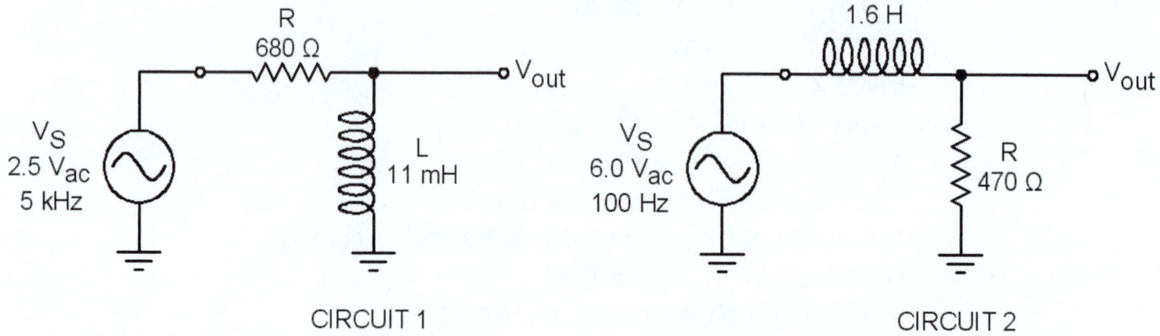

Figure 17-9: Series *RL* Circuits for Problem 5

Solution: For each circuit, first calculate the reactance $\boldsymbol{X_L}$. Then use Ohm's Law to calculate the total current $\boldsymbol{I_T}$ and circuit voltages $\boldsymbol{V_R}$ and $\boldsymbol{V_L}$.

For Circuit 1:

$\boldsymbol{X_L} = 2\pi(5\ kHz)(11\ mF)$

$= 2 \times 3.14 \times (5 \times 10^3\ Hz) \times (11 \times 10^{-3}\ H)$

$= 346\ \Omega$

For Circuit 2:

$\boldsymbol{X_L} = 2\pi(100\ Hz)(1.6\ H)$

$= 2 \times 3.14 \times (100\ Hz) \times (1.6\ H)]$

$= 1.01\ k\Omega$

The equivalent complex circuits are shown in Figure 17-10.

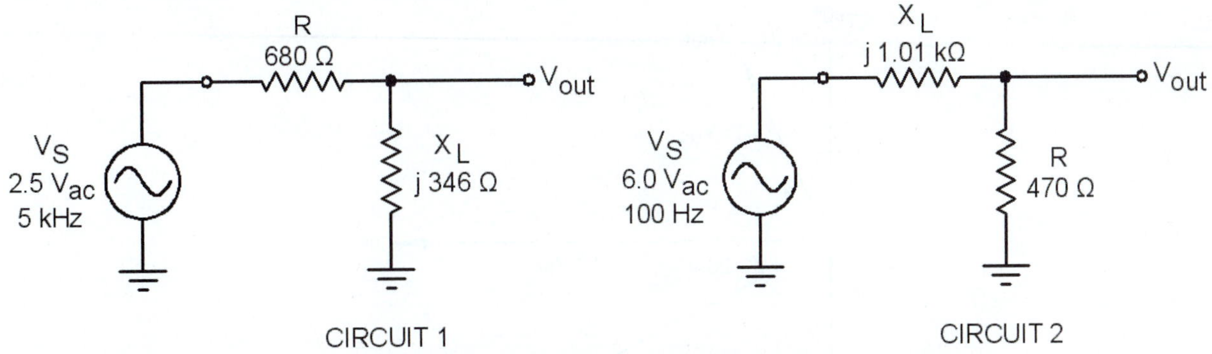

Figure 17-10: Equivalent Complex Phase Circuit for Problem 5

For Circuit 1:

$\boldsymbol{Z_T} = 680\ \Omega + \mathrm{j}\ 346\ \Omega$ in rectangular form

$= 763\ \Omega \angle 26.9°$ in polar form

Use Ohm's Law to calculate the total current and voltage drop across each component.

$\boldsymbol{I_T} = \boldsymbol{V_S} / \boldsymbol{Z_T}$

$= (2.5\ \mathrm{V_{ac}} \angle 0°) / (763\ \Omega \angle 26.9°)$

$= (2.5\ \mathrm{V_{ac}} / 763\ \Omega) \angle (0° - 26.9°)$

$= 3.28\ \mathrm{mA_{ac}} \angle -26.9°$

$\boldsymbol{V_R} = \boldsymbol{I_T} \times \boldsymbol{R}$

$= (3.28\ \mathrm{mA_{ac}} \angle -26.9°)\ (680\ \Omega)$

$= (3.28\ \mathrm{mA_{ac}} \angle -26.9°)\ (680\ \Omega \angle 0°)$

$= (3.28\ \mathrm{mA_{ac}} \times 680\ \Omega) \angle (-26.9° + 0°)$

$= 2.23\ \mathrm{V_{ac}} \angle -26.9°$

$\boldsymbol{V_{out}} = \boldsymbol{V_L}$

$= \boldsymbol{I_T} \times \mathrm{j}\ \boldsymbol{X_L}$

$= (3.28\ \mathrm{mA_{ac}} \angle -26.9°)\ (\mathrm{j}\ 346\ \Omega)$

$= (3.28\ \mathrm{mA_{ac}} \angle -26.9°)\ (346\ \Omega \angle 90°)$

$= (3.28\ \mathrm{mA_{ac}} \times 346\ \Omega) \angle (-26.9° + 90°)$

$= 1.13\ \mathrm{V_{ac}} \angle 63.1°$

Figure 17-11 shows the voltage phasor diagram for the circuit. As the diagram shows, $\boldsymbol{V_{out}}$ leads $\boldsymbol{V_S}$ by 63.1° (i.e., $\varphi_{out} = 63.1°$), so the circuit is a ***phase lead*** circuit.

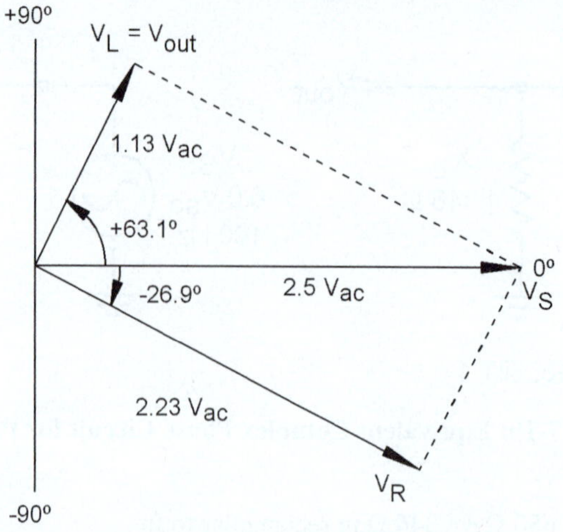

Figure 17-11: Voltage Phasor Diagram for Circuit 1

For Circuit 2:

$\boldsymbol{Z_T}$ = 470 Ω + j 1.01 kΩ in rectangular form
= 1.11 kΩ ∠ 64.9° in polar form

Use Ohm's Law to calculate the total current and voltage drop across each component.

$\boldsymbol{I_T} = \boldsymbol{V_S} / \boldsymbol{Z_T}$
$= (6.0\ V_{ac} \angle 0°) / (1.11\ k\Omega \angle 64.9°)$
$= (6.0\ V_{ac} / 1.11\ k\Omega) \angle (0° + 64.9°)$
$= 5.41\ mA_{ac} \angle -64.9°$

$\boldsymbol{V_L} = \boldsymbol{I_T} \times j\,\boldsymbol{X_L}$
$= (5.41\ mA_{ac} \angle -64.9°)\ (j\ 1.01\ k\Omega)$
$= (5.41\ mA_{ac} \angle -64.9°)\ (1.01\ k\Omega \angle 90°)$
$= (5.41\ mA_{ac} \times 1.01\ k\Omega) \angle (-64.9° + 90°)$
$=5.44\ V_{ac} \angle 25.1°$

$\boldsymbol{V_{out}} = \boldsymbol{V_R}$
$= \boldsymbol{I_T} \times \boldsymbol{R}$
$= (5.41\ mA_{ac} \angle -64.9°)\ (470\ \Omega)$
$= (5.41\ mA_{ac} \angle -64.9°)\ (470\ \Omega \angle 0°)$
$= (5.41\ mA_{ac} \times 470\ \Omega) \angle (-64.9° + 0°)$
$= 2.54\ mV_{ac} \angle -64.9°$

Figure 17-12 shows the voltage phasor diagram for the circuit. As the diagram shows, $\boldsymbol{V_{out}}$ lags $\boldsymbol{V_S}$ by 64.9° (i.e., φ_{out} =–64.9°), so the circuit is a ***phase lag*** circuit.

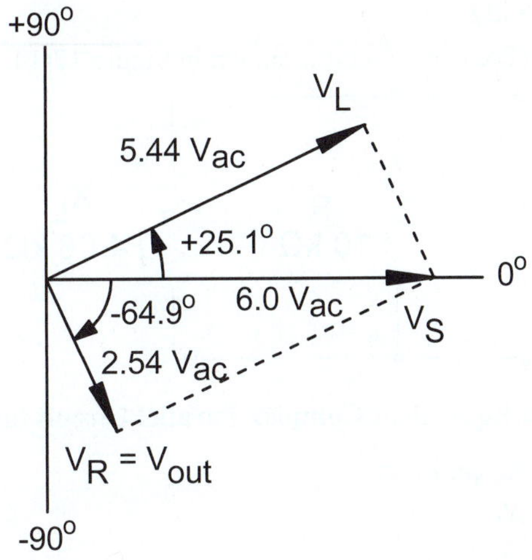

Figure 17-12: Voltage Phasor Diagram for Circuit 2

17.1.3 Parallel Circuits

1. Problem: Expand the product over sum method for parallel components to prove that the total impedance of a parallel ***RL*** circuit is

$$Z_T = [RX_L^2 / (R^2 + X_L^2)] + j\,[R^2X_L / (R^2 + X_L^2)]$$

Solution: Use the product over sum method to calculate the total impedance of parallel components.

$$\begin{aligned}
Z_T &= R \parallel j\,X_L \\
&= (j\,RX_L) / (R + j\,X_L) \\
&= [(j\,RX_L) / (R + j\,X_L)]\,[(R - j\,X_L) / (R - j\,X_L)] \\
&= [(j\,RX_L)\,(R - j\,X_L)] / [(R + j\,X_L)\,(R - j\,X_L)] \\
&= [(j\,RRX_L) - (j^2\,RX_LX_L)] / (RR + j\,RX_L - j\,RX_L - j^2\,X_LX_L) \\
&= \{(j\,R^2X_L) - [(-1)\,RX_L^2)]\} / [R^2 + j\,(RX_L - RX_L) - (-1)\,X_L^2] \\
&= (j\,R^2X_L + RX_L^2) / (R^2 + j\,0 + X_L^2) \\
&= (RX_L^2 + j\,R^2X_L) / (R^2 + X_L^2) \\
&= [RX_L^2 / (R^2 + X_L^2)] + j\,[R^2X_L / (R^2 + X_L^2)]
\end{aligned}$$

2. Problem: Refer to the circuit in Figure 17-13. Calculate the total impedance of the circuit for ***f*** = 5.0 kHz and 25 kHz. Express the answer in both rectangular and polar forms.

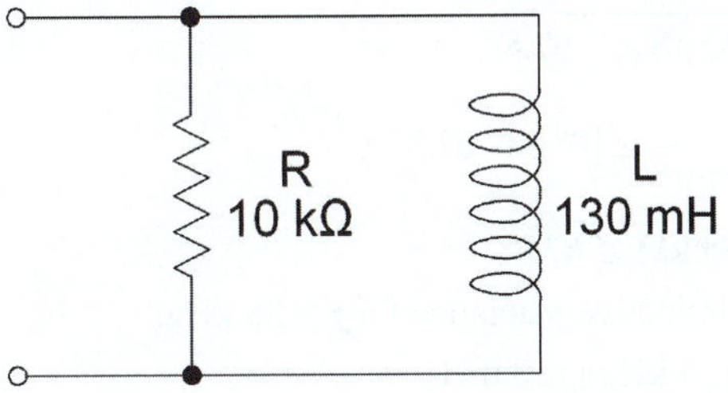

Figure 17-13: Parallel *RL* Circuit for Problem 2

Solution: First, calculate the reactance X_L. Then use complex arithmetic to find the total impedance $Z_T = R \parallel j\,X_L$ of the circuit.

Calculate the inductive reactance for ***f*** = 5.0 kHz.

$$\begin{aligned}
X_L &= 2\pi(5.0\text{ kHz})(130\text{ mH}) \\
&= 2 \times 3.14 \times (5.0 \times 10^3\text{ Hz}) \times (130 \times 10^{-3}\text{ H})]
\end{aligned}$$

$= 4.08\ \text{k}\Omega$

The equivalent complex circuit is shown in Figure 17-14.

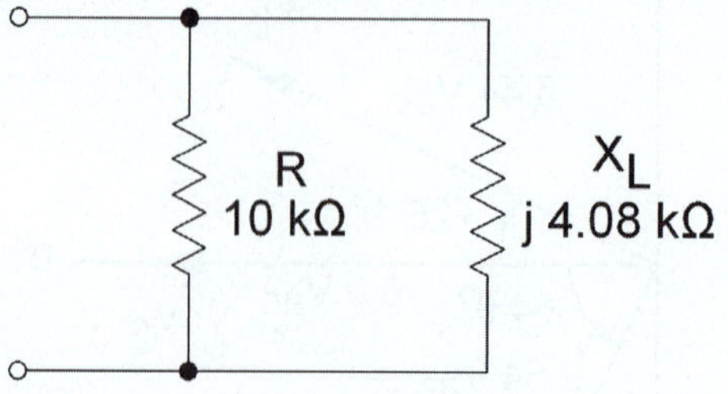

Figure 17-14: Equivalent Complex Parallel Circuit for *f* = 5.0 kHz

The total impedance, Z_T, is

$$\boldsymbol{Z_T} = \boldsymbol{R} \parallel \text{j}\,\boldsymbol{X_L}$$

$$= \frac{1}{\dfrac{1}{\boldsymbol{R}} + \dfrac{1}{\text{j}\boldsymbol{X_L}}}$$

$$= \frac{1\angle 0°}{\dfrac{1\angle 0°}{10\ \text{k}\Omega\ \angle 0°} + \dfrac{1\angle 0°}{4.08\ \text{k}\Omega\ \angle 90°}}$$

$$= \frac{1\angle 0°}{\dfrac{1}{10\ \text{k}\Omega}\angle(0° - 0°) + \dfrac{1}{4.08\ \text{k}\Omega}\angle(0° - 90°)}$$

$$= \frac{1\angle 0°}{(100\ \mu\text{S}\angle 0°) + (245\ \mu\text{S}\angle -90°)}$$

$$= \frac{1\angle 0°}{(100\ \mu\text{S} + \text{j}\,0) + (0 - \text{j}\,245\ \mu\text{S})}$$

$$= \frac{1\angle 0°}{(100\ \mu\text{S} + 0) + \text{j}\,(0 - 245\ \mu\text{S})}$$

$$= \frac{1\angle 0°}{100\ \mu\text{S} - \text{j}\,245\ \mu\text{S}}$$

$$= \frac{1\angle 0°}{265\ \mu\text{S}\angle -67.8°}$$

$$= \frac{1}{265\ \mu\text{S}}\angle[0° - (-67.8°)]$$

$$= \mathbf{3.78\ k\Omega\ \angle\ 67.8°}$$

Calculate the inductive reactance for $f = 25$ kHz.

$$X_L = 2\pi(25\ \text{kHz})(130\ \text{mH})$$

$$= 2 \times 3.14 \times (25 \times 10^3\ \text{Hz}) \times (130 \times 10^{-3}\ \text{H})$$

$$= 20.4\ \text{k}\Omega$$

The equivalent complex circuit is shown in Figure 17-15.

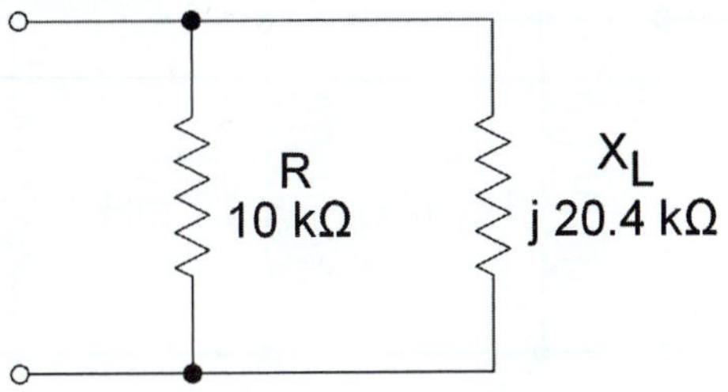

Figure 17-15: Equivalent Complex Parallel Circuit for *f* = 25 kHz

From this,

$$\begin{aligned}
\boldsymbol{Z_T} &= \boldsymbol{R} \parallel \text{j}\,\boldsymbol{X_L} \\
&= \frac{1}{\dfrac{1}{\boldsymbol{R}} + \dfrac{1}{\text{j}\boldsymbol{X_L}}} \\
&= \frac{1}{\dfrac{1}{10\,\text{k}\Omega} + \dfrac{1}{\text{j}\,20.4\,\text{k}\Omega}} \\
&= \frac{1\angle 0^\circ}{\dfrac{1\angle 0^\circ}{10\,\text{k}\Omega\angle 0^\circ} + \dfrac{1\angle 0^\circ}{20.4\,\text{k}\Omega\angle 90^\circ}} \\
&= \frac{1\angle 0^\circ}{\dfrac{1}{10\,\text{k}\Omega}\angle(0^\circ - 0^\circ) + \dfrac{1}{20.4\,\text{k}\Omega}\angle(0^\circ - 90^\circ)} \\
&= \frac{1\angle 0^\circ}{(100\,\mu\text{S}\angle 0^\circ) + (49.0\,\mu\text{S}\angle -90^\circ)} \\
&= \frac{1\angle 0^\circ}{(100\,\mu\text{S} + \text{j}\,0) + (0 - \text{j}\,49.0\,\mu\text{S})} \\
&= \frac{1\angle 0^\circ}{(100\,\mu\text{S} + 0) + \text{j}\,(0 - 49.0\,\mu\text{S})} \\
&= \frac{1\angle 0^\circ}{100\,\mu\text{S} - \text{j}\,49.0\,\mu\text{S}} \\
&= \frac{1\angle 0^\circ}{111\,\mu\text{S}\angle -26.1^\circ} \\
&= \frac{1}{111\,\mu\text{S}}\angle[0^\circ - (-26.1^\circ)] \\
&= \mathbf{8.98\,k\Omega\angle 26.1^\circ}
\end{aligned}$$

3. Problem: Refer to the circuit in Figure 17-16. Calculate the total impedance, total current, and currents through each component at *f* = 3.0 kHz. Express your answers in polar form.

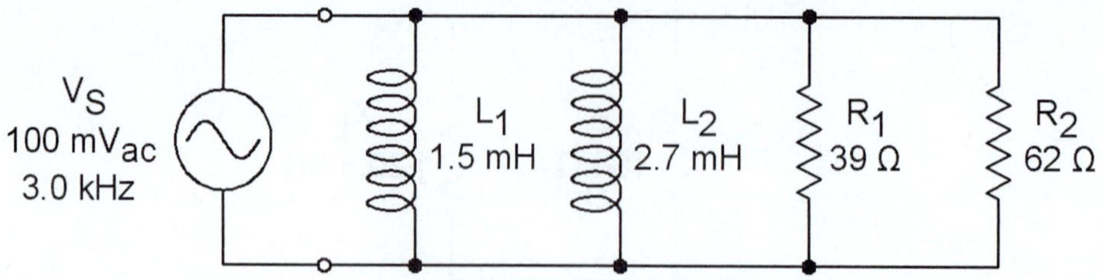

Figure 17-16: Parallel *RC* Circuit for Problem 3

Solution: First, calculate the reactance, X_L, for each inductor. Then, use Ohm's Law to calculate the current through each branch, sum the branch currents to find the total current, I_T, and use Ohm's Law to find the total impedance, Z_T.

For L_1,

$$X_{L1} = 2\pi(3.0\text{ kHz})(1.5\text{ mH})$$
$$= 2 \times 3.14 \times (3.0 \times 10^3\text{ Hz}) \times (1.5 \times 10^{-3}\text{ F})$$
$$= 28.3\ \Omega$$

For L_2,

$$X_{L2} = 2\pi(3.0\text{ kHz})(2.7\text{ mH})$$
$$= 2 \times 3.14 \times (3.0 \times 10^3\text{ Hz}) \times (2.7 \times 10^{-3}\text{ H})$$
$$= 50.9\ \Omega$$

The equivalent complex circuit is shown in Figure 17-17.

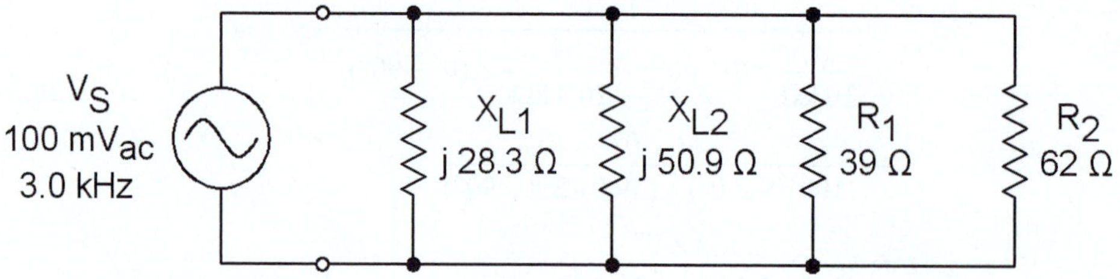

Figure 17-17: Equivalent Complex Parallel Circuit for Problem 3

Because all the components are in parallel, the voltage across each is equal to V_S. Use Ohm's Law to calculate the current through each component.

$$I_{L1} = V_S / \text{j}\,X_{L1}$$
$$= (100\text{ mV}_{ac} \angle 0°) / (28.3\ \Omega \angle 90°)$$
$$= (100\text{ mV}_{ac} / 28.3\ \Omega) \angle (0° - 90°)$$
$$= \mathbf{3.54\ mA_{ac} \angle -90°}$$

$$I_{L2} = V_S / \text{j}\,X_{L2}$$
$$= (100\text{ mV}_{ac} \angle 0°) / (50.9\ \Omega \angle 90°)$$
$$= (100\text{ mV}_{ac} / 50.9\ \Omega) \angle (0° - 90°)$$
$$= \mathbf{1.97\ mA_{ac} \angle -90°}$$

$$I_{R1} = V_S / R_1$$
$$= (100\text{ mV}_{ac} \angle 0°) / (39\ \Omega \angle 0°)$$
$$= (100\text{ mV}_{ac} / 180\ \Omega) \angle (0° - 0°)$$
$$= \mathbf{2.56\ mA_{ac} \angle 0°}$$

$$I_{R2} = V_S / R_2$$

$= (100 \text{ mV}_{ac} \angle 0°) / (62\ \Omega \angle 0°)$

$= (100 \text{ mV}_{ac} / 62\ \Omega) \angle (0° - 0°)$

$= \mathbf{1.61\ mA_{ac} \angle 0°}$

The total current I_T is the sum of these branch currents.

$I_T = I_{L1} + I_{L2} + I_{R1} + I_{R2}$

$= (3.54 \text{ mA}_{ac} \angle -90°) + (1.97 \text{ mA}_{ac} \angle -90°) + (2.56 \text{ mA}_{ac} \angle 0°) + (1.61 \text{ mA}_{ac} \angle 0°)$

$= (0 \text{ A}_{ac} - j\ 3.54 \text{ mA}_{ac}) + (0 \text{ A}_{ac} - j\ 1.97 \text{ A}_{ac})$
$\quad + (2.56 \text{ mA}_{ac} + j\ 0 \text{ A}_{ac}) + (1.61 \text{ mA}_{ac} + j\ 0 \text{ A}_{ac})$

$= (0 \text{ A}_{ac} + 0 \text{ A}_{ac} + 2.56 \text{ mA}_{ac} + 1.61 \text{ A}_{ac})$
$\quad - j\ (3.54 \text{ mA}_{ac} + 1.97 \text{ mA}_{ac} + 0 \text{ A}_{ac} + 0 \text{ A}_{ac})$

$= 4.18 \text{ mA}_{ac} - j\ 5.50 \text{ mA}_{ac}$

$= \mathbf{6.91\ mA_{ac} \angle -52.8°}$

Finally, use Ohm's Law to calculate the total impedance from V_S and I_T.

$Z_T = V_S / I_T$

$= (100 \text{ mV}_{ac} \angle 0°) / (6.91 \text{ mA}_{ac} \angle -52.8°)$

$= (100 \text{ mV}_{ac} / 6.91 \text{ mA}_{ac}) \angle [0° - (-52.8°)]$

$= \mathbf{14.5\ \Omega \angle 52.8°}$

4. Problem: For the circuit in Figure 17-18, for what value of L is $I_R = 3.54 \text{ mA}_{ac}$? What is the value of I_L for this value of L?

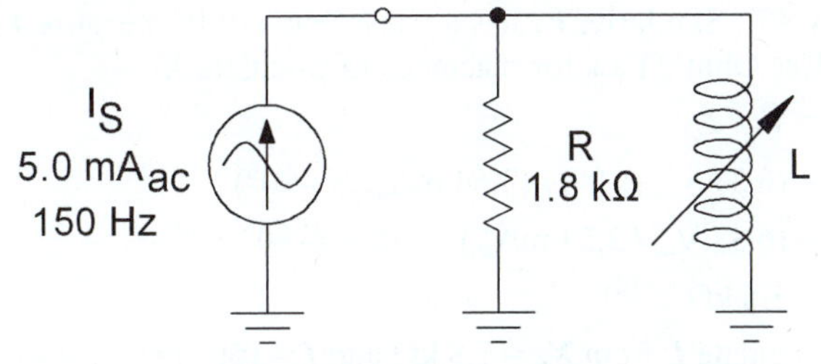

Figure 17-18: Parallel *RL* Circuit for Problem 4

Solution: First, use Ohm's Law to determine the value of V_R when $I_R = 3.54 \text{ mA}_{ac}$.

$V_R = I_R R$

$= (3.54 \text{ mA}_{ac})\ (1.8 \text{ k}\Omega)$

$= (3.54 \times 10^{-3} \text{ A}_{ac})\ (1.8 \times 10^{-3}\ \Omega)$

$= 6.37 \text{ V}_{ac}$

Because R and L are in parallel, $V_R = V_L$. The current through R is in phase with the source voltage, and the current through L lags the source voltage by 90°. Choose I_R to be the reference current, so that $\varphi_R = 0°$ and $\varphi_L = -90°$. Because the currents through R and L are 90° out of phase, the total current, I_T, is the phasor sum of I_R and I_L, as shown in Figure 17-19.

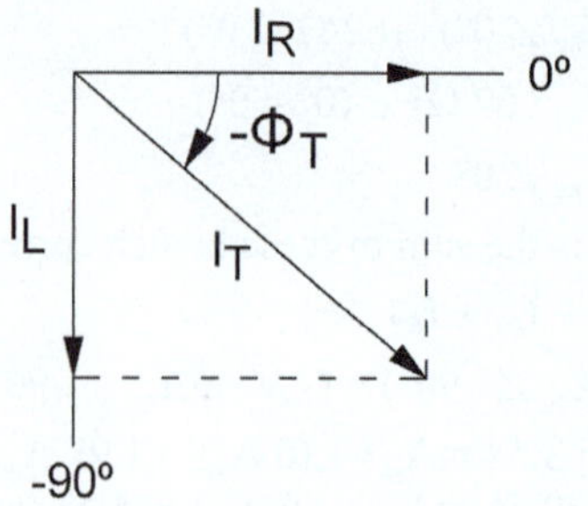

Figure 17-19: Phase Relationships of Currents for Problem 4

From the Pythagorean Theorem for the diagram in Figure 17-19, $\boldsymbol{I_T}^2 = \boldsymbol{I_R}^2 + \boldsymbol{I_L}^2$.

$$\boldsymbol{I_R}^2 + \boldsymbol{I_L}^2 = \boldsymbol{I_T}^2$$
$$(3.54\ \text{mA}_{ac})^2 + \boldsymbol{I_L}^2 = (5.0\ \text{mA}_{ac})^2$$
$$12.5\ \mu\text{A}_{ac}^{\ 2} + \boldsymbol{I_L}^2 = 25.0\ \mu\text{A}_{ac}^{\ 2}$$
$$12.5\ \mu\text{A}_{ac}^{\ 2} + \boldsymbol{I_L}^2 - 12.5\ \mu\text{A}_{ac}^{\ 2} = 25.0\ \mu\text{A}_{ac}^{\ 2} - 12.5\ \mu\text{A}_{ac}^{\ 2}$$
$$\boldsymbol{I_L}^2 = 12.5\ \mu\text{A}_{ac}^{\ 2}$$
$$\sqrt{\boldsymbol{I_L}^2} = \sqrt{12.5\ \mu\text{A}_{ac}^{\ 2}}$$
$$\boldsymbol{I_L} = 3.54\ \text{mA}_{ac} \angle -90°$$

Note that $\boldsymbol{I_L}$ has a phase angle of –90° because $\boldsymbol{I_R}$ was chosen to be the reference current and $\boldsymbol{I_L}$ lags $\boldsymbol{I_R}$ by 90°. Similarly, $\boldsymbol{V_L}$ has a phase angle of 0°, because $\boldsymbol{V_L}$ is equal to $\boldsymbol{V_R}$ and $\boldsymbol{V_R}$ is in phase with $\boldsymbol{I_R}$. Use Ohm's Law for reactance to calculate $\boldsymbol{X_L}$.

$$\boldsymbol{X_L} = \boldsymbol{V_L} / \boldsymbol{I_L}$$
$$= (6.37\ \text{V}_{ac} \angle 0°) / (3.54\ \text{mA}_{ac} \angle -90°)$$
$$= (6.37\ \text{V}_{ac} / 3.54\ \text{mA}_{ac}) \angle [0° + (-90°)]$$
$$= 1.8\ \text{k}\Omega \angle 90°$$

Finally, calculate $\boldsymbol{L}$ from $\boldsymbol{X_L} = 1.8\ \text{k}\Omega$ and $\boldsymbol{f} = 150$ Hz.

$$\boldsymbol{L} = \boldsymbol{X_L} / (2\pi \boldsymbol{f})$$
$$= 1.8\ \text{k}\Omega / [2\pi(150\ \text{Hz})]$$
$$= (1.8 \times 10^3\ \Omega) / (2 \times 3.14 \times 150\ \text{Hz})$$
$$= (1.8 \times 10^3\ \Omega) / (943\ \text{Hz})$$
$$= \mathbf{1.91\ mH}$$

17.1.4 Series-Parallel Circuits

1. Problem: Refer to the circuit in Figure 17-20. Calculate the total impedance of the circuit for $\boldsymbol{f} = 100$ Hz and $\boldsymbol{f} = 1.0$ kHz. Express the answers in both rectangular and polar forms.

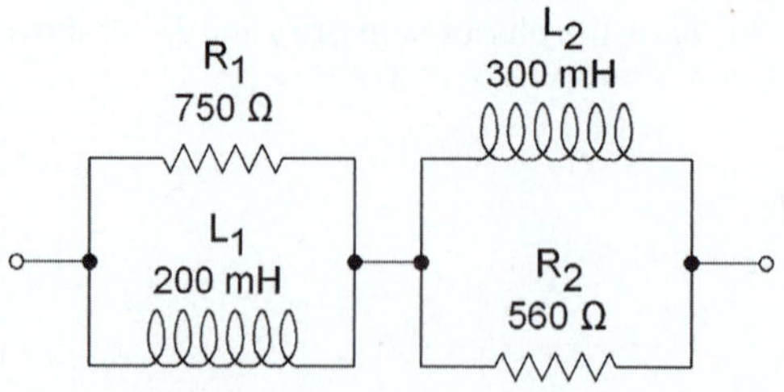

Figure 17-20: Series-Parallel *RL* Circuit for Problem 1

Solution: First, calculate the reactance, X_L, for each inductor in the circuit. Then use the techniques for finding the impedance of series and parallel **RL** circuits to find the total reactance, Z_T.

For $f = 100$ Hz:

$$\begin{aligned} X_{L1} &= 2\pi(100\text{ Hz})(200\text{ mH}) \\ &= 2 \times 3.14 \times (100\text{ Hz}) \times (200 \times 10^{-3}\text{ H}) \\ &= 126\ \Omega \end{aligned}$$

$$\begin{aligned} X_{L2} &= 2\pi(100\text{ Hz})(300\text{ mH}) \\ &= 2 \times 3.14 \times (100\text{ Hz}) \times (300 \times 10^{-3}\text{ H}) \\ &= 189\ \Omega \end{aligned}$$

The equivalent complex circuit is shown in Figure 17-21.

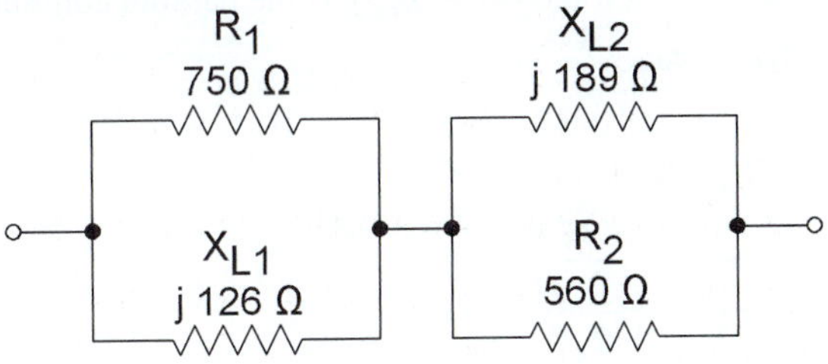

Figure 17-21: Equivalent Complex Series-Parallel Circuit for $f = 100$ Hz

For a complex parallel circuit, the total impedance is calculated in the same manner as the total resistance for a resistive series-parallel circuit, but using complex arithmetic. From standard analysis of series-parallel circuits,

$$Z_T = (R_1 \parallel X_{L1}) + (X_{L2} \parallel R_2)$$

From the standard order of operations, the first equivalent impedance, Z_{EQ1}, is the parallel combination $R_1 \parallel X_{L1}$.

$$Z_T = Z_{EQ1} + (X_{L2} \parallel R_2)$$

where

$$\begin{aligned} Z_{EQ1} &= R_1 \parallel X_{L1} \\ &= \frac{1}{\frac{1}{R_1} + \frac{1}{\mathrm{j}X_{L1}}} \\ &= \frac{1}{\frac{1}{750\,\Omega} + \frac{1}{\mathrm{j}126\,\Omega}} \\ &= \frac{1\angle 0^\circ}{\frac{1\angle 0^\circ}{750\,\Omega\angle 0^\circ} + \frac{1\angle 0^\circ}{126\,\Omega\angle 90^\circ}} \\ &= \frac{1\angle 0^\circ}{\frac{1}{750\,\Omega}\angle(0^\circ - 0^\circ) + \frac{1}{126\,\Omega}\angle(0^\circ - 90^\circ)} \\ &= \frac{1\angle 0^\circ}{(1.33\text{ mS}\angle 0^\circ) + (7.96\text{ mS}\angle -90^\circ)} \\ &= \frac{1\angle 0^\circ}{(1.33\text{ mS} + \mathrm{j}\,0\text{ S}) + (0\text{ S} - \mathrm{j}\,7.96\text{ mS})} \end{aligned}$$

$$= \frac{1\angle 0°}{(1.33\text{ mS} + 0\text{ S}) + j(0\text{ S} - 7.96\text{ mS})}$$

$$= \frac{1\angle 0°}{1.33\text{ mS} - j7.96\text{ mS}}$$

$$= \frac{1\angle 0°}{8.06\text{ mS}\angle -80.5°}$$

$$= \frac{1}{8.06\text{ mS}}\angle[0° - (-80.5°)]$$

$$= 124\ \Omega\angle 80.5°$$

The second equivalent impedance, $\boldsymbol{Z_{EQ2}}$, is the parallel combination $\boldsymbol{X_{L2}} \parallel \boldsymbol{R_2}$.

$$\boldsymbol{Z_T} = \boldsymbol{Z_{EQ1}} + \boldsymbol{Z_{EQ2}}$$

where

$$\boldsymbol{Z_{EQ2}} = \boldsymbol{X_{L2}} \parallel \boldsymbol{R_2}$$

$$= \frac{1}{\frac{1}{j\boldsymbol{X_{L2}}} + \frac{1}{\boldsymbol{R_2}}}$$

$$= \frac{1}{\frac{1}{j189\ \Omega} + \frac{1}{560\ \Omega}}$$

$$= \frac{1\angle 0°}{\frac{1\angle 0°}{189\ \Omega\angle 90°} + \frac{1\angle 0°}{560\ \Omega\angle 0°}}$$

$$= \frac{1\angle 0°}{\frac{1}{189\ \Omega}\angle(0° - 90°) + \frac{1}{560\ \Omega}\angle(0° - 0°)}$$

$$= \frac{1\angle 0°}{(5.31\text{ mS}\angle -90°) + (1.79\text{ mS}\angle 0°)}$$

$$= \frac{1\angle 0°}{(0\text{ S} - j5.31\text{ mS}) + (1.79\text{ mS} + j0\text{ S})}$$

$$= \frac{1\angle 0°}{(0\text{ S} + 1.79\text{ mS}) - j(5.31\text{ mS} + 0\text{ S})}$$

$$= \frac{1\angle 0°}{1.79\text{ mS} - j5.31\text{ mS}}$$

$$= \frac{1\angle 0°}{5.60\text{ mS}\angle -71.4°}$$

$$= \frac{1}{5.60\text{ mS}}\angle[0° - (-71.4°)]$$

$$= 179\ \Omega\angle 71.4°$$

The third (and final) equivalent impedance, $\boldsymbol{Z_{EQ3}}$, is the series combination $\boldsymbol{Z_{EQ1}} + \boldsymbol{Z_{EQ2}}$.

$$\boldsymbol{Z_T} = \boldsymbol{Z_{EQ3}}$$

where

$$\begin{aligned}
\boldsymbol{Z_{EQ3}} &= \boldsymbol{Z_{EQ1}} + \boldsymbol{Z_{EQ2}} \\
&= (124\ \Omega \angle 80.5°) + (179\ \Omega \angle 71.4°) \\
&= (20.4\ \Omega + \text{j}\ 122\ \Omega) + (57.0\ \Omega + \text{j}\ 169\ \Omega) \\
&= (20.4\ \Omega + 57.0\ \Omega) + \text{j}\ (122\ \Omega + 169\ \Omega) \\
&= \mathbf{77.5\ \Omega + j\ 292\ \Omega} \text{ in rectangular form} \\
&= \mathbf{302\ \Omega \angle 75.1°} \text{ in polar form}
\end{aligned}$$

For f = 1.0 kHz:

$$\begin{aligned}
\boldsymbol{X_{L1}} &= 2\pi(1.0\text{ kHz})(200\text{ mH}) \\
&= 2 \times 3.14 \times (1.0\text{ kHz}) \times (200 \times 10^{-3}\text{ H}) \\
&= 1.26\text{ k}\Omega
\end{aligned}$$

$$\begin{aligned}
\boldsymbol{X_{L2}} &= 2\pi(1.0\text{ kHz})(300\text{ mH}) \\
&= 2 \times 3.14 \times (1.0\text{ kHz}) \times (300 \times 10^{-3}\text{ H}) \\
&= 1.89\text{ k}\Omega
\end{aligned}$$

The equivalent complex circuit is shown in Figure 17-22.

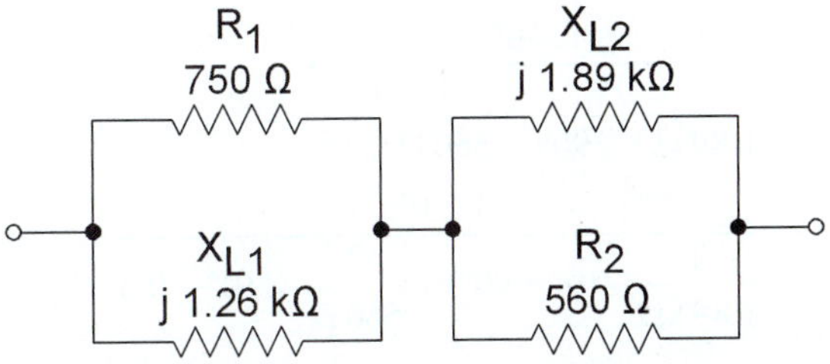

Figure 17-22: Equivalent Complex Series-Parallel Circuit for *f* = 1.0 kHz

Repeat the analysis process for f = 1.0 kHz as for f = 100 Hz.

$$\boldsymbol{Z_T} = \boldsymbol{Z_{EQ1}} + \boldsymbol{Z_{EQ2}}$$

$$\begin{aligned}
\boldsymbol{Z_{EQ1}} &= \boldsymbol{R}_1 \parallel \boldsymbol{X_{L1}} \\
&= \frac{1}{\dfrac{1}{\boldsymbol{R}_1} + \dfrac{1}{\text{j}\boldsymbol{X}_{L1}}} \\
&= \frac{1}{\dfrac{1}{750\ \Omega} + \dfrac{1}{\text{j}1.26\text{ k}\Omega}} \\
&= \frac{1\angle 0°}{\dfrac{1\angle 0°}{750\ \Omega\angle 0°} + \dfrac{1\angle 0°}{1.26\text{ k}\Omega\angle 90°}} \\
&= \frac{1\angle 0°}{\dfrac{1}{750\ \Omega}\angle(0° - 0°) + \dfrac{1}{1.26\text{ k}\Omega}\angle(0° - 90°)} \\
&= \frac{1\angle 0°}{(1.33\text{ mS}\angle 0°) + (796\ \mu\text{S}\angle -90°)} \\
&= \frac{1\angle 0°}{(1.33\text{ mS} + \text{j}\,0\text{ S}) + (0\text{ S} - \text{j}\,796\ \mu\text{S})} \\
&= \frac{1\angle 0°}{(1.33\text{ mS} + 0\text{ S}) + \text{j}\,(0\text{ S} - 796\ \mu\text{S})}
\end{aligned}$$

$$= \frac{1\angle 0°}{1.33\text{ mS} - \text{j}796\,\mu\text{S}}$$

$$= \frac{1\angle 0°}{1.55\text{ mS}\angle -30.8°}$$

$$= \frac{1}{1.55\text{ mS}}\angle[0° - (-30.8°)]$$

$$= 644\ \Omega \angle 30.8°$$

$$\boldsymbol{Z_{EQ2}} = \boldsymbol{X_{L2}} \,||\, \boldsymbol{R_2}$$

$$= \frac{1}{\frac{1}{\text{j}\boldsymbol{X_{L2}}} + \frac{1}{\boldsymbol{R_2}}}$$

$$= \frac{1}{\frac{1}{\text{j}1.89\text{ k}\Omega} + \frac{1}{560\ \Omega}}$$

$$= \frac{1\angle 0°}{\frac{1\angle 0°}{1.89\text{ k}\Omega\angle 90°} + \frac{1\angle 0°}{560\ \Omega\angle 0°}}$$

$$= \frac{1\angle 0°}{\frac{1}{1.89\text{ k}\Omega}\angle(0° - 90°) + \frac{1}{560\ \Omega}\angle(0° - 0°)}$$

$$= \frac{1\angle 0°}{(531\,\mu\text{S}\angle -90°) + (1.79\text{ mS}\angle 0°)}$$

$$= \frac{1\angle 0°}{(0\text{ S} - \text{j}531\,\mu\text{S}) + (1.79\text{ mS} + \text{j}0\text{ S})}$$

$$= \frac{1\angle 0°}{(0\text{ S} + 1.79\text{ mS}) - \text{j}(531\,\mu\text{S} + 0\text{ S})}$$

$$= \frac{1\angle 0°}{1.79\text{ mS} - \text{j}531\,\mu\text{S}}$$

$$= \frac{1\angle 0°}{1.86\text{ mS}\angle -16.6°}$$

$$= \frac{1}{1.86\text{ mS}}\angle[0° - (-16.6°)]$$

$$= 537\ \Omega \angle 16.6°$$

The total impedance is the sum of the series equivalent impedances.

$$\boldsymbol{Z_T} = \boldsymbol{Z_{EQ1}} + \boldsymbol{Z_{EQ2}}$$

$$= (644\ \Omega \angle 30.8°) + (537\ \Omega \angle 16.6°)$$

$$= (553\ \Omega + \text{j}\ 330\ \Omega) + (515\ \Omega + \text{j}\ 153\ \Omega)$$

$$= (553\ \Omega + 515\ \Omega) + \text{j}\ (330\ \Omega + 153\ \Omega)$$

= **1.07 kΩ + j 483 Ω** in rectangular form

= **1.17 kΩ ∠ 24.3°** in polar form

2. Problem: Refer to the circuit in Figure 17-23. Calculate the total impedance, total current, and the voltages across and currents through each component at f = 1.5 kHz. Express all answers in polar form.

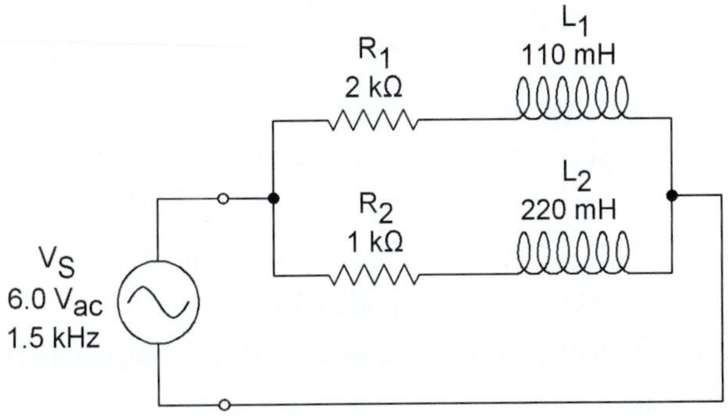

Figure 17-23: Series-Parallel *RL* Circuit for Problem 2

Solution: First, calculate the reactance, X_L, for each inductor in the circuit. Then use the techniques for analyzing series and parallel ***RL*** circuits to find the impedances, currents, and voltages of the circuit.

For $f = 1.5$ kHz:

$$X_{L1} = 2\pi(1.5 \text{ kHz})(110 \text{ mH})$$
$$= 2 \times 3.14 \times (1.5 \times 10^3 \text{ Hz}) \times (110 \times 10^{-3} \text{ H})]$$
$$= 1.04 \text{ k}\Omega$$

$$X_{L2} = 2\pi(1.5 \text{ kHz})(220 \text{ mH})$$
$$= 2 \times 3.14 \times (1.5 \times 10^3 \text{ Hz}) \times (220 \times 10^{-3} \text{ H})$$
$$= 2.07 \text{ k}\Omega$$

The equivalent complex circuit is shown in Figure 16-24.

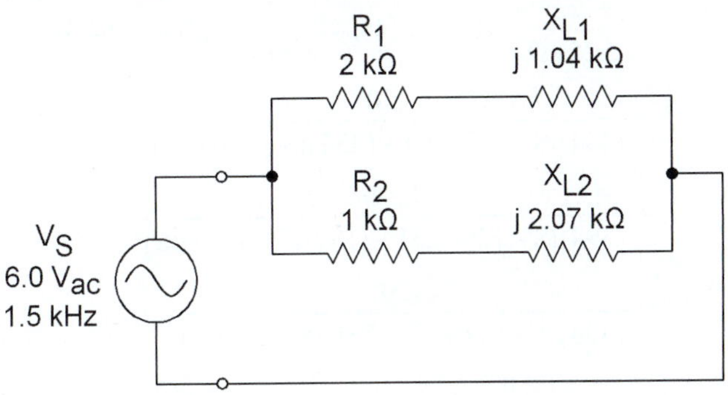

Figure 17-24: Equivalent Complex Series-Parallel Circuit for Problem 2

From standard analysis of series-parallel circuits,

$$Z_T = (R_1 + X_{L1}) \parallel (R_2 + X_{L2})$$

From the standard order of operations, the first equivalent impedance, Z_{EQ1}, is the series combination $R_1 + X_{L1}$. From this,

$$Z_T = Z_{EQ1} \parallel (R_2 + X_{L2})$$

where

$$Z_{EQ1} = R_1 + X_{L1}$$
$$= (2 \text{ k}\Omega + \text{j } 0 \ \Omega) + (0 \ \Omega + \text{j } 1.04 \text{ k}\Omega)$$
$$= (2 \text{ k}\Omega + 0 \ \Omega) + \text{j } (0 \ \Omega + 1.04 \text{ k}\Omega)$$
$$= 2 \text{ k}\Omega + \text{j } 1.04 \text{ k}\Omega \text{ in rectangular form}$$
$$= 2.25 \text{ k}\Omega \angle 27.4° \text{ in polar form}$$

The second equivalent impedance, Z_{EQ2}, is the series combination $R_2 + X_{L2}$.

$$\boldsymbol{Z_T} = \boldsymbol{Z_{EQ1}} \parallel \boldsymbol{Z_{EQ2}}$$

where

$$\begin{aligned}
\boldsymbol{Z_{EQ2}} &= \boldsymbol{R_2} + \boldsymbol{X_{L2}} \\
&= (1\text{ k}\Omega + \text{j } 0\ \Omega) + (0\ \Omega + \text{j } 2.07\text{ k}\Omega) \\
&= (1\text{ k}\Omega + 0\ \Omega) + \text{j } (0\ \Omega + 2.07\text{ k}\Omega) \\
&= 1\text{ k}\Omega + \text{j } 2.07\text{ k}\Omega \text{ in rectangular form} \\
&= 2.30\text{ k}\Omega \angle 64.3^\circ \text{ in polar form}
\end{aligned}$$

The third (and final) equivalent impedance, $\boldsymbol{Z_{EQ3}}$, is the parallel combination $\boldsymbol{Z_{EQ1}} \parallel \boldsymbol{Z_{EQ2}}$.

$$\boldsymbol{Z_T} = \boldsymbol{Z_{EQ3}}$$

where

$$\begin{aligned}
\boldsymbol{Z_{EQ3}} &= \boldsymbol{Z_{EQ1}} \parallel \boldsymbol{Z_{EQ2}} \\
&= \frac{1}{\dfrac{1}{Z_{EQ1}} + \dfrac{1}{Z_{EQ2}}} \\
&= \frac{1}{\dfrac{1}{2.25\text{ k}\Omega \angle 27.4^\circ} + \dfrac{1}{2.30\text{ k}\Omega \angle 64.3^\circ}} \\
&= \frac{1\angle 0^\circ}{\dfrac{1\angle 0^\circ}{2.25\text{ k}\Omega \angle 27.4^\circ} + \dfrac{1\angle 0^\circ}{2.30\text{ k}\Omega \angle 64.3^\circ}} \\
&= \frac{1\angle 0^\circ}{\dfrac{1}{2.25\text{ k}\Omega}\angle(0^\circ - 27.4^\circ) + \dfrac{1}{2.30\text{ k}\Omega}\angle(0^\circ - 64.3)} \\
&= \frac{1\angle 0^\circ}{(444\ \mu\text{S} \angle -27.4^\circ) + (434\ \mu\text{S} \angle -64.3^\circ)} \\
&= \frac{1\angle 0^\circ}{(394\ \mu\text{S} - \text{j}\,204\ \mu\text{S}) + (189\ \mu\text{S} - \text{j}\,391\ \mu\text{S})} \\
&= \frac{1\angle 0^\circ}{(394\ \mu\text{S} + 189\text{ S}) - \text{j}\,(204\ \mu\text{S} + 391\ \mu\text{S})} \\
&= \frac{1\angle 0^\circ}{583\ \mu\text{S} - \text{j}\,596\ \mu\text{S}} \\
&= \frac{1\angle 0^\circ}{833\ \mu\text{S} \angle -45.6^\circ} \\
&= \frac{1}{833\ \mu\text{S}}\angle[0^\circ - (-45.6^\circ)] \\
&= \mathbf{1.2\text{ k}\Omega \angle 45.6^\circ}
\end{aligned}$$

Use Ohm's Law for current to calculate the total current.

$$\begin{aligned}
\boldsymbol{I_T} &= \boldsymbol{V_S} / \boldsymbol{Z_T} \\
&= (6.0\text{ V}_{ac} \angle 0^\circ) / (1.2\text{ k}\Omega \angle 45.6^\circ) \\
&= (6.0\text{ V}_{ac} / 1.2\text{ k}\Omega) \angle (0^\circ - 45.6^\circ) \\
&= \mathbf{5.0\text{ mA}_{ac} \angle -45.6^\circ}
\end{aligned}$$

Because Z_{EQ1} and Z_{EQ2} are in parallel with V_S, $V_{ZEQ1} = V_{ZEQ2} = V_S = 6.0$ Vac. Use Ohm's Law for current to calculate the currents through the parallel branches.

$I_{ZEQ1} = V_S / Z_{EQ1}$

$= (6.0\ \text{V}_{ac} \angle 0°) / (2.25\ \text{k}\Omega \angle 27.4°)$

$= (6.0\ \text{V}_{ac} / 2.25\ \text{k}\Omega) \angle (0° - 27.4°)$

$= \mathbf{2.66\ mA_{ac} \angle -27.4°}$

$I_{ZEQ2} = V_S / Z_{EQ2}$

$= (6.0\ \text{V}_{ac} \angle 0°) / (2.30\ \text{k}\Omega \angle 64.3°)$

$= (6.0\ \text{V}_{ac} / 2.30\ \text{k}\Omega) \angle (0° - 64.3°)$

$= \mathbf{2.61\ mA_{ac} \angle -64.3°}$

Z_{EQ1} is the series combination of R_1 and X_{L1}.

$I_{R1} = I_{ZEQ1} = \mathbf{2.66\ mA_{ac} \angle -27.4°}$

$I_{XL1} = I_{ZEQ1} = \mathbf{2.66\ mA_{ac} \angle -27.4°}$

Use Ohm's Law for voltage to calculate the voltages across R_1 and X_{L1}.

$V_{R1} = I_{R1} R_1$

$= (2.66\ \text{mA}_{ac} \angle -27.4°)(2\ \text{k}\Omega \angle 0°)$

$= (2.66\ \text{mA}_{ac} \times 2\ \text{k}\Omega) \angle (-27.4° + 0°)$

$= \mathbf{5.32\ V_{ac} \angle -27.4°}$

$V_{XL1} = I_{R1} X_{L1}$

$= (2.66\ \text{mA}_{ac} \angle -27.4°)(1.04\ \text{k}\Omega \angle 90°)$

$= (2.66\ \text{mA}_{ac} \times 1.04\ \text{k}\Omega) \angle (-27.4° + 90°)$

$= \mathbf{2.76\ V_{ac} \angle 62.6°}$

Z_{EQ2} is the series combination of R_2 and X_{L2}, so

$I_{R2} = I_{ZEQ2} = \mathbf{2.61\ mA_{ac} \angle -64.3°}$

$I_{XL2} = I_{ZEQ2} = \mathbf{2.61\ mA_{ac} \angle -64.3°}$

Use Ohm's Law for voltage to calculate the voltages across R_2 and X_{L2}.

$V_{R2} = I_{R2} R_2$

$= (2.61\ \text{mA}_{ac} \angle -64.3°)(1\ \text{k}\Omega \angle 0°)$

$= (2.61\ \text{mA}_{ac} \times 1\ \text{k}\Omega) \angle (-64.3° + 0°)$

$= \mathbf{2.61\ V_{ac} \angle -64.3°}$

$V_{XL2} = I_{R2} X_{L2}$

$= (2.61\ \text{mA}_{ac} \angle -64.3°)(2.07\ \text{k}\Omega \angle 90°)$

$= (2.61\ \text{mA}_{ac} \times 2.07\ \text{k}\Omega) \angle (-64.3° + 90°)$

$= \mathbf{5.40\ V_{ac} \angle 25.8°}$

3. Problem: Refer to the circuit in Figure 17-25. Determine V_{out} for $f = 5$ kHz and $f = 25$ kHz. Express the answers in polar form.

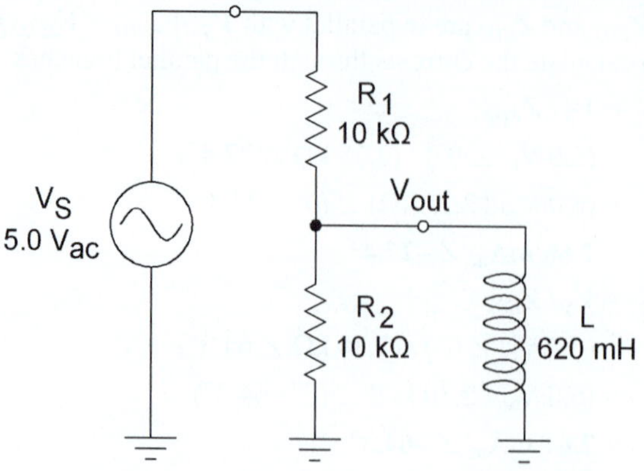

Figure 17-25: Series-Parallel *RL* Circuit for Problem 3

Solution: First, calculate the reactances, X_L, for the inductor. Then use the techniques for analyzing series and parallel ***RL*** circuits to find the circuit impedances, currents, and voltages to determine V_{out}.

For $f = 5.0$ kHz:

$$X_L = 2\pi(5.0 \text{ kHz})(620 \text{ mH})$$
$$= 2 \times 3.14 \times (5.0 \times 10^3 \text{ Hz}) \times (620 \times 10^{-3} \text{ H})$$
$$= 19.5 \text{ k}\Omega$$

For $f = 25$ kHz:

$$X_L = 2\pi(25 \text{ kHz})(620 \text{ mH})$$
$$= 2 \times 3.14 \times (25 \times 10^3 \text{ Hz}) \times (620 \times 10^{-3} \text{ H})$$
$$= 97.4 \text{ k}\Omega$$

The equivalent complex circuits are shown in Figure 17-26.

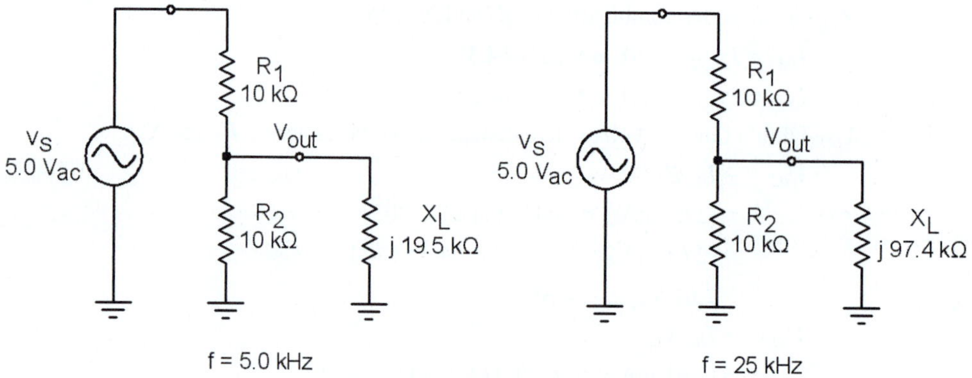

Figure 17-26: Equivalent Complex Circuit for Problem 3

From standard analysis of series-parallel circuits,

$$\boldsymbol{Z_T} = \boldsymbol{R_1} + \boldsymbol{R_2} \parallel \boldsymbol{X_L}$$

The first equivalent impedance, $\boldsymbol{Z_{EQ1}}$, is the parallel combination $\boldsymbol{R_2} \parallel \boldsymbol{X_L}$.

$$\boldsymbol{Z_T} = \boldsymbol{R_1} + \boldsymbol{Z_{EQ1}}$$

where

$$\boldsymbol{Z_{EQ1}} = \boldsymbol{R_2} \parallel \boldsymbol{X_L}$$

For $f = 5.0$ kHz:

$$\boldsymbol{Z_{EQ1}} = \boldsymbol{R_2} \parallel \boldsymbol{X_L}$$

$$= \frac{1}{\frac{1}{\boldsymbol{R_2}} + \frac{1}{\mathrm{j}\boldsymbol{X_L}}}$$

$$= \frac{1}{\frac{1}{10\ \mathrm{k}\Omega} + \frac{1}{\mathrm{j}19.5\ \mathrm{k}\Omega}}$$

$$= \frac{1\angle 0°}{\frac{1\angle 0°}{10\ \mathrm{k}\Omega\angle 0°} + \frac{1\angle 0°}{19.5\ \mathrm{k}\Omega\angle 90°}}$$

$$= \frac{1\angle 0°}{\frac{1}{10\ \mathrm{k}\Omega}\angle(0° - 0°) + \frac{1}{19.5\ \mathrm{k}\Omega}\angle(0° - 90°)}$$

$$= \frac{1\angle 0°}{(100\ \mu\mathrm{S}\angle 0°) + (51.3\ \mu\mathrm{S}\angle -90°)}$$

$$= \frac{1\angle 0°}{(100\ \mu\mathrm{S} + \mathrm{j}0\ \mathrm{S}) + (0\ \mathrm{S} - \mathrm{j}51.3\ \mu\mathrm{S})}$$

$$= \frac{1\angle 0°}{(100\ \mu\mathrm{S} + 0\ \mathrm{S}) - \mathrm{j}(0\ \mathrm{S} + 51.3\ \mu\mathrm{S})}$$

$$= \frac{1\angle 0°}{100\ \mu\mathrm{S} - \mathrm{j}51.3\ \mu\mathrm{S}}$$

$$= \frac{1\angle 0°}{112\ \mu\mathrm{S}\angle -27.2°}$$

$$= \frac{1}{112\ \mu\mathrm{S}}\angle[0° - (-27.2°)]$$

$$= 8.90\ \mathrm{k}\Omega\angle 27.2°$$

Use $\boldsymbol{Z_{EQ1}}$ to calculate the total impedance.

$$\begin{aligned}
\boldsymbol{Z_T} &= \boldsymbol{R_1} + \boldsymbol{Z_{EQ1}} \\
&= 10\ \mathrm{k}\Omega + (8.90\ \mathrm{k}\Omega\angle 27.2°) \\
&= (10\ \mathrm{k}\Omega + \mathrm{j}\,0\ \Omega) + (7.91\ \mathrm{k}\Omega + \mathrm{j}\,4.06\ \mathrm{k}\Omega) \\
&= (10\ \mathrm{k}\Omega + 7.91\ \mathrm{k}\Omega) + \mathrm{j}\,(0\ \Omega + 4.06\ \mathrm{k}\Omega) \\
&= 17.9\ \mathrm{k}\Omega + \mathrm{j}\,4.06\ \mathrm{k}\Omega \\
&= 18.4\ \mathrm{k}\Omega\angle 12.8°
\end{aligned}$$

Use the voltage divider theorem to calculate $\boldsymbol{V_{out}}$ from $\boldsymbol{Z_{EQ1}}$ and $\boldsymbol{Z_T}$.

$$\begin{aligned}
\boldsymbol{V_{out}} &= \boldsymbol{V_S}\,(\boldsymbol{Z_{EQ1}} / \boldsymbol{Z_T}) \\
&= (5.0\ \mathrm{V_{ac}}\angle 0°)\,[(8.90\ \mathrm{k}\Omega\angle 27.2°) / (18.4\ \mathrm{k}\Omega\angle 12.8°)] \\
&= (5.0\ \mathrm{V_{ac}}\angle 0°)\,[(8.90\ \mathrm{k}\Omega / 18.4\ \mathrm{k}\Omega)\angle(27.2° - 12.8°)] \\
&= (5.0\ \mathrm{V_{ac}}\angle 0°)\,(0.484\angle 14.4°) \\
&= [(5.0\ \mathrm{V_{ac}})\,(0.484)]\angle(0° + 14.4°) \\
&= \mathbf{2.42\ V_{ac}\angle 14.4°}
\end{aligned}$$

Repeat the process for $\boldsymbol{f} = 25$ kHz.

$$\boldsymbol{Z_{EQ1}} = \boldsymbol{R_2} \parallel \boldsymbol{X_L}$$

$$= \frac{1}{\dfrac{1}{R_2} + \dfrac{1}{jX_L}}$$

$$= \frac{1}{\dfrac{1}{10\ \text{k}\Omega} + \dfrac{1}{j97.4\ \text{k}\Omega}}$$

$$= \frac{1\angle 0°}{\dfrac{1\angle 0°}{10\ \text{k}\Omega \angle 0°} + \dfrac{1\angle 0°}{97.4\ \text{k}\Omega \angle 90°}}$$

$$= \frac{1\angle 0°}{\dfrac{1}{10\ \text{k}\Omega}\angle(0° - 0°) + \dfrac{1}{97.4\ \text{k}\Omega}\angle(0° - 90°)}$$

$$= \frac{1\angle 0°}{(100\ \mu\text{S}\angle 0°) + (10.3\ \mu\text{S}\angle -90°)}$$

$$= \frac{1\angle 0°}{(100\ \mu\text{S} + j0\ \text{S}) + (0\ \text{S} - j10.3\ \mu\text{S})}$$

$$= \frac{1\angle 0°}{(100\ \mu\text{S} + 0\ \text{S}) - j(0\ \text{S} + 10.3\ \mu\text{S})}$$

$$= \frac{1\angle 0°}{100\ \mu\text{S} - j10.3\ \mu\text{S}}$$

$$= \frac{1\angle 0°}{101\ \mu\text{S}\angle -5.86°}$$

$$= \frac{1}{101\ \mu\text{S}}\angle[0° - (-5.86°)]$$

$$= 9.95\ \text{k}\Omega \angle 5.86°$$

Use Z_{EQ1} to calculate the total impedance.

$$\begin{aligned}
Z_T &= R + Z_{EQ1} \\
&= 10\ \text{k}\Omega + (9.95\ \text{k}\Omega \angle 5.86°) \\
&= (10\ \text{k}\Omega + j\,0\ \Omega) + (9.90\ \text{k}\Omega + j\,1.02\ \text{k}\Omega) \\
&= (10\ \text{k}\Omega + 9.90\ \text{k}\Omega) + j\,(0\ \Omega + 1.02\ \text{k}\Omega) \\
&= 19.9\ \text{k}\Omega + j\,1.02\ \text{k}\Omega \\
&= 19.9\ \text{k}\Omega \angle 2.92°
\end{aligned}$$

Use the voltage divider theorem to calculate V_{out} from Z_{EQ1} and Z_T.

$$\begin{aligned}
V_{out} &= V_S\,(Z_{EQ1} / Z_T) \\
&= (5.0\ \text{V}_{ac} \angle 0°)\,[(9.95\ \text{k}\Omega \angle 5.86°) / (19.9\ \text{k}\Omega \angle 2.92°)] \\
&= (5.0\ \text{V}_{ac} \angle 0°)\,[(9.95\ \text{k}\Omega / 19.9\ \text{k}\Omega) \angle (5.86° - 2.92°)] \\
&= (5.0\ \text{V}_{ac} \angle 0°)\,(0.499 \angle 2.94°) \\
&= [(5.0\ \text{V}_{ac})\,(0.499)] \angle (0° + 2.94°) \\
&= \mathbf{2.50\ V_{ac} \angle 2.94°}
\end{aligned}$$

17.1.5 Power in *RL* Circuits

1. Problem: Determine the total real, reactive, and apparent power and power factor for the circuit in Figure 17-27.

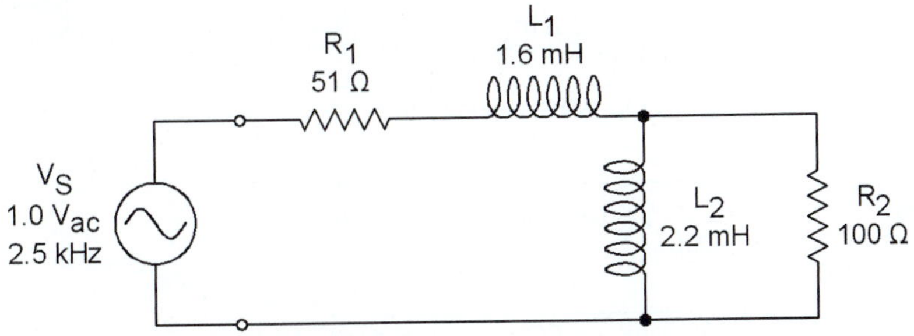

Figure 17-27: Power in *RL* Circuits Example for Problem 1

Solution: First, calculate the reactance, X_L, for each inductor. Then determine the total impedance for the circuit and use Watt's Law and the power triangle to calculate the powers in the circuit.

For $f = 2.5$ kHz,

$$X_{L1} = 2\pi(2.5\text{ kHz})(1.6\text{ mH})$$
$$= 2 \times 3.14 \times (2.5 \times 10^3\text{ Hz}) \times (1.6 \times 10^{-3}\text{ H})]$$
$$= 25.1\ \Omega$$

$$X_{L2} = 1 / [2\pi(2.5\text{ kHz})(2.2\text{ mH})]$$
$$= 1 / [2 \times 3.14 \times (2.5 \times 10^3\text{ Hz}) \times (2.2 \times 10^{-3}\text{ H})]$$
$$= 34.6\ \Omega$$

The equivalent complex circuit is shown in Figure 17-28.

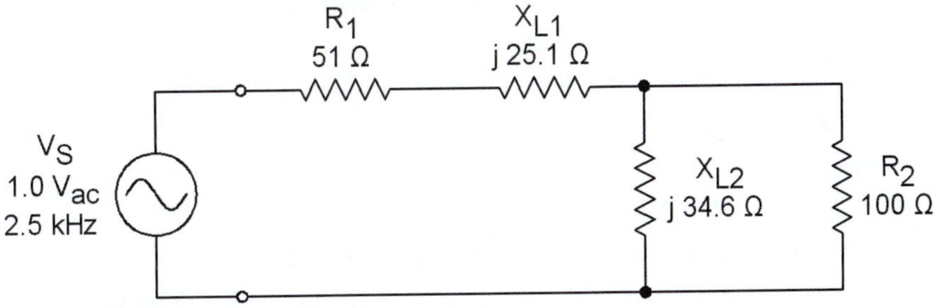

Figure 17-28: Equivalent Complex Circuit for Problem 1

From standard analysis of series-parallel circuits,

$$Z_T = R_1 + X_{L1} + X_{L2} \parallel R_2$$

From the standard order of operations, the first equivalent impedance, Z_{EQ1}, is the parallel combination $X_{L2} \parallel R_2$.

$$Z_T = X_{L1} + Z_{EQ1}$$

where

$$Z_{EQ1} = X_{L2} \parallel R_2$$
$$= \frac{1}{\dfrac{1}{jX_{L2}} + \dfrac{1}{R_2}}$$

$$= \frac{1}{\frac{1}{j34.6\,\Omega} + \frac{1}{100\,\Omega}}$$

$$= \frac{1\angle 0°}{\frac{1\angle 0°}{j34.6\,\Omega\angle 90°} + \frac{1\angle 0°}{100\,\Omega\angle 90°}}$$

$$= \frac{1\angle 0°}{\frac{1}{34.6\,\Omega}\angle(0° - 90°) + \frac{1}{10\,\Omega}\angle(0° - 0°)}$$

$$= \frac{1\angle 0°}{(28.9\text{ mS}\angle -90°) + (10\text{ mS}\angle 0°)}$$

$$= \frac{1\angle 0°}{(0\text{ S} - j\,28.9\text{ mS}) + (10\text{ mS} + j\,0\text{ S})}$$

$$= \frac{1\angle 0°}{(0\text{ S} + 10\text{ mS}) - j(28.9\text{ mS} + 0\text{ S})}$$

$$= \frac{1\angle 0°}{10\text{ mS} - j\,28.9\text{ mS}}$$

$$= \frac{1\angle 0°}{30.6\text{ mS}\angle -70.9°}$$

$$= \frac{1}{30.6\text{ mS}}\angle[0° - (-70.9°)]$$

$= 32.7\ \Omega \angle 70.9°$ in rectangular form

$= 10.7\ \Omega + j\ 30.9\ \Omega$

The second (and last) equivalent impedance, $\boldsymbol{Z_{EQ2}}$, is the series combination $\boldsymbol{R_1} + \boldsymbol{X_{L1}} + \boldsymbol{Z_{EQ1}}$

$\boldsymbol{Z_T} = \boldsymbol{Z_{EQ2}}$

where

$\boldsymbol{Z_{EQ2}} = \boldsymbol{R_1} + \boldsymbol{X_{L1}} + \boldsymbol{Z_{EQ1}}$

$= 51\ \Omega + (j\ 25.1\ \Omega) + (10.7\ \Omega + j\ 30.9\ \Omega)$

$= (51\ \Omega + j\ 0\ \Omega) + (0\ \Omega + j\ 25.1\ \Omega) + (10.7\ \Omega + j\ 30.9\ \Omega)$

$= (51\ \Omega + 0\ \Omega + 10.7\ \Omega) + j\ (0\ \Omega + 25.1\ \Omega + 30.9\ \Omega)$

$= 61.7\ \Omega + j\ 56.0\ \Omega$ in rectangular form

$= 83.3\ \Omega \angle 42.2°$

The supply voltage is in rms units, so use Watt's Law for voltage and impedance to calculate the apparent power, $\boldsymbol{P_a}$.

$\boldsymbol{P_a} = \boldsymbol{V_S}^2 / \boldsymbol{Z_T}$

$= (1.0\ V_{ac})^2 / 83.3\ \Omega$

$= 1.0\ V_{ac}^2 / 83.3\ \Omega$

= 12.0 mVA

From the total impedance the phase angle for the circuit is $\varphi = 42.2°$. Use the power triangle relating real power, $\boldsymbol{P_r}$, reactive power, $\boldsymbol{P_{reac}}$, and apparent power, $\boldsymbol{P_a}$, to calculate $\boldsymbol{P_r}$ and $\boldsymbol{P_{reac}}$.

$\boldsymbol{P_r} = \boldsymbol{P_a} \cos \boldsymbol{\varphi}$

$= (12.0\text{ mVA})[\cos(42.2°)]$

$= (12.0\ \text{mVA})(0.740)$

$= \mathbf{8.89\ mW}$

and

$P_{reac} = P_a \sin\varphi$

$= (12.0\ \text{mVA})[\sin(42.2°)]$

$= (12.0\ \text{mVA})(0.672)$

$= \mathbf{8.07\ mVAR}$

By definition the power factor, ***PF***, is

$PF = \cos\varphi$

$= \cos 42.2°$

$= \mathbf{0.740}$

2. Problem: For the circuit in Figure 17-27, determine the power for each component. Use the calculations from Problem 1 to show that the total real power is equal to the sum of the real power values and the total reactive power is the sum of the reactive power values.

Solution: From the answers to Problem 1, the total impedance, Z_T, is

$Z_T = 83.3\ \Omega \angle 42.2°$

From Ohm's Law, the total current, I_T, is

$I_T = V_S / Z_T$

$= (1.0\ V_{ac} \angle 0°) / (83.3\ \Omega \angle 42.2°)$

$= (1.0\ V_{ac} / 83.3\ \Omega) \angle (0° - 42.2°)$

$= 12.0\ mA_{ac} \angle -42.2°$

The total impedance is equal to the series combination $R_1 + X_{L1} + Z_{EQ1}$ (where $Z_{EQ1} = R_2 \parallel X_{L2}$) so the total current, I_T, flows through R_1, X_{L1}, and Z_{EQ1}. Use Watt's Law to calculate P_{R1} and P_{XL1}.

$P_{R1} = I_T^2 R_1$

$= (12.0\ mA_{ac})^2\ (51\ \Omega)$

$= (144\ \mu A^2)\ (51\ \Omega)$

$= \mathbf{7.35\ mW}$

$P_{XL1} = I_T^2 X_{L1}$

$= (12.0\ mA_{ac})^2\ (j\ 25.1\ \Omega)$

$= (144\ \mu A^2)\ (j\ 25.1\ \Omega)$

$= \mathbf{3.62\ mVAR}$

Use Ohm's Law for current and impedance to calculate the voltage across Z_{EQ1}.

$V_{ZEQ1} = I_T Z_{EQ1}$

$= (12.0\ mA_{ac} \angle -42.2°)\ (32.7\ \Omega \angle 70.9°)$

$= (12.0\ mA_{ac} \times 32.7\ \Omega) \angle (-42.2° + 70.9°)$

$= 392\ mVac \angle 28.7°$

Because X_{L2} and R_2 are in parallel, V_{ZEQ2} appears across both of them. Use Watt's Law to calculate P_{XL2} and P_{R2}.

$P_{XL2} = V_{XL2}^2 / X_{L2}$

$= V_{ZEQ1}^2 / X_{L2}$

$= (392\ mV_{ac})^2 / (j\ 34.6\ \Omega)$

$= (154\ mV_{ac}^2) / (j\ 34.6\ \Omega)$

$= \mathbf{4.45\ mVAR}$

$P_{R2} = V_{XL2}^2 / R_2$

$= \boldsymbol{V_{ZEQ1}}^2 / \boldsymbol{R_2}$

$= (392 \text{ mV}_{ac})^2 / (100\ \Omega)$

$= (154 \text{ mV}_{ac}{}^2) / (100\ \Omega)$

$= \mathbf{1.54\ mW}$

From these calculated values, the total real power, $\boldsymbol{P_R}$, is

$\boldsymbol{P_R} = \boldsymbol{P_{R1}} + \boldsymbol{P_{R2}}$

$= 7.35 \text{ mW} + 1.54 \text{ mW}$

$= \mathbf{8.89\ mW}$

and the total reactive power is

$\boldsymbol{P_{reac}} = \boldsymbol{P_{XL1}} + \boldsymbol{P_{XL2}}$

$= 3.62 \text{ mVAR} + 4.45 \text{ mVAR}$

$= \mathbf{8.07\ mVAR}$

These values match the calculated values from Problem 1.

17.2 Advanced

17.2.1 Series Circuits

1. Problem: Prove that the total inductive reactance for $\boldsymbol{N}$ inductors in series is

$\boldsymbol{X_{LT}} = \boldsymbol{X_{L1}} + \boldsymbol{X_{L2}} + \ldots + \boldsymbol{X_{LN}}$

Solution: For $\boldsymbol{N}$ inductors in series, the total inductance $\boldsymbol{L_T}$ is

$\boldsymbol{L_T} = L_1 + \boldsymbol{L_2} + \ldots + \boldsymbol{L_N}]$

The inductive reactance, $\boldsymbol{X_L}$, for a inductance, $\boldsymbol{L}$, at frequency, $\boldsymbol{f}$, is

$\boldsymbol{X_L} = 2\pi \boldsymbol{fL}$

From this, the total inductive reactance, $\boldsymbol{X_{LT}}$, for total inductance, $\boldsymbol{L_T}$, is

$\boldsymbol{X_{LT}} = 2\pi \boldsymbol{fL_T}$

$= 2\pi \boldsymbol{f}(\boldsymbol{L_1} + \boldsymbol{L_2} + \ldots + \boldsymbol{L_N})$

$= 2\pi \boldsymbol{fL_1} + 2\pi \boldsymbol{fL_2} + \ldots + 2\pi \boldsymbol{fL_N}$

$= \boldsymbol{X_{L1}} + \boldsymbol{X_{L2}} + \ldots + \boldsymbol{X_{LN}}$

2. Problem: For a series $\boldsymbol{RL}$ circuit connected across voltage source $\boldsymbol{V_S}$, prove that the voltage divider theorem is valid so that the voltage across component $\boldsymbol{K}$ is

$\boldsymbol{V_{RK}} = \boldsymbol{V_S}\,[\boldsymbol{R_K} / \boldsymbol{Z_T}]$ for a resistor

$\boldsymbol{V_{LK}} = \boldsymbol{V_S}\,[\boldsymbol{X_{LK}} / \boldsymbol{Z_T}]$ for an inductor

Solution: For a series $\boldsymbol{RL}$ circuit with $\boldsymbol{M}$ resistors and $\boldsymbol{N}$ inductors, the total impedance is $\boldsymbol{Z_T}$. Use Ohm's Law for current to calculate the total current, $\boldsymbol{I_T}$.

$\boldsymbol{I_T} = \boldsymbol{V_S} / \boldsymbol{Z_T}$

For a series circuit, the current through any component is equal to $\boldsymbol{I_T}$. From Ohm's Law, the voltage drop across resistor $\boldsymbol{R_K}$ is

$\boldsymbol{V_{RK}} = \boldsymbol{I_T}\,\boldsymbol{R_K}$

$= (\boldsymbol{V_S} / \boldsymbol{Z_T})\,\boldsymbol{R_K}$

$= (\boldsymbol{V_S}\,\boldsymbol{R_K} / \boldsymbol{Z_T})$

$= \boldsymbol{V_S}\,(\boldsymbol{R_K} / \boldsymbol{Z_T})$

Similarly, the voltage drop across inductor $\boldsymbol{L_K}$ is

$\boldsymbol{V_{LK}} = \boldsymbol{I_T}\,\boldsymbol{X_{LK}}$

$= (V_S / Z_T)\, X_{LK}$
$= (V_S X_{LK}) / Z_T)$
$= V_S (X_{LK} / Z_T)$

17.2.2 Parallel Circuits

1. Problem: Prove that the total reactance of N inductors in parallel is

$$X_{LT} = 1 / [(1 / X_{L1}) + (1 / X_{L2}) + ... + (1 / X_{LN})]$$

Solution: For N inductors in parallel, the total inductance L_T is

$$L_T = 1 / [(1 / L_1) + (1 / L_2) + ... + (1 / L_N)]$$
$$1 / L_T = (1 / L_1) + (1 / L_2) + ... + (1 / L_N)$$

The inductive reactance, X_L, for an inductor, L, at frequency, f, is

$$X_L = 2\pi f L$$

From this, the total inductive reactance, X_{LT}, for total parallel inductance, L_T, is

$$X_{LT} = 2\pi f L_T$$
$$1 / X_{LT} = 1 / (2\pi f L_T)$$
$$= [1 / (2\pi f)]\,(1 / L_T)$$
$$= [1 / 2\pi f]\,[(1 / L_1) + (1 / L_2) + ... + (1 / L_N)]$$
$$= (1 / 2\pi f L_1) + (1 / 2\pi f L_2) + ... + (1 / 2\pi f L_N)$$
$$= (1 / X_{L1}) + (1 / X_{L2}) + ... + (1 / X_{LN})$$
$$1 / (1 / X_{LT}) = 1 / [(1 / X_{L1}) + (1 / X_{L2}) + ... + (1 / X_{LN})]$$
$$X_{LT} = 1 / [(1 / X_{L1}) + (1 / X_{L2}) + ... + (1 / X_{LN})]$$

2. Problem: For a parallel ***RL*** circuit with total current I_T, prove that the current divider theorem is valid so that the current through component K is

$I_{RK} = I_T [Z_T / R_K]$ for a resistor

$I_{LK} = I_T [Z_T / X_{LK}]$ for an inductor

Solution: From Ohm's Law,

$$V_T = I_T Z_T$$

For a parallel circuit, the voltage across any component is equal to V_T. From Ohm's Law, the current I_{RK} through resistor R_K is

$$I_{RK} = V_T / R_K$$
$$= (I_T Z_T) / R_K$$
$$= I_T (Z_T / R_K)$$

and the current through inductor L_K is

$$I_{LK} = V_T / X_{LK}$$
$$= (I_T Z_T) / X_{LK}$$
$$= I_T (Z_T / X_{LK})$$

17.2.3 Series-Parallel Circuits

1. Problem: For the ***RL*** bridge circuit in Figure 17-29, the wiper setting, k, of the potentiometer R can vary from $k = 0$ (the wiper is full to the left) to $k = 1$ (the wiper is fully to the right), with $k = 0.5$ when the wiper is in the middle. If the values of R and L are known and the wiper is adjusted so that $V_A = V_B$, what is the value of the unknown inductance, L_X?

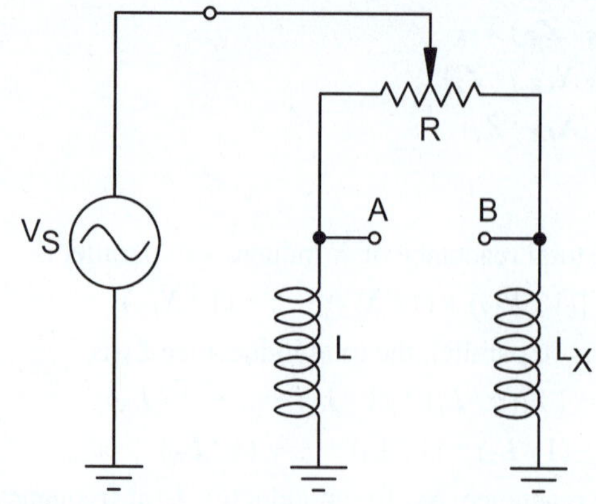

Figure 17-29: *RC* Bridge Circuit for Problem 1

Solution: When the wiper setting is k, the resistance between the wiper and L is kR and the resistance between the wiper and L_X is $(1 - k)$. From the complex voltage divider theorem,

$V_A = \mathrm{j}\,X_L / (kR + \mathrm{j}\,X_L)$

$V_B = \mathrm{j}\,X_{LX} / [(1 - k)R + \mathrm{j}\,X_{LX}]$

Set these terms equal and solve for X_{LX}.

V_A	$= V_B$
$\mathrm{j}\,X_L / (kR + \mathrm{j}\,X_L)$	$= \mathrm{j}\,X_{LX} / [(1 - k)R + \mathrm{j}\,X_{LX}]$
$[\mathrm{j}\,X_L / (kR + \mathrm{j}\,X_L)]\,(kR + \mathrm{j}\,X_L)$	$= \{\mathrm{j}\,X_{LX} / [(1 - k)R + \mathrm{j}\,X_{LX}]\}\,(kR + \mathrm{j}\,X_L)$
$\mathrm{j}\,X_L$	$= [\mathrm{j}\,X_{LX}\,(kR + \mathrm{j}\,X_L)] / [(1 - k)R + \mathrm{j}\,X_{LX}]\}$
$(\mathrm{j}\,X_L)\,[(1 - k)R + \mathrm{j}\,X_{LX}]$	$= [\mathrm{j}\,X_{LX}\,(kR + \mathrm{j}\,X_L)] / [(1 - k)R + \mathrm{j}\,X_{LX}]\}\,[(1 - k)R + \mathrm{j}\,X_{LX}]$
$\mathrm{j}\,[(1 - k)RX_L] + \mathrm{j}^2\,(X_L X_{LX})$	$= \mathrm{j}\,(kRX_{LX}) + \mathrm{j}^2\,(X_L\,X_{LX})$
$\mathrm{j}\,[(1 - k)RX_L] + (-1)(X_L X_{LX})$	$= \mathrm{j}\,(kRX_{LX}) + (-1)(X_L X_{LX})$
$\mathrm{j}\,[(1 - k)RX_L] - X_L X_{LX} + X_L X_{LX}$	$= \mathrm{j}\,(kRX_{LX}) - X_L X_{LX} + X_L X_{LX}$
$\mathrm{j}\,[(1 - k)RX_L] / (\mathrm{j}\,R)$	$= \mathrm{j}\,(kRX_{LX}) / (\mathrm{j}\,R)$
$[(1 - k)\,X_L] / k$	$= (kX_{LX}) / k$
$[(1 - k) / k]\,X_L$	$= X_{LX}$

The value of the unknown inductor L_X when $V_A = V_B$ is $\mathbf{[(1 - k) / k]\,X_L}$.

2. Problem: For the series-parallel *RL* circuit in Figure 17-30, what is V_{out}? What is V_{out} if $L_X = L$? What is V_{out} if $X_{LX} = X_L = R$?

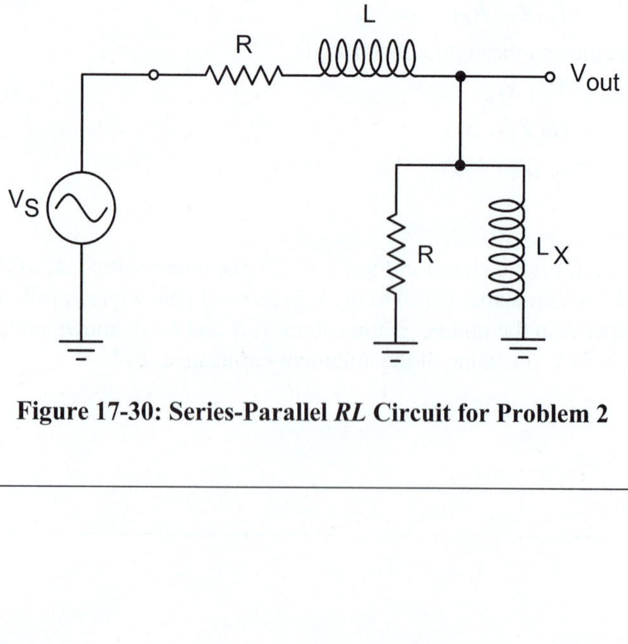

Figure 17-30: Series-Parallel *RL* Circuit for Problem 2

Solution: First, determine the values of inductive reactance

$$X_L = 2\pi f L$$

$$X_{LX} = 2\pi f L_X$$

The equivalent complex circuit is shown in Figure 17-31.

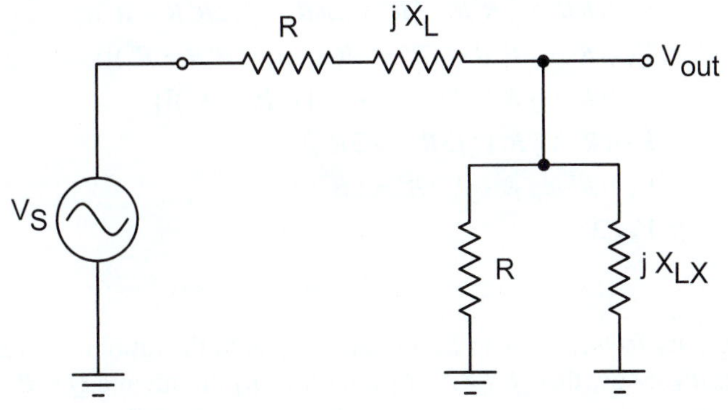

Figure 17-31: Equivalent Complex Circuit for Problem 2

From standard analysis of a series-parallel circuit, the total impedance Z_T is

$$Z_T = (R + X_L) + R \parallel X_{LX}$$

Let Z_{EQ} be the equivalent impedance of the parallel combination $R \parallel X_{LX}$.

$$Z_T = (R + X_L) + Z_{EQ}$$

where

$$\begin{aligned}
Z_{EQ} &= R \parallel X_{LX} \\
&= [(R)\,(\mathrm{j}\,X_{LX})] / (R + \mathrm{j}\,X_{LX}) \\
&= [(\mathrm{j}\,RX_{LX}) / (R + \mathrm{j}\,X_{LX})]\,[(R - \mathrm{j}\,X_{LX}) / (R - \mathrm{j}\,X_{LX})] \\
&= [(\mathrm{j}\,RX_{LX})\,(R - \mathrm{j}\,X_{LX})] / [(R + \mathrm{j}\,X_{LX})\,(R - \mathrm{j}\,X_{LX})] \\
&= (\mathrm{j}\,RRX_{LX} - \mathrm{j}^2\,RX_{LX}X_{LX}) / (RR - \mathrm{j}\,RX_{LX} + \mathrm{j}\,RX_{LX} - \mathrm{j}^2\,X_{LX}X_{LX}) \\
&= [\mathrm{j}\,R^2X_{LX} - (-1)RX_{LX}^2] / [R^2 + \mathrm{j}\,(RX_{LX} - RX_{LX}) - (-1)X_{LX}^2] \\
&= (RX_{LX}^2 + \mathrm{j}\,R^2X_{LX}) / (R^2 + X_{LX}^2)
\end{aligned}$$

Use Z_{EQ} to calculate the total impedance Z_T.

$$\begin{aligned}
Z_T &= (R + X_L) + Z_{EQ} \\
&= (R + \mathrm{j}\,X_L) + [(RX_{LX}^2 + \mathrm{j}\,R^2X_{LX}) / (R^2 + X_{LX}^2)] \\
&= (R + \mathrm{j}\,X_L)\,[(R^2 + X_{LX}^2) / (R^2 + X_{LX}^2)] + [(RX_{LX}^2 + \mathrm{j}\,R^2X_{LX}) / (R^2 + X_{LX}^2)] \\
&= \{[(R + \mathrm{j}\,X_L)\,(R^2 + X_{LX}^2)] / (R^2 + X_{LX}^2)\} + [(RX_{LX}^2 + \mathrm{j}\,R^2X_{LX}) / (R^2 + X_{LX}^2)] \\
&= \{[(R + \mathrm{j}\,X_L)\,(R^2 + X_{LX}^2)] + (RX_{LX}^2 + \mathrm{j}\,R^2X_{LX})\} / (R^2 + X_{LX}^2) \\
&= [(RR^2 + RX_{LX}^2 + \mathrm{j}\,R^2X_L + \mathrm{j}\,X_LX_{LX}^2) + (RX_{LX}^2 + \mathrm{j}\,R^2X_{LX})] / (R^2 + X_{LX}^2) \\
&= (R^3 + RX_{LX}^2 + \mathrm{j}\,R^2X_L + \mathrm{j}\,X_LX_{LX}^2 + RX_{LX}^2 + \mathrm{j}\,R^2X_{LX}) / (R^2 + X_{LX}^2) \\
&= (R^3 + 2RX_{LX}^2 + \mathrm{j}\,R^2X_L + \mathrm{j}\,X_LX_{LX}^2 + \mathrm{j}\,R^2X_{LX}) / (R^2 + X_{LX}^2) \\
&= [R^3 + 2RX_{LX}^2 + \mathrm{j}\,(R^2X_L + X_LX_{LX}^2 + R^2X_{LX})] / (R^2 + X_{LX}^2)
\end{aligned}$$

Use the complex voltage divider theorem to calculate V_{out}.

$$\begin{aligned}
V_{out} &= V_S\,(Z_{EQ} / Z_T) \\
&= V_S\,\{[(RX_{LX}^2 + \mathrm{j}\,R^2X_{LX}) / (R^2 + X_{LX}^2)] \\
&\qquad / [R^3 + 2RX_{LX}^2 + \mathrm{j}\,(R^2X_L + X_LX_{LX}^2 + R^2X_{LX})] / (R^2 + X_{LX}^2)\} \\
&= V_S\,\{(RX_{LX}^2 + \mathrm{j}\,R^2X_{LX}) / [R^3 + 2RX_{LX}^2 + \mathrm{j}\,(R^2X_L + X_LX_{LX}^2 + R^2X_{LX})]\}
\end{aligned}$$

If $L_X = L$, then $X_{LX} = X_L$.

$$V_{out} = V_S\,\{(RX_{LX}^2 + \mathrm{j}\,R^2X_{LX}) / [R^3 + 2RX_{LX}^2 + \mathrm{j}\,(R^2X_L + X_LX_{LX}^2 + R^2X_{LX})]\}$$

$= V_S\,\{(RX_L^2 + \mathrm{j}\,R^2X_L) / [R^3 + 2RX_L^2 + \mathrm{j}\,(R^2X_L + X_LX_L^2 + R^2X_L)]\}$

$= V_S\,\{(RX_L^2 + \mathrm{j}\,R^2X_L) / [R^3 + 2RX_L^2 + \mathrm{j}\,(2R^2X_L + X_L^3)]\}$

If $X_{LX} = X_L = R$,

$V_{out} = V_S\,\{(RX_L^2 + \mathrm{j}\,R^2X_L) / [R^3 + 2RX_L^2 + \mathrm{j}\,(2R^2X_L + X_L^3)]\}$

$= V_S\,\{(RR^2 + \mathrm{j}\,R^2R) / [R^3 + 2RR^2 + \mathrm{j}\,(2R^2R + R^3)]\}$

$= V_S\,\{(R^3 + \mathrm{j}\,R^2R) / [R^3 + 2RR^2 + \mathrm{j}\,(2R^2R + R^3)]\}$

$= V_S\,\{(R^3 + \mathrm{j}\,R^3) / [R^3 + 2R^3 + \mathrm{j}\,(2R^3 + R^3)]\}$

$= V_S\,[(R^3 + \mathrm{j}\,R^3) / (3R^3 + \mathrm{j}\,3R^3)]$

$= V_S\,\{(R^3 + \mathrm{j}\,R^3) / [3(R^3 + \mathrm{j}\,R^3)]\}$

$= V_S / 3$

17.2.4 Power in *RL* Circuits

1. Problem: The quality factor, Q, of an RL circuit is equal to the ratio of the reactive power, P_{VAR} to the real power, P_R. Show that $Q = X_L / R$ for a series R_L circuit and $Q = R / X_L$ for a parallel RL circuit.

   Solution: From Watt's Law, $P_{VAR} = I_L^2\, X_L = V_L^2 / X_L$ and $P_R = I_R^2\, R = V_R^2 / R$. For a series RL circuit, the inductor current, I_L, and resistor current, I_R, is equal to the total series current, I_T. From the definition of Q,

   $Q = P_{VAR} / P_R$

   $= (I_L^2\, X_L) / (I_R^2\, R)$

   $= (I_T^2\, X_L) / (I_T^2\, R)$

   $= X_L / R$

   For a parallel RL circuit, the inductor voltage, V_L, and resistor voltage, V_R, is equal to the applied source voltage, V_S. From the definition of Q,

   $Q = P_{VAR} / P_R$

   $= (V_L^2 / X_L) / (V_R^2 / R)$

   $= (V_S^2 / X_L) / (V_S^2 / R)$

   $= (V_S^2 / X_L) / (R / V_S^2)$

   $= R / X_L$

2. Problem: Prove that the total real and reactive powers in a series RL circuit containing M resistors and N inductors are equal to the sum of the real and reactive powers of the individual components, respectively.

   Solution: For a series RL circuit with total resistance R_T and reactance X_{LT}, the total impedance Z_T is

   $Z_T = R_T + \mathrm{j}\, X_{LT}$

   Use Watt's Law for complex circuits to calculate the total real and reactive powers $P_{r(T)}$ and $P_{reac(T)}$.

   $P_{r(T)} = I_T^2\, R_T$

   $P_{reac(T)} = I_T^2\, X_{LT}$

   For a series circuit, the total resistance, R_T, and total reactance, X_T, are equal to

   $R_T = R_1 + R_2 + \ldots + R_M$

   $X_T = X_{L1} + X_{L2} + \ldots + X_{LN}$

   Substitute these values into the expressions for $P_{r(T)}$ and $P_{reac(T)}$.

   $P_{r(T)} = I_T^2 R_T$

   $= I_T^2\, (R_1 + R_2 + \ldots + R_M)$

   $= I_T^2 R_1 + I_T^2 R_2 + \ldots + I_T^2 R_M$

$$\begin{aligned} \boldsymbol{P}_{reac(T)} &= \boldsymbol{I}_T^2\boldsymbol{X}_T \\ &= \boldsymbol{I}_T^2(\boldsymbol{X}_{L1} + \boldsymbol{X}_{L2} + \ldots + \boldsymbol{X}_{LN}) \\ &= \boldsymbol{I}_T^2\boldsymbol{X}_{L1} + \boldsymbol{I}_T^2\boldsymbol{X}_{L2} + \ldots + \boldsymbol{I}_T^2\boldsymbol{X}_{LN} \end{aligned}$$

Because the current through each component in a series circuit is equal to the total series current, the power for component $\boldsymbol{K}$ in a series ***RL*** circuit is

$\boldsymbol{P}_{r(K)} = \boldsymbol{I}_T^2\boldsymbol{R}_K$ for a resistor

$\boldsymbol{P}_{reac(K)} = \boldsymbol{I}_T^2\boldsymbol{X}_{LK}$ for an inductor

Substitute these values into the expressions for $\boldsymbol{P}_{r(T)}$ and $\boldsymbol{P}_{reac(T)}$.

$$\begin{aligned} \boldsymbol{P}_{r(T)} &= \boldsymbol{I}_T^2\boldsymbol{R}_1 + \boldsymbol{I}_T^2\boldsymbol{R}_2 + \ldots + \boldsymbol{I}_T^2\boldsymbol{R}_M \\ &= \boldsymbol{P}_{r1} + \boldsymbol{P}_{r2} + \ldots + \boldsymbol{P}_{rM} \\ \boldsymbol{P}_{reac(T)} &= \boldsymbol{I}_T^2\boldsymbol{X}_{L1} + \boldsymbol{I}_T^2\boldsymbol{X}_{L2} + \ldots + \boldsymbol{I}_T^2\boldsymbol{X}_{LN} \\ &= \boldsymbol{P}_{reac1} + \boldsymbol{P}_{reac2} + \ldots + \boldsymbol{P}_{reacN} \end{aligned}$$

Therefore, in a series ***RL*** circuit, the total real power is the sum of the real powers for the individual resistors and the total reactive power is the sum of the reactive powers for the individual inductors.

3. Problem: Prove that the total real and reactive powers in a parallel RL circuit are equal to the sum of the real and reactive powers of the individual components, respectively.

Solution: For a parallel ***RL*** circuit with total resistance $\boldsymbol{R}_T$ and reactance $\boldsymbol{X}_{LT}$, the total impedance $\boldsymbol{Z}_T$ is

$$\boldsymbol{Z}_T = \boldsymbol{R}_T + \mathrm{j}\,\boldsymbol{X}_{LT}$$

Use Watt's Law for complex circuits to calculate the total real and reactive powers $\boldsymbol{P}_{r(T)}$ and $\boldsymbol{P}_{reac(T)}$.

$$\begin{aligned} \boldsymbol{P}_{r(T)} &= \boldsymbol{V}_S^2 / \boldsymbol{R}_T \\ \boldsymbol{P}_{reac(T)} &= \boldsymbol{V}_S^2 / \boldsymbol{X}_{LT} \end{aligned}$$

For a parallel circuit, the total resistance, $\boldsymbol{R}_T$, and total reactance, $\boldsymbol{X}_T$, are equal to

$$\begin{aligned} \boldsymbol{R}_T &= 1 / [(1 / \boldsymbol{R}_1) + (1 / \boldsymbol{R}_2) + \ldots + (1 / \boldsymbol{R}_M)] \\ \boldsymbol{X}_T &= 1 / [(1 / \boldsymbol{X}_{L1}) + (1 / \boldsymbol{X}_{L2}) + \ldots + (1 / \boldsymbol{X}_{LN})] \end{aligned}$$

Substitute these values into the expressions for $\boldsymbol{P}_{r(T)}$ and $\boldsymbol{P}_{reac(T)}$.

$$\begin{aligned} \boldsymbol{P}_{r(T)} &= \boldsymbol{V}_S^2 / \boldsymbol{R}_T \\ &= \boldsymbol{V}_S^2 / \{1 / [(1 / \boldsymbol{R}_1) + (1 / \boldsymbol{R}_2) + \ldots + (1 / \boldsymbol{R}_M)]\} \\ &= \boldsymbol{V}_S^2 [(1 / \boldsymbol{R}_1) + (1 / \boldsymbol{R}_2) + \ldots + (1 / \boldsymbol{R}_M)] \\ &= (\boldsymbol{V}_S^2 / \boldsymbol{R}_1) + (\boldsymbol{V}_S^2 / \boldsymbol{R}_2) + \ldots + (\boldsymbol{V}_S^2 / \boldsymbol{R}_M)] \\ \boldsymbol{P}_{reac(T)} &= \boldsymbol{V}_S^2 / \boldsymbol{X}_T \\ &= \boldsymbol{V}_S^2 / \{1 / [(1 / \boldsymbol{X}_{L1}) + (1 / \boldsymbol{X}_{L2}) + \ldots + (1 / \boldsymbol{X}_{LN})]\} \\ &= \boldsymbol{V}_S^2 [(1 / \boldsymbol{X}_{L1}) + (1 / \boldsymbol{X}_{L2}) + \ldots + (1 / \boldsymbol{X}_{LN})] \\ &= (\boldsymbol{V}_S^2 / \boldsymbol{X}_{L1}) + (\boldsymbol{V}_S^2 / \boldsymbol{X}_{L2}) + \ldots + (\boldsymbol{V}_S^2 / \boldsymbol{X}_{LN})] \end{aligned}$$

Because the voltage across each component in a parallel circuit is equal to the applied source voltage, the power for component $\boldsymbol{K}$ in a parallel ***RL*** circuit is

$\boldsymbol{P}_{r(K)} = \boldsymbol{V}_S^2 / \boldsymbol{R}_K$ for a resistor

$\boldsymbol{P}_{reac(K)} = \boldsymbol{V}_S^2 / \boldsymbol{X}_{LK}$ for an inductor

Substitute these values into the expressions for $\boldsymbol{P}_{r(T)}$ and $\boldsymbol{P}_{reac(T)}$.

$$\begin{aligned} \boldsymbol{P}_{r(T)} &= (\boldsymbol{V}_S^2 / \boldsymbol{R}_1) + (\boldsymbol{V}_S^2 / \boldsymbol{R}_2) + \ldots + (\boldsymbol{V}_S^2 / \boldsymbol{R}_M)] \\ &= \boldsymbol{P}_{r1} + \boldsymbol{P}_{r2} + \ldots + \boldsymbol{P}_{rM} \\ \boldsymbol{P}_{reac(T)} &= (\boldsymbol{V}_S^2 / \boldsymbol{X}_{L1}) + (\boldsymbol{V}_S^2 / \boldsymbol{X}_{L2}) + \ldots + (\boldsymbol{V}_S^2 / \boldsymbol{X}_{LN})] \\ &= \boldsymbol{P}_{reac1} + \boldsymbol{P}_{reac2} + \ldots + \boldsymbol{P}_{reacN} \end{aligned}$$

Therefore, in a parallel ***RL*** circuit, the total real power is the sum of the real powers for the individual resistors and the total reactive power is the sum of the reactive powers for the individual inductors.

17.3 Just For Fun

1. Problem: It is sometimes convenient to use an equivalent series circuit to represent a parallel circuit. For a specific frequency f, what values of resistance R_S and inductance L_S in a series ***RL*** circuit will give the same impedance Z_T as the values of resistance R_P and inductance L_P in a parallel ***RL*** circuit?

   Solution: At a specific frequency f, the reactance X_{CS} and X_{LP} of the inductors L_S and L_P are

$$X_{LS} = 2\pi f L_S$$

$$X_{LP} = 2\pi f L_P$$

The total parallel impedance Z_T for R_P and L_P is

$$\begin{aligned}
Z_T &= R_P \,||\, \mathrm{j}\, X_{LP} \\
&= [(R_P + \mathrm{j}\,0)\,(0 + \mathrm{j}\,X_{LP})] \,/\, (R_P + \mathrm{j}\,X_{LP}) \\
&= [(R_P)(0) + (R_P)(\mathrm{j}\,X_{LP}) + (\mathrm{j}\,0)(0) + (\mathrm{j}\,0)(\mathrm{j}\,X_P)] \,/\, (R_P + \mathrm{j}\,X_{LP}) \\
&= \mathrm{j}\,R_P X_{LP} \,/\, (R_P + \mathrm{j}\,X_{LP}) \\
&= [\mathrm{j}\,R_P X_{LP} \,/\, (R_P + \mathrm{j}\,X_{LP})][(R_P - \mathrm{j}\,X_{LP}) \,/\, (R_P - \mathrm{j}\,X_{LP})] \\
&= [(\mathrm{j}\,R_P X_{LP})\,(R_P - \mathrm{j}X_{LP})][(R_P + \mathrm{j}\,X_{LP})\,(R_P - \mathrm{j}\,X_{LP})] \\
&= [\mathrm{j}\,R_P X_{LP} R_P - (\mathrm{j}^2\,R_P X_{LP} X_{LP})] \,/\, (R_P R_P - \mathrm{j}\,X_{LP} R_P + \mathrm{j}\,X_{LP} R_P - \mathrm{j}^2\,X_{LP} X_{LP}) \\
&= (\mathrm{j}\,R_P^2 X_{LP} - (-1)\,R_P X_{LP}^2) \,/\, (R_P^2 + \mathrm{j}\,0 - (-1)\,X_{LP}^2) \\
&= (R_P X_{LP}^2 + \mathrm{j}\,R_P^2 X_{LP}) \,/\, (R_P^2 + X_{LP}^2) \\
&= [R_P X_{LP}^2 \,/\, (R_P^2 + X_{LP}^2)] + \mathrm{j}\,[R_P^2 X_{LP} \,/\, (R_P^2 + X_{LP}^2)]
\end{aligned}$$

The series resistance R_S corresponds to the real term and series inductance L_S corresponds to the imaginary term.

$$\begin{aligned}
R_S &= R_P X_{LP}^2 \,/\, (R_P^2 + X_{LP}^2) \\
&= R_P\,(2\pi f L_P)^2 \,/\, [R_P^2 + (2\pi f L_P)^2]
\end{aligned}$$

and

$$\begin{aligned}
L_S &= R_P^2 X_{LP} \,/\, (R_P^2 + X_{LP}^2) \\
&= R_P^2\,(2\pi f L_P) \,/\, [R_P^2 + (2\pi f L_P)^2]
\end{aligned}$$

2. Problem: It is also sometimes convenient to use an equivalent parallel circuit to represent a series circuit. For a specific frequency f, What values of resistance R_P and inductance L_P in a parallel ***RL*** circuit will give the same impedance Z_T as the values of resistance R_S and inductance L_S in a series ***RL*** circuit?

   Solution: At a specific frequency f, the reactance X_{LS} and X_{LP} of the inductors L_S and L_P are

$$X_{LS} = 1 \,/\, (2\pi f L_S)$$

$$X_{LP} = 1 \,/\, (2\pi f L_P)$$

The total series impedance Z_T for R_S and L_S is

$$Z_T = R_S + \mathrm{j}\,X_{LS}$$

The total series admittance, Y_T, is then

$$\begin{aligned}
Y_{ST} &= 1 \,/\, (R_S + \mathrm{j}\,X_{LS}) \\
&= [1 \,/\, (R_S + \mathrm{j}\,X_{LS})]\;[(R_S - \mathrm{j}\,X_{LS})\,(R_S - \mathrm{j}\,X_{LS})] \\
&= (R_S - \mathrm{j}\,X_{LS}) \,/\, [(R_S + \mathrm{j}\,X_{LS})\,(R_S - \mathrm{j}\,X_{LS})] \\
&= (R_S - \mathrm{j}\,X_{LS}) \,/\, (R_S R_S - \mathrm{j}\,R_S X_{LS} + \mathrm{j}\,R_S X_{LS} - \mathrm{j}^2\,X_{LS} X_{LS}) \\
&= (R_S - \mathrm{j}\,X_{LS}) \,/\, (R_S^2 + \mathrm{j}\,0 - (-1)\,X_{LS}^2)
\end{aligned}$$

$= (R_S - j X_{LS}) / (R_S^2 + X_{LS}^2)$

$= [R_S / (R_S^2 + X_{LS}^2)] - j [X_S / (R_S^2 + X_{LS}^2)]$

The total parallel impedance Z_T for R_P and L_P is

$Z_{PT} = [(1 / R_P) + (1/ j X_{LP})]^{-1}$

The total parallel admittance, Y_T, is then

$Y_{PT} = (1 / R_P) + (1/ j X_{LP})$

$= (1 / R_P) + (1 / j X_{LP}) (j / j)$

$= (1 / R_P) + [j / (j^2 X_{LP})]$

$= (1 / R_P) + \{j / [(-1) X_{LP})]\}$

$= (1 / R_P) - j (1 / X_{LP})$

The parallel resistance R_P corresponds to the real term and parallel inductance L_P corresponds to the imaginary term.

$1 / R_P = [R_S / (R_S^2 + X_S^2)]$

$1 / (1 / R_P) = 1 / [R_S / (R_S^2 + X_S^2)]$

$R_P = (R_S^2 + X_S^2) / R_S$

$= [R_S^2 + (2\pi f L_S)^2] / R_S$

and

$1 / X_{LP} = [X_{LS} / (R_S^2 + X_{LS}^2)]$

$1 / (2\pi f L_P) = (2\pi f L_S) / [R_S^2 + (2\pi f L_S)^2]$

$2\pi f L_P = [R_S^2 + (2\pi f L_S)^2] / (2\pi f L_S)$

$2\pi f L_P / 2\pi f = \{[R_S^2 + (2\pi f L_S)^2] / (2\pi f L_S)\} / 2\pi f$

$L_P = [R_S^2 + (2\pi f L_S)^2] / [(2\pi f)^2 L_S]$

3. Problem: The circuit analysis techniques for resistive circuits can also be applied to ***RL*** circuits once the inductors are replaced with inductive reactances. For the ladder circuit shown in Figure 17-32, use node voltage analysis to determine the output voltage V_{out} when $R = X_L$. Use the reference directions shown for the currents and assume currents into a node are positive.

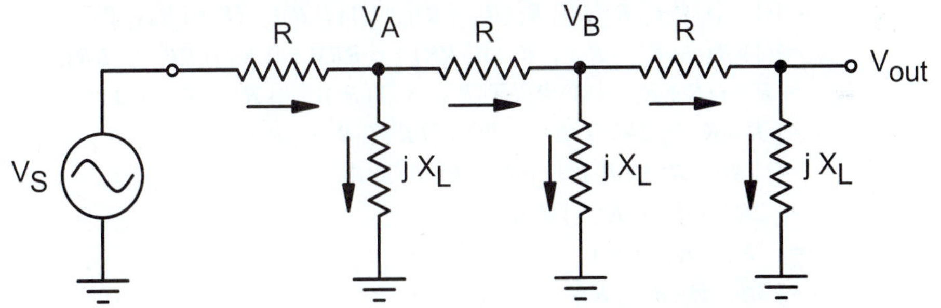

Figure 17-32: *RL* Ladder Network for Problem 3

Solution: The voltage V_{out} can be found using the complex voltage divider theorem once V_B is found. The node voltage equations and subsequent rearrangements are:

$[(V_S - V_A) / R] - (V_A / j X_L) - [(V_A - V_B) / R] = 0$

$[(V_A - V_B) / R] - (V_B / j X_L) - [V_B / (R + j X_L)] = 0$

$(V_S / R) - (V_A / R) - (V_A / j X_L) - (V_A / R) + (V_B / R) = 0$

$(V_A / R) - (V_B / R) - (V_B / j X_L) - [V_B / (R + j X_L)] = 0$

$(V_S / R) - V_A [(1 / R) + (1 / j X_L) + (1 / R)] + V_B (1 / R) = 0$

$$V_A\,(1 / R) - V_B\,\{(1 / R) + (1 / \mathrm{j}\,X_L) + [1 / (R + \mathrm{j}\,X_L)]\} = 0$$

$$(V_S / R) + V_A\,[(2 / R) + (1 / \mathrm{j}\,X_L)] + V_B\,(1 / R) - (V_S / R) = 0 - (V_S / R)$$
$$V_A\,(1 / R) - V_B\,\{(1 / R) + (1 / \mathrm{j}\,X_L) + [1 / (R + \mathrm{j}\,X_L)]\} = 0$$

$$V_A\,[(2 / R) + (1 / \mathrm{j}\,X_L)] + V_B\,(1 / R) = -(V_S / R)$$
$$V_A\,(1 / R) - V_B\,\{(1 / R) + (1 / \mathrm{j}\,X_L) + [1 / (R + \mathrm{j}\,X_L)]\} = 0$$

From the problem statement, $R = X_L$.

$$V_A\,[(2 / R) + (1 / \mathrm{j}\,R)] + V_B\,(1 / R) = -(V_S / R)$$
$$V_A\,(1 / R) - V_B\,\{(1 / R) + (1 / \mathrm{j}\,R) + [1 / (R + \mathrm{j}\,X_L)]\} = 0$$

Use a common denominator to simplify the expression $(2 / R) + (1 / \mathrm{j}\,X_L)$.

$$(2 / R) + (1 / \mathrm{j}\,R)$$
$$= (2 / R)\,(\mathrm{j} / \mathrm{j}) + (1 / \mathrm{j}\,R)$$
$$= (\mathrm{j}\,2 / \mathrm{j}\,R) + (1 / \mathrm{j}\,R)$$
$$= (1 + \mathrm{j}\,2) / \mathrm{j}\,R$$
$$= [(1 + \mathrm{j}\,2) / \mathrm{j}\,R]\,(\mathrm{j} / \mathrm{j})$$
$$= (\mathrm{j} + \mathrm{j}^2\,2) / \mathrm{j}^2\,R$$
$$= [\mathrm{j} + (-1)(2)] / [(-1)\,R]$$
$$= (-2 + \mathrm{j}) / {-R}$$
$$= (2 - \mathrm{j}) / R$$

Use a common denominator to simplify the expression $(1 / R) + (1 / \mathrm{j}\,R) + [1 / (R + \mathrm{j}\,R)]$.

$$(1 / R) + (1 / \mathrm{j}\,R) + [1 / (R + \mathrm{j}\,R)]$$
$$= [(1 / R)\,(\mathrm{j} / \mathrm{j})] + (1 / \mathrm{j}\,R) + [1 / (R + \mathrm{j}\,R)]$$
$$= (\mathrm{j} / \mathrm{j}\,R) + (1 / \mathrm{j}\,R) + [1 / (R + \mathrm{j}\,R)]$$
$$= [(\mathrm{j} + 1) / \mathrm{j}\,R]\,[(R + \mathrm{j}\,R) / (R + \mathrm{j}\,R)] + [1 / (R + \mathrm{j}\,R)]\,[(\mathrm{j}\,R) / (\mathrm{j}\,R)]$$
$$= [(1 + \mathrm{j})\,(R + \mathrm{j}\,R)] / [(\mathrm{j}\,R)\,(R + \mathrm{j}\,R)] + [(1)\,(\mathrm{j}\,R)] / [(R + \mathrm{j}\,R)\,(\mathrm{j}\,R)]$$
$$= [(1)(R) + \mathrm{j}\,R + \mathrm{j}\,R + \mathrm{j}^2\,R] / [(\mathrm{j}\,RR) + \mathrm{j}^2\,RR)] + [\mathrm{j}\,R / (\mathrm{j}\,RR + \mathrm{j}^2\,RR)]$$
$$= [R + \mathrm{j}\,(R + R) + (-1)\,R] / [(\mathrm{j}\,R^2) + (-1)\,R^2)] + \{\mathrm{j}\,R / [\mathrm{j}\,R^2 + (-1)\,R^2]\}$$
$$= [(R - R) + \mathrm{j}\,2R] / (-R^2 + \mathrm{j}\,R^2) + [\mathrm{j}\,R / (-R^2 + \mathrm{j}\,R^2)]$$
$$= [\mathrm{j}\,2R / (-R^2 + \mathrm{j}\,R^2)] + [\mathrm{j}\,R / (-R^2 + \mathrm{j}\,R^2)]$$
$$= (\mathrm{j}\,2R + \mathrm{j}\,R) / (-R^2 + \mathrm{j}\,R^2)]$$
$$= \mathrm{j}\,3R / (-R^2 + \mathrm{j}\,R^2)$$
$$= \mathrm{j}\,3R / [R(-R + \mathrm{j}\,R)]$$
$$= \mathrm{j}\,3 / (-R + \mathrm{j}\,R)$$

Substitute these simplified terms back into the node voltage equation.

$$V_A\,[(2 - \mathrm{j}) / R] + V_B\,(1 / R) = -(V_S / R)$$
$$V_A\,(1 / R) - V_B\,[\mathrm{j}\,3 / (-R + \mathrm{j}\,R)] = 0$$

The matrix-vector form of this linear system of equations is

$$\begin{vmatrix} (2 - \mathrm{j}) / R & 1 / R \\ 1 / R & \mathrm{j}\,3 / (-R + \mathrm{j}\,R) \end{vmatrix} \bullet \begin{vmatrix} V_A \\ V_B \end{vmatrix} = \begin{vmatrix} -(V_S / R) \\ 0 \end{vmatrix}$$

Use characteristic matrices to solve this system for $\boldsymbol{V_B}$. The matrix for $\boldsymbol{V_B}$ is

$$\begin{vmatrix} (2-j)/\boldsymbol{R} & -(\boldsymbol{V_S}/\boldsymbol{R}) \\ 1/\boldsymbol{R} & 0 \end{vmatrix}$$

with the determinant

$$\begin{aligned}[(2-j)/\boldsymbol{R}]\,(0)-(1/\boldsymbol{R})[-(\boldsymbol{V_S}/\boldsymbol{R})] &= (-1/\boldsymbol{R})[-(\boldsymbol{V_S}/\boldsymbol{R})] \\ &= \boldsymbol{V_S}/\boldsymbol{RR} \\ &= \boldsymbol{V_S}/\boldsymbol{R}^2\end{aligned}$$

The characteristic matrix is

$$\begin{vmatrix} (2-j)/\boldsymbol{R} & 1/\boldsymbol{R} \\ 1/\boldsymbol{R} & j\,3/(-\boldsymbol{R}+j\,\boldsymbol{R}) \end{vmatrix}$$

with the determinant

$$\begin{aligned} & [(2-j)/\boldsymbol{R}]\,[j\,3/(-\boldsymbol{R}+j\,\boldsymbol{R})]-(1/\boldsymbol{R})(1/\boldsymbol{R}) \\ = \; & [(2-j)\,(j\,3)]/[\boldsymbol{R}\,(-\boldsymbol{R}+j\,\boldsymbol{R})]-(1/\boldsymbol{R})(1/\boldsymbol{R}) \\ = \; & (j\,6-j^2\,3)/(-\boldsymbol{RR}+j\,\boldsymbol{RR})-(1/\boldsymbol{RR}) \\ = \; & [j\,6-(-1)(3)]/(-\boldsymbol{R}^2+j\,\boldsymbol{R}^2)-(1/\boldsymbol{R}^2) \\ = \; & (3+j\,6)/(-\boldsymbol{R}^2+j\,\boldsymbol{R}^2)-(1/\boldsymbol{R}^2) \\ = \; & (3+j\,6)/[\boldsymbol{R}^2\,(-1+j)]-(1/\boldsymbol{R}^2)\,[(-1+j)/(-1+j)] \\ = \; & (3+j\,6)/[\boldsymbol{R}^2\,(-1+j)]-(-1+j)/[\boldsymbol{R}^2\,(-1+j)] \\ = \; & [(3+j\,6)+(1-j)]/[\boldsymbol{R}^2\,(-1+j)] \\ = \; & [(3+1)+j\,(6-1)]/[\boldsymbol{R}^2\,(-1+j)] \\ = \; & (4+j\,5)/[\boldsymbol{R}^2\,(-1+j)] \end{aligned}$$

$\boldsymbol{V_B}$ is equal to the quotient of determinants for the $\boldsymbol{I_2}$ matrix and characteristic matrix.

$$\begin{aligned}\boldsymbol{V_B} &= (\boldsymbol{V_S}/\boldsymbol{R}^2)/\{(4+j\,5)/[\boldsymbol{R}^2\,(-1+j)]\} \\ &= (\boldsymbol{V_S}/\boldsymbol{R}^2)\times\{[\boldsymbol{R}^2\,(-1+j)]/(4+j\,5)\} \\ &= \boldsymbol{V_S}\,[\,(-1+j)/(4+j\,5)]\end{aligned}$$

Use the complex voltage divider theorem to calculate $\boldsymbol{V_{out}}$.

$$\begin{aligned}\boldsymbol{V_{out}} &= \boldsymbol{V_B}\,[j\,\boldsymbol{X_L}/(\boldsymbol{R}+j\,\boldsymbol{X_L})] \\ &= \boldsymbol{V_B}\,[j\,\boldsymbol{R}/(\boldsymbol{R}+j\,\boldsymbol{R})] \\ &= \boldsymbol{V_B}\,\{(j\,\boldsymbol{R})/[\boldsymbol{R}\,(1+j)]\} \\ &= \boldsymbol{V_B}\,[j/(1+j)]\end{aligned}$$

Substitute in the value of $\boldsymbol{V_B}$ and simplify.

$$\begin{aligned}\boldsymbol{V_{out}} &= \boldsymbol{V_B}\,[j/(1+j)] \\ &= \boldsymbol{V_S}\,\{[(-1+j)/(4+j\,5)]\,[j/(1+j)]\} \\ &= \boldsymbol{V_S}\,\{[(-1+j)\,(j)]/[(4+j\,5)\,(1+j)]\} \\ &= \boldsymbol{V_S}\,[(-j+j^2)/(4)(1)+j\,4+j\,5+j^2\,5)] \\ &= \boldsymbol{V_S}\,[(-j-1)/4+j\,(4+5)+(-1)(5)] \\ &= \boldsymbol{V_S}\,[(-1-j)/(4+j\,9+-5)] \\ &= \boldsymbol{V_S}\,[(-1-j)/(-1+j\,9)]\end{aligned}$$

Convert to polar form and simplify.

$$\begin{aligned}\boldsymbol{V_{out}} &= \boldsymbol{V_S}\,[(-1-j)/(-1+j\,9)] \\ &= \boldsymbol{V_S}\,[(1.414\angle -135°)/(9.055\angle 96.3°)] \\ &= \boldsymbol{V_S}\,[(1.414/9.055)\angle(-135° - 96.3°)]\end{aligned}$$

$= V_S\,(0.156 \angle -231°)$

$= \mathbf{V_S\,(0.156 \angle 129°)}$

The output voltage amplitude is approximately one-sixth the amplitude of the input voltage and leads the input voltage by 129°.

18. *RLC* Circuits and Resonance

Note: For all problems in this section, the phase of ac sources is 0° unless otherwise stated.

18.1 Basic

18.1.1 Analysis of Series *RLC* Circuits

For Problems 1 through 3, refer to the circuit in Figure 18-1.

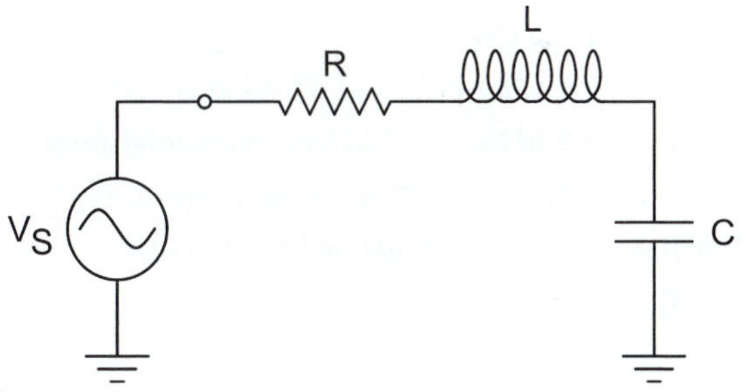

Figure 18-1: Series *RLC* Circuit for Problems 1 through 3

1. Problem: Calculate the total impedance Z_T for each of the following values of R, L, C, and f.

   a. $R = 100\ \Omega$, $L = 22$ mH, $C = 56\ \mu$F, and $f = 500$ Hz

   b. $R = 1.2$ kΩ, $L = 150$ mH, $C = 390$ nF, and $f = 75$ Hz

   c. $R = 15$ kΩ, $L = 7.5\ \mu$H, $C = 22$ pF, and $f = 400$ kHz

   d. $R = 6.8\ \Omega$, $L = 1.5$ H, $C = 1.2$ F, and $f = 10$ mHz

   Solution: For each set of circuit values, first calculate the inductive and capacitive reactances and then calculate the total series impedance using complex arithmetic.

   a. Calculate X_L for $L = 22$ mH and $f = 500$ Hz.

   $$X_L = 2\pi f L$$
   $$= 2\pi(500\text{ Hz})\ (22\text{ mH})$$
   $$= 2\pi(500\text{ Hz})\ (22 \times 10^{-3}\text{ H})$$
   $$= 69.1\ \Omega$$

   Calculate X_C for $C = 56\ \mu$F and $f = 500$ Hz.

   $$X_C = 1\ /\ (2\pi f C)$$
   $$= 1\ /\ [2\pi(500\text{ Hz})\ (56\ \mu\text{F})]$$
   $$= 1\ /\ [2\pi(500\text{ Hz})\ (56 \times 10^{-6}\text{ F})]$$
   $$= 5.68\ \Omega$$

   Calculate the total impedance Z_T from R, X_L, and X_C.

   $$Z_T = R + \text{j}\,X_L - \text{j}\,X_C$$
   $$= R + \text{j}\,(X_L - X_C)$$
   $$= 100\ \Omega + \text{j}\,(69.1\ \Omega - 5.68\ \Omega)$$
   $$= \mathbf{100\ \Omega + j\ 63.4\ \Omega}\text{ in rectangular form}$$
   $$= \mathbf{118\ \Omega\ \angle\ 32.4°}\text{ in polar form}$$

   b. Calculate X_L for $L = 150$ mH and $f = 75$ Hz.

   $$X_L = 2\pi f L$$
   $$= 2\pi(75\text{ Hz})\ (150\text{ mH})$$
   $$= 2\pi(75\text{ Hz})\ (150 \times 10^{-3}\text{ H})$$

$= 70.7\ \Omega$

Calculate $\boldsymbol{X_C}$ for $\boldsymbol{C} = 390$ nF and $\boldsymbol{f} = 75$ Hz.

$\boldsymbol{X_C} = 1 / (2\pi \boldsymbol{fC})$

$= 1 / [2\pi(75\text{ Hz})\ (390\text{ nF})]$

$= 1 / [2\pi(75\text{ Hz})\ (390 \times 10^{-9}\text{ F})]$

$= 5.44\text{ k}\Omega$

Calculate the total impedance $\boldsymbol{Z_T}$ from $\boldsymbol{R}$, $\boldsymbol{X_L}$, and $\boldsymbol{X_C}$.

$\boldsymbol{Z_T} = \boldsymbol{R} + \text{j}\, \boldsymbol{X_L} - \text{j}\, \boldsymbol{X_C}$

$= \boldsymbol{R} + \text{j}\, (\boldsymbol{X_L} - \boldsymbol{X_C})$

$= 1.2\text{ k}\Omega + \text{j}\ (70.7\ \Omega - 5.44\text{ k}\Omega)$

$=$ **1.2 kΩ – j 5.37 kΩ** in rectangular form

$=$ **5.50 kΩ ∠ –77.4°** in polar form

c. Calculate $\boldsymbol{X_L}$ for $\boldsymbol{L} = 7.5\ \mu$H and $\boldsymbol{f} = 400$ kHz.

$\boldsymbol{X_L} = 2\pi \boldsymbol{fL}$

$= 2\pi(400\text{ kHz})\ (7.5\ \mu\text{H})$

$= 2\pi(400\text{ kHz})\ (7.5 \times 10^{-6}\text{ H})$

$= 18.9\text{ k}\Omega$

Calculate $\boldsymbol{X_C}$ for $\boldsymbol{C} = 22$ pF and $\boldsymbol{f} = 400$ kHz.

$\boldsymbol{X_C} = 1 / (2\pi \boldsymbol{fC})$

$= 1 / [2\pi(400\text{ kHz})\ (22\text{ pF})]$

$= 1 / [2\pi(400\text{ kHz})\ (22 \times 10^{-12}\text{ F})]$

$= 18.1\text{ k}\Omega$

Calculate the total impedance $\boldsymbol{Z_T}$ from $\boldsymbol{R}$, $\boldsymbol{X_L}$, and $\boldsymbol{X_C}$.

$\boldsymbol{Z_T} = \boldsymbol{R} + \text{j}\, \boldsymbol{X_L} - \text{j}\, \boldsymbol{X_C}$

$= \boldsymbol{R} + \text{j}\, (\boldsymbol{X_L} - \boldsymbol{X_C})$

$= 15\text{ k}\Omega + \text{j}\ (18.9\text{ k}\Omega - 18.1\text{ k}\Omega)$

$=$ **15 kΩ – j 764 Ω** in rectangular form

$=$ **15.0 kΩ ∠ –2.92°** in polar form

d. Calculate $\boldsymbol{X_L}$ for $\boldsymbol{L} = 1.5$ H and $\boldsymbol{f} = 10$ mHz.

$\boldsymbol{X_L} = 2\pi \boldsymbol{fL}$

$= 2\pi(10\text{ mHz})\ (1.5\text{ H})$

$= 2\pi(10 \times 10^{-3}\text{ Hz})\ (1.5\text{ H})$

$= 94.3\text{ m}\Omega$

Calculate $\boldsymbol{X_C}$ for $\boldsymbol{C} = 1.2$ F and $\boldsymbol{f} = 10$ mHz.

$\boldsymbol{X_C} = 1 / (2\pi \boldsymbol{fC})$

$= 1 / [2\pi(10\text{ mHz})\ (1.2\text{ F})]$

$= 1 / [2\pi(10 \times 10^{-3}\text{ Hz})\ (1.2\text{ F})]$

$= 13.3\ \Omega$

Calculate the total impedance $\boldsymbol{Z_T}$ from $\boldsymbol{R}$, $\boldsymbol{X_L}$, and $\boldsymbol{X_C}$.

$\boldsymbol{Z_T} = \boldsymbol{R} + \text{j}\, \boldsymbol{X_L} - \text{j}\, \boldsymbol{X_C}$

$= \boldsymbol{R} + \text{j}\, (\boldsymbol{X_L} - \boldsymbol{X_C})$

$= 6.8\ \Omega + \text{j}\ (94.3\text{ m}\Omega - 13.3\ \Omega)$

$= 6.8\ \Omega + \text{j}\ (94.3 \times 10^{-3}\ \Omega - 13.3\ \Omega)$

$= \mathbf{6.8\ \Omega - j\ 13.2\ \Omega}$ in rectangular form

$= \mathbf{14.8\ \Omega \angle -62.7°}$ in polar form

2. Problem: For the each given set of circuit values, determine the unknown circuit value.

a. For $\boldsymbol{R} = 510\ \Omega$, $\boldsymbol{L} = 1.5$ mH, $\boldsymbol{f} = 25$ kHz, and $\boldsymbol{Z_T} = 510\ \Omega - \text{j}\ 2.2\ \text{k}\Omega$, what is $\boldsymbol{C}$?

b. For $\boldsymbol{R} = 2.2\ \text{k}\Omega$, $\boldsymbol{C} = 47$ nF, $\boldsymbol{f} = 5$ kHz, and $\boldsymbol{Z_T} = 2.2\ \text{k}\Omega + \text{j}\ 4.7\ \text{k}\Omega$, what is $\boldsymbol{L}$?

c. For $\boldsymbol{L} = 39$ mH, $\boldsymbol{C} = 100$ pF, $\boldsymbol{f} = 100$ kHz, and $\boldsymbol{Z_T} = 21.77\ \text{k}\Omega \angle 23.24°$, what is $\boldsymbol{R}$?

Solution: For each set of circuit values, isolate and solve for the unknown circuit value.

a. The unknown circuit value is $\boldsymbol{C}$. Begin with the equation for $\boldsymbol{Z_T}$ and isolate and solve for $\boldsymbol{X_C}$.

$$\boldsymbol{Z_T} = \boldsymbol{R} + \text{j}\,\boldsymbol{X_L} - \text{j}\,\boldsymbol{X_C}$$
$$\boldsymbol{Z_T} + \text{j}\,\boldsymbol{X_C} = \boldsymbol{R} + \text{j}\,\boldsymbol{X_L} - \text{j}\,\boldsymbol{X_C} + \text{j}\,\boldsymbol{X_C}$$
$$\boldsymbol{Z_T} + \text{j}\,\boldsymbol{X_C} - \boldsymbol{Z_T} = \boldsymbol{R} + \text{j}\,\boldsymbol{X_L} - \boldsymbol{Z_T}$$
$$\text{j}\,\boldsymbol{X_C} = \boldsymbol{R} + \text{j}\,\boldsymbol{X_L} - \boldsymbol{Z_T}$$

Calculate $\boldsymbol{X_L}$ for $\boldsymbol{L} = 1.5$ mH and $\boldsymbol{f} = 25$ kHz.

$$\boldsymbol{X_L} = 2\pi \boldsymbol{fL}$$
$$= 2\pi(25\ \text{kHz})\ (1.5\ \text{mH})$$
$$= 2\pi(25 \times 10^3\ \text{Hz})\ (1.5 \times 10^{-3}\ \text{H})$$
$$= 236\ \Omega$$

Substitute in the values for $\boldsymbol{R}$, $\boldsymbol{X_L}$, and $\boldsymbol{Z_T}$ to calculate $\boldsymbol{X_C}$.

$$\text{j}\,\boldsymbol{X_C} = \boldsymbol{R} + \text{j}\,\boldsymbol{X_L} - \boldsymbol{Z_T}$$
$$= 510\ \Omega + \text{j}\ 236\ \Omega - (510\ \Omega - \text{j}\ 2.20\ \text{k}\Omega)$$
$$= (510\ \Omega - 510\ \Omega) + \text{j}\ (236\ \Omega - 2.20\ \text{k}\Omega)$$
$$= 0\ \Omega - \text{j}\ 1.96\ \text{k}\Omega$$

The negative imaginary value verifies that the reactance $\boldsymbol{X_C}$ is capacitive reactance. Use the formula for capacitive reactance to solve for $\boldsymbol{C}$.

$$\boldsymbol{X_C} = 1\ /\ (2\pi \boldsymbol{fC})$$
$$(\boldsymbol{X_C})(\boldsymbol{C}\ /\ \boldsymbol{X_C}) = [1\ /\ (2\pi \boldsymbol{fC})](\boldsymbol{C}\ /\ \boldsymbol{X_C})$$
$$\boldsymbol{C} = [1\ /\ (2\pi \boldsymbol{f})]\ /\ \boldsymbol{X_C}$$
$$= 1\ /\ (2\pi \boldsymbol{f}\boldsymbol{X_C})$$
$$= 1\ /\ [2\pi\ (25\ \text{kHz})(1.96\ \text{k}\Omega)]$$
$$= 1\ /\ [2\pi\ (25 \times 10^3\ \text{Hz})(1.96 \times 10^3\ \Omega)]$$
$$= (1\ /\ 309 \times 10^6)\ \text{F}$$
$$= 3.24 \times 10^{-9}\ \text{F}$$
$$= \mathbf{3.24\ nF}$$

b. The unknown circuit value is $\boldsymbol{L}$. Begin with the equation for $\boldsymbol{Z_T}$ and isolate and solve for $\boldsymbol{X_L}$.

$$\boldsymbol{Z_T} = \boldsymbol{R} + \text{j}\,\boldsymbol{X_L} - \text{j}\,\boldsymbol{X_C}$$
$$\boldsymbol{Z_T} - \text{j}\,\boldsymbol{X_L} = \boldsymbol{R} + \text{j}\,\boldsymbol{X_L} - \text{j}\,\boldsymbol{X_C} - \text{j}\,\boldsymbol{X_L}$$
$$\boldsymbol{Z_T} - \text{j}\,\boldsymbol{X_C} - \boldsymbol{Z_T} = \boldsymbol{R} - \text{j}\,\boldsymbol{X_C} - \boldsymbol{Z_T}$$
$$(-\text{j}\,\boldsymbol{X_L})(-1) = (\boldsymbol{R} + \text{j}\,\boldsymbol{X_C} - \boldsymbol{Z_T})(-1)$$
$$\text{j}\,\boldsymbol{X_L} = -\boldsymbol{R} - \text{j}\,\boldsymbol{X_C} + \boldsymbol{Z_T}$$
$$= \boldsymbol{Z_T} - \boldsymbol{R} - \text{j}\,\boldsymbol{X_C}$$

Calculate $\boldsymbol{X_C}$ for $\boldsymbol{C} = 47$ nF and $\boldsymbol{f} = 5$ kHz.

$X_C = 1 / (2\pi fC)$
$= 1 / [2\pi(47 \text{ nF})(5 \text{ kHz})]$
$= 1 / [2\pi(47 \times 10^{-9} \text{ F})(5 \times 10^3 \text{ Hz})$
$= 677\ \Omega$

Substitute in the values for $\boldsymbol{R}$, $\boldsymbol{X_L}$, and $\boldsymbol{Z_T}$ to calculate $\boldsymbol{X_L}$.

$\text{j}\,X_L = Z_T - R - \text{j}\,X_C$
$= (2.2 \text{ k}\Omega + \text{j}\,4.7 \text{ k}\Omega) - 2.2 \text{ k}\Omega - \text{j}\,677\ \Omega$
$= (2.2 \text{ k}\Omega - 2.2 \text{ k}\Omega) + \text{j}\,(4.7 \text{ k}\Omega - 677\ \Omega)$
$= 0\ \Omega + \text{j}\,4.02 \text{ k}\Omega$

The negative imaginary value verifies that the reactance $\boldsymbol{X_L}$ is inductive reactance. Use the formula for inductive reactance to solve for $\boldsymbol{L}$.

$X_L = 2\pi fL$
$X_L / 2\pi f = (2\pi fL)] / (2\pi f)$
$X_L / 2\pi f = L$
$L = X_L / 2\pi f$
$= 4.02 \text{ k}\Omega / [2\pi\,(5 \text{ kHz})]$
$= (4.02 \times 10^3\ \Omega) / [2\pi\,(5 \times 10^3 \text{ Hz})]$
$= (4.02 \times 10^3\ \Omega) / (31.4 \times 10^3 \text{ Hz})$
$= 128 \times 10^{-3} \text{ H}$
$= \mathbf{128\ mH}$

c. The unknown circuit value is $\boldsymbol{R}$. Begin with the equation for $\boldsymbol{Z_T}$ and isolate and solve for $\boldsymbol{R}$.

$Z_T = R + \text{j}\,X_L - \text{j}\,X_C$
$Z_T - R = R + \text{j}\,X_L - \text{j}\,X_C - R$
$Z_T - R - Z_T = \text{j}\,X_L - \text{j}\,X_C - Z_T$
$(-R)(-1) = (\text{j}\,X_L - \text{j}\,X_C - Z_T)(-1)$
$R = -\text{j}\,X_L + \text{j}\,X_C + Z_T$
$= Z_T + \text{j}\,X_C - \text{j}\,X_L$

Calculate $\boldsymbol{X_L}$ for $\boldsymbol{L} = 39$ mH and $\boldsymbol{f} = 100$ kHz.

$X_L = 2\pi fL$
$= 2\pi(100 \text{ kHz})\,(39 \text{ mH})$
$= 2\pi(100 \times 10^3 \text{ Hz})\,(39 \times 10^{-3} \text{ H})$
$= 24.5 \text{ k}\Omega$

Calculate $\boldsymbol{X_C}$ for $\boldsymbol{C} = 100$ pF and $\boldsymbol{f} = 100$ kHz.

$X_C = 1 / (2\pi fC)$
$= 1 / [2\pi(100 \text{ kHz})\,(100 \text{ pF})]$
$= 1 / [2\pi(100 \times 10^3 \text{ Hz})\,(100 \times 10^{-12} \text{ F})]$
$= 15.9 \text{ k}\Omega$

Convert $\boldsymbol{Z_T}$ from polar to rectangular to simplify the calculation.

$Z_T = 21.77 \text{ k}\Omega\ \angle 23.24°$
$= 20.0 \text{ k}\Omega + \text{j}\,8.59 \text{ k}\Omega$

Substitute in the values for $\boldsymbol{X_L}$, $\boldsymbol{X_C}$, and $\boldsymbol{Z_T}$ to calculate $\boldsymbol{R}$.

$R = Z_T + X_C - \text{j}\,X_L$

$= (20.0\ \text{k}\Omega + \text{j}\ 8.59\ \text{k}\Omega) + \text{j}\ 15.9\ \text{k}\Omega - \text{j}\ 24.5\ \text{k}\Omega$

$= 20.0\ \text{k}\Omega + \text{j}\ (8.59\ \text{k}\Omega + 15.9\ \text{k}\Omega - 24.5\ \text{k}\Omega)$

$= 20.0\ \text{k}\Omega + \text{j}\ 0\ \Omega$

$= \mathbf{20.0\ k\Omega}$

The zero imaginary value verifies that the value $\boldsymbol{R}$ is purely resistive.

3. Problem: For each given set of circuit values, calculate the total current, $\boldsymbol{I_T}$, and the resistor, inductor, and capacitor voltages, $\boldsymbol{V_R}$, $\boldsymbol{V_L}$, and $\boldsymbol{V_C}$. Express the answers in polar form.

a. $\boldsymbol{V_S} = 1.0\ \text{V}_{\text{ac}}$, $\boldsymbol{R} = 220\ \Omega$, $\boldsymbol{L} = 120$ mH, $\boldsymbol{C} = 1.6\ \mu$F, and $\boldsymbol{f} = 500$ Hz

b. $\boldsymbol{V_S} = 50\ \text{mV}_{\text{ac}}$, $\boldsymbol{R} = 10\ \Omega$, $\boldsymbol{L} = 39$ mH, $\boldsymbol{C} = 100$ nF, and $\boldsymbol{f} = 2.5$ kHz

c. $\boldsymbol{V_S} = 10\ \text{V}_{\text{ac}}$, $\boldsymbol{R} = 56\ \text{k}\Omega$, $\boldsymbol{L} = 5.1$ mH, $\boldsymbol{C} = 4.3$ pF, and $\boldsymbol{f} = 750$ kHz

Solution: For each set of circuit values, first calculate the inductive and capacitive reactances. Next, calculate the total impedance using complex arithmetic. Finally apply Ohm's law to calculate the circuit current and voltages.

a. Calculate $\boldsymbol{X_L}$ for $\boldsymbol{L} = 120$ mH and $\boldsymbol{f} = 500$ Hz.

$\boldsymbol{X_L} = 2\pi \boldsymbol{fL}$

$= 2\pi(500\ \text{Hz})\ (120\ \text{mH})$

$= 2\pi(500\ \text{Hz})\ (120 \times 10^{-3}\ \text{H})$

$= 377\ \Omega$

Calculate $\boldsymbol{X_C}$ for $\boldsymbol{C} = 1.6\ \mu$F and $\boldsymbol{f} = 500$ Hz.

$\boldsymbol{X_C} = 1 / (2\pi \boldsymbol{fC})$

$= 1 / [2\pi(500\ \text{Hz})\ (1.6\ \mu\text{F})]$

$= 1 / [2\pi(500\ \text{Hz})\ (1.6 \times 10^{-6}\ \text{F})]$

$= 199\ \Omega$

Calculate the total impedance $\boldsymbol{Z_T}$ from $\boldsymbol{R}$, $\boldsymbol{X_L}$, and $\boldsymbol{X_C}$.

$\boldsymbol{Z_T} = \boldsymbol{R} + \text{j}\ \boldsymbol{X_L} - \text{j}\ \boldsymbol{X_C}$

$= \boldsymbol{R} + \text{j}\ (\boldsymbol{X_L} - \boldsymbol{X_C})$

$= 220\ \Omega + \text{j}\ (377\ \Omega - 199\ \Omega)$

$= 220\ \Omega + \text{j}\ 178\ \Omega$ in rectangular form

$= 283\ \Omega \angle 39.0°$ in polar form

Use Ohm's Law for current to calculate the current, $\boldsymbol{I_T}$.

$\boldsymbol{I_T} = \boldsymbol{V_S} / \boldsymbol{Z_T}$

$= (1.0\ \text{V}_{\text{ac}} \angle 0°) / (283\ \Omega \angle 39.0°)$

$= (1.0\ \text{V}_{\text{ac}} / 283\ \Omega) \angle (0° - 39.0°)$

$= \mathbf{3.53\ mA_{ac} \angle -39.0°}$

For a series circuit the current through each component is the same as the total current. Use Ohm's Law for voltage to find the voltage across each component.

$\boldsymbol{V_R} = \boldsymbol{I_R} \times \boldsymbol{R}$

$= \boldsymbol{I_T} \times \boldsymbol{R}$

$= (3.53\ \text{mA}_{\text{ac}} \angle -39.0°)(220\ \Omega \angle 0°)$

$= (3.53\ \text{mA}_{\text{ac}} \times 220\ \Omega) \angle (-39.0° + 0°)$

$= \mathbf{777\ mV_{ac} \angle -39.0°}$

$\boldsymbol{V_L} = \boldsymbol{I_L} \times \boldsymbol{X_L}$

$= \boldsymbol{I_T} \times \boldsymbol{X_L}$

$= (3.53\ \text{mA}_{\text{ac}} \angle -39.0°)(377\ \Omega \angle 90°)$

$= (3.53 \text{ mA}_{ac} \times 377\ \Omega) \angle (-39.0° + 90°)$

$= \mathbf{1.33\ V_{ac} \angle 51.0°}$

$V_C = I_C \times X_C$

$= I_T \times X_C$

$= (3.53 \text{ mA}_{ac} \angle -39.0°)(199\ \Omega \angle -90°)$

$= (3.53 \text{ mA}_{ac} \times 199\ \Omega) \angle (-39.0° + -90°)$

$= \mathbf{703\ mV_{ac} \angle -129°}$

b. Calculate X_L for $L = 39$ mH and $f = 2.5$ kHz.

$X_L = 2\pi fL$

$= 2\pi(2.5 \text{ kHz}) (39 \text{ mH})$

$= 2\pi(2.5 \times 10^3 \text{ Hz}) (39 \times 10^{-3} \text{ H})$

$= 613\ \Omega$

Calculate X_C for $C = 100$ nF and $f = 2.5$ kHz.

$X_C = 1 / (2\pi fC)$

$= 1 / [2\pi(2.5 \text{ kHz}) (100 \text{ nF})]$

$= 1 / [2\pi(2.5 \times 10^3 \text{ Hz}) (100 \times 10^{-9} \text{ F})]$

$= 637\ \Omega$

Calculate the total impedance Z_T from R, X_L, and X_C.

$Z_T = R + \text{j}\, X_L - \text{j}\, X_C$

$= R + \text{j}\, (X_L - X_C)$

$= 10\ \Omega + \text{j}\, (613\ \Omega - 637\ \Omega)$

$= 10.0\ \Omega - \text{j}\, 24.0\ \Omega$ in rectangular form

$= 26.0\ \Omega \angle -67.4°$ in polar form

Use Ohm's Law for current to calculate the current, I_T.

$I_T = V_S / Z_T$

$= (50 \text{ mV}_{ac} \angle 0°) / (26.0\ \Omega \angle -67.4°)$

$= (50 \text{ V}_{ac} / 26.0\ \Omega) \angle [0° - (-67.4°)]$

$= \mathbf{1.92\ mA_{ac} \angle 67.4°}$

For a series circuit the current through each component is the same as the total current. Use Ohm's Law for voltage to find the voltage across each component.

$V_R = I_R \times R$

$= I_T \times R$

$= (1.92 \text{ mA}_{ac} \angle 67.4°)(10\ \Omega \angle 0°)$

$= (1.92 \text{ mA}_{ac} \times 10\ \Omega) \angle (67.4° + 0°)$

$= \mathbf{19.2\ mV_{ac} \angle 67.4°}$

$V_L = I_L \times X_L$

$= I_T \times X_L$

$= (1.92 \text{ mA}_{ac} \angle 67.4°)(613\ \Omega \angle 90°)$

$= (1.92 \text{ mA}_{ac} \times 613\ \Omega) \angle (67.4° + 90°)$

$= \mathbf{1.18\ V_{ac} \angle 157°}$

$V_C = I_C \times X_C$

$= I_T \times X_C$

$= (1.92\ \text{mA}_{ac} \angle 67.4°)(637\ \Omega \angle -90°)$

$= (1.92\ \text{mA}_{ac} \times 637\ \Omega) \angle (67.4° + -90°)$

$= \mathbf{1.22\ V_{ac} \angle -22.6°}$

c. Calculate X_L for L = 5.1 mH and f = 750 kHz.

$X_L = 2\pi fL$

$= 2\pi(750\ \text{kHz})\ (5.1\ \text{mH})$

$= 2\pi(750 \times 10^3\ \text{Hz})\ (5.1 \times 10^{-3}\ \text{H})$

$= 24.0\ \text{k}\Omega$

Calculate X_C for C = 4.3 pF and f = 750 kHz.

$X_C = 1 / (2\pi fC)$

$= 1 / [2\pi(750\ \text{kHz})\ (4.3\ \text{pF})]$

$= 1 / [2\pi(750 \times 10^3\ \text{Hz})\ (4.3 \times 10^{-12}\ \text{F})]$

$= 49.4\ \text{k}\Omega$

Calculate the total impedance Z_T from R, X_L, and X_C.

$Z_T = R + \text{j}\,X_L - \text{j}\,X_C$

$= R + \text{j}\,(X_L - X_C)$

$= 56\ \text{k}\Omega + \text{j}\ (24.0\ \text{k}\Omega - 49.4\ \text{k}\Omega)$

$= 56.0\ \text{k}\Omega - \text{j}\ 25.3\ \text{k}\Omega$ in rectangular form

$= 61.5\ \text{k}\Omega \angle -24.3°$ in polar form

Use Ohm's Law for current to calculate the current, I_T.

$I_T = V_S / Z_T$

$= (10\ \text{V}_{ac} \angle 0°) / (61.5\ \text{k}\Omega \angle -24.3°)$

$= (10\ \text{V}_{ac} / 61.5\ \text{k}\Omega) \angle [0° - (-24.3°)]$

$= \mathbf{163\ \mu A_{ac} \angle 24.3°}$

For a series circuit the current through each component is the same as the total current. Use Ohm's Law for voltage to find the voltage across each component.

$V_R = I_R \times R$

$= I_T \times R$

$= (163\ \mu\text{A}_{ac} \angle 24.3°)(56\ \text{k}\Omega \angle 0°)$

$= (163\ \mu\text{A}_{ac} \times 56\ \text{k}\Omega) \angle (24.3° + 0°)$

$= \mathbf{9.11\ V_{ac} \angle 24.3°}$

$V_L = I_L \times X_L$

$= I_T \times X_L$

$= (163\ \mu\text{A}_{ac} \angle 24.3°)(24.0\ \text{k}\Omega \angle 90°)$

$= (163\ \mu\text{A}_{ac} \times 24.0\ \text{k}\Omega) \angle (67.4° + 90°)$

$= \mathbf{3.91\ V_{ac} \angle 114°}$

$V_C = I_C \times X_C$

$= I_T \times X_C$

$= (163\ \mu\text{A}_{ac} \angle 24.3°)(49.4\ \text{k}\Omega \angle -90°)$

$= (163\ \mu\text{A}_{ac} \times 49.4\ \text{k}\Omega) \angle (24.3° + -90°)$

$= \mathbf{8.03\ V_{ac} \angle -65.7°}$

18.1.2 Resonance of Series *RLC* Circuits

For Problems 1 through 3, refer to the circuit in Figure 18-2.

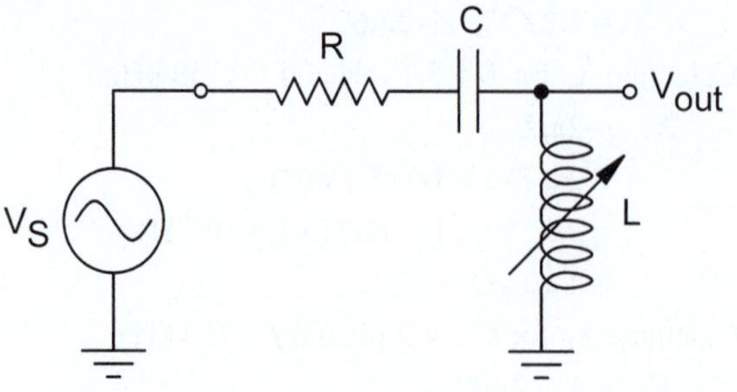

Figure 18-2: Series Resonance *RLC* Circuit for Problems 1 through 3

1. Problem: For $C = 1.0\ \mu F$, calculate the resonant frequency, f_r, for each of the following values for L: 10 mH, 20 mH, 51 mH, and 100 mH.

Solution: At the resonant frequency, f_r, the inductive reactance, X_L, and capacitive reactance, X_C, are equal. Use the equations for X_L and X_C to determine f_r in terms of L and C.

$$X_L = X_C$$

$$2\pi f_r L = 1 / (2\pi f_r C)$$

$$2\pi f_r L / 2\pi L = [1 / (2\pi f_r C)] / 2\pi L$$

$$f_r = 1 / [(2\pi f_r C)(2\pi L)]$$

$$f_r = 1 / (4\pi^2 f_r LC)$$

$$f_r \times f_r = [1 / (4\pi^2 f_r LC)] \times f_r$$

$$f_r^2 = 1 / (4\pi^2 LC)$$

$$\sqrt{f_r^{\,2}} = \sqrt{1/4\pi^2 LC}$$

$$f_r = \sqrt{1} / \sqrt{4\pi^2 LC}$$

$$= 1 / (2\pi\sqrt{LC})$$

For $C = 1.0\ \mu F$ and $L = 10$ mH,

$$f_r = 1 / [2\pi\sqrt{(10\text{ mH})(1.0\ \mu\text{F})}]$$

$$= 1 / [2\pi\sqrt{(10 \times 10^{-3}\text{ H})(1.0 \times 10^{-6}\text{ F})}]$$

$$= 1 / (628 \times 10^{-6}\text{ s})$$

$$= \mathbf{1.59\ kHz}$$

For $C = 1.0\ \mu F$ and $L = 20$ mH,

$$f_r = 1 / [2\pi\sqrt{(20\text{ mH})(1.0\ \mu\text{F})}]$$

$$= 1 / [2\pi\sqrt{(20 \times 10^{-3}\text{ H})(1.0 \times 10^{-6}\text{ F})}]$$

$$= 1 / (889 \times 10^{-6}\text{ s})$$

$$= \mathbf{1.13\ kHz}$$

For $C = 1.0\ \mu F$ and $L = 51$ mH,

$$f_r = 1 / [2\pi\sqrt{(51\text{ mH})(1.0\ \mu\text{F})}]$$

$= 1 / [2\pi\sqrt{(51 \times 10^{-3}\text{ H})(1.0 \times 10^{-6}\text{ F})}]$

$= 1 / (1.42 \times 10^{-3}\text{ s})$

$=$ **705 Hz**

For $C = 1.0\ \mu\text{F}$ and $L = 100\text{ mH}$,

$f_r = 1 / [2\pi\sqrt{(100\text{ mH})(1.0\ \mu\text{F})}]$

$= 1 / [2\pi\sqrt{(100 \times 10^{-3}\text{ H})(1.0 \times 10^{-6}\text{ F})}]$

$= 1 / (1.99 \times 10^{-3}\text{ s})$

$=$ **503 Hz**

2. Problem: For $C = 4.7\ \mu\text{F}$, determine the value of L for each of the following values of resonant frequency, f_r: 100 Hz, 200 Hz, 500 Hz, and 1.0 kHz.

Solution: At the resonant frequency, f_r, the inductive reactance, X_L, and capacitive reactance, X_C, are equal. Use the equations for calculating X_L and X_C to determine L in terms of f_r and C.

$X_L = X_C$

$2\pi f_r L = 1 / (2\pi f_r C)$

$2\pi f_r L / 2\pi f_r = [1 / (2\pi f_r C)] / 2\pi f_r$

$L = 1 / [(2\pi f_r C)(2\pi f_r)]$

$= 1 / (4\pi^2 f_r^2 C)$

For $C = 4.7\ \mu\text{F}$ and $f_r = 100\text{ Hz}$,

$L = 1 / (4\pi^2 f_r^2 C)$

$= 1 / [4\pi^2(100\text{ Hz})^2(4.7\ \mu\text{F})]$

$= 1 / [4\pi^2(10 \times 10^3\text{ Hz}^2)(4.7 \times 10^{-6}\text{ F})]$

$= 1 / (1.86\text{ H}^{-1})$

$= 539\text{ mH}$

For $C = 4.7\ \mu\text{F}$ and $f_r = 200\text{ Hz}$,

$L = 1 / (4\pi^2 f_r^2 C)$

$= 1 / [4\pi^2(200\text{ Hz})^2(4.7\ \mu\text{F})]$

$= 1 / [4\pi^2(40 \times 10^3\text{ Hz}^2)(4.7 \times 10^{-6}\text{ F})]$

$= 1 / (7.42\text{ H}^{-1})$

$= 135\text{ mH}$

For $C = 4.7\ \mu\text{F}$ and $f_r = 500\text{ Hz}$,

$L = 1 / (4\pi^2 f_r^2 C)$

$= 1 / [4\pi^2(500\text{ Hz})^2(4.7\ \mu\text{F})]$

$= 1 / [4\pi^2(250 \times 10^3\text{ Hz}^2)(4.7 \times 10^{-6}\text{ F})]$

$= 1 / (46.4\text{ H}^{-1})$

$= 21.6\text{ mH}$

For $C = 4.7\ \mu\text{F}$ and $f_r = 1.0\text{ kHz}$,

$L = 1 / (4\pi^2 f_r^2 C)$

$= 1 / [4\pi^2(1.0\text{ kHz})^2(4.7\ \mu\text{F})]$

$= 1 / [4\pi^2(1.0 \times 10^6\text{ Hz}^2)(4.7 \times 10^{-6}\text{ F})]$

$= 1 / (186\text{ H}^{-1})$

$= 5.39\text{ mH}$

3. Problem: What is V_{out} at resonance for $V_S = 5.0\text{ V}_{ac}$, $R = 120\ \Omega$, $L = 15\ \mu\text{H}$, and $C = 68\text{ pF}$?

Solution: First, use the equation from Problem 1 to calculate the resonant frequency, f_r.

$$f_r = 1/(2\pi\sqrt{LC})$$
$$= 1/[2\pi\sqrt{(15\,\mu H)(68\,pF)}]$$
$$= 1/[2\pi\sqrt{(15\times10^{-6}\,H)(68\times10^{-12}\,F)}]$$
$$= 1/(201\times10^{-9}\,s)$$
$$= 4.98\,MHz$$

At the resonant frequency, f_r, the inductive and capacitive reactances cancel, so that $Z_T = R$. Use Ohm's Law to calculate the total current, I_T.

$$I_T = V_S / Z_T$$
$$= V_S / R$$
$$= (5.0\,V_{ac}\angle 0°)/(120\,\Omega\angle 0°)$$
$$= (5.0\,V_{ac}/120\,\Omega)\angle(0° - 0°)$$
$$= 41.7\,mA_{ac}\angle 0°$$

The output voltage, V_{out}, appears across L. For a series circuit, the current through L, I_L, is equal to the total current, I_T. Calculate the value of inductive reactance, X_L, at the resonant frequency, f_r, and then use Ohm's Law to calculate V_{out}.

$$X_L = 2\pi f_r L$$
$$= 2\pi(4.98\,MHz)(15\,\mu H)$$
$$= 2\pi(4.98\times10^{6})(15\times10^{-6})$$
$$= 470\,\Omega$$

$$V_{out} = V_{XL}$$
$$= I_L X_L$$
$$= I_T X_L$$
$$= (41.7\,mA_{ac}\angle 0°)(470\,\Omega\angle 90°)$$
$$= (41.7\,mA_{ac}\times 470\,\Omega)\angle(0° + 90°)$$
$$= \mathbf{19.6\,V_{ac}\angle 90°}$$

Note that the magnitude of V_L, (which is also equal to V_C) is almost four times larger than V_S.

For Problems 4 through 6, refer to the circuit in Figure 18-3.

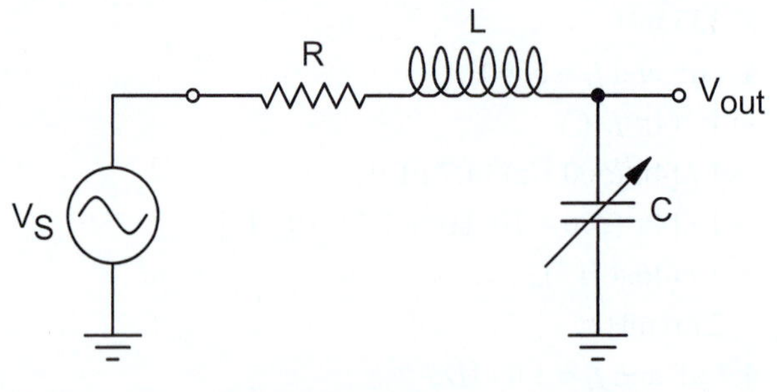

Figure 18-3: Series Resonance Circuit for Problems 4 through 6

4. Problem: Using the formula for f_r from Problem 1, calculate f_r for L = 27 mH and each of the following values of C: 1.0 pF, 2.2 pF, 4.3 pF, and 10 pF.

Solution: From the results of Problem 1 above, f_r in terms of L and C is

$$f_r = 1/(2\pi\sqrt{LC})$$

For $\boldsymbol{L}$ = 27 mH and $\boldsymbol{C}$ = 1.0 pF,

$$f_r = 1 / [\,2\pi\sqrt{(27\text{ mH})(1.0\text{ pF})}\,]$$
$$= 1 / [\,2\pi\sqrt{(27\times 10^{-3}\text{ F})(1.0\times 10^{-12}\text{ H})}\,]$$
$$= 1 / (1.03\times 10^{-6}\text{ s})$$
$$= \mathbf{969\ kHz}$$

For $\boldsymbol{L}$ = 27 mH and $\boldsymbol{C}$ = 2.2 pF,

$$f_r = 1 / [\,2\pi\sqrt{(27\text{ mH})(2.2\text{ pF})}\,]$$
$$= 1 / [\,2\pi\sqrt{(27\times 10^{-3}\text{ F})(2.2\times 10^{-12}\text{ H})}\,]$$
$$= 1 / (1.53\times 10^{-6}\text{ s})$$
$$= \mathbf{653\ kHz}$$

For $\boldsymbol{L}$ = 27 mH and $\boldsymbol{C}$ = 4.3 pF,

$$f_r = 1 / [\,2\pi\sqrt{(27\text{ mH})(4.3\text{ pF})}\,]$$
$$= 1 / [\,2\pi\sqrt{(27\times 10^{-3}\text{ F})(4.3\times 10^{-12}\text{ H})}\,]$$
$$= 1 / (2.14\times 10^{-6}\text{ s})$$
$$= \mathbf{467\ kHz}$$

For $\boldsymbol{L}$ = 27 mH and $\boldsymbol{C}$ = 10 pF,

$$f_r = 1 / [\,2\pi\sqrt{(27\text{ mH})(10\text{ pF})}\,]$$
$$= 1 / [\,2\pi\sqrt{(27\times 10^{-3}\text{ F})(10\times 10^{-12}\text{ H})}\,]$$
$$= 1 / (3.27\times 10^{-6}\text{ s})$$
$$= \mathbf{306\ kHz}$$

5. Problem: Using the formula for $\boldsymbol{f_r}$ from Problem 1, calculate $\boldsymbol{C}$ for $\boldsymbol{L}$ = 6.2 μH and each of the following values of $\boldsymbol{f_r}$: 10 kHz, 20 kHz, 50 kHz, and 100 kHz.

Solution: At the resonant frequency, $\boldsymbol{f_r}$, the inductive reactance, $\boldsymbol{X_L}$, and capacitive reactance, $\boldsymbol{X_C}$, are equal. Use the equations for calculating $\boldsymbol{X_L}$ and $\boldsymbol{X_C}$ to determine $\boldsymbol{C}$ in terms of $\boldsymbol{f_r}$ and $\boldsymbol{L}$.

$$X_L = X_C$$
$$2\pi f_r L = 1 / (2\pi f_r C)$$
$$2\pi f_r L \times C = [1 / (2\pi f_r C)] \times C$$
$$2\pi f_r L C / 2\pi f_r L = [1 / (2\pi f_r)] / 2\pi f_r L$$
$$C = 1 / [(2\pi f_r)(2\pi f_r L)]$$
$$= 1 / (4\pi^2 f_r^2 L)$$

For $\boldsymbol{L}$ = 6.2 μH and $\boldsymbol{f_r}$ = 10 kHz,

$$C = 1 / (4\pi^2 f_r^2 L)$$
$$= 1 / [4\pi^2(10\text{ kHz})^2(6.2\text{ μH})]$$
$$= 1 / [4\pi^2(100\times 10^6\text{ Hz}^2)(6.2\times 10^{-6}\text{ H})]$$
$$= 1 / (24.5\text{ kF}^{-1})$$
$$= \mathbf{40.9\ μF}$$

For $\boldsymbol{L}$ = 6.2 μH and $\boldsymbol{f_r}$ = 20 kHz,

$$C = 1 / (4\pi^2 f_r^2 L)$$
$$= 1 / [4\pi^2(20\text{ kHz})^2(6.2\text{ μH})]$$

$= 1 / [4\pi^2(400 \times 10^6 \text{ Hz}^2)(6.2 \times 10^{-6} \text{ H})]$
$= 1 / (97.9 \text{ kF}^{-1})$
$= \mathbf{10.2\ \mu F}$

For $\boldsymbol{L}$ = 6.2 μH and $\boldsymbol{f_r}$ = 50 kHz,

$C = 1 / (4\pi^2 f_r^2 L)$
$= 1 / [4\pi^2(50 \text{ kHz})^2(6.2\ \mu\text{H})]$
$= 1 / [4\pi^2(2.5 \times 10^9 \text{ Hz}^2)(6.2 \times 10^{-6} \text{ H})]$
$= 1 / (612 \text{ kF}^{-1})$
$= \mathbf{1.63\ \mu F}$

For $\boldsymbol{L}$ = 6.2 μH and $\boldsymbol{f_r}$ = 100 kHz,

$C = 1 / (4\pi^2 f_r^2 L)$
$= 1 / [4\pi^2(100 \text{ kHz})^2(6.2\ \mu\text{H})]$
$= 1 / [4\pi^2(10 \times 10^9 \text{ Hz}^2)(6.2 \times 10^{-6} \text{ H})]$
$= 1 / (2.45 \text{ MF}^{-1})$
$= \mathbf{409\ nF}$

6. Problem: The value of $\boldsymbol{V_{out}}$ at resonance is 2.0 V_{ac} ∠ –90° for $\boldsymbol{V_S}$ = 1.0 V_{ac} ∠ 0°, $\boldsymbol{R}$ = 180 Ω, and $\boldsymbol{L}$ = 39 μH. What is the value of $\boldsymbol{C}$?

Solution: At the resonant frequency, $\boldsymbol{f_r}$, the inductive and capacitive reactances, $\boldsymbol{X_L}$ and $\boldsymbol{X_C}$, are equal and opposite and cancel, so that $\boldsymbol{Z_T} = \boldsymbol{R}$. Use Ohm's Law for current to calculate the total current, $\boldsymbol{I_T}$.

$I_T = V_S / Z_T$
$= V_S / R$
$= (1.0 \text{ V}_{ac} \angle 0°) / (180\ \Omega \angle 0°)$
$= (1.0 \text{ V}_{ac} / 180\ \Omega) \angle (0° - 0°)$
$= 5.56 \text{ mA}_{ac} \angle 0°$

The output voltage, $\boldsymbol{V_{out}}$, appears across $\boldsymbol{C}$. At the resonant frequency, the inductor voltage, $\boldsymbol{V_L}$, is equal and opposite to the capacitor voltage, $\boldsymbol{V_C}$, so that $\boldsymbol{V_L} = \boldsymbol{V_{out}}$. For a series circuit, the current through $\boldsymbol{L}$, $\boldsymbol{I_L}$, is equal to the total current, $\boldsymbol{I_T}$. Use Ohm's Law for reactance to calculate the value of inductive reactance, $\boldsymbol{X_L}$.

$X_L = V_L / I_L$
$= -V_{out} / I_T$
$= (-2.0 \text{ V}_{ac} \angle -90°) / (5.56 \text{ mA}_{ac} \angle 0°)$
$= (-2.0 \text{ V}_{ac} / 5.56 \text{ mA}_{ac}) \angle (-90° - 0°)$
$= -360\ \Omega \angle -90°$
$= (-1) \times (360\ \Omega \angle -90°)$
$= (1 \angle 180°) \times (360\ \Omega \angle -90°)$
$= (1 \times 360\ \Omega) \angle (180° + -90°)$
$= 360\ \Omega \angle 90°$

Use the formula for inductive reactance to calculate the resonant frequency, $\boldsymbol{f_r}$.

$2\pi f_r L = X_L$
$2\pi f_r L / 2\pi L = X_L / 2\pi L$
$f_r = X_L / 2\pi L$
$= 360\ \Omega / [2\pi(39\ \mu\text{H})]$
$= 360\ \Omega / [2\pi(39 \times 10^{-6} \text{ H})]$

$= 1.47\text{ MHz}$

The magnitude of capacitive reactance is equal to that of the inductive reactance. Use the formula for capacitive reactance to calculate the capacitor value, $\boldsymbol{C}$.

$$\begin{aligned}
X_C &= 1 / (2\pi f_r C) \\
X_C \times C &= [1 / (2\pi f_r C)] \times C \\
[X_C C] / X_C &= [1 / (2\pi f_r)] / X_C \\
C &= 1 / (2\pi f_r X_C) \\
&= 1 / [2\pi(1.47\text{ MHz})(360\ \Omega)] \\
&= 1 / [2\pi(1.47 \times 10^6\text{ Hz})(360\ \Omega)] \\
&= 1 / (3.32 \times 10^9\text{ F}^{-1}) \\
&= \mathbf{301\ pF}
\end{aligned}$$

18.1.3 Analysis of Parallel *RLC* Circuits

For Problems 1 through 3, refer to the circuit in Figure 18-4.

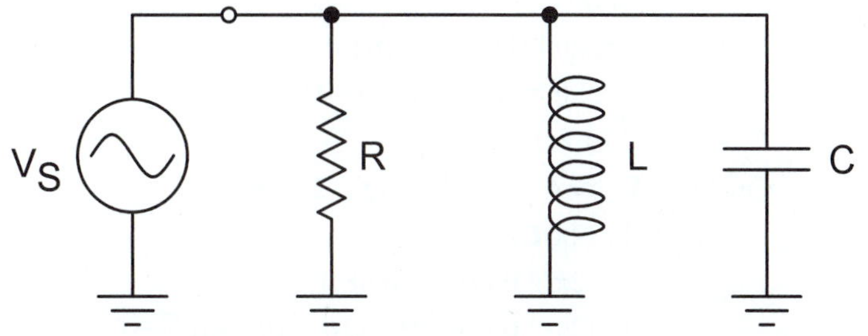

Figure 18-4: Parallel *RLC* Circuit for Problems 1 through 3

1. Problem: Calculate the total impedance Z_T for each of the following values of R, L, C, and f.

   a. $R = 130\ \Omega$, $L = 150$ mH, $C = 1.5\ \mu$F, and $f = 250$ Hz

   b. $R = 5.1$ kΩ, $L = 160$ mH, $C = 16$ nF, and $f = 2.0$ kHz

   c. $R = 68$ kΩ, $L = 130$ mH, $C = 39$ pF, and $f = 125$ kHz

   d. $R = 8.2\ \Omega$, $L = 1.2$ H, $C = 18$ mF, and $f = 1.5$ Hz

   Solution: For each set of circuit values, first calculate the inductive and capacitive reactances and then calculate the total series impedance using complex arithmetic.

   a. Calculate X_L for $L = 150$ mH and $f = 250$ Hz.

$$\begin{aligned}
X_L &= 2\pi f L \\
&= 2\pi(150\text{ mH})(250\text{ Hz}) \\
&= 2\pi(150 \times 10^{-3}\text{ H})(250\text{ Hz}) \\
&= 236\ \Omega
\end{aligned}$$

   Calculate X_C for $C = 1.5\ \mu$F and $f = 250$ Hz.

$$\begin{aligned}
X_C &= 1 / (2\pi f C) \\
&= 1 / [2\pi(250\text{ Hz})\ (1.5\ \mu\text{F})] \\
&= 1 / [2\pi(250\text{ Hz})\ (1.5 \times 10^{-6}\text{ F})] \\
&= 424\ \Omega
\end{aligned}$$

   Calculate the total impedance Z_T from R, X_L, and X_C.

$$\begin{aligned}
Z_T &= 1 / [(1 / R) + (1 / \text{j}\ X_L) + (1 / {-\text{j}}\ X_C)] \\
&= 1 / [(1 / R) + (\text{j} / \text{j})(1 / \text{j}\ X_L) + (\text{j} / \text{j})(1 / {-\text{j}}\ X_C)] \\
&= 1 / [(1 / R) + (\text{j} / \text{j}^2\ X_L) + (\text{j} / {-\text{j}^2}\ X_C)]
\end{aligned}$$

$= 1 / [(1 / \boldsymbol{R}) + (\mathrm{j} / -\boldsymbol{X_L}) + (\mathrm{j} / \boldsymbol{X_C})]$

$= 1 / \{(1 / \boldsymbol{R}) + \mathrm{j}\,[(1 / -\boldsymbol{X_L}) + (1 / \boldsymbol{X_C})]\}$

$= 1 / \{(1 / \boldsymbol{R}) + \mathrm{j}\,[(1 / \boldsymbol{X_C}) - (1 / \boldsymbol{X_L})]\}$

$= 1 / \{(1 / 130\ \Omega) + \mathrm{j}\,[(1 / 424\ \Omega) - (1 / 236\ \Omega)]\}$

$= 1 / [7.69\text{ mS} + \mathrm{j}\,(2.36\text{ mS} - 4.24\text{ mS})]$

$= 1 / (7.69\text{ mS} - \mathrm{j}\,1.89\text{ mS})$

Convert the rectangular quantity to polar form to simplify the division.

$\boldsymbol{Z_T} = 1 / (7.69\text{ mS} - \mathrm{j}\,1.89\text{ mS})$

$= (1 \angle 0°) / (7.92\text{ mS} \angle -13.8°)$

$= (1 / 7.92\text{ mS}) \angle [0° - (-13.8°)]$

= **126 Ω ∠ 13.8°** in polar form

= **123 Ω + j 30.1 Ω** in rectangular form

b. Calculate $\boldsymbol{X_L}$ for $\boldsymbol{L}$ = 160 mH and $\boldsymbol{f}$ = 2.0 kHz.

$\boldsymbol{X_L} = 2\pi \boldsymbol{fL}$

$= 2\pi(2.0\text{ kHz})\,(160\text{ mH})$

$= 2\pi(2.0 \times 10^3\text{ Hz})\,(160 \times 10^{-3}\text{ H})$

$= 2.01\text{ k}\Omega$

Calculate $\boldsymbol{X_C}$ for $\boldsymbol{C}$ = 16 nF and $\boldsymbol{f}$ = 2.0 kHz.

$\boldsymbol{X_C} = 1 / (2\pi \boldsymbol{fC})$

$= 1 / [2\pi(2.0\text{ kHz})\,(16\text{ nF})]$

$= 1 / [2\pi(2.0 \times 10^3\text{ Hz})\,(16 \times 10^{-9}\text{ F})]$

$= 4.97\text{ k}\Omega$

Calculate the total impedance $\boldsymbol{Z_T}$ from $\boldsymbol{R}$, $\boldsymbol{X_L}$, and $\boldsymbol{X_C}$.

$\boldsymbol{Z_T} = 1 / [(1 / \boldsymbol{R}) + (1 / \mathrm{j}\,\boldsymbol{X_L}) + (1 / -\mathrm{j}\,\boldsymbol{X_C})]$

$= 1 / [(1 / \boldsymbol{R}) + (\mathrm{j} / \mathrm{j})(1 / \mathrm{j}\,\boldsymbol{X_L}) + (\mathrm{j} / \mathrm{j})(1 / -\mathrm{j}\,\boldsymbol{X_C})]$

$= 1 / [(1 / \boldsymbol{R}) + (\mathrm{j} / \mathrm{j}^2\,\boldsymbol{X_L}) + (\mathrm{j} / -\mathrm{j}^2\,\boldsymbol{X_C})]$

$= 1 / [(1 / \boldsymbol{R}) + (\mathrm{j} / -\boldsymbol{X_L}) + (\mathrm{j} / \boldsymbol{X_C})]$

$= 1 / \{(1 / \boldsymbol{R}) + \mathrm{j}\,[(1 / -\boldsymbol{X_L}) + (1 / \boldsymbol{X_C})]\}$

$= 1 / \{(1 / \boldsymbol{R}) + \mathrm{j}\,[(1 / \boldsymbol{X_C}) - (1 / \boldsymbol{X_L})]\}$

$= 1 / \{(1 / 5.1\text{ k}\Omega) + \mathrm{j}\,[(1 / 4.97\text{ k}\Omega) - (1 / 2.01\text{ k}\Omega)]\}$

$= 1 / [196\ \mu\text{S} + \mathrm{j}\,(201\ \mu\text{S} - 497\ \mu\text{S})]$

$= 1 / (196\ \mu\text{S} - \mathrm{j}\,296\ \mu\text{S})$

Convert the rectangular quantity to polar form to simplify the division.

$\boldsymbol{Z_T} = 1 / (196\ \mu\text{S} - \mathrm{j}\,296\ \mu\text{S})$

$= (1 \angle 0°) / (355\ \mu\text{S} \angle -56.5°)$

$= (1 / 355\ \mu\text{S}) \angle [0° - (-56.5°)]$

= **2.82 kΩ ∠ 56.5°** in polar form

= **1.55 kΩ + j 2.35 kΩ** in rectangular form

c. Calculate $\boldsymbol{X_L}$ for $\boldsymbol{L}$ = 130 mH and $\boldsymbol{f}$ = 125 kHz.

$\boldsymbol{X_L} = 2\pi \boldsymbol{fL}$

$= 2\pi(125\text{ kHz})\,(130\text{ mH})$

$= 2\pi(125 \times 10^3\text{Hz})\,(130 \times 10^{-3}\text{ H})$

$= 102\ \text{k}\Omega$

Calculate X_C for $C = 39$ pF and $f = 125$ kHz.

$X_C = 1 / (2\pi fC)$

$= 1 / [2\pi(125\ \text{kHz})\ (39\ \text{pF})]$

$= 1 / [2\pi(125 \times 10^3\ \text{Hz})\ (39 \times 10^{-12}\ \text{F})]$

$= 32.7\ \text{k}\Omega$

Calculate the total impedance Z_T from R, X_L, and X_C.

$Z_T = 1 / [(1 / R) + (1 / \text{j}\,X_L) + (1 / -\text{j}\,X_C)]$

$= 1 / [(1 / R) + (\text{j} / \text{j})(1 / \text{j}\,X_L) + (\text{j} / \text{j})(1 / -\text{j}\,X_C)]$

$= 1 / [(1 / R) + (\text{j} / \text{j}^2\,X_L) + (\text{j} / -\text{j}^2\,X_C)]$

$= 1 / [(1 / R) + (\text{j} / -X_L) + (\text{j} / X_C)]$

$= 1 / \{(1 / R) + \text{j}\,[(1 / -X_L) + (1 / X_C)]\}$

$= 1 / \{(1 / R) + \text{j}\,[(1 / X_C) - (1 / X_L)]\}$

$= 1 / \{(1 / 68\ \text{k}\Omega) + \text{j}\,[(1 / 32.7\ \text{k}\Omega) - (1 / 102\ \text{k}\Omega)]\}$

$= 1 / [14.7\ \mu\text{S} + \text{j}\,(30.6\ \mu\text{S} - 9.79\ \mu\text{S})]$

$= 1 / (14.7\ \mu\text{S} + \text{j}\,20.8\ \mu\text{S})$

Convert the rectangular quantity to polar form to simplify the division.

$Z_T = 1 / (14.7\ \mu\text{S} + \text{j}\,20.8\ \mu\text{S})$

$= (1 \angle 0°) / (25.5\ \mu\text{S} \angle 54.8°)$

$= (1 / 25.5\ \mu\text{S}) \angle (0° - 54.8°)$

$= \mathbf{39.2\ k\Omega \angle -54.8°}$ in polar form

$= \mathbf{22.6\ k\Omega - j\,32.0\ k\Omega}$ in rectangular form

d. Calculate X_L for $L = 1.2$ H and $f = 1.5$ Hz.

$X_L = 2\pi fL$

$= 2\pi(1.5\ \text{Hz})\ (1.2\ \text{H})$

$= 11.3\ \Omega$

Calculate X_C for $C = 18$ mF and $f = 1.5$ Hz.

$X_C = 1 / (2\pi fC)$

$= 1 / [2\pi(1.5\ \text{Hz})\ (18\ \text{mF})]$

$= 1 / [2\pi(1.5\ \text{Hz})\ (18 \times 10^{-3}\ \text{F})]$

$= 5.90\ \Omega$

Calculate the total impedance Z_T from R, X_L, and X_C.

$Z_T = 1 / [(1 / R) + (1 / \text{j}\,X_L) + (1 / -\text{j}\,X_C)]$

$= 1 / [(1 / R) + (\text{j} / \text{j})(1 / \text{j}\,X_L) + (\text{j} / \text{j})(1 / -\text{j}\,X_C)]$

$= 1 / [(1 / R) + (\text{j} / \text{j}^2\,X_L) + (\text{j} / -\text{j}^2\,X_C)]$

$= 1 / [(1 / R) + (\text{j} / -X_L) + (\text{j} / X_C)]$

$= 1 / \{(1 / R) + \text{j}\,[(1 / -X_L) + (1 / X_C)]\}$

$= 1 / \{(1 / R) + \text{j}\,[(1 / X_C) - (1 / X_L)]\}$

$= 1 / \{(1 / 8.2\ \Omega) + \text{j}\,[(1 / 5.90\ \Omega) - (1 / 11.3\ \Omega)]\}$

$= 1 / [122\ \text{mS} + \text{j}\,(170\ \text{mS} - 88.4\ \text{mS})]$

$= 1 / (122\ \text{mS} + \text{j}\,81.2\ \text{mS})$

Convert the rectangular quantity to polar form to simplify the division.

$Z_T = 1 / (122\ \text{mS} + \text{j}\,81.2\ \text{mS})$

$= (1 \angle 0°) / (147\text{ mS} \angle 33.7°)$

$= (1 / 147\text{ mS}) \angle (0° - 33.7°)$

$= \mathbf{6.83\ \Omega \angle -33.7°}$ in polar form

$= \mathbf{5.68\ \Omega - j\ 3.78\ \Omega}$ in rectangular form

2. Problem: For the each given set of circuit values, determine the unknown circuit value.

a. For $\boldsymbol{R} = 62\ \Omega$, $\boldsymbol{L} = 330\ \mu\text{H}$, $\boldsymbol{f} = 50$ kHz, and $\boldsymbol{Z_T} = 61.1\ \Omega \angle 9.62°$, what is $\boldsymbol{C}$?

b. For $\boldsymbol{R} = 15\ \text{k}\Omega$, $\boldsymbol{C} = 2.2$ nF, $\boldsymbol{f} = 10$ kHz, and $\boldsymbol{Z_T} = 8.93\ \text{k}\Omega \angle -53.5°$, what is $\boldsymbol{L}$?

c. For $\boldsymbol{L} = 2.4$ H, $\boldsymbol{C} = 13\ \mu\text{F}$, $\boldsymbol{f} = 60$ Hz, and $\boldsymbol{Z_T} = 130\ \Omega \angle -29.7°$, what is $\boldsymbol{R}$?

Solution: For each set of circuit values, isolate and solve for the unknown circuit value using the total impedance, $\boldsymbol{Z_T}$, for a parallel ***RLC*** circuit:

$\boldsymbol{Z_T} = 1 / [(1 / \boldsymbol{R}) + (1 / \text{j}\ \boldsymbol{X_L}) + (1 / -\text{j}\ \boldsymbol{X_C})]$

$\boldsymbol{Z_T} = 1 / [(1 / \boldsymbol{R}) + (1 / \text{j}\ \boldsymbol{X_L}) - (1 / \text{j}\ \boldsymbol{X_C})]$

a. To solve for $\boldsymbol{C}$, first isolate $\boldsymbol{X_C}$.

$\boldsymbol{Z_T} = 1 / [(1 / \boldsymbol{R}) + (1 / \text{j}\ \boldsymbol{X_L}) - (1 / \text{j}\ \boldsymbol{X_C})]$

$1 / \boldsymbol{Z_T} = 1 / \{1 / [(1 / \boldsymbol{R}) + (1 / \text{j}\ \boldsymbol{X_L}) - (1 / \text{j}\ \boldsymbol{X_C})]\}$

$(1 / \boldsymbol{Z_T}) + (1 / \text{j}\ \boldsymbol{X_C}) = (1 / \boldsymbol{R}) + (1 / \text{j}\ \boldsymbol{X_L}) - (1 / \text{j}\ \boldsymbol{X_C}) + (1 / \text{j}\ \boldsymbol{X_C})$

$(1 / \boldsymbol{Z_T}) + (1 / \text{j}\ \boldsymbol{X_C}) - (1 / \boldsymbol{Z_T}) = (1 / \boldsymbol{R}) + (1 / \text{j}\ \boldsymbol{X_L}) - (1 / \boldsymbol{Z_T})$

$1 / (1 / \text{j}\ \boldsymbol{X_C}) = 1 / [(1 / \boldsymbol{R}) + (1 / \text{j}\ \boldsymbol{X_L}) - (1 / \boldsymbol{Z_T})]$

$\text{j}\ \boldsymbol{X_C} = 1 / [(1 / \boldsymbol{R}) + (1 / \text{j}\ \boldsymbol{X_L}) - (1 / \boldsymbol{Z_T})]$

Calculate $\boldsymbol{X_L}$ for $\boldsymbol{L} = 330\ \mu\text{H}$ and $\boldsymbol{f} = 50$ kHz.

$\boldsymbol{X_L} = 2\pi \boldsymbol{fL}$

$= 2\pi(330\ \mu\text{H})(50\text{ kHz})$

$= 2\pi(330 \times 10^{-6}\text{ H})(50 \times 10^3\text{ Hz})$

$= 104\ \Omega$

Substitute in the values for $\boldsymbol{R}$, $\boldsymbol{X_L}$, and $\boldsymbol{Z_T}$.

$\text{j}\ \boldsymbol{X_C} = 1 / [(1 / \boldsymbol{R}) + (1 / \text{j}\ \boldsymbol{X_L}) - (1 / \boldsymbol{Z_T})]$

$= 1 / \{(1 / 62\ \Omega) + (1 / \text{j}\ 104\ \Omega) - [1 / (61.1\ \Omega \angle 9.62°)]\}$

$= 1 / \{[(1 \angle 0°) / (62\ \Omega \angle 0°)] + [(1 \angle 0°) / (104\ \Omega \angle 90°)] - [(1 \angle 0°) / (61.1\ \Omega \angle 9.62°)]\}$

$= 1 / \{[(1 / 62\ \Omega) \angle (0° - 0°)] + [(1 / 104\ \Omega) \angle (0° - 90°)] - [(1 / 61.1\ \Omega) \angle (0° - 9.62°)]\}$

$= 1 / [(16.1\text{ mS} \angle 0°) + (9.65\text{ mS} \angle -90°) - (16.4\text{ mS} \angle -9.62°)]$

Convert to rectangular form to simplify the addition.

$\text{j}\ \boldsymbol{X_C} = 1 / [(16.1\text{ mS} \angle 0°) + (9.65\text{ mS} \angle -90°) - (16.4\text{ mS} \angle -9.62°)]$

$= 1 / [(16.1\text{ mS} + \text{j}\ 0\text{ S}) + (0\text{ S} - \text{j}\ 9.65\text{ mS}) - (16.1\text{ mS} - \text{j}\ 2.74\text{ mS})]$

$= 1 / (0\text{ S} - \text{j}\ 6.91\text{ mS})$

Convert to polar form to simplify the division.

$\text{j}\ \boldsymbol{X_C} = 1 / (0\text{ S} - \text{j}\ 6.91\text{ mS})$

$= (1 \angle 0°) / (6.91\text{ mS} \angle -90°)$

$= (1 / 6.91\text{ mS}) \angle [0° - (-90°)]$

$= 145\ \Omega \angle 90°$ in polar form

$= 0\ \Omega + \text{j}\ 145\ \Omega$ in rectangular form

From the above, $X_C = 145\ \Omega$. Use the equation for capacitive reactance to calculate C.

$$X_C = 1 / (2\pi f C)$$
$$X_C C = [1 / (2\pi f C)] C$$
$$(X_C C) / X_C = [1 / (2\pi f)] / X_C$$
$$C = 1 / (2\pi f X_C)$$
$$= 1 / [2\pi(50\ \text{kHz})(145\ \Omega)]$$
$$= 1 / [2\pi(50 \times 10^3\ \text{Hz})(145\ \Omega)]$$
$$= \mathbf{22\ nF}$$

b. To solve for L, first isolate X_L.

$$Z_T = 1 / [(1 / R) + (1 / \text{j}\,X_L) - (1 / \text{j}\,X_C)]$$
$$1 / Z_T = 1 / \{1 / [(1 / R) + (1 / \text{j}\,X_L) - (1 / \text{j}\,X_C)]\}$$
$$(1 / Z_T) - (1 / \text{j}\,X_L) = (1 / R) + (1 / \text{j}\,X_L) - (1 / \text{j}\,X_C) - (1 / \text{j}\,X_L)$$
$$(1 / Z_T) - (1 / \text{j}\,X_C) - (1 / Z_T) = (1 / R) - (1 / \text{j}\,X_C) - (1 / Z_T)$$
$$1 / [-(1 / \text{j}\,X_L)] = 1 / [(1 / R) - (1 / \text{j}\,X_C) - (1 / Z_T)]$$
$$-\text{j}\,X_L = 1 / [(1 / R) - (1 / \text{j}\,X_C) - (1 / Z_T)]$$

Calculate X_C for $C = 2.2$ nF and $f = 10$ kHz.

$$X_C = 1 / (2\pi f C)$$
$$= 1 / [2\pi(10\ \text{kHz})\ (2.2\ \text{nF})]$$
$$= 1 / [2\pi(10 \times 10^3\ \text{Hz})\ (2.2 \times 10^{-9}\ \text{F})]$$
$$= 7.23\ \text{k}\Omega$$

Substitute in the values for R, X_C, and Z_T.

$$-\text{j}\,X_L = 1 / [(1 / R) - (1 / \text{j}\,X_C) - (1 / Z_T)]$$
$$= 1 / \{(1 / 15\ \text{k}\Omega) - (1 / \text{j}\ 7.23\ \text{k}\Omega) - [1 / (8.93\ \text{k}\Omega \angle -53.5^\circ)]\}$$
$$= 1 / \{[(1 \angle 0^\circ) / (15\ \text{k}\Omega \angle 0^\circ)] - [(1 \angle 0^\circ) / (7.23\ \text{k}\Omega \angle 90^\circ)] - [(1 \angle 0^\circ) / (8.93\ \text{k}\Omega \angle -53.5^\circ)]\}$$
$$= 1 / ([(1 / 15\ \text{k}\Omega) \angle (0^\circ - 0^\circ)] - [(1 / 7.23\ \text{k}\Omega) \angle (0^\circ - 90^\circ)] - \{(1 / 8.93\ \text{k}\Omega) \angle [0^\circ - (-53.5^\circ)]\})$$
$$= 1 / [(66.7\ \mu\text{S} \angle 0^\circ) - (138\ \mu\text{S} \angle -90^\circ) - (112\ \mu\text{S} \angle 53.5^\circ)]$$

Convert to rectangular form to simplify the addition.

$$-\text{j}\,X_L = 1 / [(66.7\ \mu\text{S} \angle 0^\circ) - (138\ \mu\text{S} \angle -90^\circ) - (112\ \mu\text{S} \angle 53.5^\circ)]$$
$$= 1 / [(66.7\ \mu\text{S} + \text{j}\ 0\ \text{S}) - (0\ \text{S} - \text{j}\ 138\ \mu\text{S}) - (66.7\ \mu\text{S} + \text{j}\ 90.0\ \mu\text{S})]$$
$$= 1 / (0\ \text{S} + \text{j}\ 48.2\ \mu\text{S})$$

Convert to polar form to simplify the division.

$$-\text{j}\,X_L = 1 / (0\ \text{S} + \text{j}\ 48.2\ \mu\text{S})$$
$$= (1 \angle 0^\circ) / (48.2\ \mu\text{S} \angle 90^\circ)$$
$$= (1 / 48.2\ \mu\text{S}) \angle [0^\circ - (90^\circ)]$$
$$= 20.7\ \text{k}\Omega \angle -90^\circ \text{ in polar form}$$
$$= 0\ \Omega - \text{j}\ 20.7\ \text{k}\Omega \text{ in rectangular form}$$

From the above, $X_L = 20.7\ \text{k}\Omega$. Use the equation for inductive reactance to calculate L.

$$2\pi f L = X_L$$
$$2\pi f L / 2\pi f = X_L / 2\pi f$$
$$L = X_L / 2\pi f$$
$$L = 20.7\ \text{k}\Omega / [2\pi(10\ \text{kHz})]$$

$= 20.7\ \text{k}\Omega / [2\pi(10 \times 10^3\ \text{Hz})]$

$= \mathbf{330\ mH}$

c. To solve for $\boldsymbol{R}$, first isolate $\boldsymbol{R}$.

$$\boldsymbol{Z_T} = 1 / [(1 / \boldsymbol{R}) + (1 / \text{j}\, \boldsymbol{X_L}) - (1 / \text{j}\, \boldsymbol{X_C})]$$

$$1 / \boldsymbol{Z_T} = 1 / \{1 / [(1 / \boldsymbol{R}) + (1 / \text{j}\, \boldsymbol{X_L}) - (1 / \text{j}\, \boldsymbol{X_C})]\}$$

$$(1 / \boldsymbol{Z_T}) - (1 / \boldsymbol{R}) = (1 / \boldsymbol{R}) + (1 / \text{j}\, \boldsymbol{X_L}) - (1 / \text{j}\, \boldsymbol{X_C}) - (1 / \boldsymbol{R})$$

$$(1 / \boldsymbol{Z_T}) - (1 / \boldsymbol{R}) - (1 / \boldsymbol{Z_T}) = (1 / \text{j}\, \boldsymbol{X_L}) - (1 / \text{j}\, \boldsymbol{X_C}) - (1 / \boldsymbol{Z_T})$$

$$1 / [-(1 / \boldsymbol{R})] = 1 / [(1 / \text{j}\, \boldsymbol{X_L}) - (1 / \text{j}\, \boldsymbol{X_C}) - (1 / \boldsymbol{Z_T})]$$

$$-\boldsymbol{R} = 1 / [(1 / \text{j}\, \boldsymbol{X_L}) - (1 / \text{j}\, \boldsymbol{X_C}) - (1 / \boldsymbol{Z_T})]$$

$$-(-\boldsymbol{R}) = -\{1 / [(1 / \text{j}\, \boldsymbol{X_L}) - (1 / \text{j}\, \boldsymbol{X_C}) - (1 / \boldsymbol{Z_T})]\}$$

$$\boldsymbol{R} = -1 / [(1 / \text{j}\, \boldsymbol{X_L}) - (1 / \text{j}\, \boldsymbol{X_C}) - (1 / \boldsymbol{Z_T})]\}$$

Calculate $\boldsymbol{X_L}$ for $\boldsymbol{L} = 2.4$ H and $\boldsymbol{f} = 60$ Hz.

$$\begin{aligned} \boldsymbol{X_L} &= 2\pi \boldsymbol{fL} \\ &= 2\pi(60\ \text{Hz})\ (2.4\ \text{H}) \\ &= 905\ \Omega \end{aligned}$$

Calculate $\boldsymbol{X_C}$ for $\boldsymbol{C} = 13\ \mu$F and $\boldsymbol{f} = 60$ Hz.

$$\begin{aligned} \boldsymbol{X_C} &= 1 / (2\pi \boldsymbol{fC}) \\ &= 1/ [2\pi(60\ \text{Hz})\ (13\ \mu\text{F})] \\ &= 1 / [2\pi(60\ \text{Hz})(13 \times 10^{-6}\ \text{F})] \\ &= 204\ \Omega \end{aligned}$$

Substitute in the values for $\boldsymbol{X_L}$, $\boldsymbol{X_C}$, and $\boldsymbol{Z_T}$.

$$\begin{aligned} \boldsymbol{R} &= -1 / [(1 / \boldsymbol{X_L}) - (1 / \text{j}\, \boldsymbol{X_C}) - (1 / \boldsymbol{Z_T})] \\ &= -1 / \{(1 / \text{j}\ 905\ \Omega) - (1 / \text{j}\ 204\ \Omega) - [1 / (130\ \Omega \angle -29.7°)]\} \\ &= -1 / \{[(1 \angle 0°) / (905 \angle 90°)] - [(1 \angle 0°) / (204\ \Omega \angle 90°)] \\ &\quad - [(1 \angle 0°) / (130\ \Omega \angle -29.7°)]\} \\ &= -1 / ([(1 / 905\ \Omega) \angle (0° - 90°)] - [(1 / 204\ \Omega) \angle (0° - 90°)] \\ &\quad - \{(1 / 130\ \Omega) \angle [0° - (-29.7°)]\}) \\ &= -1 / [(1.11\ \text{mS} \angle -90°) - (4.90\ \text{mS} \angle -90°) - (7.69\ \text{mS} \angle 29.7°)] \end{aligned}$$

Convert to rectangular form to simplify the addition.

$$\begin{aligned} \boldsymbol{R} &= -1 / [(1.11\ \text{mS} \angle -90°) - (4.90\ \text{mS} \angle -90°) - (7.69\ \text{mS} \angle 29.7°)] \\ &= -1 / [(0\ \text{S} - \text{j}\ 1.11\ \text{mS}) - (0\ \text{S} - \text{j}\ 4.90\ \text{mS}) - (6.68\ \text{mS} + \text{j}\ 3.81\ \text{mS})] \\ &= -1 / (-6.68\ \text{mS} + \text{j}\ 0\ \text{S}) \end{aligned}$$

Convert to polar form to simplify the division.

$$\begin{aligned} \boldsymbol{R} &= -1 / (-6.68\ \text{mS} + \text{j}\ 0\ \text{S}) \\ &= (-1 \angle 0°) / (-6.68\ \text{mS} \angle 90°) \\ &= (-1 / -6.68\ \text{mS}) \angle [0° - (90°)] \\ &= 150\ \Omega \angle 0° \text{ in polar form} \\ &= 150\ \Omega - \text{j}\ 0\ \Omega \text{ in rectangular form} \end{aligned}$$

From the result of the division, $\boldsymbol{R} = \mathbf{150\ \Omega}$.

3. Problem: For each given set of circuit values, calculate the total current, $\boldsymbol{I_T}$, and the resistor, inductor, and capacitor currents, $\boldsymbol{I_R}$, $\boldsymbol{I_L}$, and $\boldsymbol{I_C}$. Express the answers in polar form.

a. $\boldsymbol{V_S} = 2.5\ \text{V}_{\text{ac}}$, $\boldsymbol{R} = 180\ \Omega$, $\boldsymbol{L} = 120$ mH, $\boldsymbol{C} = 560$ nF, and $\boldsymbol{f} = 440$ Hz

b. V_S = 100 mV$_{ac}$, R = 3.9 Ω, L = 62 mH, C = 820 µF, and f = 30 Hz

c. V_S = 120 V$_{ac}$, R = 600 Ω, L = 820 mH, C = 6.8 µF, and f = 60 Hz

Solution: Because the source voltage appears across each component in the parallel ***RLC*** circuit, it is generally easiest to use Ohm's Law to determine the current through each branch, sum the branch currents to find the total current, I_T, and then use Ohm's Law to calculate the total impedance Z_T.

a. Calculate X_L for L = 120 mH and f = 440 Hz.

$$X_L = 2\pi f L$$
$$= 2\pi(440 \text{ Hz})(120 \text{ mH})$$
$$= 2\pi(440 \text{ Hz})(120 \times 10^{-3} \text{ H})$$
$$= 332\ \Omega$$

Calculate X_C for C = 560 nF and f = 440 Hz.

$$X_C = 1/(2\pi f C)$$
$$= 1/[2\pi(440 \text{ Hz})(560 \text{ nF})]$$
$$= 1/[2\pi(440 \text{ Hz})(560 \times 10^{-9} \text{ F})]$$
$$= 646\ \Omega$$

Use Ohm's Law for current to calculate the current in each parallel branch.

$$I_R = V_S / R$$
$$= (2.5 \text{ V}_{ac} \angle 0°) / (180\ \Omega \angle 0°)$$
$$= (2.5 \text{ V}_{ac} / 180\ \Omega) \angle (0° - 0°)$$
$$= \mathbf{13.9\ mA_{ac} \angle 0°}$$

$$I_L = V_S / \text{j}\, X_L$$
$$= V_S / (X_L \angle 90°)$$
$$= (2.5 \text{ V}_{ac} \angle 0°) / (332\ \Omega \angle 90°)$$
$$= (2.5 \text{ V}_{ac} / 332\ \Omega) \angle (0° - 90°)$$
$$= \mathbf{7.54\ mA_{ac} \angle -90°}$$

$$I_C = V_S / -\text{j}\, X_C$$
$$= V_S / (X_C \angle -90°)$$
$$= (2.5 \text{ V}_{ac} \angle 0°) / (646\ \Omega \angle -90°)$$
$$= (2.5 \text{ V}_{ac} / 646\ \Omega) \angle [0° - (-90°)]$$
$$= \mathbf{3.87\ mA_{ac} \angle 90°}$$

From Kirchhoff's Current Law, the total current, I_T, is the sum of the branch currents.

$$I_T = I_R + I_L + I_C$$
$$= (13.9 \text{ mA}_{ac} \angle 0°) + (7.54 \text{ mA}_{ac} \angle -90°) + (3.87 \text{ mA}_{ac} \angle -90°)$$
$$= (13.9 \text{ mA}_{ac} + \text{j}\,0 \text{ A}_{ac}) + (0 \text{ A}_{ac} - \text{j}\,7.54 \text{ mA}_{ac}) + (0 \text{ A}_{ac} + \text{j}\,3.87 \text{ mA}_{ac})$$
$$= 13.9 \text{ mA}_{ac} - \text{j}\,3.67 \text{ mA}_{ac} \text{ in rectangular form}$$
$$= \mathbf{14.4\ mA_{ac} \angle -14.8°} \text{ in polar form}$$

Finally, use Ohm's Law for impedance to calculate the total impedance Z_T.

$$Z_T = V_S / I_T$$
$$= (2.5 \text{ V}_{ac} \angle 0°) / (14.4 \text{ mA}_{ac} \angle -14.8°)$$
$$= (2.5 \text{ V}_{ac} / 14.4 \text{ mA}_{ac}) \angle [0° - (-14.8°)]$$
$$= \mathbf{478\ \Omega \angle 14.8°}$$

b. Calculate X_L for L = 62 mH and f = 30 Hz.

$$X_L = 2\pi f L$$

$= 2\pi(30\text{ Hz})\ (62\text{ mH})$
$= 2\pi(30\text{ Hz})\ (62 \times 10^{-3}\text{ H})$
$= 11.7\ \Omega$

Calculate X_C for $C = 820\ \mu\text{F}$ and $f = 30$ Hz.

$X_C = 1\ /\ (2\pi fC)$
$= 1\ /\ [2\pi(30\text{ Hz})\ (820\ \mu\text{F})]$
$= 1\ /\ [2\pi(30\text{ Hz})(820 \times 10^{-6}\text{ F})]$
$= 6.47\ \Omega$

Use Ohm's Law for current to calculate current in each parallel branch.

$I_R = V_S\ /\ R$
$= (100\text{ mV}_{ac} \angle 0°)\ /\ (3.9\ \Omega \angle 0°)$
$= (100\text{ mV}_{ac}\ /\ 3.9\ \Omega) \angle (0° - 0°)$
$= \mathbf{25.6\ mA_{ac} \angle 0°}$

$I_L = V_S\ /\ \text{j}\,X_L$
$= V_S\ /\ (X_L \angle 90°)$
$= (100\text{ mV}_{ac} \angle 0°)\ /\ (11.7\ \Omega \angle 90°)$
$= (100\text{ mV}_{ac}\ /\ 11.7\ \Omega) \angle (0° - 90°)$
$= \mathbf{8.56\ mA_{ac} \angle -90°}$

$I_C = V_S\ /\ -\text{j}\,X_C$
$= V_S\ /\ (X_C \angle -90°)$
$= (100\text{ mV}_{ac} \angle 0°)\ /\ (6.47\ \Omega \angle -90°)$
$= (100\text{ mV}_{ac}\ /\ 646\ \Omega) \angle [0° - (-90°)]$
$= \mathbf{15.5\ mA_{ac} \angle 90°}$

From Kirchhoff's Current Law, the total current, I_T, is the sum of the branch currents.

$I_T = I_R + I_L + I_C$
$= (25.6\text{ mA}_{ac} \angle 0°) + (8.56\text{ mA}_{ac} \angle -90°) + (15.5\text{ mA}_{ac} \angle 90°)$
$= (25.6\text{ mA}_{ac} + \text{j}\ 0\text{ A}_{ac}) + (0\text{ A}_{ac} - \text{j}\ 8.56\text{ mA}_{ac}) + (0\text{ A}_{ac} + \text{j}\ 15.5\text{ mA}_{ac})$
$= 25.6\text{ mA}_{ac} + \text{j}\ 6.90\text{ mA}_{ac}$ in rectangular form
$= \mathbf{26.6\ mA_{ac} \angle 15.1°}$ in polar form

Use Ohm's Law for impedance to calculate the total impedance Z_T.

$Z_T = V_S\ /\ I_T$
$= (100\text{ mV}_{ac} \angle 0°)\ /\ (26.6\text{ mA}_{ac} \angle 15.1°)$
$= (100\text{ mV}_{ac}\ /\ 26.6\text{ mA}_{ac}) \angle (0° - 15.1°)$
$= \mathbf{3.77\ \Omega \angle -15.1°}$

c. Calculate X_L for $L = 820$ mH and $f = 60$ Hz.

$X_L = 2\pi fL$
$= 2\pi(60\text{ Hz})\ (820\text{ mH})$
$= 2\pi(60\text{ Hz})\ (820 \times 10^{-3}\text{ H})$
$= 309\ \Omega$

Calculate X_C for $C = 6.8\ \mu\text{F}$ and $f = 60$ Hz.

$X_C = 1\ /\ (2\pi fC)$

$= 1 / [2\pi(60\text{ Hz})\ (6.8\ \mu\text{F})]$

$= 1 / [2\pi(60\text{ Hz})(6.8 \times 10^{-6}\text{ F})]$

$= 390\ \Omega$

Use Ohm's Law to calculate the current in each parallel branch.

$\boldsymbol{I_R} = \boldsymbol{V_S} / \boldsymbol{R}$

$= (120\text{ V}_{ac} \angle 0°) / (600\ \Omega \angle 0°)$

$= (120\text{ mV}_{ac} / 600\ \Omega) \angle (0° - 0°)$

$= \mathbf{200\ mA_{ac} \angle 0°}$

$\boldsymbol{I_L} = \boldsymbol{V_S} / \text{j}\ \boldsymbol{X_L}$

$= \boldsymbol{V_S} / (\boldsymbol{X_L} \angle 90°)$

$= (120\text{ V}_{ac} \angle 0°) / (309\ \Omega \angle 90°)$

$= (120\text{ V}_{ac} / 309\ \Omega) \angle (0° - 90°)$

$= \mathbf{388\ mA_{ac} \angle -90°}$

$\boldsymbol{I_C} = \boldsymbol{V_S} / -\text{j}\ \boldsymbol{X_C}$

$= \boldsymbol{V_S} / (\boldsymbol{X_C} \angle -90°)$

$= (120\text{ V}_{ac} \angle 0°) / (390\ \Omega \angle -90°)$

$= (120\text{ V}_{ac} / 390\ \Omega) \angle [0° - (-90°)]$

$= \mathbf{308\ mA_{ac} \angle 90°}$

From Kirchhoff's Current Law, the total current, $\boldsymbol{I_T}$ is the sum of the branch currents.

$\boldsymbol{I_T} = \boldsymbol{I_R} + \boldsymbol{I_L} + \boldsymbol{I_C}$

$= (200\text{ mA}_{ac} \angle 0°) + (388\text{ mA}_{ac} \angle -90°) + (308\text{ mA}_{ac} \angle 90°)$

$= (200\text{ mA}_{ac} + \text{j}\ 0\text{ A}_{ac}) + (0\text{ A}_{ac} - \text{j}\ 388\text{ mA}_{ac}) + (0\text{ A}_{ac} + \text{j}\ 308\text{ mA}_{ac})$

$= 200\text{ mA}_{ac} - \text{j}\ 80.6\text{ mA}_{ac}$ in rectangular form

$= \mathbf{216\ mA_{ac} \angle -21.9°}$ in polar form

Use Ohm's Law for impedance to calculate the total impedance $\boldsymbol{Z_T}$.

$\boldsymbol{Z_T} = \boldsymbol{V_S} / \boldsymbol{I_T}$

$= (120\text{ V}_{ac} \angle 0°) / (216\text{ mA}_{ac} \angle -21.9°)$

$= (120\text{ V}_{ac} / 216\text{ mA}_{ac}) \angle [0° - (-21.9°)]$

$= \mathbf{557\ \Omega \angle 21.9°}$

18.1.4 Resonance of Parallel *RLC* Circuits

For Problems 1 through 3, refer to the circuit in Figure 18-5.

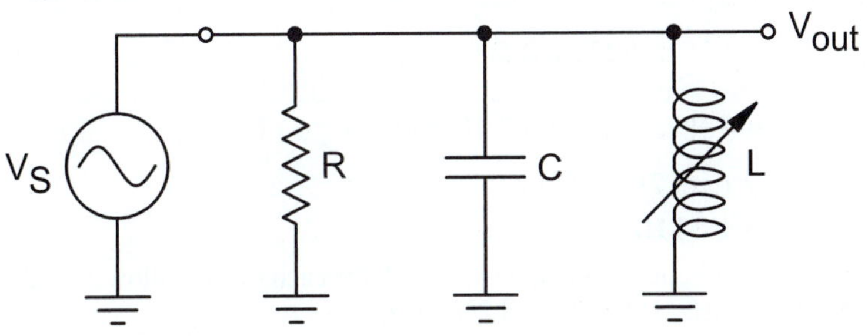

Figure 18-5: Parallel Resonance *RLC* Circuit for Problems 1 through 3

1. Problem: For $\boldsymbol{C}$ = 560 pF, calculate the resonant frequency, $\boldsymbol{f_r}$, for each of the following values for $\boldsymbol{L}$: 27 μH, 56 μH, 110 μH, and 220 μH.

Solution: At the resonant frequency, f_r, the inductive reactance, X_L, and capacitive reactance, X_C, are equal. Use the equations for X_L and X_C to determine f_r in terms of L and C.

$$X_L = X_C$$
$$2\pi f_r L = 1 / (2\pi f_r C)$$
$$2\pi f_r L / 2\pi L = [1 / (2\pi f_r C)] / 2\pi L$$
$$f_r = 1 / [(2\pi f_r C)(2\pi L)]$$
$$f_r = 1 / (4\pi^2 f_r LC)$$
$$f_r \times f_r = [1 / (4\pi^2 f_r LC)] \times f_r$$
$$f_r^2 = 1 / (4\pi^2 LC)$$
$$\sqrt{f_r^2} = \sqrt{1/4\pi^2 LC}$$
$$f_r = \sqrt{1} / \sqrt{4\pi^2 LC}$$
$$= 1 / (2\pi\sqrt{LC})$$

For $C = 560$ pF and $L = 27$ µH,

$$f_r = 1 / [2\pi\sqrt{(27\ \mu\text{H})(560\ \text{pF})}]$$
$$= 1 / [2\pi\sqrt{(27 \times 10^{-6}\ \text{H})(560 \times 10^{-12}\ \text{F})}]$$
$$= 1 / (773 \times 10^{-9}\ \text{s})$$
$$= \mathbf{1.29\ MHz}$$

For $C = 560$ pF and $L = 56$ µH,

$$f_r = 1 / [2\pi\sqrt{(56\ \mu\text{H})(560\ \text{pF})}]$$
$$= 1 / [2\pi\sqrt{(56 \times 10^{-6}\ \text{H})(560 \times 10^{-12}\ \text{F})}]$$
$$= 1 / (1.11 \times 10^{-6}\ \text{s})$$
$$= \mathbf{899\ kHz}$$

For $C = 560$ pF and $L = 110$ µH,

$$f_r = 1 / [2\pi\sqrt{(110\ \mu\text{H})(560\ \text{pF})}]$$
$$= 1 / [2\pi\sqrt{(110 \times 10^{-6}\ \text{H})(560 \times 10^{-12}\ \text{F})}]$$
$$= 1 / (1.56 \times 10^{-6}\ \text{s})$$
$$= \mathbf{641\ kHz}$$

For $C = 560$ pF and $L = 220$ µH,

$$f_r = 1 / [2\pi\sqrt{(220\ \mu\text{H})(560\ \text{pF})}]$$
$$= 1 / [2\pi\sqrt{(220 \times 10^{-6}\ \text{H})(560 \times 10^{-12}\ \text{F})}]$$
$$= 1 / (2.21 \times 10^{-6}\ \text{s})$$
$$= \mathbf{453\ kHz}$$

2. Problem: For $C = 1.5$ nF, determine the value of L for each of the following values of resonant frequency, f_r: 25 kHz, 50 kHz, 100 kHz, and 200 kHz.

Solution: At the resonant frequency, f_r, the inductive reactance, X_L, and capacitive reactance, X_C, are equal. Use the equations for calculating X_L and X_C to determine L in terms of f_r and C.

$$X_L = X_C$$

$2\pi f_r L = 1 / (2\pi f_r C)$

$2\pi f_r L / 2\pi f_r = [1 / (2\pi f_r C)] / 2\pi f_r$

$L = 1 / [(2\pi f_r C)(2\pi f_r)]$

$= 1 / (4\pi^2 f_r^2 C)$

For $C = 1.5$ nF and $f_r = 25$ kHz,

$L = 1 / (4\pi^2 f_r^2 C)$

$= 1 / [4\pi^2(25 \text{ kHz})^2(1.5 \text{ nF})]$

$= 1 / [4\pi^2(625 \times 10^6 \text{ Hz}^2)(1.5 \times 10^{-9} \text{ F})]$

$= 1 / (37.0 \text{ H}^{-1})$

$= \mathbf{27.0 \text{ mH}}$

For $C = 1.5$ nF and $f_r = 50$ kHz,

$L = 1 / (4\pi^2 f_r^2 C)$

$= 1 / [4\pi^2(50 \text{ kHz})^2(1.5 \text{ nF})]$

$= 1 / [4\pi^2(2.5 \times 10^9 \text{ Hz}^2)(1.5 \times 10^{-9} \text{ F})]$

$= 1 / (148 \text{ H}^{-1})$

$= \mathbf{6.76 \text{ mH}}$

For $C = 1.5$ nF and $f_r = 100$ kHz,

$L = 1 / (4\pi^2 f_r^2 C)$

$= 1 / [4\pi^2(100 \text{ kHz})^2(1.5 \text{ nF})]$

$= 1 / [4\pi^2(10 \times 10^9 \text{ Hz}^2)(1.5 \times 10^{-9} \text{ F})]$

$= 1 / (592 \text{ H}^{-1})$

$= \mathbf{1.69 \text{ mH}}$

For $C = 1.5$ nF and $f_r = 200$ kHz,

$L = 1 / (4\pi^2 f_r^2 C)$

$= 1 / [4\pi^2(200 \text{ kHz})^2(1.5 \text{ nF})]$

$= 1 / [4\pi^2(40 \times 10^9 \text{ Hz}^2)(1.5 \times 10^{-9} \text{ F})]$

$= 1 / (2369 \text{ H}^{-1})$

$= \mathbf{422\ \mu H}$

3. Problem: What is I_L at resonance for $V_S = 1.0$ Vac, $R = 3.9$ kΩ, $L = 75$ µH, and $C = 47$ µF?

Solution: First, use the equation from Problem 1 to calculate the resonant frequency, f_r.

$f_r = 1 / (2\pi\sqrt{LC})$

$= 1 / [2\pi\sqrt{(75\ \mu\text{H})(47\ \mu\text{F})}]$

$= 1 / [2\pi\sqrt{(75 \times 10^{-6} \text{ H})(47 \times 10^{-6} \text{ F})}]$

$= 1 / (373 \times 10^{-6} \text{ s})$

$= 2.68 \text{ kHz}$

Calculate the inductive reactance at the resonant frequency, f_r.

$X_L = 2\pi f L$

$= 2\pi(2.68 \text{ kHz})(75\ \mu\text{H})$

$= 2\pi(2.68 \times 10^3 \text{ Hz})(75 \times 10^{-6} \text{ H})$

$= 1.26\ \Omega$

The source voltage, V_S, appears across the parallel components R, C, and L. Use Ohm's Law for current to calculate I_L.

$$I_L = V_L / X_L$$
$$= V_S / X_L$$
$$= 1.0\ \mathrm{V_{ac}} / 1.26\ \Omega$$
$$= \mathbf{792\ mA_{ac}}$$

For Problems 4 through 6, refer to the circuit in Figure 18-6.

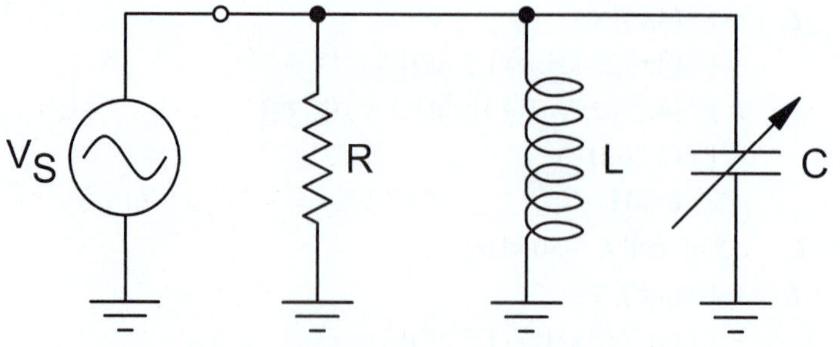

Figure 18-6: Parallel Resonance *RLC* Circuit for Problems 4 through 6

4. Problem: For $\boldsymbol{L}$ = 91 μH, calculate $\boldsymbol{f_r}$ for each of the following values of $\boldsymbol{C}$: 75 nF, 150 nF, 300 nF, and 620 nF.

Solution: From the results of Problem 1 above, $\boldsymbol{f_r}$ in terms of $\boldsymbol{L}$ and $\boldsymbol{C}$ is

$$f_r = 1 / (2\pi\sqrt{LC})$$

For $\boldsymbol{L}$ = 91 μH and $\boldsymbol{C}$ = 75 nF,

$$f_r = 1 / [\,2\pi\sqrt{(91\ \mu\mathrm{H})(75\ \mathrm{nF})}\,]$$
$$= 1 / [\,2\pi\sqrt{(91\times10^{-6}\ \mathrm{F})(75\times10^{-9}\ \mathrm{H})}\,]$$
$$= 1 / (16.4\times10^{-6}\ \mathrm{s})$$
$$= \mathbf{60.9\ kHz}$$

For $\boldsymbol{L}$ = 91 μH and $\boldsymbol{C}$ = 150 nF,

$$f_r = 1 / [\,2\pi\sqrt{(91\ \mu\mathrm{H})(150\ \mathrm{nF})}\,]$$
$$= 1 / [\,2\pi\sqrt{(91\times10^{-6}\ \mathrm{F})(150\times10^{-9}\ \mathrm{H})}\,]$$
$$= 1 / (23.2\times10^{-6}\ \mathrm{s})$$
$$= \mathbf{43.1\ kHz}$$

For $\boldsymbol{L}$ = 91 μH and $\boldsymbol{C}$ = 300 nF,

$$f_r = 1 / [\,2\pi\sqrt{(91\ \mu\mathrm{H})(300\ \mathrm{nF})}\,]$$
$$= 1 / [\,2\pi\sqrt{(91\times10^{-6}\ \mathrm{F})(300\times10^{-9}\ \mathrm{H})}\,]$$
$$= 1 / (32.8\times10^{-6}\ \mathrm{s})$$
$$= \mathbf{30.5\ kHz}$$

For $\boldsymbol{L}$ = 91 μH and $\boldsymbol{C}$ = 620 nF,

$$f_r = 1 / [\,2\pi\sqrt{(91\ \mu\mathrm{H})(620\ \mathrm{nF})}\,]$$
$$= 1 / [\,2\pi\sqrt{(91\times10^{-6}\ \mathrm{F})(620\times10^{-9}\ \mathrm{H})}\,]$$
$$= 1 / (47.2\times10^{-6}\ \mathrm{s})$$

= **21.2 kHz**

5. Problem: For $\boldsymbol{L}$ = 510 mH, calculate $\boldsymbol{C}$ for each of the following values of $\boldsymbol{f_r}$: 1.25 Hz, 2.5 Hz, 5.0 Hz, and 10 Hz.

Solution: At the resonant frequency, $\boldsymbol{f_r}$, the inductive reactance, $\boldsymbol{X_L}$, and capacitive reactance, $\boldsymbol{X_C}$, are equal. Use the equations for calculating $\boldsymbol{X_L}$ and $\boldsymbol{X_C}$ to determine $\boldsymbol{C}$ in terms of $\boldsymbol{f_r}$ and $\boldsymbol{L}$.

$$X_L = X_C$$
$$2\pi f_r L = 1 / (2\pi f_r C)$$
$$2\pi f_r L \times C = [1 / (2\pi f_r C)] \times C$$
$$2\pi f_r LC / 2\pi f_r L = [1 / (2\pi f_r)] / 2\pi f_r L$$
$$C = 1 / [(2\pi f_r)(2\pi f_r L)]$$
$$= 1 / (4\pi^2 f_r^2 L)$$

For $\boldsymbol{L}$ = 510 mH and $\boldsymbol{f_r}$ = 1.25 Hz,

$$C = 1 / (4\pi^2 f_r^2 L)$$
$$= 1 / [4\pi^2(1.25 \text{ Hz})^2(510 \text{ mH})]$$
$$= 1 / [4\pi^2(1.56 \text{ Hz}^2)(510 \times 10^{-3} \text{ H})]$$
$$= 1 / (31.5 \text{ F}^{-1})$$
$$= \mathbf{31.8\ mF}$$

For $\boldsymbol{L}$ = 510 mH and $\boldsymbol{f_r}$ = 2.5 Hz,

$$C = 1 / (4\pi^2 f_r^2 L)$$
$$= 1 / [4\pi^2(2.5 \text{ Hz})^2(510 \text{ mH})]$$
$$= 1 / [4\pi^2(6.25 \text{ Hz}^2)(510 \times 10^{-3} \text{ H})]$$
$$= 1 / (126 \text{ F}^{-1})$$
$$= \mathbf{7.95\ mF}$$

For $\boldsymbol{L}$ = 510 mH and $\boldsymbol{f_r}$ = 5.0 Hz,

$$C = 1 / (4\pi^2 f_r^2 L)$$
$$= 1 / [4\pi^2(5.0 \text{ Hz})^2(510 \text{ mH})]$$
$$= 1 / [4\pi^2(25.0 \text{ Hz}^2)(510 \times 10^{-3} \text{ H})]$$
$$= 1 / (503 \text{ F}^{-1})$$
$$= \mathbf{1.99\ mF}$$

For $\boldsymbol{L}$ = 510 mH and $\boldsymbol{f_r}$ = 10 Hz,

$$C = 1 / (4\pi^2 f_r^2 L)$$
$$= 1 / [4\pi^2(10 \text{ Hz})^2(510 \text{ mH})]$$
$$= 1 / [4\pi^2(100 \text{ Hz}^2)(510 \times 10^{-3} \text{ H})]$$
$$= 1 / (2.01 \text{ kF}^{-1})$$
$$= \mathbf{497\ \mu F}$$

6. Problem: The value of $\boldsymbol{I_C}$ at resonance is 10.0 $\text{mA}_{\text{ac}} \angle 90°$ for $\boldsymbol{V_S}$ = 250 $\text{mV}_{\text{ac}} \angle 0°$, $\boldsymbol{R}$ = 180 Ω, and $\boldsymbol{L}$ = 220 μH. What is the value of $\boldsymbol{C}$?

Solution: For the parallel ***RLC*** circuit the source voltage, $\boldsymbol{V_S}$, appears across $\boldsymbol{R}$, $\boldsymbol{L}$, and $\boldsymbol{C}$. At the resonant frequency, $\boldsymbol{f_r}$, the inductor current, $\boldsymbol{I_L}$, and capacitor current, $\boldsymbol{I_C}$, are equal and opposite. Use Ohm's Law for reactance to determine the inductive reactance at resonance.

$$X_L = V_L / I_L$$
$$= V_S / -I_C$$
$$= (250 \text{ mV}_{\text{ac}} \angle 0°) / -(10.0 \text{ mA}_{\text{ac}} \angle 90°)$$
$$= (250 \text{ mV}_{\text{ac}} \angle 0°) / [10.0 \text{ mA}_{\text{ac}} \angle (90° - 180°)]$$

$= (250\ \text{mV}_{ac} \angle 0°) / (10.0\ \text{mA}_{ac} \angle -90°)$
$= (250\ \text{mV}_{ac} / 10.0\ \text{mA}_{ac}) \angle [0° - (-90°)]$
$= 25\ \Omega \angle 90°$

Use the equation for inductive reactance to calculate resonant frequency, f_r.

$$2\pi f_r L = X_L$$
$$2\pi f_r L / 2\pi L = X_L / 2\pi L$$
$$f_r = 25\ \Omega / [2\pi\,(220\ \mu\text{H})]$$
$$= 25\ \Omega / [2\pi\,(220 \times 10^{-6}\ \text{H})]$$
$$= 18.1\ \text{kHz}$$

At resonance the magnitude of inductive reactance, X_L, and capacitive reactance, X_C, are equal. Use the equation for capacitive reactance to calculate the capacitance, C.

$$X_C = 1 / (2\pi f_r C)$$
$$X_C\,C = [1 / (2\pi f_r C)]\,C$$
$$(X_C\,C) / X_C = [1 / (2\pi f_r)] / X_C$$
$$C = 1 / (2\pi f_r X_C)$$
$$= 1 / [2\pi(18.1\ \text{kHz})(25\ \Omega)]$$
$$= 1 / 2.84\ \text{MF}^{-1}$$
$$= \mathbf{352\ nF}$$

Alternatively, the value of C could be calculated by using the equation $f_r = 1 / (2\pi\sqrt{LC}\,)$.

18.1.5 Analysis of Series-Parallel *RLC* Circuits

1. Problem: Refer to the circuit in Figure 18-7.

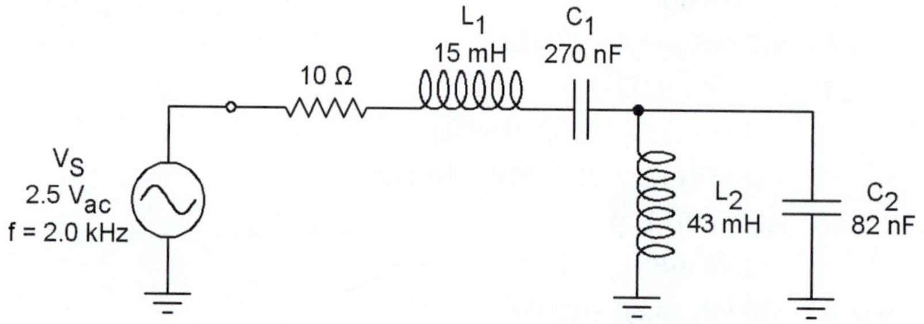

Figure 18-7: Series-Parallel *RLC* Circuit for Problem 1

Calculate the total impedance, Z_T, total current, I_T, and currents and voltages for each component. Express all answers in polar form.

Solution: First, calculate the inductive and capacitive reactances for the circuit. Then, use standard circuit analysis techniques for series-parallel circuits using complex arithmetic.

Calculate X_{L1} for $L_1 = 15$ mH and $f = 2.0$ kHz.

$$X_{L1} = 2\pi f L_1$$
$$= 2\pi(2.0\ \text{kHz})\,(15\ \text{mH})$$
$$= 2\pi(2.0 \times 10^3\ \text{Hz})\,(15 \times 10^{-3}\ \text{H})$$
$$= 189\ \Omega$$

Calculate X_{C1} for $C_1 = 270$ nF and $f = 2.0$ kHz.

$$
\begin{aligned}
X_{C1} &= 1 / (2\pi f C_1) \\
&= 1 / [2\pi(2.0 \text{ kHz})(270 \text{ nF})] \\
&= 1 / [2\pi(2.0 \times 10^3 \text{ Hz})(270 \times 10^{-9} \text{ F})] \\
&= 1 / 3.39 \text{ mS} \\
&= 295\ \Omega
\end{aligned}
$$

Calculate X_{L2} for $L_2 = 43$ mH and $f = 2.0$ kHz.

$$
\begin{aligned}
X_{L2} &= 2\pi f L_2 \\
&= 2\pi(2.0 \text{ kHz})(43 \text{ mH}) \\
&= 2\pi(2.0 \times 10^3 \text{ Hz})(43 \times 10^{-3} \text{ H}) \\
&= 540\ \Omega
\end{aligned}
$$

Calculate X_{C2} for $C_2 = 82$ nF and $f = 2.0$ kHz.

$$
\begin{aligned}
X_{C2} &= 1 / (2\pi f C_2) \\
&= 1 / [2\pi(2.0 \text{ kHz})(82 \text{ nF})] \\
&= 1 / [2\pi(2.0 \times 10^3 \text{ Hz})(82 \times 10^{-9} \text{ F})] \\
&= 1 / 1.03 \text{ mS} \\
&= 971\ \Omega
\end{aligned}
$$

Figure 18-8 shows the equivalent circuit for Figure 18-7.

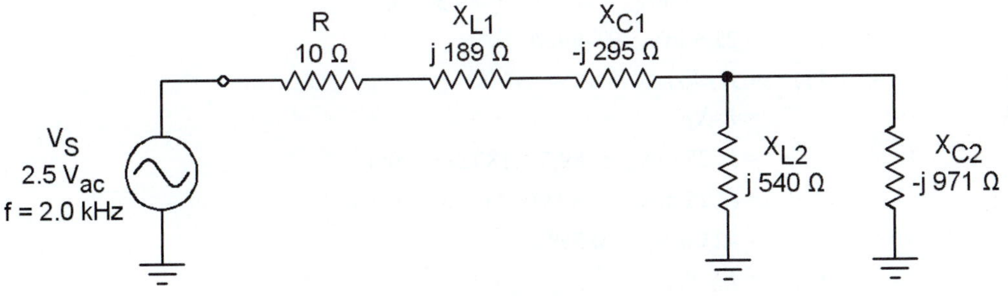

Figure 18-8: Equivalent Circuit for Figure 18-7

From standard circuit analysis, $Z_T = R + X_{L1} + X_{C1} + X_{L2} \parallel X_{C2}$. The first equivalent impedance Z_{EQ1} is the parallel combination $X_{L2} \parallel X_{C2}$.

$$Z_T = R + X_{L1} + X_{C1} + Z_{EQ1}$$

where

$$
\begin{aligned}
Z_{EQ1} &= X_{L2} \parallel X_{C2} \\
&= (\text{j } 540\ \Omega) \parallel (-\text{j } 971\ \Omega) \\
&= [(\text{j } 540\ \Omega)(-\text{j } 971\ \Omega)] / [(\text{j } 540\ \Omega) + (-\text{j } 971\ \Omega)] \\
&= [-\text{j}^2 (540\ \Omega)(971\ \Omega)] / [\text{j } (540\ \Omega - 971\ \Omega)] \\
&= [-(-1)(5.23 \times 10^3\ \Omega^2)] / [\text{j } (-431\ \Omega)] \\
&= (5.23 \times 10^3\ \Omega^2) / (-\text{j } 431\ \Omega) \\
&= [(5.23 \times 10^3\ \Omega^2) / (-\text{j } 431\ \Omega)]\ (\text{j} / \text{j}) \\
&= (\text{j } 5.23 \times 10^3\ \Omega^2) / (-\text{j}^2\ 431\ \Omega) \\
&= (\text{j } 5.23 \times 10^3\ \Omega^2) / [-(-1)\ 431\ \Omega] \\
&= \text{j } [(5.23 \times 10^3\ \Omega^2) / 431\ \Omega] \\
&= \text{j } 1.22 \text{ k}\Omega
\end{aligned}
$$

The second (and last) equivalent impedance Z_{EQ2} is the series combination $R + X_{L1} + X_{C1} + Z_{EQ1}$.

$$Z_T = Z_{EQ2}$$

where

$$\begin{aligned}
\boldsymbol{Z_{EQ2}} &= \boldsymbol{R} + \boldsymbol{X_{L1}} + \boldsymbol{X_{C1}} + \boldsymbol{Z_{EQ1}} \\
&= 10\ \Omega + j\ 189\ \Omega - j\ 295\ \Omega + j\ 1.22\ k\Omega \\
&= (10\ \Omega + j\ 0\ \Omega) + (0\ \Omega + j\ 189\ \Omega) + (0\ \Omega - j\ 295\ \Omega) + (0\ \Omega + j\ 1.22\ k\Omega) \\
&= (10\ \Omega + 0\ \Omega + 0\ \Omega + 0\ \Omega) + j\ (0\ \Omega + 189\ \Omega - 295\ \Omega + 1.22\ k\Omega) \\
&= 10\ \Omega + 1.11\ k\Omega \text{ in rectangular form} \\
&= \mathbf{1.11\ k\Omega \angle 89.5°} \text{ in polar form}
\end{aligned}$$

Use Ohm's Law for current to find the total current, $\boldsymbol{I_T}$.

$$\begin{aligned}
\boldsymbol{I_T} &= \boldsymbol{V_S} / \boldsymbol{Z_T} \\
&= (2.5\ V_{ac} \angle 0°) / (1.11\ k\Omega \angle 89.5°) \\
&= (2.5\ V_{ac} / 1.11\ k\Omega) \angle (0° - 89.5°) \\
&= \mathbf{2.25\ mA_{ac} \angle{-89.5°}}
\end{aligned}$$

The total impedance $\boldsymbol{Z_T}$ consists of $\boldsymbol{R}$, $\boldsymbol{X_{L1}}$, $\boldsymbol{X_{C1}}$, and $\boldsymbol{Z_{EQ1}}$ in series, so $\boldsymbol{I_R} = \boldsymbol{I_{L1}} = \boldsymbol{I_{C1}} = \boldsymbol{I_{ZEQ1}} = \boldsymbol{I_T} =$ **2.25 mA**$_{\mathbf{ac}}$ $\mathbf{\angle{-89.5°}}$. Use Ohm's Law for voltage to find the voltage across each series component.

$$\begin{aligned}
\boldsymbol{V_R} &= \boldsymbol{I_R}\ \boldsymbol{R} \\
&= \boldsymbol{I_T}\ \boldsymbol{R} \\
&= (2.25\ mA_{ac} \angle{-89.5°})(10\ \Omega \angle 0°) \\
&= (2.25\ mA_{ac} \times 10\ \Omega) \angle (-89.5° + 0°) \\
&= \mathbf{22.5\ mV_{ac} \angle{-89.5°}}
\end{aligned}$$

$$\begin{aligned}
\boldsymbol{V_{L1}} &= \boldsymbol{I_{L1}}\ \boldsymbol{X_{L1}} \\
&= \boldsymbol{I_T}\ \boldsymbol{X_{L1}} \\
&= (2.25\ mA_{ac} \angle{-89.5°})(189\ \Omega \angle 90°) \\
&= (2.25\ mA_{ac} \times 189\ \Omega) \angle (-89.5° + 0°) \\
&= \mathbf{423\ mV_{ac} \angle 0.51°}
\end{aligned}$$

$$\begin{aligned}
\boldsymbol{V_{C1}} &= \boldsymbol{I_{C1}}\ \boldsymbol{X_{C1}} \\
&= \boldsymbol{I_T}\ \boldsymbol{X_{C1}} \\
&= (2.25\ mA_{ac} \angle{-89.5°})(295\ \Omega \angle{-90°}) \\
&= (2.25\ mA_{ac} \times 295\ \Omega) \angle [-89.5° + (-90°)] \\
&= \mathbf{662\ mV_{ac} \angle{-179.5°}}
\end{aligned}$$

$$\begin{aligned}
\boldsymbol{V_{ZEQ1}} &= \boldsymbol{I_{ZEQ1}}\ \boldsymbol{Z_{EQ1}} \\
&= \boldsymbol{I_T}\ \boldsymbol{Z_{EQ1}} \\
&= (2.25\ mA_{ac} \angle{-89.5°})(1.22\ k\Omega \angle 90°) \\
&= (2.25\ mA_{ac} \times 1.22\ k\Omega) \angle (-89.5° + 90°) \\
&= 2.74\ V_{ac} \angle 0.51°
\end{aligned}$$

$\boldsymbol{Z_{EQ1}}$ is $\boldsymbol{X_{L2}}$ and $\boldsymbol{X_{C2}}$ in parallel, so $\boldsymbol{V_{L2}} = \boldsymbol{V_{C2}} = \boldsymbol{V_{ZEQ1}} =$ **2.74 V**$_{\mathbf{ac}}$ $\mathbf{\angle 0.51°}$. Use Ohm's Law to find the current through each component.

$$\begin{aligned}
\boldsymbol{I_{L2}} &= \boldsymbol{V_{L2}} / \boldsymbol{X_{L2}} \\
&= \boldsymbol{V_{ZEQ1}} / \boldsymbol{X_{L2}} \\
&= (2.74\ V_{ac} \angle 0.51°) / (540\ \Omega \angle 90°) \\
&= (2.74\ V_{ac} / 540\ \Omega) \angle (0.51° - 90°) \\
&= \mathbf{5.07\ mA_{ac} \angle{-89.5°}}
\end{aligned}$$

$$\boldsymbol{I_{C2}} = \boldsymbol{V_{C2}} / \boldsymbol{X_{C2}}$$

$= \boldsymbol{V_{ZEQ1}} / \boldsymbol{X_{C2}}$

$= (2.74\ \text{V}_{ac} \angle 0.51°) / (971\ \Omega \angle -90°)$

$= (2.74\ \text{V}_{ac} / 971\ \Omega) \angle [0.51° - (-90°)]$

$= \mathbf{2.82\ mA_{ac} \angle 90.5°}$

2. Problem: Refer to the circuit in Figure 18-9.

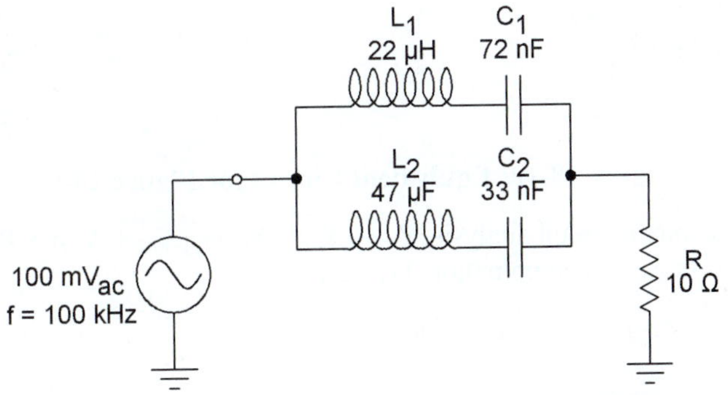

Figure 18-9: Series-Parallel *RLC* Circuit for Problem 2

Calculate the total impedance, $\boldsymbol{Z_T}$, total current, $\boldsymbol{I_T}$, and currents and voltages for each component. Express all answers in polar form.

Solution: First, calculate the inductive and capacitive reactances for the circuit. Then, use standard circuit analysis techniques for series-parallel circuits using complex arithmetic.

Calculate $\boldsymbol{X_{L1}}$ for $\boldsymbol{L}_1 = 22\ \mu\text{H}$ and $\boldsymbol{f} = 100\ \text{kHz}$.

$\boldsymbol{X_{L1}} = 2\pi \boldsymbol{f} \boldsymbol{L}_1$

$= 2\pi(100\ \text{kHz})\ (22\ \mu\text{H})$

$= 2\pi(100 \times 10^3\ \text{Hz})\ (22 \times 10^{-6}\ \text{H})$

$= 13.8\ \Omega$

Calculate $\boldsymbol{X_{C1}}$ for $\boldsymbol{C}_1 = 72\ \text{nF}$ and $\boldsymbol{f} = 100\ \text{kHz}$.

$\boldsymbol{X_{C1}} = 1 / (2\pi \boldsymbol{f} \boldsymbol{C}_1)$

$= 1 / [2\pi(100\ \text{kHz})\ (72\ \text{nF})]$

$= 1 / [2\pi(100 \times 10^3\ \text{Hz})(72 \times 10^{-9}\ \text{F})]$

$= 1 / 45.2\ \text{mS}$

$= 22.1\ \Omega$

Calculate $\boldsymbol{X_{L2}}$ for $\boldsymbol{L}_2 = 47\ \mu\text{H}$ and $\boldsymbol{f} = 100\ \text{kHz}$.

$\boldsymbol{X_{L2}} = 2\pi \boldsymbol{f} \boldsymbol{L}_2$

$= 2\pi(100\ \text{kHz})\ (47\ \text{mH})$

$= 2\pi(100 \times 10^3\ \text{Hz})\ (47 \times 10^{-6}\ \text{H})$

$= 29.5\ \Omega$

Calculate $\boldsymbol{X_{C2}}$ for $\boldsymbol{C}_2 = 33\ \text{nF}$ and $\boldsymbol{f} = 100\ \text{kHz}$.

$\boldsymbol{X_{C2}} = 1 / (2\pi \boldsymbol{f} \boldsymbol{C}_2)$

$= 1 / [2\pi(100\ \text{kHz})\ (33\ \text{nF})]$

$= 1 / [2\pi(100 \times 10^3\ \text{Hz})(33 \times 10^{-9}\ \text{F})]$

$= 1 / 20.7\ \text{mS}$

$= 48.2\ \Omega$

Figure 18-10 shows the equivalent circuit for Figure 18-9.

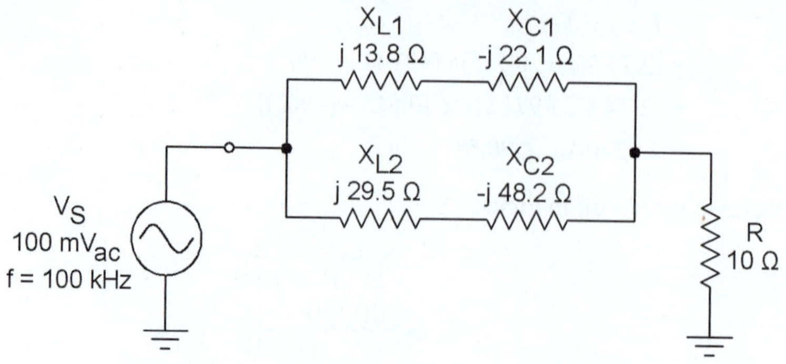

Figure 18-10: Equivalent Circuit for Figure 18-9

From standard circuit analysis, $\boldsymbol{Z_T} = (\boldsymbol{X_{L1}} + \boldsymbol{X_{C1}}) \parallel (\boldsymbol{X_{L2}} + \boldsymbol{X_{C2}}) + \boldsymbol{R}$. The first equivalent impedance, $\boldsymbol{Z_{EQ1}}$, is the series combination $\boldsymbol{X_{L1}} + \boldsymbol{X_{C1}}$.

$$\boldsymbol{Z_T} = \boldsymbol{Z_{EQ1}} \parallel (\boldsymbol{X_{L2}} + \boldsymbol{X_{C2}}) + \boldsymbol{R}$$

where

$$\begin{aligned}
\boldsymbol{Z_{EQ1}} &= \boldsymbol{X_{L1}} + \boldsymbol{X_{C1}} \\
&= (\text{j } 13.8\ \Omega) + (-\text{j } 22.1\ \Omega) \\
&= \text{j } (13.8\ \Omega - 22.1\ \Omega) \\
&= \text{j } (-8.28\ \Omega) \\
&= -\text{j } 8.28\ \Omega \text{ in rectangular form} \\
&= 8.28\ \Omega \angle -90^\circ \text{ in polar form}
\end{aligned}$$

The second equivalent impedance, $\boldsymbol{Z_{EQ2}}$, is the series combination $\boldsymbol{X_{L2}} + \boldsymbol{X_{C2}}$.

$$\boldsymbol{Z_T} = \boldsymbol{Z_{EQ1}} \parallel \boldsymbol{Z_{EQ2}} + \boldsymbol{R}$$

where

$$\begin{aligned}
\boldsymbol{Z_{EQ2}} &= \boldsymbol{X_{L2}} + \boldsymbol{X_{C2}} \\
&= (\text{j } 29.5\ \Omega) + (-\text{j } 48.2\ \Omega) \\
&= \text{j } (29.5\ \Omega - 48.2\ \Omega) \\
&= \text{j } (-18.7\ \Omega) \\
&= -\text{j } 18.7\ \Omega \text{ in rectangular form} \\
&= 18.7\ \Omega \angle -90^\circ \text{ in polar form}
\end{aligned}$$

The third equivalent impedance, $\boldsymbol{Z_{EQ3}}$, is the parallel combination $\boldsymbol{Z_{EQ1}} \parallel \boldsymbol{Z_{EQ2}}$.

$$\boldsymbol{Z_T} = \boldsymbol{Z_{EQ3}} + \boldsymbol{R}$$

where

$$\begin{aligned}
\boldsymbol{Z_{EQ3}} &= \boldsymbol{Z_{EQ1}} \parallel \boldsymbol{Z_{EQ2}} \\
&= (-\text{j } 8.28\ \Omega) \parallel (-\text{j } 18.7\ \Omega) \\
&= 1 / \{[1 / (-\text{j } 8.28\ \Omega)] + [1 / (-\text{j } 18.7\ \Omega)]\} \\
&= 1 / \{[1 / (8.28\ \Omega \angle -90^\circ)] + [1 / (18.7\ \Omega \angle -90^\circ)]\} \\
&= 1 / \{[(1 \angle 0^\circ) / (8.28\ \Omega \angle -90^\circ)] + [(1 \angle 0^\circ) / (18.7\ \Omega \angle -90^\circ)]\} \\
&= 1 / (\{(1 / 8.28\ \Omega) \angle [0^\circ - (-90^\circ)]\} + \{(1 / 18.7\ \Omega) \angle [0^\circ - (-90^\circ)]\}) \\
&= 1 / [(121\text{ mS} \angle 90^\circ) + (53.5\text{ mS} \angle 90^\circ)] \\
&= 1 / [(\text{j } 121\text{ mS}) + (\text{j } 53.5\text{ mS})] \\
&= 1 / [\text{j } (121\text{ mS} + 53.5\text{ mS})]
\end{aligned}$$

$= 1 / (j\ 174\ mS)$

$= 1 / (174\ mS \angle 90°)$

$= (1 \angle 0°) / (174\ mS \angle 90°)$

$= (1 / 174\ mS) \angle (0° - 90°)$

$= 5.74\ \Omega \angle -90°$ in polar form

$= -j\ 5.74\ \Omega$ in rectangular form

The fourth (and final) equivalent impedance, $\boldsymbol{Z_{EQ4}}$, is the series combination $\boldsymbol{Z_{EQ3}} + \boldsymbol{R}$.

$\boldsymbol{Z_T} = \boldsymbol{Z_{EQ4}}$

where

$\boldsymbol{Z_{EQ4}} = \boldsymbol{Z_{EQ3}} + \boldsymbol{R}$

$= (-j\ 5.74\ \Omega) + (10\ \Omega)$

$= 10\ \Omega + (-j\ 5.74\ \Omega)$

$= 10\ \Omega - j\ 5.74\ \Omega$ in rectangular form

$= \mathbf{11.5\ \Omega \angle -29.9°}$ in polar form

Use Ohm's Law for current to find the total current, $\boldsymbol{I_T}$.

$\boldsymbol{I_T} = \boldsymbol{V_S} / \boldsymbol{Z_T}$

$= (100\ mV_{ac} \angle 0°) / (11.5\ \Omega \angle -29.9°)$

$= (100\ mV_{ac} / 11.5\ \Omega) \angle [0° -(-29.9°)]$

$= \mathbf{8.67\ mA_{ac} \angle 29.9°}$

The total impedance $\boldsymbol{Z_T}$ consists of $\boldsymbol{R}$ and $\boldsymbol{Z_{EQ3}}$ in series, so $\boldsymbol{I_R} = \boldsymbol{I_{ZEQ3}} = \boldsymbol{I_T} = \mathbf{8.67\ mA_{ac} \angle 29.9°}$. Use Ohm's Law for voltage to find the voltage across each series component.

$\boldsymbol{V_R} = \boldsymbol{I_R}\,\boldsymbol{R}$

$= \boldsymbol{I_T}\,\boldsymbol{R}$

$= (8.67\ mA_{ac} \angle 29.9°)(10\ \Omega \angle 0°)$

$= (8.67\ mA_{ac} \times 10\ \Omega) \angle (29.9° + 0°)$

$= \mathbf{86.7\ mV_{ac} \angle 29.9°}$

$\boldsymbol{V_{ZEQ3}} = \boldsymbol{I_{ZEQ3}}\,\boldsymbol{Z_{EQ3}}$

$= \boldsymbol{I_T}\,\boldsymbol{Z_{EQ3}}$

$= (8.67\ mA_{ac} \angle 29.9°)(5.74\ \Omega \angle -90°)$

$= (8.67\ mA_{ac} \times 5.74\ \Omega) \angle [29.9° - (-90°)]$

$= 49.8\ mV_{ac} \angle -60.2°$

$\boldsymbol{Z_{EQ3}}$ is $\boldsymbol{Z_{EQ1}}$ and $\boldsymbol{Z_{EQ2}}$ in parallel, so $\boldsymbol{V_{ZEQ1}} = \boldsymbol{V_{ZEQ2}} = \boldsymbol{V_{ZEQ3}} = 49.8\ mV_{ac} \angle -60.2°$. Use Ohm's Law for current to find the current through each equivalent impedance.

$\boldsymbol{I_{ZEQ1}} = \boldsymbol{V_{ZEQ1}} / \boldsymbol{Z_{EQ1}}$

$= \boldsymbol{V_{ZEQ3}} / \boldsymbol{Z_{EQ1}}$

$= (49.8\ mV_{ac} \angle -60.2°) / (8.28\ \Omega \angle -90°)$

$= (49.8\ mV_{ac} / 8.28\ \Omega) \angle [(-60.2°) - (-90°)]$

$= 6.01\ mA_{ac} \angle 29.9°$

$\boldsymbol{I_{ZEQ2}} = \boldsymbol{V_{ZEQ2}} / \boldsymbol{Z_{EQ2}}$

$= \boldsymbol{V_{ZEQ3}} / \boldsymbol{Z_{EQ2}}$

$= (49.8\ mV_{ac} \angle -60.2°) / (18.7\ \Omega \angle -90°)$

$= (49.8\ mV_{ac} / 18.7\ \Omega) \angle [(-60.2°) - (-90°)]$

$= 2.66\ mA_{ac} \angle 29.9°$

Z_{EQ1} is X_{L1} and X_{C1} in series, so $I_{L1} = I_{C1} = I_{ZEQ1} = \mathbf{6.01\ mA_{ac} \angle 29.9°}$. Use Ohm's Law to find the voltage across each series component.

$$\begin{aligned} V_{L1} &= I_{L1}X_{L1} \\ &= (6.01\ \text{mA}_{ac} \angle 29.9°)(13.8\ \Omega \angle 90°) \\ &= [(6.01\ \text{mA}_{ac})(13.8\ \Omega)] \angle (29.9° + 90°) \\ &= \mathbf{83.1\ mV_{ac} \angle 120°} \end{aligned}$$

$$\begin{aligned} V_{C1} &= I_{C1}X_{C1} \\ &= (6.01\ \text{mA}_{ac} \angle 29.9°)(22.1\ \Omega \angle -90°) \\ &= [(6.01\ \text{mA}_{ac})(22.1\ \Omega)] \angle [29.9° + (-90°)] \\ &= \mathbf{133\ mV_{ac} \angle -60.2°} \end{aligned}$$

Z_{EQ2} is X_{L2} and X_{C2} in series, so $I_{L2} = I_{C2} = I_{ZEQ2} = \mathbf{2.66\ mA_{ac} \angle 29.9°}$. Use Ohm's Law to find the voltage across each series component.

$$\begin{aligned} V_{L2} &= I_{L2}X_{L2} \\ &= (2.66\ \text{mA}_{ac} \angle 29.9°)(29.5\ \Omega \angle 90°) \\ &= [(2.66\ \text{mA}_{ac})(29.5\ \Omega)] \angle (29.9° + 90°) \\ &= \mathbf{78.6\ mV_{ac} \angle 120°} \end{aligned}$$

$$\begin{aligned} V_{C2} &= I_{C2}X_{C2} \\ &= (2.66\ \text{mA}_{ac} \angle 29.9°)(48.2\ \Omega \angle -90°) \\ &= [(2.66\ \text{mA}_{ac})(48.2\ \Omega)] \angle [29.9° + (-90°)] \\ &= \mathbf{128\ mV_{ac} \angle -60.2°} \end{aligned}$$

18.2 Advanced

18.2.1 Series *RLC* Circuits

1. Problem: Although ideal capacitors block dc, practical capacitors will pass a small dc leakage current. The model for a practical capacitor is an ideal capacitor in parallel with a leakage resistance, R_{LEAK}. For the series ***RLC*** circuit in Figure 18-11, calculate the ideal resonant frequency, f_r, from ***L*** and ***C*** and the total impedance, Z_T, for the circuit at this frequency. From the phase angle of the impedance, is the resonant frequency for a series ***RLC*** circuit above or below the ideal value?

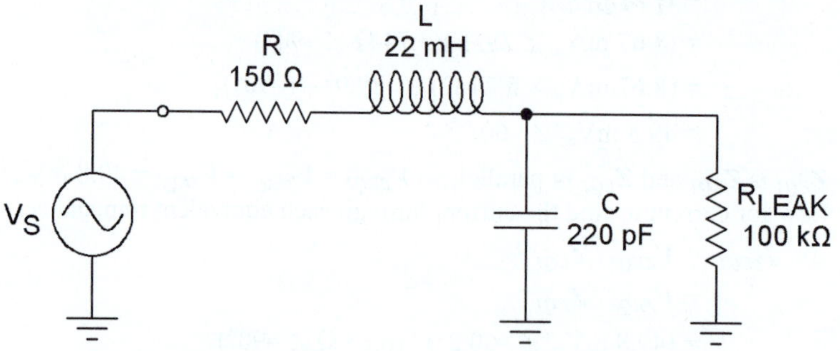

Figure 18-11: Non-ideal Series *RLC* Circuit for Problem 1

Solution: First, calculate the ideal resonant frequency, f_r, from the inductance, ***L***, and capacitance, ***C***.

$$\begin{aligned} f_r &= 1 / (2\pi\sqrt{LC}) \\ &= 1 / [2\pi\sqrt{(22\ \text{mH})(220\ \text{pF})}] \end{aligned}$$

$= 1 / [\, 2\pi\sqrt{(22\times10^{-3}\text{ H})(220\times10^{-12}\text{ F})}\,]$

$= 1 / [\, 2\pi\sqrt{4.82\times10^{-12}}\,]$

$= 1 / [2\pi(2.20 \times 10^{-6})]$

$= 1 / 13.8\ \mu s$

$= \mathbf{72.3\ kHz}$

Next, determine the values of inductive and capacitive reactances. Calculate X_L for $L = 22$ mH and $f = 72.3$ kHz.

$X_L = 2\pi fL$

$= 2\pi(72.3\text{ kHz})\,(22\text{ mH})$

$= 2\pi(72.3 \times 10^3\text{ Hz})\,(22 \times 10^{-3}\text{ H})$

$= 10.0\text{ k}\Omega$

Calculate X_C for $C = 220$ pF and $f = 72.3$ kHz.

$X_C = 1 / (2\pi fC)$

$= 1 / [2\pi(72.3\text{ kHz})\,(220\text{ pF})]$

$= 1 / [2\pi(72.3 \times 10^3\text{ Hz})(220 \times 10^{-12}\text{ F})]$

$= 1 / 100\ \mu S$

$= 10.0\text{ k}\Omega$

Figure 18-12 shows the equivalent circuit for Figure 18-11.

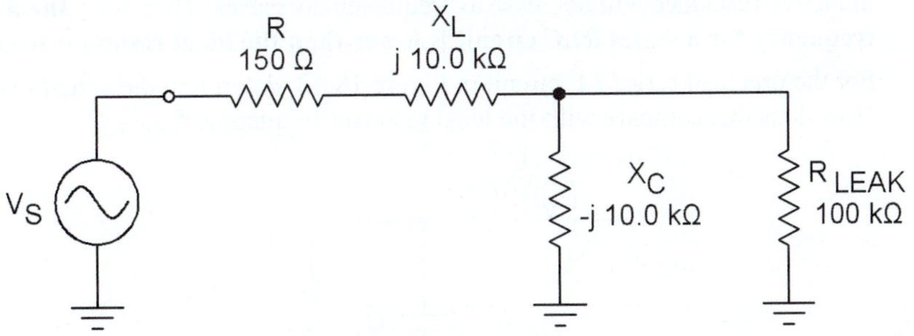

Figure 18-12: Equivalent Circuit for Figure 18-11

From standard circuit analysis, $Z_T = R + X_L + X_C \,\|\, R_{LEAK}$. The first equivalent impedance, Z_{EQ1}, is the parallel combination $X_C \,\|\, R_{LEAK}$. From this,

$Z_T = R + X_L + Z_{EQ1}$

where

$Z_{EQ1} = -j\,X_C \,\|\, R_{LEAK}$

$= (-j\ 10.0\text{ k}\Omega) \,\|\, (100\text{ k}\Omega)$

$= 1 / \{[1 / (-j\ 10.0\text{ k}\Omega)] + [1 / (100\text{ k}\Omega)]\}$

$= 1 / \{[1 / (10.0\text{ k}\Omega \angle -90°)] + [1 / (100\text{ k}\Omega \angle 0°)]\}$

$= 1 / \{[(1 \angle 0°) / (10.0\text{ k}\Omega \angle -90°)] + [(1 \angle 0°) / (100\text{ k}\Omega \angle 0°)]\}$

$= 1 / (\{(1 / 10.0\text{ k}\Omega) \angle [0° - (-90°)]\} + \{(1 / 100\text{ k}\Omega) \angle [0° - (-90°)]\})$

$= 1 / (\{(1 / 10.0\text{ k}\Omega) \angle [0° - (-90°)]\} + \{(1 / 100\text{ k}\Omega) \angle [0° - (-90°)]\})$

$= 1 / [(100\ \mu S \angle 90°) + (10\ \mu S \angle 0°)]$

$= 1 / [(0\text{ S} + j\ 100\ \mu S) + (10\ \mu S + j\ 0\text{ S})]$

$= 1 / [(0\text{ S} + 10\ \mu S) + j\,(100\ \mu S + 0\text{ S})]$

$= 1 / (10\ \mu S + j\ 100\ \mu S)$

$= 1 / (101\ \mu S \angle 84.3°)$

$= (1 \angle 0°) / (101\ \mu S \angle 84.3°)$

$= (1 / 101\ \mu S) \angle (0° - 84.3°)$

$= 9.95\ k\Omega \angle -84.3°$ in polar form

$= 990\ \Omega - j\ 9.90\ k\Omega$ in rectangular form

The second (and last) equivalent impedance, $\boldsymbol{Z_{EQ2}}$, is the series combination $\boldsymbol{R + X_L + Z_{EQ1}}$. From this,

$$Z_T = Z_{EQ2}$$

where

$$\begin{aligned} Z_{EQ2} &= R + X_L + Z_{EQ1} \\ &= 150\ \Omega + j\ 10.0\ k\Omega + (990\ \Omega - j\ 9.90\ k\Omega) \\ &= (150\ \Omega + 990\ \Omega) + (j\ 10.0\ k\Omega - j\ 9.90\ k\Omega) \\ &= 1.14\ k\Omega + j\ (10.0\ k\Omega - 9.90\ k\Omega) \\ &= 1.14\ k\Omega + j\ 99.0\ \Omega \text{ in rectangular form} \\ &= \mathbf{1.14\ k\Omega \angle 4.96°} \text{ in polar form} \end{aligned}$$

When an ideal ***RLC*** circuit is at resonance, the circuit appears purely resistive so that the phase angle should be 0°. The phase angle of the total impedance is positive, so the circuit appears inductive. To decrease the phase angle for the series ***RLC*** circuit, the capacitive reactance must increase and the inductive reactance must decrease. Capacitive reactance will increase and inductive reactance will decrease as frequency decreases. Therefore, **the actual resonant frequency for a series *RLC* circuit is lower than the ideal resonant frequency**.

2. Problem: For the practical series ***LC*** circuit in Figure 18-13, determine the actual resonant frequency, $\boldsymbol{f_r}$. How does this compare with the ideal resonant frequency, $\boldsymbol{f_{r(\text{IDEAL})}}$?

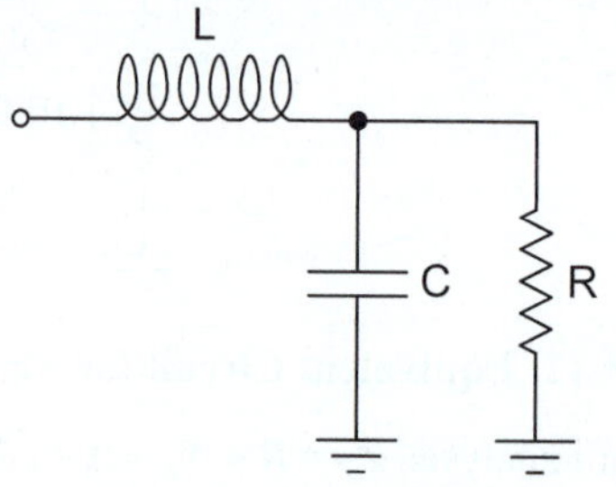

Figure 18-13: Practical Series *RLC* Circuit for Problem 2

Solution: Figure 18-13 shows the equivalent circuit for the ***RLC*** circuit.

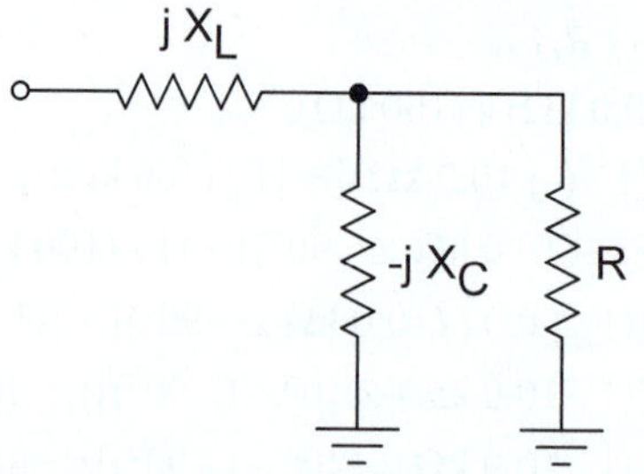

Figure 18-14: Equivalent Circuit for Figure 18-13

From standard circuit analysis, $Z_T = X_L + X_C \parallel R$. The first equivalent impedance, Z_{EQ1}, is the parallel combination $X_C \parallel R$. From this,

$$Z_T = X_L + Z_{EQ1}$$

where

$$\begin{aligned}
Z_{EQ1} &= -\mathrm{j}\,X_C \parallel R \\
&= [(-\mathrm{j}\,X_C)(R)] / [R + (-\mathrm{j}\,X_C)] \text{ using product-over-sum for parallel components} \\
&= (-\mathrm{j}\,RX_C) / (R - \mathrm{j}\,X_C) \\
&= [(-\mathrm{j}\,RX_C) / (R - \mathrm{j}\,X_C)]\,[(R + \mathrm{j}\,X_C) / (R + \mathrm{j}\,X_C)] \\
&= [(-\mathrm{j}\,RX_C)(R + \mathrm{j}\,X_C)] / [(R - \mathrm{j}\,X_C)(R + \mathrm{j}\,X_C)] \\
&= [(-\mathrm{j}\,RX_C)(R) + (-\mathrm{j}\,RX_C)(\mathrm{j}\,X_C)] / [(RR + (R)(\mathrm{j}\,X_C) + (-\mathrm{j}\,X_C)(R) + (-\mathrm{j}\,X_C)(\mathrm{j}\,X_C)] \\
&= [(-\mathrm{j}\,R^2X_C) + (-\mathrm{j}^2)(RX_C^2)] / [(R^2 + (\mathrm{j}\,RX_C) + (-\mathrm{j}\,RX_C) + (-\mathrm{j}^2)(X_C^2)] \\
&= [(-\mathrm{j}\,R^2X_C) + -(-1)(RX_C^2)] / [(R^2 + \mathrm{j}\,(RX_C - RX_C) + -(-1)(X_C^2)] \\
&= [(-\mathrm{j}\,R^2X_C) + (RX_C^2)] / [(R^2 + \mathrm{j}\,(0) + (X_C^2)] \\
&= (RX_C^2 - \mathrm{j}\,R^2X_C) / (R^2 + X_C^2)
\end{aligned}$$

The second (and last) equivalent impedance, Z_{EQ2}, is the series combination $X_L + Z_{EQ1}$. From this,

$$Z_T = Z_{EQ2}$$

where

$$\begin{aligned}
Z_{EQ2} &= \mathrm{j}\,X_L + [(RX_C^2 - \mathrm{j}\,R^2X_C) / (R^2 + X_C^2)] \\
&= (\mathrm{j}\,X_L)[(R^2 + X_C^2) / (R^2 + X_C^2)] + [(RX_C^2 - \mathrm{j}\,R^2X_C) / (R^2 + X_C^2)] \\
&= \{[(\mathrm{j}\,X_L)(R^2) + (\mathrm{j}\,X_L)(X_C^2)] / (R^2 + X_C^2)\} + [(RX_C^2 - \mathrm{j}\,R^2X_C) / (R^2 + X_C^2)] \\
&= \{[(\mathrm{j}\,X_LR^2) + (\mathrm{j}\,X_LX_C^2)] / (R^2 + X_C^2)\} + [(RX_C^2 - \mathrm{j}\,R^2X_C) / (R^2 + X_C^2)] \\
&= \{[\mathrm{j}\,(R^2X_L + X_LX_C^2)] / (R^2 + X_C^2)\} + [(RX_C^2 - \mathrm{j}\,R^2X_C) / (R^2 + X_C^2)] \\
&= \{[\mathrm{j}\,(R^2X_L + X_LX_C^2)] + (RX_C^2 - \mathrm{j}\,R^2X_C)\} / (R^2 + X_C^2) \\
&= \{RX_C^2 + [\mathrm{j}\,(R^2X_L + X_LX_C^2) + (-\mathrm{j}\,R^2X_C)]\} / (R^2 + X_C^2) \\
&= [RX_C^2 + \mathrm{j}\,(R^2X_L + X_LX_C^2 - R^2X_C)] / (R^2 + X_C^2)
\end{aligned}$$

At the resonant frequency the circuit appears purely resistive, so that the imaginary term is 0.

$$\begin{aligned}
R^2X_L + X_LX_C^2 - R^2X_C &= 0 \\
R^2X_L - R^2X_C + X_LX_C^2 &= 0 \\
R^2\,(X_L - X_C) + X_LX_C^2 &= 0
\end{aligned}$$

Use the equations for inductive and capacitive reactances to solve for the resonant frequency.

$$\begin{aligned}
R^2\,\{2\pi f_rL - [1 / (2\pi f_rC)]\} + (2\pi f_rL)[1 / (2\pi f_rC)]^2 &= 0 \\
R^2\,\{(2\pi f_rL)[(2\pi f_rC) / (2\pi f_rC)] - [1 / (2\pi f_rC)]\} + [(2\pi f_rL) / (2\pi f_rC)^2] &= 0 \\
R^2\,\{[(4\pi^2 f_r^2LC) / (2\pi f_rC)] - [1 / (2\pi f_rC)]\} + \{[(2\pi f_rL) / (2\pi f_rC)](1 / 2\pi f_rC)\} &= 0 \\
R^2\,[(4\pi^2 f_r^2LC - 1) / (2\pi f_rC)] + (L / C)](1 / 2\pi f_rC) &= 0 \\
R^2\,[(4\pi^2 f_r^2LC - 1) / (2\pi f_rC)] + [(L / C) / 2\pi f_rC] &= 0 \\
[R^2\,(4\pi^2 f_r^2LC - 1) + (L / C)] / (2\pi f_rC))(2\pi f_rC) &= (0)(2\pi f_rC) \\
R^2\,(4\pi^2 f_r^2LC - 1) + (L / C) &= 0
\end{aligned}$$

Isolate and solve for f_r.

$$\begin{aligned}
R^2\,(4\pi^2 f_r^2LC - 1) + (L / C) - (L / C) &= 0 - (L / C) \\
[R^2\,(4\pi^2 f_r^2LC - 1)] / R^2 &= (-L / C) / R^2 \\
(4\pi^2 f_r^2LC - 1) + 1 &= [-L / (R^2C)] + 1 \\
4\pi^2 f_r^2LC &= 1 - [L / (R^2C)] \\
&= (1)\,[(R^2C) / (R^2C)] - [L / (R^2C)]
\end{aligned}$$

$$= [(R^2C) / (R^2C)] - [L / (R^2C)]$$
$$= (R^2C - L) / (R^2C)$$
$$= (R^2C / R^2C) - [L / (R^2C)]$$
$$= 1 - [L / (R^2C)]$$
$$(4\pi^2 f_r^2 LC) / (LC) = \{1 - [L / (R^2C)]\} / (LC)$$
$$(4\pi^2 f_r^2) / (4\pi^2) = (\{1 - [L / (R^2C)]\} / (LC)) / (4\pi^2)$$
$$f_r^2 = \{1 - [L / (R^2C)]\} / (4\pi^2 LC)$$
$$= \{1 - [L / (R^2C)]\} [1 / (4\pi^2 LC)]$$
$$\sqrt{f_r^2} = \sqrt{1 - [L / (R^2C)][1/(4\pi^2 LC)]}$$
$$f_r = \sqrt{1 - [L / (R^2C)]}\sqrt{[1/(4\pi^2 LC)]}$$
$$= \sqrt{1 - [L / (R^2C)]}\,[1 / (2\pi\sqrt{LC})]$$

The ideal resonant frequency, $f_{r(\mathrm{IDEAL})}$, is

$$f_{r(\mathrm{IDEAL})} = 1 / (2\pi\sqrt{LC})$$

so

$$f_r = \sqrt{1 - [L / (R^2C)]} \times f_{r(\mathrm{IDEAL})}$$

Note that when R is infinite, as in the case of an ideal capacitor, (L / R^2C) is 0. Consequently, f_r and $f_{r(\mathrm{IDEAL})}$ are the same value for an ideal capacitor.

18.2.2 Parallel *RLC* Circuits

1. Problem: Although ideal inductors are shorts to dc, practical inductors have resistance. The model for a practical inductor is an ideal inductor in series with a winding resistance, R_W. For the parallel RLC circuit in Figure 18-15, calculate the ideal resonant frequency, f_r, from L and C and the total impedance, Z_T, for the circuit at this frequency. From the phase angle of the impedance, is the resonant frequency for a parallel RLC circuit above or below the ideal value?

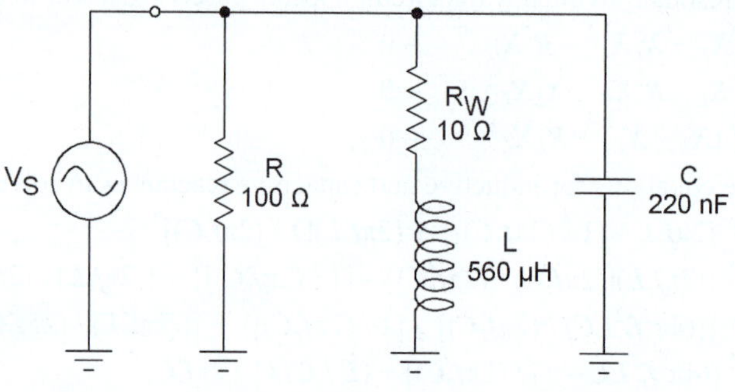

Figure 18-15: Non-Ideal Parallel *RLC* Circuit for Problem 1

Solution: First, calculate the ideal resonant frequency, f_r, from the inductance, L, and capacitance, C.

$$f_r = 1 / (2\pi\sqrt{LC})$$
$$= 1 / [2\pi\sqrt{(560\,\mu\mathrm{H})(220\,\mathrm{nF})}]$$
$$= 1 / [2\pi\sqrt{(560 \times 10^{-6}\,\mathrm{H})(220 \times 10^{-9}\,\mathrm{F})}]$$
$$= 1 / [2\pi\sqrt{123 \times 10^{-12}}]$$

$= 1 / [2\pi(11.1 \times 10^{-6})]$

$= 1 / 69.7\ \mu S$

$= \mathbf{14.3\ kHz}$

Next, determine the values of inductive and capacitive reactances. Calculate X_L for $L = 560\ \mu H$ and $f = 14.3$ kHz.

$X_L = 2\pi f L$

$= 2\pi(14.3\ kHz)\ (560\ \mu H)$

$= 2\pi(14.3 \times 10^3\ Hz)\ (560 \times 10^{-6}\ H)$

$= 50.5\ \Omega$

Calculate X_C for $C = 220$ nF and $f = 14.3$ kHz.

$X_C = 1 / (2\pi f C)$

$= 1 / [2\pi(14.3\ kHz)\ (220\ nF)]$

$= 1 / [2\pi(14.3 \times 10^3\ Hz)(220 \times 10^{-9}\ F)]$

$= 1 / 19.8\ mS$

$= 50.5\ \Omega$

Figure 18-16 shows the equivalent circuit for Figure 18-15.

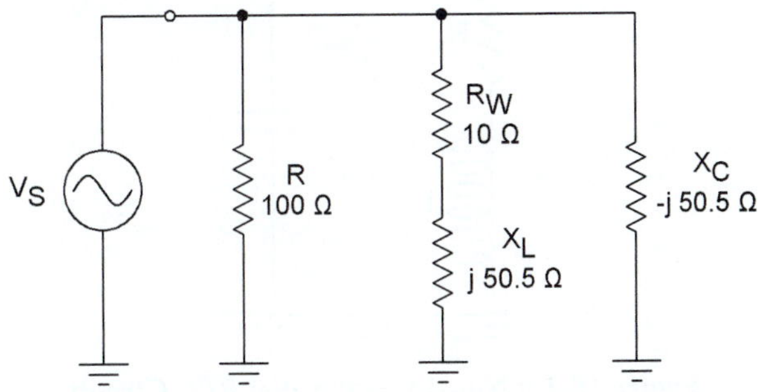

Figure 18-16: Equivalent Circuit for Figure 18-15

From standard circuit analysis, $Z_T = R \parallel (R_W + X_L) \parallel X_C$. The first equivalent impedance, Z_{EQ1}, is the parallel combination $R_W + X_L$.

$Z_T = R \parallel Z_{EQ1} \parallel X_C$

where

$Z_{EQ1} = R_W + j\,X_L$

$= 10\ \Omega + j\ 50.5\ \Omega$ in rectangular form

$= 51.4\ \Omega \angle 78.8°$ in polar form

The second (and last) equivalent impedance, Z_{EQ2}, is the parallel combination $R \parallel Z_{EQ1} \parallel X_C$.

$Z_T = Z_{EQ2}$

where

$Z_{EQ2} = R \parallel Z_{EQ1} \parallel X_C$

$= 1 / \{(1/ 100\ \Omega) + [1 / (51.4\ \Omega \angle 78.8°)] + [1 / (-j\ 50.5\ \Omega)]\}$

$= 1 / \{[(1 \angle 0°) / (100\ \Omega \angle 0°)] + [(1 \angle 0°) / (51.4\ \Omega \angle 78.8°)] + [(1 \angle 0°) / (50.5\ \Omega \angle -90°)]\}$

$= 1 / ([(1 / 100\ \Omega) \angle (0° - 0°)] + [(1 / 51.4\ \Omega) \angle (0° - 78.8°)] + \{(1 / 50.5\ \Omega) \angle [0° - (-90°)]\})$

$= 1 / [(10.0\ mS \angle 0°) + (19.4\ mS \angle -78.8°) + (19.8\ mS \angle 90°)]$

$= 1 / [(10.0\ mS \angle 0°) + (19.4\ mS \angle -78.8°) + (19.8\ mS \angle 90°)]$

$= 1 / [(10.0 \text{ mS} + j\ 0 \text{ S}) + (3.78 \text{ mS} - j\ 19.1 \text{ mS}) + (0 \text{ S} + j\ 19.8 \text{ mS})]$

$= 1 / \{(10.0 \text{ mS} + 3.78 \text{ mS} + 0 \text{ S}) + j\ [0 \text{ S} + (-19.1 \text{ mS}) + 19.8 \text{ mS}]\}$

$= 1 / (13.8 \text{ mS} + j\ 749\ \mu\text{S})$

$= 1 / (13.8 \text{ mS} \angle 3.11°)$

$= (1 \angle 0°) / (13.8 \text{ mS} \angle 3.11°)$

$= (1 / 13.8 \text{ mS}) \angle (0° - 3.11°)$

$=$ **$72.5\ \Omega \angle -3.11°$** in polar form

When an ideal ***RLC*** circuit is at resonance, the circuit appears purely resistive so that the phase angle should be 0°. The phase angle of the total impedance is negative, so the circuit appears capacitive. To increase the phase angle for the parallel ***RLC*** circuit, the capacitive reactance must increase and the inductive reactance must decrease. Capacitive reactance will increase and inductive reactance will decrease as frequency decreases. Therefore, **the actual resonant frequency for a parallel *RLC* circuit is higher than the ideal resonant frequency**.

2. Problem: For the non-ideal parallel ***LC*** circuit in Figure 18-17, determine the actual resonant frequency, f_r. How does this compare with the ideal resonant frequency, $f_{r(\text{IDEAL})}$?

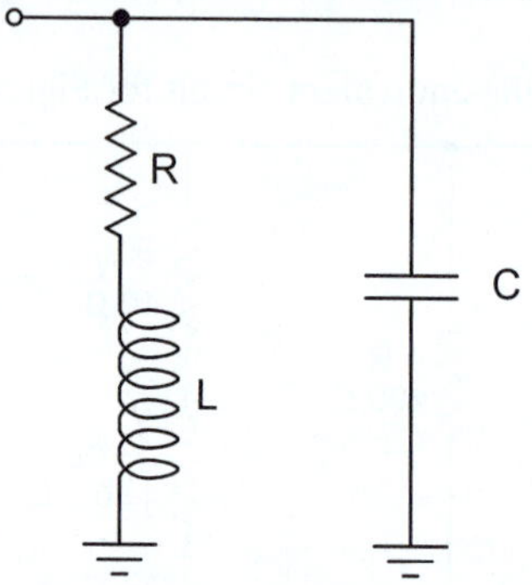

Figure 18-17: Non-Ideal Parallel *RLC* Circuit

Solution: Figure 18-18 shows the equivalent circuit for the ***RLC*** circuit.

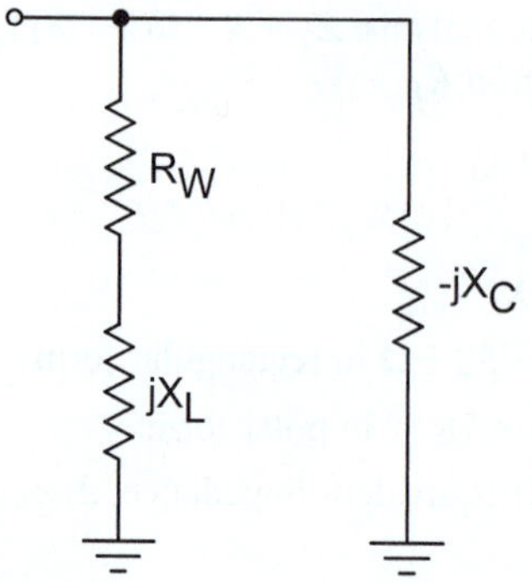

Figure 18-18: Equivalent Circuit for Figure 18-17

From standard circuit analysis, $Z_T = (R_W + X_L) \parallel X_C$. The first equivalent impedance, Z_{EQ1}, is the series combination $R_W + X_L$. From this,

$$Z_T = Z_{EQ1} + X_C$$

where

$$Z_{EQ1} = R_W + j\,X_L$$

The second (and last) equivalent impedance, $\boldsymbol{Z_{EQ2}}$, is the parallel combination $\boldsymbol{Z_{EQ1}} \parallel \boldsymbol{X_C}$. From this,

$$\boldsymbol{Z_T} = \boldsymbol{Z_{EQ2}}$$

where

$$\begin{aligned}
\boldsymbol{Z_{EQ2}} &= \boldsymbol{Z_{EQ1}} \parallel \boldsymbol{X_C} \\
&= (\boldsymbol{R_W} + \text{j}\,\boldsymbol{X_L}) \parallel (-\text{j}\,\boldsymbol{X_C}) \\
&= [(\boldsymbol{R_W} + \text{j}\,\boldsymbol{X_L})\,(-\text{j}\,\boldsymbol{X_C})] \,/\, [(\boldsymbol{R_W} + \text{j}\,\boldsymbol{X_L}) + (-\text{j}\,\boldsymbol{X_C})] \text{ from product-over sum method} \\
&= [(\boldsymbol{R_W})(-\text{j}\,\boldsymbol{X_C}) + (\text{j}\,\boldsymbol{X_L})(-\text{j}\,\boldsymbol{X_C})] \,/\, (\boldsymbol{R_W} + \text{j}\,(\boldsymbol{X_L} - \boldsymbol{X_C}) \\
&= [(-\text{j}\,\boldsymbol{R_W X_C}) + (-\text{j}^2)(\boldsymbol{X_L X_C})] \,/\, (\boldsymbol{R_W} + \text{j}\,(\boldsymbol{X_L} - \boldsymbol{X_C}) \\
&= [(-\text{j}\,\boldsymbol{R_W X_C}) + -(-1)(\boldsymbol{X_L X_C})] \,/\, (\boldsymbol{R_W} + \text{j}\,(\boldsymbol{X_L} - \boldsymbol{X_C}) \\
&= [(-\text{j}\,\boldsymbol{R_W X_C}) + (\boldsymbol{X_L X_C})] \,/\, (\boldsymbol{R_W} + \text{j}\,(\boldsymbol{X_L} - \boldsymbol{X_C}) \\
&= [(\boldsymbol{X_L X_C} - \text{j}\,\boldsymbol{R_W X_C}) \,/\, (\boldsymbol{R_W} + \text{j}\,(\boldsymbol{X_L} - \boldsymbol{X_C})]\{[\boldsymbol{R_W} - \text{j}\,(\boldsymbol{X_L} - \boldsymbol{X_C})] \,/\, [\boldsymbol{R_W} - \text{j}\,(\boldsymbol{X_L} - \boldsymbol{X_C})]\} \\
&= \{(\boldsymbol{X_L X_C} - \text{j}\,\boldsymbol{R_W X_C})\,[\boldsymbol{R_W} - \text{j}\,(\boldsymbol{X_L} - \boldsymbol{X_C})]\} \,/\, \{[\boldsymbol{R_W} + \text{j}\,(\boldsymbol{X_L} - \boldsymbol{X_C})][\boldsymbol{R_W} - \text{j}\,(\boldsymbol{X_L} - \boldsymbol{X_C})]\} \\
&= \{(\boldsymbol{X_L X_C})[\boldsymbol{R_W} - \text{j}\,(\boldsymbol{X_L} - \boldsymbol{X_C})] - (\text{j}\,\boldsymbol{R_W X_C})[\boldsymbol{R_W} - \text{j}\,(\boldsymbol{X_L} - \boldsymbol{X_C})]\} \\
&\quad / \{[\boldsymbol{R_W R_W} - \text{j}\,(\boldsymbol{R_W})(\boldsymbol{X_L} - \boldsymbol{X_C}) + \text{j}\,(\boldsymbol{R_W})(\boldsymbol{X_L} - \boldsymbol{X_C}) - \text{j}^2\,(\boldsymbol{X_L} - \boldsymbol{X_C})\,(\boldsymbol{X_L} - \boldsymbol{X_C})]\} \\
&= \{[(\boldsymbol{R_W X_L X_C}) - \text{j}\,(\boldsymbol{X_L X_C})(\boldsymbol{X_L} - \boldsymbol{X_C})] - [(\text{j}\,\boldsymbol{R_W R_W X_C}) - \text{j}^2\,(\boldsymbol{R_W X_C})(\boldsymbol{X_L} - \boldsymbol{X_C})]\} \\
&\quad / \{\boldsymbol{R_W}^2 + \text{j}\,[(\boldsymbol{R_W})(\boldsymbol{X_L} - \boldsymbol{X_C}) - (\boldsymbol{R_W})(\boldsymbol{X_L} - \boldsymbol{X_C})] - (-1)(\boldsymbol{X_L} - \boldsymbol{X_C})^2\} \\
&= \{[(\boldsymbol{R_W X_L X_C}) - \text{j}\,(\boldsymbol{X_L X_L X_C} - \boldsymbol{X_L X_C X_C})] - [(\text{j}\,\boldsymbol{R_W R_W X_C}) - (-1)(\boldsymbol{R_W X_L X_C} - \boldsymbol{R_W X_C X_C})]\} \\
&\quad / [\boldsymbol{R_W}^2 + \text{j}\,(0) + (\boldsymbol{X_L} - \boldsymbol{X_C})^2] \\
&= \{[(\boldsymbol{R_W X_L X_C}) - \text{j}\,(\boldsymbol{X_L}^2\boldsymbol{X_C} - \boldsymbol{X_L X_C}^2)] - [(\text{j}\,\boldsymbol{R_W}^2\boldsymbol{X_C}) + (\boldsymbol{R_W X_L X_C} - \boldsymbol{R_W X_C}^2)]\} \\
&\quad / [\boldsymbol{R_W}^2 + (\boldsymbol{X_L} - \boldsymbol{X_C})^2] \\
&= [(\boldsymbol{R_W X_L X_C}) + (\boldsymbol{R_W X_L X_C} - \boldsymbol{R_W X_C}^2) - \text{j}\,(\boldsymbol{X_L}^2\boldsymbol{X_C} - \boldsymbol{X_L X_C}^2) - (\text{j}\,\boldsymbol{R_W}^2\boldsymbol{X_C})] \\
&\quad / [\boldsymbol{R_W}^2 + (\boldsymbol{X_L} - \boldsymbol{X_C})^2] \\
&= [(2\boldsymbol{R_W X_L X_C} - \boldsymbol{R_W X_C}^2) - \text{j}\,(\boldsymbol{X_L}^2\boldsymbol{X_C} - \boldsymbol{X_L X_C}^2 + \boldsymbol{R_W}^2\boldsymbol{X_C})] \,/\, [\boldsymbol{R_W}^2 + (\boldsymbol{X_L} - \boldsymbol{X_C})^2] \\
&= [(2\boldsymbol{R_W X_L X_C} - \boldsymbol{R_W X_C}^2) + \text{j}\,(-\boldsymbol{X_L}^2\boldsymbol{X_C} + \boldsymbol{X_L X_C}^2 - \boldsymbol{R_W}^2\boldsymbol{X_C})] \,/\, [\boldsymbol{R_W}^2 + (\boldsymbol{X_L} - \boldsymbol{X_C})^2] \\
&= [(2\boldsymbol{R_W X_L X_C} - \boldsymbol{R_W X_C}^2) + \text{j}\,(\boldsymbol{X_L X_C}^2 - \boldsymbol{X_L}^2\boldsymbol{X_C} - \boldsymbol{R_W}^2\boldsymbol{X_C})] \,/\, [\boldsymbol{R_W}^2 + (\boldsymbol{X_L} - \boldsymbol{X_C})^2]
\end{aligned}$$

At the resonant frequency the circuit appears purely resistive so that the imaginary term is 0.

$$\begin{aligned}
\boldsymbol{X_L X_C}^2 - \boldsymbol{X_L}^2\boldsymbol{X_C} - \boldsymbol{R_W}^2\boldsymbol{X_C} &= 0 \\
[(\boldsymbol{X_L X_C} - \boldsymbol{X_L}^2 - \boldsymbol{R_W}^2)\boldsymbol{X_C}] \,/\, \boldsymbol{X_C} &= 0 \,/\, \boldsymbol{X_C} \\
\boldsymbol{X_L X_C} - \boldsymbol{X_L}^2 - \boldsymbol{R_W}^2 &= 0
\end{aligned}$$

Use the equations for inductive and capacitive reactances to solve for the resonant frequency.

$$\begin{aligned}
(2\pi \boldsymbol{f_r L})[1 \,/\, (2\pi \boldsymbol{f_r C})] - (2\pi \boldsymbol{f_r L})^2 - \boldsymbol{R_W}^2 &= 0 \\
[(2\pi \boldsymbol{f_r L}) \,/\, (2\pi \boldsymbol{f_r C})] - 4\pi^2 \boldsymbol{f_r}^2 \boldsymbol{L}^2 - \boldsymbol{R_W}^2 &= 0 \\
(\boldsymbol{L} \,/\, \boldsymbol{C}) - 4\pi^2 \boldsymbol{f_r}^2 \boldsymbol{L}^2 - \boldsymbol{R_W}^2 &= 0
\end{aligned}$$

Isolate and solve for $\boldsymbol{f_r}$.

$$\begin{aligned}
(\boldsymbol{L} \,/\, \boldsymbol{C}) - 4\pi^2 \boldsymbol{f_r}^2 \boldsymbol{L}^2 - \boldsymbol{R_W}^2 + \boldsymbol{R_W}^2 &= 0 + \boldsymbol{R_W}^2 \\
(\boldsymbol{L} \,/\, \boldsymbol{C}) - 4\pi^2 \boldsymbol{f_r}^2 \boldsymbol{L}^2 - (\boldsymbol{L} \,/\, \boldsymbol{C}) &= \boldsymbol{R_W}^2 - (\boldsymbol{L} \,/\, \boldsymbol{C}) \\
(-4\pi^2 \boldsymbol{f_r}^2 \boldsymbol{L}^2) \,/\, (-4\pi^2 \boldsymbol{L}^2) &= [\boldsymbol{R_W}^2 - (\boldsymbol{L} \,/\, \boldsymbol{C})] \,/\, (-4\pi^2 \boldsymbol{L}^2) \\
\boldsymbol{f_r}^2 &= [(\boldsymbol{R_W}^2)(\boldsymbol{C} \,/\, \boldsymbol{C}) - (\boldsymbol{L} \,/\, \boldsymbol{C})] \,/\, (-4\pi^2 \boldsymbol{L}^2) \\
&= \{[(\boldsymbol{R_W}^2\boldsymbol{C}) \,/\, \boldsymbol{C})] - (\boldsymbol{L} \,/\, \boldsymbol{C})\} \,/\, (-4\pi^2 \boldsymbol{L}^2) \\
&= [(\boldsymbol{R_W}^2\boldsymbol{C} - \boldsymbol{L}) \,/\, \boldsymbol{C}] \,/\, (-4\pi^2 \boldsymbol{L}^2) \\
&= (\boldsymbol{R_W}^2\boldsymbol{C} - \boldsymbol{L}) \,/\, (-4\pi^2 \boldsymbol{L}^2 \boldsymbol{C}) \\
&= [-(\boldsymbol{R_W}^2\boldsymbol{C} - \boldsymbol{L}) \,/\, \boldsymbol{L}] \,/\, (4\pi^2 \boldsymbol{LC}) \\
&= [(\boldsymbol{L} - \boldsymbol{R_W}^2\boldsymbol{C}) \,/\, \boldsymbol{L}] \,/\, (4\pi^2 \boldsymbol{LC}) \\
&= \{(\boldsymbol{L} \,/\, \boldsymbol{L}) - [(\boldsymbol{R_W}^2\boldsymbol{C} \,/\, \boldsymbol{L})]\} \,/\, (4\pi^2 \boldsymbol{LC})
\end{aligned}$$

$$= \{1 - [(R_W^2C) / L]\} / (4\pi^2LC)$$
$$= \{1 - [(R_W^2C) / L]\}[1 / (4\pi^2LC)]$$

$$\sqrt{f_r^2} = \sqrt{\{1-[(R_W^2C)/L]\}[1/(4\pi^2LC)]}$$

$$f_r = \sqrt{1-[(R_W^2C)/L]}\sqrt{[1/(4\pi^2LC)]}$$

$$= \sqrt{1-[(R_W^2C)/L]}[1/(2\pi\sqrt{LC})]$$

The ideal resonant frequency, $f_{r(\text{IDEAL})}$, is

$$f_{r(\text{IDEAL})} = 1/(2\pi\sqrt{LC})$$

so

$$f_r = \sqrt{1-[(R_W^2C)/L]} \times f_{r(\text{IDEAL})}$$

Note that when R_W is 0 Ω, as in the case of an ideal inductor, (R_W^2C/L) is 0. Consequently, f_r and $f_{r(\text{IDEAL})}$ are the same value for an ideal inductor.

3. Problem: The value of the quality, or Q, of a non-ideal inductor is $Q = X_L / R$. For the non-ideal RLC circuit in Figure 18-17, show that the actual resonant frequency, f_r, is

$$f_r = \sqrt{Q^2/(Q^2+1)} \times f_{r(\text{IDEAL})}$$

where $f_{r(\text{IDEAL})} = 1/(2\pi\sqrt{LC})$.

Solution: From the equivalent circuit in Figure 18-18, the total impedance, Z_T, is

$$Z_T = (R_W + X_L) \parallel X_C$$
$$= 1 / \{[1 / (R_W + \text{j}\,X_L)] + (1 / -\text{j}\,X_C)\}$$

The total admittance, Y_T, is the reciprocal of the total impedance, Z_T.

$$1 / Z_T = 1 / (1 / \{[1 / (R_W + \text{j}\,X_L)] + (1 / -\text{j}\,X_C)\})$$
$$= [1 / (R_W + \text{j}\,X_L)] + (1 / -\text{j}\,X_C)$$
$$= [1 / (R_W + \text{j}\,X_L)][(R_W - \text{j}\,X_L) / (R_W - \text{j}\,X_L)] + (1 / -\text{j}\,X_C)(\text{j} / \text{j})$$
$$= \{(R_W - \text{j}\,X_L) / [(R_W + \text{j}\,X_L)(R_W - \text{j}\,X_L)]\} + [\text{j} / -\text{j}^2 (X_C)]$$
$$= \{(R_W - \text{j}\,X_L) / [(R_W R_W - \text{j}\,(X_L)(R_W) + \text{j}\,(X_L)(R_W) - \text{j}^2 (X_L)(X_L)]\} + [\text{j} / -(-1)(X_C)]$$
$$= \{(R_W - \text{j}\,X_L) / [(R_W^2 + \text{j}\,(R_WX_L - R_WX_L) - (-1)X_L^2]\} + (\text{j} / X_C)$$
$$= \{(R_W - \text{j}\,X_L) / [(R_W^2 + \text{j}\,(0) + X_L^2]\} + (\text{j} / X_C)$$
$$= [(R_W - \text{j}\,X_L) / (R_W^2 + X_L^2)][(X_C) / (X_C)] + (\text{j} / X_C)[(R_W^2 + X_L^2) / (R_W^2 + X_L^2)]$$
$$= \{[(R_W - \text{j}\,X_L)(X_C)] / [(R_W^2 + X_L^2)(X_C)]\} + \{[\text{j}\,(R_W^2 + X_L^2)] / [X_C(R_W^2 + X_L^2)]\}$$
$$= \{[R_WX_C - \text{j}\,(X_LX_C)] / (R_W^2X_C + X_L^2X_C)\} + \{[\text{j}\,(R_W^2 + X_L^2)] / (R_W^2X_C + X_L^2X_C)\}$$
$$= [R_WX_C + \text{j}\,(R_W^2 + X_L^2) - \text{j}\,(X_LX_C)] / (R_W^2X_C + X_L^2X_C)\}$$
$$= [R_WX_C + \text{j}\,(R_W^2 + X_L^2 - X_LX_C)] / (R_W^2X_C + X_L^2X_C)\}$$

At the resonant frequency, f_r, the imaginary terms of the total impedance, Z_T, and total admittance, Y_T, are zero.

$$R_W^2 + X_L^2 - X_LX_C = 0$$
$$R_W^2 + X_L^2 - X_LX_C + X_LX_C = 0 + X_LX_C$$
$$(R_W^2 + X_L^2) = X_LX_C$$
$$(R_W^2 + X_L^2) / X_L^2 = (X_LX_C) / X_L^2$$
$$[(R_W^2 + X_L^2) / X_L^2][(1 / R_W^2) / (1 / R_W^2)] = [1 / (2\pi f_rC)] / (2\pi f_rL)$$
$$[(R_W^2 + X_L^2) / R_W^2] / (X_L^2 / R_W^2) = 1 / [(2\pi f_rC)(2\pi f_rL)]$$

$$[(R_W^2 / R_W^2) + (X_L^2 / R_W^2)] / (X_L^2 / R_W^2) = 1 / (4\pi^2 f_r^2 LC)$$
$$[(1 + Q^2) / Q^2](f_r^2) = [1 / (4\pi^2 f_r^2 LC)](f_r^2)$$
$$(f_r^2)[(Q^2 + 1) / Q^2][Q^2 / (Q^2 + 1)] = [1 / (4\pi^2 LC)][Q^2 / (Q^2 + 1)]$$
$$f_r^2 = [Q^2 / (Q^2 + 1)][1 / (4\pi^2 LC)]$$
$$\sqrt{f_r^2} = \sqrt{[Q^2 / (Q^2 + 1)][1/4\pi^2 LC]}$$
$$f_r = \sqrt{[Q^2 / (Q^2 + 1)]}\,[1/(2\pi\sqrt{LC})]$$
$$= \sqrt{[Q^2 / (Q^2 + 1)]} \times f_{r(\mathrm{IDEAL})}$$

4. Problem: For the circuit in Figure 18-19, prove that the parallel equivalent for a non-ideal inductor with inductance, L, in series with winding resistance, R_W, is an inductance, $L_{eq} = [(Q^2 + 1) / Q^2]\,L$, in parallel with a resistance, $R_{Peq} = (Q^2 + 1)\,R_W$.

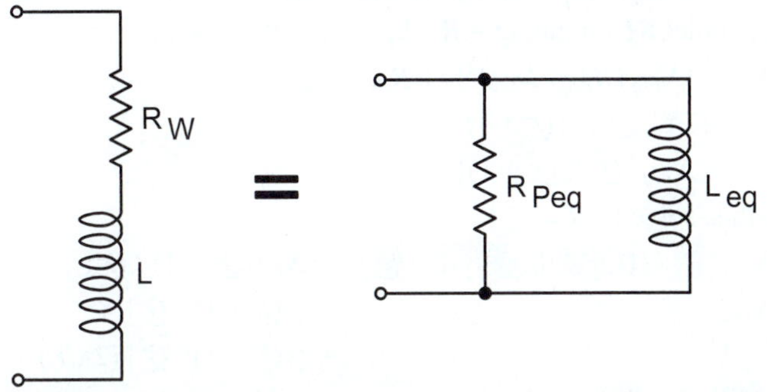

Figure 18-19: Equivalent Parallel *RL* Circuit for Non-Ideal Inductor

Solution: The total parallel impedance Z_T for R_{Peq} and L_{eq} is

$$\begin{aligned}
Z_T &= R_{Peq} \parallel \mathrm{j}\,X_{Leq} \\
&= [(R_{Peq} + \mathrm{j}\,0)(0 + \mathrm{j}\,X_{Leq})] / (R_{Peq} + \mathrm{j}\,X_{Leq}) \\
&= [(R_{Peq})(0) + (R_{Peq})(\mathrm{j}\,X_{Leq}) + (\mathrm{j}\,0)(0) + (\mathrm{j}\,0)(\mathrm{j}\,X_{Leq})] / (R_{Peq} + \mathrm{j}\,X_{Leq}) \\
&= \mathrm{j}\,R_{Peq}X_{Leq} / (R_{Peq} + \mathrm{j}\,X_{Leq}) \\
&= [\mathrm{j}\,R_{Peq}X_{Leq} / (R_{Peq} + \mathrm{j}\,X_{Leq})][(R_{Peq} - \mathrm{j}\,X_{Leq}) / (R_{Peq} - \mathrm{j}\,X_{Leq})] \\
&= [(\mathrm{j}\,R_{Peq}X_{Leq})(R_{Peq} - \mathrm{j}\,X_{Leq})][(R_P + \mathrm{j}\,X_{Leq})(R_{Peq} - \mathrm{j}\,X_{Leq})] \\
&= [\mathrm{j}\,R_{Peq}X_{Leq}R_{Peq} - (\mathrm{j}^2\,R_{Peq}X_{Leq}X_{Leq})] / (R_{Peq}R_{Peq} - \mathrm{j}\,X_{Leq}R_{Peq} + \mathrm{j}\,X_{Leq}R_{Peq} - \mathrm{j}^2\,X_{Leq}X_{Leq}) \\
&= [\mathrm{j}\,R_{Peq}^2X_{Leq} - (-1)\,R_{Peq}X_{Leq}^2] / [R_{Peq}^2 + \mathrm{j}\,(0) - (-1)\,X_{Leq}^2] \\
&= (R_{Peq}X_{Leq}^2 + \mathrm{j}\,R_{Peq}^2X_{Leq}) / (R_{Peq}^2 + X_{Leq}^2) \\
&= [R_{Peq}X_{Leq}^2 / (R_{Peq}^2 + X_{Leq}^2)] + \mathrm{j}\,[R_{Peq}^2X_{Leq} / (R_{Peq}^2 + X_{Leq}^2)]
\end{aligned}$$

The winding resistance R_W corresponds to the real term.

$$\begin{aligned}
R_W &= R_{Peq}X_{Leq}^2 / (R_{Peq}^2 + X_{Leq}^2) \\
&= [R_{Peq}X_{Leq}^2 / (R_{Peq}^2 + X_{Leq}^2)][(1 / X_{Leq}^2) / (1 / X_{Leq}^2)] \\
&= [(R_{Peq}X_{Leq}^2)(1 / X_{Leq}^2)] / [(R_{Peq}^2 + X_{Leq}^2)(1 / X_{Leq}^2)] \\
&= [(R_{Peq}X_{Leq}^2 / X_{Leq}^2)] / [(R_P^2 + X_{Leq}^2) / X_{Leq}^2] \\
&= R_{Peq} / [(R_{Peq}^2 / X_{Leq}^2) + (X_{Leq}^2 / X_{Leq}^2)] \\
&= R_{Peq} / [(R_{Peq} / X_{Leq})^2 + 1]
\end{aligned}$$

For a parallel RL circuit, $Q = R / X_L$ so $Q = R_{Peq} / X_{Leq}$.

$$\begin{aligned}
R_W &= R_{Peq} / [(R_{Peq} / X_{Leq})^2 + 1] \\
&= R_{Peq} / [(R_{Peq} / X_{Leq})^2 + 1]
\end{aligned}$$

$= R_{Peq} / (Q^2 + 1)$

From this, solve for R_{Peq}.

$[R_{Peq} / (Q^2 + 1)]\,(Q^2 + 1) = R_W (Q^2 + 1)$

$R_{Peq} = (Q^2 + 1)\, R_W$

The reactance X_L corresponds to the imaginary term.

$$
\begin{aligned}
X_L &= R_{Peq}^2 X_{Leq} / (R_{Peq}^2 + X_{Leq}^2) \\
&= [R_{Peq}^2 X_{Leq} / (R_{Peq}^2 + X_{Leq}^2)][(1 / X_{Leq}^2) / (1 / X_{Leq}^2)] \\
&= [(R_{Peq}^2 X_{Leq})(1 / X_{Leq}^2)] / [(R_{Peq}^2 + X_{Leq}^2)(1 / X_{Leq}^2)] \\
&= [(R_{Peq}^2 X_{Leq} / X_{Leq}^2)] / [(R_{Peq}^2 + X_{Leq}^2) / X_{Leq}^2] \\
&= [X_{Leq} (R_{Peq}^2 / X_{Leq}^2)] / [(R_{Peq}^2 / X_{Leq}^2) + (X_{Leq}^2 / X_{Leq}^2)] \\
&= [X_{Leq} (R_{Peq}^2 / X_{Leq}^2)] / [(R_{Peq}^2 / X_{Leq}^2) + 1] \\
&= [X_{Leq} (R_{Peq} / X_{Leq})^2] / [(R_{Peq} / X_{Leq})^2 + 1]
\end{aligned}
$$

For a parallel RL circuit, $Q = R / X_L$ so $Q = R_{Peq} / X_{Leq}$.

$$
\begin{aligned}
X_L &= [X_{Leq} (R_{Peq} / X_{Leq})^2] / [(R_{Peq} / X_{Leq})^2 + 1] \\
&= (X_{Leq} Q^2) / (Q^2 + 1) \\
&= X_{Leq} [Q^2 / (Q^2 + 1)]
\end{aligned}
$$

From this, solve for L_{eq}.

$$
\begin{aligned}
X_{Leq} [Q^2 / (Q^2 + 1)][(Q^2 + 1) / Q^2] &= X_L [(Q^2 + 1) / Q^2] \\
X_{Leq} &= [(Q^2 + 1) / Q^2]\, X_L \\
2\pi f L_{eq} &= [(Q^2 + 1) / Q^2]\,(2\pi f L) \\
(2\pi f L_{eq}) / (2\pi f) &= \{[(Q^2 + 1) / Q^2]\,(2\pi f L)\} / (2\pi f) \\
L_{eq} &= [(Q^2 + 1) / Q^2]\, L
\end{aligned}
$$

18.2.3 Series-Parallel *RLC* Circuits

1. Problem: The circuit in Figure 18-20 is a Maxwell bridge (also called a Maxwell-Wien bridge), which can determine the inductance, L_X, and winding resistance, R_X, of an inductor. To do so, the variable capacitor, C_{VAR}, and variable resistance, R_{VAR}, are adjusted so that the bridge is balanced and $V_A = V_B$. The known values of R_1, R_2, C_{VAR}, and R_{VAR} are then used to calculate the values of L_X and R_X.

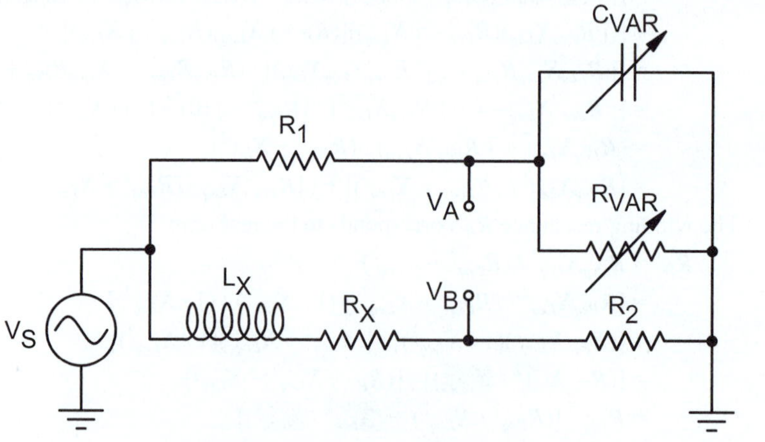

Figure 18-20: Maxwell Bridge

Determine the values of L and R_W in terms of R_1, R_2, C_{VAR}, and R_{VAR}.

Solution: Figure 18-21 shows the equivalent circuit for the Maxwell bridge when the capacitor, $\boldsymbol{C_{VAR}}$, and resistor, $\boldsymbol{R_{VAR}}$, are adjusted to $\boldsymbol{C_M}$ and $\boldsymbol{R_M}$ to balance the bridge.

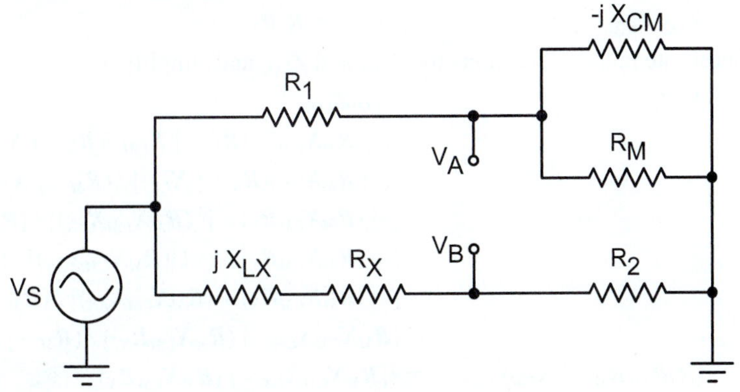

Figure 18-21: Equivalent Circuit for the Maxwell Bridge

From standard circuit analysis of the balanced bridge, the total impedance of the capacitor side, $\boldsymbol{Z_{CT}}$, is

$$Z_{CT} = R_1 + R_M \,||\, X_{CM}$$

The first equivalent impedance, $\boldsymbol{Z_{EQ1}}$, is the parallel combination $\boldsymbol{R_M \,||\, X_{CM}}$.

$$Z_{CT} = R_1 + Z_{EQ1}$$

where

$$\begin{aligned} Z_{EQ1} &= R_M \,||\, X_{CM} \\ &= R_M \,||\, -\mathrm{j}\, X_{CM} \\ &= [(R_M)(-\mathrm{j}\, X_{CM})] / (R_M - \mathrm{j}\, X_{CM}) \quad \text{using product-over-sum for parallel components} \\ &= (-\mathrm{j}\, R_M X_{CM}) / (R_M - \mathrm{j}\, X_{CM}) \end{aligned}$$

Also from standard circuit analysis of the balanced bridge, the total impedance of the inductor side of the bridge, $\boldsymbol{Z_{LT}}$, is

$$\begin{aligned} Z_{LT} &= X_{LX} + R_X + R_2 \\ &= R_2 + R_X + X_{LX} \end{aligned}$$

The first (and only) equivalent impedance, $\boldsymbol{Z_{EQ2}}$, is the series combination $\boldsymbol{R_X + X_{LX}}$.

$$Z_{LT} = Z_{EQ2} + R_2$$

where

$$\begin{aligned} Z_{EQ2} &= R_X + X_{LX} \\ &= R_X + \mathrm{j}\, X_{LX} \end{aligned}$$

When the bridge is balanced, $\boldsymbol{V_A = V_B}$. Use the voltage divider formula to find $\boldsymbol{V_A}$ and $\boldsymbol{V_B}$.

$$\begin{aligned} V_A &= V_S\,(Z_{ZEQ1} / Z_{CT}) \\ &= V_S\,[Z_{ZEQ1} / (R_1 + Z_{EQ1})] \\ V_B &= V_S\,(R_2 / Z_{LT}) \\ &= V_S\,[R_2 / (R_2 + Z_{EQ2})] \end{aligned}$$

Set $\boldsymbol{V_A}$ equal to $\boldsymbol{V_B}$ and simplify.

$$\begin{aligned} V_A &= V_B \\ V_S\,[Z_{EQ1} / (R_1 + Z_{EQ1})] &= V_S\,[R_2 / (R_2 + Z_{EQ2})] \\ \{V_S\,[Z_{EQ1} / (R_1 + Z_{EQ1})]\} / V_S &= \{V_S\,[R_2 / (R_2 + Z_{EQ2})]\} / V_S \\ [Z_{EQ1} / (R_1 + Z_{EQ1})](R_1 + Z_{ZEQ1}) &= [R_2 / (R_2 + Z_{EQ2})](R_1 + Z_{EQ1}) \\ Z_{EQ1}\,(R_2 + Z_{EQ2}) &= \{[R_2 / (R_2 + Z_{EQ2})](R_1 + Z_{EQ1})\}(R_2 + Z_{EQ2}) \end{aligned}$$

$$Z_{EQ1}(R_2 + Z_{EQ2}) = R_2(R_1 + Z_{EQ1})$$
$$R_2Z_{EQ1} + Z_{EQ1}Z_{EQ2} - R_2Z_{EQ1} = R_1R_2 + R_2Z_{EQ1} - R_2Z_{EQ1}$$
$$Z_{EQ1}Z_{EQ2} = R_1R_2$$

Substitute in the expressions for Z_{EQ1} and Z_{EQ2} and simplify.

$$R_1R_2 = Z_{EQ1}Z_{EQ2}$$
$$= [(-j\ R_MX_{CM}) / (R_M - j\ X_{CM})](R_X + j\ X_{LX})$$
$$= [(-j\ R_MX_{CM})(R_X + j\ X_{LX})] / (R_M - j\ X_{CM})$$
$$= [-j\ (R_MX_{CM}R_X) - j^2\ (R_MX_{CM}X_{LX})] / (R_M - j\ X_{CM})$$
$$= [-j\ (R_MX_{CM}R_X) - (-1)(R_MX_{CM}X_{LX})] / (R_M - j\ X_{CM})$$
$$= [-j\ (R_MX_{CM}R_X) + (R_MX_{CM}X_{LX})] / (R_M - j\ X_{CM})$$
$$= [R_MX_{CM}X_{LX} - j\ (R_MX_{CM}R_X)] / (R_M - j\ X_{CM})$$
$$(R_1R_2)(R_M - j\ X_{CM}) = \{[R_MX_{CM}X_{LX} - j\ (R_MX_{CM}R_X)] / (R_M - j\ X_{CM})\}\ (R_M - j\ X_{CM})$$
$$R_1R_2R_M - j\ (R_1R_2X_{CM}) = R_MX_{CM}X_{LX} - j\ (R_MX_{CM}R_X)$$

For two complex values to be equal, the real term of one value must equal the real term of the other, and the imaginary term of one value must equal the imaginary term of the other. Equate the real terms and solve for L_X.

$$R_MX_{CM}X_{LX} = R_1R_2R_M$$
$$(R_MX_{CM}X_{LX}) / (R_MX_{CM}) = (R_1R_2R_M) / (R_MX_{CM})$$
$$X_{LX} = (R_1R_2) / X_{CM}$$
$$2\pi fL_X = (R_1R_2) / [1 / (2\pi fC_M)]$$
$$= R_1R_2(2\pi fC_M)$$
$$(2\pi fL_X) / (2\pi f) = [R_1R_2(2\pi fC_M)] / (2\pi f)$$
$$L_X = R_1R_2C_M$$

Equate the imaginary terms and solve for R_X.

$$R_MX_{CM}R_X = R_1R_2X_{CM}$$
$$(R_MX_{CM}R_X) / (R_MX_{CM}) = (R_1R_2X_{CM}) / (R_MX_{CM})$$
$$R_X = (R_1R_2) / R_M$$

Note that frequency does not appear in the equations for L_X and R_X, so that the Maxwell bridge can theoretically operate at any frequency. Also note that, for the extreme case of $f = 0$ Hz (dc operation), L_X is a short and C_{VAR} is an open and the Maxwell bridge is essentially a Wheatstone bridge.

2. Problem: Determine the resonant frequency for the non-ideal series-parallel ***RLC*** circuit shown in Figure 18-22.

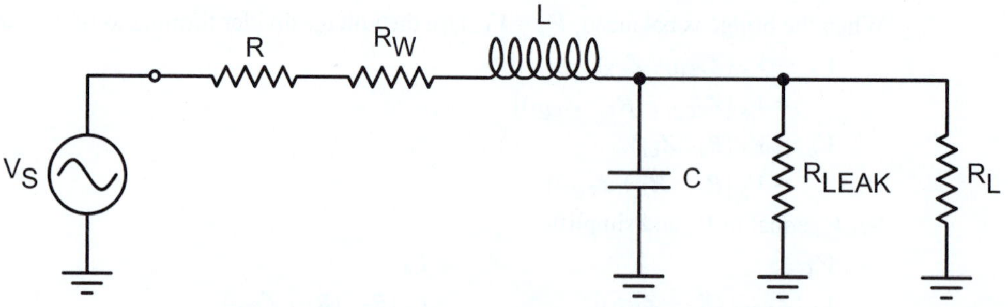

Figure 18-22: Non-Ideal Series-Parallel *RLC* Circuit

Solution: For the circuit in Figure 18-22, the series resistor, $\boldsymbol{R}$, and winding resistance, $\boldsymbol{R_W}$, can be combined into a single series resistance, $\boldsymbol{R_S}$. Similarly, the leakage resistance, $\boldsymbol{R_{LEAK}}$, and load resistor, $\boldsymbol{R_L}$, can be combined into a single parallel resistance, $\boldsymbol{R_P}$. Figure 18-23 shows the equivalent circuit with these combined resistances.

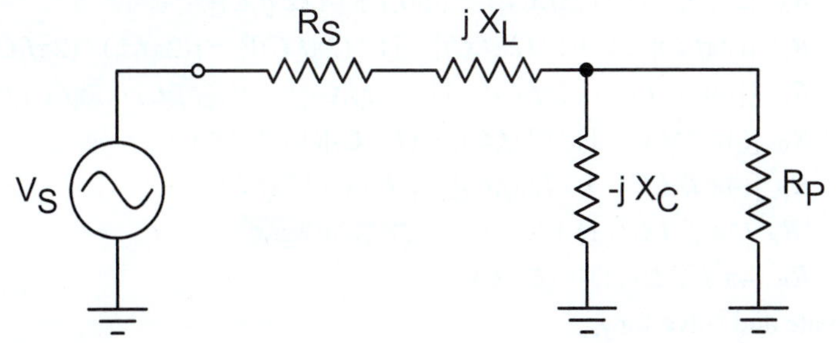

Figure 18-23: Equivalent Circuit for Non-Ideal Series-Parallel Circuit

From standard circuit analysis, $\boldsymbol{Z_T} = \boldsymbol{R_S} + \boldsymbol{X_L} + \boldsymbol{X_C} \parallel \boldsymbol{R_P}$. The first equivalent impedance, $\boldsymbol{Z_{EQ1}}$, is the parallel combination $\boldsymbol{X_C} \parallel \boldsymbol{R_P}$. From this,

$$Z_T = R_S + X_L + Z_{EQ1}$$

where

$$\begin{aligned}
Z_{EQ1} &= -\text{j}\, X_C \parallel R_P \\
&= [(-\text{j}\, X_C)(R_P)] \,/\, [R_P + (-\text{j}\, X_C)] \quad \text{using product-over-sum for parallel components} \\
&= (-\text{j}\, R_P X_C) \,/\, (R_P - \text{j}\, X_C) \\
&= [(-\text{j}\, R_P X_C) \,/\, (R_P - \text{j}\, X_C)]\, [(R_P + \text{j}\, X_C) \,/\, (R_P + \text{j}\, X_C)] \\
&= [(-\text{j}\, R_P X_C)(R_P + \text{j}\, X_C)] \,/\, [(R_P - \text{j}\, X_C)(R_P + \text{j}\, X_C)] \\
&= [(-\text{j}\, R_P X_C)(R_P) + (-\text{j}\, R_P X_C)(\text{j}\, X_C)] \\
&\quad / [(R_P R_P + (R_P)(\text{j}\, X_C) + (-\text{j}\, X_C)(R_P) + (-\text{j}\, X_C)(\text{j}\, X_C)] \\
&= [(-\text{j}\, R_P^2 X_C) + (-\text{j}^2)(R_P X_C^2)] \,/\, [(R_P^2 + (\text{j}\, R_P X_C) + (-\text{j}\, R_P X_C) + (-\text{j}^2)(X_C^2)] \\
&= [(-\text{j}\, R_P^2 X_C) + -(-1)(R_P X_C^2)] \,/\, [(R_P^2 + \text{j}\, (R_P X_C - R_P X_C) + -(-1)(X_C^2)] \\
&= [(-\text{j}\, R_P^2 X_C) + (R_P X_C^2)] \,/\, [(R_P^2 + \text{j}\, (0) + (X_C^2)] \\
&= (R_P X_C^2 - \text{j}\, R_P^2 X_C) \,/\, (R_P^2 + X_C^2)
\end{aligned}$$

The second (and last) equivalent impedance, $\boldsymbol{Z_{EQ2}}$, is the series combination $\boldsymbol{R_S} + \boldsymbol{X_L} + \boldsymbol{Z_{EQ1}}$. From this,

$$Z_T = Z_{EQ2}$$

where

$$\begin{aligned}
Z_{EQ2} &= R_S + \text{j}\, X_L + [(R_P X_C^2 - \text{j}\, R_P^2 X_C) \,/\, (R_P^2 + X_C^2)] \\
&= (R_S + \text{j}\, X_L)[(R_P^2 + X_C^2) \,/\, (R_P^2 + X_C^2)] + [(R_P X_C^2 - \text{j}\, R_P^2 X_C) \,/\, (R_P^2 + X_C^2)] \\
&= \{[(R_S + \text{j}\, X_L)(R_P^2) + (R_S + \text{j}\, X_L)(X_C^2)] \,/\, (R_P^2 + X_C^2)\} \\
&\quad + [(R_P X_C^2 - \text{j}\, R_P^2 X_C) \,/\, (R_P^2 + X_C^2)] \\
&= \{[(R_S R_P^2 + \text{j}\, X_L R_P^2) + (R_S X_C^2 + \text{j}\, X_L X_C^2)] \,/\, (R_P^2 + X_C^2)\} \\
&\quad + [(R_P X_C^2 - \text{j}\, R_P^2 X_C) \,/\, (R_P^2 + X_C^2)] \\
&= \{[(R_S R_P^2 + R_S X_C^2) + \text{j}\, (R_P^2 X_L + X_L X_C^2)] \,/\, (R_P^2 + X_C^2)\} \\
&\quad + [(R_P X_C^2 - \text{j}\, R_P^2 X_C) \,/\, (R_P^2 + X_C^2)] \\
&= \{[(R_S R_P^2 + R_S X_C^2) + \text{j}\, (R_P^2 X_L + X_L X_C^2)] + (R_P X_C^2 - \text{j}\, R_P^2 X_C)\} \,/\, (R_P^2 + X_C^2) \\
&= \{(R_S R_P^2 + R_S X_C^2 + R_P X_C^2) + [\text{j}\, (R_P^2 X_L + X_L X_C^2) + (-\text{j}\, R_P^2 X_C)]\} \,/\, (R_P^2 + X_C^2) \\
&= [(R_S R_P^2 + R_S X_C^2 + R_P X_C^2) + \text{j}\, (R_P^2 X_L + X_L X_C^2 - R_P^2 X_C)] \,/\, (R_P^2 + X_C^2)
\end{aligned}$$

At the resonant frequency the circuit appears purely resistive, so that the imaginary term is 0.

$$R_P^2 X_L + X_L X_C^2 - R_P^2 X_C = 0$$

$$R_P^2X_L - R_P^2X_C + X_LX_C^2 = 0$$
$$R_P^2 (X_L - X_C) + X_LX_C^2 = 0$$

Use the equations for inductive and capacitive reactances to solve for the resonant frequency.

$$R_P^2 \{2\pi f_rL - [1 / (2\pi f_rC)]\} + (2\pi f_rL)[1 / (2\pi f_rC)]^2 = 0$$
$$R_P^2 \{(2\pi f_rL)[(2\pi f_rC) / (2\pi f_rC)] - [1 / (2\pi f_rC)]\} + [(2\pi f_rL) / (2\pi f_rC)^2] = 0$$
$$R_P^2 \{[(4\pi^2f_r^2LC) / (2\pi f_rC)] - [1 / (2\pi f_rC)]\} + \{[(2\pi f_rL) / (2\pi f_rC)](1 / 2\pi f_rC)\} = 0$$
$$R_P^2 [(4\pi^2f_r^2LC - 1) / (2\pi f_rC)] + (L / C)(1 / 2\pi f_rC) = 0$$
$$R_P^2 [(4\pi^2f_r^2LC - 1) / (2\pi f_rC)] + [(L / C) / 2\pi f_rC] = 0$$
$$[R_P^2 (4\pi^2f_r^2LC - 1) + (L / C)] / (2\pi f_rC))(2\pi f_rC) = (0)(2\pi f_rC)$$
$$R_P^2 (4\pi^2f_r^2LC - 1) + (L / C) = 0$$

Isolate and solve for f_r.

$$RP^2 (4\pi^2f_r^2LC - 1) + (L / C) - (L / C) = 0 - (L / C)$$
$$[R_P^2 (4\pi^2f_r^2LC - 1)] / R_P^2 = (-L / C) / R_P^2$$
$$(4\pi^2f_r^2LC - 1) + 1 = [-L / (R_P^2C)] + 1$$
$$4\pi^2f_r^2LC = 1 - [L / (R_P^2C)]$$
$$= (1) [(R_P^2C) / (R_P^2C)] - [L / (R_P^2C)]$$
$$= [(R_P^2C) / (R_P^2C)] - [L / (R_P^2C)]$$
$$= (R_P^2C - L) / (R_P^2C)$$
$$= (R_P^2C / R_P^2C) - [L / (R_P^2C)]$$
$$= 1 - [L / (R_P^2C)]$$
$$(4\pi^2f_r^2LC) / (LC) = \{1 - [L / (R_P^2C)]\} / (LC)$$
$$(4\pi^2f_r^2) / (4\pi^2) = (\{1 - [L / (R_P^2C)]\} / (LC)) / (4\pi^2)$$
$$f_r^2 = \{1 - [L / (R_P^2C)]\} / (4\pi^2LC)$$
$$= \{1 - [L / (R_P^2C)]\} [1 / (4\pi^2LC)]$$
$$\sqrt{f_r^2} = \sqrt{1-[L / (R_P^2C)][1/(4\pi^2 LC)]}$$
$$f_r = \sqrt{1-[L / (R_P^2C)]}\sqrt{[1/(4\pi^2 LC)]}$$
$$= \sqrt{1-[L / (R_P^2C)]}\,[1 / (2\pi\sqrt{LC})]$$

The ideal resonant frequency, $f_{r(\text{IDEAL})}$, is

$$f_{r(\text{IDEAL})} = 1 / (2\pi\sqrt{LC})$$

so

$$f_r = \sqrt{1-[L / (R_P^2C)]} \times f_{r(\text{IDEAL})}$$

Note that the R_P, the effective resistance in parallel with C, affects the resonant frequency (especially when it is small), but that the effective series resistance, R_S, does not.

18.3 Just for Fun

The resonant frequency for an ***RLC*** circuit is the frequency at which the capacitive and inductive effects cancel and the circuit appears purely resistive (that is, the phase angle is 0°).There is, however, a special class of ***RLC*** circuit which exhibits some interesting resonance properties. The circuit in Figure 18-24 is just such a circuit. Refer to this circuit for Problems 1 and 2.

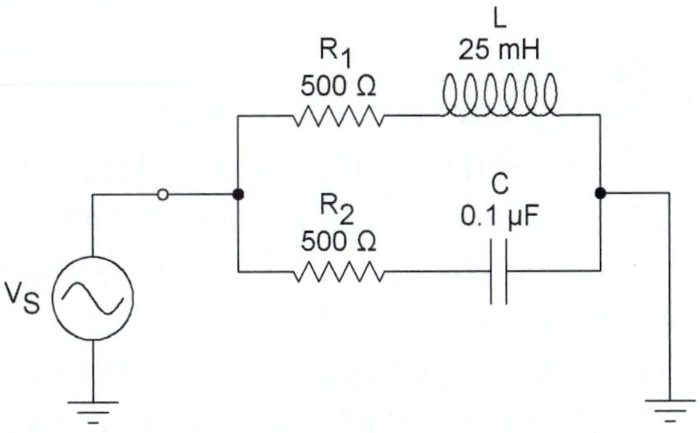

Figure 18-24: Special *RLC* Circuit

1. Problem: Determine the impedance of the circuit for $f = 0$ Hz and as $f \to \infty$.

   Solution: At dc, the 25 mH inductor appears as a short and the 0.1 μF capacitor appears as an open. Therefore, the circuit impedance is due to R_1 and equal to **500 Ω ∠ 0°**.

   As the frequency approaches infinity, the 25 mH inductor appears as an open and the 0.1 μF capacitor appears as a short. Therefore the circuit impedance is due to R_2 and equal to **500 Ω ∠ 0°**.

2. Problem: Determine the total impedance, Z_T, for $f = 100$ Hz, 1.0 kHz, and 10 kHz.

   Solution: First, calculate the value of X_L and X_C. Then use standard circuit analysis techniques to determine the total impedance.

   For $f = 100$ Hz:

$$
\begin{aligned}
X_L &= 2\pi f L \\
&= 2\pi\,(100\text{ Hz})\,(25\text{ mH}) \\
&= 2\pi\,(100\text{ Hz})\,(25 \times 10^{-3}\text{ H}) \\
&= 15.7\ \Omega
\end{aligned}
$$

$$
\begin{aligned}
X_C &= 1 / (2\pi f C) \\
&= 1 / [2\pi\,(100\text{ Hz})\,(0.1\ \mu\text{F})] \\
&= 1 / [2\pi\,(100\text{ Hz})\,(0.1 \times 10^{-6}\text{ F})] \\
&= 1 / (62.8\ \mu\text{S}) \\
&= 15.9\text{ k}\Omega
\end{aligned}
$$

From standard circuit analysis, $Z_T = (R_1 + X_L) \parallel (R_2 + X_C)$. The first equivalent impedance, Z_{EQ1}, is the series combination $R_1 + X_L$. From this,

$$Z_T = Z_{EQ1} \parallel (R_2 + X_C)$$

where

$$
\begin{aligned}
Z_{EQ1} &= R_1 + X_L \\
&= 500\ \Omega + j\,15.7\ \Omega \text{ in rectangular form} \\
&= 500\ \Omega \angle 1.80^\circ \text{ in polar form}
\end{aligned}
$$

The second equivalent impedance, Z_{EQ2}, is the series combination $R_2 + X_C$. From this,

$$Z_T = Z_{EQ1} \parallel Z_{EQ2}$$

where

$$
\begin{aligned}
Z_{EQ2} &= R_2 + X_C \\
&= 500\ \Omega - j\,15.9\text{ k}\Omega \text{ in rectangular form} \\
&= 15.9\text{ k}\Omega \angle -88.2^\circ \text{ in polar form}
\end{aligned}
$$

The third (and last) equivalent impedance, Z_{EQ3}, is the parallel combination $Z_{EQ1} \parallel Z_{EQ2}$. From this,

$$Z_T = Z_{EQ3}$$

where

$$\begin{aligned}
\boldsymbol{Z_{EQ3}} &= \boldsymbol{Z_{EQ1}} \parallel \boldsymbol{Z_{EQ2}} \\
&= (500\ \Omega \angle 1.80°) \parallel (15.9\ \text{k}\Omega \angle -88.2°) \\
&= 1 / \{[1 / (500\ \Omega \angle 1.80°)] + [1 / (15.9\ \text{k}\Omega \angle -88.2°)]\} \\
&= 1 / \{[(1 \angle 0°) / (500\ \Omega \angle 1.80°)] + [(1 \angle 0°) / (15.9\ \text{k}\Omega \angle -88.2°)]\} \\
&= 1 / (\{(1 / 500\ \Omega) \angle (0° - 1.80°)] + \{(1 / 15.9\ \text{k}\Omega) \angle [0° - (-88.2°)]\}) \\
&= 1 / [(2.00\ \text{mS} \angle -1.80°) + (62.8\ \mu\text{S} \angle 88.2°)] \\
&= 1 / [(2.00\ \text{mS} - \text{j}\ 62.8\ \mu\text{S}) + (1.97\ \mu\text{S} + \text{j}\ 62.8\ \mu\text{S})] \\
&= 1 / [(2.00\ \text{mS} + 1.97\ \mu\text{S}) + \text{j}\ (-62.8\ \mu\text{S} + 62.8\ \mu\text{S})] \\
&= 1 / (2.00\ \text{mS} + \text{j}\ 0\ \text{S}) \\
&= 1 / (2.00\ \text{mS} \angle 0°) \\
&= (1 \angle 0°) / (2.00\ \text{mS} \angle 0°) \\
&= (1 / 2.00\ \text{mS})(0° - 0°) \\
&= \mathbf{500\ \Omega \angle 0°}
\end{aligned}$$

For $\boldsymbol{f}$ = 1.0 kHz:

$$\begin{aligned}
\boldsymbol{X_L} &= 2\pi \boldsymbol{fL} \\
&= 2\pi\ (1.0\ \text{kHz})\ (25\ \text{mH}) \\
&= 2\pi\ (1.0\ \text{Hz})\ (25 \times 10^{-3}\ \text{H}) \\
&= 157\ \Omega
\end{aligned}$$

$$\begin{aligned}
\boldsymbol{X_C} &= 1 / (2\pi \boldsymbol{fC}) \\
&= 1 / [2\pi\ (1.0\ \text{Hz})\ (0.1\ \mu\text{F})] \\
&= 1 / [2\pi\ (1.0\ \text{Hz})\ (0.1 \times 10^{-6}\ \text{F})] \\
&= 1 / (628\ \mu\text{S}) \\
&= 1.59\ \text{k}\Omega
\end{aligned}$$

From standard circuit analysis, $\boldsymbol{Z_T} = (\boldsymbol{R_1} + \boldsymbol{X_L}) \parallel (\boldsymbol{R_2} + \boldsymbol{X_C})$. The first equivalent impedance, $\boldsymbol{Z_{EQ1}}$, is the series combination $\boldsymbol{R_1} + \boldsymbol{X_L}$. From this,

$$\boldsymbol{Z_T} = \boldsymbol{Z_{EQ1}} \parallel (\boldsymbol{R_2} + \boldsymbol{X_C})$$

where

$$\begin{aligned}
\boldsymbol{Z_{EQ1}} &= \boldsymbol{R_1} + \boldsymbol{X_L} \\
&= 500\ \Omega + \text{j}\ 157\ \Omega \text{ in rectangular form} \\
&= 524\ \Omega \angle 17.4° \text{ in polar form}
\end{aligned}$$

The second equivalent impedance, $\boldsymbol{Z_{EQ2}}$, is the series combination $\boldsymbol{R_2} + \boldsymbol{X_C}$. From this,

$$\boldsymbol{Z_T} = \boldsymbol{Z_{EQ1}} \parallel \boldsymbol{Z_{EQ2}}$$

where

$$\begin{aligned}
\boldsymbol{Z_{EQ2}} &= \boldsymbol{R_2} + \boldsymbol{X_C} \\
&= 500\ \Omega - \text{j}\ 1.59\ \text{k}\Omega \text{ in rectangular form} \\
&= 1.67\ \text{k}\Omega \angle -72.6° \text{ in polar form}
\end{aligned}$$

The third (and last) equivalent impedance, $\boldsymbol{Z_{EQ3}}$, is the parallel combination $\boldsymbol{Z_{EQ1}} \parallel \boldsymbol{Z_{EQ2}}$. From this,

$$\boldsymbol{Z_T} = \boldsymbol{Z_{EQ3}}$$

where

$$\boldsymbol{Z_{EQ3}} = \boldsymbol{Z_{EQ1}} \parallel \boldsymbol{Z_{EQ2}}$$

$= (524\ \Omega \angle 17.4°) \parallel (1.67\ k\Omega \angle -72.6°)$

$= 1 / \{[1 / (524\ \Omega \angle 17.4°)] + [1 / (1.67\ k\Omega \angle -72.6°)]\}$

$= 1 / \{[(1 \angle 0°) / (524\ \Omega \angle 17.4°)] + [(1 \angle 0°) / (1.67\ k\Omega \angle -72.6°)]\}$

$= 1 / ([(1 / 524\ \Omega) \angle (0° - 17.4°)] + \{(1 / 1.67\ k\Omega) \angle [0° - (-72.6°)]\})$

$= 1 / [(1.91\ mS \angle -17.4°) + (599\ \mu S \angle 72.6°)]$

$= 1 / [(1.82\ mS - j\ 572\ \mu S) + (180\ \mu S + j\ 572\ \mu S)]$

$= 1 / [(1.82\ mS + 180\ \mu S) + j\ (-572\ \mu S + 527\ \mu S)]$

$= 1 / (2.00\ mS + j\ 0\ S)$

$= 1 / (2.00\ mS \angle 0°)$

$= (1 \angle 0°) / (2.00\ mS \angle 0°)$

$= (1 / 2.00\ mS)(0° - 0°)$

$= \mathbf{500\ \Omega \angle 0°}$

For $f = 10$ kHz:

$X_L = 2\pi f L$

$= 2\pi\ (10\ kHz)\ (25\ mH)$

$= 2\pi\ (10\ kHz)\ (25 \times 10^{-3}\ H)$

$= 1.57\ k\Omega$

$X_C = 1 / (2\pi f C)$

$= 1 / [2\pi\ (10\ kHz)\ (0.1\ \mu F)]$

$= 1 / [2\pi\ (10\ kHz)\ (0.1 \times 10^{-6}\ F)]$

$= 1 / (6.28\ mS)$

$= 159\ \Omega$

From standard circuit analysis, $Z_T = (R_1 + X_L) \parallel (R_2 + X_C)$. The first equivalent impedance, Z_{EQ1}, is the series combination $R_1 + X_L$. From this,

$Z_T = Z_{EQ1} \parallel (R_2 + X_C)$

where

$Z_{EQ1} = R_1 + X_L$

$= 500\ \Omega + j\ 1.57\ k\Omega$ in rectangular form

$= 1.65\ k\Omega \angle 72.3°$ in polar form

The second equivalent impedance, Z_{EQ2}, is the series combination $R_2 + X_C$. From this,

$Z_T = Z_{EQ1} \parallel Z_{EQ2}$

where

$Z_{EQ2} = R_2 + X_C$

$= 500\ \Omega - j\ 159\ \Omega$ in rectangular form

$= 525\ \Omega \angle -17.7°$ in polar form

The third (and last) equivalent impedance, Z_{EQ3}, is the parallel combination $Z_{EQ1} \parallel Z_{EQ2}$. From this,

$Z_T = Z_{EQ3}$

where

$Z_{EQ3} = Z_{EQ1} \parallel Z_{EQ2}$

$= (1.65\ k\Omega \angle 72.3°) \parallel (525\ \Omega \angle -17.7°)$

$= 1 / \{[1 / (1.65\ k\Omega \angle 72.3°)] + [1 / (525\ \Omega \angle -17.7°)]\}$

$= 1 / \{[(1 \angle 0°) / (1.65\ k\Omega \angle 72.3°)] + [(1 \angle 0°) / (525\ \Omega \angle -17.7°)]\}$

$= 1 / ([(1 / 1.65\ \text{k}\Omega) \angle (0° - 72.3°)] + \{(1 / 525\ \Omega) \angle [0° - (-17.7°)]\})$

$= 1 / [(607\ \mu\text{S} \angle -72.3°) + (1.91\ \text{mS} \angle 17.7°)]$

$= 1 / [(184\ \mu\text{S} - \text{j}\ 578\ \mu\text{S}) + (1.82\ \text{mS} + \text{j}\ 578\ \mu\text{S})]$

$= 1 / [(184\ \mu\text{S} + 1.82\ \text{mS}) + \text{j}\ (-578\ \mu\text{S} + 578\ \mu\text{S})]$

$= 1 / (2.00\ \text{mS} + \text{j}\ 0\ \text{S})$

$= 1 / (2.00\ \text{mS} \angle 0°)$

$= (1 \angle 0°) / (2.00\ \text{mS} \angle 0°)$

$= (1 / 2.00\ \text{mS})(0° - 0°)$

$= \mathbf{500\ \Omega \angle 0°}$

As you probably suspected from this problem and Problem 1, the total impedance for this circuit will always appear to be a 500 Ω resistance because the circuit is resonant at ***all*** frequencies.

3. Problem: Determine the necessary conditions for $\boldsymbol{R}$, $\boldsymbol{L}$, and $\boldsymbol{C}$ for the circuit of Figure 18-25 to be resonant at all frequencies. Then show that the circuit in Figure 18-24 meets these conditions.

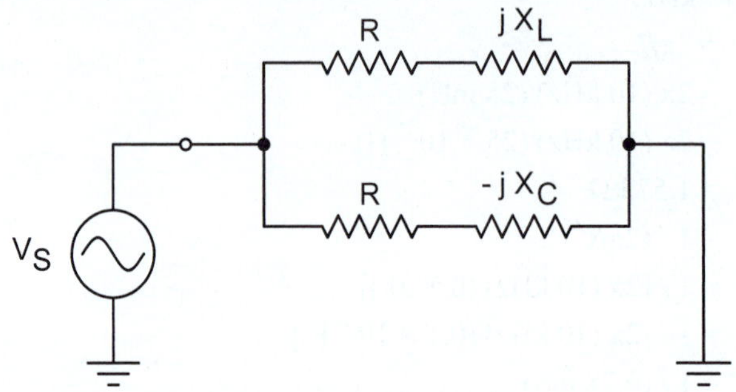

Figure 18-25: Series-Parallel *RLC* Circuit for Problem 3

Solution: From standard circuit analysis, the total impedance $\boldsymbol{Z_T} = (\boldsymbol{R} + \text{j}\ \boldsymbol{X_L}) \parallel (\boldsymbol{R} - \text{j}\ \boldsymbol{X_C})$.

$\boldsymbol{Z_T} = (\boldsymbol{R} + \text{j}\ \boldsymbol{X_L}) \parallel (\boldsymbol{R} - \text{j}\ \boldsymbol{X_C})$

$= [(\boldsymbol{R} + \text{j}\ \boldsymbol{X_L})(\boldsymbol{R} - \text{j}\ \boldsymbol{X_C})] / [(\boldsymbol{R} + \text{j}\ \boldsymbol{X_L}) + (\boldsymbol{R} - \text{j}\ \boldsymbol{X_C})]$ from product-over-sum method

$= [\boldsymbol{RR} + (\boldsymbol{R})(-\text{j}\ \boldsymbol{X_C}) + (\boldsymbol{R})(\text{j}\ \boldsymbol{X_L}) - \text{j}^2\ (\boldsymbol{X_L X_C})] / [(\boldsymbol{R} + \boldsymbol{R} + \text{j}\ \boldsymbol{X_L} - \text{j}\ \boldsymbol{X_C})]$

$= [\boldsymbol{R}^2 - \text{j}\ \boldsymbol{RX_C} + \text{j}\ \boldsymbol{RX_L} - (-1)(\boldsymbol{X_L X_C})] / (2\boldsymbol{R} + \text{j}\ \boldsymbol{X_L} - \text{j}\ \boldsymbol{X_C})$

$= [\boldsymbol{R}^2 - \text{j}\ \boldsymbol{RX_C} + \text{j}\ \boldsymbol{RX_L} + \boldsymbol{X_L X_C}] / [2\boldsymbol{R} + \text{j}\ (\boldsymbol{X_L} - \boldsymbol{X_C})]$

$= [(\boldsymbol{R}^2 + \boldsymbol{X_L X_C}) + (\text{j}\ \boldsymbol{RX_L} - \text{j}\ \boldsymbol{RX_C})] / [2\boldsymbol{R} + \text{j}\ (\boldsymbol{X_L} - \boldsymbol{X_C})]$

$= \{[(\boldsymbol{R}^2 + \boldsymbol{X_L X_C}) + (\text{j}\ \boldsymbol{RX_L} - \text{j}\ \boldsymbol{RX_C})] / [2\boldsymbol{R} + \text{j}\ (\boldsymbol{X_L} - \boldsymbol{X_C})]\}$
$\quad \times \{[2\boldsymbol{R} - \text{j}\ (\boldsymbol{X_L} - \boldsymbol{X_C})] / [2\boldsymbol{R} - \text{j}\ (\boldsymbol{X_L} - \boldsymbol{X_C})]\}$

$= \{[(\boldsymbol{R}^2 + \boldsymbol{X_L X_C}) + (\text{j}\ \boldsymbol{RX_L} - \text{j}\ \boldsymbol{RX_C})]\ [2\boldsymbol{R} - \text{j}\ (\boldsymbol{X_L} - \boldsymbol{X_C})]\}$
$\quad / \{[2\boldsymbol{R} + \text{j}\ (\boldsymbol{X_L} - \boldsymbol{X_C})][2\boldsymbol{R} - \text{j}\ (\boldsymbol{X_L} - \boldsymbol{X_C})]\}$

$= \{[(\boldsymbol{R}^2 + \boldsymbol{X_L X_C}) + (\text{j}\ \boldsymbol{RX_L} - \text{j}\ \boldsymbol{RX_C})]\ (2\boldsymbol{R} - \text{j}\ \boldsymbol{X_L} + \text{j}\ \boldsymbol{X_C})\ \}$
$\quad / \{[(2\boldsymbol{R})(2\boldsymbol{R}) + \text{j}\ (2\boldsymbol{R})(\boldsymbol{X_L} - \boldsymbol{X_C}) - \text{j}\ (2\boldsymbol{R})(\boldsymbol{X_L} - \boldsymbol{X_C}) - \text{j}^2\ (\boldsymbol{X_L} - \boldsymbol{X_C})(\ \boldsymbol{X_L} - \boldsymbol{X_C})]\}$

$= [(\boldsymbol{R}^2 + \boldsymbol{X_L X_C})(2\boldsymbol{R} - \text{j}\ \boldsymbol{X_L} + \text{j}\ \boldsymbol{X_C}) + (\text{j}\ \boldsymbol{RX_L} - \text{j}\ \boldsymbol{RX_C})(2\boldsymbol{R} - \text{j}\ \boldsymbol{X_L} + \text{j}\ \boldsymbol{X_C})]$
$\quad / \{4\boldsymbol{R}^2 + \text{j}\ [(2\boldsymbol{R})(\boldsymbol{X_L} - \boldsymbol{X_C}) - (2\boldsymbol{R})(\boldsymbol{X_L} - \boldsymbol{X_C})] - (-1)(\boldsymbol{X_L} - \boldsymbol{X_C})^2\}$

$= [(\boldsymbol{R}^2 + \boldsymbol{X_L X_C})(2\boldsymbol{R}) + (\boldsymbol{R}^2 + \boldsymbol{X_L X_C})(-\text{j}\ \boldsymbol{X_L}) + (\boldsymbol{R}^2 + \boldsymbol{X_L X_C})(\text{j}\ \boldsymbol{X_C}) +$
$\quad (\text{j}\ \boldsymbol{RX_L} - \text{j}\ \boldsymbol{RX_C})(2\boldsymbol{R}) + (\text{j}\ \boldsymbol{RX_L} - \text{j}\ \boldsymbol{RX_C})(-\text{j}\ \boldsymbol{X_L}) + (\text{j}\ \boldsymbol{RX_L} - \text{j}\ \boldsymbol{RX_C})(\text{j}\ \boldsymbol{X_C})]$
$\quad / [4\boldsymbol{R}^2 + \text{j}\ (0) + (\boldsymbol{X_L} - \boldsymbol{X_C})^2]$

$= [(2\boldsymbol{RR}^2 + 2\boldsymbol{RX_L X_C}) + (-\text{j}\ \boldsymbol{R}^2\boldsymbol{X_L} - \text{j}\ \boldsymbol{X_L X_L X_C}) + (\text{j}\ \boldsymbol{R}^2\boldsymbol{X_C} + \text{j}\ \boldsymbol{X_L X_C X_C}) +$
$\quad (\text{j}\ 2\boldsymbol{RRX_L} - \text{j}\ 2\boldsymbol{RRX_C}) + (-\text{j}^2\ \boldsymbol{RX_L X_L} + \text{j}^2\ \boldsymbol{RX_L X_C}) + (\text{j}^2\ \boldsymbol{RX_L X_C} - \text{j}^2\ \boldsymbol{RX_C X_C})]$
$\quad / [4\boldsymbol{R}^2 + (\boldsymbol{X_L} - \boldsymbol{X_C})^2]$

$$= [(2R^3 + 2RX_LX_C) + (-j\,R^2X_L - j\,X_L^2X_C) + (j\,R^2X_C + j\,X_LX_C^2) + (j\,2R^2X_L - j\,2R^2X_C) + -(-1)(RX_L^2) + (-1)(RX_LX_C) + (-1)(RX_LX_C) - (-1)(RX_C^2)] / [4R^2 + (X_L - X_C)^2]$$

$$= (2R^3 + 2RX_LX_C - j\,R^2X_L - j\,X_L^2X_C + j\,R^2X_C + j\,X_LX_C^2 + j\,2R^2X_L - j\,2R^2X_C + RX_L^2 - RX_LX_C - RX_LX_C + RX_C^2) / [4R^2 + (X_L - X_C)^2]$$

$$= (2R^3 + 2RX_LX_C + RX_L^2 - RX_LX_C - RX_LX_C + RX_C^2 + -j\,R^2X_L - j\,X_L^2X_C + j\,R^2X_C + j\,X_LX_C^2 + j\,2R^2X_L - j\,2R^2X_C) / [4R^2 + (X_L - X_C)^2]$$

$$= (2R^3 + RX_L^2 + RX_C^2 + j\,(-R^2X_L - X_L^2X_C + R^2X_C + X_LX_C^2 + 2R^2X_L - 2R^2X_C) / [4R^2 + (X_L - X_C)^2]$$

$$= [2R^3 + RX_L^2 + RX_C^2 + j\,(-X_L^2X_C + X_LX_C^2 + R^2X_L - R^2X_C] / [4R^2 + (X_L - X_C)^2]$$

$$= [2R^3 + RX_L^2 + RX_C^2 + j\,(R^2X_L - R^2X_C - X_L^2X_C + X_LX_C^2)] / [4R^2 + (X_L - X_C)^2]$$

At resonance the circuit appears purely resistive, so that the imaginary term is 0.

$$R^2X_L - R^2X_C - X_L^2X_C + X_LX_C^2 = 0$$

$$R^2(X_L - X_C) - X_LX_LX_C + X_LX_C X_C = 0$$

$$R^2(X_L - X_C) + X_LX_C(-X_L + X_C) = 0$$

$$R^2(X_L - X_C) - X_LX_C(X_L - X_C) = 0$$

$$(R^2 - X_LX_C)(X_L - X_C) = 0$$

$$(R^2 - X_LX_C)(X_L - X_C) / (X_L - X_C) = 0 / (X_L - X_C)$$

$$R^2 - X_LX_C + X_LX_C = 0 + X_LX_C$$

$$R^2 = X_LX_C$$

Substitute in the equations for inductive and capacitive reactances.

$$R^2 = (2\pi fL)[1 / (2\pi fC)]$$

$$R^2 = (2\pi fL) / (2\pi fC)$$

$$R^2 = L / C$$

$$\sqrt{R^2} = \sqrt{L / C}$$

$$R = \sqrt{L / C}$$

The series-parallel ***RLC*** circuit in Figure 18-25 will be resonant at all frequencies if the resistor value is equal to the square root of the ratio of the inductor and capacitor values. For the circuit in Figure 18-24, $R = 500\ \Omega$, $L = 25$ mH, and $C = 0.1\ \mu$F.

$$R = \sqrt{25\ \text{mH} / 0.1\ \mu\text{F}}$$

$$= \sqrt{(25 \times 10^{-3}\ \text{H}) / (0.1 \times 10^{-6}\ \text{F})}$$

$$= \sqrt{(250 \times 10^{3}\ \Omega^2}$$

$$= 500\ \Omega$$

The component values for the circuit in Figure 18-24 satisfy the necessary conditions for resonance at all frequencies.

19. Passive Filters

Note: For all problems in this section, the phase of ac sources is 0° unless otherwise stated.

19.1 Basic

19.1.1 Octaves and Decades

1. Problem: Calculate the separation, N_{10}, in decades and N_2, in octaves, between f_1 and f_2 for each frequency pair.

   a. $f_1 = 30$ Hz, $f_2 = 250$ Hz
   b. $f_1 = 500$ Hz, $f_2 = 100$ Hz
   c. $f_1 = 12.5$ kHz, $f_2 = 5.0$ kHz
   d. $f_1 = 75$ kHz, $f_2 = 7.5$ MHz

   Solution: The separation in decades, N_{10}, between two frequencies, f_1, and f_2, is log (f_2 / f_1). The separation in octaves, N_2, is [log (f_2 / f_1)] / [log (2)], where log (x) is the common (or base 10) logarithm of x. Note that $N_2 = N_{10}$ / log (2) = N_{10} / 0.301.

   a. For $f_1 = 30$ Hz and $f_2 = 250$ Hz,

      f_2 / f_1 = 250 Hz / 30 Hz
      = 8.33

      N_{10} = log (8.33) = **0.921 decades**.

      N_2 = log (8.33) / log (2) = 0.921 / 0.301 = **3.06 octaves**

   b. For $f_1 = 500$ Hz and $f_2 = 100$ Hz,

      f_2 / f_1 = 100 Hz / 500 Hz
      = 0.200

      N_{10} = log (0.200) = **–0.699 decades**.

      N_2 = log (0.200) / log (2) = –0.699 / 0.301 = **–2.32 octaves**

   c. For $f_1 = 12.5$ kHz and $f_2 = 5.0$ kHz,

      f_2 / f_1 = 5.0 kHz / 12.5 kHz
      = 0.400

      N_{10} = log (0.400) = **0.398 decades**.

      N_2 = log (0.400) / log (2) = –0.398 / 0.301 = **–1.32 octaves**

   d. For $f_1 = 75$ kHz and $f_2 = 7.5$ MHz,

      f_2 / f_1 = 7.5 MHz / 75 kHz
      = 7500 kHz / 75 kHz
      = 100

      N_{10} = log (100) = **2.00 decades**.

      N_2 = log (100) / log (2) = 2.00 / 0.301 = **6.64 octaves**

2. Problem: Calculate the frequency f_2 that is the specified number of octaves, N_2, from the indicated frequency, f_1.

   a. $f_1 = 120$ Hz, $N_2 = 2.5$ octaves
   b. $f_1 = 440$ Hz, $N_2 = -3.0$ octaves
   c. $f_1 = 10$ kHz, $N_2 = 4.75$ octaves
   d. $f_1 = 64$ kHz, $N_2 = -5.0$ octaves

   Solution: Begin with N_2 = [log (f_2 / f_1)] / [log (2)], the separation in octaves, N_2, between f_1, and f_2.

      [log (f_2 / f_1)] / [log (2)] $= N_2$

$\{[\log (f_2 / f_1)] / [\log (2)]\} \log (2) = N_2 \log (2)$

$10^{\log(f_2 / f_1)} = 10^{N_2 \log(2)}$

$f_2 / f_1 = 10^{N_2 \log(2)}$

$(f_2 / f_1) f_1 = f_1 [10^{N_2 \log(2)}]$

$f_2 = f_1 [10^{N_2 \log(2)}]$

a. For f_1 = 120 Hz and N_2 = 2.5 octaves,

$$\begin{aligned} f_2 &= f_1 [10^{N_2 \log(2)}] \\ &= (120 \text{ Hz})\{10^{(2.5)[\log (2)]}\} \\ &= (120 \text{ Hz})\{10^{(2.5)(0.301)}\} \\ &= (120 \text{ Hz})\{10^{(0.753)}\} \\ &= (120 \text{ Hz})(5.66) \\ &= \mathbf{679\ Hz} \end{aligned}$$

b. For f_1 = 440 Hz and N_2 = –3.0 octaves,

$$\begin{aligned} f_2 &= f_1 [10^{N_2 \log(2)}] \\ &= (440 \text{ Hz})\{10^{(-3.0)[\log (2)]}\} \\ &= (440 \text{ Hz})\{10^{(-3.0)(0.301)}\} \\ &= (440 \text{ Hz})\{10^{(-0.903)}\} \\ &= (440 \text{ Hz})(0.125) \\ &= \mathbf{55\ Hz} \end{aligned}$$

c. For f_1 = 10 kHz and N_2 = 4.75 octaves,

$$\begin{aligned} f_2 &= f_1 [10^{N_2 \log(2)}] \\ &= (10 \text{ kHz})\{10^{(4.75)[\log (2)]}\} \\ &= (10 \times 10^3 \text{ Hz})\{10^{(4.75)(0.301)}\} \\ &= (10 \times 10^3 \text{ Hz})\{10^{(1.43)}\} \\ &= (10 \times 10^3 \text{ Hz})(26.9) \\ &= 269 \times 10^3 \text{ Hz} \\ &= \mathbf{269\ kHz} \end{aligned}$$

d. For f_1 = 64 kHz and N_2 = –5.0 octaves,

$$\begin{aligned} f_2 &= f_1 [10^{N_2 \log(2)}] \\ &= (64 \text{ kHz})\{10^{(-5.0)[\log (2)]}\} \\ &= (64 \times 10^3 \text{ Hz})\{10^{(-5.0)(0.301)}\} \\ &= (64 \times 10^3 \text{ Hz})\{10^{(-1.51)}\} \\ &= (64 \times 10^3 \text{ Hz})(31.3 \times 10^{-3}) \\ &= 2.0 \times 10^3 \text{ Hz} \\ &= \mathbf{2.0\ kHz} \end{aligned}$$

3. Problem: Calculate the frequency f_2 that is the specified number of decades, N_{10}, from the indicated frequency, f_1.

a. f_1 = 1.0 Hz, N_{10} = –2.5 decades

b. f_1 = 250 Hz, N_{10} = 0.5 decades

c. f_1 = 50 kHz, N_{10} = 3.0 decades

d. f_1 = 3.0 MHz, N_{10} = –7.5 decades

Solution: Begin with $N_{10} = \log(f_2 / f_1)$, the separation in decades, N_{10}, between f_1 and f_2 and solve for f_2.

$$\log(f_2 / f_1) = N_{10}$$

$$10^{\log(f_2 / f_1)} = 10^{N_{10}}$$

$$f_2 / f_1 = 10^{N_{10}}$$

$$(f_2 / f_1) f_1 = f_1 (10^{N_{10}})$$

$$f_2 = f_1 (10^{N_{10}})$$

a. For f_1 = 1.0 Hz and N_{10} = –2.5 decades,

$$\begin{aligned} f_2 &= f_1 (10^{N_{10}}) \\ &= (1.0 \text{ Hz})[10^{(-2.5)}] \\ &= (1.0 \text{ Hz})(3.16 \times 10^{-3}) \\ &= (3.16 \times 10^{-3} \text{ Hz}) \\ &= \mathbf{3.16\ mHz} \end{aligned}$$

b. For f_1 = 250 Hz and N_{10} = 0.5 decades,

$$\begin{aligned} f_2 &= f_1 (10^{N_{10}}) \\ &= (250 \text{ Hz})[10^{(0.5)}] \\ &= (250 \text{ Hz})(3.16) \\ &= \mathbf{791\ Hz} \end{aligned}$$

c. For f_1 = 50 kHz and N_{10} = 3.0 decades,

$$\begin{aligned} f_2 &= f_1 (10^{N_{10}}) \\ &= (50 \text{ kHz})[10^{(3.0)}] \\ &= (50 \times 10^3 \text{ Hz})(1000) \\ &= 50 \times 10^6 \text{ Hz} \\ &= \mathbf{50\ MHz} \end{aligned}$$

d. For f_1 = 3.0 MHz and N_{10} = –7.5 decades,

$$\begin{aligned} f_2 &= f_1 (10^{N_{10}}) \\ &= (3.0 \text{ MHz})[10^{(-7.5)}] \\ &= (3.0 \times 10^6 \text{ Hz})(31.6 \times 10^{-9}) \\ &= 94.9 \times 10^{-3} \text{ Hz} \\ &= \mathbf{94.9\ mHz} \end{aligned}$$

19.1.2 Decibels

1. Problem: Convert each of the following V_{out} / V_{in} voltage ratios to decibels.
   a. V_{in} = 100 mV_{ac}, V_{out} = 2.5 V_{ac}
   b. V_{in} = 6.0 V_{ac}, V_{out} = 4.24 V_{ac}
   c. V_{in} = 7.07 V_{ac}, V_{out} = 10.0 V_{ac}
   d. V_{in} = 120 V_{ac}, V_{out} = 24.0 V_{ac}

   Solution: The voltage ratio, $A_{v(\text{dB})}$, in decibels, is $20 \log(V_{out} / V_{in})$.

   a. For V_{in} = 100 mV_{ac} and V_{out} = 2.5 V_{ac},

$$\begin{aligned} A_{v(\text{dB})} &= 20 \log(V_{out} / V_{in}) \\ &= 20 \log(2.5 \text{ V}_{ac} / 100 \text{ mV}_{ac}) \\ &= 20 \log(2.5 \text{ V}_{ac} / 100 \times 10^{-3} \text{ V}_{ac}) \end{aligned}$$

$= 20 \log (25.0)$
$= (20)(1.40)$
$= \mathbf{28.0\ dB}$

b. For $V_{in} = 6.0\ V_{ac}$ and $V_{out} = 4.24\ V_{ac}$,

$A_{v(dB)} = 20 \log (V_{out} / V_{in})$
$= 20 \log (4.24\ V_{ac} / 6.0\ V_{ac})$
$= 20 \log (0.707)$
$= (20)(-0.151)$
$= \mathbf{-3.02\ dB}$

c. For $V_{in} = 7.07\ V_{ac}$ and $V_{out} = 10.0\ V_{ac}$,

$A_{v(dB)} = 20 \log (V_{out} / V_{in})$
$= 20 \log (10.0\ V_{ac} / 7.07\ V_{ac})$
$= 20 \log (1.41)$
$= (20)(1.151)$
$= \mathbf{23.0\ dB}$

d. For $V_{in} = 120\ V_{ac}$ and $V_{out} = 24.0\ V_{ac}$,

$A_{v(dB)} = 20 \log (V_{out} / V_{in})$
$= 20 \log (24.0\ V_{ac} / 120\ V_{ac})$
$= 20 \log (0.20)$
$= (20)(-0.699)$
$= \mathbf{-14.0\ dB}$

2. Problem: Calculate the V_{out} value for each of the following V_{in} values and $A_{v(dB)}$ values in decibels.

a. $V_{in} = 150\ mV_{ac}$, $A_{v(dB)} = 50$ dB
b. $V_{in} = 5.0\ V_{ac}$, $A_{v(dB)} = 25$ dB
c. $V_{in} = 12.0\ V_{ac}$, $A_{v(dB)} = -5.0$ dB
d. $V_{in} = 120\ V_{ac}$, $A_{v(dB)} = -40.0$ dB

Solution: The voltage ratio, $A_{v\,(dB)}$, in decibels, is $20 \log (V_{out} / V_{in})$. From this,

$20 \log (V_{out} / V_{in}) = A_{v(dB)}$
$[20 \log (V_{out} / V_{in})] / 20 = A_{v(dB)} / 20$
$10^{\log (V_{out} / V_{in})} = 10^{(A_{v(dB)} / 20)}$
$(V_{out} / V_{in})\ V_{in} = V_{in}\ [10^{(A_{v(dB)} / 20)}]$
$V_{out} = V_{in}\ [10^{(A_{v(dB)} / 20)}]$

a. For $V_{in} = 150\ mV_{ac}$ and $A_{v(dB)} = 50$ dB,

$V_{out} = V_{in}\ [10^{(A_{v(dB)} / 20)}]$
$= (150\ mV_{ac})[10^{(50\ dB / 20)}]$
$= (150 \times 10^{-3}\ V_{ac})(10^{2.5})$
$= (150 \times 10^{-3}\ V_{ac})(316)$
$= \mathbf{47.4\ V_{ac}}$

b. For $V_{in} = 5.0\ V_{ac}$ and $A_{v(dB)} = 25$ dB,

$V_{out} = V_{in}\ [10^{(A_{v(db)} / 20)}]$

$= (5.0\ \text{V}_{ac})[10^{(25\ \text{dB}\ /\ 20)}]$
$= (5.0\ \text{V}_{ac})(10^{1.25})$
$= (5.0\ \text{V}_{ac})(17.8)$
$= \mathbf{88.9\ V_{ac}}$

c. For $V_{in} = 12.0\ \text{V}_{ac}$ and $A_{v(\text{dB})} = -5.0$ dB,

$V_{out} = V_{in}\,[10^{(A_{v(\text{dB})}\ /\ 20)}]$
$= (12.0\ \text{V}_{ac})[10^{(-5.0\ \text{dB}\ /\ 20)}]$
$= (12.0\ \text{V}_{ac})(10^{-0.25})$
$= (12.0\ \text{V}_{ac})(0.562)$
$= \mathbf{6.75\ V_{ac}}$

d. For $V_{in} = 120\ \text{V}_{ac}$ and $A_{v(\text{dB})} = -40.0$ dB,

$V_{out} = V_{in}\,[10^{(A_{v(\text{dB})}\ /\ 20)}]$
$= (120\ \text{V}_{ac})[10^{(-40\ \text{dB}\ /\ 20)}]$
$= (120\ \text{V}_{ac})(10^{-2.0})$
$= (120\ \text{V}_{ac})(0.01)$
$= \mathbf{1.20\ V_{ac}}$

3. Problem: Calculate the V_{in} value for each of the following V_{out} values and $A_{v(\text{dB})}$ value in decibels.

a. $V_{out} = 500\ \text{mV}_{ac}$, $A_{v(\text{dB})} = -6.0$ dB
b. $V_{out} = 2.5\ \text{V}_{ac}$, $A_{v(\text{dB})} = 12$ dB
c. $V_{out} = 10\ \text{V}_{ac}$, $A_{v(\text{dB})} = 20$ dB
d. $V_{out} = 24\ \text{V}_{ac}$, $A_{v(\text{dB})} = -33$ dB

Solution: The voltage ratio, $A_{v(\text{dB})}$, in decibels, is $20 \log (V_{out} / V_{in})$. From this,

$A_{v(\text{dB})} = 20 \log (V_{out} / V_{in})$
$A_{v(\text{dB})} / 20 = [20 \log (V_{out} / V_{in})] / 20$
$10^{(A_{v(\text{dB})}\ /\ 20)} = 10^{\log (V_{out} / V_{in})}$
$V_{in}\,[10^{(A_{v(\text{dB})}\ /\ 20)}] = (V_{out} / V_{in})\ V_{in}$
$\{V_{in}\,[10^{(A_{v(\text{dB})}\ /\ 20)}]\} / [10^{(A_{v(\text{dB})}\ /\ 20)}] = V_{out} / [10^{(A_{v(\text{dB})}\ /\ 20)}]$
$V_{in} = V_{out} / [10^{(A_{v(\text{dB})}\ /\ 20)}]$

a. For $V_{out} = 500\ \text{mV}_{ac}$ and $A_{v(\text{dB})} = -6.0$ dB,

$V_{in} = V_{out} / [10^{(A_{v(\text{dB})}\ /\ 20)}]$
$= (500\ \text{mV}_{ac}) / [10^{-6.0\ \text{dB}\ /\ 20})$
$= (500 \times 10^{-3}\ \text{V}_{ac}) / (10^{-0.30})$
$= (500 \times 10^{-3}\ \text{V}_{ac}) / 0.501$
$= 998 \times 10^{-3}\ \text{V}_{ac}$
$= \mathbf{998\ mV_{ac}}$

b. For $V_{out} = 2.5\ \text{V}_{ac}$ and $A_{v(\text{dB})} = 12$ dB,

$V_{in} = V_{out} / [10^{(A_{v(\text{dB})}\ /\ 20)}]$
$= (2.5\ \text{V}_{ac}) / [10^{12\ \text{dB}\ /\ 20})$
$= (2.5\ \text{V}_{ac}) / (10^{0.60})$
$= (2.5\ \text{V}_{ac}) / 3.98$

$= 628 \times 10^{-3}$ V

$=$ **628 mV$_{ac}$**

c. For $V_{out} = 10\ V_{ac}$ and $A_{v(dB)} = 20$ dB,

$$V_{in} = V_{out} / [10^{(A_{v(dB)}/20)}]$$
$$= (10\ V_{ac}) / [10^{20\ dB/20})$$
$$= (10\ V_{ac}) / (10^{1.0})$$
$$= (10\ V_{ac}) / 10.0$$
$$= \mathbf{1.0\ V_{ac}}$$

d. For $V_{out} = 24\ V_{ac}$ and $A_{v(dB)} = -33$ dB,

$$V_{in} = V_{out} / [10^{(A_{v(dB)}/20)}]$$
$$= (24\ V_{ac}) / [10^{-33\ dB/20})$$
$$= (24\ V_{ac}) / (10^{-1.65})$$
$$= (24\ V_{ac}) / (22.4 \times 10^{-3})$$
$$= 1.07 \times 10^{3}\ V_{ac}$$
$$= \mathbf{1.07\ kV_{ac}}$$

19.1.3 Low-Pass Filters

1. Problem: For the low-pass filter in Figure 19-1, calculate the corner frequency, f_c, and the output voltage, V_{out}, for $f = 100$ Hz and $f = 400$ Hz.

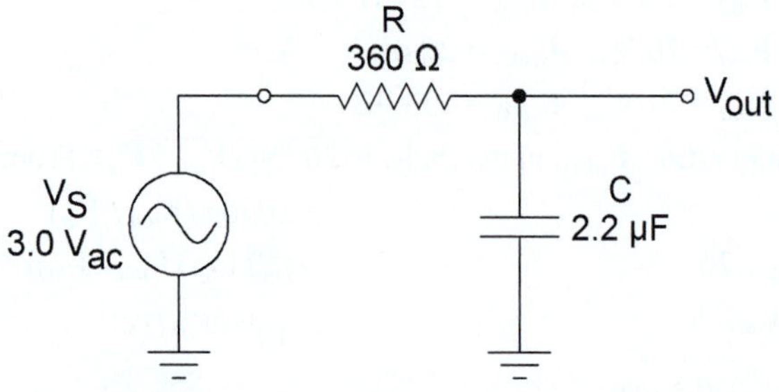

Figure 19-1: *RC* Low-Pass Filter for Problem 1

Solution: First, calculate the corner frequency, f_c. At the corner frequency, $R = X_C$.

$$R = X_C$$
$$R f_c = [1 / (2\pi f_c C)] f_c$$
$$(R f_c) / R = [1 / (2\pi C)] / R$$
$$f_c = [1 / (2\pi RC)]$$

Use the values for R and C to calculate f_c.

$$f_c = 1 / (2\pi RC)$$
$$= 1 / [2\pi(360\ \Omega)(2.2\ \mu F)]$$
$$= 1 / [2\pi(360\ \Omega)(2.2 \times 10^{-6}\ F)]$$
$$= 1 / (4.98\ ms)$$
$$= \mathbf{201\ Hz}$$

Calculate X_C for $f = 100$ Hz.

$$X_C = 1 / (2\pi f C)$$
$$= 1 / [2\pi(100\ Hz)(2.2\ \mu F)]$$

$= 1 / [2\pi(100 \text{ Hz})(2.2 \times 10^{-6} \text{ F})]$

$= 1 / (1.38 \text{ mS})$

$= 723\ \Omega$

Use the voltage divider theorem to calculate V_{out} for $X_C = 723\ \Omega$.

$V_{out} = V_S\,[X_C / Z_T]$

$= V_S\,[-j\,X_C / (R - j\,X_C)]$

$= (3.0\ V_{ac} \angle 0°)[-j\ 723\ \Omega / (360\ \Omega - j\ 723\ \Omega)]$

$= (3.0\ V_{ac} \angle 0°)[(723\ \Omega \angle -90°) / (808\ \Omega \angle -63.5°)]$

$= (3.0\ V_{ac} \angle 0°)\{(723\ \Omega / 808\ \Omega) \angle [-90° - (-63.5°)]$

$= (3.0\ V_{ac} \angle 0°)\ (0.895 \angle -26.5°)$

$= [(3.0\ V_{ac})(0.895)] \angle (0° + -26.5°)$

$= \mathbf{2.69\ Vac \angle -26.5°}$

Calculate X_C for $f = 400$ Hz.

$X_C = 1 / (2\pi f C)$

$= 1 / [2\pi(400 \text{ Hz})(2.2\ \mu\text{F})]$

$= 1 / [2\pi(400 \text{ Hz})(2.2 \times 10^{-6} \text{ F})]$

$= 1 / (5.53 \text{ mS})$

$= 181\ \Omega$

Use the voltage divider theorem to calculate V_{out} for $X_C = 181\ \Omega$.

$V_{out} = V_S\,[X_C / Z_T]$

$= V_S\,[-j\,X_C / (R - j\,X_C)]$

$= (3.0\ V_{ac} \angle 0°)[-j\ 181\ \Omega / (360\ \Omega - j\ 181\ \Omega)]$

$= (3.0\ V_{ac} \angle 0°)[(181\ \Omega \angle -90°) / (403\ \Omega \angle -26.7°)]$

$= (3.0\ V_{ac} \angle 0°)\{(181\ \Omega / 403\ \Omega) \angle [-90° - (-26.7°)]$

$= (3.0\ V_{ac} \angle 0°)\ (0.449 \angle -63.3°)$

$= [(3.0\ V_{ac})(0.449)] \angle (0° + -63.3°)$

$= \mathbf{1.35\ V_{ac} \angle -63.3°}$

2. Problem: For the low-pass filter in Figure 19-2, calculate the corner frequency, f_c, and the output voltage, V_{out}, for $f = 1.2$ Hz and $f = 2.7$ Hz.

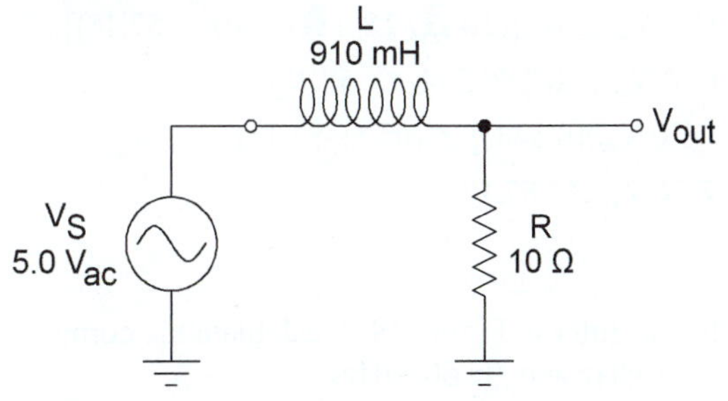

Figure 19-2: *RL* Low-Pass Filter for Problem 2

Solution: First, calculate the corner frequency, f_c. At the corner frequency, $R = X_L$.

$X_L = R$

$(2\pi f_c L)] / 2\pi L = R / (2\pi L)$

$f_c = R / (2\pi L)$

Use the values for R and L to calculate f_c.

$f_c = R / (2\pi L)$
$= (10\ \Omega) / [2\pi(910\ \text{mH})]$
$= (10\ \Omega) / [2\pi(910 \times 10^{-3}\ \text{H})]$
$= (10\ \Omega) / (5.72\ \text{H})$
$= \mathbf{1.77\ Hz}$

Calculate X_L for $f = 1.2$ Hz.

$X_L = 2\pi f L$
$= 2\pi(1.2\ \text{Hz})(910\ \text{mF})]$
$= 2\pi(1.2\ \text{Hz})(910 \times 10^{-3}\ \text{H})$
$= 6.86\ \Omega$

Use the voltage divider theorem to calculate V_{out} for $X_L = 6.86\ \Omega$.

$V_{out} = V_S\ [R / Z_T]$
$= V_S\ [R / (R + j\ X_L)]$
$= (5.0\ V_{ac} \angle 0°)[10\ \Omega / (10\ \Omega + j\ 6.86\ \Omega)]$
$= (5.0\ V_{ac} \angle 0°)[(10\ \Omega \angle 0°) / (12.1\ \Omega \angle 34.5°)]$
$= (5.0\ V_{ac} \angle 0°)[(10\ \Omega / 12.1\ \Omega) \angle (0° - 34.5.5°)]$
$= (5.0\ V_{ac} \angle 0°)\ (0.825 \angle -34.5°)$
$= [(5.0\ V_{ac})(0.825)] \angle (0° + -34.5°)$
$= \mathbf{4.12\ V_{ac} \angle -34.5°}$

Calculate X_L for $f = 2.7$ Hz.

$X_L = 2\pi f L$
$= 2\pi(2.7\ \text{Hz})(910\ \text{mF})]$
$= 2\pi(2.7\ \text{Hz})(910 \times 10^{-3}\ \text{H})$
$= 15.4\ \Omega$

Use the voltage divider theorem to calculate V_{out} for $X_L = 15.4\ \Omega$.

$V_{out} = V_S\ [R / Z_T]$
$= V_S\ [R / (R + j\ X_L)]$
$= (5.0\ V_{ac} \angle 0°)[10\ \Omega / (10\ \Omega + j\ 15.4\ \Omega)]$
$= (5.0\ V_{ac} \angle 0°)[(10\ \Omega \angle 0°) / (18.4\ \Omega \angle 57.1°)]$
$= (5.0\ V_{ac} \angle 0°)[(10\ \Omega / 18.4\ \Omega) \angle (0° - 57.1°)]$
$= (5.0\ V_{ac} \angle 0°)\ (0.544 \angle -57.1°)$
$= [(5.0\ V_{ac})(0.544)] \angle (0° + -57.1°)$
$= \mathbf{2.72\ V_{ac} \angle -57.1°}$

19.1.4 High-Pass Filters

1. Problem: For the high-pass filter in Figure 19-3, calculate the corner frequency, f_c, and the output voltage, V_{out}, for $f = 7.0$ MHz and $f = 60$ MHz.

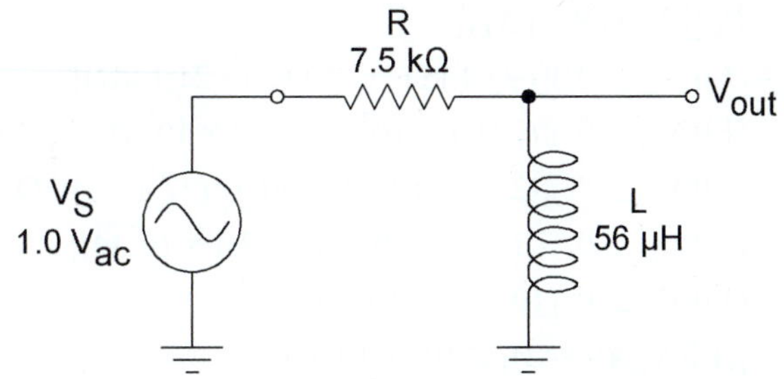

Figure 19-3: *RL* High-Pass Filter for Problem 1

Solution: First, calculate the corner frequency, f_c. At the corner frequency, $\boldsymbol{R} = \boldsymbol{X_L}$.

$$\boldsymbol{X_L} = \boldsymbol{R}$$
$$(2\pi \boldsymbol{f_c L})] / 2\pi \boldsymbol{L} = \boldsymbol{R} / (2\pi \boldsymbol{L})$$
$$\boldsymbol{f_c} = \boldsymbol{R} / (2\pi \boldsymbol{L})$$

Use the values for $\boldsymbol{R}$ and $\boldsymbol{L}$ to calculate $\boldsymbol{f_c}$.

$$\begin{aligned}
\boldsymbol{f_c} &= \boldsymbol{R} / (2\pi \boldsymbol{L}) \\
&= (7.5\ \text{k}\Omega) / [2\pi(56\ \mu\text{H})] \\
&= (7.5 \times 10^3\ \Omega) / [2\pi(56 \times 10^{-6}\ \text{H})] \\
&= (7.5 \times 10^3\ \Omega) / (352 \times 10^{-6}\ \text{H}) \\
&= 21.3 \times 10^6\ \text{Hz} \\
&= \mathbf{21.3\ MHz}
\end{aligned}$$

Calculate $\boldsymbol{X_L}$ for $\boldsymbol{f} = 7.0$ MHz.

$$\begin{aligned}
\boldsymbol{X_L} &= 2\pi \boldsymbol{fL} \\
&= 2\pi(7.0\ \text{MHz})(56\ \mu\text{F})] \\
&= 2\pi(7.0 \times 10^6\ \text{Hz})(56 \times 10^{-6}\ \text{H}) \\
&= 2.46\ \text{k}\Omega
\end{aligned}$$

Use the voltage divider theorem to calculate $\boldsymbol{V_{out}}$ for $\boldsymbol{X_L} = 2.46$ kΩ.

$$\begin{aligned}
\boldsymbol{V_{out}} &= \boldsymbol{V_S}\,[\boldsymbol{X_L} / \boldsymbol{Z_T}] \\
&= \boldsymbol{V_S}\,[\text{j}\,\boldsymbol{X_L} / (\boldsymbol{R} + \text{j}\,\boldsymbol{X_L})] \\
&= (1.0\ \text{V}_{\text{ac}} \angle 0°)[\text{j}\ 2.46\ \text{k}\Omega / (7.5\ \text{k}\Omega + \text{j}\ 2.46\ \text{k}\Omega)] \\
&= (1.0\ \text{V}_{\text{ac}} \angle 0°)[(\text{j}\ 2.46 \times 10^3\ \Omega) / (7.5 \times 10^3\ \Omega + \text{j}\ 2.46 \times 10^3\ \Omega)] \\
&= (1.0\ \text{V}_{\text{ac}} \angle 0°)[(2.46 \times 10^3\ \Omega \angle 90°) / (7.89 \times 10^3\ \Omega \angle 18.2°)] \\
&= (1.0\ \text{V}_{\text{ac}} \angle 0°)\{[(2.46 \times 10^3\ \Omega) / (7.89 \times 10^3\ \Omega)] \angle (90° - 18.2°)\} \\
&= (1.0\ \text{V}_{\text{ac}} \angle 0°)\ (0.312 \angle 71.8°) \\
&= [(1.0\ \text{V}_{\text{ac}})(0.312)] \angle (0° + 71.8°) \\
&= 0.312\ \text{V}_{\text{ac}} \angle 71.8° \\
&= \mathbf{312\ mV_{ac} \angle 71.8°}
\end{aligned}$$

Calculate $\boldsymbol{X_L}$ for $\boldsymbol{f} = 60$ MHz.

$$\begin{aligned}
\boldsymbol{X_L} &= 2\pi \boldsymbol{fL} \\
&= 2\pi(60\ \text{MHz})(56\ \mu\text{F})] \\
&= 2\pi(60 \times 10^6\ \text{Hz})(56 \times 10^{-6}\ \text{H}) \\
&= 21.1\ \text{k}\Omega
\end{aligned}$$

Use the voltage divider theorem to calculate $\boldsymbol{V_{out}}$ for $\boldsymbol{X_L} = 21.1$ kΩ.

$$\boldsymbol{V_{out}} = \boldsymbol{V_S}\,[\boldsymbol{X_L} / \boldsymbol{Z_T}]$$

$= \boldsymbol{V_S}\,[\mathrm{j}\,\boldsymbol{X_L} / (\boldsymbol{R} + \mathrm{j}\,\boldsymbol{X_L})]$
$= (1.0\ \mathrm{V_{ac}} \angle 0°)[\mathrm{j}\ 21.1\ \mathrm{k\Omega} / (7.5\ \mathrm{k\Omega} + \mathrm{j}\ 21.1\ \mathrm{k\Omega})]$
$= (1.0\ \mathrm{V_{ac}} \angle 0°)[(\mathrm{j}\ 21.1 \times 10^3\ \Omega) / (7.5 \times 10^3\ \Omega + \mathrm{j}\ 21.1 \times 10^3\ \Omega)]$
$= (1.0\ \mathrm{V_{ac}} \angle 0°)[(21.1 \times 10^3\ \Omega \angle 90°) / (22.4 \times 10^3\ \Omega \angle 70.4°)]$
$= (1.0\ \mathrm{V_{ac}} \angle 0°)\{[(21.1 \times 10^3\ \Omega) / (22.4 \times 10^3\ \Omega)] \angle (90° - 70.4°)\}$
$= (1.0\ \mathrm{V_{ac}} \angle 0°)\,(0.842 \angle 19.6°)$
$= [(1.0\ \mathrm{V_{ac}})(0.842)] \angle (0° + 19.6°)$
$= 0.842\ \mathrm{V_{ac}} \angle 19.6°$
$= \mathbf{842\ mV_{ac} \angle 19.6°}$

2. Problem: For the high-pass filter in Figure 19-4, calculate the corner frequency, $\boldsymbol{f_c}$, and the output voltage, $\boldsymbol{V_{out}}$, for $\boldsymbol{f} = 4.5$ Hz and $\boldsymbol{f} = 6.5$ Hz.

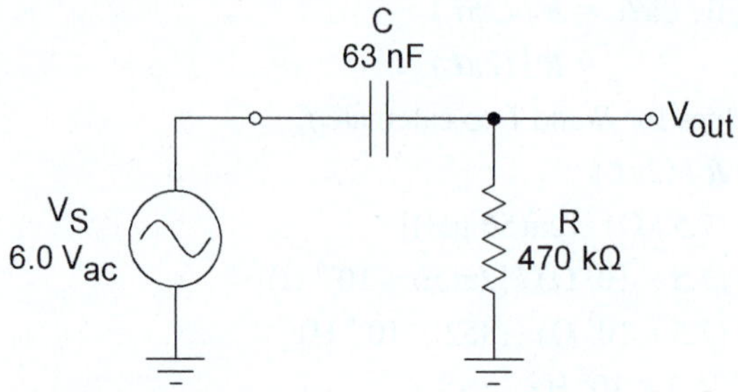

Figure 19-4: *RC* High-Pass Filter for Problem 2

Solution: First, calculate the corner frequency, $\boldsymbol{f_c}$. At the corner frequency, $\boldsymbol{R} = \boldsymbol{X_C}$.

$\boldsymbol{R} = \boldsymbol{X_C}$
$\boldsymbol{R f_c} = [1 / (2\pi \boldsymbol{f_c C})]\,\boldsymbol{f_c}$
$(\boldsymbol{R f_c}) / \boldsymbol{R} = [1 / (2\pi \boldsymbol{C})] / \boldsymbol{R}$
$\boldsymbol{f_c} = [1 / (2\pi \boldsymbol{RC})]$

Use the values for $\boldsymbol{R}$ and $\boldsymbol{C}$ to calculate $\boldsymbol{f_c}$.

$\boldsymbol{f_c} = 1 / (2\pi \boldsymbol{RC})$
$= 1 / [2\pi(470\ \mathrm{k\Omega})(63\ \mathrm{nF})]$
$= 1 / [2\pi(470 \times 10^3\ \Omega)(63 \times 10^{-9}\ \mathrm{F})]$
$= 1 / (186\ \mathrm{ms})$
$= \mathbf{5.38\ Hz}$

Calculate $\boldsymbol{X_C}$ for $\boldsymbol{f} = 4.5$ Hz.

$\boldsymbol{X_C} = 1 / (2\pi \boldsymbol{fC})$
$= 1 / [2\pi(4.5\ \mathrm{Hz})(63\ \mathrm{nF})]$
$= 1 / [2\pi(4.5\ \mathrm{Hz})(63 \times 10^{-9}\ \mathrm{F})]$
$= 1 / (1.78\ \mu\mathrm{S})$
$= 561\ \mathrm{k\Omega}$

Use the voltage divider theorem to calculate $\boldsymbol{V_{out}}$ for $\boldsymbol{X_C} = 561\ \mathrm{k\Omega}$.

$\boldsymbol{V_{out}} = \boldsymbol{V_S}\,[\boldsymbol{R} / \boldsymbol{Z_T}]$
$= \boldsymbol{V_S}\,[\boldsymbol{R} / (\boldsymbol{R} - \mathrm{j}\,\boldsymbol{X_C})]$
$= (6.0\ \mathrm{V_{ac}} \angle 0°)[470\ \mathrm{k\Omega} / (470\ \mathrm{k\Omega} - \mathrm{j}\ 561\ \mathrm{k\Omega})]$
$= (6.0\ \mathrm{V_{ac}} \angle 0°)[470 \times 10^3\ \Omega) / (470 \times 10^3\ \Omega - \mathrm{j}\ 561 \times 10^3\ \Omega)]$

$= (6.0\ V_{ac} \angle 0°)[(470 \times 10^3\ \Omega \angle 0°) / (732 \times 10^3\ \Omega \angle -50.1°)]$

$= (6.0\ V_{ac} \angle 0°)\{[(470 \times 10^3\ \Omega) / (732 \times 10^3\ \Omega)] \angle [0° - (-50.1°)]\}$

$= (6.0\ V_{ac} \angle 0°)\ (0.642 \angle 50.1°)$

$= [(6.0\ V_{ac})(0.642)] \angle (0° + 50.1°)$

$= \mathbf{3.85\ V_{ac} \angle 50.1°}$

Calculate X_C for $f = 6.5$ Hz.

$X_C = 1 / (2\pi f C)$

$= 1 / [2\pi(6.5\text{ Hz})(63\text{ nF})]$

$= 1 / [2\pi(6.5\text{ Hz})(63 \times 10^{-9}\text{ F})]$

$= 1 / (2.57\ \mu\text{S})$

$= 389\text{ k}\Omega$

Use the voltage divider theorem to calculate V_{out} for $X_C = 389\text{ k}\Omega$.

$V_{out} = V_S [R / Z_T]$

$= V_S [R / (R - \text{j}X_C)]$

$= (6.0\ V_{ac} \angle 0°)[470\text{ k}\Omega / (470\text{ k}\Omega - \text{j } 389\text{ k}\Omega)]$

$= (6.0\ V_{ac} \angle 0°)[470 \times 10^3\ \Omega) / (470 \times 10^3\ \Omega - \text{j } 389 \times 10^3\ \Omega)]$

$= (6.0\ V_{ac} \angle 0°)[(470 \times 10^3\ \Omega \angle 0°) / (610 \times 10^3\ \Omega \angle -39.6°)]$

$= (6.0\ V_{ac} \angle 0°)\{[(470 \times 10^3\ \Omega) / (610 \times 10^3\ \Omega)] \angle [0° - (-39.6°)]\}$

$= (6.0\ V_{ac} \angle 0°)\ (0.771 \angle 39.6°)$

$= [(6.0\ V_{ac})(0.771)] \angle (0° + 39.6°)$

$= \mathbf{4.62\ V_{ac} \angle 39.6°}$

19.1.5 Band-Pass Filters

1. Problem: For the cascaded ***RC*** band-pass filter in Figure 19-5, calculate the low and high corner frequencies, $f_{c(l)}$ and $f_{c(h)}$, and the bandwidth, ***BW***, of the filter.

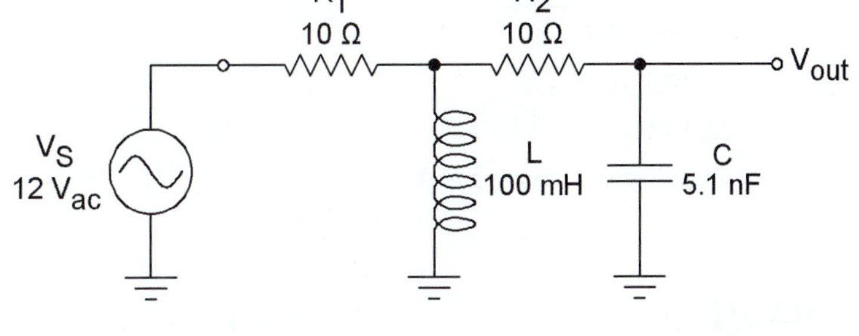

Figure 19-5: Cascaded *RLC* Band-Pass Filter for Problem 1

Solution: From the circuit configuration, R_1 and L determine the high-pass response. R_1, R_2 and C determine the low-pass response, because both R_1 and R_2 are in series with C. Calculate the corner frequency for each of these networks. For the low-pass response, calculate $f_{c(l)}$.

$f_{c(l)} = 1 / [2\pi(R_1 + R_2)C]$

$= 1 / [2\pi(10\ \Omega + 10\ \Omega)(5.1\text{ nF})]$

$= 1 / [2\pi(20\ \Omega)(5.1 \times 10^{-9}\text{ F})]$

$= 1 / (641\text{ ns})$

$= \mathbf{1.56\ MHz}$

For the high-pass response, calculate $f_{c(h)}$.

$f_{c(h)} = R_1 / (2\pi L)$

$= 10\ \Omega / [2\pi(100\text{ mH})]$

$= 10\ \Omega / [2\pi(100 \times 10^{-3}\text{ H})]$

$= 10\ \Omega / (628 \times 10^{-3}\text{ H})$

$= \mathbf{15.9\ Hz}$

The bandwidth, $\boldsymbol{BW}$, of the filter is the difference between $f_{c(h)}$ and $f_{c(l)}$.

$BW = f_{c(l)} - f_{c(h)}$

$= 1.56\text{ MHz} - 15.9\text{ Hz}$

$= \mathbf{1.56\ MHz}$

2. Problem: For the series $\boldsymbol{RLC}$ band-pass filter of Figure 19-6, calculate the center frequency, f_{mid}, the low and high corner frequencies, $f_{c(l)}$ and $f_{c(h)}$, and the $\boldsymbol{Q}$ of the filter.

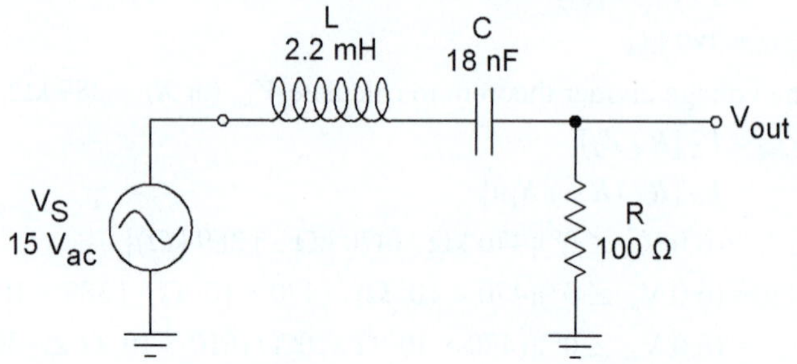

Figure 19-6: Series *RLC* Band-Pass Filter for Problem 2

Solution: The center frequency, f_{mid}, is at the resonant frequency, f_r, for the circuit.

$f_{mid} = f_r$

$= 1 / (2\pi\sqrt{LC})$

$= 1 / (2\pi\sqrt{(2.2\text{ mH})(18\text{ nF})})$

$= 1 / [2\pi\sqrt{(2.2\times10^{-3}\text{ H})(18\times10^{-9}\text{ F})}]$

$= 1 / [2\pi\sqrt{39.6\times10^{-12}\text{ s}^2}]$

$= 1 / 39.5\ \mu\text{s}$

$= \mathbf{25.3\ kHz}$

The corner frequencies occur when $X_T = R$.

$\text{j}X_T = \text{j}X_L - \text{j}X_C$

$= \text{j}(X_L - X_C)$

$(\text{j}X_T)/\text{j} = [\text{j}(X_L - X_C)]/\text{j}$

$X_T = X_L - X_C$

Substitute $\boldsymbol{R}$ for X_T and the reactance equations for X_L and X_C.

$R = (2\pi fL) - [1/(2\pi fC)]$

$= (2\pi fL) - [1/(2\pi fC)]$

$= (2\pi fL)[(2\pi fC)/(2\pi fC)] - [1/(2\pi fC)]$

$= [(2\pi fL)(2\pi fC)/(2\pi fC)] - [1/(2\pi fC)]$

$= [(4\pi^2 f^2 LC)/(2\pi fC)] - [1/(2\pi fC)]$

$= (4\pi^2 f^2 LC - 1)/(2\pi fC)$

$(R)(2\pi fC) = [(4\pi^2 f^2 LC - 1)/(2\pi fC)](2\pi fC)$

$$\begin{aligned}
2\pi f\boldsymbol{RC} &= 4\pi^2 f^2 \boldsymbol{LC} - 1 \\
(2\pi f\boldsymbol{RC}) - (2\pi f\boldsymbol{RC}) &= (4\pi^2 f^2 \boldsymbol{LC} - 1) - (2\pi f\boldsymbol{RC}) \\
0 &= 4\pi^2 f^2 \boldsymbol{LC} - 2\pi f\boldsymbol{RC} - 1 \\
&= (4\pi^2 \boldsymbol{LC})\boldsymbol{f}^2 - (2\pi\boldsymbol{RC})\boldsymbol{f} - 1
\end{aligned}$$

Substitute in the value of $\boldsymbol{R}$, $\boldsymbol{L}$, and $\boldsymbol{C}$ for this quadratic equation.

$$\begin{aligned}
(4\pi^2 \boldsymbol{LC})\boldsymbol{f}^2 - (2\pi\boldsymbol{RC})\boldsymbol{f} - 1 &= 0 \\
[4\pi^2(2.2\text{ mH})(18\text{ nF})]\boldsymbol{f}^2 - [2\pi(100\ \Omega)(18\text{ nF})]\boldsymbol{f} - 1 &= 0 \\
[4\pi^2(2.2 \times 10^{-3}\text{ H})(18 \times 10^{-9}\text{ F})]\boldsymbol{f}^2 - [2\pi(100\ \Omega)(18 \times 10^{-9}\text{ F})]\boldsymbol{f} - 1 &= 0 \\
[4\pi^2(39.6 \times 10^{-12}\text{ s}^2)]\boldsymbol{f}^2 - [2\pi(1.80 \times 10^{-6}\text{ s})]\boldsymbol{f} - 1 &= 0 \\
(1.56 \times 10^{-9}\text{ s}^2)]\boldsymbol{f}^2 - (11.3 \times 10^{-6}\text{ s})\boldsymbol{f} - 1 &= 0
\end{aligned}$$

Use the quadratic formula to solve for $\boldsymbol{f}$, where a = 1.56×10^{-9}, b = -1.13×10^{-6}, and c = -1.

$$\begin{aligned}
\boldsymbol{f} &= \frac{-\text{b} \pm \sqrt{\text{b}^2 - 4\text{ac}}}{2\text{a}} \\
&= \frac{-(-11.3\times10^{-6}) \pm \sqrt{(-11.3\times10^{-6})^2 - 4(1.56\times10^{-9})(-1)}}{(2)(1.56\times10^{-9})} \\
&= \frac{(11.3\times10^{-6}) \pm \sqrt{128\times10^{-12} + 6.25\times10^{-9}}}{3.12\times10^{-9}} \\
&= \frac{(11.3\times10^{-6}) \pm \sqrt{6.38\times10^{-9}}}{3.12\times10^{-9}} \\
&= \frac{(11.3\times10^{-6}) \pm (79.9\times10^{-6})}{3.12\times10^{-9}}
\end{aligned}$$

Evaluate this expression to find $\boldsymbol{f}_{c(l)}$ and $\boldsymbol{f}_{c(h)}$.

$$\begin{aligned}
\boldsymbol{f}_{c(l)} &= [(11.3 \times 10^{-6}) + (79.9 \times 10^{-6})] / (3.13 \times 10^{-9}) \\
&= (91.2 \times 10^{-3}) / (3.13 \times 10^{-9}) \\
&= \boldsymbol{29.2\ kHz}
\end{aligned}$$

$$\begin{aligned}
\boldsymbol{f}_{c(h)} &= [(11.3 \times 10^{-6}) - (79.9 \times 10^{-6})] / (3.13 \times 10^{-9}) \\
&= (-91.2 \times 10^{-3}) / (3.13 \times 10^{-9}) \\
&= -21.9\text{ kHz}
\end{aligned}$$

The negative sign for frequency can be ignored, so that $\boldsymbol{f}_{c(h)} = \boldsymbol{21.9\ kHz}$.

Use $\boldsymbol{f}_{c(l)}$, $\boldsymbol{f}_{c(h)}$, and $\boldsymbol{f}_{mid}$ to calculate the $\boldsymbol{Q}$ for the circuit.

$$\begin{aligned}
\boldsymbol{Q} &= \boldsymbol{f}_{mid} / \boldsymbol{BW} \\
&= \boldsymbol{f}_{mid} / (\boldsymbol{f}_{c(l)} - \boldsymbol{f}_{c(l)}) \\
&= 25.3\text{ kHz} / (29.2\text{ kHz} - 21.9\text{ kHz}) \\
&= 25.3\text{ kHz} / 7.23\text{ kHz} \\
&= \boldsymbol{3.496}
\end{aligned}$$

Note that this value of $\boldsymbol{Q}$ is identical to the value for $\boldsymbol{Q} = \boldsymbol{X}_L / \boldsymbol{R}$ at $\boldsymbol{f} = \boldsymbol{f}_{mid}$ = 25.3 kHz.

3. Problem: For the ***RLC*** band-pass filter of Figure 19-7, calculate the low and high corner frequencies, $\boldsymbol{f}_{cl}$ and $\boldsymbol{f}_{ch}$, the center frequency, $\boldsymbol{f}_{mid}$, and the $\boldsymbol{Q}$ of the filter.

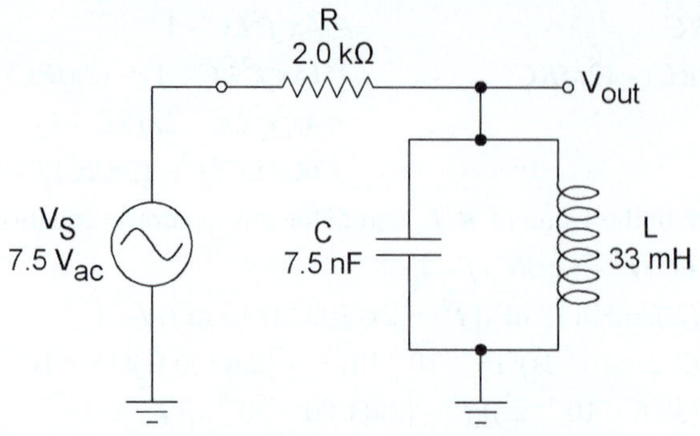

Figure 19-7: Parallel *RLC* Band-Pass Filter for Problem 3

Solution: The center frequency, f_{mid}, is at the resonant frequency, f_r, for the circuit.

$$f_{mid} = f_r$$
$$= 1 / (2\pi\sqrt{LC})$$
$$= 1 / (2\pi\sqrt{(33\text{ mH})(7.5\text{ nF})})$$
$$= 1 / [2\pi\sqrt{(33\times10^{-3}\text{ H})(7.5\times10^{-9}\text{ F})}]$$
$$= 1 / [2\pi\sqrt{248\times10^{-12}\text{ s}^2}]$$
$$= 1 / 98.8\ \mu\text{s}$$
$$= \mathbf{10.1\ kHz}$$

The corner frequencies occur when $X_T = R$.

$$\text{j}\,X_T = \text{j}\,X_L \parallel -\text{j}\,X_C$$
$$= [-\text{j}^2 (X_L X_C)] / (\text{j}\,X_L - \text{j}\,X_C)$$
$$= [-(-1)(X_L X_C)] / [\text{j}(X_L - X_C)]$$
$$= (X_L X_C) / [\text{j}(X_L - X_C)][\text{j}/\text{j}]$$
$$= (\text{j}\,X_L X_C) / [\text{j}^2 (X_L - X_C)]$$
$$= (\text{j}\,X_L X_C) / [(-1)(X_L - X_C)]$$
$$= (\text{j}\,X_L X_C) / (X_C - X_L)$$
$$(\text{j}\,X_T)/\text{j} = [(\text{j}\,X_L X_C)/(X_C - X_L)]/\text{j}$$
$$X_T = (X_L X_C)/(X_C - X_L)$$

Substitute R for X_T and the reactance equations for X_L and X_C.

$$R = \{(2\pi fL)[1/(2\pi fC)]\} / \{[1/(2\pi fC)] - (2\pi fL)\}$$
$$= (2\pi fL / 2\pi fC) / \{[1/(2\pi fC)] - [(2\pi fL)(2\pi fC / 2\pi fC)]\}$$
$$= (L/C) / \{[1/(2\pi fC)] - [(2\pi fL)(2\pi fC)/2\pi fC]\}$$
$$= (L/C) / \{[1/(2\pi fC)] - [(4\pi^2 f^2 LC)/2\pi fC]\}$$
$$= (L/C) / [(1 - 4\pi^2 f^2 LC)/2\pi fC]$$
$$= (L/C)[(2\pi fC)/(1 - 4\pi^2 f^2 LC)]$$
$$= (2\pi fL)/(1 - 4\pi^2 f^2 LC)$$
$$(R)(1 - 4\pi^2 f^2 LC) = [(2\pi fL)/(1 - 4\pi^2 f^2 LC)](1 - 4\pi^2 f^2 LC)$$
$$(R - 4\pi^2 f^2 RLC) = 2\pi fL$$
$$(R - 4\pi^2 f^2 RLC) - 2\pi fL = 2\pi fL - 2\pi fL$$
$$(-4\pi^2 f^2 RLC - 2\pi fL + R)(-1) = (0)(-1)$$

$$(4\pi^2 \boldsymbol{RLC})\boldsymbol{f}^2 + (2\pi \boldsymbol{L})\boldsymbol{f} - \boldsymbol{R} = 0$$

Substitute in the value of $\boldsymbol{R}$, $\boldsymbol{L}$, and $\boldsymbol{C}$ for this quadratic equation.

$$(4\pi^2 \boldsymbol{RLC})\boldsymbol{f}^2 + (2\pi \boldsymbol{L})\boldsymbol{f} - \boldsymbol{R} = 0$$

$$[4\pi^2(2.0\ \text{k}\Omega)(33\ \text{mH})(7.5\ \text{nF})]\boldsymbol{f}^2 + [2\pi(33\ \text{mH})]\boldsymbol{f} - (2.0\ \text{k}\Omega) = 0$$

$$[4\pi^2(2.0 \times 10^3\ \Omega)(33 \times 10^{-3}\ \text{H})(7.5 \times 10^{-9}\ \text{F})]\boldsymbol{f}^2 + [2\pi\,(33 \times 10^{-3}\ \text{H})]\boldsymbol{f} - (2.0 \times 10^3\ \Omega) = 0$$

$$[4\pi^2(495 \times 10^{-9}\ \Omega \cdot \text{s}^2)]\boldsymbol{f}^2 + [2\pi(33 \times 10^{-3}\ \text{H})]\boldsymbol{f} - (2.0 \times 10^3\ \Omega) = 0$$

$$(19.5 \times 10^{-6}\ \Omega \cdot \text{s}^2)]\boldsymbol{f}^2 + (207 \times 10^{-3}\ \text{H})\boldsymbol{f} - (2.0 \times 10^3\ \Omega) = 0$$

Use the quadratic formula to solve for $\boldsymbol{f}$, where a = 19.5×10^{-6}, b = 207×10^{-3}, and c = -2.0×10^3.

$$f = \frac{-\text{b} \pm \sqrt{\text{b}^2 - 4\text{ac}}}{2\text{a}}$$

$$= \frac{-(207 \times 10^{-3}) \pm \sqrt{(207 \times 10^{-3})^2 - 4(19.5 \times 10^{-6})(-2.0 \times 10^3)}}{(2)(19.5 \times 10^{-6})}$$

$$= \frac{(-207 \times 10^{-3}) \pm \sqrt{43.0 \times 10^{-3} + 156 \times 10^{-3}}}{39.1 \times 10^{-6}}$$

$$= \frac{(-207 \times 10^{-3}) \pm \sqrt{199 \times 10^{-3}}}{39.1 \times 10^{-6}}$$

$$= \frac{(-207 \times 10^{-3}) \pm (446 \times 10^{-3})}{39.1 \times 10^{-6}}$$

Evaluate this expression to find $\boldsymbol{f}_{c(l)}$ and $\boldsymbol{f}_{c(h)}$.

$$\boldsymbol{f}_{c(h)} = [(-207 \times 10^{-3}) + (446 \times 10^{-3})] / (39.1 \times 10^{-6})$$
$$= (239 \times 10^{-3}) / (39.1 \times 10^{-6})$$
$$= \boldsymbol{6.12\ kHz}$$

$$\boldsymbol{f}_{c(l)} = [(-207 \times 10^{-3}) - (446 \times 10^{-3})] / (39.1 \times 10^{-6})$$
$$= (-654 \times 10^{-3}) / (39.1 \times 10^{-6})$$
$$= -16.7\ \text{kHz}$$

The negative sign for frequency can be ignored, so that $\boldsymbol{f}_{c(l)} = \boldsymbol{16.7\ kHz}$.

Use $\boldsymbol{f}_{c(l)}$, $\boldsymbol{f}_{c(h)}$, and $\boldsymbol{f}_{mid}$ to calculate the $\boldsymbol{Q}$ for the circuit.

$$\boldsymbol{Q} = \boldsymbol{f}_{mid} / \boldsymbol{BW}$$
$$= \boldsymbol{f}_{mid} / (\boldsymbol{f}_{c(l)} - \boldsymbol{f}_{c(l)})$$
$$= 10.1\ \text{kHz} / (16.7\ \text{kHz} - 6.12\ \text{kHz})$$
$$= 10.1\ \text{kHz} / 10.6\ \text{kHz}$$
$$= \mathbf{0.953}$$

Note that this value of $\boldsymbol{Q}$ is identical to the value for $\boldsymbol{Q} = \boldsymbol{R} / \boldsymbol{X_L}$ at $\boldsymbol{f} = \boldsymbol{f}_{mid} = 10.1$ kHz.

19.1.6 Band-Stop Filters

1. Problem: For the parallel ***RLC*** band-stop filter in Figure 19-8, calculate the low and high corner frequencies, $\boldsymbol{f}_{cl}$ and $\boldsymbol{f}_{ch}$, the center frequency, $\boldsymbol{f}_{mid}$, and the $\boldsymbol{Q}$ of the filter.

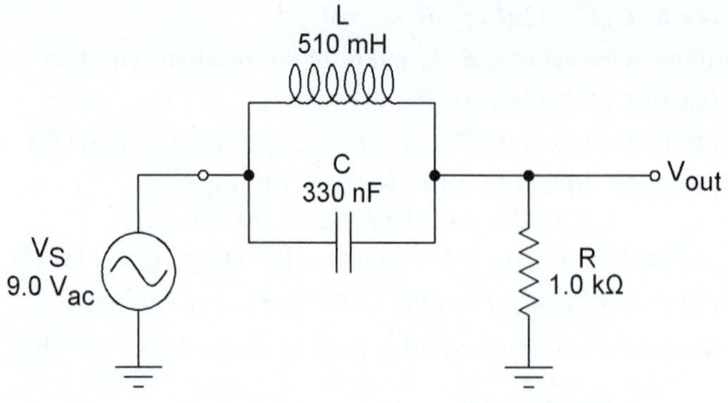

Figure 19-8: Parallel *RLC* Band-Stop Filter for Problem 1

Solution: The center frequency, f_{mid}, is at the resonant frequency, f_r, for the circuit.

$f_{mid} = f_r$

$= 1 / (2\pi\sqrt{LC})$

$= 1 / (2\pi\sqrt{(510\text{ mH})(330\text{ nF})})$

$= 1 / [2\pi\sqrt{(510 \times 10^{-3}\text{ H})(330 \times 10^{-9}\text{ F})}]$

$= 1 / [2\pi\sqrt{168 \times 10^{-9}\text{ s}^2}]$

$= 1 / 2.58\text{ ms}$

$=$ **388 Hz**

The corner frequencies occur when $X_T = R$.

$\text{j}X_T = \text{j}X_L \parallel -\text{j}X_C$

$= [-\text{j}^2 (X_L X_C)] / (\text{j}X_L - \text{j}X_C)$

$= [-(-1)(X_L X_C)] / [\text{j}(X_L - X_C)]$

$= (X_L X_C) / [\text{j}(X_L - X_C)][\text{j} / \text{j}]$

$= (\text{j}X_L X_C) / [\text{j}^2 (X_L - X_C)]$

$= (\text{j}X_L X_C) / [(-1)(X_L - X_C)]$

$= (\text{j}X_L X_C) / (X_C - X_L)$

$(\text{j}X_T) / \text{j} = [(\text{j}X_L X_C) / (X_C - X_L)] / \text{j}$

$X_T = (X_L X_C) / (X_C - X_L)$

Substitute R for X_T and the reactance equations for X_L and X_C.

$R = \{(2\pi fL)[1 / (2\pi fC)]\} / \{[1 / (2\pi fC)] - (2\pi fL)\}$

$= (2\pi fL / 2\pi fC) / \{[1 / (2\pi fC)] - [(2\pi fL)(2\pi fC / 2\pi fC)]\}$

$= (L / C) / \{[1 / (2\pi fC)] - [(2\pi fL)(2\pi fC) / 2\pi fC]\}$

$= (L / C) / \{[1 / (2\pi fC)] - [(4\pi^2 f^2 LC) / 2\pi fC]\}$

$= (L / C) / [(1 - 4\pi^2 f^2 LC) / 2\pi fC]$

$= (L / C)[(2\pi fC) / (1 - 4\pi^2 f^2 LC)]$

$= (2\pi fL) / (1 - 4\pi^2 f^2 LC)$

$(R)(1 - 4\pi^2 f^2 LC) = [(2\pi fL) / (1 - 4\pi^2 f^2 LC)](1 - 4\pi^2 f^2 LC)$

$(R - 4\pi^2 f^2 RLC) = 2\pi fL$

$(R - 4\pi^2 f^2 RLC) - 2\pi fL = 2\pi fL - 2\pi fL$

$(-4\pi^2 f^2 \boldsymbol{RLC} - 2\pi \boldsymbol{fL} + \boldsymbol{R})(-1) \quad = (0)(-1)$

$(4\pi^2 \boldsymbol{RLC})\boldsymbol{f}^2 + (2\pi \boldsymbol{L})\boldsymbol{f} - \boldsymbol{R} \quad = 0$

Substitute in the value of ***R***, ***L***, and ***C*** for this quadratic equation.

$(4\pi^2 \boldsymbol{RLC})\boldsymbol{f}^2 + (2\pi \boldsymbol{L})\boldsymbol{f} - \boldsymbol{R} = 0$

$[4\pi^2(1.0\ \text{k}\Omega)(510\ \text{mH})(330\ \text{nF})]\boldsymbol{f}^2 + [2\pi(510\ \text{mH})]\boldsymbol{f} - (1.0\ \text{k}\Omega) = 0$

$[4\pi^2(1.0 \times 10^3\ \Omega)(510 \times 10^{-3}\ \text{H})(330 \times 10^{-9}\ \text{F})]\boldsymbol{f}^2 + [2\pi\,(510 \times 10^{-3}\ \text{H})]\boldsymbol{f} - (2.0 \times 10^3\ \Omega) = 0$

$[4\pi^2(168 \times 10^{-6}\ \Omega\text{•s}^2)]\boldsymbol{f}^2 + [2\pi(510 \times 10^{-3}\ \text{H})]\boldsymbol{f} - (1.0 \times 10^3\ \Omega) = 0$

$(6.64 \times 10^{-3}\ \Omega\text{•s}^2)]\boldsymbol{f}^2 + (3.20\ \text{H})\boldsymbol{f} - (1.0 \times 10^3\ \Omega) = 0$

Use the quadratic formula to solve for ***f***, where a = 6.64 × 10^{-3}, b = 3.20, and c = –1.0 × 10^3.

$$f = \frac{-\text{b} \pm \sqrt{\text{b}^2 - 4\text{ac}}}{2\text{a}}$$

$$= \frac{-(3.20) \pm \sqrt{(3.20)^2 - 4(6.64 \times 10^{-3})(-1.0 \times 10^3)}}{(2)(6.64 \times 10^{-3})}$$

$$= \frac{(-3.20) \pm \sqrt{10.3 + 26.6}}{13.3 \times 10^{-3}}$$

$$= \frac{(-3.20) \pm \sqrt{36.8}}{13.3 \times 10^{-3}}$$

$$= \frac{(-3.20) \pm (6.07)}{13.3 \times 10^{-3}}$$

Evaluate this expression to find $f_{c(l)}$ and $f_{c(h)}$.

$f_{c(h)} = [(-3.20) + (6.07)] / (13.3 \times 10^{-3})$

$= (2.87) / (39.1 \times 10^{-6})$

$=$ **216 Hz**

$f_{c(l)} = [(-3.20) - (6.07)] / (13.3 \times 10^{-3})$

$= (-9.27) / (13.3 \times 10^{-3})$

$= -698$ Hz

The negative sign for frequency can be ignored, so that $f_{c(l)}$ = **698 Hz.**

Use $f_{c(l)}$, $f_{c(h)}$, and f_{mid} to calculate the ***Q*** for the circuit.

$\boldsymbol{Q} = f_{mid} / \boldsymbol{BW}$

$= f_{mid} / (f_{c(l)} - f_{c(l)})$

$= 388$ Hz / (698 kHz – 216 kHz)

$= 388$ Hz / 482 Hz

$=$ **0.804**

Note that this value of ***Q*** is identical to the value for $\boldsymbol{Q} = \boldsymbol{R} / \boldsymbol{X_L}$ at $f = f_{mid} = 388$ Hz.

2. Problem: For the series ***RLC*** band-stop filter in Figure 19-9, calculate the low and high corner frequencies, f_{cl} and f_{ch}, the center frequency, f_{mid}, and the ***Q*** of the filter.

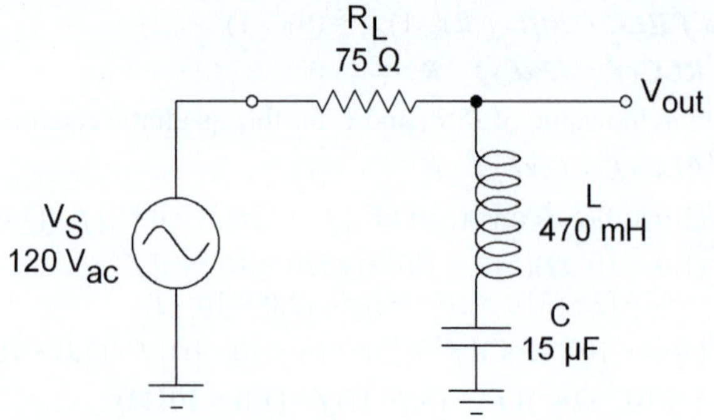

Figure 19-9: Series *RLC* Band-Stop Filter for Problem 2

Solution: The center frequency, f_{mid}, is at the resonant frequency, f_r, for the circuit.

$$f_{mid} = f_r$$
$$= 1 / (2\pi\sqrt{LC})$$
$$= 1 / (2\pi\sqrt{(470\text{ mH})(15\,\mu\text{F})})$$
$$= 1 / [2\pi\sqrt{(470\times10^{-3}\text{ H})(15\times10^{-6}\text{ F})}]$$
$$= 1 / [2\pi\sqrt{7.05\times10^{-6}\text{ s}^2}]$$
$$= 1 / 16.7\text{ ms}$$
$$= \mathbf{59.9\ Hz}$$

The corner frequencies occur when $X_T = R$.

$$\text{j}\,X_T = \text{j}\,X_L - \text{j}\,X_C$$
$$= \text{j}\,(X_L - X_C)$$
$$(\text{j}\,X_T)/\text{j} = [\text{j}\,(X_L - X_C)]/\text{j}$$
$$X_T = X_L - X_C$$

Substitute R for X_T and the reactance formulas for X_L and X_C.

$$R = (2\pi fL) - [1/(2\pi fC)]$$
$$= (2\pi fL) - [1/(2\pi fC)]$$
$$= (2\pi fL)[(2\pi fC)/(2\pi fC)] - [1/(2\pi fC)]$$
$$= [(2\pi fL)(2\pi fC)/(2\pi fC)] - [1/(2\pi fC)]$$
$$= [(4\pi^2 f^2 LC)/(2\pi fC)] - [1/(2\pi fC)]$$
$$= (4\pi^2 f^2 LC - 1)/(2\pi fC)$$
$$(R)(2\pi fC) = [(4\pi^2 f^2 LC - 1)/(2\pi fC)](2\pi fC)$$
$$2\pi fRC = 4\pi^2 f^2 LC - 1$$
$$(2\pi fRC) - (2\pi fRC) = (4\pi^2 f^2 LC - 1) - (2\pi fRC)$$
$$0 = 4\pi^2 f^2 LC - 2\pi fRC - 1$$
$$= (4\pi^2 LC) f^2 - (2\pi RC) f - 1$$

Substitute in the value of R, L, and C for this quadratic equation.

$$(4\pi^2 LC) f^2 - (2\pi RC) f - 1 = 0$$
$$[4\pi^2(470\text{ mH})(15\,\mu\text{F})] f^2 - [2\pi(75\,\Omega)(15\,\mu\text{F})] f - 1 = 0$$
$$[4\pi^2(470\times10^{-3}\text{ H})(15\times10^{-6}\text{ F})] f^2 - [2\pi(75\,\Omega)(15\times10^{-6}\text{ F})] f - 1 = 0$$
$$[4\pi^2(7.05\times10^{-6}\text{ s}^2)] f^2 - [2\pi(1.13\times10^{-3}\text{ s})] f - 1 = 0$$

$$(278 \times 10^{-6}\ s^2)]f^2 - (7.07 \times 10^{-3}\ s)f - 1 = 0$$

Use the quadratic formula to solve for f, where a = 278 × 10^{-6}, b = –7.07 × 10^{-3}, and c = –1.

$$f = \frac{-b \pm \sqrt{b^2 - 4ac}}{2a}$$

$$= \frac{-(-7.07 \times 10^{-3}) \pm \sqrt{(7.07 \times 10^{-3})^2 - 4(278 \times 10^{-6})(-1)}}{(2)(278 \times 10^{-6})}$$

$$= \frac{(7.07 \times 10^{-3}) \pm \sqrt{50.0 \times 10^{-6} + 1.11 \times 10^{-3}}}{557 \times 10^{-6}}$$

$$= \frac{(7.07 \times 10^{-3}) \pm \sqrt{1.16 \times 10^{-3}}}{557 \times 10^{-6}}$$

$$= \frac{(7.07 \times 10^{-3}) \pm (34.1^{-3})}{557 \times 10^{-6}}$$

Evaluate this expression to find $f_{c(l)}$ and $f_{c(h)}$.

$$f_{c(l)} = [(7.07 \times 10^{-3}) + (34.1 \times 10^{-3})] / (557 \times 10^{-6})$$
$$= (41.2 \times 10^{-3}) / (557 \times 10^{-6})$$
$$= \boldsymbol{74.0\ Hz}$$

$$f_{c(h)} = [(7.07 \times 10^{-3}) - (34.1 \times 10^{-3})] / (557 \times 10^{-6})$$
$$= (-27.0 \times 10^{-3}) / (557 \times 10^{-6})$$
$$= -48.6\ Hz$$

The negative sign for frequency can be ignored, so that $f_{c(h)}$ = ***48.6 Hz***.

Use $f_{c(l)}$, $f_{c(h)}$, and f_{mid} to calculate the Q for the circuit.

$$Q = f_{mid} / BW$$
$$= f_{mid} / (f_{c(l)} - f_{c(l)})$$
$$= 60.0\ Hz / (74.0\ Hz - 48.6\ Hz)$$
$$= 60\ Hz / 25.4\ Hz$$
$$= \mathbf{2.36}$$

Note that this value of Q is identical to the value for $Q = X_L / R$ at $f = f_{mid}$ = 25.3 kHz.

19.2 Advanced

19.2.1 Bode Plots

1. Problem: Sketch the idealized amplitude and phase Bode plots for an ***RC*** low-pass filter with f_c = 15 kHz. Show two decades each of the pass band and stop band.

Solution: For an ***RC*** low-pass filter, the output voltage, V_{out}, is taken across the capacitor. At frequencies below f_c, $V_{out} = V_{in}$ so that the amplitude $A_{v(dB)}$ = 0 dB. At frequencies above f_c, the output rolls off at –20 dB/decade, so at f = 150 kHz (1 decade above 15 kHz) $A_{v(dB)}$ = –20 dB, and at 1.5 MHz (2 decades above 15 kHz) $A_{v(dB)}$ = –40 dB. At f_c, $A_{v(dB)}$ is ideally 0 dB, but practically $A_{v(dB)}$ = –3.0 dB.

Below 0.1 f_c, the capacitive reactance is ten times the resistance. $Z_T \approx X_C$. $V_{out} = V_C \approx V_{in}$, so the phase angle of V_{out} is 0°.

Above 10 f_c, the capacitive reactance is less than one-tenth the resistance, so $Z_T \approx R$ and I_T is in phase with V_{in}. $I_C = I_T$ and leads V_C by 90°, so the phase angle of $V_{out} = V_C$ is –90°.

At the corner frequency, the capacitive reactance equals the resistance. The phase of the output voltage is halfway between the phases at low and high frequencies, so the phase angle of V_{out} at f_c is –45°.

Figure 19-10 shows the idealized amplitude and phase Bode plots for the ***RC*** low-pass filter.

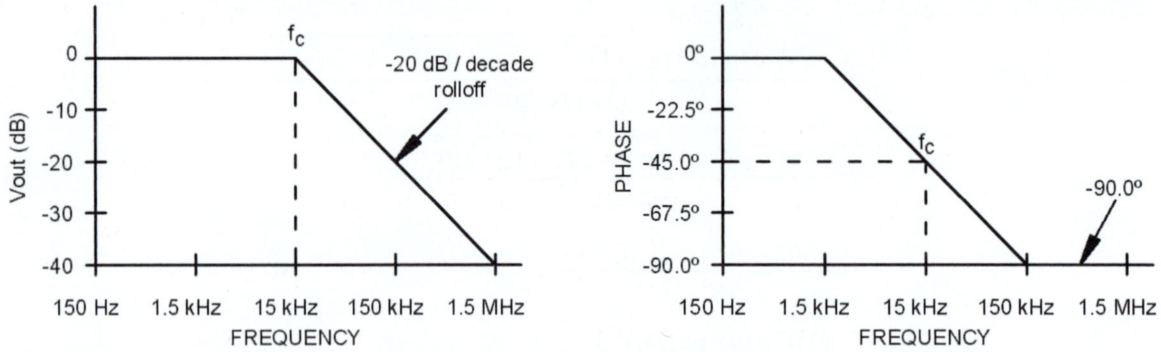

Figure 19-10: Amplitude and Phase Bode Plots for *RC* Low-Pass Filter

2. Problem: Sketch the idealized amplitude and phase Bode plots for an ***RL*** high pass filter with f_c = 60 Hz. Show two decades each of the pass band and stop band.

Solution: For an ***RL*** high-pass filter, the output voltage, V_{out}, is taken across the resistor. At frequencies above f_c, $V_{out} = V_{in}$ so that the amplitude $A_{v(dB)}$ = 0 dB. At frequencies below f_c, the output rolls off at –20 dB/decade, so at f = 6.0 Hz (1 decade below 60 Hz) $A_{v(dB)}$ = –20 dB, and at 0.6 Hz (2 decades below 60 Hz) $A_{v(dB)}$ = –40 dB. At f_c, $A_{v(dB)}$ is ideally 0 dB, but practically $A_{v(dB)}$ = –3.0 dB.

Below one-tenth the corner frequency, the inductive reactance is less than one-tenth the resistance. Because the total impedance $Z_T \approx R$, the total current, I_T, has a phase angle of 0°. The inductor current, I_L, equals I_T and the inductor voltage, V_L, leads I_L by 90°, so the phase of $V_{out} = V_L$ is 90°.

Above ten times the corner frequency, the inductive reactance is more than ten times the resistance. $Z_T \approx X_L$ so that the total current, I_T, lags V_{in} by 90°. $I_L = I_T$, and V_L leads I_L by 90°, so the phase angle of $V_{out} = V_L$ is 0°.

At the corner frequency, the inductive reactance equals the resistance. The phase of the output voltage is halfway between the phases at low and high frequencies, so the phase angle of V_{out} at f_c is 45°.

Figure 19-11 shows the idealized amplitude and phase Bode plots for the ***RL*** high-pass filter.

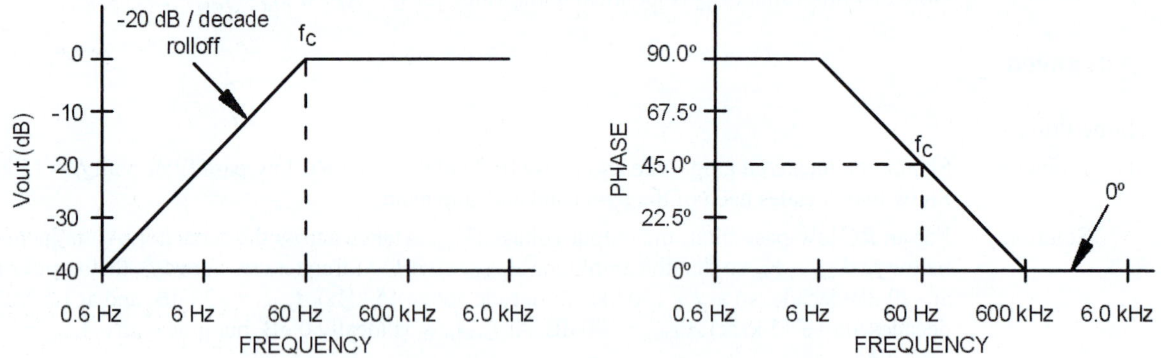

Figure 19-11: Amplitude and Phase Bode Plots for *RL* High-Pass Filter

3. Problem: Sketch the idealized amplitude and phase Bode plots for the cascaded ***RLC*** band-pass filter in Figure 19-5. Show the entire pass band and one decade for each stop band.

Solution: For the cascaded ***RLC*** band-pass filter of Figure 19-5, $f_{c(l)}$ = 1.56 MHz and $f_{c(h)}$ = 15.9 Hz. The rolloff outside the pass-band is –20 dB / decade. Because $f_{c(l)}$ and $f_{c(h)}$ are 5 decades apart, the high-pass and low-pass response can be analyzed independently.

At frequencies below 0.1 $f_{c(h)}$, the output rolls off at –20 dB/decade, so at f = 1.59 Hz (1 decade below 15.9 Hz) $A_{v(dB)}$ = –20 dB. At $f_{c(h)}$, $A_{v(dB)}$ is ideally 0 dB, but practically $A_{v(dB)}$ = –3.0 dB. At frequencies above $f_{c(l)}$, the output rolls off at –20 dB/decade, so at f = 15.6 MHz (1 decade above 1.56 MHz) $A_{v(dB)}$ = –20 dB. At $f_{c(l)}$, $A_{v(dB)}$ is ideally 0 dB, but practically $A_{v(dB)}$ = –3.0 dB.

Below 0.1 $f_{c(h)}$, the inductive reactance, X_L, is less than one-tenth the resistance, ***R***, while the capacitive reactance, X_C, is extremely large and is effectively an open. $Z_T \approx R_1$, so the phase angle of the total current, I_T, is 0°. $I_T \approx I_L$, the inductor current, and the inductor voltage, V_L, leads I_L by 90°, so the phase angle of $V_{out} \approx V_L$ is 90°.

Above 10 $f_{c(h)}$, X_L is more than ten times ***R***, and is effectively an open. X_C also is extremely large and effectively an open, so $V_{out} \approx V_{in}$ and the phase angle of V_{out} is 0°.

At $f_{c(h)}$, the phase angle is halfway between the phase angles of 0.1 $f_{c(h)}$ and 10 $f_{c(h)}$ and equal to 45°.

Below 0.1 $f_{c(l)}$, the capacitive reactance, X_C, is more than ten times the combined resistance, R_1 + R_2. $V_{out} \approx V_{in}$, so the phase angle of $V_{out} \approx$ 0°.

Above 10 $f_{c(l)}$, X_C less than one-tenth ***R***, so $Z_T \approx R$ and I_T is in phase with V_{in}. $I_C = I_T$ and leads V_C by 90° so the phase angle of $V_{out} = V_C$ is –90°..

At $f_{c(l)}$, the phase angle is halfway between the phase angles of 0.1 $f_{c(l)}$ and 10 $f_{c(l)}$ and equal to –45°.

Figure 19-12 shows the idealized amplitude and phase Bode plots for the band-pass filter.

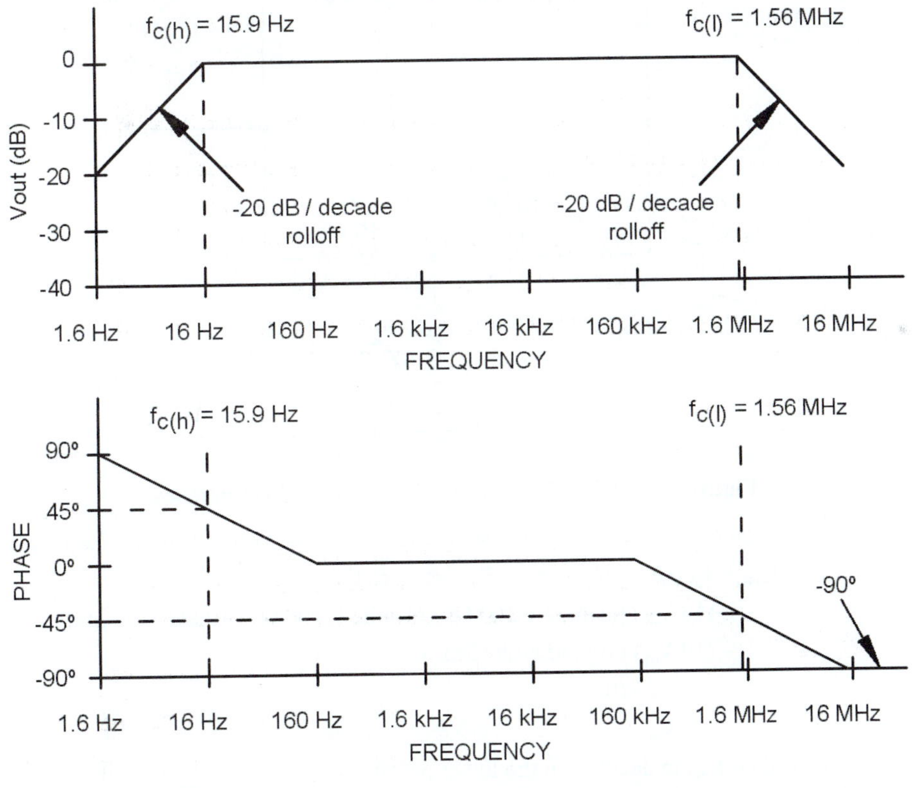

Figure 19-12: Amplitude and Phase Bode Plots for *RLC* Band-Pass Filter

4. Problem: Analyze the band-stop filter of Figure 19-13, determine the values for the low corner frequency, $f_{c(l)}$, the high corner frequency, $f_{c(h)}$, and the center frequency, f_{mid}, and sketch the idealized amplitude and phase Bode plots.

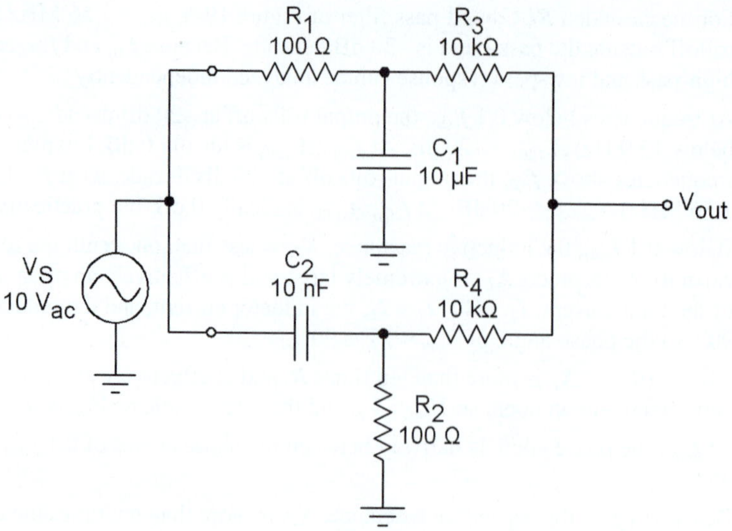

Figure 19-13: *RC* Band-Stop Filter

Solution: Figure 19-14 shows the circuit equivalent in the lower pass-band, when both C_1 and C_2 appear to be open.

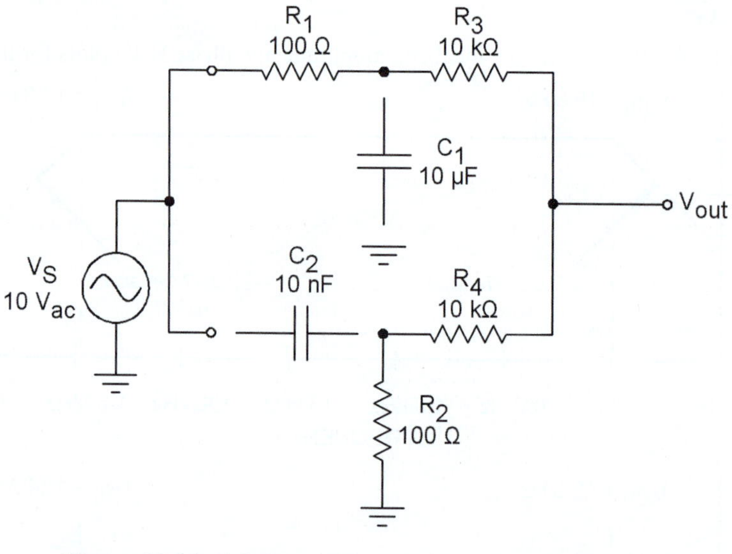

Figure 19-14: *RC* Band-Stop Filter at Low Frequencies

The circuit is a simple series resistive circuit. Use the voltage divider theorem to calculate V_{out}.

$$V_{out} = V_S\,[(R_2 + R_4) / (R_1 + R_2 + R_3 + R_4)]$$
$$= (10\ \text{V}_{ac})\,[(100\ \Omega + 10\ \text{k}\Omega) / (100\ \Omega + 100\ \Omega + 10\ \text{k}\Omega + 10\ \text{k}\Omega)]$$
$$= (10\ \text{V}_{ac})\,(10.1\ \text{k}\Omega / 20.2\ \text{k}\Omega)$$
$$= (10\ \text{V}_{ac})\,(0.5)$$
$$= 5.0\ \text{V}_{ac}$$

Calculate V_{out} in decibels in the lower pass-band.

$$A_{v(\text{dB})} = 20 \log (V_{out} / V_S)$$
$$= 20 \log (5.0\ \text{V}_{ac} / 10\ \text{V}_{ac})$$

= 20 log (0.50)
= (20)(–3.01)
= **–6.02 dB**

Because the circuit in the lower pass-band appears to be purely resistive, the phase angle of V_{out} is 0°.

At the lower corner frequency, $f_{c(l)}$, the capacitor associated with low-pass filter, C_1, will exhibit finite reactance. Because C_1 is 1000 times larger than C_2, C_2 still appears to be an open at $f_{c(l)}$. Figure 19-15 shows the circuit equivalent at these frequencies.

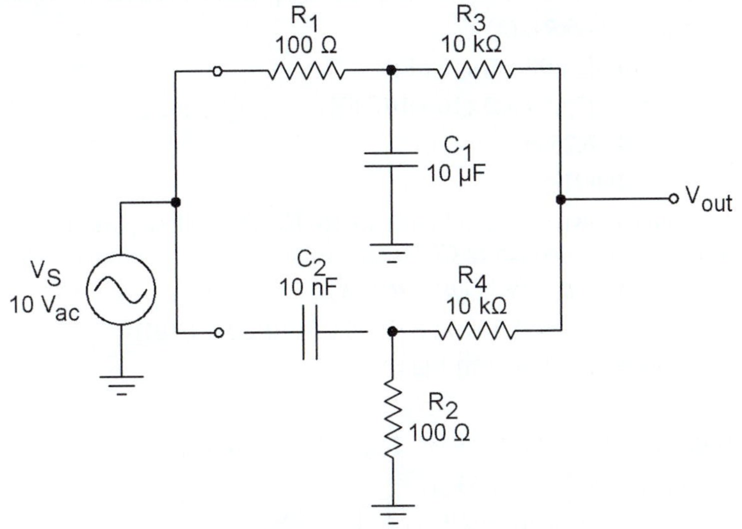

Figure 19-15: *RC* Band-Stop Filter for Finite X_{C1}

The circuit connected to C_1 in Figure 19-15 can be reduced to a Thevenin voltage, V_{TH}, and Thevenin resistance, R_{TH}, in series with C_1. To do so, redraw the circuit as shown in the left side of Figure 19-16.

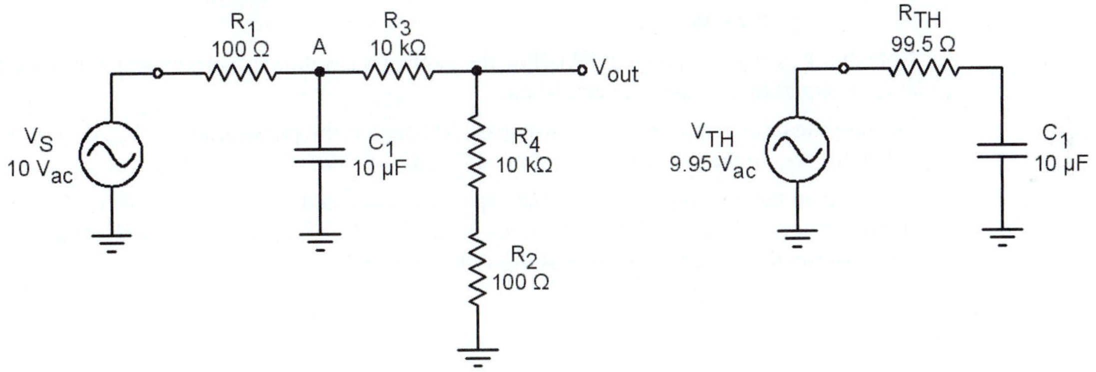

Figure 19-16: Redrawn and Thevenin Circuits for Figure 19-15

From the redrawn circuit, V_{TH} is V_A, the voltage at point A, with C_2 removed. Use the voltage divider theorem to find V_A.

$V_{TH} = V_A$

$= V_S\,[(R_2 + R_3 + R_4) / (R_1 + R_2 + R_3 + R_4)]$

$= (10\ \text{V}_{ac})\,[(100\ \Omega + 10\ \text{k}\Omega + 10\ \text{k}\Omega) / (100\ \Omega + 100\ \Omega + 10\ \text{k}\Omega + 10\ \text{k}\Omega)]$

$= (10\ \text{V}_{ac})\,(20.1\ \text{k}\Omega / 20.2\ \text{k}\Omega)$

$= (10\ \text{V}_{ac})(0.995)$

$= 9.95$ Vac

R_{TH} is the resistance seen by C_1 with V_S replaced with its internal resistance (a short).

$$R_{TH} = R_1 \parallel (R_2 + R_3 + R_4)$$
$$= 100\ \Omega \parallel (100\ \Omega + 10\ \text{k}\Omega + 10\ \text{k}\Omega)$$
$$= 100\ \Omega \parallel 20.1\ \text{k}\Omega$$
$$= 99.5\ \Omega$$

The Thevenin equivalent. shown on the right side of Figure 19-16, is a low-pass filter with a –20 dB / decade rolloff above the corner frequency. Calculate the corner frequency, $f_{c(l)}$.

$$f_{c(l)} = 1 / (2\pi R_{TH} C_1)$$
$$= 1 / [2\pi(99.5\ \Omega)(10\ \mu\text{F})]$$
$$= 1 / [2\pi(99.5\ \Omega)(10 \times 10^{-6}\ \text{F})]$$
$$= 1 / 6.25\ \text{ms}$$
$$= \mathbf{160\ Hz}$$

From the redrawn circuit in Figure 19-16, V_{out} for the low-pass circuit is developed across part of the resistor network across C_1. Use the voltage divider theorem to calculate V_{out}.

$$V_{out} = V_A\,[(R_2 + R_4) / (R_2 + R_3 + R_4)]$$
$$= V_A\,[(100\ \Omega + 10\ \text{k}\Omega) / (100\ \Omega + 10\ \text{k}\Omega + 10\ \text{k}\Omega)]$$
$$= V_A\,(10.1\ \text{k}\Omega / 20.1\ \text{k}\Omega)$$
$$= 0.502\ V_A$$

At $f_{c(l)}$, $V_A = 0.707\ V_S$. Calculate V_{out} in decibels at $f_{c(l)}$.

$$A_{v(\text{dB})} = 20 \log (V_{out} / V_S)$$
$$= 20 \log [(0.502\ V_A) / V_{in}]$$
$$= 20 \log \{[(0.502)(0.707\ V_{in})] / V_{in}\}$$
$$= 20 \log [(0.502)(0.707)]$$
$$= 20 \log (0.355)$$
$$= (20)(-0.449)$$
$$= \mathbf{-8.99\ dB}$$

Note that V_{out} at $f_{c(l)}$ is about 3 dB below the passband amplitude. For an idealized Bode plot, V_{out} at $f_{c(l)}$ is equal to the passband amplitude.

Because the circuit appears to be a low-pass ***RC*** circuit, the phase angle of V_{out} decreases from 0° below 0.1 $f_{c(l)}$, in the lower pass-band to –45° at $f_{c(l)}$, to –90° above 10 $f_{c(l)}$.

At the upper corner frequency, $f_{c(h)}$, the capacitor associated with high-pass filter, C_2, will exhibit finite reactance. Because C_1 is 1000 times larger than C_2, C_1 appears to be a short at $f_{c(h)}$. Figure 19-17 shows the circuit equivalent at these frequencies.

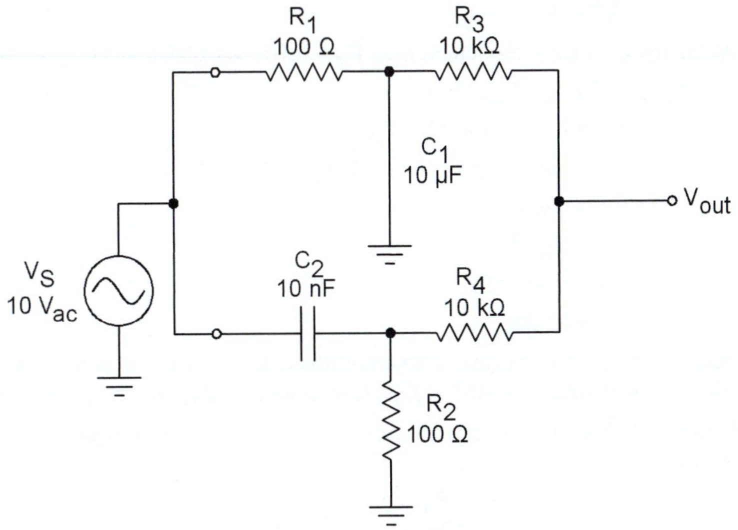

Figure 19-17: *RC* Band-Stop Filter for Finite X_{C2}

To analyze this circuit, redraw the circuit as shown in left side of Figure 19-18 and apply standard circuit analysis techniques.

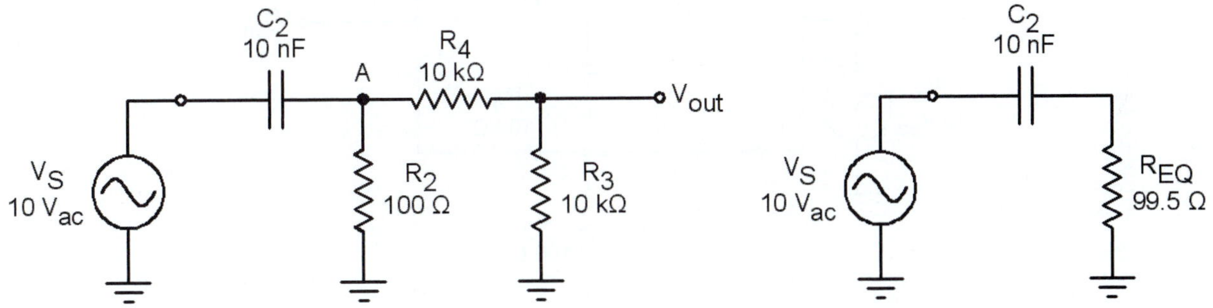

Figure 19-18: Redrawn and Equivalent Circuits for Figure 19-17

From the redrawn circuit, C_2 is in series with an equivalent series resistance $R_{EQ} = R_2 \parallel (R_3 + R_4)$.

$$
\begin{aligned}
R_{EQ} &= R_2 \parallel (R_3 + R_4) \\
&= 100\ \Omega \parallel (10\ \text{k}\Omega + 10\ \text{k}\Omega) \\
&= 100\ \Omega \parallel 20\ \text{k}\Omega \\
&= 99.5\ \Omega
\end{aligned}
$$

The circuit equivalent, shown in the right side of Figure 19-18, is a high-pass filter with a –20 dB / decade rolloff below the corner frequency. Calculate the corner frequency, $f_{c(h)}$.

$$
\begin{aligned}
f_{c(h)} &= 1 / (2\pi R_{EQ} C_2) \\
&= 1 / [2\pi(99.5\ \Omega)(10\ \text{nF})] \\
&= 1 / [2\pi(99.5\ \Omega)(10 \times 10^{-9}\ \text{F})] \\
&= 1 / 6.25\ \mu\text{s} \\
&= \mathbf{160\ kHz}
\end{aligned}
$$

From the redrawn circuit in Figure 19-18, V_{out} for the high-pass circuit is developed across part of the resistor network across C_2. Use the voltage divider theorem to calculate V_{out}.

$$
\begin{aligned}
V_{out} &= V_A\,[R_3 / (R_3 + R_4)] \\
&= V_A\,[10\ \text{k}\Omega / (10\ \text{k}\Omega + 10\ \text{k}\Omega)] \\
&= V_A\,(10\ \text{k}\Omega / 20\ \text{k}\Omega)
\end{aligned}
$$

$= 0.500\ V_A$

At the lower corner frequency, $f_{c(h)}$, $V_A = 0.707\ V_S$. Calculate V_{out} in decibels at $f_{c(h)}$.

$A_{v(\text{dB})} = 20 \log (V_{out} / V_S)$

$= 20 \log [(0.500\ V_A) / V_S]$

$= 20 \log \{[(0.500)(0.707\ V_S)] / V_S\}$

$= 20 \log [(0.500)(0.707)]$

$= 20 \log (0.354)$

$= (20)(-0.452)$

$= \mathbf{-9.03\ dB}$

Because the circuit appears to be a high-pass ***RC*** circuit, the phase angle of V_{out} increases from $-90°$ below $0.1\,f_{c(h)}$, to $-45°$ at $f_{c(h)}$, to $0°$ above $10\,f_{c(h)}$ in the upper pass-band.

Figure 19-19 shows the circuit equivalent in the upper pass-band, when both C_1 and C_2 appear to be shorts.

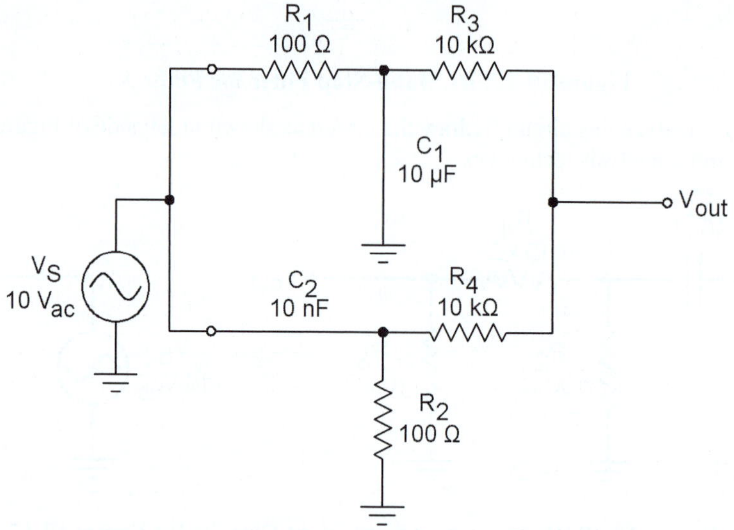

Figure 19-19: *RC* Band-Stop Filter at High Frequencies

The circuit appears to be a simple resistive circuit. C_1 shorts R_1 to ground, so that only resistors R_2, R_3, and R_4 determine V_{out}. Figure 19-20 shows the redrawn circuit.

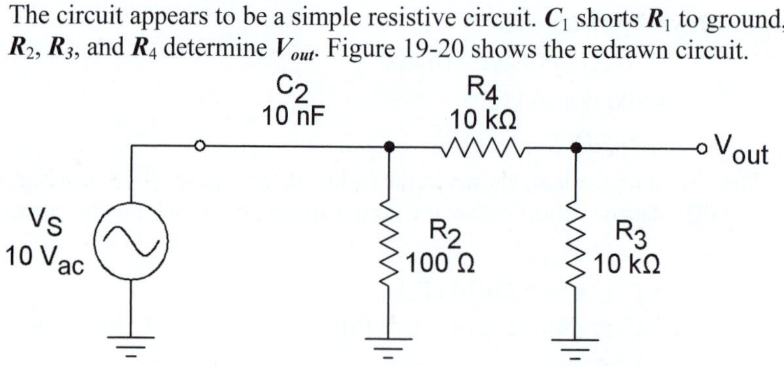

Figure 19-20: Redrawn Circuit for Figure 19-19

Both R_2 and the series combination $R_3 + R_4$ are in parallel, so that V_S appears across each. Use the voltage divider theorem to calculate V_{out}.

$V_{out} = V_S\,[R_3 / (R_3 + R_4)]$

$= (10\ V_{ac})\ [10\ k\Omega\ /\ (10\ k\Omega + 10\ k\Omega)]$

$= (10\ V_{ac})\ (10\ k\Omega\ /\ 20\ k\Omega)$

$= (10\ V_{ac})\ (0.5)$

$= 5.0\ V_{ac}$

V_{out} in decibels in the upper pass-band is

$A_{v(dB)} = 20 \log (V_{out} / V_S)$

$= 20 \log (5.0\ V_{ac} / 10\ V_{ac})$

$= 20 \log (0.50)$

$= (20)(-3.01)$

$= \mathbf{-6.02\ dB}$

Because the circuit in the lower pass-band appears to be purely resistive, the phase angle of V_{out} is 0°.

Note that the lower and upper passband amplitudes are nearly equal and can be idealized as –6 dB. Note also that V_{out} at $f_{c(l)}$ and $f_{c(h)}$ is about 3 dB below the passband amplitude and is equal to the passband amplitude for an idealized Bode plot. Figure 19-21 shows the idealized amplitude and phase Bode plots for the ***RC*** stop-band filter of Figure 19-13.

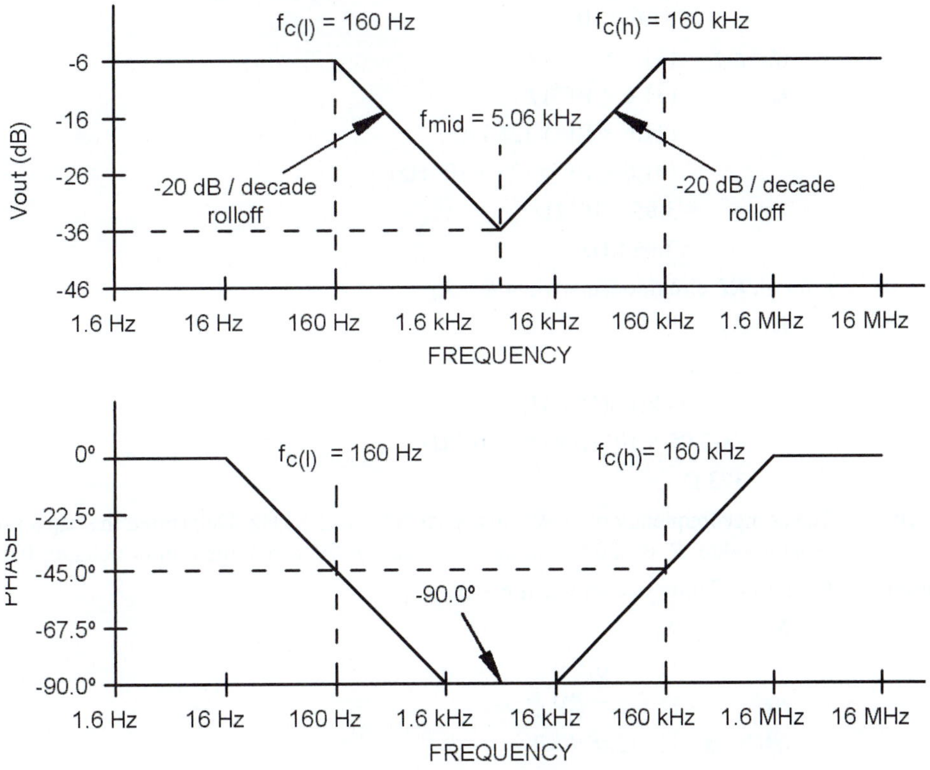

Figure 19-21: Amplitude and Phase Bode Plots for *RC* Band-Stop Filter

Because the amplitudes of the corner frequencies are equal and the roll-off for both the high- and low-pass responses are equal, the center frequency of the filter, f_{mid}, will be midway between $f_{c(l)}$ and $f_{c(h)}$, as Figure 19-21 shows. The separation between the corner frequencies is 3 decades, so f_{mid} is 1.5 decades above $f_{c(l)}$ (and 1.5 decades below $f_{c(h)}$).

$1.5 = \log [f_{mid} / f_{c(l)}]$

$10^{1.5} = 10^{\log [fmid / fc(l)]}$

$31.6 = f_{mid} / f_{c(l)}$

$31.6 f_{c(l)} = [f_{mid} / f_{c(l)}] f_{c(l)}$

$(31.6)(160 \text{ Hz}) = f_{mid}$

$\mathbf{5.06\ kHz} = f_{mid}$

19.2.2 Low-Pass Filters

1. Problem: The output voltage, V_{out}, of an ***RL*** low-pass filter is –30 dB when the frequency, f, is 125 kHz. Determine the corner frequency, f_c, and the resistor value, ***R***, if the inductor value, ***L***, is 15 mH.

   Solution: The roll-off of an ***RL*** low-pass filter is –20 dB / decade for each decade above f_c. Determine the number of decades between f_c and f.

$$
\begin{aligned}
f_c / f &= (-30 \text{ dB}) / (-20 \text{ dB / decade}) \\
&= (-30 / -20)\ [\text{dB / (dB / decade)}] \\
&= (1.5)\ [\text{dB} \times (\text{decade / dB})] \\
&= 1.5 \text{ decades}
\end{aligned}
$$

   Because the filter is a low-pass filter, f_c is 1.5 decades below f. Therefore, $f_c / f = -1.5$ decades.

$$
\begin{aligned}
f_c / f &= -1.5 \text{ decades} \\
&= 10^{-1.5} \\
&= 31.6 \times 10^{-3} \\
(f_c / f) f &= (31.6 \times 10^{-3}) f \\
f_c &= (31.6 \times 10^{-3}) f \\
&= (31.6 \times 10^{-3})(125 \text{ kHz}) \\
&= (31.6 \times 10^{-3})(125 \times 10^{3} \text{ Hz}) \\
&= 3.95 \times 10^{3} \text{ Hz} \\
&= \mathbf{3.95\ kHz}
\end{aligned}
$$

   f_c for an ***RL*** circuit occurs when $R = X_L$.

$$
\begin{aligned}
R &= X_L \\
&= 2\pi f_c L \\
&= 2\pi(3.95 \text{ kHz})(15 \text{ mH}) \\
&= 2\pi(3.95 \times 10^{3} \text{ Hz})(15 \times 10^{-3} \text{ H}) \\
&= \mathbf{373\ \Omega}
\end{aligned}
$$

2. Problem: The corner frequency of an ***RC*** low-pass filter is 2.5 kHz. Determine the capacitor value, ***C***, if the resistor value, ***R***, is 220 Ω and the frequency, f, for which the output voltage, V_{out}, is –75 dB.

   Solution: f_c for an ***RC*** filter occurs when $R = X_C$.

$$
\begin{aligned}
R &= X_C \\
R &= [1 / (2\pi f_c C)] \\
RC &= [1 / (2\pi f_c C)]\ C \\
(RC) / R &= [1 / (2\pi f_c)] / R \\
C &= 1 / (2\pi f_c R) \\
&= 1 / [2\pi(2.5 \text{ kHz})(220\ \Omega)] \\
&= 1 / [2\pi(2.5 \times 10^{3} \text{ Hz})(220\ \Omega)] \\
&= 1 / 3.46 \times 10^{6} \text{ F}^{-1} \\
&= \mathbf{289\ nF}
\end{aligned}
$$

   The roll-off of an ***RC*** low-pass filter is –20 dB / decade for each decade above f_c. Determine the number of decades between f_c and f.

N_{10} = (–75 dB) / (–20 dB / decade)
= (–75 / –20) [dB / (dB / decade)]
= (3.75) [dB × (decade / dB)]
= 3.75 decades

Because the filter is a low-pass filter, f is 3.75 decades above f_c. Therefore, f / f_c = 3.75 decades.

f / f_c = 3.75 decades
$= 10^{3.75}$
$= 5.62 \times 10^3$

$(f / f_c) f_c = (5.62 \times 10^3) f_c$
$f = (5.62 \times 10^3) f_c$
$= (5.62 \times 10^3)(2.5 \text{ kHz})$
$= (5.62 \times 10^3)(2.5 \times 10^3 \text{ Hz})$
$= 14.1 \times 10^6$ Hz
= **14.1 MHz**

19.2.3 High-Pass Filters

1. Problem: The component values of an ***RC*** high-pass filter are R = 120 kΩ and C = 510 nF. Determine the corner frequency, f_c, and the voltage output, V_{out}, in decibels for f = 0.05 Hz.

   Solution: First, calculate the corner frequency, f_c. f_c for an ***RC*** circuit occurs when $R = X_C$.

   $R = X_C$
   $R = [1 / (2\pi f_c C)]$
   $R f_c = [1 / (2\pi f_c C)] f_c$
   $(R f_c) / R = [1 / (2\pi C)] / R$
   $f_c = 1 / (2\pi RC)$
   $= 1 / [2\pi(120 \text{ k}\Omega)(510 \text{ nF})]$
   $= 1 / [2\pi(120 \times 10^3\ \Omega)(510 \times 10^{-9} \text{ F})]$
   $= 1 / 385 \times 10^{-3}$ s
   = 2.60 Hz

   Determine the number of decades, N_{10}, between f and f_c.

   $N_{10} = \log (f / f_c)$
   = log (0.05 Hz / 2.60 Hz)
   $= \log (19.2 \times 10^{-3})$
   = –1.72 decades

   The negative value confirms that f is 1.72 decades below f_c. The roll-off of a high-pass filter is –20 dB / decade for each decade below f_c.

   V_{out} = (–20 dB / decade)(1.72 decades)
   = (–20 × 1.72)[(dB / decade)(decades)]
   = **–34.3 dB**

2. Problem: The corner frequency of an ***RL*** high-pass filter is 1.5 Hz. Determine the inductor value, L, if the resistor value, R, is 1.5 Ω, and the frequency, f, for which the output voltage, V_{out}, is –50 dB.

   Solution: f_c for an ***RL*** circuit occurs when $X_L = R$.

   $X_L = R$
   $2\pi f_c L = R$
   $2\pi f_c L / 2\pi f_c = R / 2\pi f_c$

$L \quad = R / 2\pi f_c$

$= (1.5\ \Omega) / [2\pi(1.5\ \text{Hz})]$

$= (1.5\ \Omega) / (9.42\ \text{Hz})$

$= \mathbf{159\ mH}$

The roll-off of an ***RL*** high-pass filter is –20 dB / decade for each decade below f_c. Determine the number of decades between f and f_c.

$N_{10} \quad = (-50\ \text{dB}) / (-20\ \text{dB / decade})$

$= (-50 / -20)\ [\text{dB} / (\text{dB / decade})]$

$= (2.5)\ [\text{dB} \times (\text{decade / dB})]$

$= 2.5$ decades

Because the filter is a high-pass filter, f is 2.5 decades below f_c. Therefore, $f / f_c = -2.5$ decades.

$f / f_c \quad = -2.5$ decades

$= 10^{-2.5}$

$= 3.16 \times 10^{-3}$

$(f / f_c) f_c = (3.16 \times 10^{-3}) f_c$

$f_c \quad = (3.16 \times 10^{-3}) f_c$

$= (3.16 \times 10^{-3})(1.5\ \text{Hz})$

$= 4.74 \times 10^{-3}$ Hz

$= \mathbf{4.74\ mHz}$

19.2.4 Band-Pass Filters

1. Problem: Many ***RLC*** pass-band filter circuits can be Thevenized to create a circuit like that in Figure 19-22.

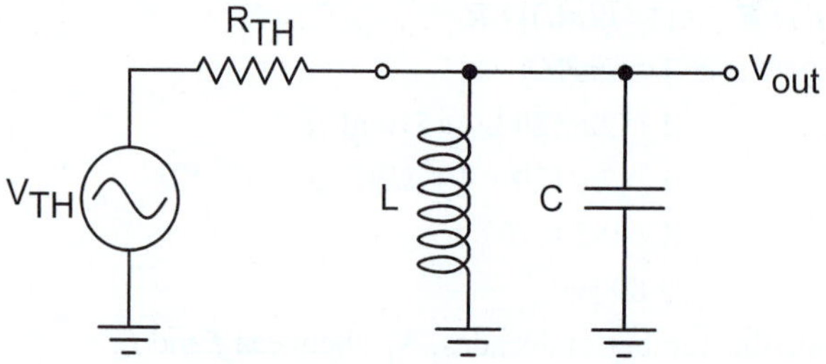

Figure 19-22: Thevenized *RLC* Pass-Band Filter

Show that the center frequency, f_0, is the geometric mean of the corner frequencies, $f_{c(h)}$ and $f_{c(l)}$ (i.e., show that $f_0 = \sqrt{f_{c(h)} f_{c(l)}}$).

Solution: For the circuit of Figure 19-22, the center frequency, f_0, is the resonant frequency of the parallel ***LC*** circuit, so that $f_0 = f_r = 1 / (2\pi\sqrt{LC})$.

The corner frequencies will occur when the magnitudes of the Thevenin resistance, R_{TH}, and Z_{EQ}, the impedance of the parallel ***LC*** circuit, are equal. First, determine the value of Z_{EQ}.

$Z_{EQ} = X_L \parallel X_C$

$= \text{j}\, X_L \parallel -\text{j}\, X_C$

$= (\text{j}\, X_L)(-\text{j}\, X_C) / [\text{j}\, X_L + (-\text{j}\, X_C)]$ using product-over-sum method

$= [-\text{j}^2 (X_L X_C)] / [\text{j}\, (X_L - X_C)]$

$= -(X_L X_C) / (X_L - X_C)$

$= (X_L X_C) / (X_C - X_L)$

$= [(2\pi f\boldsymbol{L})\,(1 / 2\pi f\boldsymbol{C})] / [(1 / 2\pi f\boldsymbol{C}) - 2\pi f\boldsymbol{L}]$

$= (\boldsymbol{L} / \boldsymbol{C}) / [(1 / 2\pi f\boldsymbol{C}) - (2\pi f\boldsymbol{L})(2\pi f\boldsymbol{C} / 2\pi f\boldsymbol{C})]$

$= (\boldsymbol{L} / \boldsymbol{C}) / \{[1 - (2\pi f\boldsymbol{L})(2\pi f\boldsymbol{C})] / (2\pi f\boldsymbol{C})\}$

$= [(\boldsymbol{L} / \boldsymbol{C})\,(2\pi f\boldsymbol{C})] / (1 - 4\pi^2 f^2\boldsymbol{LC})$

$= (2\pi f\boldsymbol{L}) / (1 - 4\pi^2 f^2\boldsymbol{LC})$

At the corner frequencies, when $f = f_C$, $R_{TH} = Z_{EQ}$.

R_{TH}	$= Z_{EQ}$
	$= (2\pi f_c L) / (1 - 4\pi^2 f_c^2 LC)$
$(R_{TH})(1 - 4\pi^2 f_c^2 LC)$	$= [(2\pi f_c L) / (1 - 4\pi^2 f_c^2 LC)](1 - 4\pi^2 f_c^2 LC)$
$R_{TH} - 4\pi^2 f_c^2 R_{TH} LC$	$= 2\pi f_c L$
$R_{TH} - 4\pi^2 f_c^2 R_{TH} LC - 2\pi f_c L$	$= 2\pi f_c L - 2\pi f_c L$
$R_{TH} - 4\pi^2 f_c^2 R_{TH} LC - 2\pi f_c L$	$= 0$
$(R_{TH} - 4\pi^2 f_c^2 R_{TH} LC - 2\pi f_c L) \times -1$	$= 0 \times -1$
$-R_{TH} + 4\pi^2 f_c^2 R_{TH} LC + 2\pi f_c L$	$= 0$
$4\pi^2 f_c^2 R_{TH} LC + 2\pi f_c L - R_{TH}$	$= 0$
$(4\pi^2 R_{TH} LC) f_c^2 + (2\pi L) f_c - R_{TH}$	$= 0$

This expression is a quadratic equation $\mathrm{a} f_c^2 + \mathrm{b} f_c + \mathrm{c} = 0$, with $\mathrm{a} = 4\pi^2 R_{TH} LC$, $\mathrm{b} = 2\pi L$, and $\mathrm{c} = -R_{TH}$. This equation has the two solutions

$$f_{c(1)} = \left|\frac{-\mathrm{b} + \sqrt{\mathrm{b}^2 - 4\mathrm{ac}}}{2\mathrm{a}}\right| \quad \text{and} \quad f_{c(2)} = \left|\frac{-\mathrm{b} - \sqrt{\mathrm{b}^2 - 4\mathrm{ac}}}{2\mathrm{a}}\right|$$

Because $f_{c(1)}$ and $f_{c(2)}$ are the magnitudes of the corner frequencies only, they are equal to the absolute values of the quadratic solutions. Start with the product of the two corner frequencies.

$$f_{c(h)} \times f_{c(l)} = \left|\frac{-\mathrm{b} + \sqrt{\mathrm{b}^2 - 4\mathrm{ac}}}{2\mathrm{a}}\right| \times \left|\frac{-\mathrm{b} - \sqrt{\mathrm{b}^2 - 4\mathrm{ac}}}{2\mathrm{a}}\right|$$

$$= \left|\frac{(-\mathrm{b})(-\mathrm{b}) + (-\mathrm{b})\left(+\sqrt{\mathrm{b}^2 - 4\mathrm{ac}}\right) + (-\mathrm{b})\left(-\sqrt{\mathrm{b}^2 - 4\mathrm{ac}}\right) + \left(+\sqrt{\mathrm{b}^2 - 4\mathrm{ac}}\right)\left(-\sqrt{\mathrm{b}^2 - 4\mathrm{ac}}\right)}{(2\mathrm{a})(2\mathrm{a})}\right|$$

$$= \left|\frac{\mathrm{b}^2 + (-\mathrm{b})\left(+\sqrt{\mathrm{b}^2 - 4\mathrm{ac}} - \sqrt{\mathrm{b}^2 - 4\mathrm{ac}}\right) - \left(\sqrt{\mathrm{b}^2 - 4\mathrm{ac}}\right)^2}{4\mathrm{a}^2}\right|$$

$$= \left|\frac{\mathrm{b}^2 + (-\mathrm{b})(0) - (\mathrm{b}^2 - 4\mathrm{ac})}{4\mathrm{a}^2}\right|$$

$$= \left|\frac{\mathrm{b}^2 - (\mathrm{b}^2 - 4\mathrm{ac})}{4\mathrm{a}^2}\right|$$

$$= \left|\frac{\mathrm{b}^2 - \mathrm{b}^2 + 4\mathrm{ac}}{4\mathrm{a}^2}\right|$$

$$= \left|\frac{4\mathrm{ac}}{4\mathrm{a}^2}\right|$$

$$= \left|\frac{c}{a}\right|$$

Substitute the values for a and c into the equation.

$$f_{c(h)} \times f_{c(l)} = \left|\frac{c}{a}\right|$$

$$= \left|\frac{-R_{TH}}{4\pi^2 R_{TH} LC}\right|$$

$$= \frac{R_{TH}}{4\pi^2 R_{TH} LC} \quad \text{(because } R_{TH}, L, \text{ and } C \text{ are all positive)}$$

$$= \frac{1}{4\pi^2 LC}$$

$$= \left(\frac{1}{2\pi\sqrt{LC}}\right)^2$$

$$= f_0^2$$

Take the square root of both sides to find f_0.

$$\sqrt{f_0^2} = \sqrt{f_{c(h)} \times f_{c(l)}}$$

$$f_0 = \sqrt{f_{c(h)} \times f_{c(l)}}$$

2. Problem: For the ***RLC*** band-pass filter circuit in Figure 19-23, find approximate values of R_W, L, and C so that $f_0 \approx 540$ kHz, $BW \approx 10.0$ kHz, and $V_{out} \approx 10.0\ \text{mV}_{ac}$ at the center frequency.

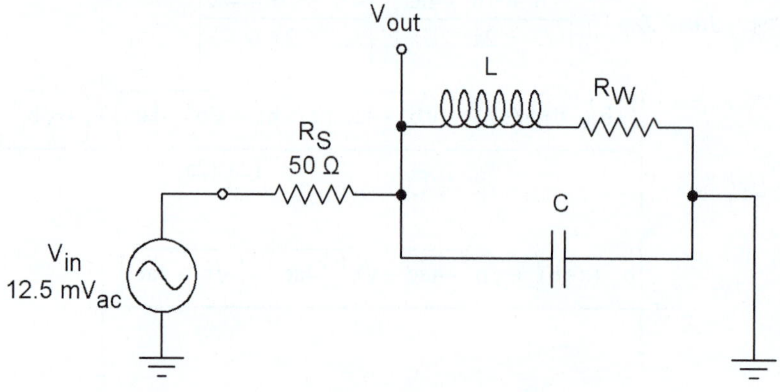

Figure 19-23: *RLC* Band-Pass Filter for Problem 2

Solution: First, redraw the circuit by replacing the series-parallel combination of R_W, L, and C with the equivalent parallel combination of R_{PEQ}, L_{EQ}, and C, as shown in Figure 19-24.

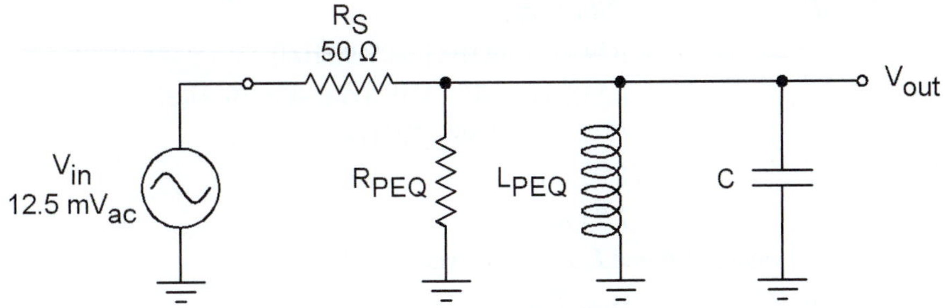

Figure 19-24: Equivalent Circuit for Figure 19-23

Q_{FILTER}, the quality factor for the filter circuit, is

$$Q_{FILTER} = f_0 / BW$$
$$= 540\ \text{kHz} / 10\ \text{kHz}$$
$$= 54$$

Because Q_{FILTER} is much larger than 10, $Q_{COIL} \approx Q_{FILTER} = Q = 54$. The equivalent parallel resistance, R_{PEQ}, for the circuit is

$$R_{PEQ} = (Q^2 + 1)\, R_W$$
$$R_{PEQ} / (Q^2 + 1) = [(Q^2 + 1)\, R_W] / (Q^2 + 1)$$
$$= R_W$$

The equivalent parallel inductance, L_{PEQ}, for the circuit is

$$L_{PEQ} = [(Q^2 + 1) / Q^2]\, L$$
$$\approx L$$

At the center frequency, f_0, the parallel combination of L_{EQ} and C in parallel appears to be an open. Use the voltage divider theorem to calculate R_{PEQ} from the given values of V_{in}, R_S, and V_{out}.

$$V_{in}\, [R_{PEQ} / (R_S + R_{PEQ})] = V_{out}$$
$$V_{in}\, [R_{PEQ} / (R_S + R_{PEQ})]\, (R_S + R_{PEQ}) = V_{out}\, (R_S + R_{PEQ})$$
$$V_{in}\, R_{PEQ} = V_{out}\, R_S + V_{out}\, R_{PEQ}$$
$$V_{in}\, R_{PEQ} - V_{out}\, R_{PEQ} = V_{out}\, R_S + V_{out}\, R_{PEQ} - V_{out}\, R_{PEQ}$$
$$(V_{in} - V_{out})\, R_{PEQ} = V_{out}\, R_S$$
$$[(V_{in} - V_{out})\, R_{PEQ}] / (V_{in} - V_{out}) = (V_{out}\, R_S) / (V_{in} - V_{out})$$
$$R_{PEQ} = (V_{out}\, R_S) / (V_{in} - V_{out})$$
$$= [(10.0\ \text{mV}_{ac})(50\ \Omega)] / (12.5\ \text{mV}_{ac} - 10.0\ \text{mV}_{ac})$$
$$= [(10.0\ \text{mV}_{ac})(50\ \Omega)] / (2.5\ \text{mV}_{ac})$$
$$= 500\ \Omega$$

Use the values of R_{PEQ} and Q to calculate the values of R_W.

$$R_W = R_{PEQ} / (Q^2 + 1)$$
$$= 500\ \Omega / [(54)^2 + 1]$$
$$= 500\ \Omega / (2916 + 1)$$
$$= 500\ \Omega / 2917$$
$$= 171 \times 10^{-3}\ \Omega$$
$$= \mathbf{171\ m\Omega}$$

Use the values of f_0, R_W, and Q to find the value of L.

$$X_L / R_W = Q$$
$$(2\pi f_0 L) / R_W = Q$$
$$[(2\pi f_0 L) / R_W]\, (R_W / 2\pi f_0) = Q\, (R_W / 2\pi f_0)$$

$L = QR_W / 2\pi f_0$
$= [(54)(171\ \text{m}\Omega)] / [2\pi(540\ \text{kHz})]$
$= [(54)(171 \times 10^{-3}\ \Omega)] / [2\pi(540 \times 10^3\ \text{Hz})]$
$= (9.26\ \Omega) / (3.39 \times 10^6\ \text{Hz})$
$= 2.73 \times 10^{-6}\ \text{H}$
$= \mathbf{2.73\ \mu H}$

Use the values of f_0 and L to find the value of C.

$f_0 = 1 / (2\pi\sqrt{LC})$
$f_0^2 = [1 / (2\pi\sqrt{LC})]^2$
$f_0^2 = 1 / (4\pi^2 LC)$
$f_0 C = [1 / (4\pi^2 LC)]\, C$
$(f_0 C) / f_0 = [1 / (4\pi^2 L)] / f_0$
$C = 1 / (4\pi^2 f_0 L)$
$= 1 / [4\pi^2 (540\ \text{kHz})^2 (2.73\ \mu\text{H})]$
$= 1 / [4\pi^2 (540 \times 10^3\ \text{Hz})^2 (2.73 \times 10^{-6}\ \text{H})]$
$= 1 / [(39.5)(292 \times 10^9\ \text{Hz}^2)(2.73 \times 10^{-6}\ \text{H})]$
$= 1 / (31.4 \times 10^6\ \text{F}^{-1})$
$= 31.8 \times 10^{-9}\ \text{F}$
$= \mathbf{31.8\ nF}$

19.2.5 Band-Stop Filters

1. Problem: For the band-stop filter in Figure 19-25, show that the center frequency, f_0, is the geometric mean of the corner frequencies, $f_{c(h)}$ and $f_{c(l)}$ (i.e., show that $f_0 = \sqrt{f_{c(h)} f_{c(l)}}$).

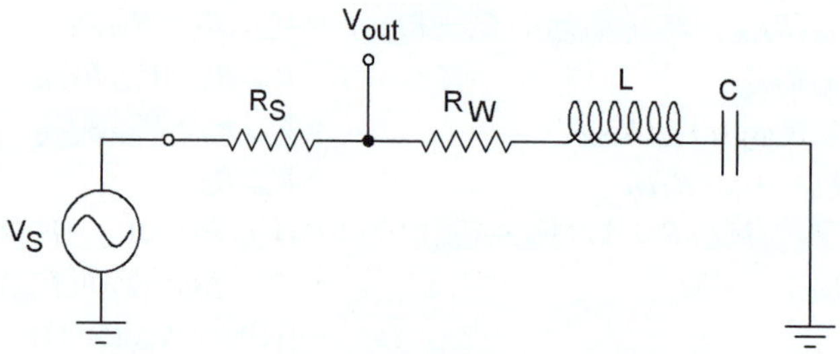

Figure 19-25: *RLC* Band-Stop Filter for Problem 1

Solution: For the circuit of Figure 19-25, the center frequency, f_0, is the resonant frequency of the parallel LC circuit, so that $f_0 = f_r = 1 / (2\pi\sqrt{LC})$.

The corner frequencies will occur when the magnitudes of the source resistance, R_S, and total impedance, Z_T (equal to the series combination of R_W, L, and C), are equal.

$R_S = Z_T$
$= R_W + X_L + X_C$
$= R_W + \mathrm{j}\,X_L - \mathrm{j}\,X_C$
$= R_W + \mathrm{j}\,(X_L - X_C)$
$= R_W + \mathrm{j}\,[2\pi f_c L - (1/2\pi f_c C)]$
$= R_W + \mathrm{j}\,[(2\pi f_c L)(2\pi f_c C / 2\pi f_c C) - (1/2\pi f_c C)]$

$$= \boldsymbol{R_W} + \mathrm{j}\ \{[(2\pi \boldsymbol{f_c L})\ (2\pi \boldsymbol{f_c C}) /\ 2\pi \boldsymbol{f_c C}] - (1 / 2\pi \boldsymbol{f_c C})\}$$
$$= \boldsymbol{R_W} + \mathrm{j}\ [(4\pi^2 \boldsymbol{f_c}^2 \boldsymbol{LC} /\ 2\pi \boldsymbol{f_c C}) - (1 / 2\pi \boldsymbol{f_c C})]$$
$$= \boldsymbol{R_W} + \mathrm{j}\ [(4\pi^2 \boldsymbol{f_c}^2 \boldsymbol{LC} - 1) / 2\pi \boldsymbol{f_c C}]$$

For a complex impedance, $\boldsymbol{Z_T} = \boldsymbol{A} + \mathrm{j}\ \boldsymbol{B}$, the magnitude, $|\boldsymbol{Z_T}|$, is $\sqrt{\boldsymbol{A}^2 + \boldsymbol{B}^2}$.

$$
\begin{aligned}
|\boldsymbol{R_S}| &= |\boldsymbol{Z_T}| \\
|\boldsymbol{R_S}|^2 &= |\boldsymbol{Z_T}|^2 \\
\boldsymbol{R_S}^2 &= \boldsymbol{Z_T}^2 \\
\boldsymbol{R_S}^2 &= \boldsymbol{R_W}^2 + [(4\pi^2 \boldsymbol{f_c}^2 \boldsymbol{LC} - 1) / 2\pi \boldsymbol{f_c C}]^2 \\
\boldsymbol{R_S}^2 - \boldsymbol{R_W}^2 &= \boldsymbol{R_W}^2 + [(4\pi^2 \boldsymbol{f_c}^2 \boldsymbol{LC} - 1) / 2\pi \boldsymbol{fC}]^2 - \boldsymbol{R_W}^2 \\
\boldsymbol{R_S}^2 - \boldsymbol{R_W}^2 &= [(4\pi^2 \boldsymbol{f_c}^2 \boldsymbol{LC} - 1) / 2\pi \boldsymbol{f_c C}]^2 \\
\boldsymbol{R_S}^2 - \boldsymbol{R_W}^2 &= (4\pi^2 \boldsymbol{f_c}^2 \boldsymbol{LC} - 1)^2 / (2\pi \boldsymbol{f_c C})^2 \\
(\boldsymbol{R_S}^2 - \boldsymbol{R_W}^2)(2\pi \boldsymbol{f_c C})^2 &= [(4\pi^2 \boldsymbol{f_c}^2 \boldsymbol{LC} - 1)^2 / (2\pi \boldsymbol{f_c C})^2](2\pi \boldsymbol{f_c C})^2 \\
(\boldsymbol{R_S}^2 - \boldsymbol{R_W}^2)(4\pi^2 \boldsymbol{f_c}^2 \boldsymbol{C}^2) &= (4\pi^2 \boldsymbol{f_c}^2 \boldsymbol{LC} - 1)^2 \\
4\pi^2 \boldsymbol{f_c}^2 (\boldsymbol{R_S}^2 - \boldsymbol{R_W}^2)\boldsymbol{C}^2 &= (4\pi^2 \boldsymbol{f_c}^2 \boldsymbol{LC})(4\pi^2 \boldsymbol{f_c}^2 \boldsymbol{LC}) - (1)(4\pi^2 \boldsymbol{f_c}^2 \boldsymbol{LC}) - (1)(4\pi^2 \boldsymbol{f_c}^2 \boldsymbol{LC}) + (1)(1) \\
&= 16\pi^4 \boldsymbol{f_c}^4 \boldsymbol{L}^2 \boldsymbol{C}^2 - 8\pi^2 \boldsymbol{f_c}^2 \boldsymbol{LC} + 1
\end{aligned}
$$

Collect and combine the terms.

$$
\begin{aligned}
&16\pi^4 \boldsymbol{f_c}^4 \boldsymbol{L}^2 \boldsymbol{C}^2 - 8\pi^2 \boldsymbol{f_c}^2 \boldsymbol{LC} + 1 = 4\pi^2 \boldsymbol{f_c}^2 (\boldsymbol{R_S}^2 - \boldsymbol{R_W}^2)\boldsymbol{C}^2 \\
&16\pi^4 \boldsymbol{f_c}^4 \boldsymbol{L}^2 \boldsymbol{C}^2 - 8\pi^2 \boldsymbol{f_c}^2 \boldsymbol{LC} + 1 - 4\pi^2 \boldsymbol{f_c}^2 (\boldsymbol{R_S}^2 - \boldsymbol{R_W}^2)\boldsymbol{C}^2 \\
&\qquad = 4\pi^2 \boldsymbol{f_c}^2 (\boldsymbol{R_S}^2 - \boldsymbol{R_W}^2)\boldsymbol{C}^2 - 4\pi^2 \boldsymbol{f_c}^2 (\boldsymbol{R_S}^2 - \boldsymbol{R_W}^2)\boldsymbol{C}^2 \\
&16\pi^4 \boldsymbol{f_c}^4 \boldsymbol{L}^2 \boldsymbol{C}^2 - 8\pi^2 \boldsymbol{f_c}^2 \boldsymbol{LC} + 1 - 4\pi^2 \boldsymbol{f_c}^2 (\boldsymbol{R_S}^2 - \boldsymbol{R_W}^2)\boldsymbol{C}^2 = 0 \\
&(16\pi^4 \boldsymbol{L}^2 \boldsymbol{C}^2)\boldsymbol{f_c}^4 - (8\pi^2 \boldsymbol{f_c}^2 \boldsymbol{LC})\,\boldsymbol{f_c}^2 + 1 - [4\pi^2 (\boldsymbol{R_S}^2 - \boldsymbol{R_W}^2)\boldsymbol{C}^2]\,\boldsymbol{f_c}^2 = 0 \\
&(16\pi^4 \boldsymbol{L}^2 \boldsymbol{C}^2)\boldsymbol{f_c}^4 - (8\pi^2 \boldsymbol{f_c}^2 \boldsymbol{LC})\,\boldsymbol{f_c}^2 - [4\pi^2 (\boldsymbol{R_S}^2 - \boldsymbol{R_W}^2)\boldsymbol{C}^2]\,\boldsymbol{f_c}^2 + 1 = 0 \\
&(16\pi^4 \boldsymbol{L}^2 \boldsymbol{C}^2)\boldsymbol{f_c}^4 - \{(8\pi^2 \boldsymbol{f_c}^2 \boldsymbol{LC}) - [4\pi^2 (\boldsymbol{R_S}^2 - \boldsymbol{R_W}^2)\boldsymbol{C}^2]\}\,\boldsymbol{f_c}^2 + 1 = 0
\end{aligned}
$$

This expression is a quadratic equation with $\mathrm{a} = 16\pi^4 \boldsymbol{L}^2 \boldsymbol{C}^2$, $\mathrm{b} = (8\pi^2 \boldsymbol{f_c}^2 \boldsymbol{LC}) - [4\pi^2 (\boldsymbol{R_S}^2 - \boldsymbol{R_W}^2)\boldsymbol{C}^2$, $\mathrm{c} = 1$, and $\boldsymbol{f} = \boldsymbol{f_c}^2$. It has the two solutions

$$\boldsymbol{f}_{(1)} = \left|\frac{-\mathrm{b} + \sqrt{\mathrm{b}^2 - 4\mathrm{ac}}}{2\mathrm{a}}\right| \quad \text{and} \quad \boldsymbol{f}_{(2)} = \left|\frac{-\mathrm{b} - \sqrt{\mathrm{b}^2 - 4\mathrm{ac}}}{2\mathrm{a}}\right|$$

The two solutions correspond to the upper and lower corner frequencies $\boldsymbol{f}_{c(l)}$ and $\boldsymbol{f}_{c(h)}$. Because $\boldsymbol{f}_{(1)}$ and $\boldsymbol{f}_{c(2)}$ are frequency magnitudes only, they are equal to the absolute values of the quadratic solutions. Solve for the product of the frequencies.

$$\boldsymbol{f}_{(l)} \times \boldsymbol{f}_{(2)} = \left|\frac{-\mathrm{b} + \sqrt{\mathrm{b}^2 - 4\mathrm{ac}}}{2\mathrm{a}}\right| \times \left|\frac{-\mathrm{b} - \sqrt{\mathrm{b}^2 - 4\mathrm{ac}}}{2\mathrm{a}}\right|$$

$$\boldsymbol{f}_{c(l)}^2 \times \boldsymbol{f}_{c(h)}^2 = \left|\frac{-\mathrm{b} + \sqrt{\mathrm{b}^2 - 4\mathrm{ac}}}{2\mathrm{a}}\right| \times \left|\frac{-\mathrm{b} - \sqrt{\mathrm{b}^2 - 4\mathrm{ac}}}{2\mathrm{a}}\right|$$

$$[\boldsymbol{f}_{c(l)} \times \boldsymbol{f}_{c(h)}]^2 = \left|\frac{-\mathrm{b} + \sqrt{\mathrm{b}^2 - 4\mathrm{ac}}}{2\mathrm{a}}\right| \times \left|\frac{-\mathrm{b} - \sqrt{\mathrm{b}^2 - 4\mathrm{ac}}}{2\mathrm{a}}\right|$$

$$= \left| \frac{(-b)(-b) + (-b)\left(+\sqrt{b^2 - 4ac}\right) + (-b)\left(-\sqrt{b^2 - 4ac}\right) + \left(+\sqrt{b^2 - 4ac}\right)\left(-\sqrt{b^2 - 4ac}\right)}{(2a)(2a)} \right|$$

$$= \left| \frac{b^2 + (-b)\left(+\sqrt{b^2 - 4ac} - \sqrt{b^2 - 4ac}\right) - \left(\sqrt{b^2 - 4ac}\right)^2}{4a^2} \right|$$

$$= \left| \frac{b^2 + (-b)(0) - (b^2 - 4ac)}{4a^2} \right|$$

$$= \left| \frac{b^2 - (b^2 - 4ac)}{4a^2} \right|$$

$$= \left| \frac{b^2 - b^2 + 4ac}{4a^2} \right|$$

$$= \left| \frac{4ac}{4a^2} \right|$$

$$= \left| \frac{c}{a} \right|$$

Substitute the values for a and c into the equation.

$$[f_{c(l)} \times f_{c(h)}]^2 = \left| \frac{c}{a} \right|$$

$$= \left| \frac{1}{16\pi^4 L^2 C^2} \right|$$

$$= \frac{1}{16\pi^4 L^2 C^2} \quad \text{(because } \boldsymbol{L} \text{ and } \boldsymbol{C} \text{ are both positive)}$$

$$= \left(\frac{1}{4\pi^2 LC} \right)^2$$

$$= \left(\frac{1}{2\pi\sqrt{LC}} \right)^4$$

$$= f_0^4$$

Finally, take the square root of both sides twice to find f_0.

$$\sqrt{f_0^4} = \sqrt{(f_{c(h)} \times f_{c(l)})^2}$$

$$f_0^2 = f_{c(l)} \times f_{c(h)}$$

$$\sqrt{f_0^2} = \sqrt{f_{c(h)} \times f_{c(l)}}$$

$$f_0 = \sqrt{f_{c(h)} \times f_{c(l)}}$$

2. Problem: For the ***RLC*** band-stop filter circuit in Figure 19-26, find values of R_W, L, and C so that $f_0 = 100$ kHz, $BW = 25.0$ kHz, and $V_{out} = 100\ \text{mV}_{ac}$ at the center frequency.

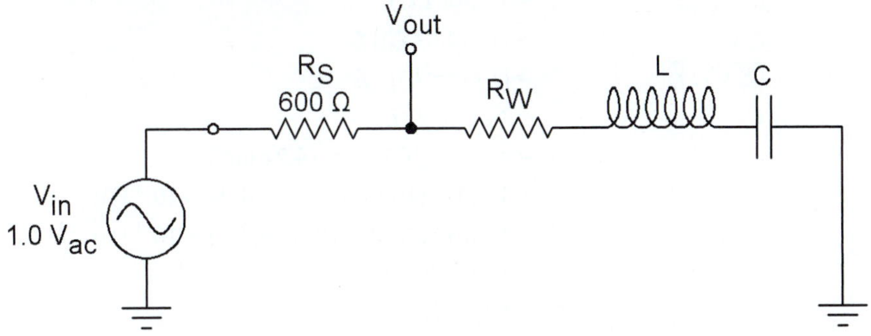

Figure 19-26: *RLC* Band-Stop Filter for Problem 2

Solution: First, find the Q of the circuit.

$$
\begin{aligned}
Q &= f_0 / BW \\
&= 100\ \text{kHz} / 25.0\ \text{kHz} \\
&= 4.0
\end{aligned}
$$

At the center frequency, f_0, the reactances of L and C cancel. Use the voltage divider theorem to calculate R_W from the given values of V_{in}, R_S, and V_{out}.

$$
\begin{aligned}
V_{in}\,[R_W / (R_S + R_W)] &= V_{out} \\
V_{in}\,[R_W / (R_S + R_W)]\,(R_S + R_W) &= V_{out}\,(R_S + R_W) \\
V_{in}\,R_W &= V_{out}\,R_S + V_{out}\,R_W \\
V_{in}\,R_W - V_{out}\,R_W &= V_{out}\,R_S + V_{out}\,R_W - V_{out}\,R_W \\
(V_{in} - V_{out})\,R_W &= V_{out}\,R_S \\
[(V_{in} - V_{out})\,R_W] / (V_{in} - V_{out}) &= (V_{out}\,R_S) / (V_{in} - V_{out}) \\
R_W &= (V_{out}\,R_S) / (V_{in} - V_{out}) \\
&= [(100\ \text{mV}_{ac})(600\ \Omega)] / (1.0\ \text{V}_{ac} - 100\ \text{mV}_{ac}) \\
&= [(0.100\ \text{V}_{ac})(600\ \Omega)] / (1.0\ \text{V}_{ac} - 0.100\ \text{V}_{ac}) \\
&= [(0.100\ \text{V}_{ac})(600\ \Omega)] / (0.90\ \text{V}_{ac}) \\
&= 60\ \Omega / 0.90 \\
&= \mathbf{66.7\ \Omega}
\end{aligned}
$$

Use the values of f_0, R_S, R_W, and Q to find the value of L.

$$
\begin{aligned}
X_L / (R_S + R_W) &= Q \\
(2\pi f_0 L) / (R_S + R_W) &= Q \\
[(2\pi f_0 L) / (R_S + R_W)]\,[(R_S + R_W) / 2\pi f_0] &= Q\,[(R_S + R_W) / 2\pi f_0] \\
L &= \{[Q(R_S + R_W)] / 2\pi f_0\} \\
&= [(4.0)(600\ \Omega + 66.7\ \Omega)] / [2\pi(100\ \text{kHz})] \\
&= [(4.0)(667\ \Omega)] / [2\pi(100 \times 10^3\ \text{Hz})] \\
&= (2667\ \Omega) / (628 \times 10^3\ \text{Hz}) \\
&= 4.24 \times 10^{-3}\ \text{H} \\
&= \mathbf{4.24\ mH}
\end{aligned}
$$

Use the values of f_0 and L to find the value of C.

$$
f_0 = 1 / (2\pi\sqrt{LC}\,)
$$

$$
\begin{aligned}
f_0^2 &= [1 / (2\pi\sqrt{LC}\,)]^2 \\
f_0^2 &= 1 / (4\pi^2 LC) \\
f_0^2 C &= [1 / (4\pi^2 LC)]\, C \\
(f_0^2 C) / f_0^2 &= [1 / (4\pi^2 L)] / f_0^2 \\
C &= 1 / (4\pi^2 f_0^2 L) \\
&= 1 / [4\pi^2 (100 \text{ kHz})^2 (4.24 \text{ mH})] \\
&= 1 / [4\pi^2 (100 \times 10^3 \text{ Hz})^2 (4.24 \times 10^{-3} \text{ H})] \\
&= 1 / [(39.5)(10 \times 10^9 \text{ Hz})(4.24 \times 10^{-3} \text{ H})] \\
&= 1 / (1.68 \times 10^9 \text{ F}^{-1}) \\
&= 597 \times 10^{-12} \text{ F} \\
&= \mathbf{597 \text{ pF}}
\end{aligned}
$$

19.3 Just for Fun

Logarithms, which are used to calculate decibels and construct Bode plots, have a number of practical applications. One application was the slide rule, a portable calculating device that was developed in the 17th century and was widely used by scientists and engineers until the introduction of low-cost electronic calculators in the early 1970s. Another application of logarithms is linearizing and analyzing exponential curves.

1. Problem: The basic slide rule, which was developed in the 17th century, consisted of two logarithmic scales that could be moved relative to each other to perform multiplication and division, just as two linear scales can be moved relative to each other to perform addition and subtraction. Refer to Figure 19-27.

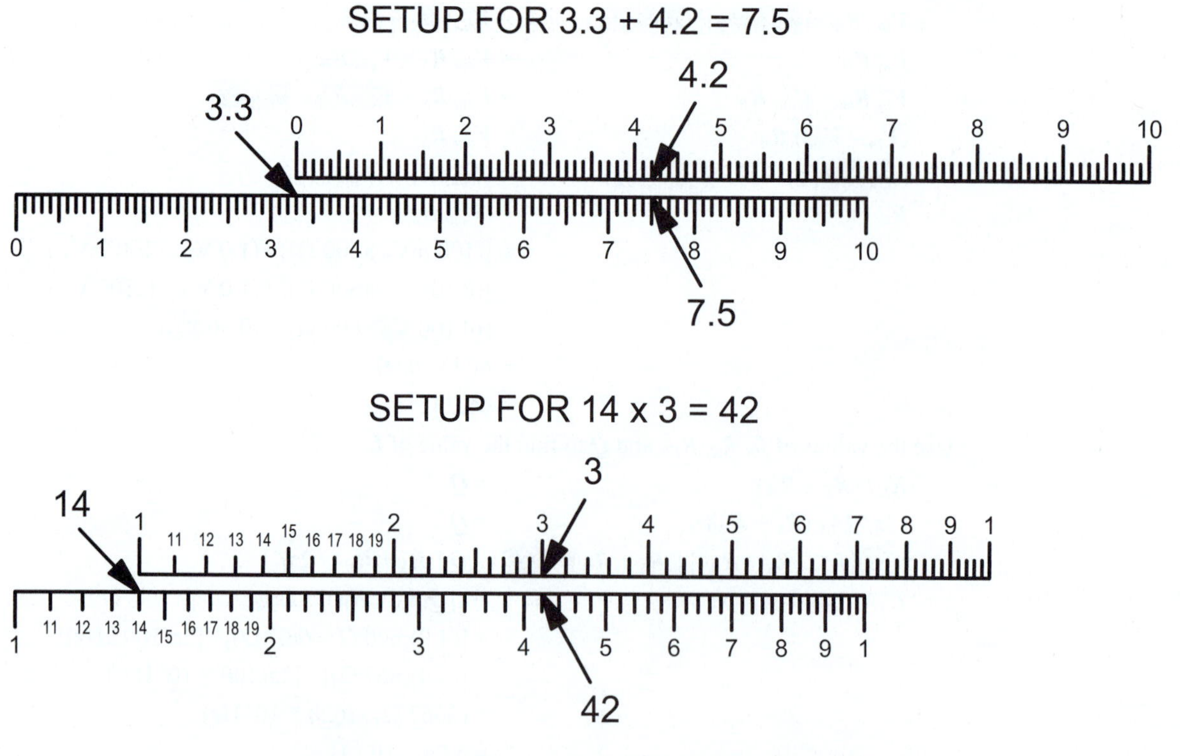

Figure 19-27: Linear and Logarithmic Arithmetic Scales

The linear scales at the top of Figure 19-27 show the setup for the addition operation "3.3 + 4.2". The leftmost mark of the top scale is set opposite the value 3.3 on the bottom scale. The answer, 7.5, is then read on the bottom scale opposite the value 4.2 on the top scale.

The logarithmic scales at the bottom of Figure 19-27 show the setup for the multiplication operation "14 × 3". The leftmost mark, or left index, of the top scale is set opposite the value 14 on the bottom scale, The answer, 42, is then read on the bottom scale opposite the value 3 on the top scale.

Note that the scales shown in Figure 19-27 do not show orders of magnitude (i.e., how "big" the numbers are). The linear scales could just as easily represent "330 + 420 = 750", just as the logarithmic scales can represent the multiplication operation "0.14 × 30". The user must determine the actual values that the marks represent and use estimation to properly interpret the result.

Adding two numbers with linear scales works because, in effect, scalar addition specifies finding a value that is some particular distance from a starting value. In this case, the top scale specifies a distance of 4.2 relative to the starting value of 3.3 on the bottom scale. Explain how using logarithmic scales can represent multiplication, and how logarithmic scales could be used to perform division.

Solution: To see why logarithmic scales can represent multiplication, use the properties of logarithms. Let $\boldsymbol{a} \times \boldsymbol{b} = \boldsymbol{c}$, $\boldsymbol{x} = \log(\boldsymbol{a})$, $\boldsymbol{y} = \log(\boldsymbol{b})$, and $z = \log(\boldsymbol{c})$.

$$
\begin{aligned}
\boldsymbol{a} \times \boldsymbol{b} &= \boldsymbol{c} \\
\log(\boldsymbol{a} \times \boldsymbol{b}) &= \log(\boldsymbol{c}) \\
\log(\boldsymbol{a}) + \log(\boldsymbol{b}) &= \log(\boldsymbol{c}) \\
\boldsymbol{x} + \boldsymbol{y} &= z
\end{aligned}
$$

Using $\boldsymbol{x}$, $\boldsymbol{y}$, and z, the logarithms of $\boldsymbol{a}$, $\boldsymbol{b}$, and $\boldsymbol{c}$, changes the multiplication operation $\boldsymbol{a} \times \boldsymbol{b}$ into the addition operation $\boldsymbol{x} + \boldsymbol{y}$. Just as using linear scales can obtain the sum of two numbers, the logarithmic scales can obtain the sum of two logarithms (which is the logarithm of the product) to represent multiplication.

Division is simply the inverse of multiplication, just as subtraction is the inverse of addition. For division, let $\boldsymbol{a} / \boldsymbol{b} = \boldsymbol{c}$, $\boldsymbol{x} = \log(\boldsymbol{a})$, $\boldsymbol{y} = \log(\boldsymbol{b})$, and $z = \log(\boldsymbol{c})$.

$$
\begin{aligned}
\boldsymbol{a} / \boldsymbol{b} &= \boldsymbol{c} \\
\boldsymbol{a} \times (1 / \boldsymbol{b}) &= \boldsymbol{c} \\
\boldsymbol{a} \times \boldsymbol{b}^{-1} &= \boldsymbol{c} \\
\log(\boldsymbol{a} \times \boldsymbol{b}^{-1}) &= \log(\boldsymbol{c}) \\
\log(\boldsymbol{a}) + \log(\boldsymbol{b}^{-1}) &= \log(\boldsymbol{c}) \\
\log(\boldsymbol{a}) + [(-1) \times \log(\boldsymbol{b})] &= \log(\boldsymbol{c}) \\
\log(\boldsymbol{a}) - \log(\boldsymbol{b}) &= \log(\boldsymbol{c}) \\
\boldsymbol{x} - \boldsymbol{y} &= z \\
\boldsymbol{x} + -\boldsymbol{y} &= z
\end{aligned}
$$

The logarithms of $\boldsymbol{a}$, $\boldsymbol{b}$, and $\boldsymbol{c}$ ($\boldsymbol{x}$, $\boldsymbol{y}$, and z, respectively) change the division operation $\boldsymbol{a} / \boldsymbol{b}$ into the subtraction operation $\boldsymbol{x} - \boldsymbol{y}$. This is the same as adding the inverse of $\boldsymbol{y}$ to $\boldsymbol{x}$. Just as linear scales can represent the difference of two numbers by specifying the distances for subtracted numbers in the direction opposite that for addition, logarithmic scales can represent the division of two logarithms (which is the logarithm of the quotient) by specifying a logarithmic distance in the direction opposite that for multiplication. One way to do so is to place the divisor value on the top scale over the dividend value on the bottom scale and find the quotient on the bottom scale below the index of the top scale. The logarithmic scales in Figure 19-27, for example, show the setup for the division operation 42 / 3 = 14.

2. Problem: A very useful property of logarithms is that they can linearize exponential functions. Let $\boldsymbol{ka}^{\boldsymbol{b}} = \boldsymbol{c}$, $\boldsymbol{x} = \log(\boldsymbol{a})$, $\boldsymbol{y} = \log(\boldsymbol{c})$, and $z = \log(\boldsymbol{k})$.

$$\boldsymbol{c} = \boldsymbol{ka}^{\boldsymbol{b}}$$

$$\begin{aligned} \log(c) &= \log(\boldsymbol{ka^b}) \\ &= \log(\boldsymbol{k}) + \log(\boldsymbol{a^b}) \\ &= \log(\boldsymbol{a^b}) + \log(\boldsymbol{k}) \\ &= \boldsymbol{b}\log(\boldsymbol{a}) + \log(\boldsymbol{k}) \\ y &= \boldsymbol{bx} + z \end{aligned}$$

The logarithmic form has the standard slope-intercept form of a straight line, in which the exponent $\boldsymbol{b}$ is the slope of the line and the logarithm of the scalar multiplier, log ($\boldsymbol{k}$), is the intercept.

A very common exponential equation is the discharging equation for ***RC*** circuits. Determine the logarithmic equations for the equation $v(t) = V_0 e^{-t/\tau}$. Then, plot a logarithmic graph of the experimental discharging data for an ***RC*** circuit in Table 19-1 and use the plot to determine the initial capacitor voltage, V_0, and time constant, τ.

Table 19-1: Experimental Discharging Data for an *RC* Circuit

t (s)	v(t) (V)	t (s)	v(t) (V)
1.5	2.037	2.0	1.843
1.6	1.997	2.1	1.807
1.7	1.957	2.2	1.771
1.8	1.919	2.3	1.736
1.9	1.881	2.4	1.702

Solution: For the discharging equation $v(t) = V_0 e^{-t/\tau}$:

$$\begin{aligned} v(t) &= V_0 e^{-t/\tau} \\ \log[v(t)] &= \log(V_0 e^{-t/\tau}) \\ &= \log(V_0) + \log(e^{-t/\tau}) \\ &= \log(e^{-t/\tau}) + \log(V_0) \\ &= (-t/\tau)\log(e) + \log(V_0) \\ &= (-t/\tau)(0.434) + \log(V_0) \\ &= [(-0.434\, t)/\tau] + \log(V_0) \\ &= (-0.434/\tau)\, t + \log(V_0) \end{aligned}$$

This equation is a linear equation of the form $y = mx + b$. To plot the data on a logarithmic graph, find the logarithmic values of $v(t)$ and plot the logarithmic values against t.

Table 19-2: Logarithmic Discharging Data for an *RC* Circuit

t (s)	*v*(*t*) (V)	log [*v*(*t*)]	*t* (s)	*v*(*t*) (V)	log [*v*(*t*)]
1.5	2.037	0.309	2.0	1.843	0.266
1.6	1.997	0.300	2.1	1.807	0.257
1.7	1.957	0.292	2.2	1.771	0.248
1.8	1.919	0.283	2.3	1.736	0.240
1.9	1.881	0.274	2.4	1.702	0.231

Plotting log [$v(t)$] vs. t produces the logarithmic graph shown in Figure 19-28.

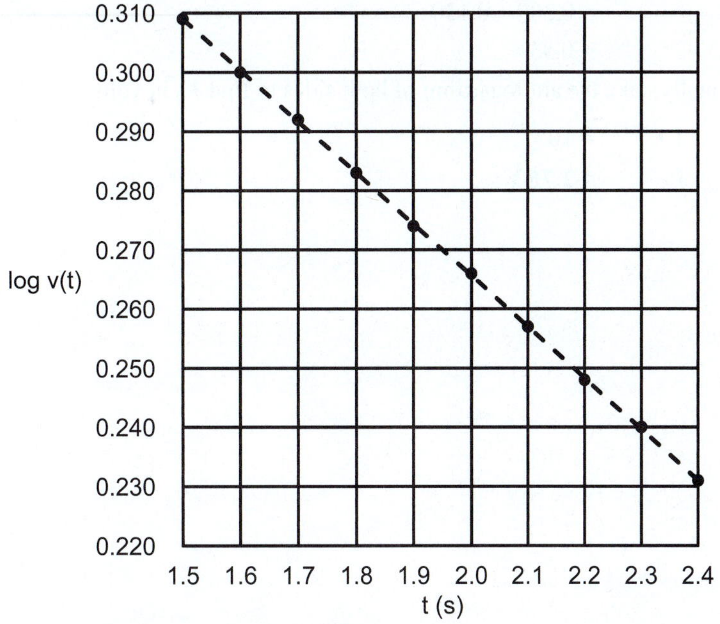

Figure 19-28: Logarithmic Graph of Discharging Data

The slope of the graph, m, is $\Delta v(t) / \Delta t$, the change in $v(t)$ divided by the change in t. Use the endpoints of the graph to calculate m.

$$\begin{aligned} m &= \Delta v(t) / \Delta t \\ &= (0.231 - 0.309) / (2.4 \text{ s} - 1.5 \text{ s}) \\ &= -0.078 / 0.9 \text{ s} \\ &= -0.0868 \text{ s}^{-1} \end{aligned}$$

From the logarithmic equation of the discharging curve, $m = -0.434 / \tau$.

$$\begin{aligned} m &= -0.434 / \tau \\ m\,\tau &= (-0.434 / \tau)\,\tau \\ (m\,\tau) / m &= -0.434 / m \\ \tau &= -0.434 / -0.0868 \text{ s}^{-1} \\ &= \mathbf{5.0 \text{ s}} \end{aligned}$$

Find the logarithm of the initial capacitor voltage, log (V_0), at time $t_0 = 0$, by using the definition of the slope for a line.

$$\begin{aligned} m &= \Delta\log[v(t)] / \Delta t \\ &= \{\log[v(t)] - \log(V_0)\} / (t - t_0) \\ &= \{\log[v(t)] - \log(V_0)\} / (t - 0) \\ &= \{\log[v(t)] - \log(V_0)\} / t \\ m\,t &= (\{\log[v(t)] - \log(V_0)\} / t)\,t \\ m\,t - \log v(t) &= \{\log[v(t)] - \log(V_0)\} - \log[v(t)] \\ (-1)\{m\,t - \log[v(t)]\} &= (-1)[-\log(V_0)] \\ \log[v(t)] - m\,t &= \log(V_0) \end{aligned}$$

Choose the first data point, for which log [$v(t)$] = 0.309 and t = 1.5 s.

$$\begin{aligned} \log(V_0) &= \log[v(t)] - m\,t \\ &= 0.309 - (-0.0868 \text{ s}^{-1})(1.5 \text{ s}) \end{aligned}$$

$= 0.309 + (0.0868\ \text{s}^{-1})(1.5\ \text{s})$

$= 0.309 + 0.130$

$= 0.439$

Finally, take the antilogarithm of both sides to find V_0 in volts.

$10^{\log(V_0)} = 10^{0.439}$

$\boldsymbol{V_0} = \mathbf{2.75\ V}$

20. AC Circuit Theorems

Note: For all problems in this section, the phase of ac sources is 0° unless otherwise stated.

20.1 Basic

20.1.1 Voltage and Current Divider Theorems

1. Problem: Use the ac voltage divider theorem to find the voltages across $\boldsymbol{R}$, $\boldsymbol{X_L}$, and $\boldsymbol{X_C}$ for the series circuit in Figure 20-1. Express all answers in polar form.

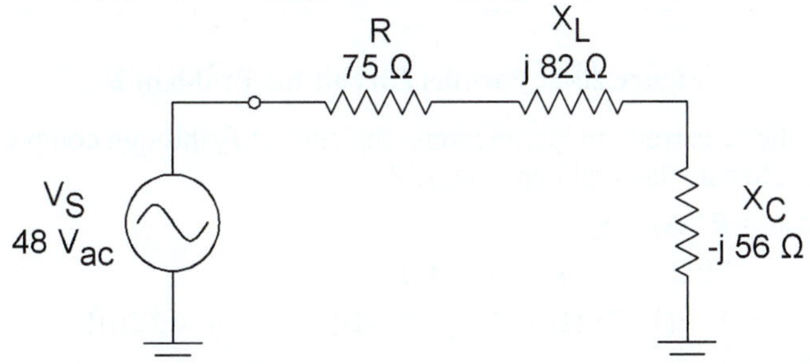

Figure 20-1: Series Circuit for Problem 1

Solution: From the ac voltage divider theorem, the voltage $\boldsymbol{V_X}$ across component $\boldsymbol{Z_X}$ is $\boldsymbol{V_X} = \boldsymbol{V_S}\,(\boldsymbol{Z_X} / \boldsymbol{Z_T})$. First, calculate the total impedance, $\boldsymbol{Z_T}$.

$$\begin{aligned}
\boldsymbol{Z_T} &= \boldsymbol{R} + \boldsymbol{X_L} + \boldsymbol{X_C} \\
&= 75\ \Omega + \mathrm{j}\ 82\ \Omega - \mathrm{j}\ 56\ \Omega \\
&= 75\ \Omega + \mathrm{j}\ (82\ \Omega - 56\ \Omega) \\
&= 75\ \Omega + \mathrm{j}\ 26\ \Omega \text{ in rectangular form} \\
&= 79.4\ \Omega \angle\ 19.1^\circ \text{ in polar form}
\end{aligned}$$

Next, use the ac voltage divider theorem to calculate the voltages for $\boldsymbol{R}$, $\boldsymbol{X_L}$, and $\boldsymbol{X_C}$.

$$\begin{aligned}
\boldsymbol{V_R} &= \boldsymbol{V_S}\,(\boldsymbol{R} / \boldsymbol{Z_T}) \\
&= (48\ \mathrm{V_{ac}} \angle\ 0^\circ)\ [(75\ \Omega \angle\ 0^\circ) / (79.4\ \Omega \angle\ 19.1^\circ)] \\
&= (48\ \mathrm{V_{ac}} \angle\ 0^\circ)\ [(75\ \Omega / 79.4\ \Omega) \angle\ (0^\circ - 19.1^\circ)] \\
&= (48\ \mathrm{V_{ac}} \angle\ 0^\circ)\ (0.945 \angle -19.1^\circ) \\
&= [(48\ \mathrm{V_{ac}})(0.945)] \angle\ [0^\circ + (-19.1^\circ)] \\
&= \mathbf{45.4\ V_{ac} \angle -19.1^\circ}
\end{aligned}$$

$$\begin{aligned}
\boldsymbol{V_{XL}} &= \boldsymbol{V_S}\,(\boldsymbol{X_L} / \boldsymbol{Z_T}) \\
&= (48\ \mathrm{V_{ac}} \angle\ 0^\circ)\ [(82\ \Omega \angle\ 90^\circ) / (79.4\ \Omega \angle\ 19.1^\circ)] \\
&= (48\ \mathrm{V_{ac}} \angle\ 0^\circ)\ [(82\ \Omega / 79.4\ \Omega) \angle\ (90^\circ - 19.1^\circ)] \\
&= (48\ \mathrm{V_{ac}} \angle\ 0^\circ)\ (1.033 \angle\ 70.9^\circ) \\
&= [(48\ \mathrm{V_{ac}})(1.033)] \angle\ (0^\circ + 70.9^\circ) \\
&= \mathbf{49.6\ V_{ac} \angle\ 70.9^\circ}
\end{aligned}$$

$$\begin{aligned}
\boldsymbol{V_{XC}} &= \boldsymbol{V_S}\,(\boldsymbol{X_C} / \boldsymbol{Z_T}) \\
&= (48\ \mathrm{V_{ac}} \angle\ 0^\circ)\ [(56\ \Omega \angle -90^\circ) / (79.4\ \Omega \angle\ 19.1^\circ)] \\
&= (48\ \mathrm{V_{ac}} \angle\ 0^\circ)\ [(56\ \Omega / 79.4\ \Omega) \angle\ (-90^\circ - 19.1^\circ)] \\
&= (48\ \mathrm{V_{ac}} \angle\ 0^\circ)\ (0.706 \angle -109^\circ) \\
&= [(48\ \mathrm{V_{ac}})(0.706)] \angle\ [0^\circ + (-109^\circ)]
\end{aligned}$$

$= \mathbf{33.9\ V_{ac} \angle -109°}$

2. Problem: Use the ac current divider theorem to find the currents through $\boldsymbol{R}$, $\boldsymbol{X_L}$, and $\boldsymbol{X_C}$ for the parallel circuit in Figure 20-2. Express all answers in polar form.

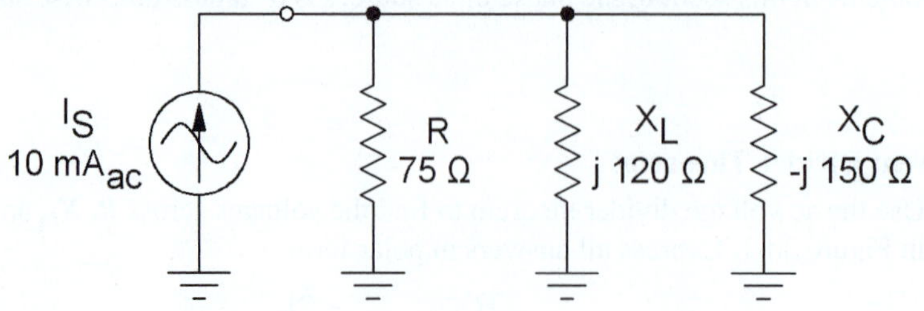

Figure 20-2: Parallel Circuit for Problem 2

Solution: From the ac current divider theorem, the current $\boldsymbol{I_X}$ through component $\boldsymbol{Z_X}$ is $\boldsymbol{I_X} = \boldsymbol{I_S}\,(\boldsymbol{Z_T} / \boldsymbol{Z_X})$. First, calculate the total impedance, $\boldsymbol{Z_T}$.

$\boldsymbol{Z_T} = \boldsymbol{R} \,||\, \boldsymbol{X_L} \,||\, \boldsymbol{X_C}$

$= 75\ \Omega \,||\, \text{j}\ 120\ \Omega \,||\, -\text{j}\ 150\ \Omega$

$= 1 / \{(1 / 75\ \Omega) + [1 / (\text{j}\ 120\ \Omega)] + [1 / (-\text{j}\ 150\ \Omega)]\}$

$\boldsymbol{Y_T} = 1 / \boldsymbol{Z_T}$

$= [(1 \angle 0°) / (75\ \Omega \angle 0°)] + [(1 \angle 0°) / (120\ \Omega \angle 90°)] + [(1 \angle 0°) / (150\ \Omega \angle -90°)]$

$= [(1 / 75\ \Omega) \angle (0° - 0°)] + [(1 / 120\ \Omega) \angle (0° - 90°)] + \{(1 / 150\ \Omega) \angle [0° - (-90°)]\}$

$= (13.3\text{ mS} \angle 0°) + (8.33\text{ mS} \angle -90°) + (6.67\text{ mS} \angle 90°)$ in polar form

$= (13.3\text{ mS} + \text{j}\ 0\text{ S}) + (0\text{ S} - \text{j}\ 8.33\text{ mS}) + (0\text{ S} + \text{j}\ 6.67\text{ mS})$ in rectangular form

$= (13.3\text{ mS} + 0\text{ S} + 0\text{ S}) + \text{j}\ (0\text{ S} - 8.33\text{ mS} + 6.67\text{ mS})$

$= 13.3\text{ mS} - \text{j}\ 1.67\text{ mS}$ in rectangular form

$= 13.4\text{ mS} \angle -7.13°$ in polar form

$\boldsymbol{Z_T} = 1 / \boldsymbol{Y_T}$

$= (1 \angle 0°) / (13.4\text{ mS} \angle -7.13°)$

$= (1 / 13.4\text{ mS}) \angle [0° - (-7.13°)]$

$= 74.4\ \Omega \angle 7.13°$

Next, use the ac current divider theorem to calculate the currents for $\boldsymbol{R}$, $\boldsymbol{X_L}$, and $\boldsymbol{X_C}$.

$\boldsymbol{I_R} = \boldsymbol{I_S}\,(\boldsymbol{Z_T} / \boldsymbol{R})$

$= (10\text{ mA}_{ac} \angle 0°)\ [(74.4\ \Omega \angle 7.13°) / (75\ \Omega \angle 0°)]$

$= (10\text{ mA}_{ac} \angle 0°)\ [(74.4\ \Omega / 75\ \Omega) \angle (7.13° - 0°)]$

$= (10\text{ mA}_{ac} \angle 0°)\ (0.992 \angle 7.13°)$

$= [(10\text{ mA}_{ac})(0.992)] \angle (0° + 7.13°)$

$= \mathbf{9.92\ mA_{ac} \angle 7.13°}$

$\boldsymbol{I_{XL}} = \boldsymbol{I_S}\,(\boldsymbol{Z_T} / \boldsymbol{X_L})$

$= (10\text{ mA}_{ac} \angle 0°)\ [(74.4\ \Omega \angle 7.13°) / (120\ \Omega \angle 90°)]$

$= (10\text{ mA}_{ac} \angle 0°)\ [(74.4\ \Omega / 120\ \Omega) \angle (7.13° - 90°)]$

$= (10\text{ mA}_{ac} \angle 0°)\ (0.620 \angle -82.9°)$

$= [(10\text{ mA}_{ac})(0.620)] \angle [0° + (-82.9°)]$

$= \mathbf{6.20\ mA_{ac} \angle -82.9°}$

$\boldsymbol{I_{XC}} = \boldsymbol{V_S}\,(\boldsymbol{Z_T} / \boldsymbol{X_C})$

$= (10\ \text{mA}_{ac} \angle 0°)\ [(74.4\ \Omega \angle 7.13°) / (150\ \Omega \angle -90°)]$

$= (10\ \text{mA}_{ac} \angle 0°)\ \{(74.4\ \Omega / 150\ \Omega) \angle [7.13° - (-90°)]\}$

$= (10\ \text{mA}_{ac} \angle 0°)\ (0.496 \angle 97.1°)$

$= [(10\ \text{mA}_{ac})(0.496)] \angle (0° + 97.1°)$

$= \mathbf{4.96\ mA_{ac} \angle 97.1°}$

20.1.2 Thevenin's Theorem

1. Problem: Calculate the voltage and impedance values, V_{TH} and Z_{TH}, of the Thevenin equivalent for the circuit in Figure 20-3 with respect to the output, V_{out}. Express the answers in polar form.

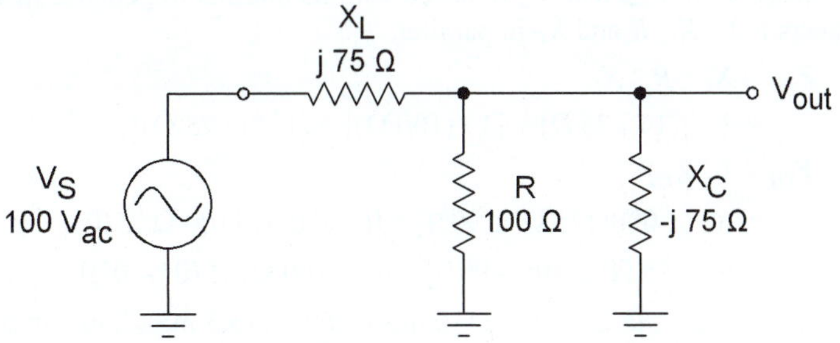

Figure 20-3: Circuit for Thevenin Problem 1

Solution: For the circuit in Figure 20-3, the Thevenin voltage, V_{TH}, is the open load voltage, V_{out}. Use the ac voltage divider theorem to find V_{TH}.

$V_{TH} = V_{out}$

$= V_S\ \{(R \,\|\, X_C) / [X_L + (R \,\|\, X_C)]\}$

First, find $Z_{EQ} = R \,\|\, X_C$.

$Z_{EQ} = R \,\|\, X_C$

$= 100\ \Omega \,\|\, -j\ 75\ \Omega$

$= 1 / [(1 / 100\ \Omega) + (1 / -j\ 75\ \Omega)]$

$Y_{EQ} = 1 / Z_{EQ}$

$= [(1 \angle 0°) / (100\ \Omega \angle 0°)] + [(1 \angle 0°) / (75\ \Omega \angle -90°)]$

$= [(1 / 100\ \Omega) \angle (0° - 0°)] + \{(1 / 75\ \Omega) \angle [0° - (-90°)]\}$

$= (10\ \text{mS} \angle 0°) + (13.3\ \text{mS} \angle 90°)$ in polar form

$= (10\ \text{mS} + j\ 0\ \text{S}) + (0\ \text{S} + j\ 13.3\ \text{mS})$ in rectangular form

$= (10\ \text{mS} + 0\ \text{S}) + j\ (0\ \text{S} + 13.3\ \text{mS})$

$= 10\ \text{mS} + j\ 13.3\ \text{mS}$ in rectangular form

$= 16.7\ \text{mS} \angle 53.1°$ in polar form

$Z_{EQ} = 1 / Y_{EQ}$

$= (1 \angle 0°) / (16.7\ \text{mS} \angle 53.1°)$

$= (1 / 16.7\ \text{mS}) \angle (0° - 53.1°)$

$= 60.0\ \Omega \angle -53.1°$ in polar form

$= 36.0\ \Omega - j\ 48.0\ \Omega$ in rectangular form

Substitute $Z_{EQ} = R \,\|\, X_C$ into the ac voltage divider equation.

$V_{TH} = V_S\ \{(R \,\|\, X_C) / [X_L + (R \,\|\, X_C)]\}$

$= V_S\ [Z_{EQ} / (X_L + Z_{EQ})]$

$= (100\ \text{V}_{ac})\ \{(60.0\ \Omega \angle -53.1°) / [j\ 75\ \Omega / (36.0\ \Omega - j\ 48.0\ \Omega)]\}$

$= (100\ V_{ac} \angle 0°)\ \{(60.0\ \Omega \angle -53.1°) / [(0\ \Omega + 36.0\ \Omega) + j\ (75\ \Omega - j\ 48.0\ \Omega)]\}$

$= (100\ V_{ac} \angle 0°)\ [(60.0\ \Omega \angle -53.1°) / (36.0\ \Omega + j\ 27\ \Omega)]$

$= (100\ V_{ac} \angle 0°)\ [(60.0\ \Omega \angle -53.1°) / (45.0\ \Omega \angle 36.9°)]$

$= (100\ V_{ac} \angle 0°)\ [(60.0\ \Omega / 45.0\ \Omega) \angle (-53.1° - 36.9°)]$

$= (100\ V_{ac} \angle 0°)\ (1.33 \angle -90°)$

$= [(100\ V_{ac})(1.33)] \angle [0° + (-90°)]$

$= \mathbf{133\ V_{ac} \angle -90°}$

For the circuit in Figure 20-3, the Thevenin impedance, Z_{TH}, is the impedance looking back into the circuit from V_{out} with V_S replaced with its internal impedance (a short). This impedance appears to be X_L, R and X_C in parallel.

$Z_{TH} = X_L \parallel R \parallel X_C$

$= 1 / \{(1 / j\ 75\ \Omega) + [1 / (100\ \Omega)] + [1 / (-j\ 75\ \Omega)]\}$

$Y_{TH} = 1 / Z_{TH}$

$= [(1 \angle 0°) / (75\ \Omega \angle 90°)] + [(1 \angle 0°) / (100\ \Omega \angle 0°)] + [(1 \angle 0°) / (75\ \Omega \angle -90°)]$

$= [(1 / 75\ \Omega) \angle (0° - 90°)] + [(1 / 100\ \Omega) \angle (0° - 0°)] + \{(1 / 75\ \Omega) \angle [0° - (-90°)]\}$

$= (13.3\ mS \angle -90°) + (10\ mS \angle 0°) + (13.3\ mS \angle 90°)$ in polar form

$= (0\ S - j\ 13.3\ mS) + (10\ mS - 0\ S) + (0\ S + j\ 13.3\ mS)$ in rectangular form

$= (0\ S + 10\ mS + 0\ S) + j\ (-13.3\ mS + 0\ S + 13.3\ mS)$

$= 10\ mS + j\ 0\ S$ in rectangular form

$= 10\ mS \angle 0°$ in polar form

$Z_{TH} = 1 / Y_{TH}$

$= (1 \angle 0°) / (10\ mS \angle 0°)$

$= (1 / 10\ mS) \angle (0° + 0°)$

$= \mathbf{100\ \Omega \angle 0°}$

2. Problem: Calculate the voltage and impedance values, V_{TH} and Z_{TH}, of the Thevenin equivalent for the circuit in Figure 20-4 with respect to the output, V_{out}.

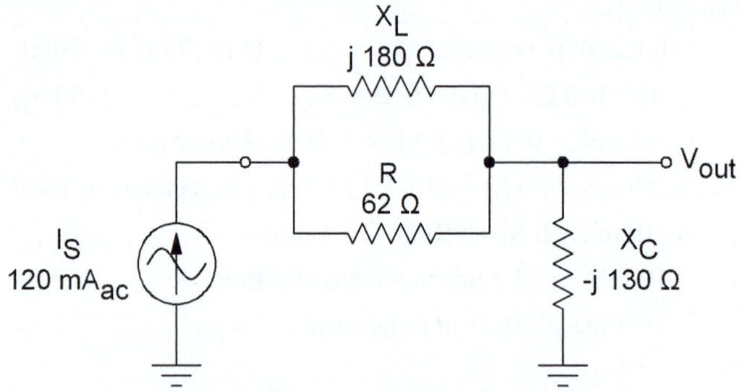

Figure 20-4: Circuit for Thevenin Problem 2

Solution: For the circuit in Figure 20-4, the Thevenin voltage, V_{TH}, is the open load voltage, V_{out}. The full source current, I_S, passes through X_C. Use Ohm's Law to find V_{TH}.

$V_{TH} = V_{out}$

$= I_S X_C$

$= (120\ mA_{ac} \angle 0°)\ (-j\ 130\ \Omega)$

$= (120\ mA_{ac} \angle 0°)\ (130\ \Omega \angle -90°)$

$= [(120\ mA_{ac})(130\ \Omega)] \angle [0° - (-90°)]$

$\mathbf{= 15.6\ V_{ac} \angle 90°}$

For the circuit in Figure 20-4, the Thevenin impedance, Z_{TH}, is the impedance looking back into the circuit from V_{out} with I_S replaced with its internal impedance (an open). This impedance appears to be X_L, R and X_C in parallel.

$Z_{TH} = X_L \parallel R \parallel X_C$

$= 1 / 1 / \{(1 / j\ 180\ \Omega) + [1 / (62\ \Omega)] + [1 / (-j\ 130\ \Omega)]\}$

$Y_{TH} = 1 / Z_{TH}$

$= [(1 \angle 0°) / (180\ \Omega \angle 90°)] + [(1 \angle 0°) / (62\ \Omega \angle 0°)] + [(1 \angle 0°) / (130\ \Omega \angle -90°)]$

$= [(1 / 180\ \Omega) \angle (0° - 90°)] + [(1 / 62\ \Omega) \angle (0° - 0°)] + \{(1 / 130\ \Omega) \angle [0° - (-90°)]\}$

$= (5.56\ mS \angle -90°) + (16.1\ mS \angle 0°) + (7.69\ mS \angle 90°)$ in polar form

$= (0\ S - j\ 5.56\ mS) + (16.1\ mS - 0\ S) + (0\ S + j\ 7.69\ mS)$ in rectangular form

$= (0\ S + 16.1\ mS + 0\ S) + j\ (-5.56\ mS + 0\ S + 7.69\ mS)$

$= 16.1\ mS + j\ 2.14\ mS$ in rectangular form

$= 16.3\ mS \angle 7.55°$ in polar form

$Z_{TH} = 1 / Y_{TH}$

$= (1 \angle 0°) / (16.3\ mS \angle 7.55°)$

$= (1 / 16.3\ mS) \angle (0° - 7.55°)$

$\mathbf{= 61.5\ \Omega \angle -7.55°}$

20.1.3 Norton's Theorem

1. Problem: Calculate the current and impedance values, I_N and Z_N, of the Norton equivalent for the circuit in Figure 20-5 with respect to the output, V_{out}.

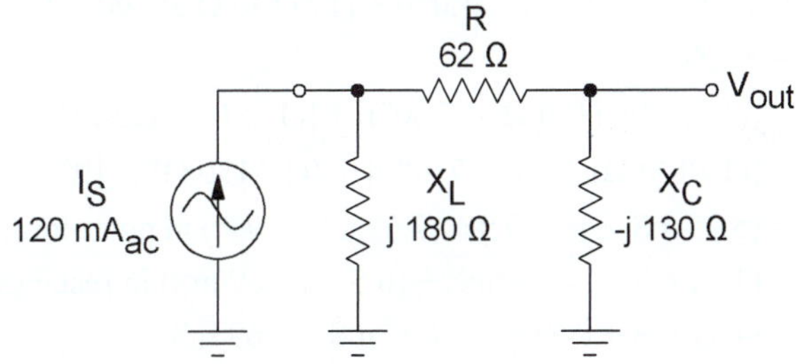

Figure 20-5: Circuit for Norton Problem 1

Solution: For the circuit in Figure 20-5, the Norton current, I_N, is the short-circuit current for V_{out}. When V_{out} is shorted to ground, X_C is bypassed so that only X_L and R offer current paths for I_S. Use the ac current divider theorem to calculate I_N.

$I_N = I_{SC}$

$= I_R$

$= I_S (Z_{EQ} / R)$

First, find $Z_{EQ} = X_L \parallel R$.

$Z_{EQ} = X_L \parallel R$

$= j\ 180\ \Omega \parallel 62\ \Omega$

$= 1 / [(1 / j\ 180\ \Omega) + (1 / 62\ \Omega)]$

$Y_{EQ} = 1 / Z_{EQ}$

$= [(1 \angle 0°) / (180\ \Omega \angle 90°)] + [(1 \angle 0°) / (62\ \Omega \angle 0°)]$

$= [(1 / 180\ \Omega) \angle (0° - 90°)] + [(1 / 62\ \Omega) \angle (0° - 0°)]$

$= (5.56\text{ mS} \angle -90°) + (16.1\text{ mS} \angle 0°)$ in polar form

$= (0\text{ S} - j\ 5.56\text{ mS}) + (16.1\text{ mS} + j\ 0\text{ S})$ in rectangular form

$= (0\text{ S} + 16.1\text{ mS}) + j\ (-5.56\text{ mS} + 0\text{ S})$

$= 16.1\text{ mS} - j\ 5.56\text{ mS}$ in rectangular form

$= 17.1\text{ mS} \angle -19.0°$ in polar form

$\boldsymbol{Z_{EQ}} = 1 / \boldsymbol{Y_{EQ}}$

$= (1 \angle 0°) / (17.1\text{ mS} \angle -19.0°)$

$= (1 \angle 17.1\text{ mS}) \angle [0° - (-19.0°)]$

$= 58.6\ \Omega \angle 19.0°$

Next, substitute $\boldsymbol{Z_{EQ}} = \boldsymbol{X_L} \parallel \boldsymbol{R}$ into the ac current divider theorem.

$\boldsymbol{I_N} = \boldsymbol{I_S} (\boldsymbol{Z_{EQ}} / \boldsymbol{R})$

$= (120\text{ mA}_{ac} \angle 0°) [(58.6\ \Omega \angle 19.0°) / (62\ \Omega \angle 0°)]$

$= (120\text{ mA}_{ac} \angle 0°) [(58.6\ \Omega / 62\ \Omega) \angle (19.0° - 0°)]$

$= (120\text{ mA}_{ac} \angle 0°) (0.946 \angle 19.0°)$

$= [(120\text{ mA}_{ac})(0.946)] \angle (0° + 19.0°)$

$= \mathbf{114\ mA_{ac} \angle 19.0°}$

For the circuit in Figure 20-5, the Norton impedance, $\boldsymbol{Z_N}$, is the impedance looking back into the circuit from $\boldsymbol{V_{out}}$ with $\boldsymbol{I_S}$ replaced with its internal impedance (an open). This impedance appears to be the series combination of $\boldsymbol{R}$ and $\boldsymbol{X_L}$ in parallel with $\boldsymbol{X_C}$.

$\boldsymbol{Z_N} = (\boldsymbol{R} + \boldsymbol{X_L}) \parallel \boldsymbol{X_C}$

$= (62\ \Omega + j\ 180\ \Omega) \parallel (-j\ 130\ \Omega)$

$= 1 / \{[1 / (62\ \Omega + j\ 180\ \Omega)] + [1 / (-j\ 130\ \Omega)]\}$

$= 1 / \{[1 / (190\ \Omega \angle 71.0°)] + [1 / (130\ \Omega \angle -90°)]\}$

$\boldsymbol{Y_N} = 1 / \boldsymbol{Z_N}$

$= [(1 \angle 0°) / (190\ \Omega \angle 71.0°)] + [(1 \angle 0°) / (130\ \Omega \angle -90°)]$

$= [(1 / 190\ \Omega) \angle (0° - 71.0°)] + \{(1 / 130\ \Omega) \angle [0° - (-90°)]\}$

$= (5.25\text{ mS} \angle -71.0°) + (7.69\text{ mS} \angle 90°)$ in polar form

$= (1.71\text{ mS} - j\ 4.97\text{ mS}) + (0\text{ S} + j\ 7.69\text{ mS})$ in rectangular form

$= (1.71\text{ mS} + 0\text{ S}) + j\ (-4.97\text{ mS} + 7.69\text{ mS})$

$= 1.71\text{ mS} + j\ 2.73\text{ mS}$ in rectangular form

$= 3.22\text{ mS} \angle 57.9°$ in polar form

$\boldsymbol{Z_N} = 1 / \boldsymbol{Y_N}$

$= (1 \angle 0°) / (3.22\text{ mS} \angle 57.9°)$

$= (1 / 3.22\text{ mS}) \angle (0° - 57.9°)$

$= \mathbf{311\ \Omega \angle -57.9°}$

2. Problem: Calculate the current and impedance values, $\boldsymbol{I_N}$ and $\boldsymbol{Z_N}$, of the Norton equivalent for the circuit in Figure 20-6 with respect to the output, $\boldsymbol{V_{out}}$.

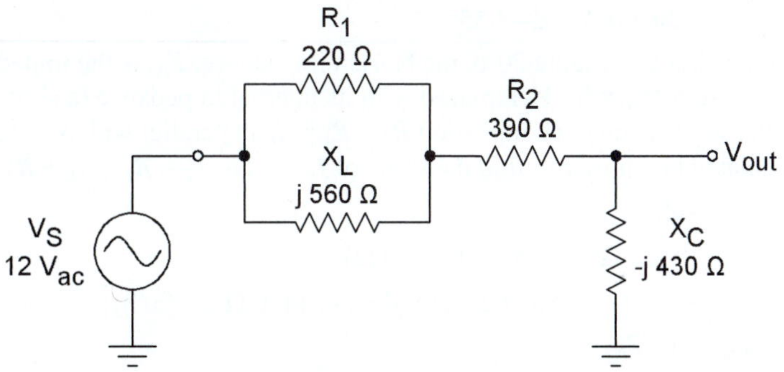

Figure 20-6: Circuit for Norton Problem 2

Solution: For the circuit in Figure 20-6, the Norton current, $\boldsymbol{I_N}$, is the short-circuit current for $\boldsymbol{V_{out}}$. When $\boldsymbol{V_{out}}$ is shorted to ground, $\boldsymbol{X_C}$ is bypassed so that the parallel combination of $\boldsymbol{X_L}$ and $\boldsymbol{R_1}$ in series with $\boldsymbol{R_2}$ is the current path for $\boldsymbol{V_S}$. First, calculate the total impedance, $\boldsymbol{Z_T}$, seen by $\boldsymbol{V_S}$ with $\boldsymbol{V_{out}}$ shorted to ground.

$\boldsymbol{Z_T} = \boldsymbol{R_1} \parallel \boldsymbol{X_L} + \boldsymbol{R_2}$

First, find $\boldsymbol{Z_{EQ}} = \boldsymbol{R_1} \parallel \boldsymbol{X_L}$.

$\boldsymbol{Z_{EQ}} = \boldsymbol{R_1} \parallel \boldsymbol{X_L}$

$= 220\ \Omega \parallel \text{j}\ 560\ \Omega$

$= 1 / [(1 / 220\ \Omega) + (1 / \text{j}\ 560\ \Omega)]$

$\boldsymbol{Y_{EQ}} = 1 / \boldsymbol{Z_{EQ}}$

$= [(1 \angle 0°) / (220\ \Omega \angle 0°)] + [(1 \angle 0°) / (560\ \Omega \angle 90°)]$

$= [(1 / 220\ \Omega) \angle (0° - 0°)] + [(1 / 560\ \Omega) \angle (0° - 90°)]$

$= (4.55\text{ mS} \angle 0°) + (1.79\text{ mS} \angle -90°)$ in polar form

$= (4.55\text{ mS} + \text{j}\ 0\text{ S}) + (0\text{ S} - \text{j}\ 1.79\text{ mS})$ in rectangular form

$= (4.55\text{ mS} + 0\text{ S}) + \text{j}\ (0\text{ S} - 1.79\text{ mS})$

$= 4.55\text{ mS} - \text{j}\ 1.79\text{ mS}$ in rectangular form

$= 4.88\text{ mS} \angle -21.5°$ in polar form

$\boldsymbol{Z_{EQ}} = 1 / \boldsymbol{Y_{EQ}}$

$= (1 \angle 0°) / (4.88\text{ mS} \angle -21.5°)$

$= (1 \angle 4.88\text{ mS}) \angle [0° - (-21.5°)]$

$= 205\ \Omega \angle 21.5°$ in polar form

$= 191\ \Omega + \text{j}\ 74.9\ \Omega$

Next, substitute $\boldsymbol{Z_{EQ}} = \boldsymbol{R_1} \parallel \boldsymbol{X_L}$ into the equation for total impedance.

$\boldsymbol{Z_T} = \boldsymbol{Z_{EQ}} + \boldsymbol{R_2}$

$= (191\ \Omega + \text{j}\ 74.9\ \Omega) + 390\ \Omega$

$= (191\ \Omega + 390\ \Omega) + \text{j}\ (74.9\ \Omega + 0\ \Omega)$

$= 581\ \Omega + \text{j}\ 74.9\ \Omega$ in rectangular form

$= 585\ \Omega \angle 7.35°$ in polar form

Use Ohm's Law to calculate $\boldsymbol{I_N}$.

$\boldsymbol{I_N} = \boldsymbol{I_{SC}}$

$= \boldsymbol{V_S} / \boldsymbol{Z_T}$

$= (12\ \text{V}_{ac} \angle 0°) / (585\ \Omega \angle 7.35°)$

$= (12\ \text{V}_{ac} / 585\ \Omega) \angle (0° - 7.35°)$

$= \mathbf{20.5\ mA_{ac} \angle -7.35°}$

For the circuit in Figure 20-6, the Norton impedance, $\boldsymbol{Z_N}$, is the impedance looking back into the circuit from $\boldsymbol{V_{out}}$ with $\boldsymbol{V_S}$ replaced with its internal impedance (a short). This impedance appears to be the series-parallel combination $\boldsymbol{R_2} + \boldsymbol{R_1} \parallel \boldsymbol{X_L}$ in parallel with $\boldsymbol{X_C}$. Part of this impedance was calculated previously to find the value of $\boldsymbol{I_N}$, and is $\boldsymbol{Z_T} = \boldsymbol{R_1} \parallel \boldsymbol{X_L} + \boldsymbol{R_2} = 585\ \Omega \angle 7.35°$.

$$\boldsymbol{Z_N} = \boldsymbol{Z_T} \parallel \boldsymbol{X_C}$$
$$= (585\ \Omega \angle 7.35°) \parallel (-\text{j}\ 430\ \Omega)$$
$$= 1 / \{[1 / (585\ \Omega \angle 7.35°)] + [1 / (430\ \Omega \angle -90°)]\}$$
$$\boldsymbol{Y_N} = 1 / \boldsymbol{Z_N}$$
$$= [(1 \angle 0°) / (585\ \Omega \angle 7.35°)] + [(1 \angle 0°) / (430\ \Omega \angle -90°)]$$
$$= [(1 / 585\ \Omega) \angle (0° - 7.35°)] + \{(1 / 430\ \Omega) \angle [0° - (-90°)]\}$$
$$= (1.71\ \text{mS} \angle -7.35°) + (2.33\ \text{mS} \angle 90°) \text{ in polar form}$$
$$= (1.69\ \text{mS} - \text{j}\ 0.219\ \text{mS}) + (0\ \text{S} + \text{j}\ 2.33\ \text{mS}) \text{ in rectangular form}$$
$$= (1.69\ \text{mS} + 0\ \text{S}) + \text{j}\ (-0.219\ \text{mS} + 2.33\ \text{mS})$$
$$= 1.69\ \text{mS} + \text{j}\ 2.11\ \text{mS} \text{ in rectangular form}$$
$$= 2.70\ \text{mS} \angle 51.2° \text{ in polar form}$$
$$\boldsymbol{Z_N} = 1 / \boldsymbol{Y_N}$$
$$= (1 \angle 0°) / (2.70\ \text{mS} \angle 51.2°)$$
$$= (1 / 2.70\ \text{mS}) \angle (0° - 51.2°)$$
$$= \mathbf{370\ \Omega \angle -51.2°}$$

20.1.4 Superposition Theorem

1. Problem: Use the ac superposition theorem to calculate the voltages across each component for the circuit in Figure 20-7. Express each answer in polar form.

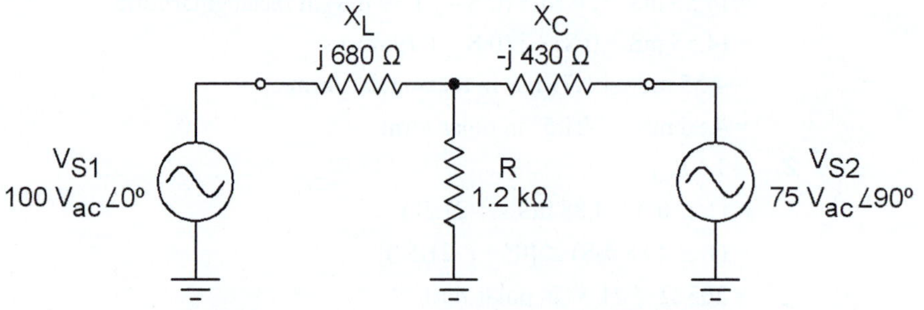

Figure 20-7: Circuit for Superposition Problem 1

Solution: First, replace $\boldsymbol{V_{S2}}$ with its internal impedance (a short) to find the voltage across each component for $\boldsymbol{V_{S1}}$. Then, replace $\boldsymbol{V_{S1}}$ with its internal impedance (a short) to find the voltage across each component for $\boldsymbol{V_{S2}}$. Finally, sum the voltages from each source to find the total voltage for each component. Figure 20-8 shows the equivalent voltage for each source.

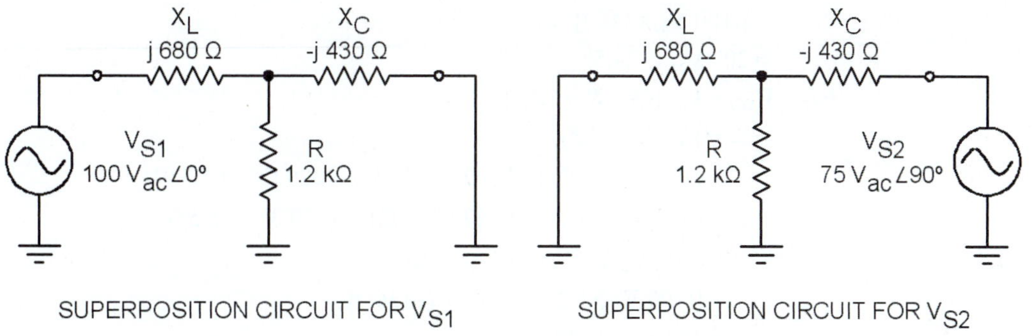

Figure 20-8: Superposition Circuits for V_{S1} and V_{S2}

The impedance of the circuit for V_{S1} is $Z_{T1} = X_L + R \parallel X_C$. Use the ac voltage divider theorem to find the voltage across each component.

$V_{XL1} = V_{S1}\,(X_L / Z_{T1})$

$V_{R1} = V_{XC1} = V_{S1}\,[(R \parallel X_C) / Z_{T1}]$

First, determine the value of $Z_{EQ1} = R \parallel X_C$.

$Z_{EQ1} = R \parallel X_C$
$= 1.2\text{ k}\Omega \parallel -\text{j }430\ \Omega$
$= 1 / [(1 / 1.2\text{ k}\Omega) + (1 / -\text{j }430\ \Omega)]$

$Y_{EQ1} = 1 / Z_{EQ1}$
$= [(1 \angle 0°) / (1.2\text{ k}\Omega \angle 0°)] + [(1 \angle 0°) / (430\ \Omega \angle -90°)]$
$= [(1 / 1.2\text{ k}\Omega) \angle (0° - 0°)] + \{(1 / 430\ \Omega) \angle [0° - (-90°)]\}$
$= (833\ \mu\text{S} \angle 0°) + (2.33\text{ mS} \angle 90°)$ in polar form
$= (833\ \mu\text{S} + \text{j }0\text{ S}) + (0\text{ S} + \text{j }2.33\text{ mS})$ in rectangular form
$= (833\ \mu\text{S} + 0\text{ S}) + \text{j }(0\text{ S} + 2.33\text{ mS})$
$= 833\ \mu\text{S} + \text{j }2.33\text{ mS}$ in rectangular form
$= 2.47\text{ mS} \angle 70.3°$ in polar form

$Z_{EQ1} = 1 / Y_{EQ1}$
$= (1 \angle 0°) / (2.47\text{ mS} \angle 70.3°)$
$= (1 / 2.47\text{ mS}) \angle (0° - 70.3°)$
$= 405\ \Omega \angle -70.3°$ in polar form
$= 137\ \Omega - \text{j }381\ \Omega$ in rectangular form

Next, substitute $Z_{EQ1} = R \parallel X_C$ into the equation for total impedance, Z_{T1}.

$Z_{T1} = X_L + Z_{EQ1}$
$= \text{j }680\ \Omega + (137\ \Omega - \text{j }381\ \Omega)$
$= (137\ \Omega + 0\ \Omega) + \text{j }(680\ \Omega - 381\ \Omega)$
$= 137\ \Omega + \text{j }299\ \Omega$ in rectangular form
$= 329\ \Omega \angle 65.5°$ in polar form

Substitute $Z_{EQ1} = R \parallel X_C$ and Z_{T1} into the appropriate ac voltage divider equations.

$V_{XL1} = V_{S1}\,(X_L / Z_{T1})$
$= (100\text{ V}_{ac} \angle 0°)\,[(\text{j }680\ \Omega) / (329\ \Omega \angle 65.5°)]$
$= (100\text{ V}_{ac} \angle 0°)\,[(680\ \Omega \angle 90°) / (329\ \Omega \angle 65.5°)]$
$= (100\text{ V}_{ac} \angle 0°)\,[(680\ \Omega / 329\ \Omega) \angle (90° - 65.5°)]$
$= (100\text{ V}_{ac} \angle 0°)\,(2.07 \angle 24.5°)$

$= [(100\ V_{ac})(2.07)] \angle (0° + 24.5°)$

$= 207\ V_{ac} \angle 24.5°$

$\boldsymbol{V_{R1}} = \boldsymbol{V_{XC1}} = \boldsymbol{V_{S1}}\ [(\boldsymbol{R} \parallel \boldsymbol{X_C}) / \boldsymbol{Z_{T1}}]$

$= \boldsymbol{V_{S1}}\ [\boldsymbol{Z_{EQ1}} / \boldsymbol{Z_{T1}}]$

$= (100\ V_{ac} \angle 0°)\ [(405\ \Omega \angle -70.3°) / (329\ \Omega \angle 65.5°)]$

$= (100\ V_{ac} \angle 0°)\ [(405\ \Omega / 329\ \Omega) \angle (-70.3° - 65.5°)]$

$= (100\ V_{ac} \angle 0°)\ (1.23 \angle -136°)$

$= [(100\ V_{ac})(1.23)] \angle [0° + (-136°)]$

$= 123\ V_{ac} \angle -136°$

The impedance of the circuit for $\boldsymbol{V_{S2}}$ is $\boldsymbol{Z_{T2}} = \boldsymbol{X_C} + \boldsymbol{R} \parallel \boldsymbol{X_L}$. Use the ac voltage divider theorem to find the voltage across each component.

$\boldsymbol{V_{XC2}} = \boldsymbol{V_{S2}}\ (\boldsymbol{X_C} / \boldsymbol{Z_{T2}})$

$\boldsymbol{V_{R2}} = \boldsymbol{V_{XL2}} = \boldsymbol{V_{S2}}\ [(\boldsymbol{R} \parallel \boldsymbol{X_L}) / \boldsymbol{Z_{T2}}]$

First, determine the value of $\boldsymbol{Z_{EQ2}} = \boldsymbol{R} \parallel \boldsymbol{X_L}$.

$\boldsymbol{Z_{EQ2}} = \boldsymbol{R} \parallel \boldsymbol{X_L}$

$= 1.2\ k\Omega \parallel j\ 680\ \Omega$

$= 1 / [(1 / 1.2\ k\Omega) + (1 / j\ 680\ \Omega)]$

$\boldsymbol{Y_{EQ2}} = 1 / \boldsymbol{Z_{EQ2}}$

$= [(1 \angle 0°) / (1.2\ k\Omega \angle 0°)] + [(1 \angle 0°) / (680\ \Omega \angle 90°)]$

$= [(1 / 1.2\ k\Omega) \angle (0° - 0°)] + [(1 / 680\ \Omega) \angle (0° - 90°)]$

$= (833\ \mu S \angle 0°) + (1.47\ mS \angle -90°)$ in polar form

$= (833\ \mu S + j\ 0\ S) + (0\ S - j\ 1.47\ mS)$ in rectangular form

$= (833\ \mu S + 0\ S) + j\ (0\ S - 1.47\ mS)$

$= 833\ \mu S - j\ 1.47\ mS$ in rectangular form

$= 1.69\ mS \angle -60.5°$ in polar form

$\boldsymbol{Z_{EQ2}} = 1 / \boldsymbol{Y_{EQ2}}$

$= (1 \angle 0°) / (1.69\ mS \angle -60.5°)$

$= (1 / 1.69\ mS) \angle [0° - (-60.5°)]$

$= 592\ \Omega \angle 60.5°$ in polar form

$= 292\ \Omega + j\ 515\ \Omega$ in rectangular form

Next, substitute $\boldsymbol{Z_{EQ2}} = \boldsymbol{R} \parallel \boldsymbol{X_L}$ into the equation for total impedance, $\boldsymbol{Z_{T2}}$.

$\boldsymbol{Z_{T2}} = \boldsymbol{X_C} + \boldsymbol{Z_{EQ2}}$

$= -j\ 430\ \Omega + (292\ \Omega + j\ 515\ \Omega)$

$= (292\ \Omega + 0\ \Omega) + j\ (-430\ \Omega + 515\ \Omega)$

$= 292\ \Omega + j\ 84.7\ \Omega$ in rectangular form

$= 304\ \Omega \angle 16.2°$ in polar form

Substitute $\boldsymbol{Z_{EQ2}} = \boldsymbol{R} \parallel \boldsymbol{X_L}$ and $\boldsymbol{Z_{T2}}$ into the ac voltage divider equations.

$\boldsymbol{V_{XC2}} = \boldsymbol{V_{S2}}\ (\boldsymbol{X_C} / \boldsymbol{Z_{T2}})$

$= (75\ V_{ac} \angle 90°)\ [(-j\ 430\ \Omega) / (304\ \Omega \angle 16.2°)]$

$= (75\ V_{ac} \angle 90°)\ [(430\ \Omega \angle -90°) / (304\ \Omega \angle 16.2°)]$

$= (75\ V_{ac} \angle 90°)\ [(430\ \Omega / 304\ \Omega) \angle (-90° - 16.2°)]$

$= (75\ V_{ac} \angle 90°)\ (1.42 \angle -106°)$

$= [(75\ V_{ac})(1.42)] \angle [90° + (-106°)]$

$= 106\ V_{ac} \angle -16.2°$

$\boldsymbol{V_{R2}} = \boldsymbol{V_{XL2}} = \boldsymbol{V_{S2}}\ [(\boldsymbol{R} \parallel \boldsymbol{X_L}) / \boldsymbol{Z_{T2}}]$

$= \boldsymbol{V_{S2}}\ [\boldsymbol{Z_{EQ2}} / \boldsymbol{Z_{T2}}]$

$= (75\ V_{ac} \angle 90°)\ [(592\ \Omega \angle 60.5°) / (304\ \Omega \angle 16.2°)]$

$= (75\ V_{ac} \angle 90°)\ [(592\ \Omega / 304\ \Omega) \angle (60.5° - 16.2°)]$

$= (75\ V_{ac} \angle 90°)\ (1.95 \angle 44.3°)$

$= [(75\ V_{ac})(1.95)] \angle [90° + 44.3°]$

$= 146\ V_{ac} \angle 134°$

Finally, sum the voltages due to $\boldsymbol{V_{S1}}$ and $\boldsymbol{V_{S2}}$ for each component.

$\boldsymbol{V_{XL}} = \boldsymbol{V_{XL1}} + \boldsymbol{V_{XL2}}$

$= (207\ V_{ac} \angle 24.5°) + (146\ V_{ac} \angle 134°)$

$= (188\ V_{ac} + j\ 86.0\ V_{ac}) + (-102\ V_{ac} + j\ 105\ V_{ac})$

$= [188\ V_{ac} + (-102\ V_{ac})] + j\ (86.0\ V_{ac} + 105\ V_{ac})$

$= 86.2\ V_{ac} + j\ 191\ V_{ac}$ in rectangular form

$=$ **$290\ V_{ac} \angle 65.7°$** in polar form

$\boldsymbol{V_R} = \boldsymbol{V_{R1}} + \boldsymbol{V_{R2}}$

$= (123\ V_{ac} \angle -136°) + (146\ V_{ac} \angle 134°)$

$= (-88.2\ V_{ac} - j\ 86.0\ V_{ac}) + (-102\ V_{ac} + j\ 105\ V_{ac})$

$= [(-88.2\ V_{ac}) + (-102\ V_{ac})] + j\ [(-86.0\ V_{ac}) + 105\ V_{ac}]$

$= -190\ V_{ac} + j\ 18.7\ V_{ac}$ in rectangular form

$=$ **$191\ V_{ac} \angle 174°$** in polar form

$\boldsymbol{V_{XC}} = \boldsymbol{V_{XC1}} + \boldsymbol{V_{XC2}}$

$= (123\ V_{ac} \angle -136°) + (106\ V_{ac} \angle -16.2°)$

$= (-88.2\ V_{ac} - j\ 86.0\ V_{ac}) + (102\ V_{ac} - j\ 29.6\ V_{ac})$

$= (-88.2\ V_{ac} + 102\ V_{ac}) + j\ [(-86.0\ V_{ac}) + (-29.6\ V_{ac})]$

$= 13.8\ V_{ac} - j\ 116\ V_{ac}$ in rectangular form

$=$ **$116\ V_{ac} \angle -83.2°$** in polar form

2. Problem: Use the ac superposition theorem to calculate the currents through each component for the circuit in Figure 20-9.

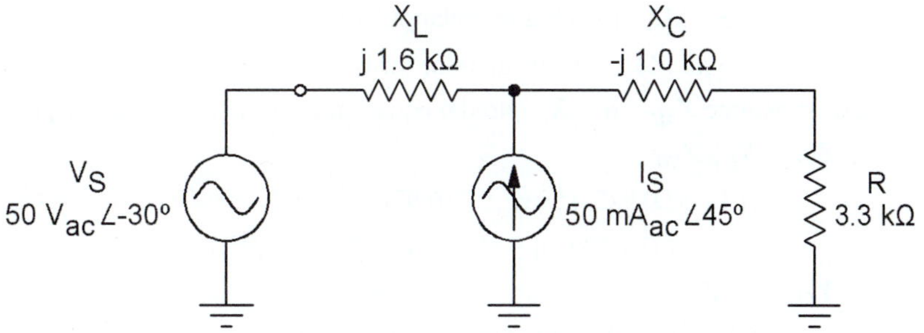

Figure 20-9: Circuit for Superposition Problem 2

Solution: First, replace $\boldsymbol{I_S}$ with its internal impedance (an open) to find the current through each component for $\boldsymbol{V_S}$. Then, replace $\boldsymbol{V_S}$ with its internal impedance (a short) to find the current through each component for I_S. Finally, sum the currents from each source to find the total current through each component. Figure 20-10 shows the equivalent voltage for each source.

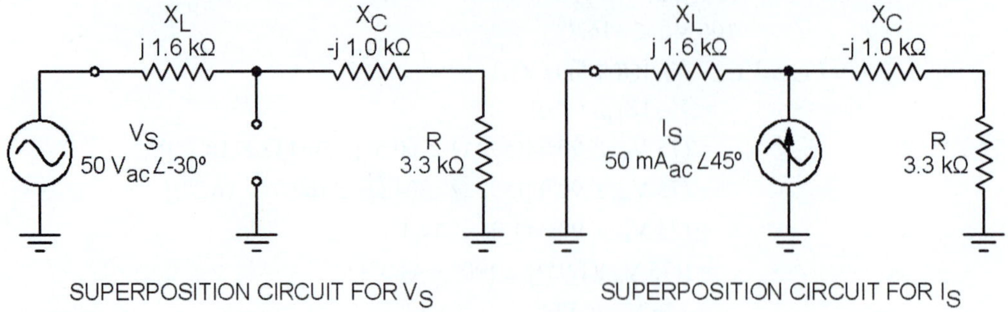

Figure 20-10: Superposition Circuits for V_S and I_S

The circuit for V_S is a series circuit. The current through each component is the same as the total current, which can be found from Ohm's Law.

$$I_{XL1} = I_{XC1} = I_{R1} = V_S / Z_{T1}$$

First, determine the value of $Z_{T1} = X_L + X_C + R$.

$$\begin{aligned} Z_{T1} &= X_L + X_C + R \\ &= (\text{j } 1.6 \text{ k}\Omega) + (-\text{j } 1.0 \text{ k}\Omega) + (3.3 \text{ k}\Omega) \\ &= 3.3 \text{ k}\Omega + \text{j } (1.6 \text{ k}\Omega - 1.0 \text{ k}\Omega) \\ &= 3.3 \text{ k}\Omega + \text{j } 600 \ \Omega \text{ in rectangular form} \\ &= 3.35 \text{ k}\Omega \angle 10.3° \text{ in polar form} \end{aligned}$$

Next, use Ohm's law to find the current for V_S through each component.

$$\begin{aligned} I_{XL1} = I_{XC1} = I_{R1} &= V_S / Z_{T1} \\ &= (50 \text{ V}_{ac} \angle -30°) / (3.35 \text{ k}\Omega \angle 10.3°) \\ &= (50 \text{ V}_{ac} / 3.35 \text{ k}\Omega) \angle (-30° - 10.3°) \\ &= 14.9 \text{ mA}_{ac} \angle -40.3° \end{aligned}$$

The impedance of the circuit for I_S is $Z_{T2} = X_L \parallel (R + X_C)$. Use the ac current divider theorem to find the current through each component.

$$I_{XL2} = I_S (Z_{T2} / X_L)$$

$$I_{R2} = I_{XC2} = I_S [Z_{T2} / (R + X_C)]$$

First, determine the value of $Z_{EQ} = R + X_C$.

$$\begin{aligned} Z_{EQ} &= R + X_C \\ &= 3.3 \text{ k}\Omega - \text{j } 1.0 \text{ k}\Omega \text{ in rectangular form} \\ &= 3.45 \text{ k}\Omega \angle -16.9° \text{ in polar form} \end{aligned}$$

Next, substitute $Z_{EQ} = R + X_C$ into the equation for total impedance, Z_{T2}.

$$\begin{aligned} Z_{T2} &= X_L \parallel Z_{EQ} \\ &= (\text{j } 1.6 \text{ k}\Omega) \parallel (3.45 \text{ k}\Omega \angle -16.9°) \\ &= 1 / \{[1 / (\text{j } 1.6 \text{ k}\Omega)] + [1 / (3.45 \text{ k}\Omega \angle -16.9°)]\} \end{aligned}$$

$$\begin{aligned} Y_{T2} &= 1 / Z_{T2} \\ &= [(1 \angle 0°) / (1.6 \text{ k}\Omega \angle 90°)] + [(1 \angle 0°) / (3.45 \text{ k}\Omega \angle -16.9°)] \\ &= [(1 / 1.6 \text{ k}\Omega) \angle (0° - 90°)] + \{(1 / 3.45 \ \Omega) \angle [0° - (-16.9°)]\} \\ &= (625 \ \mu\text{S} \angle -90°) + (290 \ \mu\text{S} \angle 16.9°) \text{ in polar form} \end{aligned}$$

$= (0\ \text{S} - j\ 625\ \mu\text{S}) + (278\ \mu\text{S} + j\ 84.1\ \mu\text{S})$ in rectangular form

$= (0\ \text{S} + 278\ \mu\text{S}) + j\ (-625\ \mu\text{S} + 84.1\ \mu\text{S})$

$= 278\ \mu\text{S} - j\ 541\ \mu\text{S}$ in rectangular form

$= 608\ \mu\text{S} \angle -62.8°$ in polar form

$\boldsymbol{Z_{T2}} = 1 / \boldsymbol{Y_{T2}}$

$= (1 \angle 0°) / (608\ \mu\text{S} \angle -62.8°)$

$= (1 / 608\ \mu\text{S}) \angle [0° - (-62.8°)]$

$= 1.65\ \text{k}\Omega \angle 62.8°$ in polar form

Substitute the values of $\boldsymbol{Z_{EQ}} = \boldsymbol{R} + \boldsymbol{X_C}$ and $\boldsymbol{Z_{T2}}$ into the ac current divider equations.

$\boldsymbol{I_{XL2}} = \boldsymbol{I_S}\ (\boldsymbol{Z_{T2}} / \boldsymbol{X_L})$

$= (50\ \text{mA}_{ac} \angle 45°)\ [(1.65\ \text{k}\Omega \angle 62.8°) / (j\ 1.6\ \text{k}\Omega)]$

$= (50\ \text{mA}_{ac} \angle 45°)\ [(1.65\ \text{k}\Omega \angle 62.8°) / (1.6\ \text{k}\Omega \angle 90°)]$

$= (50\ \text{mA}_{ac} \angle 45°)\ [(1.65\ \text{k}\Omega / 1.6\ \text{k}\Omega) \angle (62.8° - 90°)]$

$= (50\ \text{mA}_{ac} \angle 45°)\ (1.03 \angle -27.2°)$

$= [(50\ \text{mA}_{ac})(1.03)] \angle [45° + (-27.2°)]$

$= 51.4\ \text{mA}_{ac} \angle 17.8°$

$\boldsymbol{I_{R2}} = \boldsymbol{I_{XC2}} = \boldsymbol{I_S}\ [\boldsymbol{Z_{T2}} / (\boldsymbol{R} + \boldsymbol{X_C})]$

$= \boldsymbol{I_S}\ (\boldsymbol{Z_{T2}} / \boldsymbol{Z_{EQ}})$

$= (50\ \text{mA}_{ac} \angle 45°)\ [(1.65\ \text{k}\Omega \angle 62.8°) / (3.45\ \text{k}\Omega \angle -16.9°)]$

$= (50\ \text{mA}_{ac} \angle 45°)\ \{(1.65\ \text{k}\Omega / 3.45\ \text{k}\Omega) \angle [62.8° - (-16.9°)]\}$

$= (50\ \text{mA}_{ac} \angle 45°)\ (0.477 \angle 79.7°)$

$= [(50\ \text{mA}_{ac})(0.477)] \angle (45° + 79.7°)$

$= 23.9\ \text{mA}_{ac} \angle 125°$

Finally, sum the currents due to $\boldsymbol{V_S}$ and $\boldsymbol{I_S}$ for each component.

$\boldsymbol{I_L} = \boldsymbol{I_{XL1}} + \boldsymbol{I_{XL2}}$

$= (14.9\ \text{mA}_{ac} \angle -40.3°) + (51.4\ \text{mA}_{ac} \angle 17.8°)$

$= (11.4\ \text{mA}_{ac} - j\ 9.64\ \text{mA}_{ac}) + (48.9\ \text{mA}_{ac} + j\ 15.7\ \text{mA}_{ac})$

$= (11.4\ \text{mA}_{ac} + 48.9\ \text{mA}_{ac}) + j\ (-9.64\ \text{mA}_{ac} + 15.7\ \text{mA}_{ac})$

$= 60.3\ \text{mA}_{ac} + j\ 6.10\ \text{mA}_{ac}$ in rectangular form

$= \mathbf{60.6\ mA_{ac} \angle 5.78°}$ in polar form

$\boldsymbol{I_R} = \boldsymbol{I_{R1}} + \boldsymbol{I_{R2}}$

$= (14.9\ \text{mA}_{ac} \angle -40.3°) + (23.9\ \text{mA}_{ac} \angle 125°)$

$= (11.4\ \text{mA}_{ac} - j\ 9.64\ \text{mA}_{ac}) + (-13.6\ \text{mA}_{ac} + j\ 19.6\ \text{mA}_{ac})$

$= [11.4\ \text{mA}_{ac} + (-13.6\ \text{mA}_{ac}] + j\ (-9.64\ \text{mA}_{ac} + 19.6\ \text{mA}_{ac})$

$= -2.21\ \text{mA}_{ac} + j\ 9.97\ \text{mA}_{ac}$ in rectangular form

$= \mathbf{10.2\ mA_{ac} \angle 103°}$ in polar form

$\boldsymbol{I_C} = \boldsymbol{I_{XC1}} + \boldsymbol{I_{XC2}}$

$= (14.9\ \text{mA}_{ac} \angle -40.3°) + (23.9\ \text{mA}_{ac} \angle 125°)$

$= (11.4\ \text{mA}_{ac} - j\ 9.64\ \text{mA}_{ac}) + (-13.6\ \text{mA}_{ac} + j\ 19.6\ \text{mA}_{ac})$

$= [11.4\ \text{mA}_{ac} + (-13.6\ \text{mA}_{ac}] + j\ (-9.64\ \text{mA}_{ac} + 19.6\ \text{mA}_{ac})$

$= -2.21\ \text{mA}_{ac} + j\ 9.97\ \text{mA}_{ac}$ in rectangular form

$= \mathbf{10.2\ mA_{ac} \angle 103°}$ in polar form

20.1.5 Maximum Power Transfer Theorem

1. Problem: Determine the value of load impedance, Z_L, for which maximum power is transferred to the load for the circuit in Figure 20-11.

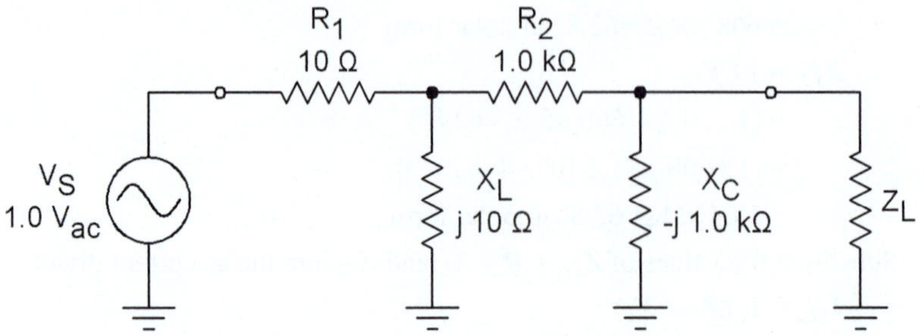

Figure 20-11: Circuit for Maximum Power Transfer Problem 1

Solution: Maximum power is transferred to Z_L when Z_L equals Z_{TH}*, the complex conjugate of the Thevenin impedance Z_{TH}. First, calculate the Thevenin impedance Z_{TH} as seen by Z_L by replacing V_S with its internal impedance (a short).

$Z_{TH} = X_C \parallel (R_2 + R_1 \parallel X_L)$

The first equivalent impedance is $Z_{EQ1} = R_1 \parallel X_L$.

$Z_{TH} = X_C \parallel (R_2 + Z_{EQ1})$

where

$Z_{EQ1} = R_1 \parallel X_L$
$= 10\ \Omega \parallel \text{j}\ 10\ \Omega$
$= [(10\ \Omega)(\text{j}10\ \Omega)] / (10\ \Omega + \text{j}\ 10\ \Omega)$ using product over sum method
$= [(\text{j}\ 100\ \Omega^2) / (10\ \Omega + \text{j}\ 10\ \Omega)]$ in rectangular form
$= (100\ \Omega^2 \angle 90°) / (14.1\ \Omega \angle 45°)$ in polar form
$= (100\ \Omega^2 / 14.1\ \Omega) \angle (90° - 45°)$
$= 7.07\ \Omega \angle 45°$ in polar form
$= 5.00\ \Omega + \text{j}\ 5.00\ \Omega$ in rectangular form

The second equivalent impedance is $Z_{EQ2} = R_2 + Z_{EQ1}$.

$Z_{TH} = X_C \parallel Z_{EQ2}$

where

$Z_{EQ2} = R_2 + Z_{EQ1}$
$= 1.0\ \text{k}\Omega + (5\ \Omega + \text{j}\ 5\ \Omega)$
$= (1.0\ \text{k}\Omega + 5\ \Omega) + \text{j}\ (0\ \Omega + 5\ \Omega)$
$= 1.005\ \text{k}\Omega + \text{j}\ 5\ \Omega$

The third and last equivalent circuit is $Z_{EQ3} = X_C \parallel Z_{EQ2}$.

$Z_{TH} = Z_{EQ3}$

where

$Z_{EQ3} = X_C \parallel Z_{EQ2}$
$= -\text{j}\ 1.0\ \text{k}\Omega \parallel (1.005\ \text{k}\Omega + \text{j}\ 5\ \Omega)$
$= [(-\text{j}\ 1.0\ \text{k}\Omega)(1.005\ \text{k}\Omega + \text{j}\ 5\ \Omega)] / [(-\text{j}\ 1.0\ \text{k}\Omega) + (1.005\ \text{k}\Omega + \text{j}\ 5\ \Omega)]$
$= [(-\text{j}\ 1.0\ \text{k}\Omega)(1.005\ \text{k}\Omega) + (-\text{j}\ 1.0\ \text{k}\Omega)(\text{j}\ 5\ \Omega)] / [1.005\ \text{k}\Omega + \text{j}\ (-1.0\ \text{k}\Omega + 5\ \Omega)]$
$= [(-\text{j}\ 1.005\ \text{M}\Omega^2) + -\text{j}^2\ (5.0\ \text{k}\Omega^2)] / (1.005\ \text{k}\Omega - \text{j}\ 995\ \Omega)$

$= [(-j\ 1.005\ M\Omega^2) + -(-1)(5.0\ k\Omega^2)] / (1.005\ k\Omega - j\ 995\ \Omega)$

$= (5.0\ k\Omega^2 - j\ 1.005\ M\Omega^2) / (1.005\ k\Omega - j\ 995\ \Omega)$ in polar form

$= (1.005\ M\Omega^2 \angle -89.7°) / (1.414\ k\Omega \angle -44.7°)$ in polar form

$= (1.005\ M\Omega^2 / 1.414\ k\Omega) \angle [-89.7° - (-44.7°)]$

$= (1.005 \times 10^6\ \Omega^2 / 1.414 \times 10^3\ \Omega) \angle -45.0°)$

$= 711\ \Omega \angle -45.0°$ in polar form

$= 503\ \Omega - j\ 503\ \Omega$ in rectangular form

The circuit transfers maximum power to $\boldsymbol{Z_L}$ when $\boldsymbol{Z_L} = \boldsymbol{Z_{TH}}^*$. The complex conjugate of a complex number $\boldsymbol{A} + j\ \boldsymbol{B} = \boldsymbol{R} \angle \varphi°$ is $= \boldsymbol{A} - j\ \boldsymbol{B} = \boldsymbol{R} \angle -\varphi°$, so the complex conjugate for $\boldsymbol{Z_{TH}} = 503\ \Omega - j\ 503\ \Omega = 711\ \Omega \angle -45.0°$ is $\boldsymbol{Z_{TH}}^* = 503\ \Omega + j\ 503\ \Omega = 711\ \Omega \angle 45.0°$. Therefore, the load for maximum power transfer is

$\boldsymbol{Z_L} = \mathbf{503\ \Omega + j\ 503\ \Omega}$ in rectangular form

$= \mathbf{711\ \Omega \angle 45.0°}$ in polar form

2. Problem: For the circuit in Figure 20-12, determine the value of load impedance, $\boldsymbol{Z_L}$, for which maximum power is transferred to the load.

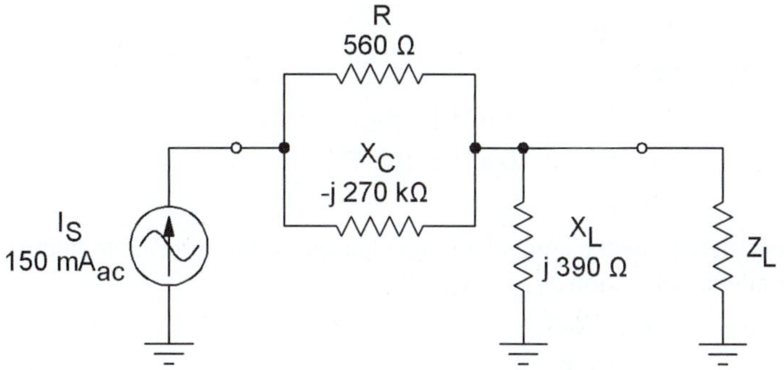

Figure 20-12: Circuit for Maximum Power Transfer Problem 2

Solution: Maximum power is transferred to $\boldsymbol{Z_L}$ when $\boldsymbol{Z_L}$ equals $\boldsymbol{Z_N}^*$, the complex conjugate of the Norton impedance $\boldsymbol{Z_N}$. First, calculate the Norton impedance $\boldsymbol{Z_N}$ as seen by $\boldsymbol{Z_L}$ by replacing $\boldsymbol{I_S}$ with its internal impedance (an open). When $\boldsymbol{I_S}$ is replaced by an open, $\boldsymbol{R}$ and $\boldsymbol{X_C}$ have no return path to ground, so that $\boldsymbol{Z_N}$ is equal to the inductive reactance, $\boldsymbol{X_L}$.

$\boldsymbol{Z_N} = \boldsymbol{X_L}$

$= j\ 390\ \Omega$

$= 0\ \Omega + j\ 390\ \Omega$ in rectangular form

$= 390\ \Omega \angle 90°$ in polar form

The circuit transfers maximum power to $\boldsymbol{Z_L}$ when $\boldsymbol{Z_L} = \boldsymbol{Z_N}^*$. The complex conjugate of a complex number $\boldsymbol{A} + j\ \boldsymbol{B} = \boldsymbol{R} \angle \varphi°$ is $= \boldsymbol{A} - j\ \boldsymbol{B} = \boldsymbol{R} \angle -\varphi°$, so the complex conjugate for $\boldsymbol{Z_N} = 0\ \Omega + j\ 390\ \Omega = 390\ \Omega \angle 90°$ is $\boldsymbol{Z_N}^* = 0\ \Omega - j\ 390\ \Omega = 390\ \Omega \angle -90°$. Therefore, the load for maximum power transfer is

$\boldsymbol{Z_L} = \mathbf{0\ \Omega - j\ 390\ \Omega}$ in rectangular form

$= \mathbf{390\ \Omega \angle -90°}$ in polar form

20.2 Advanced

20.2.1 Voltage and Current Divider Theorems

1. Problem: Use the ac voltage divider theorem to find the output voltage, $\boldsymbol{V_{out}}$, for the series-parallel circuit in Figure 20-13. Express the answer in polar form.

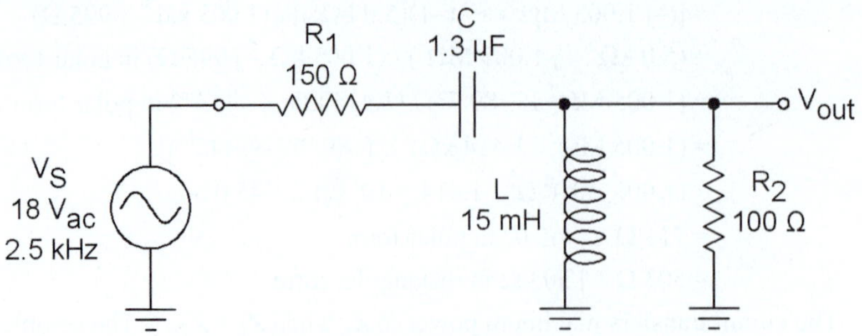

Figure 20-13: *RLC* Circuit for Voltage Divider Problem 1

Solution: First, calculate the reactances for the capacitor and inductor.

$X_C = 1 / (2\pi fC)$

$= 1 / [2\pi(2.5 \text{ kHz})(1.3 \text{ μF})]$

$= 1 / [2\pi(2.5 \times 10^3 \text{ Hz})(1.3 \times 10^{-6} \text{ F})]$

$= 1 / 20.4 \times 10^{-3} \text{ S}$

$= 49.0 \ \Omega$

$X_L = 2\pi fL$

$= 2\pi(2.5 \text{ kHz})(15 \text{ mH})]$

$= 2\pi(2.5 \times 10^3 \text{ Hz})(15 \times 10^{-3} \text{ F})]$

$= 236 \ \Omega$

Next, determine the equivalent impedances for the series combination $Z_{EQ1} = R_1 + X_C$ and the parallel combination $Z_{EQ2} = X_L \parallel R_2$.

$Z_{EQ1} = R_1 + X_C$

$= R_1 - \text{j } X_C$

$= 150 \ \Omega - \text{j } 49.0 \ \Omega$ in rectangular form

$= 158 \ \Omega \angle -18.1°$ in polar form

$Z_{EQ2} = X_L \parallel R_2$

$= \text{j } 236 \ \Omega \parallel 100 \ \Omega$

$= (\text{j } 236 \ \Omega)(100 \ \Omega) / [(\text{j } 236 \ \Omega) + 100 \ \Omega]$

$= (\text{j } 23.6 \times 10^3 \ \Omega^2) / (100 \ \Omega + \text{j } 236 \ \Omega)$ in rectangular form

$= (23.6 \times 10^3 \ \Omega^2 \angle 90°) / (256 \ \Omega \angle 67.0°)$

$= (23.6 \times 10^3 \ \Omega^2 / 256 \ \Omega) \angle (90° - 67.0°)$

$= 92.1 \ \Omega \angle 23.0°$ in polar form

$= 84.7 \ \Omega + \text{j } 36.0 \ \Omega$ in rectangular form

The total impedance, Z_T, is the series combination $Z_{EQ1} + Z_{EQ2}$.

$Z_T = Z_{EQ1} + Z_{EQ2}$

$= (150 \ \Omega - \text{j } 49.0 \ \Omega) + (84.7 \ \Omega + \text{j } 36.0 \ \Omega)$

$= (150 \ \Omega + 84.7 \ \Omega) + \text{j } (-49.0 \ \Omega + 36.0 \ \Omega)$

$= 235 \ \Omega - \text{j } 13.0 \ \Omega$ in rectangular form

$= 235 \ \Omega \angle -3.17°$ in polar form

Finally, use the ac voltage divider theorem for the equivalent impedances to calculate the output voltage, V_{out}. V_{out} is the voltage developed across the parallel combination $Z_{EQ2} = X_L \parallel R_2$.

$V_{out} = V_S (Z_{EQ2} / Z_T)$

$= (18\ \text{V}_{\text{ac}} \angle 0°)\ [(92.1\ \Omega \angle 23.0°) / (235\ \Omega \angle -3.17°)]$

$= (18\ \text{V}_{\text{ac}} \angle 0°)\ \{(92.1\ \Omega / 235\ \Omega) \angle [23.0° - (-3.17°)]\}$

$= (18\ \text{V}_{\text{ac}} \angle 0°)\ (0.392 \angle 26.2°)$

$= [(18\ \text{V}_{\text{ac}})(0.392)] \angle (0° + 26.2°)$

$= \mathbf{7.05\ V_{ac} \angle 26.2°}$

2. Problem: Use the ac current divider theorem to find the currents through each component for the series-parallel ***RLC*** circuit in Figure 20-14. Express all answers in polar form.

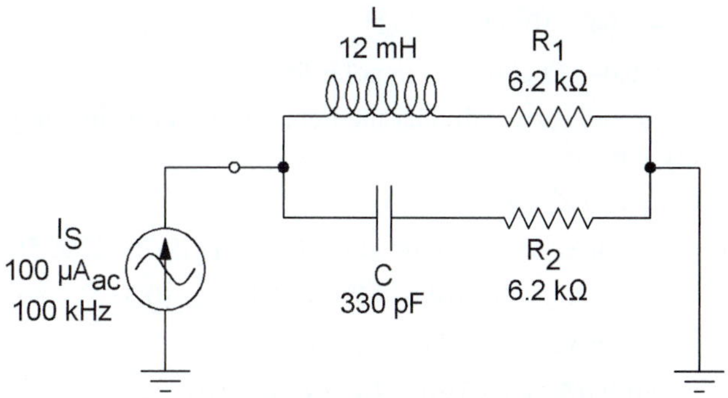

Figure 20-14: *RLC* Circuit for Current Divider Problem 2

Solution: First, calculate the reactances for the inductor and capacitor.

$\boldsymbol{X_L} = 2\pi \boldsymbol{fL}$

$= 2\pi(100\ \text{kHz})(12\ \text{mH})]$

$= 2\pi(100 \times 10^3\ \text{Hz})(12 \times 10^{-3}\ \text{F})]$

$= 7.54 \times 10^3\ \Omega$

$= 7.54\ \text{k}\Omega$

$\boldsymbol{X_C} = 1 / (2\pi \boldsymbol{fC})$

$= 1 / [2\pi(100\ \text{kHz})(330\ \text{pF})]$

$= 1 / [2\pi(100 \times 10^3\ \text{Hz})(330 \times 10^{-12}\ \text{F})]$

$= 1 / 207 \times 10^{-6}\ \text{S}$

$= 4.82 \times 10^3\ \Omega$

$= 4.82\ \text{k}\Omega$

Next, determine the equivalent impedances for the series combination $\boldsymbol{Z_{EQ1}} = \boldsymbol{X_L} + \boldsymbol{R}_1$ and the series combination $\boldsymbol{Z_{EQ2}} = \boldsymbol{X_C} + \boldsymbol{R}_2$.

$\boldsymbol{Z_{EQ1}} = \boldsymbol{X_L} + \boldsymbol{R}_1$

$= \text{j}\ \boldsymbol{X_L} + \boldsymbol{R}_1$

$= \text{j}\ 7.54\ \text{k}\Omega + 6.2\ \text{k}\Omega$

$= 6.2\ \text{k}\Omega + \text{j}\ 7.54\ \text{k}\Omega$ in rectangular form

$= 9.76\ \text{k}\Omega \angle 50.6°$ in polar form

$\boldsymbol{Z_{EQ2}} = \boldsymbol{X_C} + \boldsymbol{R}_2$

$= \text{j}\ \boldsymbol{X_C} + \boldsymbol{R}_2$

$= -\text{j}\ 4.82\ \text{k}\Omega + 6.2\ \text{k}\Omega$

$= 6.2\ \text{k}\Omega - \text{j}\ 4.82\ \text{k}\Omega$ in rectangular form

$= 7.86\ \text{k}\Omega \angle -37.9°$ in polar form

The total impedance, $\boldsymbol{Z_T}$, is the parallel combination $\boldsymbol{Z_{EQ1}} \parallel \boldsymbol{Z_{EQ2}}$.

$\boldsymbol{Z_T} = \boldsymbol{Z_{EQ1}} \parallel \boldsymbol{Z_{EQ2}}$

$= 9.76\text{ k}\Omega \angle 50.6° \parallel 7.86\text{ k}\Omega \angle -37.9°$

$= (9.76\text{ k}\Omega \angle 50.6°)(7.86\text{ k}\Omega \angle -37.9°) / [(6.2\text{ k}\Omega + j\,7.54\text{ k}\Omega) + (6.2\text{ k}\Omega - j\,4.82\text{ k}\Omega)]$

$= \{(9.76 \times 10^3\,\Omega)(7.86 \times 10^3\,\Omega) \angle [50.6° + (-37.9°)\}$
$/ \{[(6.2 \times 10^3\,\Omega) + (6.2 \times 10^3\,\Omega)] + j\,[7.54 \times 10^3\,\Omega + (-4.82 \times 10^3\,\Omega)]\}$

$= (76.7 \times 10^6\,\Omega^2 \angle 12.7°) / [(12.4 \times 10^3\,\Omega) + j\,(2.72 \times 10^3\,\Omega)]$

$= (76.7 \times 10^6\,\Omega^2 \angle 12.7°) / (12.7 \times 10^3\,\Omega \angle 12.4°)$

$= (76.7 \times 10^6\,\Omega^2 / 12.7 \times 10^3\,\Omega) \angle (12.7° - 12.4°)$

$= 6.04 \times 10^3\,\Omega \angle 0.332°$

$= 6.04\text{ k}\Omega \angle 0.332°$ in polar form

Finally, use the ac current divider theorem for the equivalent impedances to calculate the current through each branch.

$I_1 = I_S\,(Z_T / Z_{EQ1})$

$= (100\,\mu A_{ac} \angle 0°)\,[(6.04\text{ k}\Omega \angle 0.332°) / (9.76\text{ k}\Omega \angle 50.6°)]$

$= (100\,\mu A_{ac} \angle 0°)\,[(6.04\text{ k}\Omega / 9.76\text{ k}\Omega) \angle (0.332° - 50.6°)]$

$= (100\,\mu A_{ac} \angle 0°)\,(0.619 \angle -50.2°)$

$= [(100\,\mu A_{ac})(0.619)] \angle [0° + (-50.2°)]$

$= \mathbf{61.9\,\mu A_{ac} \angle -50.2°}$

$I_2 = I_S\,(Z_T / Z_{EQ2})$

$= (100\,\mu A_{ac} \angle 0°)\,[(6.04\text{ k}\Omega \angle 0.332°) / (7.86\text{ k}\Omega \angle -37.9°)]$

$= (100\,\mu A_{ac} \angle 0°)\,\{(6.04\text{ k}\Omega / 7.86\text{ k}\Omega) \angle [0.332° - (-37.9°)]\}$

$= (100\,\mu A_{ac} \angle 0°)\,(0.769 \angle 38.2°)$

$= [(100\,\mu A_{ac})(0.769)] \angle [0° + (38.2°)]$

$= \mathbf{76.9\,\mu A_{ac} \angle 38.2°}$

20.2.2 Thevenin's Theorem

1. Problem: Calculate the voltage and impedance values, V_{TH} and Z_{TH}, of the Thevenin equivalent for the circuit in Figure 20-15 with respect to the output, V_{out}.

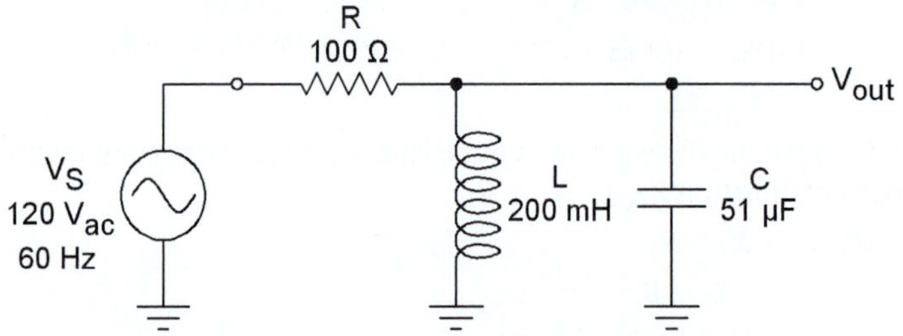

Figure 20-15: *RLC* Circuit for Thevenin Problem 1

Solution: First, calculate the reactances for the inductor and capacitor.

$X_L = 2\pi fL$

$= 2\pi(60\text{ Hz})(200\text{ mH})]$

$= 2\pi(60\text{ Hz})(200 \times 10^{-3}\text{ H})]$

$= 75.4\,\Omega$

$X_C = 1 / (2\pi fC)$

$= 1 / [2\pi(60\text{ Hz})(51\,\mu F)]$

$= 1 / [2\pi(60\text{ Hz})(51 \times 10^{-6}\text{ F})]$

$= 1 / 19.2 \times 10^{-3}\text{ S}$

$= 52.0\ \Omega$

Next, calculate the equivalent impedance of the inductor and capacitor in parallel.

$\boldsymbol{Z_{EQ}} = \boldsymbol{X_L} \parallel \boldsymbol{X_C}$

$= \text{j}\,\boldsymbol{X_L} \parallel -\text{j}\,\boldsymbol{X_C}$

$= \text{j}\ 75.4\ \Omega \parallel -\text{j}\ 52.0\ \Omega$

$= [(\text{j}\ 75.4\ \Omega)(-\text{j}\ 52.0\ \Omega)] / [(\text{j}\ 75.4\ \Omega) - (\text{j}\ 52.0\ \Omega)]$

$= [-\text{j}^2\ (75.4\ \Omega)(52.0\ \Omega)] / [\text{j}\ (75.4\ \Omega - 52.0\ \Omega)]$

$= [(1)(3.92 \times 10^3\ \Omega^2)] / (\text{j}\ 23.4\ \Omega)$

$= (3.92 \times 10^3\ \Omega^2 \angle 0°) / (23.4\ \Omega \angle 90°)$

$= (3.92 \times 10^3\ \Omega^2 / 23.4\ \Omega) \angle (0° - 90°)$

$= 168\ \Omega \angle -90°$ in polar form

$= 0\ \Omega - \text{j}\ 168\ \Omega$ in rectangular form

Use the ac voltage divider theorem with $\boldsymbol{V_S}$, $\boldsymbol{R}$, and $\boldsymbol{Z_{EQ}}$ to calculate the open load voltage, $\boldsymbol{V_{out}} = \boldsymbol{V_{TH}}$.

$\boldsymbol{V_{TH}} = \boldsymbol{V_S}\ [\boldsymbol{Z_{EQ}} / (\boldsymbol{R} + \boldsymbol{Z_{EQ}})]$

$= (120\ \text{V}_{ac} \angle 0°)\ \{(168\ \Omega \angle -90°) / [100\ \Omega + (0\ \Omega - \text{j}\ 168\ \Omega)]\}$

$= (120\ \text{V}_{ac} \angle 0°)\ \{(168\ \Omega \angle -90°) / [(100\ \Omega + \text{j}\ 0\ \Omega) + (0\ \Omega - \text{j}\ 168\ \Omega)]\}$

$= (120\ \text{V}_{ac} \angle 0°)\ \{(168\ \Omega \angle -90°) / [(100\ \Omega + 0\ \Omega) + \text{j}\ (0\ \Omega - 168\ \Omega)]\}$

$= (120\ \text{V}_{ac} \angle 0°)\ [(168\ \Omega \angle -90°) / (100\ \Omega - \text{j}\ 168\ \Omega)]$

$= (120\ \text{V}_{ac} \angle 0°)\ [(168\ \Omega \angle -90°) / (195\ \Omega \angle -59.2°)]$

$= (120\ \text{V}_{ac} \angle 0°)\ \{(168\ \Omega / 195\ \Omega) \angle [-90° - (-59.2°)]\}$

$= (120\ \text{V}_{ac} \angle 0°)\ (0.859 \angle -30.8°)$

$= [(120\ \text{V}_{ac})(0.859)] \angle [0° + (-30.8°)]$

$= \mathbf{103\ V_{ac} \angle -30.8°}$

For the circuit in Figure 20-15, the Thevenin impedance, $\boldsymbol{Z_{TH}}$, is the impedance looking back into the circuit from $\boldsymbol{V_{out}}$ with $\boldsymbol{V_S}$ replaced with its internal impedance (a short). This impedance appears to be $\boldsymbol{R}$ and $\boldsymbol{Z_{EQ}}$ in parallel.

$\boldsymbol{Z_{TH}} = \boldsymbol{R} \parallel \boldsymbol{Z_{EQ}}$

$= 1 / \{1 / 100\ \Omega) + [1 / (0\ \Omega - \text{j}\ 168\ \Omega)]\}$

$\boldsymbol{Y_{TH}} = 1 / \boldsymbol{Z_{TH}}$

$= [(1 \angle 0°) / (100\ \Omega \angle 0°)] + [(1 \angle 0°) / (168\ \Omega \angle -90°)]$

$= [(1 / 100\ \Omega) \angle (0° - 0°)] + \{(1 / 168\ \Omega) \angle [0° - (-90°)]\}$

$= (10.0\text{ mS} \angle 0°) + (5.96\text{ mS} \angle 90°)$ in polar form

$= (10.0\text{ mS} + \text{j}\ 0\text{ S}) + (0\text{ S} + \text{j}\ 5.69\text{ mS})$ in rectangular form

$= (10.0\text{ mS} + 0\text{ S}) + \text{j}\ (0\text{ S} + 5.69\text{ mS})$

$= 10.0\text{ mS} + \text{j}\ 5.69\text{ mS}$ in rectangular form

$= 11.6\text{ mS} \angle 30.8°$ in polar form

$\boldsymbol{Z_{TH}} = 1 / \boldsymbol{Y_{TH}}$

$= (1 \angle 0°) / (11.6\text{ mS} \angle 30.8°)$

$= (1 / 11.6\text{ mS}) \angle (0° - 30.8°)$

$= \mathbf{85.9\ \Omega \angle -30.8°}$

2. Problem: Calculate the voltage and impedance values, V_{TH} and Z_{TH}, of the Thevenin equivalent for the circuit in Figure 20-16 with respect to the output, V_{out}.

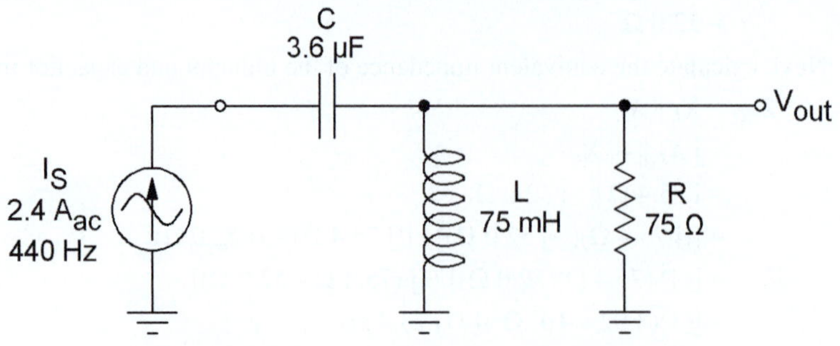

Figure 20-16: *RLC* Circuit for Thevenin Problem 2

Solution: First, calculate the reactances for the capacitor and inductor.

$X_C = 1 / (2\pi \boldsymbol{fC})$
$= 1 / [2\pi(440\text{ Hz})(3.6\ \mu\text{F})]$
$= 1 / [2\pi(440\text{ Hz})(3.6 \times 10^{-6}\text{ F})]$
$= 1 / 9.95 \times 10^{-3}\text{ S}$
$= 101\ \Omega$

$X_L = 2\pi \boldsymbol{fL}$
$= 2\pi(440\text{ Hz})(75\text{ mH})]$
$= 2\pi(440\text{ Hz})(75 \times 10^{-3}\text{ H})]$
$= 207\ \Omega$

Next, calculate the equivalent impedance of the inductor and resistor in parallel.

$Z_{EQ} = X_L \parallel R$
$= \text{j}\, X_L \parallel R$
$= \text{j}\ 207\ \Omega \parallel 75\ \Omega$
$= [(\text{j}\ 207\ \Omega)(75\ \Omega)] / [(\text{j}\ 207\ \Omega + 75\ \Omega)]$
$= [\text{j}\ (207\ \Omega)(75\ \Omega)] / (75\ \Omega + \text{j}\ 207\ \Omega)$
$= [\text{j}\ 15.5 \times 10^3\ \Omega^2)] / (75\ \Omega + \text{j}\ 207\ \Omega)$
$= (15.5 \times 10^3\ \Omega^2 \angle 90°) / (221\ \Omega \angle 70.1°)$
$= (15.5 \times 10^3\ \Omega^2 / 221\ \Omega) \angle (90° - 70.1°)$
$= 70.5\ \Omega \angle 19.9°$ in polar form
$= 66.3\ \Omega + \text{j}\ 24.0\ \Omega$ in rectangular form

Z_{EQ} is in series with I_S and X_C, so that full source current flows through Z_{EQ}. Use Ohm's Law for voltage to calculate the open load voltage, $V_{ZEQ} = V_{out} = V_{TH}$.

$V_{TH} = I_S Z_{EQ}$
$= (2.4\text{ A}_{ac} \angle 0°)\ (70.5\ \Omega \angle 19.9°)$
$= [(2.4\text{ A}_{ac})(70.5\ \Omega)] \angle [0° + 19.9°]$
$= \mathbf{169\ V_{ac} \angle 19.9°}$

For the circuit in Figure 20-16, the Thevenin impedance, Z_{TH}, is the impedance looking back into the circuit from V_{out} with V_S replaced with its internal impedance (a short). This impedance appears to be X_C and Z_{EQ} in parallel.

$Z_{TH} = X_C \parallel Z_{EQ}$

$= 1 / \{[1 / (-j\ 101\ \Omega)] + [1 / (66.3\ \Omega + j\ 24.0\ \Omega)]\}$

$\boldsymbol{Y_{TH}} = 1 / \boldsymbol{Z_{TH}}$

$= [(1 \angle 0°) / (101\ \Omega \angle -90°)] + [(1 \angle 0°) / (70.5\ \Omega \angle 19.9°)]$

$= \{(1 / 101\ \Omega) \angle [0° - (-90°)]\} + [(1 / 70.5\ \Omega) \angle (0° - 19.9°)]$

$= (9.95\text{ mS} \angle 90°) + (14.2\text{ mS} \angle -19.9°)$ in polar form

$= (0\text{ S} + j\ 9.95\text{ mS}) + (13.3\text{ mS} - j\ 4.82\text{ mS})$ in rectangular form

$= (0\text{ S} + 13.3\text{ mS}) + j\ (9.95\text{ mS} - 4.82\text{ mS})$

$= 13.3\text{ mS} + j\ 5.13\text{ mS}$ in rectangular form

$= 14.3\text{ mS} \angle 21.0°$ in polar form

$\boldsymbol{Z_{TH}} = 1 / \boldsymbol{Y_{TH}}$

$= (1 \angle 0°) / (14.3\text{ mS} \angle 21.0°)$

$= (1 / 14.3\text{ mS}) \angle (0° - 21.0°)$

$\mathbf{= 70.0\ \Omega \angle -21.0°}$

20.2.3 Norton's Theorem

1. Problem: Calculate the current and resistance values, $\boldsymbol{I_N}$ and $\boldsymbol{R_N}$, of the Norton equivalent for the circuit in Figure 20-17 with respect to the output, $\boldsymbol{V_{out}}$.

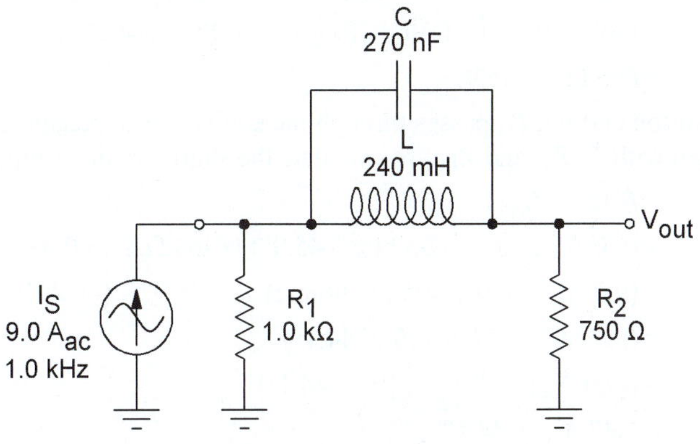

Figure 20-17: *RLC* Circuit for Norton Problem 1

Solution: First, calculate the reactances for the capacitor and inductor.

$\boldsymbol{X_C} = 1 / (2\pi \boldsymbol{fC})$

$= 1 / [2\pi(1.0\text{ kHz})(270\text{ nF})]$

$= 1 / [2\pi(1.0 \times 10^3\text{ Hz})(270 \times 10^{-9}\text{ F})]$

$= 1 / 1.70 \times 10^{-3}\text{ S}$

$= 590\ \Omega$

$\boldsymbol{X_L} = 2\pi \boldsymbol{fL}$

$= 2\pi(1.0\text{ kHz})(240\text{ mH})]$

$= 2\pi(1.0 \times 10^3\text{ Hz})(240 \times 10^{-3}\text{ H})]$

$= 1.51 \times 10^3\ \Omega$

$= 1.51\text{ k}\Omega$

Next, calculate the equivalent impedance of the capacitor and inductor in parallel.

$\boldsymbol{Z_{EQ1}} = \boldsymbol{X_C} \parallel \boldsymbol{X_L}$

$= -j\ \boldsymbol{X_C} \parallel j\ \boldsymbol{X_L}$

$= -j\ 590\ \Omega \parallel j\ 1.51\text{ k}\Omega$

$= [(-j\ 590\ \Omega)(j\ 1.51\ k\Omega)] / [(-j\ 590\ \Omega) + (j\ 1.51\ k\Omega)]$

$= [-j^2\ (590\ \Omega)(1.51 \times 10^3\ \Omega)] / [j\ (-590\ \Omega + 1.51 \times 10^3\ \Omega)]$

$= [(1)(889 \times 10^3\ \Omega^2)] / (j\ 919\ \Omega)$

$= (889 \times 10^3\ \Omega^2) / (j\ 919\ \Omega)$

$= (889 \times 10^3\ \Omega^2 \angle 0°) / (919\ \Omega \angle 90°)$

$= (889 \times 10^3\ \Omega^2 / 919\ \Omega) \angle (0° - 90°)$

$= 968\ \Omega \angle -90°$ in polar form

$= 0\ \Omega - j\ 968\ \Omega$ in rectangular form

The Norton current, $\boldsymbol{I_N}$, is equal to the current through the $\boldsymbol{V_{out}}$ terminal with $\boldsymbol{V_{out}}$ shorted to ground. When $\boldsymbol{V_{out}}$ is shorted to ground, resistor $\boldsymbol{R_2}$ is bypassed so that the total impedance seen by the source, $\boldsymbol{I_S}$, is the parallel combination of $\boldsymbol{R_1}$ and $\boldsymbol{Z_{EQ1}}$.

$\boldsymbol{Z_T} = \boldsymbol{R_1} \parallel \boldsymbol{Z_{EQ1}}$

$= 1.0\ k\Omega \parallel (968\ \Omega \angle -90°)$

$= (1.0 \times 10^3\ \Omega \angle 0°)(968\ \Omega \angle -90°) /| [(1.0 \times 10^3\ \Omega + j\ 0\ \Omega) + (0\ \Omega - j\ 968\ \Omega)]$

$= \{(1.0 \times 10^3\ \Omega)(968\ \Omega) \angle [0° + (-90°)]\} / [(1.0 \times 10^3\ \Omega + 0\ \Omega) + j\ (0\ \Omega - 968\ \Omega)]$

$= (968 \times 10^3\ \Omega^2 \angle -90°) / (1.0 \times 10^3\ \Omega - j\ 968\ \Omega)$

$= (968 \times 10^3\ \Omega^2 \angle -90°) / (1.39 \times 10^3\ \Omega \angle -44.1°)$

$= (968 \times 10^3\ \Omega^2 / 1.39 \times 10^3\ \Omega) \angle [-90° - (-44.1°)]$

$= 695\ \Omega \angle -45.9°$

The Norton current, $\boldsymbol{I_N}$, passes through the equivalent impedance, $\boldsymbol{Z_{EQ1}}$. Use the ac current divider theorem with $\boldsymbol{I_S}$, $\boldsymbol{Z_T}$, and $\boldsymbol{Z_{EQ1}}$ to calculate the short circuit current, $\boldsymbol{I_{SC}} = \boldsymbol{I_N}$.

$\boldsymbol{I_N} = \boldsymbol{I_S}\ (\boldsymbol{Z_T} / \boldsymbol{Z_{EQ1}})$

$= (9.0\ A_{ac} \angle 0°)\ [(695\ \Omega \angle -45.9°) / (968\ \Omega \angle -90°)]$

$= (9.0\ A_{ac} \angle 0°)\ \{(695\ \Omega / 968\ \Omega) \angle [-45.9° - (-90°)]\}$

$= (9.0\ A_{ac} \angle 0°)\ (0.719 \angle 44.1°)$

$= [(9.0\ A_{ac})(0.719)] \angle [0° + 44.1°]$

$\mathbf{= 6.47\ A_{ac} \angle 44.1°}$

For the circuit in Figure 20-17, the Norton impedance, $\boldsymbol{Z_N}$, is the impedance looking back into the circuit from $\boldsymbol{V_{out}}$ with $\boldsymbol{I_S}$ replaced with its internal impedance (an open). This impedance appears to be the series combination of $\boldsymbol{R_1}$ and $\boldsymbol{Z_{EQ1}}$ in parallel with $\boldsymbol{X_C}$, so that $\boldsymbol{Z_N} = (\boldsymbol{R_1} + \boldsymbol{Z_{EQ1}}) \parallel \boldsymbol{X_C}$. First, calculate the equivalent impedance, $\boldsymbol{Z_{EQ2}}$, for the series combination $\boldsymbol{R_1} + \boldsymbol{Z_{EQ1}}$.

$\boldsymbol{Z_{EQ2}} = \boldsymbol{R_1} + \boldsymbol{Z_{EQ1}}$

$= 1.0\ k\Omega + (0\ \Omega - j\ 968\ \Omega)$

$= (1.0 \times 10^3\ \Omega + j\ 0\ \Omega) + (0\ \Omega - j\ 968\ \Omega)$

$= (1.0 \times 10^3\ \Omega + 0\ \Omega) + j\ (0\ \Omega - 968\ \Omega)$

$= 1.0 \times 10^3\ \Omega - j\ 968\ \Omega$

$= 1.0\ k\Omega - j\ 968\ \Omega$ in rectangular form

$= 1.39\ k\Omega \angle -44.1°$ in polar form

Next, calculate the Norton impedance from $\boldsymbol{Z_{EQ2}}$ and $\boldsymbol{X_C}$.

$\boldsymbol{Z_N} = (\boldsymbol{R_1} + \boldsymbol{Z_{EQ1}}) \parallel \boldsymbol{X_C}$

$= \boldsymbol{Z_{EQ2}} \parallel \boldsymbol{X_C}$

$= (1.0\ k\Omega - j\ 968\ \Omega) \parallel (-j\ 590\ \Omega)$

$= (1.39\ k\Omega \angle -44.1°) \parallel (590\ \Omega \angle -90°)$

$= 1 / \{[1 / (1.39\ k\Omega \angle -44.1°)] + [1 / (590\ \Omega \angle -90°)]\}$

$\boldsymbol{Y_N} = 1 / \boldsymbol{Z_N}$

$= [(1 \angle 0°) / (1.39\ k\Omega \angle -44.1°)] + [(1 \angle 0°) / (590\ \Omega \angle -90°)]$

$= \{(1 / 1.39\ k\Omega) \angle [0° - (-44.1°)]\} + \{(1 / 590\ \Omega) \angle [0° - (-90°)]\}$

$= (719\ \mu S \angle 44.1°) + (1.70\ mS \angle 90°)$ in polar form

$= (516\ \mu S + j\ 500\ \mu S) + (0\ S + j\ 1.70\ mS)$ in rectangular form

$= (516\ \mu S + 0\ S) + j\ (500\ \mu S + 1.70\ mS)$

$= 516\ \mu S + j\ 2.20\ mS$ in rectangular form

$= 2.26\ mS \angle 76.8°$ in polar form

$\boldsymbol{Z_N} = 1 / \boldsymbol{Y_N}$

$= (1 \angle 0°) / (2.26\ mS \angle 76.8°)$

$= (1 / 2.26\ mS) \angle (0° - 76.8°)$

$= \mathbf{443\ \Omega \angle -76.7°}$

2. Problem: Calculate the current and resistance values, $\boldsymbol{I_N}$ and $\boldsymbol{R_N}$, of the Norton equivalent for the circuit in Figure 20-18 with respect to the output, $\boldsymbol{V_{out}}$.

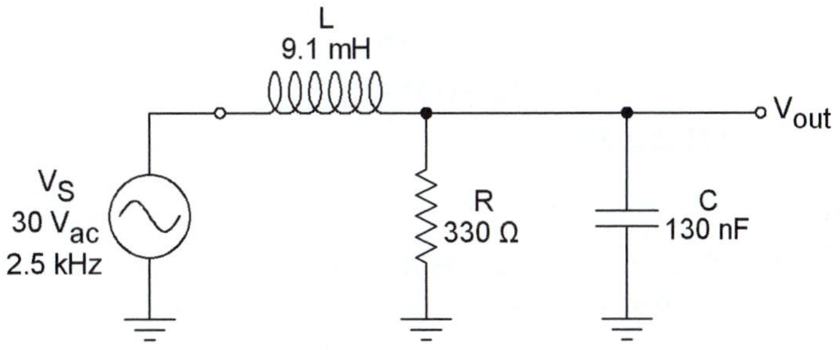

Figure 20-18: *RLC* Circuit for Norton Problem 2

Solution: First, calculate the reactances for the inductor and capacitor.

$\boldsymbol{X_L} = 2\pi \boldsymbol{fL}$

$= 2\pi(2.5\ kHz)(9.1\ mH)]$

$= 2\pi(2.5 \times 10^3\ Hz)(9.1 \times 10^{-3}\ H)]$

$= 143\ \Omega$

$\boldsymbol{X_C} = 1 / (2\pi \boldsymbol{fC})$

$= 1 / [2\pi(2.5\ kHz)(130\ nF)]$

$= 1 / [2\pi(2.5 \times 10^3\ Hz)(130 \times 10^{-9}\ F)]$

$= 1 / 2.04 \times 10^{-3}\ S$

$= 490\ \Omega$

The Norton current, $\boldsymbol{I_N}$, is equal to the current through the $\boldsymbol{V_{out}}$ terminal with $\boldsymbol{V_{out}}$ shorted to ground. When $\boldsymbol{V_{out}}$ is shorted to ground, resistor $\boldsymbol{R}$ and capacitor $\boldsymbol{C}$ are bypassed so that the total impedance seen by the source, $\boldsymbol{V_S}$, is the inductive reactance, $\boldsymbol{X_L}$.

$\boldsymbol{I_N} = \boldsymbol{V_S} / \boldsymbol{X_L}$

$= \boldsymbol{V_S} / j\ \boldsymbol{X_L}$

$= (30.0\ V_{ac} \angle 0°) / (j\ 143\ \Omega)$

$= (30.0\ V_{ac} \angle 0°) / (143\ \Omega \angle 90°)$

$= (30.0\ V_{ac} / 143\ \Omega) \angle (0° - 90°)$

$= \mathbf{210\ mA_{ac} \angle -90°}$

For the circuit in Figure 20-18, the Norton impedance, $\boldsymbol{Z_N}$, is the impedance looking back into the circuit from $\boldsymbol{V_{out}}$ with $\boldsymbol{V_S}$ replaced with its internal impedance (a short). This impedance appears to be the parallel combination of $\boldsymbol{X_L}$, $\boldsymbol{R}$, and $\boldsymbol{X_C}$. Calculate the total impedance, $\boldsymbol{Z_N}$, for the parallel combination $\boldsymbol{X_L} \parallel \boldsymbol{R} \parallel \boldsymbol{X_C}$.

$Z_N = X_L \parallel R \parallel X_C$

$= \text{j } 143\ \Omega \parallel 330\ \Omega \parallel -\text{j } 490\ \Omega$

$= 1 / \{[1 / (\text{j } 143\ \Omega)] + (1 / 330\ \Omega) + [1 / (-\text{j } 490\ \Omega)]\}$

$Y_N = 1 / Z_N$

$= [(1 \angle 0°) / (143\ \Omega \angle 90°)] + [(1 \angle 0°) / (330\ \Omega \angle 0°)] + [(1 \angle 0°) / (490\ \Omega \angle -90°)]$

$= [(1 / 143\ \Omega) \angle (0° - 90°)] + [(1 / 330\ \Omega) \angle (0° - 0°)] + \{(1 / 490\ \Omega) \angle [0° - (-90°)]\}$

$= (7.00\text{ mS} \angle -90°) + (3.03\text{ mS} \angle 0°) + (2.04\text{ mS} \angle 90°)$ in polar form

$= (0\text{ S} - \text{j } 7.00\text{ mS}) + (3.03\text{ mS} + \text{j } 0\text{ S}) + (0\text{ S} + \text{j } 2.04\text{ mS})$ in rectangular form

$= (0\text{ S} + 3.03\text{ mS} + 0\text{ S}) + \text{j } (-7.00\text{ mS} + 0\text{ S} + 2.04\text{ mS})$

$= 3.03\text{ mS} - \text{j } 4.95\text{ mS}$ in rectangular form

$= 5.81\text{ mS} \angle -58.6°$ in polar form

$Z_N = 1 / Y_N$

$= (1 \angle 0°) / (5.81\text{ mS} \angle -58.6°)$

$= (1 / 5.81\text{ mS}) \angle [0° - (-58.6°)]$

$= \mathbf{172\ \Omega \angle 58.6°}$

20.2.4 Superposition Theorem

1. Problem: The value of $\boldsymbol{V_R}$, the voltage across $\boldsymbol{R}$, for the circuit in Figure 20-19 is 7.5 $V_{ac} \angle 0°$. Use the ac superposition theorem to calculate the value of $\boldsymbol{V_{S2}}$. Express your answer in polar form.

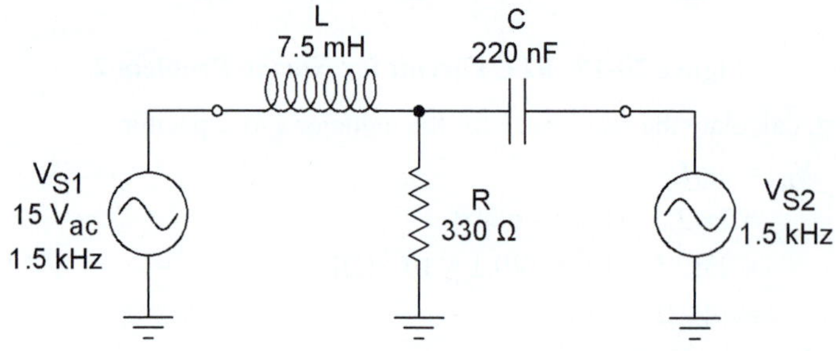

Figure 20-19: *RLC* Circuit for Superposition Problem 1

Solution: First, calculate the reactances for the inductor and capacitor.

$X_L = 2\pi f L$

$= 2\pi(1.5\text{ kHz})(7.5\text{ mH})]$

$= 2\pi(1.5 \times 10^3\text{ Hz})(7.5 \times 10^{-3}\text{ H})]$

$= 70.7\ \Omega$

$X_C = 1 / (2\pi f C)$

$= 1 / [2\pi(1.5\text{ kHz})(220\text{ nF})]$

$= 1 / [2\pi(1.5 \times 10^3\text{ Hz})(220 \times 10^{-9}\text{ F})]$

$= 1 / 2.07 \times 10^{-3}\text{ S}$

$= 482\ \Omega$

Next, determine the voltage across $\boldsymbol{R}$ due to the voltage source V_{S1} by replacing V_{S2} with its internal impedance (a short). To do so, first calculate the equivalent impedance Z_{EQ1} for the parallel combination $\boldsymbol{R} \parallel X_C$.

$$
\begin{aligned}
Z_{EQ1} &= \boldsymbol{R} \parallel X_C \\
&= \boldsymbol{R} \parallel -\mathrm{j}\, X_C \\
&= (330\ \Omega)(-\mathrm{j}\ 482\ \Omega) / (330\ \Omega - \mathrm{j}\ 482\ \Omega) \\
&= (-\mathrm{j}\ 159 \times 10^3\ \Omega^2) / (330\ \Omega - \mathrm{j}\ 482\ \Omega) \\
&= (159 \times 10^3\ \Omega^2 \angle -90°) / (584\ \Omega \angle -55.6°) \\
&= (159 \times 10^3\ \Omega^2 / 584\ \Omega) \angle [-90° - (-55.6°)] \\
&= 272\ \Omega \angle -34.4° \text{ in polar form} \\
&= 225\ \Omega - \mathrm{j}\ 154\ \Omega \text{ in rectangular form}
\end{aligned}
$$

Z_{EQ1} is the parallel combination $\boldsymbol{R} \parallel X_C$, so the voltage across $\boldsymbol{R}$ due to V_{S1} is equal to V_{ZEQ1}, the voltage across Z_{EQ1}. Use the ac voltage divider theorem to calculate V_{ZEQ1}.

$$
\begin{aligned}
V_{ZEQ1} &= V_{S1}\ [Z_{EQ1} / (X_L + Z_{EQ1})] \\
&= (15\ \mathrm{V_{ac}} \angle 0°)\ \{(272\ \Omega \angle -34.4°) / [(\mathrm{j}\ 70.7\ \Omega) + (225\ \Omega - \mathrm{j}\ 154\ \Omega)]\} \\
&= (15\ \mathrm{V_{ac}} \angle 0°)\ \{(272\ \Omega \angle -34.4°) / [(0\ \Omega + \mathrm{j}\ 70.7\ \Omega) + (225\ \Omega - \mathrm{j}\ 154\ \Omega)]\} \\
&= (15\ \mathrm{V_{ac}} \angle 0°)\ \{(272\ \Omega \angle -34.4°) / [(0\ \Omega + 225\ \Omega) + \mathrm{j}\ (70.7\ \Omega - 154\ \Omega)]\} \\
&= (15\ \mathrm{V_{ac}} \angle 0°)\ [(272\ \Omega \angle -34.4°) / (225\ \Omega - \mathrm{j}\ 83.1\ \Omega)] \\
&= (15\ \mathrm{V_{ac}} \angle 0°)\ [(272\ \Omega \angle -34.4°) / (240\ \Omega \angle -20.3°)] \\
&= (15\ \mathrm{V_{ac}} \angle 0°)\ \{(272\ \Omega / 240\ \Omega) \angle [-34.4° - (-20.3°)]\} \\
&= (15\ \mathrm{V_{ac}} \angle 0°)\ (1.14 \angle -14.1°) \\
&= [(15\ \mathrm{V_{ac}})(1.14)] \angle [0° + (-14.1°)] \\
&= 17.1\ \mathrm{V_{ac}} \angle -14.1° \text{ in polar form} \\
&= 16.5\ \mathrm{V_{ac}} - \mathrm{j}\ 4.15\ \mathrm{V_{ac}} \text{ in rectangular form}
\end{aligned}
$$

Next, determine the voltage across $\boldsymbol{R}$ due to the voltage source V_{S2} by replacing V_{S1} with its internal impedance (a short). To do so, first calculate the equivalent impedance Z_{EQ2} for the parallel combination $X_L \parallel \boldsymbol{R}$.

$$
\begin{aligned}
Z_{EQ2} &= X_L \parallel \boldsymbol{R} \\
&= \mathrm{j}\, X_L \parallel \boldsymbol{R} \\
&= (\mathrm{j}\ 70.7\ \Omega)(330\ \Omega) / (\mathrm{j}\ 70.7\ \Omega + 330\ \Omega) \\
&= (\mathrm{j}\ 23.3 \times 10^3\ \Omega^2) / (330\ \Omega + \mathrm{j}\ 70.7\ \Omega) \\
&= (23.3 \times 10^3\ \Omega^2 \angle 90°) / (338\ \Omega \angle 12.1°) \\
&= (23.3 \times 10^3\ \Omega^2 / 338\ \Omega) \angle (90° - 12.1°) \\
&= 69.1\ \Omega \angle 77.9° \text{ in polar form} \\
&= 14.5\ \Omega + \mathrm{j}\ 67.6\ \Omega \text{ in rectangular form}
\end{aligned}
$$

Z_{EQ2} is the parallel combination $X_L \parallel \boldsymbol{R}$, so the voltage across $\boldsymbol{R}$ due to V_{S2} is equal to V_{ZEQ2}, the voltage across Z_{EQ2}. Use the ac voltage divider theorem to determine the voltage, V_{ZEQ2}.

$$
\begin{aligned}
V_{ZEQ2} &= V_{S2}\ [Z_{EQ2} / (X_C + Z_{EQ2})] \\
&= V_{S2}\ \{(69.1\ \Omega \angle 77.9°) / [(-\mathrm{j}\ 482\ \Omega) + (14.5\ \Omega + \mathrm{j}\ 67.6\ \Omega)]\} \\
&= V_{S2}\ \{(69.1\ \Omega \angle 77.9°) / [(0\ \Omega - \mathrm{j}\ 482\ \Omega) + (14.5\ \Omega + \mathrm{j}\ 67.6\ \Omega)]\} \\
&= V_{S2}\ \{(69.1\ \Omega \angle 77.9°) / [(0\ \Omega + 14.5\ \Omega) + \mathrm{j}\ (-482\ \Omega + 67.6\ \Omega)]\} \\
&= V_{S2}\ [(69.1\ \Omega \angle 77.9°) / (14.5\ \Omega - \mathrm{j}\ 415\ \Omega)] \\
&= V_{S2}\ [(69.1\ \Omega \angle 77.9°) / (415\ \Omega \angle -88.0°)] \\
&= V_{S2}\ \{(69.1\ \Omega / 415\ \Omega) \angle [77.9° - (-88.0°)]\}
\end{aligned}
$$

$$= V_{S2}\ (0.167 \angle 166°) \text{ in polar form}$$

Set the sum of V_{ZEQ1} and V_{ZEQ2} equal to 7.5 $V_{ac} \angle 0°$ and solve for V_{S2}.

$$V_{ZEQ1} + V_{ZEQ2} = 7.5\ V_{ac} \angle 0°$$

$$V_{ZEQ1} + V_{ZEQ2} - V_{ZEQ1} = (7.5\ V_{ac} \angle 0°) - V_{ZEQ1}$$

$$V_{ZEQ2} = (7.5\ V_{ac} + j\ 0\ V_{ac}) - (16.5\ V_{ac} - j\ 4.15\ V_{ac})$$

$$V_{S2}\ (0.167 \angle 166°) = (7.5\ V_{ac} - 16.5\ V_{ac}) + j\ (0\ V_{ac} + j\ 4.15\ V_{ac})$$

$$= -9.03\ V_{ac} + j\ 4.15\ V_{ac} \text{ in rectangular form}$$

$$= 9.94\ V_{ac} \angle 155° \text{ in polar form}$$

$$[V_{S2}\ (0.167 \angle 166°)]\ /\ (0.167 \angle 166°) = (9.94\ V_{ac} \angle 155°)\ /\ (0.167 \angle 166°)$$

$$V_{S2} = (9.94\ V_{ac}\ /\ 0.167) \angle (155° - 166°)$$

$$= \mathbf{59.7\ V_{ac} \angle -10.6°}$$

2. Problem: The value of I_R, the current through R, for the circuit in Figure 20-20 is 12.5 $mA_{ac} \angle 0°$. Use the ac superposition theorem to calculate the value of I_S. Express your answer in polar form.

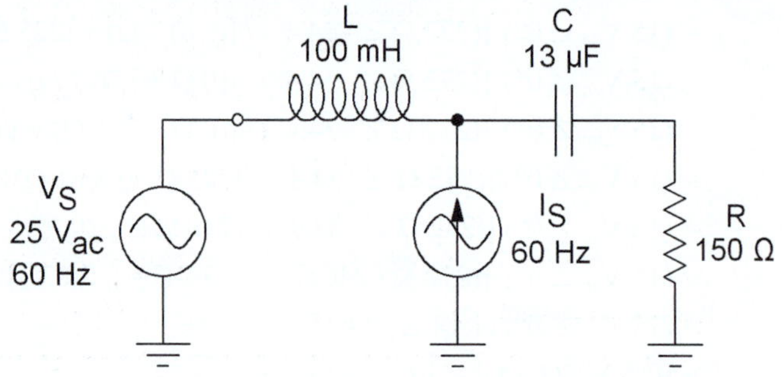

Figure 20-20: *RLC* Circuit for Superposition Problem 2

Solution: First, calculate the reactances for the inductor and capacitor.

$$X_L = 2\pi fL$$

$$= 2\pi(60\text{ Hz})(100\text{ mH})]$$

$$= 2\pi(60\text{ Hz})(100 \times 10^{-3}\text{ H})]$$

$$= 37.7\ \Omega$$

$$X_C = 1\ /\ (2\pi fC)$$

$$= 1\ /\ [2\pi(60\text{ Hz})(13\ \mu\text{F})]$$

$$= 1\ /\ [2\pi(60\text{ Hz})(0.1 \times 10^{-6}\text{ F})]$$

$$= 1\ /\ 4.90 \times 10^{-3}\text{ S}$$

$$= 204\ \Omega$$

Next, determine the current through R due to the voltage source V_S by replacing I_S with its internal impedance (an open). Because I_S appears to be an open, the equivalent impedance seen by V_S, Z_{EQ1}, is the series combination $X_L + X_C + R$.

$$Z_{EQ1} = X_L + X_C + R$$

$$= j\ X_L - j\ X_C + R$$

$$= j\ 37.7\ \Omega - j\ 204\ \Omega + 150\ \Omega$$

$$= 150\ \Omega + j\ (37.7\ \Omega - 204\ \Omega)$$

$$= 150\ \Omega - j\ 166\ \Omega \text{ in rectangular form}$$

$$= 224\ \Omega \angle -48.0° \text{ in polar form}$$

Use Ohm's Law to calculate $\boldsymbol{I_{R1}}$, the current through $\boldsymbol{R}$ due to $\boldsymbol{V_S}$.

$\boldsymbol{I_{R1}} = \boldsymbol{V_S} / \boldsymbol{Z_{EQ1}}$

$= (25\ \text{V}_{ac} \angle 0°) / (224\ \Omega \angle -48.0°)$

$= (25\ \text{V}_{ac} / 224\ \Omega) \angle [0° - (-48.0°)]$

$= 112\ \text{mA}_{ac} \angle 48.0°$ in polar form

$= 74.7\ \text{mA}_{ac} + \text{j}\ 82.9\ \text{mA}_{ac}$

Next, determine the current through $\boldsymbol{R}$ due to the current source $\boldsymbol{I_S}$ by replacing $\boldsymbol{V_S}$ with its internal impedance (a short). Because $\boldsymbol{V_S}$ appears to be a short, the impedance seen by $\boldsymbol{I_S}$ is the parallel combination $\boldsymbol{X_L} \parallel (\boldsymbol{X_C} + \boldsymbol{R})$. First, calculate the equivalent impedance $\boldsymbol{Z_{EQ2}} = \boldsymbol{X_C} + \boldsymbol{R}$.

$\boldsymbol{Z_{EQ2}} = \boldsymbol{X_C} + \boldsymbol{R}$

$= -\text{j}\ \boldsymbol{X_C} + \boldsymbol{R}$

$= -\text{j}\ 204\ \Omega + 150\ \Omega$

$= 150\ \Omega - \text{j}\ 204\ \Omega$ in rectangular form

$= 253\ \Omega \angle -53.7°$ in polar form

Next, calculate the equivalent impedance, $\boldsymbol{Z_{EQ3}}$, seen by $\boldsymbol{I_S}$.

$\boldsymbol{Z_{EQ3}} = \boldsymbol{X_L} \parallel (\boldsymbol{X_C} + \boldsymbol{R})$

$= \text{j}\ \boldsymbol{X_L} \parallel \boldsymbol{Z_{EQ2}}$

$= (\text{j}\ 37.7\ \Omega) \parallel (253\ \Omega \angle -53.7°)$

$= (37.7\ \Omega \angle 90°) \parallel (253\ \Omega \angle -53.7°)$

$= 1 / \{[1 / (37.7\ \Omega \angle 90°)] + [1 / (253\ \Omega \angle -53.7°)]\}$

$\boldsymbol{Y_{EQ3}} = 1 / \boldsymbol{Z_{EQ3}}$

$= [1 / (37.7\ \Omega \angle 90°)] + [1 / (253\ \Omega \angle -53.7°)]$

$= [(1 \angle 0°) / (37.7\ \Omega \angle 90°)] + [(1 \angle 0°) / (253\ \Omega \angle -53.7°)]$

$= [(1 / 37.7\ \Omega) \angle (0° - 90°)] + \{(1 / 253\ \Omega) \angle [0° - (-53.7°)]\}$

$= (26.5\ \text{mS} \angle -90°) + (3.95\ \text{mS} \angle 53.7°)$

$= (0\ \text{S} - \text{j}\ 26.5\ \text{mS}) + (2.34\ \text{mS} + \text{j}\ 3.18\ \text{mS})$

$= (0\ \text{S} + 2.34\ \text{mS}) + \text{j}\ (-26.5\ \text{mS} + 3.18\ \text{mS})$

$= 2.34\ \text{mS} - \text{j}\ 23.3\ \text{mS}$ in rectangular form

$= 23.5\ \text{mS} \angle -84.3°$ in polar form

$\boldsymbol{Z_{EQ3}} = 1 / \boldsymbol{Y_{EQ3}}$

$= 1 / (23.5\ \text{mS} \angle -84.3°)$

$= (1 \angle 0°) / (23.5\ \text{mS} \angle -84.3°)$

$= (1 / 23.5\ \text{mS}) \angle [0° - (-84.3°)]$

$= 42.6\ \Omega \angle 84.3°$ in polar form

$= 4.25\ \Omega + \text{j}\ 42.4\ \Omega$ in rectangular form

$\boldsymbol{Z_{EQ2}}$ is the series combination $\boldsymbol{X_C} + \boldsymbol{R}$, so that $\boldsymbol{I_{R2}}$, the current from $\boldsymbol{I_S}$ through $\boldsymbol{R}$, is the same as the current through $\boldsymbol{Z_{EQ2}}$. Use the ac current divider theorem to determine $\boldsymbol{I_{R2}}$.

$\boldsymbol{I_{R2}} = \boldsymbol{I_{ZEQ2}}$

$= \boldsymbol{I_S}\ [\boldsymbol{Z_{EQ3}} / \boldsymbol{Z_{EQ2}})]$

$= \boldsymbol{I_S}\ [(42.6\ \Omega \angle 84.3°) / (253\ \Omega \angle -53.7°)]$

$= \boldsymbol{I_S}\ \{(42.6\ \Omega / 253\ \Omega) \angle [84.3° - (-53.7°)]\}$

$= \boldsymbol{I_S}\ (0.168 \angle 138°)$

Set the sum of $\boldsymbol{I_{R1}}$ and $\boldsymbol{I_{R2}}$ equal to $12.5\ \text{mA}_{ac} \angle 0°$ and solve for $\boldsymbol{I_S}$.

$$I_{R1} + I_{R2} = 12.5\ \text{mA}_{ac} \angle 0°$$

$$I_{R1} + I_{R2} - I_{R1} = (12.5\ \text{mA}_{ac} \angle 0°) - I_{ZEQ1}$$

$$I_{R2} = (12.5\ \text{mA}_{ac} + \text{j}\ 0\ \text{mA}_{ac}) - (74.7\ \text{mA}_{ac} + \text{j}\ 82.9\ \text{mA}_{ac})$$

$$I_S\ (0.168 \angle 138°) = (12.5\ \text{mA}_{ac} - 74.6\ \text{mA}_{ac}) - \text{j}\ (74.6\ \text{mA}_{ac} + 82.9\ \text{mA}_{ac})$$

$$= -62.2\ \text{mA}_{ac} - \text{j}\ 82.9\ \text{mA}_{ac} \text{ in rectangular form}$$

$$= 104\ \text{mA}_{ac} \angle -127° \text{ in polar form}$$

$$[I_S\ (0.168 \angle 138°)] / (0.168 \angle 138°) = (104\ \text{mA}_{ac} \angle -127°) / (0.168 \angle -138°)$$

$$I_S = (104\ \text{mA}_{ac} / 0.168) \angle [-127° - (-138°)]$$

$$= \mathbf{616\ mA_{ac} \angle 95.1°}$$

20.2.5 Maximum Power Transfer Theorem

1. Problem: For the complex bridge circuit in Figure 20-21, determine the impedance for a load between points A and B that will transfer maximum power to the load.

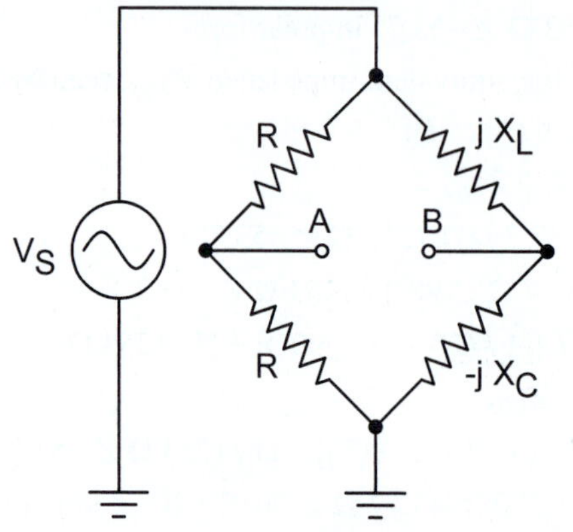

Figure 20-21: Bridge Circuit for Maximum Power Transfer Problem 1

Solution: To determine the load impedance for maximum transfer, determine the Thevenin impedance seen by the load across points A and B. To find the Thevenin impedance, replace V_S with its internal impedance (a short). The equivalent impedance, Z_{TH}, seen between A and B is the series-parallel combination $R \parallel R + X_L \parallel X_C$.

$$Z_{TH} = R \parallel R + X_L \parallel X_C$$

$$= (R \parallel R) + (\text{j}\ X_L \parallel -\text{j}\ X_C)$$

$$= [(R)(R) / (R + R)] + [(\text{j}\ X_L)(-\text{j}\ X_C) / (\text{j}\ X_L - \text{j}\ X_C)]$$

$$= (R^2 / 2R) + \{(-\text{j}^2\ X_L X_C) / [\text{j}\ (X_L - X_C)]\}$$

$$= (R / 2) + \{-(-1) X_L X_C) / [\text{j}\ (X_L - X_C)]\}$$

$$= (R / 2) + \{(X_L X_C) / [\text{j}\ (X_L - X_C)]\}$$

$$= \{(R / 2)[(X_L - X_C) / (X_L - X_C)]\} + \{(X_L X_C) / [\text{j}\ (X_L - X_C)]\}[(\text{j}\ 2) / (\text{j}\ 2)]$$

$$= \{[R\ (X_L - X_C)] / [2(X_L - X_C)]\} + \{[\text{j}\ (2X_L X_C)] / [\text{j}^2\ 2(X_L - X_C)]\}$$

$$= [R\ (X_L - RX_C) / (2X_L - 2X_C)] + \{[\text{j}\ (2X_L X_C)] / [(-1)(2)(X_L - X_C)]\}$$

$$= [R\ (X_L - X_C) / (2X_L - 2X_C)] + \{[-\text{j}\ (2X_L X_C)] / [2(X_L - X_C)]\}$$

$$= [R\ (X_L - X_C) - \text{j}\ 2X_L X_C] / [2(X_L - X_C)]$$

The bridge transfers maximum power to the load when the load impedance, Z_L, is $Z_{TH}^* = \mathbf{[R\ (X_L - X_C) + j\ 2X_L X_C] / [2(X_L - X_C)]}$, the complex conjugate of the Thevenin impedance Z_{TH}.

2. Problem: For the complex bridge circuit in Figure 20-22, determine the impedance for a load between points A and B that will transfer maximum power to the load for a frequency $f = 1 / 2\pi\sqrt{LC}$.

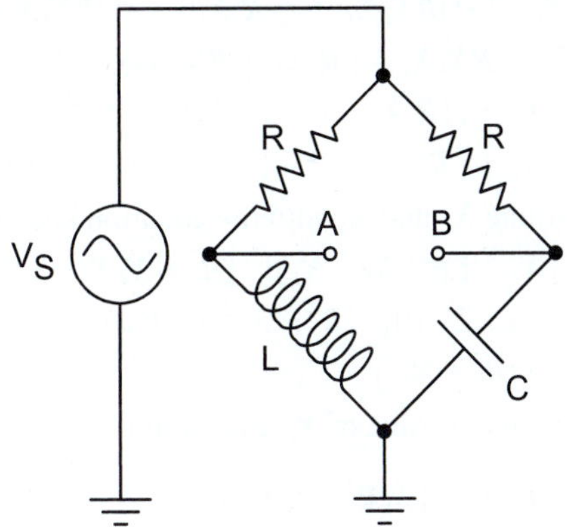

Figure 20-22: Bridge Circuit for Maximum Power Transfer Problem 2

Solution: To determine the load impedance for maximum transfer, determine the Thevenin impedance seen by the load across points A and B. To find the Thevenin impedance, replace V_S with its internal impedance (a short). The equivalent impedance, Z_{TH}, seen between A and B is the series-parallel combination $R \parallel X_L + R \parallel X_C$.

First, calculate the reactances for the inductance and capacitance.

$$
\begin{aligned}
X_L &= 2\pi f L \\
&= 2\pi\,[1 / (2\pi\sqrt{LC})]L \\
&= L / \sqrt{LC} \\
&= \sqrt{L^2} / \sqrt{LC} \\
&= \sqrt{L^2 / LC} \\
&= \sqrt{L / C}
\end{aligned}
$$

$$
\begin{aligned}
X_C &= 1 / (2\pi f C) \\
&= 1 / \{2\pi[1 / (2\pi\sqrt{LC})]C\} \\
&= 1 / (C / \sqrt{LC}) \\
&= \sqrt{LC} / C \\
&= \sqrt{LC / C^2} \\
&= \sqrt{L / C}
\end{aligned}
$$

Substitute the values of X_L and X_C into the expression for equivalent impedance Z_{TH}.

$$
\begin{aligned}
Z_{TH} &= R \parallel X_L + R \parallel X_C \\
&= (R \parallel \mathrm{j}\,X_L) + (R \parallel -\mathrm{j}\,X_C) \\
&= [(R)(\mathrm{j}\,X_L) / (R + \mathrm{j}\,X_L)] + [(R)(-\mathrm{j}\,X_C) / (R - \mathrm{j}\,X_C)] \\
&= [(\mathrm{j}\,RX_L) / (R + \mathrm{j}\,X_L)] + [(-\mathrm{j}\,RX_C) / (R - \mathrm{j}\,X_C)] \\
&= [(\mathrm{j}\,RX_L) / (R + \mathrm{j}\,X_L)][(R - \mathrm{j}\,X_C) / (R - \mathrm{j}\,X_C)] \\
&\qquad + [(-\mathrm{j}\,RX_C) / (R - \mathrm{j}\,X_C)]\,[(R + \mathrm{j}\,X_L) / (R + \mathrm{j}\,X_L)] \\
&= [(\mathrm{j}\,RX_L)(R - \mathrm{j}\,X_C)] / [(R + \mathrm{j}\,X_L)(R - \mathrm{j}\,X_C)] \\
&\qquad + [(-\mathrm{j}\,RX_C)(R + \mathrm{j}\,X_L)] / [(R - \mathrm{j}\,X_C)(R + \mathrm{j}\,X_L)]
\end{aligned}
$$

$= (j\,R^2X_L - j^2\,RX_LX_C) / (R^2 - j\,RX_C + j\,RX_L - j^2\,X_LX_C)$
$\quad + (-j\,R^2X_C - j^2\,RX_LX_C) / (R^2 - j\,RX_C + j\,RX_L - j^2\,X_LX_C)$
$= [j\,R^2X_L - (-1)RX_LX_C + -j\,R^2X_C - (-1)RX_LX_C] / [R^2 - j\,RX_C + j\,RX_L - (-1)X_LX_C]$
$= (j\,R^2X_L + RX_LX_C - j\,R^2X_C + RX_LX_C) / (R^2 - j\,RX_C + j\,RX_L + X_LX_C)$
$= [2RX_LX_C + j\,(R^2X_L - R^2X_C)] / [R^2 + X_LX_C + j\,(RX_L - RX_C)]$
$= [2RX_LX_C + j\,R^2\,(X_L - X_C)] / [R^2 + X_LX_C + j\,R\,(X_L - X_C)]$

$X_L = X_C$, so substitute X_L and X_C into the equation and simplify.

$Z_{TH} = [2RX_LX_L + j\,R^2\,(X_L - X_L)] / [R^2 + X_LX_L + j\,R\,(X_L - X_L)]$
$= [2RX_L^2 + j\,R^2\,(0)] / [R^2 + X_L^2 + j\,R\,(0)]$
$= 2RX_L^2 / (R^2 + X_L^2)$

Finally, substitute in the value of X_L and simplify.

$Z_{TH} = [2R(\sqrt{L/C}^{\,2})] / (R^2 + \sqrt{L/C}^{\,2})$
$= [2R(L / C)] / [R^2 + (L / C)]$
$= (2RL) / \{[R^2 + (L / C)]\,C\}$
$= (2RL) / [R^2C + (L / C)\,C]$
$= (2RL) / (R^2C + L)$

Z_{TH} is purely resistive (there is no imaginary component to Z_{TH}), so the bridge transfers maximum power to the load for $f = 1 / (2\pi\sqrt{LC})$ when $Z_L = Z_{TH} = (2RL) / (R^2C + L)$.

20.3 Just for Fun

Because circuit reactances and impedances can be treated mathematically in the same manner as circuit resistances, many dc theorems can be adapted for ac circuits. Two of these are the ac voltage and current divider theorems. The proofs for these theorems follow the same procedures as the proofs for the corresponding dc theorems. The following problems demonstrate the proofs for the ac voltage and current divider theorems. Similarly, proofs for ac versions of Millman's Theorem, delta-wye and wye-delta conversions, and other theorems can be derived using the proofs for dc resistive circuits as models.

1. Problem: For a series ac circuit with source voltage V_S and total impedance Z_T, prove that the voltage V_X across impedance Z_X is $V_X = V_S\,(Z_X / Z_T)$.

   Solution: From Ohm's Law for current, the total current I_T for the series circuit is

   $I_T = V_S / Z_T$

   From Ohm's Law for voltage, the voltage across Z_X, V_X, is

   $V_X = I_X Z_X$

   where I_X is the current through Z_X. Because the n impedances are connected in series, I_X, the current through Z_X, is equal to the total current, $I_T = V_S / Z_T$.

   $V_X = I_X Z_X$
   $= I_T Z_X$
   $= (V_S / Z_T)\,Z_X$
   $= [(V_S Z_X) / Z_T]$
   $= V_S\,(Z_X / Z_T)$

2. Problem: For a parallel ac circuit with source voltage V_S, source current I_S, and total impedance Z_T, prove that the current I_X through impedance Z_X is $I_X = I_S\,(Z_T / Z_X)$.

   Solution: From Ohm's Law for voltage, the source current, I_S, for the parallel circuit with source voltage V_S is

   $V_S = I_S Z_T$

From Ohm's Law for current, the current through $\boldsymbol{Z_X}$, $\boldsymbol{I_{ZX}}$, is

$$\boldsymbol{I_X} = \boldsymbol{V_{ZX}} / \boldsymbol{Z_X}$$

Because the impedances are connected in parallel, $\boldsymbol{V_X}$, the voltage across $\boldsymbol{Z_X}$, is equal to the source voltage, $\boldsymbol{V_S} = \boldsymbol{I_S}\,\boldsymbol{Z_T}$.

$$\begin{aligned}\boldsymbol{I_X} &= \boldsymbol{V_X} / \boldsymbol{Z_X} \\ &= \boldsymbol{V_S} / \boldsymbol{Z_X} \\ &= (\boldsymbol{I_S}\,\boldsymbol{Z_T}) / \boldsymbol{Z_X} \\ &= [(\boldsymbol{I_S}\,\boldsymbol{Z_T}) / \boldsymbol{Z_X}] \\ &= \boldsymbol{I_S}\,(\boldsymbol{Z_T} / \boldsymbol{Z_X})\end{aligned}$$

21. Time Response of Reactive Circuits

21.1 Basic

21.1.1 Integrators and Differentiators

1. Problem: Identify each of the following circuits shown in Figure 21-1 as an integrator or differentiator.

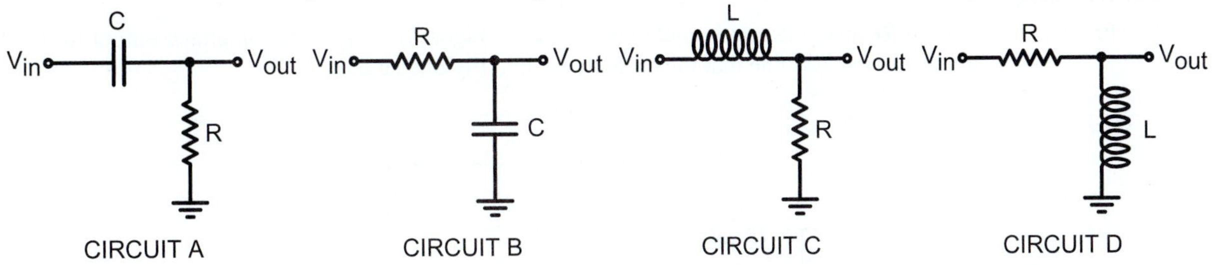

Figure 21-1: *RC* and *RL* Circuits for Problem 1

Solution: When V_{in} for Circuit A is a constant (dc) voltage, ***C*** will block it so that $V_{out} = V_R = 0$ V. Conversely, V_R must react instantly to a sudden change in V_{in}, because V_C cannot change instantly and Kirchhoff's Voltage Law requires that $V_R = -(V_{IN} + V_C)$. V_{out} responds to changes in V_{in} and not to the value of V_{in} itself, so Circuit A is a **differentiator**.

When V_{in} for Circuit B is a constant (dc) voltage, the capacitor ***C*** will charge (or discharge) through ***R*** towards the value of V_{in}, so that $V_{out} = V_C$ will increase (or decrease) proportionally. Conversely, V_C will not react instantly to a sudden change in V_{in}, so that V_{out} is not immediately affected by the change. V_{out} responds to the value of V_{in} and not to changes in V_{in}, so Circuit B is an **integrator**.

When V_{in} for Circuit C is a constant (dc) voltage, the current through ***L*** will increase (or decrease) so that voltage across ***R*** will increase (or decrease) towards the value of V_{in}. Conversely, the inductor will block any sudden change in V_{in}, so that V_{out} is not immediately affected by the change. V_{out} responds to the value of V_{in} and not to changes in V_{in}, so Circuit C is an **integrator**.

When V_{in} for Circuit D is a constant (dc) voltage, ***L*** appears to be a short so that $V_{out} = V_L = 0$ V. Conversely, V_L must react instantly to a sudden change in V_{in}, because current through ***L*** cannot change instantly and Kirchhoff's Voltage Law requires that $V_L = -(V_{IN} + V_R)$. V_{out} responds to changes in V_{in} and not to the value of V_{in} itself, so Circuit D is a **differentiator**.

2. Problem: Identify each V_{out} waveform shown in Figure 21-2 as the response of an integrator or differentiator to the indicated V_{in} waveform.

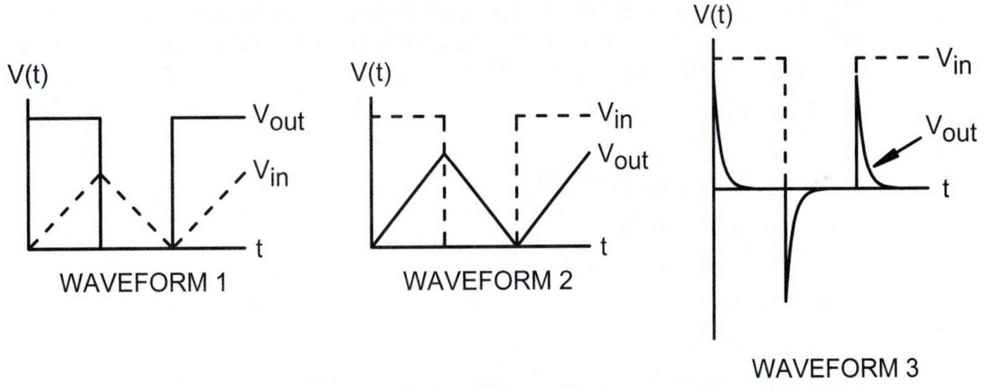

Figure 21-2: Waveforms for Problem 2

Solution: For Waveform 1, V_{out} primarily responds to changes in V_{in} rather than to the value of V_{in} itself. The waveform is the response of a **differentiator**.

For Waveform 2, V_{out} primarily responds to the value of V_{in} rather than to changes in V_{in}. The waveform is the response of an **integrator**.

For Waveform 3, V_{out} primarily responds to changes in V_{in} rather than to the value of V_{in} itself. The waveform is the response of an **differentiator**.

21.1.2 The *RC* Integrator

1. Problem: For the ***RC*** integrator and input pulse shown in Figure 21-3, plot the integrated output for $t = 0$ ms to $t = 100$ ms. Assume that the capacitor is initially uncharged.

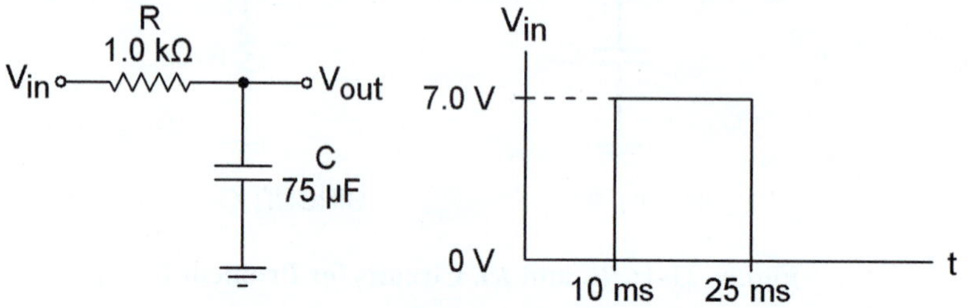

Figure 21-3: *RC* Integrator and Input Pulse for Problem 1

Solution: First, calculate the time constant, τ, for the ***RC*** circuit.

$$\tau = RC$$
$$= (1.0\ \text{k}\Omega)(75\ \mu\text{F})$$
$$= 75\ \text{ms}$$

When V_{in} changes from 0 V to 7.0 V at $t = 10$ ms, V_C, the voltage across C, cannot change instantaneously and begins to exponentially charge from 0 V towards 7.0 V. At time $t = 25$ ms, just before V_{in} changes from 7.0 V to 0 V, the capacitor has been charging for 25 ms – 10 ms = 15 ms.

$$V_{out} = V_C$$
$$= V_f[1 - e^{-(t/\tau)}]$$
$$= (7.0\ \text{V})[1 - e^{-(15\ \text{ms}/75\ \text{ms})}]$$
$$= (7.0\ \text{V})(1 - e^{-0.2})$$
$$= (7.0\ \text{V})(1 - 0.819)$$
$$= (7.0\ \text{V})(0.181)$$
$$= 1.27\ \text{V}$$

When V_{in} changes from 7.0 V to 0 V at $t = 25$ ms, V_C cannot change instantaneously and begins to exponentially discharge from 1.27 V towards 0 V. At $t = 100$ ms, the capacitor has been discharging for 100 ms – 25 ms = 75 ms = 1 τ.

$$V_{out} = V_C$$
$$= V_i e^{-(t/\tau)}]$$
$$= (1.27\ \text{V})[e^{-(75\ \text{ms}/75\ \text{ms})}]$$
$$= (1.27\ \text{V})(e^{-1})$$
$$= (1.27\ \text{V})(0.368)$$
$$= 0.467\ \text{V}$$
$$= 467\ \text{mV}$$

Figure 21-4 shows the integrated output from $t = 10$ ms to $t = 100$ ms.

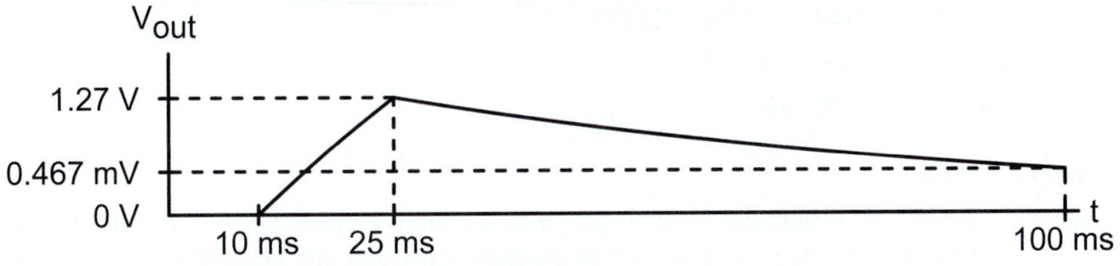

Figure 21-4: *RC* Integrator Output for Problem 1

2. Problem: For the pulse input and integrated output shown in Figure 21-5, calculate the value of ***C*** for the ***RC*** integrator.

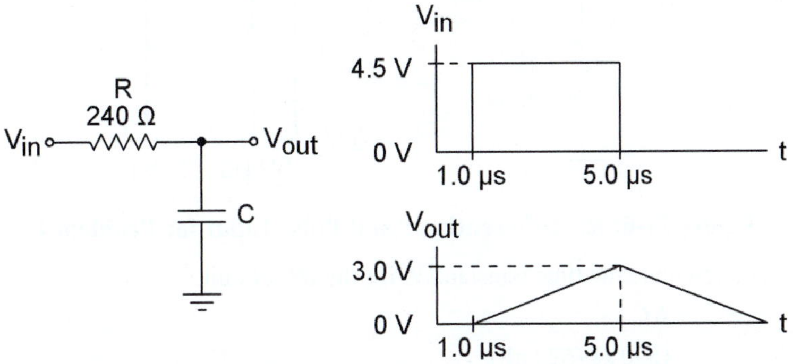

Figure 21-5: *RC* Integrator and Voltages for Problem 2

Solution: To solve for C, first solve for the time constant, τ, from the exponential charging equation.

$$V(t) = V_f[1 - e^{-(t/\tau)}]$$
$$V(t) / V_f = \{V_f[1 - e^{-(t/\tau)}]\} / V_f$$
$$[V(t) / V_f] - 1 = [1 - e^{-(t/\tau)}] - 1$$
$$-[(V(t) / V_f) - 1] = -[-e^{-(t/\tau)}]$$
$$1 - [V(t) / V_f] = e^{-(t/\tau)}$$
$$\ln[1 - (V(t) / V_f)] = \ln[e^{-(t/\tau)}]$$
$$\ln[1 - (V(t) / V_f)] = -(t/\tau)$$
$$-\ln[1 - (V(t) / V_f)] = -[-(t/\tau)]$$
$$\{-\ln[1 - (V(t) / V_f)]\} / t = (t/\tau) / t$$
$$1 / (\{-\ln[1 - (V(t) / V_f)]\} / t) = 1 / (1/\tau)$$
$$t / \{-\ln[1 - (V(t) / V_f)]\} = \tau$$

The time constant, τ, for an ***RC*** circuit is $\tau = RC$.

$$\tau = t / \{-\ln[1 - (V(t) / V_f)]\}$$
$$RC = t / \{-\ln[1 - (V(t) / V_f)]\}$$
$$(RC) / R = (t / \{-\ln[1 - (V(t) / V_f)]\}) / R$$
$$C = t / \{-R \ln[1 - (V(t) / V_f)]\}$$

From the waveforms for V_{in} and V_{out}, the capacitor attempts to charge to 4.5 V and reaches 3.0 V after charging for 4.0 μs. Therefore, $V_f = 4.5$ V, $V(t) = 3.0$ V, and $t = 4.0$ μs.

$$C = t / \{-R \ln[1 - (V(t) / V_f)]\}$$
$$= (4.0\ \mu s) / ((-240\ \Omega) \ln\{1 - [(3.0\ V) / (4.5\ V)]\})$$
$$= (4.0 \times 10^{-6}\ s) / [(-240\ \Omega) \ln(1 - 0.667)]$$
$$= (4.0 \times 10^{-6}\ s) / [(-240\ \Omega) \ln(0.333)]$$
$$= (4.0 \times 10^{-6}\ s) / [(-240\ \Omega)(-1.099)]$$

$= (4.0 \times 10^{-6}\text{ s}) / (264\ \Omega)$
$= 15.2 \times 10^{-6}\text{ F}$
$= \mathbf{15.2\ \mu F}$

21.1.3 The *RC* Differentiator

1. Problem: For the ***RC*** differentiator and input pulse shown in Figure 21-6, plot the differentiated output from $t = 0\ \mu\text{s}$ to $t = 425\ \mu\text{s}$. Assume that the capacitor is initially uncharged.

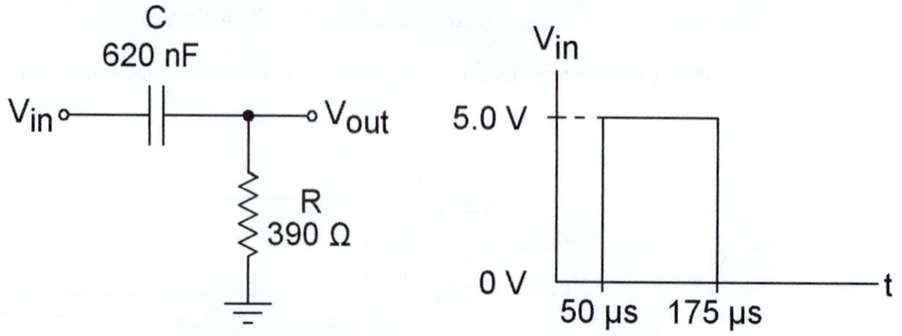

Figure 21-6: *RC* Differentiator and Pulse Input for Problem 1

Solution: First, calculate the time constant, τ, for the ***RC*** circuit.

$\tau = \mathbf{RC}$
$= (390\ \Omega)(620\text{ nF})$
$= 242\ \mu\text{s}$

From Kirchhoff's Voltage Law, the input voltage, V_{in}, must equal the voltage drops, V_C and V_R, across ***C*** and ***R***.

$V_{in} - V_C - V_R = 0\text{ V}$
$V_{in} - V_C - V_{out} = 0\text{ V}$
$5.0\text{ V} - V_C - V_{out} + V_{out} = 0\text{ V} + V_{out}$
$5.0\text{ V} - V_C = V_{out}$

When V_{in} changes from 0 V to 5.0 V at $t = 50\ \mu\text{s}$, V_C, the voltage across ***C***, cannot change instantaneously, so that $V_C = 0$ V and $V_{out} = V_{in} = 5.0$ V. From $t = 50\ \mu\text{s}$ to $t = 175\ \mu\text{s}$, V_C increases exponentially from 0 V towards 5.0 V, and V_{out} must exponentially decrease by the same amount to satisfy Kirchhoff's Voltage Law. At time $t = 175\ \mu\text{s}$, just before V_{in} changes from 5.0 V to 0 V, the capacitor has been charging for 175 μs – 50 μs = 125 μs.

$V_{out} = V_{in} - V_C$
$= 5.0\text{ V} - V_f[1 - e^{-(t/\tau)}]$
$= 5.0\text{ V} - (5.0\text{ V})[1 - e^{-(125\ \mu\text{s} / 242\ \mu\text{s})}]$
$= 5.0\text{ V} - (5.0\text{ V})(1 - e^{-0.517})$
$= 5.0\text{ V} - (5.0\text{ V})(1 - 0.596)$
$= 5.0\text{ V} - (5.0\text{ V})(0.404)$
$= 5.0\text{ V} - 2.02\text{ V}$
$= 2.98\text{ V}$

At $t = 175\ \mu\text{s}$, when V_{in} changes from 5.0 V to 0 V, V_C cannot change instantaneously. Therefore, V_{out} must change instantaneously to satisfy Kirchhoff's Voltage Law.

$V_{out} = V_{in} - V_C$
$= 0\text{ V} - V_C$
$= -V_C$

At $t = 175\ \mu s$, $V_C = 2.02$ V, so $V_R = -V_C = -2.02$ V. In effect, when V_{in} drops 5.0 V (from 5.0 V to 0 V), V_R also drops by 5.0 V (from 2.98 V to –2.02 V).

From $t = 175\ \mu s$ to $t = 425\ \mu s$, V_C exponentially discharges from 2.02 V towards 0 V so that V_{out} exponentially increases by the same amount to satisfy Kirchhoff's Voltage Law. At $t = 425\ \mu s$, the capacitor has been discharging for 425 μs – 175 μs = 250 μs.

$$\begin{aligned} V_{out} &= -V_C \\ &= -V_i\, e^{-(t/\tau)}] \\ &= -(2.02\text{ V})[e^{-(250\ \mu s / 242\ \mu s)}] \\ &= -(2.02\text{ V})(e^{-1.03}) \\ &= -(2.02\text{ V})(0.356) \\ &= -0.718\text{ V} \\ &= -718\text{ mV} \end{aligned}$$

Figure 21-7 shows the differentiated output from $t = 0\ \mu s$ to $t = 425\ \mu s$.

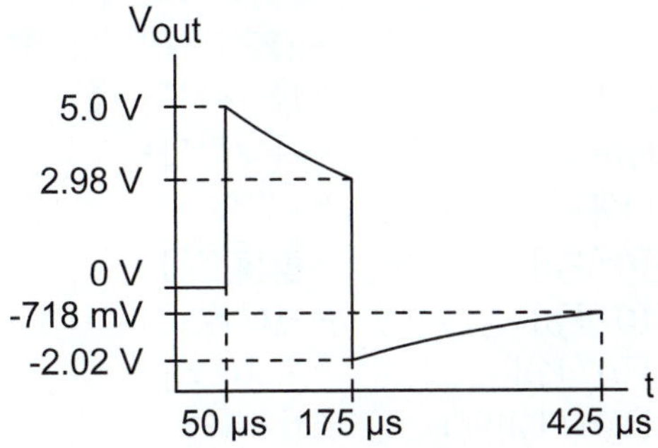

Figure 21-7: *RC* Differentiator Output for Problem 1

2. Problem: For the pulse input and differentiated output shown in Figure 21-8, calculate the required value of ***R*** for the ***RC*** differentiator.

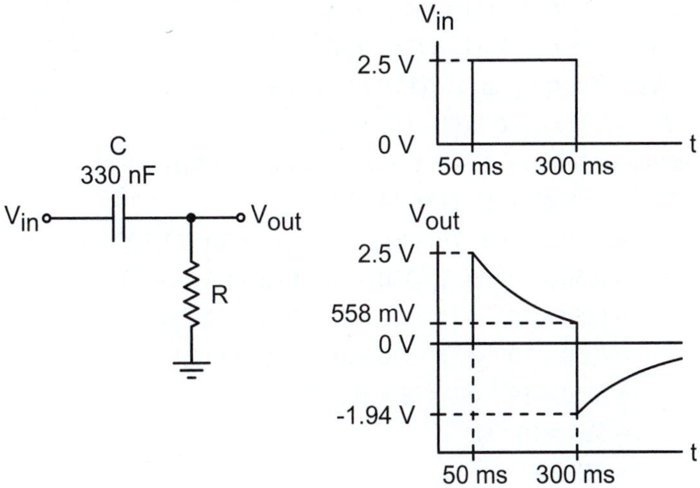

Figure 21-8: *RC* Differentiator and Voltages for Problem 2

Solution: At time $t = 50$ ms, the capacitor begins charging to $V_{in} = 2.5$ V. From Kirchhoff's Voltage Law, the input voltage, V_{in}, must equal the voltage drops, V_C and V_R, across ***C*** and ***R***.

$$\begin{aligned} V_{in} - V_C - V_R &= 0\text{ V} \\ V_{in} - V_C - V_{out} &= 0\text{ V} \\ V_{in} - V_C - V_{out} + V_C &= 0\text{ V} + V_C \end{aligned}$$

$V_{in} - V_{out} \quad = V_C$

From the waveforms for V_{in} and V_{out}, $V_{in} = 2.5$ V and V_{out} is 2.5 V at $t = 50$ ms.

$$V_C = V_{in} - V_{out}$$
$$= 2.5\text{ V} - 2.5\text{ V}$$
$$= 0\text{ V}$$

Also from the waveforms for V_{in} and V_{out}, $V_{in} = 2.5$ V and V_{out} is 588 mV at $t = 300$ ms.

$$V_C = V_{in} - V_{out}$$
$$= 2.5\text{ V} - 588\text{ mV}$$
$$= 2.5\text{ V} - 0.588\text{ V}$$
$$= 1.91\text{ V}$$

The capacitor therefore exponentially charges from 0 V to 1.91 V in 300 ms – 50 ms = 250 ms. To solve for R, first solve for the time constant, τ, from the exponential charging equation.

$$V(t) = V_f[1 - e^{-(t/\tau)}]$$
$$V(t) / V_f = \{V_f[1 - e^{-(t/\tau)}]\} / V_f$$
$$[V(t) / V_f] - 1 = [1 - e^{-(t/\tau)}] - 1$$
$$-[(V(t) / V_f) - 1] = -[-e^{-(t/\tau)}]$$
$$1 - [V(t) / V_f] = e^{-(t/\tau)}$$
$$\ln[1 - (V(t) / V_f)] = \ln[e^{-(t/\tau)}]$$
$$\ln[1 - (V(t) / V_f)] = -(t/\tau)$$
$$-\ln[1 - (V(t) / V_f)] = -[-(t/\tau)]$$
$$\{-\ln[1 - (V(t) / V_f)]\} / t = (t/\tau) / t$$
$$1 / (\{-\ln[1 - (V(t) / V_f)]\} / t) = 1 / (1/\tau)$$
$$t / \{-\ln[1 - (V(t) / V_f)]\} = \tau$$

The time constant, τ, for an RC circuit is $\tau = RC$.

$$\tau = t / \{-\ln[1 - (V(t) / V_f)]\}$$
$$RC = t / \{-\ln[1 - (V(t) / V_f)]\}$$
$$(RC) / C = (t / \{-\ln[1 - (V(t) / V_f)]\}) / C$$
$$R = t / \{-C\ln[1 - (V(t) / V_f)]\}$$

From the waveforms, $V_f = 2.5$ V, $t = 250$ ms, and $V(t) = 1.91$ V.

$$R = t / \{-C\ln[1 - (V(t) / V_f)]\}$$
$$= (250\text{ ms}) / ((-330\text{ nF})\ln\{1 - [(1.91\text{ V}) / (2.5\text{ V})]\})$$
$$= (250 \times 10^{-3}\text{ s}) / [(-330 \times 10^{-9}\text{ F})\ln(1 - 0.765)]$$
$$= (250 \times 10^{-3}\text{ s}) / [(-330 \times 10^{-9}\text{ F})\ln(0.235)]$$
$$= (250 \times 10^{-3}\text{ s}) / [(-330 \times 10^{-9}\text{ F})(-1.447)]$$
$$= (250 \times 10^{-3}\text{ s}) / (478 \times 10^{-9}\text{ F})$$
$$= 523 \times 10^{-3}\ \Omega$$
$$= \mathbf{523\ k\Omega}$$

21.1.4 The *RL* Integrator

1. Problem: For the RL integrator and pulse input shown in Figure 21-9, plot the integrated output for $t = 0$ ms to $t = 10$ ms. Assume that the initial current through L is 0 A.

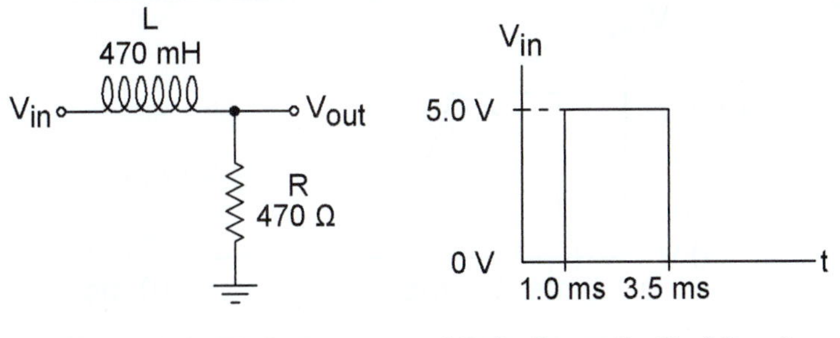

Figure 21-9: *RL* Integrator and Pulse Input for Problem 1

Solution: First, calculate the time constant, τ, for the ***RL*** circuit.

$$\tau = L / R$$
$$= (470 \text{ mH}) / (470\ \Omega)$$
$$= 1.0 \text{ ms}$$

Initially the series current $I_T = I_L = I_R = 0$ A so that $V_{out} = V_R = I_R \times R = 0$ V. When V_{in} changes from 0 V to 5.0 V at $t = 1.0$ ms, I_L, the current through L, cannot change instantaneously and begins to exponentially increase from 0 A towards its final value. Because the inductor and resistor currents are the same, the output voltage, $V_{out} = I_R \times R = I_L \times R$, will also exponentially increase towards its final value. The final values of inductor current and output voltage occur when the inductor is fully energized and $V_L = 0$ V. From Kirchhoff's Voltage Law,

$$V_{in} - V_L - V_R = 0 \text{ V}$$
$$V_{in} - 0 \text{ V} - V_{out} = 0 \text{ V}$$
$$V_{in} - V_{out} + V_{out} = 0 \text{ V} + V_{out}$$
$$V_{in} = V_{out}$$

At time $t = 3.5$ ms, just before V_{in} changes from 5.0 V to 0 V, the inductor has been energizing from 0 V towards 5.0 V for 3.5 ms – 1.0 ms = 2.5 ms. From the exponential charging equation,

$$V_{out} = V_R = V_f[1 - e^{-(t/\tau)}]$$
$$= V_{in}[1 - e^{-(2.5 \text{ ms} / 1.0 \text{ ms})}]$$
$$= (5.0 \text{ V})(1 - e^{-2.5})$$
$$= (5.0 \text{ V})(1 - 0.0821)$$
$$= (5.0 \text{ V})(0.918)$$
$$= 4.59 \text{ V}$$

When V_{in} changes from 5.0 V to 0 V at t = 3.5 ms, I_L cannot change instantaneously, so I_L begins to decrease exponentially from V_{out} towards 0 A as the inductor de-energizes. At $t = 10$ ms, the inductor has been de-energizing from 4.59 V towards 0 V for 10 ms – 3.5 ms = 6.5 ms.

$$V_{out} = V_R = V_i e^{-(t/\tau)}]$$
$$= (4.59 \text{ V})[e^{-(6.5 \text{ ms} / 1.0 \text{ ms})}]$$
$$= (4.59 \text{ V})(e^{-6.5})$$
$$= (4.59 \text{ V})(1.50 \times 10^{-3})$$
$$= 6.90 \times 10^{-3} \text{ V}$$
$$= 6.90 \text{ mV}$$

Figure 21-10 shows the integrated output from $t = 1.0$ ms to $t = 10$ ms.

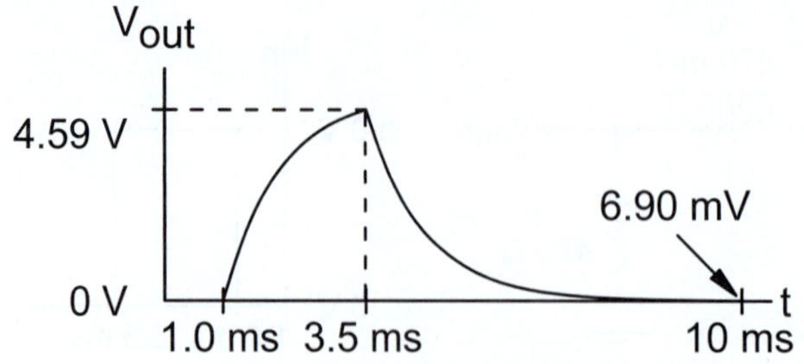

Figure 21-10: *RL* Integrator Output for Problem 1

2. Problem: For the pulse input and integrated output shown in Figure 21-11, calculate the required value of ***R*** for the ***RL*** integrator.

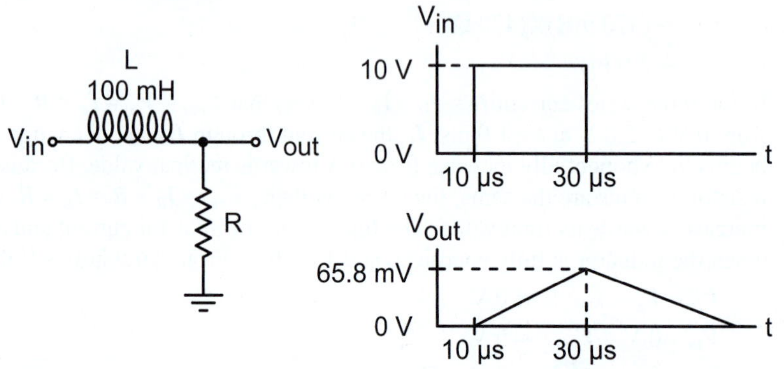

Figure 21-11: *RL* Integrator and Voltage Waveforms for Problem 2

Solution: At time $t = 10\ \mu s$, the output voltage begins to increase to $V_{in} = 2.5$ V. From Kirchhoff's Voltage Law, the input voltage, V_{in}, must equal the voltage drops, V_L and V_R, across ***L*** and ***R***.

$$V_{in} - V_L - V_R = 0\ V$$

$$V_{in} - V_L - V_{out} = 0\ V$$

$$V_{in} - V_L - V_{out} + V_L = 0\ V + V_L$$

$$V_{in} - V_{out} = V_L$$

From the waveforms for V_{in} and V_{out}, $V_{in} = 10$ V and $V_{out} = 0$ V at $t = 10\ \mu s$.

$$V_L = V_{in} - V_{out}$$

$$= 10\ V - 0\ V$$

$$= 10\ V$$

From the waveforms for V_{in} and V_{out}, $V_{in} = 10$ V and $V_{out} = 65.8$ mV at $t = 30\ \mu s$.

$$V_L = V_{in} - V_{out}$$

$$= 10\ V - 65.8\ mV$$

$$= 10\ V - 0.0658\ V$$

$$= 9.9342\ V$$

The inductor therefore exponentially de-energizes from 0 V to 9.9342 V in 20 µs. To solve for ***R***, first solve for the time constant, τ, from the exponential discharging equation.

$$V(t) = V_i\, e^{-(t/\tau)}$$

$$V(t) / V_i = [V_i\, e^{-(t/\tau)}] / V_i$$

$$\ln [V(t) / V_i] = \ln [e^{-(t/\tau)}]$$

$$\ln [V(t) / V_i] = -(t / \tau)$$

$$
\begin{aligned}
-\ln [V(t) / V_i] &= -[-(t / \tau)] \\
\{-\ln [V(t) / V_i]\} / t &= (t / \tau) / t \\
1 / (\{-\ln [V(t) / V_i]\} / t) &= 1 / (1 / \tau) \\
t / \{-\ln [V(t) / V_i]\} &= \tau
\end{aligned}
$$

The time constant, τ, for an ***RL*** circuit is $\tau = L / R$.

$$
\begin{aligned}
\tau &= t / \{-\ln [V(t) / V_i]\} \\
L / R &= t / \{-\ln [V(t) / V_i]\} \\
(L / R) / L &= (t / \{-\ln [V(t) / V_i]\}) / L \\
1 / (1 / R) &= 1 / (t / \{-L \ln [V(t) / V_i]\}) \\
R &= \{-L \ln [V(t) / V_i]\} / t \\
&= (-L / t) \ln [V(t) / V_i]
\end{aligned}
$$

From the waveforms, $V_i = 10$ V, $t = 20$ μs, and $V(t) = 65.8$ mV

$$
\begin{aligned}
R &= (-L / t) \ln [V(t) / V_f] \\
&= (-100 \text{ mH} / 20 \text{ μs}) \ln [65.8 \text{ mV} / 10 \text{ V})] \\
&= (-5.0 \times 10^3 \ \Omega) \ln (65.8 \times 10^{-3} \text{ V} / 10 \text{ V}) \\
&= (-5.0 \times 10^3 \ \Omega) \ln (6.58 \times 10^{-3}) \\
&= (-5.0 \times 10^3 \ \Omega)(-5.02) \\
&= 25.1 \times 10^3 \ \Omega \\
&= \mathbf{25.1 \ k\Omega}
\end{aligned}
$$

21.1.5 The *RL* Differentiator

1. Problem: For the ***RL*** differentiator and input pulse shown in Figure 21-12, plot the differentiated output for $t = 0$ μs to $t = 120$ μs. Assume that the initial current through ***L*** is 0 A.

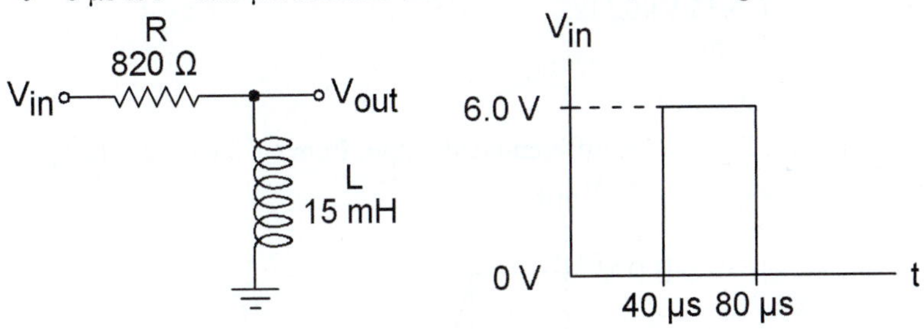

Figure 21-12: *RL* Differentiator and Input Pulse for Problem 1

Solution: First, calculate the time constant, τ, for the ***RL*** circuit.

$$
\begin{aligned}
\tau &= L / R \\
&= (15 \text{ mH})(820 \ \Omega) \\
&= 18.3 \text{ μs}
\end{aligned}
$$

From Kirchhoff's Voltage Law, the input voltage V_{in} must equal the voltage drops V_L and V_R across ***L*** and ***R***.

$$
\begin{aligned}
V_{in} - V_L - V_R &= 0 \text{ V} \\
V_{in} - V_{out} - V_R &= 0 \text{ V} \\
6.0 \text{ V} - V_{out} - V_R + V_{out} &= 0 \text{ V} + V_{out} \\
6.0 \text{ V} - V_R &= V_{out}
\end{aligned}
$$

Initially the series current $I_T = I_L = I_R = 0$ A, so that $V_R = I_R \times R = 0$ V. When V_{in} changes from 0 V to 6.0 V at t = 40 μs, the I_L cannot change instantaneously, so that $V_R = 0$ V.

$$V_{out} = 6.0 \text{ V} - V_R$$

$= 6.0\ V - 0\ V$

$= 6.0\ V$

From $t = 40\ \mu s$ to $t = 80\ \mu s$, V_L decreases exponentially from 6.0 V towards 0 V as the inductor energizes. At time $t = 80\ \mu s$, just before the V_{in} changes from 6.0 V to 0 V, the inductor has been energizing and V_L has been exponentially decreasing for $80\ \mu s - 40\ \mu s = 40\ \mu s$.

$$V_{out} = V_i\,[e^{-(t/\tau)}]$$
$$= (6.0\ V)[e^{-(40\ \mu s / 18.3\ \mu s)}]$$
$$= (6.0\ V)(e^{-2.19})$$
$$= (6.0\ V)(0.112)$$
$$= 0.674\ V$$
$$= 674\ mV$$

When V_{in} changes from 6.0 V to 0 V, $I_L = I_R$ cannot change instantaneously, so that V_R is also unchanged. Therefore, V_{out} must change instantaneously to satisfy Kirchhoff's Voltage Law.

$$V_{out} = V_{in} - V_R$$
$$= 0\ V - V_R$$
$$= -V_R$$

At $t = 80\ \mu s$, $V_R = V_{in} - V_{out} = 6.0\ V - 0.674\ V = 5.33\ V$, so $V_{out} = -5.33\ V$. In effect, when V_{in} drops 6.0 V (from 6.0 V to 0 V), V_L also drops by 6.0 V (from 0.674 V to –5.33 V).

From $t = 80\ \mu s$ to $t = 120\ \mu s$, V_L exponentially discharges from –5.33 V towards 0 V. At $t = 120\ \mu s$, the capacitor has been discharging for $120\ \mu s - 80\ \mu s = 40\ \mu s$.

$$V_{out} = V_i\,e^{-(t/\tau)}]$$
$$= (-5.33\ V)[e^{-(40\ \mu s / 18.3\ \mu s)}]$$
$$= (-5.33\ V)(e^{-2.19})$$
$$= (-5.33\ V)(0.112)$$
$$= -0.598\ V$$
$$= -598\ mV$$

Figure 21-13 shows the differentiated output from $t = 0\ \mu s$ to $t = 120\ \mu s$.

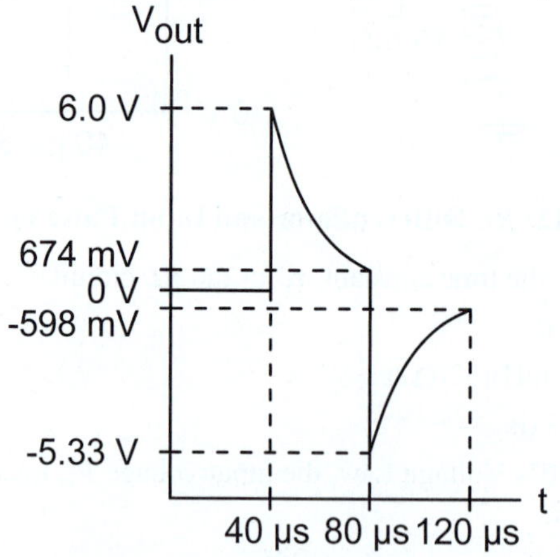

Figure 21-13: *RL* Differentiator Output for Problem 1

2. Problem: For the input pulse waveform and differentiated output shown in Figure 21-14, calculate the required value of L for the RL differentiator.

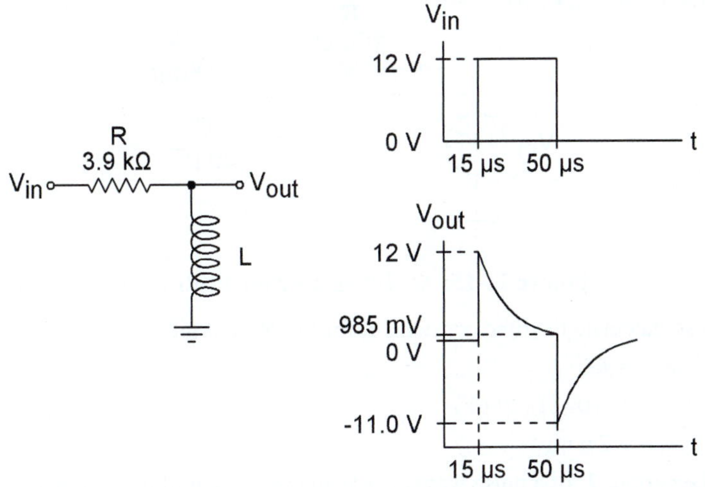

Figure 21-14: *RL* Differentiator and Voltage Waveforms for Problem 2

Solution: The inductor voltage exponentially decreases from 12 V to 985 mV in 35 µs as the inductor energizes. To solve for L, first solve for the time constant, τ, from the exponential discharging equation.

$$V(t) = V_i e^{-(t/\tau)}$$
$$V(t) / V_i = [V_i e^{-(t/\tau)}] / V_i$$
$$\ln [V(t) / V_i] = \ln [e^{-(t/\tau)}]$$
$$\ln [V(t) / V_i] = -(t / \tau)$$
$$-\ln [V(t) / V_i] = -[-(t / \tau)]$$
$$\{-\ln [V(t) / V_i]\} / t = (t / \tau) / t$$
$$1 / (\{-\ln [V(t) / V_i]\} / t) = 1 / (1 / \tau)$$
$$t / \{-\ln [V(t) / V_i]\} = \tau$$

The time constant, τ, for an RL circuit is $\tau = L / R$.

$$\tau = t / \{-\ln [V(t) / V_i]\}$$
$$L / R = t / \{-\ln [V(t) / V_i]\}$$
$$(L / R) \times R = (t / \{-\ln [V(t) / V_i]\}) \times R$$
$$L = (R\, t) / \{-\ln [V(t) / V_i]\}$$

From the waveforms, $V_i = 12$ V, $t = 35$ µs, and $V(t) = 985$ mV

$$L = (R\, t) / \{-\ln [V(t) / V_i]\}$$
$$= [(3.9\text{ k}\Omega)(35\text{ µs})] / -\ln [985\text{ mV} / 12\text{ V})]$$
$$= [(3.9 \times 10^{3}\ \Omega)(35 \times 10^{-6}\text{ s})] [-\ln (985 \times 10^{-3}\text{ V} / 12\text{ V})]$$
$$= (137 \times 10^{-3}\text{ H}) [-\ln (82.1 \times 10^{-3})]$$
$$= (137\text{ mH})[-(-2.50)]$$
$$= (137\text{ mH})(2.50)$$
$$= \mathbf{341\text{ mH}}$$

21.2 Advanced

21.2.1 Periodic Repetitive Pulses

1. Problem: For the ***RC*** integrator in Figure 21-15, calculate the output voltage, V_{out}, at the beginning and end of the first five pulses if V_{in} is a 0 V to 5 V 1.0 kHz periodic pulse waveform with a 50% duty cycle. Assume that C is initially uncharged and $V_{out} = 0$ V.

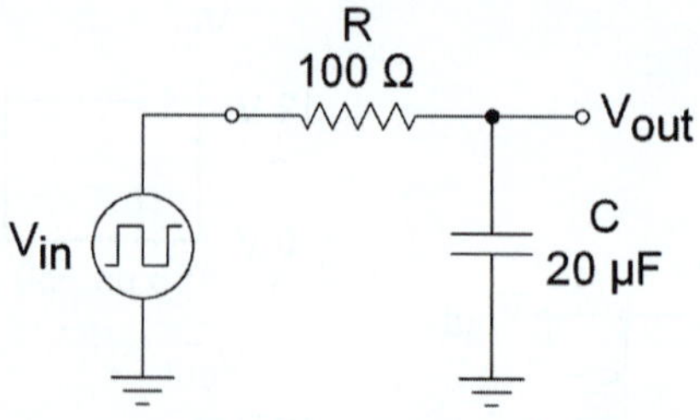

Figure 21-15: *RC* Integrator for Problem 1

Solution: First, calculate the time constant, τ, for the ***RC*** circuit.

$$\tau = RC$$
$$= (100\ \Omega)(20\ \mu\text{F})$$
$$= 2.0\ \text{ms}$$

The period, ***T***, of a pulse waveform is the reciprocal of the frequency.

$$T = 1 / f$$
$$= 1 / (1\ \text{kHz})$$
$$= 1.0\ \text{ms}$$

By definition, the duty cycle, DC, of a pulse waveform is the high time, t_H, divided by the period.

$$t_H / T = \text{DC}$$
$$(t_H / T) \times T = \text{DC} \times T$$
$$t_H = (0.5)(1.0\ \text{ms})$$
$$= 500\ \mu\text{s}$$

The low time, t_L, is the difference between ***T*** and t_H.

$$t_L = T - t_H$$
$$= (1.0\ \text{ms}) - (500\ \mu\text{s})$$
$$= 500\ \mu\text{s}$$

Therefore, the input voltage, V_{in}, alternates between 0 V for 500 µs and 5 V for 500 µs.

Before the first pulse, $V_{in} = 0$ V so that, at the start of Pulse 1, ***C*** remains uncharged and $V_{out}(0) = 0$ V. When V_{in} changes from 0 V to 5 V for 500 µs, the integrator output V_{out} rejects (does not respond to) the change in V_{in} and charges towards 5 V for 500 µs. From the general exponential equation, the output voltage at the end of Pulse 1 is

$$V_{out}(1) = V_f + (V_i - V_f)\, e^{-t/\tau}$$

where $V_f = V_{in} = 5$ V, $V_i = V_{out}(0) = 0$ V, $t = t_H = 500$ µs, and $\tau = 2.0$ ms.

$$V_{out}(1) = 5\ \text{V} + [0\ \text{V} - 5\ \text{V}]\, e^{-(500\ \mu\text{s} / 2\ \text{ms})}$$
$$= 5\ \text{V} + (-5)(e^{-0.25})$$
$$= 5\ \text{V} + (-5)(0.779)$$
$$= 5\ \text{V} + (-3.89)$$
$$= \mathbf{1.11\ V}$$

At the end of Pulse 1, $V_{out}(1) = 1.11$ V. When V_{in} changes from 5 V to 0 V for 500 µs, the integrator output V_{out} rejects the change in V_{in} and discharges towards 0 V for 500 µs. From the general exponential equation, the output voltage at the beginning of Pulse 2 is

$$V_{out}(2) = V_f + (V_i - V_f)\, e^{-t/\tau}$$

where $V_f = V_{in} = 0$ V, $V_i = V_{out}(1) = 1.11$ V, $t = t_L = 500$ µs, and $\tau = 2.0$ ms.

$$V_{out}(2) = 0\ \text{V} + [1.11\ \text{V} - 0\ \text{V}]\, e^{-(500\ \mu\text{s} / 2\ \text{ms})}$$

$= 0\text{ V} + (1.11\text{ V})(e^{-0.25})$
$= (1.11\text{ V})(0.779)$
$= \mathbf{0.861\text{ V}}$

At the start of Pulse 2, $V_{out}(2) = 0.861$ V. When V_{in} changes from 0 V to 5 V for 500 μs, the integrator output V_{out} rejects the change in V_{in} and charges towards 5 V for 500 μs. From the general exponential equation, the output voltage at the end of Pulse 2 is

$$V_{out}(3) = V_f + (V_i - V_f)\, e^{-t/\tau}$$

where $V_f = V_{in} = 5$ V, $V_i = V_{out}(2) = 0.861$ V, $t = t_H = 500$ μs, and $\tau = 2.0$ ms.

$V_{out}(3) = 5\text{ V} + [0.861\text{ V} - 5\text{ V}]\, e^{-(500\ \mu\text{s}/2\text{ ms})}$
$= 5\text{ V} + (-4.14)(e^{-0.25})$
$= 5\text{ V} + (-4.14)(0.779)$
$= 5\text{ V} + (-3.22)$
$= \mathbf{1.78\text{ V}}$

At the end of Pulse 2, $V_{out}(3) = 1.78$ V. When V_{in} changes from 5 V to 0 V for 500 μs, the integrator output V_{out} rejects the change in V_{in} and discharges towards 0 V for 500 μs. From the general exponential equation, the output voltage at the beginning of Pulse 3 is

$$V_{out}(4) = V_f + (V_i - V_f)\, e^{-t/\tau}$$

where $V_f = V_{in} = 0$ V, $V_i = V_{out}(3) = 1.78$ V, $t = t_L = 500$ μs, and $\tau = 2.0$ ms.

$V_{out}(4) = 0\text{ V} + [1.78\text{ V} - 0\text{ V}]\, e^{-(500\ \mu\text{s}/2\text{ ms})}$
$= 0\text{ V} + (1.78\text{ V})(e^{-0.25})$
$= (1.78\text{ V})(0.779)$
$= \mathbf{1.38\text{ V}}$

At the start of Pulse 3, $V_{out}(4) = 1.38$ V. When V_{in} changes from 0 V to 5 V for 500 μs, the integrator output V_{out} rejects the change in V_{in} and charges towards 5 V for 500 μs. From the general exponential equation, the output voltage at the end of Pulse 3 is

$$V_{out}(5) = V_f + (V_i - V_f)\, e^{-t/\tau}$$

where $V_f = V_{in} = 5$ V, $V_i = V_{out}(4) = 1.38$ V, t $= t_H = 500$ μs, and $\tau = 2.0$ ms.

$V_{out}(5) = 5\text{ V} + [1.38\text{ V} - 5\text{ V}]\, e^{-(500\ \mu\text{s}/2\text{ ms})}$
$= 5\text{ V} + (-3.62)(e^{-0.25})$
$= 5\text{ V} + (-3.62)(0.779)$
$= 5\text{ V} + (-2.82)$
$= \mathbf{2.18\text{ V}}$

At the end of Pulse 3, $V_{out}(5) = 2.18$ V. When V_{in} changes from 5 V to 0 V for 500 μs, the integrator output V_{out} rejects the change in V_{in} and discharges towards 0 V for 500 μs. From the general exponential equation, the output voltage at the beginning of Pulse 4 is

$$V_{out}(6) = V_f + (V_i - V_f)\, e^{-t/\tau}$$

where $V_f = V_{in} = 0$ V, $V_i = V_{out}(5) = 2.18$ V, $t = t_L = 500$ μs, and $\tau = 2.0$ ms.

$V_{out}(6) = 0\text{ V} + [2.18\text{ V} - 0\text{ V}]\, e^{-(500\ \mu\text{s}/2\text{ ms})}$
$= 0\text{ V} + (2.18\text{ V})(e^{-0.25})$
$= (2.18\text{ V})(0.779)$
$= \mathbf{1.70\text{ V}}$

At the start of Pulse 4, $V_{out}(6) = 1.70$ V. When V_{in} changes from 0 V to 5 V for 500 μs, the integrator output V_{out} rejects the change in V_{in} and charges towards 5 V for 500 μs. From the general exponential equation, the output voltage at the end of Pulse 4 is

$$V_{out}(7) = V_f + (V_i - V_f)\, e^{-t/\tau}$$

where $V_f = V_{in} = 5$ V, $V_i = V_{out}(6) = 1.70$ V, $t = t_H = 500$ μs, and $\tau = 2.0$ ms.

$$V_{out}(7) = 5\text{ V} + [1.70\text{ V} - 5\text{ V}]\, e^{-(500\ \mu\text{s} / 2\text{ ms})}$$
$$= 5\text{ V} + (-3.30)(e^{-0.25})$$
$$= 5\text{ V} + (-3.30)(0.779)$$
$$= 5\text{ V} + (-2.57)$$
$$= \mathbf{2.43\text{ V}}$$

At the end of Pulse 4, $V_{out}(7) = 2.43$ V. When V_{in} changes from 5 V to 0 V for 500 μs, the integrator output V_{out} rejects the change in V_{in} and discharges towards 0 V for 500 μs. From the general exponential equation, the output voltage at the beginning of Pulse 5 is

$$V_{out}(8) = V_f + (V_i - V_f)\, e^{-t/\tau}$$

where $V_f = V_{in} = 0$ V, $V_i = V_{out}(7) = 2.43$ V, $t = t_L = 500$ μs, and $\tau = 2.0$ ms.

$$V_{out}(8) = 0\text{ V} + [2.43\text{ V} - 0\text{ V}]\, e^{-(500\ \mu\text{s} / 2\text{ ms})}$$
$$= 0\text{ V} + (2.43\text{ V})(e^{-0.25})$$
$$= (2.43\text{ V})(0.779)$$
$$= \mathbf{1.89\text{ V}}$$

At the start of Pulse 5, $V_{out}(8) = 1.89$ V. When V_{in} changes from 0 V to 5 V for 500 μs, the integrator output V_{out} rejects the change in V_{in} and charges towards 5 V for 500 μs. From the general exponential equation, the output voltage at the end of Pulse 5 is

$$V_{out}(9) = V_f + (V_i - V_f)\, e^{-t/\tau}$$

where $V_f = V_{in} = 5$ V, $V_i = V_{out}(8) = 1.89$ V, $t = t_H = 500$ μs, and $\tau = 2.0$ ms.

$$V_{out}(9) = 5\text{ V} + [1.89\text{ V} - 5\text{ V}]\, e^{-(500\ \mu\text{s} / 2\text{ ms})}$$
$$= 5\text{ V} + (-3.11)(e^{-0.25})$$
$$= 5\text{ V} + (-3.11)(0.779)$$
$$= 5\text{ V} + (-2.42)$$
$$= \mathbf{2.58\text{ V}}$$

Figure 21-16 shows the input and output waveforms for the first five pulses.

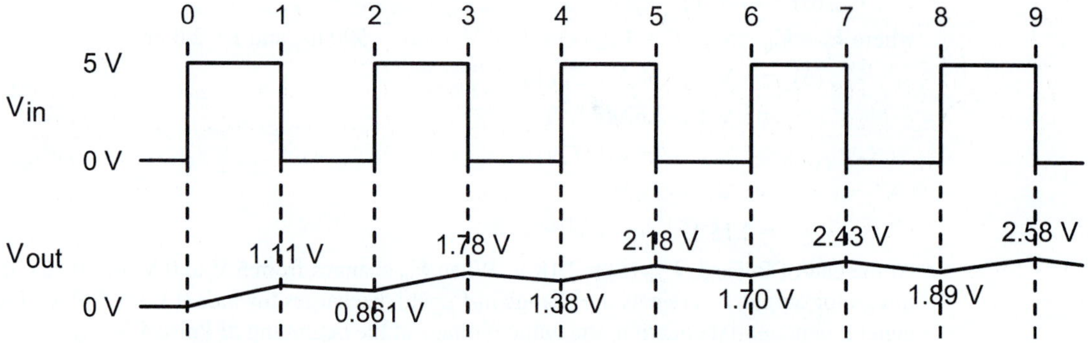

Figure 21-16: Input and Output Waveforms for *RC* Integrator

2. Problem: For the ***RL*** integrator in Figure 21-17, calculate the output voltage, V_{out}, at the beginning and end of the first five pulses if V_{in} is a 0 V to 10 V, 625 Hz periodic pulse waveform with a 40% duty cycle. Assume that L is initially de-energized and $V_{out} = 0$ V.

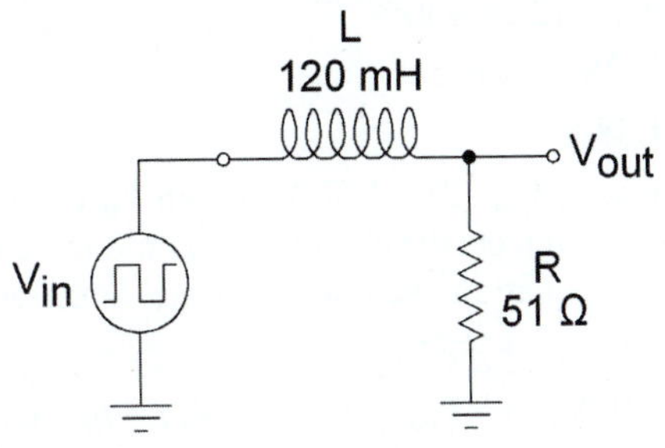

Figure 21-17: *RL* Integrator for Problem 2

Solution: First, calculate the time constant, τ, for the ***RL*** circuit.

$$\tau = L / R$$
$$= (120 \text{ mH})(51\ \Omega)$$
$$= 2.35 \text{ ms}$$

The period, T, of a pulse waveform is the reciprocal of the frequency.

$$T = 1 / f$$
$$= 1 / (625 \text{ Hz})$$
$$= 1.6 \text{ ms}$$

By definition, the duty cycle, DC, of a pulse waveform is the high time, t_H, divided by the period.

$$t_H / T = \text{DC}$$
$$(t_H / T) \times T = \text{DC} \times T$$
$$t_H = (0.4)(1.6 \text{ ms})$$
$$= 640\ \mu\text{s}$$

The low time, t_L, is the difference between T and t_H.

$$t_L = T - t_H$$
$$= (1.6 \text{ ms}) - (640)\ \mu\text{s}$$
$$= 960\ \mu\text{s}$$

Therefore, the input voltage, V_{in}, alternates between 0 V for 960 μs and 10 V for 640 μs.

Before the first pulse, V_{in} = 0 V and L is de-energized so that no current flows and V_{out} (0) = 0 V. When V_{in} changes from 0 V to 10 V for 640 μs, the integrator output V_{out} rejects (does not respond to) the change in V_{in} and increases towards 10 V for 640 μs. From the general exponential equation, the output voltage at the end of Pulse 1 is

$$V_{out}(1) = V_f + (V_i - V_f)\, e^{-t/\tau}$$

where $V_f = V_{in}$ = 10 V, $V_i = V_{out}$ (0) = 0 V, $t = t_H$ = 640 μs, and τ = 2.35 ms.

$$V_{out}(1) = 10 \text{ V} + [0 \text{ V} - 10 \text{ V}]\, e^{-(640\ \mu\text{s} / 2.35 \text{ ms})}$$
$$= 10 \text{ V} + (-10)(e^{-0.272})$$
$$= 10 \text{ V} + (-10)(0.762)$$
$$= 10 \text{ V} + (-7.62)$$
$$= \mathbf{2.38\ V}$$

At the end of Pulse 1, $V_R = V_{out}$ (1) = 2.38 V. When V_{in} changes from 10 V to 0 V for 960 μs, the integrator output V_{out} rejects the change in V_{in} and decreases towards 0 V for 960 μs. From the general exponential equation, the output voltage at the beginning of Pulse 2 is

$$V_{out}(2) = V_f + (V_i - V_f)\, e^{-t/\tau}$$

where $V_f = V_{in}$ = 0 V, $V_i = V_{out}$ (1) = 2.38 V, $t = t_L$ = 960 μs, and τ = 2.35 ms.

$$V_{out}(2) = 0 \text{ V} + [2.38 \text{ V} - 0 \text{ V}]\, e^{-(960\ \mu\text{s} / 2.35 \text{ ms})}$$

$= 0\text{ V} + (2.38\text{ V})(e^{-0.408})$

$= (2.38\text{ V})(0.665)$

$= \mathbf{1.58\text{ V}}$

At the start of Pulse 2, $V_{out}(2) = 1.58$ V. When V_{in} changes from 0 V to 10 V for 640 μs, the integrator output V_{out} rejects the change in V_{in} and increases towards 10 V for 640 μs. From the general exponential equation, the output voltage at the end of Pulse 2 is

$$V_{out}(3) = V_f + (V_i - V_f)\,e^{-t/\tau}$$

where $V_f = V_{in} = 10$ V, $V_i = V_{out}(2) = 1.58$ V, $t = t_H = 640$ μs, and $\tau = 2.35$ ms.

$V_{out}(3) = 10\text{ V} + [1.58\text{ V} - 10\text{ V}]\,e^{-(640\,\mu s\,/\,2.35\text{ ms})}$

$= 10\text{ V} + (-8.42)(e^{-0.272})$

$= 10\text{ V} + (-8.42)(0.762)$

$= 10\text{ V} + (-6.41)$

$= \mathbf{3.59\text{ V}}$

At the end of Pulse 2, $V_C = V_{out}(3) = 3.59$ V. When V_{in} changes from 10 V to 0 V for 960 μs, the integrator output V_{out} rejects the change in V_{in} and decreases towards 0 V for 960 μs. From the general exponential equation, the output voltage at the beginning of Pulse 3 is

$$V_{out}(4) = V_f + (V_i - V_f)\,e^{-t/\tau}$$

where $V_f = V_{in} = 0$ V, $V_i = V_{out}(3) = 3.59$ V, $t = t_L = 960$ μs, and $\tau = 2.35$ ms.

$V_{out}(4) = 0\text{ V} + [3.59\text{ V} - 0\text{ V}]\,e^{-(960\,\mu s\,/\,2.35\text{ ms})}$

$= 0\text{ V} + (3.59\text{ V})(e^{-0.408})$

$= (3.59\text{ V})(0.665)$

$= \mathbf{2.39\text{ V}}$

At the start of Pulse 3, $V_{out}(4) = 2.39$ V. When V_{in} changes from 0 V to 10 V for 640 μs, the integrator output V_{out} rejects the change in V_{in} and increases towards 10 V for 640 μs. From the general exponential equation, the output voltage at the end of Pulse 3 is

$$V_{out}(5) = V_f + (V_i - V_f)\,e^{-t/\tau}$$

where $V_f = V_{in} = 10$ V, $V_i = V_{out}(4) = 2.39$ V, $t = t_H = 640$ μs, and $\tau = 2.35$ ms.

$V_{out}(5) = 10\text{ V} + [2.39\text{ V} - 10\text{ V}]\,e^{-(640\,\mu s\,/\,2.35\text{ ms})}$

$= 10\text{ V} + (-7.61)(e^{-0.272})$

$= 10\text{ V} + (-7.61)(0.762)$

$= 10\text{ V} + (-5.80)$

$= \mathbf{4.20\text{ V}}$

At the end of Pulse 3, $V_C = V_{out}(5) = 4.20$ V. When V_{in} changes from 10 V to 0 V for 960 μs, the integrator output V_{out} rejects the change in V_{in} and decreases towards 0 V for 960 μs. From the general exponential equation, the output voltage at the beginning of Pulse 4 is

$$V_{out}(6) = V_f + (V_i - V_f)\,e^{-t/\tau}$$

where $V_f = V_{in} = 0$ V, $V_i = V_{out}(5) = 4.20$ V, $t = t_L = 960$ μs, and $\tau = 2.35$ ms.

$V_{out}(6) = 0\text{ V} + [4.20\text{ V} - 0\text{ V}]\,e^{-(960\,\mu s\,/\,2.35\text{ ms})}$

$= 0\text{ V} + (4.20\text{ V})(e^{-0.408})$

$= (4.20\text{ V})(0.665)$

$= \mathbf{2.79\text{ V}}$

At the start of Pulse 4, $V_{out}(6) = 2.79$ V. When V_{in} changes from 0 V to 10 V for 640 μs, the integrator output V_{out} rejects the change in V_{in} and increases towards 10 V for 640 μs. From the general exponential equation, the output voltage at the end of Pulse 4 is

$V_{out}(7) = V_f + (V_i - V_f)\, e^{-t/\tau}$

where $V_f = V_{in} = 10$ V, $V_i = V_{out}(6) = 2.79$ V, $t = t_H = 640$ μs, and $\tau = 2.35$ ms.

$$\begin{aligned} V_{out}(7) &= 10\text{ V} + [2.79\text{ V} - 10\text{ V}]\, e^{-(640\ \mu s / 2.35\ ms)} \\ &= 10\text{ V} + (-7.21)(e^{-0.272}) \\ &= 10\text{ V} + (-7.21)(0.762) \\ &= 10\text{ V} + (-5.49) \\ &= \mathbf{4.51\ V} \end{aligned}$$

At the end of Pulse 4, $V_C = V_{out}(7) = 4.51$ V. When V_{in} changes from 10 V to 0 V for 960 μs, the integrator output V_{out} rejects the change in V_{in} and decreases towards 0 V for 960 μs. From the general exponential equation, the output voltage at the beginning of Pulse 5 is

$V_{out}(8) = V_f + (V_i - V_f)\, e^{-t/\tau}$

where $V_f = V_{in} = 0$ V, $V_i = V_{out}(7) = 4.51$ V, $t = t_L = 960$ μs, and $\tau = 2.35$ ms.

$$\begin{aligned} V_{out}(8) &= 0\text{ V} + [4.51\text{ V} - 0\text{ V}]\, e^{-(960\ \mu s / 2.35\ ms)} \\ &= 0\text{ V} + (4.51\text{ V})(e^{-0.408}) \\ &= (4.51\text{ V})(0.665) \\ &= \mathbf{3.00\ V} \end{aligned}$$

At the start of Pulse 5, $V_{out}(8) = 3.00$ V. When V_{in} changes from 0 V to 5 V for 640 μs, the integrator output V_{out} rejects the change in V_{in} and increases towards 5 V for 640 μs. From the general exponential equation, the output voltage at the end of Pulse 5 is

$V_{out}(9) = V_f + (V_i - V_f)\, e^{-t/\tau}$

where $V_f = V_{in} = 10$ V, $V_i = V_{out}(8) = 3.00$ V, $t = t_H = 640$ μs, and $\tau = 2.35$ ms.

$$\begin{aligned} V_{out}(9) &= 10\text{ V} + [3.00\text{ V} - 10\text{ V}]\, e^{-(640\ \mu s / 2.35\ ms)} \\ &= 10\text{ V} + (-7.00)(e^{-0.272}) \\ &= 10\text{ V} + (-7.00)(0.762) \\ &= 10\text{ V} + (-5.33) \\ &= \mathbf{4.67\ V} \end{aligned}$$

Figure 21-18 shows the input and output waveforms for the first five pulses.

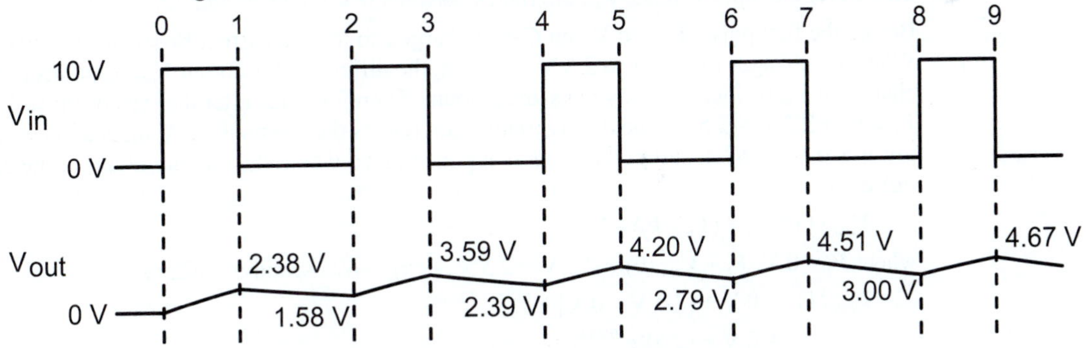

Figure 21-18: Input and Output Waveforms for *RL* Integrator

3. Problem: For the ***RC*** differentiator in Figure 21-19, calculate the output voltage, V_{out}, at the beginning and end of the first five pulses if V_{in} is a 0 V to 2.5 V, 10 MHz periodic pulse waveform with a 67% duty cycle. Assume that C is initially uncharged and $V_{out} = 0$ V.

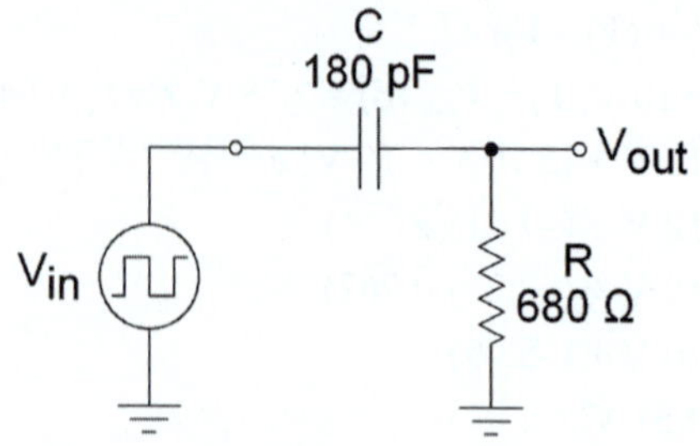

Figure 21-19: *RC* Differentiator for Problem 3

Solution: First, calculate the time constant, τ, for the ***RC*** circuit.

$\tau = RC$
$= (680\ \Omega)(180\ \text{pF})$
$= 122\ \text{ns}$

The period, T, of a pulse waveform is the reciprocal of the frequency.

$T = 1/f$
$= 1/(10\ \text{MHz})$
$= 100\ \text{ns}$

By definition, the duty cycle, DC, of a pulse waveform is the high time, t_H, divided by the period.

$t_H / T = \text{DC}$
$(t_H / T) \times T = \text{DC} \times T$
$t_H = (0.67)(100\ \text{ns})$
$= 67\ \text{ns}$

The low time, t_L, is the difference between T and t_H.

$t_L = T - t_H$
$= (100\ \text{ns}) - (67\ \text{ns})$
$= 33\ \text{ns}$

Therefore, the input voltage, V_{in}, alternates between 0 V for 33 ns and 2.5 V for 67 ns.

Before the first pulse, $V_{in} = 0$ V and C is uncharged so that no current flows and $V_{out}(0) = 0$ V. When V_{in} changes from 0 V to 2.5 V for 67 ns, the differentiator output V_{out} responds to the change in V_{in} by increasing by the same amount. Therefore, just after the start of Pulse 1, $V_{out} = V_{out}(0) + 2.5\ \text{V} = 2.5\ \text{V}$. The differentiator then rejects the constant 2.5 V input and decreases towards 0 V for 67 ns. From the general exponential equation, the output voltage at the end of Pulse 1 is

$V_{out}(1) = V_f + (V_i - V_f)\, e^{-t/\tau}$

where $V_f = 0$ V, $V_i = V_{out}(0) + 2.5\ \text{V} = 2.5$ V, $t = t_H = 67$ ns, and $\tau = 122$ ns.

$V_{out}(1) = 0\ \text{V} + [2.5\ \text{V} - 0\ \text{V}]\, e^{-(67\ \text{ns}\,/\,122\ \text{ns})}$
$= 0\ \text{V} + (2.5)(e^{-0.547})$
$= (2.5)(0.579)$
$= \mathbf{1.45\ V}$

At the end of Pulse 1, $V_R = V_{out}(1) = 1.45$ V. When V_{in} changes from 2.5 V to 0 V for 33 ns, the differentiator output V_{out} responds to the change in V_{in} by decreasing by the same amount. Therefore, just after the end of Pulse 1, $V_{out} = V_{out}(1) - 2.5\ \text{V} = 1.45\ \text{V} - 2.5\ \text{V} = \mathbf{-1.05\ V}$. The differentiator then rejects the constant 0 V input and decreases towards 0 V for 33 ns. From the general exponential equation, the output voltage at the start of Pulse 2 is

$V_{out}(2) = V_f + (V_i - V_f)\, e^{-t/\tau}$

where $V_f = V_{in} = 0$ V, $V_i = V_{out}(1) - 2.5$ V $= -1.05$ V, $t = t_L = 33$ ns, and $\tau = 122$ ns.

$V_{out}(2) = 0\text{ V} + [-1.05\text{ V} - 0\text{ V}]\, e^{-(33\text{ ns}/122\text{ ns})}$
$= 0\text{ V} + (-1.05\text{ V})(e^{-0.270})$
$= (-1.05\text{ V})(0.764)$
$= \mathbf{-0.805\ V}$

At the start of Pulse 2, $V_{out}(2) = -0.805$ V. When V_{in} changes from 0 V to 2.5 V for 67 ns, the differentiator output V_{out} responds to the change in V_{in} by increasing by the same amount, Therefore, just after the start of Pulse 2, $V_{out} = V_{out}(2) + 2.5$ V $= -0.805$ V $+ 2.5$ V $=$ **1.70 V**. The differentiator then rejects the constant 2.5 V input and decreases towards 0 V for 67 ns. From the general exponential equation, the output voltage at the end of Pulse 2 is

$V_{out}(3) = V_f + (V_i - V_f)\, e^{-t/\tau}$

where $V_f = V_{in} = 0$ V, $V_i = V_{out}(2) + 2.5$ V $= 1.70$ V, $t = t_H = 67$ ns, and $\tau = 122$ ns.

$V_{out}(3) = 0\text{ V} + [1.70\text{ V} - 0\text{ V}]\, e^{-(67\text{ ns}/122\text{ ms})}$
$= 0\text{ V} + (1.70)(e^{-0.547})$
$= (1.70)(0.579)$
$= \mathbf{0.981\ V}$

At the end of Pulse 2, $V_C = V_{out}(3) = 0.981$ V. When V_{in} changes from 2.5 V to 0 V for 33 ns, the differentiator output V_{out} responds to the change in V_{in} by decreasing by the same amount. Therefore, just after the end of Pulse 2, $V_{out} = V_{out}(3) - 2.5$ V $= 0.981$ V $- 2.5$ V $=$ **–1.52 V**. The differentiator then rejects the constant 0 V input and decreases towards 0 V for 33 ns. From the general exponential equation, the output voltage at the beginning of Pulse 3 is

$V_{out}(4) = V_f + (V_i - V_f)\, e^{-t/\tau}$

where $V_f = V_{in} = 0$ V, $V_i = V_{out}(3) - 2.5$ V $= -1.52$ V, $t = t_L = 33$ ns, and $\tau = 122$ ns.

$V_{out}(4) = 0\text{ V} + [-1.52\text{ V} - 0\text{ V}]\, e^{-(33\text{ ns}/122\text{ ns})}$
$= 0\text{ V} + (-1.52\text{ V})(e^{-0.270})$
$= (-1.52\text{ V})(0.764)$
$= \mathbf{-1.16\ V}$

At the start of Pulse 3, $V_{out}(4) = -1.16$ V. When V_{in} changes from 0 V to 2.5 V for 67 ns, the differentiator output V_{out} responds to the change in V_{in} by increasing by the same amount. Therefore, just after the start of Pulse 3, $V_{out} = V_{out}(4) + 2.5$ V $= -1.16$ V $+ 2.5$ V $=$ **1.34 V**. The differentiator then rejects the constant 2.5 V input and decreases towards 0 V for 67 ns. From the general exponential equation, the output voltage at the end of Pulse 3 is

$V_{out}(5) = V_f + (V_i - V_f)\, e^{-t/\tau}$

where $V_f = V_{in} = 0$ V, $V_i = V_{out}(4) + 2.5$ V $= 1.34$ V, $t = t_H = 67$ ns, and $\tau = 122$ ns.

$V_{out}(5) = 0\text{ V} + [1.34\text{ V} - 0\text{ V}]\, e^{-(67\text{ ns}/122\text{ ns})}$
$= 0\text{ V} + (1.34)(e^{-0.547})$
$= (1.34)(0.579)$
$= \mathbf{0.775\ V}$

At the end of Pulse 3, $V_C = V_{out}(5) = 0.775$ V. When V_{in} changes from 2.5 V to 0 V for 33 ns, the differentiator output V_{out} responds to the change in V_{in} by decreasing by the same amount. Therefore, just after the end of Pulse 3, $V_{out} = V_{out}(5) - 2.5$ V $= 0.775$ V $- 2.5$ V $=$ **–1.73 V**. The differentiator then rejects the constant 0 V input and decreases towards 0 V for 33 ns. From the general exponential equation, the output voltage at the beginning of Pulse 4 is

$V_{out}(6) = V_f + (V_i - V_f)\, e^{-t/\tau}$

where $V_f = V_{in} = 0$ V, $V_i = V_{out}(5) - 2.5$ V $= -1.73$ V, $t = t_L = 33$ ns, and $\tau = 122$ ns.

$V_{out}(6) = 0\text{ V} + [-1.73\text{ V} - 0\text{ V}]\, e^{-(33\text{ ns}/122\text{ ns})}$

$= 0\text{ V} + (-1.73\text{ V})(e^{-0.270})$
$= (-1.73\text{ V})(0.764)$
$= \mathbf{-1.32\text{ V}}$

At the start of Pulse 4, V_{out} (6) = –1.32 V. When V_{in} changes from 0 V to 2.5 V for 67 ns, the differentiator output V_{out} responds to the change in V_{in} by increasing by the same amount. Therefore, just after the start of Pulse 4, $V_{out} = V_{out}$ (6) + 2.5 V = –1.32 V + 2.5 V = **1.18 V**. The differentiator then rejects the constant 2.5 V input and decreases towards 0 V for 67 ns. From the general exponential equation, the output voltage at the end of Pulse 4 is

$$V_{out}(7) = V_f + (V_i - V_f)\, e^{-t/\tau}$$

where $V_f = V_{in} = 0$ V, $V_i = V_{out}$ (6) + 2.5 V = 1.18 V, $t = t_H = 67$ ns, and $\tau = 122$ ns.

$V_{out}(7) = 0\text{ V} + [1.18\text{ V} - 0\text{ V}]\, e^{-(67\text{ ns}/122\text{ ns})}$
$= 0\text{ V} + (1.18)(e^{-0.547})$
$= (1.18)(0.579)$
$= \mathbf{0.684\text{ V}}$

At the end of Pulse 4, V_{out} (7) = 0.684 V. When V_{in} changes from 2.5 V to 0 V for 33 ns, the differentiator output V_{out} responds to the change in V_{in} by decreasing by the same amount. Therefore, just after the end of Pulse 4, $V_{out} = V_{out}$ (7) – 2.5 V = 0.684 V – 2.5 V = **–1.82 V**. The differentiator then rejects the constant 0 V input and decreases towards 0 V for 33 ns. From the general exponential equation, the output voltage at the beginning of Pulse 5 is

$$V_{out}(8) = V_f + (V_i - V_f)\, e^{-t/\tau}$$

where $V_f = V_{in} = 0$ V, $V_i = V_{out}$ (7) – 2.5 V = –1.82 V, $t = t_L = 33$ ns, and $\tau = 122$ ns.

$V_{out}(8) = 0\text{ V} + [-1.82\text{ V} - 0\text{ V}]\, e^{-(33\text{ ns}/122\text{ ns})}$
$= 0\text{ V} + (-1.82\text{ V})(e^{-0.270})$
$= (-1.82\text{ V})(0.764)$
$= \mathbf{-1.39\text{ V}}$

At the start of Pulse 5, V_{out} (8) = –1.39 V. When V_{in} changes from 0 V to 2.5 V for 67 ns, the differentiator output V_{out} responds to the change in V_{in} by increasing by the same amount. Therefore, just after the start of Pulse 5, $V_{out} = V_{out}$ (8) + 2.5 V = –1.39 V + 2.5 V = **1.11 V**. The differentiator then rejects the constant 2.5 V input and decreases towards 0 V for 67 ns. From the general exponential equation, the output voltage at the end of Pulse 5 is

$$V_{out}(9) = V_f + (V_i - V_f)\, e^{-t/\tau}$$

where $V_f = V_{in} = 0$ V, $V_i = V_{out}$ (8) + 2.5 V = 1.11 V, $t = t_H = 67$ ns, and $\tau = 122$ ns.

$V_{out}(9) = 0\text{ V} + [1.11\text{ V} - 0\text{ V}]\, e^{-(67\text{ ns}/122\text{ ms})}$
$= 0\text{ V} + (1.11)(e^{-0.547})$
$= (1.11)(0.579)$
$= \mathbf{0.644\text{ V}}$

At the end of Pulse 5, V_{out} (9) = 0.644 V. When V_{in} changes from 2.5 V to 0 V for 33 ns, the differentiator output V_{out} responds to the change in V_{in} by decreasing by the same amount. Therefore, just after the end of Pulse 5, $V_{out} = V_{out}$ (9) – 2.5 V = 0.644 V – 2.5 V = **–1.86 V**.

Figure 21-20 shows the input and output waveforms for the first five pulses.

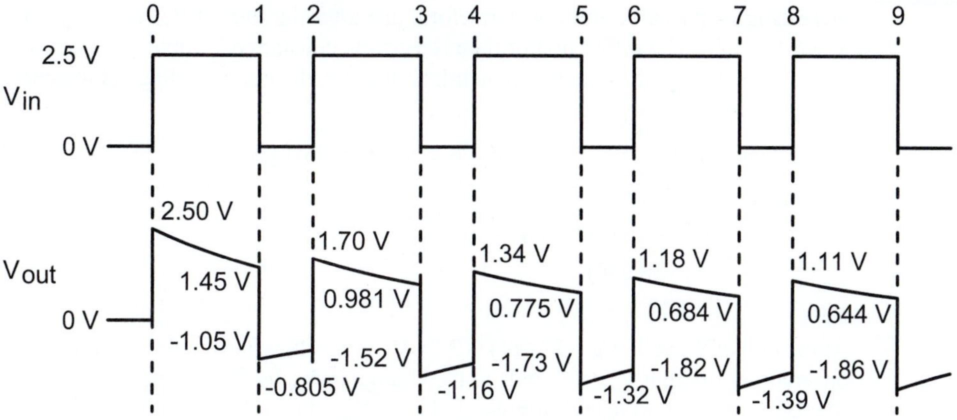

Figure 21-20: Input and Output Waveforms for *RC* Differentiator

4. Problem: For the ***RL*** differentiator in Figure 21-21, calculate the output voltage, V_{out}, at the beginning and end of the first five pulses if V_{in} is a 0 V to 6 V, 3.33 kHz periodic pulse waveform with a 25% duty cycle. Assume that L is initially de-energized so that V_{out} = 0 V.

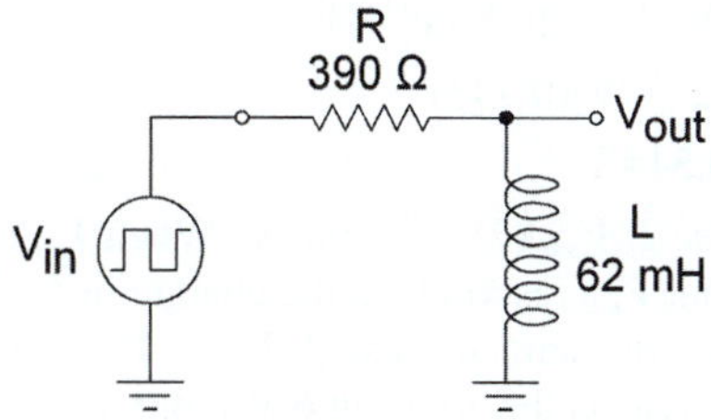

Figure 21-21: *RL* Differentiator for Problem 4

Solution: First, calculate the time constant, τ, for the ***RL*** circuit.

$\tau = L / R$

$= (62 \text{ mH}) / (390\ \Omega)$

$= 159\ \mu\text{s}$

The period, T, of a pulse waveform is the reciprocal of the frequency.

$T = 1 / f$

$= 1 / (3.33 \text{ kHz})$

$= 300\ \mu\text{s}$

By definition, the duty cycle, DC, of a pulse waveform is the high time, t_H, divided by the period.

$t_H / T = \text{DC}$

$(t_H / T) \times T = \text{DC} \times \tau$

$t_H = (0.25)(300\ \mu\text{s})$

$= 75.1\ \mu\text{s}$

The low time, t_L, is the difference between T and t_H.

$t_L = T - t_H$

$= (300\ \mu\text{s}) - (75.1\ \mu\text{s})$

$= 225\ \mu\text{s}$

Therefore, the input voltage, V_{in}, alternates between 0 V for 225 μs and 6 V for 75.1 μs.

Before the first pulse, V_{in} = 0 V and L is de-energized so that $V_{out}(0)$ = 0 V. When V_{in} changes from 0 V to 6 V for 75.1 μs, the differentiator output V_{out} responds to the change in V_{in} by increasing by the same amount. Therefore, just after the start of Pulse 1, $V_{out} = V_{out}(0) + 6\text{ V} = 0\text{ V} + 6\text{ V} = 6\text{ V}$. The differentiator then rejects the constant 6 V input and decreases towards 0 V for 75.1 μs. From the general exponential equation, the output voltage at the end of Pulse 1 is

$$V_{out}(1) = V_f + (V_i - V_f)\, e^{-t/\tau}$$

where $V_f = 0$ V, $V_i = V_{out}(0) + 6\text{ V} = 6$ V, $t = t_H = 75.1$ μs, and $\tau = 159$ μs.

$$\begin{aligned} V_{out}(1) &= 0\text{ V} + [6\text{ V} - 0\text{ V}]\, e^{-(75.1\ \mu s\,/\,159\ \mu s)} \\ &= 0\text{ V} + (6)(e^{-0.472}) \\ &= (6)(0.624) \\ &= \mathbf{3.74\ V} \end{aligned}$$

At the end of Pulse 1, $V_R = V_{out}(1) = 3.74$ V. When V_{in} changes from 6 V to 0 V for 225 μs, the differentiator output V_{out} responds to the change in V_{in} by decreasing by the same amount. Therefore, just after the end of Pulse 1, $V_{out} = V_{out}(1) - 6\text{ V} = 3.74\text{ V} - 6\text{ V} = $ **–2.26 V**. The differentiator then rejects the constant 0 V input and decreases towards 0 V for 225 μs. From the general exponential equation, the output voltage at the start of Pulse 2 is

$$V_{out}(2) = V_f + (V_i - V_f)\, e^{-t/\tau}$$

where $V_f = V_{in} = 0$ V, $V_i = V_{out}(1) - 6\text{ V} = -2.26$ V, $t = t_L = 225$ μs, and $\tau = 159$ μs.

$$\begin{aligned} V_{out}(2) &= 0\text{ V} + [-2.26\text{ V} - 0\text{ V}]\, e^{-(225\ \mu s\,/\,159\ \mu s)} \\ &= 0\text{ V} + (-2.26\text{ V})(e^{-1.42}) \\ &= (-2.26\text{ V})(0.243) \\ &= \mathbf{-0.548\ V} \end{aligned}$$

At the start of Pulse 2, $V_{out}(2) = -0.548$ V. When V_{in} changes from 0 V to 6 V for 75.1 μs, the differentiator output V_{out} responds to the change in V_{in} by increasing by the same amount. Therefore, just after the start of Pulse 2, $V_{out} = V_{out}(2) + 6\text{ V} = -0.548\text{ V} + 6\text{ V} = $ **5.45 V**. The differentiator then rejects the constant 6 V input and decreases towards 0 V for 75.1 μs. From the general exponential equation, the output voltage at the end of Pulse 2 is

$$V_{out}(3) = V_f + (V_i - V_f)\, e^{-t/\tau}$$

where $V_f = V_{in} = 0$ V, $V_i = V_{out}(2) + 6\text{ V} = 5.45$ V, $t = t_H = 75.1$ μs, and $\tau = 159$ μs.

$$\begin{aligned} V_{out}(3) &= 0\text{ V} + [5.45\text{ V} - 0\text{ V}]\, e^{-(75.1\ \mu s\,/\,159\ \mu s)} \\ &= 0\text{ V} + (5.45)(e^{-0.472}) \\ &= (5.45)(0.624) \\ &= \mathbf{3.40\ V} \end{aligned}$$

At the end of Pulse 2, $V_C = V_{out}(3) = 3.40$ V. When V_{in} changes from 6 V to 0 V for 225 μs, the differentiator output V_{out} responds to the change in V_{in} by decreasing by the same amount. Therefore, just after the end of Pulse 2, $V_{out} = V_{out}(3) - 6\text{ V} = 3.40\text{ V} - 6\text{ V} = $ **–2.60 V**. The differentiator then rejects the constant 0 V input and decreases towards 0 V for 225 μs. From the general exponential equation, the output voltage at the beginning of Pulse 3 is

$$V_{out}(4) = V_f + (V_i - V_f)\, e^{-t/\tau}$$

where $V_f = V_{in} = 0$ V, $V_i = V_{out}(3) - 6\text{ V} = -2.60$ V, $t = t_L = 225$ μs, and $\tau = 159$ μs.

$$\begin{aligned} V_{out}(4) &= 0\text{ V} + [-2.60\text{ V} - 0\text{ V}]\, e^{-(225\ \mu s\,/\,159\ \mu s)} \\ &= 0\text{ V} + (-2.60\text{ V})(e^{-1.42}) \\ &= (-2.60\text{ V})(0.243) \\ &= \mathbf{-0.631\ V} \end{aligned}$$

At the start of Pulse 3, $V_{out}(4) = -0.631$ V. When V_{in} changes from 0 V to 6 V for 75.1 μs, the differentiator output V_{out} responds to the change in V_{in} by increasing by the same amount. Therefore, just after the start of Pulse 3, $V_{out} = V_{out}(4) + 6\text{ V} = -0.631\text{ V} + 6\text{ V} = $ **5.37 V**. The differentiator then rejects the constant 6 V input and decreases towards 0 V for 75.1 μs. From the general exponential equation, the output voltage at the end of Pulse 3 is

$$V_{out}(5) = V_f + (V_i - V_f)\, e^{-t/\tau}$$

where $V_f = V_{in} = 0$ V, $V_i = V_{out}(4) + 6$ V $= 5.37$ V, $t = t_H = 75.1$ μs, and $\tau = 159$ μs.

$$\begin{aligned} V_{out}(5) &= 0\text{ V} + [5.37\text{ V} - 0\text{ V}]\, e^{-(75.1\ \mu\text{s}/159\ \mu\text{s})} \\ &= 0\text{ V} + (5.37)(e^{-0.472}) \\ &= (5.37)(0.624) \\ &= \mathbf{3.35\ V} \end{aligned}$$

At the end of Pulse 3, $V_C = V_{out}(5) = 3.35$ V. When V_{in} changes from 6 V to 0 V for 225 μs, the differentiator output V_{out} responds to the change in V_{in} by decreasing by the same amount. Therefore, just after the end of Pulse 3, $V_{out} = V_{out}(5) - 6$ V $= 3.35$ V $- 6$ V $=$ **–2.65 V**. The differentiator then rejects the constant 0 V input and decreases towards 0 V for 225 μs. From the general exponential equation, the output voltage at the beginning of Pulse 4 is

$$V_{out}(6) = V_f + (V_i - V_f)\, e^{-t/\tau}$$

where $V_f = V_{in} = 0$ V, $V_i = V_{out}(5) - 6$ V $= -2.65$ V, $t = t_L = 225$ μs, and $\tau = 159$ μs.

$$\begin{aligned} V_{out}(6) &= 0\text{ V} + [-2.65\text{ V} - 0\text{ V}]\, e^{-(225\ \mu\text{s}/159\ \mu\text{s})} \\ &= 0\text{ V} + (-2.65\text{ V})(e^{-1.42}) \\ &= (-2.65\text{ V})(0.243) \\ &= \mathbf{-0.643\ V} \end{aligned}$$

At the start of Pulse 4, $V_{out}(6) = -0.643$ V. When V_{in} changes from 0 V to 6 V for 75.1 μs, the differentiator output V_{out} responds to the change in V_{in} by increasing by the same amount. Therefore, just after the start of Pulse 4, $V_{out} = V_{out}(6) + 6$ V $= -0.643$ V $+ 6$ V $=$ **5.36 V**. The differentiator then rejects the constant 6 V input and decreases towards 0 V for 75.1 μs. From the general exponential equation, the output voltage at the end of Pulse 4 is

$$V_{out}(7) = V_f + (V_i - V_f)\, e^{-t/\tau}$$

where $V_f = V_{in} = 0$ V, $V_i = V_{out}(6) + 6$ V $= 5.36$ V, $t = t_H = 75.1$ μs, and $\tau = 159$ μs.

$$\begin{aligned} V_{out}(7) &= 0\text{ V} + [5.36\text{ V} - 0\text{ V}]\, e^{-(75.1\ \mu\text{s}/159\ \mu\text{s})} \\ &= 0\text{ V} + (5.36)(e^{-0.472}) \\ &= (5.36)(0.624) \\ &= \mathbf{3.34\ V} \end{aligned}$$

At the end of Pulse 4, $V_{out}(7) = 3.34$ V. When V_{in} changes from 6 V to 0 V for 225 μs, the differentiator output V_{out} responds to the change in V_{in} by decreasing by the same amount. Therefore, just after the end of Pulse 4, $V_{out} = V_{out}(7) - 6$ V $= 3.34$ V $- 6$ V $=$ **–2.66 V**. The differentiator then rejects the constant 0 V input and decreases towards 0 V for 225 μs. From the general exponential equation, the output voltage at the beginning of Pulse 5 is

$$V_{out}(8) = V_f + (V_i - V_f)\, e^{-t/\tau}$$

where $V_f = V_{in} = 0$ V, $V_i = V_{out}(7) - 6$ V $= -2.66$ V, $t = t_L = 225$ μs, and $\tau = 159$ μs.

$$\begin{aligned} V_{out}(8) &= 0\text{ V} + [-2.66\text{ V} - 0\text{ V}]\, e^{-(225\ \mu\text{s}/159\ \mu\text{s})} \\ &= 0\text{ V} + (-2.66\text{ V})(e^{-1.42}) \\ &= (-2.65\text{ V})(0.243) \\ &= \mathbf{-0.645\ V} \end{aligned}$$

At the start of Pulse 5, $V_{out}(8) = -0.645$ V. When V_{in} changes from 0 V to 6 V for 75.1 μs, the differentiator output V_{out} responds to the change in V_{in} by increasing by the same amount. Therefore, just after the start of Pulse 5, $V_{out} = V_{out}(8) + 6$ V $= -0.645$ V $+ 6$ V $=$ **5.36 V**. The differentiator then rejects the constant 6 V input and decreases towards 0 V for 75.1 μs. From the general exponential equation, the output voltage at the end of Pulse 5 is

$$V_{out}(9) = V_f + (V_i - V_f)\, e^{-t/\tau}$$

where $V_f = V_{in} = 0$ V, $V_i = V_{out}$ (8) + 6 V = 5.36 V, $t = t_H = 75.1$ ns, and $\tau = 122$ ns.

$$V_{out}(9) = 0\text{ V} + [5.36\text{ V} - 0\text{ V}]\, e^{-(75.1\ \mu s / 159\ \mu s)}$$
$$= 0\text{ V} + (5.36)(e^{-0.472})$$
$$= (5.36)(0.624)$$
$$= \mathbf{3.34\ V}$$

At the end of Pulse 5, V_{out} (9) = 0.3.34 V. When V_{in} changes from 6 V to 0 V for 225 ns, the differentiator output V_{out} responds to the change in V_{in} by decreasing by the same amount. Therefore, just after the end of Pulse 5, $V_{out} = V_{out}$ (9) – 6 V = 3.34 V – 6 V = **–2.66 V**.

Figure 21-22 shows the input and output waveforms for the first five pulses.

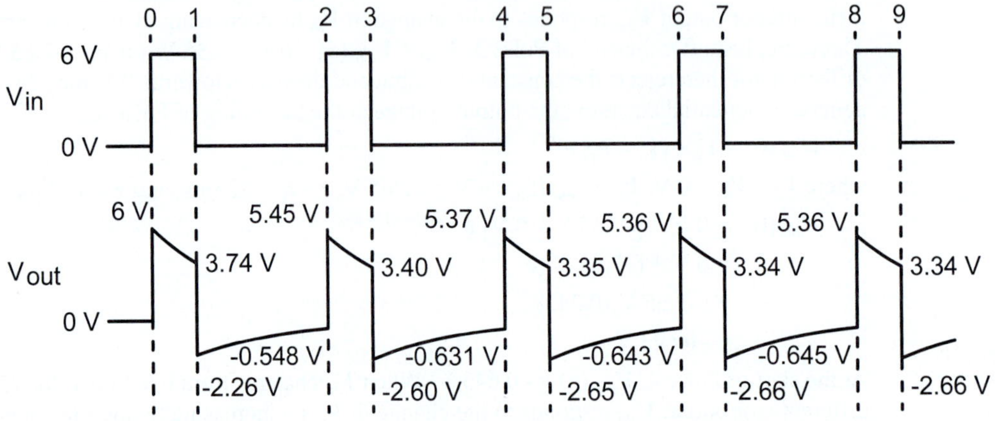

Figure 21-22: Input and Output Waveforms for *RL* Differentiator

21.2.2 Nonperiodic Repetitive Pulses

For Problems 1 and 2, refer to the nonperiodic pulse waveform in Figure 21-23.

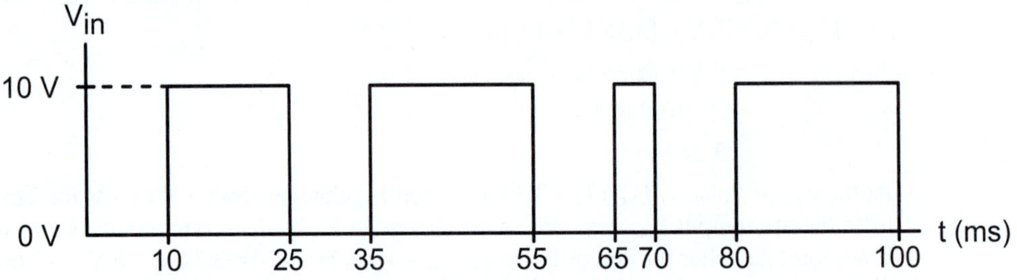

Figure 21-23: Nonperiodic Pulse Waveform for Problems 1 and 2

1. Problem: The nonperiodic pulse waveform in Figure 21-23, V_{in}, is applied to an integrator with a time constant $\tau = 20$ ms. What is V_{out}, the output voltage of the integrator, at $t = 100$ ms? Assume that $V_{out} = 0$ V at $t = 0$ ms.

   Solution: To determine the value of V_{out} at $t = 100$ ms, successively apply the general exponential equation to each section of the input waveform to determine V_{out} at the start and end of each pulse.

   From the problem statement, at $t = 0$ ms, $V_{out} = 0$ V. V_{in} remains at 0 V until $t = 10$ ms, so that V_{out} remains at 0 V.

   At $t = 10$ ms, V_{in} changes from 0 V to 10 V and remains at 10 V for 25 ms – 10 ms = 15 ms. The integrator rejects the change in V_{in} and increases towards 10 V for 15 ms. From the general exponential equation, V_{out} (25 ms) = $V_f + (V_i - V_f)\, e^{-t/\tau}$, where $V_i = 0$ V, $V_f = 10$ V, $t = 15$ ms, and $\tau = 20$ ms.

$$
\begin{aligned}
V_{out}\,(25\text{ ms}) &= 10\text{ V} + [0\text{ V} - 10\text{ V}]\; e^{-(15\text{ ms}/20\text{ ms})} \\
&= 10\text{ V} + (-10\text{ V})(e^{-0.75}) \\
&= 10\text{ V} + (-10\text{ V})(0.472) \\
&= 10\text{ V} + (-4.72\text{ V}) \\
&= 5.28\text{ V}
\end{aligned}
$$

At $t = 25$ ms, V_{in} changes from 10 V to 0 V and remains at 0 V for 35 ms – 25 ms = 10 ms. The integrator rejects the change in V_{in} and decreases towards 0 V for 10 ms. From the general exponential equation, V_{out} (35 ms) = $V_f + (V_i - V_f)\, e^{-t/\tau}$, where $V_i = 5.28$ V, $V_f = 0$ V, $t = 10$ ms, and $\tau = 20$ ms.

$$
\begin{aligned}
V_{out}\,(35\text{ ms}) &= 0\text{ V} + [5.28\text{ V} - 0\text{ V}]\; e^{-(10\text{ ms}/20\text{ ms})} \\
&= 0\text{ V} + (5.28\text{ V})(e^{-0.50}) \\
&= (5.28\text{ V})(0.607) \\
&= 3.20\text{ V}
\end{aligned}
$$

At $t = 35$ ms, V_{in} changes from 0 V to 10 V and remains at 10 V for 55 ms – 35 ms = 20 ms. The integrator rejects the change in V_{in} and increases towards 10 V for 20 ms. From the general exponential equation, V_{out} (55 ms) = $V_f + (V_i - V_f)\, e^{-t/\tau}$, where $V_i = 3.20$ V, $V_f = 10$ V, $t = 20$ ms, and $\tau = 20$ ms.

$$
\begin{aligned}
V_{out}\,(55\text{ ms}) &= 10\text{ V} + [3.20\text{ V} - 10\text{ V}]\; e^{-(20\text{ ms}/20\text{ ms})} \\
&= 10\text{ V} + (-6.80\text{ V})(e^{-1.0}) \\
&= 10\text{ V} + (-6.80\text{ V})(0.368) \\
&= 10\text{ V} + (-2.50\text{ V}) \\
&= 7.50\text{ V}
\end{aligned}
$$

At $t = 55$ ms, V_{in} changes from 10 V to 0 V and remains at 0 V for 65 ms – 55 ms = 10 ms. The integrator rejects the change in V_{in} and decreases towards 0 V for 10 ms. From the general exponential equation, V_{out} (65 ms) = $V_f + (V_i - V_f)\, e^{-t/\tau}$, where $V_i = 7.50$ V, $V_f = 0$ V, $t = 10$ ms, and $\tau = 20$ ms.

$$
\begin{aligned}
V_{out}\,(65\text{ ms}) &= 0\text{ V} + [7.50\text{ V} - 0\text{ V}]\; e^{-(10\text{ ms}/20\text{ ms})} \\
&= 0\text{ V} + (7.50\text{ V})(e^{-0.50}) \\
&= (7.50\text{ V})(0.607) \\
&= 4.55\text{ V}
\end{aligned}
$$

At $t = 65$ ms, V_{in} changes from 0 V to 10 V and remains at 10 V for 70 ms – 65 ms = 5 ms. The integrator rejects the change in V_{in} and increases towards 10 V for 5 ms. From the general exponential equation, V_{out} (70 ms) = $V_f + (V_i - V_f)\, e^{-t/\tau}$, where $V_i = 4.55$ V, $V_f = 10$ V, $t = 5$ ms, and $\tau = 20$ ms.

$$
\begin{aligned}
V_{out}\,(70\text{ ms}) &= 10\text{ V} + [4.55\text{ V} - 10\text{ V}]\; e^{-(5\text{ ms}/20\text{ ms})} \\
&= 10\text{ V} + (-5.45\text{ V})(e^{-0.25}) \\
&= 10\text{ V} + (-5.45\text{ V})(0.779) \\
&= 10\text{ V} + (-4.25\text{ V}) \\
&= 5.75\text{ V}
\end{aligned}
$$

At $t = 70$ ms, V_{in} changes from 10 V to 0 V and remains at 0 V for 80 ms – 70 ms = 10 ms. The integrator rejects the change in V_{in} and decreases towards 0 V for 10 ms. From the general exponential equation, V_{out} (80 ms) = $V_f + (V_i - V_f)\, e^{-t/\tau}$, where $V_i = 5.75$ V, $V_f = 0$ V, $t = 10$ ms, and $\tau = 20$ ms.

$$
\begin{aligned}
V_{out}\,(80\text{ ms}) &= 0\text{ V} + [5.75\text{ V} - 0\text{ V}]\; e^{-(10\text{ ms}/20\text{ ms})} \\
&= 0\text{ V} + (5.75\text{ V})(e^{-0.50}) \\
&= (5.75\text{ V})(0.607) \\
&= 3.49\text{ V}
\end{aligned}
$$

At $t = 80$ ms, V_{in} changes from 0 V to 10 V and remains at 10 V for 100 ms – 80 ms = 20 ms. The integrator rejects the change in V_{in} and increases towards 10 V for 20 ms. From the general exponential equation, V_{out} (100 ms) = $V_f + (V_i - V_f)\, e^{-t/\tau}$, where $V_i = 3.49$ V, $V_f = 10$ V, $t = 20$ ms, and $\tau = 20$ ms.

$$\begin{aligned} V_{out}\text{ (100 ms)} &= 10\text{ V} + [3.49\text{ V} - 10\text{ V}]\, e^{-(20\text{ ms} / 20\text{ ms})} \\ &= 10\text{ V} + (-6.51\text{ V})(e^{-1.0}) \\ &= 10\text{ V} + (-6.51\text{ V})(0.368) \\ &= 10\text{ V} + (-2.40\text{ V}) \\ &= \mathbf{7.60\ V} \end{aligned}$$

Figure 21-24 shows the integrated output for the input waveform in Figure 21-23.

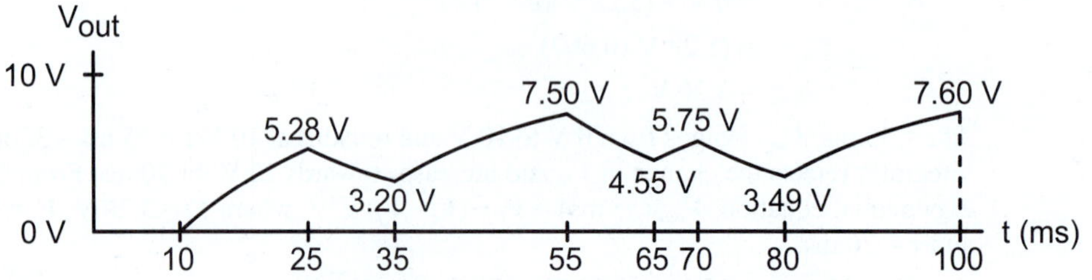

Figure 21-24: Integrated Output for Figure 21-23

2. Problem: The nonperiodic pulse waveform Figure 21-23, V_{in}, is applied to a differentiator with a time constant $\tau = 20$ ms. What is V_{out}, the output voltage of the differentiator, at $t = 100$ ms? Assume that $V_{out} = 0$ V at $t = 0$ ms.

Solution: To determine the value of V_{out} at $t = 100$ ms, successively apply the general exponential equation to each section of the input waveform to determine V_{out} at the start and end of each pulse.

From the problem statement, at $t = 0$ ms, $V_{out} = 0$ V. V_{in} remains at 0 V until $t = 10$ ms, so that V_{out} remains at 0 V.

At $t = 10$ ms, V_{in} changes from 0 V to 10 V and remains at 10 V for 25 ms – 10 ms = 15 ms. The differentiator responds to the 10 V increase in V_{in} by increasing to 0 V + 10 V = 10 V. It then rejects the constant 10 V input and decreases towards 0 V. From the general exponential equation, V_{out} (25 ms) = $V_f + (V_i - V_f)\, e^{-t/\tau}$, where $V_i = 10$ V, $V_f = 0$ V, $t = 15$ ms, and $\tau = 20$ ms.

$$\begin{aligned} V_{out}\text{ (25 ms)} &= 0\text{ V} + [10\text{ V} - 0\text{ V}]\, e^{-(15\text{ ms} / 20\text{ ms})} \\ &= 0\text{ V} + (10\text{ V})(e^{-0.75}) \\ &= (10\text{ V})(0.472) \\ &= 4.72\text{ V} \end{aligned}$$

At $t = 25$ ms, V_{in} changes from 10 V to 0 V and remains at 0 V for 35 ms – 20 ms = 10 ms. The differentiator responds to the 10 V decrease in V_{in} by decreasing to 4.72 V – 10 V = –5.28 V. It then rejects the constant 0 V input and increases towards 0 V. From the general exponential equation, V_{out} (35 ms) = $V_f + (V_i - V_f)\, e^{-t/\tau}$, where $V_i = -5.28$ V, $V_f = 0$ V, $t = 10$ ms, and $\tau =$ 20 ms.

$$\begin{aligned} V_{out}\text{ (35 ms)} &= 0\text{ V} + [-5.28\text{ V} - 0\text{ V}]\, e^{-(10\text{ ms} / 20\text{ ms})} \\ &= 0\text{ V} + (-5.28\text{ V})(e^{-0.50}) \\ &= (-5.28\text{ V})(0.607) \\ &= -3.20\text{ V} \end{aligned}$$

At $t = 35$ ms, V_{in} changes from 0 V to 10 V and remains at 10 V for 55 ms – 35 ms = 20 ms. The differentiator responds to the 10 V increase in V_{in} by increasing to –3.20 V + 10 V = 6.80 V. It then rejects the constant 10 V input and decreases towards 0 V. From the general exponential equation, V_{out} (55 ms) = $V_f + (V_i - V_f)\, e^{-t/\tau}$, where $V_i = 6.80$ V, $V_f = 0$ V, $t = 20$ ms, and $\tau = 20$ ms.

$$\begin{aligned} V_{out}(55\text{ ms}) &= 0\text{ V} + [6.80\text{ V} - 0\text{ V}]\, e^{-(20\text{ ms}/20\text{ ms})} \\ &= 0\text{ V} + (6.80\text{ V})(e^{-1.0}) \\ &= 0\text{ V} + (6.80\text{ V})(0.368) \\ &= 0\text{ V} + (2.50\text{ V}) \\ &= 2.50\text{ V} \end{aligned}$$

At t = 55 ms, V_{in} changes from 10 V to 0 V and remains at 0 V for 65 ms – 55 ms = 10 ms. The differentiator responds to the 10 V decrease in V_{in} by decreasing to 2.50 V – 10 V = –7.50 V. It then rejects the constant 0 V input and increases towards 0 V. From the general exponential equation, V_{out} (65 ms) = $V_f + (V_i - V_f)\, e^{-t/\tau}$, where V_i = –7.50 V, V_f = 0 V, t = 10 ms, and τ = 20 ms.

$$\begin{aligned} V_{out}(65\text{ ms}) &= 0\text{ V} + [-7.50\text{ V} - 0\text{ V}]\, e^{-(10\text{ ms}/20\text{ ms})} \\ &= 0\text{ V} + (-7.50\text{ V})(e^{-0.50}) \\ &= (-7.50\text{ V})(0.607) \\ &= -4.55\text{ V} \end{aligned}$$

At t = 65 ms, V_{in} changes from 0 V to 10 V and remains at 10 V for 70 ms – 65 ms = 5 ms. The differentiator responds to the 10 V increase in V_{in} by increasing to –4.55 V + 10 V = 5.45 V. It then rejects the constant 10 V input and decreases towards 0 V. From the general exponential equation, V_{out} (70 ms) = $V_f + (V_i - V_f)\, e^{-t/\tau}$, where V_i = 5.45 V, V_f = 0 V, t = 5 ms, and τ = 20 ms.

$$\begin{aligned} V_{out}(70\text{ ms}) &= 0\text{ V} + [5.45\text{ V} - 0\text{ V}]\, e^{-(5\text{ ms}/20\text{ ms})} \\ &= 0\text{ V} + (5.45\text{ V})(e^{-0.25}) \\ &= (5.45\text{ V})(0.779) \\ &= 4.25\text{ V} \end{aligned}$$

At t = 70 ms, V_{in} changes from 10 V to 0 V and remains at 0 V for 80 ms – 70 ms = 10 ms. The differentiator responds to the 10 V decrease in V_{in} by decreasing to 4.25 V – 10 V = –5.75 V. It then rejects the constant 0 V input and increases towards 0 V. From the general exponential equation, V_{out} (80 ms) = $V_f + (V_i - V_f)\, e^{-t/\tau}$, where V_i = –5.75 V, V_f = 0 V, t = 10 ms, and τ = 20 ms.

$$\begin{aligned} V_{out}(80\text{ ms}) &= 0\text{ V} + [-5.75\text{ V} - 0\text{ V}]\, e^{-(10\text{ ms}/20\text{ ms})} \\ &= 0\text{ V} + (5.75\text{ V})(e^{-0.50}) \\ &= (-5.75\text{ V})(0.607) \\ &= -3.49\text{ V} \end{aligned}$$

At t = 80 ms, V_{in} changes from 0 V to 10 V and remains at 10 V for 100 ms – 80 ms = 20 ms. The differentiator responds to the 10 V increase in V_{in} by increasing to –3.49 V + 10 V = 6.51 V. It then rejects the constant 10 V input and decreases towards 0 V. From the general exponential equation, V_{out} (100 ms) = $V_f + (V_i - V_f)\, e^{-t/\tau}$, where V_i = 6.51 V, V_f = 0 V, t = 20 ms, and τ = 20 ms.

$$\begin{aligned} V_{out}(100\text{ ms}) &= 0\text{ V} + [6.51\text{ V} - 0\text{ V}]\, e^{-(20\text{ ms}/20\text{ ms})} \\ &= 0\text{ V} + (6.51\text{ V})(e^{-1.0}) \\ &= (6.51\text{ V})(0.368) \\ &= 2.40\text{ V} \end{aligned}$$

At t = 100 ms, V_{in} changes from 10 V to 0 V. The differentiator responds to the change in V_{in} by decreasing by 10 V to 2.40 V – 10 V = –7.60 V. **At t = 100 ms, V_{out} is a transition from 2.40 V (just before V_{in} changes from 10 V to 0 V) to –7.60 V (just before V_{in} changes from 10 V to 0 V).**

Figure 21-25 shows the integrated output for the input waveform in Figure 21-23.

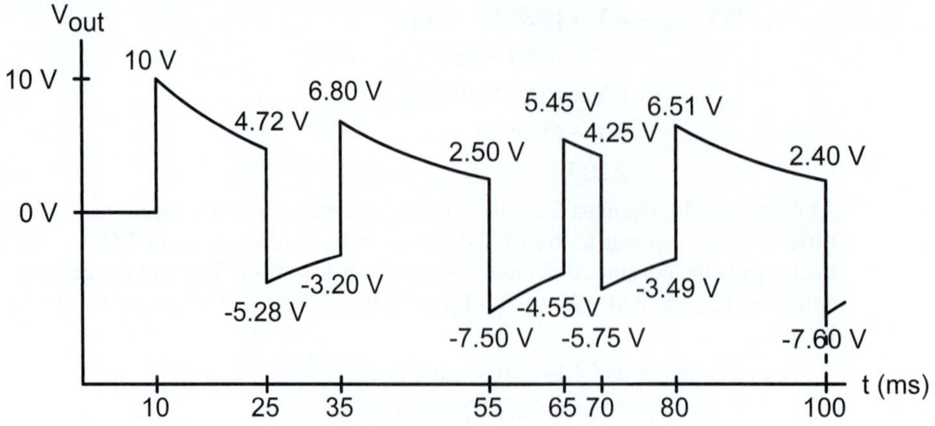

Figure 21-25: Differentiated Output for Figure 21-23

For Problems 3 and 4, refer to the nonperiodic pulse waveform in Figure 21-26.

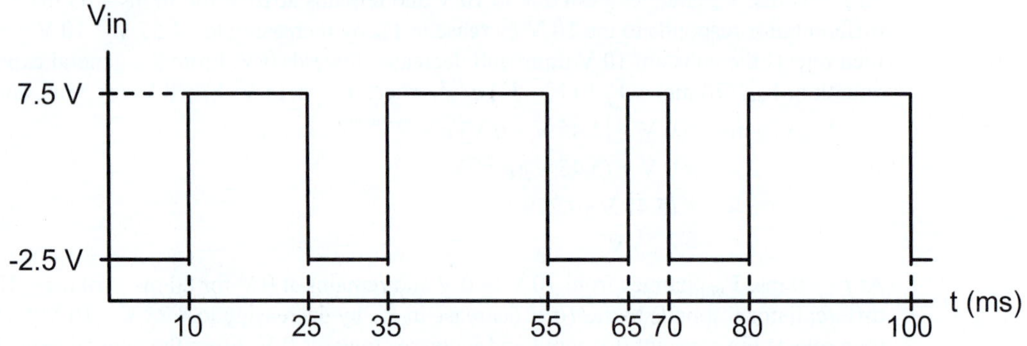

Figure 21-26: Nonperiodic Pulse Waveform for Problems 3 and 4

3. Problem: The nonperiodic pulse waveform in Figure 21-26, V_{in}, is applied to an integrator with a time constant $\tau = 20$ ms. What is V_{out}, the output voltage of the integrator, at $t = 100$ ms? Assume that $V_{out} = 0$ V at $t = 0$ ms.

Solution: To determine the value of V_{out} at $t = 100$ ms, successively apply the general exponential equation to each section of the input waveform to determine V_{out} at the start and end of each pulse.

From the problem statement, at $t = 0$ ms, $V_{out} = 0$ V and V_{in} remains at –2.5 V until $t = 10$ ms. Because $V_{out} > V_{in}$, V_{out} decreases towards –2.5 V for 10 ms – 0 ms = 10 ms. From the general exponential equation, $V_{out}\,(10\text{ ms}) = V_f + (V_i - V_f)\,e^{-t/\tau}$, where $V_i = 0$ V, $V_f = -2.5$ V, $t = 10$ ms, and $\tau = 20$ ms.

$$
\begin{aligned}
V_{out}\,(10\text{ ms}) &= -2.5\text{ V} + [0\text{ V} - (-2.5\text{ V})]\,e^{-(10\text{ ms}/20\text{ ms})}\\
&= -2.5\text{ V} + (2.5\text{ V})(e^{-0.50})\\
&= -2.5\text{ V} + (2.5\text{ V})(0.607)\\
&= -2.5\text{ V} + (1.52\text{ V})\\
&= -0.984\text{ V}
\end{aligned}
$$

At $t = 10$ ms, V_{in} changes from –2.5 V to 7.5 V and remains at 7.5 V for 25 ms – 10 ms = 15 ms. The integrator rejects the change in V_{in} and increases towards 10 V for 15 ms. From the general exponential equation, $V_{out}\,(25\text{ ms}) = V_f + (V_i - V_f)\,e^{-t/\tau}$, where $V_i = -0.984$ V, $V_f = 7.5$ V, $t = 15$ ms, and $\tau = 20$ ms.

$$
\begin{aligned}
V_{out}\,(25\text{ ms}) &= 7.5\text{ V} + [-0.984\text{ V} - 7.5\text{ V}]\,e^{-(15\text{ ms}/20\text{ ms})}\\
&= 7.5\text{ V} + (-8.48\text{ V})(e^{-0.75})
\end{aligned}
$$

$= 7.5\text{ V} + (-8.48\text{ V})(0.472)$

$= 7.5\text{ V} + (-4.01\text{ V})$

$= 3.49\text{ V}$

At $t = 25$ ms, V_{in} changes from 7.5 V to –2.5 V and remains at –2.5 V for 35 ms – 20 ms = 10 ms. The integrator rejects the change in V_{in} and decreases towards –2.5 V for 10 ms. From the general exponential equation, V_{out} (35 ms) = $V_f + (V_i - V_f)\, e^{-t/\tau}$, where $V_i = 3.49$ V, $V_f = -2.5$ V, $t = 10$ ms, and $\tau = 20$ ms.

$$V_{out}\text{ (35 ms)} = -2.5\text{ V} + [3.49\text{ V} - (-2.5\text{ V})]\, e^{-(10\text{ ms}/20\text{ ms})}$$
$$= -2.5\text{ V} + (5.99\text{ V})(e^{-0.50})$$
$$= -2.5\text{ V} + (5.99\text{ V})(0.607)$$
$$= -2.5\text{ V} + 3.64\text{ V}$$
$$= 1.14\text{ V}$$

At $t = 35$ ms, V_{in} changes from –2.5 V to 7.5 V and remains at 7.5 V for 55 ms – 35 ms = 20 ms. The integrator rejects the change in V_{in} and increases towards 7.5 V for 20 ms. From the general exponential equation, V_{out} (55 ms) = $V_f + (V_i - V_f)\, e^{-t/\tau}$, where $V_i = 1.14$ V, $V_f = 7.5$ V, $t = 20$ ms, and $\tau = 20$ ms.

$$V_{out}\text{ (55 ms)} = 7.5\text{ V} + [1.14\text{ V} - 7.5\text{ V}]\, e^{-(20\text{ ms}/20\text{ ms})}$$
$$= 7.5\text{ V} + (-6.37\text{ V})(e^{-1.0})$$
$$= 7.5\text{ V} + (-6.37\text{ V})(0.368)$$
$$= 7.5\text{ V} + (-2.34\text{ V})$$
$$= 5.16\text{ V}$$

At $t = 55$ ms, V_{in} changes from 7.5 V to –2.5 V and remains at –2.5 V for 65 ms – 55 ms = 10 ms. The integrator rejects the change in V_{in} and decreases towards –2.5 V for 10 ms. From the general exponential equation, V_{out} (65 ms) = $V_f + (V_i - V_f)\, e^{-t/\tau}$, where $V_i = 5.16$ V, $V_f = -2.5$ V, $t = 10$ ms, and $\tau = 20$ ms.

$$V_{out}\text{ (65 ms)} = -2.5\text{ V} + [5.16\text{ V} - (-2.5\text{ V})]\, e^{-(10\text{ ms}/20\text{ ms})}$$
$$= -2.5\text{ V} + (7.66\text{ V})(e^{-0.50})$$
$$= -2.5\text{ V} + (7.66\text{ V})(0.607)$$
$$= -2.5\text{ V} + 4.65\text{ V}$$
$$= 2.15\text{ V}$$

At $t = 65$ ms, V_{in} changes from –2.5 V to 7.5 V and remains at 7.5 V for 70 ms – 65 ms = 5 ms. The integrator rejects the change in V_{in} and increases towards 7.5 V for 5 ms. From the general exponential equation, V_{out} (70 ms) = $V_f + (V_i - V_f)\, e^{-t/\tau}$, where $V_i = 2.15$ V, $V_f = 7.5$ V, $t = 5$ ms, and $\tau = 20$ ms.

$$V_{out}\text{ (70 ms)} = 7.5\text{ V} + [2.15\text{ V} - 7.5\text{ V}]\, e^{-(5\text{ ms}/20\text{ ms})}$$
$$= 7.5\text{ V} + (-5.36\text{ V})(e^{-0.25})$$
$$= 7.5\text{ V} + (-5.36\text{ V})(0.779)$$
$$= 7.5\text{ V} + (-4.17\text{ V})$$
$$= 3.33\text{ V}$$

At $t = 70$ ms, V_{in} changes from 7.5 V to –2.5 V and remains at –2.5 V for 80 ms – 70 ms = 10 ms. The integrator rejects the change in V_{in} and decreases towards –2.5 V for 10 ms. From the general exponential equation, V_{out} (80 ms) = $V_f + (V_i - V_f)\, e^{-t/\tau}$, where $V_i = 3.33$ V, $V_f = -2.5$ V, $t = 10$ ms, and $\tau = 20$ ms.

$$V_{out}\text{ (65 ms)} = -2.5\text{ V} + [3.33\text{ V} - (-2.5\text{ V})]\, e^{-(10\text{ ms}/20\text{ ms})}$$
$$= -2.5\text{ V} + (5.83\text{ V})(e^{-0.50})$$
$$= -2.5\text{ V} + (5.83\text{ V})(0.607)$$
$$= -2.5\text{ V} + 3.54\text{ V}$$

$= 1.04\text{ V}$

At $t = 80$ ms, V_{in} changes from –2.5 V to 7.5 V and remains at 7.5 V for 100 ms – 80 ms = 20 ms. The integrator rejects the change in V_{in} and increases towards 7.5 V for 20 ms. From the general exponential equation, $V_{out}\,(100\text{ ms}) = V_f + (V_i - V_f)\,e^{-t/\tau}$, where $V_i = 1.04$ V, $V_f = 7.5$ V, $t = 20$ ms, and $\tau = 20$ ms.

$$V_{out}\,(100\text{ ms}) = 7.5\text{ V} + [1.04\text{ V} - 7.5\text{ V}]\,e^{-(20\text{ ms}/20\text{ ms})}$$
$$= 7.5\text{ V} + (-6.47\text{ V})(e^{-1.0})$$
$$= 7.5\text{ V} + (-6.47\text{ V})(0.368)$$
$$= 7.5\text{ V} + (-2.38\text{ V})$$
$$= \mathbf{5.12\text{ V}}$$

Figure 21-27 shows the integrated output for the input waveform in Figure 21-26.

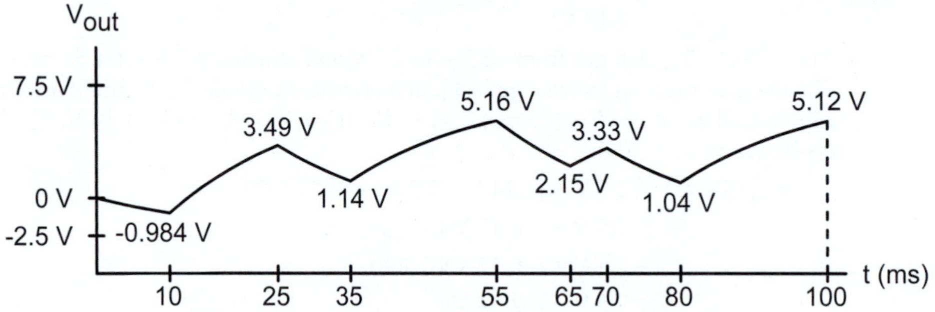

Figure 21-27: Integrated Output for Figure 21-26

4. Problem: Assume that the nonperiodic pulse waveform Figure 21-26 is applied to a differentiator with a time constant $\tau = 20$ ms. What is V_{out}, the output voltage of the differentiator, at $t = 100$ ms? Assume that $V_{out} = 0$ V at $t = 0$ ms.

Solution: To determine the value of V_{out} at $t = 100$ ms, successively apply the general exponential equation to each section of the input waveform to determine V_{out} at the start and end of each pulse.

From the problem statement, at $t = 0$ ms, $V_{out} = 0$ V and V_{in} remains at –2.5 V until $t = 10$ ms. The differentiator rejects the constant –2.5 V input, so that V_{out} remains at 0 V.

At $t = 10$ ms, V_{in} changes from –2.5 V to 7.5 V and remains at 7.5 V for 25 ms – 10 ms = 15 ms. The differentiator responds to the 10 V increase in V_{in} by increasing to 0 V + 10 V = 10 V. It then rejects the constant 7.5 V input and decreases towards 0 V. From the general exponential equation, $V_{out}\,(25\text{ ms}) = V_f + (V_i - V_f)\,e^{-t/\tau}$, where $V_i = 10$ V, $V_f = 0$ V, $t = 15$ ms, and $\tau = 20$ ms.

$$V_{out}\,(25\text{ ms}) = 0\text{ V} + [10\text{ V} - 0\text{ V}]\,e^{-(15\text{ ms}/20\text{ ms})}$$
$$= 0\text{ V} + (10\text{ V})(e^{-0.75})$$
$$= (10\text{ V})(0.472)$$
$$= 4.72\text{ V}$$

At $t = 25$ ms, V_{in} changes from 7.5 V to –2.5 V and remains at –2.5 V for 35 ms – 20 ms = 10 ms. The differentiator responds to the 10 V decrease in V_{in} by decreasing to 4.72 V – 10 V = –5.28 V. It then rejects the constant –2.5 V input and increases towards 0 V. From the general exponential equation, $V_{out}\,(35\text{ ms}) = V_f + (V_i - V_f)\,e^{-t/\tau}$, where $V_i = -5.28$ V, $V_f = 0$ V, $t = 10$ ms, and $\tau = 20$ ms.

$$V_{out}\,(35\text{ ms}) = 0\text{ V} + [-5.28\text{ V} - 0\text{ V}]\,e^{-(10\text{ ms}/20\text{ ms})}$$
$$= 0\text{ V} + (-5.28\text{ V})(e^{-0.50})$$
$$= (-5.28\text{ V})(0.607)$$
$$= -3.20\text{ V}$$

At $t = 35$ ms, V_{in} changes from –2.5 V to 7.5 V and remains at 7.5 V for 55 ms – 35 ms = 20 ms. The differentiator responds to the 10 V increase in V_{in} by increasing to –3.20 V + 10 V = 6.80 V. It then rejects the constant 7.5 V input and decreases towards 0 V. From the general exponential equation, V_{out} (55 ms) = $V_f + (V_i - V_f)\, e^{-t/\tau}$, where $V_i = 6.80$ V, $V_f = 0$ V, $t = 20$ ms, and $\tau = 20$ ms.

$$\begin{aligned} V_{out}\,(55\text{ ms}) &= 0\text{ V} + [6.80\text{ V} - 0\text{ V}]\, e^{-(20\text{ ms}/20\text{ ms})} \\ &= 0\text{ V} + (6.80\text{ V})(e^{-1.0}) \\ &= 0\text{ V} + (6.80\text{ V})(0.368) \\ &= 0\text{ V} + (2.50\text{ V}) \\ &= 2.50\text{ V} \end{aligned}$$

At $t = 55$ ms, V_{in} changes from 7.5 V to –2.5 V and remains at –2.5 V for 65 ms – 55 ms = 10 ms. The differentiator responds to the 10 V decrease in V_{in} by decreasing to 2.50 V – 10 V = –7.50 V. It then rejects the constant –2.5 V input and increases towards 0 V. From the general exponential equation, V_{out} (65 ms) = $V_f + (V_i - V_f)\, e^{-t/\tau}$, where $V_i = -7.50$ V, $V_f = 0$ V, $t = 10$ ms, and $\tau = 20$ ms.

$$\begin{aligned} V_{out}\,(65\text{ ms}) &= 0\text{ V} + [-7.50\text{ V} - 0\text{ V}]\, e^{-(10\text{ ms}/20\text{ ms})} \\ &= 0\text{ V} + (-7.50\text{ V})(e^{-0.50}) \\ &= (-7.50\text{ V})(0.607) \\ &= -4.55\text{ V} \end{aligned}$$

At $t = 65$ ms, V_{in} changes from –2.5 V to 7.5 V and remains at 7.5 V for 70 ms – 65 ms = 5 ms. The differentiator responds to the 10 V increase in V_{in} by increasing to –4.55 V + 10 V = 5.45 V. It then rejects the constant 7.5 V input and decreases towards 0 V. From the general exponential equation, V_{out} (70 ms) = $V_f + (V_i - V_f)\, e^{-t/\tau}$, where $V_i = 5.45$ V, $V_f = 0$ V, $t = 5$ ms, and $\tau = 20$ ms.

$$\begin{aligned} V_{out}\,(70\text{ ms}) &= 0\text{ V} + [5.45\text{ V} - 0\text{ V}]\, e^{-(5\text{ ms}/20\text{ ms})} \\ &= 0\text{ V} + (5.45\text{ V})(e^{-0.25}) \\ &= (5.45\text{ V})(0.779) \\ &= 4.25\text{ V} \end{aligned}$$

At $t = 70$ ms, V_{in} changes from 7.5 V to –2.5 V and remains at –2.5 V for 80 ms – 70 ms = 10 ms. The differentiator responds to the 10 V decrease in V_{in} by decreasing to 4.25 V – 10 V = –5.75 V. It then rejects the constant –2.5 V input and increases towards 0 V. From the general exponential equation, V_{out} (80 ms) = $V_f + (V_i - V_f)\, e^{-t/\tau}$, where $V_i = -5.75$ V, $V_f = 0$ V, $t = 10$ ms, and $\tau = 20$ ms.

$$\begin{aligned} V_{out}\,(80\text{ ms}) &= 0\text{ V} + [-5.75\text{ V} - 0\text{ V}]\, e^{-(10\text{ ms}/20\text{ ms})} \\ &= 0\text{ V} + (5.75\text{ V})(e^{-0.50}) \\ &= (-5.75\text{ V})(0.607) \\ &= -3.49\text{ V} \end{aligned}$$

At $t = 80$ ms, V_{in} changes from –2.5 V to 7.5 V and remains at 7.5 V for 100 ms – 80 ms = 20 ms. The differentiator responds to the 10 V increase in V_{in} by increasing to –3.49 V + 10 V = 6.51 V. It then rejects the constant 7.5 V input and decreases toward 0 V. From the general exponential equation, V_{out} (100 ms) = $V_f + (V_i - V_f)\, e^{-t/\tau}$, where $V_i = 6.51$ V, $V_f = 0$ V, $t = 20$ ms, and $\tau = 20$ ms.

$$\begin{aligned} V_{out}\,(100\text{ ms}) &= 0\text{ V} + [6.51\text{ V} - 0\text{ V}]\, e^{-(20\text{ ms}/20\text{ ms})} \\ &= 0\text{ V} + (6.51\text{ V})(e^{-1.0}) \\ &= (6.51\text{ V})(0.368) \\ &= 2.40\text{ V} \end{aligned}$$

At $t = 100$ ms, V_{in} changes from 7.5 V to –2.5 V. The differentiator responds to the 10 V decrease in V_{in} by decreasing to 2.40 V – 10 V = –7.60 V. **At $t = 100$ ms, V_{out} is a transition from 2.40 V (just before V_{in} changes from 10 V to 0 V) to –7.60 V (just after V_{in} changes from 10 V to 0 V).** Note that output values are the same as those for the input waveform in Figure 21-23, as the only difference for the input waveform in Figure 21-26 is a –2.5 V dc offset and differentiators reject constant (dc) inputs.

Figure 21-28 shows the differentiated output for the input waveform in Figure 21-26.

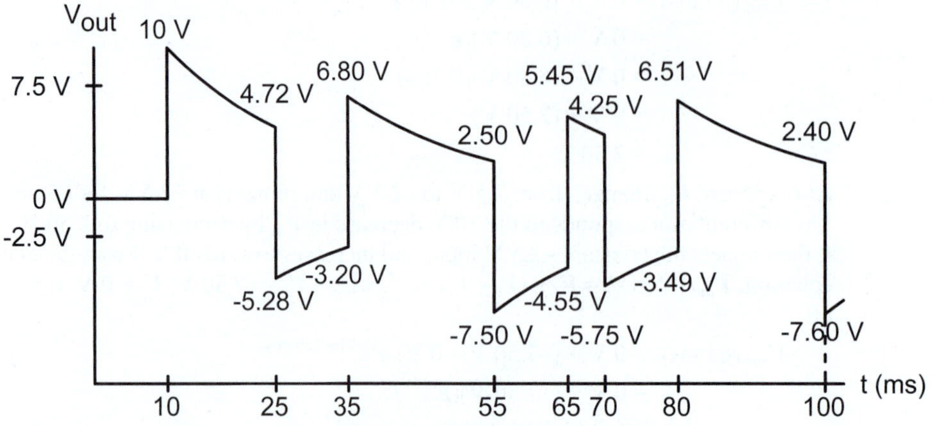

Figure 21-28: Differentiated Output for Figure 21-26

For Problems 5 and 6, refer to the nonperiodic pulse waveform in Figure 21-29.

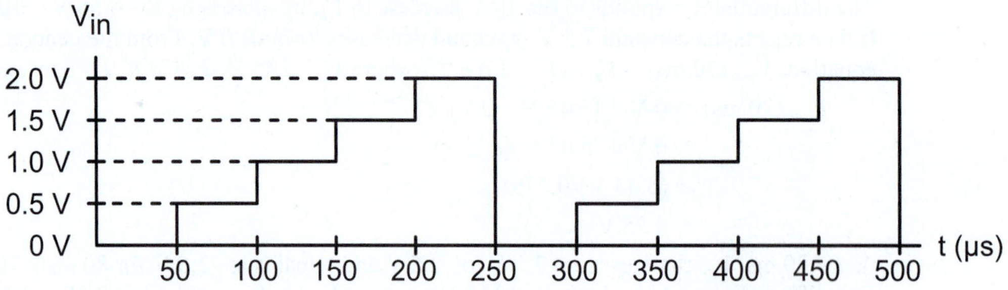

Figure 21-29: Nonperiodic Pulse Waveform for Problems 5 and 6

5. Problem: The nonperiodic pulse waveform in Figure 21-29, V_{in}, is applied to an integrator with a time constant τ = 100 µs. What is V_{out}, the output voltage of the integrator, at t = 500 µs? Assume that V_{out} = 0 V at t = 0 µs.

Solution: To determine the value of V_{out} at t = 500 µs, successively apply the general exponential equation to each section of the input waveform to determine V_{out} at the start and end of each pulse.

From the problem statement, at t = 0 µs, V_{out} = 0 V. V_{in} remains at 0 V until t = 50 µs, so that V_{out} remains at 0 V.

At t = 50 µs, V_{in} changes from 0 V to 0.5 V and remains at 0.5 V for 100 µs – 50 µs = 50 µs. The integrator rejects the change in V_{in} and increases towards 0.5 V for 50 µs. From the general exponential equation, $V_{out}(100\ \mu s) = V_f + (V_i - V_f)\,e^{-t/\tau}$, where V_i = 0 V, V_f = 0.5 V, t = 50 µs, and τ = 100 µs.

$$\begin{aligned} V_{out}(100\ \mu s) &= 0.5\text{ V} + [0\text{ V} - 0.5\text{ V}]\,e^{-(50\ \mu s\,/\,100\ \mu s)} \\ &= 0.5\text{ V} + (-0.5\text{ V})(e^{-0.50}) \\ &= 0.5\text{ V} + (-0.5\text{ V})(0.607) \\ &= 0.5\text{ V} + (-0.303\text{ V}) \\ &= 0.197\text{ V} \end{aligned}$$

At $t = 100\ \mu s$, V_{in} changes from 0.5 V to 1.0 V and remains at 1.0 V for 150 μs – 100 μs = 50 μs. The integrator rejects the change in V_{in} and increases towards 1.0 V for 50 μs. From the general exponential equation, $V_{out}\,(150\ \mu s) = V_f + (V_i - V_f)\,e^{-t/\tau}$, where $V_i = 0.197$ V, $V_f = 1.0$ V, $t = 50\ \mu s$, and $\tau = 100\ \mu s$.

$$\begin{aligned}V_{out}\,(150\ \mu s) &= 1.0\text{ V} + [0.197\text{ V} - 1.0\text{ V}]\,e^{-(50\ \mu s\,/\,100\ \mu s)}\\ &= 1.0\text{ V} + (-0.803\text{ V})(e^{-0.50})\\ &= 1.0\text{ V} + (-0.803\text{ V})(0.607)\\ &= 1.0\text{ V} + (-0.487\text{ V})\\ &= 0.513\text{ V}\end{aligned}$$

At $t = 150\ \mu s$, V_{in} changes from 1.0 V to 1.5 V and remains at 1.5 V for 200 μs – 150 μs = 50 μs. The integrator rejects the change in V_{in} and increases towards 0.5 V for 50 μs. From the general exponential equation, $V_{out}\,(200\ \mu s) = V_f + (V_i - V_f)\,e^{-t/\tau}$, where $V_i = 0.513$ V, $V_f = 1.5$ V, $t = 50\ \mu s$, and $\tau = 100\ \mu s$.

$$\begin{aligned}V_{out}\,(200\ \mu s) &= 1.5\text{ V} + [0.513\text{ V} - 1.5\text{ V}]\,e^{-(50\ \mu s\,/\,100\ \mu s)}\\ &= 1.5\text{ V} + (-0.987\text{ V})(e^{-0.50})\\ &= 1.5\text{ V} + (-0.987\text{ V})(0.607)\\ &= 1.5\text{ V} + (-0.599\text{ V})\\ &= 0.901\text{ V}\end{aligned}$$

At $t = 200\ \mu s$, V_{in} changes from 1.5 V to 2.0 V and remains at 2.0 V for 250 μs – 200 μs = 50 μs. The integrator rejects the change in V_{in} and increases towards 2.0 V for 50 μs. From the general exponential equation, $V_{out}\,(250\ \mu s) = V_f + (V_i - V_f)\,e^{-t/\tau}$, where $V_i = 0.901$ V, $V_f = 2.0$ V, $t = 50\ \mu s$, and $\tau = 100\ \mu s$.

$$\begin{aligned}V_{out}\,(250\ \mu s) &= 2.0\text{ V} + [0.901\text{ V} - 2.0\text{ V}]\,e^{-(50\ \mu s\,/\,100\ \mu s)}\\ &= 2.0\text{ V} + (-1.10\text{ V})(e^{-0.50})\\ &= 2.0\text{ V} + (-1.10\text{ V})(0.607)\\ &= 2.0\text{ V} + (-0.666\text{ V})\\ &= 1.33\text{ V}\end{aligned}$$

At $t = 250\ \mu s$, V_{in} changes from 2.0 V to 0 V and remains at 0 V for 300 μs – 250 μs = 50 μs. The integrator rejects the change in V_{in} and decreases towards 0 V for 50 μs. From the general exponential equation, $V_{out}\,(300\ \mu s) = V_f + (V_i - V_f)\,e^{-t/\tau}$, where $V_i = 1.33$ V, $V_f = 0$ V, $t = 50\ \mu s$, and $\tau = 100\ \mu s$.

$$\begin{aligned}V_{out}\,(300\ \mu s) &= 0\text{ V} + [1.33\text{ V} - 0\text{ V}]\,e^{-(50\ \mu s\,/\,100\ \mu s)}\\ &= 0\text{ V} + (1.33\text{ V})(e^{-0.50})\\ &= (1.33\text{ V})(0.607)\\ &= 0.809\text{ V}\end{aligned}$$

At $t = 300\ \mu s$, V_{in} changes from 0 V to 0.5 V and remains at 0.5 V for 350 μs – 300 μs = 50 μs. The integrator rejects the change in V_{in} and increases towards 0.5 V for 50 μs. From the general exponential equation, $V_{out}\,(350\ \mu s) = V_f + (V_i - V_f)\,e^{-t/\tau}$, where $V_i = 0.809$ V, $V_f = 0.5$ V, $t = 50\ \mu s$, and $\tau = 100\ \mu s$.

$$\begin{aligned}V_{out}\,(350\ \mu s) &= 0.5\text{ V} + [0.809\text{ V} - 0.5\text{ V}]\,e^{-(50\ \mu s\,/\,100\ \mu s)}\\ &= 0.5\text{ V} + (0.309\text{ V})(e^{-0.50})\\ &= 0.5\text{ V} + (0.309\text{ V})(0.607)\\ &= 0.5\text{ V} + (0.187\text{ V})\\ &= 0.687\text{ V}\end{aligned}$$

At $t = 350\ \mu s$, V_{in} changes from 0.5 V to 1.0 V and remains at 1.0 V for 400 μs – 350 μs = 50 μs. The integrator rejects the change in V_{in} and increases towards 1.0 V for 50 μs. From the general exponential equation, V_{out} (150 μs) = $V_f + (V_i - V_f)\, e^{-t/\tau}$, where V_i = 0.687 V, V_f = 1.0 V, t = 50 μs, and τ = 100 μs.

$$
\begin{aligned}
V_{out}\,(400\ \mu s) &= 1.0\ V + [0.687\ V - 1.0\ V]\ e^{-(50\ \mu s\ /\ 100\ \mu s)} \\
&= 1.0\ V + (-0.313\ V)(e^{-0.50}) \\
&= 1.0\ V + (-0.313\ V)(0.607) \\
&= 1.0\ V + (-0.190\ V) \\
&= 0.810\ V
\end{aligned}
$$

At t = 400 μs, V_{in} changes from 1.0 V to 1.5 V and remains at 1.5 V for 450 μs – 400 μs = 50 μs. The integrator rejects the change in V_{in} and increases towards 0.5 V for 50 μs. From the general exponential equation, V_{out} (450 μs) = $V_f + (V_i - V_f)\, e^{-t/\tau}$, where V_i = 0.810 V, V_f = 1.5 V, t = 50 μs, and τ = 100 μs.

$$
\begin{aligned}
V_{out}\,(450\ \mu s) &= 1.5\ V + [0.810\ V - 1.5\ V]\ e^{-(50\ \mu s\ /\ 100\ \mu s)} \\
&= 1.5\ V + (-0.690\ V)(e^{-0.50}) \\
&= 1.5\ V + (-0.690\ V)(0.607) \\
&= 1.5\ V + (-0.418\ V) \\
&= 1.08\ V
\end{aligned}
$$

At t = 450 μs, V_{in} changes from 1.5 V to 2.0 V and remains at 2.0 V for 500 μs – 450 μs = 50 μs. The integrator rejects the change in V_{in} and increases towards 2.0 V for 50 μs. From the general exponential equation, V_{out} (500 μs) = $V_f + (V_i - V_f)\, e^{-t/\tau}$, where V_i = 1.08 V, V_f = 2.0 V, t = 50 μs, and τ = 100 μs.

$$
\begin{aligned}
V_{out}\,(500\ \mu s) &= 2.0\ V + [1.08\ V - 2.0\ V]\ e^{-(50\ \mu s\ /\ 100\ \mu s)} \\
&= 2.0\ V + (-0.918\ V)(e^{-0.50}) \\
&= 2.0\ V + (-0.918\ V)(0.607) \\
&= 2.0\ V + (-0.557\ V) \\
&= \mathbf{1.44\ V}
\end{aligned}
$$

Figure 21-30 shows the integrated output for the input waveform in Figure 21-29.

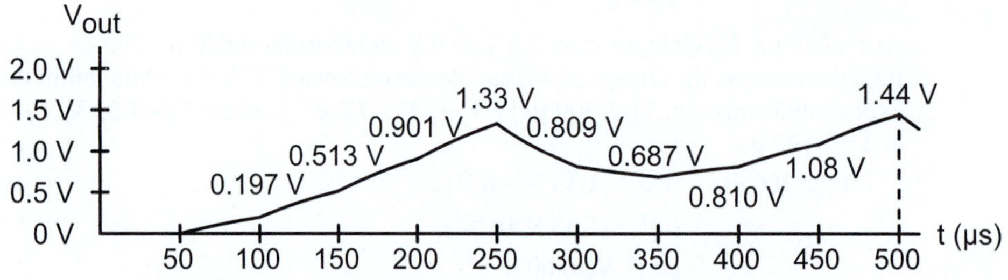

Figure 21-30: Integrated Output for Figure 21-29

6. Problem: The nonperiodic pulse waveform in Figure 21-29, V_{in}, is applied to a differentiator with a time constant τ = 100 μs. What is V_{out}, the output voltage of the differentiator, at t = 500 μs? Assume that V_{out} = 0 V at t = 0.

Solution: To determine the value of V_{out} at t = 500 μs, successively apply the general exponential equation to each section of the input waveform to determine V_{out} at the start and end of each pulse.

From the problem statement, at t = 0 μs, V_{out} = 0 V. V_{in} remains at 0 V until t = 50 μs, so that V_{out} remains at 0 V.

At t = 50 μs, V_{in} changes from 0 V to 0.5 V and remains at 0.5 V for 100 μs – 50 μs = 50 μs. The differentiator responds to the 0.5 V increase in V_{in} by increasing to 0 V + 0.5 V = 0.5 V. It then rejects the constant 0.5 V input and decreases towards 0 V for 50 μs. From the general exponential equation, V_{out} (100 μs) = $V_f + (V_i - V_f)\, e^{-t/\tau}$, where V_i = 0.5 V, V_f = 0 V, t = 50 μs, and τ = 100 μs.

$$V_{out}\,(100\ \mu s) = 0\ V + [0.5\ V - 0\ V]\ e^{-(50\ \mu s\ /\ 100\ \mu s)}$$
$$= 0\ V + (0.5\ V)(e^{-0.50})$$
$$= (0.5\ V)(0.607)$$
$$= 0.303\ V$$

At $t = 100\ \mu s$, V_{in} changes from 0.5 V to 1.0 V and remains at 1.0 V for 150 μs – 50 μs = 50 μs. The differentiator responds to the 0.5 V increase in V_{in} by increasing to 0.303 V + 0.5 V = 0.803 V. It then rejects the constant 1.0 V input and decreases towards 0 V for 50 μs. From the general exponential equation, $V_{out}\,(150\ \mu s) = V_f + (V_i - V_f)\,e^{-t/\tau}$, where $V_i = 0.803$ V, $V_f = 0$ V, $t = 50$ μs, and $\tau = 100$ μs.

$$V_{out}\,(150\ \mu s) = 0\ V + [0.803\ V - 0\ V]\ e^{-(50\ \mu s\ /\ 100\ \mu s)}$$
$$= 0\ V + (0.803\ V)(e^{-0.50})$$
$$= (0.803\ V)(0.607)$$
$$= 0.487\ V$$

At $t = 150\ \mu s$, V_{in} changes from 1.0 V to 1.5 V and remains at 1.5 V for 200 μs – 150 μs = 50 μs. The differentiator responds to the 0.5 V increase in V_{in} by increasing to 0.487 V + 0.5 V = 0.987 V. It then rejects the constant 1.5 V input and decreases towards 0 V for 50 μs. From the general exponential equation, $V_{out}\,(200\ \mu s) = V_f + (V_i - V_f)\,e^{-t/\tau}$, where $V_i = 0.987$ V, $V_f = 0$ V, $t = 50$ μs, and $\tau = 100$ μs.

$$V_{out}\,(200\ \mu s) = 0\ V + [0.987\ V - 0\ V]\ e^{-(50\ \mu s\ /\ 100\ \mu s)}$$
$$= 0\ V + (0.987\ V)(e^{-0.50})$$
$$= (0.987\ V)(0.607)$$
$$= 0.599\ V$$

At $t = 200\ \mu s$, V_{in} changes from 1.5 V to 2.0 V and remains at 2.0 V for 250 μs – 200 μs = 50 μs. The differentiator responds to the 0.5 V increase in V_{in} by increasing to 0.599 V + 0.5 V = 1.10 V. It then rejects the constant 2.0 V input and decreases towards 0 V for 50 μs. From the general exponential equation, $V_{out}\,(250\ \mu s) = V_f + (V_i - V_f)\,e^{-t/\tau}$, where $V_i = 1.10$ V, $V_f = 0$ V, $t = 50$ μs, and $\tau = 100$ μs.

$$V_{out}\,(250\ \mu s) = 0\ V + [1.10\ V - 0\ V]\ e^{-(50\ \mu s\ /\ 100\ \mu s)}$$
$$= 0\ V + (1.10\ V)(e^{-0.50})$$
$$= (1.10\ V)(0.607)$$
$$= 0.666\ V$$

At $t = 250\ \mu s$, V_{in} changes from 2.0 V to 0 V and remains at 0 V for 300 μs – 250 μs = 50 μs. The differentiator responds to the 2.0 V decrease in V_{in} by decreasing to 0.666 V – 2.0 V = 1.33 V. It then rejects the constant 0 V input and decreases towards 0 V for 50 μs. From the general exponential equation, $V_{out}\,(300\ \mu s) = V_f + (V_i - V_f)\,e^{-t/\tau}$, where $V_i = -1.33$ V, $V_f = 0$ V, $t = 50$ μs, and $\tau = 100$ μs.

$$V_{out}\,(300\ \mu s) = 0\ V + [-1.33\ V - 0\ V]\ e^{-(50\ \mu s\ /\ 100\ \mu s)}$$
$$= 0\ V + (-1.33\ V)(e^{-0.50})$$
$$= (-1.33\ V)(0.607)$$
$$= -0.809\ V$$

At $t = 300\ \mu s$, V_{in} changes from 0 V to 0.5 V and remains at 0.5 V for 350 μs – 300 μs = 50 μs. The differentiator responds to the 0.5 V increase in V_{in} by increasing to –0.809 V + 0.5 V = –0.309 V. It then rejects the constant 0.5 V input and decreases towards 0 V for 50 μs. From the general exponential equation, $V_{out}\,(350\ \mu s) = V_f + (V_i - V_f)\,e^{-t/\tau}$, where $V_i = -0.309$ V, $V_f = 0$ V, $t = 50$ μs, and $\tau = 100$ μs.

$$V_{out}\,(350\ \mu s) = 0\ V + [-0.309\ V - 0\ V]\ e^{-(50\ \mu s\ /\ 100\ \mu s)}$$
$$= 0\ V + (-0.309\ V)(e^{-0.50})$$
$$= (-0.309\ V)(0.607)$$

$= -0.187\ \text{V}$

At $\boldsymbol{t}$ = 350 μs, V_{in} changes from 0.5 V to 1.0 V and remains at 1.0 V for 400 μs – 350 μs = 50 μs. The differentiator responds to the 0.5 V increase in V_{in} by increasing to –0.187 V + 0.5 V = 0.313 V. It then rejects the constant 1.0 V input and decreases towards 0 V for 50 μs. From the general exponential equation, $V_{out}\,(400\ \mu s) = V_f + (V_i - V_f)\, e^{-t/\tau}$, where $V_i = 0.313$ V, $V_f = 0$ V, t = 50 μs, and τ = 100 μs.

$$V_{out}\,(400\ \mu\text{s}) = 0\ \text{V} + [0.313\ \text{V} - 0\ \text{V}]\ e^{-(50\ \mu\text{s}\,/\,100\ \mu\text{s})}$$
$$= 0\ \text{V} + (0.313\ \text{V})(e^{-0.50})$$
$$= (0.313\ \text{V})(0.607)$$
$$= 0.190\ \text{V}$$

At $\boldsymbol{t}$ = 400 μs, V_{in} changes from 1.0 V to 1.5 V and remains at 1.5 V for 450 μs – 400 μs = 50 μs. The differentiator responds to the 0.5 V increase in V_{in} by increasing to 0.190 V + 0.5 V = 0.690 V. It then rejects the constant 1.5 V input and decreases towards 0 V for 50 μs. From the general exponential equation, $V_{out}\,(450\ \mu s) = V_f + (V_i - V_f)\, e^{-t/\tau}$, where $V_i = 0.690$ V, $V_f = 0$ V, t = 50 μs, and τ = 100 μs.

$$V_{out}\,(450\ \mu\text{s}) = 0\ \text{V} + [0.690\ \text{V} - 0\ \text{V}]\ e^{-(50\ \mu\text{s}\,/\,100\ \mu\text{s})}$$
$$= 0\ \text{V} + (0.690\ \text{V})(e^{-0.50})$$
$$= (0.690\ \text{V})(0.607)$$
$$= 0.418\ \text{V}$$

At $\boldsymbol{t}$ = 450 μs, V_{in} changes from 1.5 V to 2.0 V and remains at 2.0 V for 500 μs – 450 μs = 50 μs. The differentiator responds to the 0.5 V increase in V_{in} by increasing to 0.418 V + 0.5 V = 0.918 V. It then rejects the constant 2.0 V input and decreases towards 0 V for 50 μs. From the general exponential equation, $V_{out}\,(500\ \mu s) = V_f + (V_i - V_f)\, e^{-t/\tau}$, where $V_i = 0.918$ V, $V_f = 0$ V, t = 50 μs, and τ = 100 μs.

$$V_{out}\,(500\ \mu\text{s}) = 0\ \text{V} + [0.918\ \text{V} - 0\ \text{V}]\ e^{-(50\ \mu\text{s}\,/\,100\ \mu\text{s})}$$
$$= 0\ \text{V} + (0.918\ \text{V})(e^{-0.50})$$
$$= (0.918\ \text{V})(0.607)$$
$$= 0.557\ \text{V}$$

At $\boldsymbol{t}$ = 500 μs, V_{in} changes from 2.0 V to 0 V. The differentiator responds to the 2.0 V decrease in V_{in} by decreasing to 0.557 V – 2.0 V = –1.44 V. **At t = 500 μs, V_{out} is a transition from 0.557 V (just before V_{in} changes from 2.0 V to 0 V) to –1.44 V (just after V_{in} changes from 2.0 V to 0 V).**

Figure 21-31 shows the differentiated output for the input waveform in Figure 21-29.

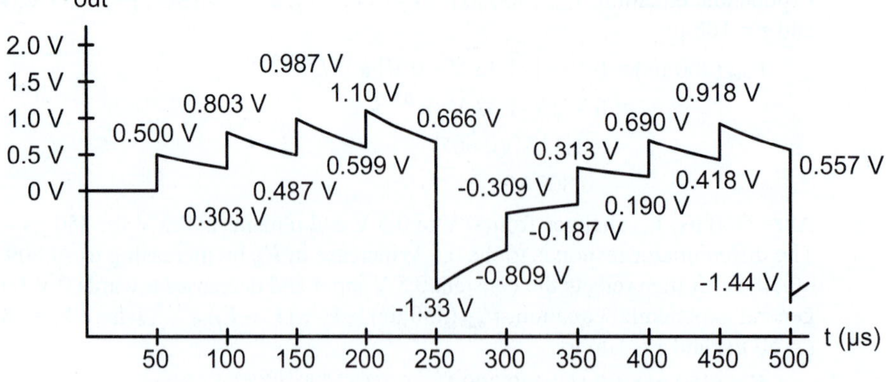

Figure 21-31: Differentiated Output for Figure 21-29

21.3 Just for Fun

1. Problem: The rise time of a pulse, t_r, is defined as the time for a pulse to change from 10% to 90% (V_{10} to V_{90}) of the voltage between the low voltage, V_L, and high voltage, V_H. For an ideal pulse, $t_r = 0$ s, but the circuitry of practical oscilloscopes limits the rise time that the oscilloscope can measure. Figure 21-32 shows the equivalent circuit for a typical oscilloscope probe and V_{osc}, the measured value of V_{ckt}, an ideal pulse input. If the RC circuit of the oscilloscope probe determines the response of the oscilloscope to a pulse input, prove that $t_r = 2.2\ RC$.

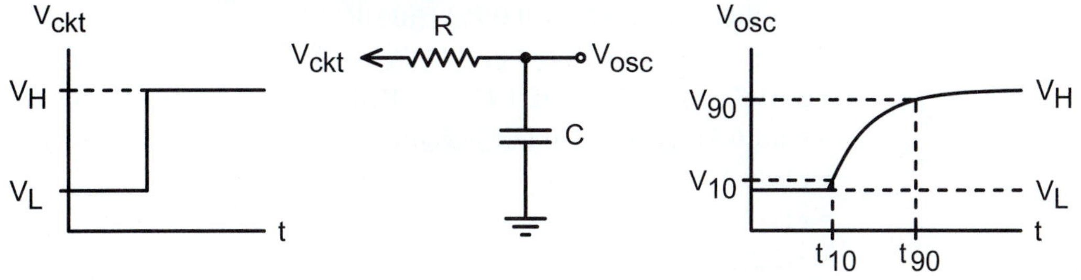

Figure 21-32: Oscilloscope Response to Ideal Pulse

Solution: If the RC circuit of the oscilloscope probe determines the response of the oscilloscope to a pulse input, then V_{osc} is the exponential response given by the general exponential equation. Solve for the time, t, for a specific value of V_{osc}.

$$V_{osc} = V_H + (V_L - V_H)\, e^{-t/\tau}$$

$$V_{osc} - V_H = V_H + (V_L - V_H)\, e^{-t/\tau} - V_H$$

$$(V_{osc} - V_H) / (V_L - V_H) = [(V_L - V_H)\, e^{-t/\tau}] / (V_L - V_H)$$

$$\ln [(V_{osc} - V_H) / (V_L - V_H)] = \ln (e^{-t/\tau})$$

$$\{\ln [(V_{osc} - V_H) / (V_L - V_H)]\}\ (-\tau) = (-t / \tau)\ (-\tau)$$

$$-\tau \ln [(V_{osc} - V_H) / (V_L - V_H)] = t$$

From the V_{osc} waveform in Figure 21-32, the 10% and 90% voltages, V_{10} and V_{90}, occur at times $t = t_{10}$ and $t = t_{90}$.

$$t_{10} = -\tau \ln [(V_{10} - V_H) / (V_L - V_H)]$$

$$t_{90} = -\tau \ln [(V_{90} - V_H) / (V_L - V_H)]$$

The rise time, t_r, is the difference between t_{90} and t_{90}.

$$\begin{aligned}
t_r &= t_{90} - t_{10} \\
&= \{-\tau \ln [(V_{90} - V_H) / (V_L - V_H)]\} - \{-\tau \ln [(V_{10} - V_H) / (V_L - V_H)]\} \\
&= -\tau \ln [(V_{90} - V_H) / (V_L - V_H)] - \ln [(V_{10} - V_H) / (V_L - V_H)] \\
&= -\tau \ln \{[(V_{90} - V_H) / (V_L - V_H)] / [(V_{10} - V_H) / (V_L - V_H)]\} \\
&= -\tau \ln \{[(V_{90} - V_H) / (V_L - V_H)]\ [(V_L - V_H) / (V_{10} - V_H)]\} \\
&= -\tau \ln [(V_{90} - V_H) / (V_{10} - V_H)] \\
&= \tau \ln [(V_{10} - V_H) / (V_{90} - V_H)]
\end{aligned}$$

Express V_{10} and V_{90} in terms of V_L and V_H.

$$\begin{aligned}
V_{10} &= V_L + 0.1\ (V_H - V_L) \\
&= V_L + 0.1\ V_H - 0.1\ V_L \\
&= 1.0\ V_L - 0.1\ V_L + 0.1\ V_H \\
&= (1.0 - 0.1)\ V_L + 0.1\ V_H \\
&= 0.9\ V_L + 0.1\ V_H
\end{aligned}$$

$$\begin{aligned}
V_{90} &= V_L + 0.9\ (V_H - V_L) \\
&= V_L + 0.9\ V_H - 0.9\ V_L
\end{aligned}$$

$= 1.0\ V_L - 0.9\ V_L + 0.9\ V_H$
$= (1.0 - 0.9)\ V_L + 0.9\ V_H$
$= 0.1\ V_L + 0.9\ V_H$

Substitute the value of V_{10} and V_{90} into the expression for t_r.

$t_r = \tau \ln [(V_{10} - V_H) / (V_{90} - V_H)]$
$= \tau \ln \{[(0.9\ V_L + 0.1\ V_H) - V_H] / [(0.1\ V_L + 0.9\ V_H) - V_H]\}$
$= \tau \ln \{[(0.9\ V_L + 0.1\ V_H) - 1.0\ V_H] / [(0.1\ V_L + 0.9\ V_H) - 1.0\ V_H]\}$
$= \tau \ln \{[0.9\ V_L + (0.1 - 1.0)\ V_H] / [0.1\ V_L + (0.9 - 1.0)\ V_H]\}$
$= \tau \ln [(0.9\ V_L - 0.9\ V_H) / (0.1\ V_L - 0.1\ V_H)]$
$= \tau \ln \{[0.9\ (V_L - V_H)] / [0.1\ (V_L - V_H)]\}$
$= \tau \ln (0.9 / 0.1)$
$= \tau \ln (9)$
$= \tau\ (2.2)$
$= 2.2\ \tau$

For an RC circuit, $\tau = RC$.

$t_r = 2.2\ \tau$
$= 2.2\ RC$

2. Problem: An oscilloscope has a bandwidth specification that indicates the frequency response of the oscilloscope and the range of frequencies for which it can accurately reproduce voltage waveforms, such as pulses. Because oscilloscopes can measure down to dc, the bandwidth is equivalent to the upper corner frequency, f_H, of the oscilloscope. A 225 MHz oscilloscope, for example, has an upper corner frequency of 225 MHz. Use $t_r = 2.2\ RC$ from Problem 1 to prove that the upper corner frequency of an oscilloscope with rise time t_r is $f_H = 0.35 / t_r$.

Solution: The equivalent circuit shown in Figure 21-32 is an RC low-pass filter. Use the equation for the upper corner frequency, f_H, to solve for RC.

$f_H = 1 / (2\pi RC)$
$(f_H)(RC) = [1 / (2\pi RC)]\ (RC)$
$[(f_H)(RC)] / f_H = [1 / (2\pi)] / (f_H)$
$RC = 1 / (2\pi f_H)$

Substitute the expression for RC into the equation for t_r.

$t_r = 2.2\ RC$
$= 2.2\ [1 / (2\pi f_H)]$
$(t_r)(f_H) = [2.2 / (2\pi f_H)]\ (f_H)$
$[(t_r)(f_H)] / t_r = [2.2 / (2\pi)] / t_r$
$= (2.2 / 6.28) / t_r$
$= 0.35 / t_r$

3. Problem: Refer to the RC integrator and waveforms shown in Figure 21-33.

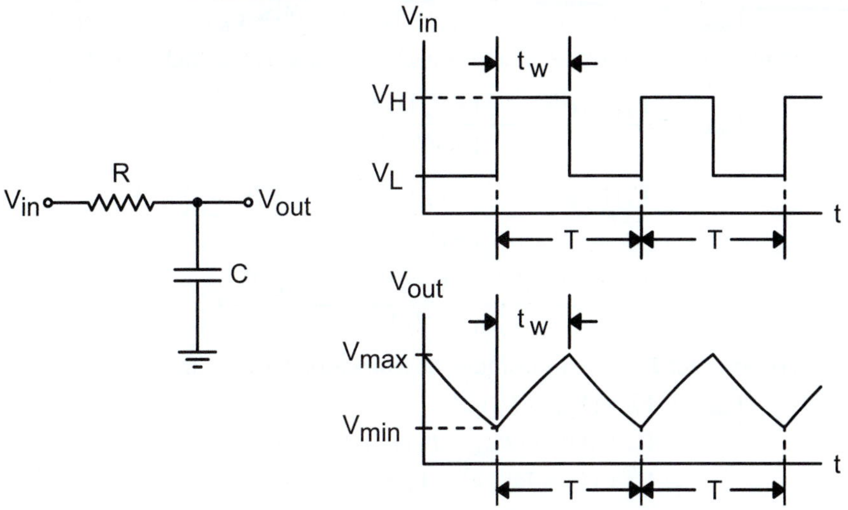

Figure 21-33: *RC* Integrator and Voltage Waveforms for Problem 3

Prove that the steady-state output waveform, V_{out}, has minimum and maximum amplitudes

$$V_{min} = \frac{V_L[1-e^{(t_w-T)/\tau}]+V_H[e^{(t_w-T)/\tau}-e^{-T/\tau}]}{1-e^{-T/\tau}}$$

$$V_{max} = \frac{V_H(1-e^{-t_w/\tau})+V_L(e^{-t_w/\tau}-e^{-T/\tau})}{1-e^{-T/\tau}}$$

Solution: At steady state, V_{out} will charge from V_{min} to V_{max} for $t = t_w$, and discharge from V_{max} to V_{min} for $t = T - t_w$ From the general exponential equation $V_{out} = V_f + (V_i - V_f)\, e^{-t/\tau}$. For the discharging portion of the waveform, $V_{out} = V_{min}$, $V_f = V_L$, $V_i = V_{max}$, and $t = T - t_w$.

$$V_{min} = V_L + (V_{max} - V_L)\, e^{-(T-t_w)/\tau}$$

For the charging portion of the waveform, $V_{out} = V_{max}$, $V_f = V_H$, $V_i = V_{min}$, and $t = t_w$.

$$V_{max} = V_H + (V_{min} - V_H)\, e^{-t_w/\tau}$$

To simplify these expressions, let $\mathrm{M} = -(T - t_w) / \tau = (t_w - T) / \tau$ and $\mathrm{N} = -t_w / \tau$.

$$V_{min} = V_L + (V_{max} - V_L)\, e^{\mathrm{M}}$$

$$V_{max} = V_H + (V_{min} - V_H)\, e^{\mathrm{N}}$$

To solve for V_{min}, first substitute the expression for V_{max} into the expression for V_{min}.

$$\begin{aligned}
V_{min} &= V_L + (V_{max} - V_L)\, e^{\mathrm{M}} \\
&= V_L + \{[V_H + (V_{min} - V_H)\, e^{\mathrm{N}}] - V_L\}\, e^{\mathrm{M}} \\
&= V_L + [V_H + V_{min}\, e^{\mathrm{N}} - V_H\, e^{\mathrm{N}} - V_L]\, e^{\mathrm{M}} \\
&= V_L + V_H\, e^{\mathrm{M}} + V_{min}\, e^{\mathrm{N}}\, e^{\mathrm{M}} - V_H\, e^{\mathrm{N}}\, e^{\mathrm{M}} - V_L\, e^{\mathrm{M}} \\
&= V_L + V_H\, e^{\mathrm{M}} + V_{min}\, e^{\mathrm{M+N}} - V_H\, e^{\mathrm{M+N}} - V_L\, e^{\mathrm{M}} \\
&= V_L - V_L\, e^{\mathrm{M}} + V_H\, e^{\mathrm{M}} - V_H\, e^{\mathrm{M+N}} + V_{min}\, e^{\mathrm{M+N}} \\
&= V_L\, (1 - e^{\mathrm{M}}) + V_H\, (e^{\mathrm{M}} - e^{\mathrm{M+N}}) + V_{min}\, e^{\mathrm{M+N}}
\end{aligned}$$

Collect the V_{min} terms on one side and solve for V_{min}.

$$\begin{aligned}
V_{min} &= V_L\, (1 - e^{\mathrm{M}}) + V_H\, (e^{\mathrm{M}} - e^{\mathrm{M+N}}) + V_{min}\, e^{\mathrm{M+N}} \\
V_{min} - V_{min}\, e^{\mathrm{M+N}} &= V_L\, (1 - e^{\mathrm{M}}) + V_H\, (e^{\mathrm{M}} - e^{\mathrm{M+N}}) + V_{min}\, e^{\mathrm{M+N}} - V_{min}\, e^{\mathrm{M+N}} \\
V_{min}\, (1 - e^{\mathrm{M+N}}) &= V_L\, (1 - e^{\mathrm{M}}) + V_H\, (e^{\mathrm{M}} - e^{\mathrm{M+N}}) \\
[V_{min}\, (1 - e^{\mathrm{M+N}})] / (1 - e^{\mathrm{M+N}}) &= [V_L\, (1 - e^{\mathrm{M}}) + V_H\, (e^{\mathrm{M}} - e^{\mathrm{M+N}})] / (1 - e^{\mathrm{M+N}})
\end{aligned}$$

$$V_{min} = [V_L(1 - e^{M}) + V_H(e^{M} - e^{M+N})] / (1 - e^{M+N})$$

Substitute the values for M and N back into the equation and simplify.

$$V_{min} = \frac{V_L[1-e^{(t_w-T)/\tau}]+V_H[e^{(t_w-T)/\tau}-e^{[(t_w-T)/\tau]+(-t_w/\tau)}]}{1-e^{[(t_w-T)/\tau]+(-t_w/\tau)}}$$

$$= \frac{V_L[1-e^{(t_w-T)/\tau}]+V_H[e^{(t_w-T)/\tau}-e^{[(t_w-T-t_w)/\tau}]}{1-e^{[(t_w-T-t_w)/\tau}}$$

$$= \frac{V_L[1-e^{(t_w-T)/\tau}]+V_H[e^{(t_w-T)/\tau}-e^{-T/\tau}]}{1-e^{-T/\tau}}$$

To solve for V_{max}, first substitute the expression for V_{min} into the expression for V_{max}.

$$V_{max} = V_H + (V_{min} - V_H)\,e^{N}$$
$$= V_H + \{[V_L + (V_{max} - V_L)\,e^{M}] - V_H\}\,e^{N}$$
$$= V_H + [V_L + V_{max}\,e^{M} - V_L\,e^{M} - V_H]\,e^{N}$$
$$= V_H + V_L\,e^{N} + V_{max}\,e^{N}\,e^{M} - V_L\,e^{N}\,e^{M} - V_H\,e^{N}$$
$$= V_H + V_L\,e^{N} + V_{max}\,e^{M+N} - V_L\,e^{M+N} - V_H\,e^{N}$$
$$= V_H - V_H\,e^{N} + V_L\,e^{N} - V_L\,e^{M+N} + V_{max}\,e^{M+N}$$
$$= V_H(1 - e^{N}) + V_L(e^{N} - e^{M+N}) + V_{max}\,e^{M+N}$$

Collect the V_{max} terms on one side and solve for V_{max}.

$$V_{max} = V_H(1 - e^{N}) + V_L(e^{N} - e^{M+N}) + V_{max}\,e^{M+N}$$
$$V_{max} - V_{max}\,e^{M+N} = V_H(1 - e^{N}) + V_L(e^{N} - e^{M+N}) + V_{max}\,e^{M+N} - V_{max}\,e^{M+N}$$
$$V_{max}(1 - e^{M+N}) = V_H(1 - e^{N}) + V_L(e^{N} - e^{M+N})$$
$$[V_{max}(1 - e^{M+N})] / (1 - e^{M+N}) = [V_H(1 - e^{N}) + V_L(e^{N} - e^{M+N})] / (1 - e^{M+N})$$
$$V_{max} = [V_H(1 - e^{N}) + V_L(e^{N} - e^{M+N})] / (1 - e^{M+N})$$

Substitute the values for M and N back into the equation and simplify.

$$V_{max} = \frac{V_H(1-e^{-t_w/\tau})+V_L[e^{-t_w/\tau}-e^{[(t_w-T)/\tau]+(-t_w/\tau)}]}{1-e^{[(t_w-T)/\tau]+(-t_w/\tau)}}$$

$$= \frac{V_H(1-e^{-t_w/\tau})+V_L[e^{-t_w/\tau}-e^{[(t_w-T-t_w)/\tau}]}{1-e^{[(t_w-T-t_w)/\tau}}$$

$$= \frac{V_H(1-e^{-t_w/\tau})+V_L(e^{-t_w/\tau}-e^{-T/\tau})}{1-e^{-T/\tau}}$$

22. Three-Phase Circuits

22.1 Basic

22.1.1 Characteristics of Three-Phase Circuits

1. Problem: Determine $V_{L(ab)}$, $V_{L(bc)}$, and $V_{L(ca)}$, the line voltages of a three-phase wye generator relative to $V_{\varphi a}$, $V_{\varphi b}$, and $V_{\varphi c}$, the phase voltages. Use $V_{\varphi a}$ for the reference angle.

   Solution: Refer to the phasor diagrams in Figure 22-1.

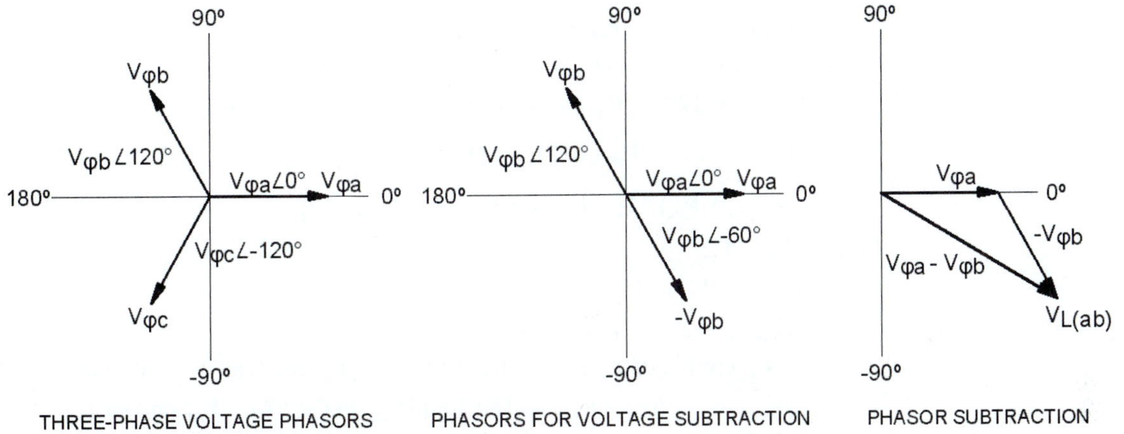

Figure 22-1: Phasor Diagrams for Three-Phase Circuit

As Figure 22-1 shows, the phase voltages, $V_{\varphi a}$, $V_{\varphi b}$, and $V_{\varphi c}$ of a three-phase wye generator are separated by 120°. The line voltages, $V_{L(ab)}$, $V_{L(bc)}$, and $V_{L(ca)}$, are the differences between the phase voltages.

$$V_{L(ab)} = V_{\varphi a} - V_{\varphi b}$$
$$V_{L(bc)} = V_{\varphi b} - V_{\varphi c}$$
$$V_{L(ca)} = V_{\varphi c} - V_{\varphi a}$$

Subtracting a phasor is the same as adding its opposite.

$$\begin{aligned} V_{L(ab)} &= V_{\varphi a} - V_{\varphi b} \\ &= V_{\varphi a} + (-V_{\varphi b}) \\ V_{L(bc)} &= V_{\varphi b} - V_{\varphi c} \\ &= V_{\varphi b} + (-V_{\varphi c}) \\ V_{L(ca)} &= V_{\varphi c} - V_{\varphi a} \\ &= V_{\varphi c} + (-V_{\varphi a}) \end{aligned}$$

The opposite of a phasor is the phasor rotated by 180°. Figure 22-1 shows the relationship between $V_{\varphi b}$ and $-V_{\varphi b}$.

$$\begin{aligned} -V_{\varphi b} &= -(V_{\varphi b} \angle 120°) \\ &= V_{\varphi b} \angle (120° - 180°) \\ &= V_{\varphi b} \angle{-60°} \end{aligned}$$

Similarly, $-V_{\varphi a} = V_{\varphi a} \angle 180°$ and $-V_{\varphi c} = V_{\varphi c} \angle 60°$. Use these values to calculate the line voltages. Figure 22-1 shows the phasor subtraction between the phase voltages $V_{\varphi a}$ and $V_{\varphi b}$ that results in the line voltage V_{ab}.

$$\begin{aligned} V_{L(ab)} &= V_{\varphi a} - V_{\varphi b} \\ &= V_{\varphi a} + (-V_{\varphi b}) \end{aligned}$$

$$= (V_{\varphi a} \angle 0°) + (V_{\varphi b} \angle -60°)$$
$$= [V_{\varphi a} \cos 0° + \mathrm{j}\,(V_{\varphi a} \sin 0°)] + \{ V_{\varphi b} \cos(-60°) + \mathrm{j}\,[V_{\varphi b} \sin(-60°)]\}$$
$$= [V_{\varphi a} \cos 0° + V_{\varphi b} \cos(-60°)] + \mathrm{j}\,[V_{\varphi a} \sin 0° + V_{\varphi b} \sin(-60°)]$$
$$= [V_{\varphi a}(1) + V_{\varphi b}(1/2)] + \mathrm{j}\,[V_{\varphi a}(0) + V_{\varphi b}(-\sqrt{3}/2)]$$
$$= \mathbf{[V_{\varphi a} + (V_{\varphi b}/2)] + j\,[(-\sqrt{3}\,V_{\varphi b})/2]}$$

Apply the same procedure for V_{bc} and V_{ca} using the values for $V_{\varphi a}$, $V_{\varphi b}$, and $V_{\varphi c}$.

$$V_{L(bc)} = V_{\varphi b} - V_{\varphi c}$$
$$= V_{\varphi b} + (-V_{\varphi c})$$
$$= (V_{\varphi b} \angle 120°) + (V_{\varphi c} \angle 60°)$$
$$= [V_{\varphi b} \cos 120° + \mathrm{j}\,(V_{\varphi b} \sin 120°)] + [V_{\varphi c} \cos 60° + \mathrm{j}\,(V_{\varphi c} \sin 60°)]$$
$$= (V_{\varphi b} \cos 120° + V_{\varphi c} \cos 60°) + \mathrm{j}\,(V_{\varphi b} \sin 120° + V_{\varphi c} \sin 60°)$$
$$= [V_{\varphi b}(-1/2) + V_{\varphi c}(1/2)] + \mathrm{j}\,[V_{\varphi b}(-\sqrt{3}/2) + V_{\varphi c}(-\sqrt{3}/2)]$$
$$= \mathbf{[(-V_{\varphi b} + V_{\varphi c})/2] + j\,[(-\sqrt{3}\,V_{\varphi b} + -\sqrt{3}\,V_{\varphi c})/2]}$$

$$V_{L(ca)} = V_{\varphi c} - V_{\varphi a}$$
$$= V_{\varphi c} + (-V_{\varphi a})$$
$$= (V_{\varphi c} \angle -120°) + (V_{\varphi a} \angle 180°)$$
$$= \{ V_{\varphi c} \cos(-120°) + \mathrm{j}\,[V_{\varphi c} \sin(-120°)\} + \{ V_{\varphi a} \cos(180°) + \mathrm{j}\,[V_{\varphi a} \sin(180°)]\}$$
$$= [V_{\varphi c} \cos(-120°) + V_{\varphi a} \cos 180°] + \mathrm{j}\,[V_{\varphi c} \sin(-120°) + V_{\varphi a} \sin 180°]$$
$$= [V_{\varphi c}(-1/2) + V_{\varphi a}(-1)] + \mathrm{j}\,[V_{\varphi c}(-\sqrt{3}/2) + V_{\varphi a}(0)]$$
$$= \mathbf{[(-V_{\varphi c}/2) - V_{\varphi a}] + j\,[(-\sqrt{3}\,V_{\varphi c})/2]}$$

2. Problem: Use the results of Problem 1 to calculate the line voltages of a balanced three-phase wye generator relative to the phase voltages.

Solution: For a balanced three-phase wye generator, $V_{\varphi a} = V_{\varphi b} = V_{\varphi c} = V_S$. Substitute this into the expressions for $V_{L(ab)}$, $V_{L(bc)}$, and $V_{L(ca)}$ and simplify.

$$V_{L(ab)} = [V_{\varphi a} + (V_{\varphi b}/2)] + \mathrm{j}\,[(-\sqrt{3}\,V_{\varphi b})/2]$$
$$= [V_S + (V_S/2)] + \mathrm{j}\,[(-\sqrt{3}\,V_S)/2]$$
$$= [V_S(2/2) + V_S(1/2)] + \mathrm{j}\,[V_S(-\sqrt{3}/2)]$$
$$= \{V_S[(2/2) + (1/2)]\} + \mathrm{j}\,[V_S(-\sqrt{3}/2)]$$
$$= [V_S(3/2)] + \mathrm{j}\,[V_S(-\sqrt{3}/2)]$$
$$= \{V_S[(-\sqrt{3} \times -\sqrt{3})/2]\} + \mathrm{j}\,[V_S(-\sqrt{3}/2)]$$
$$= [-\sqrt{3}\;V_S(-\sqrt{3}/2)] + \mathrm{j}\,[-\sqrt{3}\;V_S(-1/2)]$$
$$= -\sqrt{3}\;V_S[(-\sqrt{3}/2) + \mathrm{j}\,(-1/2)]$$
$$= -\sqrt{3}\;V_S\,\mathbf{[\cos(-30°) + j\sin(-30°)]} \text{ in rectangular form}$$
$$= -\sqrt{3}\;V_S\,\angle\mathbf{-30°} \text{ in polar form}$$

$$V_{L(bc)} = [(-V_{\varphi b} + V_{\varphi c})/2] + \mathrm{j}\,[(\sqrt{3}\,V_{\varphi b} + \sqrt{3}\,V_{\varphi c})/2]$$
$$= [(-V_S + V_S)/2] + \mathrm{j}\,[(\sqrt{3}\,V_S + \sqrt{3}\,V_S)/2]$$

$= \{V_S [(-1 + 1) / 2]\} + j\ \{V_S [(\sqrt{3} + \sqrt{3}) / 2)]\}$

$= [V_S (0 / 2)] + j\ [V_S (2\sqrt{3} / 2)]$

$= [V_S (0)] + j\ [V_S (\sqrt{3})]$

$= [V_S (0 \times \sqrt{3})] + j\ [\sqrt{3}\ V_S (1)]$

$= \sqrt{3}\ V_S [0 + j\ (1)]$

$= \sqrt{3}\ V_S$ **[cos 90° + j sin (90°)]** in rectangular form

$= \sqrt{3}\ V_S \angle$ **90°** in polar form

$V_{L(ca)} = [(-V_{\varphi c} / 2) - V_{\varphi a}] + j\ [(-\sqrt{3}\ V_{\varphi c}) / 2]$

$= [(-V_S / 2) - V_S] + j\ [(-\sqrt{3}\ V_S) / 2]$

$= [V_S (-1 / 2) + V_S (-2 / 2)] + j\ [V_S (-\sqrt{3} / 2)]$

$= \{V_S [(-1 / 2) + (-2 / 2)]\} + j\ [V_S (-\sqrt{3} / 2)]$

$= (V_S \{[(-1) + (-2)] / 2\} + j\ [V_S (-\sqrt{3} / 2)]$

$= [V_S (-3 / 2)] + j\ [V_S (-\sqrt{3} / 2)]$

$= \{V_S [(\sqrt{3} \times \sqrt{3}) / 2]\} + j\ [V_S (-\sqrt{3} / 2)]$

$= [\sqrt{3}\ 3\ V_S (\sqrt{3} / 2)] + j\ [\sqrt{3}\ V_S (-1 / 2)]$

$= \sqrt{3}\ V_S [(\sqrt{3} / 2) + j\ (-1 / 2)]$

$= \sqrt{3}\ V_S$ **[cos (–150°) + j sin (–150°)]** in rectangular form

$= \sqrt{3}\ V_S \angle$ **–150°** in polar form

Figure 22-2 shows the phasor diagrams of the line voltages relative to the phase voltages.

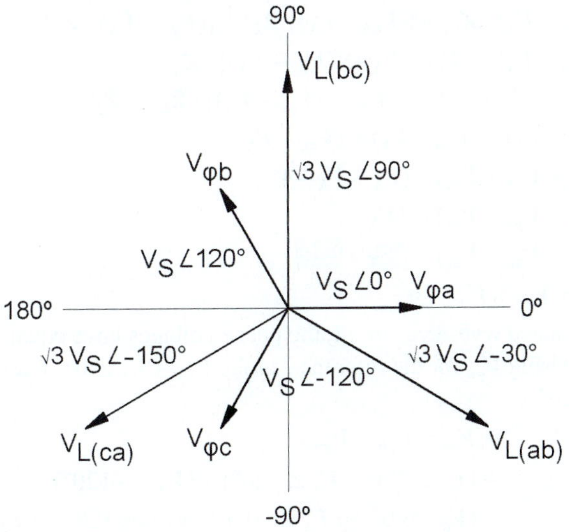

Figure 22-2: Phase and Line Voltage Phasors for Balanced Three-Phase Wye Generator

3. Problem: Prove that V_X for the balanced wye-wye circuit in Figure 22-3 is 0 V. Use $V_{\varphi a}$ for the reference angle.

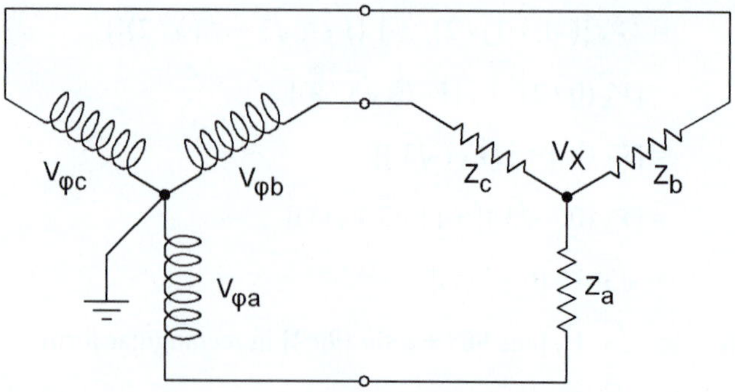

Figure 22-3: Balance Wye-Wye Circuit for Problem 3

Solution: Refer to the node voltage analysis of the balanced wye load in Figure 22-4.

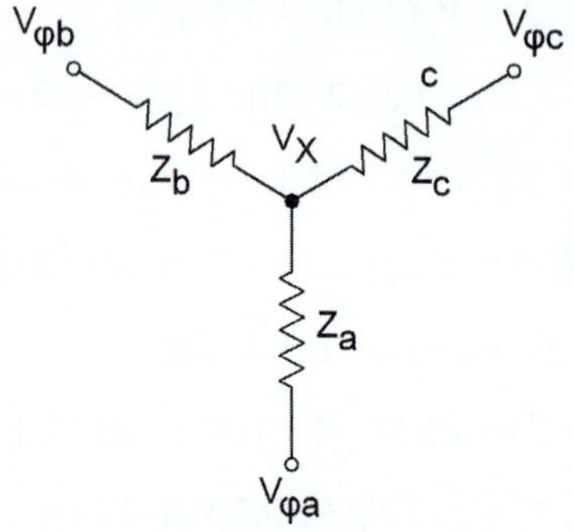

Figure 22-4: Analysis of Balanced Wye Load for Problem 3

Let V_X be the reference voltage for the node voltage analysis equation.

$$[(V_{\varphi a} - V_X) / Z_a] + [(V_{\varphi b} - V_X) / Z_b] + [(V_{\varphi c} - V_X) / Z_c] \quad = 0$$

For a balanced wye-wye circuit, $Z_a = Z_b = Z_c = Z_L$.

$$[(V_{\varphi a} - V_X) / Z_a] + [(V_{\varphi b} - V_X) / Z_b] + [(V_{\varphi c} - V_X) / Z_c] \quad = 0$$

$$[(V_{\varphi a} - V_X) / Z_L] + [(V_{\varphi b} - V_X) / Z_L] + [(V_{\varphi c} - V_X) / Z_L] \quad = 0$$

$$[(V_{\varphi a} - V_X) + (V_{\varphi b} - V_X) + (V_{\varphi c} - V_X)] / Z_L \quad = 0$$

$$\{[(V_{\varphi a} - V_X) + (V_{\varphi b} - V_X) + (V_{\varphi c} - V_X)] / Z_L\} \times Z_L \quad = 0 \times Z_L$$

$$(V_{\varphi a} - V_X) + (V_{\varphi b} - V_X) + (V_{\varphi c} - V_X) \quad = 0$$

$$(V_{\varphi a} + V_{\varphi b} + V_{\varphi c}) - (V_X - V_X - V_X) \quad = 0$$

$$(V_{\varphi a} + V_{\varphi b} + V_{\varphi c}) - 3V_X \quad = 0$$

$$(V_{\varphi a} + V_{\varphi b} + V_{\varphi c}) - 3V_X + 3\,V_X \quad = 0 + 3\,V_X$$

$$V_{\varphi a} + V_{\varphi b} + V_{\varphi c} \quad = 3\,V_X$$

For a balanced wye-wye circuit, the phase voltages have equal magnitude and a phase separation of 120°. Using $V_{\varphi a}$ for the reference angle, $V_{\varphi a} = V_S \angle 0°$, $V_{\varphi b} = V_S \angle 120°$, and $V_{\varphi c} = V_S \angle -120°$.

$$\begin{aligned}
3\,V_X &= V_{\varphi a} + V_{\varphi b} + V_{\varphi c} \\
&= (V_S \angle 0°) + (V_S \angle 120°) + (V_S \angle -120°) \\
&= (V_S \cos 0° + j\,V_S \sin 0°) + (V_S \cos 120° + j\,V_S \sin 120°) \\
&\qquad + [V_S \cos(-120°) + j\,V_S \sin(-120°)] \\
&= [V_S(1) + j\,V_S(0)] + [V_S(-1/2) + j\,V_S(\sqrt{3}/2)] + [V_S(-1/2) + j\,V_S(-\sqrt{3}/2)]
\end{aligned}$$

$= [V_S(1) + V_S(-1/2) + V_S(-1/2)] + [\mathrm{j}\ V_S(0) + \mathrm{j}\ V_S(\sqrt{3}/2) + \mathrm{j}\ V_S(-\sqrt{3}/2)]$

$= \{V_S[1 + (-1/2) + (-1/2)]\} + \mathrm{j}\ \{V_S[0 + (\sqrt{3}/2) + (-\sqrt{3}/2)]\}$

$= V_S(0) + \mathrm{j}\ [V_S(0)]$

$= 0 + \mathrm{j}\ 0\ \mathrm{V}$

$= 0\ \mathrm{V}$

$(3\ V_X)/3 = 0\ \mathrm{V}/3$

$V_X = 0\ \mathrm{V}$

22.1.2 Balanced Wye-Wye Circuits

For Problems 1 through 3, refer to the balanced wye-wye circuit in Figure 22-5.

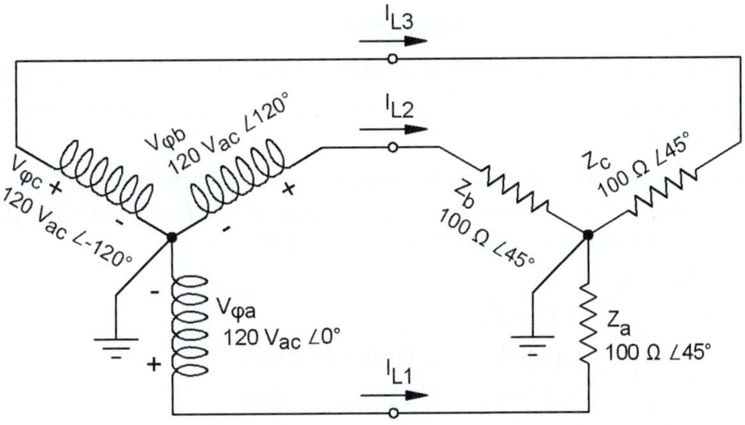

Figure 22-5: Balanced Wye-Wye Circuit

1. Problem: For the circuit in Figure 22-5, determine the phase, line, and load voltages. Express all answers in polar form.

Solution: The phase voltages are the voltages generated by each phase of the wye generator.

$\boldsymbol{V_{\varphi a} = 120\ \mathrm{Vac} \angle 0°}$

$\boldsymbol{V_{\varphi b} = 120\ \mathrm{Vac} \angle 120°}$

$\boldsymbol{V_{\varphi c} = 120\ \mathrm{Vac} \angle -120°}$

The line voltages are the voltages between the phases of the wye generator.

$V_{L(ab)} = V_{\varphi a} - V_{\varphi b}$

$= (120\ \mathrm{V_{ac}} \angle 0°) - (120\ \mathrm{V_{ac}} \angle 120°)$

$= (120\ \mathrm{V_{ac}} + \mathrm{j}\ 0\ \mathrm{V_{ac}}) - (-60\ \mathrm{V_{ac}} + \mathrm{j}\ 104\ \mathrm{V_{ac}})$

$= [120\ \mathrm{V_{ac}} + 60\ \mathrm{V_{ac}})] - \mathrm{j}\ (0\ \mathrm{V_{ac}} + 104\ \mathrm{V_{ac}})$

$= 180\ \mathrm{V_{ac}} - \mathrm{j}\ 104\ \mathrm{V_{ac}}$ in rectangular form

$= \mathbf{208\ V_{ac} \angle -30°}$ in polar form

$V_{L(bc)} = V_{\varphi b} - V_{\varphi c}$

$= (120\ \mathrm{V_{ac}} \angle 120°) - (120\ \mathrm{V_{ac}} \angle -120°)$

$= (-60\ \mathrm{V_{ac}} + \mathrm{j}\ 104\ \mathrm{V_{ac}}) - (-60\ \mathrm{V_{ac}} - \mathrm{j}\ 104\ \mathrm{V_{ac}})$

$= (-60\ \mathrm{V_{ac}} + 60\ \mathrm{V_{ac}}) + \mathrm{j}\ (104\ \mathrm{V_{ac}} + 104\ \mathrm{V_{ac}})$

$= 0\ \mathrm{V_{ac}} + \mathrm{j}\ 208\ \mathrm{V_{ac}}$ in rectangular form

$= \mathbf{208\ V_{ac} \angle 90°}$ in polar form

$V_{L(ca)} = V_{\varphi c} - V_{\varphi a}$

$= (120\ V_{ac} \angle -120°) - (120\ V_{ac} \angle 0°)$

$= (-60\ V_{ac} - j\ 104\ V_{ac}) - (120\ V_{ac} + j\ 0\ V_{ac})$

$= (-60\ V_{ac} - 120\ V_{ac}) - j\ (104\ V_{ac} + 0\ V_{ac})$

$= -180\ V_{ac} - j\ 104\ V_{ac}$ in rectangular form

$= \mathbf{208\ V_{ac} \angle -150°}$

The central nodes of the wye generator and load both are connected to ground, so use Kirchhoff's Voltage Law to determine the load voltage V_{Za} for load Z_a.

$$V_{\varphi a} - V_{Za} = 0\ \text{V}$$

$$V_{\varphi a} - V_{Za} - V_{\varphi a} = 0\ \text{V} - V_{\varphi a}$$

$$-V_{Za} \times -1 = -V_{\varphi a} \times -1$$

$$V_{Za} = V_{\varphi a}$$

$$V_{Za} = \mathbf{120\ Vac \angle 0°}$$

Similarly, $V_{Zb} = V_{\varphi b} = \mathbf{120\ Vac \angle 120°}$ and $V_{Zc} = V_{\varphi c} = \mathbf{120\ Vac \angle -120°}$.

2. Problem: For the circuit in Figure 22-5, determine the phase, line, and load currents. Express all answers in polar form.

Solution: Each phase, line, and load in the circuit forms a series path from the central node of the wye generator to the central node of the wye load, so that $I_\varphi = I_L = I_Z$. Use the load voltages and Ohm's Law for current to calculate each set of currents.

$I_{\varphi a} = I_{L1} = I_{Za}$

$= V_{Za} / Z_a$

$= (120\ V_{ac} \angle 0°) / (100\ \Omega \angle 45°)$

$= (120\ V_{ac} / 100\ \Omega) \angle (0° - 45°)$

$= \mathbf{1.2\ A_{ac} \angle -45°}$

$I_{\varphi b} = I_{L2} = I_{Zb}$

$= V_{Zb} / Z_b$

$= (120\ V_{ac} \angle 120°) / (100\ \Omega \angle 45°)$

$= (120\ V_{ac} / 100\ \Omega) \angle (120° - 45°)$

$= \mathbf{1.2\ A_{ac} \angle 75°}$

$I_{\varphi c} = I_{L3} = I_{Zc}$

$= V_{Zc} / Z_c$

$= (120\ V_{ac} \angle -120°) / (100\ \Omega \angle 45°)$

$= (120\ V_{ac} / 100\ \Omega) \angle [(-120°) - 45°]$

$= \mathbf{1.2\ A_{ac} \angle -165°}$

3. Problem: For the circuit in Figure 22-5, determine the real, reactive, and apparent power for each load.

Solution: For a balanced wye-wye circuit, the power in each load is identical. First, use Watt's Law for voltage and current to calculate P_{app}, the apparent power. Then, use the phase angle of the load to calculate P_r and P_{reac}, the real and reactive powers.

$P_{app} = V_Z I_Z$

$= (120\ V_{ac})(1.2\ A_{ac})$

$= \mathbf{144\ VA}$

The phase angle, φ, of each load is 45°.

$P_r = P_{app} \cos |\varphi|$

$= (144\ \text{VA})(\cos 45°)$

$= (144\text{ VA})(0.707)$

$= \mathbf{102\ W}$

$\boldsymbol{P_{reac}} = \boldsymbol{P_{app}} \sin|\varphi|$

$= (144\text{ VA})(\sin 45°)$

$= (144\text{ VA})(0.707)$

$= \mathbf{102\ VAR}$

22.1.3 Balanced Wye-Delta Circuits

For Problems 1 through 3, refer to the balanced wye-delta circuit in Figure 22-6.

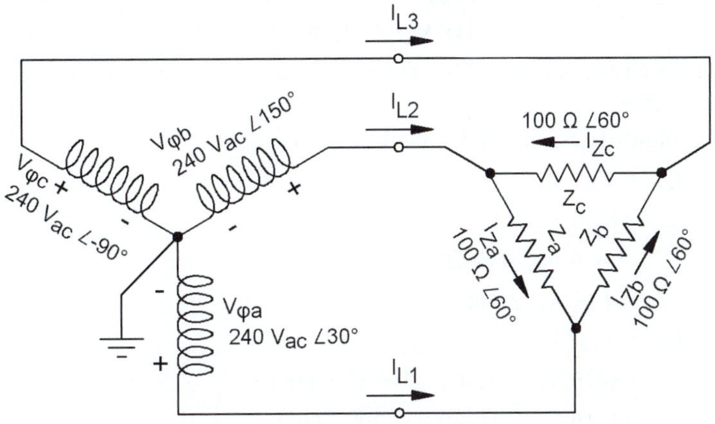

Figure 22-6: Balanced Wye-Delta Circuit

1. Problem: For the circuit in Figure 22-6, determine the phase, line, and load voltages. Express all answers in polar form.

Solution: The phase voltages are generated by each phase of the wye generator.

$\boldsymbol{V_{\varphi a}} = \mathbf{240\ Vac} \angle \mathbf{30°}$

$\boldsymbol{V_{\varphi b}} = \mathbf{240\ Vac} \angle \mathbf{150°}$

$\boldsymbol{V_{\varphi c}} = \mathbf{240\ Vac} \angle \mathbf{-90°}$

The line voltages are the voltages between the phases of the wye generator.

$\boldsymbol{V_{L(ab)}} = \boldsymbol{V_{\varphi a}} - \boldsymbol{V_{\varphi b}}$

$= (240\text{ V}_{ac} \angle 30°) - (240\text{ V}_{ac} \angle 150°)$

$= (208\text{ V}_{ac} + j\,120\text{ V}_{ac}) - (-208\text{ V}_{ac} + j\,120\text{ V}_{ac})$

$= (208\text{ V}_{ac} + 208\text{ V}_{ac}) + j\,(120\text{ V}_{ac} - 120\text{ V}_{ac})$

$= 416\text{ V}_{ac} + j\,0\text{ V}_{ac}$ in rectangular form

$= \mathbf{416\ V_{ac}} \angle \mathbf{0°}$ in polar form

$\boldsymbol{V_{L(bc)}} = \boldsymbol{V_{\varphi b}} - \boldsymbol{V_{\varphi c}}$

$= (240\text{ V}_{ac} \angle 150°) - (240\text{ V}_{ac} \angle -90°)$

$= (-208\text{ V}_{ac} + j\,120\text{ V}_{ac}) - (0\text{ V}_{ac} - j\,240\text{ V}_{ac})$

$= (-208\text{ V}_{ac} - 0\text{ V}_{ac}) + j\,(120\text{ V}_{ac} + 240\text{ V}_{ac})$

$= -208\text{ V}_{ac} + j\,360\text{ V}_{ac}$ in rectangular form

$= \mathbf{416\ V_{ac}} \angle \mathbf{120°}$ in polar form

$\boldsymbol{V_{L(ca)}} = \boldsymbol{V_{\varphi c}} - \boldsymbol{V_{\varphi a}}$

$= (240\text{ V}_{ac} \angle -90°) - (240\text{ V}_{ac} \angle 30°)$

$= (0\text{ V}_{ac} - j\,240\text{ V}_{ac}) - (208\text{ V}_{ac} + j\,120\text{ V}_{ac})$

$= (0\ V_{ac} - 208\ V_{ac}) - j\ (240\ V_{ac} + 120\ V_{ac})$

$= -208\ V_{ac} - j\ 360\ V_{ac}$ in rectangular form

$= \mathbf{416\ V_{ac} \angle -120°}$ in polar form

Each load is connected directly across a line voltage of the wye generator.

$V_{Za} = V_{L(ab)}$

$= \mathbf{416\ V_{ac} \angle 0°}$

$V_{Zb} = V_{L(bc)}$

$= \mathbf{416\ V_{ac} \angle 120°}$

$V_{Zc} = V_{L(ab)}$

$= \mathbf{416\ V_{ac} \angle -120°}$

2. Problem: For the circuit in Figure 22-6, determine the phase, line, and load currents. Express all answers in polar form.

Solution: Use Ohm's law to determine the load currents.

$I_{Za} = V_{Za} / Z_a$

$= (416\ V_{ac} \angle 0°) / (100\ \Omega \angle 60°)$

$= (416\ V_{ac} / 100\ \Omega) \angle (0° - 60°)$

$= \mathbf{4.16\ A_{ac} \angle -60°}$

$I_{Zb} = V_{Zb} / Z_b$

$= (416\ V_{ac} \angle 120°) / (100\ \Omega \angle 60°)$

$= (416\ V_{ac} / 100\ \Omega) \angle (120° - 60°)$

$= \mathbf{4.16\ A_{ac} \angle 60°}$

$I_{Zc} = V_{Zc} / Z_c$

$= (416\ V_{ac} \angle -120°) / (100\ \Omega \angle 60°)$

$= (416\ V_{ac} / 100\ \Omega) \angle (-120° - 60°)$

$= \mathbf{4.16\ A_{ac} \angle -180° = 4.16\ A_{ac} \angle 180°}$

The relative direction of the currents in a three-phase delta load are as shown in Figure 22-6, so that each line current is the difference between two load currents. Because the line currents are shown entering the delta load, define currents entering a node to be positive and currents leaving a node to be negative. Then use Kirchhoff's Current Law to determine the line currents.

$I_{L1} + I_{Za} - I_{Zb} = 0\ A$

$I_{L1} + I_{Za} - I_{Zb} - I_{Za} + I_{Zb} = 0\ A - I_{Za} + I_{Zb}$

$I_{L1} = I_{Zb} - I_{Za}$

$= (4.16\ A_{ac} \angle 60°) - (4.16\ A_{ac} \angle -60°)$

$= (2.08\ A_{ac} + j\ 3.60\ A_{ac}) - (2.08\ A_{ac} - j\ 3.60\ A_{ac})$

$= (2.08\ A_{ac} - 2.08\ A_{ac}) + j\ (3.60\ A_{ac} + 3.60\ A_{ac})$

$= 0\ A_{ac} + j\ 7.20\ A_{ac}$ in rectangular form

$= \mathbf{7.20\ A_{ac} \angle 90°}$ in polar form

$I_{L2} - I_{Za} + I_{Za} = 0\ A$

$I_{L2} - I_{Za} + I_{Zc} + I_{Za} - I_{Zc} = 0\ A + I_{Za} - I_{Zc}$

$I_{L2} = I_{Za} - I_{Zc}$

$= (4.16\ A_{ac} \angle -60°) - (4.16\ A_{ac} \angle 180°)$

$= (2.08\ A_{ac} - j\ 3.60\ A_{ac}) - (-4.16\ A_{ac} + j\ 0\ A_{ac})$

$= (2.08\ A_{ac} + 4.16\ A_{ac}) - j\ (3.60\ A_{ac} + 0\ A_{ac})$

$= 6.24\ A_{ac} - j\ 3.60\ A_{ac}$ in rectangular form

$= \mathbf{7.20\ A_{ac} \angle -30°}$ in polar form

$\boldsymbol{I_{L3}} + \boldsymbol{I_{Zb}} - \boldsymbol{I_{Zc}} = 0\ A$

$\boldsymbol{I_{L3}} + \boldsymbol{I_{Zb}} - \boldsymbol{I_{Zc}} - \boldsymbol{I_{Zb}} + \boldsymbol{I_{Zc}} = 0\ A - \boldsymbol{I_{Zb}} + \boldsymbol{I_{Zc}}$

$\boldsymbol{I_{L3}} = \boldsymbol{I_{Zc}} - \boldsymbol{I_{Zb}}$

$= (4.16\ A_{ac} \angle 180°) - (4.16\ A_{ac} \angle 60°)$

$= (-4.16\ A_{ac} + j\ 0\ A_{ac}) - (2.08\ A_{ac} + j\ 3.60\ A_{ac})$

$= (-4.16\ A_{ac} - 2.08\ A_{ac}) + j\ (0\ A_{ac} - 3.60\ A_{ac})$

$= -6.24\ A_{ac} - j\ 3.60\ A_{ac}$ in rectangular form

$= \mathbf{7.20\ A_{ac} \angle 210°}$ in polar form

The phases of the wye generator are in series with the lines, so that the phase currents are equal to the line currents.

$\boldsymbol{I_{\varphi a}} = \boldsymbol{I_{L1}}$

$= \mathbf{7.20\ A_{ac} \angle 90°}$

$\boldsymbol{I_{\varphi b}} = \boldsymbol{I_{L2}}$

$= \mathbf{7.20\ A_{ac} \angle -30°}$

$\boldsymbol{I_{\varphi c}} = \boldsymbol{I_{L3}}$

$= \mathbf{7.20\ A_{ac} \angle 210°}$

3. Problem: For the circuit in Figure 22-6, determine the real, reactive, and apparent power for each load.

Solution: For a balanced wye-delta circuit, the power in each load is identical. First, use Watt's Law for voltage and current to calculate $\boldsymbol{P_{app}}$, the apparent power. Then, use the phase angle of the load to calculate $\boldsymbol{P_r}$ and $\boldsymbol{P_{reac}}$, the real and reactive powers.

$\boldsymbol{P_{app}} = V_Z I_Z$

$= (416\ V_{ac})(4.16\ A_{ac})$

$= \mathbf{1730\ VA}$

The phase angle, φ, of each load is 60°.

$\boldsymbol{P_r} = \boldsymbol{P_{app}} \cos |\varphi|$

$= (1730\ VA)(\cos 60°)$

$= (1730\ VA)(0.500)$

$= \mathbf{865\ W}$

$\boldsymbol{P_{reac}} = \boldsymbol{P_{app}} \sin |\varphi|$

$= (1730\ VA)(\sin 60°)$

$= (1730\ VA)(0.866)$

$= \mathbf{1500\ VAR}$

22.1.4 Balanced Delta-Wye Circuits

For Problems 1 through 3, refer to the balanced delta-wye circuit in Figure 22-7.

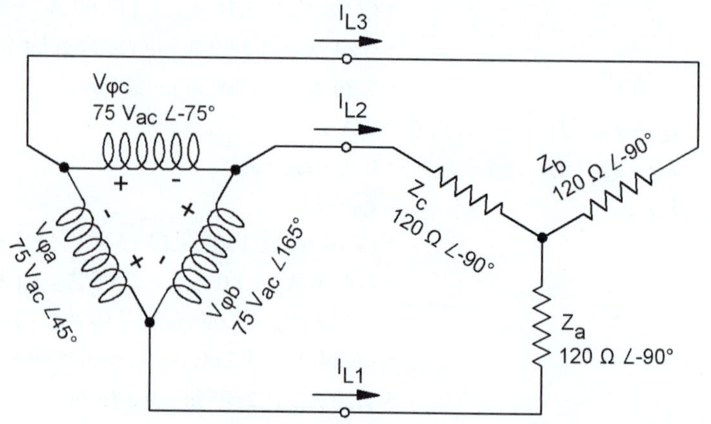

Figure 22-7: Balanced Delta-Wye Circuit

1. Problem: For the circuit in Figure 22-7, determine the line and load currents. Express all answers in polar form.

Solution: To determine the line and load currents, first convert the wye load to a delta load, as shown in Figure 22-8, using the delta-to-wye conversion formulas.

$$\boldsymbol{Z_1} = (\boldsymbol{Z_a Z_b} + \boldsymbol{Z_b Z_c} + \boldsymbol{Z_a Z_c}) / \boldsymbol{Z_b}$$
$$\boldsymbol{Z_2} = (\boldsymbol{Z_a Z_b} + \boldsymbol{Z_b Z_c} + \boldsymbol{Z_a Z_c}) / \boldsymbol{Z_c}$$
$$\boldsymbol{Z_3} = (\boldsymbol{Z_a Z_b} + \boldsymbol{Z_b Z_c} + \boldsymbol{Z_a Z_c}) / \boldsymbol{Z_a}$$

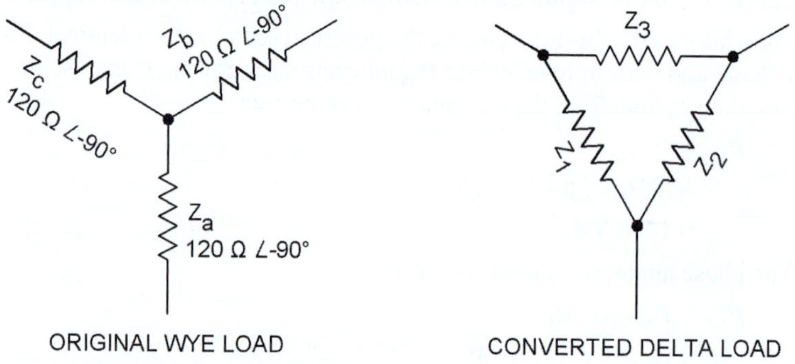

Figure 22-8: Conversion of Wye Load to Delta Load

For the balanced wye load, $\boldsymbol{Z_a} = \boldsymbol{Z_b} = \boldsymbol{Z_c} = \boldsymbol{Z_Y}$, so the equation for each delta load, $\boldsymbol{Z_1} = \boldsymbol{Z_2} = \boldsymbol{Z_3} = \boldsymbol{Z_\Delta}$, simplifies to

$$
\begin{aligned}
\boldsymbol{Z_\Delta} &= (\boldsymbol{Z_Y Z_Y} + \boldsymbol{Z_Y Z_Y} + \boldsymbol{Z_Y Z_Y}) / \boldsymbol{Z_Y} \\
&= (\boldsymbol{Z_Y}^2 + \boldsymbol{Z_Y}^2 + \boldsymbol{Z_Y}^2) / \boldsymbol{Z_Y} \\
&= (3\,\boldsymbol{Z_Y}^2) / \boldsymbol{Z_Y} \\
&= 3\,\boldsymbol{Z_Y} \\
&= (3)(120\ \Omega \angle -90°) \\
&= (3 \angle 0°)(120\ \Omega \angle -90°) \\
&= [(3)(120\ \Omega)] \angle (0° + -90°) \\
&= 360\ \Omega \angle -90°
\end{aligned}
$$

Figure 22-9 shows the converted circuit.

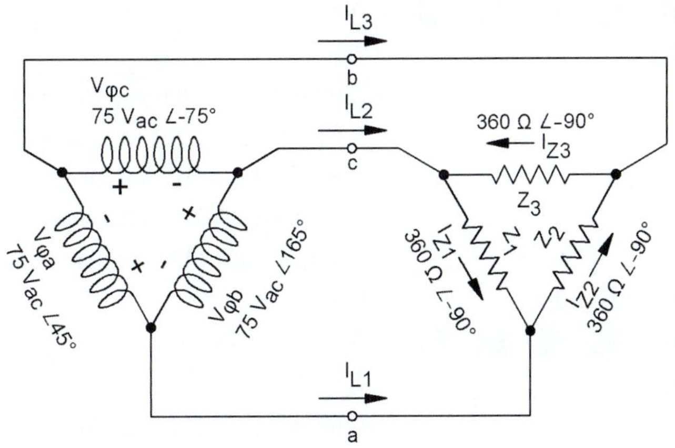

Figure 22-9: Delta-Wye Circuit Converted to Delta-Delta Circuit

In the delta-delta circuit of Figure 22-9, the delta phase voltages are applied directly across the delta loads. Use Ohm's Law for current to calculate the delta phase currents.

$$\begin{aligned}
\boldsymbol{I_{Z1}} &= \boldsymbol{V_{Z1}} / \boldsymbol{Z_1} \\
&= \boldsymbol{V_{\varphi b}} / \boldsymbol{Z_1} \\
&= (75\ \mathrm{V_{ac}} \angle 165°) / (360\ \Omega \angle -90°) \\
&= (75\ \mathrm{V_{ac}} / 360\ \Omega) \angle [165° - (-90°)] \\
&= 208\ \mathrm{mA_{ac}} \angle 255° \\
&= 208\ \mathrm{mA_{ac}} \angle -105°
\end{aligned}$$

$$\begin{aligned}
\boldsymbol{I_{Z2}} &= \boldsymbol{V_{Z2}} / \boldsymbol{Z_2} \\
&= \boldsymbol{V_{\varphi a}} / \boldsymbol{Z_2} \\
&= (75\ \mathrm{V_{ac}} \angle 45°) / (360\ \Omega \angle -90°) \\
&= (75\ \mathrm{V_{ac}} / 360\ \Omega) \angle [45° - (-90°)] \\
&= 208\ \mathrm{mA_{ac}} \angle 135°
\end{aligned}$$

$$\begin{aligned}
\boldsymbol{I_{Z3}} &= \boldsymbol{V_{Z3}} / \boldsymbol{Z_3} \\
&= \boldsymbol{V_{\varphi c}} / \boldsymbol{Z_3} \\
&= (75\ \mathrm{V_{ac}} \angle -75°) / (360\ \Omega \angle -90°) \\
&= (75\ \mathrm{V_{ac}} / 360\ \Omega) \angle [(-75°) - (-90°)] \\
&= 208\ \mathrm{mA_{ac}} \angle 15°
\end{aligned}$$

For the direction of the load currents as shown in Figure 22-9, each line current is the difference between two load currents. Because the line currents are shown entering the delta load, define currents entering a node to be positive and currents leaving a node to be negative. Then use Kirchhoff's Current Law to determine the line currents.

$$\begin{aligned}
\boldsymbol{I_{L1}} + \boldsymbol{I_{Z1}} - \boldsymbol{I_{Z2}} &= 0\ \mathrm{A} \\
\boldsymbol{I_{L1}} + \boldsymbol{I_{Z1}} - \boldsymbol{I_{Z2}} - \boldsymbol{I_{Z1}} + \boldsymbol{I_{Z2}} &= 0\ \mathrm{A} - \boldsymbol{I_{Z1}} + \boldsymbol{I_{Z2}} \\
\boldsymbol{I_{L1}} &= \boldsymbol{I_{Z2}} - \boldsymbol{I_{Z1}} \\
&= (208\ \mathrm{mA_{ac}} \angle 135°) - (208\ \mathrm{mA_{ac}} \angle -105°) \\
&= (-147\ \mathrm{mA_{ac}} + j\ 147\ \mathrm{mA_{ac}}) - (-53.9\ \mathrm{mA_{ac}} - j\ 201\ \mathrm{mA_{ac}}) \\
&= (-147\ \mathrm{mA_{ac}} + 53.9\ \mathrm{mA_{ac}}) + j\ (147\ \mathrm{mA_{ac}} + 201\ \mathrm{mA_{ac}}) \\
&= -93.4\ \mathrm{mA_{ac}} + j\ 349\ \mathrm{mA_{ac}} \text{ in rectangular form} \\
&= \mathbf{361\ mA_{ac} \angle 105°} \text{ in polar form}
\end{aligned}$$

$$\boldsymbol{I_{L2}} - \boldsymbol{I_{Z1}} + \boldsymbol{I_{Z3}} = 0\ \mathrm{A}$$

$$I_{L2} - I_{Z1} + I_{Z3} + I_{Z1} - I_{Z3} = 0\text{ A} + I_{Z1} - I_{Z3}$$

$$I_{L2} = I_{Z1} - I_{Z3}$$

$= (208\text{ mA}_{ac} \angle -105°) - (208\text{ mA}_{ac} \angle 15°)$

$= (-53.9\text{ mA}_{ac} - j\ 201\text{ mA}_{ac}) - (201\text{ mA}_{ac} + j\ 53.9\text{ mA}_{ac})$

$= (-53.9\text{ mA}_{ac} - 201\text{ mA}_{ac}) - j\ (201\text{ mA}_{ac} + 53.9\text{ mA}_{ac})$

$= -255\text{ mA}_{ac} - j\ 255\text{ mA}_{ac}$ in rectangular form

$= \mathbf{361\ mA_{ac} \angle -135°}$ in polar form

$$I_{L3} + I_{Z2} - I_{Z3} = 0\text{ A}$$

$$I_{L3} + I_{Z2} - I_{Z3} - I_{Z2} + I_{Z3} = 0\text{ A} - I_{Z2} + I_{Z3}$$

$$I_{L3} = I_{Z3} - I_{Z2}$$

$= (208\text{ mA}_{ac} \angle 15°) - (208\text{ mA}_{ac} \angle 135°)$

$= (201\text{ mA}_{ac} + j\ 53.9\text{ mA}_{ac}) - (-147\text{ mA}_{ac} + j\ 147\text{ mA}_{ac})$

$= (201\text{ mA}_{ac} + 147\text{ mA}_{ac}) + j\ (53.9\text{ mA}_{ac} - 147\text{ mA}_{ac})$

$= 349\text{ mA}_{ac} - j\ 93.4\text{ mA}_{ac}$ in rectangular form

$= \mathbf{361\ mA_{ac} \angle -15°}$ in polar form

Referring back to Figure 22-7, wye loads are in series with the lines, so that the load currents are equal to the line currents.

$I_{Za} = I_{L1}$

$= \mathbf{361\ mA_{ac} \angle 105°}$

$I_{Zb} = I_{L3}$

$= \mathbf{361\ mA_{ac} \angle -15°}$

$I_{Zc} = I_{L2}$

$= \mathbf{361\ mA_{ac} \angle -135°}$

2. Problem: For the circuit in Figure 22-7, determine the phase, line, and load voltages.

Solution: For a delta source, the phase voltages and line voltages are equal.

$V_{\varphi a} = V_{L(ab)} = \mathbf{75\ V_{ac} \angle 45°}$

$V_{\varphi b} = V_{L(ca)} = \mathbf{75\ V_{ac} \angle 165°}$

$V_{\varphi c} = V_{L(bc)} = \mathbf{75\ V_{ac} \angle -75°}$

Use Ohm's Law for voltage to calculate the load voltages.

$V_{Za} = I_{Za} Z_a$

$= (361\text{ mA}_{ac} \angle 105°)(120\ \Omega \angle -90°)$

$= [(361\text{ mA}_{ac})(120\ \Omega)] \angle [105° + (-90°)]$

$= \mathbf{43.3\ V_{ac} \angle 15°}$

$V_{Zb} = I_{Zb} Z_b$

$= (361\text{ mA}_{ac} \angle -15°)(120\ \Omega \angle -90°)$

$= [(361\text{ mA}_{ac})(120\ \Omega)] \angle [(-15°) + (-90°)]$

$= \mathbf{43.3\ V_{ac} \angle -105°}$

$V_{Zc} = I_{Zc} Z_c$

$= (361\text{ mA}_{ac} \angle -135°)(120\ \Omega \angle -90°)$

$= [(361\text{ mA}_{ac})(120\ \Omega)] \angle [(-135°) + (-90°)]$

$= \mathbf{43.3\ V_{ac} \angle -225° = 43.3\ V \angle 135°}$

3. Problem: For the circuit in Figure 22-7, determine the real, reactive, and apparent power for each load.

Solution: For a balanced delta-wye circuit, the power in each load is identical. First, use Watt's Law for voltage and current to calculate $\boldsymbol{P_{app}}$, the apparent power. Then, use the phase angle of the load to calculate $\boldsymbol{P_r}$ and $\boldsymbol{P_{reac}}$, the real and reactive powers.

$$\boldsymbol{P_{app}} = \boldsymbol{V_Z I_Z}$$
$$= (43.3\ \mathrm{V_{ac}})(361\ \mathrm{mA_{ac}})$$
$$= \mathbf{15.6\ VA}$$

The phase angle, φ, of each load is –90°.

$$\boldsymbol{P_r} = \boldsymbol{P_{app}} \cos|\varphi|$$
$$= (15.6\ \mathrm{VA})(\cos 90°)$$
$$= (15.6\ \mathrm{VA})(0)$$
$$= \mathbf{0\ W}$$

$$\boldsymbol{P_{reac}} = \boldsymbol{P_{app}} \sin|\varphi|$$
$$= (15.6\ \mathrm{VA})(\sin 90°)$$
$$= (15.6\ \mathrm{VA})(1)$$
$$= \mathbf{15.6\ VAR}$$

22.1.5 Balanced Delta-Delta Circuits

For Problems 1 through 3, refer to the balanced delta-delta circuit in Figure 22-10.

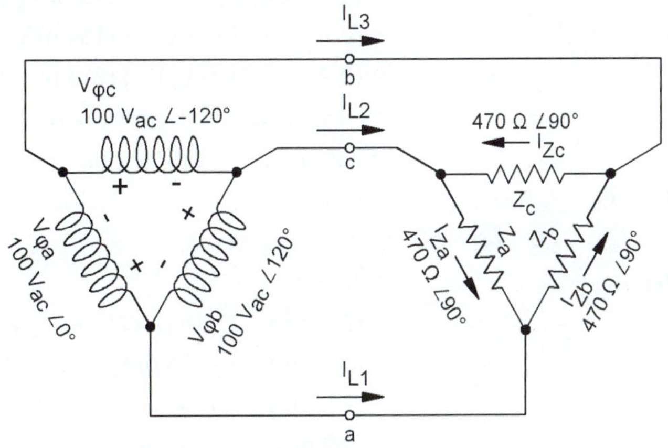

Figure 22-10: Balanced Delta-Delta Circuit

1. Problem: For the circuit in Figure 22-10, determine the phase, line, and load voltages.

Solution: For a delta-delta circuit, the phase and load voltages are equal.

$$\boldsymbol{V_{\varphi a}} = \boldsymbol{V_{L(ab)}} = \boldsymbol{V_{Zb}} = \mathbf{100\ V_{ac}\ \angle\ 120°}$$
$$\boldsymbol{V_{\varphi b}} = \boldsymbol{V_{L(ca)}} = \boldsymbol{V_{Za}} = \mathbf{100\ V_{ac}\ \angle\ 0°}$$
$$\boldsymbol{V_{\varphi c}} = \boldsymbol{V_{L(bc)}} = \boldsymbol{V_{Zc}} = \mathbf{100\ V_{ac}\ \angle\ {-120°}}$$

2. Problem: For the circuit in Figure 22-10, determine the line and load currents.

Solution: In the delta-delta circuit of Figure 22-10, the delta phase voltages are applied directly across the delta loads. Use Ohm's Law for current to calculate the delta load currents.

$$\boldsymbol{I_{Za}} = \boldsymbol{V_{Za}} / \boldsymbol{Z_a}$$
$$= \boldsymbol{V_{\varphi b}} / \boldsymbol{Z_a}$$
$$= (100\ \mathrm{V_{ac}} \angle\ 120°) / (470\ \Omega \angle\ 90°)$$
$$= (100\ \mathrm{V_{ac}} / 470\ \Omega) \angle\ (120° - 90°)$$

$= \mathbf{213\ mA_{ac} \angle 30°}$

$\boldsymbol{I_{Zb}} = \boldsymbol{V_{Zb}} / \boldsymbol{Z_b}$

$= \boldsymbol{V_{\varphi a}} / \boldsymbol{Z_b}$

$= (100\ V_{ac} \angle 0°) / (470\ \Omega \angle 90°)$

$= (100\ V_{ac} / 470\ \Omega) \angle (0° - 90°)$

$= \mathbf{213\ mA_{ac} \angle -90°}$

$\boldsymbol{I_{Zc}} = \boldsymbol{V_{Zc}} / \boldsymbol{Z_c}$

$= \boldsymbol{V_{\varphi c}} / \boldsymbol{Z_c}$

$= (100\ V_{ac} \angle -120°) / (470\ \Omega \angle 90°)$

$= (100\ V_{ac} / 470\ \Omega) \angle [(-120°) - 90°]$

$= \mathbf{213\ mA_{ac} \angle -210° = 213\ mA_{ac} \angle 150°}$

The relative direction of the currents in the delta load are as shown in Figure 22-10, so that each line current is the difference between two load currents. Because the line currents are shown entering the converted delta load, define currents entering a node to be positive and currents leaving a node to be negative. Then use Kirchhoff's Current Law to determine the line currents.

$\boldsymbol{I_{L1}} + \boldsymbol{I_{Za}} - \boldsymbol{I_{Zb}} = 0\ A$

$\boldsymbol{I_{L1}} + \boldsymbol{I_{Za}} - \boldsymbol{I_{Zb}} - \boldsymbol{I_{Za}} + \boldsymbol{I_{Zb}} = 0\ A - \boldsymbol{I_{Za}} + \boldsymbol{I_{Zb}}$

$\boldsymbol{I_{L1}} = \boldsymbol{I_{Zb}} - \boldsymbol{I_{Za}}$

$= (213\ mA_{ac} \angle -90°) - (213\ mA_{ac} \angle 30°)$

$= (0\ A_{ac} - j\ 213\ mA_{ac}) - (184\ mA_{ac} + j\ 106\ mA_{ac})$

$= (0\ A_{ac} - 184\ mA_{ac}) - j\ (213\ mA_{ac} + 106\ mA_{ac})$

$= -184\ mA_{ac} - j\ 319\ mA_{ac}$ in rectangular form

$= \mathbf{369\ mA_{ac} \angle -120°}$ in polar form

$\boldsymbol{I_{L2}} - \boldsymbol{I_{Za}} + \boldsymbol{I_{Zc}} = 0\ A$

$\boldsymbol{I_{L2}} - \boldsymbol{I_{Za}} + \boldsymbol{I_{Zc}} + \boldsymbol{I_{Za}} - \boldsymbol{I_{Zc}} = 0\ A + \boldsymbol{I_{Za}} - \boldsymbol{I_{Zc}}$

$\boldsymbol{I_{L2}} = \boldsymbol{I_{Za}} - \boldsymbol{I_{Zc}}$

$= (213\ mA_{ac} \angle 30°) - (213\ mA_{ac} \angle 150°)$

$= (184\ mA_{ac} + j\ 106\ mA_{ac}) - (-184\ mA_{ac} + j\ 106\ mA_{ac})$

$= (184\ mA_{ac} + 184\ mA_{ac}) + j\ (106\ mA_{ac} - 106\ mA_{ac})$

$= 369\ mA_{ac} + j\ 0\ A_{ac}$ in rectangular form

$= \mathbf{361\ mA_{ac} \angle 0°}$ in polar form

$\boldsymbol{I_{L3}} + \boldsymbol{I_{Zb}} - \boldsymbol{I_{Zc}} = 0\ A$

$\boldsymbol{I_{L3}} + \boldsymbol{I_{Zb}} - \boldsymbol{I_{Zc}} - \boldsymbol{I_{Zb}} + \boldsymbol{I_{Zc}} = 0\ A - \boldsymbol{I_{Zb}} + \boldsymbol{I_{Zc}}$

$\boldsymbol{I_{L3}} = \boldsymbol{I_{Zc}} - \boldsymbol{I_{Zb}}$

$= (213\ mA_{ac} \angle 150°) - (213\ mA_{ac} \angle -90°)$

$= (-184\ mA_{ac} + j\ 106\ mA_{ac}) - (0\ A_{ac} - j\ 213\ mA_{ac})$

$= (-184\ mA_{ac} + 0\ A_{ac}) + j\ (106\ mA_{ac} + 213\ mA_{ac})$

$= -184\ mA_{ac} + j\ 319\ mA_{ac}$ in rectangular form

$= \mathbf{369\ mA_{ac} \angle 120°}$ in polar form

3. Problem: For the circuit in Figure 22-10, determine the real, reactive, and apparent power for each load.

Solution: For a balanced delta-delta circuit, the power in each load is identical. First, use Watt's Law for voltage and current to calculate $\boldsymbol{P_{app}}$, the apparent power. Then, use the phase angle of the load to calculate $\boldsymbol{P_r}$ and $\boldsymbol{P_{reac}}$, the real and reactive powers.

$$\boldsymbol{P_{app}} = \boldsymbol{V_Z I_Z}$$
$$= (100\ \text{V}_{ac})(213\ \text{mA}_{ac})$$
$$= \mathbf{21.3\ VA}$$

The phase angle, φ, of each load is 90°.

$$\boldsymbol{P_r} = \boldsymbol{P_{app}} \cos|\varphi|$$
$$= (21.3\ \text{VA})(\cos 90°)$$
$$= (21.3\ \text{VA})(0)$$
$$= \mathbf{0\ W}$$

$$\boldsymbol{P_{reac}} = \boldsymbol{P_{app}} \sin|\varphi|$$
$$= (21.3\ \text{VA})(\sin 90°)$$
$$= (21.3\ \text{VA})(1)$$
$$= \mathbf{21.3\ VAR}$$

22.2 Advanced

22.2.1 Unbalanced Wye-Wye Circuits

1. Problem: Determine the line and neutral currents for the unbalanced wye-wye circuit shown in Figure 22-11. Express all answers in polar form.

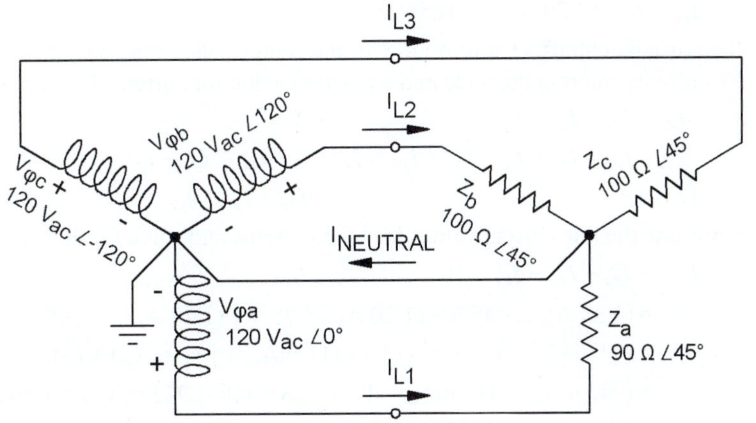

Figure 22-11: Unbalanced Wye-Wye Circuit from Problem 1

Solution: The center nodes of the wye source and wye load are connected, so use Kirchhoff's Voltage Law and Ohm's Law for voltage to determine the line currents.

$$\boldsymbol{V_{\varphi a}} - \boldsymbol{V_{Za}} = 0$$
$$\boldsymbol{V_{\varphi a}} - (\boldsymbol{I_{L1}})(\boldsymbol{Z_a}) = 0$$
$$\boldsymbol{V_{\varphi a}} - (\boldsymbol{I_{L1}})(\boldsymbol{Z_a}) - \boldsymbol{V_{\varphi a}} = 0 - \boldsymbol{V_{\varphi a}}$$
$$-(\boldsymbol{I_{L1}})(\boldsymbol{Z_a}) / -\boldsymbol{Z_a} = -\boldsymbol{V_{\varphi a}} / -\boldsymbol{Z_a}$$
$$\boldsymbol{I_{L1}} = \boldsymbol{V_{\varphi a}} / \boldsymbol{Z_a}$$
$$= (120\ \text{V}_{ac} \angle 0°) / (90\ \Omega \angle 45°)$$
$$= (120\ \text{V}_{ac} / 90\ \Omega) \angle (0° - 45°)$$
$$= \mathbf{1.33\ A_{ac} \angle -45°}$$

$$\boldsymbol{V_{\varphi b}} - \boldsymbol{V_{Zb}} = 0$$
$$\boldsymbol{V_{\varphi b}} - (\boldsymbol{I_{L2}})(\boldsymbol{Z_b}) = 0$$
$$\boldsymbol{V_{\varphi b}} - (\boldsymbol{I_{L2}})(\boldsymbol{Z_b}) - \boldsymbol{V_{\varphi b}} = 0 - \boldsymbol{V_{\varphi b}}$$

$-(\boldsymbol{I_{L2}})(\boldsymbol{Z_b}) / -\boldsymbol{Z_b} = -\boldsymbol{V_{\varphi b}} / -\boldsymbol{Z_b}$

$\boldsymbol{I_{L2}} = \boldsymbol{V_{\varphi b}} / \boldsymbol{Z_b}$

$= (120\ \text{V}_{\text{ac}} \angle 120°) / (100\ \Omega \angle 45°)$

$= (120\ \text{V}_{\text{ac}} / 100\ \Omega) \angle (120° - 45°)$

$= \mathbf{1.20\ A_{ac} \angle 75°}$

$\boldsymbol{V_{\varphi c}} - \boldsymbol{V_{Zc}} = 0$

$\boldsymbol{V_{\varphi c}} - (\boldsymbol{I_{L3}})(\boldsymbol{Z_c}) = 0$

$\boldsymbol{V_{\varphi c}} - (\boldsymbol{I_{L3}})(\boldsymbol{Z_c}) - \boldsymbol{V_{\varphi c}} = 0 - \boldsymbol{V_{\varphi c}}$

$-(\boldsymbol{I_{L3}})(\boldsymbol{Z_c}) / -\boldsymbol{Z_c} = -\boldsymbol{V_{\varphi c}} / -\boldsymbol{Z_c}$

$\boldsymbol{I_{L3}} = \boldsymbol{V_{\varphi c}} / \boldsymbol{Z_c}$

$= (120\ \text{V}_{\text{ac}} \angle -120°) / (100\ \Omega \angle 45°)$

$= (120\ \text{V}_{\text{ac}} / 100\ \Omega) \angle (-120° - 45°)$

$= \mathbf{1.20\ A_{ac} \angle -165°}$

Because the wye loads are in series with the line currents, the load currents are equal to the line currents.

$\boldsymbol{I_{Za}} = \boldsymbol{I_{L1}} = 1.33\ \text{A}_{\text{ac}} \angle -45°$

$\boldsymbol{I_{Zb}} = \boldsymbol{I_{L2}} = 1.20\ \text{A}_{\text{ac}} \angle 75°$

$\boldsymbol{I_{Zc}} = \boldsymbol{I_{L3}} = 1.20\ \text{A}_{\text{ac}} \angle -165°$

Next, use Kirchhoff's Current Law for the center node of the wye load. Assign a negative value for currents entering the node and a positive value for currents leaving the node.

$-\boldsymbol{I_{Za}} - \boldsymbol{I_{Zb}} - \boldsymbol{I_{Zc}} + \boldsymbol{I_N} = 0$

$-\boldsymbol{I_{Za}} - \boldsymbol{I_{Zb}} - \boldsymbol{I_{Zc}} + \boldsymbol{I_N} + \boldsymbol{I_{Za}} + \boldsymbol{I_{Zb}} + \boldsymbol{I_{Zc}} = 0 + \boldsymbol{I_{Za}} + \boldsymbol{I_{Zb}} + \boldsymbol{I_{Zc}}$

$\boldsymbol{I_N} = \boldsymbol{I_{Za}} + \boldsymbol{I_{Zb}} + \boldsymbol{I_{Zc}}$

Substitute the line current values for load currents and solve for the value of neutral current.

$\boldsymbol{I_N} = \boldsymbol{I_{Za}} + \boldsymbol{I_{Zb}} + \boldsymbol{I_{Zc}}$

$= (1.33\ \text{A}_{\text{ac}} \angle -45°) + (1.20\ \text{A}_{\text{ac}} \angle 75°) + (1.20\ \text{A}_{\text{ac}} \angle -165°)$

$= (943\ \text{mA}_{\text{ac}} - j\ 943\ \text{mA}_{\text{ac}}) + (311\ \text{mA}_{\text{ac}} + j\ 1.16\ \text{A}_{\text{ac}}) + (-1.16\ \text{A}_{\text{ac}} - j\ 311\ \text{mA}_{\text{ac}})$

$= (943\ \text{mA}_{\text{ac}} + 311\ \text{mA}_{\text{ac}} - 1.16\ \text{A}_{\text{ac}}) + j\ [(-j\ 943\ \text{mA}_{\text{ac}}) + 1.16\ \text{A}_{\text{ac}} + (-311\ \text{mA}_{\text{ac}})]$

$= 94.3\ \text{mA}_{\text{ac}} - j\ 94.3\ \text{mA}_{\text{ac}}$ in rectangular form

$= \mathbf{113\ mA_{ac} \angle -45°}$

2. Problem: Determine the center node voltage, $\boldsymbol{V_X}$, relative to ground for the unbalanced wye-wye circuit shown in Figure 22-12. Express all answers in polar form.

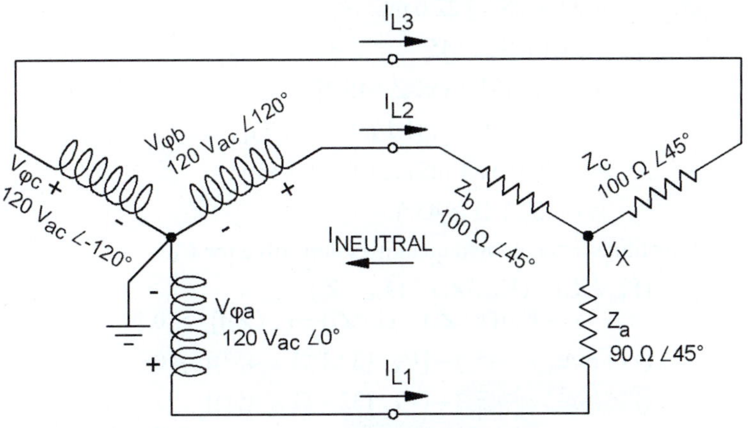

Figure 22-12: Unbalanced Wye-Wye Circuit for Problem 2

Solution: The phase voltages of the wye generator are referenced to ground, so use node voltage method to determine the value of V_X relative to ground.

$$[(V_{\varphi a} - V_X) / Z_a] + [(V_{\varphi b} - V_X) / Z_b] + [(V_{\varphi c} - V_X) / Z_c] = 0$$
$$(V_{\varphi a} / Z_a) - (V_X / Z_a) + (V_{\varphi b} / Z_b) - (V_X / Z_b) + (V_{\varphi c} / Z_c) - (V_X / Z_c) = 0$$
$$[(V_{\varphi a} - V_X) / Z_a] + [(V_{\varphi b} - V_X) / Z_b] + [(V_{\varphi c} - V_X) / Z_c] = 0$$
$$(V_{\varphi a} / Z_a) + (V_{\varphi b} / Z_b) + (V_{\varphi c} / Z_c) - (V_X / Z_a) - (V_X / Z_b) - (V_X / Z_c) = 0$$
$$(V_{\varphi a} / Z_a) + (V_{\varphi b} / Z_b) + (V_{\varphi c} / Z_c) - V_X [(1 / Z_a) + (1 / Z_b) + (1 / Z_c)] = 0$$

Solve for the known terms.

$$V_{\varphi a} / Z_a = (120\ V_{ac} \angle 0°) / (90\ \Omega \angle 45°)$$
$$= (120\ V_{ac} / 90\ \Omega) \angle (0° - 45°)$$
$$= 1.33\ mA_{ac} \angle -45°$$

$$V_{\varphi b} / Z_b = (120\ V_{ac} \angle 120°) / (100\ \Omega \angle 45°)$$
$$= (120\ V_{ac} / 100\ \Omega) \angle (120° - 45°)$$
$$= 1.20\ mA_{ac} \angle 75°$$

$$V_{\varphi c} / Z_c = (120\ V_{ac} \angle -120°) / (100\ \Omega \angle 45°)$$
$$= (120\ V_{ac} / 100\ \Omega) \angle (-120° - 45°)$$
$$= 1.20\ mA_{ac} \angle -165°$$

$$(V_{\varphi a} / Z_a) + (V_{\varphi b} / Z_b) + (V_{\varphi c} / Z_c)$$
$$= (1.33\ mA_{ac} \angle -45°) + (1.20\ mA_{ac} \angle 75°) + (1.20\ mA_{ac} \angle -165°)$$
$$= (943\ mA_{ac} - j\ 943\ mA_{ac}) + (311\ mA_{ac} + j\ 1.16\ A_{ac}) + (-1.16\ A_{ac} - j\ 311\ mA_{ac})$$
$$= (943\ mA_{ac} + 311\ mA_{ac} - 1.16\ A_{ac}) + j\ [(-j\ 943\ mA_{ac}) + 1.16\ A_{ac} + (-311\ mA_{ac})]$$
$$= 94.3\ mA_{ac} - j\ 94.3\ mA_{ac} \text{ in rectangular form}$$
$$= 113\ mA_{ac} \angle -45°$$

$$(1 / Z_a) + (1 / Z_b) + (1 / Z_c)$$
$$= [1 / (90\ \Omega \angle 45°)] + [1 / (100\ \Omega \angle 45°)] + [1 / (100\ \Omega \angle 45°)]$$
$$= [(1 \angle 0°)/(90\ \Omega \angle 45°)] + [(1 \angle 0°)/(100\ \Omega \angle 45°)] + [(1 \angle 0°)/(100\ \Omega \angle 45°)]$$
$$= [(1 / 90\ \Omega) \angle (0° - 45°)] + [(1 / 100\ \Omega) \angle (0° - 45°)] + [(1 / 100\ \Omega) \angle (0° - 45°)]$$
$$= (11.1\ mS \angle -45°) + (10.0\ mS \angle -45°) + (10.0\ mS \angle -45°)$$
$$= (7.86\ mS - j\ 7.86\ ms) + (7.07\ mS - j\ 7.07\ ms) + (7.07\ mS - j\ 7.07\ ms)$$
$$= (7.86\ mS + 7.07\ mS + 7.07\ mS) - j\ (7.86\ ms + 7.07\ mS + 7.07\ mS)$$

$= 22.0 \text{ mS} - j\ 22.0 \text{ mS}$

$= 31.1 \text{ mS} \angle -45°$

$= 1 / [1 / (31.1 \text{ mS} \angle -45°)]$

$= 1 / [(1 \angle 0°) / (31.1 \text{ mS} \angle -45°)]$

$= 1 / [(1 / 31.1 \text{ mS}) \angle [0° - (-45°)]$

$= 1 / (32.1\ \Omega \angle 45°)$

Substitute in the known quantities and solve for V_X.

$$(V_{\varphi a} / Z_a) + (V_{\varphi b} / Z_b) + (V_{\varphi c} / Z_c) - V_X [(1 / Z_a) + (1 / Z_b) + (1 / Z_c)] = 0$$

$$(113 \text{ mA}_{ac} \angle -45°) - [V_X / (32.1\ \Omega \angle 45°)] = 0$$

$$(113 \text{ mA}_{ac} \angle -45°) - [V_X / (32.1\ \Omega \angle 45°)] - (113 \text{ mA}_{ac} \angle -45°) = 0 - (113 \text{ mA}_{ac} \angle -45°)$$

$$[V_X / (32.1\ \Omega \angle 45°)](32.1\ \Omega \angle 45°) = (113 \text{ mA}_{ac} \angle -45°)(32.1\ \Omega \angle 45°)$$

$$V_X = [(113 \text{ mA}_{ac})(32.1\ \Omega) \angle (-45° + 45°)$$

$$= \mathbf{4.29\ V_{ac} \angle 0°}$$

22.2.2 Unbalanced Wye-Delta Circuits

1. Problem: Determine the line and load currents for the unbalanced wye-delta circuit shown in Figure 22-13. Express all answers in polar form.

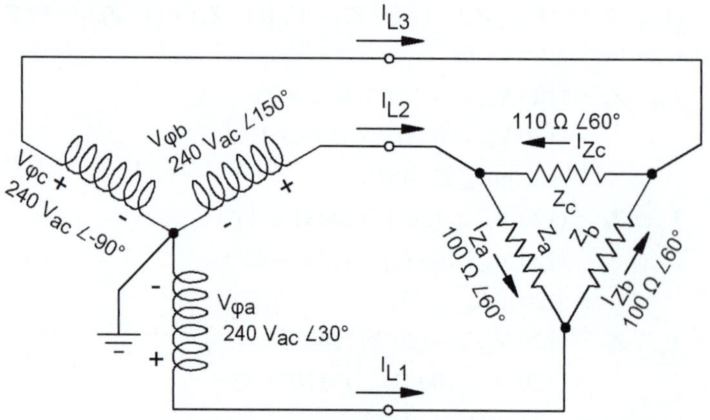

Figure 22-13: Unbalanced Wye-Delta Circuit for Problem 1

Solution: The delta loads are connected directly across the line voltages. First solve for the line voltages and then use Ohm's Law for current to calculate the load currents.

$V_{L(ba)} = V_{\varphi b} - V_{\varphi a}$

$= (240 \text{ V}_{ac} \angle 150°) - (240 \text{ V}_{ac} \angle 30°)$

$= (-208 \text{ V}_{ac} + j\ 120 \text{ V}_{ac}) - (208 \text{ V}_{ac} + j\ 120 \text{ V}_{ac})$

$= (-208 \text{ V}_{ac} - 208 \text{ V}_{ac}) + j\ (120 \text{ V}_{ac} - 120 \text{ V}_{ac})$

$= -416 \text{ V}_{ac} + j\ 0 \text{ V}_{ac}$

$= 416 \text{ V}_{ac} \angle 180°$

$V_{L(cb)} = V_{\varphi c} - V_{\varphi b}$

$= (240 \text{ V}_{ac} \angle -90°) - (240 \text{ V}_{ac} \angle 150°)$

$= (0 \text{ V}_{ac} - j\ 240 \text{ V}_{ac}) - (-208 \text{ V}_{ac} + j\ 120 \text{ V}_{ac})$

$= (0\ V_{ac} + 208\ V_{ac}) - j\ (240\ V_{ac} + 120\ V_{ac})$

$= 208\ V_{ac} - j\ 360\ V_{ac}$

$= 416\ V_{ac} \angle -60°$

$\boldsymbol{V_{L(ac)}} = \boldsymbol{V_{\varphi a}} - \boldsymbol{V_{\varphi c}}$

$= (240\ V_{ac} \angle 30°) - (240\ V_{ac} \angle -90°)$

$= (208\ V_{ac} + j\ 120\ V_{ac}) - (0\ V_{ac} - j\ 240\ V_{ac})$

$= (208\ V_{ac} - 0\ V_{ac}) + j\ (120\ V_{ac} + 240\ V_{ac})$

$= 208\ V_{ac} + j\ 360\ V_{ac}$

$= 416\ V_{ac} \angle 60°$

$\boldsymbol{I_{Za}} = \boldsymbol{V_{L(ba)}} / \boldsymbol{Z_a}$

$= (416\ V_{ac} \angle 180°) / (100\ \Omega \angle 60°)$

$= (416\ V_{ac} / 100\ \Omega) \angle (180° - 60°)$

$= \mathbf{4.16\ A_{ac} \angle 120°}$

$\boldsymbol{I_{Zb}} = \boldsymbol{V_{L(ac)}} / \boldsymbol{Z_b}$

$= (416\ V_{ac} \angle 60°) / (100\ \Omega \angle 60°)$

$= (416\ V_{ac} / 100\ \Omega) \angle (60° - 60°)$

$= \mathbf{4.16\ A_{ac} \angle 0°}$

$\boldsymbol{I_{Zc}} = \boldsymbol{V_{L(cb)}} / \boldsymbol{Z_c}$

$= (416\ V_{ac} \angle -60°) / (110\ \Omega \angle 60°)$

$= (416\ V_{ac} / 110\ \Omega) \angle [(-60°) - 60°)$

$= \mathbf{3.78\ A_{ac} \angle -120°}$

Use Kirchhoff's Current Law to calculate the line currents from the load currents. Because the line currents are shown entering the nodes, define currents entering a node to be positive and currents leaving a node to be negative.

$\boldsymbol{I_{L1}} + \boldsymbol{I_{Za}} - \boldsymbol{I_{Zb}} = 0$

$\boldsymbol{I_{L1}} + \boldsymbol{I_{Za}} - \boldsymbol{I_{Zb}} - \boldsymbol{I_{Za}} + \boldsymbol{I_{Zb}} = 0 - \boldsymbol{I_{Za}} + \boldsymbol{I_{Zb}}$

$\boldsymbol{I_{L1}} = \boldsymbol{I_{Zb}} - \boldsymbol{I_{Za}}$

$= (4.16\ A_{ac} \angle 0°) - (4.16\ A_{ac} \angle 120°)$

$= (4.16\ A_{ac} + j\ 0\ A_{ac}) - (-2.08\ A_{ac} + j\ 3.60\ A_{ac})$

$= (4.16\ A_{ac} + 2.08\ A_{ac}) + j\ (0\ A_{ac} - 3.60\ A_{ac})$

$= 6.24\ A_{ac} - j\ 3.60\ A_{ac}$ in rectangular form

$= \mathbf{7.20\ A_{ac} \angle -30°}$ in polar form

$\boldsymbol{I_{L2}} + \boldsymbol{I_{Zc}} - \boldsymbol{I_{Za}} = 0$

$\boldsymbol{I_{L1}} + \boldsymbol{I_{Zc}} - \boldsymbol{I_{Za}} - \boldsymbol{I_{Zc}} + \boldsymbol{I_{Za}} = 0 - \boldsymbol{I_{Zc}} + \boldsymbol{I_{Za}}$

$\boldsymbol{I_{L1}} = \boldsymbol{I_{Za}} - \boldsymbol{I_{Zc}}$

$= (4.16\ A_{ac} \angle 120°) - (3.78\ A_{ac} \angle -120°)$

$= (-2.08\ A_{ac} + j\ 3.60\ A_{ac}) - (-1.89\ A_{ac} - j\ 3.27\ A_{ac})$

$= (-2.08\ A_{ac} + 1.89\ A_{ac}) + j\ (3.60\ A_{ac} + 3.27\ A_{ac})$

$= -189\ mA_{ac} + j\ 6.87\ A_{ac}$ in rectangular form

$= \mathbf{6.88\ A_{ac} \angle 91.6°}$ in polar form

$\boldsymbol{I_{L3}} + \boldsymbol{I_{Zb}} - \boldsymbol{I_{Zc}} = 0$

$\boldsymbol{I_{L3}} + \boldsymbol{I_{Zb}} - \boldsymbol{I_{Zc}} - \boldsymbol{I_{Zb}} + \boldsymbol{I_{Zc}} = 0 - \boldsymbol{I_{Zb}} + \boldsymbol{I_{Zc}}$

$\boldsymbol{I_{L3}} = \boldsymbol{I_{Zc}} - \boldsymbol{I_{Zb}}$

$= (3.78\ A_{ac} \angle -120°) - (4.16\ A_{ac} \angle 0°)$
$= (-1.89\ A_{ac} - j\ 3.27\ A_{ac}) - (4.16\ A_{ac} + j\ 0\ A_{ac})$
$= (-1.89\ A_{ac} - 4.16\ A_{ac}) - j\ (3.27\ A_{ac} + 0\ A_{ac})$
$= -6.05\ A_{ac} - j\ 3.27\ A_{ac}$ in rectangular form
$= \mathbf{6.88\ A_{ac} \angle -152°}$ in polar form

2. Problem: Determine the line and load currents for the unbalanced wye-delta circuit shown in Figure 22-14. Express all answers in polar form.

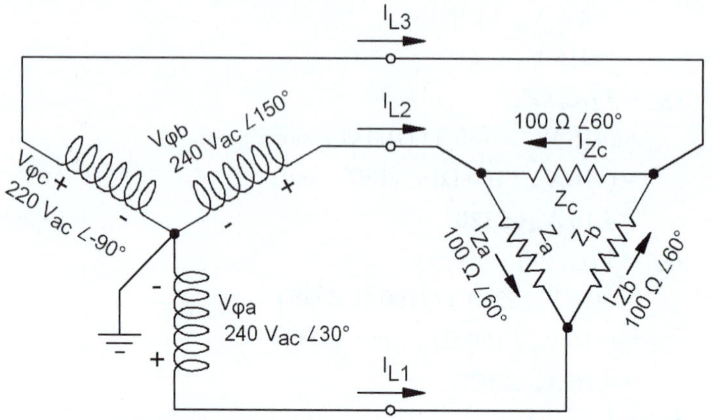

Figure 22-14: Unbalanced Wye-Delta Circuit for Problem 2

Solution: The delta loads are connected directly across the line voltages. First solve for the line voltages and then use Ohm's Law for current to calculate the load currents.

$\boldsymbol{V_{L(ba)}} = \boldsymbol{V_{\varphi b}} - \boldsymbol{V_{\varphi a}}$
$= (240\ V_{ac} \angle 150°) - (240\ V_{ac} \angle 30°)$
$= (-208\ V_{ac} + j\ 120\ V_{ac}) - (208\ V_{ac} + j\ 120\ V_{ac})$
$= (-208\ V_{ac} - 208\ V_{ac}) + j\ (120\ V_{ac} - 120\ V_{ac})$
$= -416\ V_{ac} + j\ 0\ V_{ac}$
$= 416\ V_{ac} \angle 180°$

$\boldsymbol{V_{L(cb)}} = \boldsymbol{V_{\varphi c}} - \boldsymbol{V_{\varphi b}}$
$= (220\ V_{ac} \angle -90°) - (240\ V_{ac} \angle 150°)$
$= (0\ V_{ac} - j\ 220\ V_{ac}) - (-208\ V_{ac} + j\ 120\ V_{ac})$
$= (0\ V_{ac} + 208\ V_{ac}) - j\ (220\ V_{ac} + 120\ V_{ac})$
$= 208\ V_{ac} - j\ 340\ V_{ac}$
$= 398\ V_{ac} \angle -58.6°$

$\boldsymbol{V_{L(ac)}} = \boldsymbol{V_{\varphi a}} - \boldsymbol{V_{\varphi c}}$
$= (240\ V_{ac} \angle 30°) - (220\ V_{ac} \angle -90°)$
$= (208\ V_{ac} + j\ 120\ V_{ac}) - (0\ V_{ac} - j\ 220\ V_{ac})$
$= (208\ V_{ac} - 0\ V_{ac}) + j\ (120\ V_{ac} + 220\ V_{ac})$
$= 208\ V_{ac} + j\ 340\ V_{ac}$
$= 398\ V_{ac} \angle 58.6°$

$\boldsymbol{I_{Za}} = \boldsymbol{V_{L(ba)}} / \boldsymbol{Z_a}$
$= (416\ V_{ac} \angle 180°) / (100\ \Omega \angle 60°)$

$= (416\ V_{ac} / 100\ \Omega) \angle (180° - 60°)$

$= \mathbf{4.16\ A_{ac} \angle 120°}$

$\boldsymbol{I_{Zb}} = \boldsymbol{V_{L(ac)}} / \boldsymbol{Z_b}$

$= (398\ V_{ac} \angle 58.6°) / (100\ \Omega \angle 60°)$

$= (398\ V_{ac} / 100\ \Omega) \angle (58.6° - 60°)$

$= \mathbf{3.98\ A_{ac} \angle -1.44°}$

$\boldsymbol{I_{Zc}} = \boldsymbol{V_{L(cb)}} / \boldsymbol{Z_c}$

$= (398\ V_{ac} \angle -58.6°) / (110\ \Omega \angle 60°)$

$= (398\ V_{ac} / 100\ \Omega) \angle [(-58.6°) - 60°)$

$= \mathbf{3.98\ A_{ac} \angle -119°}$

Use Kirchhoff's Current Law to calculate the line currents from the load currents. Because the line currents are shown entering the nodes, define currents entering a node to be positive and currents leaving a node to be negative.

$\boldsymbol{I_{L1}} + \boldsymbol{I_{Za}} - \boldsymbol{I_{Zb}} = 0$

$\boldsymbol{I_{L1}} + \boldsymbol{I_{Za}} - \boldsymbol{I_{Zb}} - \boldsymbol{I_{Za}} + \boldsymbol{I_{Zb}} = 0 - \boldsymbol{I_{Za}} + \boldsymbol{I_{Zb}}$

$\boldsymbol{I_{L1}} = \boldsymbol{I_{Zb}} - \boldsymbol{I_{Za}}$

$= (3.98\ A_{ac} \angle -1.44°) - (4.16\ A_{ac} \angle 120°)$

$= (3.98\ A_{ac} - j\ 100\ mA_{ac}) - (-2.08\ A_{ac} + j\ 3.60\ A_{ac})$

$= (3.98\ A_{ac} + 2.08\ A_{ac}) - j\ (100\ mA_{ac} + 3.60\ A_{ac})$

$= 6.06\ A_{ac} - j\ 3.70\ A_{ac}$ in rectangular form

$= \mathbf{7.10\ A_{ac} \angle -31.4°}$ in polar form

$\boldsymbol{I_{L2}} + \boldsymbol{I_{Zc}} - \boldsymbol{I_{Za}} = 0$

$\boldsymbol{I_{L1}} + \boldsymbol{I_{Zc}} - \boldsymbol{I_{Za}} - \boldsymbol{I_{Zc}} + \boldsymbol{I_{Za}} = 0 - \boldsymbol{I_{Zc}} + \boldsymbol{I_{Za}}$

$\boldsymbol{I_{L1}} = \boldsymbol{I_{Za}} - \boldsymbol{I_{Zc}}$

$= (4.16\ A_{ac} \angle 120°) - (3.98\ A_{ac} \angle -119°)$

$= (-2.08\ A_{ac} + j\ 3.60\ A_{ac}) - (-1.91\ A_{ac} - j\ 3.50\ A_{ac})$

$= (-2.08\ A_{ac} + 1.91\ A_{ac}) + j\ (3.60\ A_{ac} + 3.50\ A_{ac})$

$= -173\ mA_{ac} + j\ 7.10\ A_{ac}$ in rectangular form

$= \mathbf{7.10\ A_{ac} \angle 91.4°}$ in polar form

$\boldsymbol{I_{L3}} + \boldsymbol{I_{Zb}} - \boldsymbol{I_{Zc}} = 0$

$\boldsymbol{I_{L3}} + \boldsymbol{I_{Zb}} - \boldsymbol{I_{Zc}} - \boldsymbol{I_{Zb}} + \boldsymbol{I_{Zc}} = 0 - \boldsymbol{I_{Zb}} + \boldsymbol{I_{Zc}}$

$\boldsymbol{I_{L3}} = \boldsymbol{I_{Zc}} - \boldsymbol{I_{Zb}}$

$= (3.98\ A_{ac} \angle -119°) - (3.98\ A_{ac} \angle -1.44°)$

$= (-1.91\ A_{ac} - j\ 3.50\ A_{ac}) - (3.98\ A_{ac} + j\ 100\ mA_{ac})$

$= (-1.91\ A_{ac} - 3.98\ A_{ac}) - j\ (3.50\ A_{ac} + 100\ mA_{ac})$

$= -5.89\ A_{ac} - j\ 3.40\ A_{ac}$ in rectangular form

$= \mathbf{6.80\ A_{ac} \angle -150°}$ in polar form

22.2.3 Unbalanced Delta-Wye Circuits

1. Problem: Determine the line and load currents for the unbalanced delta-wye circuit shown in Figure 22-15. Express all answers in polar form.

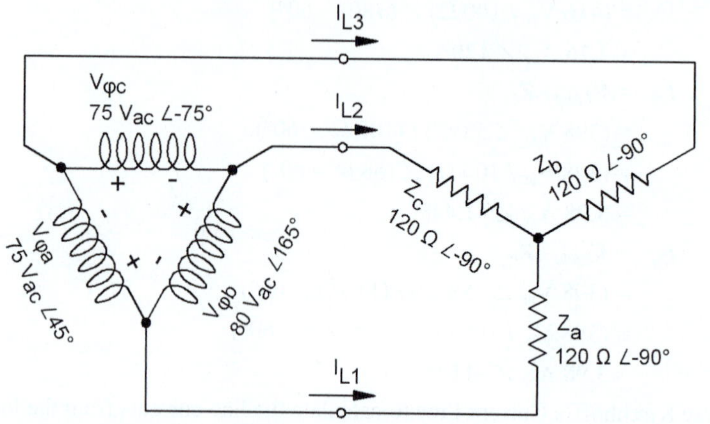

Figure 22-15: Unbalanced Delta-Wye Circuit for Problem 1

Solution: To determine the line and load currents, first convert the wye load to a delta load, as shown in Figure 22-18, using the delta-to-wye conversion formulas.

$$\boldsymbol{Z_1} = (\boldsymbol{Z_a Z_b} + \boldsymbol{Z_b Z_c} + \boldsymbol{Z_a Z_c}) / \boldsymbol{Z_b}$$
$$\boldsymbol{Z_2} = (\boldsymbol{Z_a Z_b} + \boldsymbol{Z_b Z_c} + \boldsymbol{Z_a Z_c}) / \boldsymbol{Z_c}$$
$$\boldsymbol{Z_3} = (\boldsymbol{Z_a Z_b} + \boldsymbol{Z_b Z_c} + \boldsymbol{Z_a Z_c}) / \boldsymbol{Z_a}$$

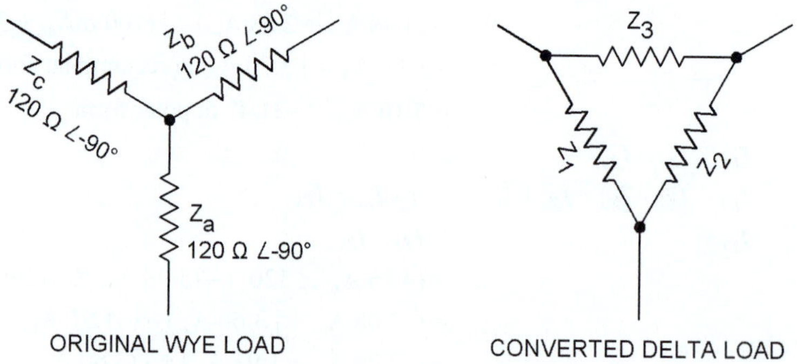

Figure 22-16: Conversion of Wye Load to Delta Load

For the balanced wye load, $\boldsymbol{Z_a} = \boldsymbol{Z_b} = \boldsymbol{Z_c} = \boldsymbol{Z_Y}$, so the equation for each delta load, $\boldsymbol{Z_1} = \boldsymbol{Z_2} = \boldsymbol{Z_3} = \boldsymbol{Z_\Delta}$, simplifies to

$$\begin{aligned}
\boldsymbol{Z_\Delta} &= (\boldsymbol{Z_Y Z_Y} + \boldsymbol{Z_Y Z_Y} + \boldsymbol{Z_Y Z_Y}) / \boldsymbol{Z_Y} \\
&= (\boldsymbol{Z_Y}^2 + \boldsymbol{Z_Y}^2 + \boldsymbol{Z_Y}^2) / \boldsymbol{Z_Y} \\
&= (3\,\boldsymbol{Z_Y}^2) / \boldsymbol{Z_Y} \\
&= 3\,\boldsymbol{Z_Y} \\
&= (3)(120\ \Omega \angle -90^\circ) \\
&= (3 \angle 0^\circ)(120\ \Omega \angle -90^\circ) \\
&= [(3)(120\ \Omega)] \angle (0^\circ + -90^\circ) \\
&= 360\ \Omega \angle -90^\circ
\end{aligned}$$

Figure 22-17 shows the converted circuit.

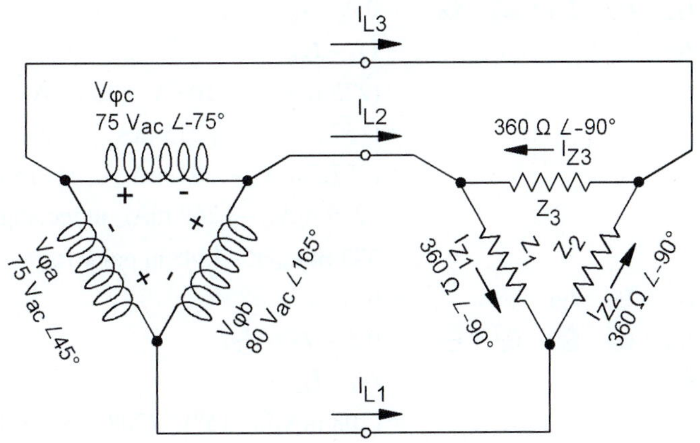

Figure 22-17: Delta-Wye Circuit Converted to Delta-Delta Circuit

In the delta-delta circuit of Figure 22-17, the delta phase voltages are applied directly across the delta loads. Use Ohm's Law for current to calculate the delta phase currents.

$\boldsymbol{I_{Z1}} = \boldsymbol{V_{Z1}} / \boldsymbol{Z_1}$

$= \boldsymbol{V_{\varphi b}} / \boldsymbol{Z_1}$

$= (80\ V_{ac} \angle\ 165°) / (360\ \Omega \angle -90°)$

$= (80\ V_{ac} / 360\ \Omega) \angle\ [165° - (-90°)]$

$= 222\ mA_{ac} \angle\ 255°$

$= 222\ mA_{ac} \angle -105°$

$\boldsymbol{I_{Z2}} = \boldsymbol{V_{Z2}} / \boldsymbol{Z_2}$

$= \boldsymbol{V_{\varphi a}} / \boldsymbol{Z_2}$

$= (75\ V_{ac} \angle\ 45°) / (360\ \Omega \angle -90°)$

$= (75\ V_{ac} / 360\ \Omega) \angle\ [45° - (-90°)]$

$= 208\ mA_{ac} \angle\ 135°$

$\boldsymbol{I_{Z3}} = \boldsymbol{V_{Z3}} / \boldsymbol{Z_3}$

$= \boldsymbol{V_{\varphi c}} / \boldsymbol{Z_3}$

$= (75\ V_{ac} \angle -75°) / (360\ \Omega \angle -90°)$

$= (75\ V_{ac} / 360\ \Omega) \angle\ [(-75°) - (-90°)]$

$= 208\ mA_{ac} \angle\ 15°$

For the directions of the load currents as shown in Figure 22-17, each line current is the difference between two load currents. Because the line currents are shown entering the converted delta load, define currents entering a node to be positive and currents leaving a node to be negative. Then use Kirchhoff's Current Law to determine the line currents.

$\boldsymbol{I_{L1}} + \boldsymbol{I_{Z1}} - \boldsymbol{I_{Z2}} = 0\ A$

$\boldsymbol{I_{L1}} + \boldsymbol{I_{Z1}} - \boldsymbol{I_{Z2}} - \boldsymbol{I_{Z1}} + \boldsymbol{I_{Z2}} = 0\ A - \boldsymbol{I_{Z1}} + \boldsymbol{I_{Z2}}$

$\boldsymbol{I_{L1}} = \boldsymbol{I_{Z2}} - \boldsymbol{I_{Z1}}$

$= (208\ mA_{ac} \angle\ 135°) - (222\ mA_{ac} \angle -105°)$

$= (-147\ mA_{ac} + j\ 147\ mA_{ac}) - (-57.5\ mA_{ac} - j\ 215\ mA_{ac})$

$= (-147\ mA_{ac} + 57.5\ mA_{ac}) + j\ (147\ mA_{ac} + 215\ mA_{ac})$

$= -89.8\ mA_{ac} + j\ 362\ mA_{ac}$ in rectangular form

$=$ **$373\ mA_{ac} \angle\ 104°$** in polar form

$\boldsymbol{I_{L2}} - \boldsymbol{I_{Z1}} + \boldsymbol{I_{Z3}} = 0\ A$

$$I_{L2} - I_{Z1} + I_{Z3} + I_{Z1} - I_{Z3} = 0\text{ A} + I_{Z1} - I_{Z3}$$

$$I_{L2} = I_{Z1} - I_{Z3}$$

$$= (222\text{ mA}_{ac} \angle -105°) - (208\text{ mA}_{ac} \angle 15°)$$

$$= (-57.5\text{ mA}_{ac} - j\ 215\text{ mA}_{ac}) - (201\text{ mA}_{ac} + j\ 53.9\text{ mA}_{ac})$$

$$= (-57.5\text{ mA}_{ac} - 201\text{ mA}_{ac}) - j\ (215\text{ mA}_{ac} + 53.9\text{ mA}_{ac})$$

$$= -259\text{ mA}_{ac} - j\ 269\text{ mA}_{ac} \text{ in rectangular form}$$

$$= \mathbf{373\ mA_{ac} \angle -134°} \text{ in polar form}$$

$$I_{L3} + I_{Z2} - I_{Z3} = 0\text{ A}$$

$$I_{L3} + I_{Z2} - I_{Z3} - I_{Z2} + I_{Z3} = 0\text{ A} - I_{Z2} + I_{Z3}$$

$$I_{L3} = I_{Z3} - I_{Z2}$$

$$= (208\text{ mA}_{ac} \angle 15°) - (208\text{ mA}_{ac} \angle 135°)$$

$$= (201\text{ mA}_{ac} + j\ 53.9\text{ mA}_{ac}) - (-147\text{ mA}_{ac} + j\ 147\text{ mA}_{ac})$$

$$= (201\text{ mA}_{ac} + 147\text{ mA}_{ac}) + j\ (53.9\text{ mA}_{ac} - 147\text{ mA}_{ac})$$

$$= 349\text{ mA}_{ac} - j\ 93.4\text{ mA}_{ac} \text{ in rectangular form}$$

$$= \mathbf{361\ mA_{ac} \angle -15°} \text{ in polar form}$$

Referring back to Figure 22-15, the wye loads are in series with the lines, so that the load currents are equal to the line currents.

$$I_{Za} = I_{L1}$$

$$= \mathbf{373\ mA_{ac} \angle 104°}$$

$$I_{Zb} = I_{L3}$$

$$= \mathbf{361\ mA_{ac} \angle -15°}$$

$$I_{Zc} = I_{L2}$$

$$= \mathbf{373\ mA_{ac} \angle -134°}$$

2. Problem: Determine the line and load currents for the unbalanced delta-wye circuit shown in Figure 22-18. Express all answers in polar form.

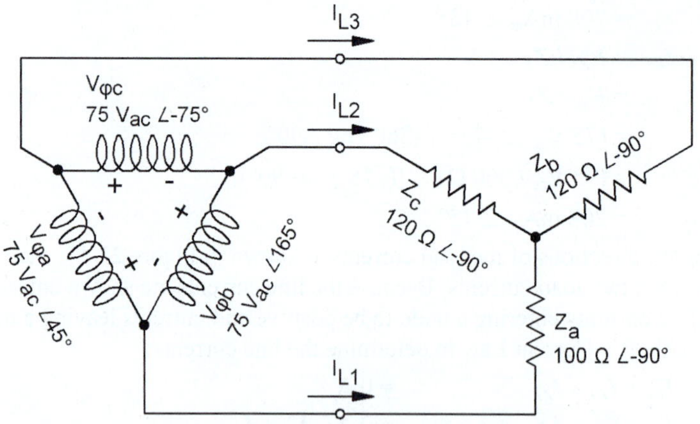

Figure 22-18: Unbalanced Delta-Wye Circuit for Problem 2

Solution: To determine the line and load currents, first convert the wye load to a delta load, as shown in Figure 22-19, using the delta-to-wye conversion formulas.

$$Z_1 = (Z_a Z_b + Z_b Z_c + Z_a Z_c) / Z_b$$

$$Z_2 = (Z_a Z_b + Z_b Z_c + Z_a Z_c) / Z_c$$

$Z_3 \ = (Z_a Z_b + Z_b Z_c + Z_a Z_c) / Z_a$

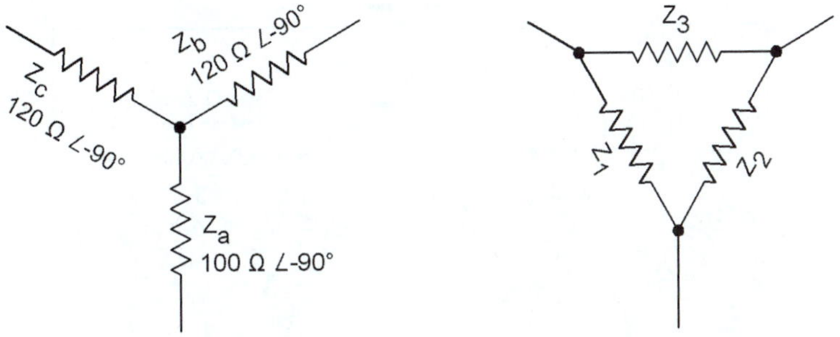

Figure 22-19: Conversion of Wye Load to Delta Load

First, calculate Z_{NUM}, the common numerator for all three converted impedances. Then calculate the loads for the converted delta load.

$$\begin{aligned}
Z_{NUM} &= Z_a Z_b + Z_b Z_c + Z_a Z_c \\
&= (100\ \Omega \angle -90°)(120\ \Omega \angle -90°) + (120\ \Omega \angle -90°)(120\ \Omega \angle -90°) \\
&\qquad + (100\ \Omega \angle -90°)(120\ \Omega \angle -90°) \\
&= [(100\ \Omega)(120\ \Omega)] \angle [(-90°) + (-90°)] + [(120\ \Omega)(120\ \Omega)] \angle [(-90°) + (-90°)] \\
&\qquad + [(100\ \Omega)(120\ \Omega)] \angle [(-90°) + (-90°)] \\
&= (12.0\ \text{k}\Omega^2 \angle -180°) + (14.4\ \text{k}\Omega^2 \angle -180°) + (12.0\ \text{k}\Omega^2 \angle -180°) \\
&= (-12.0\ \text{k}\Omega^2 + j\ 0\ \Omega^2) + (-14.4\ \text{k}\Omega^2 + j\ 0\ \Omega^2) + (-12.0\ \text{k}\Omega^2 + j\ 0\ \Omega^2) \\
&= (-12.0\ \text{k}\Omega^2 - 14.4\ \text{k}\Omega^2 - 12.0\ \text{k}\Omega^2) + j\ (0\ \Omega^2 + 0\ \Omega^2 + 0\ \Omega^2) \\
&= -38.4\ \text{k}\Omega^2 + j\ 0\ \Omega^2 \\
&= 38.4\ \text{k}\Omega^2 \angle 180°
\end{aligned}$$

$$\begin{aligned}
Z_1 &= Z_{NUM} / Z_b \\
&= (38.4\ \text{k}\Omega^2 \angle 180°) / (120\ \Omega \angle -90°) \\
&= (38.4\ \text{k}\Omega^2 / 120\ \Omega) \angle [180° - (-90°)] \\
&= (38400\ \Omega^2 / 120\ \Omega) \angle [180° - (-90°)] \\
&= 320\ \Omega \angle 270° = 320\ \Omega \angle -90°
\end{aligned}$$

$$\begin{aligned}
Z_2 &= Z_{NUM} / Z_c \\
&= (38.4\ \text{k}\Omega^2 \angle 180°) / (120\ \Omega \angle -90°) \\
&= (38.4\ \text{k}\Omega^2 / 120\ \Omega) \angle [180° - (-90°)] \\
&= (38.4\ \Omega^2 / 120\ \Omega) \angle [180° - (-90°)] \\
&= 320\ \Omega \angle 270° = 320\ \Omega \angle -90°
\end{aligned}$$

$$\begin{aligned}
Z_3 &= Z_{NUM} / Z_a \\
&= (38.4\ \text{k}\Omega^2 \angle 180°) / (100\ \Omega \angle -90°) \\
&= (38.4\ \text{k}\Omega^2 / 100\ \Omega) \angle [180° - (-90°)] \\
&= (38400\ \Omega^2 / 100\ \Omega) \angle [180° - (-90°)] \\
&= 384\ \Omega \angle 270° = 384\ \Omega \angle -90°
\end{aligned}$$

Figure 22-20 shows the converted circuit.

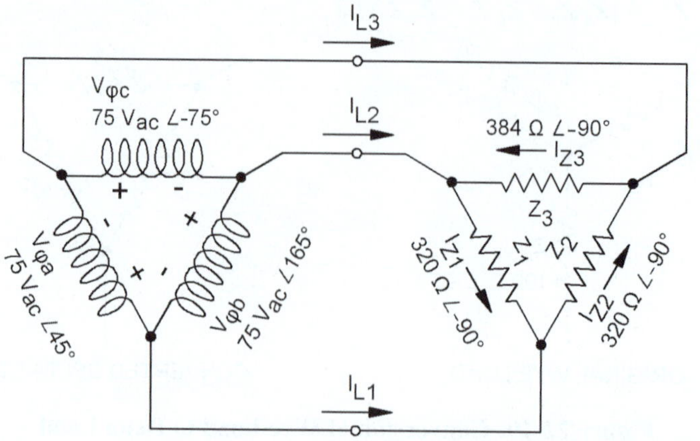

Figure 22-20: Delta-Wye Circuit Converted to Delta-Delta Circuit

In the delta-delta circuit of Figure 22-20, the delta phase voltages are applied directly across the delta loads. Use Ohm's Law for current to calculate the delta phase currents.

$$
\begin{aligned}
\boldsymbol{I_{Z1}} &= \boldsymbol{V_{Z1}} / \boldsymbol{Z_1} \\
&= \boldsymbol{V_{\varphi b}} / \boldsymbol{Z_1} \\
&= (75\ \mathrm{V_{ac}} \angle 165°) / (320\ \Omega \angle -90°) \\
&= (75\ \mathrm{V_{ac}} / 320\ \Omega) \angle [165° - (-90°)] \\
&= 234\ \mathrm{mA_{ac}} \angle 255° \\
&= 234\ \mathrm{mA_{ac}} \angle -105°
\end{aligned}
$$

$$
\begin{aligned}
\boldsymbol{I_{Z2}} &= \boldsymbol{V_{Z2}} / \boldsymbol{Z_2} \\
&= \boldsymbol{V_{\varphi a}} / \boldsymbol{Z_2} \\
&= (75\ \mathrm{V_{ac}} \angle 45°) / (320\ \Omega \angle -90°) \\
&= (75\ \mathrm{V_{ac}} / 320\ \Omega) \angle [45° - (-90°)] \\
&= 234\ \mathrm{mA_{ac}} \angle 135°
\end{aligned}
$$

$$
\begin{aligned}
\boldsymbol{I_{Z3}} &= \boldsymbol{V_{Z3}} / \boldsymbol{Z_3} \\
&= \boldsymbol{V_{\varphi c}} / \boldsymbol{Z_3} \\
&= (75\ \mathrm{V_{ac}} \angle -75°) / (384\ \Omega \angle -90°) \\
&= (75\ \mathrm{V_{ac}} / 384\ \Omega) \angle [(-75°) - (-90°)] \\
&= 195\ \mathrm{mA_{ac}} \angle 15°
\end{aligned}
$$

For the directions of the load currents as shown in Figure 22-20, each line current is the difference between two load currents. Because the line currents are shown entering the converted delta load, define currents entering a node to be positive and currents leaving a node to be negative. Then use Kirchhoff's Current Law to determine the line currents.

$$
\begin{aligned}
\boldsymbol{I_{L1}} + \boldsymbol{I_{Z1}} - \boldsymbol{I_{Z2}} &= 0\ \mathrm{A} \\
\boldsymbol{I_{L1}} + \boldsymbol{I_{Z1}} - \boldsymbol{I_{Z2}} - \boldsymbol{I_{Z1}} + \boldsymbol{I_{Z2}} &= 0\ \mathrm{A} - \boldsymbol{I_{Z1}} + \boldsymbol{I_{Z2}} \\
\boldsymbol{I_{L1}} &= \boldsymbol{I_{Z2}} - \boldsymbol{I_{Z1}} \\
&= (234\ \mathrm{mA_{ac}} \angle 135°) - (234\ \mathrm{mA_{ac}} \angle -105°) \\
&= (-166\ \mathrm{mA_{ac}} + j\ 166\ \mathrm{mA_{ac}}) - (-60.7\ \mathrm{mA_{ac}} - j\ 226\ \mathrm{mA_{ac}}) \\
&= (-166\ \mathrm{mA_{ac}} + 60.7\ \mathrm{mA_{ac}}) + j\ (166\ \mathrm{mA_{ac}} + 226\ \mathrm{mA_{ac}}) \\
&= -105\ \mathrm{mA_{ac}} + j\ 392\ \mathrm{mA_{ac}} \text{ in rectangular form}
\end{aligned}
$$

$$
\begin{aligned}
& = \mathbf{406\ mA_{ac} \angle 105^\circ} \text{ in polar form} \\
\boldsymbol{I_{L2}} - \boldsymbol{I_{Z1}} + \boldsymbol{I_{Z3}} & = 0\text{ A} \\
\boldsymbol{I_{L2}} - \boldsymbol{I_{Z1}} + \boldsymbol{I_{Z3}} + \boldsymbol{I_{Z1}} - \boldsymbol{I_{Z3}} & = 0\text{ A} + \boldsymbol{I_{Z1}} - \boldsymbol{I_{Z3}} \\
\boldsymbol{I_{L2}} & = \boldsymbol{I_{Z1}} - \boldsymbol{I_{Z3}} \\
& = (234\text{ mA}_{ac} \angle -105^\circ) - (195\text{ mA}_{ac} \angle 15^\circ) \\
& = (-60.7\text{ mA}_{ac} - \text{j } 226\text{ mA}_{ac}) - (189\text{ mA}_{ac} + \text{j } 50.6\text{ mA}_{ac}) \\
& = (-60.7\text{ mA}_{ac} - 189\text{ mA}_{ac}) - \text{j } (226\text{ mA}_{ac} + 50.6\text{ mA}_{ac}) \\
& = -249\text{ mA}_{ac} - \text{j } 277\text{ mA}_{ac} \text{ in rectangular form} \\
& = \mathbf{373\ mA_{ac} \angle -132^\circ} \text{ in polar form} \\
\boldsymbol{I_{L3}} + \boldsymbol{I_{Z2}} - \boldsymbol{I_{Z3}} & = 0\text{ A} \\
\boldsymbol{I_{L3}} + \boldsymbol{I_{Z2}} - \boldsymbol{I_{Z3}} - \boldsymbol{I_{Z2}} + \boldsymbol{I_{Z3}} & = 0\text{ A} - \boldsymbol{I_{Z2}} + \boldsymbol{I_{Z3}} \\
\boldsymbol{I_{L3}} & = \boldsymbol{I_{Z3}} - \boldsymbol{I_{Z2}} \\
& = (195\text{ mA}_{ac} \angle 15^\circ) - (234\text{ mA}_{ac} \angle 135^\circ) \\
& = (189\text{ mA}_{ac} + \text{j } 50.6\text{ mA}_{ac}) - (-166\text{ mA}_{ac} + \text{j } 166\text{ mA}_{ac}) \\
& = (189\text{ mA}_{ac} + 166\text{ mA}_{ac}) + \text{j } (50.6\text{ mA}_{ac} - 166\text{ mA}_{ac}) \\
& = 354\text{ mA}_{ac} - \text{j } 115\text{ mA}_{ac} \text{ in rectangular form} \\
& = \mathbf{373\ mA_{ac} \angle -18^\circ} \text{ in polar form}
\end{aligned}
$$

Referring back to Figure 22-18, wye loads are in series with the lines, so that the load currents are equal to the line currents.

$$
\begin{aligned}
\boldsymbol{I_{Za}} & = \boldsymbol{I_{L1}} \\
& = \mathbf{406\ mA_{ac} \angle 105^\circ} \\
\boldsymbol{I_{Zb}} & = \boldsymbol{I_{L3}} \\
& = \mathbf{373\ mA_{ac} \angle -18^\circ} \\
\boldsymbol{I_{Zc}} & = \boldsymbol{I_{L2}} \\
& = \mathbf{373\ mA_{ac} \angle -132^\circ}
\end{aligned}
$$

22.2.4 Unbalanced Delta-Delta Circuits

1. Problem: Determine the current and the apparent, real, and reactive power for each load for the unbalanced delta-delta circuit shown in Figure 22-21. Express all answers in polar form.

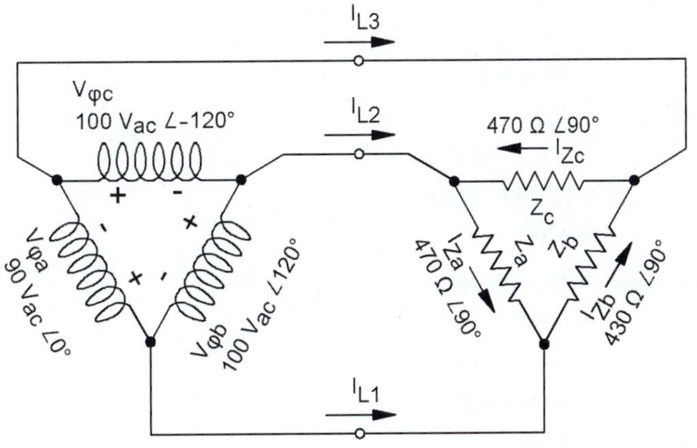

Figure 22-21: Unbalanced Delta-Delta Circuit for Problem 1

Solution: In the delta-delta circuit of Figure 22-21, the delta phase voltages are applied directly across the delta loads. Use Ohm's Law for current to calculate the delta load currents.

$\boldsymbol{I_{Za}} = \boldsymbol{V_{Za} / Z_a}$

$= \boldsymbol{V_{\varphi b} / Z_a}$

$= (100\ \text{V}_{ac} \angle 120°) / (470\ \Omega \angle 90°)$

$= (100\ \text{V}_{ac} / 470\ \Omega) \angle (120° - 90°)$

$= \mathbf{213\ mA_{ac} \angle 30°}$

$\boldsymbol{I_{Zb}} = \boldsymbol{V_{Zb} / Z_b}$

$= \boldsymbol{V_{\varphi a} / Z_b}$

$= (90\ \text{V}_{ac} \angle 0°) / (430\ \Omega \angle 90°)$

$= (100\ \text{V}_{ac} / 470\ \Omega) \angle (0° - 90°)$

$= \mathbf{209\ mA_{ac} \angle -90°}$

$\boldsymbol{I_{Zc}} = \boldsymbol{V_{Zc} / Z_c}$

$= \boldsymbol{V_{\varphi c} / Z_c}$

$= (100\ \text{V}_{ac} \angle -120°) / (470\ \Omega \angle 90°)$

$= (100\ \text{V}_{ac} / 470\ \Omega) \angle [(-120°) - 90°]$

$= \mathbf{213\ mA_{ac} \angle -210° = 213\ mA_{ac} \angle 150°}$

First, use Watt's Law for voltage and current to calculate $\boldsymbol{P_{app}}$, the apparent power for each load. Then, use the phase angle of the load to calculate $\boldsymbol{P_r}$ and $\boldsymbol{P_{reac}}$, the real and reactive powers.

$\boldsymbol{P_{app(Za)}} = \boldsymbol{V_{Za} I_{Za}}$

$= (100\ \text{V}_{ac})(213\ \text{mA}_{ac})$

$= \mathbf{21.3\ VA}$

$\boldsymbol{P_{r(Za)}} = \boldsymbol{P_{app}} \cos |\varphi|$

$= (21.3\ \text{VA})(\cos 90°)$

$= (21.3\ \text{VA})(0)$

$= \mathbf{0\ W}$

$\boldsymbol{P_{reac(Za)}} = \boldsymbol{P_{app}} \sin |\ \varphi\ |$

$= (21.3\ \text{VA})(\sin 90°)$

$= (21.3\ \text{VA})(1)$

$= \mathbf{21.3\ VAR}$

$\boldsymbol{P_{app(Zb)}} = \boldsymbol{V_{Zb} I_{Zb}}$

$= (90\ \text{V}_{ac})(209\ \text{mA}_{ac})$

$= \mathbf{18.8\ VA}$

$\boldsymbol{P_{r(Zb)}} = \boldsymbol{P_{app}} \cos |\varphi|$

$= (18.8\ \text{VA})(\cos 90°)$

$= (18.83\ \text{VA})(0)$

$= \mathbf{0\ W}$

$\boldsymbol{P_{reac(Zb)}} = \boldsymbol{P_{app}} \sin |\ \varphi\ |$

$= (18.8\ \text{VA})(\sin 90°)$

$= (18.8\ \text{VA})(1)$

$= \mathbf{18.8\ VAR}$

$\boldsymbol{P_{app(Zc)}} = \boldsymbol{V_{Zc} I_{Zc}}$

$= (100\ \text{V}_{ac})(213\ \text{mA}_{ac})$

$= \mathbf{21.3\ VA}$

$\boldsymbol{P_{r(Zc)}} = \boldsymbol{P_{app}} \cos|\varphi|$

$= (21.3\ \text{VA})(\cos 90°)$

$= (21.3\ \text{VA})(0)$

$= \mathbf{0\ W}$

$\boldsymbol{P_{reac(Zc)}} = \boldsymbol{P_{app}} \sin|\varphi|$

$= (21.3\ \text{VA})(\sin 90°)$

$= (21.3\ \text{VA})(1)$

$= \mathbf{21.3\ VAR}$

2. Problem: Determine the current and power for each load for the unbalanced delta-delta circuit shown in Figure 22-22. Express all answers in polar form.

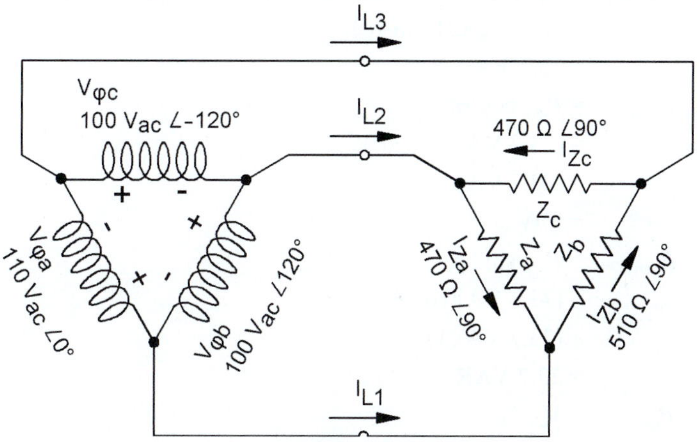

Figure 22-22: Unbalanced Delta-Delta Circuit for Problem 2

Solution: In the delta-delta circuit of Figure 22-22, the delta phase voltages are applied directly across the delta loads. Use Ohm's Law for current to calculate the delta load currents.

$\boldsymbol{I_{Za}} = \boldsymbol{V_{Za}} / \boldsymbol{Z_a}$

$= \boldsymbol{V_{\varphi b}} / \boldsymbol{Z_a}$

$= (100\ \text{V}_{ac} \angle 120°) / (470\ \Omega \angle 90°)$

$= (100\ \text{V}_{ac} / 470\ \Omega) \angle (120° - 90°)$

$= \mathbf{213\ mA_{ac} \angle 30°}$

$\boldsymbol{I_{Zb}} = \boldsymbol{V_{Zb}} / \boldsymbol{Z_b}$

$= \boldsymbol{V_{\varphi a}} / \boldsymbol{Z_b}$

$= (110\ \text{V}_{ac} \angle 0°) / (510\ \Omega \angle 90°)$

$= (110\ \text{V}_{ac} / 510\ \Omega) \angle (0° - 90°)$

$= \mathbf{216\ mA_{ac} \angle -90°}$

$\boldsymbol{I_{Zc}} = \boldsymbol{V_{Zc}} / \boldsymbol{Z_c}$

$= \boldsymbol{V_{\varphi c}} / \boldsymbol{Z_c}$

$= (100\ \text{V}_{ac} \angle -120°) / (470\ \Omega \angle 90°)$

$= (100\ \text{V}_{ac} / 470\ \Omega) \angle [(-120°) - 90°]$

$= \mathbf{213\ mA_{ac} \angle -210° = 213\ mA_{ac} \angle 150°}$

First, use Watt's Law for voltage and current to calculate $\boldsymbol{P_{app}}$, the apparent power for each load. Then, use the phase angle of the load to calculate $\boldsymbol{P_r}$ and $\boldsymbol{P_{reac}}$, the real and reactive powers.

$\boldsymbol{P_{app(Za)}} = \boldsymbol{V_{Za}} \boldsymbol{I_{Za}}$

$= (100\ \text{V}_{ac})(213\ \text{mA}_{ac})$

$= \mathbf{21.3\ VA}$

$P_{r(Za)} = P_{app} \cos|\varphi|$
$= (21.3\ VA)(\cos 90°)$
$= (21.3\ VA)(0)$
$= \mathbf{0\ W}$

$P_{reac(Za)} = P_{app} \sin|\varphi|$
$= (21.3\ VA)(\sin 90°)$
$= (21.3\ VA)(1)$
$= \mathbf{21.3\ VAR}$

$P_{app(Zb)} = V_{Zb} I_{Zb}$
$= (110\ V_{ac})(216\ mA_{ac})$
$= \mathbf{23.7\ VA}$

$P_{r(Zb)} = P_{app} \cos|\varphi|$
$= (23.7\ VA)(\cos 90°)$
$= (23.7\ VA)(0)$
$= \mathbf{0\ W}$

$P_{reac(Zb)} = P_{app} \sin|\varphi|$
$= (23.7\ VA)(\sin 90°)$
$= (23.7\ VA)(1)$
$= \mathbf{23.7\ VAR}$

$P_{app(Zc)} = V_{Zc} I_{Zc}$
$= (100\ V_{ac})(213\ mA_{ac})$
$= \mathbf{21.3\ VA}$

$P_{r(Zc)} = P_{app} \cos|\varphi|$
$= (21.3\ VA)(\cos 90°)$
$= (21.3\ VA)(0)$
$= \mathbf{0\ W}$

$P_{reac(Zc)} = P_{app} \sin|\varphi|$
$= (21.3\ VA)(\sin 90°)$
$= (21.3\ VA)(1)$
$= \mathbf{21.3\ VAR}$

22.3 Just for Fun

1. Problem: Prove that the total power, P_T, developed by the loads, Z_a, Z_b, and Z_c, in a balanced three-phase ac system is constant.

   Solution: From Watt's Law for voltage and impedance, the total power, P_T, developed by the loads is

   $$P_T = (V_{Za}^2 / Z_a) + (V_{Zb}^2 / Z_b) + (V_{Zc}^2 / Z_c)$$

   In a balanced three-phase system, $Z_a = Z_b = Z_c = Z$.

   $$P_T = (V_{Za}^2 / Z) + (V_{Zb}^2 / Z) + (V_{Zc}^2 / Z)$$

   The amplitudes of the voltages across the loads are identical and differ only in a phase separation of 120°. For an ac load voltage with amplitude V_Z and angular frequency $\omega = 2\pi f$, the amplitude of the voltages at any time t (where V_{Za} is the phase reference) is

   $$V_{Za} = V_Z \sin(2\pi f t) = V_S \sin(\omega t)$$

$V_{Zb} = V_Z \sin(2\pi ft + 120°) = V_S \sin(\omega t + 120°)$

$V_{Zc} = V_Z \sin(2\pi ft - 120°) = V_S \sin(\omega t - 120°)$

Substitute this into the equation for P_T and factor out the common terms.

$$P_T = (V_{Za}^2 / Z) + (V_{Zb}^2 / Z) + (V_{Zc}^2 / Z)$$
$$= \{[V_Z \sin(\omega t)]^2 / Z\} + \{[V_Z \sin(\omega t + 120°)]^2 / Z\} + \{[V_Z \sin(\omega t - 120°)]^2 / Z\}$$
$$= \{[V_Z^2 \sin^2(\omega t)] / Z\} + \{[V_Z^2 \sin^2(\omega t + 120°)] / Z\} + \{[V_Z^2 \sin^2(\omega t - 120°)] / Z\}$$
$$= (V_Z^2 / Z)\,[\sin^2(\omega t) + \sin^2(\omega t + 120°) + \sin^2(\omega t - 120°)]$$

From trigonometry, $\sin(\alpha + \beta) = (\sin\alpha)(\cos\beta) + (\sin\beta)(\cos\alpha)$.

$$\sin^2(\omega t + 120°) = \{[\sin(\omega t)](\cos 120°) + (\sin 120°)[\cos(\omega t)]\}^2$$
$$= \{[\sin(\omega t)(1/2) + (\sqrt{3}/2)[\cos(\omega t)]\}^2$$
$$= (\{[\sin(\omega t)] / 2\} + \{[\sqrt{3}\cos(\omega t)] / 2\})^2$$
$$= \{[\sin(\omega t)] / 2)\}^2 + \{[\sin(\omega t)] / 2)\}\{[\sqrt{3}\cos(\omega t)] / 2\}$$
$$+ \{[\sin(\omega t)] / 2)\}\{[\sqrt{3}\cos(\omega t)] / 2\} + \{[\sqrt{3}\cos(\omega t)] / 2]^2$$
$$= \{[\sin^2(\omega t)] / 2^2)\} + (\{[\sin(\omega t)][\sqrt{3}\cos(\omega t)]\} / [(2)(2)])$$
$$+ (\{[\sin(\omega t)][\sqrt{3}\cos(\omega t)]\} / [(2)(2)]) + \{[\sqrt{3}^{\,2}\cos^2(\omega t)] / 2^2\}$$
$$= \{\sin^2(\omega t)] / 4\} + 2\,(\{[(\sqrt{3}\sin(\omega t)][\cos(\omega t)]\} / [(2)(2)])$$
$$+ \{[3\cos^2(\omega t)] / 4\}$$
$$= \{[\sin^2(\omega t)] / 4\} + \{[\sqrt{3}\sin(\omega t)][\cos(\omega t)] / 2\} + \{[3\cos^2(\omega t)] / 4\}$$

$$\sin^2(\omega t - 120°) = \sin^2[\omega t + (-120°)]$$
$$= \{[\sin(\omega t)][\cos(-120°)] + [\sin(-120°)][\cos(\omega t)]\}^2$$
$$= \{[\sin(\omega t)(-1/2) + (\sqrt{3}/2)[\cos(\omega t)]\}^2$$
$$= (\{[-\sin(\omega t)] / 2\} + \{[\sqrt{3}\cos(\omega t)] / 2\})^2$$
$$= \{[-\sin(\omega t)] / 2)\}^2 + \{[-\sin(\omega t)] / 2)\}\{[\sqrt{3}\cos(\omega t)] / 2\}$$
$$+ \{[-\sin(\omega t)] / 2)\}\{[\sqrt{3}\cos(\omega t)] / 2\} + \{[\sqrt{3}\cos(\omega t)] / 2]^2$$
$$= \{[\sin^2(\omega t)] / 2^2)\} - (\{[\sin(\omega t)][\sqrt{3}\cos(\omega t)]\} / [(2)(2)])$$
$$- (\{[\sin(\omega t)][\sqrt{3}\cos(\omega t)]\} / [(2)(2)]) + \{[\sqrt{3}^{\,2}\cos^2(\omega t)] / 2^2\}$$
$$= \{\sin^2(\omega t)] / 4\} - 2\,(\{[(\sqrt{3}\sin(\omega t)][\cos(\omega t)]\} / [(2)(2)])$$
$$+ \{[3\cos^2(\omega t)] / 4\}$$
$$= \{[\sin^2(\omega t)] / 4\} - \{[\sqrt{3}\sin(\omega t)][\cos(\omega t)] / 2\} + \{[3\cos^2(\omega t)] / 4\}$$

Substitute the expressions for $\sin^2(\omega t + 120°)$ and $\sin^2(\omega t - 120°)$ into the equation for P_T and simplify.

$$P_T = (V_Z^2 / Z)\,[\sin^2(\omega t) + \sin^2(\omega t + 120°) + \sin^2(\omega t - 120°)]$$
$$= (V_Z^2 / Z)\,(\sin^2(\omega t)$$
$$+ \{[\sin^2(\omega t)] / 4\} + \{[\sqrt{3}\sin(\omega t)][\cos(\omega t)] / 2\} + \{[3\cos^2(\omega t)] / 4\}$$
$$+ \{[\sin^2(\omega t)] / 4\} - \{[\sqrt{3}\sin(\omega t)][\cos(\omega t)] / 2\} + \{[3\cos^2(\omega t)] / 4\})$$
$$= (V_Z^2 / Z)\,(\sin^2(\omega t)$$
$$+ \{[\sin^2(\omega t)] / 4\} + \{[3\cos^2(\omega t)] / 4\}$$
$$+ \{[\sin^2(\omega t)] / 4\} + \{[3\cos^2(\omega t)] / 4\})$$
$$= (V_Z^2 / Z)\,(\sin^2(\omega t) + 2\,\{[\sin^2(\omega t)] / 4\} + 2\,\{[3\cos^2(\omega t)] / 4]\})$$

$= (V_Z^2 / Z)$ $(\sin^2(\omega t) + \{[2\sin^2(\omega t)] / 4\} + \{[6\cos^2(\omega t)] / 4]\})$

$= (V_Z^2 / Z)$ $([4\sin^2(\omega t)] / 4\} + \{[2\sin^2(\omega t)] / 4\} + [(6\cos^2 \omega t) / 4]\})$

$= (V_Z^2 / Z)$ $(\{[6\sin^2(\omega t) / 4] + \{[6\cos^2(\omega t)] / 4\})$

$= (V_Z^2 / Z)$ $\{[(3 / 2)\sin^2(\omega t)] + [(3 / 2)\cos^2(\omega t)]\}$

$= (3 / 2)(V_Z^2 / Z)[\sin^2(\omega t) + \cos^2(\omega t)]$

From trigonometry, $\sin^2(\omega t) + \cos^2(\omega t) = 1$.

$P_T = (3 / 2)(V_Z^2 / Z)[\sin^2(\omega t) + \cos^2(\omega t)]$

$= (3 / 2)(V_Z^2 / Z)(1)$

$= (3 / 2)(V_Z^2 / Z)$

Because the time t used to calculate the three-phase power in the load was arbitrary and both V_Z and Z are constant, the total power delivered to the load for a balanced three-phase circuit is always the same and therefore constant.

2. Problem: Three-phase circuits are one type of a general class of circuits known as *polyphase* circuits, which are circuits that have more than one phase. The analysis of these circuits is very similar to that for three-phase circuits. Determine the line and load voltages and currents for the four-phase circuit shown in Figure 22-23. Express all answers in polar form.

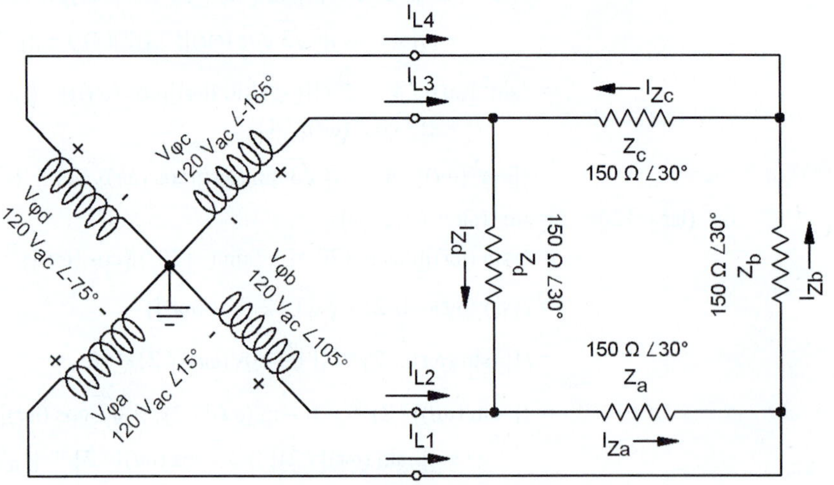

Figure 22-23: Four-Phase Circuit

Solution: To determine the line and load voltages and currents, first calculate the line voltages.

$V_{ba} = V_b - V_a$

$= (120\ V_{ac} \angle 105°) - (120\ V_{ac} \angle 15°)$

$= (-31.1\ V_{ac} + j\ 116\ V_{ac}) - (116\ V_{ac} + j\ 31.1\ V_{ac})$

$= (-31.1\ V_{ac} - 116\ V_{ac}) + j\ (116\ V_{ac} - 31.1\ V_{ac})$

$= -147\ V_{ac} + j\ 84.9\ V_{ac}$ in rectangular form

$=$ **$170\ V_{ac} \angle 150°$** in polar form

$V_{cb} = V_c - V_b$

$= (120\ V_{ac} \angle -165°) - (120\ V_{ac} \angle 105°)$

$= (-116\ V_{ac} - 31.1\ V_{ac}) - (-31.1\ V_{ac} + j\ 116\ V_{ac})$

$= (-116\ V_{ac} + 31.1\ V_{ac}) - j\ (31.1\ V_{ac} - 116\ V_{ac})$

$= -84.9\ V_{ac} - j\ 147\ V_{ac}$ in rectangular form

$= \mathbf{170\ V_{ac} \angle -120°}$ in polar form

$V_{dc} = V_d - V_c$

$= (120\ V_{ac} \angle -75°) - (120\ V_{ac} \angle -165°)$

$= (31.1\ V_{ac} - j\ 116\ V_{ac}) - (-116\ V_{ac} - j\ 31.1\ V_{ac})$

$= (31.1\ V_{ac} + 116\ V_{ac}) - j\ (116\ V_{ac} - 31.1\ V_{ac})$

$= 147\ V_{ac} - j\ 84.9\ V_{ac}$ in rectangular form

$= \mathbf{170\ V_{ac} \angle -30°}$ in polar form

$V_{ad} = V_a - V_d$

$= (120\ V_{ac} \angle 15°) - (120\ V_{ac} \angle -75°)$

$= (116\ V_{ac} + 31.1\ V_{ac}) - (31.1\ V_{ac} - j\ 116\ V_{ac})$

$= (116\ V_{ac} - 31.1\ V_{ac}) + j\ (31.1\ V_{ac} + 116\ V_{ac})$

$= 84.9\ V_{ac} + j\ 147\ V_{ac}$ in rectangular form

$= \mathbf{170\ V_{ac} \angle 60°}$ in polar form

The line voltages are applied directly across the loads, so the load voltages are equal to the line voltages.

$V_{Za} = V_{ba}$

$= \mathbf{170\ V_{ac} \angle 150°}$

$V_{Zb} = V_{ad}$

$= \mathbf{170\ V_{ac} \angle 60°}$

$V_{Zc} = V_{dc}$

$= \mathbf{170\ V_{ac} \angle -30°}$

$V_{Zd} = V_{cb}$

$= \mathbf{170\ V_{ac} \angle -120°}$

Use Ohm's Law for current to calculate the load currents.

$I_{Za} = V_{Za} / Z_a$

$= (170\ V_{ac} \angle 150°) / (150\ \Omega \angle 30°)$

$= (170\ V_{ac} / 150\ \Omega) \angle (150° - 30°)$

$= \mathbf{1.13\ A_{ac} \angle 120°}$

$I_{Zb} = V_{Zb} / Z_a$

$= (170\ V_{ac} \angle 60°) / (150\ \Omega \angle 30°)$

$= (170\ V_{ac} / 150\ \Omega) \angle (60° - 30°)$

$= \mathbf{1.13\ A_{ac} \angle 30°}$

$I_{Zc} = V_{Zc} / Z_c$

$= (170\ V_{ac} \angle -30°) / (150\ \Omega \angle 30°)$

$= (170\ V_{ac} / 150\ \Omega) \angle [(-30°) - 30°]$

$= \mathbf{1.13\ A_{ac} \angle -60°}$

$I_{Zd} = V_{Zd} / Z_d$

$= (170\ V_{ac} \angle -120°) / (150\ \Omega \angle 30°)$

$= (170\ V_{ac} / 150\ \Omega) \angle [(-120°) - 30°]$

$= \mathbf{1.13\ A_{ac} \angle -150°}$

For the directions of the load currents as shown in Figure 22-23, each line current is the difference between two load currents. Because the figure shows the line currents entering the four-phase load, define currents entering a node to be positive and currents leaving a node to be negative. Then use Kirchhoff's Current Law to determine the line currents.

$\boldsymbol{I_{L1}} = \boldsymbol{I_{Za}} - \boldsymbol{I_{Zb}}$
$= (1.13\ A_{ac} \angle 120°) - (1.13\ A_{ac} \angle 30°)$
$= (-566\ mA_{ac} + j\ 980\ mA_{ac}) - (980\ mA_{ac} + j\ 566\ mA_{ac})$
$= (-566\ mA_{ac} - j\ 980\ mA_{ac}) + (980\ mA_{ac} - j\ 566\ mA_{ac})$
$= -1.55\ A_{ac} + j\ 414\ mA_{ac}$ in rectangular form
$= \mathbf{1.60\ A_{ac} \angle 165°}$ in polar form

$\boldsymbol{I_{L2}} = \boldsymbol{I_{Zd}} - \boldsymbol{I_{Za}}$
$= (1.13\ A_{ac} \angle -150°) - (1.13\ A_{ac} \angle 120°)$
$= (-980\ mA_{ac} - j\ 566\ mA_{ac}) - (-566\ mA_{ac} + j\ 980\ mA_{ac})$
$= (-980\ mA_{ac} + j\ 566\ mA_{ac}) - (566\ mA_{ac} + j\ 566\ mA_{ac})$
$= -414\ mA_{ac} - j\ 1.55\ A_{ac}$ in rectangular form
$= \mathbf{1.60\ A_{ac} \angle -105°}$ in polar form

$\boldsymbol{I_{L3}} = \boldsymbol{I_{Zc}} - \boldsymbol{I_{Zd}}$
$= (1.13\ A_{ac} \angle -60°) - (1.13\ A_{ac} \angle -150°)$
$= (566\ mA_{ac} - j\ 980\ mA_{ac}) - (-980\ mA_{ac} - j\ 566\ mA_{ac})$
$= (566\ mA_{ac} + j\ 980\ mA_{ac}) - (980\ mA_{ac} - j\ 566\ mA_{ac})$
$= 1.55\ A_{ac} - j\ 414\ mA_{ac}$ in rectangular form
$= \mathbf{1.60\ A_{ac} \angle -15°}$ in polar form

$\boldsymbol{I_{L4}} = \boldsymbol{I_{Zb}} - \boldsymbol{I_{Zc}}$
$= (1.13\ A_{ac} \angle 30°) - (1.13\ A_{ac} \angle -60°)$
$= (980\ mA_{ac} + j\ 566\ mA_{ac}) - (566\ mA_{ac} - j\ 980\ mA_{ac})$
$= (980\ mA_{ac} - j\ 566\ mA_{ac}) + (566\ mA_{ac} + j\ 980\ mA_{ac})$
$= 414\ mA_{ac} + j\ 1.55\ A_{ac}$ in rectangular form
$= \mathbf{1.60\ A_{ac} \angle 75°}$ in polar form

3. Problem: Prove that the total power, $\boldsymbol{P_T}$, developed by the loads, $\boldsymbol{Z_a}$, $\boldsymbol{Z_b}$, $\boldsymbol{Z_c}$, and $\boldsymbol{Z_d}$, in a balanced four-phase ac system is constant.

Solution: From Watt's Law for voltage and impedance, the total power, $\boldsymbol{P_T}$, developed by the loads is

$$\boldsymbol{P_T} = (\boldsymbol{V_{Za}}^2 / \boldsymbol{Z_a}) + (\boldsymbol{V_{Zb}}^2 / \boldsymbol{Z_b}) + (\boldsymbol{V_{Zc}}^2 / \boldsymbol{Z_c}) + (\boldsymbol{V_{Zd}}^2 / \boldsymbol{Z_d})$$

In a balanced four-phase system, $\boldsymbol{Z_a} = \boldsymbol{Z_b} = \boldsymbol{Z_c} = \boldsymbol{Z_d} = \boldsymbol{Z}$.

$$\boldsymbol{P_T} = (\boldsymbol{V_{Za}}^2 / \boldsymbol{Z}) + (\boldsymbol{V_{Zb}}^2 / \boldsymbol{Z}) + (\boldsymbol{V_{Zc}}^2 / \boldsymbol{Z}) + (\boldsymbol{V_{Zd}}^2 / \boldsymbol{Z})$$

The amplitudes of the voltages across the loads are identical and differ only in a phase separation of 90°. For an ac load voltage with amplitude $\boldsymbol{V_Z}$ and angular frequency $\omega = 2\pi\boldsymbol{f}$, the amplitude of the voltages at any time $\boldsymbol{t}$ (where $\boldsymbol{V_{Za}}$ is the phase reference) is

$\boldsymbol{V_{Za}} = \boldsymbol{V_Z} \sin (2\pi\boldsymbol{ft}) = \boldsymbol{V_S} \sin (\omega\boldsymbol{t})$
$\boldsymbol{V_{Zb}} = \boldsymbol{V_Z} \sin (2\pi\boldsymbol{ft} + 90°) = \boldsymbol{V_S} \sin (\omega\boldsymbol{t} + 90°)$
$\boldsymbol{V_{Zc}} = \boldsymbol{V_Z} \sin (2\pi\boldsymbol{ft} + 180°) = \boldsymbol{V_S} \sin (\omega\boldsymbol{t} + 180°)$
$\boldsymbol{V_{Zd}} = \boldsymbol{V_Z} \sin (2\pi\boldsymbol{ft} - 90°) = \boldsymbol{V_S} \sin (\omega\boldsymbol{t} - 90°)$

Substitute this into the equation for $\boldsymbol{P_T}$ and factor out the common terms.

$$\boldsymbol{P_T} = (\boldsymbol{V_{Za}}^2 / \boldsymbol{Z}) + (\boldsymbol{V_{Zb}}^2 / \boldsymbol{Z}) + (\boldsymbol{V_{Zc}}^2 / \boldsymbol{Z}) + (\boldsymbol{V_{Zd}}^2 / \boldsymbol{Z})$$
$$= \{[\boldsymbol{V_Z} \sin (\omega\boldsymbol{t})]^2 / \boldsymbol{Z}\} + \{[\boldsymbol{V_Z} \sin (\omega\boldsymbol{t} + 90°)]^2 / \boldsymbol{Z}\} + \{[\boldsymbol{V_Z} \sin (\omega\boldsymbol{t} + 180°)]^2 / \boldsymbol{Z}\} + \{[\boldsymbol{V_Z} \sin (\omega\boldsymbol{t} - 90°)]^2 / \boldsymbol{Z}\}$$

$= \{[\boldsymbol{V_Z}^2 \sin^2 (\omega \boldsymbol{t})] / \boldsymbol{Z}\} + \{[\boldsymbol{V_Z}^2 \sin^2 (\omega \boldsymbol{t} + 90°)]^2 / \boldsymbol{Z}\} + \{[\boldsymbol{V_Z}^2 \sin^2 (\omega \boldsymbol{t} + 180°)]^2 / \boldsymbol{Z}\}$
$+ \{[\boldsymbol{V_Z}^2 \sin^2 (\omega \boldsymbol{t} - 90°)]^2 / \boldsymbol{Z}\}$

$= (\boldsymbol{V_Z}^2 / \boldsymbol{Z}) [\sin^2 (\omega \boldsymbol{t}) + \sin^2 (\omega \boldsymbol{t} + 90°) + \sin^2 (\omega \boldsymbol{t} + 180°) + \sin^2 (\omega \boldsymbol{t} - 90°)]$

From trigonometry, sin (α + 90°) = cos α, sin (α – 90°) = –cos α, and sin (α + 180°) = –sin α for 0° ≤ α ≤ 180°. Substitute these expressions into the equation for $\boldsymbol{P_T}$ and simplify.

$\boldsymbol{P_T} = (\boldsymbol{V_Z}^2 / \boldsymbol{Z}) [\sin^2 (\omega \boldsymbol{t}) + \sin^2 (\omega \boldsymbol{t} + 90°) + \sin^2 (\omega \boldsymbol{t} + 180°) + \sin^2 (\omega \boldsymbol{t} - 90°)]$

$= (\boldsymbol{V_Z}^2 / \boldsymbol{Z}) \{\sin^2 (\omega \boldsymbol{t}) + \cos^2 (\omega \boldsymbol{t}) + [-\cos (\omega \boldsymbol{t})]^2 + [-\sin (\omega \boldsymbol{t})]^2\}$

$= (\boldsymbol{V_Z}^2 / \boldsymbol{Z}) [\sin^2 (\omega \boldsymbol{t}) + \cos^2 (\omega \boldsymbol{t}) + \cos^2 (\omega \boldsymbol{t}) + \sin^2 (\omega \boldsymbol{t})]$

$= (\boldsymbol{V_Z}^2 / \boldsymbol{Z}) [2 \sin^2 (\omega \boldsymbol{t}) + 2 \cos^2 (\omega \boldsymbol{t})]$

$= (2) (\boldsymbol{V_Z}^2 / \boldsymbol{Z}) [\sin^2 (\omega \boldsymbol{t}) + \cos^2 (\omega \boldsymbol{t})]$

From trigonometry, $\sin^2 (\omega \boldsymbol{t}) + \cos^2 (\omega \boldsymbol{t}) = 1$.

$\boldsymbol{P_T} = (2) (\boldsymbol{V_Z}^2 / \boldsymbol{Z}) [\sin^2 (\omega \boldsymbol{t}) + \cos^2 (\omega \boldsymbol{t})]$

$= (2) (\boldsymbol{V_Z}^2 / \boldsymbol{Z}) (1)$

$= (2) (\boldsymbol{V_Z}^2 / \boldsymbol{Z})$

$= (2\, \boldsymbol{V_Z}^2) / \boldsymbol{Z}$

Because the time $\boldsymbol{t}$ used to calculate the four-phase power in the load was arbitrary and both $\boldsymbol{V_Z}$ and $\boldsymbol{Z}$ are constant, the total power delivered to the load for a balanced four-phase circuit is always the same and therefore constant.